Meaning of Life, Human Nature, and Delusions

Rui Diogo

Meaning of Life, Human Nature, and Delusions

How Tales about Love, Sex, Races, Gods and Progress Affect Our Lives and Earth's Splendor

Rui Diogo
Department of Anatomy
Howard University College of Medicine
Washington, DC, USA

ISBN 978-3-319-70400-5 ISBN 978-3-319-70401-2 (eBook)
https://doi.org/10.1007/978-3-319-70401-2

This Springer imprint is published by the registered company Springer Nature Switzerland AG
The registered company address is: Gewerbestrasse 11, 6330 Cham, Switzerland

Love does not consist of gazing at each other, but in looking outward together in the same direction.
(Antoine de Saint-Exupéry)

I dedicated my last book—which was about how evolution is mainly driven by organismal behavior—to Tots, an exceptional member of the species Canis lupus that was crucial for my understanding of biological evolution and was a true companion, 24/7, until her very last breath. Therefore, I dedicate the present book particularly to Alejandra—a truly fulfilled person with a broad interest in knowledge, arts, people, life, and the pursuit of happiness—and to our naturally "wild" Salehe, a name that in Swahili and Arabic basically means, in a very simple but also profoundly deep way, "Good man." My hope is that my passion and enthusiasm for life in all its diverse, beautiful, and amazingly fascinating combination of chaotic, contingent, and random events will be passed to him and to the broader public, particularly the new generations, and will give them the necessary tools to help them escape from the chains of Neverland once for all, and live in a world of reality with much less hate, oppression, discrimination, inequalities, wars, famines, animal abuse, ecological disasters, and delusions.

Praise for the Book

Rui Diogo is becoming the Slavoj Zizek of evolutionary biology (Marcelo Sanchez-Villagra, Director of the Paleontological Institute and Museum of the University of Zurich).

I applaud the enormous work that Diogo has invested in this follow-up to his widely acclaimed Evolution driven by organismal behavior book, and the challenge of getting people to think beyond and outside of our usual set of definitions and expectations. The case-studies provided in the book are fascinating and insightful (Drew Noden, Emeritus Professor, Cornell University).

Standing on the Shoulders of Farsighted Humans

One glance at a book and you hear the voice of another person, perhaps someone dead for 1,000 years. To read is to voyage through time. (Carl Sagan)

Why should I fear death? If I am, then death is not. If Death is, then I am not. Why should I fear that which can only exist when I do not? Long time men lay oppressed with slavish fear. Religious tyranny did domineer. (Epicurus)

For fools admire and love those things they see hidden in verses turned all upside down, and take for truth what sweetly strokes the ears and comes with sound of phrases fine imbued. (Lucretius)

Sublime Lucretius' work will not die, until the day the world itself passes away. (Ovid)

I have never thought, for my part, that man's freedom consists in his being able to do whatever he wills, but that he should not, by any human power, be forced to do what is against his will. (Jean-Jacques Rousseau)

Those who can make you believe absurdities, can make you commit atrocities. (Voltaire)

I am in this endless lack of solitude an animal of light corralled by his mistakes and by his foliage. (Pablo Neruda)

Two types of choices seem to me to have been crucial in tipping the outcomes [of the various societies' histories] towards success or failure: long-term planning and willingness to reconsider core values. On reflection we can also recognize the crucial role of these same two choices for the outcomes of our individual lives. (Jared Diamond)

A bank is a place where they lend you an umbrella in fair weather and ask for it back when it begins to rain. (Robert Frost)

When the last tree is cut down, the last fish eaten and the last stream poisoned, you will realize that you cannot eat money. (Prophecy of the Native American Cree people)

White people cannot live with the idea of living aimlessly. They think that work is the reason for their existence. They enslaved so much 'others', that now they need to enslave themselves.. as if becoming 'civilized' was our destiny. They can't stop and experience life as something that is simply part of a marvelous world. This is their religion: the religion of civilization. (Ailton Krenak, leader of an indigenous movement in Brazil)

The Earth is 4.6 billion years old...let's scale that down to 46 years...we've been here for 4 hours...our industrial revolution began 1 minute ago...in that time we've destroyed more than 50% of the World's Rain Forests...this isn't sustainable. (Greenpeace)

Preface

Before the creation of the universe, God did nothing, apparently...suddenly, one day he decided to create the Universe, we don't known why or for what...according to the Bible, he then made the Universe in six days, six days only, six days...then he rested on the seventh day, until today. He did nothing, ever again. Does this make any sense?! (Jose Saramago)

Our tendency to wonder for "why" life is as it is and what is its "purpose" is often considered to be among the most "noble" features of humanity. This book is the first to provide a multidisciplinary account showing that while this propensity does play crucial functions such as help coping with death and a plethora of societal troubles, thus decreasing depression, it is also profoundly linked with some of the darker moments in our history, including atrocious wars, animal abuse, colonialism, slavery, misogyny, and racism. The central topics discussed here are beautifully shown by **Jacopo Bassano**'s (ca. 1510–1592)—an Italian painter also known as **Jacopo dal Ponte**—stunning painting displayed in Fig. 1. **Saint Jerome** (340–420 AD)—a monk who lived for 4 years in the Syrian desert where he devoted himself to the practice of penance and the study of Hebrew and the Scriptures—is often shown in paintings contemplating the skull as a reminder of the inevitability of death and the vanity of worldly events. The dark tone of the painting and the skull at the bottom show very well the obscurity of death and the anguish of St. Jerome—and of humans in general—when confronted with death's inexorableness. This point is also stressed by the saint's body showing its marks of age, and the menacing sky suggesting that death might be near. In this book I defend that our awareness of the certainty of death and randomness of both our lives and demise—an aptitude that seems to be uniquely found in humans—is profoundly related to our peculiar, compulsive tendency to create, and believe in, complex teleological imaginary narratives. Such tales are based on the notion that everything has a "special" cosmic purpose or use, or meaning. These imaginary stories can therefore be either religious or non-religious, as are for example tales about a purposeful "Mother Nature" or many narratives characteristic of atheistic humanism.

Accordingly, on the right portion of the painting one can see clear symbols—including Christ himself, crucified on the cross—of complex teleological narratives: in this case, those of Christianity. That is, within the main ideas discussed in the

Fig. 1 Jacobo da Ponte's splendid painting of St. Jerome

present book—and not necessarily within those defended by Jacoppo when he did this painting as they are still controversial—moving from left to right, one sees the inevitability of death (left), how it leads St. Jerome—and humans in general—to be profoundly anguished when confronted with it (center), and consequently to create teleological narratives (right) about a cosmic purpose of life: that our existence is "meant to be." Lastly, reflecting perhaps the most peculiar and in this sense essential take-home message of this book is the idea that this quest for a purpose of life actually leads to some of the darker moments of our human history. This point is reflected in the painting by details suggesting the self-flagellation of St. Jerome. Using this as metaphor for the ideas of this book, it can remain ambiguous if the flagellation was knowingly done by him or unconsciously self-imposed during a delusional dream about supernatural deities. Within the context of this book, it is particularly relevant that the flagellation shown in the painting is very likely related to the popular story about St. Jerome's dream, in which he saw himself being flagellated by angels of God *because* he studied classical literature—often symbolizing knowledge about the *world of reality*, including historical texts and scientific ones such as those of Aristotle—more than the Bible—one of the most emblematic examples of complex imaginary narratives created by humans. That is, one could say that the flagellation of St. Jerome represents both the conscious and unconscious self-flagellations that humans imposed, and continue to inflict, to themselves over and over again by creating, believing in, and blindly acting—or being obliged to act, either by force or due to social norms or peer pressure—according to such imaginary tales.

A critical point is that my aim is not to present a detailed, encyclopedic account of all the diverse subjects discussed, and references mentioned, in this book. Most of these issues have been widely analyzed separately in numerous publications and by countless thinkers. As an avid reader I could not resist to read, in most occasions, the originals, even if they were written centuries, or even millennia, ago. However, I try to not provide too many details in many cases, as they can be fascinating to some readers, but can bore others, or at least distract them from the main ideas elaborated in this book. Therefore, I often provide an introduction to each of the many topics analyzed, so readers can understand what is being discussed and where they can then find more information, in case they are particularly interested in knowing more about a specific topic. This is because the main aim of the book is precisely to put together all these topics in a broader way than what is often provided in other works and in particular within what can be defined as empirical or scientific philosophy. That is, using scientific—including experimental—data to discuss in an holistic way broader questions that have fascinated humans for a long time. In other words, I do not want to provide details about each and every small branch of the tree, but instead to provide a uniquely multidisciplinary, far-reaching analysis of the tree as a whole. In that sense, this book literally contains "a thousand different voices," including empirical data, historical accounts, and points of view from authors from numerous different backgrounds, places, cultures, and times, from the first epic stories and religious texts written millennia ago to current discussions on artificial intelligence (AI) and virtual reality. Similarly, as I intend to have the ideas analyzed in this book read, discussed, and hopefully taken into account by a wide audience across the globe, I also had to make the difficult decision to leave out numerous references and fascinating case studies, and reduced the use of jargon to the maximum. At least, I can surely say, humbly and respectfully, that I really tried to do my best in order to do so: this is a book directed to everybody—whatever you are from, or how many years you were at school, or if you are religious or not, this book is for you. So, what I can indeed guarantee is that whatever is your background or interests, there will be something in the book that will engage you, make you learn something new, and more importantly, lead you to think deeply about critical aspects of your daily life, from how you have sex to how you love, from what you eat to the physical activities you do, from how you see other people to how you deal with them, with other animals, and with the planet in general.

Washington, DC, USA Rui Diogo

Acknowledgments

If our brains were simple enough for us to understand them, we'd be so simple that we couldn't. (Ian Stewart)

I would like to acknowledge the hundreds of colleagues, students, and friends with whom I had the privilege to collaborate and discuss numerous topics covered in this book. I have been very lucky to meet and interact with people interested in so many different fascinating issues, and this book would have never been possible without them. Among all those people, my most profound admiration, and gratitude, is to Michel Chardon, former professor of the University of Liege and advisor of my first PhD thesis. As I always told him, for me he was in a way the last emblematic example of the so-called Old Europe: a true humanist, savant, and caring person and sage. We desperately need more people like him—he was and still is the main inspiration for my academic life. I also want to thank my parents, Valter and Fatima—since I was a small child they emphasized the importance of reading, traveling, and thinking "outside the box," which was further stimulated by my brothers, Luis and Hugo. I also want to thank Howard University, particularly its mission. Being part of one of the most renowned historically Black Universities clearly made me more aware of, and dedicated to address, a plethora of societal problems and the long-standing and still prevailing fictional stories that have chiefly contributed to them.

Contents

List of Boxes

I think that God, in creating man, somewhat overestimated His ability. (Oscar Wilde)

Chapter 1
Introduction

> *We are like dwarfs sitting on the shoulders of giants...we see more, and things that are more distant, than they did, not because our sight is superior or because we are taller than they, but because they raise us up, and by their great stature add to ours.*
>
> (John of Salisbury, excerpt from 1159 treatise Metalogicon)

1.1 Standing on the Shoulders of Others

> *What Descartes did was a good step...you have added much several ways, and especially in taking the colours of thin plates into philosophical consideration...if I have seen a little further it is by standing on the shoulders of Giants. (Isaac Newton, letter to Robert Hooke in 1676)*

Let's first talk about what this book is about, and what it *does* not aim to be. The book indeed stands on the shoulder of not only giants, but of all those other scholars, thinkers, historians, and lay people that have been interested on the broader topics discussed here, which are those that have basically fascinated the most people since times immemorial. Why are we here? What is the purpose of life? What is its meaning? Are we progressing towards any direction? Will we thrive? Therefore, this is a "popular book" in which I compile the information provided in previous works by other authors in a relatively simple way so the broader public can understand it, think about it, discuss it, and most importantly take it into account in their daily lives. Accordingly, I avoided the excessive use of scientific and philosophical jargon, and when I do so, particularly when I cite the works of others, I usually provide brief definitions of terms that might not be so familiar to readers. A major, and somewhat peculiar, characteristic of this book, which is in a way a logical follow-up from what I did in my 2017 book *Evolution Driven by Organismal Behavior*, is that I often cite brief excerpts of the original texts of other authors.

I consider that we, scientific writers, should not only clearly mention, but also actually pay a direct tribute to authors who have influenced our ideas or a large number of people, either because we agree with them or because by not agreeing

R. Diogo, *Meaning of Life, Human Nature, and Delusions*,
https://doi.org/10.1007/978-3-319-70401-2_1

with them, they make us re-think and analyze a topic from a different perspective. My viewpoint is that we should be humble, and not pretend that we invented the wheel: we do need to recognize, unambiguously, that *we do stand on the shoulders of others*, and that is why I opted to refer *directly* to their original works—particularly those that don't agree with some of the ideas defended in the current volume—so that, in all fairness, you can make an impartial judgment about which ideas you agree with. It is just too easy to criticize ideas or authors by misrepresenting them, and I don't want to do so. I recently read a book review that commented that one of the major strengths of the book that was being reviewed is that it *did* include numerous direct citations of other works, and that, by doing so, it literally contained "*a thousand different voices.*" This applies to the present volume: voices from authors from very different fields of science or knowledge in general, from very different times and places and cultures, and with very different perspectives on life, politics, love, sex, "races," and another number of greater-than-life topics, from texts written in ancient "civilizations" to works published about Covid-19 or Artificial Intelligence just a few weeks ago, at the end of 2020.

Regarding the use of brief definitions, I need to explain what I mean, when I use the terms "**cosmic meaning of life**" and "**cosmic purpose of life**." Obviously, we all often have some aims, or "purposes" that we want to fulfill when we wake up, either the same day or week or year or during our life in general. This morning I had the aim of spending a few hours writing this book. But in the way the terms are used in this book—and commonly in philosophical or biology works and broader discussions about the "meaning of life"—such specific aims refer to our ***purpose in life***. That is completely different from the discussions about and quests to find a "*cosmic meaning of life*" or "*cosmic purpose of life*," which have nothing to do with specific tasks such as going shopping or cleaning a bathroom or writing an article, that is, to purposes *we set up* to be fulfilled *in* our lives. Instead, they refer to much broader, transcendent "purposes" *of* life in the cosmos as a whole that *a supernatural deity such as **God** or **Mother Nature** or another type of agent set up* for you, or others, or for humans or life in this planet in general. Theists usually argue that God created the universe as well as life, and that this means we are here for a reason, as part of **God's purposeful "masterplan."** But, importantly—and too often neglected—the quest for the "cosmic purpose of life," or the "meaning of life," is not only undertaken by theists: many atheists follow ways of thinking that are actually very similar to those followed by theists and talk about cosmic "purposes of life." This is done by **humanist atheists**, which often argue that humans have some kind of "purpose" or "duty" to "make a better world," or "protect the planet," or the living beings that live in it. As we will see in this book, basically the vast majority of people from our species *believe*, and have always *believed*, that there is some kind of cosmic purpose of life.

Regarding specifically the two terms with "of," that is *meaning of life* and *purpose of life*, some authors argue that the former is a psychological concept that is focused on the significance of life, while the latter is a spiritual concept of life. Other authors argue that the latter is related to a belief that everything has a use or a reason for being, while the former refers to the value or values that are assigned to that belief. However, for the context of the present book what is important to stress is that both terms are similar in the sense that they refer to something more abstract and transcendent than the mere specific short- or long-term aims or goals, that is the *purposes in life*, that *we* set up to accomplish in a material universe. In other words, in a very simplified way *purpose or meaning of life* would refer to something as saying "the reason I am here, as a human being living in this planet, is to change the world by being a science communicator, this is what I was meant to be by God, Mother Nature, or the interactive energy of the Cosmos." Accordingly, in this book I commonly use the term *purpose of life* because the word "purpose" is more directly related to the term ***teleological narratives,*** which are precisely often related to the notion that things have a "special" *purpose* or *use*, or "**telos**." Specifically, the word *teleology* builds on the Greek *telos* ("purpose," or "end") and *logia*, which refers to "a branch of learning."

Similarly, I should make another distinction: in biology and various other scientific areas, "**how questions**" and "**why questions**" are normally associated with, respectively, "proximate" and "ultimate" causes. However, when I refer in this book to "why-questions," I am doing this more as it is often done by scholars that discuss these types of "broader-than-life" issues, such as philosophers or theologians: that is, to refer to ***teleological why-questions***. A biologist could answer a ***scientific why-question*** such as "why are humans in this planet" with a naturalistic answer like "because our ancestors were apes and then there were ecological changes and consequently there were anatomical and genetic changes in the ancestors of the lineage that lead to our species." Such an answer would not invoke any cosmic "purpose" or "masterplan." However, within the sense applied in the present book, a typical answer to the ***teleological why-question*** "why are humans in this planet" would involve something about a cosmic purpose or goal, for example imaginary tales such as "because God wanted so," or "because Mother Earth wanted a better world and needed humans to do so" or, alternatively, "wanted a species that would destroy numerous others so that there would be a renewal of life in the planet."

As it might already be obvious to you, one of the main take-home messages from the compilation of the extensive amount of scientific data that is discussed in this book is that there is *no* cosmic *purpose of life.* This includes the clear inconsistencies that exist between the different cosmic teleological narratives created by different groups of people, between them and the available scientific evidence, between the different parts of a single tale or single religious monograph, or cross-cultural studies about what people from different groups believe, or how differently they "feel" or "see" "death" and "afterlife" in near-death experiences. Many events in life including its very end—death—are mainly arbitrary, but our brains often try very hard to attribute them a "meaning," a "purpose," or connect them within inexistent patterns. Within an endless number of examples available in the literature and in the media, some that most shockingly illustrate how life can be so empathetically absurd and ephemeral are the cell-phone "**selfies**" that were taken just seconds or minutes before the people that are displayed in them died, almost always for reasons that were completely unpredictable just days before they happened.

One emblematic example of this is provided in a dignifying article from the *Cosmopolitan*, which was published with the approval, and including interviews with, the family of **Collette Morenos**, and was accordingly entitled "*The real story of Collette Moreno's viral 'selfie death'.*" Collette's family was particularly disappointed that the media, and the broader public in general, tried to quickly conclude that Collette's death occurred "*because*" she and her friend took a "selfie," while actually the car crash that led to Collette's death occurred after a few minutes and at a different place in which that selfie was taken. Therefore, apart from the dignifying mission of telling the truth about what really happened and of not blaming the selfie for Collette's death, I refer to this specific case because it does emphasize a critical point of this book: how random and ephemeral life is. As noted in the *Cosmopolitan* article:

> Collette Moreno was on the way to her own bachelorette party, and she was choking. The truck in front of her and her best friend, Ashley Theobald, was spewing fumes that were aggravating Moreno's asthma, but Missouri Highway 5 had a double yellow line; they couldn't pass. Her friend tearing up, Theobald craned her head to the left. The coast looked clear. She tilted the wheel, guiding the Chevy Malibu across the lines, speeding up to make the pass quick. But as the Malibu sped forward, a Dodge Ram came cruising up a slight hill that neither of them had seen. Theobald swerved, but the Dodge swerved with them. The cars collided head-on. On June 20, 2014, 26-year-old Collette Moreno died five weeks before her wedding, leaving her 5-year-old son motherless.

Few cases reflect how transient our lives are, and amplify our discomfort in recognizing this reality, as this one, particularly when we know that College was going to her bachelorette party and was mother of a boy who has lived only half of a decade. That is precisely why most people, and the media, quickly tried so hard to create an alternative reality. In this era of "alternative facts," the "fake news" that so quickly became so viral about this case were chiefly due to that huge discomfort we feel when facing the absurdity of death: the need to create a posteriori imaginary narratives about "why" people die, what is the "*purpose*." Humans love to talk about "purposes," to seek for **causality**, a ***causal*** **chain of events** in which any one event causes the next, and hate to recognize that most of the events are actually not causal but *casual*. In this specific case, such human tendencies lead the media and broader public to quickly *believe*—by following the premises of such a "causal chain of events"—that Collette's death was causally connected with the last previous moment of Collette's life they knew about: the moment she and her friend took the selfie. Of course, there are cases where the act of taking a selfie does *lead* directly to the death of the person or persons taking it, further reinforcing the absurdity of life. However, as it will be shown below, such cases are in *reality* extremely rare: in the vast majority of other cases, including this one, there is no causal chain at all, no *pattern* connecting the dots, and that is what makes humans feel so uneasily and create "why" teleological narratives that assign a *purpose* to a completely unpurposeful natural event, death. In this case, the "purpose" story that so quickly became so viral—without most people and the media even caring to search for the true facts—is that Collette's story was "teaching us a lesson": karma, or "*narcissism*" was surely the "*cosmic reason*" for this death. As explained in the *Cosmopolitan* article:

> Collette Moreno died five weeks before her wedding, leaving her 5-year-old son motherless…but that, according to the internet, wasn't the worst thing that happened that day: eight minutes before the fumes and the double yellow lines, Moreno took a selfie – grinning

from the passenger seat, with Theobald in the background driving in shiny heart-shaped sunglasses. '*Dying in a car crash...but first, LEMME TAKE A SELFIE!*' an anonymous commenter wrote.... '*That's natural selection – idiots die*' wrote another.... '*With great selfies must also come great stupidity*' commented a third....

'*We were just beginning to learn about everything ourselves*' Moreno's sister Samantha says over the phone. '*It was just completely overwhelming and heartbreaking that people that didn't even know the situation were saying things*'. Within a year of Moreno's death, the internet's selfie-death obsession peaked.... *The Guardian* called 2015 the year of the '*dangerous selfie*'. The world had started seeing selfies not as a novelty, but rather caught squarely in the crosshairs of the cruel repercussions of a narcissistic culture. The names of people that accumulated on the Wikipedia page for '*List of selfie-related injuries and deaths*' became punching bags. Moreno's photo, for example, had absolutely nothing to do with her death. But that didn't matter to the internet. Moreno's relationship with Brayden's father fell apart, and while she was happy as a single mom, she hoped one day to fall in love again. A few years later, she met Jesse Arcobasso at a party, and they started dating. He was 25, she was 23, and the young couple was goofy and carefree. After three years together, Arcobasso and Moreno stopped by a mall caricature artist, who drew a cartoonish Arcobasso holding a diamond ring and asking, "Will you marry me?" They set the date – July 26, 2014 – and decided to marry in Jamaica. Moreno never made it down the aisle.

After the crash, Theobald...told Fox News that Moreno didn't appear terribly injured. "I was talking to her. She couldn't talk back but she was nodding at me", Theobald says through tears.... An ambulance took Moreno to the hospital, where she would die a few hours later from injuries sustained in the crash. Most of the news coverage about Moreno stated at least in the body of the article that she had snapped the photo a full eight minutes before the crash, but many readers missed this from the headlines like '*Collette Moreno Killed En Route To Her Bachelorette Party Moments After Taking Selfie*' and '*Bride-to-be, 26, killed in head-on car crash as she and best friend drove to her bachelorette party moments after pair snapped this selfie*'. Unfortunately, 55 percent of readers spend fewer than 15 seconds actually reading an article, according to founding CEO of Chartbeat Tony Haile. Moreno isn't the only one who has been wrongly lumped into the selfie-death craze. In September 2015, the family of Kristi Kafcaloudis, a student who fell to her death from a cliff in Norway, came forward to clear their daughter's name after a similar internet mob. '*It was an accident. She was nowhere near the end of the rock, and not taking a selfie*' her mother Milli Kafcaloudis said.... Data journalism site Priceonomics estimated in January that, of the 49 people who have reportedly died while taking a selfie since 2014, '*not a single death was caused by the selfie itself*'.

Of the 49 reported "**death selfies**," "not a single death was caused by the selfie itself": wow dangerous is the combination between our biological tendencies to look for causal connections and for transcendental **karma** or the **punishment of sins** and for a *purpose of life and death*, and the current **sensationalism of the media** and the related fast public consumption and network dissemination of **fake news**, even when it involves wrongly blaming others for their own tragic death in order to give meaningful "cosmic purposes" to our lives.

1.2 Notes on Interdisciplinarity

History makes little sense without prehistory, and prehistory makes little sense without biology...knowledge of prehistory and biology is increasing rapidly, bringing into focus how humanity originated and why a species like our own exists on the planet. (Edward Wilson)

When I discuss the issues covered in this book with friends and colleagues, they often say: "you seem to be moving from the natural sciences to philosophy." However, while some of the issues are indeed frequently discussed by philosophers, I answer them—and I want to make this very clear here—that this book is not at all a philosophical essay. This point leads us to a note about *interdisciplinarity*: this book includes data and discussions about topics that refer to areas as diverse as biological anthropology, cultural anthropology, history and philosophy of sciences, neurobiology, philosophy, genetics, behavioral sciences, theology, psychology, sociology, social psychology, and evolutionary biology, among many others. And this is precisely because I want to discuss broader issues that have fascinated humans from times immemorial *within the realm of empirical scientific data*, so at the maximum the type of discussions I am including in this book could be classified as "**experimental philosophy**" or "**scientific philosophy.**" Namely, this is an interdisciplinary work that focuses on human evolution, biology and diversity by using current ideas and paradigms of biological evolution to understand human history and specifically how we think and behave, including the fascinating fact that what we "want" or "desire" is often very different from what we "do" in society. Somewhat strangely, traditionally there has been a huge disconnect between the study of biological evolution and of human history, as if humans, and particularly our mental capacities, were somehow not part—or, accordingly to longstanding and still prevailing teleological narratives, "above those"—of the natural world. Remarkably, this tendency is also seen among scientists, and even among many evolutionary biologists. As noted in Peterson's 2001 book *Being Human*, the evolutionist **Alfred Russel Wallace**, a contemporary of Darwin, stated that "the body of man is indeed a biological structure, clearly descended from the apes, but his culture, which stems from his extraordinary and unique mind, is on a new, higher hierarchical level of its own; evolutionary biology has nothing to tell us about this higher level." As she further emphasized, "this idea still holds sway among many scholars in the social sciences and humanities." Some of the factors leading to this tendency were briefly summarized by Van Arsdale in a 2017 book chapter:

> Today, it is unusual to find a researcher whose specialization is the behavioral and morphological evolution of humanity in a biology department (human genetics as a focus in molecular biology being a notable exception). Instead, researchers who focus on human evolution are more often found in departments of anthropology, anatomy divisions of medical schools, or more recent incarnations such as departments of human evolutionary biology. **Stephen Jay Gould**'s voluminous *The Structure of Evolutionary Theory* (2002), as one example of this trend, only makes passing reference to hominid evolution in its more than 1300 pages. The drift of human evolutionary studies away from mainstream evolutionary research, or vice versa, in the period after World War II is understandable. The revelations of the atrocities of science engineered under the National Socialist regime of Germany, especially those focused on human subjects, made public by the **Nuremberg Trials** were a watershed moment in twentieth-century human biology. Given the historical focus of anthropology on race, and the prominence of race-based perspectives on human evolution within **anthropology** prior to WWII, it is easy to understand the movement away from studies focused on humans in mainstream biology. The resistance to E. O. Wilson's *Sociobiology: The New Synthesis* (1975) from areas of the social sciences critical of any hint of biological determinism only furthered this trend. Humans are too complex to distinguish between genetic

and environmental ("cultural") effects. We live too long to look at trans-generational changes in allele frequencies. The data needed to study evolution for humans is too messy. Humans, quite simply, are not a good model organism for the study of evolution. Or so the logic went. Despite the sidelining of humans within evolutionary studies, humans remain a major focus of the public facing side of evolutionary studies. Major fossils relevant for human evolution are disproportionately represented on the covers of *Science* and *Nature*. Documentaries on evolution rarely bypass, and more often than not highlight as a central topic, human evolution. While often devoting entire spacious halls to narratives of human evolution, organisms like *Drosophila* [flies]...or even *Mus musculus* (mice)...rarely get the public coverage warranted by their importance within the scientific process itself.

On the more controversial side of things, it is the evolution of humans, rather than evolutionary theory more broadly, which often raises legal and political challenges to the teaching of evolution or public acceptance of evolution. Likewise, the acceptance of scientific knowledge itself, regardless of its evolutionary content, often is strongly correlated with one's understanding of the application of evolution to humans. In the time period that human evolution has drifted away from the center of evolutionary studies, traditional biological sciences have also been involved in critical self-examination of its foundational framework. These debates encompass a broad range of topics and developments within the fields of evolutionary studies but can be summarized as discontentment (or a lack of discontentment) with the traditional gene-centric view of the **Modern Synthesis**. For some researchers, the major developments within evolutionary theory over the past 80 years – neutral theory, renewed engagement with evolutionary perspectives on development, epigenetics and complex genomic structure, and hierarchically structured plasticity – have shifted the main focus of evolutionary causation away from natural selection and raised questions as to whether the traditional neo-Darwinian framework remains the best approach to understanding and presenting the action of evolution. These are not, it should be pointed out, arguments that "evolution is wrong" but instead are arguments about where the focus is placed on the processes of evolutionary change through time.

This issue was also discussed by one of the most prominent biologists in the last decades, **Edward Wilson**—who was cited in the above excerpt and wrote, in his 2014 book *The Meaning of Human Existence*:

Studying the relation between science and the humanities should be at the heart of liberal education everywhere, for students of science and the humanities alike. That's not going to be easy to achieve, of course. Among the fiefdoms of academia and punditry there exists a great variation in acceptable ideology and procedure. Western intellectual life is ruled by hard-core specialists. At Harvard University, for example, where I taught for four decades, the dominant criterion in the selection of new faculty was preeminence or the promise of preeminence in a *specialty*. Starting with the deliberations of department-level search committees, then recommendations to the dean of the faculty of arts and sciences, and at last the final decision by the president of Harvard, who was assisted by an ad hoc committee drawn from both within and outside the university, the pivotal question asked was, 'Is the candidate the best in the world in his research specialty'? The guiding philosophy overall was that the assembly of a sufficient number of such world-class *specialists* would somehow coalesce into an intellectual superorganism attractive to both students and financial backers. [However] the early stages of a creative thought, the ones that count, do not arise from jigsaw puzzles of specialization. The most successful scientist thinks like a poet – wide-ranging... – and works like a bookkeeper. It is the latter role that the world sees.... Science and technology reveal with increasing precision the place of humanity, here on Earth and beyond in the cosmos as a whole...[but the specialists and/or those within humanities] don't even pose the question in a manner that can be answered. Confined to a small box of awareness, they celebrate the tiny segments of the continua they know, in minute detail and over and over again in endless permutations. These segments alone do not address the origins of the traits we fundamentally possess – our overbearing instincts, our moderate intelligence, our dangerously limited wisdom, even, critics will insist, the hubris of our science.

There is indeed a major problem created by the disengagement between what we now know about biological evolution and the way human evolution and history is portrayed not only by many social scientists but also by numerous authors from areas such as biological anthropology, evolutionary psychology, and evolutionary medicine. The use of outdated evolutionary ideas to discuss the evolution of our lineage is particularly prominent within evolutionary psychology and evolutionary medicine, which include a substantial portion of scholars that are among the most extreme **adaptationists** and/or gene-centered **Neo-Darwinists** (see Chap. 6). It is therefore very important, and imperative, to use sound empirical evolutionary data—instead of **just-so-scientific-stories** or **philosophical theoretical speculations**—to undertake a re-examination of the evolution of our lineage, and of our beliefs, sexuality, racism, misogyny, and other prejudices, and our tendency to create and believe in fictional and often highly irrational narratives and to seek for a "cosmic purpose of life." As explained in Smith's 2016 paper on Freud and his "**just-so-stories**," this term refers to the fairytale-like creations of **Kipling's *Just so stories for little children***: they are mostly unfalsifiable ad hoc stories based on little or no empirical evidence.

I have nothing against non-empirical and non-evolutionary philosophical works; they were actually often the ones I most liked to read at school because they often engage us in fascinating profound reflections, with their theoretical case studies and thought experiments. Accordingly, since then I have read numerous philosophy books, particularly when I was writing the present monograph. However, I do have to admit that a few of them, particularly those including *exclusively* non-empirical-based discussions, can feel a bit vague, and even empty. Sometimes, their authors act as if historians and scientists have not gathered any new empirical data since the epoch of Socrates and Plato or, in a better scenario, since the epoch of Kant and Nietzsche: yes, we can and often should refer to those authors and their ideas—as I do in this book—or use ***thought experiments***, but why should one completely disregard the results of several real scientific experiments that directly address the issues being discussed as, let's say, near-death experiences, or the loss or gain of consciousness in studies involving mice or humans? One can understand that the ideas of **Socrates** were not rooted on an extensive, interdisciplinary review of empirical data because in his epoch the knowledge about the origin of the planets and stars, or about human evolution and our closest living relatives—the apes—or about human development from embryos to adulthood, or about consciousness, was extremely scarce. However, when philosophers nowadays discuss such topics it does seem rather odd if they don't include in their discussions the empirical data accumulated since Socrates was alive, such as information about the Big Ban, the age of our sun and our planet earth, and so on. Of course, this is *not* a criticism to philosophy per se, nor to all current philosophers, well on the contrary: some current philosophers provide admirable examples of interdisciplinary, in which empirical data from various fields are used in their reflections, as is the case with those I met, for example, in a meeting organized by the Philosophy Department of the

University of Lisbon about Human Enhancement some years ago. The meeting included researchers from natural sciences, such as biologists—that is why I was invited to be there and give a talk—and philosophers and bioethicists, which included detailed information, in their talks, about genome editing tools, the evolution of the human genome and behavior, and so on.

I will therefore provide here just an example to be contrasted with the type of cases I encountered in that meeting, that is, an example of "non-empirical-based philosophical discussions": Sehon's 2005 book *Teleological Realism – Mind, Agency and Explanation.* It is important to note that my aim is *not* to use here a "straw man" or criticize a specific book, because Sehon is otherwise an excellent scholar, in my opinion. In that book he discusses the "mystery" of how, seemingly paradoxically, **materialism** tries to explain natural organisms using a **reductionist approach** by merely using physical language, but then when one refers to organisms such as humans, notions like *purpose* and *morality* "appear to have no role in purely physical descriptions of the world." As he notes, "we would never say that an asteroid was morally responsible for its motion, even if it crashed into earth." I completely agree with this point, and this is precisely an example of how there is *no* cosmic purpose or meaning *of* life: the asteroid just crashed into earth, without any designed goal or cosmic purpose, there is no "masterplan." In the *reality* of the natural world "purpose" can only be factually applied within the term "**purpose *in* life**," and this obviously only applies to organisms that have the intellectual capacities to elaborate ***conscious purposeful behaviors***, such as humans, primates, and many other animals (see Box 1.1). Sehon notes that many authors explain this seemingly paradoxical dichotomy—how can some organisms have a purpose *in* life within an unpurposeful world—by recurring to **dualism**. In other words, they argue that there is both a material world—including our bodies—and nonphysical *souls*. Sehon does not accept dualism—rightly so, as this view has been shown to be factually inaccurate (see Box 1.4)—nor materialism, defending instead what he defines as a "third option": **teleological realism**. Namely, although he agrees "that human beings are composed of physical particles," he claims that "the facts about the mind are not ultimately a species of physical fact" and that "they are not going to be subsumed within physical science." He defines teleological realism as a version of "**weak naturalism**," as occupying a middle ground between supernaturalism and strong naturalism, because the latter defends that natural sciences are able to completely explain the existence of a purpose *in* life.

The problem illustrated with the case study provided by Sehon's book is that throughout the whole book there are almost no references to any type of empirical scientific data: it mainly includes theoretical philosophical discussions and speculations. However, when an author such as Sehon discusses topics such as consciousness, or the existence of "souls," he should at least refer to at least some available empirical data obtained in the last decades in areas such as neurobiology and **systems biology** that *do* show that natural sciences are actually fully able to explain the existence of purposeful organisms and of consciousness, through a natural evolutionary phenomenon known as **emergence** (see Box 1.1).

Box 1.1: Behavior, Behavioral Choices, Intentionality, and Emergence
This box is mainly extracted from my 2017 book *Evolution Driven by Organismal Behavior*: when I use the term **behavior** I am referring to a very simple, and broad, definition often seen in the literature – a response of an organism to stimuli or inputs, whether they are conscious or unconscious. Accordingly, **behavioral choices** refer to cases in which at least more than one potential choice is possible. These are considered behavioral choices no matter if in organisms such as bacteria they are likely often unconscious while in organisms such as humans and chimpanzees they are usually conscious. If we think about a bird in the air, it feels the effect of the same force of gravity than an object does, but there are many possible outcomes, which are thus behavioral choices: the bird can let itself passively go down, towards the center of the earth, but it can instead fly to counterbalance the force of gravity, staying at about the same altitude, or decide to even fly to a higher altitude, and so on. In this sense, behavioral choices are always undertaken by the organism as a whole, i.e., they are **organismal behaviors**. A crucial concept is thus **emergence**, in which the organism can display a behavioral choice as a single unit, no matter whether it has a central nervous system or any type of consciousness. The dichotomy between organismal behavioral choices versus other types of behaviors can thus match the dichotomy about having or not "intentionality," but only if "intentionality" refers to the drive that the whole organism has to undertake certain behavioral choices, and not necessarily to consciousness nor to any teleological concept related to "evolutionary purpose" or "evolutionary goal" or "cosmic purpose of life." That is, in the present monograph such "intentionality" refers exclusively to the purposeful actions of organisms that are in the realm of a *purpose in life*, not of a *cosmic purpose of life*.

Therefore, as put by Lindholm in a 2015 paper, behavioral choices cannot be reduced to genetics – or, I would add, to mere automatic, physiological, and/or localized epigenetic reactions to external stimuli or other factors – because they require a *subject* to take *choices* and have the drive to undertake them, which is the whole organism. This capacity and drive to undertake behavioral choices obviously depends on intrinsic genetic/genomic and epigenetic (for instance, hormonal/physiological) features linked with external factors, but is ultimately indeed mainly related to a phenomenon that is now becoming more and more prominent in biology, particularly due to the rise of **systems biology**: **emergence**. That is, a strikingly high number of complex factors, intrinsic and extrinsic, including the complex network connections made by the parts (for instance, neuronal networks) are combined in a way in which the overall outcome is *more* than just the sum of the part: the capacity to take a behavioral choice, and having the drive to undertake it. Contrary to mechanistic, reductionist, and atomistic views that have prevailed for a long time in the history of biology, this capacity does not apply to any of the organismal subunits or regional parts/organs – for example, individual atoms, or electrons, or neurons do not walk bipedally as we do, nor can they choose to do so. This capacity only applies to the *whole organism*, thence the term "**organismal behavior**." Within the context of the present monograph, purposeful actions of living organisms refer exclusively to those instances of emergent organismal behavior in which those organisms display **intentionality**, that is, to cases of a purpose *in* life. Accordingly, such purposeful actions and such intentionality can never apply to non-living objects, or to supernatural agents, or to any other entity that is not a living organism with the capacity to undertake – through emergence – purposeful thoughts and/or actions.

I also want to stress that the present book is *not*—and surely it was never my intention for it to be—a nihilist manifesto. I am saying this because you could have this idea, as some of the points that I already made about the inexistence of a cosmic purpose or meaning of life do match some key aspects of **nihilism**, which is a philosophical doctrine that, when presented in the form of **existential nihilism**, defends that life has no objective cosmic meaning or purpose. Or, that, when presented in the form of **moral nihilism**, argues that there is no **intrinsic human morality**, as explained in Stevens' 2016 book *Nihilism – A Philosophy Based on Nothingness and Eternity*. I do empirically show that the features that we tend to accept as **moral values** are abstractly contrived and are in no way **universal moral truths**: they are just dynamic social constructions that often differ depending on the time and geographical region or culture to which they refer to. However, nihilism often takes also other forms—metaphysical, epistemological, ontological, and so on—that stand for the idea that in some way reality does not exist at all, or that no form of knowledge is truly achievable, what is exactly the opposite of what I defend in all my books and other publications, which strictly follow the root of **scientific empiricism**. The earth moving around the sun is a physical reality, at least it seems to be in the sense that it is supported by a huge amount of empirical data and was never contradicted so far by any scientific empirical study.

Moreover, nihilists often—but of course not always—express an unease or despair related to the awareness that there is no higher, transcendental meaning or purpose of life, and some authors therefore consider that **postmodernity** is in a way a reflection of nihilism. Stevens' 2016 book is, in fact, an illustrative example of this attitude: in some ways it does have that "feeling" of despair and negativity that is exactly the opposite of what I show in the present book about the *reality of the natural world*. This planet and the millions of species that live and have lived in it are fascinating, despite the fact—or better say, chiefly because—there is no cosmic purpose of life nor cosmic progress: all of them, including our own species, are just the result of an aimless, mesmerizing combination of mainly chaotic, random, and contingent phenomena. *Things just happen, life is just as it is*, and that is what makes it particularly fascinating.

It is interesting that, within all the books and papers I read in order to write the present volume, there was not even a single one that specifically, directly related the darkest events of human history with the human tendency to formulate ***teleological why-questions*** and build imaginary narratives to answer them. An illustrative example is the superb book *Behave – The Biology of Humans at Our Best and Worst*, written by **Robert Sapolsky**. This is one of the most outstanding and integrative books I have ever read, full of case studies from a wide range of scientific disciplines to exemplify how complex and often incongruent human behaviors are, particularly concerning subjects such as aggression, hate, discrimination, and war. Still, the word teleology is almost never used in that 790-pages book, including its whole index, despite the fact that teleological narratives are profoundly connected to the subjects discussed in it, such as the "**nature *versus* nurture**" debate and the related question on whether humans are "naturally good or bad." To put those questions in context, I should explain that they are linked to the ideas of **Hobbes**—in a very simplistic way, he defended that "in a state of nature, life is solitary, poor, nasty brutish and short" and that a strong government is thus crucial to impose law and order—and of **Rousseau**, who, also in a very crude way, defended that humans are "born free" and are mostly "good"—the so-called **Noble Savage**—and that "everywhere they are in chains" because of things such as strong governments. The reality of what these two authors truly defended is obviously much more complex, as we will see.

Within this introductory chapter, it suffices to say that there is often a gap between what we define as "nature"—for instance, there is no "cosmic purpose" in the natural cosmos—versus "nurture"—for example, most organized religions desperately try to find, describe, or use the notion of a "cosmic meaning of life" in their own way. But the division clearly cannot be absolute, because humans are highly social animals, and moreover they tend to construct such imaginary narratives because their evolution leads to a natural propensity for them to do so, as explained in the 2018 book *Why We Disagree About Human Nature*, edited by Elizabeth Hannon and Tim Lewens. So, the typical "ideal" study of what humans "truly are naturally" that many talk about – that is, having a child completely alone in an island to then check if he would be "naturally" violent or not, or monogamous or not, or religious or not – would not only be unethical but also completely flawed scientifically. This is because such a study would force humans to lose any type of social interaction, for instance we could not check if that person would be "naturally monogamous or not" because there would be nobody to copulate with anyway—see also Boxes 1.2 and 1.3 about studies of deaf-mute children and of "feral" children. Similarly, the typical divisions between "**heart**" **versus** "**brain**" or "**body**" **versus** "**mind**," or "**emotion**" **versus** "**reason**," which are often discussed together with the "nature *versus* nurture" debate, are also clearly *Neverland* constructions that don't correspond at all to the reality of the natural world. Both the heart and the brain are simply internal body organs, and of course without a human body there is no human mind or "soul." And, as noted by **Damasio** in his elegant 1994 book *Descartes' Error – Emotion, Reason and the Human Brain*, many scientific case studies, including about people with major brain injuries, have shown that so-called emotions often play a crucial role in decision-making and in what we often call "rationality," contrary to **Descartes**' dualistic separation of mind and body, and emotion and rationality. In fact, erroneous dualistic ideas have been around much before Descartes, since thousands of years ago, being for instance defended by **Plato**, as pointed out in Malik's 2014 book *The Quest for a Moral Compass*:

> In Plato's eyes…the appetitive part of the soul is linked to bodily desires, such as the yearning for food or pleasure. The spirited is concerned with honour, and with anger and indignation. The rational is driven by a desire for knowledge and truth. This division, especially between the appetites, or bodily desires, and reason, or the mind, was to exert enormous pressure upon subsequent ethical thinking. For Plato, and for many of those who followed in his footsteps, reason and desire, the body and the mind, the ego and the id, were locked in mortal combat. Humans, according to Plato, fall into one of three categories depending on which part of their soul is dominant, three categories that correspond, of course, to the three social roles necessary for the healthy functioning of the state. The common people are driven by base desires, soldiers by a yearning for honour, while rulers look to reason. Upbringing may help an individual regulate his soul and thereby change the group to which he should belong. Mostly, though, it is a matter of birth – we are born to be blacksmiths or soldiers or philosopher kings.

Box 1.2: Deaf-Mute Children, Animism, and "Nature Versus Nature"

Bering, in his 2011 book *The Belief Instinct*, argues that his own empirical studies contradict the idea, defended by many authors, that kids born isolated in an secluded island would probably have some kind of belief. However, the kind of evidence provided in his studies does not encompass at all the very different aspects of belief: they focus more on aspects such as the **theory of mind**, which is seemingly related to some kind of beliefs, for instance about **supernatural beings**, but not necessarily with others, such as the typical **animistic ideas** that are over and over reported in **deaf-mute children**, as well as in most kids from a very young age. Anyway, some parts of Bering's book are very interesting and relate directly with issues discussed in the present book:

> Scientists would be hard-pressed to find and interview feral children who've been reared in a cultural vacuum to probe for aspects of quasi-religious thinking. In reality, the closest we may ever get to conducting this type of thought experiment is to study the few accounts of deaf-mutes who, allegedly at least, spontaneously invented their own cosmologies during their **prelinguistic childhoods**. In his book *The child's religion* (1928), the Swiss educator Pierre Bovet recounted that even Helen Keller, who went deaf and blind at nineteen months of age from an undiagnosed illness, was said to have instinctively asked herself, 'who made the sky, the sea, everything?' Such rare accounts of deaf-mute children pontificating about Creation through some sort of internal monologue of **nonverbal thought** – thought far removed from any known cultural iterations or socially communicated tales of Genesis – are useful to us because they represent the unadulterated mind at work on the problem of origins. If we take these accounts at face value, the basic existential problem of reasoning about our purpose and origins would appear not to be the mental poison of religion, society, or education, but rather an insuppressible eruption of our innate human minds. We're preoccupied with why things are. Unlike most people, these deaf-mute children – most of whom grew up before the invention of a standardized symbolic communication system of gestures, such as **American Sign Language (ASL)** – had no access to the typical explanatory balms of science and religion in calming these bothersome riddles. Without language, one can't easily share the idea of a purposeful, monotheistic God with a naive child. And the theory of natural selection is dif-

ficult enough to convey to a normal speaking and hearing child, let alone one who can do neither. These special children were therefore left to their own devices in making sense of how the world came to be and, more intriguingly, in weaving their own existence into the narrative fabric of this grand cosmology.

In an 1892 issue of *The Philosophical Review*, William James, brother to the novelist Henry James and himself arguably the world's most famous psychologist of his era, penned an introduction to the autobiographical account of one such deaf-mute, Theophilus Hope d'Estrella. Born in 1851 in San Francisco to a French-Swiss father he never met and a Mexican mother who died when he was five years old, D'Estrella grew up as an orphan raised by his mother's short-tempered best friend – another Mexican woman who, judging by her fondness for whipping him over the slightest misdeeds, apparently felt burdened by his frustratingly incommunicative presence. With no one to talk to otherwise, and only wordless observations and inborn powers of discernment to guide his naive theories of the world, D'Estrella retreated into his own imagination to make sense of what must have been a very confusing existential situation. For example, he developed an animistic theory of the moon that hints at the egocentric nature of children's minds, particularly with respect to morality: *"He wondered why the moon appeared so regularly…so he thought that she must have come out to see him alone…then he talked to her in gestures, and fancied that he saw her smile or frown…[he] found out that he had been whipped oftener when the moon was visible…it was as though she were watching him and telling his guardian (he being an orphan boy) all about his bad capers"*….

What type of mind does it take to be superstitious, and how can one investigate this in the laboratory? In the summer of 2005, my University of Arkansas colleague Becky Parker and I began the first study ever to investigate the psychology underlying the human capacity to see messages – signs or omens – in unexpected natural events. We knew that theory of mind was involved, because again such a capacity requires sleuthing out the mental reasons for the supernatural agent to have acted in such a manner. But because previous research had shown that a fully developed **theory of mind** does not appear in children's thinking until about four years of age (before this, children still mind-read, but they're just not as good at taking the perspectives of others and they tend to make frequent egocentric errors), we suspected there might be subtle, age-related differences in children's ability to engage in the divination of everyday events. In these initial experiments, which have come to be known among my students as the '**Princess Alice studies**', we invited a group of three- to nine-year-old children into our lab and told them they were about to play a fun guessing game. It was a simple game in which each child was tested individually. The child was asked to go to the corner of the room and to cover his or her eyes before coming back and guessing which of two large boxes contained a hidden ball. All the child had to do was place a hand on the box that he or she believed contained the ball. A short time was allowed for the decision to be made but, importantly, during that time the children were allowed to change their mind at any time by moving their hand to the other box. The final answer on each of the four trials was reflected simply by where the child's hand was when the experimenter said, 'Time's up!' Children who guessed right won a sticker prize. In reality, the game was a little more complicated than this.

There were secretly two balls, one in each box, and we had decided in advance whether the children were going to get it 'right' or 'wrong' on each of the four guessing trials. At the conclusion of each trial, the child was shown the contents of only one of the boxes. The other box remained closed. For example, for 'wrong' guesses, only the unselected box was opened, and the child was told to look inside ('Aw, too bad. The ball was in the other box this time. See?'). Children who had been randomly assigned to the control condition were told that they had been successful on a random two of the four trials. Children assigned to the experimental condition received some additional information before starting the game. These children were told that there was a friendly magic princess in the room, 'Princess Alice', who had made herself invisible. We showed them a picture of Princess Alice hanging against the door inside the room (an image that looked remarkably like Barbie), and we gave them the following information: 'Princess Alice really likes you, and she's going to help you play this game. She's going to tell you, somehow, when you pick the wrong box.' We repeated this information right before each of the four trials, in case the children had forgotten. For every child in the study, whether assigned to the standard control condition ('No Princess Alice') or to the experimental condition ('Princess Alice'), we engineered the room such that a spontaneous and unexpected event would occur just as the child placed a hand on one of the boxes. For example, in one case, the picture of Princess Alice came crashing to the floor as soon as the child made a decision, and in another case a table lamp flickered on and off.

The predictions were clear: if the children in the experimental condition interpreted the picture falling and the light flashing as a sign from Princess Alice that they had chosen the wrong box, they would move their hand to the other box. What we found was rather surprising, even to us. Only the oldest children, the seven- to nine-year-olds, from the experimental (Princess Alice) condition, moved their hands to the other box in response to the unexpected events. By contrast, their same-aged peers from the control condition failed to move their hands. This finding told us that the explicit concept of a specific supernatural agent – likely acquired from and reinforced by cultural sources – is needed for people to see communicative messages in natural events. In other words, children, at least, don't automatically infer meaning in natural events without first being primed somehow with the idea of an identifiable supernatural agent such as Princess Alice (or God, one's dead mother, or perhaps a member of Doreen Virtue's variegated flock of angels). More curious, though, was the fact that the slightly younger children in the study, even those who had been told about Princess Alice, apparently failed to see any communicative message in the light-flashing or picture-falling events. These children kept their hands just where they were. When we asked them later why these things happened, these five- and six-year-olds said that Princess Alice had caused them, but they saw her as simply an eccentric, invisible woman running around the room knocking pictures off the wall and causing the lights to flicker. To them, Princess Alice was like a mischievous poltergeist with attention deficit disorder: she did things because she wanted to, and that's that. One of these children answered that Princess Alice had knocked the picture off the wall because she thought it looked better on the ground. In other words, they completely failed to see her 'behavior' as having any meaningful connection with the decision they had just made on the guessing game; they saw no 'signs' there. The youngest children in the study, the three- and four-year-olds in both conditions, only shrugged their shoulders or gave physical explanations for the events, such as the picture not being sticky enough to stay on the wall or the light being broken. Ironically, these youngest children were actually the most scientific of the bunch, perhaps because they interpreted 'invisible' to mean simply 'not present in the room' rather than 'transparent'.

Contrary to the common assumption that superstitious beliefs represent a childish mode of sloppy and undeveloped thinking, therefore, the ability to be superstitious actually demands some mental sophistication. At the very least, it's an acquired cognitive skill. Still, the real puzzle to our findings was to be found in the reactions of the five- and six-year-olds from the Princess Alice condition. Clearly they possessed the same understanding of invisibility as did the older children, because they also believed Princess Alice caused these spooky things to happen in the lab. Yet although we reminded these children repeatedly that Princess Alice would tell them, somehow, if they chose the wrong box, they failed to put two and two together. So what is the critical change between the ages of about six and seven that allows older children to perceive natural events as being communicative messages about their own behaviors (in this case, their choice of box) rather than simply the capricious, arbitrary actions of some invisible or otherwise supernatural entity? The answer probably lies in the maturation of children's theory-of-mind abilities in this critical period of brain development. Research by University of Salzburg psychologist Josef Perner, for instance, has revealed that it's not until about the age of seven that children are first able to reason about 'multiple orders' of mental states. This is the type of everyday, grown-up social cognition whereby theory of mind becomes effortlessly layered in complex, soap opera–style interactions with other people. Not only do we reason about what's going on inside someone else's head, but we also reason about what other people are reasoning is happening inside still other people's heads!

Box 1.3: Human Smiling, Animal Laughter, Feral Children, and Social Isolation

In Hood's superb 2013 book *The Self Illusion – How the Social Brain Creates Identity,* he explains how both "nature" and "nurture" are crucial to create a story of our *self*, that is, *the constructed narrative that our brain creates* depends obviously on the biological attributes of the brain and its development *and* on the environment and context in which that development takes place (see also Boxes 9.2, 9.4–9.6 and Fig. 9.14). About racism, he notes that "unlike most adults who think members of other ethnic groups look very similar, babies initially have no problem…they can tell everyone apart…it is only after exposure to lots of faces from the same race that our **discrimination** kicks in." Therefore, "you can train babies not to become tuned into their own race if you keep exposing them to faces from other races…so the next time you think that other races all look alike, don't worry, it isn't **racism** – it's your lack of brain plasticity." Similarly, notions of **happiness**, often associated with smiling and laughter, also show a very interesting mix between what some people call "natural" behaviors, as seen in human fetuses and other animals, and the social context surrounding us since we are born. He explains:

> We tend to only smile when there are others around. In one study, players in a tenpin bowling alley were found to smile only 4% of the time after a good score if they were facing away from their friends but this increased to 42% when they turned round to face them, indicating that this expression is primarily a signal to others. Smiling is linked to the development of the brain regions that support social behaviour, which are located towards the front of the brain in a cortical area known as the orbital cortex because it sits over the orbits of the eye sockets. Although smiling has been observed using ultrasound in unborn babies, indicating that it is a **hard-wired behaviour**, at around two months it operates in combination with the higher order centres of the brain that are recruited for social interaction. At two months, the baby is already using a smile to control others. The built-in capacity for smiling is proven by the remarkable observation that babies who are congenitally both deaf and blind, who have never seen a human face, also start to smile around two months. However, smiling in blind babies eventually disappears if nothing is done to reinforce it.

Without the right feedback smiling dies out, just like the following instinct does in goslings. But here's a fascinating fact: blind babies will continue to smile if they are cuddled, bounced, nudged and tickled by the adult – anything to let them know that they are not alone and that someone cares about them. This social feedback encourages the baby to continue smiling. In this way, early experience operates with our biology to establish social behaviours. **Animal laughter** has been a controversial claim. Until fairly recently, laughter was considered uniquely human. However, most human behaviours have evolved and so we should not be too surprised to find primitive versions in other species.

Puppies and kittens seem to engage in behaviour that has no obvious rewards other than the joy of play. Initially it was argued that these behaviours were precursors to adult aggression – a means of developing survival skills for hunting. Even the interpretation of animal behaviour was misguided. For example, chimpanzees who bare their teeth in a smile are generally regarded as displaying a threat or fear response. However, animal laughter during play had to be rethought when Jaak Panksepp made an amazing discovery with rats. First, he noticed that rats that had been deafened for experiments on hearing did not engage in as much rough-and-tumble play as normal rats. There was something missing in these **deaf rats**. It turns out that it was the squeals of delight. When Panksepp placed a sensitive microphone in the cage that makes high-frequency sound audible to human hearing, they discovered a cacophony of 50 kHz chirping during the play sessions – the rat equivalent of laughing. He soon discovered that rats were also ticklish and would chase the experimenter's hand until they were tickled. Apparently, rats are most ticklish at the nape of the neck. They would play chase with the hand and all the other familiar baby tickling games like 'coochie-coo'. Baby pup rats laughed the most, and as the play activity declined with age so did the laughing….

Feral children have been sparking the imagination of intellectuals interested in nature and nurture for centuries. In one of the better-documented cases from the 1970s, psychologists studied 'Genie', a fourteen-year-old girl who had been kept in social isolation from infancy in the backroom of her psychotic grandfather's condo in Los Angeles…she had limited communication and understanding, despite the concerted attempts of speech therapists and child psychologists to rehabilitate her. The case of Genie has been used as evidence to support the critical period of social development, but without knowing the initial state of these children, it is still difficult to draw firm conclusions. Maybe they were abandoned because they were already brain-damaged. We also do not know whether and to what extent early malnourishment of feral children contributes to potential brain damage. Maybe it was not the lack of social interaction so much as the damaging consequences of not being cared for by others who provide the necessary nutrition to develop normally. However, the fall of a Romanian dictator in 1989 would reveal that both physical and psychological nurturing is essential for long-term social development. When the plight of the orphans came to light, the world descended on Romania to rescue these children. Families determined to give them a better start in life brought around 300 **orphans** to the United Kingdom. In the United States, Nelson and his colleagues studied 136 of them. How would they fare? British psychiatrist Sir Michael Rutter led a team that would study 111 of these children who were less than two years of age when they first came to the UK. There were no medical records for these orphans and there is always the problem of knowing if an individual child suffered from congenital disorders, but the research revealed some amazingly consistent findings. When they arrived, the orphans were **mentally retarded** and physically stunted with significantly smaller heads than normal children. However, by four years of age, most of this impairment had gone. Their IQs were below the average for other four-year-olds, but within the normal range that could be expected. These children seemed to be largely rehabilitated. Some had done much better than others.

Orphans who were younger than six months of age when they arrived were indistinguishable from other normal British children of the same age. They made a full recovery. Their window of opportunity had not yet closed when they arrived in the UK. The longer they had been in the orphanage after six months of age, the more impaired their recovery was despite the best efforts of their adopted families. The orphans were followed up again at six, eleven and fifteen years of age. Again as a group they fared much better than expected, given their poor start, but not all was well. Those who had spent the longest time in the orphanage were beginning to show disturbed behaviour with problems forming relationships and hyperactivity. Just as Bowlby and others had predicted, the absence of a normal social attachment during infancy had left a legacy of poor social attachment as an adult. Rutter concluded that infants younger than six months recovered fully from social deprivation, but older infants were increasingly at risk of later problems in life. While malnutrition played some role in their impaired development, it could not be the only reason. When they looked at the weight of babies when they entered the UK, this did not predict their development. Rather it was the amount of time that they had been socially isolated that played a greater role. Their ability to fit in socially had been irrevocably ruined by their isolation as infants.

The **Romanian orphans** responded similarly to the rhesus monkeys in Harry Harlow's infamous **isolation studies** during the 1960s. Harlow had been inspired by Bowlby's theory of why children raised in orphanages develop antisocial behaviour, but he wanted to rule out the alternative explanations that these were children from poorer backgrounds or that poor nutrition in the institutions had led to these effects. To test this, he raised infant rhesus monkeys in total social isolation for varying amounts of time (these studies would never be approved today now that we know how similar monkeys are to humans). Despite feeding them and keeping them warm, those monkeys that spent at least the first six months of life in total isolation developed abnormally. They compulsively rocked back and forth while biting themselves and found it difficult to interact with other monkeys. When they became mothers themselves, they ignored or sometimes attacked their own babies. The **social deprivation** they had experienced as infants had left them as socially retarded adults. If they were introduced to the rest of the monkeys before the six months was up, then they recovered more social behaviours. **Monkeys** that were only isolated after the first six months were not affected. Clearly monkeys and humans from birth require something more than sustenance. It isn't food and warmth they need, it is love – without the love of others, we are lost as individuals, unable to form the social behaviours that are so necessary to becoming a normal social animal. What is it about **social isolation** that is so destructive for the developing primate? There is no simple answer and one can speculate about different mechanisms. For example, babies who are born extremely prematurely can spend several weeks isolated in an incubator to provide a suitable breathing and sterile environment for their immature lungs. Not only are they born too early, but they are also very small and have a low birthweight. However, if you interact with them by stroking them and massaging them while they are still inside the incubator, this minimal contact significantly improves their physical development. They grow and put on weight much faster than premature babies left alone. The most likely explanation comes from animal studies that show that grooming and tactile contact stimulate the release of growth hormones in the brain. These growth hormones affect metabolism and the calorific uptake so that these little guys can absorb more from their food.

At a time in which there are many new interdisciplinary scientific journals, and in which multidisciplinarity is so in vogue within most funding institutions, one should think about the many ways in which we can truly be interdisciplinary. In fact, there are still too many cases in which, if a certain author tries to link research from different fields, the "specialists" of those fields attack her/him for not "knowing all the details" of and works published within a certain field. As it is of course impossible for any author in the twenty-first century to know everything published by all the scholars working in all fields of knowledge, such an attitude—which is, above all, both arrogant and defensive, a reflection of how many specialists work in a certain field "feel" that they are the only ones "entitled" to work on that field—does not help at all the efforts to enhance interdisciplinarity. I remember when I was doing my Bachelor at the University of Aveiro, and I explained to a biology professor that I choose to study biology mainly because I read the divulgation books of the late evolutionary biologist **Stephen Jay Gould**. My professor said that I "made a good choice of studying biology," but that Gould was in fact "not a true biologist and researcher, but *merely* a science disseminator." When I then read, various years later, the more specialized papers of Gould about snails, and in particular his last, *opus magnum* 2002 book *The Structure of Evolutionary Theory*, I realized how wrong my professor was. In fact, that 2002 book is written in such a dense, encyclopedic, detailed way that clearly seems a way of Gould—who knew this would be his last major work, before dying of cancer—proving wrong all those that accused him of being "just a disseminator."

This story does not illustrate a problem specifically about my biology professor, or about Gould, per se: it mainly concerns the scientific system as a whole, a system in which, contrary to the good intentions of the "*interdisciplinarity*" *motto* so much in vogue nowadays, researchers that are specialized on a small field of science, or on a very specific group of animals or objects, or a specific geological time, feel highly unease about researchers that are willing or able to try to link several fields and/or items of knowledge in order to comprehend the bigger picture and explain it to the broader public, such as Gould did for biology or **Hawkins** or **Sagan** did for physics. That is, a system that is, unfortunately, too often plagued by jealousy and hypocrisy: do as I say—be innovative, be interdisciplinary!—but not as I do. In fact, I am writing these lines just after I saw a documentary about the life of **Carl Sagan**, which clearly illustrates this point. Sagan was excellent at communicating with the broader public and discussing the validity, strengths, and wonders of science, particularly in his series *Cosmos*, which is one of the most watched TV shows in human history. However, instead of having the vast majority of ultra-specialized scientists being proud of this, as the broader public would know more, and be more interested, in science overall, and therefore in the specialized work they are doing as well, many of those scientists were moved by jealously and even hate towards Sagan. As explained in the documentary, many of them criticized him for "being too broad," as so many biologists did with Gould. Notably, after Sagan was already highly renowned for both his scientific merits—including his participation in various missions of the **National Aeronautics and Space Administration (NASA)**—and scientific dissemination, he was refused a tenure at **Harvard University**, by his colleagues and peers. Not only that, even much later, after doing still more top-quality scientific work, being a key player in many NASA missions, and being the most successful scientific disseminator in the globe, when Sagan was nominated for membership in the National Academy of Sciences, the nomination was refused. As clearly—and extremely sadly, for Sagan and for science in general—pointed out by his former student David Morrison, in an article about Sagan's life:

> Sagan was nominated for membership in the National Academy of Sciences. Academy membership requires distinguished research scholarship, but that is rarely sufficient to ensure membership. Considerable weight is also given to public service, as well as more political factors such as where a nominee works and whom he or she knows. Most colleagues agreed that Sagan's research record was more than adequate, and that his additional journal editorship, government service, and contributions to public understanding of science should have ensured his election. But Sagan was blackballed in the first voting round, requiring a full debate and vote by the Academy membership. In the final vote he barely received 50 percent yes votes, far short of the two-thirds majority required for election to membership. Two years later, the National Academy awarded Sagan its prestigious Public Welfare Medal, perhaps in partial compensation for his earlier rejection. The damage was done, however: not only a stinging personal blow, but also an attack on his credibility as a spokesperson for science. For all his accomplishments – or perhaps because of some of them – influential members of the academic "old boys" network never accepted him.

Box 1.4: Descartes, "Mind" Versus "Body," Immune System, and Mindfulness

In Davis's 2018 book *The Beautiful Cure,* he explains how scientific data gathered in the last decades has contradicted **Descartes**' ideas about the separation of "**mind**" and "body," and moreover also stresses the crucial power of "narratives," even concerning the fights against **disease**:

> As well as being one of the world's most important medicines, the discovery of cortisol opened up the molecular basis for how our mind and body are connected. 350 years after Descartes theorised the separation of mind and body, **cortisol** brought them together, showing how a mental experience – stress – results in physiological effects. Understanding the full implications of this connection between our mental state and our **immune system** is an especially fascinating but controversial subject of ongoing enquiry.... There is evidence that people who are stressed for prolonged periods of time suffer worse from viral infections, take longer to heal wounds, and respond less well to vaccination. All kinds of stresses have been linked with diminished immune responses, from burnout at work to unemployment. Even natural disasters like a hurricane can alter the state of people's immune system. Well over a hundred clinical studies have reported that stress can contribute to poor health, which leads many to suppose that a super-charged lifestyle perhaps increases our risk of all kinds of illnesses, from **autoimmune disease** to cancer. The topic remains controversial, however, because so many factors affect our ability to fight disease that it is difficult to assess the effect of any one.... In humans, elderly people stressed by caring for a spouse with dementia have a reduced response to a shot of flu vaccine. There is also evidence that stress can affect our response to HIV. Our immune system can keep the virus in check before eventually AIDS develops, but the length

of time it can do so varies between people. Over a five-and-a-half-year period of study, it was found that the probability of men infected with HIV developing AIDS increased two to three times if they had higher than average stress, or less social support. A separate study of homosexual men came to the conclusion that **AIDS** advanced more rapidly in men who conceal their sexuality, although the reasons for this were not established. Many other studies have found that stressed individuals are more prone to reactivation of herpes. Overall, the bad effect of stress on health is probably the best-established link between lifestyle and the immune system....

Practitioners of **t'ai chi**, developed as a **martial art** in China, or related exercises such as qigong, perform a slow meditative choreography of movements. There is good evidence that t'ai chi can help improve pain and physical mobility for elderly arthritis patients. Whether or not t'ai chi impacts the immune system, however, is controversial. In one study, a t'ai chi class taken for an hour three times a week led to elderly adults responding better to a flu virus vaccine. This is an interesting result but this type of research is often less definitive than it might seem at first. One problem is that studies such as this often involve only small numbers of people. In this particular study, only fifty people were tested – twenty-seven people who took t'ai chi classes were compared to twenty-three who didn't. Other studies testing for a link between practising t'ai chi and health test similarly small numbers of people. This is something like the number of people who would be enrolled in the first phase of a clinical trial for a new pharmaceutical drug, merely to test the safety of the drug, not whether it works. For a drug to be approved as a new medicine, it is usually tested in thousands of people, and compared to other interventions. A second problem is bias. In around half the trials testing the effects of t'ai chi on immune defense, it's not clear whether those who took the t'ai chi classes, as opposed to those who did not, were selected randomly. If those who took the classes had already been doing so before the study began, and were merely picked out as a group of people practising t'ai chi, then there is no way of knowing if the effects observed are owing to the classes themselves or to some other shared characteristic that also happens to result in people taking up t'ai chi. More subtly, the control group – those subjects who don't take the class – should be given another activity to perform to replicate the possible benefits of a t'ai chi class that are not actually to do with t'ai chi itself, such as the contact time with a social group.... What sets t'ai chi apart from (for instance) bouncing on a trampoline is that it provides not just a method of exercise but a narrative for health. There is a story to the movements of t'ai chi; practitioners talk about moving energy around the body to balance one's chi. The power of story is often part of a cure. It's why naming a condition is important, and why a physician's bedside manner, their description of an illness and how they intend to deal with it, can have such a major impact on how patients respond. This power of t'ai chi – the power of its narrative – is hard to quantify.

To take another example, there has recently been great interest in using **mindfulness**, a non-religious form of meditation, to improve health. Developed in 1979 by Jon Kabat-Zinn – the son of an immunologist – at the University of Massachusetts Medical School, mindfulness uses attention-focusing techniques to instil a moment-to-moment awareness. As Ruby Wax, comedian, writer and mindfulness practitioner, puts it: 'Mindfulness is a way of exercising your ability to pay attention: when you can bring focus to something, the critical thoughts quieten down'. A review of forty-seven trials testing a total of 3515 participants concluded that mindfulness can indeed ward off the negative effects of **stress**, **anxiety**, **depression** and pain. The effect is small but similar to what is often achieved with an antidepressant drug. In one clinical trial directly comparing mindfulness with **antidepressants**, both improved the well-being of patients with recurrent depression to a similar extent. As well as helping people cope with depression or anxiety, mindfulness is practised more widely as a way of dealing with everyday stress. To enthusiasts, mindfulness is the ideal antidote to the pre-eminent problem of our age: distraction. It might be assumed that, like t'ai chi, reducing stress by practising mindfulness could lower a person's cortisol levels and, in turn, boost the immune system. And in 2016, an analysis of twenty trials with a total of 1602 participants tested this exact idea. It was found that mindfulness could indeed lower some markers of inflammation and increase numbers of particular T cells in HIV-diagnosed individuals, but other measures – levels of **cytokines** or antibodies in blood, for example – were affected in some trials and unaffected in others. The authors concluded: 'we caution against exaggerating the positive effects of mindfulness meditation on immune system dynamics until these effects are further replicated and additional studies are performed'. In fact, it's not actually clear whether mindfulness impacts cortisol levels at all; different trials come to different conclusions. Unsatisfactory as it is, all we know is that mindfulness may help. One of the reasons we don't know for sure if t'ai chi or mindfulness can boost the immune system is that the cost of finding out is prohibitively high. In general, a clinical trial big enough that it would lead to FDA approval if the results were positive costs around $40 million. Eyeing up the possible profits, pharmaceutical companies are willing to pay such sums to test their novel compounds. But who would, should, or could pay to test an unpatented practice like t'ai chi?

We will never fully know why Sagan was blackballed, because the details about such meetings and decisions in academia are often not disclosed—for instance, they are not transparent at all in the sense that it is not reported who votes "yes" or "no" and for what reason, as it also happens in decisions about one of the most crucial moments in academia: receiving tenure, or not. In this sense, such a nontransparent process precisely gives a lot of power—by using their anonymity and therefore removing them for any responsibility in wrongdoing—to the ones that are often the most mediocre scientists or teachers, which are often, not surprisingly, the ones that dedicate more time to "corridor politics," "gossiping," and creating intrigues. These are obviously things that people as occupied and highly successful as Sagan or Gould did not have the time to do, and in general would *not* like spending their whole time doing, anyway. In fact, in that specific documentary about Sagan at least one person that was present at those meetings claimed—unofficially—that this decision was in part, or mainly, because Sagan had done "so much television work," which many ultra-specialized scientists predictably considered as something "fluff," including the series *Cosmos*. When asked about this particular case, Sagan famously answered that he was, indeed, not surprised at all. In summary, for many scientists, the fact that Sagan did *Cosmos*, one of the most successful series ever in terms of how it disseminated science to the broader public of so many regions of the globe, and literally leading a huge number of kids to be interested, and even wanting to do a career, in science, is something that should be "penalized," instead of gratified.

And moreover this does not apply only to the series *Cosmos*, because this was just part of the extraordinary work that Carl did as a science disseminator, including many books, among them one of the most passionate defenses of science ever written, his 1997 masterpiece *The Demon-Haunted World – Science as a Candle in the Dark*, which was also a best seller within the broader public.

It is important to point out that such "penalizations" were not always the prevailing attitude among scientists. The history of sciences shows us that there have been repeating cycles of holism and reductionism, and accordingly since the eighteenth century a huge number of scientists have embraced a more reductionist approach to science, as explained in Daston and Park's excellent 1998 book *Wonders and the Order of Nature, 1150–1750*. As they point out, European naturalists from the High Middle Ages through the **Enlightenment** often used wonder and wonders to envision themselves and the natural world: the history of wonders as objects of natural inquiry was simultaneously an *intellectual history* of the orders of nature. Of course, I am not defending a return to all those aspects that permeated such a scientific history of wonders, because as it will be seen in the present volume, many of those aspects were related with beliefs in imaginary narratives. As recognized by Daston and Park: "monsters, gems that shone in the dark, petrifying springs, celestial apparitions – these were the marvels that adorned romances, puzzled philosophers, lured collectors, and frightened the devout…the two sides of knowledge, objective order and subjective sensibility, were obverse and reverse of the same coin rather than opposed to one another." In fact, within the field of biology this continued to be a prevalent way of doing science for many authors until much later.

As it will be explained in the present volume, **German romanticism** and ***Naturphilosophie*** have influenced many prominent biologists up to the first decades of the twentieth century and continue to influence some ideas defended by some biologists nowadays. The type of reductionism so prevalent in Neo-Darwinism was in part precisely the result of a reaction—rightly so—against such a way of doing science. However, the problem was that such a reaction against a type of scientific methodology not entirely based on "objectivity" went too far, leading to the wide spread and acceptance, in the last decades of the twentieth century, of extreme reductionist ideas such as those defended by Dawkins about "selfish genes," which are also subjective, not testable and mostly factually wrong. As a famous saying goes, "if you go more and more to the east, you end up reaching the west." In other words, both extreme reductionist ideas such as Dawkins' "selfish genes" and extreme holistic ideas such as **Goethe**'s "romantic ideal archetypal plant" are not a reflection of the reality of the natural world: they are both plagued by narratives from *Neverland*'s world of unreality. Any type of extremism, in any aspect of life, including science, is almost always contraproductive because, using another old saying, "the truth is often somewhere in the middle."

Indeed, the reason why we now know that the notion of "**selfish genes**" does not correspond at all to true complexity of life is precisely because we are seeing again a move towards the "middle," that is towards a somewhat more holistic approach within the field of biology—as well as of many other fields of science. One of the most damaging legacies of the type of **reductionism** that was so prevalent in sciences in general since the eighteenth century and in particular in the second half of the twentieth century is unfortunately still very prevalent: the "*shut up and calculate*" type of mentality that was precisely used to attack so harshly Sagan and Gould. That is: focus on your specific subject, don't talk about other stuff and above all don't be too broad, as that is something restricted to "philosophers." This type of mentality was indeed not only damaging for fields such as biology and anthropology, but also for many other fields of science such as physics, and therefore for our understanding of the origin of life and of the cosmos as a whole. In Kaiser's 2011 book *How the Hippies Saved Physics – Science, Counterculture, and the Quantum Revival*—from which I coined the phrase "*shut up and calculate*"—it is explained:

> The woolly pursuits of the 1970s hearkened back to an earlier way of doing physics and of being a physicist. The roots of **quantum information science** stretch all the way back to the golden age of theoretical physics of the 1920s and 1930s, when giants like **Albert Einstein**, **Niels Bohr**, **Werner Heisenberg**, and **Erwin Schrödinger** cobbled quantum mechanics together. From their earliest wranglings, they found themselves tangled up with all sorts of strange, counterintuitive notions. Many have become well-known catchphrases like 'wave-particle duality', 'Heisenberg's uncertainty principle' and 'Schrödinger's cat'. Each signaled that atom-sized objects could behave fantastically different from what our usual experience would suggest. To Einstein, Bohr, and the rest, it seemed axiomatic that progress could only be made by tackling these *philosophical challenges* head on. Manipulating equations for their own sake would never be enough. That style of doing physics did not last long. The clouds of fascism gathered quickly across Europe, scattering a once-tight community. The ensuing war engulfed physicists around the world. Torn from their prewar routines and thrust into projects of immediate, worldly significance – radar, the atomic bomb, and dozens of lesser-known gadgets – physicists' day-to-day activities in 1945 bore little resemblance to those of 1925. Over the next quarter century, Cold War imperatives shaped not just who received grants to pursue this or that problem; they left an indelible mark on the world of ideas, on what counted as 'real' physics. Physicists in the United States adopted an aggressively pragmatic attitude. The equations of quantum mechanics had long since lost their novelty, even if their ultimate meaning still remained obscure. The pressing challenge became to put those equations to work. How much radiation would be emitted from a particular nuclear reaction? How would electric current flow through a transistor or a superconductor? As far as the postwar generation of physicists was concerned, their business was to *calculate*, not to daydream about philosophical chestnuts.
>
> Before the war, Einstein, Bohr, Heisenberg, and Schrödinger had held one model in mind for the aspiring physicist. A physicist should aim, above all, to be a Kulturträger – *a bearer of culture* – as comfortable reciting passages of Goethe's Faust from memory or admiring a Mozart sonata as jousting over the strange world of the quantum. The physicists who came of age during and after World War II crafted a rather different identity for themselves. Watching their mentors stride through the corridors of power, advising generals, lecturing politicians, and consulting for major industries, few sought to mimic the otherworldly, detached demeanor of the prewar days. Philosophical engagement with **quantum theory**, which had once seemed inseparable from working on quantum theory itself, rapidly fell out of fashion. Those few physicists who continued to wrestle with the seemingly outlandish features of quantum mechanics found their activity shoved ever more sharply to the margins. Before there could be a field like quantum information science a critical mass of researchers needed to embrace a different mode of doing physics once more. They had to incorporate philosophy, interpretation, even bald speculation back into their daily routine. **Quantum physicists** needed to daydream again.

Such a type of "shut up and calculate" mindset has contributed to the type of skepticism, suspicion, jealousy, and "corridor intrigues" against authors such as Sagan and Gould, and numerous others top-quality researchers. The combination of all these factors therefore made much more difficult not only the work of such disseminators in particular, but also any attempt to promote science within the broader public in general. This is particularly disturbing when we now see some political parties, TV chains, and so many people around the world displaying an increasing skepticism about science and scientists. Scientists do not seem to be winning the "war of ideas" in these particular sensitive times of "fake news" and "alternative facts," but still many of them continue to give themselves a shoot in the foot when they harshly criticize those few that actually have a main interest in, and are good at, promoting science to the broader public. On the other hand, the combination of the "shut up" mentality with scientific jealousy and "corridor intrigues" also makes much more difficult the type of interdisciplinary, synthetic scientific work that is so badly needed in science, particularly because we now have literally millions of specialized scientists producing data and writing papers within their specific fields: someone needs to try to make sense of all these data, to try to see the whole tree—as authors such as Sagan and Gould did so amazingly well—instead of just focusing on each of its numerous branches.

With respect to this, it is interesting to note that—precisely contradicting such a "shut up and calculate" type of narrative—numerous scientists that made key contributions to science, particularly in terms of changing the mindsets and leading to completely new avenues of research, including many Nobel Prize winners, moved from a scientific field to another, or approached several of them at the same type. This makes sense, because often when one becomes too specialized and focused on a single small area of research, one tends to think only "inside the box." If someone comes from another field, particularly those that have already a rich and more holistic scientific background and a tendency to try to put things together within a truly interdisciplinary context, it can therefore be easier to think "outside-the-box" within—and in particular to apply new ideas from other areas to—a certain field of science. This is therefore, in a way, also one of the goals of this book. That is, to apply an empiricist mindset and combine my background in natural sciences—I did a PhD at a biology department and later another PhD at an anthropology department—with an extensive overview of empirical data obtained from many other fields of knowledge from both natural and social sciences in order to discuss ideas that are traditionally more discussed in areas that are considered part of the humanities, such as history or philosophy.

Chapter 2
Death and Cosmic Purpose of Life

More than well-being and pleasure, pain, suffering and the perception of death are particularly empowering. I imagine that religions have developed around this conscience. To a certain extent, from a historical and evolutionary point of view, consciousness was a forbidden fruit that once tasted left us vulnerable to pain and suffering and, ultimately, exposed to the tragic confrontation with death.... Death as a source of tragedy is very present in biblical narratives and in Greek theater, continuing to impose itself on current artistic creations.... We are puppets in the hands of pain and pleasure, occasionally liberated by our creativity.

(Antonio Damasio)

2.1 Quests to Find a Cosmic Purpose of Life

Since we cannot change reality, let us change the eyes which see reality. (Nikos Kazantzakis)

The quest for a **cosmic meaning/purpose of life** has been a constant trend in humans since times immemorial. The oldest written epic tale—the fascinating ***Epic of Gilgamesh*** (Fig. 2.1), composed earlier than c. 2150 BC—specifically deals with **Gilgamesh**'s—"the great king of Uruk"—quest for the meaning of life. The tale includes enthralling details about the myths and beliefs of the people of **Mesopotamia**, their Gods and heroes, and their history. Importantly, *The Epic of Gilgamesh* was one of the sources from which the **Bible** absorbed many of its stories, including the ones related to the **garden of Eden** and the **genesis flood narrative**. Such examples are often used by historians as empirical chronological evidence that it was not God that created humans, but humans that created—better said, that recycled ideas about—God (Box 2.1). Another famous story, *The Appointment in Samarra*, modified in 1933 by W. Somerset Maugham from an ancient **Mesopotamian tale** recorded in the **Babylonian Talmud**—Sukkah 53a—in which the narrator is death itself, also clearly illustrates our obsession with the inevitability

R. Diogo, *Meaning of Life, Human Nature, and Delusions*,
https://doi.org/10.1007/978-3-319-70401-2_2

Fig. 2.1 The "Flood Tablet," a Neo-Assyrian clay tablet, part of the oldest epic tale, the *Epic of Gilgamesh*, which focuses on the quest for the meaning of life undertaken by "Gilgamesh, the great king of Uruk"

of **death**, and its links to the building of teleological narratives and to the notion of "**fate**" and "**meant to be**":

> There was a merchant in Bagdad who sent his servant to market to buy provisions and in a little while the servant came back, white and trembling, and said, "Master, just now when I was in the marketplace I was jostled by a woman in the crowd and when I turned I saw it was Death that jostled me…she looked at me and made a threatening gesture, now, lend me your horse, and I will ride away from this city and avoid my fate…. I will go to Samarra and there Death will not find me." The merchant lent him his horse, and the servant mounted it, and he dug his spurs in its flanks and as fast as the horse could gallop he went. Then the merchant went down to the marketplace and he saw me standing in the crowd and he came to me and said, "why did you make a threatening gesture to my servant when you saw him this morning?". "That was not a threatening gesture", I said, "it was only a start of surprise…. I was astonished to see him in Bagdad, for I had an appointment with him tonight in Samarra".

Most researchers defend that the human quest for the cosmic meaning of life is very likely universal among our species *Homo sapiens*—which originated more than 200,000 years ago—and that **religion** *sensu lato*—that is, defined in a broader way—likely appeared earlier than the Upper Paleolithic 30,000 years ago. Within our species, the earliest undisputed **human burial** dates back to about 100,000 years ago. As put by **Philip Lieberman** in his influential 1991 book *Uniquely Human*, such burials are commonly associated with some kind of religious thinking and/or with a quest for the meaning of life because they may signify "a concern for the dead that transcends daily life." In a 2009 book chapter Lahti claimed that burial practices may originally have protected the group from disease or wild animals, but when burials started to include the use of red ochre and other **ritualistic burial practices** they almost surely involved a ***belief in an afterlife*** (see Box 2.9). Importantly, there is increasing evidence that burials were also done by another species of humans, **Neanderthals**—see the 2014 paper by Rendu and colleagues—although some authors question the evidence for *intentional* Neanderthal burials—see the 2015 paper by Dibble and colleagues.

Box 2.1 Links Between the Epic of Gilgamesh and the Bible

In their exceptional 2016 book *The Good Book of Human Nature – An Evolutionary Reading of the Bible*, Van Schaik and Michel provide several examples showing that the **Bible** absorbed many stories disseminated by previous and/or surrounding cultures. An illustrative example is the people of Mesopotamia and their *Epic of Gilgamesh*, which was composed at least two millennia BC (Fig. 2.1):

> Probably the closest religious-historical parallel to the story of the **Garden of Eden** appears in an early Babylonian version of the *Epic of Gilgamesh* – in the biography of Gilgamesh's friend Enkidu, to be precise. After Enkidu was formed from clay, he first lived like a wild man together with the animals. It was not the gods, however, who civilized him, but rather a harlot skilled in all the womanly arts of seduction. Although all of the animals left Enkidu and his ability to run was compromised, his intellect expanded, and for the first time he understood her words. The harlot commented on his transformation as follows: "You are wise, Enkidu, and now you have become like a god." That sounds familiar – "and you shall be as gods," or so reads the serpent's well-known promise to **Adam and Eve**. The similarities are striking. In the Gilgamesh epic, the civilizing act has equally negative repercussions: **Enkidu** loses his innocence and his feeling of security among the animals. Now that he is a civilized human being, his thirst for action and glory will harm the animals and offend the gods. As a result he brings the gods' wrath upon himself and suffers a tragic end. Henrik Pfeiffer describes these stories so typical of the ancient world of the Middle East as "two-staged anthropogonies." Man's transformation from an evidently wild creature into a cultural being led to a life of backbreaking labor.... The flood narrative also found its way into the...saga of the feats of the Sumerian king Gilgamesh. Here, too, we find motifs familiar to readers of the Bible. For example, Gilgamesh finds the herb of life, only to have it stolen by a snake. Gilgamesh also encounters a second snake, which hides in the roots of the huluppu tree planted in a holy garden.

Many authors argue that we can actually see evidence of symbolic and/or religious thinking also in **chimpanzees**, which share a common ancestor with us that dates back to at least 6 million years ago. In a 2016 study Kühl and colleagues reported **stone tool-use behavior** and stone accumulation sites in wild chimpanzees that are somewhat evocative of some human **ritual sites** (Fig. 2.2). Namely, in some West African chimpanzee populations individuals routinely bang and throw rocks against trees, or toss them into the cavities of the trees, leading to conspicuous stone accumulations at these sites. That is, such stone tool use does not seem to be aimed to extractive foraging at what seems to be targeted trees, being instead apparently an instance of **ritualized behavioral** display, which can perhaps help us to understand the origins of ritual sites. Various other types of ritualized behavior have been described in chimpanzees. *In the shadow of man*, the pioneer and charismatic researcher **Jane Goodall** beautifully described such behaviors, as for instance what she called "**rain dance**." When it rains, sometimes chimpanzees perform energetic physical and auditory displays, charging through the forest and rhythmically swaggering, drumming their feet on the buttresses of the trees, and slapping the ground. Jane wrote about these displays: "my enthusiasm was not merely scientific as I watched…. I could only watch, and marvel at the magnificence of these splendid creatures…with a display of strength and vigor such as this, primitive man himself might have challenged the elements." More recently other researchers have described

Fig. 2.2 Photographs and stills of accumulative stone throwing behavior and sites. Above: Adult male chimpanzee tossing a stone; hurling a stone (Boé, Guinea-Bissau); and banging a stone (Comoé GEPRENAF, Côte d'Ivoire). Below: Boé, Guinea-Bissau landscape: stones accumulated in a hollow tree; a chimpanzee accumulative stone throwing site; and stones accumulated in-between buttress roots

similar ritualized chimpanzee "dances" when they encountered wildfire, as noted in a 2010 paper by Pruetz and LaDuke: instead of running away, some chimpanzees serenely monitored the wildfires and changed their behavior, anticipating the fire's movement. These authors argued that "the ability to conceptualize the 'behavior' of fire may be a...trait characterizing the human-chimpanzee clade...if the cognitive underpinnings of fire conceptualization are [such] a primitive...trait, hypotheses concerning the origins of the control and use of fire may need revision."

As I have argued in my 2017 book *Evolution Driven by Organismal Behavior*, there is indeed strong empirical evidence supporting the idea that, both in terms of anatomy and behavior, chimpanzees are much more similar to us than it has been assumed for centuries in the scientific literature. Therefore it is completely plausible that chimpanzees might display true examples of ritualized behaviors that have a direct connection with religious experiences in humans. However, in the present book I argue that there is very likely a main, crucial behavioral difference between chimpanzees and humans, which might explain why, contrary to them, we are particularly prone to create and *believe in complex teleological narratives*. Namely, we don't have any type of direct evidence showing that they, or any other non-human animals, fully understand both the inevitability and randomness of death, as we do: we know not only that we will surely die one day, but also that both our life and death are moreover highly dependent on chance and randomness, although we try so hard to create narratives to think that this is not the case. This is "our blessing and our curse."

2.2 Beliefs, Religion, and Evolution

God made Man because He loves stories. (Elie Wiesel)

Among the works that try to search for a scientific reason for how we evolved **beliefs** and a tendency to ask **teleological-why-questions** and to "find" nonexistent patterns and **causal chains of events**, one that does a good job at summarizing the scientific data available is Shermer's 2011 *The Believing Brain*. According to Shermer:

> The brain is a belief engine. From sensory data flowing in through the senses the brain naturally begins to look for and find **patterns**, and then infuses those patterns with meaning. The first process I call ***patternicity***: the tendency to find meaningful patterns in both meaningful and meaningless data. The second process I call ***agenticity***: the tendency to infuse patterns with meaning, **intention** and **agency**. We can't help it. Our brains evolved to connect the dots of our world into meaningful patterns that explain *why* things happen. These meaningful patterns become beliefs, and these beliefs shape our understanding of reality. Once beliefs are formed, the brain begins to look for and find confirmatory evidence in support of those beliefs, which adds an emotional boost of further confidence in the beliefs and thereby accelerates the process of reinforcing them, and round and round the process goes in a positive feedback loop of **belief confirmation**.

It is interesting to note that **animism**—which, in a Western-centric way, is often viewed as a "precursor of religion" or a "basal religion"—includes the belief that inanimate objects possess life or a **spirit** and even intentions or purposes: that is, it is directly related to **agenticity**. A powerful example provided by Shermer about humans, beliefs, and rationalization is a 2009 Harris Poll of 2303 adults living in the U.S. in which people were asked to "indicate for each category below if you believe in it, or not." The results are striking: *God: 82%; Miracles: 76%; Heaven: 75%; Angels: 72%; Hell: 61%; The virgin birth of Jesus: 61%; The devil: 60%; Darwin' theory of evolution: 45%; Ghosts: 42%*. More people living in the U.S. believe in **angels** and the **devil** than in the reality of biological evolution—an example of how most humans are chiefly immersed in *Neverland*'s world of *unreality*. Even more striking is the fact that a 2002 article in the *Skeptic* magazine has reported that there is no negative correlation between scientific knowledge and **paranormal beliefs**, stressing that "students are taught *what* to think but not *how* to think." According to Shermer, part of the problem may be that 70% of Americans still do not understand the scientific process, and a solution would thus be to teach *how science works* in addition to *what science knows*. However, teaching how science works is only one—although a very important one—of many steps that have to be taken to liberate humans from *Neverland*. Actually, as we will see, one of the main points of religion is precisely to accept as true stories that are scientifically completely illogical, such as a virgin having a child: that is ***faith***, precisely. In this sense, if one thinks deeply, the story about angels and the devil also suffers from major logical inconsistencies: if the devil wants people to do "bad" things while alive, would it not be more logical to think that he should reward them in hell, instead of punishing them? The ultimate implication of this would be that for us it would always be a win-win situation: *if we are "good," we will go to heaven and be rewarded there by God; if we are "bad" we will be sent to hell and rewarded there by the Devil*. The point of this aside divagation is just to stress that teleological stories in general and religious tales in particular are not logical at all, that is exactly one of their key features. Shermer refers to a *Science* paper titled *On being sane in insane places* by Stanford University psychologist David Rosenhan:

> Rosenhan and his associates…entered a dozen mental hospitals in five different states on the East and West (USA) coasts, reporting having had a brief auditory **hallucination**. All eight (three psychologists, a psychiatrist, a pediatrician, a housewife, a painter and a psychologist student, none of whom had any history of **mental illness**) were admitted, seven of them diagnosed as schizophrenic and one as manic-depressive. Outside of the faux auditory hallucination and false names, they were instructed to tell the truth after admission, act normally, and claim that the **hallucinations** had stopped and that they now felt perfectly fine. Despite the fact that the nurses reported the patients as 'friendly' and 'cooperative' and said they 'exhibited no abnormal indications', none of the hospital psychiatrists or staff caught on the experiment, consistently treating these normals as abnormals. After an average stay of nineteen days…all of Rosenhan's shills were discharged with a diagnosis of schizophrenia 'in remission'. The power of the diagnostic belief engine was striking…. To pass the time they (the researchers) kept detailed notes of their experiences. In one poignant descriptor, the staff reported that 'patient engages in writing behavior' on a list of signs of pathology. Tellingly, the real patients suspected something was up right away. Of the 118 patients whose remarks were recorded, 25 of them indicated that they knew what was really

> going on. As one exclaimed "you're note crazy…you're a journalist, or a professor…you're checking up on the hospital'. Of course. Who else would be sitting around a mental hospital taking copious notes? What you believe is what you see. The label is the behavior. Theory molds data. Concepts determine percepts. Belief-dependent realism".

This example reinforces the lack of a clear negative correlation between so-called scientific knowledge—for instance, the number of years you studied science in school to be a psychiatrist in a mental hospital—and the influence of wrong beliefs and/or of **confirmation biases**. Most of the patients of the mental institution, which are generally considered to be "insane", were actually more efficient at seeing what was really going on than were the hospital psychiatrists, because they had no reason, and thus no confirmation biases, that would lead them to try/want to think otherwise, contrary to the psychiatrists. The same applies to so-called **intelligence**, as measured by **IQ tests**. There is also no negative correlation between belief in **superstition** or **magic** and a higher IQ, particularly as you move up the IQ spectrum. As noted by Shermer, this is because when people commit to a belief, the smarter they are the better they are at rationalizing those beliefs:

> *Smart people believe weird things because they are skilled at defending beliefs they arrived at for nonsmart reasons*. Most people, most of the time, arrive at their beliefs for a host of reasons involving personality and temperament, family dynamics and cultural background, parents and siblings, peer groups and teachers, education and books, mentors and heroes, and various life experiences, very few of which have anything at all to do with intelligence. The **Enlightenment** idea of *Homo rationalis* has us sitting down before a table of facts, weighing them in the balance pro and con, and then employing logic and reason to determine which set of facts best supports this or that theory. *This is not at all how we form beliefs*. What happens is that facts of the world are filtered by our brains through the colored lenses of worldviews, paradigms, theories, hypotheses, conjectures, hunches, **biases**, and **prejudices** we have accumulated through living. We then sort through the facts and select those that confirm what we already believe and ignore or rationalize away those that contradict our beliefs.

So, how does Shermer try to explain these differences between humans and other organisms, particularly concerning the two most crucial topics of his book, ***patternicity*** and ***agenticity*** (Box 2.3)? About patternicity, he states:

> Imagine that you are a hominid walking along the Savanna of an African valley three million years ago. You hear a rustle in the grass. It is just the wind or it is a dangerous predator? Your answer could mean life or death. If you assume that the rustle in the grass is a dangerous predator but it turns out that it is just the wind, you have made what is called a ***Type I error*** in **cognition**, also known as a ***false positive***, or believing something is real when it is not. That is, you have found a nonexistent pattern. No harm. You move away from the rustling sound, become more alert and cautious, and find another path to your destination. If you assume that the rustle in the grass is just the wind but it turns out that it is a dangerous predator, you have made what is called a ***Type II error*** in cognition, also known as a **false negative**, or believing something is not real when it is. That is, you have missed a real pattern. You're lunch. Congratulations, you have won a Darwin Award. You are no longer a member of the hominid gene pool (if you did not have kids yet). Our brains are belief engines, evolved pattern-recognition machines that connect the dots and create meaning out of patterns that we think we see in nature.

About **agenticity**, he explains:

Let us return to our erstwhile hominid on the planes of Africa who hears a rustle in the grass, and the crucial matter of whether the sound represents a dangerous predator or just the wind: 'wind' represents an ***inanimate force*** whereas 'dangerous predator' indicates an ***intentional agent***. There is a big difference between an inanimate force and an intentional agent. Most animals can make this distinction on the superficial (but vital) life-or-death level, but we do something other animals do not do. As large-brained hominids with a developed cortex and a '**theory of mind**' – the capacity to be aware of such mental states as desires and intentions in both ourselves and others – we practice what I call *agenticity*: the tendency to infuse patterns of meaning, intention, and agency. That is, we often impart the patterns we find with agency and intention, and believe that these intentional agents control the world, sometimes invisibly from the top down, instead of bottom-up causal laws and randomness that makes up much of the world. Souls, **spirits**, **ghosts**, **gods**, **demons**, **angels**, **aliens**, **intelligent designers**, **government conspiracists**, and all manner of invisible agents with power and intention are believed to haunt our world and control our lives. Combined with our propensity to find meaningful patterns in both meaningful and meaningless noise, patternicity and agencity form the cognitive basis of **shamanism**, **paganism**, **animism**, **polytheism**, **monotheism**, and all modes of Old and New Age **spiritualism**.

In his book *Supersense*, University of Bristol psychologist documented the growing body of data that demonstrates our tendency not only to infuse patterns with agency and intention, but to also believe that objects, animals, and people contain essence and that this essence may be transmitted from objects to people, and from people to people. There are evolutionary reasons for this ***essentialism***, rooted in fears about diseases and contagions that contain all-too-natural essences that can be deadly, and thus there was a natural selection for those who avoided deadly diseases by following their instincts about essence avoidance. But we also generalize these essence emotions to both natural and supernatural beings, to any and all objects and people, and to things seen and unseen. Children believe that the sun can think and follows them around, and when asked to draw a picture of the sun they often add a smiley face to give agency to it. A third of transplant (adult) patients believe that the donor's personality or essence is transplanted with the organ.

Box 2.2 Neural Correlates of Belief and Dopamine

Shermer's 2011 *The Believing Brain* provides a detailed review of empirical examples that help us to understand how our brains are somewhat "wired" to believe and to build teleological narratives. The book refers to the ideas of neuroscientist Michael Gazzaniga, who argues that we have a **neural network** that coordinates all the other neural networks and weaves them together into a whole. That is, this network—named ***left-hemisphere interpreter*** by Gazzaniga—is like a **brain storyteller** that puts together countless inputs into a meaningful narrative story. Fascinatingly, Gazzaniga reported the existence of this network when he was studying **split-brain patients** whose hemispheres have been separated to stop the spread of **epileptic seizures**. In one experiment, he presented the word "walk" to only the right hemisphere of one of these patients, who promptly got up and stated walking. When asked why he did so, the *left hemisphere interpreter* of the patient made up an imaginary story to explain this behavior: "I wanted to go get a coke." Within other

neural correlates of belief, one that is particularly important is **dopamine**, as noted by Shermer:

> Dopamine, in fact, is critical in association learning and the reward system of the brain…whereby any behavior that is reinforced tends to be repeated. In the divided brain stem – one of the most evolutionarily ancient parts of the brain shared by all vertebrates – there are pockets of roughly fifteen thousand to twenty-five thousand dopamine-producing neurons on each side that shoot out long axons connecting to other parts of the brain. These neurons stimulate the release of dopamine whenever it is determined that a received reward is more than expected, which causes the individual to repeat this behavior. The release of dopamine is a form of information, a message that tells the organism "do that again". You get a hit (reinforcement) and your brain gets a hit of dopamine: *behavior-reinforcement-behavior; repeat sequence*. In fact, dopamine appears to fuel…[a]…so-called pleasure center of the brain (***nucleus accumbens***) that has been implicated in the 'high' derived from cocaine and orgasms…addictive drugs take over the role of reward signals that feed into the dopamine neurons. So, too, do addictive ideas, most notably bad ideas, such as those propagated by cults that lead to mass suicides…or those propagated by religions that lead to suicide bombing…dopamine reinforces behaviors and beliefs and patternicity, and thus is one of the primary belief drugs.
>
> The connection between dopamine and belief was established by experiments conducted by Peter Brugger and his colleague Christine Mohr at the University of Bristol in England…[whom]…found that people with high levels of dopamine are more likely to find significance in coincidences and pick out meaning and patterns where there are none. In one study…they compared 20 self-professed believers in **ghosts**, **gods**, **spirits**, and **conspiracies** to 20 self-professed skeptics of such claims. They showed all subjects a series of slides consisting of people's faces, some of which were normal while others had their parts scrambled, such as swapping out eyes or ears or noses from different faces. In general, the scientists found that the believers were much more likely than the skeptics to mistakenly assess a scrambled face as real, and to read a scrambled word as normal. In the second part of the experiment, Brugger and Mohr gave all forty subjects L-dopa, the drug used for **Parkinson's disease** patients that increases the levels of dopamine in the brain. They then repeated the slide show with the scrambled or real faces and words. The boost of dopamine caused both believers and skeptics to identify scrambled faces and real and jumbled words as normal…[supporting the idea]…that that patternicity may be associated with high levels of dopamine in the brain. One theory – promulgated by Mohr, Brugger, and their colleagues – is that dopamine increases the signal-to-noise ratio, that is, the amount of signal that your brain will detect in background noise – this is the **error-detention problem** associated with patternicity. Dopamine…increases the rate of neural firing in association with pattern recognition, which means that synaptic connections between neurons are likely to increase to a perceived pattern, thereby cementing those perceived patterns into long-term memory through the actual physical growth of neural connections and the reinforcement of old synaptic skills.

In a very interesting 2009 book chapter, Vaas noted that although there is no consensus yet, there are indeed various candidates for **neural correlates of religious experiences** based on empirical data, such as:

a) **Superstition** has a physiological basis: proneness to gullibility, belief in **paranormal phenomena**, and to "seeing" things, faces, for example, in random patterns are tendencies that are associated with higher levels of dopamine in the brain or can be increased by the intake of L-dopa, a dopamine precursor…it is also a matter of debate whether there is a connection between superstitious behavior, scrupulosity, religious rituals, and the neurological symptoms of **obsessive-compulsive disorder**.

b) Some **drugs** create spiritual experiences with long-lasting effects. **Psychotropic substances** have been used for religious purposes probably since pre-historic times and still influence **ceremonial practice** and spiritual orientations. **Visual hallucinations** are especially intense and can also be triggered by endogenous neurophysiological processes, which are the cause, for example, of the aura preceding a **migraine attack**.

c) Spiritual experiences during **meditation** are associated with an increased activity of the prefrontal cortex and a decreased activity of the object-association area in the parietal lobe and other areas. **Brain activity** becomes more synchronous, **empathy** is enhanced (**insula activation**, etc.).

d) **Hearing voices**, often interpreted as messages from God, are common in **schizophrenia**.

e) **Temporal lobe epilepsies** are frequently associated with **hyper-religiosity** to the point of extremism. Many founders of religion could have been temporal lobe personalities.

f) Artificially created "**micro-seizures**" (temporal lobe transients) produced by **transcerebral magnetic stimulation** cause a "sensed presence" of God, an **angel** or an **alter ego**.

g) **Out-of-body experiences** can be triggered experimentally via electrical stimulation or **virtual reality illusions** as can the "appearance" of **ghosts** (temporo-parietal junction). They all disrupt the body representation in the brain.

h) There is an association between pictorial representations and concepts of God and the reading/writing directions of **holy scriptures** with or without vowels asymmetrically represented in the brain's hemispheres.

i) Strong religious convictions and associations, for example of **evangelical believers**, correlate with frontal and parietal lobe activation.

j) Frontal areas are also involved with decision-making, emotional evaluations, **moral judgments**, and **altruistic behavior**. Experiments in **cognitive psychology** have shown that religious attitudes have some influence here.

In fact, an interesting point about both patternicity and agenticity, as well as teleological thinking, is that these features are universally present in humans not only in the sense that they are found in adults of each extant human group, but also in the sense that they are present in kids since a very young age. Voland, in a chapter of the 2009 book *The Biological Evolution of Religious Mind and Behavior* that was edited by him and Schiefenhovel, reviews various different empirical studies published in the literature and stated that:

> Children under the age of 5 years attribute omniscience to all the persons in their immediate environment. Only with the development of a '**theory of mind**' do children begin to understand that different sets of knowledge are at home in different brains. Children under the age of 5 years think teleologically: there are clouds so that it rains; and it rains, so that flowers can thrive. Finally, they not only think dualistically, but at the same time, they store the assumption of a **life after death**. Interestingly enough, these cognitive basic attitudes of **early childhood**, namely the assumption of omniscient persons and a teleological and dual-

istic way of thinking, also form the basis of crucial theoretical assumptions in many theistic systems of beliefs…children appear [to be] born to believe.

In a chapter from the same book, Rossano provides some examples of data obtained in **developmental psychology studies** that also indicate that children have a natural inclination to think supernaturally. According to him this tendency appears to support the idea that belief has been **adaptative**—that is, evolutionarily advantageous—within human evolution, a subject that I will discuss in further detail below. Namely, Rossano proposed:

This inclination…helps them [the children] hone the social reasoning skills essential for successful functioning as an adult. Childhood **supernatural thinking** became selectively advantageous as the complexity of the hominin social world increased beginning about 100,000 years ago. Among our ancestors, those adults who were the most socially skilled were the ones who as children tended to think supernaturally. Developmental research has documented evidence of childhood supernatural thinking in a number of different forms. Twelve-month-olds treat computer-animated images as intentional objects with desires and goals; six-month-olds make primitive 'moral' judgments based on the inferred intentions of those objects. Older children [also] readily link natural events with **moral behavior**, such as bridge collapsing, because the children on it behave badly earlier. Cross-culturally, children prefer God as the cause of the existence of animate and inanimate objects over other causes. Children form a theory of mind about God prior to forming one about other humans.

Also in the same book, a chapter by Richert and Smith is mainly focused on **the cognitive foundations of the development of belief** and of a **religious mind**. They review psychological studies that indicate that belief in the supernatural, and particularly in the existence of a **soul**, is very likely related to the awareness, and fear, of the inevitability and randomness of death, as I am arguing in the present volume. They note that in such studies adults who did affirm the existence of the soul were likely to claim that it would continue to exist after death—that is, contrary to the brain the soul was dissociated from the cycle of conception, life, and death. Moreover, they also review various developmental studies that provide support to the idea—also defended in the present volume—that an awareness of the inexorableness of death probably leads to an increased tendency to believe in teleological narratives during human evolution, as a very similar pattern is seen during human development as well. They noted:

Seven- and 11-year-old children from public schools in Spain were given two stories about a grandparent that died. The language in one story was designed to evoke religious reasoning; the language in the other story was designed to reflect a biological (secular) stance to **death**. The results indicated that the older children were less likely to claim processes stopped after death. This finding was especially strong in the context of religious narrative and for questions related to body processes. Furthermore, children's justifications, coded into biological and metaphysical responses, further supported these findings. Biological justifications were more frequent in the secular story, for body questions and among younger children. Metaphysical justifications, on the other hand, were used equally for body and mind questions, but were more common in the religious story and among older children. This is consistent with the authors' interpretation that *as children mature in their understanding of the **finality and completeness of biological death**, the religious concept of death and the **after-life** is increasingly persuasive*.

Further exploring the **teleological roots of children's ideas**, they wrote:

Past research has suggested children have a tendency to prefer creation rather than evolutionary explanations for origins…about the origins of plants, animals, the sky, the earth, and large rocks…children were asked to choose from three possible creators: people, God, or nobody knows/unknown power. The preschoolers in this study were about seven times more likely to attribute responsibility for the natural world to God, and not to people. Furthermore…regardless of religious affiliation (fundamentalist Christian communities *versus* non-fundamentalist communities) a large majority of 5- to 8-year-old children preferred **creationist accounts** for the origins of the natural world to either evolutionary, artificialist (created by humans), or **emergentist accounts**. Young children have strong inclinations to understand both living and non-living things as purposeful. They see living and non-living things as possessing attributes purposefully designed to help them or serve themselves or other things. For example, 4- and 5-year-old children often claim mountains are "for climbing" or clouds are "for raining". Kelemen has suggested the possibility that children naturally develop as "intuitive theists," and religious instruction merely fills in the forms that already exist in children's minds. Kelemen and DiYanni specifically examined whether children's assumptions about the purposeful nature of natural phenomena are related to their **intelligent design** reasoning. They interviewed 6- through 10-year-old children for their intuitions about the origins of artifacts, animals, natural events, and natural objects. Regardless of age, children tended to prefer teleological explanations for the origins of artifacts, animals, and natural objects. In addition, **children's teleofunctional intuitions** about the origins of artifacts, animals, natural events, and natural objects were significantly correlated with their claims about whether the first things (i.e., artifacts, animals, natural events, and natural objects) just appeared or were made by someone or something. A conclusion from this research might be that developing a concept of a **purposeful Creator God** is not cognitively demanding for children because this concept builds on their intuitive assumptions about the teleological nature of the world.

Box 2.3 Believing Brain, Patternicity, Madness, and Creativity

Another point worthy of note discussed by Shermer in his 2011 book is the connection between teleological narratives, belief, **dopamine levels, creativity,** and **madness**: as explained by him, the empirical data available further supports the idea that believers don't necessarily "know less" or are "less intelligent" than non-believers. All humans, to an extent, have some kind of belief in fictional teleological narratives, and Nobel Prize winners and geniuses are not an exception, sometimes well on the contrary. Following up with what was explained in Box 2.2 about the neural correlates of belief and dopamine, Shermer notes:

Increasing dopamine increases pattern detection…dopamine agonists not only enhance learning but in higher doses can also trigger symptoms of **psychosis**, such as **hallucinations**, which may be related to the fine line between creativity (discriminate patternicity) and madness (indiscriminate patternicity). The dose is the key: too much of it and you are likely to be making type I errors – false positives – in which you find connections that are not really there. Too little and you make type II errors – false negatives – in which you miss connections that are real. The ***signal-to-noise ratio*** is everything. People on the **schizophrenic spectrum** tend to have an all-inclusive thinking style, which means they see patterns where no meaningful patterns exist, and cannot tell the difference between a meaningful or a non-meaningful pattern. Psychologist Hans Eysenck…was the first to suggest a possible correlation of ***psychotism*** with creativity. This is, in fact, what was found by Max Planck Institute cognitive neuroscientist Anna Abraham and her colleagues, in a 2005 study designed to explore…[this]…link: subjects with higher level of psychotism were more creative but in less practical ways.

In a way, there's a fine line between the creative genius of finding novel patterns that change the world and the madness or paranoia of seeing patterns everywhere and being unable to pick up the important ones. An instructive example would be a comparison between Nobel Prize-winning physicist **Richard Feynman**, who did top-secret government work on the Manhattan Project to build an atomic bomb (and whose quirkiness extended no further than playing bonobo drums, sketching nudes, and cracking safes), and the Nobel Prize-winning mathematician **John Nash**, who was diagnosed schizophrenic and portrayed in the film *Beautiful Mind* as a man struggling with **paranoid delusions** about top-secret government work on a code-breaking project to detect enemy information patterns. Both Feynman and Nash were creative geniuses who made novel discoveries and unique patterns worthy of a Nobel Prize – Feynman in **quantum physics** and Nash in game theory – but Nash's cognitive style was all inclusive: he saw patterns everywhere, including complex **conspiracies** with nonexistent government agents and no basis on reality.

In a 2009 book chapter, Brune also provided a mesmerizing discussion on the links between belief, religion, delusion, and clinical medicine:

Ever since **psychiatry** has emerged as a branch of clinical medicine, religiosity and religiousness have been intimately interwoven with shifting concepts of **psychopathology**. Pre-scientific theories proposed that mental illness was a consequence of personal failure and punishment inflicted by God. These untenable speculations prevailed well until the middle of the nineteenth century, and contributed to the maltreatment of many psychiatrically ill in **asylums** and **mental hospitals**. One of the reasons for linking psychopathology with religious matters could have been that psychotic experiences frequently involve ideas of being influenced by supernatural powers, including God or the devil. Even though the prevalence of such ideas may wax and wane with cultural attitudes toward religiosity, they represent an important aspect of subjective experience during **psychotic states** to the present day. Recently, the role of religiousness in **psychiatric disorders** has received renewed interest, and the number of papers published in prestigious psychiatric journals on religious issues in relation to psychiatry has grown substantially. For example, several studies found evidence in support of both protective and negative effects on suicide risk in patients with **schizophrenia** and moderate effects in terms of strengthening resiliency in suicidal adolescents. Moreover, in widowed individuals religiousness was associated with decreased grief but did not influence depression. One of the least well understood phenomena is, however, the relationship between religiousness and **delusional beliefs**. According to a recent report, 57% of patients with schizophrenia believed that their illness was in one or the other way influenced by their religious convictions, with 31% believing in a positive (e.g., "a test sent by God to put me on the right path") and 26% in a negative impact (e.g., "my illness is a punishment sent by God for my sins") of religious matters on their illnesses.

About studies done in babies and infants about agency, Bering, in his 2011 book *The Belief Instinct*, explains (see also Boxes 2.4 and 2.5):

> Some developmental psychologists even believe that this **cognitive bias** to see intentions in inanimate objects – and thus a very basic **theory of mind** – can be found in babies just a few months out of the womb. For example, Hungarian psychologists György Gergely and Gergely Csibra from the Central European University in Budapest have shown in their work that babies, on the basis of their staring response, act surprised when a dot on a computer screen continues to butt up against an empty space on the screen after a computerized barrier blocking its path has been deleted. It's as if the baby is staring at the dot trying to figure out why the dot is acting as though it 'thinks' the barrier is still there. By contrast, the infants are not especially interested – that is, they don't stare in surprise – when the dot stops in front of the block, or when the dot continues along its path in the absence of the barrier. The most famous example of this cognitive phenomenon of seeing minds in nonliving objects, however, is a 1944 *American Journal of Psychology* study by Austrian researchers Fritz Heider and Mary-Ann Simmel. In this very early study, the scientists put together a simplistic animated film depicting three moving, black-and-white figures: a large triangle, a small triangle, and a small circle. Participants watched the figures moving about on the screen for a while and then were asked to describe what they had just seen. Most reported using a human social behavioral narrative – for example, seeing the large triangle as 'bullying' the 'timid' smaller triangle, both of 'whom' were 'seeking' the 'affections' of the 'female' circle.

Box 2.4 Why-Questions, Agency, Beliefs, and Creationism

Bering's 2011 book *The Belief Instinct* provides several other examples of interesting experimental studies done in children about agency and beliefs (see also Boxes 1.2 and 1.3):

> As Boston University psychologist Deborah Kelemen has found in study after study, young children erroneously endow…natural, inanimate entities – waterfalls, clouds, rocks, and so on – with their own teleo-functional purposes…. Because of this tendency to over-attribute reason and purpose to aspects of the natural world, Kelemen refers to young children as 'promiscuous teleologists'. For example, Kelemen and her colleagues find that seven-and eight-year-olds who are asked why mountains exist overwhelmingly prefer, regardless of their parents' religiosity or irreligiosity, **teleo-functional explanations** ('to give animals a place to climb') over mechanistic, or physical, causal explanations ('because volcanoes cooled into lumps'). It's only around fourth or fifth grade that children begin abandoning these incorrect teleo-functional answers in favor of scientifically accurate accounts. And without a basic science education, **promiscuous teleology** remains a fixture of adult thought. In studies with uneducated Romany adults, Kelemen and psychologist Krista Casler revealed the same preference for teleo-functional reasoning that is seen in young children; it also appears in **Alzheimer's patients**, presumably because their scientific knowledge has been eaten away by disease, thus allowing the unaffected teleo-functional bias to recrudesce. So with artifacts and some biological features (those modified by human beings), we're on solid ground using teleo-functional reasoning. Again, however, young children and adults lacking a basic scientific education overdo it; they're promiscuously teleological when reasoning about happenstance properties of nonbiological, inanimate objects. For example, when asked why rocks are pointy, the seven- and eight-year-olds in Kelemen's studies endorse teleo-functional accounts, treating rocks as something like artifacts ('so that animals could scratch on them when they get itchy') or as though the rocks were organisms themselves with evolved adaptations ('so that animals wouldn't sit on them and smash them').

If you think this type of response is just the result of what kids hear on television or from their parents, Kelemen is one step ahead of you, at least with respect to parental input. In looking at spontaneous dialogues occurring between preschoolers and their parents – particularly with respect to 'why' and 'what's that for' questions – Kelemen and her colleagues showed that parents generally reply with naturalistic causal answers (that is, scientific) rather than teleo-functional explanations. And even when they're given a choice and told that all-important adults prefer non-functional explanations over teleo-functional ones, children still opt strongly for the latter. 'So current evidence suggests the answer does not lie there', says Kelemen. At least, not in any straightforward sense. Furthermore, not only do children err teleologically about inanimate natural entities like mountains, or about the physical features of inorganic objects like the shapes of rocks; they even display **teleo-functional reasoning** when it comes to the existence of whole organisms. One wouldn't (at least, one shouldn't) say that turkey vultures as a whole exist 'for' cleaning up roadkill-splattered interstates. Dogs, as a domesticated species, may have been designed for human purposes, but, like buzzards, canines as a group aren't 'for' anything either. Rather, they simply are; they've come to exist; they've evolved. And yet, again, Kelemen has found that when children are asked why, say, lions exist, they prefer teleo-functional explanations ('to go in the zoo').

All of this may sound silly to you, but such findings, and the distorting lens of our species' theory of mind more generally, have obvious implications for our ability to ever truly grasp the completely mindless principles of evolution by random mutation and natural selection. In fact, for the past decade University of Michigan psychologist Margaret Evans has been investigating why creationist thinking comes more easily to the human mind than does evolutionary thinking. 'Persistence [of creationist beliefs] is not simply the result of fundamentalist politics and socialization', writes Evans. 'Rather, these forces themselves depend on certain propensities of the human mind'. According to Evans, the stubborn preponderance of creationist beliefs is due in large part to the way our cognitive systems have, interestingly enough, evolved. Like Kelemen, Evans has discovered that irrespective of their parents' beliefs or whether they attend religious or secular school, when asked where the first member of a particular animal species came from, five- to seven-year-old children give either spontaneous generationist ('it got born there') or **creationist** ('God made it') responses. By eight years of age, however, children from both secular and religious backgrounds give more or less exclusively creationist answers. Usually these answers predictably manifest as 'God made it', but otherwise Nature is personified, seen as a deliberate agent that intentionally made the animal for its own ends.

It's at eight years or so, then, that **teleo-functional reasoning** seems to turn into a full-blown 'design stance', in which children envisage an actual being as intentionally creating the entity in question for its own personal reasons. Only among the oldest children she has studied, the ten- to twelve-year-olds, has Evans uncovered an effect of developmental experience, with children of evolutionary-minded parents finally giving evolutionary responses and those of evangelical parents giving creationist answers to the question of species origins. And even the 'evolutionary' responses are often corrupted by culturally based misunderstandings. For example, Japanese fifth-graders tend to believe that human beings evolved directly from monkeys, probably because macaque monkeys are prodigious in Japan. In other words, all of this suggests that thinking like an evolutionist is hard work because, ironically, our psychological development – and, in particular, our theory of mind – strongly favors the purposeful-design framework. Evolutionists will probably never outnumber creationists, because the latter have a paradoxical ally in the way natural selection has lent itself to our species' untutored penchant for reasoning about its own origins.

2.3 Fear of Death, Purpose, and Human Evolution

> *'tis too horrible...the weariest and most loathed worldly life...that age, ache, penury, and imprisonment...can lay on nature is a paradise...to what we fear of death. (Shakespeare, in his "Measure for Measure")*

It is important to note that, as Shermer himself recognizes in his 2011 book, the process of **patternicity**—the tendency to find meaningful patterns in both meaningful and meaningless noise—is shared by many animals, "from *C. elegans* (a roundworm) to *H. sapiens*":

> B. F. Skinner was the first scientist to systematically study **superstitious behavior in animals**, noting that when food was presented to pigeons at random intervals instead of more predictable schedules of reinforcement – for which pecking a key inside a box in which the pigeon was placed would result in delivery of the food trough a small food hopper – the pigeons exhibited an odd assortment of behaviors, such as side-to-side hopping or twirling around counterclockwise before pecking the key. It was an avian rain dance of sorts. The pigeons did this because they were put on something called a variable interval schedule of reinforcement, in which the time interval between getting the food reward for pecking a key varied. In that interval of time between pecking the key and the hopper delivering the food, whatever the pigeons happened to be doing was scored in their little brains as a pattern.
>
> Inspired by Skinner's classic experiments, Koichi Ono of Komazawa University in Japan ran human subjects trough the equivalent of a Skinner box...after minute nine of the thirty-minute session, subject 5 had his ritual down pat. Subject 15 developed the stranger rite of all. Five minutes into her session a point was delivered the moment she happened to touch the point counter. Thereafter she started touching anything and everything within reach, and, of course, since the points continued to be delivered, this odd touching behavior was reinforced. At the ten-minute mark she got a point just as she happened to jump on the floor, whereby she promptly abandoned touching and took up jumping as her new strategy, climaxing in a point being scored when she touched the ceiling, leading her to end the session early from ceiling touching exhaustion. Technically speaking, in Ono's words 'superstitious behavior is defined as behavior produced by response independent schedules of reinforced delivery, in which only an accidental relation exists between responses and delivery of reinforcers'. That's a fancy way of saying that **superstitions** are just an accidental form of learning.

Therefore, it seems difficult to argue that the main difference between humans and other organisms that probably lead to our tendency to build, and believe in, complex fictional teleological narratives is merely related to this process of patternicity, when such a process is shared with many other animals. Shermer argues that the difference is not so much the process itself, but our greater cognitive capacity for holding associations in our memory, compared to birds and other animals. That is, he states that our enhanced capacities are "a double-edged sword; our greater capacity for learning is often offset by our greater capacity for **magical thinking**: **superstition in pigeons** can be easily extinguished; in humans it is much more difficult." I do agree with this point, and I would point out that this is one of the main reasons why we are indeed, *Homo irrationalis*—that is, we are mainly an irrational being, as other animals are, but contrary to them we are much more immersed in *Neverland*'s world of unreality. Do dogs believe in angels, or ghosts? We do not have enough evidence to completely discard this possibility, but the direct evidence we have so far by observing their behaviors, and the scarce studies on animal beliefs available so far, indicate that they probably don't have complex teleological narratives and don't fear ghosts or dream with angels, and one of the reasons for this is precisely that humans can more easily create imaginary stories and defend them a posteriori. But this is clearly not the *only* reason, very far from it.

So, another reason that Shermer and various other authors propose is the following: humans are particularly prone to build and believe in teleological narratives because of the profound connection between **agency** and the processed called ***theory of mind***—that is, that we are self-aware of our own beliefs, desires, and intentions, as well as aware that others also have them. According to Shermer, theory of mind probably evolved from various preexisting **neural networks** used for activities such as the ability to distinguish between living and non-living objects, to hold the attention of other beings, and to represent actions that are goal directed. Shermer relates a theory of mind with activities as important as imitation, anticipation, and **empathy**, and gives the example of **mirror neurons**—which are specialized to "mirror" the actions of others—and were discovered by neuroscientist Giacomo Rizzolatti and his colleagues at the University of Parma in macaque monkeys. In one of the experiments a member of Rizzolatti's team reached in and grabbed one of the peanuts that was being used for an experiment with monkeys, causing the motor neurons of a monkey to fire, as detected by hair-thin electrodes applied to the monkeys' heads. This and other subsequent experimental studies, done in non-human animals such as monkeys and also in humans, have revealed that mirror neurons are important in both predicting others' actions and inferring their actions, which is the very foundation of **agenticity**. As non-human primates reveal similar patterns concerning mirror neurons, it seems difficult to argue that such neurons, and the theory of mind in general, are what makes humans so prone to build and believe in complex teleological narratives. In fact, it is now becoming more and more accepted that at least some non-human animals, including our most close living relatives—common chimpanzees and bonobos—also have a theory of mind, as pointed out in De Wall's *Are We Smart Enough to Know How Smart Animals Are?*

So, apart from the fact that we can more easily create imaginary stories and also a posteriori ways to reinforce our beliefs in them, what is truly the main driver of our *unique* trend to become so immersed in *Neverland*, contrary to other animals? Could it be because our peculiar brains and mental capacities made us *more aware about the inevitability of death, and the crucial role played by random, chaotic, and contingent events in our lives*, from their very beginning to the very last breath, as I suggested above? Apart from trying to escape from death by avoiding accidents and predators as many other animals do, chimpanzees and many other primates, as well as elephants, do seem to display another level of behavioral complexity in the sense that they seemingly have a deeper knowledge about death, as indicated for instance by the rituals elephants do when they pass near the bones of their dead relatives. Such **ritual behaviors** seem to indicate that elephants somehow know that those bones were a physical part of the lives of their relatives and companions, whom are therefore clearly *no longer alive*: they seem to be aware of, and even grief about, this. As pointed out in an article published online in August 24th, 2018, by Pierce in

Fig. 2.3 Segasira, a juvenile gorilla, was seen making a night nest and staying close to the dead mother until the morning, grooming her, resting against her and attempting to move her head

the *Smithsonian Magazine*, entitled "*Animals experience grief?*," many scientists still have a prejudice against the idea that animals "feel **grief**" or respond in complex ways to death. However, things are changing quickly regarding the scientific recognition of such complex behaviors within non-human animals. Namely, there are various reports, including video footages—including some made by scientific researchers, such as the one done in 2016 by Shifra Goldenberg, who was at that time a Colorado State University doctoral student studying **elephants** in Africa—suggesting that members of elephant families visit the corpses of deceased relatives, repeatedly smelling, touching, and passing by those corpses. Another case study concerned chimpanzees, which have also been repeatedly observed engaging in **death-related behaviors**: in one case, a small group of captive **chimpanzees** was carefully observed after one of their members, an elderly female named Pansy, died. The chimpanzees checked her body for signs of life and cleaned bits of straw from her fur, and then refused to go to the place where she had died for several days afterwards.

Two Portuguese scholars, Gonçalves and Carvalho, discussed this fascinating topic extensively in their superb 2019 paper *Death Among Primate – A Critical Review of Non-human Primate Interactions Towards Their Dead and Dying*. For instance, they show how Segasira, a juvenile gorilla made a night nest and stayed close to the dead mother until the morning, grooming her and resting against her and attempting to move her head (Fig. 2.3). These authors refer to the common visitations, by not only non-human apes but also by other primates such as monkeys, to the place of death, something that remind us so much of our own species:

> Visitations are defined as returns to the place where death ensued or the corpse was last seen. Such places may hold residual information about the event which can arouse curiosity or emotional distress. Smuts describes how, in the weeks following the **infanticide** of a

yellow baboon, the bereaved mother (Zandra) became extremely agitated and called when passing the site of death, apparently initiating a search for her dead infant. In captive pottos, Cowgill reported a surviving couple searching for a dead male in its usual sleeping site following its removal from the cage, and leaving portions of food, presumably for the absent male (according to the author) – a behaviour maintained even when the portion size was reduced. Following the cagemate's removal, the surviving pottos may have suffered a decrease in appetite, suggesting a grief-like response. Similar searches have been described in chimpanzees when no corpse was visible. Perry and Manson describe capuchins, after the removal of a dead infant, alarm calling at site where the corpse was previously seen. Chimpanzees, gorillas, long-tailed macaques and hanuman langurs have been observed returning to the place where a body was last seen and inspecting the ground.... If the corpse is not removed, chimpanzees may revisit it the following day.... Returning to a corpse has been recorded for wild lowland and mountain gorillas at three different sites and captive marmosets. The chimpanzee Flint, soon after the death of his mother Flo, spent two minutes staring at a nest they had shared prior to her death. Later, he returned to the place where Flo had died and 'sank deeper into depression', before his final excursion to the site, where he 'curled up' and died. Patricia Wright reports on a family of sifakas that, after predation of the adult male, gave out lost calls and visited the corpse 14 times in five days. While some of these events may simply indicate curiosity and an attempt to obtain information on the death event, others illustrate continuation of emotional bonds after death that were maintained during life.

It might well be that some animals are even aware that their own death is near, when they are very old, sick, or weak. Also, it is clear that animals such as gazelles seem terrified when they are hunted by carnivore animals such as lions, indicating that they perceive that something that they consider to be "very bad" will happen if they are hunted. However, that is completely different from humans that, *at the prime, strongest time of their existence, and not facing any eminent danger, are fully aware not only that they will die—no matter what they might do to avoid it—but also that both their death and life are dependent on events that they cannot control or predict*. Although it is difficult to be sure that no other animals are aware of this—and I would be very glad to change my position about this idea, if future empirical studies contradict this idea—there is indeed no direct evidence that chimpanzees and gorillas, or elephants, or any other non-human animals, have such an awareness. Examples of the very few, anecdotal, accounts about such a potential awareness in other animals are those done within the framework of ***The American Sign Language* (ASL)** projects, as explained by Gonçalves and Carvalho:

The American Sign Language (ASL) projects...with great apes that emerged in the 1970s also contributed to our knowledge of primate thanatology. Despite no formal tests being done with regards to communicating **concepts of mortality**, and the available data on this matter remaining anecdotal, attempts by researchers to communicate about death for both for the western **lowland gorilla Koko**, and the **chimpanzee Washoe**, had inconclusive results. Koko, when she was seven, was asked a series of questions relating to death, such as when do gorillas die, to which she signed 'Trouble, old', where do they go after death, signing 'Comfortable Hole Bye', and how they feel upon death, signing 'Sleep'. When told that her cat was killed by a speeding car she 'cried'; three days after she was questioned again about the cat, signing 'Sleep'. On one occasion she saw a picture of a similar cat pointing and signing to it 'Cry, Sad, Frown'. A second case involves, Washoe, whose infant had died. Immediately after being told the news, Washoe dropped her cradling arms and 'moved over to a far corner and looked away her eyes vacant'. Others did not follow this

> line of inquiry, such as David Premack, renowned for his research in chimpanzee cognition, who stated: 'Until I can suggest concrete steps in teaching the concept of death without fear I have no intention of imparting the knowledge of mortality to the ape'. Gordon Gallup, whose insightful experiments with mirror recognition suggested that great apes possess an awareness of self, claimed that apes could very well also have an awareness of death.

Of course, another question that is related to the study of biology—that is, the study of *life*—is why do we, and other animals, age and die. This is of course a very interesting question, but I will not discuss it in this book. For readers interested to know more about it, I recommend Klarsfeld and Revah's 2003 book *The Biology of Death – Origins of Mortality*, and many of the fascinating biological, anthropological, and philosophical references cited therein.

In summary, we can see that the only indication, of any kind, that non-human animals might be aware of the inevitability of their own death when they are *not* old, sick, weak, or attacked would be Koko's answer to the question "when do gorillas die?": "Trouble, old." That is, this answer, if it eventually truly reflects what Koko meant, would point out that some non-human animals *might eventually* know they would surely—or just probably? We can't really discern that with Koko's "answer"—die when they will be "old." However, this is far too anecdotal, and concerns a study that is too controversial to be considered solid scientific evidence. Indeed, Koko's study is often criticized within the field of anthropology, and many researchers have serious doubts about its methodology and "results". Scientists, and the broader public in general, would want so much to know what other animals truly think about such issues, but so far unfortunately we were not able to do this in a sound way. Even if Koko really meant what she is reported to have said, we don't know if her ideas would also apply to the randomness of death, because clearly a huge number of gorillas don't die of "old age" but instead due to many other, often completely random, factors, such as being arbitrarily infected by a parasite, virus, or bacteria. In fact, the idea that only humans seem to have a deep understanding about the inevitability and randomness of death has been defended in many works, as in the superb 2020 review paper by Monsó and Osuna-Mascaró, "*Death is common, so is understanding it: the concept of death in other species,*" These authors argue that the number of species that have some kind of knowledge about, or understanding of, death is likely much higher than most lay people and even scientists have admitted, but that this is very different from our **cognitive awareness of death**, that is to think of our death in abstract terms "as something that we know will inevitably befall all of us":

> Brosnan and Vonk use the distinction between, on the one hand, the process of dying and the resulting state of being dead and, on the other hand, death itself, which they consider to be a hypothetical construct, to argue that only animals who can reason about unobservables can acquire a ***concept of death***. However, death is only an abstract concept when one has the human perspective in mind. When we speak of humanity's **fear of death**, for instance, we are thinking of death in abstract terms, as something that we know will inevitably befall all of us, but which we cannot point to or perceive with any of our senses. Depictions of death as a hooded figure with a scythe are attempts to make concrete this unobservable entity that haunts our lives. While death in this sense is 'unobservable', we disagree with Brosnan and Vonk's claim that this warrants us considering death as a hypothetical construct. The process

> of dying and the state of being dead are both very concrete and perceptually accessible entities. Once they occur they are neither 'hypothetical' nor 'constructed'...death can and does assume a physical form whenever it happens. The hypothetical and constructed nature of death only applies to it as our inevitable and not-yet-fulfilled destiny. However, it is unwarranted to assume, without further argument, that this is what we are talking about when discussing whether animals can understand death. It amounts to departing from one of the most sophisticated notions of death and asking whether animals can have *that concept of death*, i.e., *our* concept of death. Understood like this, the question becomes uninteresting: it is self-evident that creatures without a linguistic capacity that can enable an oral culture of narratives surrounding death cannot reach as sophisticated a notion of death as ours.... It seems unlikely that non-linguistic animals can develop a notion of the inevitability of their own and others' death, but **comparative thanatologists** sometimes bring this up in their discussions of what it would take for animals to understand death.... We believe that this has to do with the strong meaning that humans attach to death, which is precisely linked to its being the unavoidable fate of ourselves and those we love. This is what makes death so terrifying, and it can have such a strong influence on our lives that there is even a psychological discipline – ***terror management theory*** – devoted entirely to how humans cope with this fear.... But this gives us a reason to think that, if [other, non-human] animals can develop a *concept of death*, it is more likely not to have these...components...it has been argued that selective pressures are unlikely to have pushed for animals to learn about the **inevitability of death**. According to this view, defended, for instance, by Varki, natural selection is likely to have acted against it, given the fitness-diminishing fear that would come from the knowledge of the inevitability of one's death.

The same point applies to the behavioral studies on **superstition** done in pigeons and other animals. The pigeons display superstitious behaviors when one increases randomness and uncertainty in the system, for instance giving them food at completely random times. But they do not seem to be aware, at all, of that randomness, nor did the Japanese human subjects of Ono's study. That is, contrary to death, which is part of human oral culture since times immemorial, the pigeons and the human subjects of Ono's study had no a priori information about what the experiments done by the researchers would bring them. This is a crucial difference: there was a moment when humans actually became aware of both the *inevitability of* and *the importance of randomness in* death, and then used their linguistic capacities to enable oral cultures of narratives about it. This awareness of not having any kind of true control of many important events that happen during our lives and that lead to our death is the *key difference* between the type of teleological narratives that humans built as a response of that awareness versus the kind of superstitious behaviors displayed by the pigeons and by the humans subjects of Ono's study. Not having control over our life and death, or those of our family members or loved ones or friends, is one of the most difficult ideas for humans to handle. This point has been shown in numerous studies, and unfortunately confirmed by, and used in, some of the most atrocious acts done by humans. One of the most powerful types of mental tortures that can be done to humans is precisely to take away their **sense of control**, thus increasing the feeling of **randomness**, or of **helplessness**, as explained in Diamond's spectacular 2012 book *The World Until Yesterday*:

> Our craving for relief from **feeling helpless** is illustrated by a study of religious Israeli women, carried out by anthropologists Richard Sosis and W. Penn Handwerker. During the 2006 Lebanon War the Hezbollah launched Katyusha rockets against the Galilee region of

northern Israel, and the town of Tzfat and its environs in particular were hit by dozens of rockets daily. Although siren warnings while rockets were en route alerted Tzfat residents to protect their own lives by taking refuge in bomb shelters, they could do nothing to protect their houses. Realistically, that threat from the rockets was unpredictable and uncontrollable. Nevertheless, about two-thirds of the women interviewed by Sosis and Handwerker recited psalms every day to cope with the stress of the rocket attacks. When they were asked why they did so, a common reply was that they felt compelled "to do something" as opposed to doing nothing at all. Although reciting psalms does not actually deflect rockets, it did provide the chanters with a **sense of control** as they went through the semblance of taking action. (Of course, they themselves did not give that explanation; they did believe that reciting psalms can protect one's house from destruction by a rocket). Compared to women in the same community who did not recite psalms, the psalm reciters had less difficulty falling asleep, had less difficulty concentrating, were less inclined to bursts of anger, and felt less anxious, nervous, tense, and depressed. Thus, they really did benefit, by reducing the risk that natural anxiety over uncontrollable danger would cause them to endanger themselves in a different way by doing something foolish. As all of us who have been in situations of unpredictable and uncontrollable danger know, we do become prone to multiply our problems by thoughtlessness if we can't master our **anxiety**.

Diamond also defends the idea that humans are very likely the only extant animals that know they will surely die, no matter what:

Let's now turn to a **function of religion** that must have expanded over the last 10,000 years: to provide comfort, hope, and meaning when life is hard. A specific example is to comfort us at the prospect of our own death and at the death of a loved one. Some mammals – elephants are a striking example – appear to recognize and mourn the death of a close companion. But we have no reason to suspect that any animal except us humans understands that, one day, it too will die. We would inevitably have realized that that fate lay in store for us as we acquired self-consciousness and better reasoning power, and began to generalize from watching our fellow band members die. Almost all observed and archaeologically attested human groups demonstrate their understanding of death's significance by not just discarding their dead but somehow providing for them by burial, cremation, wrapping, mummification, cooking, or other means. This **comforting function of religion** must have emerged early in our evolutionary history, as soon as we were smart enough to realize that we'd die, and to wonder why life was often painful. Hunter-gatherers do often believe in survival after death as spirits. But this function expanded greatly later with the rise of so-called world-rejecting religions, which assert not only that there is an afterlife, but that it's even more important and long-lasting than this earthly life, and that the overriding goal of earthly life is to obtain salvation and prepare you for the afterlife. While **world rejection** is strong in Christianity, Islam, and some forms of Buddhism, it also characterizes some secular (i.e., non-religious) philosophies such as **Plato**'s. Such beliefs can be so compelling that some religious people actually reject the worldly life. Monks and nuns in residential orders do so insofar as they live, sleep, and eat separately from the secular world, although they may go out into it daily in order to minister, teach, and preach. But there are other orders that isolate themselves as completely as possible from the secular world. Among them were the **Cistercian order**, whose great monasteries at Rievaulx, Fountains Abbey, and Jerveaulx in England remain England's best-preserved monastic ruins because they were erected far from towns and hence were less subject to plunder and re-use after they were abandoned. Even more extreme was the world rejection practiced by a few Irish monks who settled as hermits in otherwise uninhabited Iceland. Small-scale societies place much less emphasis on world rejection, salvation, and the afterlife than do large-scale, more complex and recent societies.

There are at least three reasons for this trend. First, **social stratification** and **inequality** have increased, from **egalitarian small-scale societies** to large complex societies with their kings, nobles, elite, rich, and members of highly ranked clans contrasting with their mass of poor peasants and laborers. If everybody else around you is suffering as much as you are, then there is no unfairness to be explained, and no visible example of the good life to which to aspire. But the observation that some people have much more comfortable lives and can dominate you takes a lot of explaining and comforting, which religion offers. A second reason why large, complex societies emphasize comforting and the **afterlife** more than do small-scale societies is that archaeological and ethnographic evidence shows that life really did become harder as **hunter-gatherers** became **farmers** and assembled in larger societies. With the transition to agriculture, the average daily number of work hours increased, nutrition deteriorated, **infectious disease** and body wear increased, and lifespan shortened. Conditions deteriorated even further for urban proletariats during the **Industrial Revolution**, as work days lengthened, and as hygiene, health, and pleasures diminished. Finally…complex populous societies have more formalized moral codes, more black-and-white emphasis on good and evil, and bigger resulting problems of theodicy: why, if you yourself are behaving virtuously and obeying the laws, do law-breakers and the rest of the world get away with being cruel to you? All three of these reasons suggest why the comforting function of religion has increased in more populous and recent societies: it's simply that those societies inflict on us more bad things for which we crave comfort. This comforting role of religion helps explain the frequent observation that misfortune tends to make people more religious, and that poorer social strata, regions, and countries tend to be more religious than richer ones: they need more comforting. Among the world's nations today, the percentage of citizens who say that religion is an important part of their daily lives is 80%–99% for most nations with per-capita gross domestic products (GDP) under $10,000, but only 17%–43% for most nations with per-capita GDP over $30,000. (That doesn't account for high religious commitment in the rich U.S.)…. Even within just the U.S., there appear to be more churches and more church attendance in poorer areas than in richer areas, despite the greater resources and leisure time available to build and attend churches in richer areas. Within American society, the highest religious commitment and the most radical Christian branches are found among the most marginalized, underprivileged social groups.

Importantly, there are various empirical studies supporting the idea that an **awareness of death**—either as a **psychological death** or as a **cessation of bodily "agency"**—is already found in young children, as noted in Bering's 2011 book *Belief Instinct*:

In a 2004 study reported in *Developmental Psychology*, Florida Atlantic University psychologist David Bjorklund and I performed a puppet show for two hundred three- to twelve-year-olds. Every child was presented with the story of Baby Mouse, who was out strolling innocently in the woods. 'Just then', we told them, 'he notices something very strange…. The bushes are moving! An alligator jumps out of the bushes and gobbles him all up. Baby Mouse is not alive anymore'…the children were asked about the dead character's psychological functioning. 'Does Baby Mouse still want to go home?' we asked them. 'Does he still feel sick?' 'Can he still smell the flowers?' The youngest children in the study, the three- to five-year-olds, were significantly more likely to reason in terms of psychological continuity than were children from the two older age groups. But here's the really curious part. Even the preschoolers had a solid grasp of biological cessation; they knew, for example, that dead Baby Mouse didn't need food or water anymore. They knew he wouldn't grow up to be an adult mouse. Eighty-five percent of the youngest kids even told us that his brain no longer worked. Yet, in answering our specific questions, most of these very young children still attributed thoughts and emotions to dead Baby Mouse, telling us that he was hungry or thirsty, that he felt better, or that he was still angry at his brother. One couldn't say that the preschoolers lacked a concept of death, therefore, because nearly all of the kids realized that biological imperatives no longer apply after death. Rather, they seemed to have trouble using this knowledge to theorize about related mental functions.

From an evolutionary perspective, some scholars believe that a coherent theory about psychological death is not necessarily vital. Anthropologist H. Clark Barrett of the University of California at Los Angeles, argues instead that understanding the cessation of bodily 'agency' (for example, that a dead creature isn't going to suddenly leap up and bite you) is probably what saved lives (and thus genes) in the ancestral past. According to Barrett, comprehending the cessation of the mind, on the other hand, has no survival value and is, in an evolutionary sense, neither here nor there. In a 2005 study published in the journal *Cognition*, Barrett and psychologist Tanya Behne of the University of Manchester in England reported that city-dwelling four-year-olds from Berlin were just as good at distinguishing sleeping animals from dead ones as were hunter-horticulturalist children from the Shuar region of Ecuador. So even today's urban children, who generally have very little exposure to dead bodies, appear tuned in to perceptual cues signaling death. A 'violation of the body envelope' (in other words, a mutilated carcass) is a pretty good sign that one needn't worry about tiptoeing around. On the one hand, then, from a very early age, children realize that dead bodies are not coming back to life. On the other hand, also from a very early age, kids endow the dead with ongoing psychological functions, using their **theory of mind**. This conception may be at least partially why the idea of brainless zombies seems to us so implausible and the stuff of horror movies, limited to only a handful of cultures, whereas some type of **spiritual afterlife**, with the **mindful souls** of the deceased passing on to the other side, is by contrast strikingly mundane.

One illustrative and very tragic example of the randomness of both life and death that we teach our medical students at Howard University concerns **amyotrophic lateral sclerosis**, or ALS (see Fig. 2.4). In most cases, this disease occurs apparently at random with no clearly associated risk factors, and most people develop symptoms between the ages of 40 and 70 years, with an average age of 55 at the

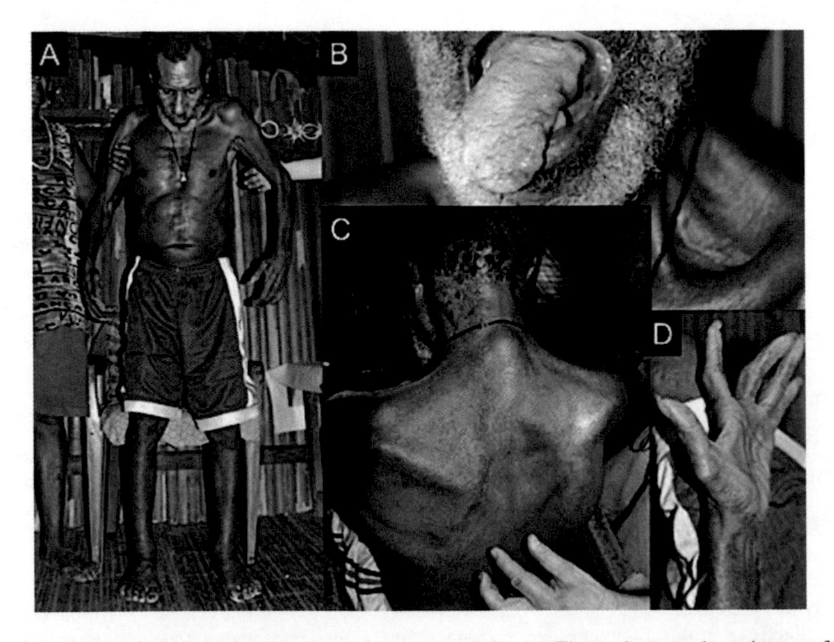

Fig. 2.4 Patient with amyotrophic lateral sclerosis (ALS). (**a**) The patient needs assistance from family members to stand. (**b**) Advanced atrophy of the tongue. (**c**) There is upper limb and truncal muscle atrophy. (**d**) Advanced atrophy of the thumb region

time of diagnosis. It is a dreadful progressive **neurodegenerative**, **motor neuron disease** with no known cure: when the motor neurons die, the ability to initiate and control muscle movement is lost, and the loss of skeletal muscle innervation causes muscle atrophy. Terribly, the life expectancy of a person with this disease averages only about 2–5 years from the time of diagnosis, although many people can live with the disease for 5 years or even more, as was the case of the theoretical physicist **Stephen Hawking**, who was the world's longest surviving ALS patient. He was diagnosed at the age of 21—a very early onset of this disease—and died in 2018 at the age of 76. Most deaths result from respiratory complications—as respiratory muscles weaken patients lose the ability to breathe on their own and must depend on ventilatory support for survival. In the U.S., over 90% of people choose to die rather than live with this disease on ventilatory support.

Of course, this disease is relatively rare—it has a prevalence of about 1 in 30,000–50,000 persons in the U.S. general population, but knowing, when you are younger than 55, that you *might* turn out, without any apparent reason, to have this disease, and then just have 2 or 3 or 4 years to live your life, under very dreadful conditions, is the type of thought that most humans cannot stand. And, in this specific case, thinking about something as terrible as this is clearly something that other animals would not do, because they obviously are not aware about the details of ALS, the average age when it normally occurs, that it is a progressive disease, what it does, and how it kills. The knowledge we have about this disease is therefore indeed an example of how our awareness of these diseases and the inexorableness of death is both a "blessing"—in the sense that it allows us to start developing some kind of medicine to try to delay the progression of the disease—and a "curse"—because we are aware of this, and when we know that we have this disease we are completely conscious of how our body will deteriorate, progressively. Faced with such a consciousness about the randomness of both life and death and the inevitability of death, it is really not a surprise that humans had to create and believe complex teleological narratives to try to "make sense" of all of this, such as to think that such horrible diseases in particular, or death in general, are part of a "cosmic purpose," that they have a "meaning" within a "masterplan" made by a "higher power."

There are many studies about the connection between **patternicity**, **randomness**, the **feeling of control**, and beliefs in such "higher powers," as reviewed by Shermer:

> Patternicites do not occur randomly but are instead related to the context and environment of the organism, to what extent it believes that it is in control of its environment. Psychologists call this ***locus of control***. People who rate high on internal locus of control tend to believe that they make things happen and that they are in control of their circumstances, whereas people who score high on external locus of control tend to think that circumstances are beyond their control and that things just happen to them.... In fact, people who consider themselves 'skeptics' about the **paranormal** and **supernatural** tend to score high in internal locus of control, whereas self-reported 'believers' in **spiritualism**, **reincarnation**, and **mystical experiences** in general tend to rate high in external locus of control. Locus of control is also mediated by levels of certainty or uncertainty in physical and social environments. Broniskaw Malinowski's famous studies of **superstitions** among the Trobriand Islanders in the South Pacific demonstrated that as the level of uncertainty in the environment increases so, too, does the level of superstitious behavior.

> In a 2008 study descriptively entitled "*Lacking control increases illusory pattern perception*" by management researchers Jennifer Whitson at the University of Texas-Austin and Adam Galisnky from Northwestern University…the researchers conducted six experiments to test the thesis that 'when individuals are unable to gain a sense of control objectively, they will try to gain in perceptually…**feelings of control** are essential for our well-being…lacking control is highly aversive, and one fundamental way we can bolster our sense of control is to understand what is going on…so we instinctively seek out patterns to regain control – even if those patterns are illusory'. This is reminiscent of a 1976 study by Harvard psychologist Ellen Langer and her colleague Judith Rodin…in a New England nursing home. Residents were given plants and the opportunity to see weekly films but with some variation of control. Residents of the fourth floor, who were in charge of watering the plants and could choose the night of the week they wanted to view the film, lived longer and healthier lives than the other residents, even those given plants that were watered by the staff. It was the **sense of control** that had the apparent effect on health and well-being.

The "curse" side of our "*blessing and curse*" awareness of the inevitability and randomness of death was elegantly summarized by Bangambiki Habyarimana: "it's better to be ignorant and live in bliss than know the truth and live in agony." So, very likely since the moment humans gained this awareness, there were two main ways of reacting to it, which are chiefly the same two ways of dealing with it nowadays for most people, with of course some exceptions characterizing a smaller amount of people. One type of reaction to this "curse," to this "agony"—which is somewhat similar to what many people nowadays call "**depressive Nihilist ideas**" (see Chap. 9)—is to basically think: *if I will surely die one day when I will be old, and my death, as well as the serious diseases/injuries I will have before it, are in great part related to random events that I can't control—a gene inherited from my mother, a drunken person crashing a car against mine—why should I make an effort to find food, take care of myself and others, learn stuff, and reproduce and put more people in such a random, meaningless world?* As noted in Thagard's 2010 book *The Brain and the Meaning of Life*: "why don't you kill yourself? **Albert Camus** began his book *The myth of Sisyphus* with the startling assertion 'there is but one truly serious philosophical problem and that is suicide'." It is important to note that Camus—a French novelist and philosopher who won the Nobel Prize for literature in 1957—is often not considered to be a "pessimist Nihilist." Instead, his "**absurdism**"—related to the "absurd" situations created for instance by the dissonance between the **pursuit of meaning** and a meaningfulness cosmos—is instead more often said to reflect an "**optimistic Nihilism**," eventually. Still, as noted by Thagard, "Camus said that judging whether life is or is not worth living amounts to answering the fundamental question of philosophy…if life is meaningless, there is no point to pursuing traditional philosophical questions about the nature of reality, knowledge, and morality." An opposite reaction is to think instead: yes, I will die, but that is precisely part of the "masterplan" designed by God or by Mother nature or another "higher power," so by knowing that "masterplan" I do have some sense of control because I know what is going on, and moreover by being part of it I am *useful*, as I am here to be *part of it, I am meant to be*. So, I will accordingly fulfill my cosmic purpose, and by doing this I will make that "higher power" proud of me, perhaps I can even be compensated for this in some kind of **afterlife**. As explained in the same book by

Thagard—who is a non-believer, it should be noted—for "most people today, religion provides a major source of answers to such questions about the meaning of life…from a religious perspective, meaning arises not from any meager aspect of our daily lives, but from our profound connections with God, who brought us into existence and who provides the possibility of eternal happiness."

This idea is supported by the obvious fact that, apart from notions of afterlife, basically all religions involve imaginary stories that make us feel as if we, humans, are special and thus have a cosmic purpose, a "duty" to be made, even within *this* life. This is a key trick of religious thought, if not most people would just do nothing, and wait for afterlife, particularly in those cases that such afterlife is supposedly to be lived in a **paradise**: why wait? It is therefore not at all a coincidence that many religions condemn suicide, and use teleological narratives about how we need to "work" for God, or "reproduce," or even "fight" for him: that is our "duty," that is why he put us here. The fact that most religions try so hard to make their followers feel "special" was extensively discussed in the 2000 book *Science and Religion in Search of Cosmic Purpose*, edited by Haught. This book documented how all major **cosmologies** share an obsession for teleological narratives, including those of **Christianity** and **Islam**, the **Indian cosmogony** and the **law of *Karma***, and even **Confucianism** and its notion of a "moral purpose in the universe." One of the obvious examples is the idea, almost consensually accepted in the West for many centuries, that it was the sun that moved around the Earth, and not vice versa. "Our planet" has to be the "center of the universe." Of course, this idea has been contradicted by science since then, but, interestingly, this does not seem to affect at all the believers of all those religions that have cosmologies related to that idea. Facts can never defeat faith: one just needs to adapt a bit the narrative, which is, now, that those cosmologies were not made to be taken "literally," but "figuratively."

Similarly, humans tended to believe—and many still do—that they, and not other organisms, are the center of "Creation," as famously illustrated in **Michelangelo**'s beautiful painting *The Creation of Adam*, depicted at the ceiling of the **Sistine Chapel** in the **Vatican** (Fig. 2.5). That is, there is a need to think that not only this planet, but the whole universe, was made *thinking about us, Homo sapiens*. Of course, once again, the available empirical data shows otherwise: if one would *truly think rationally* about life in this planet, one would realize that within a 24-h clock representing the history of planet Earth, our species actually only appeared 4 s before midnight (see Fig. 2.6). This is an extraordinary late timing, if the whole planet is supposed to exist mainly *for* us. That is, if we were to analyze this in an objective, rational way, we would easily conclude that we are not at all the "center of Creation," but instead simply a random, and very late, by-product of the particular type of biological evolution that developed in this planet. But thinking in such an objective, rational way is something extremely difficult for the self-designated "sapient being," who is instead truly a **storytelling animal** that, as will be explained in this book, is better defined as a ***Homo fictus, irrationalis, et socialis***.

So, this begs the question: which of the two main different ways to react to the randomness of life and inescapability of death mentioned above do you think was more evolutionarily successful within our human lineage? In other words, in a

Fig. 2.5 Michelangelo's "Creation of Adam," from the ceiling of the Sistine Chapel in the Vatican

context of survival and reproduction that leads to a higher number of descendants in the long term, within two groups of humans reacting in such different ways many hundreds of thousands or even millions of years ago—that is, a "believer" group and a "non-believer" one—which of them would outcompete the fewer and fewer members of the other group? Most people to whom I ask this question, both from the scientific community and the lay public, have no doubts about the answer: at that moment in time, the "believers" would normally outcompete the non-believers. Yes, this was very likely the case then, and today we continue to feel the long-term repercussions of such an evolutionary event, as most people in this planet continue to be believers, and the proportion is actually seemingly increasing as we will see below (see Figs. 2.9 and 2.10). One of the major reasons that very likely lead to the ascendancy of the believers is that that way of reacting leads to a higher *motivation* to live and survive, and thus to fight and to kill, as well as to reproduce, precisely because they are absolutely convinced that they are part of a transcendental, higher "purpose of life" or "masterplan." **Blind acceptance of beliefs**, **uncritical fanaticism, and zeal** and **obsessive enthusiasm** are among the most powerful mental weapons one can use in a fight. As a personal example, I surely would not spend months or years of my life—and eventually lose my whole life, as religious suicide bombers do—to try to get bombs, hide from the police and everybody else, then put those bombs around my body, then meticulously avoid controls in order to reach a metro and activate those bombs so that I can kill dozens or hundreds of people from a group of people that don't believe in a certain God, or that believe in another God. I would

Fig. 2.6 Life on earth as a 24-h clock

never do this, because I do *not* believe in a specific God or any other kind of particular "higher purpose" or ideology that would make me do so.

Indeed, within the type of classification shown in Fig. 2.10, which we will discuss below, I would be among a very small minority: those 18% of people—basically, less than 1 out of 5—that live in a so-called modern secular country such as the U.S. that declare themselves to be "neither religious nor spiritual." Still, that does not mean that I don't believe in anything. What make humans truly unique is that, because of our awareness of the inevitability of death, we begun to build and believe in complex narratives about a "cosmic" purpose of life, and that applies both to people that declare themselves religious—for instance, there is a meaning for our existence, which was created by God or other religious deities—or spiritual—for example, the meaning of life has to do with Mother Earth's sapience, or to the elemental energies swirling within each of us and Mother Earth, and so on. But much before humans even existed, there were already other types of beliefs, based not on complex stories about cosmic purpose but on simple phenomena such as **agenticity** and **patternicity**, which are illustrated by the superstitions of pigeons and the paranoiac behavior of animals in the Savannah when they detect any type of movement in the grass. So even within those 18% of non-religious, non-spiritual people living in the U.S., most still believe in strange stuff, other than there is a cosmic meaning for their existence. This point surely includes me: despite trying as hard to be objective, and scientifically knowing that there is no link between hearing a music about death in a plane and the engines of that plane stopping or that plane being hit by something, when I am listening to music in a plane and a song about death starts, I just immediately stop it.

The point that, in a way or another, we are basically all believers was brilliantly made by Sproul in her 1991 exceptional book *Primal Myths* compiling numerous different **creation myths** around the globe, in which she shows that many people that declare themselves as "non-believers" nowadays actually do accept narratives that are very similar to those of such "**primal myths**":

> The most profound human questions are the ones that give rise to creation myths: Who are we? Why are we here? What is the purpose of our lives and our deaths? How should we understand our place in the world, in time and space? These are central questions of value and meaning, and, while they are influenced by issues of fact, they are not in themselves factual questions; rather, they involve attitudes toward facts and reality. As such, the issues that they raise are addressed most directly by myths. Myths proclaim such attitudes toward reality. They organize the way we perceive facts and understand ourselves and the world. Whether we adhere to them consciously or not, they remain pervasively influential. Think of the power of the **first myth of Genesis** (1-2:3) in the **Old Testament**. While the scientific claims it incorporates, so obviously at odds with modern ones, may be rejected, what about the myth itself? Most Westerners, whether or not they are practicing Jews or Christians, still show themselves to be the heirs of this tradition by holding to the view that people are sacred, the creatures of God. Declared unbelievers often dispense with the frankly religious language of this assertion by renouncing God, yet even they still cherish the consequence of the myth's claim and affirm that people have inalienable rights (as if they were created by God). And, further, consider the beliefs that human beings are superior to all other creatures and are properly set above the rest of the physical world by intelligence and spirit with the obligation to govern it – these beliefs are still current and very powerful. Even the notion that time is properly organized into seven-day weeks, with one day for rest, remains widely accepted. These attitudes toward reality are all part of the first myth of Genesis. And whether people go to temple or church, whether they consider themselves religious, to the extent they reflect these attitudes in their daily behavior, they are still deeply Judeo-Christian…. It is no accident that cultures think their creation myths the most sacred, for these myths are the ground on which all later myths stand. In them members of the group (and now outsiders) can perceive the main elements of entire structures of value and meaning…. Thus, because of the way in which domestic myths are transmitted, people often never learn that they are myths; people become submerged in their viewpoints, prisoners of their own traditions.

As famously stated by Chesterton, "*when men stop believing in God they don't believe in nothing; they believe in anything*." In reality, this even applies to those that follow ideologies that are supposedly "against beliefs and religion," such as communism (see also Boxes 2.5 and 2.6). As explained in Figes' 2007 book *The Whisperers – A Private Life in Stalin's Russia*:

> The published **Gulag memoirs** influenced not only the recollection of scenes and people, but the very understanding of the experience. All the memoirs of the ***Stalin Terror*** are reconstructed narratives by survivors. The story they tell is usually one of purgatory and redemption – a journey through the 'hell' of the Gulag and back again to 'normal life' – in which the narrator transcends death and suffering. This uplifting moral helps to account for the compelling influence of these literary memoirs on the way that other Gulag survivors recalled their own stories. Ginzburg's memoirs, in particular, became a model of the survivor narrative and her literary structure was copied by countless amateur memoirists with life-stories not unlike her own. The unifying theme of **Ginzburg's memoirs** is regeneration through love – a theme which gives her writing powerful effect as a work of literature. Ginzburg explains her survival in the camps as a matter of her faith in human beings; the flashes of humanity she evokes in others, and which help her to survive, are a response to her faith in people. In the first part of her memoirs, *Into the whirlwind*, Ginzburg highlights her work in a nursery at Kolyma where caring for the children reminds her of her son and gives her the strength to go on. In the second part, *Within the whirlwind* (in 1981), Ginzburg is transferred from the nursery to a hospital, where she falls in love with a fellow prisoner serving as a doctor in the camp. Despite the anguish of repeated separations, they both survive and somehow keep in touch until **Stalin**'s death; freed but still in exile from the major Russian cities, they get married and adopt a child. This narrative trajectory is endlessly repeated in the memoir literature. The uniformity of such 'family chronicles' and 'documentary tales', which are virtually identical in their basic structure, in their form and moral tone, is remarkable and cannot be explained by literary fashion on its own. Perhaps these memoirists, who all lived such extraordinary lives, felt some need to link their destiny to that of others like themselves by recalling their life-story according to a literary prototype. **The Soviet narrative** offered a different type of consolation, assuring the victims that their *sacrifices* had been in the service of *collective goals and achievements*. The idea of a common ***Soviet purpose*** was not just a propaganda myth. It helped people to come to terms with their suffering by giving them a sense that their lives were validated by the part they had played in the struggle for the *Soviet ideal*.
>
> The collective memory of the **Great Patriotic War** was very potent in this respect. It enabled veterans to think of their pain and losses as having a larger *purpose and meaning*, represented by the victory of 1945, from which they took pride. The historian Catherine Merridale, who conducted interviews with veterans in Kursk for her book on the **Soviet army** in the war, found that they did not speak about their experiences with bitterness or self-pity, but accepted all their losses stoically, and that 'rather than trying to relive the grimmest scenes of war, they tended to adopt the language of the vanished Soviet state, talking about honour and pride, of justified revenge, of motherland, Stalin, and the absolute necessity of faith'. As Merridale explains, this identification with the **Soviet war *myth*** was a coping mechanism for these veterans, enabling them to live with their painful memories: "Back then, during the war, it would have been easy enough to break down, to feel the depth of every horror, but it would also have been fatal…the path to survival lay in stoical acceptance, a focus on the job at hand…the men's vocabulary was businesslike and optimistic, for anything else might have induced despair…sixty years later, it would have been easy again to play for sympathy or simply to command attention by telling bloodcurdling tales…but that, for these people, would have amounted to a betrayal of the values that have been their collective pride, their way of life." People who returned from the labour camps similarly found consolation in the Stalinist idea that, as Gulag labourers, they too had made a contribution to the **Soviet economy**. Many of these people later looked back with enormous pride at the factories, dams and cities they had built. This pride stemmed in part from their continued ***belief in the Soviet system and its ideology***, despite the injustices they had been dealt, and in part, perhaps, from their need to find a *larger meaning* for their suffering.

Box 2.5 Theism, Atheism, Science, and Belief as a Human Universal

In his 2018 book *Seven Types of Atheism*, Gray provides interesting reflections about some of the key issues discussed in the present book:

> Atheism has not always been like this. Along with many who have searched for a surrogate Deity to fill the hole left by the God that has departed, there have been some who stepped out of monotheism altogether and in doing so found freedom and fulfillment. Not looking for cosmic meaning, they were content with the world as they found it…. Rather than **atheism** being a worldview that recurs throughout history, there have been many atheisms with conflicting views of the world. In ancient Greece and Rome, India and China there were schools of thought that, without denying that gods existed, were convinced they were not concerned with humans. Some of these schools developed early versions of the philosophy which holds that everything in the world is composed of matter. Others held back from speculating about the nature of things. The Roman poet **Lucretius** thought the universe was composed of 'atoms and the void', whereas the Chinese mystic **Chuang Tzu** followed the (possibly mythical) Taoist sage **Lao Tzu** in thinking the world had a way of working that could not be grasped by human reason. Since their view of things did not contain a divine mind that created the universe, both were atheists. But neither of them fussed about 'the existence of God', since they had no conception of a creator-god to question or reject.
>
> Religion is universal, whereas monotheism is a local cult. Many 'primitive' cultures contain elaborate **creation myths** – stories of how the world came into being. Some tell of it emerging from a primordial chaos, others of it springing from a cosmic egg, still others of it arising from the dismembered parts of a dead god. But few of these stories feature a god that fashioned the universe. There may be gods or spirits, but they are not supernatural. In **animism**, the original religion of all humankind, the natural world is thick with spirits…. Just as not all religions contain the idea of a creator-god, there have been many without any idea of an immortal soul. In some religions – such as those that produced **Norse mythology** – the gods themselves are mortal. Greek polytheists expected an **afterlife**, but believed it would be populated by the shades of people who had once existed, not these people in a posthumous form. **Biblical Judaism** conceived of an underworld (**Sheol**) in much the same way. **Jesus** promised his disciples salvation from death, but through the resurrection of their fleshly bodies, divinely perfected. There have been atheists who believed human personality continues after bodily death. In Victorian and Edwardian times, some psychical researchers thought an afterlife meant passing into another part of the natural world.

About science, knowledge, beliefs, and atheism, Gray writes:

> The story of **Adam and Eve** eating from the **Tree of Knowledge** is a mythical imagining of the ambiguous impact of knowledge on human freedom. Rather than being inherently liberating, knowledge can be used for purposes of enslavement. That is what is meant when, having eaten the forbidden apple after the serpent promises them they will become like gods, Adam and Eve find themselves exiled from the **Garden of Eden** and condemned to a life of unceasing labour. Unlike scientific theories, myths cannot be true or false. But myths can be more or less truthful to human experience. The **Genesis myth** is a more truthful rendition of enduring human conflicts than anything in **Greek philosophy**, which is founded on the myth that knowledge and goodness are inseparably connected.

Science cannot replace a religious view of the world, since there is no such thing as 'the scientific worldview'. A method of inquiry rather than a settled body of theories, science yields different views of the world as knowledge advances. Until **Darwin** showed that species change over time, science pictured a world of fixed species. In the same way, classical physics has been followed by quantum mechanics. It is commonly assumed that science will someday yield a single unchanging view of things. Certainly some views of the world are eliminated as scientific knowledge advances. But there is no reason for supposing that the progress of science will reach a point where only one worldview is left standing. Some will stay this is tantamount to **relativism** – the claim that views of the world are only cultural constructions, none of them true or false. Against this philosophy, it is asserted that science is the exercise of discovering universal laws of nature. But unless you believe the human mind mirrors a rational cosmos – the faith of **Plato** and the **Stoics**, which helped shape **Christianity** – science can only be a tool the human animal has invented to deal with a world it cannot fully understand. No doubt our knowledge has increased, and will continue to increase. But the order that appears to prevail in our corner of the universe may be local and ephemeral, emerging randomly and then melting away. The very idea that we live in a law-governed cosmos may be not much more than a fading legacy of faith in a divine law-giver. Atheists who think of religions as erroneous theories mistake **faith** – trust in an unknown power – for belief. But if there is a problem with belief, it is not confined to religion. Much of what passes as scientific knowledge is as open to doubt as the miraculous events that feature in traditional faiths. Wander among the shelves of the social sciences stacks in university libraries, and you find yourself in a mausoleum of dead theories. These theories have not passed into the intellectual netherworld by being falsified. Most are not even false; they are too nebulous to allow empirical testing. Systems of ideas such as **Positivism** and **Marxism** that forecast the decline of religion have been confounded time and time again. Yet these cod-scientific speculations linger on in a dim afterlife in the minds of many who have never heard of the ideas from which they sprang.

Box 2.6 Communism, Science, Vavilov, Lysenko, and Stalin

A clear, and particularly sad, and outrageous, example of how political narratives and scientific biases are profoundly linked, and that further shows how science is not as "objective" as it is often portrayed, is the case of **Nikolai Vavilov**. He was one of the most bright **soviet biologists**, and many of his broader evolutionary ideas are now coming back in force, as I could attest when I gave a talk in Moscow, at the Russian Academy of Sciences, some years ago. However, for decades his work was almost forgotten, and actively persecuted, in the **Soviet Union**, while that of **Lysenko**, who was in some ways a charlatan, was increasingly seen as of a scientific genius. Worse than that, there is strong evidence that Lysenko, and particularly **Stalin**, were directly involved in the persecution against Vavilov, which ultimately resulted in his imprisonment and his death because of disease and starvation in a miserable soviet prison. This is how far **scientific biases**, and **political teleological narratives**, such as those invoked in **communism** and **Marxism**, as well as those invoked in capitalism and in basically all types of political storytelling, can go. In this particularly case, the problem for Vavilov is that his ideas about genes and their importance in evolution and for plant breeding were opposed to the Stalin's communist view that epigenetics was the key, and that genetics was mainly a field related to the West and to capitalism, as was defended by Lysenko. A book that I strongly recommend about Vavilov, Lysenko, and Stalin is Pringle's *The Murder of Nikolai Vavilov – The Story of Stalin's Persecution of One of the Great Scientists of the Twentieth Century*. Of course, as will be explained in the present book, we now know that epigenetics does play a much more important role than most Neo-Darwinists were defending at that time, so in this respect some of the biological ideas defended by Stalin's communism were not at all wrong. However, the main political narrative that the total "real theory of the knowledge of the world was given by Marx, Engels, and Lenin" is obviously wrong and, whether or not some of Vavilov's ideas about genetics might have eventually been not completely right, his personal persecution, imprisonment, and ultimately death in prison are completely unacceptable as well. As explained by Pringle:

> When Stalin praised Lysenko in a Kremlin hall packed with plant breeders and collective farm workers, he marked a new and destructive era in the history of **Soviet biology**. A jubilant Lysenko quickly claimed a series of new plant breeding achievements, each one less credible than the last, yet each one carefully couched as "practical science", and linked expressly to Stalin's demands for quicker results. The agricultural bosses followed their leader by giving Lysenkoism renewed support. Commissar Yakovlev praised Lysenko as a "practical worker whose vernalization of plants has opened a new chapter in agricultural science"...a scientist who was now "heeded by the entire agricultural world, not only here, but abroad as well". Lysenko's "people" would become the "backbone of the real Bolshevik apparatus". Vavilov did not respond, at first. It was not in his nature to engage in an academic brawl – and, even if he had wanted to, the stakes were now much higher. For a geneticist to criticize Lysenko's claims when Soviet agricultural production was failing was to seem unpatriotic at best and, at worst, to be engaging in economic sabotage. Lysenko himself encouraged such treasonable thoughts. His speeches were routinely laced with references to the nefarious activities of "bourgeois scientists", "saboteurs", and "class enemies". In the Kremlin speech to the collective workers, the passage that drew the most applause, and Stalin's extraordinary exclamation, was not about any new achievement in crop production but Lysenko's attack on bourgeois science. All the bourgeois academics ever did was "observe and explain phenomena", he said, while socialist science aimed to "alter the plant and animal world in favor of the building of socialist society".
>
> Lysenko was a fiery speaker who knew his audience. "You see, comrades, saboteur-kulaks are found not only in your kolhoz life.... They are no less dangerous, no less accursed, in science.... Comrades, was there – and is there – really no class struggle on the vernalization front?.... Indeed there was.... Instead of helping the collective farmers, they sabotaged things. Both within the scientific world and out-

side it, a class enemy is always an enemy, even if a scientist". Isaak Prezent, Lysenko's viperish political minder, wrote the speeches, but Lysenko's delivery was masterful. He knew exactly how and when to grab Stalin's attention. It was the kolhoz system, the "mass of kolhoz workers", the collective farmers, who would provide solutions to Soviet agriculture, Lysenko said, not these "so called scientists". And it was practical solutions like his own vernalization, solutions that came from ordinary, poorly educated farmers such as himself, that would give the collective workers a chance to prove their worth. Apologizing for his own lack of training, Lysenko had ended his Kremlin speech by emphasizing the distinction between himself and the academic theorists. He was not an academic, he was not an orator, he modestly declared, he was "only a vernalizer". It was at this point that Stalin jumped to his feet and shouted his approbation and the Kremlin audience broke out in wild applause….

In his new powerful position as head of the Lenin Academy, Lysenko, the poorly educated vernalizer, the barefoot scientist from Kharkov, had become a monster, an arrogant, selfimportant party hack. He called Vavilov to Moscow to give an account of the Leningrad Institute, but he had no interest in the constructive scientific dialogue Vavilov had once offered him. He had no interest in Vavilov's report of his international seed collection now including 250,000 samples and the biggest in the world. He was only interested in Vavilov's humiliation as a scientist and in his total defeat. For this odious exercise Lysenko employed one of his stooges at the Lenin Academy, a schoolyard bully named Lukyanenko. The May 1939 session was recorded by a stenographer and here is a sample of the interrogation…. "*LYSENKO: I understood from what you wrote that you came to agree with your teacher Bateson that evolution must be viewed as a process of simplification…; VAVILOV: When I studied with Bateson…; LUKYANENKO: An anti-Darwinist.; V: No. Some day I'll tell you about Bateson, a most fascinating, most interesting man.; LUKYANENKO: Couldn't you learn from Marx? Marxism is the only science. Darwinism is only a part, the real theory of the knowledge of the world was given by Marx, Engels, and Lenin*"….

[Many years later] on January 24, 1943, emaciated and suffering from a fever, he [Nikolai Vavilov] was moved to the prison hospital. As he entered he introduced himself, "You see before you, talking of the past, the Academician Vavilov, but now according to the opinion of the investigators, nothing but dung". In the hospital he complained of chest pains and shortness of breath. He had diarrhea and was hardly eating. He was put on "meals of the 2nd type" that included milk, and glass jars were placed on his chest in an effort to extract the fever that the doctor noted might be a recurrence of malaria. A committee of Saratov jail doctors examined Nikolai Ivanovich. They noted that he was complaining of overall weakness. And they saw "maceration, pale skin, and swelling in the feet". Their diagnosis was: "dystrophy from prolonged malnutrition". On January 26, 1943, at seven o'clock in the morning, Nikolai Ivanovich's heart stopped beating. The official cause of death was recorded as pneumonia, a cold caught in the prison exercise yard, perhaps. This eminent plant hunter who had a plan to feed the world had died of starvation.

In his short, but profound, 2009 paper "*Human uniqueness and the denial of death*" Varki has already proposed—as well as Brower, cited by him—that the awareness of the inevitability of death would have been an even much worse *curse* for humans—that is, concerning the evolutionary survival of our human lineage as a whole—if no groups of humans had been able to start creating *and* believing in imaginary tales to 'deny death':

> Among key features of **human uniqueness** are full self-awareness and…attributes [that] may have been positively selected because of their benefits to interpersonal communication, cooperative breeding, language and other critical human activities. However, the late Danny Brower, a geneticist from the University of Arizona, suggested to me that the real question is why they should have emerged in only one species, despite millions of years of opportunity. Here, I attempt to communicate Brower's concept. He explained that with full self-awareness and inter-subjectivity would also come awareness of [the inevitability of] death and mortality. Thus, far from being useful, the resulting overwhelming fear would be a dead-end evolutionary barrier, curbing activities and cognitive functions necessary for survival and reproductive fitness. Brower suggested that, although many species manifest features of self-awareness (including orangutans, chimpanzees, orcas, dolphins, elephants and perhaps magpies), the transition to a fully human-like phenotype was blocked for tens of millions of years of mammalian (and perhaps avian) evolution. In his view, the only way these properties could become positively selected was if they emerged simultaneously with ***neural mechanisms for denying mortality***. Although aspects such as denial of death and awareness of mortality have been discussed as contributing to human culture and behaviour…to my knowledge Brower's concept of a long-standing evolutionary barrier had not previously been entertained. Brower's contrarian view could help modify and reinvigorate ongoing debates about the origins of human uniqueness and inter-subjectivity. It could also steer discussions of other uniquely human 'universals', such as the ability to hold false beliefs, existential angst, theories of ***after-life***, religiosity, severity of grieving, importance of death rituals, risk-taking behaviour, panic attacks, suicide and martyrdom…. If this logic is correct, many warm-blooded species may have previously achieved complete self-awareness and inter-subjectivity, but then failed to survive because of the extremely negative immediate consequences. Perhaps we should be looking for the mechanisms (or loss of mechanisms) that allow us to delude ourselves and others about reality, even while realizing that both we and others are capable of such ***delusions and false beliefs***.

One superb book that focuses entirely on the history of our fear of death and our belief in afterlife was just published a few months ago—Erhman's 2020 *Heaven and Hell – A History of the Afterlife*, which explains:

> Traditional **Christian beliefs** in the **afterlife** continue to be widely held in our society. A recent Pew Research Poll showed that 72 percent of all Americans agree that there is a literal heaven where people go when they die; 58 percent believe in an actual, literal **hell**. These numbers are, of course, down seriously from previous periods, but they are still impressive. And for the historian, it is important to realize that in the Christian West prior to the modern period – think, for example, the Middle Ages or, for that matter, the 1950s – virtually everyone believed that when they died their soul would go to one place or the other (or to **Purgatory** in painful preparation for ultimate glory). Through the course of this book we will see that there was indeed a time when literally no one thought that at death their soul would go to heaven or hell. In the oldest forms of Western culture, as far back as we have written records, people believed everyone experienced the same fate after death, an uninteresting, feeble, and rather boring eternity in a place often called Hades. This is the view clearly set forth in **Homer's Odyssey**. But eventually people came to think this could not be right, largely because it was not fair. If there are gods with anything like our moral code who oversee the world, there must be justice, both in this life and the next. That must mean that faithful, well-meaning, and virtuous people in the world will be rewarded for how they live, and the wicked will be punished. This is the view that developed next, as we will see in the writings of **Plato**. A similar transformation happened in the ancient religion of Israel. Our oldest sources of the **Hebrew Bible** do not talk about "life after death" but simply the state of death, as all people, righteous and wicked, reside in their grave or in a mysterious entity called **Sheol**. The focus for these texts, therefore, is on life in the present, in particular the life of the nation Israel, chosen and called by God to be his people. He would make the nation great in

exchange for its worship and devotion. But that longheld view came to be challenged by the realities of history as tiny Israel experienced one disaster and calamity after another: economic, political, social, and military. When parts of the nation came to be destroyed, some survivors wrestled seriously with how to understand the disaster in light of God's justice. How could God allow his own chosen people to be wiped out by a foreign, pagan power?

Starting in the sixth century BCE, **Hebrew prophets** began to proclaim that the nation that had been destroyed would be restored to life by God. In a sense, it would be "raised from the dead". This was a national resurrection – not of the people who lived in the nation but a restoration of the nation Israel itself – to become, once more, a sovereign state. Toward the very end of the **Old Testament** period, some Jewish thinkers came to believe this future "resurrection" would apply not to the fortunes of the nation but to individuals. If God was just, surely he could not allow the suffering of the righteous to go unrequited. There would be a future **day of judgment**, when God would literally bring his people, each of them, back to life. This would be a resurrection of the dead: those who had sided with God would be returned to their bodies to live forevermore. **Jesus of Nazareth** inherited this view and forcefully proclaimed it. Those who did God's will would be rewarded at the end, raised from the dead to live forever in a glorious kingdom here on earth. Those opposed to God would be punished by being annihilated out of existence. For Jesus this was to happen very soon. Evil had taken control of this world and was wreaking havoc in it, especially among the people of God. But God would soon intervene to overthrow these forces of evil and establish his kingdom here on earth.

After Jesus' death, his disciples carried on his message, even as they transformed it in light of the new circumstances they came to face. Among other things, the expected end never did come, which led to a reevaluation of Jesus' original message. Some of his followers came to think that God's vindication of his followers would not be delayed until the end of human history. It would happen to each person at the point of death. Believers in Christ would be taken into the presence of Christ in heaven as they awaited the return to their bodies at the future resurrection. Those opposed to God, however, would be punished. Eventually **Christians** came to think this punishment would not entail annihilation (Jesus' view) but torment, and not just for a short day or two but forever. God is eternal; his creation is eternal; humans are eternal; and eternity will show forth God's glorious judgments: paradise for the saints and pain for the sinners. **Heaven and hell** were born. In short, the ideas of the afterlife that so many billions of people in our world have inherited emerged over a long period of time as people struggled with how this world can be fair and how God or the gods can be just. Death itself cannot be the end of the story. Surely all people will receive what they deserve. But this is not what people always thought. It was a view that Jews and Christians came up with over a long period of time as they tried to explain the injustice of this world and the ultimate triumph of good over evil.

Erhman shows that the **fear of death** was not only present since the beginning of recorded history, but was *the very central theme* in many of the first written stories, as it has been within the works of an endless number of writers, philosophers, politicians, and artists since then (Box 2.7). Namely, he provides interesting details about the **Epic of Gilgamesh** (see Box 2.1 and Fig. 2.1):

Many…believe that at death our life is extinguished and we cease to exist in every way. The idea of nonexistence itself – of not waking up, of a personal identity permanently lost, world without end – inspires not relief but horror. How can we even imagine it? At all times of our lives, since we have been able to think, we have existed. How can we think of not existing? And so it is no surprise that **death** is often lamented in the great literature of the world, including the **Bible**. As the psalmist says, praising God for saving him for the time being from death, imaged as the realm of **Sheol**: *I will give thanks to you, O Lord my God, with my wholeheart…for great is your steadfast love toward me; you have delivered my soul from the* *depths of Sheol.* In no small part the Bible's authors praise God for saving them from untimely death because they realize all too clearly that life is short and death certain. And so the psalmist laments that people "like smoke…vanish away" (Psalm 37:20); elsewhere we hear that "our days on earth are like a shadow" (Chronicles 29:15); or, as the **New Testament** book of James says, "[we] are a mist that appears for a while, and after which it disappears" (James 4:14). That is our life. Short and temporary like smoke, a shadow, or the morning mist. Once gone it will never return. And we don't have long to wait. The obsession with death and fear of what comes next extends beyond even the most ancient biblical records to the beginning of recorded history. It can be found in the ancient Mesopotamian epic known as **Gilgamesh**.

One of the goddesses creates a mortal equal to him (to Gilgamesh), named Enkidu, who begins as his adversary but after confronting him in hand-to-hand combat becomes his most beloved friend and partner in rampaging mischief. In one of their adventures, the gods send a sacred beast, the "bull of heaven," to wreak havoc in retribution for Gilgamesh's ungodly and outlandish behavior, but the two supermen kill it. The gods are incensed at this violation of their divine prerogative and decide that one of the two supermen must die. Enkidu mourns because he knows it will be he, and in expressing his grief he provides us with the earliest record in human history of the **terror of death**. He has a dream of being overwhelmed by a powerful man and recounts the nightmarish outcome in poignant terms: "*He seized me, drove me down to the dark house, dwelling of Erkalla's god, To the house which those who enter cannot leave, On the road where travelling is one way only, To the house where those who stay are deprived of light, Where dust is their food, and clay their bread. They are clothed like birds, with feathers, And they see no light, and they dwell in darkness*". Even the most powerful superhumans alive are powerless in the face of death. We all will eat dust and dwell forever in darkness. Not a happy prospect. And then Enkidu experiences it. He dies. Gilgamesh bitterly mourns his lost companion and roams the countryside, disconsolate. Most of his grief, however, is not for his friend but for himself: he too will eventually be confronted by death, and he hates the prospect: "*Shall I die too? Am I not like Enkidu? Grief has entered my innermost being, I am afraid of Death, and so I roam open country*". He decides he needs to find a path to immortality, and for that he needs advice. There is only one man in all of history who has escaped death to live life everlasting, a man named Ut-napishtim. Gilgamesh ventures on a journey to find him, to learn the secret of immortal existence.

In that realm he first comes upon a mysterious woman identified simply as an "alewife". It is not clear who she is or what she is doing there, but Gilgamesh is pleased to find a human of any sort and spills out to her the dreadful reason for his mission: "*I am afraid of Death, and so I roam open country…. How, O how could I stay silent, how, O how could I keep quiet? My friend whom I love has turned to clay: Enkidu my friend whom I love has turned to clay. Am I not like him? Must I liedown too, Never to rise, ever again*"? The alewife tells him how to find Ut-napishtim, and so he continues his journey, finally arriving to meet the one immortal human ever to have lived. At first Utnapishtim is not encouraging about Gilgamesh's hopes for immortality: "*Since [the gods] made you like your father and mother, [Death is inevitable…] at some time, both for Gilgamesh and for a fool*". It is a gloomy prospect. We do our best to accomplish things in life, but then we die without warning and our life is over, leaving everything we have done and produced in the possession of others. We have no more existence or meaning. And so Gilgamesh goes on another quest. With the boatman Ur-shanabi as a guide, he sails to the designated spot, ties stones to his feet, sinks to the bottom of the sea, and retrieves the plant of life, exclaiming: "*Ur-shanabi, this plant is a plant to cure a crisis! With it a man may win thebreath of life…. Its name shall be: An old man grows into a young man. I too shall eat it and turn into the young man that I once was.*" Anyone familiar with tales about plants that can bring eternal life – think, the **Garden of Eden** – should be braced for what is to come next. Gilgamesh's plans are tragically foiled. On his return home he comes to a calm pool of water and decides to have a dip to cool off. While he is in the water, a "*snake [smells] the fragrance of the plant*" that had been left in the boat, and it slithers to the spot and absconds with the plant. "*As it [takes] it* *away, it shed[s] its scaly skin*". More familiar resonances: immortality is lost because of the nefarious working of a sly serpent. As one can imagine, Gilgamesh is deeply distraught and weeps, having lost his one chance at immortality. His fear will be realized. Like all mortals, he has to die. So, of course, will we. We may seek immortality – in our day and age, not by finding the plant of immortality *per se*, but certainly by finding the right diet, exercise regimen, vitamin and mineral supplements, and other protocols to prolong our lives. But we too, like Gilgamesh, are mortal, and our time is short. The question is whether we stand in terror before the inevitable or have resources to deal with what is certainly to be.

Box 2.7 Fear of Death, Sex, Chastity, and How to Not "Misbehave"

In his book *Heaven and Hell*, Erhman refers to a not-so-known, but illuminating, story:

> One of the most bizarre accounts of the ***Acts of Thomas*** involves an episode of sex, mad jealousy, murder, and resurrection. The story begins with a young Christian man who has come to a worship service in Thomas's church in India, where he tries to take communion. But he is thwarted by a divine miracle: as the man brings the Eucharistic bread to his mouth, his hands wither. The parishioners who see this happen report to Thomas, who asks the man what sin he has recently committed. Underlying the man's tale is a major ideological point made repeatedly by this entire long narrative: to be a truly committed Christian means abstaining from the pleasures of the flesh. And that means not having sex. The man explains to Thomas that he had recently converted to Christianity, opting, when he did so, to go all in for the new faith and live a **life of chastity**. This was not welcome news to the woman he loved, who refused to make that kind of commitment herself. So the man flew into a fit of rage, imagining that she would become sexually involved with someone else, and murdered her with a sword. This had just happened before he arrived to take communion. Thomas responds by lamenting deeply the lust and **sexual immorality** of the world (the root of all evil, apparently) and instructs the man to wash his hands in a basin of **sacred water**. The man does so, and his hands are restored. Thomas then asks to be shown the woman's corpse, and they go off to the inn where the murder had been committed. When they find the body, the apostle prays that God will raise her from the dead. He instructs her former lover to take her by the hand, and she comes back to life. But rather than exulting in her new lease on life, she looks on them with terror, exclaiming that when she was dead she had been taken to a horrible place of immense suffering. She desperately does not want to go again. She then tells her tale.
>
> After she died, an exceedingly hateful man in filthy clothes came and took her to a place filled with deep chasms and an unbearable stench. He forced her to look into each chasm, all of which contained souls of the dead being subject to hellish torments. In the first were souls hung on wheels of fire that were running and ramming each other. These people, she was told, had "perverted the intercourse of man and wife". We're not told what exactly they had done. Committed adultery? Engaged in illicit sexual practices within the confines of marriage? Something else? Whatever it was, it involved sex and it brought eternal torment. Another chasm was filled with souls wallowing in mud and worms. These were women who had left their husbands to commit adultery. Yet another contained people hanging by various body parts: women who had gone into public without head coverings, possibly to show off their beauty, were hanging by their hair; thieves who reveled in their wealth and didn't give to the poor were hanging by their hands; those who walked in the ways of wickedness were hanging by their feet. After seeing the various chasms, she was shown a vast, dark cavern

filled with a vile stench. This was a holding pen for souls: some were there after being tortured in one chasm or another, others were those who had perished in their anguish, and yet others were waiting for tortures to come. Some of the demonic torturers who guarded the cavern asked the woman's guide to give her soul over to them to torture, but he refused. He had received strict instructions not to hand her over yet. She then was met by someone who looked like Thomas himself (presumably Jesus, his twin) who told her guide: "Take her, for she is one of the sheep that have gone astray". At that moment the woman regained consciousness, not awakening from a dream but arriving back from the reality of hell itself. When she sees Thomas, she begs him to save her from "those places of punishment which I have seen".

Thomas tells those who have come to observe her resuscitation that they need to repent or they themselves will end up in that place of torment: "You have heard what this woman has recounted. And these are not the only punishments, but there are others worse than these". Worse than these? How could they be worse than these? Apparently they are. You don't want to go there. And neither did Thomas's hearers. He tells them how to escape. They need to turn to God, believe in Christ for forgiveness, and cleanse themselves "from all your bodily desires that remain on earth". They are no longer to steal, commit adultery, covet, lie, get drunk, slander, or execute vengeance. As one would expect in a Christian text such as this, Thomas's brief sermon, backed with irrefutable visions of fire and brimstone, has its desired effect: "The whole people therefore believed and presented obedient souls to the living God and Christ Jesus". Clearly this tale of hell had, for the author of the vision, a didactic purpose: a brief life of chastity and purity is the only prophylactic for fiery punishments awaiting those who cannot control themselves. Still, the ethical function of the near-death experience does not mean that the hearers of this tale took it all to be metaphor. On the contrary, **early Christians** appear to have believed the literal truth of such grisly descriptions of what is to come. Many Christians today still do. The point may be to behave now, but it is a point rooted in the belief that there will be torment later for those who *misbehave.*

The fear of death was therefore deeply related to the rise of religion, and was accordingly then used by different organized religions in different ways to "discipline" people according to the particular stories used in those religions: an emblematic, and strange example is given in Box 2.7. **Antonio Damasio**, one of the leading world experts in consciousness and the links between emotions, reason, feelings, and knowledge, noted in his new book *Feeling & Knowing – Making Mind Conscious* that the connection between religions such as **Christianity** and human's obsession with death, and therefore with eternal life, is visibly displayed in many passages of religious texts. According to **John's Gospel**, **Jesus** asked the 12 apostles if they would also leave him after many of his disciples abandoned him because of the difficulty of accepting his teaching—Simon Peter answered: "Lord, to whom shall we go? You have the words of eternal life." Damasio's book included the beautiful quote that I used at the beginning of this chapter, which summarizes the link between our anxieties about death and the beginning of religion in a very powerful and elegant way.

However, it is crucial to point out that the idea that I present in this book differs from the ideas defended by authors such as Damasio and Erhman in a critical respect: the moment that humans became aware about the inevitability of death and the randomness of life is not necessarily the moment when religion starts. That moment is instead deeply related to the much broader origin of the unique human tendency to build and accept as true complex teleological narratives related to a "cosmic purpose of life."

That is, religion is nothing more than just *one of the many types of beliefs* related to such complex teleological narratives. A non-religious type of belief in a cosmic meaning of life involves, for instance, the spiritual credence that there are **vitalistic forces** connecting us and Mother Earth—in a very simplified way, **vitalism** holds that living entities, including their basic structures such as molecules, contain some kind of distinctive "vital spirit" (see also Box 3.6). So, when was the exact time when humans gained an awareness about the inexorableness of death and reacted to it by starting to create and believe in such "cosmic purpose" stories? The only thing that we can say about this with some certainty, based on the scientific ideas that chimpanzees are *not* aware of the inevitability of death and that **Neanderthals** had **ritualistic burial practices**, is that this happened after we split from the chimpanzee lineage more than 6 millions of years ago and before we split from the lineage of **Neanderthals** about half a million years ago. Anything else is, right now, just a guess, a speculation. I could speculate, for instance, that because chimpanzees seem to be relatively "close cognitively" to the level required to have such an awareness (see Fig. 2.2), this might well have happened during the earlier stages of human evolution, about 5 or 4 or 3 millions of years ago, but who knows?

What we *do* know, and is consensually accepted, based on a huge amount of empirical evidence, is that the rise of **organized religions** was a much later event that dates back to only some thousands of years ago, being related to the rise of so-called early civilizations that arose with **sedentism** and particularly with the advent of **agriculture** (see Box 2.9 and Chap. 4). In this sense, it is important to remember that while organized religions—also known as **institutional religions**, including **belief systems and rituals** that are systematically arranged, formally established, typically based on an official doctrine or dogma, and that involve a hierarchical or bureaucratic leadership structure and various rules—are "seen" as something "normal" nowadays, they only existed in the very last 0.16% of the about 6 millions of years since we split from the chimpanzee lineage. So, in a factual historical way, one can say that organized religions, as agriculture, are in reality anomalies, very rare exceptions within the huge number of generations of ancestors we had, during our long evolutionary history. That is why in this book I avoid the use of terms such as "**pre-agricultural**" or "**prehistorical**" or "**precivilization**"—or, when I discuss the ideas of authors that have used them, I use them in between single quotation marks, as I just did here. In fact, such terms moreover reflect a legacy of a very dark past—and, unfortunately, present—plagued by atrocious racist acts such as slavery and the genocide of "others" that were precisely often "morally" justified by imaginary tales about cosmic purpose and progress.

Specifically, the term "pre" only makes sense within a teleological view of the natural world in which there is a cosmic progress towards a "pro." This applies to a term that was and still is too often used in evolutionary biology, "**preadaptation**": as if nature could "know" that by creating something at a moment in time that would be advantageous in the future. Accordingly, the term "pre-agricultural" invokes the notion of progress by implying that there is a "natural progress" that leads to agriculture, as if hunter-gatherers are missing something, lacking the "blessing" of agriculture. Factually, that is surely not true, because although sedentism and agriculture arose in various locations of the globe independently, the vast majority of hunter-gatherers that have existed in this planet never started to domesticate animals and plants, nor did any group of non-human primates. The term "prehistorical" is even worse, because it implies not only that there is a cosmic direction towards the "development of writing"—what is, of course, also factually wrong—but also that all human groups that did not use writing were not even part of history, as if basically they never existed or were completely irrelevant. But the reality, the empirical evidence, shows us exactly the opposite: agriculture, organized religions, and writing are the ones that actually apply only to an infinitesimal portion of the history of life in this planet, a historical anomaly, an exception within not only the natural world but also human history per se, representing less than 0.2% of our evolutionary history.

Brune wrote in his 2009 book chapter: "many individuals believe that they have been 'born to do' something or are 'meant to' lead certain lives; in other words, according to a widespread belief an individual's existence is designed to serve a particular purpose…such a position could additionally be fuelled by one's existential anxiety and inconceivability of one's personal death." As noted by John Gray, author of the 2013 book *The Silence of Animals: On Progress and Other Modern Myths*, the very idea of **progress** in history is in itself a "myth created by the need for meaning". As further explained in Tallis' book *The Incurable Romantic*:

> The Greek philosopher **Epicurus** maintained that all of our anxieties and sadness can be traced back to a root cause: the fear of death…a great deal of existential psychotherapy is concerned with the **search for meaning**, which must be personal, because the universe is intrinsically meaningless…. It is natural to fear death; however, for some people this fear becomes so intense and troublesome that they can't enjoy life. When this happens, the condition attracts medical appellations as '**death anxiety**' and '**thanatophobia**'. There are a number of arguments that can be employed to help people with death anxiety. They aren't always effective, but when they do work, patients experience a change of perspective and the idea of death becomes less alien and strange. We are more intimate with oblivion than is generally acknowledged. Every night there are discontinuities in our existence during dreamless sleep. Moreover, we forget things every day, so in a sense we are constantly dissolving into nothingness. Recognizing that our nativity was preceded by aeons of oblivion can, for some individuals, turn 'the great unknown' into 'more of the same'. The chemical constituents of our bodies were assembled by exploding stars at an inconceivably distant point in the past, and these constituents continue to exist, in some shape or form, after our death. We are woven into the fabric of the universe – and will always be. There is also a kind of **afterlife** in procreation, making cultural contributions, leaving legacies, or simply being remembered by those who survive us. By merely existing, we influence an expanding web of cause and effect relationships that will continue indefinitely.

So, apart from such religious narratives and related tales about "progress," as well as the type of spiritual vitalistic tales related with a cosmic meaning of life that are now so in vogue within for instance New Age people, which other types of complex teleological narratives did humans built within this never-ending quest? This probably might surprise you, but one type that has existed since millennia ago but that has been so in vogue in the last centuries and particularly in the last decades, concerns many of the narratives often used in ***humanism***, even when they are used in an atheistic context, as was the case with those so frequently used by the previous President of the U.S., **Barack Obama** (see Chap. 7). Still another example are scientific just-so stories related to the teleological notion of ***Chain of Being***, which is itself related to the concept of cosmic progress (see Chap. 3).

One of the most known and awarded books about how the anxiety about death was hugely important and crucial in human history, but also how paradoxically the creation of just-so-stories about a cosmic purpose to reduce this anxiety by basically ***denying death***—for instance, by talking about an afterlife—often actually leads, in the long term, to a lot of suffering, pain, and social atrocities, is **Ernest Becker**'s 1973 book *The Denial of Death*. That book was the culmination of his life's work and won the Pulitzer Prize in 1974, and partially inspired the development of a whole theory, discussed in numerous specialized papers and books and by researchers of various scientific fields, about how the **awareness of death** plays a critical role in many aspects of our daily lives—even when we don't realize it consciously. Importantly, those papers and books also show, empirically, how such an **awareness and denial of death** is indeed far from being always something "positive," either at a psychological or social level, well on the contrary in many cases (see Box 2.8). This theory—called the *Terror management theory*—was discussed in detail in a 2016 book edited by Harvell and Nisbett with a title that plainly resonates with that of Becker's masterpiece, namely: *Denying Death – An Interdisciplinary Approach to Terror Management Theory*. As explained in the first chapter of that latter book:

> The problem of death has likely troubled human beings ever since awareness of its inevitability first emerged among our early ancestors. Evidence of **ritualized funerary practices**, which predate the emergence of agriculture and written language, demonstrate that death was an early concern, dating back at least 20,000 years; there are even some hints of the possibility of **ritual burial among Neanderthals** 100,000 years ago, but this evidence is less clear. Thus it is not surprising that history is full of musings about death from diverse cultural perspectives. **Cultural mythologies** portray death, or awareness of it, entering the world as a result of human caprice or **divine punishment**. In the **Judeo-Christian tradition**, eating from the tree of knowledge bestowed **Adam and Eve** with the **awareness of mortality**; in the Greek account, opening **Pandora's Box** released death and suffering into the world; and for the **Chiricahua Apache**, the Coyote tricked humankind into acknowledging death. Philosophers and theologians have continued to discuss the role that death plays in life ever since **Gilgamesh**, the first epic hero, sought immortality after the death of his close friend Enkidu at the hands of fickle gods. For all of Gilgamesh's legendary strength and bravery, it was the singular knowledge of death that terrified him. Cultural anthropologist Ernest Becker synthesized ideas about the role that death plays in life from various prominent thinkers, including Søren Kierkegaard, Otto Rank, **Sigmund Freud**, William James, **Charles Darwin**, Gregory Zilboorg, Norman Brown, Eric Fromm, and Robert Jay Lifton, in his Pulitzer Prize winning book, *The denial of death*, which was the primary inspiration for TMT (**Terror Management Theory**). Becker observed the human preoccupation with death and concluded that it stemmed from an inherent paradox unique to our species. The cognitive developments of language, future-oriented thinking, and self-awareness provide distinct adaptive advantages, increasing the flexibility of human behavior and therefore the likelihood of survival and procreation. Humans are able to use abstraction, which is manifested in the manipulation of symbols as well as long-term plan-

ning such as setting goals and imagining possible future situations in the absence of direct experience. These increased cognitive capabilities rendered our ancestors able to imagine the end of their individual existence, and thus aware of the **inevitability of death**. Awareness of certain mortality in an organism with a strong desire for life is the paradox at the core of our existence. According to Becker, this awareness creates the potential for an ever-present fear of death that must be continually repressed to make productive functioning in the world possible.

Becker theorized that one way of addressing the problem of death was through culture. According to Becker all societies and cultures are manifestations of the belief that human existence is meaningful, significant, and unending. By living up to the standards of value that our cultures provide, our existence, matters. As Becker put it, "society itself is a codified hero system, which means that society everywhere is a living myth of the significance of human life, a defiant creation of meaning". It is the shared illusion of permanence and immortality among members of all cultures that gives the **societal myth** its power. Living up to such standards provides selfesteem, and thus emotional security, while failing to do so leaves one vulnerable to fear, anxiety, and terror. This was the initial reasoning behind TMT. According to TMT, in order to provide existential protection, cultural worldviews provide a shared lens for viewing life and reality that (a) gives life meaning and significance, (b) is perceived as permanent and enduring over time, (c) establishes the standards of value for individuals within the culture to live up to, and (d) provides some hope of immortality. **Cultural worldviews** provide the hope for two kinds of immortality: literal and symbolic. Literal immortality consists of continued existence after biological death; usually this takes the form of a prescribed religious afterlife (e.g., reincarnation, heaven, Elysium, Valhalla, etc.). Symbolic immortality entails being part of something greater than oneself that endures long after one's own physical death. It involves making meaningful and valuable contributions to something greater than oneself and thereby remaining relevant and remembered after one's biological death. For example, although Gilgamesh failed to obtain the secret of literal immortality, his story is still recounted and his name remembered many thousands of years after his purported death. Once awareness of morality emerged, it could not be sealed back in Pandora's Box.

Box 2.8 Empirical Evidence Supporting the Terror Management Theory

As noted in the first chapter of the book *Denying Death – An interdisciplinary Approach to Terror Management Theory*, in the last decades hundreds of fascinating experiments have provided strong support for this theory (TMT), for instance concerning its "*mortality-salient hypothesis*":

> The first empirical support for TMT came from studies testing what has become known as the **mortality-salience hypothesis**, which states that if **cultural worldviews** provide protection from the ever-present fear of nonexistence (i.e., death), then explicitly reminding people of death (morality salience, MS) should increase defense of their worldviews, self-esteem, and attachments. To test this hypothesis Rosenblatt, Greenberg, Solomon, Pyszczynski, and Lyon (in 1989) asked municipal court judges to review a case and recommend a bond amount for a woman accused of prostitution. Judges are typically thought of as stewards of social order who are tasked with safeguarding law and order in society; prostitution violates moral values and state laws and was therefore considered to be a violation of the judges' cultural worldviews. Therefore, it was hypothesized that if MS increased the need to defend one's worldview, judges who thought about their own death would render harsher judgments (i.e., set higher bonds). The MS manipulation, which is still commonly

used, is described as a type of novel personality assessment and takes the form of two open-ended questions: (a) please briefly describe the emotions that the thought of your own death arouses in you, and (b) write-down, as specifically as you can, what you think will happen to you as you physically die and once you are physically dead. Judges reminded of their mortality levied significantly harsher penalties against the prostitute than those who did not receive these reminders ($455 and $50, respectively). Follow-up studies using undergraduates helped expand our understanding of responses to MS. The initial MS results could lead some to reasonably wonder if the observed response might be due to some general type of negative affect, or physical arousal. Because thinking about death is unpleasant and bound to put one in a bad mood, perhaps the effects are not due to fear of death, *per se*, but rather due to the arousing and unpleasant nature of the questions. The first follow-up studies indicated that MS effects are not mediated by affect and do not result in general negative affect. But they do largely depend on previously established beliefs and values: MS leads to harsher judgments of prostitutes only for those who view prostitution as wrong (Study 2). Importantly, responses to MS are not categorically negative: Individuals who are perceived to uphold societal values are rated as significantly as deserving of a greater monetary reward under MS compared to the control condition (Study 3; $3,476 and $1,112). Additionally, MS effects are different from those produced by self-awareness and negative mood (Study 4), do not increase and are not mediated by physiological arousal (Study 5), and can be induced in multiple ways (Study 6). The initial battery of experiments provided compelling evidence that thinking about death produced a unique pattern of results directed at bolstering one's worldview.

In a complementary set of studies, Greenberg *et al.* (in 1990) examined how MS affects responses to ingroup and outgroup members and to people and ideas that either directly support or challenge one's worldview. Because people require consensual validation of their worldviews, TMT suggests that reminders of death would lead to more positive evaluations of those who share one's worldview and more negative evaluations of those who hold worldviews different from one's own. To test this idea, Christian participants were given either the previously described MS induction or a parallel set of neutral questions and then asked to indicate their impressions of either a fellow Christian or a Jew (Study 1). As predicted, MS increased ingroup bias, leading to more positive evaluations of the fellow Christian and more negative evaluations of the Jewish student. Many other studies have shown that MS increases positive reactions to those who explicitly praise one's worldview and more negative reactions toward those who criticize one's worldview (e.g., Greenberg *et al.* 1990, Study 3). H. McGregor *et al.* (in 1998) investigated whether these effects would extend to actual physical aggression. Participants with strong and clear political attitudes were randomly assigned to MS or aversive thought control conditions, and then received an essay ostensibly written by a fellow student that was highly critical of either liberals or conservatives, and thus either supported or challenged participants' own **political ideologies**. They then participated in a bogus "food preference" study, in which their task was to measure out and administer a painfully mouth-burning hot sauce to their politically similar or dissimilar fellow participant. MS increased the amount of hot sauce given to the politically dissimilar other, but decreased the amount given to the politically similar other participant. Later studies showed that MS increased support for the use of violence in international conflicts. Americans, Israelis, Koreans, Palestinians, and Iranians have all been found to be more supportive of extreme military tactics, or suicide terrorism, when reminded of their mortality.

Other research has shown that thinking about death increases people's striving to maintain self-esteem. In one of the first studies to document this effect, individuals who tied a significant portion of their self-worth to their driving abilities engaged in more risky driving behavior on a driving simulator after MS. However, this MS-enhanced risky driving was eliminated if participants were given positive feedback about their driving before using the simulator. In a related vein, Peters, Greenberg, Williams, and Schneider (in 2005) found that MS increased the grip strength of people who value physical strength as a source of self-esteem, and Goldenberg, McCoy, Pyszczynski, Greenberg, and Solomon (in 2000) found that MS increased identification with one's body and sexual desire among individuals who place a high value on physical appearance as a source of self-esteem. Mikulincer and Florian (in 2002) provided further evidence of the terror management function of selfesteem by showing that MS increases self-serving attributional biases. Additionally, research has demonstrated that MS increases the desirability of close romantic relationships among securely attached individuals. Mikulincer and Florian (in 2000; Study 5) demonstrated that following the typical MS induction, securely attached participants scored significantly higher on a measure of desired intimacy following MS compared to a control condition. A battery of studies by Taubman-Ben-Ari, Findler, and Mikulincer (in 2002) also demonstrated that MS increases willingness to initiate social interaction, increases perceived interpersonal skills, and decreases sensitivity to social rejection. Taken together these studies suggest that close relationships can be sources of existential security and that individual differences in attachment moderate these tendencies. Together these studies provide general support for the MS hypothesis: that the psychological structures of cultural worldview, self-esteem, and close relationships buffer against the awareness of death, as explicit reminders of mortality increase the activation of, striving, and reliance on these structures.

The studies described in Box 2.8, as well as many other studies described in the book *Denying Death*, are truly fascinating because they reinforce the idea, key for the present volume, that the awareness of our inevitable death leads humans to create and believe in tales about, as well as to make and follow social norms that make us feel as having, a *cosmic purpose of life*. In turn, this subsequently leads humans to further create additional stories and **social norms** to distinguish members of some groups from "other" humans and thus to criticize, discriminate, oppress, attack, and often even kill those "others" that do not accept the same tales, prey to the same deities, behave in the same way, and/or follow the same social norms. So, in a way, by doing this, such tales led the storytelling animal to become immersed in the vicious cycle of *Neverland*'s biases, prejudices, stereotypes, mistrust, and hate, which then further increased our fear of death and paranoiac apprehension towards others, for instance if we label and eventually start to truly believe that those "others" are "bad people" or "infidels," or "witches," or "vampires," or "demons." In his 2006 book *Slayers and Their Vampires – A Cultural History of Killing the Dead*, McClelland states that some of the first historical mentions about "**vampires**" referred mainly to pagans that were not adhering to the social norms of the **Orthodox church** in Slavic countries such as Bulgaria. Apart from not adhering to such norms, they often performed some **pagan rituals** such as **sacrifices**, so they were seen as "others" and that is chiefly why they were labeled as "vampires" by the local **Orthodox Christian communities**. Indeed, as explained in *Denying Death*, the *Terror management theory* (TMT) links directly the fear of death, teleological narratives, and acts of discrimination, violence, and eventually terrorism against people from "other" religious groups, or so-called races, or another gender or type of sexual preferences:

> People often respond quite negatively – and at times even violently – to those who threaten their worldview, and this tendency can be clearly seen in experimental studies that heighten concerns about mortality. Not only does MS (*'mortality-salient hypothesis'*) increase negative and discriminatory reactions to those who espouse or represent different beliefs, but it also fuels overall aggression toward worldview-threatening targets as the individual seeks to demonstrate the superiority of his or her belief system. Consequently, efforts to manage concerns about mortality may contribute to even more extreme forms of aggression such as the commission of **hate crimes** and **terrorist acts**. Hate crimes are unusual offences, because attacks are typically committed in a spontaneous manner by a group of people who have no prior direct contact with the victim, and who generally do not take any items of monetary value from the victim. Thus, from a rational standpoint, hate crimes make little sense. However, regardless of whether hate crime attacks involve the commission of either violent or property crimes, they serve as an opportunity to devalue the victim, because of the victim's group identity (e.g., homosexual, Jewish, African American, etc.). By devaluing the victim, the offender is able to reinforce the perception of the superiority of his or her own cultural worldview group. Consequently, TMT can be used to explain motivations for hate crimes. Hate crimes can provide a mechanism for several avenues of worldview defense, including derogation (e.g., racial slurs combined with threats used to intimidate particular group members) and annihilation (e.g., physically attacking and even killing worldview threatening outgroup members). Similarly, TMT can explain the motivation behind terrorist acts, and why individuals have the propensity to inflict destruction and death upon innocent victims, simply because they do not share common religious, political, or cultural views. Research has shown that reminders of death lead individuals to express greater support of violent resistance against those who undermine one's ideology, increased support for military interventions and martyrdom, support for counterterrorism policies, and increased support for extreme and violent approaches to **terrorism**.
>
> The importance of upholding one's worldviews is so significant that there is an increased willingness for self-sacrifice in the name of one's beliefs or nation. Although self-sacrifice is contradictory to death avoidance, this behavior is seen as securing symbolic immortality and is overall an **honorable death**, thus bolstering one's self-esteem and personal significance. This is evident with the extreme actions of **militant Islamic groups** as well as being a cross-cultural phenomenon with occurrences in Saudi Arabia, Pakistan, Afghanistan, Britain, Egypt, and Jordan. The role of religion is particularly important when discussing terrorism as it provides a justification for many to engage in this behavior. When faced with reminders of mortality, **religious fundamentalists** have been found to have higher levels of prejudice, **militarism**, and **ethnocentrism**. Adhering to strict religious beliefs often provides a sense of serving one's God, adding pressures of a greater cause to protect one's people and culture. Scriptural depictions of violence and violence sanctioned by God have been found to produce aggressive responses to threats by religious individuals. As religion has been assumed to be the root of **Islamic terrorism**, it is no surprise that this behavior is often justified by one's religious beliefs. Acts of terrorism also have the potential to foster feelings of hostility and aggression toward outgroups. Evidence of this is shown through research examining the psychological impacts and reactions to the events of the September 11th attacks, which served as an extreme reminder of death for many Americans. After the attacks, there was a surge of intolerance, prejudice, discrimination, and violence, against

Arab Americans and Muslims, as well as a call for greater surveillance and restrictions of civil rights of these individuals…. From a TMT perspective, **sexual violence** is viewed as a coping and defense mechanism for individuals who developed feelings of vulnerability and poor self-esteem. Further, sexual attraction can also be viewed as threatening to the extent that it implicates existential concerns with the physicality and corporeality of the body. As such, when threatened with reminders of mortality and primed with lustful thoughts, men actually show greater tolerance for aggression against women.

Although there is a general tendency for death-related cognition to cause more punitive reactions and harsher judgments toward lawbreakers across a wide range of criminal offenses, research also demonstrates the potential for leniency toward defendants. The critical determining factor seems to be whether it is the defendant or the victim who poses more of a worldview threat to the individual making a judgment. For example, whereas people are generally more punitive toward hate-crime offenders, when reminded of mortality, they show greater tolerance for a hate crime offender if the victim poses a worldview threat. Similar results were obtained by Greenberg, Schimel, Martens, Solomon, and Pyszczynski (in 2001) in a study exploring perceptions of individuals expressing **racist beliefs**, that found MS White participants to be more tolerant of racist statements made by a **White supremacist**, than control participants were. Thus, it appears that reminders of death activate stereotypic perceptions of victims who may pose a worldview threat and intensify positive ingroup identification.

Hood et al.'s 2009 work *The Psychology of Religion – An Empirical Approach* also provides strong support for the idea that "the search for meaning is of central importance to human functioning and that **religion** is uniquely capable of helping in that search…the greatest uncertainty, death, is often regarded as a major stimulus for the origin of religion…people need to find their particular niche in the world…meaning helps meet perhaps an even greater underlying need for control…though the ideal in life is *actual* control, the need to perceive personal mastery is often so great that the *illusion* of control will suffice." One of the examples they mention concerns the **death of a child**. As they note, faced with the absurdity of such a death, "naturalistic explanations are unsatisfactory for most people, because they imply no future, no hope – simply complete and total termination." Therefore, humans create narratives about "why" people die, what is the cause, the "meaning," obsessively looking for causality, as well as for a way to have some hope. Teleological narratives of cosmic purpose/meaning, including religion, not only offer a cause, but even a future, not only for the child, but for a reunion between the child and the parents: the **afterlife**. As they explain, religion has three teleological ways to defend God's goodness despite the **child's death**: (1) the death of your child had a purpose (for instance, God loves her so much that He took her to heaven sooner); (2) you will meet your child again in afterlife, don't worry; or, in a more dark way (3) He did it as a punishment for the way the father, mother, or other relatives/survivors behaved.

While I completely agree with the part of making the coping of parents much easier by providing a "meaning" to the death of the child, as well as a hope regarding the future, I think that the part about "keeping God's goodness intact" is so illogical that it can only be applied—as most religious people *do* apply it in such moments of extreme sufferance—precisely because humans are so deeply immersed in *Neverland*. This is because within such an imaginary story there is no way to "keep

God's goodness intact" unless one accepts several obvious logical inconsistencies. Indeed, just imagine that any of the three teleological narratives listed above would be applied to a human being, instead: it would be immediately evident that this human being would truly be a horrible person. Within the huge number of reported cases in which a woman or a man kidnaps a child because she/he "really wants to have that child with her/him" (first narrative—"purpose"), do we "still defend the goodness" of the kidnaper? Of course not. Even in cases when the police is able to find where the child is and to give the child back to the parents (second narrative—future reunion), do we then defend the "goodness" of the kidnaper? Obviously not. Even worse, if we say that someone killed a child as a form of punishment because, let's say, her father did not pay the bills, or robbed a bank, would we defend the "goodness" of the killer? Surely not. So, why do we defend the "goodness of a God" that, moreover, by being so powerful and having "full knowledge of everything," should completely understand how all these things are so atrocious? Unless God decided to do them despite knowing that, what would be even worse.

2.4 Teleology, Morality, and Organized Religions

Those are my principles, and if you don't like them...well, I have others. (Groucho Marx)

One of the various types of evidence supporting the idea that the creation of and belief in complex teleological narratives is much more ancient than the rise of institutional religions comes from anthropological studies of extant **hunter-gatherer groups**, that is non-agricultural groups that hunt wild animals—a task often, but not always, done by men—and that gather wild plants—a task frequently done by women, but sometimes also by men (Chap. 4). All those groups studied so far indeed have *both* an awareness of the inevitability and randomness of death and a fascination for why-questions about a cosmic purpose of life and numerous imaginary tales to answer such questions. In *The Good Book of Human Nature* Van Schaik and Michel note that in "prehistorical" times there was "a belief in a natural environment full of **ghosts**, ancestors, and **animal spirits**; in the nineteenth century, social anthropologist Edward Tylor introduced the term '**animism**' to describe this phenomenon." They further note that "as studies of hunter-gatherer groups existing today have shown, these peoples have an entire cosmos of beings to explain 'why it snows, why wind blows, why clouds obscure the moon, why thunder crashes, why dreams contain dead people and so on'."

One crucial point that is often not included within discussions on the evolution of **morality** is that there is strong empirical evidence indicating that so-called "moral" features such as **fairness** and **altruism** are present in other animals such as chimpanzees, so it is very likely that such features arose much before humans did. In a 1992 paper Boesch showed that Tai chimpanzees were frequently seen to tend injured animals, even if they did not have a direct kin relationship: "Tai chimpanzees, however, totally independent of kin relationship, were regularly seen to tend wounded animals for extended periods of time…once this care was observed for more than 2 months…individual reactions tend to indicate that they are aware of the needs of the wounded, e.g., they lick the blood away and remove all dirt particles with fingers and lips, as well as preventing flies from coming near the wounds." In addition, "**empathy** for the pain resulting from such wounds was clearly demonstrated by the reaction of other group members: after having received fresh wounds from an attack of a leopard, the injured individual is constantly looked after by group members, all trying to help by grooming and tending the wounds…dominant adult males prevented other group members from disturbing the wounded chimp by chasing playing infants or noisy group members away from his vicinity." He added: "as wounds handicapped the movements of the injured animal, group members remained with him as long as he needed before he was able to begin to walk again; some just waited whereas others would return to him until he started to move (three times the group waited for four hours at the same spot)…whenever he stopped they waited for him." For more empirical examples of **fairness and altruism in chimpanzees**, I strongly recommend De Wall's exceptional book, *Are We Smart Enough to Know How Smart Animal Are?*

Another clear indication that "morality," as socially constructed by humans, cannot be seen as the product of religions, and particularly of organized religions, is the obvious fact that—as shown in many works reviewed in the 2009 book *The Psychology of Religion – an empirical approach*—the development of "morality" in children has usually nothing to do with ideas related to religious narratives about Gods, angels, or demons. At early stages of development, doing "good" or "bad" things has much more to do with simple punishments and rewards externally imposed/provided by adults. Moreover, the idea that the enforcement of morality is deeply related to the belief in teleological narratives, particularly in religious ones, is further contradicted by the fact that in general religious people don't seem at all to truly be more "moral" than non-religious people, in average. Many of the atrocities that were, and continue to be, done in human history were/are performed or at least justified by religious people, as will be seen in this book. Moreover, even less extreme, day-a-day items such as helping others or being tolerant or honest—features commonly seen as "good moral conducts"—are actually often *negatively correlated* with the acceptance of religious beliefs, as emphasized in Shermer's book *The Science of Good and Evil*:

> **Absolute morality** leads logically to **absolute intolerance**. Once it is determined that one has the absolute and final answers to moral questions, why be tolerant of those who refuse to accept the Truth? Religiously based **moral systems** apply this principle in spades. From the **medieval Crusades** and the **Spanish Inquisition** to the **Holocaust** and Bosnia, history is rife with examples of intolerance. In the name of God, religious people have sanctioned slavery, anti-Semitism, racism, homophobia, torture, genocide, ethnic cleansing, and war. In the name of their religion, people have burned women accused of witchcraft. The events of September 11, 2001, are a potent example of religious extremism gone political. "All faiths that come out of the biblical tradition – **Judaism**, **Christianity** and **Islam** – have the tendency to believe that they have the exclusive truth", writes Rabbi David Hartman of the Shalom Hartman Institute in Jerusalem. "When the **Taliban** wiped out the Buddhist statues, that's what they were saying…but others have said it too". **Religious extremism** in America

is particularly potent when gathered together under the umbrella of militia groups. An example can be seen in the life of Eric Robert Rudolph, the American terrorist charged in the 1996 Atlanta Olympics bombing that killed one person and wounded over a hundred others, as well as in the bombing of a gay nightclub and two abortion clinics. When he was captured in May 2003, it was reported that he was a member of the **Christian Identity movement** (an extremist group that believes Jews are satanic and blacks are subhuman), was known for his anti-Semitic and racist views, and that in his seven-year evasion of the FBI and law enforcement agencies, he probably had help from militia groups as well as local townspeople in Murphy, North Carolina, many of whom apparently share his views.

Not only is there no evidence that a lack of religiosity leads to less **moral behavior**, a number of studies actually support the opposite conclusion. In 1934 Abraham Franzblau found a negative correlation between acceptance of religious beliefs and three different measures of honesty. As religiosity increased, honesty decreased. In 1950 Murray Ross conducted a survey among 2,000 associates of the YMCA and discovered that agnostics and atheists were more likely to express their willingness to aid the poor than those who rated themselves as deeply religious. In 1969 sociologists Travis Hirschi and Rodney Stark reported no difference in the self-reported likelihood to commit crimes between children who attended church regularly and those who did not. In 1975 Ronald Smith, Gregory Wheeler, and Edward Diener discovered that college-age students in religious schools were no less likely to cheat on a test than their atheist and agnostic counterparts in nonreligious schools. Finally, David Wulff's comprehensive survey of correlational studies on the **psychology of religion** revealed that there is a consistent positive correlation between "religious affiliation, church attendance, doctrinal orthodoxy, rated importance of religion, and so on" with "**ethnocentrism**, **authoritarianism**, **dogmatism**, social distance, rigidity, intolerance of ambiguity, and specific forms of **prejudice**, especially against Jews and blacks". The conclusion is clear: not only does religion not necessarily make one more moral, it can lead to greater intolerance, **racism**, **sexism**, and the erosion of other values cherished in a free and democratic society. In a similar vein, pro-lifer Randall Terry, founder of Operation Rescue, succinctly summarized the absolute intolerance that is possible with absolute morality: "let a wave of intolerance wash over you…yes, hate is good…our goal is a Christian nation…we are called by God to conquer this country…we don't want pluralism". Putting an exclamation point on Terry's philosophy is the **abortion clinic bomber** John Brockhoeft: "I'm a very narrow-minded, intolerant, reactionary, bible-thumping fundamentalist…a zealot and fanatic…the reason the United States was once a great nation, besides being blessed by God, is because she was founded on truth, justice, and narrow-mindedness". There is a simple historical problem with this theory. According to the 2001 New Historical Atlas of Religion in America, it is a myth that America was once a Christian nation that has since lapsed into secular debauchery. Most revealing are the historical maps and charts that track the changing demographics of American religion. Conservative pundits who worry about the "moral fiber of America and proclaim that we need to return to the good old days when America was a Christian nation should (consider) that church membership among the U.S. population over the past century and a half has increased from 25 percent to 65 percent".

In *The Psychology of Religion – An Empirical Approach* Hood and colleagues also refer to studies showing a negative correlation between the acceptance of religious beliefs and features that are often considered to be "moral," such as honesty. They mention several examples in which the public stance of religious people about a subject is much different from what they truly do in their private lives: an emblematic case is that 70% of the Seventh-Day Adventist youngsters included in a study endorsed their church's **prohibition of premarital sex**, but 54% of them reported that they had engaged in premarital sex. At least in theory, it is indeed to be expected that dishonesty, and hypocrisy between public stance and private acts, are in general more frequent in religious people that impose themselves difficult goals such as "resisting" basic "bodily needs and desires" such as having sex. Actually, not only such religious people often fail to fulfill and have to lie about such illogical goals that make no sense at all within the reality of the natural world but, in more extreme cases in which such prohibitions apply to their lifetimes—as is the case with the **celibacy vows** of **Catholic priests**—they often end up by doing things that are much worse than just lying, such as raping or/and molesting children (see Fig. 9.2) (Box 2.9).

Box 2.9 From Belief in Afterlife to Institutional Religion
Lahti's 2009 chapter of the book *The Biological Evolution of Religious Mind and Behavior* provides a very brief account on hypotheses of how belief in **afterlife** and **ancestor worship** might have led to institutional religion:

> Ancestor worship generally evolved into or coexisted with an emerging **polytheism**, with **magic** and local spirits common. With the weakening of hereditary leadership and the increased contact between groups, the ancestors of one group would not necessarily continue to hold sway, and so they would have to gain power or else be supplemented or replaced by other powers. As ancestors "broke the ice" for belief in **supernatural persons** who were interested in human affairs and could be consulted for leadership, local and regional gods could have taken their places, or ancestors could even transform into them. **Magic**, including **shamanic medicine**, would have attracted and strengthened commitment to these **spirits**. Spirits could be appropriated for all aspects of life, as witness the proliferation of **fertility figurines**. Shared rituals between groups could invoke spirits that all participating social groups revered, cementing social relationships and a common **moral commitment**. Thus, with an increase in the extent of communication among social groups, both **morality** and religion became accordingly less parochial.
>
> The **agricultural revolutions** that began to occur about 11,000 years ago resulted in population centers with a greater diversity and number of interacting people, and a greater degree of control over nature than had ever been customary for humans. One of two alternative cultural strategies could have been adopted in these situations: isolation or integration. Smaller regional groups that were not at the helm of the new **agricultural civilizations** but lived near or within them may have chosen to preserve their traditional social and moral structures by avoiding external influence, strengthening old dispositions to mistrust outsiders and to associate preferentially with those who are closely related. A ruling state, however, would have found greater advantage to being integrationist, as its culture would spread and strengthen by influencing and incorporating more groups. As for social structure within the **agricultural society**, leaders would have been relatively ineffective if they depended on kinship ties to give weight to commands in a cosmopolitan social environment, so political expediencies were necessary to facilitate social order. A strong social hierarchy or class system appears generally to have been imposed. Moral and legal standards were more generalized and less parochial than those of tribal cultures, a trend that became especially evident upon the **development of writing**. For instance, most of the laws in the **Sumerian Hammurabi Codex** (c. 1760 BC) begin with the generality '*If any one...*' although some are class-specific ('*If a chieftain...*'). Citizens of city-states would encounter and depend upon various kinds of people, including many known only by name and reputation. Therefore, indirect reciprocity in such a cosmopolitan situation would have assumed paramount importance in the enforcement of moral standards in a community of peers. Leaders and their law codes could encourage and intensify

indirect reciprocity by enhancing public dissemination of social information, ensuring its accuracy, and formalizing reputational rewards and punishments. For instance, the **Code of the Assura** (c. 1075 BC) lists "ruining of reputation" as one of three nonlethal forms of punishment for breaking laws.

Likewise, the **Teaching For Merikare** demonstrates that indirect reciprocity was so central by 4,000 years ago in Egypt that the impacts of reputation even after one's death were considered a primary reason for being good in this life. A variety of gods associated with local cultural or natural entities gradually gave way to a more generalized pantheon where gods represented less parochial objects (such as the sun, fertility, or the harvest), although these ideas were undoubtedly present in local traditions earlier. Leaders also tended to consecrate a chief god, or a god devoted to the civilization. A culture's most significant rituals would generally be in service to this chief deity, and moral codes were most likely claimed to proceed from that deity as well. Often this deity was a glorified local god, a deified ancestor king (or living king), or a transformed favorite god in a preexisting cosmopolitan pantheon. The favoritism of a single deity above others, and the attribution of an increasing proportion of natural events and features to that deity, marks a transition from the picture of nature as chaotic and controlled by many forces (the animistic-polytheistic view) to a picture of nature as regular and reliable, as if controlled by a single personality or power. Thus a cosmopolitan social environment and increased human control of nature would have led **Neolithic agricultural societies** at least part of the way toward **monotheism** and the universalization of religious and moral ideas.

Any culture that favored the view that nature was orderly and understandable, and that humans from different lineages could (at least in principle) agree to abide by similar moral and legal guidelines in a society, might not rest comfortably at any parochial understanding of morality or religion. A favored god could only maintain an orderly universe if competing gods had no say or no existence. Divergent human groups could only be morally similar if they shared a common origin and purpose. If this is true, it is not surprising that from perhaps 4,000 years ago, the chief god of some societies began to transform into an only god, generally responsible for creating the universe and its value structure, and establishing human life and its purpose. The earliest written evidence of such monotheism is some passages from the **Bible**, the **Vedas**, and **Akhenaten's Great Hymn to the Aten**, all likely to have been written in the second millennium BC but probably stemming from older traditions. Morality would be transformed by this monotheism or **monism**; the **moral guidelines** would be perceived as created by a Supreme Being and thus inherent in the nature of the universe rather than proceeding from a social group or its leadership, whether human or supernatural. Thus, leaders in such a situation would no longer be able to assume divinity, but would have to claim a special connection to God or the universe in order to deliver moral commands.

As explained in my 2017 book *Evolution Driven by Organismal Behavior*, in the natural world evolutionary changes cannot be classified as "good" or "bad," or even as simply advantageous or detrimental, as if this was a timeless attribute. This is because the evolution of life in this planet is instead made of **evolutionary trade-offs**, that is, what might be advantageous within a particular context at a certain time, place, and ecological network might be—and often will be—detrimental in a future different context. In this respect, evolutionary biology has changed a lot in the last decades: not only traits that are advantageous only make sense when they are seen as **local adaptations** as **Darwin** suggested, but many features don't even confer any advantage at all, being for instance simply evolutionary neutral and/or the by-product of other features that are developed for a completely different reason, as emphasized by **Stephen Jay Gould**. So, taking this into account, is there empirical data to directly support the key idea that within the first groups of humans that had an awareness of the inevitability of death and randomness of life the ones that built and believed in complex teleological narratives about cosmic purpose and afterlife indeed outcompeted the non-believers? To answer this question, we should keep in mind that, at the onset of any evolutionary change, there is usually variation: Darwin's use of the term **natural selection** precisely refers to the fact that some of the existing variants, but not others, were selected at a certain time and place for reasons often concerning a plethora of different factors and particular circumstances. So, as there are nowadays humans that have "more faith" and others that have less, there were surely humans at that time in the past that had a greater predisposition to believe, either because of their genetic inheritance or because of **epigenetic factors**—"epigenetic" basically meaning "above" or "other than" genetic aspects—such as living in more unpredictable or harsh environments.

So, we can now rephrase the question: within that initial variation, does the available scientific data indicate that believers would indeed tend to survive more and/or have more children? Regarding reproduction, the empirical data available from numerous studies comparing different regions of the globe and people with different beliefs clearly show that in general *religious people have more kids*—the only major exception, interestingly, being the "Latin America-Caribbean" region (see Fig. 2.7). Moreover, several studies also show that within people that express/display/"feel" beliefs in different ways, people that do so more intensively also have, in general, more children (see Fig. 2.8). At the very moment that I write this text, an illustrative example is being highly discussed within the media, because of the political events that are occurring in Israel, concerning the discussion on whether **Haredi Jews**—sometimes called "ultra-orthodox"—should or not participate in the country's army. This is because, with a current number of children per woman of 6.2, they have passed from representing about 9.9% of the Israeli population in 2009 to 11.1% by 2014, 12% by 2017, and are projected to become one third of the whole Israeli population by 2065.

That is why even if in regions such as North America and Europe a significant part of youth—including those born in religious families—tend to be atheist, in the globe as a whole the religiously unaffiliated population is actually expected to *decline* as a percentage of the global population, from 16% in 2015 to 13% in 2060 (Fig. 2.9). This is particularly so because the higher rates of population increase are precisely seen in regions where the vast majority of people are religious, such as sub-Saharan Africa and India. Moreover, as shown in Fig. 2.10, that 13% worldwide 2060 prediction refers to people that are *not religious*—that is, it corresponds to the 45% of people in the U.S. in 2017 that are not religious (18% + 27%). So, as in the U.S. 2017 sample the vast majority of non-religious people were "spiritual" (60% of them, that is, 27 out of 45), it might well be that this will also happen with those 13% of people worldwide that will be non-religious, particularly because spiritualism is clearly on the rise nowadays as also noted above and also shown in Fig. 2.10. If this will be so, this would mean that only about 5% of people would be

Muslims and Christians have more children per woman than other religious groups

Total fertility rate, by religion, 2015-2020

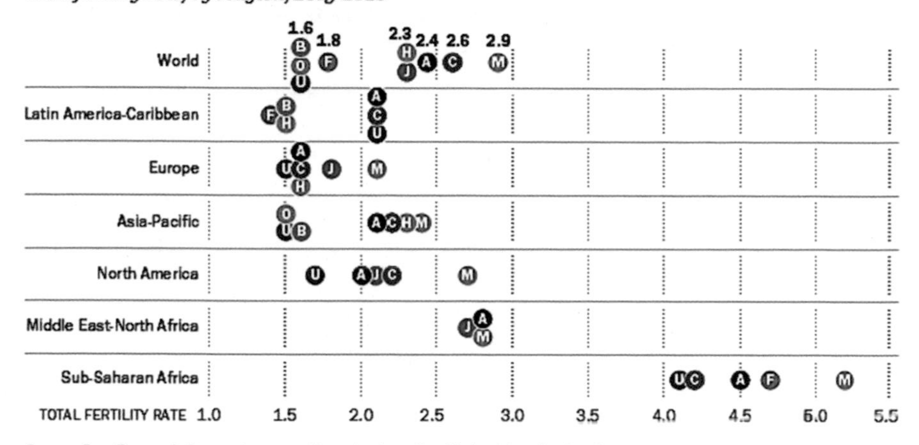

Source: Pew Research Center demographic projections. See Methodology for details.
"The Changing Global Religious Landscape"

PEW RESEARCH CENTER

KEY B: Buddhists C: Christians F: Folk religions H: Hindus J: Jews M: Muslims O: Other religions U: Unaffiliated A: All

Fig. 2.7 The changing global religious landscape

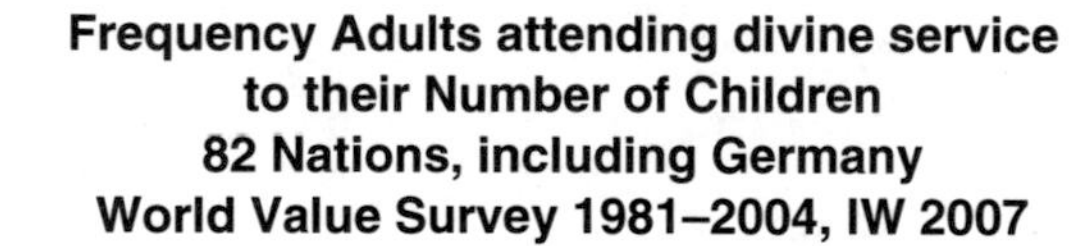

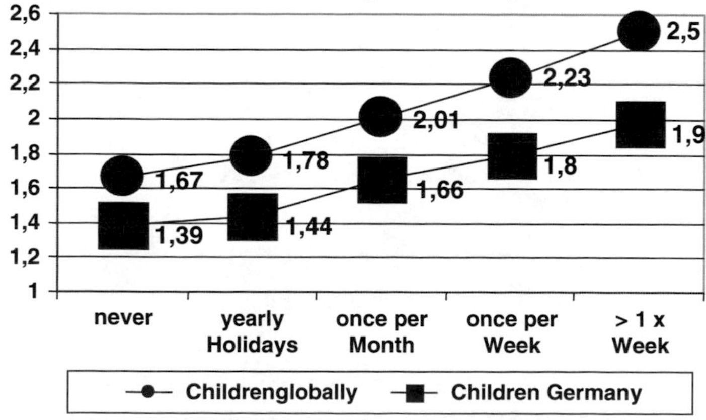

Data: Dominik Enste, Institut der deutschen Wirtschaft Köln 2007

Fig. 2.8 Global correlation of worship attendance and number of children

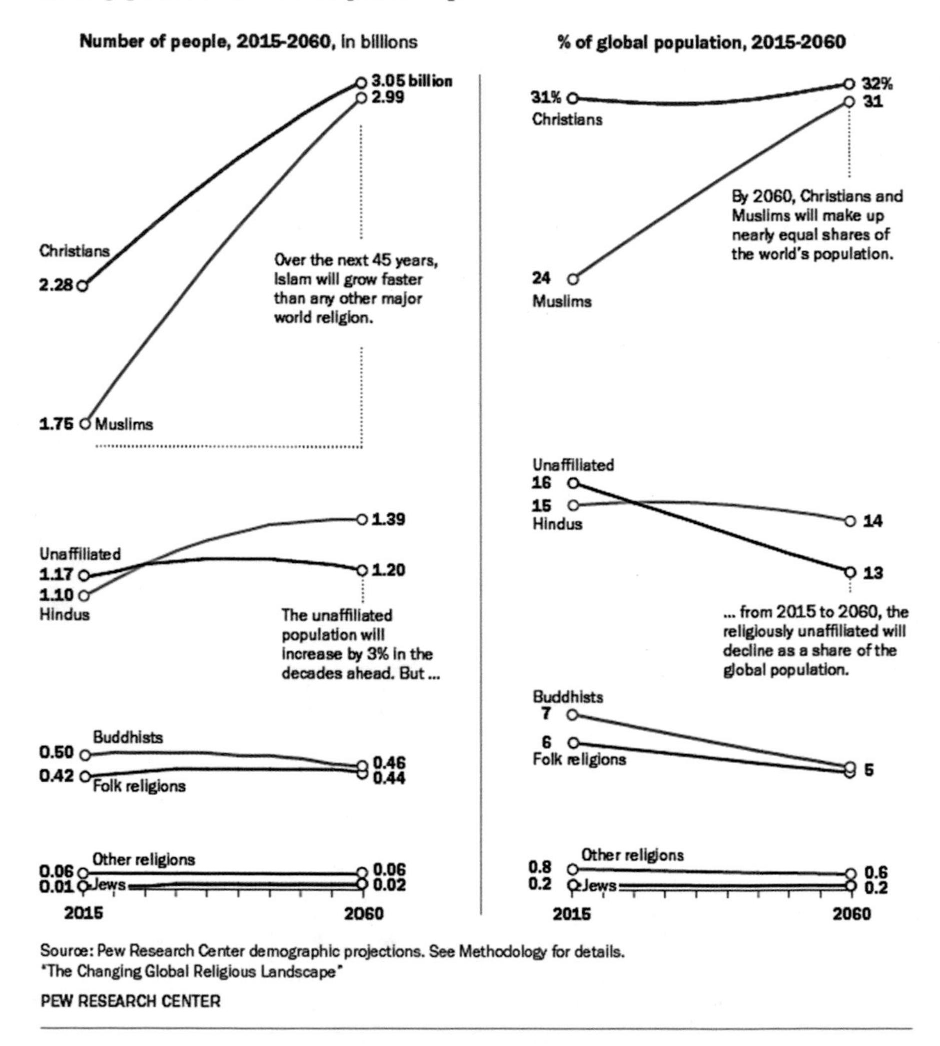

Fig. 2.9 Projected change in global population, 2015–2060

both non-religious and non-spiritual in 2060: in other words, in the second half of the twenty-first century about 19 out of 20 people would still be deeply immersed in *Neverland* and specifically believe in imaginary transcendental tales about a cosmic meaning of life, stressing how difficult it is for humans to escape the world of unreality they have entered hundreds of thousands or millions of years ago.

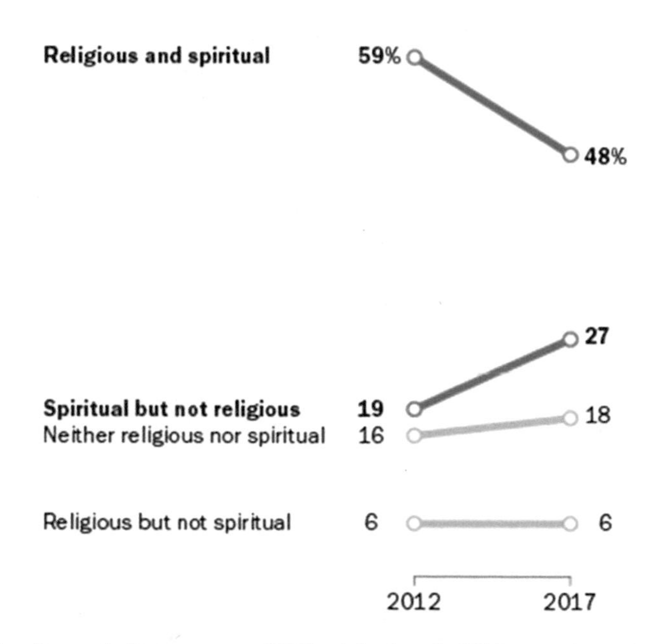

Fig. 2.10 Pew Research Center survey of U.S. adults done in 2017, reporting an increase of people who see themselves as spiritual but not religious

Indeed, these numbers show that **Peter Berger**, an influential sociologist who in 1968 famously stated that a major "collapse of religion" would occur in the twenty-first century, was correct when he later recognized, in 1999, that his 1968 prediction was completely wrong: he admitted that "the world is furiously religious as it ever was, and in some places more so than ever…a whole body of literature by historians/social scientists on 'secularization theory' was essentially mistaken." Apart from Fig. 2.10 showing that the number of people that are "neither religious nor spiritual" in countries such as the U.S. has been mainly stable in the last years because the decrease of "religious" people has been accompanied by an increase of "spiritual but not religious" people, I will refer to my own country, Portugal. In a relatively large 2018 *YouGov-Cambridge* study including 1003 adults living in Portugal, only 7% answered that they trust "a great deal" in academics to "tell the truth": 51% answered "a fair amount," 33% "not much," and 5% "not at all." In contrast, when asked how much they consider that their life is decided by a "higher force" such as God, fate, or destiny, 10% answered "everything," 12% said "most," and 33% said "some," while only 28% unambiguously said that "none" of what happens in their life is caused by such a "higher force."

Regarding conspiracy theories, 10% agreed that "humans have made contact with aliens and this fact has been deliberately hidden from the public," 8% that "the AIDS virus was created and spread around the world on purpose by a secret group or organization," 10% that "the truth about the harmful effects of vaccines is being deliberately hidden from the public," and, strikingly, 42%—yes, 42%—that "regardless of who is officially in charge of governments and other organizations, there is a single group of people who secretly control events and rule the world together." The continuous recycling of, and belief in, fairytales about the ***Illuminati***—people that supposedly conspire to control world affairs, by masterminding events and planting agents in government and corporations—as illustrated in literally thousands of books, movies, TV shows, music videos, and video games made in the last decades, is one of the most emblematic examples of the world of *Neverland*.

Interestingly, the link between religiosity and a higher fertility rate was even recognized by Fetchenhauer in the last chapter of the 2009 book *The Biological Evolution of Religious Mind and Behavior*, in which he actually argued against the notion that belief and religion might have led to an evolutionarily more successful reproductive strategy. Indeed, in his summary of the other chapters of the book he had to accept that "there is evidence that religiosity is linked to a long-term reproductive strategy and that this strategy is reproductively more successful under the conditions of modern societies." However, he then argued:

> Yet, there is no reason to believe that such a strategy has been more successful in the EEA (Environment of Evolutionary Adaptedness) and therefore it is doubtful whether the evolution of religion can be explained by a causal link between religiosity and fertility. What is missing is the proximate mechanism that would link both variables with each other. Why do religious people get more children? This question remains largely unclear. By the way, there is evidence showing that making people aware of their own mortality increases their wish to reproduce and to get offspring. Terror management theory regards this as a proof that mortality salience increases humans' need to make oneself immortal…. If this actually is the case, religious people (at least when they adhere to a variant of Islam or Christendom) should have a lower desire for getting offspring, as they do not need own children to overcome their mortality. Rather, they might try to strictly adhere to the rules of their God to make sure they actually come into heaven (e.g., by becoming a monk and totally refrain from sexuality).

The arguments of Fetchenhauer are somewhat confusing, and seem to miss the main point. Firstly, the objective, quantitative scientific data that are available about this issue do indicate that religious people have in average more children, in a wide range of geographic regions and social and geopolitical contexts, as recognized by him (see Figs. 2.7, 2.8, 2.9, and 2.10). A scientist needs to explain such objective data a posteriori, and not to argue a priori against those data just because one finds it difficult to discern "proximate mechanisms" to explain them. Of course, this is easier said than done, because we humans do have a strong "wired" tendency to create post hoc narratives that have nothing to do with the facts, as noted above and explained in detail in books such as Sapolsky's *Behave* and Taleb's *Black Swan* (see below and Box 2.10).

Box 2.10 Cognition, Brains, Confirmation Bias, and Dopamine

Discussions about **evolutionary adaptations** and advantages are particularly plagued by teleological narratives, such as those about "cosmic purpose" and "progress," and by our related "wired" tendency—of all of us, humans, including scientists such as myself—for ***post hoc rationalization***, as explained in Taleb's 2010 tour de force *The Black Swan – The Impact of the Highly Improbable*:

> Cognitive scientists have studied our natural tendency to look only for corroboration; they call this vulnerability to the corroboration error the **confirmation bias**. You can test a given rule either directly, by looking at instances where it works, or indirectly, by focusing on where it does not work…disconfirming instances are far more powerful in establishing truth…yet we tend to not be aware of this property. [One of the first experiments] concerning this phenomenon was done by the psychologist P. C. Wason. He presented subjects with the three-number sequence 2, 4, 6, and asked them to try to guess the rule generating it. Their method of guessing was to produce other three-number sequences, to which the experimenter would respond 'yes' or 'no' depending on whether the new sequences were consistent with the rule. Once confident with their answers, the subjects would formulate the rule. The correct rule was 'numbers in ascending order,' nothing more. Very few subjects discovered it because in order to do so they had to offer a series in descending order (that the experimenter would say 'no' to). Wason noticed that the subjects had a rule in mind, but gave him examples aimed at confirming it instead of trying to supply series that were inconsistent with their hypothesis. Subjects tenaciously kept trying to confirm the rules that they had made up. This experiment inspired a collection of similar tests, of which another example: subjects were asked which questions to ask to find out whether a person was extroverted or not, purportedly for another type of experiment. It was established that subjects supplied mostly questions for which a 'yes' answer would support the hypothesis.
>
> We like stories, we like to summarize, and we like to simplify, i.e., to reduce the dimension of matters…the *narrative fallacy*…is associated with our vulnerability to overinterpretation and our predilection for compact stories over raw truths. It severely distorts our mental representation of the world; it is particularly acute when it comes to the rare event. [Regarding] *post hoc rationalization*…in an experiment, psychologists asked women to select from among twelve pairs of nylon stockings the ones they preferred. The researchers then asked the women their reasons for their choices. Texture, 'feel,' and color featured among the selected reasons. All the pairs of stockings were, in fact, identical. The women supplied backfit, **post hoc explanations**. Does this suggest that we are better at explaining than at understanding? Let us see. A series of famous experiments on **split-brain patients** gives us convincing physical – that is, biological – evidence of the automatic aspect of the act of interpretation. There appears to be a sense-making organ in us – though it may not be easy to zoom in on it with any precision. Let us see how it is detected. Split-brain patients have no connection between the left and the right sides of their brains, which prevents information from being shared between the two cerebral hemispheres. These patients are jewels, rare and invaluable for researchers.
>
> You literally have two different persons, and you can communicate with each one of them separately; the differences between the two individuals give you some indication about the specialization of each of the hemispheres. This splitting is usually the result of surgery to remedy more serious conditions like severe epilepsy; no, scientists in Western countries (and most Eastern ones) are no longer allowed to cut human brains in half, even if it is for the pursuit of knowledge and wisdom. Now, say

that you induced such a person to perform an act – raise his finger, laugh, or grab a shovel – in order to ascertain how he ascribes a reason to his act (when in fact you know that there is no reason for it other than your inducing it). If you ask the right hemisphere, here isolated from the left side, to perform the action, then ask the other hemisphere for an explanation, the patient will invariably offer some interpretation: 'I was pointing at the ceiling in order to'.... 'I saw something interesting on the wall'. In addition to the story of the left-brain interpreter, we have more physiological evidence of our ingrained pattern seeking, thanks to our growing knowledge of the role of neurotransmitters, the chemicals that are assumed to transport signals between different parts of the brain. It appears that pattern perception increases along with the concentration in the brain of the chemical **dopamine**. *Dopamine* also regulates moods and supplies an internal reward system in the brain (not surprisingly, it is found in slightly higher concentrations in the left side of the brains of right-handed persons than on the right side). A higher concentration of dopamine appears to lower skepticism and result in greater vulnerability to pattern detection; an injection of L-dopa, a substance used to treat patients with Parkinson's disease, seems to increase such activity and lowers one's suspension of belief. The person becomes vulnerable to all manner of fads, such as **astrology**, **superstitions**, economics, and **tarot-card reading**.

[In general] by finding the pattern, the logic of the series, you no longer need to memorize it all. You just store the pattern. And, as we can see here, a pattern is obviously more compact than raw information. It is along these lines that the great probabilist Andrey Nikolayevich Kolmogorov defined the degree of randomness; it is called '**Kolmogorov complexity**'. We, members of the human variety of primates, have a hunger for rules because we need to reduce the dimension of matters so they can get into our heads. Or, rather, sadly, so we can squeeze them into our heads. The more random information is, the greater the dimensionality, and thus the more difficult to summarize. The more you summarize, the more order you put in, the less randomness. Hence the same condition that makes us simplify pushes us to think that the world is less random than it actually is. And the ***Black Swan*** is what we leave out of simplification. Both the artistic and scientific enterprises are the product of our need to reduce dimensions and inflict some order on things. Think of the world around you, laden with trillions of details. Try to describe it and you will find yourself tempted to weave a thread into what you are saying. A novel, a story, a myth, or a tale, all have the same function: they spare us from the complexity of the world and shield us from its randomness. Myths impart order to the disorder of human perception and the perceived 'chaos of human experience'. Now, if you think that science is an abstract subject free of sensationalism and distortions, I have some sobering news. Empirical researchers have found evidence that scientists too are vulnerable to narratives, emphasizing titles and 'sexy' attention-grabbing punch lines over more substantive matters. They too are human and get their attention from sensational matters. The way to remedy this is through meta-analyses of scientific studies, in which an überresearcher peruses the entire literature, which includes the less-advertised articles, and produces a synthesis. Now, I do not disagree with those recommending the use of a narrative to get attention. Indeed, our consciousness may be linked to our ability to concoct some form of story about ourselves. It is just that narrative can be lethal when used in the wrong places.

Contrary to the statements of authors such as Fetchenhauer, some authors, including myself in the present book, have actually proposed specific evolutionary mechanisms to link belief in a cosmic meaning of life and an increase in the number of people that first started to have that belief, by increasing both (a) the motivation to live and fight for one's life and faith and (b) the levels of fertility. Regarding the rise in the levels of fertility, empirical data show that this is commonly indeed the case, as seen above. In this sense, the argument about monks not having children, used by Fetchenhauer, is completely flawed. Firstly, this would be as like arguing that termites can't really be evolutionary successful, or increase their numbers, because their "workers" and "soldiers" are usually sterile: such an assertion is obviously factually wrong because a single termite colony can sometimes have up to millions of individuals. Moreover, compared to termite "workers" and "soldiers," the proportion of monks within the whole population of believers is extremely small. Thirdly, it has been shown empirically that by having a very small subset of people within a religion that express their beliefs—and thus send a signal to that broader religious community—in an extremely strong way, as do monks or Catholic priests when they do **celibacy vows**, that community tends to become even more unified as a whole, thus often becoming more successful overall both in terms of overall survival and of reproduction due to group selection (see Box 2.11). In this sense, in an evolutionary context one can say that monks somehow "sacrifice" themselves for their religious community.

That is, the mistake done by Fetchehauer in this specific example about monks was to focus on **individual selection** and neglect the importance of **group selection**, as many other authors unfortunately continue to do, wrongly, about a wide range of other evolutionary topics. The power of religions is mainly related with group selection, and that is precisely why religious groups often can afford to have members such as monks, or suicide bombers, or religious warriors that "sacrifice themselves" evolutionarily for the sake of the whole religious community. So, let's now discuss in more detail the empirical evidence available relating religiosity to the higher fertility rates, and particularly to the higher survival rates, of a religious group as a whole. One work that provides an illuminating analysis of these issues is the 2009 book *The Biological Evolution of Religious Mind and Behavior*, edited by Voland and Schiefenhovel. In his chapter Voland writes:

> Owing to the close connection between therapy and **mysticism**, it is debatable as to whether or not **shamanism** belongs to the history of medicine or to the history of religion. In any case, mystic elements in day-to-day living can improve one's physical and mental well-being and thus provide for improved mastery of contingencies. This connection has an interesting evolutionary feedback: to the degree that shamanism was therapeutically successful, it selected genotypes which tended to accept suggestions and precisely for this reason, they were also open to unusual experiences that we call religious. There is an extensive literature on the correlation between religious practice and mastery of life events; not only have interesting single studies repeatedly found a positive correlation, but so have statistically reliable meta-analyses. Of course, there is also the "dark side" of religious fears and obsessions, which are definitely associated with significant health risks. Overall, however, the positive effects clearly predominate, which is why religion proves to be extremely functional from a biological standpoint. Therefore, a first biological benefit function for religious behavior is described: self-preservation through an improved mastery of contingencies. The function of spirituality is not limited to personal benefits, however. Joint participation in rituals lends it a social dimension. Not infrequently, ritual performances are very rigid, redundant, compulsory and oriented towards "useless" behavioral goals. The

whole process is frequently supported by rhythms and ends in a kind of "emotional synchronization" of the participants. Without rituals that have an emotional impact, religions would lack both an emotional depth and a motivating power. This means that rituals are used in particular when the intent is to demand collective efforts or special altruistic services from the faithful (**war**, **competition** or **solidarity**). Psychologically, this is done by a form of the loss of self, by the feeling of being at one with the universe. Individuality and egocentrism are displaced in favour of collectiveness. Accordingly, collective rituals have a lot to do with social coordination and cohesion, with the bundling of forces and with enabling gains through cooperation. Various empirical studies show a clear correlation in migrant groups (e.g. Moslem youths in the Netherlands) between finding one's personal identity through group cohesion, personal well-being and religious practice. Thus, there appears to be a second biological benefit function of religiosity: strengthening the community by obligating its members to work towards common goals.

Voland also refers to a third biological "beneficial function" of belief and religiosity, which is a recurrent theme in many chapters published in the same book: "religion survives because it produces children, not because it is true, [as noted by] Economics Nobel Prize winner Friedrich-August von Hayek." Voland refers to various studies that showed, regarding Spain and the U.S., that **religious commitment** correlates with fertility, noting that "it appears that religious people, even in modern, enlightened societies, are more successful than others in overcoming the personal barriers in having children; however the correlation between religiosity and differential reproduction might have come about, religious persons are often observed to over reproduce." Importantly, it should be noted that Vaas proposes that "it is quite possible that the beginning of *Homo religiosus* is an outcome of the development of I-consciousness (verbalizable, reflective self-consciousness) and death awareness," an idea similar to that defended in the present book. Moreover, as also proposed in this volume, he directly relates belief in religious teleologies with evolutionary **reproductive success**:

Be "fruitful and multiply" were God's first words to mankind according to the **book of Genesis**. And indeed, religiosity is correlated with reproductive benefits all over the world (including nomadic **hunter-gatherers** like the !Kung San). Adults have more children if they confess strong faith, if they pray more often, attend religious services more frequently, or have very religious parents; and having children is more important for religious than for less religious persons. This at least has been the case in recent decades and cannot be explained by other factors alone, such as wealth or education, though they have a much bigger effect. The larger number of children is due to the religious doctrines themselves, as well as social and psychological factors (family bonding, social support, better coping with stress, a more trusted mate).

Interestingly, and also providing an indirect link between belief in teleological narratives, evolutionary selection, and heritability, Vaas further notes:

There are indications that religiosity is genetically determined to a significant degree, up to about 40–60%, according to twin studies. First, there is quite a strong correlation between religiosity, **authoritarianism**, and **conservatism** – that is certain views about the organization of nature, family, and society – which greatly influences **mate choice**. This refers both to extrinsic and intrinsic religiosity. **Extrinsic religiosity** is mostly socially determined (parent's religion is the strongest determinant of child's) and more about rules than feelings. But there is also a high correlation, about 50%, for **intrinsic religiosity**, especially **spirituality**. Spirituality is a basic personality trait. It has to do with **mysticism**: **self-**

transcendence – that is creative self-forgetfulness (transcend self-boundary when deeply involved in work or relationship; frequent "flow state" or "peak experiences"; **creativity**), transpersonal identification (feeling of strong connections to the entire world and everything in it; **idealism**), and spiritual acceptance (e.g., of **miracles**, extrasensory perception, **telepathy**, **vitalization**). There is even a candidate gene identified, the VMAT2 gene on chromosome 10: people with at least one C (cytosine) at nucleotide position 33,050 instead of A (adenine) seem to be significantly more spiritual. The gene encodes for the VMAT2 Vesicular Monoamine Transporter. This protein puts monoamines such as dopamine into synaptic vesicles, making them available for neural processing, including emotional (and mystical) states. It has been pointed out, however, that at least 50 more genes with the same effect would be required to explain the **twin studies data**. The search for "God genes", a very misleading term, is only at the beginning, but the **twin studies** demonstrate a surprisingly strong inheritance. However, **genes** correlated with religiosity are, by themselves, not sufficient for proving adaptivity (even fully inherited traits like eye colors could be selectively neutral). But the data are consistent with a weak adaptation at least (for a strong adaptation the variability might be too large, because strong selection often reduces it). And religiosity could be just a by-product of inheritable traits like authoritarianism and conservatism while spirituality may even decrease reproductive interests.

Box 2.11 Potential Evolutionary Advantages of Religious Narratives

The 2009 book chapter written by Vaas provides a succinct but very informative discussion of various hypotheses that have been proposed in the literature to argue for a selective advantage of believing in teleological, and specifically in religious, narratives:

> How could religiosity be adaptive? Many hypotheses try to answer this question, and most are compatible with one another. However, the situation is quite confusing, because speculations are numerous whereas convincing data are still rare. One can distinguish between advantages for individuals and for groups, but even these are not mutually exclusive. And the advantages under consideration – many of which are controversial – are not helpful for each group member:
>
> • *Explanations*: One function of **myths** (and some rites) was – and for some people still is – to understand things and events in nature. Causal attributions are a powerful way to cope with a world that displays a degree of predictability, and **magical thinking** as well as **superstitious beliefs** and behavior are a spillover of this advantageous cognitive stance. Though meanwhile science can do much better regarding phenomena like earthquakes or thunderstorms, there remain questions which cannot be explained scientifically. For example: Why is there something rather than nothing? Why is the world the way it is? What is the meaning of everything (if there is one)? Believers think that transcendent entities might be helpful in this respect. But more important for most, is the following aspect [mentioned in the next item].
>
> • *Meaning and consolation*: Religion promises the mastering or acceptance of **contingency**: helping, for instance, to cope with death, illness, injustice...religion postulates meaning, order, orientation (as opposed to blind chance or fate) and, thus, relief and distraction, a protection from **absurdity** and an unpleasant reality.
>
> • *Happiness and health*: There are indications that religious contingency mastering, a kind of psychic placebo, and the social bonding in religious communities often increase psychological and **physical health**, **life-expectancy** and **happiness**, while

decreasing **depression**, **drug abuse**, **divorces**, and **suicide rates**. But **rigid religiosity** can also have negative effects like depression.

• *Shaping of behavior*: Rulers can gain, justify, and keep their **power** (note: more powerful people have, statistically speaking, more children). Sometimes this even transcends their death, and **ancestor worship** probably played a major role in the origin of religions. **Moral rules** can be more easily justified and enforced if they come along with religion. People can be motivated and manipulated – even as far as **martyrdom** (which, if there were no **group selection** advantage, would appear to be a dysfunctional extreme) and "**holy wars**" (which might be eerily advantageous as a result of the **looting** of resources and **rape**).

• *Group stabilization*: Intra-group conformism and inter-group demarcation.

• *Cooperation*: Increase of reciprocal **altruism**, which is beneficial (for instance, in **food sharing**, trade, hunting, **warfare**, defense, division of labor) as long as it is not exploited by free riders (even **celibacy** could be adaptive in some way, for instance due to inclusive fitness or **group selection**). Thus religion might be good for **intra-group loyalty**, strengthening commitments between members, because public religious activities (for example, **excessive prayer**, **food taboos**, **abstinence**, **pilgrimage**, **flagellations**, **circumcision**) serve as costly or **hard-to-fake signals**. They can help…deterring free riders, making it too expensive for them to fake the signals just to attain the group privileges. Public display of costly signals increases trustfulness and reciprocal altruism by helping to identify credible partners for cooperation (so do not costly, but otherwise hard-to-fake signals: displays of emotions). Moreover, it has been shown that religious groups are more stable and long-lasting than secular ones, especially those that make great demands on their members, that is insist on costly signals. Also the widespread belief in an omniscient, watching and punishing God or in still present ancestors reinforces reliability of partners via fear and remorse.

• *Sexual selection*: This is a special case of cooperation and another kind of selective factor in addition to natural selection. It is mainly driven by female choice. While natural selection is about survival, **sexual selection** is directly about reproduction. In human evolution the importance of paternal investment increased – together with the development of bigger brains, higher intelligence, and an ever more complex culture – because of the "physiological prematurity" of infants and the extended early childhood period. Therefore, it has been suggested that women use religion for manipulation of males: food-sharing, decrease of intra-sexual competition, and an increase of **sexual fidelity**. Also, religiosity can be seen as a fitness indicator…empirical data support the role of religiosity for sexual selection…there are more believing women than men; religion is more often practiced in public by men but more appreciated and experienced by women; there are more women in religious communities than men, and women gravitate more strongly toward religious groups that emphasize **family values** and **faithfulness**; furthermore, religious women have more children, often a religious spouse and there is a lower probability of them being single mothers.

However, I would like to make some brief comments about one of the specific hypotheses listed by Vaas and shown in Box 2.11 about a potential selective advantage of believing in teleological, and specifically in religious, narratives: the one concerning the "shaping of behavior and cooperation." These comments mainly have to do with three points that are often given as examples of the advantages of not only belief and religion per se, but of organized religion: cooperation/altruism, morality, and following rules. Referring specifically to organized religions, Shermer wrote in his 2004 book *The Science of Good and Evil*:

> When bands and tribes gave way to chiefdoms and states, religion developed as the principal social institution to facilitate cooperation and goodwill. It did so by encouraging altruism and selflessness, discouraging excessive greed and selfishness, promoting cooperation over competition, and revealing the level of commitment to the group through social events and religious rituals. If I see you every week in church, mosque, or synagogue, consistently participating in our religion's activities and following the prescribed rituals and customs, it is a positive indication that you can be trusted and you are a reliable member of our group that I can count on. As an organization with codified moral rules, with a hierarchical structure so well suited for hierarchical social primates like humans, and with a higher power to enforce the rules and punish their transgressors, religion responded to a need. In this social and moral mode I define religion as a social institution that evolved as an integral mechanism of human culture to encourage altruism and reciprocal altruism, to discourage selfishness and greed, and to reveal the level of commitment to cooperate and reciprocate among members of the community.

But such an idea clearly does not hold against the available empirical data, because even works that attack Rousseau's "Noble Savage," such as Pinker's *The Better Angels of Our Nature*, usually agree that in "pre-agriculture," and thus before the rise of organized religions, there was in general much more social and economic equality and sharing of food and other items than in agricultural societies with organized religion. Moreover, even when authors such as Pinker argue that in agricultural societies the rates of violent crimes are lower than in "pre-agricultural" ones, they chiefly use that argument to defend the Hobbesian idea that this is mainly due to the existence of powerful, efficient central governments in agricultural "civilizations," and not of organized religion per se. In fact, countries in which there is not only a huge proportion of religious people but also a very powerful church, and in which there is a week central government that is not able to enforce the rule of law, such as Mexico, Honduras, and more recently Venezuela, are among those with higher rates of violent crimes—clearly much higher than countries in which much more people are non-religious and in which the church is much less powerful, such as Finland, Denmark, or Norway. Actually, contrary to the Hobbesian ideas of authors such as Pinker, the empirical data available indicate that violence and particularly warfare *increased* with the rise of "early civilizations" and that organized religion played a critical role in that increase by helping particularly aggressive rulers to gain, validate, use, and keep their enormous power, including by undertaking atrocious acts of violence that were often precisely "morally" justified by teleological narratives.

2.5 Beliefs, Atheism, and Spirituality

When men stop believing in God they don't believe in nothing; they believe in anything. (Gilbert Chesterton)

In *The Psychology of Religion – An Empirical Approach* Hood and colleagues review several studies that strongly support the idea that practices such as meditation provide benefits for **well-being** and health, and show that religion is in general associated with lower consumption of illegal substances, tobacco, and alcohol, less anxiety, stress, and depression, better health in general, higher life satisfaction, optimism and self-esteem, and a more optimistic way of coping with chronic illness or with the death of loved ones. The authors provide various reasons for such trends. Some religions sponsor a wide variety of **healthful practices** that are adopted by many believers, such as condemning the use of alcohol in Islam. Religions often also provide social connection and support for believers, including the elders, which are particularly at risk of depression or suicide due to isolation. And, because religions provide a "meaning," a "cosmic purpose of life," being religious per se probably also contributes to many of these items *as a* ***placebo effect***. This point is supported by empirical studies that show that even religious patients with conditions that resist improvement and that isolate them tend to adapt better to such difficult situations and often resolve their depression faster than those low in religiosity. Furthermore, they cite several studies that support the idea that "faith is associated with lower reported cases of cardiovascular conditions, hypertension, stroke, and different forms of cancer, all of which are concentrated among the elderly."

Still another possible reason, in some way related to a kind of placebo effect, is that some studies suggest that people tend to "feel" what the others around them are *apparently* feeling. That is, if the other people from your religion or cult seem to be "happy," then probably you will also tend to "feel happy." Hood and colleagues cite a 1962 experiment in which Schachter and Singer injected with **epinephrine**—a hormone that produces arousal, including increased heartbeat, trembling, and rapid breathing—a group of 184 male participants. All of them were told that they were being injected with a new drug to test their eyesight, but one subgroup of participants was informed of the possible side effects that the injection could cause; the other participants were not told so. Participants were then placed in a room with another participant that was actually a "stooge," who either acted as euphoric or as angry. Participants that were not informed about the effects of the injection were more likely to feel happier if they were in a room with the "euphoric stooge" or angry if they were exposed to the "angry stooge." That is, these results suggest that people who had no a priori descriptions—and thus expectations—about how they would likely "feel" are more likely to then use external cues to label what emotion is occurring: in other worlds, in this case, as in many others, specific "emotions" are at least partially socially constructed, stressing once again that we are truly a peculiar combination of *Homo fictus, irrationalis, et socialis*. As put by Hood and colleagues, "despite major methodological criticisms of the Schachtner and Singer 1962 study, its importance for the theory of **religious/spiritual experience** is that physiological processes per se cannot account for emotional experiences; cognition must also occur, at least in ambiguous circumstances."

On the other hand, Hood and colleagues also cite several studies showing that certain types of religious beliefs can lead to having *more* health problems, for instance because of **inbreeding**: some **Amish groups**—often portrayed as the "best defined inbred groups" in the U.S.—have a higher incidence of **hereditary dwarfism**, **hemophilia** (**pathological bleeding**), and **muscular dystrophy**, among other inherited conditions. Studies show that Jews in the U.S. are, overall, also more prone to have a wide variety of inherited illnesses, including the development of **breast cancer** in women. They also stress the fact that within many **religious cults**, **social isolation**—for instance from the family—is a requirement, and the belief that those who commit suicide never truly die is so emphasized in such cults that they "create the condition to make suicide appear to be the only means of achieving ultimate happiness," sometimes leading to dramatic cases of **mass suicides**. They further note that some empirical studies have indicated that **family abuse** might be higher "in the strong **patriarchal family** structure espoused by some conservative religions." Moreover, "some clergy counsel women to remain with **abusive husbands** because it is their religious duty and responsibility…indeed, there is evidence that religious beliefs among **evangelical Christian women** may make them more reluctant to leave a physically abusive husband…this patriarchal issue may cross cultural boundaries…higher level of religiosity among **Arab women in Israel** were associated with the attitude that **abused women** should assume more personal responsibility for their husbands' violent behavior."

Furthermore, Hood and colleagues also refer to the problem of **self-reporting**, concerning specifically romantic love, marriage, and the teleological notion of "meant to be." First of all, they note that, because religious people tend to think they are doing "what is right"—as married people often also tend to—within the teleological narratives and social norms accepted by them, they tend to self-report that they are more happy with their lives than other indicators—for instance based on detailed social psychology studies—actually seem to indicate. Therefore, it is likely that religious people will also often tend to self-report that they are more optimistic, more relaxed, less anxious, and so on. Furthermore, either because this might be effectively true in some cases, or because religious people might start believing their own self-reports, or because they don't truly believe them but might not want to recognize that in public, they might be "less inclined to report symptoms of illness and therefore may downplay their possible significance," as noted by Hood and colleagues. This point is further complicated by the fact that it is often difficult to discern what is the true **causation** in many cases. These authors state that frequent church attendees tend to live longer than infrequent ones but, as they point out, in order to go to church one has to have at least some physical health, so this might be one example of a confounding factor: indeed, they noted that a meta-analysis that considered "some 15 possible confounding factors…found that religious involvement and longevity continued to be positively related…the association was, however, rather weak."

A point that is too often neglected in the literature—but not in the book of Hood and colleagues—and that I will discuss in detail in the chapters below when I will analyze Van Schaik and Michel's 2016 captivating book *The Good Book of Human Nature*, is that organized religions were very likely critical as "**systems of crisis management**" after the rise of agriculture, to cope with crucial problems such as the dramatic increase of diseases, famines, and plagues after the rise of agriculture (see Fig. 8.7). Organized religions proposed strict rules about **bathing**,

menstruation, nutrition, and so on, which were very likely hugely important to save many lives back then, but that are often mainly obsolete nowadays, although many of them continue to be followed dogmatically by many believers—a further example that in evolution what might be advantageous at a certain point in time might become neutral or even detrimental later. In this sense, I completely agree with Van Schaik and Michel in that most people nowadays don't realize that this might have been one of the most crucial functions of organized religion back then—in great part because most people, including many scholars, continue to *believe* in *Neverland*'s fairytales about how "good" and "noble" and "liberating" agriculture was, when the *reality* is that a huge amount of empirical data show the opposite, as stressed by Van Schaik and Michel (see Chap. 4).

However, it also needs to be noted that organized religion in particular, and teleological narratives in general, have also been a major "curse," since the rise of agriculture—also designated as "Neolithic revolution"—and of so-called "early civilizations." This is actually the *key difference* between this book and previous ones that discussed these topics: its focus not only on the mere curiosities about and the "positive" aspects of such teleological narratives, but also on the very dark aspects that are directly related to, or were at least justified by, them, such as discrimination, oppression, racism, misogyny, colonialism, slavery, atrocious wars, animal abuse, and so on. An emblematic example of how these dark aspects are indeed often neglected in works about religion in particular or beliefs in general is Shermer's 2011 book, which is one of the most cited books about these topics and which, despite being entirely focused on beliefs, provides only very few examples of some "negative" consequences to which they might lead. In fact, only two pages of that whole book are truly dedicated to discuss the "power and perils of patternicity" at a broader human scale, for instance. One of the few cases provided in those two pages concerns the belief in conspiracy theories:

> Occasionally I am challenged about the harm of people embracing superstitions, along the lines of: 'Oh, come on, Shermer, let people have their delusions. What's the harm?'. What's the harm? Ask the victims of John Patrick Bedell, the gunman who attacked guards at the entrance of the Pentagon in March 2010, who now appears to have been a right-wing extremist and 9/11 'truther'. In an internet posting, he said that he intended to expose the truth behind the 9/11 'demolitions'. Apparently the delusional Bedell intended to shoot his way into the Pentagon to find out what really happened on 9/11. Death by conspiracy.

Unfortunately, imaginary tales have in reality been directly or indirectly associated with the death of not only a few people but of literally several millions of human beings, as well as with the discrimination and oppression of, and abuses made to, billions of people and also with atrocious maltreatment of billions of non-human animals. This disturbing fact is summarized in Fig. 2.11, which includes an endless number of **conspiracy theories** that are believed by numerous people nowadays, in the twenty-first century, including many that have indeed led to the death of millions of people—for instance, anti-Semitic ones—and others that might well do the same in the near future—for example, **QAnon and anti-vax theories**. Indeed, as illustrated in that figure, the more we detach from reality and immerse ourselves in *Neverland*'s world of unreality, the more we pay an enormous price for

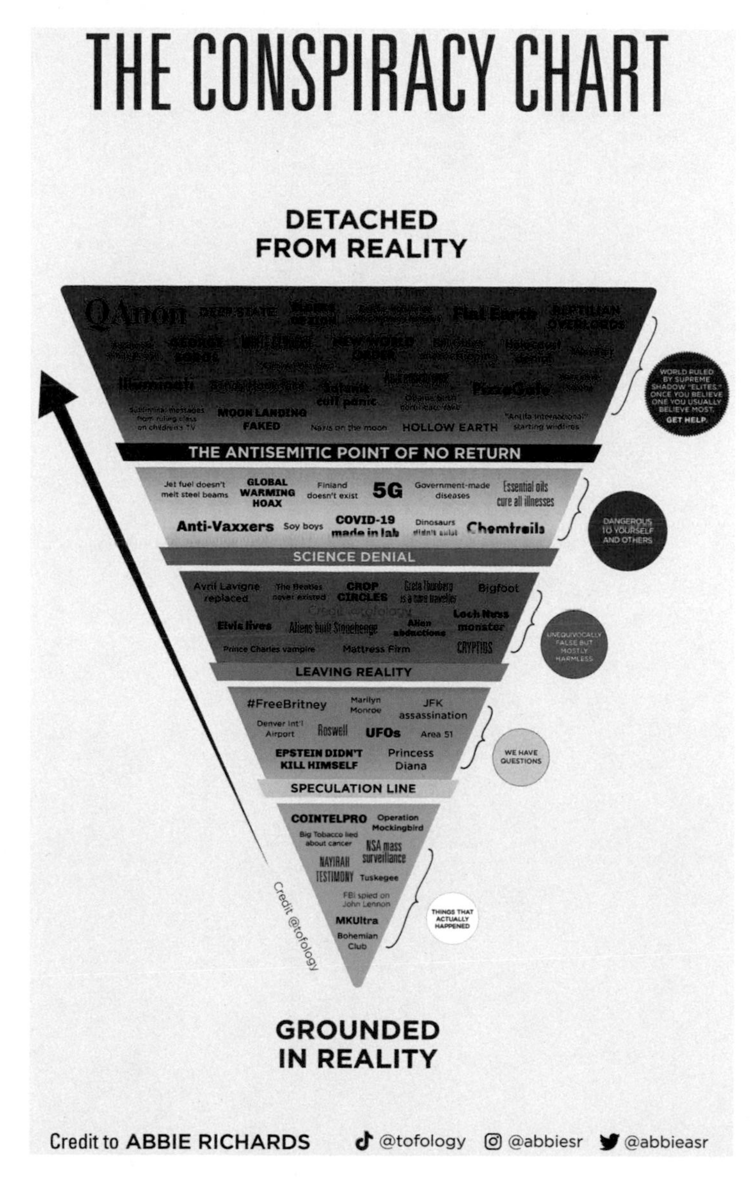

Fig. 2.11 It is disturbing to see that, in the twenty-first century, many millions of people believe in such illogical, absurd, and often dangerous *Neverland* conspiracy theories, many of—and unfortunately often the most hazardous of—them being completely detached from the world of reality

it. And, at the present moment, we are effectively very close to paying an ever higher price because if we can't escape from it we will very likely soon reach an ecological point of no return that will put at risk the existence of an endless number of species, including our own.

Chapter 3
Our Place in Nature, Progress, and Racism

(In the Holocaust) rather than producing goods, the raw material was human beings and the end product was death, so many units per day marked carefully on the manager's production charts…the chimneys, the very symbol of the modern factory system, poured forth acrid smoke produced by burning human flesh…the brilliantly organized railroad grid of modern Europe carried a new kind of raw material to the factories.

(Zygmunt Bauman)

3.1 Quests to Understand Our Place in Nature

I believe our heavenly Father invented man because he was disappointed in the monkey. (Mark Twain)

Numerous books have already been written on racism. Here, I will provide a novel way to understand the deeper basis of racism by emphasizing how teleology, and in particular the notion of a "**chain of being**" that is profoundly related to our quest to understand humankind—including its "subgroups," or imaginary "races"—and its place in nature, is profoundly related to the heartbreaking history of racism, colonialism, and slavery. "Chain of being," or "**ladder of life**," is a translation from the Latin "*scala naturae*": that there is a "progress" from "lower forms," such as plants, to non-human animals, and then to humans, which self-designated themselves as the culmination point of such a "progression" towards perfection (see Boxes 3.1 and 3.6 and Fig. 3.2).

R. Diogo, *Meaning of Life, Human Nature, and Delusions*,
https://doi.org/10.1007/978-3-319-70401-2_3

Box 3.1 Plato, Aristotle, Teleological Roots of Western Science, Causality, and Intentionality

One of the most complete and influential books about the history of the teleological notion of *scala naturae* (Fig. 3.2) is Lovejoy's book *The Great Chain of Being – A Study of the History of an Idea.* I highly recommend you to read that book. Regarding more recent books, in *The Tinkerer's Accomplice – How Design Emerges from Life Itself*, Turner provides a succinct historical background on the idea of **intentionality in nature**, referring, for example, to **platonic teleology**, in which **crabs** were seen as made by God to offer particular moral examples to teach humans. Regarding **Aristotle**, he saw **purpose** in a different, more dynamic way, focusing on physiology: for instance, an animal is supposed to do something, and when taken out of that context/niche, say a fish taken out of water, it will "intentionally" try to go back to it (water) or to adapt to the new context (e.g., to life on land). According to Turner, Aristotle used in particular two main ways to explain the phenotype (anatomy) of organisms: **teleology**—natural entities have intrinsic purposes—and "matter." It can also be said that in a sense Aristotle accepted some notions of **vitalism** (see also Box 3.6). A 2009 book chapter written by Achtner also provides a succinct, interesting take on the teleological origins of Western science and in particular of natural sciences, stating:

> The first representative of a teleological understanding of evolution is **Aristotle**, who coined the notion of **teleology** as a result of his observations in nature in general and of organic life in biology in particular. Interestingly, he had a dispute on the teleological development in nature with **Empedokles**, who claimed that all development in nature is driven by pure chance, thus foreshadowing the debate in the nineteenth century between **Darwin** and the representatives of natural theology based on teleology. In antiquity Aristotle won the battle, in the nineteenth century the revenant of Empedokles, Charles Darwin, was the winner. Due to the Aristotelian reception by theologians and philosophers in the Middle Ages teleology becomes part of the scientific canon. However, already as early as the fourteenth century teleology was severely criticized in the context of the emerging nominalistic philosophy of nature and extinguished in favor of the **causa efficiens**, the efficient causes. In the seventeenth century, teleology was completely replaced by the concept of law of nature within the **Scientific Revolution**.
>
> It was only in biology that teleology survived as a scientific concept as late as the nineteenth century. Nineteenth century famous biologists like Gerhard Oncken (1800-1884) or Johannes Müller (1801–1858) took it still seriously as scientific. As is generally known, British **natural theology** was based on the **teleological interpretation of nature**. The long-standing tradition of natural theology found its climax in the work of **William Paley** (1743–1805) and his book *Natural theology*. The studied theologian Charles Darwin owes his decisive inspiration to coin "adaptation" or "fitness" as a scientific term to him. In natural theology adaptation and fitness was interpreted as the result of divine creation. Darwin replaced this theological interpretation of adaptation as a result of the divine creator by a scientific explanation, in which the evolutionary process is governed solely by pure chance. Thus teleology was extinguished from biology and theology suffered because the argument of divine design could no longer be sustained. The problem arose whether or

not the evolutionary process of creation based on chance could be reconciled with the idea of **divine creation** and **teleological providence**.

On a similar topic, and in the same book, Frey (2009) notes:

The Aristotelian theory of **causality** corresponds to animistic beliefs, in which the subjective part of the explanation is not quite overcome. The "Impetus" theory **of Philoponus** (490-570 A.D.) and Buridan and Oresmes are the next stage toward a **mechanistic Newtonian explanation**: an object moves because a "moving force" has been transferred into it. The object stops moving, as soon as this force is used up. Interestingly, this is the top level most – even educated – people reach. Actually, most physics college students do not master **Newtonian concepts**... People in non-industrialized societies have a fascinating concept of causality. If illiterates are asked questions like "Why do sun and moon move?" or "Why does a stone fall to the ground?" the answers are usually quite uniform. Two types of answers are dominant. First, **animistic answers**: the clouds, the sun, and the moon live and thus move intentionally by themselves. Second, answers are given that require a creator: they are moved by someone else (a higher and very powerful being). Depending on the domain of the question, the number of animistic answers is quite high, ranging from 20 up to 80%. For example, 38% of Turkish illiterates are convinced that clouds have a consciousness and 55% think that the sun is alive. In general, it is possible to distinguish four basic patterns: 1) explanations show an explicit bias toward the originator/agent/cause of an action or situation; 2) in analogy to living social agents it is insinuated that inanimate objects have goals and intentions; 3) external physical forces (e.g., force of gravity) are regarded as object-internal, that is as intrinsic forces; 4) forces are always thought to be in the direction of movement; if there is no apparent external force, an internal [intrinsic] force is presumed. Causality is just one example of deeply engrained ideas about folk physics, but it shows that these ideas are very strong, persistent, and powerful...[and]...how humans handle causal events and how the combination of these patterns of thought connects to the idea of God's existence.

As noted above, teleological narratives are based on the premise that everything has a special "purpose" or use, and the word *teleology* builds on the Greek *telos* ("purpose," or "end") and *logia*, which refers to "a branch of learning." The **ancient Greeks** were in fact crucial for the widespread use and acceptance of teleological narratives in the West, particularly within sciences. Natural sciences, including biology, are an emblematic example of this, because **Aristotle**—famous for his teleological statements about the organs of organisms—is often considered to be the "first biologist" (see Boxes 3.1 and 3.2). A wonderful book that I highly recommend to those interested in knowing more about Aristotle's life and works, and their influence for biology and society in general since then, is Leroi's 2004 *The Lagoon – How Aristotle Invented Science*. Alternatively, or additionally, for someone interested to read an extensive study specifically focused on Aristotle and teleology, I recommend Johnson's 2005 book *Aristotle on Teleology*. Of course, it should be noted that the **scientific method**, specially its trial-and-error methodology—which in modern scientific terms could be termed "testing-and-falsification"—was done by Greeks before Aristotle. Also, we always should avoid the trap of **ethnocentrism**, and therefore one needs to stress that most human societies, since times immemorial, clearly had some kind of science: for instance, at least some so-called traditional medicines are clearly the product of an endless number of trials-and-errors, done during several generations, all the way until modern times, as it is clearly exemplified by several of the ones used by the so-called pygmies of Central-Africa (see Chap. 4).

However, having said this, one can say that, in a somewhat simplistic way, **Aristotle** might well be the "father," or at least the most influential "originator," of Western biology, of **Western natural sciences**, and perhaps in some way even of **Western science** as a whole, as suggested by Leroi. At least, one needs to recognize that the ideas of the vast majority of the most eminent Western scientists, including Darwin and most other scholars within the natural sciences, have been deeply influenced by Aristotle's ideas. Apart from Leroi's book, and Lovejoy's 1936 book *The Great Chain of Being* (see Box 3.1), in case you want to read more about these issues, I strongly recommend Ruse's 2003 book *Darwin and Design* and 2018 book *On Purpose*, and Reiss's 2009 book *Not by Design – Retiring Darwin's Watchmaker*. Unfortunately, some of Aristotle's societal ideas based on the notion of *scala naturae* (Boxes 3.1, 3.2, and 3.3), such as his thoughts about the inferiority of women and "superiority" of humankind, also became hugely influential within the humanities and social sciences, and within the broader public in general, as noted by Leroi:

> The history of Western thought is littered with teleologists. From fourth-century **Attica** to twenty-first-century Kansas, the ***Argument from Design*** has never lost its appeal. **Aristotle** and **Darwin**, however, share the more unusual conviction that though the organic world is filled with design there is no designer. But if the designer is dead for whose benefit is the design? It's the prosecutor's question: *cui bono*? Darwin answered that individuals benefit. Biologists have batted the question about ever since. The answers that they've essayed are: memes, genes, individuals, groups, species, some combination or all of the above. Aristotle, however, generally appears to agree with Darwin: organs exist for the sake of the survival and reproduction of individual animals. This is why so much of his biology seems so familiar. Yet there is a deep difference between Aristotle's teleology and Darwin's **adaptationism**, one that appears when we follow the chain of explanation that any theory of organic design invites. Why does the elephant have a trunk? To snorkel. Why must it snorkel? Because it's slow and lives in swamps. Why is it slow? Because it's big. Why is it big? To defend itself. Why must it defend itself? Because it wants to survive and reproduce. Why does it want to survive and reproduce? Because…Because natural selection has designed the elephant to reproduce itself. Darwin gave teleology a mechanistic explanation. He halted the march of whys. It is for this reason that Ogle celebrated Darwin as **Democritus** reincarnated. For, where Aristotle's organismal teleology is imposed upon recalcitrant matter, Darwin showed how, given a few simple conditions, it emerges from it. Darwin is an ontological reductionist; Aristotle is not.
>
> Why, then, should Aristotelian animals strive to survive and reproduce? Aristotle can hardly invoke natural selection. (He's dismissed at least one version of it.) He could have said 'they just do', and left it at that, but then he would not be Aristotle, so he does have an answer, beautiful and a little mystical. Living things, he says, desire to survive and reproduce so that they can 'participate in the eternal and the divine'. When he asserts that living things desire to participate in the eternal he means that they are designed not to become extinct. *Cui bono*? It turns out that organismal design is not, after all, for the sake of individuals, for they always die, but to ensure that their forms/kinds, their species, persist for ever. When Aristotle speaks of the divine he is not – the point must be made again – invoking a divine craftsman for none exists; rather, he is telling us that immortality is a property

of divine things and that reproduction makes animals a little bit divine...his teleology is riddled with such value judgments. He says that the position of the heart in the middle of the body is dictated by its embryonic origins. But it is also located more *above* than *below* and more *before* than *behind*, 'For nature when allocating places puts more honourable things in more honourable positions, unless something more important prevents this' – the language suggests the seating plan at a dinner.

One may wonder why, then, the human heart (actually its apex) is located on the inferior left, but Aristotle has inserted a caveat – '*when nature does nothing (in vain)...*' – and gives a patently *ad hoc* explanation that it's needed there to 'balance the cooling of things on the left'. He thinks, of course, that the right-hand side of the body, being more honourable, is hotter than the left, and that this is especially so in humans, and so the heart has to shift to compensate for the left's relative coolness. **Plato**'s influence is most obvious when Aristotle considers man. He is explicit: man is his model not only because he's the animal we know best, but also because he is the most perfect animal of all. The axes of the body are most differentiated in humans; in other animals they're present but in a confused sort of way. In the same way, the characters of animals – courage, timidity, intelligence and the like – that are regulated by the sensitive soul are better developed in humans than in any other animal...there's a sense in which a swallow shows intelligence when it builds its beautiful little nest, but human intelligence is of an altogether different kind. We are beginning to touch on Aristotle's theology, his ultimate explanation for why the cosmos is arranged the way it is and its relationship to an immortal God. Why should animal kinds be immortal? This is where we come to the end of explanation, to one of those indemonstrable axioms that lie at the bottom of every Aristotelian science, and from which all else flows, and it is simply this: it is better to exist than not to exist.

Box 3.2 Adaptationism, Aristotle's teleology, and Intelligent Design

Biological **adaptationism** is deeply rooted in **Aristotle**'s teleological thinking, in the sense that adaptationists tend to try to assign an "adaptative" cause for the origin and persistence of traits such as morphological or behavioral ones, by arguing that these traits necessarily gave a functional advantage to—and thus contributed for the evolutionary success of—the organisms possessing them. The most extreme adaptationists try as hard as they can to assign an "adaptative" current function even to **vestigial organs or traits**—that is, those that are mainly a reminiscence of past ones. In contrast, many other authors, including me, have refuted such adaptationist ideas, based on empirical data. In my 2017 book *Evolution Driven by Organismal Behavior*, I provided several factual cases of vestigial muscles that are normally present only in the early stages of human development and that become absent in later, postnatal stages. I will provide here an illustrative example of the type of circular reasoning that is typical of adaptationist works by referring to a recent, influential, and otherwise magnificent book: *Evolutionary Behavioral Ecology*, edited in 2010 by Westneat and Fox. In that book, Fox and Westneat's chapter on "*Adaptation*" is a particularly strong—and I would say extreme—defense of the **adaptationist framework**. First, they explain that in the last decades of the twentieth century, **Stephen Jay Gould** and Lewontin provided a powerful criticism of the **adaptationist paradigm** that was crucial for the beginning of the decline of adaptationism since then. As Fox and Westneat recognize, this objection to the adaptationist paradigm was even adhered to by many "behavioral ecologists," but they note that this paradigm "remains dominant in behavioral ecology because, in case after case, the focus on **adaptationist explanations** has led to new insights." However, "leading to new insights" does not mean that one is following the right scientific path. Everything, even non-scientific reasoning, tends to lead to "new insights." Surely, the biased discussions of authors defending the creationist idea designated as "**Intelligent Design**" have led to many "new insights," as evidenced by the endless number of new books and articles and blog posts they continue to write and even museums they build. They continue to do so much stuff, and to always confirm their a priori ideas, precisely because they blindly follow a path and a way of thinking that is often difficult to falsify—and, in the rare cases it can be and it indeed is, they ignore this anyway. Although I do not in any way intend to place Intelligent Design and the adaptationist paradigm on the same level, there are, unfortunately, a few striking parallels between the two. Adaptationist ideas are often very hard to falsify, as stressed by Gould and Lewontin, and it is precisely not a good sign when defenders of adaptationism argue that this is not a major reason to be careful, because it leads to "new insights." This can lead to a dangerous circular reasoning, of "predictions," "new insights," and "confirmation of (the often untestable) predictions," and so on.

When I referred to empirical examples that refute some of the most influential just-so stories created by adaptationists in a 2017 meeting on Human Enhancement organized by philosophers of the University of Lisbon, these philosophers asked me: "but one can still say that the function of the heart, its reason for existence, is to beat, thus pumping the blood that allow us to live, right"? Actually, this is *not* factually accurate: this way of talking is deeply rooted in our tendency to ask why-questions and to build teleological narratives to answer them, and in particular on the type of teleological biological thinking of **Aristotle**. The origin of the heart is not related to the goal, or "purpose," to allowing us to live. There is no "purpose" for the existence of the heart, because nobody designed it or planned it to evolve in a certain way, contrary to what creationists and/or followers of intelligent design argue. The focus on the heart as a whole, as an organ that pumps our blood to allow us to live is itself teleological. In fact, if we focus instead on each of the cells that are part of the heart, the scenario is very different: if we were to ask each of those cells "why" they do what they do, do you think they would answer "well, my whole existence, all that I do, that is my *ultimate function*, is just to allow you to live." Each cell of the heart has no idea that the organ formed by them is allowing us to live. And they were surely not "originated" with that purpose, because they derive evolutionary from other types of cells that did completely different things—what we call an **exaptation**, in contrast to an

adaptation—and that existed billions of years before the origin of humans. That is, the cells of the heart, and the heart as a whole, do not exist *to* allow us to live: instead, the fact that we can live is just a *by-product* of the fact that those cells exist and are part of a complex anatomical and physiological network of cells that we designate as the "heart." Each of the cells of our heart is just living, and evolving, as are all the other cells of our body and all the types of living beings since life began. There is no cosmic purpose for the existence of any type of living cells or organisms, they just exist and evolve, and sometimes they can be simply co-opted to be part of the body of multicellular organisms, that's all.

Box 3.3 Teleology, Aristotle, Voltaire, Kant, and Hegel

For those that might think that **teleology** is just an outdated idea, with no interest for current history, culture, or politics, I strongly recommend the 2015 book *Historical Teleologies in the Modern World*, edited by Trüper and colleagues. I will provide some excerpts from it, below. However, one needs to keep in mind that- as explained in Lenoir's 1982 book *The Strategy of Life – Teleology and Mechanics in Nineteenth-Century German Biology*—some of the ideas defended by Kant and the German researchers that were highly influenced by him, including some "romantics," were not so out-of-touch with the scientific method as suggested in that 2015 book, which states that:

> The word was a success. Invented by Christian Wolff in 1728, with **Diderot**'s *Encyclopédie* and the philosophy of **Kant** as its most powerful transmitters, 'teleology' soon projected itself into the furthest corners of philosophical discourse. Wolff's placid definition held that we 'might call Teleology' that 'part of natural philosophy which explains the ends of things, and which thus far lacks a name, even if it is most ample and useful'. Some two millennia after the **Greek philosophers** had begun to discuss the status of the explanation of something by reference to an end, goal, aim or purpose, the issue was antique. It had been overwritten by uncountable theoretical texts and traditions in various languages throughout the vast space of reception the Greek classics enjoyed. It was not, perhaps, so surprising that this palimpsest bore no definitive name when Wolff entered the scene. His seemingly minuscule intrusion into the philosophical mess of the final cause was grave in consequence, since by sleight of hand he temporalized it. As opposed to the dry technicality of the *causa finalis* or the *conatus* (inherent tendency or direction, a Leibnizian term), 'teleology' promised the future achievement of a well-ordered doctrine, a disciplined logos with a clearly and distinctly defined ambit, and a 'most ample and useful' application. In a word, **teleology** became a project. The very term expressed a conviction about what philosophy had so far failed to achieve but would, soon, amend. Even if the future course of the development of thought was not known in detail, the sense of direction was clear. In an oblique and imbalanced manner, Wolff coupled history and teleology. Crucially, teleology was to be self-reflective; it would ultimately include an account of itself since it was itself goal-directed. In this way,

history, too, would comprise its own explanation. When Wolff entered the fray, his intention was to reconcile explanation by reference to the pursuit of ends with explanation by reference to efficient causes. The protagonists of the so-called '**scientific revolution**' had almost unanimously dismissed the *causa finalis* as, in **Francis Bacon**'s words, a 'barren virgin', devoted to God in the pursuit of the inbuilt purposes of all creatures in creation, but unproductive of any such knowledge of nature that would be capable of begetting further knowledge.

Baconian, **Hobbesian**, **Cartesian** and **Spinozist** philosophies all focused on the competing notion of a comprehensive and total 'chain' of efficient, 'mechanical' causes in order to eliminate any space for teleological explanation outside – possibly – the domain of intentional human (or divine) action. Yet, the notion of the world as mechanism seemed to fall short of explaining numerous phenomena, and it laid claim to a kind of universality that was difficult to bear out. Most prominently, as Wolff believed, mechanism failed to answer the question of how mind was able to produce any effect on matter. The mechanists, he charged, had to posit a *qualitas obscura*, an inexplicable *explanans*, on at least one side of the Cartesian abyss between matter – understood as extended, inert and solid – and consciousness. More precisely, since **consciousness** seemed rather obviously capable of anticipation and directedness, for the anti-mechanists it was clear that the modification had to take place in the concept of matter. Wolff, for instance, proposed adding what he labeled a specific 'moving force' (*vis motrix*) to the understanding of matter. By contrast, the force of the conceptual debate in physics has often been, but ought not to be, underestimated. It was this debate that provided the plot; theology was already reduced to acting as *deus ex machina* providing a forced resolution if the action otherwise failed to achieve closure.... Wolff's concerns were expressed through an appropriation of **Leibnizian ideas**, the reception of which over the course of the eighteenth century, owing to the erratic posthumous publication of several of his most important works, was a convoluted affair. It seems clear that his development of the infinitesimal calculus (simultaneous to **Newton**'s) persuaded **Leibniz** to posit the existence of unextended, and therefore unqualified, atomic substances....

In the subsequent decades, cognate programmes of re-conceptualizing substance spread through European metaphysics, engendering – across emerging disciplinary fields such as chemistry and biology – a cascade of different conceptions of matter as imbued with a variety of natural forces and immanent directions. At the same time, in the distant field of historical studies, the traditional *ars historica*, the moral-educational paradigm of history as life's *magistra*, was collapsing. An opening emerged for new notions and approaches of historical writing. Into this opening intruded the new idiom that metaphysics and **natural philosophy** produced for the purpose of analysing natural reality. Still, this intrusion was not obvious and direct. Rather, it occurred by way of a bastardization that transported the ontological vocabulary of physics into the transforming field of history by way of **political philosophy**. Certainly, the convergence of natural and political philosophy in the eighteenth century is not a novel discovery. Various interwar period theorists, for instance, already agreed that the mediation of nature and politics had been an indispensable driving force behind the emergence of the famed, and defamed, philosophies of history of the **Enlightenment**. However, the eighteenth-century pioneers of philosophy of history in France, most notably **Voltaire** and, in his wake, Turgot, d'Alembert and Condorcet, all sided with the mechanists. As Voltaire had it, physics was to provide the epistemological model for a renewed understanding of history....

Kant's most momentous contribution to the vocabulary of natural philosophy was arguably his explication of the meaning of 'organism' in terms of a whole as consisting of parts that were to be understood by way of their inherent purposes for the functioning of the other parts as well as the whole. Organic reality thus required a set of purposive explanatory means categorically different from the efficient causes that alone were admitted in the mechanical nexus of causality. At the same time, teleology became constitutive, through the powerful **vitalist metaphor** of the organism, of a novel idea of self-organizing complexity. It is hardly necessary to point to the tremendously important history of **organicist perspectives** in the nineteenth and twentieth centuries. The decisive passage, in the introduction to the *Critique*, in which **Kant** stages the overarching understanding of a 'principle of purposiveness in nature' as a condition of possibility of theoretical knowledge, teems with references to mechanism and the problem of explaining change in the realm of physical bodies…it is thus the very intelligibility of natural reality that constitutes its teleological character…. One might easily be tempted to conclude that it is the progressive understanding of this reality that constitutes the realization of its inherent *telos*. Arguably, the most influential teleological philosophy of history of the period after 1800 was that of **Hegel**. In his conception, which was heavily, but not perhaps obviously, indebted to the Kantian, history was progress towards, and the realization of, freedom. This progress was to take place in the form of the objectivation of reason, the workings of the 'spirit' in time, its tireless labour, in the sequence of nations, to give itself form and in this way to know itself. For much of the nineteenth and twentieth centuries…the colonizers had legitimized their rule through the vocabulary of civilization and modernization, progress, backwardness and catching up. This vocabulary deployed the device of placement in history to work out a wealth of distinctions which often went so far as to refuse a colonized population a share in historicity altogether. Local, previously established models of staging historical time, and their respective histories, were often marginalized or submerged; and other such models, as based on objects, practices and notions otherwise excluded, appeared in response. On the colonial scene, the transition from the nineteenth to the twentieth century, supposedly effected by the *caesura* of the First World War, seems far less obvious. On the contrary, in many non-European contexts the obstinacy of teleologies of history appears only to have increased in the twentieth century, be it in alliance with anti-colonialist or anti-imperial nationalisms or emancipatory **Marxisms**, the critique of modernization theory, or that of 'development'. Even if more recent 'western' recurrences of historical teleologies, such as the perfectibility of the 'ever closer union' pursued by the European Union, or, possibly, the reassertion of autonomous historical agency in the US presidential election slogan **'Yes we can'** seem in part designed precisely to avoid reckoning with the imperial past, nonetheless they only confirm the plurality of available formulations of historicity.

In his 1980 book *History of the Idea of Progress*, Nisbet spends 357 pages exclusively focusing on the historical roots of teleological narratives about the notion of a "ladder of life" and of "cosmic progress" and on how they have deeply influenced our sciences, politics, art, and daily lives in general, in the last millennia. The book is truly impressive, as it includes a huge number of small sections discussing how a particular renowned, influential scientist, or philosopher, or politician, was deeply influenced by such narratives. In fact, basically all the Western "big names" are there, from **Aristotle** to **Darwin**, from **Kant** to **Spinoza**, from **Marx** to **Adam Smith**, or **Rousseau**. Nisbet explains:

> No single idea has been more important than, perhaps as important as, the ***idea of progress*** in Western civilization for nearly three thousand years. Other ideas will come to mind, properly: liberty, justice, equality, community and so forth.... But this must be stressed: throughout most of Western history, the substratum of even these ideas has been a philosophy of history that lends past, present, and future to their importance. Nothing gives greater importance or credibility to a moral or political value than belief that it is more than something cherish or to be cherished that it is an essential element of historical movement from past through present to future. Simply stated, the idea of progress holds that mankind has advanced in the past – from some aboriginal condition of primitiveness, barbarism, or even nullity – is now advancing, and will continue to advance through the foreseeable future. But what does 'advance' or passage from 'inferior to superior' mean in substantive terms? We shall find from the Greeks down to the twentieth century two closely related though distinguishable propositions. First slow, gradual, and cumulative improvement in knowledge, the kind of knowledge embodied in the arts and sciences in the manifold ways man has for coping with the problems presented by nature or by the sheer efforts of human beings to live with one another in groups.
>
> From **Hesiod** and, more vividly, Protagoras, through such Romans as **Lucretius** and **Seneca**, through **St. Augustine** and his descendants all the way to the seventeenth-century Puritans and beyond, down to the great prophets of progress in the nineteenth and twentieth centuries, such as Saint-Simon, **Comte**, **Hegel**, **Marx**, and **Herbert Spencer**, we find a rarely interrupted conviction that the very nature of knowledge – objective knowledge such as that in science and technology – is to advance, to improve, to become more perfect. Matters become almost hopelessly complicated and conflicting when we try to speak of progress (or regress) in reference to 'humanity', 'mankind' or 'civilization'. And yet these inherent complications, conflicts, and paradoxes notwithstanding, many very wise and eminent philosophers, scientists, historians, and statesmen have spoken of progress in these terms. To call the roll is to summon up such names as **Protagoras**, **Plato**, **Aristotle**, Lucretius, Seneca, St. Augustine, **Jean Bodin**, **Isaac Newton**, **Robert Boyle**, **Joseph Priestley**, Comte, Hegel, Darwin, Marx, Herbert Spencer, and in America, a line that commenced with **Cotton Mather** and **Jonathan Edwards**, and included **Jefferson**, **John Adams**, **Franklin**, and very nearly every major thinker and statesman in the United States who succeeded the Founding Fathers. These are but a few of the West's light and leading for whom the progress of mankind, especially in the arts and sciences, was as real and as certain as any law in physical science. Nor can we overlook the masses. From at least the early nineteenth century until a few decades ago, belief in the progress of mankind, with Western civilization in the vanguard, was virtually a *universal religion* on both sides of the Atlantic.

Indeed, the notion of a "cosmic progress" has been, and continues to be, almost a "universal religion" within humans in general—clearly not only on both sides of the Atlantic, but on all sides of all Earth's oceans, as shown by many examples given in the present volume. We will therefore now expand this fascinating historical topic to a discussion that has direct implications for a directly related issue that has also plagued, and unfortunately continues to plague, humans everywhere: **racism** (see also Box 3.4). Namely, it is important to focus on these related questions: is racism "innately" part of us, or a social construct? And is it associated to a posteriori narratives justifying—paraphrasing **Robert Sapolsky**'s fascinating book *Behave – The Biology of Humans at Our Best and Worst*—"**our worse behaviors**"? As explained in that book, humans often build a posteriori justifications for acts that they or others may deem as "wrong." Taleb's 2010 masterwork, *The Black Swan – The Impact of the Highly Improbable*, also provides numerous empirical examples of how our "wired" tendency to build a posteriori narratives applies basically to all aspects of life, and to all people, including scientists and even historians:

Our minds are wonderful explanation machines, capable of making sense out of almost anything, capable of mounting explanations for all manner of phenomena, and generally incapable of accepting the idea of unpredictability…events were [often] unexplainable, but intelligent people thought they were capable of providing convincing explanations for them – *after the fact*. Furthermore, the more intelligent the person, the better sounding the explanation. What's more worrisome is that all these *beliefs* and accounts appeared to be logically coherent and devoid of inconsistencies. (For instance) the Levant has been something of a mass producer of consequential events nobody saw coming. Who predicted the rise of **Christianity** as a dominant religion in the Mediterranean basin, and later in the Western world? The Roman chroniclers of that period did not even take note of the new religion – historians of Christianity are baffled by the absence of contemporary mentions. Apparently, few of the big guns took the ideas of a seemingly heretical Jew seriously enough to think that he would leave traces for posterity. We only have a single contemporary reference to **Jesus of Nazareth** – in The Jewish Wars of Josephus – which itself may have been added later by a devout copyist. How about the competing religion that emerged seven centuries later; who forecast that a collection of horsemen would spread their empire and Islamic law from the Indian subcontinent to Spain in just a few years? Even more than the rise of Christianity, it was the spread of **Islam** (the third edition, so to speak) that carried full unpredictability; many historians looking at the record have been taken aback by the swiftness of the change. Georges Duby, for one, expressed his amazement about how quickly close to ten centuries of Levantine Hellenism were blotted out 'with a strike of a sword'. A later holder of the same history chair at the Collège de France, Paul Veyne, aptly talked about religions spreading 'like bestsellers' – a comparison that indicates unpredictability. These kinds of discontinuities in the chronology of events did not make the historian's profession too easy: the studious examination of the past in the greatest of detail does not teach you much about the mind of History; it only gives you the illusion of understanding it. History and societies do not crawl. They make jumps. They go from fracture to fracture, with a few vibrations in between. Yet we (and historians) like to believe in the predictable, small incremental *progression*.

(I) will outline the *Black Swan problem* in its original form: how can we know the future, given knowledge of the past; or, more generally, how can we figure out properties of the (infinite) unknown based on the (finite) known? Think of the feeding again: What can a turkey learn about what is in store for it tomorrow from the events of yesterday? A lot, perhaps, but certainly a little less than it thinks, and it is just that "little less" that may make all the difference. The turkey problem can be generalized to any situation where the same hand that feeds you can be the one that wrings your neck. Let us go one step further and consider induction's most worrisome aspect: learning backward. Consider that the turkey's experience may have, rather than no value, a negative value. It learned from observation, as we are all advised to do (hey, after all, this is what is believed to be the scientific method). Its confidence increased as the number of friendly feedings grew, and it felt increasingly safe even though the slaughter was more and more imminent. Consider that the feeling of safety reached its maximum when the risk was at the highest! But the problem is even more general than that; it strikes at the nature of empirical knowledge itself. Something has worked in the past, until – well, it unexpectedly no longer does, and what we have learned from the past turns out to be at best irrelevant or false, at worst viciously misleading. Now, there are other themes arising from our blindness to the **Black Swan**: a) we focus on preselected segments of the seen and generalize from it to the unseen: *the error of confirmation*; b) we fool ourselves with stories that cater to *our Platonic thirst for distinct patterns*: *the narrative fallacy*; c) we behave as if the Black Swan does not exist: human nature is not programmed for Black Swans; d) what we see is not necessarily all that is there…history hides Black Swans from us and gives us a mistaken idea about the odds of these events: this is the distortion of silent evidence; e) we 'tunnel': that is, we focus on a few well-defined sources of uncertainty, on too specific a list of Black Swans (at the expense of the others that do not easily come to mind).

Fortunately, more and more scientists seem to be able to start deconstructing, or at least to be more aware of, our tendency to build post hoc narratives of "purpose," "design," and "progress" and consequently also of the importance of randomness and chance in nature. The 2016 book *Chance in Evolution*, edited by Ramsey and Pence, provides a detailed historical account on discussions about **contingency**, **chance** and **randomness** in ancient, medieval, and modern biology, referring to the Christian objections to the role of chance emphasized in Darwin's evolutionary works. Importantly, the book discusses a question that is crucial within the context of the present volume—"are we, humans, here by chance?"—and answers it by providing several empirical examples about the huge importance of chance in biological evolution, including obviously within our evolutionary lineage. Many other fascinating—some of them truly mind-blowing—examples are presented in Queiroz' outstanding 2014 book *The Monkey's Voyage – How Improbable Journeys Shaped the History of Life*.

Box 3.4 Noah, Christianity, Islam, and Judaism

In his 2009 book *On Monsters*, Asma explains (see also Fig. 3.1):

> In addition to…theorizing about the souls of monsters, theologians were also intrigued by the question of their genealogy: Who or what were the progenitors of these misshapen creatures? In particular, the races of monsters were difficult to square with the **biblical Table of Nations**. If they were indeed men, then we must conclude that they, like every other human race, were descendants of **Adam**. The descent of monsters was usually put in the context of *Genesis 9*. Two very important themes arise from this chapter. One theme is **Noah**'s lineage: that 'the sons of Noah, who came out of the ark, were Shem, Ham, and Japheth…and from these was all mankind spread over the whole earth' (9:18-19). Another theme is the **'curse of Ham'**. In this narrative Noah gets drunk and passes out naked in his tent. Ham accidentally witnesses his naked father and reports it to his two brothers, Shem and Japheth, who quickly walk backward (to prevent seeing Noah's nakedness) and cover him with a cloth. When Noah awakes from his drunken state and 'learns what his younger son has done to him' he curses the descendants of Ham, decreeing that subsequent generations of Ham's son, **Canaan**, will have to be the servants or slaves of Japheth's and Shem's descendants. This influential episode eventually served as a map by which Christians viewed infidel races. By the time of the **Crusades** the Table of Nations had become a handy template for metaphysically separating the 'noble' races from the ever-threatening exotic foreign hordes. Monsters, **Jews**, races of color, and **Muslims** all came to occupy a conceptual territory outside orthodoxy. The curse of Ham was just one of these many boundary inventions. Whatever the actual sin was, the cursed party was more obvious. Ham's descendants were people of color (i.e., Africans, but also Asians and eventually Americans), and their plight in life was to be subservient to the favored races (i.e., Semitic descendants of Shem and Indo-European descendants of Japheth)….
>
> The questions of race and monstrosity became even more intertwined in the age of exploration. In *The city of God* **Augustine** suggests that distant monstrous races and local individual monstrous births are closely interconnected: we have direct evidence of abnormal births, but only hearsay about faraway abnormal races…when we encounter an innocent child born with extreme physical maladies, we might naturally conclude that God is a poor craftsman. But Augustine proposes that monstrous races may exist in order to prevent us from drawing this impious conclusion and

show us instead that God knows what He's doing. When we realize that our newborn cycloptic child has some parallel with an entire race of **Cyclopes**, we cannot think of our child as a 'mistake' or a 'failure.' Moreover, he suggests that the logic works both ways. We know that the individual child is not a mistake because of the existence of monstrous races, and we know that monstrous races are not mistakes because individual monsters crop up regularly. What if God has seen fit to create some races in this way, that we 'might not suppose that the monstrous births which appear among ourselves are the failures of that wisdom whereby He fashions the human nature, as we speak of the failure of a less perfect workman?' As time passed, few followed Augustine's charitable single-ancestry theory for all races, including monsters. Most seemed to prefer the **xenophobic uses of monsters**, and by the late medieval period mainstream **Christians** were not only distancing themselves further from the legendary exotic tribes but they were also adding more proximate ethnicities (Jews, **Tartars**, **Moors**, etc.) to that reprobate category….

The xenophobic idea of dangerous monsters culminated in a popular story about Alexander's gates. The European version of the story, of a barrier erected against barbarian enemies, seems to have first appeared in sixth-century accounts of the Alexander Romance, but the legend is probably much older. Alexander supposedly chased his foreign enemies through a mountain pass in the Caucasus region and then enclosed them behind unbreachable iron gates. The details and the symbolic significance of the story changed slightly in every medieval retelling, and it was retold often, especially in the age of exploration. By the thirteenth and fourteenth centuries, the meaning of **Alexander's gates** had long since been Christianized and played an important role in both the geography of monsters and the ultimate end-time purpose of the fiends. The maps of the time, the **mappaemundi**, almost always include the gates, though their placement is not consistent. Most maps and narratives of the later medieval period agree that this prison territory, created proximately by Alexander but ultimately by God, houses the savage tribes of Gog and Magog, who are referred to with great ambiguity throughout the **Bible**, sometimes as individual monsters, sometimes as nations, sometimes as places. In the story of Alexander's gates, a kind of synthesis occurs, in which 'Gog and Magog' becomes a label for designating infidel nations and monstrous races, a monster zone, which different scribes can populate with all manner of projected fears. Mathew Paris was the chronicler of the Benedictine abbey of St. Albans in England from 1235 to 1259, and he drew up a series of influential maps, usually with Jerusalem and the Holy Land as the central focus. In his maps he placed the monster zone of Gog and Magog in northern Asia and populated it with Tartars (multiethnic Muslim populations). The British Hereford mappamundi (ca. 1300) continued the tradition of moral geography, placing Jerusalem as the righteous navel, with lesser known territories, some quite deviant, near the perimeter. In addition to the Alexander Romance, the Hereford map drew heavily for its source material on the writings of Solinus, a fourth-century author of *De Mirabilibus Mundi* (On the Wonders of the World). The Mirabilibus itself drew significantly on **Pliny's Natural History** and therefore repeats the familiar monsters of the ancient world. But now the creatures, including the **dog-headed Cynocephali**, the **Satyrs**, the Blemmyae, the **cannibal Anthropophagi**, and others, are all reconceptualized as players in the metaphysical geography of **Christianity**….

In the Zohar of the **Kabbalah** the Jewish midrash tradition further develops the Cain story, suggesting that Cain's own depravity was partly genetic. Cain's mother, Eve, fouled the bloodline by having a relationship with the serpent: 'When the serpent injected his impurity into Eve, she absorbed it and so when Adam had intercourse with her she bore two sons – one from the impure side and one from the side of Adam…. Hence it was that their ways in life were different…. From [Cain] originate all the evil habitations and demons and goblins and evil spirits in the world'.

Apocryphal scriptures, legends, and even the pictorial traditions tend to characterize Cain as misshapen, with horns and lumps on his body, and often draped in fur pelts like a feral man. But this story is important for the way that it broadcasts the tendency in ancient and medieval thought to connect sin and heredity, the tendency to explain monsters and evil generally as the result of unholy sexual union or dysgenics. In the late medieval mind, these myriad errant offspring could be localized in one contained place. The monsters' incarceration behind Alexander's gates is only temporary. They await their imminent release, the medievals believed, and will be upon us shortly. The Travels of Sir John Mandeville (published between 1357 and 1371) reveals precisely how this unleashing will finally occur. Mandeville retells the story of a monster zone full of dragons, serpents, and venomous beasts in the Caspian Mountains, but he adds another ethnic group, indeed, what he considers the main ethnic group, to the famous confinement. In chapter 29 he writes, 'Between those mountains the Jews of ten lineages be enclosed, that men call Gog and Magog and they may not go out on any side'. Here he is referring to the legendary ten lost tribes that disappeared from history after the Assyrian conquest in the eighth century BCE. These Jews, according to Mandeville, will escape during the time of the Antichrist and 'make great slaughter of Christian men. And therefore all the Jews that dwell in all lands learn always to speak Hebrew, in hope, that when the other Jews shall go out, that they may understand their speech, and to lead them into Christendom for to destroy the Christian people.'

Christian paranoia about Jews is, of course, an old story. Here in Mandeville we find a late medieval anti-Semitic maneuver that linked Jews directly with other monsters behind the gates and also gave Christians reason for increased paranoia about the local Jewry. To Christians, Jews were proximate in-house monsters (the **Diaspora**) who also had genealogical relations with the most foreign and distant of monsters. **Anti-Semitism** didn't really need help from Mandeville's like because the pious fury of the formally anti-Muslim crusades (1095–1291 and beyond) had been spilling over to include violence against local Jewry for centuries.... Muslims, like everyone else, accepted the existence of barbaric races. The historian Aziz Al-Azmeh even suggests three common markers that Muslims used to diagnose foreign peoples for barbaric status; filth, profligate sexuality (ascribed to Europeans), and unholy funerary rites. In principle, then, the idea of a great king shutting up dangerous uncivilized races behind an iron gate made sense, but the question was, Who were these brutes? Muslims could not and would not interpret the gates as enclosing themselves or the relatively more familiar peoples of the Eurasian steppes, nor did they believe Gog and Magog comprised the lost Jewish tribes. Islamic civilization of the time, unlike European Christendom, was simply too close to the region to accept any facile identification of the monstrous Gog and Magog. During the Patriarchal and Umayyad Caliphate expansions of Islam (632-750 CE), for example, the territories near the legendary gates would likely have been Muslim. When, in the ninth century, Caliph al-Wathiq-Billah sent an interpreter named Sallam to find Alexander's renowned gates, Sallam failed to discover them in the Caucasus but claimed to find them much further inside Asia. This tells us something about the human tendency to keep locating barbarism and monstrosity farther and farther away from oneself and one's own tribes. Instead of naming the ethnic groups inside Gog and Magog, Aziz Al-Azmeh claims, Arab Islamic culture left them unnameable, imaginary placeholders. These unnamed were the antithesis of civilization, and Muslims accepted the idea that their counterhumanity would strike against pious culture once the gates were breached, but the creatures themselves were more anonymous than in the European narratives. Both Christians and Muslims had deep-seated monster narratives to explain the evil and the uncivilized.

Fig. 3.1 As explained in Asma's *On Monsters*, "the Psalter *mappamundi* (ca. 1225) continued the tradition of moral geography, placing Jerusalem as the righteous navel, with lesser known monster territories near the perimeter"

In *Behave*, Sapolsky discusses some interesting neurobiological aspects related to **discrimination** and **tribalism**:

> The strength of Us/Them-ing is shown by its emergence in **kids**. By age three to four, kids already group people by race and **gender**, have more negative views of such Thems, and perceive other-race faces as being angrier than same-race **faces**. And even earlier. Infants learn same-race faces better than other-race. (How can you tell? Show an infant a picture of someone repeatedly; she looks at it less each time. Now show a different face – if she can't

tell the two apart, she barely glances at it. But if it's recognized as being new, there's excitement, and longer looking). Four important thoughts about kids dichotomizing: 1) Are children learning these prejudices from their parents? Not necessarily. Kids grow in environments whose nonrandom stimuli tacitly pave the way for dichotomizing. If an infant sees faces of only one **skin color**, the salient thing about the first face with a different skin color will be the skin color. 2) **Racial dichotomies** are formed during a crucial developmental period. As evidence, children adopted before age eight by someone of a different race develop the expertise at **face recognition** of the adoptive parent's race. 3) Kids learn dichotomies in the absence of any ill intent. When a kindergarten teacher says, "Good morning, boys and girls", the kids are being taught that dividing the world that way is more meaningful than saying, "Good morning, those of you who have lost a tooth and those of you who haven't yet". It's everywhere, from "she" and "he" meaning different things to those languages so taken with gender dichotomizing that inanimate objects are given honorary gonads. 4) Racial Us/Them-ing can seem indelibly entrenched in kids because the parents most intent on preventing it are often lousy at it. As shown in studies, liberals are typically uncomfortable discussing race with their **children**. Instead they counter the lure of Us/Them-ing with abstractions that mean squat to kids – "It's wonderful that everyone can be friends" or "Barney is purple, and we love Barney". Thus, the strength of Us/Them-ing is shown by: (a) the speed and minimal **sensory stimuli** required for the brain to process group differences; (b) the **unconscious automaticity** of such processes; (c) its presence in other **primates** and very young humans; and (d) the tendency to group according to arbitrary differences, and to then imbue those markers with power.

Sapolsky then expands his analysis to the links between discrimination, racism, beliefs, the physiology of **disgust**, and our thoughts and feelings:

By age three Being disgusted by another group's abstract **beliefs** isn't naturally the role of the **insula**, which evolved to care about disgusting tastes and smells. Us/Them markers provide a stepping-stone. Feeling disgusted by Them because they eat repulsive, sacred, or adorable things, slather themselves with rancid scents, dress in scandalous ways – these are things the insula can sink its teeth into. In the words of the psychologist Paul Rozin… 'Disgust serves as an ethnic or out-group marker'. Establishing that They eat disgusting things provides momentum for deciding that They also have disgusting ideas about, say, deontological ethics. The role of disgust in Them-ing explains some individual differences in its magnitude. Specifically, people with the strongest negative attitudes toward **immigrants**, **foreigners**, and **socially deviant groups** tend to have low thresholds for interpersonal disgust (e.g., are resistant to wearing a stranger's clothes or sitting in a warm seat just vacated). Thus Us/Them-ing can arise from cognitive capacities to generalize, imagine the future, infer hidden motivations, and use **language** to align these cognitions with other Us-es. As we saw, other primates not only kill individuals because they are Thems but have negative associations about them as well. The automaticity of Us/Them-ing is shown by the speed of the amygdala and insula in making such dichotomies – the brain weighing in affectively precedes conscious awareness, or there never is **conscious awareness**, as with **subliminal stimuli**. Another measure of the **affective core** of Them-ing is when no one even knows the basis of a **prejudice**.

Consider the Cagots, a minority in France whose persecution began in the eleventh century and continued well into the last one. **Cagots** were required to live outside villages, dress distinctively, sit separately in church, and do menial jobs. Yet they didn't differ in appearance, religion, accent, or names, and no one knows why they were pariahs. They may have descended from Moorish soldiers in the Islamic invasion of Spain and thus were discriminated against by Christians. Or they might have been early **Christians**, and discrimination against them was started by *non*-Christians. No one knew the sins of ancestral Cagots or how to recognize Cagots beyond community knowledge. During the **French Revolution**, Cagots burned birth certificates in government offices to destroy proof of their status. How's

> this for a fascinating influence on Us/Them-ing, way below the level of awareness…when women are ovulating, their fusiform face areas respond more to faces…. Carlos Navarrete at Michigan State University has shown that white women, when ovulating, have more negative attitudes toward African American men. Thus the intensity of Us/Them-ing is being modulated by hormones. Our feelings about Thems can be shaped by subterranean forces we haven't a clue about. Our cognitions run to catch up with our affective selves, searching for the minute factoid or plausible fabrication that explains why we hate Them.

So, racism seems to be linked—in a very complex, interactive way—to both **innate tendencies** (see also Box 3.5) and **social constructions**, although in the last sentences of the text above, Sapolsky seems to suggest that discrimination might be mainly related to innate, physiological features such as disgust and that then we try to rationalize such "gut feelings" a posteriori, often using teleological narratives such as "God prefers us." However, things are probably even more complicated than this: our tendency to build and believe in such narratives per se is at least partially "natural." That is, we seem to have a tendency to do so, which is present already in young children, as the tendencies to discriminate seem to be. Sapolsky suggests that such tendencies to discriminate are more "basal" than those to build and believe in teleological narratives, because as he pointed out, the former are present in other primates. That is true: tribalism is present in almost all social mammals, while the belief in imaginary stories is not. However, as he also pointed out, "nonetheless, no other primate kills over ideology, theology, or aesthetics," three items that are often precisely deeply related to teleological narratives. Moreover he also noted that teleological myths of "No one has it all" can reinforce the *status quo*: the "cultural trope of "poor but happy"—the poor are more carefree, more in touch with and able to enjoy the simple things in life—and the myth of the rich as unhappy, stressed, and burdened with responsibility (think of miserable, miserly Scrooge and those warm, loving Cratchits) are great ways to keep things from changing... the trope of "poor but honest" by throwing a sop of prestige to Thems is another great means of rationalizing the (discriminatory) system".

Box 3.5 Neurobiological Cues, Fear, Racism, and Stereotypes

Sapolsky's book *Behave* provides some fascinating case studies about the links between neurobiological cues, fear, racism, and stereotypes:

> Over the course of seconds **sensory cues** can shape your behavior unconsciously. A hugely unsettling sensory cue concerns race. Our brains are incredibly attuned to skin color. Flash a face for less than a tenth of a second (one hundred milliseconds), so short a time that people aren't even sure they've seen something. Have them guess the race of the pictured face, and there's a better-than-even chance of accuracy. We may claim to judge someone by the content of their character rather than by the color of their skin. But our brains sure as hell note the color, real fast. By one hundred milliseconds, brain function already differs in two depressing ways, depending on the race of the face (as shown with neuroimaging). First, in a widely replicated finding, the **amygdala** activates. Moreover, the more racist someone is in an implicit test of race bias, the more activation there is. Similarly, repeatedly show subjects a picture of a face accompanied by a shock; soon, seeing the face alone activates the amygdala. As shown by Elizabeth Phelps of NYU, such "fear conditioning" occurs

faster for other-race than same-race faces. Amygdalae are prepared to learn to associate something bad with Them. Moreover, people judge neutral other-race faces as angrier than neutral same-race faces. So if whites see a black face shown at a subliminal speed, the amygdala activates. But if the face is shown long enough for conscious processing, the **anterior cingulate** and the "cognitive" dlPFC (**dorsolateral prefrontal cortex**) then activate and inhibit the amygdala. It's the frontal cortex exerting executive control over the deeper, darker amygdaloid response. Second depressing finding: subliminal signaling of race also affects the **fusiform face area**, the cortical region that specializes in **facial recognition**. Damaging the fusiform, for example, selectively produces "face blindness" (aka **prosopagnosia**), an inability to recognize faces. Work by John Gabrieli at MIT demonstrates less fusiform activation for other-race faces, with the effect strongest in the most implicitly racist subjects. This isn't about novelty – show a face with purple skin and the fusiform responds as if it's same-race. The fusiform isn't fooled – "That's not an Other; it's just a 'normal' Photoshopped face". In accord with that, white Americans remember white better than black faces; moreover, mixed-race faces are remembered better if described as being of a white rather than a black person. Remarkably, if mixed-race subjects are told they've been assigned to one of the two races for the study, they show less fusiform response to faces of the arbitrarily designated "other" race. Our attunement to race is shown in another way, too. Show a video of someone's hand being poked with a needle, and subjects have an "isomorphic sensorimotor" response – hands tense in **empathy**. Among both whites and blacks, the response is blunted for other-race hands; the more the implicit racism, the more blunting. Similarly, among subjects of both races, there's more activation of the (emotional) medial PFC when considering misfortune befalling a member of their own race than of another race.

This has major implications. In work by Joshua Correll at the University of Colorado, subjects were rapidly shown pictures of people holding either a gun or a cell phone and were told to shoot (only) gun toters. This is painfully reminiscent of the 1999 killing of **Amadou Diallo**. Diallo, a West African immigrant in New York, matched a description of a rapist. Four white officers questioned him, and when the unarmed Diallo started to pull out his wallet, they decided it was a gun and fired forty-one shots. The underlying neurobiology concerns "event-related potentials" (ERPs), which are stimulus-induced changes in electrical activity of the brain (as assessed by EEG – electroencephalography). Threatening faces produce a distinctive change (called the P200 component) in the ERP waveform in under two hundred milliseconds. Among white subjects, viewing someone black evokes a stronger P200 waveform than viewing someone white, regardless of whether the person is armed. Then, a few milliseconds later, a second, inhibitory waveform (the N200 component) appears, originating from the frontal cortex- "Let's think a sec about what we're seeing before we shoot". Viewing a black individual evokes less of an N200 waveform than does seeing someone white. The greater the P200/N200 ratio (i.e., the greater the ratio of I'm-feeling-threatened to Hold-on-a-sec), the greater the likelihood of shooting an unarmed black individual. In another study subjects had to identify fragmented pictures of objects. Priming white subjects with subliminal views of black (but not white) faces made them better at detecting pictures of weapons (but not cameras or books). Finally, for the same criminal conviction, the more stereotypically African a black individual's facial features, the longer the sentence. In contrast, juries view black (but not white) male defendants more favorably if they're wearing big, clunky glasses; some defense attorneys even exploit this "nerd defense" by accessorizing their clients with fake glasses, and prosecuting attorneys ask whether those dorky glasses are real. In other words, when blind, impartial justice is supposedly being administered, jurors are unconsciously biased by racial stereotypes of someone's face. There's also subliminal cuing about beauty. From an early age, in both sexes and across cultures, attractive people are judged to be smarter, kinder, and more honest. We're more likely to vote for attractive people or hire them, less likely to convict them of crimes, and, if they are convicted, more likely to dole out shorter sentences. Remarkably, the **medial orbitofrontal cortex** assesses both the beauty of a face and the goodness of a behavior, and its level of activity during one of those tasks predicts the level during the other. The brain does similar things when contemplating beautiful minds, hearts, and cheekbones. And assumes that cheekbones tell something about minds and hearts.

3.2 Cultural and Innate Notions of Race

The Negro 'with us' is not an actual physical being of flesh and bones and blood, but a hideous monster of the mind, ugly beyond all physical portraying, so utterly and ineffably monstrous as to frighten reason from its throne, and justice from its balance, and mercy from its hallowed temple, and to blot out shame and probity, and the eternal sympathies of nature, so far as these things have presence in the breasts or being of American republicans! No sir! It is a constructive Negro – a John Roe and Richard Doe Negro, that haunts with grim presence the precincts of this republic, shaking his gory locks over legislative halls and family prayers. (James McCune Smith)

The "basal," physiological, tribal "disgust" type of "racism"/discrimination against the other to which Sapolsky refers (see Box 3.5) is indeed very old, and this makes completely sense evolutionarily, because one of the most crucial aspects of evolution is **speciation**, in which members of a taxon start to no longer reproduce with other members of that taxon. This can happen for many reasons, such as a physical barrier that is formed between the two subgroups, such as a river or mountain. This is called **allopatric speciation**. But it is well known that within **sympatric speciation**—that is, speciation that happens in a same region, without such physical barriers—there are many cases in which the beginning of the reproductive/biological separation is related to **behavioral choices**. That is, for some behavioral reasons, the individuals of a subgroup start copulating less and less with those of the other subgroups. I provided several examples in my 2017 book *Evolution Driven by Organismal Behavior*, and I have seen myself other examples while traveling around the globe. In the Galapagos islands, a tour conservation guide explained me how turtles from a same species are seemingly slowly starting to form new subspecies. This is because researchers working with that guide have studied the behavior and DNA of the turtles and concluded that some subgroups of turtles that live nearby certain human villages that surround one of the volcanoes of the Galapagos no longer interbreed with other subgroups living nearby, around the same volcano. There is no physical barrier, or sexual anatomical differences between any of the subgroups: as the guide told me, they "simply seem to have decide to not 'do it' with the *other* turtles." This case, which probably is similar to early stages of many other cases of sympatric speciation, therefore seems mainly related to what can be called a type of discrimination against "*the other*," in a way. I am not sure, of course, if this is because those turtles acquired a physiological "gut feeling" of disgust—*sensu*

Sapolsky—against all the *others*, or if their behavioral choice is due to any other factor, but what is clear is that, as suggested by Sapolsky, such a type of discrimination against "*the other*" is far from being a uniquely human feature.

However, it is important to note that in humans there are mainly two types of racism, that is two ways of seeing the "other," which I will call "type A/epigenetic racism" and "type B/genetic/innate racism." **Type A/epigenetic racism** was the most prevalent one in human history until the end of the seventeenth century. Then, particularly in Western countries, there was a rise of **type B/innate racism**, which led to some of the atrocious acts done by some eugenic movements and particularly by the Nazis: this is the type of racism nowadays occurring within **Neo-Nazis** and/or **white supremacy groups** (see Figs. 4.5 and 9.17). Although the difference between these two types of racism is widely consensual in the specialized literature—it was explained in detail in Gould's excellent 1981 book *The Mismeasure of Man*—it is unfortunately too often neglected in discussions about racism in many textbooks, and particularly within the media and social media. However, the fact is that, as explained by Gould, for the ancient Egyptians, Greeks, and Romans, as well as for most European colonizers until the end of the seventeenth century, there were clearly "others" that were seen as "inferior," for instance "blacks," or "Native Americans," but the explanations for why they were "inferior" were mainly related to **epigenetic factors**. For instance, it could be because of the "way they lived," or the environment where they lived, the air they inhaled, the food they ate, and so on. One example, among many others, concerns the writings of the ancient Greek historian **Herodotus**—the so-called father of Western history—which reflect some of the views more commonly defended in **ancient Greece**. As pointed out in Malik's 2014 book *The Quest for a Moral Compass*:

> Herodotus examines the customs, beliefs and institutions not just of the Greeks but also of **Persians**, **Egyptians**, Libyans, **Scythians** and **Arabs**. Differences, he insists, are neither accidental nor the result of divine intervention but derive from material, earthly causes. The Egyptians have unusual customs because of their need to deal with their unusual climate. The natural poverty of Greece encouraged its inhabitants to develop appropriate laws and institutions to overcome it. The success of the Athenians was rooted not simply in the endeavours of great individuals but also in a democratic system that had nurtured a sense of common responsibility. Herodotus attempted to use rational explanations to understand the social and cultural differences between cities and nations, peoples and ages; he also believed that such differences helped in turn to explain the movement of history. The **Trojan War**, the **rise of Athenian democracy**, the **Persian invasion of Greece**, the **conflict between Athens and Sparta** – none could be explained by appealing simply to individual decisions or whims, whether human or divine Each was also the result of the way human or divine. Each was also the result of the way in which people in a given society with particular customs could be expected to act in certain circumstances.

One could think that—as this is the most prevalent type of racism since humans have written historical records until the end of the seventeenth century—such type A of racism could thus correspond to the "basal" type of racism found in other animals. Clearly, as Sapolsky suggests, this does seem to be the case in the sense that both the human type A/epigenetic racism and the discrimination of others done by other animals seem to be related, at least in part, to the physiological feeling of "disgust." However, the reality is that such a feeling of "disgust" is very likely associated with both the type A/epigenetic racism and the type B/genetic racism, the main difference between the two being the type of a posteriori rationalization of this "disgust" feeling: both see the other as inferior, but in the former the inferiority is due to something that him/her did *during his/her life* while in the latter it is due to the "way things are," that is due to the "genes" or "skin color" of that "race" and so on. So, we don't know if when chimpanzees are tribal they do so by invoking that "the others" are "disgusting" because the way they live—type A of racism—or instead because they are "innately inferior"—type B of racism. Many would say that none of these apply, as chimpanzees cannot rationalize, but that is not sure at all as explained above. So, in this sense, the timeline for the origin of racism in humans seems to be that it is just a by-product of the type of "disgust" type of tribalism that other animals have, but then humans begun to rationalize it in more and more complex ways a posteriori, mainly using a "they don't live as us" type A of racist narratives, and then more recently also/instead using a "they are inferior since they are born" type B of racist imaginary stories. This timeline also makes sense because the beginning of the sympatric speciation cases I mentioned just above in other animals seem to be mainly done through behavioral choices, so it would make sense that the discrimination itself would be related to some type of *different behaviors* displayed by "the others." That is, the subgroup of turtles that lives in a specific place and that no longer reproduces with other turtles that they often encounter might do so because they perceive something different/"disgusting" within the behavior of those other turtles.

This timeline also makes sense because racism type B in humans is often against groups that don't display a certain type of homogeneous behavior that can truly differentiate them from others—except in cases such as against a certain specific religious group, as in those cases there might be some distinctive religious practices such as using a type of dressing, or going to synagogues, which could be "identifiable behaviors" in the point of view of the racist people that hate them. But that clearly does not apply to "blacks," or "Africans," whom as groups such as Jews or Muslims or Christians are *not* a true **monophyletic (separate) group** biologically but, contrarily to the people from those religious groups, do not have any homogeneous cultural tradition or belief that can make them "behaviorally identifiable," as Africans have a plethora of different religions, beliefs, and so on. Instead, what groups that racist people define as "**Africans**" or "**blacks**" have is that their superficial anatomical features such as the color of the skin are *perceived*, within *Neverland*'s racist world of unreality, as if they mean that the "other" is part of a true separate group. This is in a way similar to what Nazis tried to do with Jews, saying that they had ways of "measure" and "identify" them, anatomically, what is obviously factually wrong. As it is obviously also wrong with "blacks" or "Africans": **Neonazis** clearly include **African albinos** in such a group, but obviously their skin is not dark at all, being actually in many cases even lighter than that of the vast majority of neonazis themselves. Biologically, "blacks" could never be a true separate group because a monophyletic group, by definition, needs to include *all* the descendants of its common ancestors, as do true biological races of non-human taxa: this clearly does not apply to "blacks" because "whites" are derived from "blacks"/Africans, that is, they came from a subset of so-called Africans that left Africa various dozens of thousands of years ago. That is, the grouping of "blacks" is a social construction, a **paraphyletic group** such as are "dinosaurs"—because we have to artificially exclude birds from them. Or "apes"—we have to exclude humans, which derived from them and are more closely related to chimpanzees than chimpanzees are to gorillas and orangutans.

Indeed, the key reason why people from different regions of the globe have a darker color of skin has nothing to do with being from a separate biological group, but instead because the color of the skin is clearly associated with life in different parts of the planet exposed to different levels of ultraviolet rays. People tend to have a very dark **color of skin** if their ancestors have lived for millennia in equatorial South America—for example, indigenous populations of Ecuador—or equatorial Africa, or equatorial Asia—for instance, many parts of Bangladesh—or Australia—such as many groups of Australian aborigines. So, this is the result of *independent* adaptations to the exposure to sunlight in those regions, for example, a lighter skin in such regions would tend to lead to more skin cancers, as it does within albinos in equatorial Africa. So, suggesting that "blacks" form a "true biological race" is as absurd as suggesting that populations that have lived for a long time in the Andes, Alps, and Himalayas form a "true biological race"—a "mountain race"?—because they are adapted to life in high altitudes by using the available oxygen in more efficient ways than do populations living in lower altitudes.

Amazingly, although these scientific facts are well known since many decades ago—there is a consensus within the vast majority of anthropologists about this and this was explicitly stated, officially, by the *American Anthropological Association*—most media and most lay people, and even many scientists from other areas of science, continue to confuse evolutionary terms and to argue that "**human biological races**" are biologically "real." No, they aren't: they are only "real" in *Neverland*'s world of unreality and racism. Some other scientists are a bit more "sophisticated" and recognize that "blacks" can't be a monophyletic group, but then keep talking over and over about the "biological reality of race" by referring to "clusters," or by saying that people from similar regions of the globe, such as Africa, "join up" in genetic evolutionary trees of *Homo sapiens*, as did Kirkegaard in his 2019 paper entitled "*Race differences – a very brief review*." However, "join up" has nothing to do, evolutionarily, with forming a true, natural, monophyletic group, or "race." Dinosaurs also "join up" in evolutionary trees because birds appear on the top or bottom of those trees as they precisely descended from "dinosaurs," but that does not make "dinosaurs" to be a true biological clade, in any possible way. Actually, in the case of "Africans," it is exactly the opposite. The fact that so-called whites or Asians obviously derived from them means that at a moment in time, before *Homo sapiens* left Africa, there was a group of African ancestors that gave rise to a group A, let's say of Africans that remained in Northeastern regions of Africa such as nowadays' Egypt, and a group B, which then left Africa. So, this means that me, a so-called white from group B, I am more closely related, genetically/phylogenetically to a so-called black person from group A than that person is to any other "black" person from a group other than group A. So, if you are walking on the street, just think about that: if you are "white," you are surely more closely related to some so-called blacks—that is, you share a more recent common ancestor with some of that constructed "them", a grand, grand, grand, grand father or mother, many generations ago—than they share with many other "blacks." Although it is quite possible that some biologists, and particularly non-biologist scientists, don't know these evolutionary facts, it is also quite likely that others might know but still continue to defend that there are "true biological human races" due to their own biases, prejudices, teleological narratives in which they believe, and/or agendas.

One important point is that both racist ideas *and* "gut feelings" have clearly changed during human history, being in particular deeply connected to changing teleological narratives, further supporting the notion that at least a substantial part of these ideas and "feelings" are socially constructed, although of course there is surely a complex, interactive feedback between them and our "deep physiology" *sensu* Sapolsky. This notion is also supported by the fact that in many instances, particularly in cases in which teleological narratives are contradicted by the reality when people contact/meet each other, the categorization Us-Them can be reversed. An emblematic example occurred among soldiers fighting against each other, which were acculturated to "hate" each other but realized that "Them" and "Us" were basically all the same thing and the mainly reason they were fighting was because of imaginary stories—about "patriotism," defending the "flag," and so on—created by others, as it so often happens in human history. This happened during the famous **German-British "Christmas Truce of 2014"** during World War I.

Therefore, apart from the numerous examples provided in specialized papers and books such as Gould's *The Mismeasure of Man* showing that type A of racism was more prevalent within ancient Egyptians, Greeks, and Romans, I think it is worthy to refer here to an illustrative example given in a less known, but also fascinating, book: Canizares-Esguerra's 2006 *Nature, Empire and Nation – Explorations of the History of Science in the Iberian World*. This is because he shows that, in at least some cases, type A/epigenetic racism—for instance, **Native Americans** being considered to be "inferior" because of the humid air of America—obviously is often related to teleological narratives. Moreover, the author provides some interesting exceptions to the rule that until the end of the seventeenth century—mainly as a reaction of Tyson's 1699 first detailed description of a chimpanzee, as will be explained below—there was mainly type A/epigenetic racism. Gould had also mentioned some exceptions, which as he noted are the type of exceptions that "prove the general rule," as such few cases of racism type B are much less frequent than the huge number of cases of racism type A occurring before the end of the seventeenth century. Canizares-Esguerra does confirm that ancient ideas of racism were indeed in general more related to type A racism:

As **conquistadores** discovered large river basins, lakes, and tropical forests, a sense that America was a temperate, yet humid, continent came to dominate in the imaginations of European scholars…. In 1579, the Franciscan Diego Valdez asserted in Italy that the Amerindians were 'stupid' because they were born in thick air. In 1591, Juan de Cardenas…maintained that humidity…was not only the cause of frequent earthquakes but also sapped the strength of the population…. The purported humidity of America was a claim with heavy ideological baggage. Scholarship had associated masculinity with warm, dry environments, and femininity with moist, cold ones. America posed the threat of impeding sexual transformations for colonists. Some Spanish authors contended in the course of the 16th century that America was a land where women urinated standing, while men did so seated…. The Franciscan friar Bernardo de Lizana, astonished by the beauty and grandeur of **Maya ancient buildings** and the complexity and extension of their polities, insisted that the Maya were the descendants of Carthaginians, for only a nation like the Carthaginians could have had the intellectual skills required to design such buildings…. Yet Lizana, who thought that the contemporary Maya were childish and brutish, argued that climate and total isolation from Carthage had been responsible for transforming ancient Carthaginians into barbarous and crude…the Amerindian was represented as a psychologically arrested child, whose innate natural rights had to be administered by proxy. A closer look at these views shows that nurture, not nature, was the dominant explanation offered for religious deviancy…it was the assumption that culture could become so ingrained as to become as second human nature that led 16th century Spanish scholars to argue that it would take a great many generations of hard work by missionaries to transform the psychologically arrested, childish Amerindians into full-fledged, Christian European adults. Nonetheless, Amerindian conversion and Europeanization was deemed possible. Rocha also argued that Creoles had somehow been spared from rapidly becoming effeminate, stupid Amerindians thanks, in part, to the European food they ate and to the fresh influx of European blood that arrived in the colonies with every generation…these two processes had slowed and even stopped the environmentally induced degeneration.

Creoles and long-term European residents in the Indies were left facing an extraordinary paradox: how to maintain that America was under benign, soothing cosmic influences without giving up their construct of the Amerindians as phlegmatic miscreants. The works of Leon Pinelo and Salinas de Cordoba…show the solution adapted by colonial intellectuals, that is, to postulate that Amerindians and Europeans had different sorts of bodies – racism type B *sensu* the present volume – which "would make any radical transformation of the later due to climatic or astral influences unlikely, if not impossible. This discourse permeated the works of the most important representatives of Creole patriotism in the 17th century. By the early 17th century…[the] faith in the transformation (and redeeming) power of evangelization and acculturation gave way to a marked skepticism in Creole learned circles…. Creole authors, witnessing the indigenous refusal to give up ancestral religious practices lightly, embraced racialist views to explain religious deviance among Andeans. In previous centuries Christian scholars had maintained that the devil was largely responsible for idolatrous behavior, because Satan misled worshippers through the manipulation of people's mental faculties, making them see false phantasms and religious visions…but by the 17th century, the Creole clergy…began to blame Amerindian idolatry on the flawed operation of the Amerindian body, rather than on the machination of the devil. Indigenous religious deviance thus became a psychological problem, largely attributable to the physical malfunctioning of the internal senses of Amerindians, the failure of their brains to grasp the logical, scientific structure of the universe. Creole and…European scholars, hammered out…a science of the racialized body that long predated that invented in the late 18th and early 19th century in Europe. But this challenge to ancient views" – racism type B *sensu* the present volume – "inadvertently developed by (Iberian) colonial intellectuals, did not influence, and was not even acknowledged in, later European discourses of the racialized body."

3.3 Aristotle, Galen, Monkeys, and Chain of Being

No one…saw anything wrong with the tyranny…slavery was as customary as prisons are today…few could imagine an ordered world without them. (Ibram X. Kendi)

As noted in my 2018a paper "*Links between the discovery of primates and anatomical comparisons with humans, the chain of being, our place in nature, and racism*," some works, including a few monographs, have analyzed the links between discussions on our place in nature, the teleological notion of chain of being or *scala naturae*, and the scientific descriptions of non-human primates by Westerners. Some examples included in that paper, and in the list of references I provide in the present volume, are Martin's 1984, Bowler's 1987, Groves's 2008, Corbey's 2005, Corbey and Theunissen's 1995, Delisle's 2007, Kuklick's 2008, Barsanti's 2009, Sommer's 2015, Hoßfeld's 2016, Persaud's 1984, Persaud et al.'s 2014, and Engelmeier's 2016 works. It should however be noted that those publications do not focus specifically on the links between these subjects and the history of **primate comparative anatomy**, which is truly at the heart of the explosion of **scientific racism** and the rise of **type B/innate racism** in the eighteenth century. One striking fact about primate comparative anatomy, that is neglected in most textbooks and even specialized papers, is that the first detailed anatomical studies of non-human primates done by ancient Greeks and Romans were, in reality, even older than the first comprehensive studies of human anatomy. The anatomical descriptions of **Galen—Claudius Galenus** (130 to about 210 AD), or **Aelius Galenus**, Anglicized as Galen and often known as **Galen of Pergamon**, was a surgeon, physician, and philosopher in the **Roman Empire**—that were used for centuries as the "basis of human anatomy" were actually mainly based on dissections of the "**Barbary ape**." This Old World monkey was often called an "ape" because apes have no tail, and the members of this species, *Macaca sylvanus*, display a vestigial tail. In my 2018 paper, I therefore provided a succinct but in this sense more accurate account on the links between the teleological roots of Western science—and in particular about the quests to understand our place in nature—and the history of anatomical comparisons between humans and other primates and between so-called different human groups and therefore of **racism**. A significant part of this Section is thus based on that 2018 paper, and in addition includes key new parts that are particularly relevant to a broader understanding of these complex links.

Box 3.6 Aristotle, Scala Naturae, Religion, and Design

Subjects such as the passive versus active role of organisms and related topics, such as the notion of evolutionary trends, the idea that **complexity** supposedly increases during evolution, and the **form *versus* function debate**, have been crucial within the **history of biology**—including **Aristotle**—and particularly of evolutionary biology—including authors such as **Darwin**, **Wallace** and others. These subjects are also related to the long-standing notion of *scala*

naturae (Fig. 3.2) and to associated **teleological topics** such as the notion of "**design**" or "**cosmic purpose**" in the natural world. Specifically, as noted in Turner's 2013 book chapter "*Biology's second law: homeostasis, purpose and desire*," such teleological notions are linked to the question "why are organisms 'constructed' so well to perform their functions?"—which is not only the type of why-questions that have been so damaging within natural sciences, but also that follow a "perfection of design" type of narrative that is factually inaccurate. Lovejoy's 1936 book (see Box 3.1), as well as Reiss' 2009 book *Not by Design*, provide very detailed and well-documented discussions on these topics and summaries of the history of teleological reasoning, from the Greeks to modern times. **McShea**'s 2012 paper "*Upper-directed systems: a new approach to teleology in biology*" argues that a major reason why scientists have been so interested in such topics for millennia—which, in turn, makes discussions on these issues so difficult and, often, contentious—is that "there has always been an aura of mystery, of magic, around such systems on account of their seeming future directedness." He further notes that "the three standard terms of discourse (teleology, **goal-directedness**, purpose) all imply a future object or event (a **telos**, a goal, an achieved purpose) that is in some sense explanatory of present *behavior*." In this sense, it is interesting to note that even some scientists have defended the notion of a "cosmic progress" by invoking factually inexistent forces, such as **vitalistic forces** or forces created by an "intelligent designer." I provide more details on these topics in my 2017 book *Evolution Driven by Organismal Behavior*. As noted in that book, examples of the renewed interest in these issues include the publication of several books about them in the last decades, such as Bonner's 2013 *Randomness in Evolution*, Butler's 2012 *Evolution Without Darwinism*, Minelli's 2009 *Forms of Becoming*, Vinicius' 2012 *Modular Evolution – How Natural Selection Produces Biological Complexity*, Reiss' 2009 *Not by Design – Retiring Darwin's Watchmaker*, Bonner's 2013 *Randomness in Evolution*, Odling-Smee et al.'s 2003 *Niche Construction – The Neglected Process in Evolution*, Ruse 2003's *Darwin and Design – Does Evolution Have a Purpose?* Wagner's 2014 *Arrival of the Fittest: Solving Evolution's Greatest Puzzle*, and Noble's 2006 *The Music of Life: Biology Beyond the Genome*, as well as other publications included in the list of references of the present volume, such as Johnson et al.'s 2012 work, Nee's 2005 work, Omland et al.'s 2008 work, or Rigato and Minelli's 2013 work.

How is it possible that humans knew, for many centuries, more about the **internal anatomy** of a single monkey species than about that of their own bodies? One of the main reasons is that some of the first detailed anatomical studies date back to the ancient Greeks and Romans, whom often viewed a human corpse as something "impure"—contrary to the culture of **human dissection** developed mainly in the Christian West much later. That is why **Galen** mainly based his descriptions of

Fig. 3.2 "Ascent of Life," by F. Besnier, 1886

"human anatomy" on dissections of animals such as sheep, oxen, pigs, dogs, bears, and particularly the "Barbary ape." As pointed out in Singer's 1959 book *A History of Biology to About the Year 1900* and in Cole's 1975 book *A History of Comparative Anatomy – From Aristotle to the Eighteenth Century*, it is indeed remarkable that, for a millennium—after Galen and before authors such as **Vesalius**—few scholars recognized this fact and most scholars Galen's studies to learn, and teach, human anatomy. Moreover, in a few instances, Galen inaccurately described features in

macaques, therefore contributing to further erroneous ideas about what then became to be accepted by most as the "standard human anatomy"—a recent review on these questions was published in 2017 by my former PhD student Malak Alghamdi, my colleague Janine Ziermann, and me. This is one of the most striking examples of how science can be, and often is, biased, and also of the power of **authoritarianism** and **idealization** within sciences, because while the internal anatomy of monkeys in general is somewhat similar to that of humans, there are still numerous specific anatomical differences between them. For more details about such differences, principally concerning soft tissues such as muscles, you can refer to the various books and papers published by Bernard Wood and me, for instance in 2011, 2012, 2013, and 2016.

The works, and errors, of Galen had crucial repercussions for anatomy in particular, and biology and science in general, because Galen so impressed the people of his time and of succeeding ages that for centuries his works were regarded as almost infallible. This reverence for Galen is partially related to teleological narratives. In particular, to the fact that, although he remained a pagan, he believed in one God and developed the idea that every organ in the human body was created by a God in the best possible form and for its perfect use, an idea that fitted in well with that of **Christianity**, as explained in Mayr's 1976 book *Evolution and the Diversity of Life: Selected Essays* and Cunningham's 1997 book *The Anatomical Renaissance: The Resurrection of the Anatomical Projects of the Ancients*. In fact, only very few "pre-Vesalius" authors—including some Muslim scholars, contrary to the narratives presented in most Western textbooks, as explained in our Alghamdi et al. 2017 paper—realized and/or were brave enough to state that Galen's descriptions did not match human anatomy. That is why the blind acceptance of Galen's works mostly started to be challenged, at a broader level, only after **Vesalius** dissected both monkeys and humans and conclusively showed that Galen's descriptions were mainly based on **monkey anatomy**—see Lagerkvist's splendid 2005 book *The Enigma of Ferment – From the Philosopher's Stone to the First Biochemical Nobel Prize*. Still, the 1543 ***Fabrica* of Vesalius** is often seen by historians as a "corrected and expanded version" of the ***Corpus Galenicum***, a view that does not reflect the reality, as this is comparing oranges to apples. The *Corpus Galenicum* is the first detailed, but partially inaccurate, anatomical description of non-human primates. In contrast, the *Fabrica* is the first comprehensive Western report of human anatomy based on numerous actual **human dissections** and done specifically to set the record straight and provide accurate information concerning these topics, despite the fact that Vesalius was a Galenist nonetheless and that his descriptions also contain some factual errors.

But perhaps the most critical question, for the purpose of the present volume and to truly understand the history of human racism and its deep links to teleological narratives, is: why did Galen extrapolate that the anatomical features of the monkey *Macaca sylvanus* would apply to humans, when faced with the difficulty of dissecting human bodies that was so characteristic of his epoch? Part of the answer is that he considered that monkeys are, in general, essentially similar, internally, to us. Interestingly, this idea, which is correct anatomically—with some exceptions as noted above—was influenced by the human–animal continuity implied by the Greek teleological notion of the "great chain of being," or *scala naturae* ("ladder of being") (Fig. 3.2). In his treatise *On Anatomical Procedure*, Galen wrote: "of all other animals, **monkeys** are most like humans in viscera, muscles, arteries, veins, and nerves…because of this they walk on two legs and use their forelimbs as hands…the more human sort have a nearly erect posture." As explained in Lovejoy's 1936 book, this notion of *scala naturae*—which dates back to **Plato**, and was then further developed by **Aristotle** and other **Greek naturalists** and philosophers (Boxes 3.1 and 3.6)—was a crucial aspect of the Greco-Roman way of seeing our place in nature by Galen's time. Aristotle actually did not hold that all organisms can be arranged in one ascending sequence of forms, but he introduced the idea of continuity that was destined to fuse with the Platonic doctrine of the necessary "fullness" of the world. He stated that "nature passes so gradually from the inanimate to the animate that their continuity renders the boundary between them indistinguishable…and the transition from plants to animals is continuous." Specifically about **primates**, Aristotle noted that it cannot be said that mammals are either quadrupeds or bipeds, the latter being solely represented by humans, for "participating in the nature of both man and quadrupeds is the ape" belonging to neither group or both—note that until the seventeenth century, the name "ape" mainly referred to monkeys (see below). In particular, the hierarchical arrangement of all organisms in Aristotle's *De Anima* paved the way for later naturalists and philosophers to arrange them in a single graded *scala naturae* leading to "perfection." This is because Aristotle referred to the "powers of soul" from the "nutritive"—typical of plants—to the "rational"—characteristic of "man, and possibly another kind superior to his"—each higher order possessing all the powers of those below it in the scale, and an additional differentiating one of its own.

But not all Greek and Roman authors defended the notion of a *scala naturae*. As brilliantly shown in Greenblatt's 2011 book *The Swerve – How the World Became Modern*, the Roman poet and philosopher **Lucretius** (c. 99 BC to c. 55 BC) wrote, more than two millennia ago, *De rerum natura* ("*On the nature of things*"), in part to explain to Romans the philosophy of **Epicurus** (c. 341 BC to c. 270 BC) (Fig. 9.21). Defending **Epicurean philosophy**, Lucretius argued that the universe was composed of "atoms and the void," a naturalistic—and in this specific example, non-teleological—view of the universe and life that is in some way very similar to the idea I defend in the present book: that life is just what it is, that "things" just happen, without any cosmic purpose or designed masterplan or goal (see Boxes 5.1 and 9.10). However, as Greenblatt notes, Lucretius' *De rerum natura* became mainly a forgotten masterpiece, until the winter of 1417, when "an unemployed papal secretary turned book hunter named **Poggio Bracciolini**" discovered a copy of it in a remote monastery. Together with many other factors including the predisposition of Christians to be much more in line with the teleological notion of *scala naturae*, during the many centuries before that discovery the work of Epicurus and Lucretius was in general much less influential than that of Greek and Roman scholars defending a teleological view of nature, such as **Aristotle**. That is why the notion of *scala naturae* has been so influential in the history of religion, philosophy, sciences, and arts since then. Moreover, it is important to point out that despite Lucretius' non-teleological assertion that the "universe was composed of atoms and the void," he did defend an idea of humanity that was in general characterized by "progress," as noted in **Stephanie Moser**'s superb, and beautifully illustrated, 1998 book *Ancestral Images – The Iconography of Human Origins* (see Box 3.7).

Box 3.7 Hesiod and Lucretius, Rousseau and Hobbes, Adam and Eve

Moser's 1998 book *Ancestral Images – The Iconography of Human Origins* has not only beautiful illustrations: its text is highly informative and interesting. It shows that ancient Greeks and Romans already had the two major types of views about human nature that remain until today. For instance, the Greek poet **Hesiod** defended a **primitivist framework** that shared some similarities with **Rousseau**'s "**noble savage**," while the Roman poet **Lucretius** defended a more **anti-primitivist framework** that resembles in some ways the ideas of **Hobbes** about "**brutish savages**." Moreover, the book brilliantly connects such views with the stories about, and iconography displaying, **Adam** and **Eve**. This, in turn, is very interesting because it also shows a paradox: although Christians in general accepted the *scala naturae* view of life and "progress" defended by most ancient Greeks, they tended to see the lives of Adam and Eve before eating the apple as a paradisiacal example of existence in the **Garden of Eden**. That is, within this view, before eating the apple Adam and Eve were mainly "noble savages": in a sense, this view is thus more in agreement with the primitivist ideas of writers such as Hesiod, than with the notion of *scala naturae* and its related notion of "progress." Moser explains:

> The poet Hesiod was one of the first writers of the western tradition to provide an explanation of our human beginnings. This was outlined in *Works and days*, written around the eighth century BC, which was an account of Greek cultural history. Here Hesiod presented the legend of the *Five Races* or generations of the world, in which the successive races of humankind are described. First was the Golden race, who were peaceful and happy, living in a natural state without hardship. They were succeeded by the Silver race (**Fig. 3.3**), who represented a deterioration in that they were inferior physically and morally and they began to fight with one another. Third was the Bronze race, who represented a further decline in that they became violent and began to eat flesh. Fourth was the race of Heroes, who did not represent a further deterioration in culture, being noble and righteous. They were followed by the Iron race, who were savage, libidinous and cruel. With this final race of humanity there was a complete physical and moral decline. The Five Races scheme, with its metallic ages, incorporated elements from other creation myths, especially Babylonian, Persian and Asiatic ones. The symbolism of the metals is important in that it embodies a sense of cultural evolution, later to become a fundamental feature of scientific schemes of prehistory. However, cultural evolution was certainly not the underlying theme of Hesiod's scheme. His scheme was one of degeneration, in which the most distant past was conceived of in the most positive light and the most recent past was considered to represent a decayed state. Thus, in the Five Races legend, history was characterized not by progress, but rather by decline.

An alternative to Hesiod's scheme was outlined by those who argued that human evolution was characterized by progress rather than decline. The major proponent of this view was the Greek [actually, Roman] poet and philosopher Lucretius, who, in his *De rerum natura* of the first century BC, developed the idea that history was characterized by gradual social and technological advance. Emphasizing the brutal and primitive nature of our beginnings, Lucretius informs us that the first humans were little different from beasts…thus we learn that the first ancestors were born of the earth, and that they were large, muscular and strong. They did not live as couples, but rather lived like animals in a pack. The skills of working the land were unknown to them, as was the knowledge of metals and the ability to grow and maintain crops. Individuals thought only of themselves; they did not share or give to others. Male and female were united not by love but by lust.

These two different perspectives have been described as **primitivism** and **anti-primitivism**. While the primitivist view was nostalgic and glorified a simple idyllic existence, the anti-primitivist view appreciated the advances technology had brought. It is the latter that is most closely aligned with the development of scientific views of prehistory. In many ways the life of Adam and Eve in the Garden of Eden resembled the descriptions of the life of the Golden race in the Five Ages of Hesiod. Like the people of the Golden race, Adam and Eve did not need clothes, shelter or weapons because there were no threats to their life. While the Creation and the Golden Age were both very popular as visual themes, the biblical story was more prolifically illustrated throughout the early Christian and medieval periods. Similarities between the classical and biblical schemes can also be seen in the story of the expulsion from Eden, which led to a degradation in the life of the first couple, and the model of Hesiod, which embodied the concept of regression from an early idyllic state. The scenes of Adam and Eve labouring after their expulsion [**Fig. 3.4**] are also important. It is here that Adam is represented as a digger toiling on the land and Eve as a nurturer raising infants. The other point to note in the labouring imagery is the standardization of gender roles. The representation of the division of labour in this way was to become a dominant theme in the imagery of human ancestry.

It is also important to note that the notion of *scala naturae* per se has been applied in many different ways during the last two millennia. In the middle ages and the centuries that followed, there were various philosophical conflicting views about it. Although originally implying **animal–human continuity**, as emphasized in the terms "chain or ladder of being," many medieval authors defended that humans were essentially different from animals. They argued that humans were made "in the image of God" and could therefore use all non-human organisms—including primates—as they pleased. An example is **Francis Bacon**'s 1609 *De sapientia veterum*, which stated: "man…may be regarded as the centre of the world; insomuch that if man were taken away from the world, the rest would seem to be all astray, without aim or purpose…and leading to nothing…for the whole world works together in the service of man." This view was beautifully illustrated in a monograph edited by Perrault in 1676—that is, 67 years after the publication of Bacon's *Sapientia*—which was one of the first works illustrating in detail the internal

Fig. 3.3 "The Fruits of Jealousy"—"The Close of the Silver Age," by Lucas Cranash the Elder, circa 1530

anatomy, including the internal organs and brain, of both New Word and Old World **monkeys**. Perrault's monograph is one of the numerous works published on the seventeenth century on **comparative anatomy**, which truly started as a discipline in about 1600. Notable examples of early seventeenth century works are those of **Fabricius**—for instance, his famous 1600s volume *De formato foetu*—and his student **Casserius**, including his famous work of 1600–1601 that included textual and visual descriptions of the skull and laryngeal region of several animals including monkeys (Fig. 3.5), as explained in Riva et al.'s 2001 paper *Iulius Casserius (1552-1616) – the self-made anatomist of Padua's golden age*. Both Fabricius and

Fig. 3.4 "Adam and Eve," Woodcut from *Nuremburg Chronicle* 1483, pl. IX

Casserius were influenced by **Vesalius**' 1543 study of human anatomy and Belon's previous comparisons of humans and birds. Casserius in particular is renowned for his detailed anatomical studies of animals as diverse as fish, insects, and mammals, and for giving private courses in his house in which he dissected humans, dogs, and monkeys.

French scholars followed and further developed this tradition of comparative anatomy, and the 1676 volume *Memoires pour servir a l'histoire naturelle des animaux* edited by Perrault is an emblematic example of this. As explained in Guerrini's 2015 book *The Courtiers' Anatomists – Animals and Humans in Loius XIV's Paris*, the beautiful figure that shows the "sapajou," or capuchin monkey from South America, and the "guenon," or *Cercopithecus* from Africa, and their internal anatomy (Fig. 3.6) is a powerful example of the "**man-the-master**" teleological view defended by authors such as **Francis Bacon**, for several reasons. At the time, it was usual for painters, including the court painter **Lebrun**, to show human buildings in the background, even if humans were not present, to reinforce the idea that what

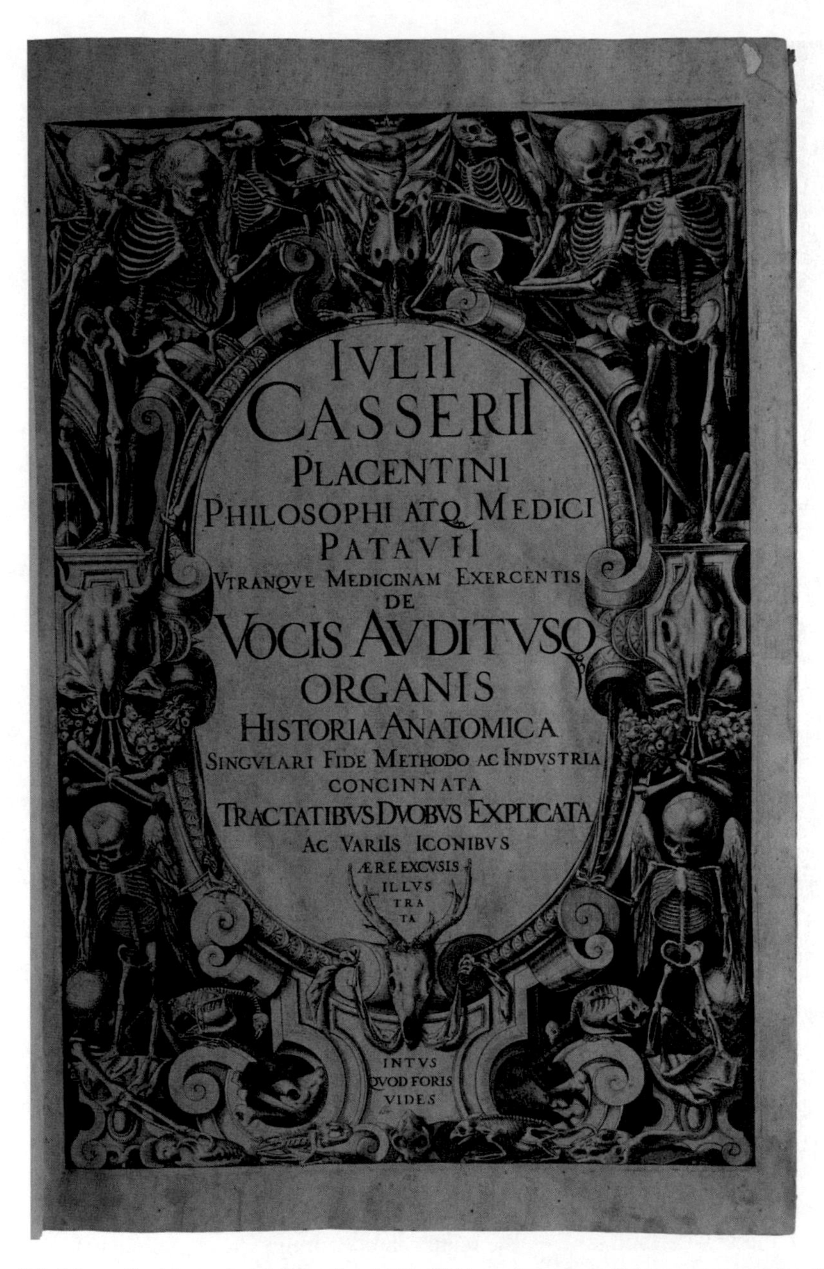

IVLII
CASSERII
PLACENTINI
PHILOSOPHI ATQ. MEDICI
PATAVII
VTRANQVE MEDICINAM EXERCENTIS
DE
VOCIS AVDITVSQ.
ORGANIS
HISTORIA ANATOMICA
SINGVLARI FIDE METHODO AC INDVSTRIA
CONCINNATA
TRACTATIBVS DVOBVS EXPLICATA
AC VARIIS ICONIBVS
ÆRE EXCVSIS
ILLVS
TRA
TA

INTVS
QVOD FORIS
VIDES

Fig. 3.5 Ligozzi's title page of Casserius 1600–1601 work is one of the most dramatic of the baroque period, depicting whole or partial skeletons of several animals, including skulls of oxen, birds, dogs, and deers and tailless monkeys

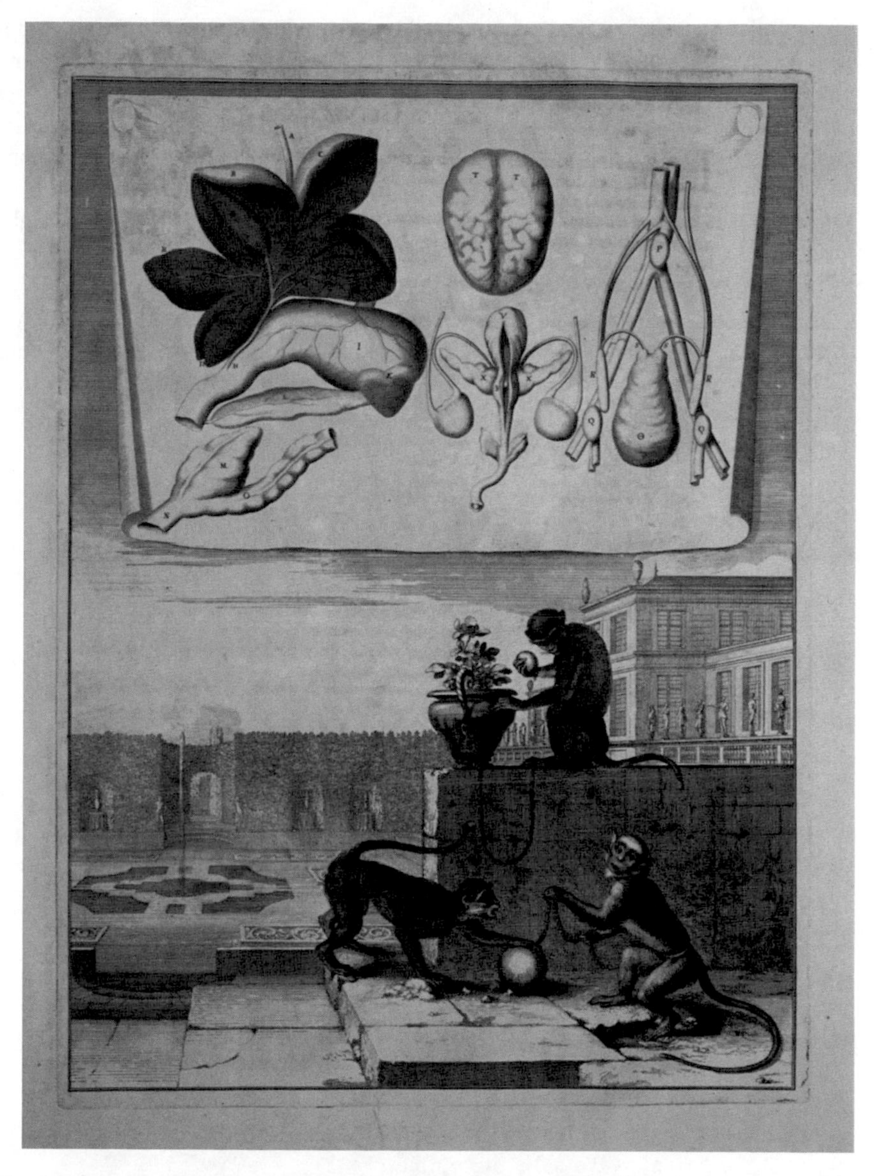

Fig. 3.6 "Sapajou et Guenon," showing the internal organs of these monkeys on top and the monkeys on a domesticated background on the bottom to reinforce the idea of human hegemony that was shared by many scholars and painters in the seventeenth century

was being displayed was not the natural habitat of the animals shown, but instead their *human ownership*. That is, although the monkeys were depicted outside the buildings where they spent their last days, "these animals nonetheless were shown in domesticated landscapes for denatured animals, human settings rather than animal settings," as noted by Guerrini. The inclusion of dissected animal parts and of the potted plant and the monkey chains further asserted "human hegemony" and a "nature completely subsumed to human desires." Such teleological view of "**humans as the masters of nature**"—strongly criticized by authors such as **Descartes**—was related to another conflict regarding the exact place of humans in the "ladder of being." Those scholars that defended the teleological idea that all non-human organisms were made by God to be merely used by humans usually emphasized the very prominent position of humans in the ladder, while their opponents typically highlighted instead the "infinite" distance between humans and God in that ladder, as noted in Lovejoy's 1936 book. He explained that in the seventeenth century **John Locke**—who defended the idea of human–animal continuity—stated that "there are far more species of creatures above us (e.g. angels), than there are beneath; we being in degrees of perfection much more remote from the infinite Being of God, than we are from the lowest state of being." (see als Box 3.8)

Box 3.8 Comparative Anatomy, a Field Deeply Influenced by Teleology
Comparative anatomy is a field that has been historically highly influenced by teleology, as explained in **Schmitt**'s 2004 outstanding book *Histoire d'une question anatomique: la repetition des parties*, a book that is unfortunately often neglected as it was written in French. A clear example of such teleological influence is given in an excerpt of that book that refers to three of the most influential comparative anatomists, **Geoffroy Saint-Hillaire**, **Cuvier**, and **Owen**: "the work of Owen…tries…to combine the formal and functional approaches of anatomy, that is the conceptions of Geoffroy Saint-Hilaire and of Cuvier: '*one organ is never really known until one knows perfectly both its teleological and morphological relationships*', wrote Owen in 1841." One can see how comparative morphologists—particularly **adaptationists**, following the tradition of Cuvier—often confused functional explanations with teleological explanations, as if functionality and teleology were mere synonyms. That is, as if searching for the *function* that an organ might perform in a certain animal is equivalent to search for the *reason why* the organ appeared/was evolved. We now know very well that this is not the same, as explained above. An additional example that further illustrates this point concerns the wings of birds, which now are mainly functionally related to flying but which probably were ancestrally mainly related with thermoregulation: so, saying that wings evolved "for flying" is factually inaccurate.

Interestingly, in contrast to the highly different and conflicting ideas about the place of humans in nature, the popular views about non-human primates in Europe from the rise of Christianity until the last centuries of the middle ages were relative constant. Namely, they were generally more negative than they had been before, and than they were in most other regions of the globe. Fascinating details about this subject can be read in Morris's 2013 book *Monkey*, Sorenson's 2009 book *Ape*, and Veracini and

Teixeira's 2016 paper *Perception and description of New World non-human primates in the travel literature of the fifteenth and sixteenth centuries*. As often happens in cases of discrimination and **racism**—see Bancel et al.'s 2014 book *The invention of race – scientific and popular representations*—such **negative views of monkeys** by Europeans were in great part due to the fact that almost no Europeans had physically seen monkeys until then. With the exception of *Macaca sylvanus*—the 'Barbary ape'—from the small island of Gibraltar—in the Mediterranean sea between Western Europe and Africa—there are no native non-human primates in Europe. In other words, apart from very few scholars such as Galen, in general Europeans had no knowledge about non-human primate biology and behavior. So, the authors of such negative ideas against monkeys—in great part, theologians—mainly referred to exaggerations of some old, mainly fantasized, stories written about monkeys by the ancient Greeks, or even to completely made-up stories about mythological human-like creatures (see Fig. 3.12). Actually, it should be noted that—contrary to **Aristotle** and **Galen**—even some Greek and Roman scholars had negative view of monkeys, as noted by Morris. In the seventh century BC, the Greek poet **Simonides** identified, based on external anatomical comparisons, "the very worst kind of woman" as descending from monkeys: "she is short, in the neck, hardly moves, has no buttocks, is withered of limb…and she knows all the intrigues and tricks like a monkey." But such rare negative views of ancient Greeks were often not related directly to real monkeys, as happens in this example, but instead to "half human-half animal" beings, or even with true apes, as explained in Janson's edited 1952 book *Apes and Ape Lore in the Middle Ages and the Renaissance* (note that primates include humans, apes, monkeys, and non-anthropoids).

No other culture and/or region of the globe displayed the consistent highly negative views of monkeys typical of the **European Christian Middle Ages**. There are a few exceptions, such as in Sub-Saharan African regions where there is an actual competition for crops between humans and large-sized monkeys such as baboons: in such few cases, the views tend to be negative or at least ambiguous. This is the case of the ***Dogon society***. However, most non-European cultures have in general a positive view of monkeys, in particular those in which people have a closer physical proximity with, and thus more knowledge about the behavior of, other primates. One illustrative example is the **God-monkey of Egyptians**, who commonly used monkeys as pets and knew several aspects of their biology, including the reactions of **baboons** to the sunrise and their relatively lengthy penis and the fact that they spent a great deal of time sitting in the squatting posture (Fig. 3.7). In fact, in one more interesting link between **comparative anatomy** and the human views on other primates and their relations and value to us, Morris noted that "compared with the human penis, that of the hamadryas (baboons) appears to lack a foreskin and the animal was therefore thought to be born circumcised." He added that "it has been suggested that the Egyptian priests who attended the sacred baboons honored them by imitating this condition…in this way the ritual of **human circumcision** is thought to have arisen, spreading later to nearby tribes who wished to emulate the advanced Egyptians" (Fig. 3.7). Other examples include the **Indian monkey God Hanuman**, the **sacred forest monkey of Bali**, the **monkey King Sun Wukong of China**, the

Fig. 3.7 Wall painting of sacred baboons from the tomb of Tutankhamen. Although Egyptians did seemingly not provide detailed reports on the internal anatomy of baboons, they had a relatively good knowledge on their external anatomy. Some authors argue that it was their detailed anatomical comparisons with humans that actually lead to the origin and spread of the ritual of human circumcision in at least some African regions

three wise monkeys of Japan, the **Aztec monkey God Ozmatli**, and the **Mayan monkey deity Batz**, among many others. In contrast, the teleological stories created by European Christians about monkeys were markedly negative, in great part because these stories had clear strategic purposes for **Christianity**, such as to directly criticize the previous **non-monotheistic religions** and/or "**paganism**." As noted in Morris's 2013 book *Monkeys*, "from the fall of the **Roman empire** until the late Middle Ages the official view of the **Christian Church** was that the monkey was a **diabolical beast**." For instance, "in the 4th century, when early Christian zealots were eagerly setting about the destruction of **Egyptian idols** in Alexandria, their leader ordered that one statue should be preserved as a monument to heathen depravity…needless to say, that statue was one of a sacred baboon." That is, "the

monkey-god of ancient Egyptians had, in one powerful gesture, become the **monkey-devil of Christianity**…the **Devil** himself became known as ***Simia Dei*** or **God's Monkey**."

The popular views of Europeans about monkeys only started to be more positive when they started to physically see, interact, and directly study non-human primates more frequently, particularly from the fourteenth century on when they traveled more often to, traded with, and received monkeys from other regions of the globe. This change of view was also deeply related with the **Renaissance**, the period of European history from the fourteenth to the seventeenth century that is often said to be the cultural bridge between the **Middle Ages** and the so-called modern history, in which observation and direct study of nature—supposedly less influenced from teleological narratives, but see Box 3.2—was largely promoted. In his 2008 book *Extended Family: Long Lost Cousins*, Groves explains how Europeans started to get monkeys from Sub-Saharan Africa well before the first European navigators—the Portuguese—reached even Cape Bojador in 1434. That is probably also why Europeans such as the ancient Greeks—including Galen—were able to dissect them much before the rise of Christianity. He refers to the **Trans-Saharan caravan**

Fig. 3.8 From *Six Monkeys and a Sturgeon*, 1430s Pisanello sketchbook—note that Groves (2008) stated that the monkey species is Campbell's mona, but in reality there are doubts about its true identity, with some scholars considering that it is very likely a *Chlorocebus* species, possibly *Chlorocebus aethiops* (Veracini, pers. comm.)

routes that passed through **Mali**. The drawing shown in Fig. 3.8 is an illustrative example of how, a few years before 1434, **Renaissance artists** were already painting monkeys more realistically, showing their true proportions, postures, and gestures. This contrasts markedly with the unrealistic and mainly religious symbolic manner in which they were almost always depicted during the Middle Ages. These points, together with the fact that monkeys started to be commonly used as pets in the Renaissance, led to a critical change in mindset, in which monkeys became to be less and less seen as terrifying evil creatures: they started to be seen as "funny"—for instance, as "clumsy human imitators"—as they are usually often portrayed nowadays in many movies and books, particularly those for younger kids.

3.4 Apes and Rise of Innate Notions of "Race"

We cannot differ from animals in general any more than a squirrel or a salmon can. (Anna Peterson)

The idea that ancient Greeks could have access to reports about true **apes**—that is, about non-human **hominoids** according to current scientific terminology—is based on the interpretation of some current authors about some ancient written accounts. For instance, those of the Greek **Herodotus**, whom supposedly referred, in the fifth century BC, to African apes in *The Histories*, following the description by the Phoenician Hanno in his *Periplus or Circumnavigation of Africa*: a strange creature inhabiting a fabulous land (far-) west of Cyrenaica (probably coinciding with what is today Cameroon, according to my colleague Marco Masseti). This might explain the common references, in the Greco-Roman literature, to "pygmies" and other ***semi-human creatures***, known for their lasciviousness. One example is **Pliny**'s 77 AD *Natural History*, which was a "beautiful mixture of accurate information, acute observation, and credulity" as put by Groves. Actually, Groves agrees with **Tyson**'s 1699 work (see below) in that many of these "semi-human creatures" do not seem to refer to true apes: for instance, Pliny's "satyrs" generally had hooves and tails while true apes are tailless. But in some cases they probably *did* refer to **chimpanzees**—such as the sixth century BC "**onocentaura**" quoted by Aelian and reported by **Pythagoras**—and possibly, but less likely, even gorillas—regarding the sixth century BC quote by Hanno.

Groves notes that in the first confirmed written report of a direct, close contact between an ape and an European, Purchas' 1625 *Purchas His Pilgrimes* tells a story of a sailor that was captive by the Portuguese in Angola before 1610 and that referred to an ape that was "very tall…his face and ears are without hair, and his hands also…his body is full of hair, but not very thick…they feed on fruit they find in the woods." This story, which on the one hand still retains some old habits of exaggerating up to the point of even making up certain features, but on the other hand also incorporates the more sober type of naturalistic facts that were more and more typical in that epoch, clearly seems to refer to a **Gorilla**. As put by Groves, "and so the

Fig. 3.9 Depiction of a great ape in Tulpius' 1641 *Observationes medicae*

last of the Great Apes to be described scientifically was the first to be described popularly." Just 16 years later, **Tulpius**—a physician and anatomist immortalized in **Rembrandt**'s painting *The Anatomical Lesson*—published the first relatively realistic external anatomical depiction of a great ape (Fig. 3.9), which was likely a common chimpanzee, a bonobo, or even an orangutan. Be that as it may, what is truly important is to note how even great apes were generally depicted in a rather positive, even docile, way in the seventeenth century, in contrast to the way apes begun to be

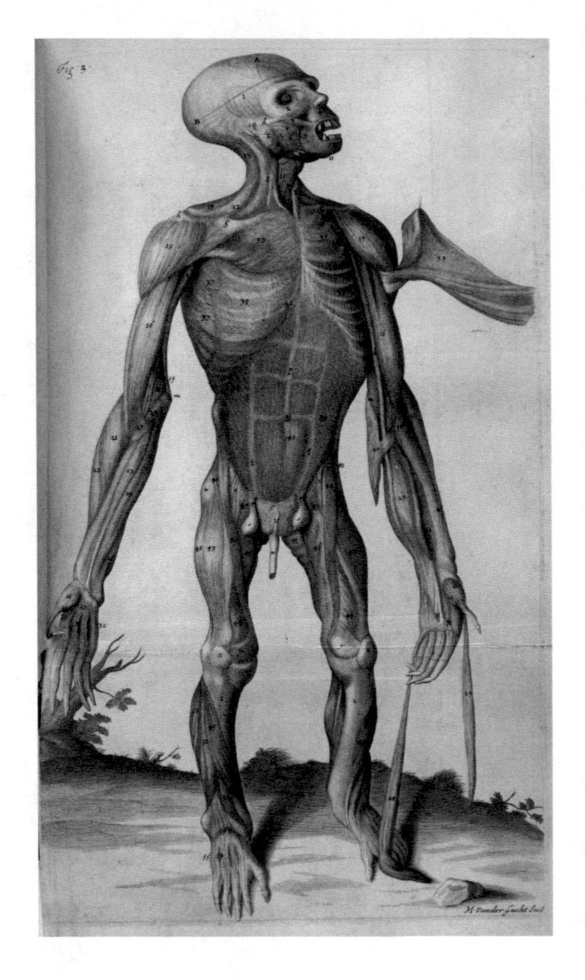

Fig. 3.10 One of the depictions of the muscles of the immature common chimp (*Pan troglodytes*) dissected in Tyson's 1699 work

represented after the rise of scientific racism type B—the innate, segregationist type—in the eighteenth and nineteenth centuries.

This marked change of attitude towards apes is primarily related to the publication of a landmark work that literally changed human history and our perception of our place in nature: **Tyson**'s 1699 monograph *Orang-Outang Sive Homo Sylvestris, or the Anatomy of a Pygmie Compared to That of a Monkey, an Ape and a Man*, which was the very first comprehensive description of the ***internal anatomy*** of an ape (Fig. 3.10). Tyson's 1699 work—which is, in a nutshell, the culmination of the trend to depict non-human primates in a more naturalistic, realistic, and also positive way that emerged in the fourteenth century (see also Fig. 3.11)—is indeed considered to be one of the most outstanding breakthroughs not only in the **history of comparative anatomy** but also of biology and even of science as a whole. In the preface of his 1943 volume entitled *Edward Tyson, M.D., F.R.S., 1650-1708*, Montagu noted that Tyson "did not discover the theory of evolution…but he accomplished in a modest and honest way a goodly share of the (anatomical) analytical work without which the scientific formulation of that theory would have remained impossible", being in this sense "a forerunner…of [**Darwin**'s] the *Origin of species* (1859) and the *Descent of man* (1871)". However, as also noted by Montagu, most authors "entirely overlooked…Tyson's (transitional) gradational view of the relation of animals in general, and of the special kind of relationship of…the chimpanzee to man in particular". Tyson stated "we may better observe Nature's Gradation in the Formation of Animal Bodies, and the Transitions made from one to another…the animal of which I have given the anatomy, coming nearest to Mankind, seems the nexus of the Animal and Rational…in the *Chain of the Creation*, as an intermediate link between an ape [meaning monkey] and a man, I would place our pygmie."

The ape studied by Tyson came from Angola and died a few months after its arrival in London, before being dissected, being probably a juvenile "a little over 2 years old," since it "had only its milk dentition, and since none of its permanent teeth had erupted," as noted by Montagu. Therefore, the name "pygmie" used by Tyson probably refers to both the small size of the individual and the fact that the ancients, including **Homer**, used the name "pygmie" to refer to semi-human creatures. Tyson concluded that the "pygmies" of ancients probably referred to indirect, or at least not close, contacts with true apes and argued that the "satyrs," "cynocephaly," or "sphinges" probably referred to monkeys or were merely **mythological creatures**. In this sense, comparative anatomy also played a crucial role in the **history of sciences** and biology and of broader discussions on our place in nature. This is because it was critical to end with the confusion, speculation, and imaginary creation myths about human-like creatures that had been so deeply immersed in the minds of thinkers, philosophers, scientists, and the broader public for millennia. Similarly, the detailed anatomical studies of human fossils that started to be found by Europeans in the nineteenth century—the **first Neanderthal remains** were discovered in Belgium, Gibraltar, and Germany in 1829, 1848, and 1856, respectively—were also crucial to inform and clarify discussions on those subjects, as noted in Corbey's 2005 book *The Metaphysics of Apes: Negotiating the Animal-Human Boundary*.

It is important to recognize that there were some erroneous assertions in Tyson's 1699 monograph. These mainly concerned the muscles, which were in reality dissected and described by William Cowper, who also assisted Tyson by doing the anatomical figures for that 1699 monograph (Fig. 3.10). Tyson incorrectly suggested that this chimpanzee was "in all respects designed by Nature, to walk erect," although he never actually stated that it *did* walk bipedally. However, despite those few errors, the central conclusion of Tyson's work—that he found more anatomical features shared by chimpanzees and humans (48 according to him) than by chimpanzees and monkeys (34 according to his comparison)—is completely right in view of current knowledge and is a key moment in our history. Namely, it was the first scientific publication to provide empirical data to break the ***humans versus other animals dichotomist narrative*** that is so predominant in organized religions, and the most critical teleological tale to attribute us a "special cosmic purpose," a transcendental "meaning of life." *Tyson's data showed, empirically, that humans are not special at all, that there is no* *scala naturae, because there is not only a continuity between humans and other animals, but humans are actually anatomically—and behaviorally as seen in Chap. 2—more similar to other animals such as chimpanzees, than those animals are to any other living animals*. This fact was later confirmed by other anatomical studies, and particularly by more recent genetic studies: chimpanzees are indeed more closely related to us than they are to any other living beings.

So, terms that we all use—as I use in this book—such as **"other animals," or "apes," or "non-human primates"** don't correspond at all to true biological groups, they are, using biological terms, **paraphyletic groups**, as are terms used for humans such as **"blacks" or "Africans" or "whites" or "Asians,"** because they all have to exclude some of its own members, as explained above. "Apes" cannot be a true group because it does not include humans but humans came from apes. "Africans" neither, because "whites," "Asians," etc., came from Africans. "Whites" neither, because "Asians" came from them: again, this means, pragmatically, that some "whites" are more closely related to some "Asians" then they are to other "whites." So, similarly, "Asians" is not a true group because **Native Americans** derived from them, and so on. So, while we can use these old terms, because most people understand them and it is thus easier to communicate, we need to understand that they don't refer to the reality of the natural world, they are just a relic of our distorter *Neverland* way to interpret it, they are part of our *world of unreality*. In other words, we can use them in the sense that we can also use the term **Mickey Mouse**, because everybody knows what we mean by it *and* everybody—hopefully—knows that Mickey Mouse is not really a true living organism. This just needs to be understood, once for all: not just by a few scientists, but by all people, as this aspect of the natural world completely destroys the vast majority of anthropocentric and racist ideas that are still so prevalent nowadays—thence, again, the huge importance of *scientific communication to the broader public*.

As noted in Lovejoy's 1936 book, in the eighteenth century decades after Tyson's 1699 work, "the sense of the separation between man from the rest of the animal creation was beginning to break down." **Rousseau** asserted in 1753 that humans and great apes—orangutans and chimpanzees, as gorillas were still not described scientifically, then—should be included in a same species and that language is not natural to humans but instead "an art which one variety of this species (humans) has gradually developed." And in 1781, **Bonnet** stated that great apes have the size, members, carriage, and "upright posture" of humans, having a "true face" and being "entirely destitute of a tail" and "susceptible of education," to the point of acquiring even a sort of politeness: whether we compare their minds or bodies with ours, "we are astonished to see how slight and how few are the differences, and how manifold and how marked are the resemblances." In fact, it was the discomfort caused by such **human–chimpanzee similarities** that mainly led several thinkers and scholars to *re-emphasize* an **animal–human discontinuity** in which the "true" gap was no longer between humans versus other animals but instead between "civilized" European humans versus non-European humans plus other primates. In 1714, just 15 years after Tyson's work, Blackmore and Hughes, noting how "surprising and delightful it is" to trace "the scale or gradual ascent from minerals to man," placed the **African**

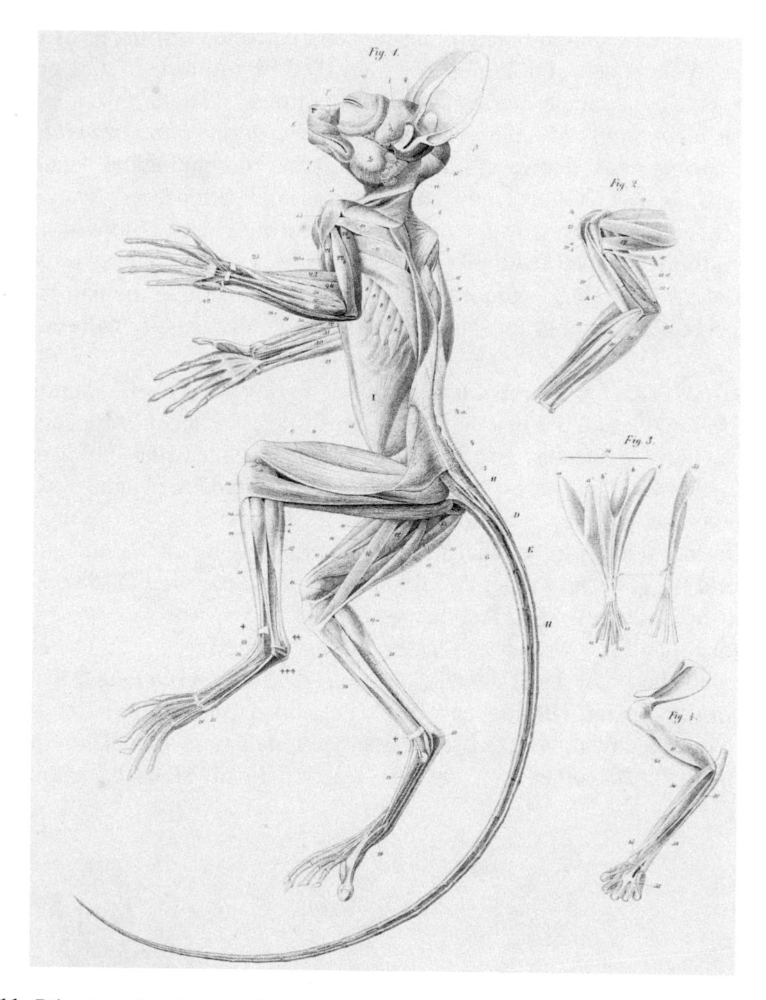

Fig. 3.11 Primates other than monkeys and apes were the last ones to have their internal anatomy described in detail, with meticulous gross anatomical descriptions of tarsiers and strepsirhines mainly starting to be published in the nineteenth century

Hottentots (see Fig. 3.14) between "humans"—including mainly people from "civilized" countries—and apes. They wrote: "the ape or the monkey that bears the greatest similitude to man, is the next order of animals below him…as the Hottentot, or stupid native of Nova Zembla."

Of course, there had been comparisons between non-human primates and non-European humans much before the eighteenth century, including, strikingly, even comparisons between small-sized New World monkeys such as spider monkeys and human ethnic groups native to the West Coast of Africa, for instance, by Oviedo in the sixteenth century, as noted in a 2016 paper by Veracini and Teixeira. Moreover, although this is often neglected in the literature, during the Middle Ages various Muslim scholars have also compared certain human groups with non-human primates and even explicitly stated that humans derived from them, as noted by Montagu and explained in a 2017 paper published by my former student Aamina Malik, my colleague Janine Ziermann, and me, entitled "*An untold story in biology: the historical continuity of evolutionary ideas of Muslim scholars from the 8th century to Darwin's time*." However, the profusion and morphological detail of such comparisons, as well as their type B/innate racist implications, only truly emerged in the eighteenth century with the so-called **scientific racial studies**—often also called "**anatomical racial studies**." These studies became particularly prominent in the fields of **anthropology** and **anatomy** in the nineteenth and the first half of the twentieth centuries, and became widely used to "morally" justify **colonialism** and **slavery**.

Linnaeus' 1735 *Systema Naturae* is a landmark example of the influence of Tyson's 1699 work and the reactionary responses to it by most European thinkers and researchers. Linnaeus' work profoundly influenced the publications of **Blumenbach**'s and **Camper**, whom are usually considered the "fathers of physical anthropology" and of "racial anatomical studies." On the one hand, Linnaeus went one step further than Tyson did because in the great scheme of classification of the living world proposed in the tenth edition of *Systema Naturae* (1758), he further developed his previous classifications and listed two *Homo* species: *H. sapiens* ("*H. diurnus*") and *H. troglodytes* ("*H. nocturnus*") (see Fig. 3.12). The latter species included the great apes, namely "*H. sylvestris Orang-Outang*," that is both Bontius' orangutan and **Tulpius**' and Tyson's chimpanzees, as gorillas had not been officially described by then. However, this suggested human–animal continuity was accompanied, on the other hand, by the reactionary divisions between humans

Fig. 3.12 The "Anthropomorpha" of Linnaeus (1935): *Troglodyta*, *Lucifer*, *Satyrus* and *Pygmaeus*, which are based on a mixture between scientific descriptions such as those of Tyson 1699 and imaginary human-like creatures mentioned by earlier authors

themselves, as Linnaeus divided humans into different "variations" that were defined by both anatomical and social/moral traits.

In this sense, Linnaeus work was also a landmark publication that changed human history, but in a very dark, factually inaccurate way that was completely the opposite of what Tyson's revolutionary, open-minded, empirical work did: his *Systema Naturae* was the *first* influential scientific work by a renowned biologist to imply that different "groups" of humans were defined by *fixed, unchangeable, morphological, and mental traits*. That is, to construct, based on data that are factually wrong, the basis of ***scientific type B/innate racism***. Specifically, Linnaeus defined four different human "groups" or "variants"—which later became commonly called "**races**" and which mainly prevail until today, as attested by the common use of the terms "blacks"—"Africans" and their descendants—"whites"—"Europeans" and their descendants—"yellows"—or "pales," that is, "Asians"—and "reds"—Native Americans. Namely, he defined: (1) (Native) Americans: red, bilious, straight—governed by customs; (2) Europeans: white, sanguine, muscular—governed by customs; (3) Asians: sallow (pale), melancholic, stiff—governed by opinion; and (4) Africans: black, phlegmatic, stiff—governed by chance (for more details, see Box 3.9). So, using Linnaeus' definitions, a "black" person is *always* governed by "chance." There is no "escape" to it: "black" people were born like that, and would remain in that way during their whole lives, this was "innate" or, using current terms, a genetically imprinted characteristic. This is exactly the way people defending type B/innate racist ideas continue to think: **Nazis**, **Neonazis**, **White supremacists**, and so on consider that blacks are, were, and always will be mentally "inferior," so they have the "moral" right—and the only solution is—to enslave them, segregate them, or even lynch/kill them.

Numerous authors have discussed the links between Linnaeus work and the development of the term "race" and of racism in the eighteenth and nineteenth centuries. However, surprisingly—or maybe not, as those authors were mainly "white"—most of them have completely neglected their connections with Tyson's 1699 work, leading to a rather incomplete, and often incorrect, account of the history of racism, principally of type B/innate racism. The 2014 book *The Invention of Race*, edited by Bancel and colleagues, provides a comprehensive, updated discussion on some of these topics, but fails to connect them directly with Tyson's 1699 masterpiece. Still, despite this weakness, some chapters of that book, in particular the one by Hoquet, do provide some interesting and pertinent facts. Namely, Hoquet states that despite the numerous nuances concerning this topic, there is no doubt that "Linnaeus description, due to the strict union of the physical and the moral, goes beyond a physical characterization of race: it includes the moral character of peoples." That is why for Hoquet, more than the works and terms used by other authors that are often said to have invented the modern sense of the word "race," such as Bernier and **Buffon**, "the categories developed by Linnaeus ultimately do indeed correspond to what we call '**races**': they are *physical and moral categories* that divide humans by color and by continental zones, and are unified by Hippocratic temperaments."

This is because for most scholars previous to, or living in, Linnaeus's epoch—including those that, as Buffon, defended that the differences between human "groups" were the result of a "degeneration" from an "ideal type"—such dissimilarities were mainly related to climatic/environmental differences. That is, they were grounded in **humoralism** and were therefore in theory *reversible*. Kendi, in his superb 2016 book *Stamped from the Beginning*, states that in 1684—that is, just 15 years before Tyson's 1699 work—the "French physician and travel writer **François Bernier**, a friend of John Locke's, anonymously crafted a 'new division of the earth'…through this essay, Bernier became the first popular classifier of all humans into races, which he differentiated fundamentally by their phenotypic characteristics." As noted by Kendi, "to Bernier, there existed 'races of men so notably differing from each other that this may serve as the just foundation of a new division of the world'…as a **monogenesist**, he held that 'all men are descended from one individual,' and described four races: the 'first' race, which included Europeans, was the original humans; then there were the Africans, the East Asians, and the 'quite frightful' people of northern Finland, 'the Lapps'." As put by Kendi, Bernier thus "gave future taxonomists some revisionist work to do when he lumped with Europeans in the 'first' race the people of North Africa, the Middle East, India, the Americas, and Southeast Asia…the notion of Europeans – save the Lapps – as being in the 'first' race was part of Western thought almost from the beginning of racist ideas." That is, as previous scholars, Bernier was setting such ideas "in the conceptual core of climate theory: Africans darkened by the sun could return to their original White complexion by living in cooler Europe…in advancing White originality and normality, Bernier positioned the 'first' race as the 'yardstick against which the others are measured', as historian Siep Stuurman later explained." In other worlds, Bernier's displayed the—although a bit more extreme, and systematized—type A/epigenetic racism that was prevalent before the type B/innate racism exploded in early eighteenth century works such as those of Linnaeus as a reaction to Tyson's 1699 monograph.

The term **monogenesist** refers to two main ideas concerning the number of species of humans living today: **polygenism** and **monogenism**. **Polygenism** is the idea that the different color of skin or other traits seen in human individuals is due to the fact that such individuals belong to different human species: this idea was very prominent in the nineteenth century—particularly after the publication of Darwin's evolutionary works—and the first half of the twentieth century. As an example, many of these scholars defended that "blacks" are part of a "species" that in many ways is more similar, and closely related, to chimpanzees than to the "species" including "whites." This contrasts with **monogenism**—the idea that there is a single human species—which was defended by most scientists before the eighteenth century and also by creationists: because God created only one **Adam and Eve**, there can be only a species of humans. However, it should be noted that monogenesists were not necessarily less racist than polygenesists: simply, they were racist in different ways—in this specific case, they would argue, for instance, that "blacks" were inferior because they "degenerated" from "whites" such as Adam and Eve. In clear contrast to the "climate theories" of the monogenesist Bernier, Linnaeus paved the way—by instead connecting in an *irreversible*, *innate way* the social/moral with anatomical traits—for authors such as Meiners, **Blumenbach**, **Camper**, and **Broca** and their "anatomical race studies," which undertook a "**biologization of the social**," as put by Reynaud-Paligot in *The Invention of Race*. Or, I would say, truly an **anatomization of the social/moral** (see also Box 3.9).

The particularly dark history of the fields of anthropology—particularly biological and physical anthropology—and biology—particularly anatomy and evolutionary biology—is illustrated by the fact that **Camper**, one of the main promoters of scientific type B/"innate" racism, is celebrated by many scholars as the "father of physical anthropology." This Dutch anatomist provided, in 1772, the first detailed anatomical description of **orangutans** and, importantly, also published the first detailed "human anatomical racial study" using "**craniology**"—a so-called scientific area that is actually a *Neverland* mixture of fictional gibberish defending that one can "identify" different human "races" by "measuring" the shape and size of skulls. As noted by Guedron in 2014 book *The Invention of Race*, Camper started a tradition in which the design of human anatomical plates and illustrations became to be used within the broader discourse of **racial hierarchization**. Some authors, such as Meijer in the same 2014 book, defend that Camper's engravings that initiated **craniology** (see Fig. 3.13) should not be isolated "from the rest of Camper's work, the original context about organic interconnectiveness, human head shapes' plasticity and their mutual reciprocity." Meijer accurately emphasized that Camper stated that "prioritizing whiteness was narcissism, for those who gave precedence to whites were always white themselves." However, many if not most authors consider that Camper clearly contributed to the rise of scientific racism in the eighteenth and nineteenth centuries and paved the way for the creation of the "negro." An interesting overview on this topic is given by Panese in that 2014 book, in which he explains that Camper used the "facial angle" to differentiate "between the Negro, the 'Calmouque', and the European, noting 'analogies' between the head of the Negro and that of the monkey" (Fig. 3.13) and that Camper specifically stated that "the

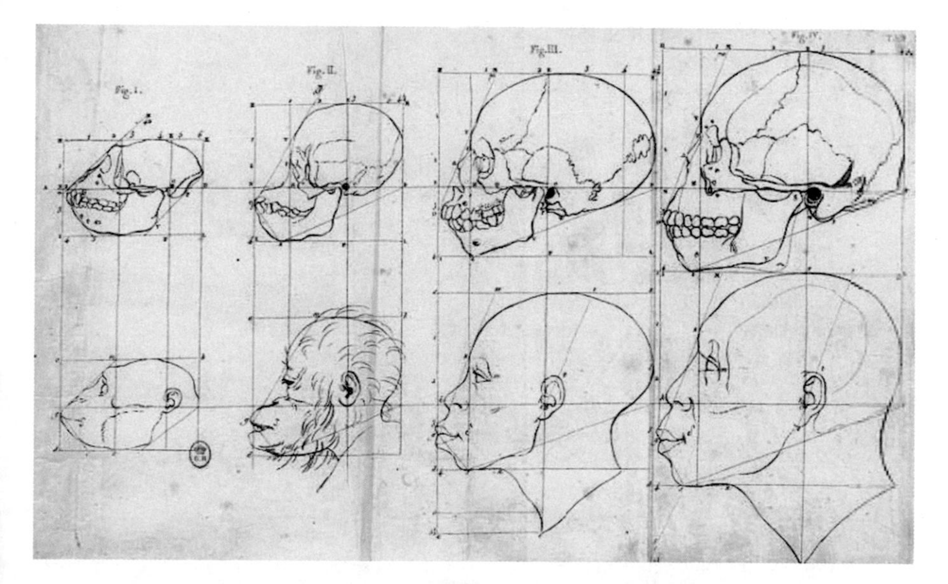

Fig. 3.13 Skulls and facial angles of a monkey, an orangutan, an "African moor" and an "Asian" as shown in a plate of 1791s French translation of Camper's work on the facial angles

Box 3.9 Scientific Revolution, Enlightenment, and the American Revolution

Kendi, in his 2016 book *Stamped from the Beginning*, fully recognizes the critical contribution of comparative human and primate anatomy, and of **Linnaeus** work, for the rise of scientific racism, and for the spread of racist notions within the broader population in general:

> The **scientific revolution** of the 1600s had given way to a greater intellectual movement in the 1700s. Secular knowledge, and notions of the propensity for universal **human progress**, had long been distrusted in Christian Europe. That changed with the dawn of an age that came to be known as les Lumières in France, Aufklärung in Germany, Illuminismo in Italy, and the **Enlightenment** in Great Britain and America. For Enlightenment intellectuals, the metaphor of light typically had a double meaning. Europeans had rediscovered learning after a thousand years in religious darkness, and their bright continental beacon of insight existed in the midst of a 'dark' world not yet touched by light. Light, then, became a metaphor for Europeanness, and therefore Whiteness, a notion that **Benjamin Franklin** and his philosophical society eagerly embraced and imported to the colonies. White colonists, Franklin alleged in *Observations Concerning the Increase of Mankind* (1751), were "making this side of our Globe reflect a brighter Light". Let us bar uneconomic slavery and Black people, Franklin suggested. "But perhaps", he thought, "I am partial to the complexion of my Country, for such kind of partiality is natural to Mankind". Enlightenment ideas gave legitimacy to this long-held racist "partiality", the connection between lightness and Whiteness and reason, on the one hand, and between darkness and Blackness and ignorance, on the other. These Enlightenment counterpoints arose, conveniently, at a time when Western Europe's triangular transatlantic trade was flourishing. Great Britain, France, and colonial America principally furnished ships and manufactured goods. The ships sailed to West Africa, and traders exchanged these goods, at a profit, for human merchandise.
>
> **Carl Linnaeus**, the progenitor of Sweden's Enlightenment, followed in the footsteps of **François Bernier** and took the lead classifying humanity into a **racial hierarchy** for the new intellectual and commercial age. In *Systema Naturae*, first published in 1735, Linnaeus placed humans at the pinnacle of the animal kingdom. He sliced the genus *Homo* into *Homo sapiens* (humans) and *Homo troglodytes* (ape), and so on, and further divided the single *Homo sapiens* species into four varieties. At the pinnacle of his human kingdom reigned *H. sapiens europaeus*: "Very smart, inventive. Covered by tight clothing. Ruled by law". Then came *H. sapiens americanus* ("Ruled by custom") and *H. sapiens asiaticus* ("Ruled by opinion"). He relegated humanity's nadir, *H. sapiens afer*, to the bottom, calling this group "sluggish, lazy…[c]rafty, slow, careless. Covered by grease. Ruled by caprice", describing, in particular, the "females with genital flap and elongated breasts". Carl Linnaeus created a hierarchy within the animal kingdom and a hierarchy within the human kingdom, and this human hierarchy was based on race. His "enlightened" peers were also creating human hierarchies; within the European kingdom, they placed Irish people, Jews, Romani, and southern and eastern Europeans at the bottom. Enslavers and slave traders were creating similar ethnic hierarchies within the African kingdom. Enslaved Africans in North America were coming mainly from seven cultural-geopolitical regions: Angola (26 percent), Senegambia (20 percent), Nigeria (17 percent), Sierra Leone (11 percent), Ghana (11 percent), Ivory Coast (6 percent), and Benin (3 percent). Since the hierarchies were usually based on which ancestral groups were thought to make the best slaves, or whose ways most resembled those

of Europeans, different enslavers with different needs and different cultures had different hierarchies. Generally, Angolans were classed as the most inferior Africans, since they were priced so cheaply in slave markets (due to their greater supply)....

Voltaire, France's Enlightenment guru, used Linnaeus's racist ladder in the book of additions that supplemented his half-million-word Essay on Universal History in 1756. He agreed there was a permanent natural order of the species. He asked, "were the flowers, fruits, trees, and animals with which nature covers the face of the earth, planted by her at first only in one spot, in order that they might be spread over the rest of the world"? No, he boldly declared. "The negro race is a species of men as different from ours as the breed of spaniels is from that of greyhound.... If their understanding is not of a different nature from ours it is at least greatly inferior". The African people were like animals, he added, merely living to satisfy "bodily wants". However, as a "warlike, hardy, and cruel people", they were "superior" soldiers. With the publication of *Essay on Universal History*, Voltaire became the first prominent writer in almost a century daring enough to suggest polygenesis. The theory of separately created races was a contrast to the **assimilationist idea** of **monogenesis**, that is, of all humans as descendants of a White Adam and Eve. Voltaire emerged as the eighteenth century's chief arbiter of segregationist thought, promoting the idea that the races were fundamentally separate, that the separation was immutable, and that the inferior Black race had no capability to assimilate, to be normal, or to be civilized and White. The Enlightenment shift to secular thought had thus opened the door to the production of more segregationist ideas. And segregationist ideas of permanent Black inferiority appealed to enslavers, because they bolstered their defense of the permanent enslavement of Black people.... The argument over Voltaire's multiple human species *versus* **Buffon**'s single human species was one aspect of a larger scientific divide during the Enlightenment era. Their beloved Sir **Isaac Newton** envisioned the natural world as an assembled machine running on "natural laws"....

Most of the leading Enlightenment intellectuals were producers of racist ideas and abolitionist thought. Buffon defined a species as "a constant succession of similar individuals that can reproduce together". And since different races could reproduce together, they must be of the same species, he argued. Buffon was responding to some of the first segregationist denigrations of biracial people. **Polygenesists** were questioning or rejecting the reproductive capability of biracial people in order to substantiate their arguments for racial groups being separate species. If Blacks and Whites were separate species, then their offspring would be infertile. And so the word **mulatto**, which came from "mule", came into being, because mules were the infertile offspring of horses and donkeys. In the eighteenth century, the adage "black as the devil" battled for popularity in the English-speaking world with "God made the white man, the devil made the mulatto".... The foundational thinker of modern European art history, Johann Joachim Winckelmann of Germany [said]: "a beautiful body will be all the more beautiful the whiter it is"...in his disciplinary classic, *Geschichte der Kunst des Alterthums* (*History of the Art of Antiquity*) in 1764. These were the "enlightened" ideas on race that Benjamin Franklin's American Philosophical Society and a young **Thomas Jefferson** were consuming and importing to America on the eve of the **American Revolution**.... The United States [later] joined the growing band of nations seeking to colonize Africa. By 1824, American settlers had built fortifications there. They renamed the settlement "**Liberia**", and its capital "**Monrovia**", after the US president [at that time]. Between 1820 and 1830, only 154 Black northerners out of more than 100,000 sailed to Liberia.

whitest" humans were "also the most beautiful and well-proportioned in the known world."

Guedron's 2014 chapter in *The Invention of Race* shows how comparative anatomical illustrations, in particular those including human sequential images such as that shown in Fig. 3.13, were indeed crucial for the development of the **concept of "race"** and of **scientific racism**. This is because almost all figures used in subsequent "racial" studies were based on Camper's writings and in the sequential illustrations published in the French translation of those writings. Guedron explains that "at first glance there is no narrative link between the different skulls…yet the horizontal organization of the first two plates suggests to the reader that he must read left to right". He notes that "Camper is careful to warn us that we must not jump to hasty conclusions (based on these images)…still, the sequence implies evolution, with the white race nearing perfection – Europeans were placed next to **Greek Gods** – and the African in proximity with the monkey." Therefore, "it is not difficult to understand how such images came to be interpreted as a kind of *teleological demonstration, from the beast to the divine*…[Camper's] words could not have been also clearer: the upper and lower jaws protrude in the same way in all black people…in this way, they are more like monkeys than us, or the faces from antiquity."

Guedron wisely points out that such illustrations are indeed a crucial part of the **biologization of the social**, as they recall old conceptions of the human body—the split between the spiritual and the material, the **soul** and the beast. He writes: "the top of the head was considered to be more developed among whites…meanwhile, the inferior part of the face (considered to be more developed in non-Europeans)…was related to…violence and instinct, which were associated with primitive peoples." Sequential illustrations similar to those provided in Camper's 1791 work appeared in countless publications that followed, well before Darwin proposed his evolutionary ideas in his 1859 book *The Origin of Species*. A few relevant examples are Cuvier's 1797 very influential *Tableau elementaire de 'histoire naturelle des animaux*, White's 1799 *An Account of the Regular Gradation in Man*, Blumenbach's 1804 *De l'unite du genre humain*, and Cloquet's 1821–1831 *Anatomie de 'homme*. Such works provided a crucial contribution to the **racial determinism**—that is, type B/"innate" racism—that became so widespread among scholars in the nineteenth century. Such determinism was related to the idea of biological transmission, through blood and heredity, of not only anatomical but also intellectual and moral attributes within a given people, as further explained by Reynaud-Paligot in *The Invention of Race*.

So, what were the broader societal implications of scientific racism? Its implications and consequences were huge, as after it emerged in the beginning of the eighteenth century, it quickly became widespread to the broader society, as it continues to be, today. An emblematic example of this, and of the fundamental—but too often neglected—role played by comparative anatomy in discussions about our place in nature, human "races," and the chain of being, is the life of **Sarah Baartman**, nicknamed the "**Hottentot Venus**" (Fig. 3.14; see also Box 3.10 for other examples). She was taken on board of a British ship in South Africa by a surgeon of the Royal Navy in 1810 and then arrived in London before being sent to Paris. As noted by Boetsch and Blanchard 2014 in *The Invention of Race*, although she was not the first

Fig. 3.14 "A Pair of Broad Bottoms," a caricature of Sarah Baartman by William Heath from 1810

person to be displayed in exhibitions in Europe, by quickly starting to be an object, all at once, of entertainment, media interest, "sexual fantasy," and science, she marked in a sense the beginning of a new way of thinking about "the Other" in Western countries. The obsession with her is at least partly explained by the fact that she was preceded by a mythical figure of the "**African Hottentot**" in the European imagination: "Hottentots were a source of fascination, and were ear-marked to fill the role of a 'missing link' [in the chain of being]…a symbol of an 'intermediate race' between human and animal…or at the very least act as proof of degeneration within the human species." The shape of the body of "Venus" was considered peculiar, because of her **steatopygia**—buttock and hip hypertrophy—and **macronymphia**—protruding sexual organs (Fig. 3.14).

In 1815 **Saint-Hillaire** published a report in which he compared her face with that on an orangutan and her posterior with that of a female mandrill monkey, while **Cuvier** stated that he had never "seen a human head that more closely resembled that of a monkey." Just 24 hours after her death on the December 29, 1815, Cuvier dissected her body, removing her sexual organs and anus for preservation and then having her other anatomical parts removed, including the brain, and putting them in

Fig. 3.15 Ota Benga (second from the left, with monkey), a widower from what is now the Democratic Republic of Congo, was exhibited in New York's Bronx Zoo in 1910, and is here shown with other African men who were exhibited to the public and studied by psychologists at the 1904 St. Louis World's Fair

jars to be kept in the museum archives as a "reference anatomical specimen," as was commonly done for non-human animals. Taking her sexual anatomical features as proof of a **"primitive" sexual appetite** in African women, Cuvier concluded that "races with depressed, compressed skulls are *forever* condemned to inferiority." In a nutshell, just 116 years after Tyson's revolutionary 1699 work, and just eight decades after Linnaeus' 1735 reactionary response to it in *Systema Naturae*, type B/ innate racism was indeed spread everywhere, from lay people to the most prominent scientists, from colonizers to slave-owners, from politicians to religious leaders, from museum collections to world's fairs and zoo displays. Indeed, another truly tragic, atrocious example of this, among an endless number of them, is that of **Ota Benga**, whom was "acquired" and "displayed" by the Bronx Zoological Gardens in 1906 as if he was literally an ape or a monkey (see Fig. 3.15 and Box 3.10).

In *Stamped from the Beginning*, Kendi provides further—and mostly horrifying—details about **Sarah Baartman**'s life:

> London was blitzed with a broadsheet picturing a seminude African woman standing sideways to the viewer, her oversized buttocks exposed on one side, the unseen side draped in animal skin. A headband wraps her forehead, and she holds a body-sized stick. Whitening Blacks, Black exhibits, and "converted Hottentots", sharing their supposed journeys from savagery to civilization, were becoming less remarkable with each passing year. But Londoners were captivated by Sarah Baartman, or rather, her enormous buttocks and genitalia. Baartman's **Khoi people** of southern Africa had been classified as the lowest Africans, the closest to animals, for more than a century. Baartman's buttocks and genitals were

irregularly large among her fellow Khoi women, not to mention African women across the continent, or across the Atlantic on Jefferson's plantation. And yet Baartman's enormous buttocks and genitals were presented as regular and authentically African. She was billed on stage in the fashionable West End of London as the "Hottentot Venus", which tightened the bolt on the racist stereotype linking Black women to big buttocks. Polygenesist Charles White had already tightened the bolt linking Black men with big genitalia. Retiring colonial official Alexander Dunlop and Baartman's South African master Hendrik Cesars brought Baartman to London in July 1810. Upon Dunlop's death in 1814, exhibiter Henry Taylor brought the thirty-six or thirty-seven-year-old Baartman to Paris for another round of shows. Papers rejoiced over her arrival. She appeared in the grand Palais-Royal, the centerfold of Parisian debauchery, where prostitutes mixed with printers, restaurants with gambling houses, coffee gossipers with drunk dancers, beggars with elites.

On November 19, 1814, Parisians strolled into the Vaudeville Theater across from the Palais-Royal to view the opening of *La Venus Hottentote*, ou *Haine aux Francais* (or the Hatred of French Women). In the opera's plot, a young Frenchman does not find his suitor sufficiently exotic. When she appears disguised as the "Hottentot Venus", he falls in love. Secure in his attraction, she drops the disguise. The Frenchman drops the ridiculous attraction to the Hottentot Venus, comes to his senses, and the couple marries. The opera revealed Europeans' ideas about Black women. After all, when Frenchmen are seduced by the Hottentot Venus, they are acting like animals. When Frenchmen are attracted to Frenchwomen, they are acting rationally. While hypersexual Black women are worthy of sexual attraction, asexual Frenchwomen are worthy of love and marriage. In January 1815, animal showman S. Reaux obtained Baartman from Henry Taylor. Reaux paraded her, sometimes with a collar around her neck, at cafés, at restaurants, and in soirées for Parisian elites – wherever there was money. One day in March 1815, Reaux shepherded Baartman to the Museum of Natural History in Paris, which housed the world's greatest collection of natural objects. They had a meeting with Europe's most distinguished intellectual, the comparative anatomist **Georges Cuvier**.

That rare segregationist who rejected **polygenesis**, Cuvier believed that all humans descended from Europe's Garden of Eden. A catastrophic event 5,000 years earlier had sent the survivors fleeing to Asia and Africa; three races had emerged and had started passing on unchangeable hereditary traits. "The white race" was the "most beautiful of all" and was "superior", according to Cuvier. The African's physical features "approximate[d] it to the monkey tribe". In his lab, Cuvier asked Baartman to take off her long skirt and shawl, which she had worn to ward off the March wind. Baartman refused. Startled, Cuvier did all he could to document her with her clothes on over the next three days, measuring and drawing her body. Sometime in late December 1815, Baartman died, perhaps of pneumonia. No Black woman was the subject of more obituaries in Parisian newspapers in the nineteenth century than Sarah Baartman. Cuvier secured her corpse and brought her to his laboratory. He removed her clothes, cracked open her chest wall, removed and studied all of her major organs. Cuvier spread her legs, studied her buttocks, and cut out her genitals, setting them aside for preservation. After Cuvier and his team of scientists finished their scientific rape, they boiled off the rest of Baartman's flesh. They reassembled the bones into a skeleton. Cuvier then added her remains to his world-famous collection. In his report, he claimed to have "never seen a human head more resembling a monkey's than hers".

The **Khoi people** of South Africa, he concluded, were more closely related to the ape than to the human. Parisians displayed Baartman's skeleton, genitals, and brain until 1974. When President **Nelson Mandela** took office in 1994, he renewed South Africans' calls for Baartman's return home. France returned her remains to her homeland in 2002. After a life and afterlife of unceasing exhibitions, Baartman finally rested in peace. Baartman's fate was particularly horrific in the early 1810s, and Cuvier's conclusions about Black bodies were consumed with little hesitation by those seeking evidence of Black inferiority to justify their commerce on both sides of the Atlantic, a commerce taking root in the wombs of Black women.

Box 3.10 Circus Africanus, Human Zoos, World's Fairs, and Western "Progress"

The story of **Sarah Baartman**, nicknamed the "Hottentot Venus," has some resemblances, particularly concerning its horrendous aspects, with that of **Ota Benda**, whom was "acquired" by the Bronx Zoological Gardens in 1906, after being displayed in the 1904 St. Louis World's Fair (Fig. 3.15). This and other horrific cases were described in detail in the chapter entitled '*Circus Africanus*' of Washington's outstanding 2006 book *Medical Apartheid*:

> By 1904, swashbuckling missionary-explorer Samuel Phillips Verner had acquired a veritable Noah's Ark of exotic fauna during three trips to the interior of the Dark Continent. The last expedition was commissioned in 1903 by the St. Louis Exposition Company, which paid the South Carolina–born Verner to hunt men instead of monkeys: He was to bring **African Pygmies** to America for display at the **St. Louis World's Fair**. Upon his return to America, Verner found himself romanticized as a reincarnation of Dr. David Livingstone, whom he claimed as his "posthumous mentor". As an ordained minister in the Presbyterian Church, Verner was also lionized in church circles as an imparter of morality to the Congo natives he doggedly hectored at the Southern Presbyterian Missionary House in Luebo, chiding them for their immodest dress and sexual behavior. His American admirers did not know that between 1895 and 1899, Verner had fathered a daughter and son on an African orphan girl there. By 1906, the World's Fair was over and the cash-strapped Verner was selling off his animals, artifacts, and more. Upon the receipt of a financial gift, he bestowed a prized equatorial specimen upon William T. Hornaday, director of the **Bronx Zoological Gardens**. Verner's present was twenty-three-year old **Ota Benga**, an Mbuti widower from southern Africa, in what is now the Democratic Republic of Congo. Around 1903, Benga had returned from a hunting trip, only to find his village in smoking ruins and his wife, children, and entire tribe slaughtered by Force Publique thugs supported by the Belgian government. Benga himself was seized and sold into Verner's hands. Hornaday's views about the natives of sub-Saharan Africa mirrored Verner's own, conscripting Darwin in the service of racism: He told the New York Times that there exists "a close analogy of the African savage to the apes". *Scientific American* agreed: "The Congo pygmies [are] small, apelike, elfish creatures, furtive and mischievous, they closely parallel the brownies and goblins of our fairy tales. They live in the dense tangled forests in absolute savagery…while they exhibit many ape-like features in their bodies".
>
> But Hornaday espoused a more progressive vision as a scientific artist, and we have him to thank for the modern American zoo. As chief taxidermist of the National Museum (the Smithsonian), a position he held until 1890, he had inherited a static mausoleum of tatty taxidermy enshrined on plaster pedestals with only laconic placards to suggest what the animal had been like in life. In 1888, Hornaday persuaded the museum to add a wing of living animals in lifelike settings, which proved so popular a revolution that it became the National Zoological Gardens. He resigned over differences of vision, but in 1896 he reemerged as the first director of the New York Zoological Gardens (known as the Bronx Zoo), the world's largest, lushest, and most varied zoo. Hornaday's passion was for colorful verisimilitude in the re-creation of his animals' natural habitats. With a verdant Bronx park as his canvas, Hormaday installed colorful exotic animals of every genus grouped with their natural companions amid native vegetation. So when Benga was locked in the monkey house, before the staring crowd and with keepers always nearby, he was given a bow

and arrow to brandish, his cage was littered with bones, and his two cage mates were Dinah, a gorilla, and an orangutan called Dohung. The placard on Benga's enclosure read, "The African Pygmy, 'Ota Benga'...height 4 feet 11 inches...weight 103 pounds...brought from the Kasai River, Congo Free State, South Central Africa by Dr. Samuel P. Verner...exhibited each afternoon during September". *New York Times* headline trumpeted, "*BUSHMAN SHARES A CAGE WITH THE BRONX PARK APES*". Black New Yorkers were incensed, and representatives of the clergy, led by the Reverend Dr. MacArthur, pressed Mayor George B. McClellan to withdraw the city's support from the exhibit. As another minister, a Reverend Gordon, told the New York Times, "Our race...depressed enough without exhibiting one of us with the apes...we think we are worthy of being considered human beings, with souls". The Times turned an unsympathetic ear to African American objections: "One reverend colored brother objects to the curious exhibition on the grounds that it is an impious effort to lend credibility to Darwin's dreadful theories...the reverend colored brother should be told that evolution...is now taught in the textbooks of all the schools, and that it is no more debatable than the multiplication table". The swipe at creationism did not address Gordon's immediate concerns but did hit a nerve among many whites who shared Gordon's outrage. Some were angered by this inhumane insult to blacks, and others, who opposed the teaching of Darwin's theory of evolution, were afraid that Benga's dramatic presence would offer a powerful plebeian argument for the theory of evolution. The entertainment of a "monkey-man" might persuade people who were untouched by the theory's scientific merits. Mayor McClellan snubbed the black delegation, referring them to the Parks Department, and another *Times* account hinted that Benga differed little from the zoo's animals: "Ota Benga...is a normal specimen of his race or tribe, with a brain as much developed as are those of its other members...and can be studied with profit...the pygmies are an efficient people in their native forests...but they are very low in the human scale, and the suggestion that Benga should be in a school instead of a cage ignores the high probability that school would be a place of torture to him and one from which he could draw no advantage whatever".

A lively epistolatory debate ensued in the pages of the *Times*, heavily weighted in favor of retaining Benga, and many of the letters were signed by respondents with M.D. and Ph.D. degrees. One doctor suggested, "It is a pity that Dr. Hornaday does not introduce the system of short lectures or talks in connection with such exhibitions...[to] help our clergymen to familiarize themselves with the scientific point of view so foreign to many of them". *Times* journalists agreed that Benga provided a valuable tool for illustrating basic evolutionary precepts: To oppose his internment was to oppose science. These precepts included physical similarities to the lower primates that scientific racism was beginning to popularize widely. Anthropometric portraits of blacks and apes demonstrated how blacks' facial angles, stature, stance, and gait resembled those of monkeys, chimpanzees, and orangutans. Blacks' hair, or "wool", was compared to animal pelts. Such uncomplimentary images were published in scientific journals and would soon adorn children's textbooks. Scientists alleged that apes preferred to mate with black women, just as black men lusted after white women, their own evolutionary "betters". At the zoo, the *Times* revealed that Benga's situation was escalating: "There were 40,000 visitors to the park on Sunday...nearly every man, woman and child of this crowd made for the monkey house to see the star attraction in the park, the wild man from Africa...they chased him about the grounds all day, howling, jeering, and yelling...some of them poked him in the ribs, others tripped him up, all laughed at him". Finally, Benga retaliated by attacking visitors with a knife and a bow and arrows, and the zoo ejected him.

Black New Yorkers organized a collection, which was insufficient to return him home, as he wished, but provided enough to cap his filed teeth and send him to the Virginia Theological Seminary and College, where he proved himself an able student. Benga then found work in a Lynchburg, Virginia, tobacco factory, where he fit in well as an efficient worker and a beloved Pied Piper who taught local children to fish and hunt. But he spoke often and tearfully of wishing to return home to the Congo, and when he realized he could never save enough for passage, his depression became profound. In 1916, Benga committed suicide with that ubiquitous icon of Western technological achievement, a handgun.

3.5 Apish Humans, Malthus, and Darwin

Philosophies that capture the imagination never wholly fade. From Animism to Zoroastrianism, every view known to man retains at least a few devotees. There might always be Freudians, and there will always be admirers of Freud's great imaginative and literary powers; these two, as the foregoing remarks suggest, are intimately linked. But as to Freud's claims upon truth, the judgment of time seems to be running against him. (AC Grayling, The Guardian)

Although many authors, including Muslim scholars, defended—well before **Darwin** and **Wallace**—that organisms can change during time and even that humans "derived" from non-human primates—see Malik et al. 2017 paper—it is evident that Darwin's 1859 *On the Origin of Species* was a landmark work that had an astonishing impact on discussions about our place in nature and that literally changed human history. In the present chapter, I just want to emphasize the historical links between Tyson's 1699 monograph, Darwin's 1859 book, and the history of racism, as these two books were crucial to destroy, scientifically, teleological narratives about how humans were "special" or "made in the image" of God. While most people knows about Darwin's book and is not familiar with Tyson's monograph, the reality is that Tyson's detailed anatomical comparisons between humans, apes and monkeys were crucial to pave the way for Darwin's publications. Grounded in this foundational knowledge, Darwin then expanded the comprehension about our place in nature, by providing a naturalistic, rational way—instead of supernatural or vitalistic ideas—to explain how evolution occurs, including how humans evolved from non-human primates, through **natural selection**. However, because of Darwin's obsession with the notion of a "struggle for life" based on the Malthusian theory of population growth, unfortunately his works provided, in at least some cases, easy ammunition for authors defending extremist racist ideas. It is well known how Darwin's ideas—in particular, about humans being derived from apes—provoked passionate and even violent reactions from not only theologians but also the broader public, as well as from researchers from various fields of science, including the most renowned comparative anatomist of that time, **Richard Owen**. The discomfort created by Darwin's ideas was not merely due to the fact that one would no longer need to evoke **supernatural beings** to discuss and/or explain our place in Nature. It was also due to the fact that his ideas about **evolutionary randomness** and chance put in question our longstanding tendency to create teleological narratives about the "cosmic purpose" and "specialness" of human beings. The idea that we are merely one of many primate species and that our evolution was not related to a cosmic purpose or a noble "masterplan," was inconceivable for a huge portion of the broader public—as it continues to be today—and the scientific community at that time. This is one of the reasons why Darwin's opponents were particularly furious with his 1871 book *The Descent of Man*, and with **Huxley**'s 1863 *Evidence as to Man's Place in Nature*.

This was mainly because, as it had happened with Tyson's 1699 book, Darwin's and Huxley's works had the double effect of leading many scholars to defend even more vigorously either the continuity between humans and other animals—for instance, those opposing his evolutionary theory—or the discontinuity between Europeans and other humans—for example, those accepting, or misinterpreting, his theory. Indeed, within the latter, many started to defend **human polygenism** by using evolutionary trees to "explain" for example how "blacks" are part of a "species" that is more closely related to chimpanzees than to the "species" including "whites." It is important to note that Darwin never defended such an absurd idea. Concerning **eugenics** and its links to the ideas of Darwin and his followers, it should also be clarified that in a wide sense eugenics is usually broadly defined as "a set of beliefs and/or practices that aims at improving the genetic quality of a human population" and that countless biologists could be somehow considered to be eugenicists in the early 1900s. So, not all **eugenicists** were necessarily racist, although most obviously were, as were the vast majority of biologists, and scientists in general, by then. But it is true that many eugenicists did not necessarily approve negative measures such as **sterilization** or murder of "others," being instead more focused—or at least they said they were—on using science to "improve" qualities via, for instance, the production of "positive traits," as explained in Buklijas and Gluckman's 2013 book chapter "*From evolution and medicine to evolutionary medicine*."

What can be said for sure is that, by stressing over and over the notion of "struggle for life," Darwin's did unfortunately provide a line of argument to those members of **eugenics movements** that became so highly influential in the U.S. and then in other countries such as Germany, before and during the **Second World War**. In fact, nowadays when scholars refer to Darwin's 1859 book, they almost always refer to its title as *On the Origin of Species by Means of Natural Selection*. Most are unaware—or don't want to make others aware—that its full title was actually *On the Origin of Species by Means of Natural Selection, or, the Preservation of Favored Races in the Struggle for Life*. As put by Corbey in his 2005 book *The Metaphysics of Apes*, "the Darwinian perception of nature as competition provided new support to the age-old icon of a beastly, humanlike, and now preferably apish Other." It is therefore not a coincidence that particularly in the end of the nineteenth century, many ethnographical books started to portray non-European men carrying weapons, to emphasize the "aggressive practices of savages," such as warfare and hunting. Many scholars begun to see different so-called races or other "groups" competing against one another in a selfish, desperate way, an idea that is related to what is nowadays often designated as "racial Darwinism" or "social Darwinism." As explained by Andreassen in *The Invention of Race*, at the end of the nineteenth century, "indigenous people were literally being exterminated by white colonizers in Australia, but their extermination was not understood as a result of the atrocities being committed against them but rather as a result of biological determinism that mandated that the stronger (white) race survive while the weaker race (of color) disappeared."

It is thus also not a coincidence that at the same time that non-Europeans were increasingly represented as aggressive and as carrying weapons in the Western media and literature, the exact same pattern was applied to great apes in the end of the nineteenth and first half of the twentieth century, as emblematically illustrated by the 1933 movie ***King Kong***. A very interesting book about this movie, and its links with, and influence on, popular culture is Erb's 1998 *Tracking King Kong – A Hollywood Icon of World Culture*. The image of King Kong shown in Fig. 3.16 is not at all a mere coincidence: the message is that the "other"—clearly, here King Kong is representing not only apes but also "blacks"—cannot just be ignored, or even segregated. This is because even if you do so, he will not just stay "there," in the segregated place that you have designated for "others." No, he will leave that place and aggressively come to "your" cities, to "your" neighborhoods, and will destroy "your" civilization—represented by a plane, one of the main symbols of Western "civilization" in the 1930s—and take and eventually even rape "your" women—represented not simply by a Western woman, but by a *Western blonde* woman, in case there were still some doubts about the meaning of this key message. So, there is only a thing you can do: the "other" has to be exterminated, for instance, by sterilization as proposed by many followers of the **eugenics movement** at that time, or even by direct murder and **genocide**, as was done just a few years later by the **Nazis**. Indeed, particularly during **colonial expansion** in the second half of the nineteenth and the first half of the twentieth centuries, there was a tendency to use overlapping textual and visual representations of both apes and so-called lower human races conceived as living ancestral forms that were wild, savage, and *aggressive*. Such a view clearly contrasts with the views of earlier writers such as **Rousseau** who tended to emphasize the peaceful behavior of both apes and "Noble savages" and with the type of scientific illustrations of apes provided by earlier scholars such **Tulp** in 1641 and **Tyson** in 1699 (Figs. 3.9 and 3.10).

In summary, it was precisely the reactionary response of humans, and in particular of so-called civilized Westerners, to the revolutionary way in which these illustrations portrayed apes, and particularly to Tyson's groundbreaking conclusions that apes are more similar anatomically to humans then they are to other animals that in great part lead to the very aggressive way in which King Kong in particular and "others" in general—both apes and "non-white" humans—were portrayed more than two centuries after Tulp's and Tyson's works (Fig. 3.16, compare with Figs. 3.9 and 3.10). As put by Corbey in *The Metaphysics of Apes*, as happened with the so-called lower human races, great apes and particularly gorillas "came to be seen as powerful personifications of wildernesses to be fought heroically and conquered by civilized Westerners." One of the most influential channels through which the "**beast-in-man stereotype**" spread from the nineteenth into the twentieth century

Fig. 3.16 Austrian poster to advertise the 1933 movie *King Kong*

scientific and cultural discourse was the **psychoanalysis Sigmund Freud**. In fact, Freud provides an emblematic illustration of the type of scholars that became so influential at that time: a "white" entitled man that was both profoundly racist and misogynist and that subscribed to the imaginary teleological tales about "progress" in human evolution. In that sense, calling Freud a "scientist" as most of the media and textbooks continue to do, or at least his "**psychoanalytic theories**," clearly is too far-stretched, as those theories were not only mostly factually wrong and based in such imaginary tales, but were not tested and supported using the scientific method: they were mostly fictional, non-testable, just-so-stories, as more and more psychologists finally started to publicly recognize in the last decades.

This tendency was summarized in a short article published in *The Guardian* in 2002, entitled "*Scientist or storyteller?*" which referred to the publication by the publisher company Penguin of the major translations of Freud's work for over 30 years under the general editorship of Adam Phillips, which prompted "serious questions about the nature of Freud's contribution and his legacy." As correctly put in the last paragraph of the article, Freud's "psychoanalytic theories" are not chiefly only based on, but are in themselves, nothing more than imaginary tales: "*Philosophies that capture the imagination never wholly fade. From Animism to Zoarastrianism, every view known to man retains at least a few devotees. There might always be Freudians, and there will always be admirers of Freud's great imaginative and literary powers; these two, as the foregoing remarks suggest, are intimately linked. But as to Freud's claims upon truth, the judgment of time seems to be running against him.*" Among the huge number of recent books—finally—recognizing that Freud was not only "not scientific" but also, in many instances, "a liar and a fraud" and "putting the final nail in his coffin," I recommend Frederick Crews' 2017 book *Freud: The Making of an Illusion*. This is because that book directly connects these topics with a key issue discussed in the present volume: how **Freud**'s huge success and influence were not only due to his use of very appealing imaginary teleological stories but also due to the fact that many scientists and most lay people, so immersed in *Neverland*'s illusions and delusions, were so eager and even anxious to read about and believe in such fairytales.

The overlap between narratives about apes and "non-white" humans is clearly evidenced by the common use of terms such as "**apish humans**" in a huge number of scientific and literary works in the last two centuries and continues to have atrocious consequences at numerous levels, including in American popular culture. Detailed archival content analyses reveal that news articles keep on creating implicit associations between "black criminals" and apes and that those identified as more "**ape-like**" are more likely to be executed, as reported in Sorenson's 2009 book *Ape*. A particularly illustrative and upsetting example of the links between the discovery of non-human primates by Westerners, science, **comparative anatomy**, **colonialism**, and also animal abuse concerns the **colonial propaganda** film made in the 1950s in the **Belgian Congo**. The film was made on behalf of the Belgian government and circulated broadly in Belgian cinemas, programmed on Sunday afternoons for families with children. As described by Corbey, "the footage shows in great and, by present-day standards, shocking detail how scientists of the **Royal Belgian Institute of Natural Sciences** shoot and kill and adult female gorilla carrying young; subsequently the body is skinned and washed in a nearby stream, with the distressed youngster sitting next to it; the adult's skeleton, skin and other body parts were collected for scientific [anatomical] study and conservation, while the live young gorilla was sent to the **Antwerp zoo**."

In his 2016 book *Stamped from the Beginning*, Kendi shows how the history of **colonialism** and the **Trans-Atlantic slave trade**, as well as of **American slavery**, **oppression**, and **discrimination**, was deeply related with the two types of racism discussed above: **type B/"innate" racism** was defended by what he calls the "**segregationists**," while the more ancient **type A/"cultural-epigenetic" racism** was defended by what he designates as the "**assimilitationists**." One of the most original and remarkable aspects of his book is that it reveals that even those scholars, politicians, and activists that history books tend to praise by being more "progressive" and "humanist" actually tend to be, at least in some respects, assimilationists. That is, according to him, they are in fact not able to escape, themselves, from the more ancestral type of cultural racism: for him, this includes even some of the narratives employed or defended by people like **Barack Obama**. In this sense, Kendi's book allows us to understand the history of racism and "racial progress" in a completely different, more holistic and profound, way:

> The title *Stamped from the Beginning* comes from a speech that Mississippi senator Jefferson Davis gave on the floor of the US Senate on April 12, 1860. This future president of the **Confederacy** objected to a bill funding Black education in Washington, DC. "This Government was not founded by negroes nor for negroes", but "by white men for white men", Davis lectured his colleagues. The bill was based on the false notion of racial equality, he declared. The "inequality of the white and black races" was "stamped from the beginning". It may not be surprising that **Jefferson Davis** regarded Black people as biologically distinct and inferior to White people – and Black skin as an ugly stamp on the beautiful White canvas of normal human skin – and this Black stamp as a signifier of the Negro's everlasting inferiority. This kind of **segregationist thinking** is perhaps easier to identify – and easier to condemn – as obviously racist…. Historically, there have been three sides to this heated argument. A group we can call **segregationists** has blamed Black people themselves for the racial disparities. A group we can call antiracists has pointed to racial discrimination. A group we can call **assimilationists** has tried to argue for both, saying that Black people and racial discrimination were to blame for racial disparities. And yet so many prominent Americans, many of whom we celebrate for their progressive ideas and activism, many of whom had very good intentions, subscribed to assimilationist thinking that also served up racist beliefs about Black inferiority. We have remembered assimilationists' glorious struggle against racial discrimination, and tucked away their inglorious partial blaming of inferior Black behavior for racial disparities. In embracing biological racial equality, assimilationists point to environment – hot climates, discrimination, culture, and poverty – as the creators of inferior Black behaviors. For solutions, they maintain that the ugly Black stamp can be erased – that inferior Black behaviors can be developed, given the proper environment. As such, assimilationists constantly encourage Black adoption of White cultural traits and/or physical ideals. In his landmark 1944 study of race relations, a study widely regarded as one of the instigators of the civil rights movement, Swedish economist and Nobel Laureate Gunnar Myrdal wrote, "It is to the advantage of American Negroes as individuals and as a group to become assimilated into American culture, to acquire the traits held in esteem by the dominant white Americans". He had also claimed, in *An American Dilemma*, that "in practically all its divergences, American Negro culture is…a distorted development, or a pathological condition, of the general American culture".
>
> There was simple or straightforward or predictable about racist ideas, and thus their history. Frankly speaking, for generations of Americans, racist ideas have been their common sense. The simple logic of racist ideas has manipulated millions over the years, muf-

fling the more complex antiracist reality again and again. And so, this history could not be made for readers in an easy-to-predict narrative of absurd racists clashing with reasonable antiracists. This history could not be made for readers in an easy-to-predict, two-sided Hollywood battle of obvious good *versus* obvious evil, with good triumphing in the end. From the beginning, it has been a three-sided battle, a battle of antiracist ideas being pitted against two kinds of racist ideas at the same time, with evil and good failing and triumphing in the end. Both segregationist and assimilationist ideas have been wrapped up in attractive arguments to seem good, and both have made sure to re-wrap antiracist ideas as evil…. In colonizing Virginia (and later New England), the British had already begun to conceive of distinct races. The word race first appeared in Frenchman Jacques de Brézé's 1481 poem "The Hunt", where it referred to hunting dogs. As the term expanded to include humans over the next century, it was used primarily to identify and differentiate and animalize African people. The term did not appear in a dictionary until 1606, when French diplomat **Jean Nicot** included an entry for it. "Race…means descent", he explained, and "it is said that a man, a horse, a dog or another animal is from good or bad race"…. In 1683, Increase and **Cotton Mather** founded colonial America's first formal intellectual group, the *Boston Philosophical Society*. Modeled after London's Royal Society, the Boston Society lasted only four years. The Mathers never published a journal, but if they had, they might have modeled it after the Royal Society's Philosophical Transactions, or the Journal des Savants in Paris. These were the organs of Western Europe's scientific revolution, and new ideas on race were a part of that revolution….

One of the early leaders of the Royal Society was one of England's most celebrated young scholars, the author of *The Sceptical Chymist* (1661) and the father of English chemistry – **Robert Boyle**. In 1665, Boyle urged his European peers to compile more "natural" histories of foreign lands and peoples, with Richard Ligon's *Historie of Barbados* serving as the racist prototype. The year before, Boyle had jumped into the ring of the racial debate with *Of the Nature of Whiteness and Blackness*. He rejected both curse and climate theorists and knocked up a foundational antiracist idea: "The Seat" of human pigmentation "seems to be but the thin Epidermes, or outward Skin," he wrote. And yet, this antiracist idea of skin color being only skin deep did not stop Boyle from judging different colors. Black skin, he maintained, was an "ugly" deformity of normal Whiteness. The physics of light, Boyle argued, showed that Whiteness was "the chiefest color". He claimed to have ignored his personal "opinions" and "clearly and faithfully" presented the truth, as his Royal Society deeded. As Boyle and the Royal Society promoted the innovation and circulation of racist ideas, they promoted objectivity in all their writings. Intellectuals from Geneva to Boston, including Richard Mather's youngest son, Increase Mather, carefully read and loudly hailed Boyle's work in 1664. A twenty-two-year-old unremarkable Cambridge student from a farming family copied full quotations. As he rose in stature over the next forty years to become one of the most influential scientists of all time, **Isaac Newton** took it upon himself to substantiate Boyle's color law: light is white is standard. In 1704, a year after he assumed the presidency of the Royal Society, Newton released one of the most eminent books of the modern era, *Opticks*. "Whiteness is produced by the Convention of all Colors", he wrote. Newton created a color wheel to illustrate his thesis. "The center" was "white of the first order", and all the other colors were positioned in relation to their "distance from Whiteness". In one of the foundational books of the upcoming European intellectual renaissance, Newton imaged "perfect whiteness." Thanks to this malleable concept in Western Europe, the British were free to lump the multiethnic Native Americans and the multiethnic Africans into the same racial groups. [Centuries later] assimilationists first used and defined and popularized the term "racism" during the 1940s. All the while, they refused to define their own assimilationist ideas of Black behavioral inferiority as racist. These assimilationists defined only segregationist ideas of Black biological inferiority as racist. And segregationists, too, have always resisted the label of "racist". They have claimed instead that they were merely articulating God's word, nature's design, science's plan, or plain old common sense.

Fig. 3.17 Drawing of the facial muscles by Huber, showing what he considered to be "significant racial differences" between a so-called adult male negro (on the left) and a so-called adult male Australian (on the right)

In a 2010 paper entitled "*Comparative anatomy, anthropology and archaeology as case studies on the influence of human biases in natural sciences: the origin of 'humans', of 'behaviorally modern humans' and of 'fully civilized humans'*," I discuss how the complex interplay between the agendas and **biases** of scientists, politicians, financial markets, and the media and broader public played a key role in the rise of racism and eugenics in the last half of the nineteenth and first half of the twentieth centuries. One powerful example is the text and figures of **Huber**'s 1931 book *Evolution of Facial Musculature and Expression*, which he used to defend that "blacks" and Australian natives are more similar to chimpanzees than to "whites" not only morphologically but also "mentally" (see Fig. 3.17). As noted by him, the "first three decades of the twentieth century" were a particularly "active period of racial anatomical research on **facial musculature**," because the expressions of the face were said to be a window to our "emotional and mental" states. He stated that "the facial musculature of the adult American Negro is generally composed of bundles which are much coarser and also darker in color than are those of the White…in a prevailing percentage of White cases, the zygomaticus musculature has reached a *higher stage* in evolution."

In a particularly appalling example of how **racial determinism** and the deep links between anatomy and the "social/moral" proposed by **Linnaeus** were so prevalent in the first half of the twentieth century, Huber stated: "in the responsive faces of Whites we notice, especially in the upper region of the face and about the mouth, a great range of varied expressions with many modulations; the mouth, even closed, may serve as an admirable index of character or mental state through a slightly increased tonus of its musculature…the smile turns into a happy, hearty laugh." In contrast, "apparently nerve impulses that are less finely graded reach the respective mimetic muscle groups, thus setting them into sudden, strong contraction which rather suggests more primitive muscle actions [of the 'negroes']; the expression is characteristic…the large white teeth show in vivid contrast to the dark face; instead of grades laugh typical of the white we notice *the characteristic grinning of the negro, and through sounds, often simultaneously uttered, which differ in tone of voice from those of the white, the negro's grinning becomes even more characteristic*." What is most striking about such racist quotes is that they were not written by only a few, unknown researchers, but instead by the most eminent, respected, and influential scientists of the first decades of the twentieth century. That is, although almost all the biologists of that time accepted Darwin's evolutionary ideas, they were *in reality* more interested in stressing the "astonishing" differences between the "civilized" Europeans versus the "savage" non-European humans and non-human primates than to focus on the biological continuity between humans and other primates. And what is also extremely disturbing is that scientists were so biased by their prejudices and a priori teleological narratives about "progress" that they could inclusively provide wrong data to support their ideas, either unconsciously because of their biases or even intentionally, by fabricating "evidence" or alternatively by ignoring empirical data, as explained in my 2010 paper and in the powerful example about inoculation discussed in Box 3.11.

Box 3.11 Medicine, Inoculation, African Medical Practices, and Scientific Racism

In Kendi's 2016 book *Stamped from the Beginning*, he provides a very powerful example of the link between racist narratives and stereotypes and **scientific biases**:

> On April 21, 1721, the HMS Seahorse sailed into Boston Harbor from Barbados. A month later, **Cotton Mather** logged in his journal, "the grievous calamity of the smallpox has now entered the town". One thousand Bostonians, nearly 10 percent of the town, fled to the countryside to escape the judgment of the Almighty. Fifteen years prior, Mather had asked **Onesimus** one of the standard questions that Boston

slaveholders asked new house slaves – Have you had **smallpox**? "Yes and no", Onesimus answered. He explained how in Africa before his enslavement, a tiny amount of pus from a smallpox victim had been scraped into his skin with a thorn, following a practice hundreds of years old that resulted in building up healthy recipients' immunities to the disease. This form of **inoculation** – a precursor to **modern vaccination** – was an innovative practice that prevented untold numbers of deaths in West Africa and on disease-ridden slave ships to ports throughout the Atlantic. Racist European scientists at first refused to recognize that African physicians could have made such advances. Indeed, it would take several decades and many more deaths before British physician Edward Jenner, the so-called father of **immunology,** validated inoculation. Cotton Mather, however, became an early believer when he read an essay on inoculation in the *Royal Society's Philosophical Transactions* in 1714. He then interviewed Africans around Boston to be sure. Sharing their inoculation stories, they gave him a window into the intellectual culture of West Africa. He had trouble grasping it, instead complaining about how "brokenly and blunderingly and like Idiots they tell the Story".

On June 6, 1721, Mather calmly composed an "Address to the Physicians of Boston", respectfully requesting that they consider inoculation. If anyone had the credibility to suggest something so new in a time of peril it was Cotton Mather, the first American-born fellow in London's Royal Society, which was still headed by Isaac Newton. Mather had released fifteen to twenty books and pamphlets a year since the 1690s, and he was nearing his mammoth career total of 388 – probably more than the rest of his entire generation of New England ministers combined. The only doctor who responded to Mather was Zabadiel Boylston, President John Adams's great-uncle. When Boylston announced his successful inoculation of his six-year-old son and two enslaved Africans on July 15, 1721, area doctors and councilmen were horrified. It made no sense that people should inject themselves with a disease to save themselves from the disease. Boston's only holder of a medical degree, a physician pressing to maintain his professional legitimacy, fanned the city's flames of fear. Dr. **William Douglass** concocted a conspiracy theory, saying there was a grand plot afoot among African people, who had agreed to kill their masters by convincing them to be inoculated. "There is not a Race of Men on Earth more False Liars" than Africans, Douglass barked.

Fortunately, since the 1950s, there has been a dramatic change of mindset in a substantial part of the scientific community and a still small but increasing proportion of the broader public, towards a vision of continuity and unity both between all extant human groups and between them and other primates. This change was in great part influenced by the aim of not repeating the errors that lead to ideas that were used to justify the atrocities committed during **World War II**. The **UNESCO Statement on Race**—an official declaration against racism that attempted to break the connection between "race" and **biological determinism**—was published in 1950, becoming an extremely important turning point, as explained in Brattain's 2007 paper "*Race, racism, and anti-racism: UNESCO and the politics of presenting science to the postwar public*." In fact, in many countries, **eugenics** was already losing some ground as early as the 1930s because geneticists could not scientifically support any of the key conceptual tenets defended by eugenicists, and also because of the socioeconomic crisis that affected many of these countries in that decade, as explained in Kevles' 1998 book *In the Name of Eugenics*. This change of attitude also led to—and then was interactively further expanded by—new and less biased comparative anatomical, behavioral, and genetic works on primates, such as the groundbreaking behavioral works of apes done by researchers such as **Dian Fossey**, **Birute Galdikas**, and **Jane Goodall** in the second half of the twentieth century.

One crucial consequence of those changes is that many researchers—such as **De Waal**, as emphasized in his 2016 book *Are We Smart Enough to Know How Smart Animal Are?*—are now recognizing that a huge historical problem of behavioral studies has been not so much that they tended to **anthropomorphize non-human primates**, as previously thought, but instead the non-acceptance that these primates are in fact so much like us: that is, that we are not so "special" or "unique" after all. Indeed, we now know that apes also use tools, display highly complex behaviors including some related to "fairness," "morality," and altruism, have similar emotions and facial expressions, plan tasks in advance, are able to deal with abstract concepts, and so on. A broader reflection of this change of mindset is ***The Great Ape Project***, which calls for great apes to be accorded the same basic rights as humans. Accordingly, more and more authors are calling for humans and chimpanzees to be placed in the same genus, as reflected by the provocative title of Diamond's excellent 1991 book about some of these topics, *The Rise and Fall of the Third Chimpanzee*.

However, this surely does not mean that "all is well now." Racism is still prevalent in most countries, and it is alarming that there is a re-appearance of attitudes and patterns—across the globe, including the country where I currently live, the U.S.—that are in some ways so disturbingly similar to those that were commonly seen in the first half of the twentieth century. Indeed, as shown in Kendi's 2016 *Stamped from the Beginning*, while there are some positive aspects in the fight against racism and discrimination in the U.S., as illustrated by **Obama's presidential victory**, there are also some disturbingly negative ones such as the rise of racist reactions already before, and in particular after, that victory, which likely contributed in great part to the subsequent re-emergence of white supremacy groups and victory of Trump. As explained by him, everything is not yet well, very far from it:

> As the economic and racial disparities grew and middle-class incomes became more unstable in the late 1970s and early 1980s, old segregationist fields – like **evolutionary psychology**, preaching genetic intellectual hierarchies, and **physical anthropology**, preaching biological racial distinctions – and new fields, like sociobiology, all seemed to grow in popularity. After all, new racist ideas were needed to rationalize the newly growing disparities. Harvard biologist **Edward Osborne Wilson**, who was trained in the dual-evolution theory, published *Sociobiology: The New Synthesis* in 1975. Wilson more or less called on American scholars to find "the biological basis of all forms of social behavior in all kinds of organisms, including man". Though most **sociobiologists** did not apply **sociobiology** directly to race, the unproven theory underlying sociobiology itself allowed believers to apply the field's principles to racial disparities and arrive at racist ideas that blamed Blacks' social behavior for their plight. It was the first great academic theory in the post-1960s era whose producers tried to avoid the label "racist". Intellectuals and politicians were producing theories – like welfare recipients are lazy, or inner cities are dangerous, or poor people are ignorant, or one-parent households are immoral – that allowed Americans to call Black people lazy, dangerous, and immoral without ever saying "Black people", which allowed them to deflect charges of racism. Assimilationists and antiracists, realizing the implica-

tions of Sociobiology, mounted a spirited reproach, which led to a spirited academic and popular debate over its merits and political significance during the late 1970s and early 1980s. Harvard evolutionary biologist **Stephen Jay Gould**, who released *The Mismeasure of Man* in 1981, led the reproach in the biological sciences against segregationist ideas. Edward Osborne Wilson, not to be deterred, emerged as a public intellectual. He no doubt enjoyed hearing Americans say unproven statements that showed how popular his theories had become, such as when someone quips that a particular behavior "is in my DNA". He no doubt enjoyed, as well, taking home two Pulitzer Prizes for his books and a National Medal of Science from President **Jimmy Carter**. Wilson's sociobiology promoted but never proved the existence of genes for behaviors like meanness, aggression, conformity, homosexuality, and even xenophobia and racism....

Weeks after passing the most antiracist bill of the decade over **Reagan**'s veto – the Comprehensive Anti-Apartheid Act with its strict economic sanctions – Congress passed the most racist bill of the decade. On October 27, 1986, Reagan, "with great pleasure", signed the **Anti-Drug Abuse Act**, supported by both Republicans and Democrats. "The American people want their government to get tough and to go on the offensive", Reagan commented. By signing the bill, he put the presidential seal on the "Just say no" campaign and on the "tough laws" that would now supposedly deter drug abuse. While the AntiDrug Abuse Act prescribed a minimum five-year sentence for a dealer or user caught with five grams of crack, the amount typically handled by Blacks and poor people, the mostly White and rich users and dealers of powder cocaine – who operated in neighborhoods with fewer police – had to be caught with five hundred grams to receive the same five-year minimum sentence. Racist ideas then defended this racist and elitist policy. The bipartisan act led to the mass incarceration of Americans.

The prison population quadrupled between 1980 and 2000 due entirely to stiffer sentencing policies, not more crime. Between 1985 and 2000, **drug offenses** accounted for two-thirds of the spike in the inmate population. By 2000, Blacks comprised 62.7 percent and Whites 36.7 percent of all drug offenders in state prisons – and not because they were selling or using more drugs. That year, the National Household Survey on **Drug Abuse** reported that 6.4 percent of Whites and 6.4 percent of Blacks were using illegal drugs. Racial studies on drug dealers usually found similar rates. One 2012 analysis, the National Survey on Drug Use and Health, found that White youths (6.6 percent) were 32 percent more likely than Black youths (5 percent) to sell drugs. But Black youths were far more likely to get arrested for it. During the crack craze in the late 1980s and early 1990s, the situation was the same. Whites and Blacks were selling and consuming illegal drugs at similar rates, but the Black users and dealers were getting arrested and convicted much more. In 1996, when two-thirds of the crack users were White or Latina/o, 84.5 percent of the defendants convicted of crack possession were Black. Even without the crucial factor of racial profiling of Blacks as drug dealers and users by the police, a general rule applied that still applies today: wherever there are more police, there are more arrests, and wherever there are more arrests, people perceive there is more crime, which then justifies more police, and more arrests, and supposedly more crime.... After all, African Americans possessed 1 percent of the national wealth in 1990, after holding 0.5 percent in 1865, even as the Black population remained at around 10 to 14 percent during that period.

Fourteen years after 1990, the situation was as shown in Fig. 3.18. And 30 years after 1990, while I am writing these lines, the situation is indeed far from being acceptable, contrary to reactionary narratives about how we now mainly live in a "post-racial" time and how one should no longer find "excuses" for the **inequality between "whites" and African-Americans** in the U.S. As explained in a report published by the Brookings Institution in February 27, 2020—based on a detailed analysis of empirical data, not on "gut feelings" or wishful thinking: "a close

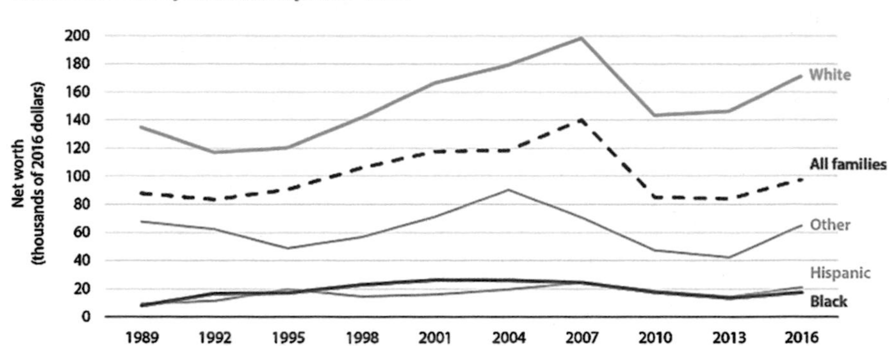

Fig. 3.18 Inequality in the U.S., in 2004

examination of wealth in the U.S. finds evidence of staggering racial disparities…at $171,000, the net worth of a typical white family is nearly ten times greater than that of a Black family ($17,150) in 2016." The report plainly states that: "gaps in wealth between Black and white households reveal the effects of accumulated **inequality and discrimination**, as well as differences in power and opportunity that can be traced back to this nation's inception…the **Black-white wealth gap** reflects a society that has not and does not afford equality of opportunity to all its citizens." Unfortunately, contrary to the erroneous notion of "linear cosmic progress," history tends instead to repeat itself. That is why historical discussions such as the ones provided in the present volume, and in works such as Kendi's splendid 2016 book, are urgently needed.

3.6 Medical Experimentation, Eugenics, and Genocide

When I began working at the institute, I recalled my adolescent dream of becoming a medical research worker…daily I saw young…[white] boys and girls receiving instruction in chemistry and medicine that the average black boy or girl could never receive…when I was alone, I wandered and poked my fingers into strange chemicals, watched intricate machines trace red and black lines upon ruled paper…at times I paused and stared at the walls of the rooms, at the floors, at the wide desks at which the white doctors sat; and I realized – with a feeling that I could never quite get used to – that I was looking at the world of another race. (Richard Wright, in 1994)

The history of racism and of teleological narratives is profoundly linked to the history of forced **medical experimentation** and to **genocides**. Many details about these links are particularly atrocious, concerning some of the most horrible things ever done by the self-proclaimed "sapient being," and rising very difficult philosophical and ethical questions. While we all—hopefully—agree that forced medical experimentations should never be done, should the information obtained from dreadful medical experiments *that were already done*, such as those forcibly done

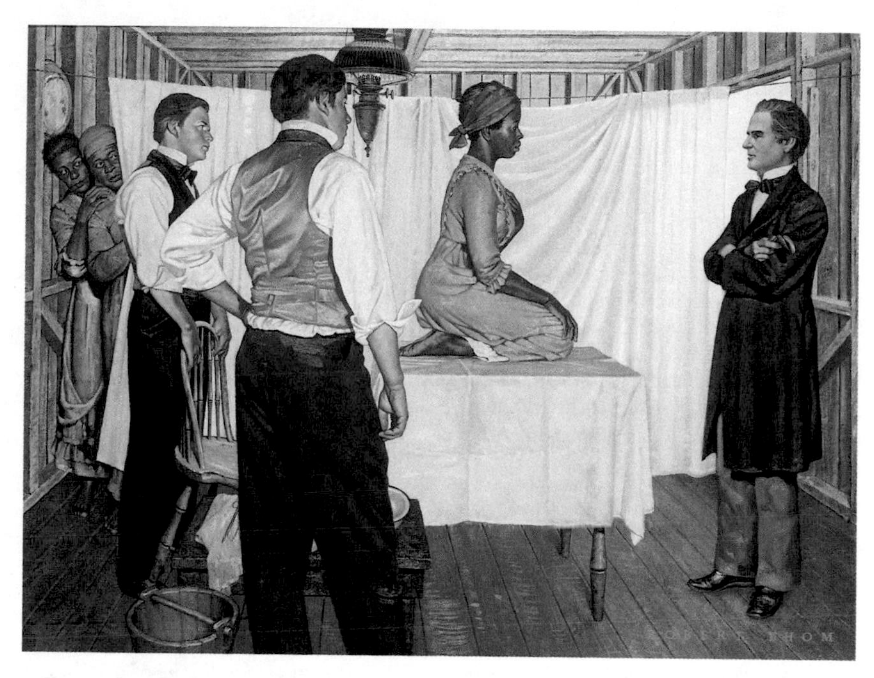

Fig. 3.19 Robert Thom's "J. Marion Sims: Gynecologic Surgeon," an oil representation of an experimental surgery upon a powerless slave, from Thom's *The history of medicine*, circa 1952, archived at the University of Michigan

with African slaves in the U.S. or Jews and Gypsies in Nazi Germany—often without any type of anesthesia or respect for human lives and bodies—be used to potentially save lives, or not? As will be seen below, information obtained from doing horrible experiments on "others" has indeed contributed to save lives not only among the "ingroup" oppressing population but also, more recently, among groups of "others," increasing even more the complexity and difficulty of such questions. As pointed out by Washington in her excellent 2006 book *Medical Apartheid*, there are several books that discuss such highly sensitive and disturbing issues, but strangely—or maybe predictably, because of the discomfort caused by them in scholars and historians from the "ingroup"—there are very few that actually discuss them in a broader way and within a comprehensive historical context. Although Washington's book was published 14 years ago, it directly relates with events that occurred while I was writing these lines in 2020, concerning the ***Black Lives Matter*** movement and the removal or destruction, by some people, of statues of so-called historical personalities that are seen by many among the "in-group majority" as "heroes," such as James Marion Sims (see Fig. 3.19):

> On a sylvan stretch of New York's patrician upper Fifth Avenue, just across from the New York Academy of Medicine, a colossus in marble, august inscriptions, and a bas-relief caduceus grace a memorial bordering Central Park. These laurels venerate the surgeon **James Marion Sims**, M.D., as a selfless benefactor of women. Nor is this the only statuary erected in honor of Dr. Sims. Marble monuments to his skill, benevolence, and humanity

> guard his native South Carolina's statehouse, its medical school, the Alabama capitol grounds, and a French hospital. In the mid-nineteenth century, Dr. Sims dedicated his career to the care and cure of women's disorders and opened the nation's first hospital for women in New York City. He attended French royalty, his Grecian visage inspired oil portraits, and in 1875, he was elected president of the American Medical Association. Hospitals still bear his name, including a West African hospital that utilizes the eponymous gynecological instruments that he first invented for surgeries upon black female slaves in the 1840s. But this benevolent image vies with the detached Marion Sims portrayed in Robert Thom's J. Marion Sims: Gynecologic Surgeon, an oil representation of an experimental surgery upon his powerless slave Betsey [**Fig. 3.19**]. Sims stands aloof, arms folded, one hand holding a metroscope (the forerunner of the speculum) as he regards the kneeling woman in a coolly evaluative medical gaze. His tie and morning coat contrast with her simple servants' dress, head rag, and bare feet. The painting, commissioned and distributed by the Parke-Davis pharmaceutical house more than a century after the surgeries as one of its *A History of Medicine in Pictures* series, takes telling liberties with the historical facts. Thom portrays Betsey as a fully clothed, calm slave woman who kneels complacently on a small table, hand modestly raised to her breast, before a trio of white male physicians. Two other slave women peer around a sheet, apparently hung for modesty's sake, …each woman's body was a bloodied battleground. Each naked, unanesthetized slave woman had to be forcibly restrained by the other physicians through her shrieks of agony as Sims determinedly sliced, then sutured her genitalia. The other doctors, who could, fled when they could bear the horrific scenes no longer. It then fell to the women to restrain one another….
>
> Betsey's voice has been silenced by history, but as one reads Sims's biographers and his own memoirs, a haughty, self-absorbed researcher emerges, a man who bought black women slaves and addicted them to morphine in order to perform dozens of exquisitely painful, distressingly intimate vaginal surgeries. Not until he had experimented with his surgeries on Betsey and her fellow slaves for years did Sims essay to cure white women. Was Sims a savior or a sadist? *It depends, I suppose, on the color of the women you ask.* Marion Sims epitomizes the two faces – one benign, one malevolent – of American medical research…a doctor could be open about buying slaves for experiments, or locating or moving hospitals to areas where blacks furnished bodies for experimentation and dissection. Public Health Service physician Thomas Murrell could brashly insist in the 1940s, "The future of the Negro lies more in the research laboratory than in the schools…. When diseased, he should be registered and forced to take treatment before he offers his diseased mind and body on the altar of academic and professional education." Even more recently, the segregated nature of U.S. medical training emboldened some physicians to speak with candor of misusing black subjects. "[It was] cheaper to use Niggers than cats because they were everywhere and cheap experimental animals", neurosurgeon Harry Bailey, M.D., reminisced in a 1960s speech he delivered while at Tulane Medical School. But as societal attitudes changed, so did physician reticence, and most became more circumspect. However, as late as 1995, radiation scientist Clarence Lushbaugh, M.D., explained that he and his partner, Eugene Saenger, M.D., chose "slum" patients as radiation subjects because "these persons don't have any money and they're black and they're poorly washed"…[there were] numerous instances of such shocking frankness on the part of white researchers and physicians when they thought that nobody outside of their peer group was listening.

In this specific case, and many others that will be discussed below, Slims surely is not a "hero" when seen from the perspective of "others," particularly the slaves that he abused, or their descendants. The fact that he has been idealized by numerous "whites" for his "humanity"—by using information gained from those atrocious experiments done to slaves to help the life of "white" woman—can only make sense in a *Neverland* world where the lives of "others" are less important than those of the "ingroup," or simply do not matter at all. In a *word of reality* where all lives matter, clearly it makes no sense at all to have such a statue being displayed in public spaces and particularly in front or across public buildings such as the New York Academy of Medicine, where African Americans that are descendants of those slaves that were obliged to undertake such atrocious experiments pass by, on a daily basis. This is because such statues are sending a powerful message: that they were done by, and *for*, "whites," and that "others" don't matter at all. The following excerpts of Washington's book also directly connect past events to what is occurring right now, during the **Covid-19** pandemic, which affected a much higher proportion of African-Americans than of the rest of the population living in the U.S., stressing that health and economic disparities are still so profound in this country:

> "Of all the forms of **inequality**, injustice in health is the most shocking and the most inhumane". In 1965, **Martin Luther King**, Jr., spoke these words in Montgomery, Alabama, at the end of the ***Selma to Montgomery* march** that had been attended by the black and white physicians of the Medical Committee for Human Rights. King had invited the doctors not only to give medical succor to injured marchers but also to witness the abuse suffered at the hands of segregationists. With these almost unnoticed words, King ushered in a new era in civil rights, because as Delegate to Congress Donna Christian-Christensen, M.D., chair of the Congressional Black Caucus Health Braintrust, has declared, "Health disparities are the civil rights issue of the 21st century". Thus Dr. King's alarm over racial health injustice was prescient, and were he alive today, his concern would be redoubled. Mounting evidence of the racial health divide confronts us everywhere we look, from doubled black infant death rates to **African American life expectancies** that fall years behind whites. **Infant mortality** of African Americans is twice that of whites, and black babies born in more racially segregated cities have higher rates of mortality. The life expectancy of African Americans is as much as six years less than that of whites. Old measures of health not only have failed to improve significantly but have stayed the same: some have even worsened. Mainstream newspapers and magazines often report disease in an ethnocentric manner that shrouds its true cost among African Americans.
>
> For example, despite the heavy emphasis on genetic ailments among blacks, fewer than 0.5 percent of black deaths – that's less than one death in two hundred – can be attributed to hereditary disorders such as **sickle-cell anemia**. A closer look at the troubling numbers reveals that blacks are dying not of exotic, incurable, poorly understood illnesses nor of genetic diseases that target only them, but rather from common ailments that are more often prevented and treated among whites than among blacks. Three times as many African Americans were diagnosed with **diabetes** in 1993 as in 1963. This rate is nearly twice that of white Americans and is sorely underestimated: the real black diabetes rate is probably double that of whites. As with most **chronic diseases**, African Americans suffer more complications, **including limb loss**, **blindness**, **kidney disease**, and **terminal heart disease**. **Cancer**, the nation's second greatest killer, is diagnosed later in blacks and carries off proportionately more African Americans than whites. African Americans suffer the nation's highest rate of cancer and **cancer deaths**.
>
> The distortion of African American death rates is illustrated by the common dismissal of black women's **breast-cancer** risks as "lower than white women's". This characterization implies that black women are at low risk from **breast cancer**, but their risk is only slightly lower, because the estimated lifetime risk of developing breast cancer is ten per one hundred for white women born in 1980, and seven per one hundred for black women born that year. Moreover, this lower risk of developing breast cancer is overshadowed by blacks' much higher risk of dying from it: eighty-six percent of white women with breast cancer are alive five years later; only 71 percent of black women survive that long. A black woman is 2.2 times as likely as a white woman to die of breast cancer. Black women have been under-

going mammograms at the same rate as white women but are more likely to receive poorer quality screening, which may not detect a cancer in time for a cure. A black woman is also more likely to develop her cancer before age forty, too early for recommended mammograms to catch it, and black women are diagnosed at a more advanced stage than either Hispanic or white breast-cancer patients. Black breast-cancer patients have a worse overall prognosis, and a worse prognosis at each stage. Black men have the nation's highest rates of developing and of dying from prostate and lung cancers. Despite its image as a disease that affects middle-aged white men, **heart disease** claims 50 percent more African Americans than whites and African Americans die from heart attacks at a higher rate than whites. African Americans are more likely to develop serious liver ailments such as **hepatitis C**, the chief cause of liver transplants. They are also more likely to die from **liver disease**, not because of any inherent racial susceptibility, but because blacks are less likely to receive aggressive treatment with drugs such as interferon or lifesaving liver transplants.

Washington further notes:

Even the legion of newest illnesses – emerging disease such as **HIV/AIDS** and **hepatitis C** – kills blacks at much higher rates than whites. AIDS, the scourge of our time, has become a disease of people of color here and abroad: forty-nine percent of HIV infected Americans are African Americans and 86 percent of children with AIDS are African American or Hispanic. Blacks are ten times as likely to develop AIDS as whites. **Mental ailments** are destroying blacks, as well: black women suffer the highest rates of stress and major depression in the nation and suicide rates soared 200 percent among young black men within just twenty years. These are dire statistics, born of complex interactions among unhealthy environments, social pressures and limitations, lifestyle factors, and limited access to health care, including very limited access to cutting-edge therapeutic medical research that is meant to help treat or cure a patient with a disorder. But this dearth of therapeutic research is accompanied by a plethora of nontherapeutic research with African Americans, which is meant to investigate medical issues for the benefit of future patients or of medical knowledge. And this brings us to…a peculiar type of injustice in health: the troubled history of medical experimentation with African Americans – and the resulting behavioral fallout that causes researchers and African Americans to view each other through jaundiced eyes. In his 1909 preface to *The doctor's dilemma*, George Bernard Shaw scathingly observed, "the tragedy of illness at present is that it delivers you helplessly into the hands of a profession which you deeply mistrust". He could have been speaking for contemporary African Americans, because studies and surveys repeatedly confirm that no other group as deeply mistrusts the American medical system, especially medical research. The problem is growing.

As the *Wall Street Journal* observed several years back, "it hasn't been a good time for scientists who experiment on people – or the people they experiment on". This is a masterpiece of understatement, especially if you consider the recent history of medical research with African Americans. The *Office for Protection from Research Risks (OPRR)* has been busily investigating abuses at more than sixty research centers, including experimentation-related deaths at premier universities, from Columbia to California. Another important subset of human subject abuse has been scientific fraud, wherein scientists from the University of South Carolina to MIT have also been found to have lied through falsified data or fictitious research agendas, often in the service of research that abused black Americans. Within recent years, the OPRR has also suspended research at such revered universities as Alabama, Pennsylvania, Duke, Yale, and even Johns Hopkins. Many studies enrolled only or principally African Americans, although some included a smattering of Hispanics. Some research studies specifically excluded white subjects according to the terms of their official protocols, the federally required plans that detail how research studies are conducted. However, in other human medical experiments, the recruitment of blacks and the poor is a tacit feature of the study because they recruit subjects from heavily black inner-city areas that tend to surround American teaching hospitals. American university research centers have historically been located in inner-city areas, and accordingly, a disproportionate number of these abuses have involved experiments with African Americans.

Washington's book is crucial to understand the direct links between the so-called racial anatomical studies and the medical *beliefs* about, and experimentation forced upon, Africa-Americans, during and after slavery:

In 1839, **Morton** published *Crania Americana*, a book written to demonstrate how human skull measurements indicated a **hierarchy of racial types**. Morton determined that Caucasians had the largest skulls, and therefore the largest brains, and blacks the smallest. His tests were the forerunner of phrenology, which sought to determine character and intelligence by interpreting the shape of the skull. By 1848, Louisiana's Samuel A. Cartwright, M.D., had gained renown by publishing a plethora of articles on Negro medicine in southern medical journals, leading the Medical Association of Louisiana to appoint him chair of its committee to investigate black health and physiology. That same year, Cartwright published his paper "*The Diseases and Physical Peculiarities of the Negro Race*".... By 1851, Cartwright had also discovered and described a host of **imaginary "black" diseases**, whose principal symptoms seemed to be a lack of enthusiasm for slavery. Escape might have seemed normal behavior for a slave in ancient Greece or Rome, but Cartwright medically condemned such behavior in American blacks, offering a diagnosis of **drapetomania**, from the Greek words for flight and insanity. Hebetude was a singular laziness or shiftlessness that caused slaves to mishandle and abuse their owners' property. **Dysthesia Aethiopica** was another black behavioral malady, which was characterized by a desire to destroy the property of white slave owners. Cartwright claimed that it "differs from every other species of mental disease, as it is accompanied with physical signs or lesions of the body discoverable to the medical observer". **Struma Africana** was a form of tuberculosis that physicians misdiagnosed as a peculiarly African disease. **Cachexia Africana** referred to blacks' supposed propensity for eating nonfood substances such as clay, chalk, and dirt. Actually, this disorder, which is called pica today, is not racially specific and the cravings it inspires were probably related to the rampant malnutrition among slaves.

Tellingly, Dr. Cartwright recommended that these ailments be treated with corporal punishment or with internment in "work camps": "put the patient to some hard kind of work in the open air and sunshine...the compulsory power of the white man, by making the slothful negro take active exercise, puts into active play the lungs, through whose agency the vitalized blood is sent to the brain to give liberty to the mind". Other medical disorders were thought to manifest differently, usually less severely, in blacks. **Syphilis**, for example, was held to be racially dimorphic. Physicians believed it worked its most feared damage within the neurological system of whites but that the less evolved nervous system of blacks was left relatively unimpaired. In blacks, syphilis was thought to attack the muscles, including the heart. This belief that syphilis in blacks differed dramatically from the disease in whites provided a rationale for the infamous U.S. Public Health Service's (PHS) ***Tuskegee Study of Syphilis in the Untreated Negro Male***. Between 1932 and 1972, six hundred black men, their wives, and their children were deceived into participating in a research study that denied them treatment, so that PHS scientists could trace the progress of the disease in blacks. Allegedly inferior cognition was only the tip of the iceberg. In 1854, several years after Cartwright published "*Report on the Diseases and Physical Peculiarities of the Negro Race*", and five years before Darwin published *On the Origin of Species*.... Alabama, physician Josiah Nott, M.D., and George R. Gliddon produced an equally popular screed entitled *Types of Mankind*. In it, they claimed that blacks' physical and mental differences signaled their polygenic origins and proved **black inferiority**. For example, Nott theorized that the distinctive knee joint and "long heel" of the black man proved he had been created as a "submissive knee-bender" – a servant to whites. Scientists adjudged the dark skin of Africans as a biblical curse that set them aside as eternal servants to other men.... As late as

1903, Dr. W. T. English observed, "a careful inspection reveals the body of the negro a mass of imperfections from the crown of the head to the soles of the feet". Even biological advantages were cast as racial flaws: in discussing the tendency of blacks to survive yellow fever epidemics that killed whites, one physician denounced the "inferior susceptibility" of black slaves.

This last sentence is a powerful illustration of a crucial aspect of *Homo fictus et irrationalis*, the **confirmation bias**: even when empirical data show that "others" tend to survive more a certain disease, that information is transformed in order to "confirm" the "inferiority" of those others, in this specific case by arguing that they have an "inferior susceptibility." How can one argue against such *Neverland*'s irrational, illogical, and circular types of reasoning? Washington's book provides many examples in which Westerns applied such a type of reasoning, in order to discard relevant or potential contributions of "Africans" or African-Americans to medicine and to other important aspects of daily life (see Box 3.11):

Interestingly, the contradiction of the black slave as both "riddled with imperfections from head to toe" and as a hardy laborer who was impervious to most illness escaped the scientific racists. Scientists expressed whichever opinion fit their political needs at the moment, as abolitionist **Frederick Douglass** suggested when he observed that ninety-nine of one hundred polygenists were Anglo-Saxon slave owners. Scientists also claimed that the primitive nervous systems of blacks were "immune" to physical and emotional pain and to mental illness. This belief, which will be discussed at greater length in the next chapter, released physicians and owners from the responsibility of shielding black slaves from painful medical procedures and justified torture such as branding, whipping, hobbling, and maiming. All these precepts of **scientific racism**, although convenient for the slave owner and physician, were highly illogical articles of faith. So was the supposed inferior intelligence of blacks, because planters and doctors behaved in many contexts as though they held the abilities and judgment of blacks in high regard, employing slaves in responsible positions as nurses, cooks, herbalists, midwives, overseers, leaders of work gangs, accountants, and operators of farm and factory implements. Owners reaped profits from the many patents on slave inventions, and physicians used slaves as skilled apprentices, who often went on to practice independently. White households depended upon the specialized skills and discernment of slaves, not the other way around….

[Yes], appeals to God, the importance of moral fitness, and enlisting the help of departed spirits, especially the intercession of ancestors, were all key to the African-based healing process. Ancestors who were angered by disrespect or neglect could cause illness, alienation, and other troubles for the living. This is one reason the respectful ritual treatment of the dead was so important to slaves and why they reviled **Western medicine** when they discovered that physicians appropriated the bodies of dead slaves for display and dissection. Western medicine was thought ineffective against spirit-caused illness, and slaves often lacked confidence in a Western doctor's ability to cure them: if a doctor did not believe that one could be cursed or "conjured", how could he remove the threat? This is a wide generalization, because some slaves mistrusted African practitioners, who sometimes used their skills to harm as well as to heal. But in planters' farm books and in medical journals, physicians and slave owners repeatedly berated the ignorance and superstition that led slaves to conceal illness and to shrink from "scientific" Western medicine in favor of conjure women and witch doctoring. However, whites had no monopoly on science.

The **African tradition** involved physiological as well as spiritual approaches to healing, including an encyclopedic knowledge of herbs, roots, and other natural medicaments. This detailed knowledge was continually passed down along lines of apprenticeship from wise women and male herb doctors to gifted young members of the community. Despite their characterization as primitive, African healers first employed citrus juice for **scurvy** and **inoculation for smallpox** and other viral illnesses; midwives used African techniques, herbs, and medicines so successfully – without dangerous tools of the day, such as forceps – that many white women called them to attend births. Some whites were impressed by the success rate of Negro doctors and "doctresses", consulted them, and placed their medication recipes in the family book, and Western doctors faced brisk competition from black herb doctors. In an 1855 journal article, Dr. R. H. Whitfield of Alabama railed against "unscientific" midwives: "*[There are no practices wherein which] the female practitioners are less educated, being chiefly negresses or mulatresses, or foreigners without anatomical, physiological and obstetrical education…that such uneducated persons should be generally successful is owing to the fact that [in] a great majority of cases no scientific skill is required, and thus a lucky negress become[s] the rival of the most learned obstetrician*".

For all their complaints, physicians in the early to midnineteenth century were happy to leave the business of birthing in the hands of **black midwives**. However, physicians wanted black healers under the scrutiny and supervision of white physicians. White doctors denigrated black midwives and healers, calling them "uneducated", but white physicians themselves usually had no academic preparation beyond a few months in proprietary medical school or a few years of apprenticeship, which many blacks also shared. So until the mid-1800s, such claims of superior education rang hollow. Also, regular medicine embraced no consistent curriculum, but roped in a motley association of disciplines. At the bedside, healers practiced a variety of Western fads such as hydrotherapy, which utilized harmless but ineffectual "water cures" for many ailments, and **Thomsonism**, followed by disciples of New Hampshire farmer **Samuel Thomson**, who, like black healers, advocated the use of milder herbal and vegetable remedies and emetics. The constant friction between white physicians and enslaved healers sometimes erupted into open hostility. Physicians denigrated black medical practice and imposed punishment, including execution, upon black healers, on the punishment, including execution, upon black healers, on the pretext of protecting the larger community from poisoning and from the evil machinations of occultists who, doctors claimed, controlled the minds and actions of superstitious blacks.

Black contributions to early American medicine included research. In fact, slave doctors sometimes developed medications that were so highly prized as to garner them fame, fortune, and their freedom. In 1729, Lieutenant Governor Gooch of Virginia authorized the payment of sixty pounds to manumit an unnamed "negro man". Gooch declared that his mixture of pharmacologically active roots and bark had proved an effective syphilis remedy. "It is well worth the price of the negro's freedom", wrote Gooch, "since it is now known how to cure negroes without mercury". In 1751, a South Carolina slave doctor named Cesar developed several medical innovations, including an almost foolproof snakebite antidote. The cure featured a shrub called plantane and horehound, a plant that derived its name from Egyptian priests who called it the "seed of Horus" (the Egyptian god of the sun and virtue), mixed with sassafras, wood ashes, and tobacco. On February 25 of that year, the *South Carolina Gazette* published the recipe as a public service, and demand ran so high that it was reprinted widely and published as a monograph in 1789. In 1799, it was mentioned in the text *Domestick Medicine*. Cesar's medical acumen earned him his freedom from the South Carolina General Assembly, which also granted him an annual pension of one hundred pounds. Primus was another slave who won fame for medical achievements, which included a rabies treatment. The medical career of Wilcie Elfe of Charleston, South Carolina, benefited rather than suffered when the white pharmacist to whom he was apprenticed turned out to be incapacitated by alcoholism. Left to his own devices, Elfe formulated new medications, which proved so effective that his patent drugs became popular across the state. Meanwhile, Western doctors complained that overseers resorted to a standard remedy for every complaint: "an emetic followed by calomel and oil".

Another very interesting aspect of Washington's book is that she refers to **Georgia Dunston**, a geneticist that worked at the same university in which I research and teach currently—**Howard University**, one of the most renown historical "Black" universities in the USA. Namely, she explains that Dunston claimed in the mid-1990s that of the more than 60 families whose genes were analyzed by the **Human Genome Project**, there were no people of African descent. She "lamented that severing the African branch of the family tree is a critical error because African gene pools are the oldest and consequently the most diverse on the planet, due to human life's having evolved in Africa." Dunston asked, "what picture of humankind can emerge without Africa"?

Black's 2003 book *War Against the Weak* provides a superb historical context of the rise of **eugenics** and explains its profound and disturbing—and often untold—direct links to the subsequent rise of the area of biology that is now known as "genetics." In particular, the book shows that not only a huge number of biologists but also a vast number of famous U.S. philanthropists—such as **Carnegie** and **Rockefeller**—as well as the enormously renowned and influential institutions that they created—as the **Carnegie Institution** and **Rockefeller Foundation**—were deeply involved in the eugenic enterprise. Together with what we have seen above about "highly respected" physicians such as Sims and scientists such as Aristotle and Cuvier, and what will be seen in the chapters below about "discoverers" such as Columbus, "explorers" such as Charles Darwin, "thinkers" such as Kant, or "noble politicians" such as Thomas Jefferson, what is clear is that while *Neverland* enculturated us to *believe* that Western "civilizations" are mostly built "on the shoulder of noble giants," this is not truly so. Instead, in the *world of reality*, almost all those "noble giants" that are displayed as "heroes" in thousands of statues around the globe and in an endless number of books, TV series and movies, were actually almost always extremely racist or misogynistic, often both. Black wrote:

> The victims of **eugenics** [in the first decades of the 20th century within the USA] were poor urban dwellers and rural 'white trash' from New England to California, immigrants from across Europe, Blacks, Jews, Mexicans, Native Americans, epileptics, alcoholics, petty criminals, the mentally ill and anyone else who did not resemble the blond and blue-eyed Nordic ideal the eugenics movement glorified. Eugenics contaminated many otherwise worthy social, medical and educational causes from the birth control movement to the development of psychology to urban sanitation. Psychologists persecuted their patients. Teachers stigmatized their students. Charitable associations clamored to send those in need of help to lethal chambers they hoped would be constructed. Immigration assistance bureaus connived to send the most needy to **sterilization mills**. Leaders of the ophthalmology profession conducted a long and chilling political campaign to round up and coercively sterilize every relative of every American with a vision problem. All of this churned throughout America years before the **Third Reich** rose in Germany. Eugenics targeted all mankind, so of course its scope was global. **American eugenic evangelists** spawned similar movements and practices throughout Europe, Latin America and Asia. **Forced sterilization laws** and regimens took root on every continent. Each local American eugenic ordinance or statute – from Virginia to Oregon – was promoted internationally as yet another precedent to be emulated by the international movement. A tightly-knit network of mainstream medical and eugenical journals, international meetings and conferences kept the generals and soldiers of eugenics up to date and armed for their nation's next legislative opportunity.

Eventually, **America's eugenic movement** spread to Germany as well, where it caught the fascination of Adolf Hitler and the Nazi movement. Under **Hitler**, eugenics careened beyond any American eugenicist's dream. National Socialism transduced America's quest for a "superior Nordic race" into Hitler's drive for an "**Aryan master race**". The **Nazis** were fond of saying "National Socialism is nothing but applied biology", and in 1934 the *Richmond Times-Dispatch* quoted a prominent American eugenicist as saying, "the Germans are beating us at our own game". **Nazi eugenics** quickly outpaced American eugenics in both velocity and ferocity. In the 1930s, Germany assumed the lead in the international movement. Hitler's eugenics was backed by brutal decrees, custom-designed IBM data processing machines, eugenical courts, mass sterilization mills, concentration camps, and virulent biological anti-Semitism – all of which enjoyed the open approval of leading American eugenicists and their institutions. The cheering quieted, but only reluctantly, when the United States entered the war in December of 1941. Then, out of sight of the world, Germany's eugenic warriors operated extermination centers. Eventually, Germany's eugenic madness led to the Holocaust, the destruction of the Gypsies, the rape of Poland and the decimation of all Europe. But none of America's far-reaching scientific racism would have risen above ignorant rants without the backing of corporate philanthropic largess. Within these pages you will discover the sad truth of how the scientific rationales that drove killer doctors at **Auschwitz** were first concocted on Long Island at the **Carnegie Institution**'s eugenic enterprise at **Cold Spring Harbor**...during the prewar Hitler regime, the Carnegie Institution, through its Cold Spring Harbor complex, enthusiastically propagandized for the **Nazi regime** and even distributed **anti-Semitic Nazi Party films** to American high schools...[there were] links between the Rockefeller Foundation's massive financial grants and the German scientific establishment that began the eugenic programs that were finished by **Mengele** at **Auschwitz**. Only after the truth about **Nazi extermination** became known did the American eugenics movement fade. **American eugenic institutions** rushed to change their names from eugenics to genetics. With its new identity, the remnant eugenics movement reinvented itself and helped establish the modem, enlightened human genetic revolution. Although the rhetoric and the organizational names had changed, the laws and mindsets were left in place. So for decades after Nuremberg labeled eugenic methods genocide and **crimes against humanity**, America continued to forcibly sterilize and prohibit eugenically undesirable marriages.

The involvement of the Carnegie Institution in the eugenic enterprise, in particular through one of the most prominent and racist promoters of eugenism in the U.S., the zoologist **Charles Davenport**, is a clear example of how racist teleological narratives were widely accepted among and used by scientists, politicians, and other powerful figures during the first decades of the twentieth century. As noted by Black:

Fresh from his European travels [at the beginning of the 20th century], and fortified with the latest international views on eugenics, Davenport dispatched to the Carnegie Institution a more detailed letter plus a lengthy report on the state of human evolution studies to date. The documents made clear that far-reaching American race policy could not be directed without supportive scientific data based on breeding experiments with lower species. The results of those experiments would be applied in broad strokes to humans. "Improvement of the human race can probably be effected only by understanding and applying these methods", he argued. "How appalling is our ignorance, for example, concerning the effect of a mixture of races as contrasted with pure breeding; a matter of infinite importance in a country like ours containing numerous races and subspecies of men". Davenport hoped to craft a **super race of Nordics**. "Can we build a wall high enough around this country", he asked his colleagues, "so as to keep out these cheaper races, or will it be a feeble dam...leaving it to our descendants to abandon the country to the blacks, browns and yellows and seek and an asylum in New Zealand". Man was still evolving, he reasoned, and that evolution could and should be to a higher plane. Carnegie funds could accelerate and direct that process. "But what are these processes by which man has evolved", posited Davenport, "and which we should know…in hastening his further evolution". He disputed the value of improved conditions for those considered genetically inferior. He readily admitted that with schooling, training and social benefits, "a person born in the slums can be made a useful man". But that usefulness was limited in the evolutionary scheme of things.

No amount of book learning, "finer mental stuff" or "intellectual accumulation" would transfer to the next generation, he insisted, adding that "permanent improvement of the race can only be brought about by breeding the best". Drawing on his belief in **raceology**, Davenport offered the Carnegie trustees an example he knew would resonate: "We have in this country the grave problem of the negro", he wrote, "a race whose mental development is, on the average, far below the average of the Caucasian. Is there a prospect that we may through the education of the individual produce an improved race so that we may hope at last that the negro mind shall be as teachable, as elastic, as original, and as fruitful as the Caucasian's? Or must future generations, indefinitely, start from the same low plane and yield the same meager results? We do not know; we have no data. Prevailing 'opinion' says we must face the latter alternative. If this were so, it would be best to export the black race at once". Proof was needed to fuel the social plans the eugenicists and their allies championed. Davenport was sure he could deliver the proof. "As to a person to carry out the proposed work", he wrote Carnegie, "I am ready at the present moment to abandon all other plans for this". To dispel any doubt of his devotion, Davenport told the institution, "I propose to give the rest of my life unreservedly to this work". The men of Carnegie were impressed. They said yes….

It didn't matter that the majority of the American people opposed sterilization and the eugenics movement's other draconian solutions. It didn't matter that the underlying science was a fiction, that the intelligence measurements were fallacious, that the Constitutionality was tenuous, or that the whole idea was roundly condemned by so many. None of that mattered because Davenport, Laughlin and their eugenic constellation were not interested in furthering a democracy-they were creating a *supremacy*. Of course, American eugenicists did not seek the approbation of the masses whose defective germ plasm they sought to wipe away. Instead, they relied upon the *powerful, the wealthy and the influential* to make their war against the weak a conflict fought not in public, but in the administrative and bureaucratic foxholes of America. A phalanx of shock troops sallied forth from obscure state agencies and special committees-everyone from the elite of the academic world to sympathetic legislators who sought to shroud their racist beliefs under the protective canopy of science. In tandem, they would hunt, identify, label and take control of those deemed unfit to populate the earth.

The "powerful, wealthy, and influential" figures that supported the eugenic enterprise included even people that were, or had been, U.S. presidents, as explained by Black:

Eventually, the eugenics movement and its supporters began to speak a common language that crept into the general mindset of many of America's most influential thinkers. On January 3, 1913, former President **Theodore Roosevelt** wrote Davenport, "I agree with you…that society has no business to permit degenerates to reproduce their kind…some day, we will realize that the prime duty, the inescapable duty, of the good citizen of the right type, is to leave his or her blood behind him in the world; and that we have no business to permit the perpetuation of citizens of the wrong type". Episcopalian Bishop John T. Dallas of Concord, New Hampshire, issued a public statement: "eugenics is one of the very most important subjects that the present generation has to consider". Episcopalian Bishop Thomas F. Gailor of Memphis, Tennessee, issued a similar statement: "The science of eugenics…by devising methods for the prevention of the propagation of the feebleminded, criminal and unfit members of the community, is…one of the most important and valuable contributions to civilization". Dr. Ada Comstock, president of Radcliffe College, declared publicly, "eugenics is 'the greatest concern of the human race'. The development of civilization depends upon it". Dr. Albert Wiggam, an author and a leading member of the *American Association for the Advancement of Science*, pronounced his belief: "Had Jesus been among us, he would have been president of the First Eugenic Congress".

While many of America's elite exalted eugenics, the original **Galtonian eugenicists** in Britain were horrified by the sham science they saw thriving in the United States and taking root in their own country. In a merciless 1913 scientific paper written on behalf of the Galton Laboratory, British scientist **David Heron** publicly excoriated the American eugenics of Davenport, Laughlin, and the Eugenics Record Office. Using the harshest possible language, Heron warned against "certain recent American work which has been welcomed in this country as of first-class importance, but the teaching of which we hold to be fallacious and indeed actually dangerous to social welfare". His accusations: "careless presentation of data, inaccurate methods of analysis, irresponsible expression of conclusions, and rapid change of opinion". Heron lamented further, "those of us who have the highest hopes for the new science of Eugenics in the future are not a little alarmed by many of the recent contributions to the subject which threaten to place Eugenics…entirely outside the pale of true science…when we find such teaching-based on the flimsiest of theories and on the most superficial of inquiries-proclaimed in the name of Eugenics, and spoken of as 'entirely splendid work', we feel that it is not possible to use criticism too harsh, nor words too strong in repudiation of advice which, if accepted, must mean the death of Eugenics as a science". Heron emphasized "that the material has been collected in a most unsatisfactory manner, that the data have been tabled in a most slipshod fashion, and that the Mendelian conclusions drawn have no justification whatever…" He went so far as to say the data had been deliberately skewed. As an example, he observed that "a family containing a large number of defectives is more likely to be recorded than a family containing a small number of defectives". In sum, he called American eugenics rubbish. Davenport exploded.

However, unfortunately, the eugenics movement thriving in the U.S. did export itself very successfully to many other countries, including to the UK and politicians such as **Winston Churchill**, and Germany and ultimately to **Hitler**'s Third Reich, as noted by Black:

By 1912, America's negative eugenics had been purveyed to like-minded social engineers throughout Europe, especially in Germany and the Scandinavian nations, where **theories of Nordic superiority** were well received. Hence the *First International Congress of Eugenics* attracted several hundred delegates and speakers from the United States, Belgium, England, France, Germany, Italy, Japan, Spain and Norway…[for instance] the ambitious British eugenic plans encompassed not just those who seemed mentally inferior, but also criminals, debtors, paupers, alcoholics, recipients of charity and "other parasites". In 1909 and 1910, other so-called welfare societies for the feebleminded, such as the ***Cambridge Association for the Care of the Feebleminded***, contacted the **Eugenics Education Society** to urge more joint lobbying of the government to sanction forced sterilization. Mass letter-writing campaigns began. Every candidate for Parliament was sent a letter demanding they "support measures…that tend to discourage parenthood on the part of the feebleminded and other degenerate types". As in America, sterilization advocacy focused first and foremost on the most obviously impaired, in this case, the feeble-minded, but then escalated to include "other degenerate types". Seeking support for the **Mental Deficiency Act**, society members mailed letters to every sitting member of Parliament, long lists of social welfare officials, and virtually every education committee in England. When preliminary governmental committees shrank from support, the society simply redoubled its letter-writing campaign. Finally the government agreed to consider the legislation. Home Secretary **Winston Churchill**, an enthusiastic supporter of eugenics, reassured one group of eugenicists that

Britain's 120,000 feebleminded persons "should, if possible, be segregated under proper conditions so that their curse died with them and was not transmitted to future generations". The plan called for the creation of vast colonies. Thousands of Britain's unfit would be moved into these colonies to live out their days....

Germany was no exception. **German eugenicists** had formed academic and personal relationships with Davenport and the American eugenic establishment from the turn of the century. Even after World War I, when Germany would not cooperate with the **International Federation of Eugenic Organizations** because of French, English and Belgian involvement, its bonds with Davenport and the rest of the U.S. movement remained strong. American foundations such as the **Carnegie Institution** and the **Rockefeller Foundation** generously funded German race biology with hundreds of thousands of dollars, even as Americans stood in breadlines. Germany had certainly developed its own body of eugenic knowledge and library of publications. Yet German readers still closely followed American eugenic accomplishments as the model: **biological courts**, **forced sterilization**, detention for the socially inadequate, debates on euthanasia. As America's elite were describing the socially worthless and the ancestrally unfit as "bacteria", "vermin", "mongrels" and "subhuman", a superior race of Nordics was increasingly seen as the final solution to the globe's eugenic problems. America had established the value of race and blood. In Germany, the concept was known as ***Rasse und Blut***.

U.S. proposals, laws, eugenic investigations and ideology were not undertaken quietly out of sight of Gennan activists. They became inspirational blueprints for Germany's rising tide of race biologists and **race-based hatemongers**, be they white-coated doctors studying *Eugenical News* and attending congresses in New York, or brown-shirted agitators waving banners and screaming for social upheaval in the streets of Munich. One such agitator was a disgruntled corporal in the German army. He was an extreme nationalist who also considered himself a race biologist and an advocate of a **master race**. He was willing to use force to achieve his nationalist racial goals. His inner circle included Germany's most prominent eugenic publisher. In 1924, he was serving time in prison for mob action. While in prison, he spent his time poring over eugenic textbooks, which extensively quoted Davenport, Popenoe and other American raceo-logical stalwarts. Moreover, he closely followed the writings of Leon Whitney, president of the American Eugenics Society, and Madison Grant, who extolled the Nordic race and bemoaned its corruption by Jews, Negroes, Slavs and others who did not possess blond hair and blue eyes. The young Gennan corporal even wrote one of them fan mail...[he] would soon burn and gas his name into the blackest corner of history. He would duplicate the American eugenic program-both that which was legislated and that which was only brashly advocated-and his group would consistently point to the United States as setting the precedents for Germany's actions. And then this man would go further than any American eugenicist ever dreamed, further than the world would ever tolerate, further than humanity will ever forget. The man who sent those letters was **Adolf Hitler**.

As pointed out by Black, the links between the U.S. eugenics movement, **Hitler**, and teleological narratives are indeed disturbing:

Where did Hitler develop his racist and **anti-Semitic views**? Certainly not from anything he read or heard from America. Hitler became a mad racist dictator based solely on his own inner monstrosity, with no assistance from anything written or spoken in English. But like many rabid racists, from Plecker in Virginia to Rentoul in England, Hitler preferred to legitimize his race hatred by medicalizing it, and wrapping it in a more palatable **pseudo-scientific facade-eugenics**. Indeed, Hitler was able to recruit more followers among reasonable Germans by claiming that science was on his side. The intellectual outlines of the eugenics Hitler adopted in 1924 were strictly American. He merely compounded all the virulence of long established American race science with his fanatic anti-Jewish rage. Hitler's extremist eugenic science, which in many ways seemed like the logical extension of America's own entrenched programs and advocacy, eventually helped shape the institutions and even the machinery of the **Third Reich's genocide**. By the time Hitler's concept of **Aryan superiority** emerged, his politics had completely fused into a biological and eugenic mindset. When Hitler used the term master race, he meant just that, a biological "master race". America crusaded for a biologically superior race, which would gradually wipe away the existence of all inferior strains. Hitler would crusade for a master race to quickly dominate all others. In Hitler's view, eugenically inferior groups, such as Poles and Russians, would be permitted to exist but were destined to serve **Germany's master race**. Hitler demonized the Jewish community as social, political and racial poison, that is, a biological menace. He vowed that the Jewish community would be neutralized, dismantled and removed from Europe. Nazi eugenics would ultimately dictate who would be persecuted, how people would live, and how they would die. **Nazi doctors** would become the unseen generals in Hitler's war against the Jews and other Europeans deemed inferior. Doctors would create the science, devise the eugenic formulas, write the legislation, and even hand-select the victims for sterilization, euthanasia and mass extermination. Hitler's deputy, **Rudolf Hess**, coined a popular adage in the Reich, "National Socialism is nothing but applied biology".

In page after page of **Mein Kampfs** rantings, Hitler recited social Darwinian imperatives, condemned the concept of charity, and praised the policies of the United States and its quest for Nordic purity. Perhaps no passage better summarized Hitler's views than this from chapter 11: "the Germanic inhabitant of the American continent, who has remained racially pure and unmixed, rose to be master of the continent; he will remain the master as long as he does not fall a victim to defilement of the blood"…. Moreover, as Hitler's knowledge of **American pedigree techniques** broadened, he came to realize that even he might have been eugenically excluded. In later years, he conceded at a dinner engagement, "I was shown a questionnaire drawn up by the Ministry of the Interior, which it was proposed to put to people whom it was deemed desirable to sterilize. At least three-quarters of the questions asked would have defeated my own good mother. If this system had been introduced before my birth, I am pretty sure I should never have been born at all"…. On January 30, 1933, Adolf Hitler seized power following an inconclusive election. During the twelve-year Reich, he never varied from the eugenic doctrines of identification, segregation, sterilization, euthanasia, eugenic courts and eventually mass termination of germ plasm in lethal chambers. During the Reich's first ten years, eugenicists across America welcomed Hitler's plans as the logical fulfillment of their own decades of research and effort. Indeed, they were envious as Hitler rapidly began sterilizing hundreds of thousands and systematically eliminating non-Aryans from German society. This included the Jews. Ten years after Virginia passed its **1924 sterilization act**, Joseph Dejarnette, superintendent of Virginia's Western State Hospital, complained in the *Richmond TimesDispatch*, "the Germans are beating us at our own game".

Most of all, **American raceologists** were intensely proud to have inspired the purely eugenic state the Nazis were constructing. In those early years of the Third Reich, Hitler and his race hygienists carefully crafted eugenic legislation modeled on laws already introduced across America, upheld by the Supreme Court and routinely enforced. **Nazi doctors** and even Hitler himself regularly communicated with American eugenicists from New York to California, ensuring that Germany would scrupulously follow the path blazed by the United States. American eugenicists were eager to assist. As they followed the day-to-day progress of the **Third Reich**, American eugenicists clearly understood their continuing role. This was particularly true of California's eugenicists, who led the nation in sterilization and provided the most scientific support for Hitler's regime. In 1934, as Germany's sterilizations were accelerating beyond five thousand per month, the California eugenic leader and immigration activist C. M. Goethe was ebullient in congratulating E. S. Gosney of the San Diego-based **Human Betterment Foundation** for his impact on Hitler's work. Upon his return in 1934 from a eugenic fact-finding mission in Germany, Goethe wrote Gosney a letter of praise.

> The *Human Betterment Foundation* was so proud of Goethe's letter that they reprinted it in their 1935 Annual Report. "You will be interested to know", Goethe's letter proclaimed, "that your work has played a powerful part in shaping the opinions of the group of intellectuals who are behind Hitler in this epoch-making program. Everywhere I sensed that their opinions have been tremendously stimulated by American thought, and particularly by the work of the Human Betterment Foundation. I want you, my dear friend, to carry this thought with you for the rest of your life, that you have really jolted into action a great government of 60 million people".... While much of the world recoiled in revulsion, American eugenicists covered eugenic developments in Germany with pride and excitement. By the summer of 1933...Cold Spring Harbor quickly obtained a full copy of the eighteen-paragraph **Nazi sterilization law** from German Consul Otto Kiep, and rushed a verbatim translation into the next issue as its lead item. In accompanying commentary, Eugenical News declared: "Germany is the first of the world's major nations to enact a modern eugenical sterilization law for the nation as a unit...the law recently promulgated by the **Nazi Government** marks several substantial advances...doubtless the legislative and court history of the experimental sterilization laws in 27 states of the American union provided the experience, which Germany used in writing her new national sterilization statute. To one versed in the history of **eugenical sterilization in America**, the text of the German statute reads almost like the 'American model sterilization law'." Proudly pointing out the American origins of the Nazi statute, the article continued, "in the meantime it is announced that the Reich will secure data on prospective sterilization cases, that it will, in fact, in accordance with 'the American model sterilization law', work out a census of its socially inadequate human stocks".

The examples provided by Black about how many Nazis continued to practice medicine after the Second World War, either in Germany or, commonly, in countries such as the U.S., bring us back to the very difficult ethical question on whether the medical data obtained by the Nazis in their horrible experiments with living humans should be used or not to potentially save lives:

> As the ashes of Jews and Gypsies wafted into the air of Europe and were dumped into the Vistula River coursing through the heart of Europe, so their victimization flowed into the mainstream of modern medical literature. Medical literature evolves from decade to decade. As **American eugenic pseudoscience** thoroughly infused the scientific journals of the first three decades of the twentieth century, Nazi-era eugenics placed its unmistakable stamp on the medical literature of the twenties, thirties and forties. The writings of Nazi doctors not only permeated the spectrum of German medical journals, they also appeared prominently in American medical literature. These writings included the results of war crime experimentation at concentration camps. Verschuer's own bibliographies, circa 1939, enumerated a long list of Nazi scientific discoveries, authored by him, his colleagues and assistants, including **Mengele**. Such scientific publication continued right through the last days of the **Third Reich**. The topics included everything from rheumatism, heart disease, eye pathology, blood studies, brain function, tuberculosis, and the gastric system to endless permutations of hereditary pathology. Much of it was sham science. Some of it was astute. Both types found their way into the medical literature of the fifties and sixties. Hence, Nazi victimization contributed significantly to many of the modem medical advances of the postwar period. For example, the Nazis at Dachau, using ice water tests, were the first to experimentally lower human body temperature to 79.7 degrees Fahrenheit-this to discover the best means of reviving Luftwaf e pilots downed over the North Sea. Nazi scientists learned that the most effective method was rapid rewarming in hot water. Nuremberg testimony revealed that Dr. Sigmund Rascher, who oversaw these heinous hypothermia tests, prominently reported his breakthroughs at a 1942 medical symposium with a paper entitled "*Medical Problems Arising from Sea and Winter*". After the war, Rascher's conclusions were gleaned

from Nazi reports and reluctantly adopted by British and American air-sea rescue services. A Nuremberg war crimes report on Nazi medicine summed up the extreme discomfort of Allied military doctors: "Dr. Rascher, although he wallowed in blood…and in obscenity…nevertheless appears to have settled the question of what to do for people in shock from exposure to cold…the method of rapid and intensive rewarming in hot water…should be immediately adopted as the treatment of choice by the Air-Sea Rescue Services of the United States Armed Forces".

Rascher reported to Hubertus Strughold, director of the Luftwaffe Institute for Aviation Medicine. Strughold attended the Berlin medical conference that reviewed Rascher's revelations. A Nazi scientist wrote at the time that there were no "objections whatsoever to the experiments requested by the Chief of the Medical Service of the Luftwaffe to be conducted at the Rascher experimental station in the Dachau concentration camp. If possible, Jews or prisoners held in quarantine are to be used". After the war, Strughold was smuggled into the United States under the infamous Operation Paperclip project, which offered Nazi scientists refuge and immunity in exchange for their scientific expertise. Once in the U.S., **Strughold** became the leader in American aviation medicine. His work was directly and indirectly responsible for numerous aeromedical advances, including the ability to walk effortlessly in a pressurized air cabin-now taken for granted-but which was also developed as a result of Dachau experiments. He was called "the father of U.S. Space Medicine", and Brooks Air Force Base in Texas named its Aeromedical Library in his honor. A celebratory mural picturing Strughold was commissioned by Ohio State University.

As explained by Black:

When Jewish and Holocaust-survivor groups, led by the **AntiDefamation League**, discovered the honors extended to Strughold, they objected. Ohio State University removed its mural in 1993. The U.S. Air Force changed its library's name in 1995. In 2003, the state of New Mexico still listed Strughold as a member of its International Space Hall of Fame. But on February 13,2003, when this reporter asked about their honoree's Nazi connection, a startled museum official declared, "if he was doing experiments at Dachau, it would give one pause why anyone would ever nominate him in the first place". Museum officials added they would immediately look into removing his name. Another case involved Nazi doctors Hallervorden and Spatz. In 1922, the two had successfully identified a rare and devastating brain disease caused by a genetic mutation. The disease came to be known as **Hallervorden-Spatz Syndrome** in their honor. During the Hitler era, while working at the Kaiser WIlhelm Institute for Brain Research, Hallervorden and Spatz furthered their research by utilizing hundreds of brains harvested from T-4 victims. Right through the 1960s, Hallervorden authored numerous influential scientific papers on the subject. For decades, the name Hallervorden-Spatz has been used by the leading medical institutions in the world, honoring the two Nazis who discovered the disorder. Thousands of articles and presentations have been made on the topic, using the name HallervordenSpatz. Medical investigators created an "International Registry of Patients with Hallervorden-Spatz Syndrome and Related Disorders".

Leading family support groups involved with the disorder have also taken their organizational names from the two Nazi doctors. But the news about Hallervorden and Spatz's Nazi past recently became known to many in the field. In 1993, two doctors expressed the view of many in a letter to the editor of the journal *Neurology*. "It is also time to stop using the term, 'Hallervorden-Spatz disease' whose only purpose is to honor Hallervorden by using his name". Another journal, Lancet, expressed a similar view in 1996, describing the continued honorary use of the name "Hallervorden-Spatz" as "indefensible" because "both Hallervorden and Spatz were closely associated with the Nazi extermination policies". In January of 2003, the Hallervorden-Spatz Syndrome Association renamed itself the NBIA Disorders Association; the acronym was derived from "neurodegeneration with brain iron accumulation". Just after the announcement, the newly-renamed association's president,

Patricia Wood, told this reporter that the name change was certainly due to the legacy of Nazi experiments attached to Hallervorden and Spatz.

The association's website confirmed that the name change was driven by "concerns about the unethical activities of Dr. Hallervorden (and perhaps also Dr. Spatz) involving euthanasia of mentally ill patients during **World War II**". Nazi medical victims suffered torture to substantially advance Reich scientific knowledge and modern medicine. Then the murdered specimens were delivered to the likes of Verschuer and Hallervorden and their eugenic institutions. But then what? After the war, victims' remains were transferred to or maintained by some of Germany's leading medical research facilities. Hence the exterminated continued to provide organic service to German medicine. In 1989, the **Max Planck Institute for Brain Research**, the successor to Hallervorden's center, admitted that it still possessed thirty tissue samples in its files. That same year, tissue samples and skeletons were also found in universities in Tübingen and Heidelberg. In 1997, investigators confirmed that the University of Vienna's Institute of Neurobiology still housed four hundred Holocaust victims' brains. The **University of Vienna** had functioned as part of the Reich after Austria's union with Germany in 1938. Similar discoveries have been made elsewhere in former Nazi occupied Europe.

One of the most disconcerting parts of Black's *War Against the Weak* concerns the resurgence, in the last decades, of many of the troubling aspects of eugenics, which in some cases is now often re-branded under the name of "genetics" or, as Black calls it, "**newgenics**":

Insurance companies vigorously claim they do not seek ancestral or genetic information. This is not true. In fact, the international insurance field considers ancestral and genetic information its newest high priority. The industry is now grappling with the notion of underwriting not only the individual applicant, but his family history as well. Insurers increasingly consider genetic traits "preexisting conditions" that should either be excluded or factored into premiums. A healthy individual may be without symptoms, or asymptomatic, but descend from a family with a history of a disease. In the industry's view, that individual presumably knows his family history; the insurance company doesn't. Insurers call this disparity "asymmetrical information", and it is hotly discussed at numerous industry symposiums and in professional papers. Governments and privacy groups worldwide want to prohibit the acquisition and use of genetic testing. Many in the insurance world, however, argue that their industry cannot survive without such information, and the resulting coverage restrictions, exclusions and denials that would protect company liquidity. Insurance discrimination based on genetics has already become the subject of an active debate in Great Britain. British insurers were widely employing predictive genetic testing by the late 1990s to underwrite life and medical insurance, and utilizing the results to increase premiums and deny coverage. The science of such testing is by no means authoritative or even reliable, but it allows insurers to justify higher prices and exclusions. Complaints of genetic discrimination have already become widespread. A third of those polled from genetic disorder support groups in Britain reported difficulties obtaining insurance, compared to just 5 percent from a general population survey. Similarly, a U.S. study cited by the American insurance publication Risk Management found that 22 percent of nearly one thousand individuals reported genetic discrimination. A British Medical Journal study paper asserted, "our findings suggest that in less clear cut instances, where genes confer an increased susceptibility rather than 100% or zero probability, some people might be charged high premiums that cannot be justified on the actuarial risk they present"....

British insurers began widely utilizing genetic tests after a leading geneticist consulting for the industry's trade association recommended the action, a Norwich Union executive explained. The widespread concern in England is generation-to-generation discrimination pivoting not on race, color or religion, but on genetic caste. "We are concerned, of course", warned Dr. Michael Wilks, of the British Medical Association's Medical Ethics Committee, "that the more we go down the road of precision testing for specific patients for specific insurance policies the more likely we are to create a group who simply will not be insurable". Wilks called such a group a genetic "underclass". A member of Parliament characterized Norwich Union's actions as an attempt to construct a "**genetic ghetto**". Prominent voices in the genetic technology field believe that mankind is destined for a genetic divide that will yield a superior race or species to exercise dominion over an inferior subset of humanity. They speak of "self-directed evolution" in which genetic technology is harnessed to immeasurably correct humanity, and then immeasurably enhance it. Correction is already underway. So much is possible: genetic therapies, embryo screening in cases of inherited disease and even modification of the genes responsible for adverse behaviors, such as aggression and gambling addiction. Even more exotic technologies will permit healthier babies and stronger, more capable individuals in ways society never dreamed of before the **Human Genome Project** was completed. These improvements are coming this decade.

Some are available now. But correction will not be cheap. Only the affluent who can today afford personalized elective health care will be able to afford expensive genetic correction. Hence, economic class is destined to be associated with genetic improvement. If the genetically "corrected" and endowed are favored for employment, insurance, credit and the other benefits of society, then that will only increase their advantages. But over whom will these advantages be gained? Those who worry about "gene lining", "genetic ghettos" and a "genetic underclass" see a sharp societal gulf looming ahead to rival the current inequities of the health care and judicial systems. The vogue term designer babies itself connotes wealth. The term designer babies is by and large just emblematic of the idea that genetic technology can do more than merely correct the frail aspects of human existence. It can redress nature's essential randomness.

Purely elective changes are in the offing. The industry argues over the details, but many assure that within our decade, depending upon the family and the circumstances, height, weight and even eye color will become elective. Gender selection has been a fact of birth for years with a success rate of up to 91 percent for those who use it. It goes further – much further. A deaf lesbian couple in the Washington, D.C., area sought sperm from a deaf man determined to produce a deaf baby because they felt better equipped to parent such a child. A child was indeed born and the couple rejoiced when an audiology test showed that the baby was deaf. A dwarf couple reportedly wants to design a dwarf child. A Texas couple reportedly wants to engineer a baby who will grow up to be a large football player. One West Coast sperm bank caters exclusively to Americans who desire Scandinavian sperm from select and screened Nordics…. It will be an international challenge to successfully regulate such genetic tampering and the permutations possible because few can keep up with the moment-to-moment technology. It goes much further than designer babies. **Mass social engineering** is still being advocated by eminent voices in the genetics community. Celebrated geneticist **James Watson**, co-discoverer of the double helix and president of **Cold Spring Harbor Laboratories**, told a British film crew in 2003, "if you are really stupid, I would call that a disease. The lower 10 per cent who really have difficulty, even in elementary school, what's the cause of it? A lot of people would like to say, 'Well, poverty, things like that'. It probably isn't. So I'd like to get rid of that, to help the lower 10 per cent". For the first half of the twentieth century, Cold Spring Harbor focused on the "submerged tenth"; apparently, the passion has not completely dissipated.

Sadly, within the self-designated species *Homo sapiens*—which, if the name was indeed correct, should be able to learn with its mistakes—history and imaginary narratives tend, in fact, to repeat itself, over and over. The Nobel Laureate **James Watson** (Fig. 3.20), one of the so-called fathers of DNA, is indeed one of the most emblematic examples showing that many current so-called geneticists, including the most renowned ones, defend exactly the same atrocious racist—and

Fig. 3.20 James Watson, Nobel Laureate: one of the most emblematic examples of how even people that are capable of doing an important scientific discovery are not able to escape the heavy chains of *Neverland*'s longstanding completely unscientific racist and misogynistic imaginary tales. Such people are particularly dangerous because they precisely use their very specific discoveries to give the appearance that they have a deep knowledge about issues they know little about, in order to better "sell" such fake tales to the broader community, thus dragging many people down with them, to the deep dungeons of *Neverland*

misogynistic—ideas that eugenicists defended a century ago, based on the very same absurd, factually wrong *Neverland* fairytales. Basically, people like Watson use their scientific expertise on a very specific topic to give the appearance that they know about issues they know little about and, worse, they pretend to be experts in order to better "sell" such false tales to the broader community—in other words, *intellectual dishonesty*. As explained in an article published in January 2019, in the website *Vox*, entitled "*DNA scientist James Watson has a remarkably long history of sexist, racist public comments*":

> The legacy of James Watson – who discovered DNA along with Francis Crick, Maurice Wilkins, and Rosalind Franklin – has once again been tarnished by the American biologist's offensive, baseless comments. In a new PBS documentary, Watson, now 90, affirms his previously stated view that black people are intellectually inferior to white people. "There's a difference on the average between blacks and whites in IQ tests", Watson said in the film *American Masters: Decoding Watson*. "I would say the difference is genetic". The remarks prompted the Cold Spring Harbor Laboratory in New York, where Watson was a director from 1968 to 1994, to sever its ties with the Nobel Prize winner…saying his views are "reprehensible, unsupported by science, and in no way represent the views of [the lab]." But this is far from the first public embarrassment over comments he's made. The famed scientist has a long history of provocation with racist, sexist, homophobic, anti-Semitic, and even

fat-shaming remarks. When you look at them in their totality, it's amazing it took this long for Cold Spring Harbor to fully sever ties. Watson began to recede from public life in 2007, after he told a British reporter that he was "gloomy" about Africa because "all our social policies are based on the fact that their intelligence is the same as ours – whereas all the testing says not really…while people might wish all humans were equal, people who have to deal with black employees find this not true." In 2014, Watson became the first Nobel winner to sell his prize because, he said, the race remarks made him an "unperson". The new documentary was a chance for Watson to redeem his public image and walk back from the offensive comments. But he instead confirmed them…. It's a deeply troubling pattern, particularly because he's a scientist with great influence and authority. By misrepresenting science to belittle minority groups and women, he can easily mislead people. Here's a sampling of some of his most egregious public comments made over the decades:

Watson's 1968 book *The Double Helix* includes a sexist depiction of Rosalind Franklin, the British chemist whose work on X-ray crystallography enabled Watson and his DNA co-discoverer Francis Crick to actually see the structure of DNA…. At *Boing Boing*, Maggie Koerth-Baker outlines how Watson repeatedly refers to her as "Rosy", a nickname Franklin didn't use, undermines her contributions to science, and criticizes her appearance…. In the years since these quips were made, journalists and historians have noted that Franklin's contributions to the discovery of DNA were initially overlooked. *1997*: Watson reportedly argued in a Sunday Telegraph interview that women should be allowed to abort fetuses that carried a **"gay gene"**, should one ever be discovered…. During a 2000 guest lecture at the University of California Berkeley, Watson shared his belief that thin people are unhappier than larger people, and therefore harder-working. He also said: "Whenever you interview fat people, you feel bad, because you know you're not going to hire them". In that same lecture, the *Chronicle* reported, Watson commented on the (nonexistent) link between sun exposure (and darker skin color) and sexual prowess: "That's why you have Latin Lovers. You've never heard of an English lover. Only an English patient". In a 2003 documentary interview called *DNA*, which aired in 2003 on Channel 4 in the UK, Watson delivered a zinger on gene editing for beauty: "People say it would be terrible if we made all girls pretty. I think it would be great." In the same documentary, he suggests stupidity is a disease to be abolished. In a 2007 interview with Esquire, he asked, "Why isn't everyone as intelligent as Ashkenazi Jews?" and suggested that rich people should be paid to have children because "[i]f there is any correlation between success and genes, IQ will fall if the successful people don't have children." Of women in science, he said at the 2012 EuroScience Open Forum in Dublin, "I think having all these women around makes it more fun for the men but they're probably less effective."…*In other words, Watson isn't being persecuted for unpopular scientific views; his views just aren't scientific at all. They're hurtful and dangerous – and are fuel for bigots and white supremacists to draw on to justify their views.*

Chapter 4
Myths and Reality About "Savages" and "Civilization"

> *"Writing appears to be necessary for the centralized, stratified state to reproduce itself...has accompanied...the integration...of a considerable number of people into a hierarchy of castes and classes...it seems to favor rather the exploitation than the enlightenment of mankind"*
>
> (Claude Levi-Strauss)

4.1 "Savages," "Civilization," and Inequality

Americans wanted to believe that the racial, ethnic, class, and gender hierarchies in the United States were natural and normal...they wanted to believe that they were passing their traits on to their children (Ibram X Kendi)

Among the many works that discuss the links between the notion of "savages," the lifestyle of hunter-gatherers, "civilization" and inequality, Kelly's 2013 book *The Lifeways of Hunter-Gatherers* is probably the one that provides the most clear, succinct, empirically based, and unbiased account on both the myths and facts about the changes that normally occur from nomadic hunter-gatherer groups to sedentary ones and then eventually to agricultural groups. Indeed, Kelly has a much more neutral and pragmatic take of the long-standing **"Hobbes *versus* Rousseau debate"** than most other authors have. As noted by him, the fact that for six millions of years there were—and there still are—hunter-gatherer groups, particularly nomadic ones, raises an important question that has too often been addressed merely by using imaginary ethnocentric teleological narratives about "cosmic purpose" and principally about "progress":

> If everyone has been on earth for the same amount of time, why have some peoples made more 'progress' than others? The Enlightenment paradigm provided the answer: variability among the world's peoples was attributed to variability in the tempo of mental improvement. Some people moved ("progressed") up the evolutionary ladder more quickly than others. Handily enough, this meant that the evolutionists could see less-advanced societies as relics of an earlier age, "monuments of the past". By placing the world's peoples into a ranked sequence, human prehistory could be reconstructed – and without dirtying one's

R. Diogo, *Meaning of Life, Human Nature, and Delusions*,
https://doi.org/10.1007/978-3-319-70401-2_4

hands in archaeological sites! The criteria for constructing evolutionary sequences were various and included technological, social, political, intellectual, and moral factors. These criteria exposed the **ethnocentrism** of the comparative method, for invariably *Western scholars judged other societies against the standard of European society*. Monogamy was superior to polygamy, patrilineal descent was better than matrilineal descent, monotheism was morally superior to ancestor worship, and science was the successor to magic and religious superstition. Rankings also had a strongly racialist basis, with people of color at the bottom and Europeans (and especially northwestern, light-skinned Europeans) at the top of the sequence.

Two factors helped place hunter-gatherers near the bottom of the evolutionary scale. First, they had few belongings. It might have been obvious that material goods were a hindrance to nomadic peoples, but nineteenth-century European scholars reversed the causal arrow: hunter-gatherers were nomadic because they were intellectually incapable of developing the technology needed to permit a sedentary existence – agricultural implements, storage facilities, houses, ceramics, and the like. Were their moral and intellectual character to be raised, hunter-gatherers would settle down and reap the material rewards of **progress**. Second, because many were nomadic, hunter-gatherers had concepts of private property quite different from those of Europeans. Although it is incorrect to say that there are no territorial boundaries among hunter-gatherers, the subtlety of the ways in which huntergatherers relate people to geography was lost on European explorers and colonizers. To them, hunter-gatherers had no concept of **private property**, a sure sign of arrested development. Early twentieth-century descriptions of foragers were often so bleak that they left students wondering 'not only how hunters managed to make a living, but whether, after all, this was living'.

However, the Second World War and subsequent events led to a major change of mindset, as noted above. In some cases, such a change lead to the opposite extreme, that is, to a notion of a completely "noble," "good," "pure," altruistic "savage" that is far more extremist—and biased—than anything that **Rousseau** ever defended. As noted by Kelly:

In 1966, seventy-five scholars from around the world met in Chicago to discuss the state of knowledge about hunter-gatherers. Organized by **Richard Lee** and **Irven DeVore**...the ***Man the Hunter*** **conference** proved to be the twentieth-century's watershed for knowledge about **foragers...**foragers of the Kalahari Desert, and especially the **Ju/'hoansi**, came to be the model **hunter-gatherers**. And not just a model but a model we should emulate. Dissatisfaction with modern life had been growing since World War I, and it came to a head in the 1960s and 1970s, with the grinding war of attrition in Vietnam, political assassinations and corruption, and widespread environmental degradation. Nineteenth-century notions of progress collapsed and, instead of an inexorable climb upward, social evolution now seemed to be a long fall from Eden. Increasingly dissatisfied, many rejected the **materialism** of Western society and searched for an alternative way of life in which material possessions meant little, people lived in harmony with nature, and there were no national boundaries to contest. It was the context for **John Lennon's song, *Imagine***, and for the numerous **hippie communes**. Hunting and gathering had kept humanity alive for 99 percent of its history; what could we learn from it? **Marshall Sahlins** answered this question with his eloquent formulation of the "**original affluent society**", perhaps the most enduring legacy of *Man the Hunter*. Prior to the conference, many social scientists saw foraging as a perpetual and barely adequate search for food. Paleolithic hunters, the argument went, adopted **agriculture** and **animal domestication** to relieve themselves of the time-consuming burden of hunting and gathering. They were evolution's success stories. Living hunter-gatherers, on the other hand, were the unfortunates who had been pushed into environments hostile to agriculture.

Spending all of their waking hours in the food quest, hunter-gatherers could not develop elaborate culture because they did not have the spare time to build irrigation systems, bake ceramics, invent complex rituals, or erect pyramids. Inspired by economist John Kenneth Galbraith's *The affluent society*, Sahlins sought to overturn this misconception with "the most shocking terms possible." He argued that ethnographic data actually painted the opposite picture: hunter-gatherers spent relatively little time working, had all the food they needed, and spent leisure hours sleeping or socializing. Their devil-may-care attitude toward the future, which many explorers interpreted as stupidity or foolishness, Sahlins claimed was an expression of self-confidence and assurance that nature would meet one's needs. The carelessness with which hunter-gatherers treated material goods, previously interpreted as an inability to recognize personal property, was, Sahlins argued, a response to a **nomadic lifestyle** in which material goods are a hindrance. In Sahlins's memorable phrase, the foraging economy was a Zen economy: wanting little, hunter-gatherers had all they wanted. He dramatized the fact that **Australian Aborigines** and the **Ju/'hoansi** work only a few hours a day, yet they did not develop civilization. The development of writing, arts, architecture, and the like required something more than just free time. Sahlins's idea of hunter-gatherers as "affluent" captured wide attention (and continues to do so).

Anthropologists' view of foraging societies became myopic, and they excluded matrilineal, sedentary, territorial, warring, and ranked foraging societies (e.g., those of North America's Northwest Coast). In archaeology, the concept of affluence had a particularly dramatic effect on explaining the origins of agriculture. Although archaeologists had long seen agriculture as a great improvement in human life, in the 1960s, we saw it as a lifeway adopted only under dire circumstances. Theories explaining the origin of agriculture focused on how population growth and migration to environmentally marginal areas forced hunter-gatherers to leave their life of leisure behind, become agriculturalists, and work for a living. A question that goes to the heart of the generalized foraging model's concept of original affluence is: how much do hunter-gatherers work, and why? Reexaminations of Ju/'hoansi and Australian work effort do not support Sahlins's claim. Kristen Hawkes and James O'Connell found a major discrepancy between the **Paraguayan Ache**'s nearly seventy-hour work week and the Ju/'hoansi's reportedly twelve- to nineteen-hour week. The discrepancy, they discovered, lay in Lee's definition of work.

Lee counted as work only the time spent in the bush searching for and procuring food, not the labor needed to process food resources in camp. Add in the time it takes to manufacture and maintain tools, carry water, care for children, process nuts and game, gather firewood, and clean habitations, and the Ju/'hoansi work well over a forty-hour week. In addition, one of Sahlins's Australian datasets was generated from a foraging experiment of only a few days' duration, performed by nine adults with no dependents. There was little incentive for these adults to forage much (and apparently they were none too keen on participating). More accurate estimates of the time hunter-gatherers spend foraging and performing other chores demonstrate that some hunter-gatherers work hard, foraging for eight or more hours a day. But many hunter-gatherers do not spend much time foraging, and some only forage every other day or so. Why don't they forage more? Do they intend to have an affluent life of leisure? At *Man the Hunter*, Lorna Marshall pointed out that Ju/'hoansi women may not work as hard as they could because, in gathering more than needed, a woman would soon be confronted by demands to share the fruits of her extra efforts and face accusations of stinginess if she refused.

As it often happens in the *world of reality*, the truth if probably in between the two extremes, as noted by Kelly:

The concept of **original affluence** cannot account for variability in forager work effort and reproduction – or for conditions that lead to increased work effort and population growth. In addition, many hunter-gatherers are also chronically undernourished and undergo dramatic seasonal fluctuations in weight and nutritional status that, for women, affect fecundity and the welfare of nursing offspring. Members of that original affluent society, the Ju/'hoansi, "are very thin and complain often of hunger, at all times of the year…it is likely that hunger is a contributing cause to many deaths which are immediately caused by infectious and parasitic diseases, even though it is rare for anyone simply to starve to death". In fact, pregnant and lactating Ju/'hoan women have a body mass index (weight/height) of 18.5, a value usually associated with chronic energy deficiency (Howell 2010). This is not just a product of contact. Archaeological data also demonstrate that prehistoric hunter-gatherers in a variety of environments lived physically demanding lives and witnessed seasonal food shortages. Life among some hunter-gatherers may also be more violent than previously thought. Per capita **homicide rates** among some hunter-gatherers, including the Ju/'hoansi, are quite high, rivaling those of large Western cities. The North American rates are higher if we take deaths due to warfare into account, and some violence results when nomadic foragers are forced into large settlements with no dispute-managing apparatus, or when alcohol becomes easily available.

Nonetheless, the Ju/'hoansi do experience violence, and many other hunter-gatherers fought and raided one another for **revenge**, food, and slaves. Other hunter-gatherers are quite territorial, including some in the Kalahari Desert and vigorously defend their territories, sometimes violently. The emphasis on plant food – and women's labor – in the generalized foraging model also does not apply to all hunter-gatherers. It is obviously not true of **Arctic foragers**, but it is also untrue for many who live at lower latitudes. Using Murdock's 1967 ethnographic atlas, Carol Ember showed that as a simple statistical percentage, meat was more important than plant food and, not surprisingly, that men contributed more to subsistence than women in the majority of foraging societies. Brian Hayden also found that whereas hunted food provides a mean of only 35 percent by weight in a sample of forager diets, it provides at least half of many groups' total caloric needs. Others have found that the alleged egalitarian relations of hunter-gatherers are pervaded by inequality, if only between the young and the old and between men and women. Food is not shared equally, and women may eat less meat than do men. Archaeologists find more and more evidence of **nonegalitarian hunter-gatherers** in a variety of different environments, most of whom lived under **high population densities** and stored food on a large scale. Put simply, we cannot equate foraging with egalitarianism.

In fact, the most complete and fascinating book that I have ever read, and that specifically addresses the broader question raised by Kelly—"if everyone has been on earth for the same amount of time, why have some peoples made more 'progress' than others?"—, and provides a huge amount of evidence, from several fields of knowledge, to support the idea that the truth lies indeed between the "enormously brutish savage" and "amazingly noble savage" ideas about nomadic hunter-gatherers is **Jared Diamond's** magnificent 1999 *Guns, Germs and Steel*. The main aim of that book was precisely to answer exactly the same question, which Diamond designated as "***Yali question***" because his New Guinean friend Yali once asked him: "why is it that you white people developed so much cargo…but we black people had little cargo of our own?" As very briefly summarized by James Clear—author of the New York Times bestseller, *Atomic Habits*—Diamond answered that question in his 1999 book by discussing in detail a plethora of fascinating issues concerning the evolution of germs, of writing, of religion, of government, of technology, and so on, and putting them all together, concluding that:

Some environments provide more starting materials and more favorable conditions for utilizing inventions and building societies than other environments. This is particularly notable in the rise of European peoples, which occurred because of **environmental differences**

> and not because of **biological differences** in the people themselves. There are four primary reasons Europeans rose to power and conquered the natives of North and South America [and New Guinea and many other places], and not the other way around: 1) the continental differences in the plants and animals available for domestication, which led to more food and larger populations in Europe and Asia, 2) the **rate of diffusion of agriculture, technology and innovation** due to the geographic orientation of Europe and Asia (east-west) compared to the Americas (north-south) [and Africa], 3) the ease of intercontinental diffusion between Europe, Asia, and [North] Africa, and 4) the differences in continental size, which led to differences in total population size and technology diffusion.... The primary geographic axis of North and South America is north-south. That is, the land mass is more longitudinal than latitudinal. The same for Africa. But for Europe and Asia, the primary axis is east-west. Interestingly, this positioning and shape matters greatly because it appears that agriculture and innovations spread more rapidly along east-west axes than along north-south axes...[because] locations along the same east-west axis share similar latitudes and thus have similar day lengths, seasons, climate, rainfalls, and biomes...all of which increase the speed of innovation relative to north-south axes. Geographic location is a key determinant in the pace of technological innovation and acceleration because a centrally located society will not only accumulate knowledge and technology from their own inventions, but also from neighboring societies. In the case of a particularly large land mass like Eurasia, technologies can spread from one culture to another and continue to do so along the entire span of the continent. This spread occurs much more quickly in these locations than it would to, say, aboriginal cultures in Tasmania, which did not receive outside contact from other civilizations for over 10,000 years. One collection of evidence for the difference in spread along geographic axes is the spread of domesticated crops. Many crops spread across Asia with one domestication, while crops like cotton or squash were domesticated in multiple individual areas throughout Mesoamerica. This is because the crop spread too slowly for one domestication to takeover the region.
>
> [Related to these differences in geography leading to dramatic regarding the evolution of agriculture, technology and innovation, the evolution of *germs* was also crucial...for instance] North America was populated by about 20 million Native Americans when Columbus landed in 1492...within two centuries, 95 percent of the native population had died, most of them from infectious diseases. One reason farming communities developed immunity to diseases that wiped out hunter gatherer populations is that some diseases ([in particular **infectious diseases**] like **measles**) are “**crowd diseases**”.... They require a large population to sustain themselves because they act quickly: you either die or develop immunity. In order for the disease to sustain itself there must be enough new babies born to contract the disease from those who have already developed immunity. Only agricultural communities could grow to the required population size.... [Lastly, *guns* were also crucial because] agriculture allowed food production per unit area to increase, which meant a given area could support a larger population...this allowed farming cultures to defeat hunter gatherer cultures by sheer force due to larger populations. This, in turn, led to the spread of more **agricultural societies** across the globe. [Importantly, this means that] **agriculture** did not lead to an unequivocally better lifestyle. In fact, for those who actually grow food life tends to be worse than it would be as a hunter gatherer...if this is true, and the evidence seems to point that way, then it means that advancement of civilization has essentially happened on the backs of **society’s have-nots**. In other words, the entire system we live within – agriculture, **capitalism**, etc. – **requires inequality** to function.

Coming back to Kelly’s book, for me the most effective aspect of that book is that it clearly deconstructs the narrative that hunter-gatherers have a single “way of life.” He makes a superb job in distinguishing hunter-gatherers that tend to live in smaller, nomadic groups from those that tend to live in bigger and more sedentary groups, and notes that even within each of those two main divisions, there is a huge

Fig. 4.1 Figure adapted from Diamond's 2012 book, who wrote about it: "First contact: a New Guinea Highlander weeps in terror at his first sight of a European, during the 1933 Leahy Expedition"

diversity. Moreover, he emphasized that it is extremely difficult to know exactly how the **lifeways of hunter-gatherers** truly were before they were firstly contacted by agricultural societies because "long before anthropologists arrived on the scene, hunter-gatherers had already been given diseases, shot at, traded with, employed, and exploited by colonial powers or agricultural neighbors" (see Fig. 4.1):

> The result in many cases (some would say all) was dramatic alterations in huntergatherers' livelihoods. Family trapping territories among the Canadian **Naskapi** and **Montagnais** were probably adaptations to the fur trade rather than precontact forms of land ownership. The **Micmac** division of labor shifted, moving men but not women into the more public and prestigious arena of trade with Europeans. Virtually no hunter-gatherer in the tropical forest today lives without trading heavily with horticulturalists for carbohydrates or eating government or missionary rations. Some foragers retreated into forests or deserts to avoid conscription, taxes, and the administrative arms of colonial powers. **Madagascar's Mikea**, for example, retreated to the forest to avoid slavers in the nineteenth century and again, in the 1960s, to avoid taxation. Yet other foragers today forage as a way to affirm their cultural worth, as a political message that only makes sense in a world of enclaved minorities. There

can be little doubt that all ethnographically known hunter-gatherers are tied into the world economic system in one way or another; and, in some cases, they have been linked to it for hundreds of years. Foragers are not evolutionary relics. The concern with contact-induced change threatens to reduce analysis of variability among **hunter-gatherers** to yet another stereotype, one that focuses on issues of power and control, that treats modern hunter-gatherers as disenfranchised rural proletariat, and that ultimately denies the usefulness of the study of modern hunter-gatherers for understanding prehistory.

This is as much an oversimplification as was the generalized foraging model. And it is as much an overstatement to claim that modern ethnography is useless to prehistory as it is naive to suppose that the effects of contact can be easily subtracted from living foragers. When defined in terms of social relations, hunter-gatherers are often divided into two types, egalitarian and nonegalitarian, or what Woodburn labels ***immediate-return* and *delayed-return* hunter-gatherers**. In immediate-return systems, no surplus is created and resources, especially food, are consumed on a daily basis. These are egalitarian hunter-gatherers and include groups such as the **Hadza**, **Mbuti**, and **Ju/'hoansi**. Delayed-return hunter-gatherers, conversely, are those who reap the benefits of their labor some time after investing it. This category includes hunter-gatherers who store food for later consumption. But, in Woodburn's view, it also includes **Australian Aborigines** because adult men give kinswomen away as brides in the expectation that their patrilineage will receive a bride back in the future; thus, men store obligations in the form of women. Extensive food storage does appear to be associated with nonegalitarian sociopolitical organizations among foragers, although it is not clear how (or even if) storage itself necessarily results in exploitation. Delayed-return or storing hunter-gatherers do not fit the model of '**primitive communism**'.

There are a number of difficulties with **Marxism** as it is applied to hunter-gatherers. Since Marx's social analysis was designed with class societies in mind, one can question its applicability to many classless hunter-gatherers. Many Marxists, however, argue that classes are not necessary for a Marxist analysis to proceed since all societies contain contradictions and exploitation at some level between groups that theoretically approach classes. For huntergatherers, the two most obvious categories are those of gender and age. Among some Australian Aborigines, for example, old men control the distribution of women as wives; young men acquire wives by obeying older men, hunting for them, and allowing them to distribute the product of the hunt. Likewise, men who have received wives are in debt to the older men who gave them wives until they are able to return a woman as a wife. Woodburn sees this as establishing inequality and exploitation between men and women, as well as between older and younger men (although, unlike true social classes where there is no or limited social mobility, all surviving young men in a group eventually become older men). Throughout the history of anthropological thought, the stereotypes of hunter-gatherers have changed from one extreme to another: from lives that are nasty, brutish, and short to ones of affluence; from a diet of meat to a diet of plant food; from egalitarianism to inequality; from isolated relic to rural proletariat.

I completely agree with Kelly's point that researchers, and others, talk too often about hunter-gatherers as a compact, homogeneous "group" mainly to support their a priori, biased ideas about other issues such as **capitalism**—for instance to criticize it, thus describing them as "noble primitive communists," or to praise it, calling them "savage brutish egalitarians." By doing this, both the people that are supposedly "praising" and "criticizing" their constructed notion of "hunter-gatherers" are doing something typical of discrimination, racism and tribalism. That is, firstly, they refer to the "others" as a single, homogeneous group, without diversity or complexities, as it is often done in imaginary racist narratives, and secondly they take out the agency of these "others," as if they are just "primitive passive slaves of nature," what is completely false as noted by Kelly:

> The term 'egalitarian' does not mean that all members have the same of everything – goods, food, prestige, or authority. Not everyone is equal in **egalitarian societies**, but everyone has (or is alleged to have) equal access to food, to the technology needed to acquire resources, and to the paths leading to status and prestige. Even in this regard, the **inheritance of material wealth** (especially productive land) and **relational wealth** (political connections) give some individuals a head start in life. For these reasons, the key property of egalitarianism is not **material equality** (although that may result) but rather an ethos and practice of ***individual autonomy***. Many hunter-gatherers emphasize autonomy in their everyday lives. They describe their societies as those in which each person "is headman over himself". But **egalitarianism** is not simply the absence of **hierarchy**. Egalitarianism is not human nature but is itself an adaptation. Indeed, Christopher Boehm argues that human egalitarianism arose in the distant past from some kind of social hierarchy that characterizes many nonhuman primates today. Hunter-gatherers are sometimes described as being "fiercely egalitarian", not because they routinely take up arms to protect their way of life (although some might be willing to do so) but instead because *the maintenance of an egalitarian society requires effort*. Egalitarian relations do not come easily; they are not "natural" in that they are not what is left in the absence of stratification.
>
> There are people in every society who will try to lord it over others, but egalitarian cultures contain ways to level individuals, to "cool their hearts" as the **Ju/'hoansi** say. Humor is used to belittle the successful but boastful Ju/'hoan hunter; if that fails, he will be shamed with the label "far-hearted", meaning mean or stingy. The **Martu** berate such people with warnings that they are "like rocks", with no compassion. Wives use sexual humor to keep a husband in line; and gambling, accusations of stinginess, or demand-sharing maintain a constant circulation of goods and prevent hoarding. Many foraging societies contain ritualized means of defusing tensions and ending feuds (e.g., the **Australian Aborigine** penis-holding ritual, **Inuit** song duels, or the Selk'nam's wrestling matches). And, in nomadic societies, a family can simply pack up and move away from belligerent individuals. Mobility, in fact, is often what allows foragers to maintain an egalitarian ethos and practice because it permits autonomy. Sharing helps even out the variability inherent in foraging returns (especially the hunting of large game). One might think that would be a good thing, and it is. But sharing can also create tension because it establishes debts and proclaims differences in ability, and so self-effacing behavior makes sharing easier. A hunter who acknowledges his worthlessness as his wife distributes meat from a fat antelope he has just killed relieves the tension of sharing. He is saying, "I know I'm a good hunter…. I know you owe me…but I'm not going to use that against you". And that behavior creates and is created by a culture that is assertively egalitarian, one in which the open hoarding of goods or the imposition of one's will on another is at odds with cultural norms.

Another crucial point made in Kelly's book is that, contrary to the views that are often portrayed in popular culture, the origin of major **inequalities** as well as of so-called civilization does not necessarily relate to the **rise of agriculture**, being instead also very often linked to the transitions from so-called simple to so-called complex hunter-gatherer groups. Although I consider that Kelly's use of the terms "simple" and "complex" is unfortunate—and inaccurate, as these terms seem to reinforce the notion of "progress" that Kelly is precisely criticizing in his book—, his table (shown in Fig. 4.2) is a powerful reminder that there are many more key items shared by **agricultural societies** and **"complex" sedentary hunter-gatherer groups**—such as large settlement sites, **sedentism**, **higher population density**, **more inequalities** and **warfare**, **use of slavery**—than by the latter groups and **"simple" nomadic hunter-gatherer groups**. The main difference between sedentary hunter-gatherer groups and agricultural societies is, as

	Simple	Complex
Environment	Unpredictable and/or variable	Highly predictable, less variable
Diet	Terrestrial game, or game/plant food mix	Marine or plant foods
Settlement size	Small	Large
Residential mobility	Medium to high	Low to sedentary
Demography	Low population density relative to food supply	High population density relative to food supply
Food storage	Little to no dependence	Medium to high dependence
Social organization	No corporate groups	Corporate descent groups (e.g., lineages)
Political organization	Egalitarian	Hierarchical; classes based on wealth and/or descent
Occupational specialization	Only for elderly	Common
Land tenure	Social boundary defense	Perimeter defense
Warfare	Rare	Common
Slavery	Absent	Frequent
Ethic of competition	Not tolerated	Encouraged
Resource ownership	Diffuse	Tightly controlled
Exchange	Generalized reciprocity	Wealth objects, competitive feasts

Fig. 4.2 So-called simple *versus* complex hunter-gatherers, showing again that even the most careful, and knowledgeable scientists often recur to use terms based on factually wrong teleological narratives about "progress" and "complexity"

their names indicate, that the former don't use domesticated animals and plants, instead hunting wild animals and gathering wild plants. But that does not mean they cannot have high agglomerations of people, or live in houses or have statues: many sedentary hunter-gatherer societies do, as shown in Fig. 4.3. But this has nothing to do with having more or less "complexity" than nomadic hunter-gatherer groups. As will be explained below, more and more authors are actually recognizing that a more sedentary type of life, in particular one in which there is an agricultural mass production of crops, or an industrial mass manufacture of items such as cars or bottles that require the reiteration of mainly repetitive, "no-brain" tasks for many hours, every single day, probably leads to a much more monotonous, less mentally challenging, and thus in this sense "less complex and diverse" daily life.

These topics are further discussed in Kelly's excerpts included in Box 4.1. Firstly, he explains one of the biggest paradoxes analyzed in the present chapter, that is, why so-called technological revolutions and progress are *in reality* often associated to people having to "work" in average *more hours* per day and thus *less time for leisure* (see also Box 4.3). In addition, he explains that women, and children, were in general the most affected ones by such "revolutions," a subject also discussed in Box 4.4. Having said that, Kelly, always careful to avoid generalizations, also points

Fig. 4.3 An house, but not of an agricultural society: the interior of a Nootka (Nuuchahnulth) house, Vancouver Island, displaying the daily-lives of Northwest Coast's hunter-gatherers—pen-and-ink drawing by John Webber, April 1778, photographed by Hillel Burger. Adapted from Kelly (2013), who wrote about it: "Note stored, dried fish hanging from ceiling; the cedar boxes for ceremonial paraphernalia on shelves; the decorated whale dorsal fin on the bench to the left…to the right, low plank walls separate family units in the house…the women in the center are roasting fish and heating water with stones from the fire"

out that this does not necessarily mean that there was/is a strict male-female egalitarianism in all nomadic hunter-gatherers societies. He explains:

> Prior to *Man the hunter*, women in foraging societies were often seen as chattel and slaves, dominated by male authority in the realms of subsistence, **marriage**, religion, and sex. After *Man the hunter*, however, anthropologists portrayed hunter-gatherers as useful role models for a Western society striving for **gender equality**. The argument was that female foragers provide as much (if not more) food as men do and therefore have a status equal to that of men. However…this is almost never actually true…**Ju/'hoan men** do about two-thirds of the talking at public meetings and act as group spokespersons more frequently than do women…. In domestic conflicts, **Ju/'hoan women** are more often the victims than are men; the same holds true for **Australian Aboriginal women**. And yet, even where women have less public authority than men, they can still exert power, often using a culture's own precepts…still, differences in power can result in real impacts in well-being. Some studies suggest that foraging women eat less meat than do men…. **Aka women** have significantly more caries than **Aka men**, suggesting that they eat more carbohydrate and less meat than do men. And, in a number of societies, women are forbidden to eat fat during pregnancy and lactation, just when they could use the extra calories, fat-soluble vitamins, and fatty acids. Sanday also found that colonialism increased male dominance in indigenous societies, although she saw this as a function of changes in resource availability and men's *versus* women's tasks (e.g., **warfare** and rebellion) rather than a straightforward imposition of European customs.

Box 4.1 Sedentism, Leaders, Inequality, Slavery, and "Civilization"
Kelly's 2013 book *The Lifeways of Hunter-Gatherers* states that:

> Through analysis of thirty-three foraging societies, Keeley found that **sedentism** (defined as a stay of longer than five months in one village), food storage, and population pressure were all correlated with **nonegalitarian organization**. In fact, Keeley concluded that population pressure "fits very well the expectations for a necessary and sufficient condition for and the efficient cause of complexity among hunter-gatherers". Foragers who are sedentary, store food, and have a **nonegalitarian sociopolitical organization** live under **high population pressure**. How does population pressure produce nonegalitarian organizations? From my perspective, a reduction in residential mobility due to increasing population density, which eventually results in sedentism, is the "kick" that sets sociopolitical changes in motion…sedentism results from the interplay between the distribution of food across a landscape and population density…the process will happen more or less quickly depending on the "patchiness" of the environment and on the cost of residential movement compared to the foraging return rate. Where resources are localized and defensible and travel is difficult, residential mobility will decrease quickly as population density increases. Where foragers can store food in quantity and/or rely on a food base – such as marine resources – that rebounds more quickly than terrestrial ones, we can expect a forager population to grow and to reach carrying capacity fairly quickly in archaeological time. Although standard ecological theory suggests that the growth rate should slow as foragers reach carrying capacity, as they close in on it, the diet-breadth model leads us to conclude that foragers would still have to use lower ranked resources with *high processing costs*; they would probably do so with ***technological innovations*** (e.g., acorn leaching, fish nets, boats). The result is that foragers would have to *work longer hours to achieve the same foraging return rate they experienced under a lower population density*; this work will probably *fall heavily on women* since they are generally the ones who harvest more reliable, lower-ranked foods. Working more will reduce on-demand breastfeeding but it probably also reduces aerobic work by exchanging time spent foraging for time spent processing food. At the same time, the use of stored foods may increase long-term energy balance. All of these may conspire to reduce women's energy flux and increase fecundity. On the other side of the demography coin, sedentism and storage may reduce child mortality, and the increased need for labor and the potential for peer-rearing in sedentary communities may decrease the perceived cost of children and lower the frequency of birth-spacing **infanticide**. Population size, food storage, and decreasing residential mobility are linked in a self-reinforcing cycle.
>
> Storage creates a second problem if the resource can only be gathered in large quantities for a short period of time. This appears to be the case with many stored foods (e.g., anadromous fish or migratory herd animals). The problem is that gathering the resource in bulk may *require considerable labor*. Fish weirs, for example, can require the coordinated effort of many workers, as can the rapid spearing/netting and processing of fish. The hunting of sea mammals, especially whales, requires the effort of a dozen hunters in a large boat (and someone has to make the boat); the hunting of a herd of bison also may require the efforts of many people. Making a storage economy work, then, may require that someone coordinate and/or control the efforts of some number of foragers. Hayden, for example, argues that the limiting factor on the Northwest Coast "was not the salmon, but the labor required to procure and above all process, dry, and store the salmon". Here is where the cost of **slavery** – raiding other villages and risking your own life – becomes worth the benefit (I do not mean to ignore the morality of slavery, but our perspective here is

evolutionary and thus focuses on the costs and benefits of any particular behavior). This is also where men might seek to control the labor of their sons- or brothers-in-law. This discussion suggests that the course to **inequality** is charted by *a leader's ability to organize labor*. Jeanne Arnold sees the "sustained or on-demand control over nonkin labor" as an essential element of **nonegalitarian societies**. How might this happen? One way is through the benefit foragers gain if they relinquish some of their autonomy to a leader. One of the problems of cooperative groups is free-riders, people who gain from a group's efforts without contributing their share…group leaders can arise where an economy of scale means that the per capita return rate increases with a larger group size. An example might be whale-boat crews among the **Inuit** or on the Northwest Coast. It is easy to imagine that a single forager might have a rough time (to put it mildly) taking down a whale from his personal kayak, and he would quickly see that having a crew of eight to twelve men in a large umiak or ocean-going canoe would increase his own return rate. Why would a leader be needed in such cases? For a share of the return, a leader ensures that each whaler contributes his share of labor and coordinates efforts to avoid duplication and inefficiency. Thus, leaders become more important the larger the cooperating group since (a) they take on the cost of ejecting the free-riders who can invade large groups if no one is watching, and (b) they reduce the potential for inefficiency that increases with larger groups (and this becomes especially important with **warfare**).

If new members join a group that is already at optimal size, they lower per capita returns for everyone. An answer to this problem is to enlist their participation in activities that deliver prestige to the ranking individual or family. These activities are costly signals and usually come in the form of what Paul Roscoe calls *conspicuous distribution, conspicuous performance, and conspicuous construction* – feasts, artistry, and public works. These are designed to communicate the numerical strength of a population and, importantly, the capacity of its leaders to mobilize that strength. Trade goods figure into this process as well. Hxaro goods among the **Bushmen** are visible evidence of social connections beyond the immediate group. But in the context of nonegalitarian societies, elaborate, nonfunctional goods or esoteric knowledge and immaterial goods (e.g., dances, songs) communicate elite's connections to other elites and the power they can draw on should their position be challenged. When a Northwest Coast chief held a potlatch, he was communicating to both his guests and his constituency 'this is how powerful I am; this is how many people stand behind me; this is how much labor I command; crossing me would be foolish'. The **leadership** in nonegalitarian societies must provide an explanation for why some members receive more than others, why some live in larger houses, have more stored food, or can command labor – and why others cannot adopt the tactics of the elite. And that explanation is invariably an ideology (e.g., my forefathers founded this village, so you are permitted here by the good graces of my family). Costly displays, such as feasts, are visual demonstrations of the "honesty" of this claim. And warfare is perhaps the most definitive and most costly of all possible displays…leaders in **nonegalitarian societies** control access to resources either by controlling physical access to key resource-extraction localities or by controlling the necessary technology…in so doing, they can exchange access for the labor of others – and hence *control* that labor. Woodburn notes that certain **Hadza** sought to acquire control over other Hadza by virtue of what they were able to acquire through their contacts with outsiders, which gave them access to valuable goods and associated them with intimidating Europeans: "in every instance…in which a particular Hadza has been said to be any sort of figure of authority…he was someone with contacts with outsiders who was attempting to use these contacts to acquire power over other Hadza". But these individuals were not accorded prestige. Instead, they were treated by other Hadza as nothing more than "rather predatory entrepreneurs". Why don't similar attitudes and behavior prevent the formation of nonegalitarian communities?

We know that **nonegalitarian hunter-gatherers** are sedentary…a shift toward sedentism may precipitate changes in the structure of foraging activities, which can alter childrearing methods from parent-reared to peer-reared and change the modal personality. Some of this shift entails changes in how people perceive individual autonomy and gender relations. Peer-reared children tend to display greater gender differentiation and to manipulate the world through social relations rather than through technology. Sanday's 1981 cross-cultural study pointed to an association of large-game hunting, a perception that the environment is hostile, and **segregation of the sexes** in work and childrearing with a predisposition for **competition** to be culturally endorsed, and for men to see women as potentially dangerous. As we have seen, sedentism establishes structural conditions that encourage men's absence from a village (long distance hunting or fishing). Eventually, as population grows, some men devote time to prestige-seeking (or -giving) activities (including trade of wealth objects and warfare) and thus remove themselves further from their wives and children. Therefore, by changing the nature of the **enculturative process**, the advent of sedentism may, after several generations, alter a population's modal personality toward one that sees social manipulation – *the control of another's labor* – and competition as the primary way of achieving goals. Peggy Sanday's 1981 study suggests that this may be especially true for men, and thus it sets the stage for the **manipulation of women by men**. Variability in **inculturation** within a community could also promote inequality. Children of high-ranking families will learn a different set of values and expectations than the children of low-ranking families. If high-ranking men invest time in prestige-seeking activities and have additional wives or slaves to care for children, then they may spend little time with their children. As a result, children in high-ranking families may be more heavily impacted by the general enculturative process we have associated with sedentism, and sons would see competition and **social manipulation** as the keys to success. And, through inheritance, they would have the capacity to do so. Low-ranking men, being limited in their resource-acquiring potential, may devote more time to childcare and raise children who are less inclined to competition. This leaves children of low-ranking families open to **exploitation** by the competitive attitudes and greater resources of children of high-ranking men. If true, this would help account for Collier's observation that in unequal-bridewealth societies, people see their fortunes in *life as being controlled primarily by birth.*

The last part of Box 4.1, in which Kelly associates **schooling** and **education** with **inculturation** and the promotion and maintenance of **inequality** can be mind-blowing for most people at first sight, as schooling is often seen as one of the most "noble" features of, and within, Western "civilization." In this sense, Kelly's remarks are very similar to a point famously made by **Claude Levi-Strauss**—which is probably also mind-blowing for many readers, because writing is also commonly see as a landmark of "civilization": "*writing appears to be necessary for the centralized, stratified state to reproduce itself…has accompanied…the integration…of a considerable number of people into a hierarchy of castes and classes…it seems to favor rather the exploitation than the enlightenment of mankind.*" Of course, I am not saying that schools and education are just "bad" and should be terminated, or that we should not promote education for children in let's say poor rural communities of the U.S. or African countries, because, within the current sociopolitical system, that is probably one of the very few chances they will have to be able to take part of the "cake" within their countries' economies. But if we step back and think about this, it is obvious that this is so precisely due to the fact that these economies are now part of a widespread dominant socioeconomic system that we are enculturated to see as "normal" or even as the "only possible way." But that is clearly not the case, because for 99% of the six millions of years of our human lineage, as well as for 100% of the billions of years of life in this planet before humans existed, there were no "modern schools." So, what we tend to see as "normal" today is often just a *Neverland* construction, the result of inculturation, that does not reflect at all—and it's frequently actually the opposite of, as in this specific case—what is truly "normal"—that is, the norm—within the *reality of the natural world.*

For, let's say, a **Caduveo indigenous girl** (see Fig. 9.4), it would surely not seem "normal" at all if someone would tell her that she needs—actually, in many countries, that she is obliged—to go to a "school" and sit down for hours and hours, standing still, to learn for instance the importance of things like "money" or of "working" 8 hours often doing tasks one does not like to do, or of using a seat belt. Things are only important, or "necessary," or seen as "normal" within a certain specific socioeconomic system, at a certain moment in time and geographical place. Unfortunately, we tend to forget this. In fact, as noted by Kelly and by Levi-Strauss, this is precisely one of the main goals of inculturation: to forget such facts. That is why inculturation has, historically, been so successful in maintaining inequalities in not just the ways mentioned by Kelly, but also in many other ways, including being part of a system that is ruled by using another long-standing habit in politics: the "*Divide et Impera*" (see Figs. 4.4 and 9.5). This Latin phrase—an approach attributed to the Roman emperor **Julius Caesar**—, means divide and conquer, or better said, **divide to conquer**.

Such a "divide to conquer" attitude continues to be as prevalent today as millennia ago. In the last months of 2019 we have assisted to mass protests in a huge number of very diverse regions: Bolivia, Chile, Colombia, Lebanon, Hong Kong, Iraq, Egypt, and so on. And what is amazing is that most people—either that are in those regions or that are not and that see or read about them—continue to fall into a trap that does not allow them to realize who are those that are truly creating the problems—for instance by being corrupt—and that are therefore enriching with those problems and increasing the inequality (see Figs. 4.4 and 9.5). That is, most people continue to be divided into narratives related to **left wing *versus* right wing**, **communism *versus* capitalism**, and so on, while it is obvious that people from both the left and right wing can be corrupt, and benefit from a system of inequality, or oppress people. Not only is this obvious at a logical level: it has been also demonstrated over and over in history. As an example, some media state—and many people from the left agree—that the protests in Chile announce the end of **neoliberalism**, while other media defend—and many people from the right concur—that the protests in Bolivia against Evo Morales announce the end of the **"left wave" in Latin America**, particularly because one of the most decadent countries in the region right now is Venezuela, a so-called socialist

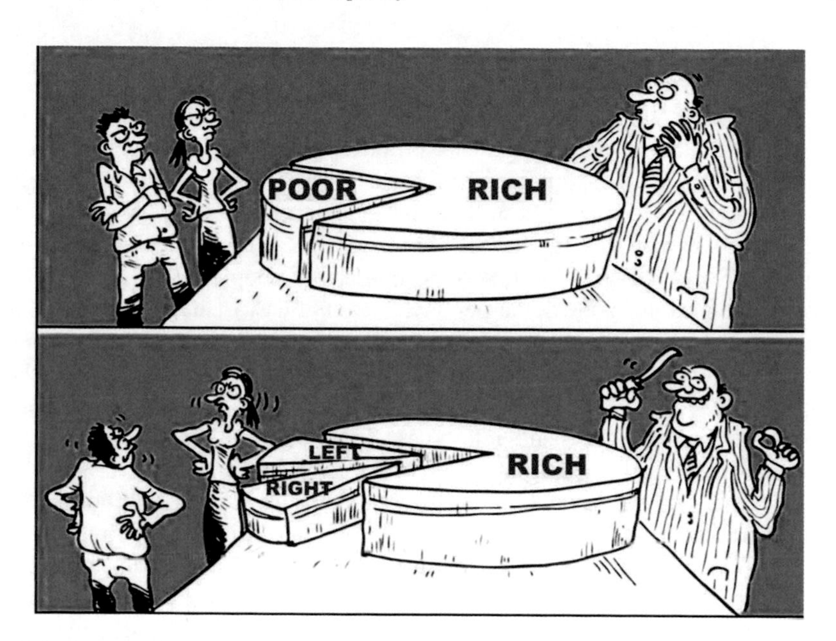

Fig. 4.4 The *"Divide et Impera"* approach attributed to the Roman emperor Julius Caesar, which means divide and conquer, keeps being commonly used nowadays to help maintaining the *status-quo*: the only difference is that the narratives used to do so are recycled, from time to time, so people continue to be distracted—for instance trough political partisanship—about what is truly happening around them in the *world of reality*, including the continuation of huge social inequalities

country. But when **Evo Morales** then left Bolivia, there were huge protests in Bolivia, but they were instead against the "right-wing politicians" that gained power in that country, and accordingly when there were new elections after that, the left wing politicians won again. Then when many see the protests in Lebanon, they say that those people "are just Arabs, it's their normal chaos." What most people seem to be unable to realize is that the common pattern is not left *versus* right, communism *versus* capitalism, Arabs *versus* others, and so on. It is instead people that are, and that want to continue to be, in power, mainly because they benefit from it, either because of corruption or because politicians in general live a type of life that that most people—including those that voted for them—would never be able to live: it is the **cycle of inequality**, as brilliantly shown in the cartoon of Fig. 4.4.

Obviously, some leaders might have, at least at the beginning of their mandates, some noble intentions. I am an optimist, not a cynic. If I tell you the name **Barack Obama**, most of you, even those that might normally not vote in, or appreciate ideas of, the U.S. Democratic party, would probably not think that he tried to be President just to become richer. However, this does not mean that, by being President, he did not become richer. Clearly, he did: he had, and continues to have, a type of life that 99%, or even 99.9% of the 8 billion people alive nowadays can't have even in their dreams. U.S. Presidents have a huge number of privileges even when they leave office, such as a monthly pension for life at the rate of the head of an executive department; an office staff, with an aggregate salary up to $96,000 per year, office space appropriately furnished and equipped; and up to $1000,000 per year for security and travel related expenses, including secret service protection. In addition, their spouses are also protected, as are their children until they are 16. Former U.S. presidents, their spouses and their children are also entitled to have free health treatment at a military hospital for their lifetimes, and former presidents are furthermore granted state funerals with military honors, depending on the wishes of their family. Moreover, after leaving office, Obama gave numerous speeches, some earning him up to $400,000 an hour. A huge number of people living in agricultural states live with less than one dollar per day, so even if they manage to live until they are 70 under such a huge poverty, this means that they would have less than $25,000 during their whole lives, that is, 16 times less than Obama can make in a single hour.

So, we can be optimistic, but should not be naive. This *is* the system, a system plagued by huge inequalities and unfairness. And we are talking about Obama, who represents a very peculiar case, particularly concerning the history of "civilizations" as a whole, because most people of the globe considers that he did not try to be President just to make money or to escape justice: that is precisely why I used his example. Moreover, this example concerns the U.S., which is one of the most democratic countries in the planet, and has been a democracy for more than two centuries. That is, this obscene example of privilege and inequalities is nothing compared to what most political leaders or kings had *and* were able to do during the history of "civilizations," including having total control of a whole country, or being in power for decades until they died, or leading to wars that have killed millions of people, or having harems, or simply raping or forcing thousands of women to have sex with them. To give just an example, **Genghis Khan's** conquests are said to have led to about 40 million casualties, and to a total of about 12 million people that are his descendants nowadays.

Right now, in the planet, there is a huge number of dictators, as well as of so-called democratic authoritarian leaders such as Putin or Duterte, or religious leaders that control entire countries such as Iran's ayatollahs, or kings such as those of Thailand. Let's see what the newspaper I read everyday says *today*, 20th January 2021, *on a single page of the International section*: a woman has been condemned to spend 43 years—yes, 43 years—in jail merely because she criticized the king of Thailand; Navalny, after being poisoned by men obeying to Russia's current President Putin, tried to return to Russia to just be directly put in jail; in Uganda, after the very dubious re-election of Museveni, who has been already in power for 35 years, the leader of the opposition Wine was also arrested, in his own home. Fortunately, this does not happen in every country as it did in "early civilizations," but a huge number of so-called modern countries don't differ so much from what happened in those earlier states, for instance concerning social inequities. History and imaginary narratives just repeat themselves over and over, in the world of *Neverland*. Even in "developed" countries that supposedly have a "functional" democratic process, there are numerous political leaders such as **Netanyahu**, who desperately wants to continue to be a leader in order to not face charges of fraud, bribery and breach of trust. That leaders can escape from such charges *because* they are in power shows how turned-up-side-down this *Neverland* system is: leaders should be subject to higher standards, not lower, than most people, if they are supposed to *lead by example*. But still, just some weeks ago, the 27th of November 2019, Netanyahu supporters attacked the justice system, defending Netanyahu against the "leftists," the "media" and so on.

It is however important to emphasize that authors such as Kelly and Levi-Strauss are not simply stating that hunter-gatherers, even those that live in small and more mobile groups, are all noble "natural conservationists." It is true that being sedentary per se, particularly when associated to agriculture, is obviously related, in general, to more changes within and damage of the natural environments, but this does not mean that nomadic hunter-gatherer groups never damage the environment. Moreover, even those nomadic groups that don't do so, it is not necessarily *always*—although in many cases it is, indeed, as will be seen below—a matter of being "consciously noble": it may instead be mainly related to the specific type of life, mobility, and diet displayed by those more mobile groups, for instance. In *The Lifeways of Hunter-Gatherers*, Kelly explains:

> Prior to *Man the hunter*, foragers were seen as giving no thought to the future, unconcerned with the impact today's actions might have on tomorrow. A significant result of Sahlins' portrayal of hunter-gatherers as "affluent", however, was to overturn this perception. By the late 1960s, anthropologists saw foragers as intentionally managing their resources. Optimal foraging theory acknowledges and can accommodate foragers' intentional or unintentional modifications of their environment. **Conservation ethics** are reflected in spiritual beliefs as well. The **G/wi** believe that N!adima (God) will be angered if they do not leave enough plants behind for regeneration. The **Waswanipi Cree** look upon animals as *chimiikonow*, or gifts, because animals are "like persons", who act willfully and intelligently, and who give themselves over to a hunter who has lived up to standards of reciprocity. The Cree and **Naskapi** believe that improper acts, such as killing more animals than needed, ignoring obligations to share, or treating the remains inappropriately, result in retaliation by game, who might not allow themselves to be captured. At the same time, however, we can point to instances in which foragers did indeed overhunt their prey. For example, after **Alaska's Inupiat** exterminated caribou and mountain sheep in one part of their range, they simply moved to another part, where they did it again. Foragers nearly hunted beaver to extinction in eastern Canada during the seventeenth-century fur-trade frenzy. When the **Nuvugmiut** of northern Alaska would drive molting birds into nets, they would fill their umiaks with the adults while 'thousands of downy young are…thrown away'. And, finally: foragers periodically burn land, ostensibly to attract game by promoting young growth.
>
> The **Alaskan Tanana**, for example, burned hillsides to promote the growth of willow shoots as forage for moose. Likewise, **native Californians** claimed that burning increased the size and abundance of tubers. The question is whether these behaviors that allegedly conserve resources are *intentionally* directed at resource conservation or whether conservation is an unintentional (although real) consequence of optimal foraging by low-density human societies. Shifting hunting territories may permit game to rebound, and burning may promote growth and **biodiversity**, but is that why foragers shift their hunting territories or burn their land? The **Piro** are not "natural" conservationists; in fact, no one is. But it is also clear that many former foragers do indeed think about the connection between today's actions and tomorrow's consequences. One has only to look at their feelings toward the development projects on their ancestral lands to understand this. Seeing hunting territory disappear beneath mammoth dams and reservoirs, **Canada's Cree** were puzzled and saddened by what they see as the wanton and irreversible destruction of resources that belong to future generations. Robert Brightman suggests that a widespread ethic of resource conservation among **North American boreal forest peoples** is a post-fur-trade phenomenon,

replacing an ethic in which to not kill an animal when encountered would result in a lack of game in times of need – since the hunter is "refusing" the animal's gift of itself. But it is unlikely that conservation was an ethic imported from Western society (the culture that brought us Love Canal, global warming, and extinction of the passenger pigeon).

One of the most weird theories defended by some of those who have defended and continue to defend that hunter-gatherers are just "brute savages," is that hunter-gatherers cannot even survive by themselves on rainforests. Such imaginary stories are typical of an "agriculture-centered" view of human evolution. For instance, the idea that the so-called central **African forest pygmies** needed the "help" of "**Bantu farming populations** with superior technology…and a reliable subsistence base – domestic plants and animals" to survive was discussed by Lupo and colleagues in the 2014 book *Hunter-Gatherers of the Congo Basin*:

> Despite the widespread acceptance of the view that rainforests lacked sufficient wild starches to support full-time foragers, a number of scholars challenged this basic assumption. Botanical surveys in concert with ethnobotanical studies conducted in forested west central and central Africa identified high densities of edible wild yams (*Dioscorea* spp.) in some areas. At the same time, ethnographic studies show that some forest foragers practice a type of paracultivation or incipient management of wild yams by replanting stems after harvesting. Management practices, ritual beliefs and traditional knowledge of wild yams imply an established long-time depth of exploitation. Experimental studies conducted among forest foragers demonstrate that foraging for **wild yams** can seasonally yield high returns that rival those derived from other **domesticated starches**. Although domesticated foods are clearly important, wild yams comprised a consistent proportion of the diet for many forest foragers and in some locations were seasonally the primary starch source. For example, Kitanishi (1995) found that the proportion of wild yams consumed by **Aka** in northeastern Congo varied seasonally but during the late wet season greatly exceeded the amount of domesticated foods exploited. More recently, Yasouka (2006) documented forest foraging trips (so-called molongo's) spanning several months during which **Baka foragers** in Cameroun subsisted entirely on forest products including wild yams. In addition to these studies, archaeological research conducted in the central African rainforest and in tropical forested regions throughout the world shows occupation by human foragers that predates the emergence of domesticated food sources. Cumulatively, these archaeological and ethnographic studies weaken the underlying assumption of a nutritional basis for these relationships and reopen lingering questions about the timing, nature and catalysts that gave rise to these inter-relationships.

It is truly amazing how self-evident facts—humans were foragers for at least millions of years before agriculture arose some millennia ago—can be neglected, or forgotten, or simply ignored, to create such "progress," or "agriculture-centered," or "civilization" narratives. In the same book, a chapter by Lewis provides a very good account on the **egalitarian social base** of the so-called **forest pygmies** that goes in line with—and sometimes expands—the ideas of Kelly. Lewis notes that "while both delayed and immediate return societies exist among hunter-gatherers, only delayed return societies exist among non-hunter-gatherers." Importantly—and this is one of the key points that I want to stress in the present volume—, he explains that in hunter-gatherer societies that are politically egalitarian "*no one can force others to do their will* – people who brag or try to assert their wishes or views on others are mercilessly teased, fought, avoided and if they persist even exiled." This point contrasts with a central aspect of **hierarchical societies**: that many people within the 99%, particularly those that are "at the bottom," mainly work—and sometimes even "live"—for the *will and/or dreams of others*. This reminds us of a famous quote "*life is too short to be living somebody else's dream,*" which was not said by a philosopher or scholar, but by **Hugh Hefner**, the founder and editor-in-chief of **Playboy magazine**—and one thing is sure, in this regard, he did apply his quote during his lifetime.

When people work on building construction, or even in a bank, 8 or more hours per day, for years, basically they are working for *the dreams of others*: for instance, the people that drove Hefner from place to place, or that cleaned his mansions, or that helped to produce the Playboy magazine, most of them were working for Hefner's dream, as the people that are building the Trump towers are doing that not for their dreams but for someone that has the dream of seeing his name in big edifices across the planet. The fact that we simply tend to accept such **hierarchically organized societies** as a given is a fascinating, and troubling, fact, particularly when we realize how recent—a mere thousands of years—such a system is. As noted by Lewis, egalitarian "societies are indeed rare today…[they] include some Pygmy groups in Central Africa (**Aka**, **Baka**, **Efe**, **Mbendjele**, **Mbuti**); **Hadza** in Tanzania, some **San** groups in Namibia and Botswana; several groups in India such as the **Jarawa** and **Ongee Andaman Islanders**, **Hill Pandaram** and **Nayaka**; and in south-east Asia, the **Agta**, **Batek**, **Maniq**, **Penan**, and others." Although "numerically insignificant today, these societies are hugely significant for anthropology since their **egalitarian immediate-return orientation** represents such a radically different mode of social organization to the numerous hierarchically organized delayed-return systems that currently dominate human societies." He further notes:

> Reviewing how his typology had stood up to the evidence from 30 years of new ethnography Woodburn (2005) noted that **immediate-return societies** have shown remarkable resilience over time. They are stable and enduring systems, internally coherent and meaningful to those who live in them. Despite the combined forces of government **sedentarization and assimilationist policies, agricultural expansion, industrial exploitation** and fortress conservation all putting huge pressures on these societies, they tenaciously cling to their immediate-return lifestyle. **Demand-sharing** is the core practice that ensures egalitarian economic relations. In contrast to the **donor organized sharing** familiar to most people, where the person owning the resource dispenses it according to their whim, demand-sharing is recipient controlled. Potential recipients constantly demand shares of things they suspect may be around. It is the donor's duty to give whatever they are requested, refusal is impolite even offensive. This is crucial to prevent sharing being manipulated to the donor's advantage. For most material items need determines who can claim the item, especially when they are consumable. In this context, possessing something is more like a guardianship or caretaker role until someone else needs it. Certain personal possessions, such as a woman's basket, her cooking pots, and machete, and a man's bag, his spear, knife and axe, are recognized as belonging to named individuals, often the person who made, found, took or bought the item. These individuals have priority over others' claims to the item. But when not in use by them, any of these objects will be shared on demand with someone who needs it.

> ***Ekila* taboos** [for example] serve to enforce and define proper sharing: by not sharing animals and meat properly among all present a hunter's *ekila* is ruined so that he is unsuccessful. If parents of infants eat *ekila* animals it can provoke illness and even death in their children. If either husband or wife inappropriately share their sexuality with others outside their marriage, both partners have their *ekila* ruined. A menstruating woman is *ekila* and must share her menstrual blood (also *ekila*) with spirits so that her male relatives continue to find food. Even laughter should be shared properly. Laughter shared between people in camp during the evening makes the forest rejoice, whereas laughing at hunted animals ruins the hunter's *ekila*. Everyone is encouraged to share according to ability, but if you are old, physically or mentally challenged in some way and only rarely contribute, your entitlement is not diminished. The principle is that if someone has something that you need just ask them for it; and, as **Mbendjele** often say '*since we have easy hands we just give it*'. Mbendjele adults should epitomize this quality by being generous to a fault – so they will give away all that is asked of them even when this results in them having nothing left for themselves. They contrast this behaviour with the 'hard hands' of **Bilo** (**Bantu**) villagers. In an egalitarian society no one can play the role of 'judge' since this would imply status or authority. Occasionally people may discuss a recurring problem and collectively suggest a solution, but no-one has the authority to impose it. Often one party simply moves away, without even acknowledging the dispute. This doesn't mean that Mbendjele do not recognise individual skill or expertise, but rather that such recognition is not associated with any special advantage or privilege. Specialist roles are held by both men and women, except for men's role as tuma (elephant hunter). While the titles konja mokondi (spirit guardian), kombo (song composer), lipwete (speaker) and nganga (healer) recognize that certain people are particularly skilled or knowledgeable about a particular activity, they get no privilege or special treatment from this recognition. Rather each role is recognized because the activities that they are associated with are potentially dangerous or stressful to the community as a whole. People described by one of these titles are expected to manage these stressful situations well for the benefit of all, so that they have a positive outcome.

In his 2012 book *The world until yesterday*, **Diamond** provides various examples illustrating how the education and play of children in hunter-gatherer groups often involves cooperation and sharing, rather than winning and losing as it often does in so-called developed countries (see also Box 4.2):

> For instance, the anthropologist Jane Goodale watched a group of children (the **Kaulong people** of New Britain) who had been given a bunch of bananas sufficient to provide one banana for each child. The children proceeded to play a game. Instead of a contest in which each child sought to win the biggest banana, each child cut his/her banana into two equal halves, ate one half, offered the other half to another child, and in turn received half of that child's banana. Then each child proceeded to cut that uneaten half of the banana into two equal quarters, ate one of the quarters, offered the other quarter to another child, and received another child's uneaten quarter banana in return. The game went on for five cycles, as the residual piece of banana was broken into equal eighths, then into equal sixteenths, until finally each child ate the stub representing one-thirty-second of the original banana, gave the other thirty-second to another child to eat, and received and ate the last thirty-second of another banana from still another child. That whole play ritual was part of the practicing by which **New Guinea children** learn to share, and not to seek an advantage for themselves.

Diamond gives several other examples of crucial differences between the **education** of children in small nomadic groups *versus* more sedentary/agricultural ones, including **physical punishment** and **child autonomy**, although of course one should be careful to not over generalize, as there are huge differences within each of these major types of societies:

Similarly, in modern Africa the **Aka Pygmies** never beat or even scold their children, and they consider horrible and abusive the **child-rearing practices** of neighboring **Ngandu farmers**, who do beat their children. However, there does seem to be a broad trend: most **hunter-gatherer bands** do minimal physical punishment of young children, many **farming societies** do some punishment, and **herders** are especially likely to punish. One contributing explanation is that misbehavior by a hunter-gatherer child will probably hurt only the child and not anyone or anything else, because hunter-gatherers tend to have few valuable physical possessions. But many farmers, and especially herders, do have valuable material things, especially valuable livestock, so herders punish children to prevent serious consequences to the whole family – e.g., if a child fails to close the pasture gates, valuable cows and sheep can run away. More generally, compared to mobile societies of **egalitarian hunter-gatherers**, **sedentary societies** (e.g., most farmers and herders) have more **power differences**, more gender-based and age-based and **individual inequality**, more emphasis on learning deference and respect – and hence more **punishment of children**. Similar attitudes prevail among most other hunter-gatherer groups studied. If one **Aka Pygmy** parent hits an infant, the other parent considers that ground for divorce. The **!Kung** explain their policy of not punishing children by saying that children have no wits and are not responsible for their actions. Instead, !Kung and Aka children are permitted to slap and insult parents. The **Siriono** practice mild punishment of a child that eats dirt or a taboo animal, by roughly picking up the child, but they never beat a child, whereas children are allowed to have temper tantrums in which they beat their father or mother as hard as possible. In addition, how much freedom children enjoy seems to depend partly on how dangerous the environment is, or is perceived to be. Some environments are relatively safe for children, but others are dangerous because of either environmental hazards or else dangers from people. Consider the following spectrum of environments, from the most dangerous to the least dangerous, paralleled by a range of child-rearing practices from adults severely restricting the freedom of young children to adults permitting young children to wander.

Among the most dangerous environments are the **New World's tropical rainforests**, which teem with biting, stinging, poisonous insects (army ants, bees, scorpions, spiders, and wasps), dangerous mammals (jaguars, peccaries, and pumas), large poisonous snakes (fer-de-lance and bushmasters), and stinging plants. No infant or small child left alone would survive for long in the Amazon rainforest. Hence, Kim Hill and A. Magdalena Hurtado write, "**[Ache] infants** under one year of age spend about 93% of their daylight time in tactile contact with a mother or father, and they are never set down on the ground or left alone for more than a few seconds…it is not until about three years of age that Ache children begin to spend significant amounts of time more than one meter from their mother. Even still, Ache children between three and four years of age spend 76% of their daylight time less than one meter away from their mother and are monitored almost constantly". As a result, Hill and Hurtado commented, Ache children don't learn to walk independently until they are 21 to 23 months old, 9 months later than American children. Ache children between three and five years of age are often carried piggyback in the forest by an adult, rather than being allowed to walk. Only when an Ache child is five years old does it begin to explore the forest on its own legs, but even then Ache children remain within 50 meters of an adult for most of the time.

Diamond then clarified:

Naturally, I'm not saying that we should emulate all child-rearing practices of hunter-gatherers. I don't recommend that we return to the hunter-gatherer practices of **selective**

infanticide, **high risk of death in childbirth**, and letting infants play with knives and get burned by fires. Some other features of hunter-gatherer childhoods, like the **permissiveness of child sex play**, feel uncomfortable to many of us, even though it may be hard to demonstrate that they really are harmful to children. Still other practices are now adopted by some citizens of state societies, but make others of us uncomfortable – such as having infants sleep in the same bedroom or in the same bed as parents, nursing children until age three or four, and avoiding physical punishment of children. But some other hunter-gatherer child-rearing practices may fit readily into modern state societies. It's perfectly feasible for us to transport our infants vertically upright and facing forward, rather than horizontally in a pram or vertically upright but facing backwards in a pack. We could respond quickly and consistently to an infant's crying, practice much more extensive **allo-parenting**, and have far more physical contact between infants and care-givers. We could encourage **selfinvented play of children**, rather than discourage it by constantly providing complicated so-called educational toys. We could arrange for multi-age child playgroups, rather than playgroups consisting of a uniform age cohort. We could maximize a **child's freedom to explore**, insofar as it is safe to do so. I find myself thinking a lot about the New Guinea people with whom I have been working for the last 49 years, and about the comments of Westerners who have lived for years in hunter-gatherer societies and watched children grow up there. A recurring theme is that the other Westerners and I are struck by the **emotional security**, **self-confidence**, **curiosity**, and **autonomy** of members of small-scale societies, not only as adults but already as children. We see that people in small-scale societies spend far more time talking to each other than we do, and they spend no time at all on passive entertainment supplied by outsiders, such as television, video games, and books.

We are struck by the precocious development of social skills in their children. These are qualities that most of us admire, and would like to see in our own children, but we discourage development of those qualities by ranking and grading our children and constantly telling them what to do. The Westerners who have lived with hunter-gatherers and other small-scale societies speculate that these admirable qualities develop because of the way in which their children are brought up: namely, with constant security and stimulation, as a result of the long nursing period, sleeping near parents for several years, far more social models available to children through allo-parenting, far more social stimulation through constant physical contact and proximity of caretakers, instant caretaker responses to a child's crying, and the minimal amount of physical punishment. But our impressions of greater adult security, autonomy, and social skills in small-scale societies are just impressions: they are hard to measure and to prove. Even if these impressions are real, it's difficult to establish that they are the result of a long nursing period, allo-parenting, and so on. At minimum, though, one can say that hunter-gatherer rearing practices that seem so foreign to us aren't disastrous, and they don't produce societies of obvious sociopaths. Instead, they produce individuals capable of coping with big challenges and dangers while still enjoying their lives. The hunter-gatherer lifestyle worked at least tolerably well for the nearly 100,000-year history of behaviorally modern humans. Everybody in the world was a hunter-gatherer until the local origins of agriculture around 11,000 years ago, and nobody in the world lived under a state government until 5,400 years ago. The lessons from all those experiments in child-rearing that lasted for such a long time are worth considering seriously.

Box 4.2 How Does One Learn to Be Egalitarian?

In the 2014 book, *Hunter-Gatherers of the Congo Basin*, Hewlett, who is the main editor of the whole book, provides a superb summary on the links between **education**, **egalitarianism**, **childhood**, and **gender**, based on cases studies from different groups of so-called central **African forest pygmies**:

> Men and women, young and old, are viewed as relatively equal and have similar access to resources. Respect for an **individual's autonomy** is also a foundational schema among foragers. One does not coerce others, including children. Men and women, young and old, are generally free to do what they want. If an infant wants to play with a machete, she is allowed to do so. **Congo Basin (Bantu) farmers** have foundational schema that are relatively distinct from those of the foragers: gender and age hierarchy, communalism, and material/economic dimensions to social relations. Village women are expected to defer to the requests of men, and the young should be respectful of elders, whether they are older siblings or parents. **Communalism** refers to the cultural value placed on putting the needs of the group, generally clan members or the extended family, over the needs of an individual and the importance of relying upon and expecting support from these specific others.... The importance of physical as well as **emotional proximity** to others is illustrated in two studies. In a study of conflicts between toddlers and older juveniles among **Bofi hunter-gatherers and farmers**, Fouts and Lamb found that **Bofi forager toddlers** were substantially more likely to have conflicts over staying close to juveniles, while **Bofi farmer toddlers** were more likely to have conflicts with juveniles over competition for objects or over the juvenile hitting the toddler, which never occurred among the Bofi hunter-gatherer toddlers. This study illustrates early acquisition and manifestation of foundational schema – emotional proximity to others among the forest hunter-gatherers and the **economic-material dimensions of social relations** among the farmers. In another study, **Aka forager** and **Ngandu farmer adolescents** were asked about their experiences and feelings about the death and loss of friends and relatives. Forager expressions of grief emphasized their love and emotional connections to the person, while farmer adolescents' expressions of grief focused on the material objects the child received upon the death of a relative.
>
> **Congo Basin hunter-gatherer children** are granted autonomy in their daily lives, while farmer children are subject to the control of parents and older children. For instance, Hewlett found that **Aka forager three- and four month-old infants** took the breast on their own to nurse during 58 percent of feeding bouts by comparison to only 2 percent of feeding bouts among farmers. **Ngandu farmer mothers** decided when to nurse, not the infant. At weaning, Bofi forager mothers said the child decided when she or he wanted to wean, while Bofi farmer mothers said they decided when to wean the child. The forager mothers said that if they initiated the weaning it would cause the child to get sick, whereas the farmers said nursing too long causes the child to become lazy. In a study of cosleeping among the Aka foragers and Ngandu farmers, forager parents indicated their children slept wherever they wanted, whereas farmer parents said they told their children where to sleep. Play is highly valued and occurs at all ages. The frequency declines from middle childhood to adolescence, but it is a regular feature of both child and adult life among Congo Basin hunter-gatherers. Several researchers have reported that hunter-gatherer children spend most of the day playing and are not expected to contribute much to subsistence or maintenance. By comparison, **children in farming communities** are more likely to be given responsibilities for childcare and other tasks. Studies have

shown that forest foragers learn how to share, take care of infants, and hunt and gather by age ten…. Features of the forager learning environment that enhance early and rapid learning include security and trust in others, tolerant instructors (i.e., adults do not push children away if they want to learn something), freedom to explore the natural and social environment, highly self-motivated learners, learning through highly valued play content, easy access to material artifacts and multiple skilled models, collaborative learning in multiage child groups from infancy through childhood, and plenty of time to practice and innovate.

Studies that have compared Congo Basin forager and farmer infant and early childhood consistently indicate that the foragers are more giving and responsive to their infants and young children than their neighboring farmers. The Western child development literature sometimes refers to this high responsiveness as "indulgence", which tends to imply that parents are "childcentered" and indulge (and potentially spoil) the child in whatever he or she wants. Congo Basin forager childcare is not child-focused, and parents are not trying to spoil the child. The term "giving" is used instead of "indulgence" because childcare is similar to giving and sharing food. Caregivers give to infants or young children who need or request care just as they would give and share food with others. One measure of giving and responsiveness is frequency and nature of **breastfeeding**. Aka infants and young children were breastfed on demand about four times per hour, whereas farmers averaged about two times per hour. Fouts *et al.* examined breastfeeding among Aka and Bofi forager and Ngandu and Bofi farmer three- and four-month-olds, nine- and ten-month-olds, and one- to four-year-olds and found that at all ages forest foragers breastfed more frequently, had more breast-feeding bouts per hour, and were more likely to be holding infants when nursing than did neighboring farmers.

Aka and Baka forager caregivers are also significantly more likely than Ngandu and **Bombong farmer caregivers** to respond to infant crying and fussing. Farmer infants cried significantly longer and more frequently than did forager infants in both groups. Forager caregivers responded to fussy or crying infants much more quickly than farmer caregivers and responded to most every fuss and cry event, while farmer caregivers did not respond to about one-third of the fuss and cry events, and it took them much longer to respond when they did. **Efe caregivers** also respond to infants' cries or fussing rapidly – within ten seconds of a fuss over 85 percent of the time at three and seven weeks of age and over 75 percent of the time at eighteen weeks of age. Foragers give and are responsive when infants are hungry, desire physical contact, want help walking, want attention, do not feel well, and so on. Foragers tend to feel secure and trust their social environments because responses to distress are rapid and come from many individuals in the camp. Tronick *et al.* were the first to provide quantitative evidence of **allomaternal care in Congo Basin forager early infancy** (one to four months of age). They found that **Efe mothers** were not the first to nurse a newborn; four month-olds spent only 40 percent of their time with their mothers, infants were transferred to alternative caregivers 8.3 times per hour on average, and infants were cared for by 14.2 different people on average during an eight-hour period. Farming mothers assign middle-aged girls more childcare tasks than boys, and cultural models expect girls to help more with childcare than boys.

This gender bias with middle-childhood caregivers does not exist with Congo Basin foragers. Systematic research with Efe and Aka foragers indicates both males and females in middle childhood contribute similar amounts of caregiving to infants and young children. Congo Basin hunter-gatherers are relatively distinct from neighboring farmers in that both male and female adolescents are initiated into forest spirit associations. Farmers, by contrast, often have adolescent initiation ceremonies for males, generally associated with circumcision, but seldom for females. Several studies indicate that hunter-gatherers are generally more peaceful, nonviolent or nonaggressive by comparison to farmers and peoples in other modes of production. Second, children learn how to avoid conflict and aggression by moving away from contentious situations. A quantitative study of aggression among Aka children found no sex differences in physical aggression between adult men and women, but did find that male children and adolescents were generally more aggressive than females. Several researchers have indicated that Congo Basin forager parents rarely if ever use corporal punishment to discipline children, whereas it is common among farming populations in the Congo Basin. Cross-cultural studies indicate a statistical relationship between corporal punishment and the frequency of violence and social stratification in a society.

4.2 Agriculture, Labor, Slavery, and "Progress"

Forward us a model of the [cotton] gin and you will receive your patent immediately, [Thomas] Jefferson wrote...Cotton became America's leading export, exceeding in dollar value all exports, helping...helping to expand the factory system in the North, and helping to power the Industrial Revolution in the United States.... Cotton – more than anyone or anything else – economically freed American enslavers from England and tightened the chains of African people in American slavery. (Ibram X Kendi)

The rise of sedentism and of agriculture, which occurred independently within various groups of people, was often accompanied by a *worsening* of the quality of life experienced by the vast majority of the members of those groups. This includes health issues—for instance concerning famines and pandemics (see Fig. 8.7)—as well as the rise of inequality and associated subjugation and oppression of certain subgroups, such as children, women and "others," including those used as slaves (see Boxes 4.3 and 4.4, 8.3). In *Against the Grain – A Deep History of the Earliest States*, Scott explains that the recognition of such "dark" aspects of agriculture only started to be widespread among scholars within the last decades—being more and more consensual within historians and anthropologists—because of the prevalence, for two millennia, of very influential imaginary just-so-stories about "civilization" and "progress." However, due to the typical time gap between what scholars *know versus* what the media disseminate and thus what the broader public tends to *believe*, and also due to the prevailing influence of such just-so-stories in many components of our "developed" countries—including via direct *inculturation*—, most people in such countries continue to accept such imaginary stories. Indeed, scientific facts can take decades, or even centuries, to permeate—and even more time to then be accepted by—the broader public, and sometimes might never be broadly accepted by most people. As noted by Scott in the Preface of his book:

> The astonishing advances in our understanding over the past decades have served to radically revise or totally reverse what we thought we knew about the first '**civilizations**' in the **Mesopotamian alluvium** and elsewhere. We thought (most of us anyway)

that the **domestication** of plants and animals led directly to **sedentism** and fixed-field **agriculture**. It turns out that sedentism long preceded evidence of plant and animal domestication and that both sedentism and domestication were in place at least four millennia before anything like agricultural villages appeared. Sedentism and the first appearance of towns were typically seen to be the effect of irrigation and of states. It turns out that both are, instead, usually the product of wetland abundance. We thought that sedentism and cultivation led directly to **state formation**, yet states pop up only long after fixed-field agriculture appears. Agriculture, it was assumed, was a great step forward in human well-being, nutrition, and leisure. Something like the opposite was initially the case. The state and early civilizations were often seen as attractive magnets, drawing people in by virtue of their luxury, culture, and opportunities. In fact, the early states had to capture and hold much of their population by forms of bondage and were plagued by the **epidemics** of crowding. The early states were fragile and liable to collapse, but the ensuing 'dark ages' may often have marked an actual improvement in human welfare. Finally, there is a strong case to be made that life outside the state – life as a '**barbarian**' – may often have been materially easier, freer, and healthier than life at least for nonelites inside civilization.... I suggest that the broadest understanding of domestication as control over reproduction might be applied not only to fire, plants, and animals but also to slaves, state subjects, and women in the patriarchal family. I propose that the cereal grains have unique characteristics such that they would be, virtually everywhere, the major **tax commodity** essential to early state building.... We may have grossly underestimated the importance of the [infectious] diseases of crowding in the demographic fragility of the early state. Unlike many historians, I wonder whether the frequent abandonment of early state centers might often have been a boon to the health and safety of their populations rather than a "dark age" signaling the collapse of a civilization. And finally, I ask whether those populations that remained outside state centers for millennia after the first states were established may not have remained there (or fled there) because they found conditions better.

Box 4.3 Work, Leisure, Affluence, and Myths About the Agricultural Revolution

The book *Sex at Dawn* provides a relevant compilation of ideas advanced by various authors about several topics discussed in this Section—some of these ideas might be biased, but do offer a worthy basis for further discussions on these issues:

> Prehistoric humans did not habitually store food, but this doesn't mean they lived in chronic hunger. Studies of prehistoric human bones and teeth show ancient human life was marked by episodic fasts and feasts, but prolonged periods of starvation were rare. How do we know our ancestors weren't living at the brink of starvation? When children and adolescents don't get adequate nutrition for as little as a week, growth slows in the long bones in their arms and legs. When their nutritional intake recovers and the bones begin to grow again, the density of the new bone growth differs from before the interruption. X-rays reveal these telltale lines in ancient bones, known as Harris lines. Periods of more prolonged malnutrition leave signs on the teeth known as **hypoplasias** – discolored bands and small pits in the enamel surface, which can still be seen

many centuries later in fossilized remains. Archaeologists find fewer Harris lines and dental hypoplasias in the remains of prehistoric hunter-gatherer populations than they do in the skeletons of settled populations who lived in villages dependent on cultivation for their food supply. Being highly mobile, hunter-gatherers were unlikely to suffer from prolonged starvation since in most cases, they could simply move to areas where conditions were better. Approximately eight hundred skeletons from the Dickson Mounds in the lower Illinois Valley have been analyzed. They reveal a clear picture of the health changes that accompanied the shift from foraging to corn farming around 1200 AD. Archaeologist George Armelagos and his colleagues reported that the farmers' remains show a 50 percent increase in **chronic malnutrition**, and three times the incidence of **infectious diseases** (indicated by bone lesions) compared with the foragers who preceded them. Furthermore, they found evidence of **increased infant mortality**, **delayed skeletal growth** in adults, and a fourfold increase in **porotic hyperostosis**, indicating **iron-deficiency anemia** in more than half the population.

Many have noted the strangely cavalier approach to food among foragers, who have nothing in the freezer. French Jesuit missionary Paul Le Jeune, who spent some six months among the Montagnais in present-day Quebec, was exasperated by the natives' generosity. "If my host took two, three, or four Beavers", wrote Le Jeune, "whether it was day or night, they had a feast for all neighboring Savages…and if those people had captured something, they had one also at the same time; so that, on emerging from one feast, you went to another, and sometimes even to a third and a fourth". When Le Jeune tried to explain the advantages of saving some of their food, "they laughed at me. 'Tomorrow' (they said) 'we shall make another feast with what we shall capture'". Israeli anthropologist Nurit Bird-David explains, "just as Westerners' behaviour is understandable in relation to their assumption of shortage, so **hunter-gatherers**' behaviour is understandable in relation to their assumption of affluence…moreover, just as we analyze, even predict, Westerners' behavior by presuming that they behave as if they did not have enough, so we can analyze, even predict, huntergatherers' behaviour by presuming that they behave as if they had it made". While **farmers** toil to grow rice, potatoes, wheat, or corn, a forager's diet is characterized by a variety of nutritious plants and critters.

But how much work is foraging? Is it an efficient way to get a meal? Archaeologist David Madsen investigated the energy efficiency of foraging for Mormon crickets, which had been on the menu of the local native people in present-day Utah. His group collected crickets at a rate of about eighteen crunchy pounds per hour. At that rate, Madsen calculated that in just an hour's work, a forager could collect the caloric equivalent of eighty-seven chili dogs, forty-nine slices of pizza, or forty-three Big Macs – without all the heart-clogging fats and additives. Another study found that the**!Kung San** (in the Kalahari desert, mind you) had an average daily intake (in a good month) of 2140 calories and ninety-three grams of protein. Marvin Harris puts it simply: "stone age populations lived healthier lives than did most of the people who came immediately after them". And maybe healthier than people who came long after them, too…. Skeletons dug up in Greece and Turkey show that pre-agricultural men in those areas were about five foot nine on average, with women being about five foot five. But with the **adoption of agriculture**, average height plummeted. Modern Greeks and Turks still aren't as tall, on average, as their ancient ancestors. Throughout the world, the shift to agriculture accompanied a dramatic drop in the quality of most people's diets and overall health. Describing what he terms "the worst mistake in human history", **Jared Diamond** writes, "hunter-gatherers practiced the most successful and longest-lasting life style in human history…in contrast…we're still struggling with the mess into which agriculture has tumbled us, and it's unclear whether we can solve it".

In his provocative essay *The original affluent society*, **Sahlins** notes that among foraging people, "the food quest is so successful that half the time the people do not seem to know what to do with themselves". Even **Australian Aborigines** living in apparently unforgiving and empty country had no trouble finding enough to eat (as well as sleeping about three hours per afternoon in addition to a full night's rest). Richard Lee's research with **!Kung San** bushmen of the Kalahari Desert in Botswana indicates that they spend only about fifteen hours per week getting food. A woman gathers on one day enough food to feed her family for three days, and spends the rest of her time resting in camp, doing embroidery, visiting other camps, or entertaining visitors from other camps. For each day at home, kitchen routines, such as cooking, nut cracking, collecting firewood, and fetching water, occupy one to three hours of her time. This rhythm of steady work and steady leisure is maintained throughout the year. In *Hierarchy in the forest*, primatologist Christopher Boehm argues that **egalitarianism** is an eminently rational, even hierarchical political system, writing, "individuals who otherwise would be subordinated are clever enough to form a large and united political coalition, and they do so for the express purpose of keeping the strong from dominating the weak". According to Boehm, foragers are downright feline in refusing to follow orders, writing, "nomadic foragers are universally – and all but obsessively – concerned with being free of the authority of others". Prehistory must have been a frustrating time for megalomaniacs. "An individual endowed with the passion for control", writes psychologist Erich Fromm, "would have been a social failure and without influence".

Difficult as it may be for some to accept, skeletal evidence clearly shows that our ancestors didn't experience widespread, chronic scarcity until the advent of agriculture. Chronic **food shortages** and **scarcity-based economies** are artifacts of social systems that arose with f**arming**. In his introduction to *Limited wants, unlimited means*, Gowdy points to the central irony: "hunter gatherers…spent their abundant leisure time eating, drinking, playing, socializing – in short, doing the very things we associate with affluence"…. Numbers (about life expectancy)…are, in fact, based upon the same erroneous calculation distorted by **high infant mortality rates**. When this factor is eliminated, we see that prehistoric humans who survived beyond childhood typically lived from sixty-six to ninety-one years, with higher levels of overall health and mobility than we find in most Western societies today. It's a game of averages, you see. While it's true that many infants and small children died in prehistoric populations – as indicated by the larger numbers of infant skeletons in most burial sites – these skeletons tell us nothing about what constituted a 'ripe old age'. Life expectancy at birth, which is the measure generally cited, is far from an accurate measure of the typical life span. When you read, "at the beginning of the 20th century, life expectancy at birth was around 45 years…it has risen to about 75 thanks to the advent of antibiotics and public health measures that allow people to survive or avoid infectious diseases", keep in mind that this dramatic increase is much more a reflection of increased infant survival than of adults living longer.

Box 4.4 Sedentism, Agriculture, Colonialism and Gender Inequality

Suzman's 2017 book *Affluence without Abundance* provides an interesting discussion on **sedentism**, **agriculture**, **gender inequality** and **colonialism**:

When they still foraged independently, **Ju/'hoansi**, like most other well-documented **hunter-gatherers**, defined gender roles very clearly. But they were adamant that **gender differences** were no grounds to assert the superiority of one gender over the other. Individual charisma, strength of character, persuasiveness, common sense, and humility were much more important factors in an individual's influence within a band than his or her genitals. Both men and women could be healers and both could be *n!orekxausi*, the holders of inheritance rights to any particular territory. Gender equality in hunter-gatherer societies also just made sense in terms of the practicalities of making a living. Men and women both played important roles in food provision, and it was plainly obvious that when making decisions about when and where to camp, both men and women offered perspectives that needed to be taken into consideration.

Traditional Ju/'hoan views on gender relations, however, would not endure long after the arrival of white farmers. White farmers in the Omaheke desperately needed Ju/'hoan labor, and the kind of laborer they wanted was male. A Ju/'hoan laborer's female dependents might be given some domestic tasks to do, but more often than not they were simply expected to sit in the workers' compounds and not make trouble. In the early days of white settlement, Ju/'hoan women occupied themselves by gathering. But since they were not able to move between camps on the farms, all bush foods within a reasonable distance of the workers' compounds were quickly exhausted. And, just as importantly, farmers now provided them with bags of cheap and easy-to-make carbohydrate-rich corn porridge, so they often had just about enough to eat. With no productive roles on the farm other than making babies, minding children, and preparing meals for their families, Ju/'hoan women found themselves suddenly marginalized even within their families. More than this, they became dependent on males for food in their bellies and places to stay. Farmers had no reason to accommodate "stray" women with no clear links to their laborers.

Another problem was that white farmers, like the Herero, had their own particular views about appropriate roles for Ju/'hoan women that were often spiced up by fantasies regarding their sexual availability. Several farmers in the Omaheke informed me without batting an eyelid that "Bushman girls" were incapable of resisting a male's sexual advances when they were "on heat" or that they were required to make themselves permanently sexually available to Ju/'hoan men. Herero and white farmers' views on gender roles were also a product of the **Neolithic Revolution**, a process that sowed the seeds of modern patriarchy. With the shift to **agriculture**, gender roles changed dramatically. Tasks immediately related to production probably played only a minor role in this transition. In productive terms alone, subsistence farming does place a slightly higher value on male brawn simply because of the energy input involved in some farming-related tasks. Thus, societies with roots in "plow agriculture", which required significant upper-body strength, are typically more patriarchal than those that prepared their fields with equipment like hoes that were wielded just as effectively by women as men. But making a living from farming involved much more than hefting heavy rocks and digging furrows. It involved many tasks in which strength did not really make a meaningful difference – tasks like toolmaking, food storage and preparation, livestock herding and rearing, sowing seeds, and harvesting. There was plenty of work for everyone regardless of gender.

With an increased workload came an increased emphasis on making babies, who in time could also be put to work. The increased demands for infant care meant that women focused on jobs that could be undertaken while nursing at or near the home, like processing cereals, making clothing, preparing food, and mending tools. This resulted in the development of the almost universal association of women with domestic spaces and men with outside "public" spaces in agricultural societies. In and of itself, this wouldn't make a difference if it were not for the increased productivity that farming created and the risks and fears that haunted farmers' dreams. Increased productivity also meant larger, denser communities. And larger, denser communities enabled more complex, often hierarchical social institutions to manage behavior, distribute resources, and manage risks. Tethered to their domestic duties, women typically had fewer opportunities to participate in ever more important public spaces. At the same time competition between households, villages, or even coalitions of villages for resources like land placed a far higher premium on the distribution of resources and correspondingly the ability to organize, lead, and project influence whether by means of persuasion or, as often as not, the ability to make **war** – a task to which men were far better suited. And the locus for the evolution of social life to do this was inevitably the male, public world. As a result, over time, while the male-dominated public realm grew in importance and complexity, women remained largely confined to their homes, where they had no option but to exercise their influence indirectly through their male kin. In all **farming societies**, social and political power was ultimately shaped by the flow of goods and resources between people. Among **Africa's pastoralists**, like the Herero, cattle became the primary form of wealth and the number of wives a man had become the primary expression of that wealth, with women being exchanged for cattle in the form of a bride-price. Among European and Asian societies, exchange relationships became increasingly characterized by more symbolic tokens representing credit and debt – which, by the time da Gama and Dias (two Portuguese sailors) sailed around the Cape – had taken the form of coinage.

Scott further wrote:

Historical humankind has been mesmerized by the narrative of **progress** and **civilization** as codified by the first great **agrarian kingdoms**. As new and powerful societies, they were determined to distinguish themselves as sharply as possible from the populations from which they sprang and that still beckoned and threatened at their fringes. In its essentials, it was an "ascent of man" story. Agriculture, it held, replaced the savage, wild, primitive, lawless, and violent world of hunter-gatherers and nomads. Fixed-field crops, on the other hand, were the origin and guarantor of the settled life, of **formal religion**, of society, and of government by laws. Those who refused to take up **agriculture** did so out of ignorance or a refusal to adapt. In virtually all early agricultural settings the superiority of farming was underwritten by an elaborate mythology recounting how a powerful god or goddess entrusted the sacred grain to a chosen people. Once the basic assumption of the superiority and attraction of fixed-field farming over all previous forms of subsistence is questioned, it becomes clear that this assumption itself rests on a deeper and more embedded assumption that is virtually never questioned. And that assumption is that sedentary life itself is superior to and more attractive than mobile forms of subsistence. The place of the domus and of fixed residence in the civilizational narrative is so deep as to be invisible; fish don't talk about water! It is simply assumed that weary *Homo sapiens* couldn't wait to finally settle down permanently, could not wait to end hundreds of millennia of mobility and seasonal movement.

Yet there is massive evidence of determined resistance by mobile peoples everywhere to permanent settlement, even under relatively favorable circumstances. Pastoralists and hunting-and-gathering populations have fought against permanent settlement, associating it, often correctly, with disease and state control. Many Native American peoples were confined to reservations only on the heels of military defeat. Others seized historic opportunities presented by European contact to increase their mobility, the Sioux and Comanche becoming horseback hunters, traders, and raiders, and the Navajo becoming sheep-based pastoralists. Most peoples practicing mobile forms of subsistence – herding, foraging, hunting, marine collecting, and even shifting cultivation – while adapting to modern trade with alacrity, have bitterly fought permanent settlement.

It turns out that the greater part of what we might call the standard narrative has had to be abandoned once confronted with accumulating archaeological evidence. Contrary to earlier assumptions, hunters and gatherers – even today in the marginal refugia they inhabit – are nothing like the famished, one-day-away-from-starvation desperados of folklore. Hunters and gathers have, in fact, never looked so good…agriculturalists, on the contrary, have never looked so bad – in terms of their diet, their health, and their leisure…. For example, it has been assumed that fixed residence – **sedentism** – was a consequence of crop-field agriculture. **Crops** allowed populations to concentrate and settle, providing a necessary condition for state formation. Inconveniently for the narrative, sedentism is actually quite common in ecologically rich and varied, preagricultural settings – especially wetlands bordering the seasonal migration routes of fish, birds, and larger game. There, in ancient southern Mesopotamia (Greek for "between the rivers"), one encounters sedentary populations, even towns, of up to five thousand inhabitants with little or no agriculture. The opposite anomaly is also encountered: crop planting associated with mobility and dispersal except for a brief harvest period.

This last paradox alerts us again to the fact that the implicit assumption of the standard narrative – namely that people couldn't wait to abandon mobility altogether and "settle down" – may also be mistaken. On a generous reading, until the past four hundred years, one-third of the globe was still occupied by hunter-gatherers, shifting cultivators, pastoralists, and independent horticulturalists, while states, being essentially agrarian, were confined largely to that small portion of the globe suitable for cultivation. Much of the world's population might never have met that hallmark of the state: a tax collector. Many, perhaps a majority, were able to move in and out of state space and to shift modes of subsistence; they had a sporting chance of evading the heavy hand of the state. If, then, we locate the era of definitive state hegemony as beginning about 1600 CE, the state can be said to dominate only the last two-tenths of one percent of our species' political life. In focusing our attention on the exceptional places where the earliest states appeared, we risk missing the key fact that in much of the world there was no state at all until quite recently. The classical states of Southeast Asia are roughly contemporaneous with Charlemagne's reign, more than six thousand years after the "invention" of farming. Those of the New World, with the exception of the Mayan Empire, are even more recent creations. They too were territorially quite small. Outside their reach were great congeries of "unadministered" peoples assembled in what historians might call tribes, chiefdoms, and bands. They inhabited zones of no sovereignty or vanishingly weak, nominal sovereignty. The states in question were only rarely and then quite briefly the formidable Leviathans that a description of their most powerful reign tends to convey. In most cases, interregna, fragmentation, and "dark ages" were more common than consolidated, effective rule.

Importantly, Scott astutely points out that:

Here again, we – and the historians as well – are likely to be mesmerized by the records of a dynasty's founding or its classical period, while periods of disintegration and disorder leave little or nothing in the way of records. Greece's four-century-long "**Dark Age**", when literacy was apparently lost, is nearly a blank page compared with the vast literature on the plays and philosophy of the Classical Age. This is entirely understandable if the purpose of a history is to examine the cultural achievements that we revere, but it overlooks the brittleness and fragility of state forms. In a good part of the world, the state, even when it was robust, was a seasonal institution. Until very recently, during the annual monsoon rains in Southeast Asia, the state's ability to project its power shrank back virtually to its palace walls. For example, it appears that flight from the early state domains to the periphery was quite common, but, as it contradicts the narrative of the state as a civilizing benefactor of its subjects, it is relegated to obscure legal codes. By comparing the life world of agriculture – strapped as it is to the metronome of a major cereal grain – with the life world of the hunter-gatherer, I make the case that the life of farming is comparatively far narrower experientially and, in both a cultural and a ritual sense, more impoverished. If the formation of the earliest states were shown to be largely a coercive enterprise, the vision of the state, one dear to the heart of such social-contract theorists as Hobbes and Locke, as a magnet of civil peace, social order, and freedom from fear, drawing people in by its charisma, would have to be reexamined. The early state, in fact…often failed to hold its population; it was exceptionally fragile epidemiologically, ecologically, and politically and prone to collapse or fragmentation. If, however, the state often broke up, it was not for lack of exercising whatever coercive powers it could muster. Evidence for the extensive use of unfree labor – war captives, indentured servitude, temple **slavery**, slave markets, forced resettlement in labor colonies, convict labor, and communal slavery (for example, Sparta's helots) – is overwhelming. Unfree labor was particularly important in building city walls and roads, digging canals, mining, quarrying, logging, monumental construction, wool textile weaving, and of course agricultural labor.

The attention to "husbanding" the subject population, including women, as a form of wealth, like livestock, in which fertility and high rates of reproduction were encouraged, is apparent. The ancient world clearly shared Aristotle's judgment that the slave was, like a plough animal, a "tool for work". Even before one encounters terms for slaves in the early written records, the archaeological record speaks volumes with its bas relief depictions of ragged captive slaves being led back from the field of victory and, in **Mesopotamia**, thousands of identical, small, beveled bowls used, in all likelihood, for barley or beer rations for gang labor. Formal slavery in the ancient world reaches its apotheosis in **classical Greece** and early **imperial Rome**, which were slave states in the full sense one applies to the antebellum South in the United States. Chattel slavery on this order, though not absent in Mesopotamia and early Egypt, was less dominant than other forms of unfree labor, such as the thousands of women in large workshops in Ur making textiles for export. That a good share of the population in Greece and Roman Italy was being held against its will is testified to by slave rebellions in Roman Italy and Sicily, by the wartime offers of freedom – by **Sparta** to Athenian slaves and by the Athenians to Sparta's helots – and by the frequent references to fleeing and absconding populations in Mesopotamia. One is reminded in this context of Owen Lattimore's admonition that the great walls of China were built as much to keep Chinese taxpayers in as to keep the barbarians out. Early states surely did not invent the institution of slavery, but they did codify and organize it as a state project.

And what about these **barbarians** who, in the epoch of the early states, are massively more numerous than state subjects and, though dispersed, occupy most of the earth's habitable surface? The term "barbarian", we know, was originally applied by the Greeks to all non–Greek speakers – captured slaves as well as quite "civilized" neighbors such as the Egyptians, the Persians, and the Phoenicians. "Ba-ba" was meant to be a parody of the sound of non-Greek speech. In one form or another the term was reinvented by all early states to distinguish themselves from those outside the state. I will continue to use the term "barbarian" – with tongue planted firmly in cheek – in part because I want to argue that the era of the earliest and fragile states was a time when it was good to be a barbarian. The length of this period varied from place to place depending on state strength and military technology; while it lasted it might be called the golden age of barbarians. The barbarian zone, as it were, is essentially the mirror image of the agro-ecology of the state. It is a zone of hunting, slash-and-burn cultivation, shellfish collection, foraging, **pastoralism**, roots and tubers, and few if any standing grain crops. It is a zone of physical mobility, mixed and shifting subsistence strategies: in a word, "illegible" production. If the barbarian realm is one of diversity and complexity, the state realm is, agro-economically speaking, one of relative simplicity. Barbarians are not essentially a cultural category; they are a political category to designate populations not (yet?) administered by the state. The line on the frontier where the barbarians begin is that line where taxes and grain end. The Chinese used the terms "raw" and "cooked" to distinguish between barbarians. Among groups with the same language, culture, and kinship systems, the "cooked" or more "evolved" segment comprised those whose households had been registered and who were, however nominally, ruled by Chinese magistrates. They were said to 'have entered the map'.

The part about China—"*the great walls of China were built as much to keep Chinese taxpayers in as to keep the barbarians out…early states surely did not invent the institution of slavery, but they did codify and organize it as a state project*"—is particularly powerful as it resonates so profoundly with what is happening today. When I talk about these subjects with both laypeople and scientists, I often hear answers such as "come on, everybody knows that nomads and 'barbarians' would like to be part of 'civilization' as they *should* know that they *would* be much 'better of'…so obviously walls are to not allow 'others' to come in, not to forbid state people to go out." Of course, this is part of the problem of thinking that what is happening *now* is what is "normal," what has always occurred—"Mexicans want to come to the U.S., Egyptians to Europe, always the same story." No, history is dynamic: using the terms often used by reactionary and racist people, before the arrival of Europeans the region that is now Mexico had much more "civilized" societies than that which is now the U.S., and the same applies to Egypt versus Europe, some millennia ago. In fact, what is remarkable about such *Neverland* racist stories is that they even forget examples that are historically very recent, or that are happening right now before our eyes: the **Soviet Union**, or **East Germany**—officially the **German Democratic Republic**—had walls and guards to *not* allow people to *go out* of those countries, not the other way round, and this continues to happen with countries such as **North Korea**. Moreover, regarding so-called barbarians and nomads, right now **China** is applying policies to "tame them and round them up for *military*-style classes and taught *work discipline*" that are strikingly similar to those that have been used by "early civilizations" millennia ago. This was reported in a New York Times article published when I was writing these lines, entitled "*China has a new plan to tame Tibet*":

> Before Xinjiang, there was **Tibet**. **Repressive policies** tested there between 2012 and 2016 were then applied to the **Uighurs** and other **ethnic minorities** in northwestern China: entire cities covered in surveillance cameras, ubiquitous neighborhood police stations, residents made to report on one another. Now that process also works the other way around. **Xinjiang's coercive labor program** – which includes mandatory training for farmers and herders in centralized vocational facilities and their reassignment to state-assigned jobs, some far away – is being applied to Tibet. (Not the internment camps, though.) Call this a feedback loop of **forcible assimilation**. It certainly is evidence of the scale of Beijing's ruthless campaign to suppress cultural and ethnic differences – and not just in Tibet and **Xinjiang**. I analyzed more than 100 policy papers and documents from the Tibetan authorities and state-media reports for a study published with the Jamestown Foundation this week. Photos show Tibetans training, wearing fatigues. Official documents outline how Beijing is rolling out for them a militarized labor program much like the one in place in Xinjiang: Tibetan **nomads** and farmers are being rounded up for military-style classes and taught work discipline, "gratitude" for the **Chinese Communist Party** and Chinese-language skills.
>
> More than half a million workers have been trained under this policy during the first seven months of the year, according to official documents…. The people of what the Chinese government refers to as the Tibet Autonomous Region – about 3.5 million, mostly nomads and farmers scattered throughout the vast Himalayan plateau – have resisted its encroachment for decades. Notably, riots broke out in the capital, Lhasa, in 2008, just weeks before the Olympic Games in Beijing, following years of tightening restrictions on cultural and religious freedoms. There reportedly have been more than 150 cases of self-

immolation carried out in protest since 2011.... This is but one of the many ways in which Beijing has been doubling down on imposing state controls over Tibetan traditional ways of life. Tibet, like Xinjiang, nominally is an autonomous region, yet in 2019, its government mandated that all Tibetan nomads and farmers be subjected to what some government directives call "military-style" training for vocational skills and then be assigned low-skilled jobs, for example in manufacturing or the services sector.

Some of the reports...including one by Tibet's Ethnic Affairs Commission, claim that Tibetans' religion cultivates "backward thinking". The city of Chamdo claims to have "carried out the transfer of surplus labor force in agricultural and pastoral areas" in order to overcome Tibetans' purportedly "poor organizational skills". According to a major policy paper by the Tibetan regional government, "The 2019-2020 Farmer and Pastoralist Training and Labor Transfer Action Plan", the military drill-style skills training, coupled with what the government calls "thought education", will supposedly compel Tibetans to voluntarily participate in the poverty alleviation efforts prescribed by the state. As of this year, Tibet's labor plan has explicitly included the transfer of Tibetan workers to other parts of China, with target quotas for each Tibetan region. Local officials who fail to meet those quotas are subject to punishment.... Many of the program's main features, and objectives, bear a striking similarity with the plan in place in Xinjiang. So do other measures designed to marginalize Tibetan culture. For example, Beijing has drastically accelerated in recent years its efforts to minimize the teaching of the Tibetan language, including outside Tibet. In late 2015, Tashi Wangchuk, a Tibetan from the remote nomadic region of Yulshul in Qinghai Province, tried to sue his local government over the curtailment of Tibetan language education. In 2018, he was sentenced to five years in prison for "inciting separatism"...

This strategy has old roots. Back in 1989, the eminent Chinese anthropologist Fei Xiaotong wrote that through a long process of "mixing and melding", the Han majority and other ethnic groups in China would eventually combine into a single entity: the **Chinese nation-race**. In Fei's view, the Han would be at the center of this fusion, because they were the superior culture into which so-called backward minority groups would inevitably assimilate. The Chinese government adopted Fei's vision, and for a time tried to help it along with a large dose of top-down economic development. In 2000, President Jiang Zemin launched the Great Western Development Campaign, bringing infrastructure – and numerous Han – to the western part of China. Local ethnic minorities would benefit from the new economic activity and employment opportunities so long as they were willing to assimilate culturally and linguistically. Many resisted. Local expressions of ethnic identities flourished. Tibetans and members of other minority groups flocked to schools that taught their languages, and kept their distinct religions alive.... The purpose of these policies is clear, as are the stakes, and targeted groups are trying to push back. The central government's recent efforts to replace Mongolian with Chinese as the main teaching language in schools in Inner Mongolia has triggered major protests there.... In Fei's vision, ethnic fusion would happen slowly, naturally. That has failed. In Mr. Xi's vision, the assimilation of minority groups must be coerced by the state. That, too, will fail.

I will focus here, just briefly, on some of the fascinating data provided in Scott's *Against the Grain*, particularly those that clearly contradict the longstanding myths linking sedentism and agriculture with aspects of "progress," such as those related to health and the complexity and diversity of daily-life and the knowledge about the *reality of the natural world*:

Just as human **sedentism** represents a reduction in mobility and increased crowding in the village and domus, so the relative confinement and crowding of **domestic animals** has immediate consequences for health. The stress and physical trauma of confinement, together with a narrower spectrum diet and the ease with which infections can spread among individuals of the same species packed together, make for a variety of **pathologies**.

Bone pathologies due to repeated infection, relative inactivity, and a poorer diet are particularly common. Archaeologists have come to expect cases of **chronic arthritis**, evidence of **gum disease**, and bone signatures of confinement in analyzing the remains of archaic domestic animals. The result is also far higher **mortality rates** among newborn domesticates. Among confined llamas, for example, the mortality rate for newborns approaches 50 percent, far higher than among **wild llamas (guanacos)**. The difference can be largely attributed to the effects of confinement – muddy, feces-rich corrals in which virulent clostridium **bacteria**, among others, thrives and, like other **parasites**, finds an abundant supply of hosts close at hand. The high rates of **mortality** for newborn domesticates would seem to defeat the purpose of human management, which is largely to maximize the reproduction of animal protein as one maximizes one's crop of grain. It appears, however, that the rates of fertility may increase so dramatically as to more than offset the losses through mortality. The reasons are not entirely clear, but domesticated animals generally reach reproductive age earlier, ovulate and conceive more frequently, and have longer reproductive lives. **Tame silver foxes** in the Russian experiment came into heat twice a year compared with once a year for undomesticated foxes. The pattern for rats is more striking, although as commensals even in their wild state, they allow only speculative inferences to other domesticates. Captured wild rats have quite low rates of fertility, but after only eight (short!) generations of captivity, their rate of fertility was found to increase from 64 percent to 94 percent and by the twenty-fifth generation, the reproductive life of captive rats was twice as long as "noncaptives." They were, overall, nearly three times as fecund. The paradox of relative ill health and high newborn mortality on the one hand, coupled with more-than-compensating increases in fertility on the other, is one to which we shall return, as it bears directly on the **demographic explosion** of agricultural peoples at the expense of hunters and gatherers.

One way of determining whether a woman who died nine thousand years ago was living in a sedentary, grain-growing community as compared with a foraging band was simply to examine the bones of her back, toes, and knees. Women in grain villages had characteristic bent-under toes and deformed knees that came from long hours kneeling and rocking back and forth grinding grain. It was a small but telling way that that new subsistence routines – what today would be called a repetitive stress injury – shaped our bodies to new purposes, much as the work animals domesticated later – cattle, horses, and donkeys – bore skeletal signature of their work routines. No wonder then that the archaeological signs for a life lived largely in the domus are strikingly similar for man and beast. "Domiciled" sheep, for example, are generally smaller than their wild ancestors; they bear telltale signs of domesticate life: bone pathologies typical of crowding and a narrow diet with distinctive deficiencies. The bones of "domiciled" *Homo sapiens* compared with those of hunter-gatherers are also distinctive: they are smaller; the bones and teeth often bear the signature of **nutritional distress**, in particular, an **iron-deficiency anemia** marked above all in women of reproductive age whose diets consist increasingly of grains. The parallel, of course, arises from a common environment of more restricted mobility, crowding and the cross-infection opportunities it presents, a narrower diet (less variety for herbivores, less variety and less protein for omnivores like *Homo sapiens*), and relaxation of some of the selection pressures from predators lurking outside the domus. In the case of *Homo sapiens*, however, the process of **self-domestication** had begun long before (some of it even before "sapiens") with the **use of fire**, **cooking**, and the **domestication of grain**. Thus declining tooth size, facial shortening, a reduction in stature and skeletal robustness and less sexual dimorphism were evolutionary effects that had a far longer history than the Neolithic alone. Nevertheless, sedentism, crowding, and a diet increasingly dominated by cereals were revolutionary changes that left an immediate and legible mark on the archaeological record.

To return to the concept of tempo, one might think of hunters and gatherers as attentive to the distinct metronome of a great diversity of natural rhythms. Farmers, especially fixed-field, cereal-grain farmers, are largely confined to a single food web, and their routines are geared to its particular tempo. Bringing a handful of crops successfully to harvest is to be

> sure a demanding and complex activity, but it is usually dominated by the requirements of one dominant starch plant. It is no exaggeration to say that hunting and foraging are, in terms of complexity, as different from cereal-grain farming as cereal-grain farming is, in turn, removed from repetitive work on a modern assembly line. Each step represents a substantial narrowing of focus and a simplification of tasks. The metaphors with which people reasoned were increasingly dominated by domesticated grains and domesticated animals: "a time to sow and a time to reap", being "a good shepherd." There is hardly a passage in the **Old Testament** that fails to make use of such imagery. This codification of subsistence and ritual life around the domus was powerful evidence that, with domestication, *Homo sapiens* had traded a wide spectrum of wild flora for a handful of cereals and a wide spectrum of wild fauna for a handful of livestock. I am tempted to see the late Neolithic revolution, for all its contributions to large-scale societies, as something of a deskilling. Adam Smith's iconic example of the productivity gains achievable through the division of labor was the pin factory, where each minute step of pin making was broken down into a task carried out by a different worker. **Alexis de Tocqueville** read *The wealth of nations* sympathetically but asked, "what can be expected of a man who has spent twenty years of his life putting heads on pins"? If this is a too bleak view of a breakthrough credited with making civilization possible, let us at least say that it represented a contraction of our species' attention to and practical knowledge of the natural world, a contraction of diet, a contraction of space, and perhaps a contraction, as well, in the breadth of ritual life.

I consider the last lines to be particularly insightful, profoundly deep, and disturbingly accurate. This is because they not only go against what the majority of people learned in school and then believed in, during most of their lives, but also make us re-think key aspects that we tend to see as "self-evident" and "normal" in our current societies. Two of the so-called major revolutions in human history—the "agricultural" and "industrial" ones—have actually led to a *decrease* of the complexity of daily-life of most people and of their contact with and knowledge about the natural world. For most people, they indeed led to a life plagued by a huge number of tedious, repetitive, and often relatively simple tasks, as well as by more diseases, less diversity of food, and so on. Of course, it is true that this applies particularly to the periods that occurred *just after* the occurrence of such "revolutions": most people living in Western countries *nowadays* don't spend decades of their lives "putting heads on pins" and have, in average, a higher life expectancy than most people living in hunter-gatherer societies, lower child mortality rates, and so on. In addition, one can argue that some Western biologists know, in the overall—let's say, if they have studied in detail the fauna and flora of several different regions, and types of habitats, of the planet—more about the natural life of Earth as a whole than a single hunter-gatherer person that lives in a specific region of the globe does.

However, we—the authors of books such as this one, or their readers, or the students of schools and universities, and so on—tend to forget that we are part of a very small and lucky minority of people, even within Western countries. A majority of people living in these countries—including many that are often said to be "well-off," such as for instance bankers—still have jobs that involve doing very repetitive, tedious tasks, during 8, or often even more, hours every single weekday. The type of life of a huge number of Western bankers would hardly be considered to be as "enjoyable" for nomadic hunter-gatherers, and probably even less as an indicator of "progress." This is not merely philosophical speculation, nor "wishful thinking":

there are numerous empirical case studies of hunter-gatherers that *were forced* to live so-called Westerner-lifestyles for some time and that, when they were finally able to subsequently choose what *they wanted to do*, utterly refused to continue living in Western countries (see Chap. 6 and Fig. 6.1). Moreover, it is obvious that apart from a few scientists, the *vast majority of people* living in our countries clearly know much less about the natural world than most hunter-gatherers did/do. It suffices to say that even the most prominent biologists often need—when they travel in research trips to remote places such as Amazonia or the Congo basin—local indigenous people to help them to be able to help them survive under tropical conditions, recognize dangers and even identify plants and animals and so on. Within lay people living in big Western cities the loss of contact with and knowledge about the natural work is particularly devastating: many kids have never seen a living chicken, and don't even know that there is a link between that animal and the **"Chicken McNuggets"** they eat at **McDonald's**. There is no doubt that, in average, the so-called highly educated children of so-called developed civilized countries just know an infinitesimal part of the natural world, compared with most kids of nomadic hunter-gatherer societies. This makes us think, again, about what Levi-Strauss wrote about "education," writing, inculturation, "development," and "civilization."

One example that illustrates how ideas about these topics started to change in the second half of the twentieth century is McKeown's 1988 book, *The Origins of Human Disease*. This is because McKeown provides empirical data that actually forced him to recognize that **infectious diseases** in particular, and **diseases** in general, became more prevalent in people that started to live in larger sedentary groups, principally in agricultural societies. However, McKeown was not able to completely abandon the longstanding—and still prevailing, then—*Neverland* stories about how the life of nomadic hunter-gatherers was always, and still is, "nasty, brutish and short." So, he spends dozens of pages speculating about what type of "evidence" could support such imaginary stories, despite the fact that the data compiled in his own book clearly contradicted them. As usually happens in the world of unreality, he was able to "find" an a posteriori way to "solve this paradox": for him, the "solution" was that nomadic people were overall "worse-off" anyway, because they "failed" to increase their numbers, contrary to agricultural societies.

This is a common obsession—and fallacy—of so-called Darwinism, based on Malthus' ideas, which McKeown indeed cited an endless number of times: that somehow a group is only "successful" when it increases the number of individuals included in it. According to that *Neverland* idea, the life of most animals in the planet is surely "unsuccessful" and "nasty, brutish, and short": poor lions, poor gorillas, poor chimpanzees, poor wolves, poor bears, poor eagles, they were not able to increase their numbers to match the "successful" number of eight billion people living right now in planet Earth. But within the *reality of the natural world* the opposite is often true: empirical data shows that, in general, wild animals normally have a much better quality of life, less diseases, eat food with more quality and diversity, and obviously have more freedom and mobility, than the huge number of domesticated animals that we use and abuse (see Fig. 9.6). If in a "factory farm" with let's say 1000 pigs—all with horrible lives without any mobility and with their bodies full of germs and antibiotics to fight them—100 die, this clearly does not mean that because there are 900 left one can say that they are "better off," or "more developed," or "more successful" than a group of only 90 wild pigs living in a certain forest just because the domesticated ones have a higher rate of reproduction. So, in a similar way, the fact that so-called **civilizations** are able to "produce" such a huge number of human individuals—many of them to work, or fight, or die, for the dream of others, many times even being literally slaves, as domesticated non-human animals tend to be—does not mean at all that those individuals are "better off" or are more "developed" or more "successful" than are the people living in small nomadic hunter-gatherer societies (see Figs. 8.1, 8.6, 9.24).

Having said that, one needs to recognize that, in a way, McKeown already has a lot of merit to have been brave enough to go against *some of the myths* of the Hobbesian and "progressive" imaginary stories that were still so dominant when he wrote his 1988 book—even if he was only able to go halfway, while doing so. Within the context of the 1980s his book was indeed very courageous, as is attested by the fact that it was indeed highly criticized by many Hobbesians, principally because it moreover argued that the available data also indicated that there was a huge increase of non-communicable diseases after the "industrial revolution":

> **Neanderthal man** is believed to have suffered frequently from **osteoarthritis** – arthritic changes were found in seventeen of twenty-seven skeletons from a site in California…in the same series there was also evidence of osteomyelitis, sinusitis and alveolar abscesses, although dental caries is rare in hunter-gatherers…[but] **traumatic lesions** are quite common in archaeological material, the site of the injury varying, as might be expected, with the nature of the habitat. In **Palaeolithic times** population densities must have varied considerably, but they were always low, rarely above a few persons per square mile…it would be difficult to describe conditions less suitable for the establishment of human infectious organisms…. By 10,000 BC, just before the **Agricultural Revolution**, it is thought [the population of humans in the planet] to have been about 4 million…. [After the rise of agriculture] **tuberculosis** probably can be identified from about 3000 BC, **leprosy** from AD 500 and **syphilis** from AD 1500. Under such conditions many of the **diseases** which were prevalent in the historical period could not have existed: **measles**, **smallpox**, **whooping cough**, **poliomyelitis** and most enteric and **respiratory infections** are examples. These diseases are characteristic of generalized human **viral infections**, for which there is no other animal host and in which latency and recurrent infection does not occur. The only 'specifically human' **viral diseases** that we could expect to survive in primitive man are those marked by latency and recurrent disease. **Herpes simplex** and **chickenpox virus**, for example, could survive even in isolated family units, because of this characteristic. Among bacterial and protozoal infections, tuberculosis, leprosy and **treponematosis** are also diseases from which early man might have suffered, since they are characterized by chronicity and recurrent excretion…. Wild animal reservoirs would have existed for diseases such as **brucellosis**, **leptospirosis**, **relapsing fever**, **salmonelloses**, **tularaemia**, **mite-born typhus**, the **rickettsioses** and **plague**. In northern regions wolves become infected with **rabies** in times of stress when they will attack human beings. In warmer climates man would be exposed to infection by **arboviruses** carried by primates. Indeed Fenner suggested that jungle **yellow fever**, mainly a disease of the adult male whose work brings him into close contact with the forest, is a prototype of the sort of viral infection to which Palaeolithic man was exposed.
>
> **Falciparum malaria** is believed to have emerged as a human disease in the tropics during the Palaeolithic, but although it may have caused considerable mortality, it is unlikely

to have been serious until the development of **agriculture** led to large populations and conditions suitable for the breeding of mosquito vectors. Fenner considered that **cholera** was unknown to nomadic Palaeolithic man, but developed when villages and village water supplies were established. Finally, hunter-gatherers would have been infected by a group of pathogens referred to as commensals, of which the **helminths** and other pathogens of the bowel are typical. Most of these organisms are usually harmless, and cause disease only under conditions of stress due to influences such as overcrowding (not a problem for early man) and food-shortage. I conclude that in spite of occasional dramatic exceptions, early man, like other animals, generally 'lived his life in balance with his pathogens, which only have serious effects at time of ecological imbalance'.

Lacking data for truly isolated communities, observers have used serological and clinical methods to assess the infections [of hunter-gatherers]...on the basis of immunological evidence from several investigations, Black concluded that the disease of Brazilian Indian tribes fall into four groups: endemic diseases of high incidence and low morbidity (such as **herpes** and **hepatitis B**); diseases of low prevalence (such as yellow fever) acquired from other animals; explosive but transient diseases (such as measles and influenza) introduced from outside the hunter-gatherer communities; persistent diseases (such as tuberculosis and malaria) also introduced from outside. But perhaps the most inconsistent findings are in respect of food and nutrition. In his appraisal of health and disease in hunter-gatherers, Dunn concluded that malnutrition is rare and starvation occurs only infrequently. **Kung bushmen**, for example, are reported to live well by a few hours of daily hunting and gathering, and the Hadza of Tanzania are said to be better off than their agricultural neighbours: 'for a **Hadza** to die of hunger, or even to fail to satisfy his hunger for more than a day or two is almost inconceivable'. But there are exceptions: for some **Alaskan Eskimos** 'everything is focused entirely and absolutely upon the requirements that the increasing search for meat necessitates'; the **Birhor of India** not only work hard for their food but often go hungry; and among the **Siriono**, in the Bolivian rain forest, there appears to be a good deal of hunger which leads to widespread anxiety about food and often to aggressive behaviour. But while recognizing such exceptions, recent writers have emphasized the adequate, sometimes abundant food supplies of hunter-gatherers, and have concluded that the frequency of famine and malnutrition have been exaggerated. Diseases such as **cancer**, **obesity**, **diabetes**, **hypertension** and **heart disease** are rare or absent in other primates in their natural habitats; they are very uncommon in present-day hunter-gatherers and peasant agriculturists and only begin to appear when traditional ways of life are abandoned.

Indeed it is the observation that primitive societies are largely free from non-communicable diseases which has led to the conclusion that so long as they are undisturbed by external influences, hunter-gatherers remain essentially healthy. We (thus) must leave our present-day perspective and recognize the considerable attractions of the hunting and gathering life. In a symposium on hunting people it was said to be 'the most successful and persistent adaptation man has ever achieved'. Unquestionably it was the most persistent, since it lasted for a few million years; and in many ways it was successful, particularly from the viewpoint of hunters unaware of the attractions of a more sophisticated life. Outside the Arctic, a wide range of plant and animal foods is thought to have provided a diet rich in vitamins, minerals and proteins, sufficient in composition and amount to maintain a healthy population; and because of its variety and inclusion of substantial amounts of meat, the food was more attractive than that of a farmer, based on one or a few cereals. Moreover, there was a degree of security and continuity in the supply that was lost with the advent of agriculture, cultivated crops being more susceptible to climatic influences such as drought. Finally, if experience of present-day hunters can be taken as a guide, it was easier and pleasanter to obtain food by hunting and gathering than by agriculture: 'there is nothing to suggest that people would shift to agriculture to save labour or gain leisure time'...against the attractions of the hunter's life, agriculture offered only one advantage, but it was an enormous one: by providing a greater amount of food per unit area of land it could feed a larger population.

So, what about health and disease *after* the rise of agriculture? McKeown wrote:

A plausible explanation was proposed by Cohen…for the appearance of **agriculture** within a short period in many different parts of the world…. 'The pattern of events for the various regions is consistent with a picture of continuous (although not necessarily steady or consistent) population growth and population pressure'. He suggested that the human population had been growing throughout its history, and that this expansion was a cause rather than merely a result of technological change. For although hunting and gathering is a successful form of life for small groups, it is not well suited to the support of large or dense populations, and eventually our ancestors 'were forced to adjust to further increases in population by artificially increasing, not those resources which they preferred to eat, but those which responded well to human attention and could be made to produce the greatest number of edible calories per unit of land'. [Regarding agriculture] the best evidence on non-communicable diseases comes from a number of studies of peasant agriculturalists who have retained, or only recently changed, their traditional ways of life. In Papua New Guinea, for example, **arteriosclerosis** and its various manifestations, including **coronary heart disease**, **cerebro-vascular disease** and **peripheral vascular disease**, were rarely seen; **obesity**, **diabetes**, **hypertension**, **carcinoma** of the bowel and **varicose veins** were all uncommon. In West Nile Ugandans, blood pressure did not rise with age and essential hypertension and stroke were virtually unknown. **Diabetes mellitus** was rare in the **Bantu rural areas** of South Africa and **acute appendicitis** was not observed in the first thousand Kenyan autopsies. Similar results were obtained in Zimbabwe where **coronary thrombosis** has only recently appeared in Africans and **angina** is still a rare disorder. From these and other observations Trowell and Burkitt constructed a list of non-communicable diseases which appear to be as rare in primitive **agriculturalists** as in **hunter-gatherers**…they included hypertension, obesity, diabetes, **gall stones**, **renal stones** and coronary heart disease…appendicitis, **haemorrhoids**, varicose veins, **colo-rectal cancer**, **hiatus hernia** and **diverticular disease**. Trowell and Burkitt attributed the rarity of these diseases in agriculturalists mainly to their conditions of life, and their increase since the eighteenth century to adoption of the western life-style.

It is in keeping with this interpretation that most of the changes in ways of life from hunting and gathering to agriculture would not be expected to lead to the occurrence of non-communicable diseases. Most people still lived an active rural life, and even the minority living in towns were not exposed to many of the hazards of the present day, such as **atmospheric pollution**, widespread **use of chemicals**, adverse working conditions, **road traffic, tobacco** and **drug abuse**. The most significant change was in respect of food, which Trowell and Burkitt considered to be the most important of the multiple influences responsible for western diseases in the industrial period. Under agriculture, there were two important changes in the types of food. Hunter-gatherers lived on meat, fish, fruit and vegetables, and although the proportions of the different foods varied from one population to another, on the average about two-thirds of the diet came from plant sources. They were not often able to have cereals and they had almost no dairy products. Under agriculture man's diet also consisted essentially of vegetable foods, supplemented with meat and fish where these were available. But the common vegetables were cultivated **cereals**, particularly **wheat**, **rice** and **maize**; wheat accounted for 50 to 70 per cent of the food of those who ate it and rice for 80 to 90 per cent. For some populations dairy products provided a significant part of the diet.

The consumption of cereals did not lead to a substantial increase in the frequency of non-communicable diseases. The vegetables of hunter-gatherers were not limited to those of present-day supermarkets, but included a wide variety of plant foods which contained starch – leaves, fleshy roots, wild berries, wild grass seeds and wild grasses, the precursors of modern wheats. The human body, therefore, had no difficulty in accepting planted and cultivated cereals, since it was well adapted to vegetables with a high proportion of carbohydrates. The problems were to arise later from the introduction of refined and processed foods in which the **starch** was separated from fibre. Nevertheless a diet composed mainly of one or a few cereals did expose people from time to time to disease due to deficiencies of proteins, vitamins and minerals. Maize, for example, is short of nicotinic acid which is needed to prevent **pellagra**; all cereals contain phytate which contributes to **rickets** because it interferes with the absorption of calcium; and deficiencies may be caused by the preparation of cereals, as in the polishing or rice which reduces the amount of thiamine needed to prevent **beri-beri**. According to Yudkin, beri-beri, pellagra, **riboflavin deficiency** and rickets all resulted largely from the dietary changes brought about by the **Neolithic revolution**.

As noted by McKeown, regarding infectious diseases there was indeed a dramatic change, together with the rise of agriculture:

There were four main influences which led to the predominance of **infectious diseases** as causes of sickness and death (in **agricultural societies**): the existence, probably for the first time, of populations large enough to enable some human infections to become established and others to be amplified; defective hygiene and crowding, which further increased exposure to communicable diseases; insufficient food which lowered resistance to infection; and close contact with domesticated and other animals which were the probable source of many micro-organisms. Remarkably, we owe the origin of most serious infectious diseases to the conditions which led to our cultural heritage, the city states made possible by the planting of crops in the flood plains of **Mesopotamia**, **Egypt** and the **Indus Valley**. The conditions which resulted from agriculture had a profound effect on the frequency of exposure to infectious disease. The most important influence was the proximity of large numbers of people, which facilitated the spread of airborne and other infections. But hygienic conditions were also significant, particularly those determined by methods of handling food and water and disposing of excreta and waste. **Cholera**...appeared when villages and village water supplies were established, and **malaria** became serious when 'the size of human populations and the opportunities for breeding of vectors increased with advances in agricultural practices'. **Tuberculosis**, possibly an ancient disease, could almost be described as a disease of cities, for it became a common cause of death under the conditions prevailing in large towns. The spread of **intestinal infections** – **typhoid**, **dysentery**, **tuberculosis**, **salmonella** and the like – resulted from contamination of food and water. So the hygienic conditions which followed the introduction of agriculture made it possible for new diseases to appear, and for some diseases already present to become more serious...

Contact with other animals increased under agriculture, with **domesticated animals** such as cattle, sheep, goats, pigs, horses, cats and dogs, and with unwanted intruders attracted to human settlements, such as **rats**, **mice**, sparrows, **ticks**, **fleas** and mosquitoes. The human type of the **tubercle bacillus** which causes **respiratory tuberculosis** probably resulted from mutation by the bovine type, which was transferred to man from wild cattle. Fiennes suggested that the water buffalo is possibly the original source of **leprosy**, the cow of **diphtheria** and the monkey of **syphilis**. Many other diseases such as **mumps** and **small-pox** are believed to have arisen from related conditions in other animals, although the original hosts are not certainly known. Fiennes also reviewed the origins of the few diseases which are capable of spreading as global pandemics. In order to do this they must: 'be capable of rapid transmission from one host to another'; 'find susceptible hosts in sufficient numbers to maintain the momentum'; and (to fulfill the first criterion) 'must be spread by airborne infection'. Only two diseases – **plague** and **influenza** – have met these exacting requirements. Plague is believed to have originated as an unapparent infection of **gerbils** in eastern Asia, and the first great pandemic occurred in the years AD 592–594. The human **virus of influenza** appears to be identified with the virus of **swine influenza**, and it is thought that new pandemic strains of influenza A viruses arise as **zoonoses**.

While in some respects a number of the earlier writings of McKeown were seen as controversial, his 1988 book, and in particular these excerpts comparing nomadic hunter-gatherers and sedentary agriculturalists, are completely in line with what is now usually accepted by many anthropologists and historians. I will finish this Section by adding two examples, one of them being from Fabregas' 1997 *Evolution of Sickness and Healing*:

> The kinds of **zoonotic diseases** that might have affected early humans (and do affect contemporary hunter-gatherers) are protean in their physiological effects, long term and short term. These diseases include the following: **rabies**, **tularemia**, **toxoplasmosis**, **brucellosis**, **salmonellosis**, **trichinosis**, **tapeworm infestations**, **typhus**, **yellow fever**, **malaria**, and **encephalitis**. As outlined by Mark N. Cohen, the zoonotic and soil-borne diseases have the following characteristics: (1) they do not claim many victims; (2) they can have a severe impact on the body; (3) their rarity precludes individual immunity or resistance; (4) their independence from humans means that there has taken place no selection for less virulent forms compatible with human life; (5) they are more likely to strike adults and productive members (as compared to children) who venture into the wild, away from the camp, with severe economic and demographic consequences for the group; and (6) they have limited geographic distribution. There are a number of theoretical and informative summaries of the ecological balances that have existed between early human (and, later, fully human) communities and parasites and microorganisms of environments located in different regions of the world. Early family-level foragers and hunters were less vulnerable to infections that were based on a fecal-oral mode of transmission since they moved frequently and accumulated less human wastes at any one site.
>
> A figure of around 200 deaths per 1,000 infants is cited as the average rate of survival among family-level subsistence foragers [by Cohen]. This means that approximately 50 to 65 percent of all babies are reared to adulthood. These figures naturally vary in relation to the physical environment. Although the figures do not compare favorably with those of contemporary industrial societies (**infant mortality** averages around 10 deaths per 1,000 infants), they match and in some instances are better than levels achieved in the historical past, especially in urban settings of civilizations. The pattern of mortality in hunter-gatherers varies in relation to the physical habitat. It is appropriate to here quote Cohen: "[There is] a changing distribution of causes of death with latitude. In the tropics, indigenous infections are a significant source of mortality, but by most accounts starvation is rarely a cause of death, and accidents are relatively unimportant. **Malaria** and other diseases of greater antiquity account for a significant fraction of deaths in some societies. Hunting accidents appear surprisingly unimportant as causes of death. Such accidents as falls, burns, and (more rarely) snakebites are mentioned more frequently. In high latitudes, in contrast, famine and accidental death are significant sources of mortality". Studies of contemporary family-level foragers indicate that so-called **degenerative diseases** (**heart diseases**, **cancers**, **hypertension**, **diabetes**, and **bowel disorders** such as **peptic ulcer** and **diverticulitis**) are comparatively rare, as are also **epidemic virus diseases** (**influenza**, **parainfluenza**, **mumps**, **measles**, **polio**, **whooping cough**, **rubella**, **tuberculosis**, and **smallpox**); the former largely a result of diet and the latter of the small size, isolation, and migratory lifestyle of the group. Studies of contemporary hunter-gatherer societies (which rely on some agriculture to be sure) strongly suggest that **skin diseases**, either due to insect bites or to infected injuries with resulting abscess and ulcer formation and chronicity, were an important consideration.

The other example is from a book that also provides an interesting, and very original, take on these subjects: Green's 2019 *A Fistful of Shells – West Africa from the Rise of the Slave Trade to the Age of Revolution*:

> By the early eighteenth century, rituals surrounding the installation of kings had become highly elaborate in many parts of West Africa.... What were the major factors at play in the

growing gulf between rulers and subjects in West Africa, one that in many ways continues to the present day? As we have seen, the increasing currency imports allowed the fiscalization of the state and the concentration of **military power**. By controlling currency imports and production, rulers enhanced their control of society. Growing state power both arose with the imports of currency, and fomented the place of 'real' money in society. Newly powerful states such as **Asante** and **Dahomey**, with their treasuries of cowries and pouches of gold dust, were underwriters of the new social and economic dispensations. These transformations prompted fierce struggles against them. A core dynamic to emerge was the struggle between civil society and state power. While warrior elites and their chosen rulers benefited from the rise of the fiscal-military state, it was commoners who became expendable as troops and potential captives. Pushback followed. By the middle of the seventeenth century, rulers recognized that in order to consolidate power in the hands of the king and local elites, they would need to develop systems to control people's time and labour. Correspondingly, religious practice itself changed to reflect the new priorities and experiences that had come to be normal…

Changes in agricultural productivity and in the organization of labour and the trade all required a new understanding of land use. But this was difficult, since, as the historian Assan Sarr has brilliantly shown, land was itself seen as sacred, much of it occupied by spirits. These beliefs meant that large tracts of land were seen by outside observers as 'unoccupied', though to the people who lived there they were fully occupied by potentially malevolent forces; and, indeed, this system of belief allowed often delicate ecologies to be strengthened, since 'empty land' produced wildlife havens, so that here was a sort of ecophilosophy long before such ideas had been considered by Western thinkers. With the erosion of these ideas in the nineteenth century, the land was put to more utilitarian and productive economic uses, and the jeeri became drier…. This spiritual power of land is essential to understanding the political and cultural transformations we have examined in this chapter. As material value and currency equivalence became ever more important, land, too, adopted its own monetary power. Indeed, English traders on the Gold Coast were so well aware of the priceless nature of these beliefs that they were prepared to accept religious objects (which they called fetishes) as pawns, with one factor writing in 1693 that he had been brought one such weighing more than nine ounces, 'which I have taken in pawne for 35 English carpets'. For these European traders, the relative nature of economic value was a given in the decisions they took when it came to trade. Through such eyes it is the idea of universal value, rationally calculated according to numerical equivalence, that seems like a myth. This different approach to land must also be seen as a cause of the economic divergence that opened up between West Africa and the world in this era.

In the worldview of Western economics, land was a material good to be used as a basis for **capital accumulation**. This was an approach deriving from the land security that came with the ruthless process of land enclosures, something that its advocates always called an 'improvement' – a teleological herald of '**progress**'. Some modern economists still see lack of titled access to land as a core driver of poverty in the Global South. That in West and West-Central Africa land had a spiritual value above its material use meant that societies there would always be at a 'competitive disadvantage' in a global economic system in which 'rational' exploitation of land was essential for **economic growth**. The importance of this relationship of spiritual and political power to land is, then, hard to overstate. In the previous chapter, we saw how the ability of hunters to vanquish spirits in the bush was associated with their hold over power in new states; and this spiritual power was something that the rulers who followed them retained, and, with it, their control over the land. In time, though, spiritual practices altered to accommodate the new parallel worldviews. Though land retained its spiritual meaning, new shrines and **secret societies** emerged in which trade goods associated with the Atlantic world became valuable, and, indeed, in which trade goods could become part of the offerings to the deities of a shrine – 'commodity fetishes' a century before Marx invented the term. In order to gain entry to the Ékpè masquerade of

Calabar, or the Hupila Hudjenk shrine of the Diola of Casamance in **Senegambia**, it was necessary to 'pay' material goods that it was possible to obtain only through the Atlantic trade, such as copper rods or slave fetters. Yorùbá proverbs describe how slaves could not wear masks and pawns were banned from parading in the Otomporo masquerade. Thus, new religious traditions enforced growing hierarchies, representing the power of the new worldview that had intruded into West and West-Central Africa.

This perspective is not somehow to barbarize or primitivize such a complex history, but rather to take seriously the perspectives of the actors themselves. The idea that taking African spiritual beliefs seriously is to 'exoticize' and 'primitivize' them begins in the long history of ridiculing **African spiritual beliefs** and denying the religious practice of Africans, which are instead seen as 'rituals', a perspective that goes back to European travel accounts of the sixteenth and seventeenth centuries. It starts from the view that Western economic rationality is some kind of universal given to which all societies ought to aspire, and that **Judaeo-Christian monotheism** offers universalism, whereas everything else is a particularistic 'cult'. Yet the reality that 'rational' exploitation of land is not some sort of universal value is becoming more and more apparent in the light of ecological degradation. Alternative perspectives are of potential significance in themselves. Certainly, it is impossible to understand the history of West Africa without acknowledging the place of spiritual beliefs in shaping social realities.

About the links between agriculture, capitalism, **religion**, "progress," and **misogyny**, Green explains:

Prior to the arrival of **Christian missionaries** and the **Islamic reform movements**, in many **African societies** women had substantial autonomy, which was by no means limited to trade. Women were active warriors in states as distant from each other as **Dahomey** and **Ndongo**, where women held positions of military command in the armies led by Nzinga in the 1640s and 1650s against the Portuguese in Luanda. In both Dahomey and Ndongo, too, women possessed religious power and could act as priests. Women rulers existed as well, in Ndongo, and also in some smaller decentralized states in Senegambia and in the Niger Delta. Power was becoming masculinized in many parts of West Africa, as we shall see, but this process was in some places overturning earlier structures that could be more equitable. Beyond trade, the other major arena in which women had the chance to gain autonomy was within the royal palaces. As more and more men either died in warfare or were exported as captives into the trans-Atlantic trade, gender imbalances grew. This increased **polygamy**; and in royal palaces such as Dahomey's this offered royal wives the opportunity for substantial influence. Even the Yovogan of Ouidah had several hundred wives in the 1730s, despite being a eunuch, showing that the huge number of royal wives did not always have sexual connotations. The extent of this female power is expressed well by the account of the Brazilian priest Vicente Pires, who described, of his 1797 to Dahomey visit, how the migan (or prime minister) was the most powerful of all the king's councillors – except for the Council of Wives, who could overrule him on all affairs. According to Pires, the king's wives all had different jobs, with some being barbers, others key bearers and others supplicating the deities in the shrines. This pattern is well documented at Dahomey. All residents of the palace were called ahosi, or royal wives, and they were often titled officers of the state. Princesses, the ahovi, also lived in the palace, and had a central role in selecting dadás. The ahosi farmed selected lands, and could trade and sell goods in their own right. No man was allowed within twenty feet of the dadá, not even the most senior military commander, and instead had to communicate through the dakhlo, or queen mother. Meanwhile, in Oyo (to which Dahomey was subject), when the alafin was no longer popular with his subjects, it was his wives who strangled him…

In sum, the growing impact of social transformations had made for an increasing division between and within the sexes. Men ruled, and in some places their wives governed, while at the same time the labour demands on poorer women became ever harsher. One of the driving features of this nascent process was, indeed, the growing burden of the export agricultural trade, and the increasing shortage of men. This heavy agricultural burden – 'it is women who do the agriculture', one 1770s account of Cacongo, north of Kongo, put it – made the lot of many women hard. Though women could gain power and better their lot, this was often through masculine structures of kingship or trade. In the nineteenth century, these patterns would then be exacerbated by the appearance of Christian missionaries from a Victorian Britain regressing in terms of gender roles, and of Islamic scholars fresh from the reaction to the Salafiya revival movement, which advocated a return to patriarchal values of early Islam. The rise of **global capitalism**, it becomes clear, was also a conservative male retrenchment to bolster patriarchy: that this was the case in Africa, Europe and the Islamic world shows again how interconnected the rising global value systems were, and how closely the emergence of **capitalism** relied on squeezing female power. There were exceptions, such as among the Igbo of the Niger Delta, where women could hold political power. In Dahomey, too, women remained key traders of textiles and ceramics, and retained their military functions. But, by the later eighteenth century, it was often hard for women to gain political power on their own account, as the role of a masquerade such as Mama-Jori in enforcing male choices shows. And today, in southern Senegal, when the Kankurang masquerade, seen by many as a descendant of Mama-Jori appears, it is women who wait up all night, playing percussive rhythms on sharp sticks, in anticipation and fear of the morning.

4.3 Hobbes, Rousseau's "Noble Savage," and Violence

The first person who, having enclosed a plot of land, took it into his head to say this is mine and found people simple enough to believe him was the true founder of civil society...what crimes, wars, murders, what miseries and horrors would the human race have been spared, had some one pulled up the stakes or filled in the ditch and cried out to his fellow men: "do not listen to this imposter...you are lost if you forget that the fruits of the earth belong to all and the earth to no one! (Jean-Jacques Rousseau)

Let's now focus on the questions about levels of **violence** in nomadic and sedentary hunter-gatherer societies. As discussed in Lee's 2018 paper "*Hunter-Gatherers and human evolution: new light on old debates,*" such questions have animated philosophical discussions since a long time, and nowadays are often simplified as two fundamental different ways of seeing human history—the so-called Rousseau *versus* Hobbes debate:

In **Thomas Hobbes**'s social evolutionary view, life in the "state of nature" was "nasty, brutish, and short", while **Jean-Jacques Rousseau** launched humanity's trajectory from a baseline of the "**noble savage**". Despite the publication of much more accurate data from twentieth-century archaeology and ethnography, the underlying debate has remained. In a recent book, *The better angels of our nature*, psychologist **Steven Pinker**, an avowed **Hobbesian**, added a new twist to the debate. Despite humanity's deep flaws, he argues, there is reason for hope – things are getting better. Like the famous figure of Dr. Pangloss, in Voltaire's eighteenth-century classic *Candide*, Pinker sought to affirm that **civilization**, if not the best-of-all-possible-worlds, is at least vastly superior to the state of humanity during its long history of hunting and gathering. In *The Better Angels* and elsewhere, Pinker draws on recent studies that assert a baseline of **primordial violence** by prestate peoples. The HNF – **Historically Nomadic Foragers** (HNF), small in scale, mobile, and egalitarian...reflect most closely the characteristics of ancient foragers...[and] are [indeed] not nonviolent. They fight and sometimes kill...but there is an enormous distance between that statement and the canonical assertion of the bellicose school that 5%, 15%, or even 50% of all hunter-gatherer deaths are due to interpersonal violence. In *The Better Angels of Our Nature* Pinker examined the **!Kung** data specifically and set the !Kung homicide death rate at 40.0/100,000; these levels are comparable to the high US urban homicide rates, which, for 1972, were 36.8 for Baltimore and 40.1 for Detroit. Despite the apparent magnitude of the **Ju/'hoan**/!Kung homicide rate, these still represent only 1.0-1.6% of overall deaths, compared to the 8-58% figure referenced in Pinker's TED Talk. There are crucial differences to consider. First is the question of US assault victims – unlike the !Kung – having access to excellent emergency room and trauma center facilities. Beckett recently asked, "While the number of gun murders has decreased in recent years, there's a debate over whether this reflects a drop in the total number of shootings, or an improvement in how many lives emergency room doctors can save". Second, the 25 listed killings represented all the !Kung homicides that our research group collected. The !Kung waged no **wars** in the twentieth century, and the Americans and other modern nations did (and still do). Adding to twentieth-century totals the deaths on both sides from the World Wars, Korean War, Vietnam War, and many other smaller conflicts more than triples the modern violent death rates, which I estimate for Europeans in the period 1914–1945 at close to 100 per 100,000 population, 2.5 times that of the !Kung.

Pinker conflates all **prestate societies** under a general heading and glosses over a very well documented and durable tenet of anthropology, namely that, with a few exceptions, **warfare**, as commonly understood, is rare or uncommon in many hunting and gathering societies. Evidence for it and its dire effects becomes prevalent only with the dramatic changes brought about by the **Neolithic Revolution**. The domestication of plants and animals, the transition from nomadic to sedentary living, and the subsequent growth of population and of fixed property brought profound changes to human societies, including rising rates of intergroup conflict and its deadly consequences. What sets foragers apart from farmers? In marked contrast to early farmers, their foraging predecessors lived more lightly on the land, and, although violence was present, they had other ways of resolving conflict. Living at very low densities, foragers had fewer things to fight over and, with little or no fixed **property**, could easily vote with their feet and disperse to diffuse conflict.

In fact, comparing the number of crimes in major cities—as for instance a rate of X crimes within 100,000 people—does make sense in most cases, but *surely not* concerning cases involving for example violence in very small nomadic hunter-gatherer communities *versus* major **genocides**, as Pinker does. Within any type of "morality" that we might socially construct, it does seem very hard to argue that the execution of a meticulous plan to kill six million Jews—estimates vary, but this is one of the most consensual numbers—during 5 years (1941–1945) within a globe population of about 3000 million people is really "morally" equivalent to having just a single person killed within a band of 100 hunter-gatherers during 25 years. The total ratio of persons killed by 100,000 people is exactly the same in the two cases, but the *reality* is that the first case involves the detailed planning made by the **Nazis**, the construction of concentration camps, of train lines to reach them, of killing centers, the acquisition of the gas used in those centers to suffocate people, the design of some of the most cruel, painful medical experiments ever done with humans, and so on (**see** Figs. 4.5 and 9.17).

In contrast, the killing of a single person in the second case could be an isolated/spontaneous act of revenge/jealousy, or even be merely related to an incidental killing due to a lost arrow being used in hunting other animals, as such incidents are not unlikely to happen over a long period of 25 years. Would someone that is *not* blinded

Fig. 4.5 How to *not* compare one killing with a genocide

by *Neverland* a priori assumptions and/or an a priori agenda, and that is not—in the specific case of Pinker—an avowed Hobbesian, really defend that an occurrence of a single lost arrow every 25 years, or an attack of jealousy, within a hunter-gatherer group of 100 people, is "equivalent" to the whole **Holocaust**? The problem is that these are precisely the type of flawed arguments—which are moreover often based on biased data, for instance only considering the number of people killed in crimes, but not in wars, within countries such as the U.S., as noted by Lee—that have been and continue to be used to support narratives about hunter-gatherers being particularly "violent," or "uncivilized," or "brute." So, while it is important to emphasize that **Pinker** as least had the merit to *try* to use quantitative data to address such questions, contrary to most authors in the past—including Hobbes—, his books are, in reality, full of these types of odd comparisons, biased data and even factually inaccurate items, unfortunately.

Interestingly, months after I wrote the above paragraph, I read Malesevic's 2017 book *The Rise of Organized Brutality – A Historical Sociology of Violence*, which basically defended, point by point, the very same line of argumentation that I followed above. Actually, when Pinker recognizes, at the very beginning of *Better Angels*, that he is an avowed Hobbesian, the whole premise and goal of his book seems clear: to "prove" the a priori *conclusion*—rather than to test the hypothesis—that Hobbes was indeed right. One wonders, if Pinker's goal was instead *to test this hypothesis*, and the data compiled by him would contradict it, would he really still publish a whole book about those data and its anti-Hobbesian implications? So far, all of Pinker's books and his renowned talks/speeches always have the very same conclusions: they all "prove" that Hobbes was mainly right, and that there is a "progress" in humanity, towards things such as an "expanding moral circle." Basically, they all defend the *status-quo*: accordingly, it is not a surprise that the richer people in the globe, such as Bill Gates, Mark Zuckerberg, and so on, are so in love with Pinker's books. So, it is indeed a valid question to wonder what is the true goal, or agenda, of what Pinker has been saying and publishing. This contrasts with the books of authors such as **Jared Diamond**, which cover a much wider range of subjects in a much more nuanced way, being very difficult to discern a "goal" that is common in all of his publications, except what should be the main aim of a scientist, ideally: to gather data, testing hypotheses with them, either supporting or contradicting them, and providing broader discussions based on those results. As noted by **Malesevic**:

> Pinker's data indicate that all regions of the world have followed similar patterns of decline in **violence** with dramatic decrease in the number of **war casualties**, **homicides**, acts of revolutionary violence and **terrorism** and so on. In contrast, studies that empirically analyse patterns of structural violence find continuous increase in the scale of structural and **symbolic violence** in different parts of the world. For example, using the ***Human Development Index*** as a comparative measure of structural violence, Iadicola and Shupe (in 2012) show that many underdeveloped regions of the world have experienced a rise in different forms of violence: from the decrease in **life expectancy** (in **DC Congo** and **Somalia**, only forty-four years), to the increased number of stunted development of children due to **malnutrition** (58 per cent in **Yemen** and 54 per cent in **Guatemala**), to the substantial rise of serious **diseases** and deaths caused by pollution (where the underdeveloped countries experience thirty-three times more deaths than those in the developed world)…. In Pinker's strange and deeply misleading system of calculation…non-war-related population increases were used to offset the casualties taking place elsewhere…. (For instance) that the overall violence has decreased because the relative death rates are lower has nothing to do with the populations of the world which did not actively participate in the two world wars. The fact that such populations might have experienced a demographic boom during the two world wars has very little to do with the bloody realities of the war taking place elsewhere. What does a substantial population increase in 1939-1945 Brazil or India have to do with the mass slaughters on the battlefields of **Stalingrad**, Verdun and Kursk?
>
> [Contrary to Pinker's ideas] the historical expansion of violence follows the expansion of organisational structure. Simple foragers lack organization and are generally egalitarian and less violent. Chiefdoms possess contours of complex organizations and are also deeply hierarchical and violence prone. For example, Genghis Khan's conquests (1206–1227) are often considered to be among the most violent episodes in human history, amounting to as many as 40 million casualties. Obviously, this is no coincidence. It is one of the main principles that can be encountered throughout history: the better, more complex and durable social organizations possess greater coercive capacities which often generate more violence. Most scholars of early history agree that the emergence of **civilizations** was paralleled with the substantial increase in the scale and scope of violence. There is wealth of evidence that early empires, city-states and other forms of statehood were much more belligerent than those of their nomadic foraging predecessors. The establishment of complex **hierarchical polities** with permanent, usually densely populated urban settlements, elaborate division of labour, established patterns of **social stratification** and **power centralization**,

was crucial for the increase in the **coercive capacity** of early civilizations. These organisational advancements often developed together with elaborate proto-ideological belief systems, ceremonial centres of worship and the practice of the elite-level writing that helped justify the use of violence on a large scale.

Malesevic's book carefully deconstructs the highly biased Hobbesian ideas of Pinker, as well as the ideas of another author that is far more respected and influential within the scientific community—not within the broader public—than Pinker is: **Norbert Elias**, and his ideas about the "**civilizing process**":

The relationship between **organized violence** and civilisation is something that has puzzled analysts since antiquity. From **Confucius** and Mosi to **Plato**, **Aristotle** and **Ibn Khaldun**, scholars have attempted to identify whether civilisations tame or ignite violence. Norbert Elias develops an original theory that centres on the interdependence of the civilising process and violence. For Elias, violence and civilisation have a complex relationship. In some respects, Elias follows **Weber** arguing that the monopoly on the use of violence is a precondition of civilisational advancement, as it pacifies the social order. However, in other, more pronounced, respects, for Elias violence is the exact opposite of civilisation. More specifically, his most celebrated book, *The Civilising Process*, traces the steady decrease in individual and collective forms of violence to lasting processes of expanding external social control, coupled with a gradual internalisation of self-restraint. In Elias' writings, the civilising process is understood as a dual phenomenon through which individuals learn how to constrain their own 'natural' violent impulses and through which entire social orders become more pacified. However, not only is it the case that civilisation and violent action are fully congruent, as all coordinated collective violence requires a substantial degree of self-restraint, but more importantly, *civilisation is the cradle of organized violence*. Despite the popular view that human beings have engaged in **warfare** since time immemorial, numerous archaeological and anthropological studies have shown that organized violence emerged only in the last twelve thousand years, and large-scale warfare only in the last three thousand years of human existence. Organized violence appears on the historical stage, together with **sedentary cultures** – with the **domestication** of plants and animals, **organized farming**, **land ownership**, **fortified towns**, **institutionalised religions**, political orders and elaborate forms of **social stratification**. In a word: **civilisation**. What distinguished the first known civilisations – **Sumer**, **ancient Egypt**, **Shang China**, **Harrapan India** and later **Mesoamerican worlds** – from the earlier social formations was their ability to use organized violence and fight wars of conquest. The pristine states of early civilisations were created through warfare, and distinct civilisations have expanded through organized violence. *Hence violence is not the 'Other' of civilisation but one of its most important components*.

[Indeed] the advancements in science, technology and administration on the one hand and the staggering increase in the population size on the other have fostered the emergence and proliferation of specialised social organizations responsible for violent action as well as for coercive coordination of the large numbers of human beings (i.e. military, police, private security companies, armed militias, etc.).... Whereas a handful of individuals living in a flexible foraging band could freely roam African savannahs and engage occasionally in violent interpersonal disputes, millions of individuals inhabiting Ptolemaic Egypt could not survive without the presence of the well-established polity able to generate enough food and establish internal order and external security, including periodic warfare with its neighbours. As recent macrosociological research demonstrates convincingly, **ethnic cleansing** and **genocide** are modern phenomena inspired by modern ideological blueprints, modern means of organization, modern and mutually exclusive state-building projects, and conflicting visions of modernity. While the modern subject might avoid spitting or blowing her nose in the table cloth, the populations of modern states are complicit in many episodes of mass violence, whether detonating atomic bombs, perpetrating 'targeted assassinations' or launching 'preemptive' and 'surgical strikes', and the reality of these actions is often sanitised through the language of 'collateral damage'.... The military ideologies and strategies behind these two wars [WW1 and WW2] were conceived and implemented by highly refined and selfdisciplined gentlemen bent on implementing Clausewitz's Dictum of absolute war as a realm of 'utmost violence' where one side is determined to annihilate the other.

When [these] facts fly in the face of his theory, Elias utilises concepts such as 'decivilising spurt' to rescue his explanatory model. An example is how Elias accounts for **Nazism** and the **Holocaust**, where he argues that the civilising process can occasionally go into reverse. So **concentration camps**, **gas chambers**, extensive systems of **torture** and **genocide** are understood as no more than a 'deepest regression into barbarism' whereby war removes all internal and external constraints and individuals revert to their 'animalistic selves'. In particular, Elias emphasises the role of specific social agents wedded to irrationally held belief systems with 'high fantasy content' that provided them with 'a high degree of immediate emotional satisfaction'. In other words, a decivilising spurt strips away the civilising benefits of detached thinking and marks the return of emotionally charged communal fantasies: [Elias wrote that] '[the] national Socialist movement was mainly led by half-educated men'; 'the Nazi belief system with its pseudo-scientific varnish spread thinly over a primitive, barbaric national mythology…that it could not withstand the judgment of more educated people'. As most recent studies of the Nazi movement show, much of its leadership as well as its support base were very well educated. Many German intellectuals, university professors and broader cultural elite were sympathetic to National Socialist ideas, and its core constituency was much more educated than the rest of German society. For example…41 per cent of SD [**Nazi intelligence service**] had higher education at the time when national average was 2 or 3 per cent; the SS recruits and officers were highly educated; majority of doctors, judges and solicitors were members of NSDAP. As Müller-Hill shows the majority of the commanders of **Einsatzkommandos** (mobile killing squads), who were the main protagonists of genocide, were highly educated individuals: economists, solicitors, academics. More than two-thirds of these commanders had higher education and one-third had doctorates. In a similar vein, half of the German students were Nazi sympathisers by 1930; university-trained professionals (i.e. "academic professionals") were overrepresented in the NSDAP and in the SA and SS officer corps. While National Socialist ideology did attract many social strata, some of which had little or no education, its core ideological support base were young and educated males: (as noted by Mann) 'fascism was capturing the young and educated males because it was the latest wisdom of half a continent…its ideological resonance in its era…was the main reason it was a generational movement'.

These last lines about **Nazism** being mainly promoted by more educated people remind us of something that unfortunately continues to be mainly neglected within the media and broader public in general, and even among the scientific community. Namely, that smarter people tend to be particularly dangerous because, when they commit to a belief, they are better at rationalizing that belief a posteriori—or as nicely put by Shermer, "*smart people believe weird things because they are skilled at defending beliefs they arrived at for nonsmart reasons.*" However, one important point that needs to be stressed is that I am not defending here—nor did authors such as Diamond or Malesevic—the notion of a "noble savage" that is never violent. As clearly explained in Diamond's 2012 book *The world until today*, traditional societies also perform acts that can be designated, in some ways, as "wars" (see Fig. 4.6). In that book—which, it should be noted, has been accused by some researchers to actually overemphasize the violence of such hunter-gatherer societies—Diamond argues that:

Fig. 4.6 Adapted from Diamond's 2012 book, who wrote: "Traditional warfare: Dani tribesmen fighting with spears in the Baliem Valley of the New Guinea Highlands…the highest one-day death toll in those wars occurred on June 4, 1966, when northern Dani killed face-to-face 125 southern Dani, many of whom the attackers would personally have known (or known of)…the death toll constituted 5% of the southerners' population"

State armies spare and take prisoners because they are able to feed them, guard them, put them to work, and prevent them from running away. Traditional "armies" do not take enemy warriors as prisoners, because they cannot do any of those things to make use of prisoners. Surrounded or defeated traditional warriors do not surrender, because they know that they would be killed anyway. The earliest historical or archaeological evidence of states taking prisoners is not until the time of **Mesopotamian states** of about 5,000 years ago, which solved the practical problems of getting use out of prisoners by gouging out their eyes to blind them so that they could not run away, then putting them to work at tasks that could be carried out by the sense of touch alone, such as spinning and some gardening chores. A few large, sedentary, economically specialized tribes and chiefdoms of hunter-gatherers, such as coastal **Pacific Northwest Indians** and **Florida's Calusa Indians**, were also able routinely to enslave, maintain, and make use of captives. However, for societies simpler than Mesopotamian states, Pacific Northwest Indians, and the Calusa, defeated enemies were of no value alive. War's goal among the **Dani**, **Fore**, **Northwest Alaskan Inuit**, **Andaman Islanders**, and many other tribes was to take over the enemy's land and to exterminate the enemy of both sexes and all ages, including the dozens of Dani women and children killed in the June 4, 1966, massacre. Other traditional societies, such as the **Nuer** raiding the **Dinka**, were more selective, in that they killed Dinka men and clubbed to death Dinka babies and older women but brought home Dinka women of marriageable age to force-marry to Nuer men, and also brought home Dinka weaned children to rear as Nuer. The **Yanomamo** similarly spared enemy women in order to use them as mates. **Total warfare** among traditional societies also means mobilizing all men, including the Dani boys down to age six who fought in the battle of August 6, 1961. State war, however, is usually fought with proportionally tiny professional armies of adult men.

> The remaining big difference between tribal and **state warfare**, after that distinction between total and limited warfare, involves the differing ease of ending war and maintaining peace…wars of small-scale societies often involve cycles of **revenge killings**. A death suffered by side A demands that side A take vengeance by killing someone from side B, whose members now in turn demand vengeance of their own against side A. Those cycles end only when one side has been exterminated or driven out, or else when both sides are exhausted, both have suffered many deaths, and neither side foresees the likelihood of being able to exterminate or drive out the other. While analogous considerations apply to ending state warfare, states and large chiefdoms go to war with much more limited goals than do bands and tribes: at most, just to conquer all of the enemy's territory. But it's much harder for a tribe than for a state (and a large centralized chiefdom) to reach a decision to seek an end to fighting, and to negotiate a truce with the enemy – because a state has centralized decision-making and negotiators, while a tribe lacks centralized leadership and everyone has his say. It's even harder for a tribe than for a state to maintain peace, once a truce has been negotiated. In any society, whether a tribe or a state, there will be some individuals who are dissatisfied with any peace agreement, and who want to attack some enemy for their own private reasons and to provoke a new outbreak of fighting. A state government that asserts a **centralized monopoly** on the use of power and force can usually restrain those hotheads; a weak tribal leader can't. Hence tribal peaces are fragile and quickly deteriorate to yet another cycle of war. That difference between states and small centralized societies is a major reason why states exist at all.
>
> There has been a long-standing debate among political scientists about how states arise, and why the governed masses tolerate kings and congressmen and their bureaucrats. Full-time political leaders don't grow their own food, but they live off of food raised by us peasants. How did our leaders convince or force us to feed them, and why do we let them remain in power? The French philosopher **Jean-Jacques Rousseau** speculated, without any evidence to back up his speculations, that governments arise as the result of rational decisions by the masses who recognize that their own interests will be better served under a leader and bureaucrats. In all the cases of state formation now known to historians, no such farsighted calculation has ever been observed. Instead, states arise from chiefdoms through competition, conquest, or external pressure: the chiefdom with the most effective decision-making is better able to resist conquest or to outcompete other chiefdoms. For example, between 1807 and 1817 the dozens of separate chiefdoms of southeastern **Africa's Zulu people**, traditionally warring with each other, became amalgamated into one state under one of the chiefs, named Dingiswayo, who conquered all the competing chiefs by proving more successful at figuring out how best to recruit an army, settle disputes, incorporate defeated chiefdoms, and administer his territory. Thus, it could not be claimed that some societies are inherently or genetically peaceful, while others are inherently warlike. Instead, it appears that societies do or don't resort to war, depending on whether it might be profitable for them to initiate war and/or necessary for them to defend themselves against wars initiated by others. Most societies have indeed participated in wars, but a few have not, for good reasons. While those societies that have not are sometimes claimed to be inherently gentle (e.g., the **Semang**, **!Kung**, and **African Pygmies**), those gentle people do have intragroup violence ("murder"); they merely have reasons for lacking organized inter-group violence that would fit a definition of war. When the normally gentle Semang were enlisted by the British army in the 1950s to scout and kill Communist rebels in Malaya, the Semang killed enthusiastically. It is equally fruitless to debate whether humans are intrinsically violent or else intrinsically cooperative. All human societies practise both violence and cooperation; which trait appears to predominate depends on the circumstances.

This last point is indeed one of the main take-home messages of the present book as well: as Diamond, I am clearly *not* arguing that nomadic hunter-gatherers are "good" and so-called civilized people are "bad" but instead that it is in great part the *circumstance created by the specific social organization* of nomadic hunter-gatherers that leads them to behave in a more "noble" way, in general, with some exceptions of course. So, the question is—as pointed out in the cover of Diamond's book—, *what can we learn—and use in our "developed" societies—from their social organization and way of life*? Malesevic further discusses these key topics with his characteristic deep and very perceptive type of argumentation in *The Rise of Organized Brutality*:

> The tribal groupings of the Great Plains, such as the Blackfoot, Arapaho, **Cheyenne**, **Comanche**, Crow and **Kiowa Apache**, among others, have often been described as being war prone and having strong warrior cultures. The popular images of scalped enemy heads, raped women, pillaged villages and tortured war prisoners are typically associated with many of these tribes. Nevertheless, as Farb demonstrates, rather than having an intrinsic proclivity for violence from time immemorial, these tribes were, for the most part, a historical product of the **European colonialism**, as an overwhelming majority of these groups did not exist as such before the European conquest. Simply put, the **Great Plains Indians** developed from the descendants of refugees of various precontact tribes. As European colonisation advanced, most native populations were decimated by disease and colonial expansion, while the remnants of these populations created *new social orders and cultures built around horses, guns and alcohol*, all of which were introduced by the European invasion. Hence there is nothing inherently violent about the Great Plains huntergatherers: their violent cultural practices are not ubiquitous but were a product of the specific structural transformation.
>
> Just as Hobbes and Rousseau were in some important respects representatives of the similar philosophical tradition (i.e. the social contract theory), so are contemporary Rousseauian and Hobbesian anthropologists of violence: despite their diametrically opposed views on human relations to violence, they both embrace an essentialist understanding of social relations. Simply put, instead of analysing violent interaction through the prism of changing social and structural relations, these scholars tend to focus on the spurious and sociologically meaningless question, that is, are human beings inherently violent or peaceful? Hence for Sponsel, 'nonviolence and peace are natural, ubiquitous, and normal in the human species throughout its evolution and adaptation. In fact, for well over a million years evolution has selected for a human nature that is naturally inclined toward nonviolence, peace, cooperation, reciprocity, empathy, and compassion'. In stark contrast, for Pinker…human beings are genetically wired for violence. [However] violence is first and foremost a social relation between two or more living organisms; *it is not a biological quality*. It is a particular form of social action that human beings alone categorise as violent. It is not a fixed trait but something that is historically generated, structurally shaped and ideologically framed.
>
> Once the focus of our analysis is on the social and historical context instead of biological traits, it becomes apparent that violence proliferates in specific structural conditions. More specifically, large-scale violent action necessitates the development of durable social organizations, as disorganised violence does not last and is unlikely to have any long-term impact. Therefore, unlike some scholars who attribute violent acts to individual or collective motivation and see violence as the property of specific groups, it is much more fruitful to look not at groups or individuals but at social organizations as the dominant purveyors of violent, and other, action…. It seems much more logical to assume that those individuals and small groups who initially demonstrated skill, organisational competence and coercive dominance to produce more food than others may have been just as skilled in relying on these qualities to safeguard their food surpluses. As a number of anthropological studies indicate, the nascent hierarchies in the **prepatrimonial world** were usually built around the distribution, not appropriation, of goods. The so-called Big Men tended to distribute food surpluses and protection in order to attain group support and maintain their Big Men posi-

tion. This is still evident among the existing hunter-gatherers and was particularly pronounced in the complex hunting-gathering communities (chiefdoms). What seems to be of more importance than the accumulation or distribution of economic goods is the social mechanisms through which such processes take place – the creation of social organization itself. Binding individuals together around a shared long-term project gives impetus to the development of organisational power.

Malesevic then points out:

So regardless of what a particular undertaking is (the hunting and storage of animal flash, cultivation of land, coordinated distribution or appropriation of resources, etc.), what really matters is the ever-increasing organisational capacity that such joint undertakings generate. As Michels noted long ago, organizations are the epicentres of all power, as the organized minorities generally tend to overpower the disorganised majorities. As both archaeological and anthropological records show, foragers as a rule do not like organized structures. The historical records demonstrate many instances of foraging bands being integrated into wider imperial or other state structures and their prolonged struggles to break free from such organisational shells (i.e. from the neighbouring tribes of Akkadian Sumer to the indigenous populations of Australia, South and North America, etc.). Even today, most existing hunting-gathering communities reject all attempts to bring them into the fold of the modern nation-state. Hence the shift towards social organizations was extremely slow and turbulent, involving numerous reversals and countless new beginnings.

While **foraging bands** consist of very small, fluid, nonhierarchic, mobile and generally disorganised groups, the chiefdoms are much more populous and better organized entities built around a centralised leadership and possessing distinct social hierarchies in terms of age, gender, marriage and military role. Furthermore, unlike disorganised foragers who tend to scavenge for food and run away from external attacks (by carnivorous animals or organized human groups) and are generally unable to establish collective resistance, the chiefdoms are characterised by their political and military prowess and are capable of utilising organized collective action for both attack and defence. However, unlike chiefdoms, which are able to mobilise relatively large numbers of individuals and as such can force those individuals to act against their will, the foraging bands enjoy almost unlimited individual freedom. A Microsoft employee is provided with a regular salary, pension, health insurance and social security protection as long as she abides by the strict rules of that organization and devotes eight or more hours per day to Microsoft. This historical trade-off between individual liberty and organisational security is at the heart of both social development and organisational power…

The early complex hunter-gatherers relied on the developed organisational power to generate more food, better accommodation and clothing, but this very same organisational power was also used to dominate, enslave and kill one's neighbours in the new thrust for territory, resources, status and other acquisitions. There is a tendency to keep these two sets of processes separate as the organized production is regularly opposed to the organized destruction, but in fact these two are outcomes of the same phenomenon – organisational power. Once organisational power is generated, it can be deployed in variety of ways, some of which are likely to be perceived differently by those who utilise such power and those who become objects of this power. While well established commercial fishing fleets such as American Seafoods or Trident Seafoods use their superior organisational power to catch hundreds of tonnes of fish per year which feed millions of humans, such undertakings are also premised on the destruction of particular ecosystems and the lives of millions of seabed animals and also negatively impact the livelihoods of local fishing communities. There is no social development without coercive costs…

Therefore, violence is not an unnecessary addition or an avoidable byproduct of organisational power. Instead, violence is the core of all organisational social action, and this link between organization and violence has its structural logic. The origins and development of

> **violence** are rooted in the dynamics of organisational power. If one looks only at the death casualties of wars over the past seven centuries, it is possible to see the exponential rise: for the fourteenth and fifteenth centuries, the total war death toll is less than 1 and 4 million, respectively; this jumps to 7 and 8 million for the sixteenth and seventeenth centuries and rockets to 19 million for the nineteenth century alone. However, the twentieth century surpasses all of the recorded history combined with its 135 million war deaths. This rise in death tolls is clearly linked with the increase in the organisational capacity of polities: even though some rulers of the premodern empires might have been more inclined to kill than their modern counterparts, the emperors lacked the organisational means (i.e. technology, infrastructure, communications, transport, etc.) to achieve such bloody ends.... Furthermore, as the records from the prehistory and early history demonstrate, large-scale structures do not come naturally to human beings. On the contrary, it seems that our predecessors were reluctant to embrace life under complex and durable social organisations even when this has proved beneficial in terms of security and the regular supply of food and shelter. This might be in part linked to the loss of individual and communal freedoms, sharp rise in inequalities and general worsening of quality of life. In contrast to popular perceptions, reinforced by evolutionary thinking, our late Palaeolithic predecessors had better health, were taller, had greater pelvic inlet depth index and had a slightly longer lifespan than average individuals living from the fifteenth to the nineteenth century. This nearly universal resistance towards organisational structures is much better documented with the contemporary foragers from Amazonia to Papua New Guinea, who generally tend to reject all attempts by the state authorities to bring them into the fold of the established organisational structures.

The above excerpts, from various books written by authors with very different backgrounds, do illustrate that there has been a change of mindset in the last decades and an increasing consensus, supported by a huge amount of empirical evidence compiled from different areas, regarding the idea that the rise of "civilizations" was *not* necessarily a "blessing" in human history, after all (see Box 7.12). Another illuminating example of this is Kirkham's 2019 superb *Our Shadowed World – Reflections on Civilization, Conflict, and Belief*, which moreover directly connects many atrocities done by "civilizations" plagued by teleological narratives, ideologies and beliefs:

> Civilizations are formidable and frightening creations. Civilization is often seen as the opposite of **savagery** – the living city rising above the threatening wilderness, a progressive idea leaving in its wake a more primitive state…Rather, I contend, **civilization** brings its own form of savagery; and the greater the cities and the more "advanced" the civilization, the greater the scale of savagery. It is a story that culminates in our own times with one of the most incomparable acts of savagery imaginable – the **Shoah**, or **Jewish Holocaust** of the **Second World War**. It is a story that can be perhaps best represented by the monumental concrete face, rising some fifteen meters high, which stands on a hill alongside the Road of Bones (human bones!) overlooking Magadan in the Kolyma region of Eastern Russia. Here in what has been called the capital of the **Soviet Gulag** this overpowering memorial – the Mask of Sorrow – was built in memory of all those countless millions of people who perished under the flimsiest of pretexts in the forced labor camps of **Stalin**'s regime over a period of three decades. This modern pietà has an Aztec quality to it, with its grim visage and weeping tears of skulls, which bears witness to the industrial scale of slaughter that took place among civilized people. It would need all the capabilities of a modern state to exterminate twenty or forty million (the actual numbers will never be known) of its own people as occurred in Stalin's Russia and **Mao**'s China. The rise of civilization with the city-state some five thousand years ago in **Mesopotamia**, the "land between the rivers" also called "the cradle of civilization", saw the appearance of new forms of institutionalized

savagery: organized violence or war, **slavery**, and the oppression of women, to name but three. And to support and defend the city-state there arose the organized expansion of savagery we call empire. Empire was invented by **Sargon the Akkadian**, one of the first identifiable figures of history, who also created the first legal code and thereby laid the foundations for legalized male domination, for civilizations have generally been patriarchal affairs. One of the earliest marks of civilization is the humble brick, and among other irrevocable precepts of Sargon was that a woman who dared to speak out of turn should have her teeth smashed, with a brick.

Civilization grew and proliferated largely through empire. It would be pedantic to rehearse the benefits that civilization has provided to humanity, but it is sobering to reflect on the human cost by which they were acquired. A typical boast of a victorious king and aspiring emperor, like that of the **Assyrian Shalmaneser III**, was that "I covered the wide plain with the corpses of their fighting men…. I dyed the mountains red with their blood". Another kind, **Ashurbanipal**, recorded, "I tore the tongues of those whose slanderous mouths had uttered blasphemies against my **God Ashur**…. I fed their corpses, cut into small pieces, to dogs". The invention of writing – one of the key achievements and marks of civilizations – has, ironically, enabled us to chronicle the record of such savagery. Indeed, it is the grim determination of its victims not to be forgotten that provided the motivation for writers of many eras such as Varlam Shalamov, a victim of **Stalin's Great Terror**, whose Kolyma Tales bears witness to the truth of human savagery that must never be forgotten. Nor must its cause be forgotten. Let us be clear: *it is human belief, both secular and religious, that provides the motivation and legitimation of savagery*. As one of Stalin's activists, Lev Kopelev, would write of the goal of the universal triumph of **Communism**, "in the terrible spring of 1933 I saw people dying from hunger…women and children with distended bellies, turning blue…corpses in ragged sheepskin coats…I saw all this and did not go out of my mind…nor did I lose my faith…I believed because I wanted to believe". Belief provides the numinous justification for civilization as much as bricks facilitate its construction. The terrifying judgment of the papal legate **Arnaud Amaury** on the inhabitants of Béziers in 1209 – "Kill them all…God will know his own" – reverberates across the centuries from that "Age of Faith" to the modern faith-based caliphate of **Abu Bakr al-Baghdadi** in Syria/Iraq, which also now aspires to rule the world. As I write these words contemporary jihadists are perpetrating similar atrocities in that same part of the world where civilization began in the name of their beliefs. Inshallah. "God's will". Plus ca change?

In the meantime, empires have continued to rise and fall, and civilizations have woven people's noble ideals and religious beliefs into an ever-expanding remorseless spiral of savagery. One need only consider the **Holy Roman Empire of Christendom**, baptized with blood of thousands of butchered Saxons by its founder **Charlemagne**, and expanded by the swords of those merciless holy warriors of God, the **Crusaders**. Such were the **Teutonic Knights**, whose mission was to create a **Christian civilization** among the heathen Slavonic savages of the East. Whatever noble truths they sought to establish in the name of **Christian belief**, the **Inquisition**, which was also the instrument of their implementation, was perhaps the most cruel and savage institution ever devised. Its savagery shaped the minds of millions, and its legacy poisons the present in often unnoticed ways. It is a story I will endeavor to explore as one wonders to what extent the vast ambition of Caesaropapism and the Christian civilization or **Christendom** it created reflected the teaching of the humble Nazarene carpenter. To many these are uncomfortable and indeed unwelcome thoughts, for the notions of civilization and savagery, "special providence" and spectacular destruction do not sit easily together. They are like ill-matched conjoined twins or the mythical brothers **Cain** and **Abel**, whose fratricidal enmity precedes the rise of civilization. So has it ever been. Rome, which sprang from the blood of **Remus** spilled by his brother **Romulus**, would become a mighty empire whose distinguishing features were patriarchy, slavery, hierarchy, and violent conquest of its enemies – aptly epitomized in the

> greatest popular spectacles of all, the gladiatorial games that took place in its most imposing and appropriately named monumental structure (given its colossal size), the **Colosseum**. The orgies of blood-soaked savagery that this masterpiece of civil engineering made possible – as in many other such structures across the empire – are without historical parallel. Since those ancient times an exponential growth of the possibilities for violence and savagery has occurred as the flint-tipped arrow has given way to the nuclear missile. The twentieth century became characterized not only by the scope of genocidal savagery but also by the destructive violence of its **warfare** and **weaponry**. Now humanity had the power to destroy the whole of civilization in an atomic holocaust. And in the twenty-first century this shadow over the earth has been further darkened by an even more perilous threat: **ecocide**.

In summary, the ideas of authors such as Kirkham, Diamond and Malesevic are indeed very similar to the overall ideas I am defending in the current book. That is, aspects such as violence, sexuality, egalitarianism, and so on, don't depend exclusively on different "moral values"—as some defenders of the notion of a "Noble Savage," or alternatively some reactionary racists, would argue—nor on the natural environment—as adaptationists would tend to defend—, but instead, on specific *circumstances* created by a complex combination of random, chaotic and contingent factors. In this sense, this view is also similar to that recently defended in a book written by two authors—Daron Acemoglu and James Robinson—that clearly cannot be accused to be "left-wing Rousseauians," as their book *Why Nations Fail: The Origins of Power, Prosperity, and Poverty* was considered "best book of the year" by the clearly non-left wing *Wall Street Journal*. Namely, Acemoglu and Robinson argue that it is not so much geography—as defended in Diamond's *Guns, Germs and Steel* (see above)—nor culture—as defended by most authors, and of course by reactionary and racist people—but rather human-made **political and economic institutions** that explain why some countries "fail" and others not. It should be said that, in my opinion, their book does not really contradicts directly Diamond's *Guns, Germs and Steel* because the latter book explains mainly the differences that occurred various millennia ago—for instance, why ancient Egypt was so powerful—while *Why Nations Fail* mainly focuses on the differences between later "civilizations" or countries.

In this regard, these books complement themselves, and that is probably why Diamond himself praised Acemoglu and Robinson's book, as things are indeed complicated and related to a series of interactive, random, and contingent cultural, geographical and sociopolitical factors. Lee's 2018 paper "*Hunter-Gatherers and human evolution: new light on old debates*" also notes how complicated are these topics and in particular the historical accounts about and comparisons between "civilizations" and nomadic hunter-gatherer societies by pointing out that such comparisons often include hunter-gatherer groups that were already significantly influenced by contacts with sedentary groups and particularly with colonialist powers, as noted above:

> From my own area of study, there are historic southern African **San/Bushman** groups who did wage war…[but] to these anomalous cases, some analysts of the bellicose school add the famous war-like South American **Yanomamo** and **Jivaro**, as well as the war-like pig-raising farmers of Highland New Guinea. All are included under the rubric of hunter-gatherers; all are war-like, and yet as practicing farmers (and for New Guinea, pig raisers as well), they are emphatetically not hunters and gatherers. With sampling procedures such as

these, the apparent level of warfare is artificially jacked up. In terms of the skeletal evidence, Haas and Piscitelli ambitiously surveyed 400 Paleolithic sites with 2,930 skeletons, gleaned from a review of more than 75 published sources on skeletal remains in Europe, Asia, and Africa. They report that, in a vast array of prehistoric sites, there is scant evidence of **warfare**. Clear evidence of some violence is found in two Italian and two Ukrainian sites with individual skeletons that indicated embedded points. Only a single site – the Jebel Sahaba ossuary – on the Upper Nile, with 24 of 58 skeletons showing serious evidence of violent death, supports the bellicose thesis. In marked contrast, more than 390 of the 400 sites across the Old World (97.5%) are completely lacking in such signs. The evidence indicates that early humans, rather than being "killer apes" in the Pleistocene and early Holocene, lived as relatively peaceful hunter-gathers for some 15,000 generations, from the emergence of modern *Homo sapiens* up until the invention of agriculture, roughly from 300,000 to 8,000 years BCE. Therefore, there is a major gap between the purported violence of our chimp-like ancestors and the documented violence of post-Neolithic humanity.

Historically Nomadic Forager groups rely on movement in the course of the annual round, moving 3-6 (or more) times per year. These movements take place both within the traditional "territory" and with visits to kin in adjacent localities. Most exhibit an annual cycle of dispersal and of aggregation. A key corollary of this mobility is the basic fact that groups' social world extends far beyond their home territory, and a second corollary is the necessity of maintaining a low accumulation of material property. Ease of movement is important in dealing with conflicts. Within the local group, there are strong injunctions about food sharing, a key characteristic of egalitarian societies. Most observers report the markedly **higher status of women** in hunter-gatherer society, when compared to women's status in tribal, chiefly, and state-level societies. In the latter, observers note that the rise of patriarchy and male dominance are closely associated with the post-Neolithic increases in warfare and social complexity. Physical punishment of children is very rare. One of the common threads through this literature is the phenomenon of allo-parenting, care of children by individuals other than the parents. The practice is ubiquitous, especially in the areas of provisioning and food sharing.

Chapter 5
Sex, Love, Marriage, and Misogyny

> *Love is our true destiny...we do not find the meaning of life by ourselves alone – we find it with another.*
>
> (Thomas Merton)

5.1 Homosexuality, Sex, and Romantic Love

> *Faith, in other words, no longer meant a state of surrender or openness to Revelation but rather an absolute trust in the Church hierarchy, which alone possessed the reason to discern God's meaning...faith had become the means of enforcing authority. (Kenan Malik)*

As explained in Diane Ackerman's 1994 book *Natural History of Love*, in most "post-Neolithic" (agricultural) societies, there is a clear link between the notion of **romantic love** and **teleological narratives**, including those about a **cosmic purpose of life** and the related concept of **"meant to be."** She provides several examples of this, from art in **ancient Egypt** and philosophy in **ancient Greece** to studies of **organized religions** and stories about the **troubadours** of the European Middle Ages and fascinating facts about the Victorian era. She argues that the Western vision of romantic love mainly derives from ancient Greek thought, noting that "to **Plato**, lovers are incomplete halves of a single puzzle, searching for each other in order to become whole." This is shown in Plato's *The symposium*, which includes the following famous fable: "originally there were three sexes: men, women, and a hermaphroditic combination of man and woman…these primitive beings had two heads…two sets of genitals, and so on…threatened by their potential power, **Zeus** divided each one of them in half, making individual lesbians, **homosexual** men, and heterosexuals…but each person longed for its **missing half**, which it sought out, tracked down, and embraced, so that it could become one again." She explains that this "is an amazing fable, saying, in effect, that each person has an ideal love waiting somewhere to be found…each of us has a one-and-only, and finding that person makes us whole…this **romantic ideal**…appealed so strongly to hearts and minds that people *believed* it in all the following centuries, and many still *believe* in it today."

I italicized "believed" and "believe" to emphasize that, as she is implying, this is a teleological narrative that is *not* based on empirical data or on a comprehensive analysis of our evolutionary history or biology. Such *beliefs*, as many others that humans created to "find" a cosmic meaning or purpose that is "meant to be,"

R. Diogo, *Meaning of Life, Human Nature, and Delusions*,
https://doi.org/10.1007/978-3-319-70401-2_5

continue to be—or, more accurately, are becoming increasingly—predominant in not only **Hollywood and Bollywood movies** and popular culture in general but also in many books written even by scholars that sometimes defend ideas such as "we are meant to love, and specifically to love our other half." In fact, contrary to some religious narratives, the teleological narratives about "romantic love" did not tend to lose importance within people of Western countries in the last centuries. Instead, they have become more and more extremely romanticized, such as the "Hollywood" versions that are now so much in vogue not only in those countries, but around many other countries as well. Of course, such romantic ideas clearly contrast with the *reality of biological evolution*: animals—including humans—are not "meant" to do anything—they just exist, they don't follow any fixed masterplan, there is no planned future goal to be achieved. Actually, even within the few things that *need* to—rather than are "meant" to—be done so let's say an animal species does not become extinct, love between a male and a female, or between a male and a male or a female and a female, is surely *not* one of them. Evolutionarily, it suffices that they merely survive and reproduce. Therefore, **sex**—mainly driven by ***desire*** and ***pleasure***, linked to an interplay between **hormones**, enjoyable sensations, and many other items—without "romantic love" is more than enough. Actually, we need to remember that most organisms in this planet—which is terms of individual number of organisms is mainly inhabited by bacteria and viruses—don't even have sex, reproducing asexually.

Apart from Ackerman's *Natural history of love*, many books discuss how this social construction was directly related to the use of and believe in teleological narratives, such as Stephen Greenblastt's *The Rise and Fall of Adam and Eve*, or Bruce Feiler's *The First Love Story – Adam, Eve, and Us*. In the latter book, Feiler notes how Viktor Frankl's postwar classic *Man's Search for Meaning* called love "the ultimate and the highest goal to which man can aspire," and how Erich Fromm's *The Art of Loving* described love as "an intensification of life, a completeness, a fullness, a wholeness of life," following the narrative that "being alive is too overwhelming to be done by yourself: we can only fully be ourselves when we are with another." What is interesting about all these teleological narratives, and further illustrates that they are not based at all on biological or anthropological data, is that they assume that we are "meant to be" with "another," with "the only one," with "the love of our life": that is, they imply that humans are naturally not only a **social monogamous** species, but even a **sexual monogamous** one (see Boxes 5.1 and 5.2). It is important to make a distinction between these two terms, which were recently discussed in detail in Carter and Perkeybile's in 2018 paper "*The monogamy paradox – what do love and sex have to do with it*". In the past, researchers—often based on a priori ideas about how "true monogamy" was widely seen in "nature"—*assumed* that several animal groups studied by them, such as gibbons, were sexually monogamous because they *saw* that female–male pairs were together for most of the time. Yes, but they were *not* together all the time: the reality is that subsequent DNA studies revealed that the vast majority of those groups were not sexually monogamous—instead, as it often happens in human couples, they sometimes "fooled around" and had sex with others. That is, as also noted in Kvarnemo's 2018 review on animal monogamy—sexual polygamy is actually much more common that it had been assumed for a long time, being indeed present in many of the so-called socially monogamous species. Once again, empirical facts—in this case, based on DNA/genetic data—clearly contradict longstanding *Neverland* "scientific" just-so stories that were, in turn, in great part influenced by teleological narratives (see also Box 5.4).

As noted in Feiler's *The First Love Story*, **Plato**'s old notion of romantic love is one of these narratives: it "echoes this notion…we 'find our soul mate', 'fit well together', 'feel whole', with someone we love…'you complete me'…." According to Feiler the notion of monogamic romantic love as "meant to be" is clearly seen in the religious teleological narrative about **Adam** and **Eve**: "Genesis 1 deserves credit for introducing this now universal concept…it begins the discussion of human relationships with the idea that the first man and woman weren't just thrust together because they were the only choices available: they were made for each other." Indeed, a crucial part of the common narrative about romantic love is that it is not only related to finding a "meaning" of life but also about pleasing our individual human egos by building a narrative that make each of us feel "special" and "needed" by others. However, rather than something cosmically "designed" or "meant" to be, it is actually the opposite: in most cases, couples end up by being with each other because that is with whom they *could* be within both the randomness and contingency of life, as elegantly put by **Ann Druyan**, the ex-spouse of late **Carl Sagan** (see Chap. 9). People living in small, isolated islands or villages logically mostly end up by marrying with someone from the same islands or villages, by the mere fact that they were born in, or moved to, those places: it is not at all about finding the "most special person in the world." Instead, as it is often the case in biological, including human, evolution, they have to deal with what is *available* and what is *possible*: *not the "best," not the "worse," not "meant to be," it is just what it is.*

So, what do the biological and anthropological data available truly tell us, about these topics? First, sexual monogamy is almost virtually inexistent in primates, so it would be more logical to ask, biologically speaking, if our species is "naturally" **social monogamous** or instead **polygamous**, as there is *no* empirical data, from any field of science, indicating that we are sexually monogamous. In fact, as we will see below, data from physiology, biological anthropology, evolutionary biology, sociology, neurobiology, and various other fields strongly indicate that humans have, in general, a tendency to be **polygamous**—very likely with a **multimale-multifemale type of sexuality**. A major problem related to these questions is that it is difficult to differentiate what the members of the *Homo fictus, irrationalis, socialis et servus* truly *desire and want* from what the prevailing *Neverland* teleological narratives, social constructions, cultural norms, and peer pressure forcibly impose or voluntarily lead them to *say that they want* and often—but not always, for instance, not in millions of cases of so-called infidelity—to ultimately *do*. This topic, about what we truly want to do in the world of reality versus what we say we want in Neverland's world of delusions, alternative facts, fake news, and lies—often to ourselves—is at the core of the present book. According to Feiler, the notion of **romantic love** is likely "universal" for our species: "in a detailed examination of 166 cultures around the world, anthropologists found evidence of romantic love in 151 of them, or 91%…in the remaining 15 cultures, researchers simply failed to study this aspect of people's lives…the ways we experience love also differ little across age, race, gender, religion, and sexual preference…people of all types can be just as passionate, tender, aroused, angry, and committed toward their loved ones as anyone else." However, it is important to note that such a "universal," widespread notion of "love" is *not at all* what might come to your mind when you think about "romantic love" as currently defined in most Western countries. A huge number of **non-agricultural societies** socially accept **polygamy** and, in those cases in which they don't, sex outside the couple is very frequent anyway, as it will be seen below. Moreover, almost none of those societies share some of the aspects that we are acculturated to believe to be crucial components of "romantic love," such as the so-called **romantic kiss**: as noted in Tallis' *The Incurable Romantic*, "in a study of 168 cultures, it was found that people in only 46% of them kissed romantically." So, clearly not "universal."

Tallis' book combines his own observations as a clinical psychologist with his comprehensive literature review on the subject. In the preface of the book, he writes: "**lovesickness** is not a trivial matter…unrequited love is a frequent cause of **suicide** (particularly among the young) and approximately 10 per cent of all **murders** are connected with **sexual jealousy**…moreover, there is a view that intermittently gains currency within psychiatry and psychology that **troubled close relationships** are not merely associated with **mental illness** but are a primary cause." He further notes that "men are more likely to kill women, but women also kill men…about 1/3 of women who are murdered around the globe are killed by their husbands or boyfriends—often stabbed or beaten to death…*a woman is statistically much safer getting into bed with a total stranger than with someone she knows*…although **jealousy** arises along a spectrum of severity, even mild forms can be explosive." About one of his own case studies, he explains that "even her delusional thinking was, in a sense, normal, because romantic love is often very irrational—love at first sight, ascribing chance meetings to destiny, oceanic feelings and powerful affinities that can transcend time and space are all common place." Indeed, what is particularly dazzling about that book is that it shows, with solid empirical evidence, that there is a very thin line between the currently so in vogue "Hollywood" narratives of romantic love and what is clinically considered to be **pathological behavior** (see Box 5.1).

Box 5.1: The Incurable Romantic, Epicurus, and Homosexuality

What I particularly like about Tallis' book *The Incurable Romantic* is that it shows that, as defended in the present book, teleological narratives are deeply embedded in almost everything we do, including—if not particularly—about loving someone and how we run our sexual lives. On the other hand, Tallis also explains that some of the behaviors that are nowadays considered to be "sexually pathological" might not be considered to be so, in the future. For instance,

homosexuality was long considered to be a **sexual pathology**. Then, in the 1960s and 1970s, many psychotherapists and psychiatrists started to question if **homosexuality** should really be considered a form of **mental illness**, and in 1973, it was removed from the diagnostic system. He explains that there are other so-called **paraphilic disorders** included in the diagnostic system that actually only merit professional attention when they are present in an extreme or debilitating form. Merely having a **fetish** was once regarded as clinically significant, but this is certainly not true today. He further explains:

> Atheists often attack believers for invoking 'God of the gaps'. Whenever a gap in scientific knowledge is identified, the gap is enlisted as a religious proof. We cannot explain the ultimate origin of the universe – therefore the creator must exist. We make the same thinking error when we are in love. The reasons why we fall in love with one person rather than another are simply too numerous, subtle, elusive and complicated to untangle. Many of them are the result of unconscious processes. Subsequently, there are always gaps in our understanding of love and like hopeful theists we tend to fill those gaps with supernatural explanations. We hint at strange affinities and the operation of mysterious forces…"She is unique" (said one of Tallis' patients)…"isn't everybody unique?" (asked Tallis)…"Psychoanalysts view **idealization** as a defense. It simplifies the word in order to reduce the anxiety caused by inconsistency and troublesome complexities…his love had much more to do with what he wanted her to be, rather than what she actually was. Idealization always incorporates a degree of denial, because in order to see someone as perfect we must deny the existence of their less favorable attributes. In reality, few people get to marry their ideal partner. Love involves making a series of compromises. This is no bad thing, because an idealized partner is only nominally human. "Sometimes love – however deep – just isn't returned" (said Tallis)…he (Tallis patient) raised his head from his hand and at me as though I had uttered a blasphemy. I had contradicted a sacred principle: if you love someone enough, your love will be reciprocated. It is an instance of a more generalized assumption that social psychologists call the just-world hypothesis: you deserve what you get, and you get what you deserve. The world, however, is not a fair place. There are no invisible forces at work restoring moral equity, and sincerity of feeling has never guaranteed that a declaration of love is accepted.
>
> "It feels so right…our bodies just seemed to fit together" (said Tallis patient)…the desire for completion is perhaps finally mitigated when…the Greek creation myth recounted by **Plato** (the one noted just above, about Zeus division of bodies into two parts)…and evolutionary objectives coincide with the production of children…thereafter, the wound inflicted by **Zeus** is healed and yearning is replaced by more practical necessities such as paying bills, doing housework, getting the kids to school and trying to get a good night's sleep. I had (however) been too direct, too insensitive…a broader, philosophical view offered him new hope: "maybe this has all happened for a reason. It's said that when bad things happen, they make us stronger…if we did get back together, I'd be a stronger person – a better person…and all this misery would have a point – a *purpose*" (said the patient). He was reassigning a different meaning to his separation…it wasn't an end, but a beginning – a trial that he must endure true and supreme love. His thinking was following a convention of courtly romance. Like an exiled Arthurian knight, he would face temptations and dangers, pass tests and return in triumph providing evidence of his virtue: "if we did get back together again, I'm sure we'd appreciate each other more". He was going to demonstrate the durability of his love and win his Queen's favor.

"Religion has never held any answers for me, which is a shame, because, actually, I'd like to *believe* in something" (said the patient). But he did believe in something, he believed in love. The Greek philosopher **Epicurus** maintained that all of our anxieties and sadness can be traced back to a root cause: the fear of death…a great deal of existential psychotherapy is concerned with the **search for meaning**, which must be personal, because the universe is intrinsically meaningless. We must decide what is meaningful for ourselves. Love gives us **purpose**. And sex, which promises a surrogate for of eternal life through procreation, reduces (albeit temporarily) the potency of the two great existential terrors – aloneness and mortality…the psychoactive substances released into the blood when we are aroused can take us out of time and make us feel that we are boundless and eternal. In the ecstatic delirium of **orgasm** we are beyond Death's reach. "It feels wrong to just let her go…I mean, there are so many songs and films – and they all have the same message: love will find a way, love conquers everything…that's what we *believe*…that's why they're popular…they strike a chord" (said the patient). Week after week, Paul (the patient) came to my consulting room to express his longing…sometimes, I would simply listen, and at other times I would point out how his unhappiness was being maintained by a *belief* system full of contradictions and dysfunctional assumptions.

The word **romantic** is extraordinary rich and complex, because it represents many beliefs and ideas about love that have accumulated and blended together over a period of a thousand years. The concept of romance is so much a part of our **cultural heritage** that we accept its implicit assumptions without question. In plays, operas, films and novels, anything – if it is done for love – is acceptable. Today, **Islam** is frequently characterized as an exporter of hate, however, in actuality, the Islamic world's most successful export is love. The Arab **Bedouin** composed a form of poetry that contained several motifs now familiar to a global readership: an idealized lover, thwarted passions and melancholic yearning. Building on this tradition, eleventh century Islamic authors wrote large-scale epic romances. The dissemination of Islamic love stories across Europe followed the **Moorish** conquest of Iberian Peninsula. Thereafter, the chivalric songs and verses of the **troubadours** provided the foundation for courtly adventures…during the **Renaissance**, poets such as **Petrarch** and **Dante** took the idealization to new and ecstatic heights. The word 'romance' was infused with further meanings in the late eighteenth century, when **romanticism** – a movement that valued violent passions over cold reason – found its initial impetus in a story of doomed love by **Goethe**…the fundamental problem with the notion of **romantic love** is that it is based on…unrealistic expectations…it is really the case that there is only one person (like a singular deity) with whom true love is possible? Fate (or the hand of good) does not bring people together, there are only random occurrences. Obstacles to love have no significance, they do not appear in order to test and intensify love. There is no divine plan. The romantic word view is rooted in literatures that construe love as nascent tragedy. As such, it is a potentially dangerous body of ideas. To be romantic is, for the most part, an unhappy, **hallucinatory experience**. Romantic love promises one thing, but delivers another. The trappings and accessories of romantic love have now been successfully monetized. On **Valentine's day** we celebrate romance with cards, bouquets and candle'-lit meals. But what are we celebrating, exactly?

An illustrative example of the dangerous links between **teleological narratives**—both religious and nonreligious—**sex**, and **misogyny**, concerns one of the most horrible practices: **female genital mutilation**. This practice—which is nowadays particularly common in Africa and the Middle East but that was also common until very recently in the U.S. and Europe as will be seen below—is nowadays often performed by women but was historically mainly imposed—and continues to be—by social norms, peer pressure, and narratives based on man-made imaginary stories (man-made meaning literally made by men specifically, not by humans in general). As explained in Nour's 2008 paper "*Female genital cutting: a persisting practice*":

> **Female genital cutting** (FGC), also known as **female circumcision** or female genital mutilation, is an ancient practice that predates the **Abrahamic religions**. Fraught with medical, legal, and bioethical debates, FGC is practiced in 28 African countries and some countries in Asia. In 1997, the World Health Organization, United Nations Children's Fund, and United Nations Population Fund issued a joint statement that defined FGC as "all procedures involving partial or total removal of the external female genitalia or other injury to the female genital organs whether for cultural or other non-therapeutic reasons". Approximately 3 million girls every year are at risk of undergoing FGC. The health, psychological, and sexual complications of FGC depend on the type of procedure that is performed, sterility during the procedure, the experience of the operator, and the social atmosphere at the time the cutting is performed. The origins of FGC are a mystery. It is thought to have existed in **ancient Egypt**, **Ethiopia**, and **Greece**. The practice transcends religion, geography, and socioeconomic status. Although FGC predates **Islam**, a small number of Muslims have adopted the practice as a religious requirement. As late as the 1960s, American obstetricians performed **clitoridectomies** to treat **erotomania**, **lesbianism**, **hysteria**, and **clitoral enlargement**. Girls typically undergo FGC between the ages of 6 and 12 years. It is performed on newborns, at menarche, and prior to marriage. Usually girls are aware that they will be cut some day, and some eagerly anticipate it. Villagers gather girls and celebrate the **rite of passage** with food, song, and gifts. Generally, midwives or trained circumcisers go from village to village and perform the cutting with no anesthesia, antibiotics, or sterile technique. Their instruments are knives, razors, scissors, or hot objects that are reused. After the tissue has been excised, sutures, thread, and local concoctions such as oil, honey, dough, or tree sap are used to ease bleeding. Postoperatively, wound care depends on the extent of damage. Girls who have undergone type I usually heal within a few days, whereas girls who have undergone type III require bed rest for approximately 1 week. Their thighs and legs are bound together to ensure proper healing of the infibulated scar. Some girls are unaware they will be cut. FGC is performed on these girls suddenly, without mental preparation, celebration, or fanfare. In this situation, girls can be emotionally traumatized. In other cases, nurses and physicians perform FGC in their offices under anesthesia in order "to protect" girls from complications. The international medical community strongly opposes medicalizing FGC on ethical grounds. Medical involvement is also seen as justifying and perpetuating a practice that should instead be eradicated.
>
> Parents who continue this practice are compassionate and loving. They believe that they are protecting their daughters from harm. Reasons that parents and practitioners give for the procedure include rite of passage, preserving **chastity**, ensuring **marriageability**, improving fertility, **religious requirement**, hygiene, and enhancing **sexual pleasure** for men. Parents who insist that their daughters undergo FGC are driven by a fear that their daughters may never marry. An unmarried daughter is ostracized and shunned in these societies, and may be seen as unclean, unhygienic, and perhaps even labeled as a prostitute. Some societies believe that the **clitoris** is toxic, and if during child birth the clitoris touches the baby's head, the baby will die. Some societies believe that if unchecked, the clitoris will grow until it touches the ground. Thus, removing the clitoris improves survival, ensures beauty, and preserves their daughter's reputation. Women with types I and II FGC who survive the procedure rarely have long-term complications given that they do not have an infibulated scar covering their external genitalia. Women who undergo type III FGC are at the highest risk for immediate and long-term complications.

> The most common immediate complications are uncontrolled bleeding, fever, wound infection, sepsis, and death. The most common long-term complications are **dysmenorrhea**, **dyspareunia**, recurrent vaginal and **urinary tract infections**, **infertility**, **cysts**, **abscesses**, **keloid formation**, difficult labor and delivery, and **sexual dysfunction**. Infertility is a devastating psychosocial complication to the infibulated woman. Her infertility rate can be as high as 30%. This infertility rate is secondary to both anatomic and psychologic barriers. The infibulated scar that supposedly protects girls from pregnancy out of wedlock becomes the obstacle that prevents them from getting pregnant within marriage. With multiple coital attempts over several months and using ample lubricants, the scar can stretch, but coitus is still very painful. This creates an unhealthy and distressing **sexual relationship** between husband and wife. Women fear that they may never become mothers, and husbands question their masculinity. Although some studies have demonstrated that men prefer to marry uncircumcised women, other studies have found the opposite to be true. Once pregnant, infibulated women face another daunting challenge: labor and delivery. In a large study, women with FGC were found to be at an increased risk of having adverse obstetric outcomes, including **postpartum hemorrhage**, **episiotomies**, **cesarean deliveries**, extended maternal hospital stay, **infant resuscitation**, **stillbirth**, or **neonatal death**.

In *The Fate of Gender*, Browning refers to similar practices that were done by American obstetricians just some decades ago, calling attention to the hypocrisy surrounding this topic, and also to the fact that the removal of sexual tissue is indeed related not only to religious narratives but also to non-religious ones, including some that were deemed to be "scientific" in the past. He notes that "both the United Nations and the World Health Organization have mounted vigorous public health campaigns to stop female genital cutting…they have, however, rarely acknowledged that very similar practices were undertaken by European and American doctors well into the last century to calm obstreperous and overexcited young girls who were judged to have touched themselves too often or too publicly." Doctors "urged mothers to choose undergarments that would not stimulate little girl's sexual terrain and thereby risk activating their uteruses, which could send them into the widely accepted diagnosis of '**hysteria**' (Sigmund **Freud** dismissed **clitoral orgasm** as 'infantile' and 'immature' when compared to what he called 'the mature' orgasm associated with vaginal stimulation)." He further explains that "those diagnoses and the bizarre notions of female physiology on which they were based have largely disappeared, but cutting of strange or abnormal or simply unfamiliar sexual tissue has not disappeared in American neonatal surgical wards."

Box 5.2: The Particular Obsession of Christianity with the Body, Sex, Shame, Women, and Homosexuality

Nixey explains, in her 2017 book *The Darkening Age – The Christian Destruction of the Classic World*:

> In the writings of the **Roman elite**…God's Sex ought to be contained – but it was not denied…like any other appetite it was…something to be admitted and managed rather than something to be ashamed of. The feast of the Liberalia was on 17 March; a now sadly forgotten festival at which Roman citizens celebrated a boy's first **ejacu-**

lation. In Roman medical manuals, ejaculation had been readily and openly discussed by classical doctors who advised it for health and getting rid of the seed that might otherwise cause headaches. It was thought that if athletes could abstain from sex they would be stronger. **Orgasms** and sex were even recommended for women's health. Sex, **sexual desire** and the consequences of sex were frankly discussed. Poets chastised their lovers when they had **abortions**, less for the abortion than for endangering their own health. **Ovid** professed himself furious with his lover Corinna for rashly attempting one – but less because she had committed this act than because she had taken 'that risk, and she never told me!' Others followed more laborious methods of avoiding pregnancy. When Julia, the famously racy and beautiful daughter of the emperor **Augustus**, was asked how, given her many lovers, her children all resembled her husband, she replied that she would 'take on a passenger only when the ship's hold is full'. Why not have sex? Life was short and one didn't know what was coming next. Live now, proclaimed countless mosaics, paintings and poems in the **old Roman world**. For who knows what tomorrow might bring? In a recently discovered mosaic in **Antioch**, a skeleton reclines, a cup in his hand, and an amphora of wine near by. Over his head, in clear Greek lettering, the mosaic gave an instruction to those diners above: 'Be cheerful', it reads. 'Enjoy your life'. The injunction to enjoy yourself is written in stone. One of the most famous of all classical poems had put this ideal into rather more elegant lines. *'Quam minimum credula postero'* – trust as little as possible in tomorrow – advised the poet **Horace**, and instead 'carpe diem'. Seize the day.

[However] for **St Paul** and other Christian preachers, the body and its urges were not to be celebrated but smothered. In tortuous and embarrassed circumlocutions, Paul raged at 'this body of death'. The rewards of a virgin in heaven were said to be sixty times greater. Christian writers in this period recorded the stirrings of their sexuality with great distaste – perhaps none more influentially than **Augustine**. **Sex** was, he felt, permissible if children resulted from the union but even then the action itself was lustful, evil and 'bestial', while erections were 'unseemly'. The West would reap a bitter harvest of **sexual shame** from the disgusted writings of these two men. In the earliest days of the **religion**, some **Christians** went further, arguing that there was no need for sex any more at all. A new form of creation, in the form of a great conflagration and rebirth of the godly, was imminent. What need for awkward, messy, inexact human reproduction? Eternal life rendered reproduction redundant. If the most famous non-Christian manual had been **Ovid**'s, then one of the most famous manuals by a Christian writer was a third-century tract by the theologian **Clement of Alexandria**. It is called the **Paedagogus** – the instructor – and its stated aim was to 'compendiously describe what the man who is called a Christian ought to be during the whole of his life'. Clement, in precise and authoritative paragraphs, peppered with frequent quotations and not infrequent threats from the scriptures, advised the faithful on every aspect of their day, from what they were allowed to eat and drink to what they could wear and put on their feet; from how they could style their hair to even what they were allowed to do in bed. In the three volumes of his guide he censured almost every human activity. 'Unblushing pleasure', he wrote, 'must be cut out by the roots'. He began with eating, opening with the reminder that we are all ultimately dust, before turning his attention on particular dishes. In starchy and unforgiving sentences, unseasoned by even a dash of humour, the extravagant dinner was deplored. As too was almost everything eaten at it. Overuse of the pestle and mortar was frowned upon. Condiments were considered unacceptable, as too were white bread ('emasculated') and sweetmeats, honey cakes, sugar plums, dried figs…One should not, Clement warned, be like the gourmands who source their lampreys from Sicily, their turbots in Attica, their thrushes in Daphnis…The list went on.

Male **homosexuality** was denounced; and then outlawed. By the sixth century, those who were, as one chronicler put it, 'afflicted with homosexual lust' started to live in fear. And with good reason. When a bishop called Alexander was accused of having a homosexual relationship, he and his partner were 'in accordance with a sacred ordinance…brought to Constantinople and were examined and condemned by Victor the city prefect, who punished them: he tortured Isaiah severely and exiled him and he amputated Alexander's genitals and paraded him around on a litter'. The emperor immediately decreed that those detected in pederasty should have their genitals amputated. At that time many homosexuals were arrested and died after having their genitals amputated. Sex between a husband and wife was allowed but it should not, preachers said, be enjoyed. The old merry **marriage ceremonies**, in which people had eaten, drunk and sung profane songs about sex, were bluntly deplored as the Devil's dungheap. Admiring stories of married couples who never slept with each other but spent their nights wearing hair shirts proliferated. In the glare of Christian disapproval the onceloved art of pantomime dance withered and died. Sexually explicit – and sexually joyful – poetry stopped being openly written. There were no new Ovids. **Desire** started to be called **'lust'** – and it became something shameful that was to be feared, despised, smothered and – if homosexual – punished, horribly. What was celebrated at holy days and festivals also underwent a transformation…[While] on 17 March, Roman citizens had celebrated the Liberalia…the Christian Church celebrated instead the saint's day of Ambrose of Alexandria – a pupil of Origen, the man who (allegedly) had castrated himself for the sake of heaven. What was acceptable in terms of sexuality narrowed. It would be well over a thousand years before Western civilization could come to see homosexuality as anything other than a perversion and a crime. Throughout the empire, statues were brought out of the bathhouses, their bodies mutilated, and burned as jeering crowds looked on in delight. Nipples were smashed from one naked figure of **Aphrodite**; Aphrodite herself was beheaded and left in the dirt.

As both Boxes 5.1 and 5.2 refer to **homosexuality**, and this topic will also be further discussed below (see also Box 5.7), it is perhaps opportune to provide a brief note on the occurrence of same-sex intercourse in other animals, as moreover this topic also illustrates the links between **teleological narratives**, **"purpose**," and **biases**, including **scientific biases**. This is because for a long time the evolution of homosexuality was not a main focus of scientific publications. It only begun to be the central focus of some broader synthetic reviews, including a few books, recently, as explained in one of those books, *Homosexual Behavior in Animals – An Evolutionary Perspective*, edited by Sommer and Vasey and published in 2006:

> Earlier studies of **animal behaviour** tended to dismiss occurrences of **same-sex sexual behaviour** as mere quirks or such instances were classified as pathological manifestations. The use of caged subjects was prevalent and meant that these interactions were invariably characterized as abnormal products of captivity, unlikely to be found in 'nature'. As early as the 1700s biologists such as George Edwards (1758–64) were speculating on the causes of such **behavioural 'abnormalities'**. He stated that 'three or four young [bantam] cocks remaining where they could have no communication with hens…each endeavoured to tread his fellow, though none of them seemed willing to be trodden…reflections on this odd circumstance hinted to me, why the natural appetites, in some of our own species, are diverted into wrong channels'. Research conducted throughout the 1890s purported that an absence

of opposite-sex partners and artificial confinement could 'force' individuals to choose same-sex mates. It is interesting to note that the same researchers who reported such findings showed that pigeons will participate in same-sex sexual interactions even if they are housed in mixed-sex groups. Moreover, some researchers even demonstrated that certain same-sex pairs stopped their homosexual activity upon being isolated from their opposite-sex group mates. For example, Whitman (in 1919) stated that 'a number of pairs of mature males were isolated; some of these were observed for several months, but no real matings resulted from any of these cases'. He then went on to conclude, however, that 'confinement will thus force matings which would not otherwise occur...pairings between like sexes are secured in this manner'. Clearly, such an interpretation is more reflective of the opinions of the observer than an objective observation of the behavioural phenomenon under investigation. Nevertheless, more and more detailed studies of animals in their natural environments made it increasingly difficult to discount all sexual interactions in animals among members of the same sex as exceptions, as idiosyncrasies, or as pathologies. Slowly, but steadily, a quite different picture emerged. A recent encyclopedic volume by Bruce Bagemihl (in 1999) on **animal homosexual behaviour** provides evidence that hundreds of mammals, birds, reptiles, amphibians, fishes, insects, spiders and other invertebrates engage in **same-sex sexual activity**. Clearly, what was once thought to be an aberration appears to be a behavioural pattern that is broadly, albeit unevenly, distributed across the animal kingdom. Indeed, within a select number of species, homosexual activity is widespread and occurs at levels that approach or sometimes even surpass **heterosexual** activity.

So, contrary to religious narratives, and to what many people still *believe* nowadays, homosexuality is completely "natural," in the sense that it occurs naturally in many non-human animal species: it is *not* something unique in humans, nor a "human abnormality" or an "**immoral sin**." However, a major problem that continues to plague such research on homosexuality concerns another teleological narrative—namely, the quest of so many scientists for a "purpose." This particularly applies to **adaptationists**, who want to find, at all cost, an adaptive advantage—very often a current one—for each and every trait found in living beings. However, contrary to such **adaptationist just-so-stories**, we now know that many traits can be, and remain, present in taxa without providing them any current advantage. But unfortunately, **adaptationism** continues to plague evolutionary research, particularly fields such as evolutionary psychology, and although various authors of that book's chapters do recognize the danger of falling into the "**adaptationist trap**," they often fall into that trap themselves when they *desperately try* to explain the "**purpose**" of homosexuality in the animals studied by them. Numerous "purposes" are proposed in that book based on the a priori *assumption* that homosexuality might be an **evolutionary adaptation**—a clear illustration of **circular adaptationist reasoning**—despite the obvious fact that same-sex intercourse does not lead to producing offspring. Such "purposes" include "controlling population size," "dominance expression," "social tension regulation," "reconciliation," "social bonding," "alliance formation," "acquisition of alloparental care," "mate attraction," "inhibition of competitor's reproduction," "kin selection," or "practice for heterosexual activities."

I am not saying that it is not possible that, within the huge, amazing diversity of sexual organisms in this planet, some or most of these "functions" of homosexuality do apply in at least some, or even in many, cases. We know that **Catholic priests** having no sex—well, except in cases such as those when they rape and molest children—and no progeny may actually increase the cohesion of the ingroup—in this case of **Catholicism** as a whole—and therefore lead to its evolutionary maintenance via group selection. In a chapter of that book written by Kotrschal and colleagues concerning **homosexuality in geese**, these authors argue that "male-male pairings may be seen as a tactic of securing an ally needed to maintain a high rank in the flock." This might indeed be evolutionarily sound, because as they explained, "high rank is important for access to resources, including female mates…hence, if the number of available female partners is limited, males tend to resort to pairing with another male…this idea is supported by the fact that the frequency of male-male pairs is positively correlated with the male bias in the flock." Moreover, there are anthropological studies that show that, similarly to what happens in male jails in our countries, in some non-Western societies homosexuality between men might indeed be at least partially explained by a lack of female partners rather than by a strong sexual preference per se. An illustrative example was provided in **Claude Levi-Strauss**' book *Tristes Tropiques*. About the **Nambikwara**, an indigenous group that he encountered during his travels to the Brazilian Amazon, he wrote:

> By withdrawing, as he does, a number of young girls from the normal **matrimonial cycle**, the chief creates a disequilibrium between the number of young men and the number of available girls. The young men suffer most from this, for they are condemned either to remain single for years, or to ally themselves with widows or older women whose husbands have had enough of them. The Nambikwara have, however, another way of resolving the problem, and that is by **homosexual relations** or, as they call them, *tamindige kihandige*: 'lying love'. These relations, common among the younger men, are carried on with a publicity uncommon in the case of more normal relations. The partners do not go off into the bush, as they would with a partner of the opposite sex, but get down to it beside the campfire, much to the amusement of their neighbours. The incident provokes a joke or two, on the quiet, the relations in question being regarded as childishness and of no serious account. It remains doubtful whether these exercises are carried to the point of complete satisfaction or whether, like much that goes on between husbands and wives among the Nambikwara, they are limited to sentimental out pourings and a certain amount of erotic fore-play. Homosexual relations are only allowed between adolescent boys who stand to one another in the relations of crossed cousins cases, that is to say, in which one partner would normally marry the other's sister and is taking her brother as a provisional substitute. Whenever I asked an Indian about a relationship of this sort, the answer was always the same: 'They are two cousins (or brothers-in-law) who make love together'. Even when fully grown, the brothers-in-law are still very free in their ways, and it is not unusual to see two or three men, all married and the fathers of children, walking round in die evening with their arms round one another's waists.

However, it also seems clear that in many—probably in *most*—known cases of homosexuality, the so-called purposes or adaptative functions mentioned above do not apply at all. In fact, by simply denying a priori the hypothesis that perhaps there is no adaptive function at all in those cases, that is, that **homosexuality** might be, in many instances, merely a manifestation of ***sexual desire*** related to ***pleasure***, scholars might be inadvertently adding fuel to the longstanding idea that homosexuality is somewhat an "evolutionary abnormality," by implying that it cannot be the "real thing," compared to **heterosexuality**. This is done when scholars insist, over and over, in seeing homosexuality as a *frustrated* answer to the impossibility to mate with the "truly desired sex." Yes, there are cases in which this is so, but independently of the fact that this might indeed happen in some—likely only a *few*—non-human animal species, scientists too often use unfortunate terms to talk about such cases. Such as the ones used in the title of Yamane's chapter in the book *Homosexual Behavior in Animals*: "*Frustrated Felines: Male-Male Mounting in Feral Cats*"—this title is clearly inopportune because it can be used to propagate stereotypes about homosexuality, such as human homosexuals being seen as "frustrated beings" and so on.

Let's refer for example to the singer **Elton John**: can we really say that his homosexuality is merely a "practice for heterosexual activities," which is described in that 2006 book as a "learning opportunity for immature individuals, often during play, that functions to facilitate social and motor development for adult heterosexual roles"? Right now, when I am writing these lines, Elton John is 73 years old, and he is married—"happily," according to his own words—to David Furnish [I should note that I took years to write different parts of this book, so in some cases I might be referring to March 2016, others to April 2018, others to January 2020 and so on—I only finished the book in early 2021]. So, it seems strange to argue that he might still be "practicing" with his husband to then assume a "normal adult heterosexual" role later in life, or to assume that his homosexuality is a result of his "frustration" of not being able to have sex with women—clearly, he *would* be able, if he wanted. So, to assume a priori that *all* cases of homosexuality are always explained by such **adaptationist "purposes"** is just not only logically absurd but scientifically plainly wrong.

It is indeed significant that within the 16 chapters of a 2006 book written just 15 years ago and that is completely focused on the study of—and entitled—*Homosexual Behavior in Animals*, only one chapter concluded that the most plausible explanation for the homosexuality of the non-human animal group discussed in that chapter was essentially related to *sexual preference and gratification*. Namely, Vasey, the author of that chapter—which was aptly entitled "*The pursuit of pleasure – an evolutionary history of homosexual behavior in Japanese macaque*"—concluded that "some same-sex sexual partners might be preferred over certain opposite-sex alternatives, simply because females derived more sexual gratification during interactions with them." Elton John couldn't have said it better, nor couldn't the many women that engage in same-sex relationships and do often have a very high sexual gratification by doing so, including in what regards the frequency, intensity, and number of the orgasms that they have, as we will see below.

5.2 Agriculture, "Original Sin," and Monogamy

It is indeed probable that more harm and misery has been caused by men determined to use coercion to stamp out a moral evil than by men intent on doing evil. (Friedrich Hayek)

In their 2016 book *The Good Book of Human Nature – An Evolutionary Reading of the Bible*, **Van Schaik** and Michel provide a brief introductory note about the links between **agriculture** and the **Neolithic Revolution**, the **Bible**, the evolution of human **religion**, **monogamy**—at that moment mainly only applied to, and forced upon, women—and **misogyny**:

> Poor **Eve**. Just look at all she was made to suffer. **God** cursed her with the words, "I will greatly multiply thy sorrow and thy conception; in sorrow thou shalt bring forth children". If that were not enough, he added, "and thy **desire** shall be to thy husband, and he shall rule over thee". The biblical exegetes upped the ante with their interpretation of Eve's experience. This bias against women is quite a recent phenomenon. The relationship between the sexes was actually much more balanced among hunter-gatherers. Men were the dominant sex, but women were able to return to their families at any time or even to switch partners; pair bonds were not necessarily exclusive, and in practice the notion that a woman was bound to one man for her entire life was alien. Sometimes women even maintained parallel relationships. Engaging in sexual relations with several men was in a woman's best interest, allowing her to establish a network of potential fathers for each of her children. The fact that women knew how to best use their charms was a natural part of their sexual freedom. **Sexual freedom** was effective in egalitarian groups because it brought all members a bit closer together by creating an invisible "network of love". This was all over as soon as this group's way of life fell apart and every man started worrying about his property and demanding absolute faithfulness from his wives.
>
> The new **concept of ownership** had led to a situation in which sons remained with their fathers, and daughters were married off into other families. They served to help forge alliances or were simply viewed as tradable goods. As these were arranged **marriages** and thus not generally love matches, the women did not necessarily harbor strong feelings of solidarity with their new families. Only after they had born children to the son of the household did things change, for then a shared genetic interest had taken hold that bound the entire family together. In this new state of affairs, patriarchs did all they could to prevent their wives from sleeping with other men. And where wealth and power prospered, men turned to **polygyny**. This is commonplace in the stories of the **Old Testament**; nearly everyone had more than one wife; when women become the property of men, their power has to be reined in, and a large share of this power lies in their sexual attractiveness. The Bible underscores this point. **Adam** and Eve have to reach for the fig leaf, and God himself fashions them clothes. The patriarchal world raises a woman's fidelity to her husband to the level of commandment; a woman's sexual obligation to only one man is a cultural rule, not a biological one.

As noted in Jack Holland's *A Brief History of Misogyny*, **Tertullian** (AD 160–220), one of the founding fathers of the **Catholic Church**, wrote, about women: "you are the **devil**'s gateway; you are the unsealer of that forbidden tree; you are the first deserter of **Divine law**…you are she who persuaded him whom the devil was not valiant enough to attack…you destroyed so easily God's image, man." **Christianity** is particularly obsessed with—and against—women, as well as with **sexuality** in general, and **homosexuality** in particular (see Box 5.2). However, it is important to emphasize that this does not mean that women are just passive players within, and victims of, religious narratives and activities. Yes, most **religious teleological narratives** were written by men, and they are often oppressive to women. However, apart from the typical disturbing characteristic of human oppression discussed above—the oppressed ending up by believing in the very same narratives that were historically used to oppress them (see Fig. 6.3)—in some cases women may paradoxically actually also benefit from those man-made narratives. These two factors, combined, indeed explain why women are now particularly crucial *active players* in religions such as **Christianity**, not only regarding practicing them in general but also concerning praying and attending specific ritual services, as explained in Box 5.3. About some of the possible biological benefits of social monogamy in humans, Van Schaik and Michel wrote in *The Good Book of Nature*:

> Once again, the stories of the **patriarchs** appear at just the right moment in the **Bible**. They offer impressive examples of the enormous problems resulting from the **monopolization of women** and **property**. The patriarchs thus have what it takes to make an anthropological classic. They illustrate the difficulties produced by the new institution of **polygyny** (a man with various woman) and help us understand why **monogamy** established itself in the long run. But monogamy's hegemony was by no means inevitable. In many of the world's economically advanced states, monogamy may be the standard today, but throughout most of history things have looked rather different. Here it is important to be specific, for we are talking about what we today understand by monogamy: a lasting, exclusive relationship between two partners. Most of the marriages among the hunter-gatherers were also monogamous, but they rarely lasted a lifetime. Importantly, polygyny was not forbidden – on the contrary, unusually successful hunters had two (or, very rarely, even three) wives. Anthropologist Frank Marlowe has evaluated the available ethnological data: around 85 percent of all societies ever studied by anthropologists have allowed a man to have multiple wives. In contrast, **polyandry** (one woman has two or more husbands) is found in less than 1 percent of all societies. Why does this change? The phenomenon is often described as a "puzzle": "historically, the emergence of **monogamous marriage** is particularly puzzling since the very men who most benefit from **polygynous marriage** – wealthy aristocrats – are often those most influential in setting norms and shaping laws." However, if you read the stories of the patriarchs and thus gain an impression of family reality under polygyny – even if only a literary one – you will immediately appreciate the advantages of monogamy. It reduces competition for women among men. It also reduces competition between women and among their sons. Furthermore, there are fewer unmarried men in monogamous societies, providing less fuel for violent conflict.

Box 5.3: Female Religiosity, Monogamy, Biological Evolution, Marriage Vows, and Religious Rites

Blume's 2009 chapter of the book *The Biological Evolution of Religious Mind and Behavior* discusses an interesting idea about a connection between **female religiosity** and **monogamy**:

> **God**'s first **biblical words** to mankind are "Be fruitful and multiply" (Gen. 1, 28), forming the first of all 613 commandments according to **Jewish tradition**. Religious communities preaching family values with some success might be observed in everyday as well as in political life. Higher-birth rates of religious affiliated have been empirically verified globally. For example, according to a representative survey conducted among younger people aged 16–29 years in Germany in 2006, 42% of the respondents who described themselves as non-religious said it was "important" to them to "have children". Among the religious respondents, children were valued that way by 61%. As any mammalian reproduction is associated with heavy costs and risks especially to the mothers, biologists are used to comprehend animal reproductive strategies by investment games leading retrospectively (that means: by relative success) to **Evolutionary Stable Strategies**. But as soon as humans began to decide their pairing and reproduction strategies in advance, the setting changed to the prisoner's game with uneven stakes.... Men had to struggle with a certain insecurity of whether the children they were raising were their own offspring. Wrong choices could be crucial.... For women, the risk of defecting partners seeking reproductive

chances otherwise instead of staying and contributing to the children has been a constant topic of human history.... We might remember that Emma Wedgwood probably would not have accepted Darwin if he had not been religious at the time of his courtship. **Marriage contracts** are cultural institutions to secure cooperation and to prevent betrayal. And among almost all cultures, **marriage vows** are accompanied by **religious rites**, regularly involving supernatural as well as natural witnesses. And if there are communities around whose marriage vows are blessed by all-knowing, transcendent beings like the living dead or Gods punishing **unfaithfulness** in this life or the next, the "social control" is expanded and deepened beyond secular possibilities.

Therefore, women's preferences of **religious communities** and their sexual selection towards "catechised" men turns out to be **evolutionarily adaptive**. We may test this by comparing all religious denominations not predominated by immigration in the Swiss Census 2000. Swiss women not only dominate memberships in all major denominations. They significantly prefer those communities, where living together means to marry, where pairs tend to have children together and where **divorce rates** are low. In contrast, the non-affiliated form the single category dominated by men, featuring the lowest percentage of married couples, the lowest percentage of pairs with children and (although characterized by the lowest rate of marriages and births) having the highest percentage of divorcees and mothers raising children alone. Therefore, we should not be too surprised to find women attracted to religious movements like patriarchal monotheisms, which did not and do not exactly promote self-actualization and gender mainstreaming. Men might perceive religion as a stage to earn status and therefore to win economic and reproductive chances. But women tend to prefer communities and prospective husbands sending honest signals of familial cooperation. We might be fascinated by the Kung San-Num healing dancer role only accessible to men by long, demanding and dangerous initiation. These costly requirements secure the trustworthiness of the dancers, as obligations and **rituals** secure the longevity of religious communities abroad. But we should not overlook the fact that the women behold the performers, sitting in the middle, singing and clasping the tune. Although the **Bible** clearly indicates that **Adam** "recognized" **Eve**, she thanks God for the birth of her child (Gen 4, 1). The Lord is even reported to have slain two sons of Judah before humiliating the patriarch himself while siding with a foreign woman, Tamar, for rightfully insisting on her marriage contract (Gen 38).

Moreover, one needs to keep in mind that although most religions, including **Christianity**, are mostly based on man-made tales, these tales are in general very appealing and comforting to not only males but also females because they are related to our human drive to build and believe in a cosmic purpose of life. This is why religious narratives about female monogamy, or against luxury, or even about female inferiority per se, can still be so appealing to women. In a way, this is similar to what happened to Native Americans, who embraced some of the **Catholic religious tales** that were precisely used to colonize and massacre them even more fervently than their colonizers did: in contrast with continental Europe, which has been traditionally associated with **Catholicism** but recently had a huge increase of secularism, more than 90% of Latin-Americans currently self-identify as Catholic. Therefore, that a group of people embraces some imaginary stories does not mean that that they have not paid, and don't continue to pay, a huge price for doing that, particularly when they embrace them with such a passion. Concerning teleological narratives about sexuality specifically, such as those created to enforce **social monogamy** (see Box 5.4), they indeed have caused, and continue to cause, a huge suffering, particularly to women. This is not only because monogamy was and in many regions of the globe continues to be exclusively imposed to women through the use of man-made *Neverland* tales, but also due to the enormous mismatches between these tales and the *reality* of our **evolutionary history**, **physiology**, and **sexual desires**, as will be seen below.

With regard to this, Coontz's emphasized, in her excellent 2005 book *Marriage – A History*, that "Christian [very negative] attitudes toward **marriage** and **sexuality** stood in sharp contrast with those of most ancient religions…to **Hindus** in India, marrying was a **holy act**, and a celibate or unmarried person was considered impious, or at least incomplete…the **Old Testament** and later Jewish teachings called marriage God's commandment and celebrated sexuality within marriage." In contrast, in Christianity, **Pope Gregory the Great** explained early in the sixth century that although marriage was not sinful, "conjugal union cannot take place without **carnal pleasure**, and such pleasure cannot under any circumstance be without blame." However, while it is true that **Christianity** is one of the religions that has a more negative view about sex and the naked body, as well as about other animals in general and primates in particular, all institutional religions, without exception, have a negative view of what reminds us of our animality, because that reminds us that we are nothing more than mere mortal animals. I explained this in detail in Chap. 2, providing examples of various religions, including Buddhism, and the way **Hinduism** is practiced in India, a country that nowadays has one of the highest rates of misogynistic atrocities—including a huge number of rapes and murders—and taboos about sex, female menstruation, and so on (see Fig. 7.2).

Box 5.4: Genetic, Sexual and Social Monogamy, Primates, and Oxytocin

Jon Cavanaugh's recent 2018 PhD thesis, titled *The prosocial paradox: unraveling oxytocin's role in monogamous relationships*, provides a succinct, relatively simple—for the broader public—and encompassing introduction to different types of monogamy and their links to **oxytocin**:

> Recently, the neuropeptide **oxytocin** has attracted enormous interest as a neuromodulator of social, emotional, and **cognitive processes**…it has been described as 'the great facilitator of life' for it's critical and pervasive role in reproductive physiology and behavior…. While it is clear that oxytocin has an important role in the initiation of **social behavior** (e.g., offspring care; affiliation) as well as the formation of **relationships** (e.g., offspring-caregiver; adult partners), less attention has been given to whether it regulates social behaviors that contribute to the preservation of long-lasting bonds. As early as the late-nineteenth century, the term **monogamy** has been used to describe animals that cohabitate and reproduce as a male-female pair. Yet, after several bird and primate species that were once branded as monogamous were found to have produced offspring outside of their established male-female unit, the

concept of monogamy was reevaluated and subdivided to better characterize the behavioral phenotypes of these species.... A species or individual is categorized as **genetically monogamous** when the gametes from one individual exclusively combine with the gametes from a second individual to produce offspring. A species or individual is categorized as **sexually monogamous** when partners exclusively engage in sexual interactions with each other. A species or individual is categorized as **socially monogamous** when two individuals exclusively cohabitate over an extended period of time. **Social monogamy** can also involve numerous elemental components, including the expression of high levels of mate-directed affiliation, distress upon separation, a **partner preference**, **biparental care**, low **sexual dimorphism**, and **mate-guarding behavior**. While decisions about engaging in monogamous versus nonmonogamous sociosexual relationships in humans are not limited to **heterosexual partners**, the literature on the **neurobiology of social monogamy** focuses primarily on male-female relationships. The incidence of social monogamy across vertebrates is widely variable. While approximately 90% of bird species are classified as socially monogamous, the prevalence of this trait is quite rare among mammals (<5% of species). The majority of mammalian orders (e.g., ungulates, aquatic mammals, rodents) have very few socially monogamous species. Primates, however, have a relatively high incidence of social monogamy (about 30% of species)...(particularly) New World primates (about 60%)...(which) diverged from Old World primates (e.g., macaques; baboons) and hominoids (e.g., gibbons; orangutans; gorillas; bonobos; chimpanzees; humans) about 43 million years ago.

While **social organization** (i.e., pair-living), social relationship (i.e., pair behaviors), and **mating system** (i.e., exclusivity of mating) are often overlapping in primates, they do not necessarily always covary. Recently, researchers have argued for the use of more nuanced definitions and classification of social systems in primates, and maintain that while no single social repertoire describes all monogamous primates, there are three broad categories of behavior that typically comprise monogamous relationships: **pairbonding**, **attachment**, and **mate guarding**. The most conspicuous features of primate socially monogamy are the behavioral manifestations of a pairbond between two individuals, which include high levels of mate-directed sociality and a partner preference. High-quality social interactions with a mate are, not surprisingly, critical to both the development and the preservation of an enduring bond. Mate-directed sociality is characterized by high rates of physical contact, affiliative behavior (e.g., grooming, food sharing), and **sexual behavior**. A pervasive and reciprocal preference for a long-term partner over an opposite-sex stranger (i.e., partner preference) is the second hallmark of a pairbond. Sexual exclusivity to a long-term mate is threatened when one, or both, members of a pair spend time with an opposite-sex stranger, especially if it is at the expense of spending time with a long-term mate. While a robust partner preference is important for monogamy, it is not necessarily required. Some monogamous primates (e.g., titi monkeys) consistently show a preference for their long-term mate over an opposite-sex stranger (and even over individuals they were formerly bonded with), while others (e.g., marmosets, tamarins) have more flexible partner preferences. Behavioral and physiological responses to mate separation have been used as measures of social attachment between mates, much like measures of the attachment of an infant to its mother. Individuals separated from their pairmate generally display increased vocalization rate, heart rate, hypothalamic-pituitary-adrenal-axis activity, and locomotor activity. These behavioral and physiological indicators of separation distress are indicative of an attachment between mates. The presence of a long-term mate can also serve as a powerful buffer against environmental stressors (i.e., **social buffering**).

In **monogamous primates** that display social buffering, the benefits of social support may occur either through passive presence of a pairmate or active intervention by a pairmate (e.g., vocal reassurance, physical contact) during a stressor, mitigating the physiological and behavioral stress response. Partners in a monogamous relationship can therefore utilize each other to minimize the negative impacts of stressors and can benefit from the widespread social advantages associated with social attachments. The third behavioral element that comprises primate monogamous relationships is mate-guarding behavior, which includes the expression of selective aggression toward same-sex strangers, and/or maintaining close proximity with and expressing mate-directed sociality during social intrusions. Mate-guarding behavior is an important reproductive strategy to avoid cuckoldry by a sexual rival. Mate guarding is a prevalent trait across many primate social relationships, as both non-monogamous and monogamous primates exhibit mate-guarding behaviors. For example, in non-monogamous primates including many Old-World monkeys and hominoids, **male aggression** toward other adult males are quite prevalent. This behavior facilitates **polygyny** (one male-many females) by enabling males to maintain and defend access to multiple mates, which in many primates is often determined by exhibiting dominance and aggression to maintain position in a social hierarchy. Likewise, **monogamous primates** also engage in aggression toward same-sex adults to maintain **sexual exclusivity** with their long-term partner.

The distinct behavioral responses to a same-sex rival (i.e., selective aggression or mate-directed sociality) may be regulated by independent neurobiological mechanisms. Oxytocin and the closely related **arginine-vasopressin** are evolutionary ancient molecules dating back about 600 million years ago…they are classic neuromodulatory agents in areas of the '**social brain**' that regulate attachment, parental care, reward, social memory, and aggression, including the **social behavior network** and the mesolimbic reward system; many of these effects are sex- and context-specific. In **prairie voles**, central infusion of oxytocin reduced the duration required to form a stable pair bond and blocking endogenous oxytocin activity by administering an oxytocin antagonist diminished a partner preference and inhibited pairbond formation. In the socially monogamous **marmoset monkey**, blocking endogenous oxytocin activity in both males and females reduced mate-directed sociality during the bond formation period. These results suggest that endogenous oxytocin activity has a critical role in the expression of partner-preference behavior and the formation of adult male-female bonds. There is also evidence that the oxytocin system has distinct roles in bond formation compared to bond preservation. In humans, neural activity in the anterior cingulate gyrus and the caudate nucleus, which have high densities of neuropeptide and dopamine receptors, decreased following the bond formation period. Additionally, peripheral levels of oxytocin were significantly greater in new couples than in singles and parents. These findings indicate that oxytocin-system activity changes across the development of a bond, and suggest that oxytocin may uniquely regulate the behavioral manifestations of a monogamous bond from formation to preservation.

5.3 Women, Men, Sexual Desire, and Orgasms

Everything in the world is about sex except sex…sex is about power. (Oscar Wilde)

Contrary to what people from Western countries often think—using their countries as a model of the planet or of what is "normal," that is, being Western centric—**polygamy** is still prevalent in many agricultural states. Figure 5.1 shows that there are

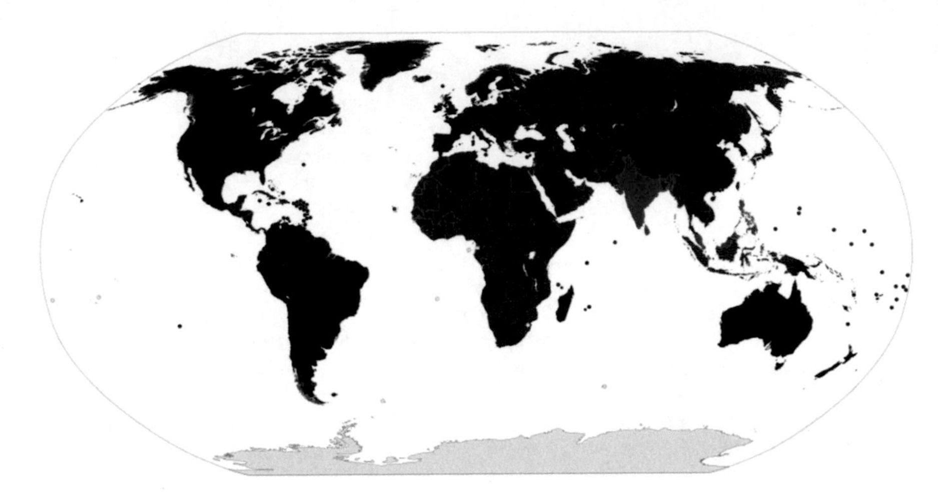

Fig. 5.1 Sexual preferences, social constructions and norms, and law. Green: Polygyny is only legal for Muslims. Blue: Polygyny is legal. Blue and red: Polygyny is legal in some regions (Indonesia). Red: Polygyny is illegal, but practice is not criminalized. Black: Polygyny is illegal and practice criminalized. Grey: Legal status unknown. Note that in Eritrea, India, Philippines, Singapore, and Sri Lanka polygamy is only legal for Muslims, in Nigeria and South Africa polygamous marriages under customary law and for Muslims are legally recognized, in Mauritius, polygamous unions have no legal recognition, and Muslim men may "marry" up to four women, but they do not have the legal status of wives

numerous countries in which **polygamous marriages** are not legally forbidden or at least not criminalized. This map emphasizes four very important points. First, it shows that individuals can't simply do *what they want to do*, even when it refers to aspects of their lives which are—or should be—as private as deciding with whom they have sex or get married with: it depends on the dominant cultures, social norms, laws, and so on. In one interview for a Portuguese newspaper, I said to a journalist that taking into account the available evidence, humans—both men and women—probably have, in general, a tendency to desire to have sex with many partners, as I will show below. The journalist looked surprised and told me: "to me it clearly seems that humans *want* to be monogamous, because I see everybody in the street with a single partner." She made a profoundly wrong assumption that basically almost never applies to humans, particularly in *Neverland*'s world of unreality: that what people *do* is what they truly *want to do*. I answered to her by pointing out that, using the same logic that she used, if she had gone to the U.S. some centuries ago, she would say that African-Americans wanted to be slaves, because that is what they *did*. And that, if she would live currently in Afghanistan instead of Portugal, she would say that "clearly people want to be polygamous," because men often *do* have more than one wife there.

What one wants to do is very different from what one often does, and even from what one tells to others about what one wants to do. This obviously applies even to a more extreme degree to people that are subjugated and oppressed by others: in the example of Afghanistan, it principally applies to women, which are often obliged, at a young age, to get married to a much older man that they don't even know, that they don't like at all, that are already married with other women, and that often will treat them badly. Basically, most of those women don't even have a word to say about what they truly want: this is a key difference between "agricultural civilizations" and nomadic hunter-gatherer societies. In the latter, people do have to conform to general social norms that are advantageous for the group as a whole, but normally are not *obliged* to do what a single individual wants them to do, such as being a slave or a sexual vassal—or to work for his/her dreams, as it happens so often nowadays in "developed" countries.

This topic is related to a second major point emphasized by Fig. 5.1: in many of those agricultural states where **polygamous marriages** are not forbidden and in which such marriages are indeed prevalent, such as Afghanistan, the type of polygamy practiced there almost exclusively refers to that preferred by men, who precisely have in general a much higher status than women in agricultural societies: **polygyny** (one male–many females). In a way, this type of polygamy reflects what most men truly tend to want, sexually, in most regions of the globe, including also those shown in black in that figure: a man wants to have sex with many partners—women or/and men—but due to jealousy, territoriality and the feeling of possession, he wants each of those many partners to have sex only with him. The problem is that, contrary to the imaginary tales created by men, and then believed even by so many women, exactly the same happens with women: in general, *a woman wants to have sex with many partners – men or/and women – but she wants that each of them just have sex with her*. What is the perfect solution, then? There is none: biological evolution and life have nothing to do with perfection, things are just what they are. What we do know, based on historical facts and what is happening in countries plagued by misogyny is that when people are in a position of power—as men have over women in places such as **Afghanistan**—the very first thing they try to do is apply a situation that is ideal for them, but very unfair for those that they oppress: they use their power to indeed have sex with many *and* oblige those many to only have sex with them. Historically, those that had even more power, such as emperor or kings that were often obviously men, were sometimes able to even go a step further in terms of their sexual fulfillment and the oppression of women: they had **harems** with hundreds of women that were supposed to only have sex with their "masters" (see below). Indeed, in agricultural societies, **polyandry** (one female–many males) is much more rare, occurring nowadays only in a few regions with particularly scarce natural resources, although it was much more prevalent before Westerns colonized the rest of globe, as will be explained below.

This leads us to a third major point stressed by Fig. 5.1: in general, nowadays, the countries shown in black tend to be agricultural states that are less misogynistic than those not shown in black. So one can argue that this is why these countries forbid polygamous marriages: to defend the rights of the oppressed ones, that is of women in this context. This might be part of the explanation, as we know that political leaders use this argument to explain why some U.S. laws are against the polygamous marriages of **Mormons**, a group in which women are indeed often subjugated by men. However, this is surely *not the whole story*, because in the past, many Western countries have not only forbidden polygamous marriages but also imposed monogamy in the most atrocious ways, even to the point of torturing and burning people—mostly the oppressed ones that they now say to defend, that is, women—for adultery, or for having sex with many partners, or with demons, or even with the Devil himself. Moreover, as we also will see below, it was precisely the colonization by Western countries, and their laws and social norms, that contributed in great part to the rise of misogyny in many of the countries not shown in black in Fig. 5.1, particularly those in Sub-Saharan Africa. Before colonization, in many of them women had a much higher social status that they have today, and there were various cases of **polyandry** that were immediately forbidden by the colonizers.

This leads us to the fourth key point stressed by Fig. 5.1. As wisely pointed out by Ryan and Jetha in the book *Sex at Dawn*, the very fact that so many countries had to create so many social norms, and even laws, to *forbid polygamy*, as well as the fact that despite all those norms and laws in most countries shown in black basically nobody adopts nowadays **sexual monogamy**, clearly show how unnatural monogamy is for humans. In fact, contrary to socially monogamous animals such as gibbons, a huge number of people in Western countries do not have stable relationships, or when they have they then break up or divorce, and many have multiple partners at once. Indeed, even the decreasing number of people that, as my parents, got married and never divorced, before getting married they practiced **serial social monogamy**, not the type of social monogamy practiced by gibbons. That is, most people that marry now in Western countries such as the U.S. or Norway have a "stable relationship" with a partner A—for instance, a boyfriend—then a partner B—for example, a second boyfriend—then a partner C, D, E, F, G, and so on, until they have partner Q that becomes a husband, and in many cases, they then divorce and have a partner R, S, and so on until they marry for a second time with a partner U, with moreover some "fooling around" during those marriages/relationships. So, even if polygamous marriages are forbidden in such countries, in reality polygamy was always adopted by many people, and is becoming particularly widespread, there.

Indeed, it is interesting to see how statistics related to monogamy have been changing dramatically in the last decades, precisely in those regions of the map shown in black. One of these statistics is shown in Fig. 5.2: although there has been, fascinatingly—and this should clearly be the subject of further study—a small reversion in the 1970s and 1980s, the percentage of Americans reporting 0–1 partners since age 18 has dramatically decreased, compared with the 1930s. Figures 5.3 and 5.4 also display remarkable graphs: they show the enormous increase of **births out of wedlock** as well as of **divorce rates** at a regional and global scale, in the last decades. But perhaps the most relevant graph is that shown in Fig. 5.5. To read that graph one needs to understand that it is about the percentage of a sample of married people from the U.S. that reported to have had sex with someone other than the spouse. So, that people under 29 years of age have a 10 or 11% does not mean they "cheat" less than people that have a 24% when they are older than 80 years, because if let's say both got married at 18 years, a 29-year-old person had only 11 years to "cheat" while married, while a 80-year old person had 62 years to do so. As expected, there is a huge discrepancy between men and women older than 80 years: those

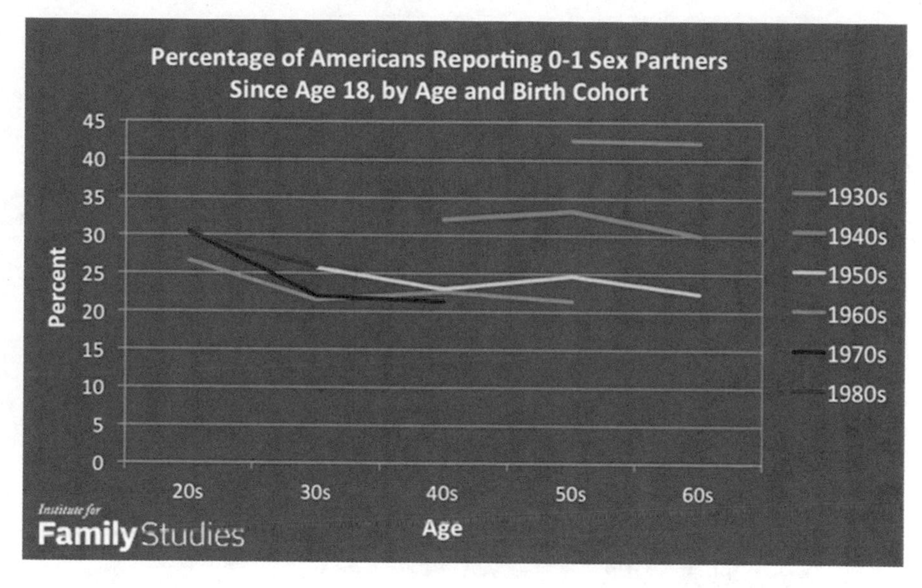

Fig. 5.2 Although there has been a small reversion in the 1970s and 1980s, the percentage of Americans reporting 0–1 partners since age 18 has dramatically decreased, compared with the 1930s

people were born in the 1930s or 1940s, so mostly got married in the 1950s or 1960s, at a time when social norms about women being with a single man were much stronger than they are today, and when it was actually much more difficult, for a woman, to physically do so even if she would want to, because she stayed at home more time, or did not have a job at all, traveled less alone, and so on.

The most amazing point of Fig. 5.5 is that, on average, married women under 29 years *already* report "cheating" more their spouses than men do: 11% versus 10%. In fact, the true difference is likely to be even higher than these self-reported estimates. This is because it is well known, based on empirical studies, that women tend to report less sexual partners/activity than they actually have, while men tend to do exactly the opposite, in great part precisely due to still-prevalent social norms and narratives about women being "supposed to be less sexual" than men. As noted by clinical psychologist **Frank Tallis**—the author of *The incurable romantic*, among various other books—in a 2018 interview for the Portuguese newspaper *O Publico*, in detailed studies about patterns of "*Google search*" use, it is actually women, more than men, that ask questions such as: why my partner does not want to have sex with me? As he explains, when they are in public, or even when they talk in private with psychologists, women tend to talk more about emotions—as they are supposed to do, within our accepted **cultural and gender stereotypes**—but when they are truly alone, they do "care a lot about sex," as evidenced by such studies. This is a further example of the huge difference between *what an individual truly wants* versus what an individual *pretends to be within the cultural context of a certain society*. This is also displayed in Fig. 5.5: it does not reflect a change about what

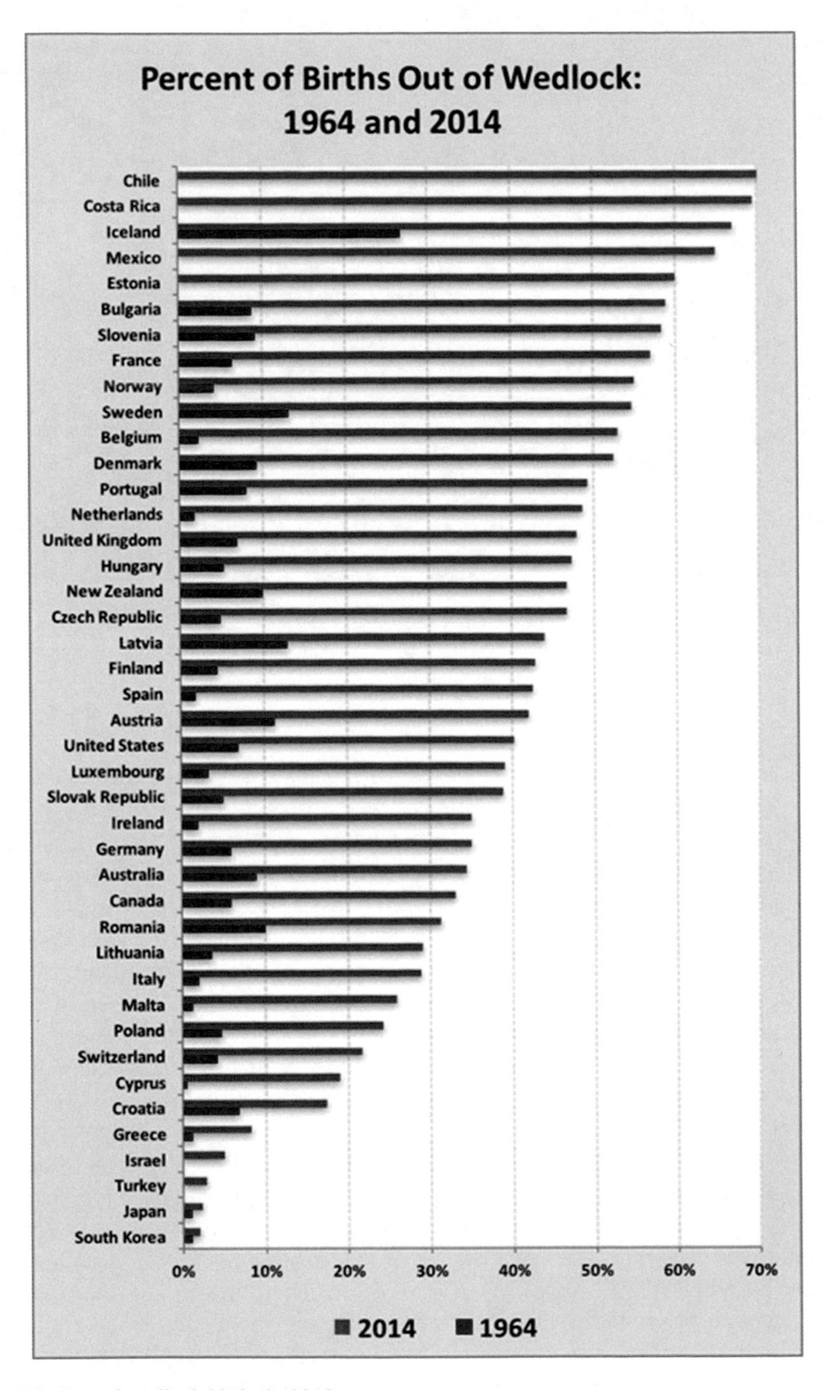

Fig. 5.3 Out of wedlock births in 2018

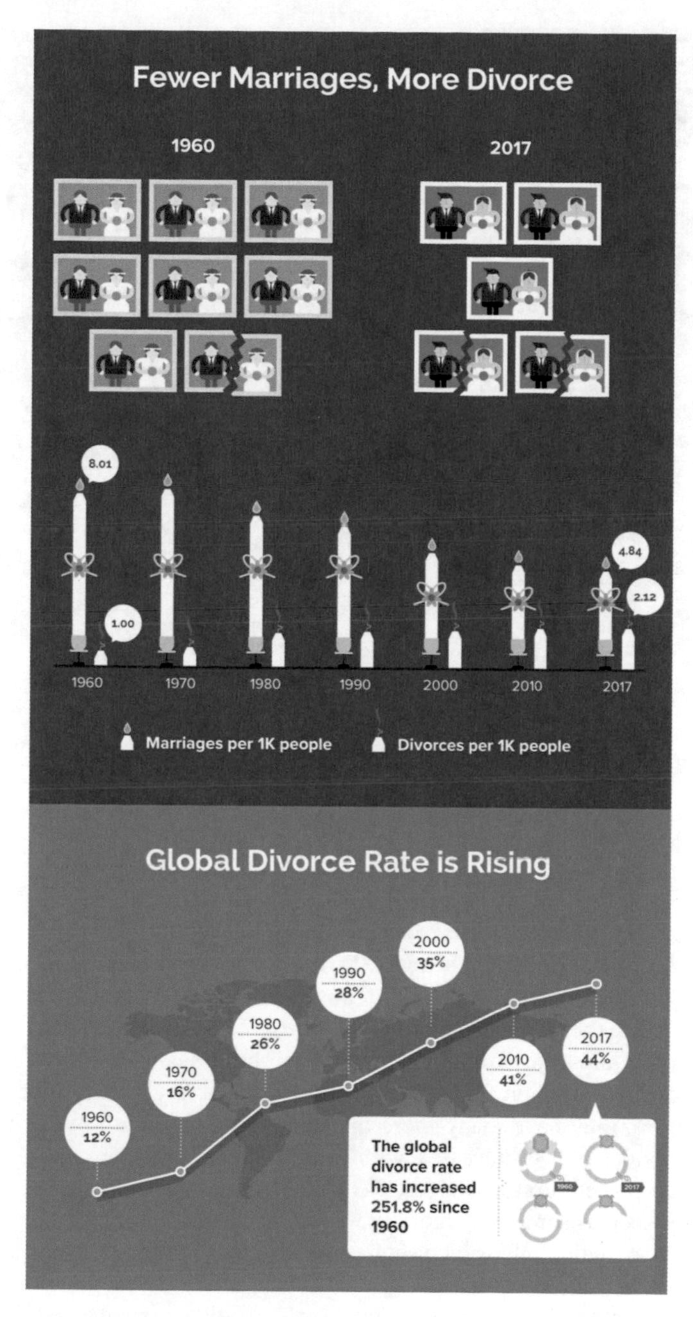

Fig. 5.4 Increase of divorce rates in the last decades

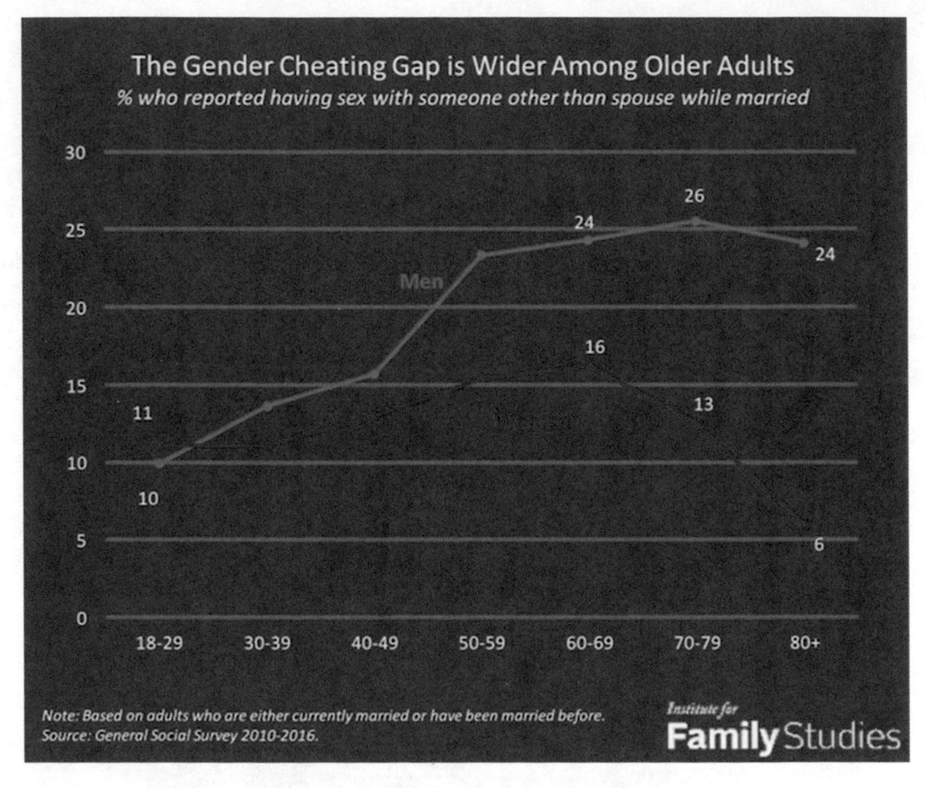

Fig. 5.5 The gender cheating gap, based on a study of adults who are either currently married or have been married before. Source: General Social Survey 2010–2016

women *desired*, but instead the fact that the social and cultural norms that have constrained them for millennia begun to be less and less pronounced/accepted.

Contrary to the "domestic model" imposed to women in the last centuries and particularly since the 1950s in countries such as the U.S., in the last decades Western women begun again to work more outside of their homes, and moreover have begun to travel alone or with their female friends more than at any period since the rise of agriculture. So, they have many more opportunities to finally do what they—and men—always desired: to have sex with more than one partner—or, as it is now called within our cultural narratives, clearly showing how within those narratives polygamy is seen as something morally wrong, to "cheat." If women are really less "sexual," if they really only want a single man within a stable "relationship" to help them to raise their kids—who are supposedly their only interest and even only aim in life, according to many of such narratives—why would young married U.S. women already "cheat" their spouses more than married men do? Clearly, something is wrong with such narratives, and it is not difficult to know what it is: they are man-made imaginary tales created to subjugate, control, and oppress women. That is, these *Neverland* just-so-stories are rather about *what men desire their spouses to be*.

Note that I use the term "spouses," and not women in general, because obviously many men do have the desire to have sex with women that desire—or that pretend to desire—sex for the sake of sex and pleasure per se, as evidenced by the huge historical demand, by men, for prostitutes, pornographic movies, and so on. The very fact that prostitution is said to be the "oldest profession in the world" clearly shows, by itself, that monogamy is not natural at all for humans.

So, digging a bit deeper, what does the data available from anthropology and evolutionary biology tell us about the "sexuality" of women versus men? Let's start by talking about the "hottest" topic: **orgasms**. **Darwin** was one of the main exponents, in the Victorian era, of the "man-made" narratives described just above. He was himself influenced by the highly misogynistic ideas that were predominant during that time in England: another illustrative example of how society, cultural narratives, and science have a very complex, interactive link. That is, he was influenced by those ideas, and then created just-so scientific stories to "support" them empirically, instead of using biological data to test them, in a way that is very similar to what authors such as Pinker do today—influenced by prevailing ideas about progress, they desperately try to find data that support them, instead of truly testing them. Namely, Darwinists often supposedly provide "objective scientific support" for the idea that men evolved to have desire for many women in order to have as many kids as possible while women mainly evolved to prefer to have a single male partner. And of course, within that idea women want to have a male partner not so much because of the pleasure of having sex with him per se, but mainly as an "exchange" for food or any other kind of help they can have to raise the only thing they care in life: their kids. In other words, *the true desire of the father is sex*, while *the true desire of the mother is to take care of her kids*. It is interesting to see that within this man-made narrative, there was no "**romantic love**" between men and women, thus keeping with a tradition that was prevalent in most cultures for millennia and just changed very recently, as explained in Coontz' 2005 book *Marriage, a History*. Darwin himself, in his writings and personal letters, was very far from being a romantic: overall, he was mainly a very misogynistic, unromantic Victorian man. In fact, as will be explain in Chap. 6, in a way Darwin tended to see living beings, including humans – and in particular women – as mostly passive, that is, subjugated to/by natural selection – mainly the external environment –, in contrast to a more modern understanding of biological evolution (see Fig. 5.6).

Strikingly, although the Victorian era finished more than a century ago, Darwin's ideas about women and in particular their sexuality are still highly prevalent within the broader public in many regions of the globe, including Western countries, being dominant even among many scientists. Since Darwin's time until this very moment, many researchers have desperately tried to answer the supposedly "puzzling" question of why women have **orgasms** at all. This is because, by assuming a priori that such man-made Darwinian narratives are right, scientists argue that women would not at all need to have orgasms, as they don't really care about pleasure: they just care about having kids, and they can have kids without having orgasms as long as men have them. Amazingly, such attempts to scientifically "explain this puzzle" are done even by scientists that deem themselves of being "progressive" and

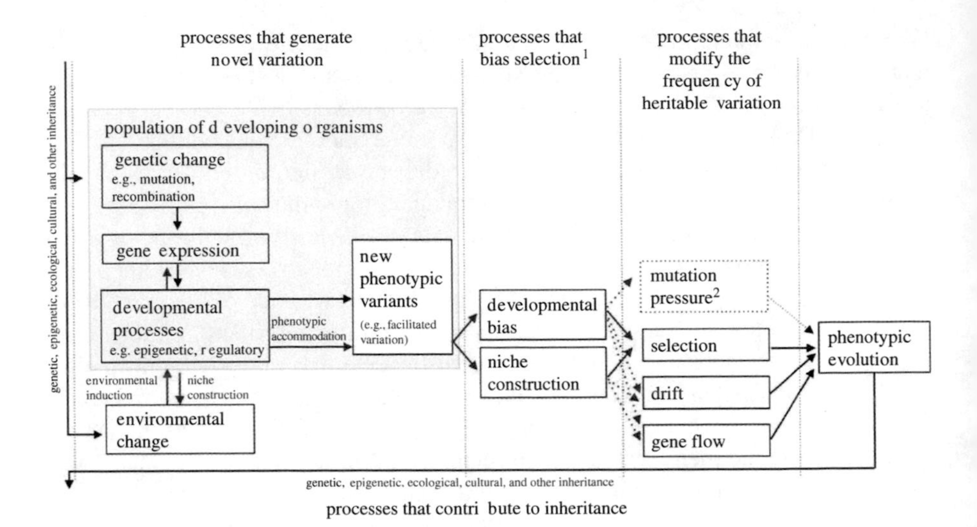

Fig. 5.6 The structure of the Extended Evolutionary Synthesis (EES) according to, and modified from, Laland et al. (2015), with arrows representing causal influences and processes shown in red being those emphasized by the EES but not by a more traditional Neo-Darwinist perspective. EES includes as evolutionary causes processes that create new variants, influence selection, change the frequency of heritable variation and contribute to inheritance. A diversity of ontogenetic processes (e.g., epigenetic effects, regulation of gene expression, construction of internal and external developmental environments) contributes to the rise of new phenotypic variation, which may be viable and adaptive (e.g., "facilitated variation"). In addition to accepted evolutionary processes that directly change gene frequencies, processes that influence the outcome of natural selection, particularly ontogenetic biases and niche construction, are also recognized in EES. A broadened notion of inheritance encompasses genetic, epigenetic, and ecological (including cultural) inheritance. "Mutation pressure" refers to the population-level consequences of repeated mutation, depicted as dashed because mutation is also shown in "processes that generate novel variation." (1) Developmental bias and niche construction can also affect other evolutionary processes, such as mutation, drift, and gene flow. (2) In EES, this category of processes will often need to be broadened to include processes that change the frequencies of other heritable resources

"non-misogynistic," many of them women. An emblematic example is a paper published just a few years ago, by one of the most renowned current developmental biologists—Gunter Wagner, who was amazingly kind with me when I gave a talk at his lab and show me Yale University—and a researcher who is often considered to be a potential rising female star within that field, Michaela Pavlicev.

At the very beginning of the abstract of their Pavlicev and Wagner 2016 paper, they summarized what are still the prevailing views on the subject within a substantial part of the scientific community: "the evolutionary explanation of female orgasm has been difficult to come by…. The **orgasm** in women does not obviously contribute to the reproductive success, and surprisingly unreliably accompanies **heterosexual intercourse**…. Two types of explanations have been proposed: one insisting on extant adaptive roles in reproduction, another explaining **female orgasm** as a byproduct of selection on **male orgasm**, which is crucial for sperm transfer." I am still amazed to see that such man-made narratives without any scientific validity are so deeply impregnated in our minds that even the ones that have been oppressed by their use—in this case, a women, who is the first author of the paper—blindly accept them. That is, within such narratives, the behavior/attributes of the "others"—either from another religion, or so-called race, or gender as in this case—are always seen as *passive*, as mainly a "by-product" of evolution or history, which of course "truly" concern only selective pressures exerted on the "ingroup," which is thus seen as the only "active player": in this case, the men. This is not at all a criticism of the authors of that specific paper, as when they do research about other subjects that are less emotionally connected to us, such as the evolution of cells, they can really do great work. So, their paper is precisely chosen to illustrate the fact that this is not a case of using a straw-man: they are rather top researchers that often defend "progressive ideas," but that—when it comes to issues that directly concern longstanding *Neverland* imaginary teleological stories about humans—use these stories as a priori dogmas in their papers about these issues.

Before accepting such misogynistic Darwinian ideas a priori as scientific dogmas, researchers should instead question them and test them empirically. A key question that researchers should try to answer in the first place, about this topic, is: why are women the ones that actually have more frequently **multiple orgasms**, if women are supposed to be "less sexual" and care less about pleasure than men? A study by Puts and colleagues in 2021 found that 43% of women reported usually experience multiple orgasms. Of course, the numbers in different studies change, depending on the geographical location and the cultural narratives commonly accepted, the context in which the questionnaires are done, and of course the type of questions. For example, when one uses the term "usually," what does this mean? Every week? Every month? Every year? But what is clear, from the vast majority of studies done so far, is that multiple orgasms are indeed much more frequent in women than in men, on average. In their 2016 review titled "*Multiple orgasms in men – what we know so far*," Wibowo and Wassersug wrote that "few men are multiorgasmic: <10% for those in their 20s, and <7% after the age of 30." Interestingly, they point out that a change of sexual partner is said to potentially increase how likely a male is to have multiple orgasms, or a similar sensation such as the so-called **Coolidge effect**: a re-arousal phenomenon where the refractory ("recovery") period from an orgasm is reduced in the presence of a *new sex partner*. As they further note, evidence of a similar phenomenon has been shown empirically in other animals: access to new or novel sexual partners may promote **sexual desire** and motivation to orgasm, truncating the refractory period. Of course, such data put into question, once again, the narratives that our truly "human nature" is to be sexually or socially monogamous, because empirical studies show, over and over, that risk and novelty—rather than having sex exclusively with the very same partner for various decades—are among the main promoters of sexual motivation and desire. As noted in *Sex at Dawn*:

> Over fifty years ago, sex researchers Clellan Ford and Frank Beach declared, "in those [hunter-gatherer] societies which have no double standard in sexual matters and in which a variety of liaisons are permitted, the women avail themselves as eagerly of their opportunity as do the men". Nor do the females of our closest primate cousins offer much reason to believe the human female should be sexually reluctant due to purely biological concerns. Instead, primatologist Meredith Small has noted that female primates are highly attracted to novelty in mating. Unfamiliar males appear to attract females more than known males with any other characteristic a male might offer (high status, large size, coloration, frequent grooming, hairy chest, gold chains, pinky ring, whatever). Small writes, "the only consistent interest seen among the general primate population is an interest in novelty and variety.... In fact," she reports, "the search for the unfamiliar is documented as a female preference more often than is any other characteristic our human eyes can perceive".

In addition, various studies indicate that women experience, on average, more complex, elaborate, and **intense orgasms** than men, as reported in a 2002 paper by Mah and Binik. **Ackerman**, in her 1994 book *Natural History of Love*, noted that "men's **oxytocin** levels quintuple during orgasm…but a Stanford University study showed that women have even higher levels of oxytocin than men do during sex, and that it takes more **oxytocin** for a woman to achieve orgasm…drenched in this spa of the chemical, women are able to have more multiple orgasms than men, as well as **full body orgasms**." Furthermore, the **clitoris** normally has about 8000 **sensory receptors**—compared to half as many in an ordinary **penis**—and its sole function is to provide ecstatic pleasure to its bearing, contrary to the penis, as explained in Browning's 2017 paper "*Survival secrets – what is about women that makes them more resilient than men.*" So, based on these data, and various other lines of evidence that will be discussed below, it seems that instead of asking "why women have orgasms," we should instead ask: with a much higher **orgasmic potential**, why in numerous regions and contexts there are less women that orgasm-as compared to men-on average, per male–female sexual act? I was actually asked this question, in an interview for a TV show about this subject, by a sex therapist, some months ago. I answered that, indeed, as explained in a 2018 review by Blair and colleagues appropriately titled "*Not all orgasms are created equal,*" according to some of the most extensive studies done on the subject, men report experiences of orgasm in 85–95% of their partnered sexual activities, while the percentage is only 40–65% in women. As those authors explain, this difference is *not* related to a lack of orgasmic potential by women—well on the contrary—but rather due to other reasons, one of them being precisely the prevailing man-made narratives about how "good" women should be more passive in, or more ashamed of, their sexuality.

This women–men gap is further reinforced by the typical "women–men sexual scripts" that are also prevalent in our societies, which tend to favor sexual activities that are aimed to give pleasure to, and thus to more likely result in orgasms for, men, such as **penile-vaginal intercourse**. In *Testosterone Rex*, Cordelia Fine discusses this topic, noting that "a large-scale study of thousands of female North American college students found that they had only 11% chance of experiencing an orgasm from a first casual 'hookup'…follow-up interviews revealed why it was that women had such slim odds of reaching a climax…students generally agreed that it was important for a man to be sexually satisfied in any context, and for women to be sexually satisfied in the context of a relationship". Indeed, "there was no perceived obligation to provide sexual satisfaction to a woman in **hookup sex**…while many men felt that bringing their girlfriend to orgasm reflected well on their masculinity, they often didn't feel the same way about hookup partners."

When I am interviewed for TVs and radios, I always use the following example: image that the opposite happens, that is, that a woman would say to a man "OK, we can have sex for hours and hours, but we can never touch, or use, your penis." Almost no men would be happy with this, as basically they would not be able to have a single orgasm, even after several hours. Well, that is precisely what many women accept, when they have sex with men that penetrate them without ever stimulating their clitoris, which is the female region that embryologically corresponds to the penis. Many women, by trying to follow the script of being a "good, respectful woman," not only accept that men do that but also don't have the courage to even stimulate their clitoris themselves when that happens, particularly during the first time they have sex with a man. So, that U.S. college students only have 11% chance of experiencing an orgasm from a first casual "hookup" with a man has nothing to do with them, and their clitoris, having less **orgasmic potential** than men and their penises, well on the contrary: it is instead because within *Neverland*'s man-made tales most women have to follow a sexual script that only cares about the pleasure of man, as it literally happens also with the scripts of so many **pornographic movies**. This point is further supported by empirical data that show that **lesbians** have significantly higher rates of orgasm than self-identified heterosexual or bisexual women, a topic that we will discuss below.

The fact that such absurd man-made scripts are still prevalent in popular culture, including in Western countries, is further illustrated by a sentence typically accepted by both men and women in those countries, a sentence we all heard several times: "men have a penis, women a vagina." In other words, this man-made narrative emphasizes the passive role that men want to assign to women: women basically have a hole to receive the active player, that is, men's penis. Strikingly-but not surprisingly-such scripts have been defended, and "allegedly scientifically supported," by so-called scientists such as **Freud**, who defended that contrary to "immature women," "mature women" are able to have vaginal orgasms. That is, only women that obey men and their factually wrong scripts can achieve the recognition of men and be called "proper mature, good women." As put by Ackerman in *Natural History of Love*:

> It's worth noting that when we talk about gender we say that a man has a **penis** and a woman has a **vagina**. This distinction, which we take for granted, hides a prejudice about the baseness of women. A man's pleasure organ is the penis, and a woman's pleasure organ is her **clitoris**, not her vagina. Even if we're talking about procreation, it's not accurate: a man's penis delivers sperm and can impregnate, and a woman's womb contains eggs, which can become fertile. Equating the man's penis with the woman's vagina says, in effect, that the natural order of things is for a man to have pleasure during sex, and for a woman to have a sleeve for man's pleasure. It perpetuates the notion that women aren't supposed to enjoy sex, that they're bucking the natural and social order if they do. I don't think this will change very soon, but it reminds me how many of our mores travel almost invisibly in the plasma of language.

As part of an effort to precisely finish once for all with such longstanding biases, prejudices, and unscientific narratives, let's continue referring to the empirical lines of evidence that contradict the fairytale that women are less "sexual" than men. One of them, related to **physiology**, concerns the fact that studies have indicated—although more studies, with more individuals, are clearly needed—that women outperform men in **smell sensitivity** tests—as noted in a 2014 paper by Oliveira-Pinto and colleagues—including those regarding scents associated to **sexual arousal**, as documented in Hirsch's 1998 book *Scentsational Sex*. Another line of evidence concerns studies on **sex fluidity** and **homosexuality** (see also Boxes 5.1, 5.2, and 5.7). As pointed out in a 2018 review of many studies done on this topic, published by Jeffery and colleagues, the data obtained in such studies indicates that "heterosexual women are more likely than heterosexual men to report **same-sex sexual attractions.**" Individuals "with high sexual fluidity experience sexual responses toward a broad and shifting range of stimuli; male sexual desire is usually considered category-specific, as it strongly favors one sex; for men, there is a negative correlation between sexual attraction to one sex and sexual attraction to the other sex, and men of all sexual orientations with higher sex drives have more sex with a single preferred sex."

As they note, "this is not true of women, who show a weak positive correlation between attraction to one sex and attraction to the other sex." Also, "women report more frequent shifts in sexual attractions than men across their lifetimes, particularly among non-heterosexuals." Importantly, such **studies on homosexuality** emphasize the complex links between "nature" and "nurture" discussed in the present book. That is, while it seems that women tend to be truly more sexually fluid by "nature"—biologically/evolutionarily—as supported by those studies, their results also reveal that there is an important social—environmental—component of **sexual orientation**. As the authors note, female prisons represent an extreme example, with "somewhere between 30 and 60% of female prison inmates engaging in same-sex sexual behaviors and relationships; most of these women did not identify as homosexual prior to incarceration and, the more time women spend in prison, the more accepting they become of sexual interactions between other inmates and of having a 'gay cellmate'."

As they also stress, various researchers argue that frustration with the typical model of "male–female intercourse" that is prevalent in most societies may contribute to homosexuality. For instance, "heterosexual men and women report that sexual relationships inflict greater costs than are reported by homosexual men and women; meanwhile, homosexual men enjoy more rapid and frequent sexual experiences than heterosexual men (on average) and homosexual women report greater sexual satisfaction, more sexual desire, more frequent orgasms, and greater satisfaction with their own bodies than heterosexual women." This point is further made in a very detailed study about Brazilian female jails done by one of the most renowned medical researchers in Latin America, **Drauzio Varella**, for more than a decade. In his 2017 book *Prisioneiras*, he argues that the only place where women nowadays have "true sexual freedom" is in the jail, because they don't live on a daily basis with their main oppressors—men—so they don't have to follow man-made scripts and to face so much discrimination and gender violence, and sex is often more consensual. Accordingly, one can see the more profound, and complex, expression of feminine sexuality, with an incredibly wide range of identities and expressions. He notes that homosexuality is much more frequent, and also more subtle, than in male prisons, and states that, although he is a professional oncologist and gynecologist, "in the female jails, I realized that I only truly knew about 10% of the variability of what **women sexuality** can be."

Many recent studies, based on anonymous online questionnaires have as well been reported as "surprising" by the media, because they go against the narrative of "man-the-sexual-being." As the number of such studies is so overwhelming, I am only referring to a few of them here. In a large anonymous online questionnaire made to 1300 persons, by the sexshop *Lovehoney*, 46% of the women recognized that they think about other people when they are having sex with their partners, while only 42% of men said the same. Moreover, studies about phenomena such as so-called **hypersexual disorders**—such as what is called "**satyriasis**" in men and "**nymphomania**" in women—consistently show that, in different contexts and countries, when compared with men, women tend to score higher on "**hypersexuality.**" Moreover, in many of such studies, women also engage more often in **risky sexual behavior** than men, contrary to what the broader public tends to *believe*, as explained in a 2017 paper by Öberg and colleagues. Another example of women sexual fluidity is given in the book *Sex at Dawn*, which uses the term "**erotic plasticity**":

> This greater erotic plasticity appears to manifest in women's more holistic responses to sexual imagery and thoughts. In 2006, psychologist Meredith Chivers set up an experiment where she showed a variety of sexual videos to men and women, both straight and gay. The videos included a wide range of possible erotic configurations: man/woman, man/man, woman/woman, lone man masturbating, lone woman masturbating, a muscular guy walking naked on a beach, and a fit woman working out in the nude. To top it all off, she also included a short film clip of bonobos mating. While her subjects were being buffeted by this onslaught of varied eroticism, they had a keypad where they could indicate how turned on they felt. In addition, their genitals were wired up to **plethysmographs** (measures blood flow to the genitals…think of it as an **erotic lie detector**). What did Chivers find? Gay or straight, the men were predictable. The things that turned them on were what you'd expect. The straight guys responded to anything involving naked women, but were left cold when only men were on display. The gay guys were similarly consistent, though at 180 degrees. And both straight and gay men indicated with the keypad what their genital blood flow was saying.
>
> As it turns out, men can think with both heads at once, as long as both are thinking the same thing. The female subjects, on the other hand, were the very picture of inscrutability. Regardless of sexual orientation, most of them had the plethysmograph's needle twitching over just about everything they saw. Whether they were watching men with men, women with women, the guy on the beach, the woman in the gym, or bonobos in the zoo, their genital blood was pumping. But unlike the men, many of the women reported (via the keypad) that they weren't turned on. As Daniel Bergner reported on the study in *The New York Times*, "With the women…mind and genitals seemed scarcely to belong to the same person". Watching both the lesbians and the gay male couple, the straight women's vaginal blood flow indicated more arousal than they confessed on the keypad. Watching good old-fashioned vanilla heterosexual couplings, everything flipped and they claimed more arousal than their bodies indicated. Straight or gay, the women reported almost no response to the

hot bonobo-on-bonobo action, though again, their bodily reactions suggested they kinda liked it. This disconnect between what these women experienced on a physical level and what they consciously registered is precisely what the theory of differential erotic plasticity predicts. It could well be that the price of women's greater erotic flexibility is more difficulty in knowing – and, depending on what cultural restrictions may be involved, in accepting – what they're feeling. This is worth keeping in mind when considering why so many women report lack of interest in sex or difficulties in reaching orgasm [particularly when having sex with men].

It is very likely that many, if not most, of the women that did not self-report to be aroused when there was a high blood flow to their genitals did that *consciously*, because a "good woman" is not supposed to be aroused, and even less to say so—emphasizing again the crucial *Neverland* dichotomy: what one *wants* versus what one is *supposed to want or say*. This idea is further supported by many other studies, one of them being reported in Browning's *The Fate of Gender*: "in a series of controlled surveys published…in 2011…men recorded thinking about sex an average of 18 times a day while women recorded thinking about sex 10 times a day…similarly, when men and women were asked orally how many partners they would like to have per year, young college-age men reported many more than young women did…however, when both were attached to lie detector machines, the number of desired partners per year turned out to be about equal". However, and perhaps even more disturbing, is that it is also likely that, within the study of blood flow cited above, at least some women probably did *not even know* that there was a high blood flood to their genitals, that is, because of the way women are bombarded with man-made narratives about sex and religion in so-called civilized countries since they are extremely young, it is very likely that many of them repress so much their sexual desires that they become completely disconnected with how their bodies are reacting physiologically. If this is so, such women would have "reached" a state that is the end goal of many religious and other teleological narratives that were mainly created, disseminated, and defended by misogynistic men to dominate them and oppress their desires: the complete separation between their "spiritual soul"—immersed in *Neverland*—and their "animal instincts," "bodily needs," and "earthly desires"—within the reality of the natural world.

5.4 Sex at Dawn, Sex at Dusk, and Scientific Biases

Love is blind, they say; sex is impervious to reason and mocks the power of all philosophers…but, in fact, a person's sexual choice is the result and sum of their fundamental convictions…tell me what a person finds sexually attractive and I will tell you their entire philosophy of life. (Any Rand)

In Van Schaik and Michel's 2016 *The Good Book of Human Nature*, it is said that only about 1% of all extant human societies have **polyandry** (one female–many males). This is an interesting point because a **multimale-multifemale** organization in found in our closest relatives, the chimpanzees, being particularly evident in the so-called highly promiscuous bonobos, and also because most extant human gathered-gathered groups don't follow at all sexual, or even social, monogamy in a strict way. These are the central topics of Ryan and Jetha's book ***Sex at Dawn***, which caused a huge controversy within the academic community: several papers, commentaries, and even a whole book—***Sex at Dusk: Lifting the Shiny Wrapping from Sex at Dawn***—were published just to attack it, an honor that very few books have, evidently. Clearly, *Sex at Dawn* hit a nerve among many people. Why? Well, when one reads most works that were produced after, and as a reaction to, *Sex at Dawn*, one can see that the major controversy about that book is not really about monogamy versus polygamy, but instead about *which kind of polygamy* is most prominent within our "**human nature**." That is, the controversy is more about our "sexual nature" being mainly **polygynous** (one male-several females) versus **multimale-multifemale** (each female and each male having several partners of the other sex) as proposed in *Sex at Dawn*. In other words, both models—and most lay people, nowadays—assume that it is mainly part of our "human nature" that males desire several sexual partners: what people often attack about *Sex at Dawn* is mainly its idea that this also applies to women. One can see this in the very aggressive tone in which many scholars criticize *Sex at Dawn*: it is not at all the typical tone that scientists often use, when they criticize a book that has some flaws. Clearly, that book provoked a huge reaction—above all, from **reactionary people**, which precisely tend to *react* when the prevailing narratives and *status quo* are put in question—because it referred to a sensitive topic that is deeply related to man-made narratives that most humans have been accepting since the rise of agriculture and of organized religions.

In fact, while *Sex at Dawn* does have some scientific flaws as most books do—I will discuss these flaws below—it does discuss a huge amount of empirical data from many fields of science, contrary to many other books written about human sexuality and particularly ***women sexuality***, which are mainly just-so stories based on "opinion" or "gut feelings" or "personal experiences" or pseudoscientific tales plagued by prejudices. David Barash, an emeritus professor of psychology at the University of Washington, wrote, in a 2012 article of the *Chronicle of Higher Education*: "A little while ago, I worried that the next time someone asked me about the book, *Sex at dawn*…I might vomit. An over-reaction? Perhaps. And one likely composed, in part, of simple envy, since their book seems to have sold a lot of copies. At least as contributory, however, is the profoundly annoying fact that *Sex at dawn* has been taken as scientifically valid by large numbers of naïve readers…whereas it is an intellectually myopic, ideologically driven, pseudo-scientific fraud." How many times have you seen an emeritus professor publicly using the word "vomit" to describe a book? An over-reaction, indeed.

Let's thus now come back to the pinnacle of the reactionary way in which *Sex at Dawn* was received by a huge number of people: Saxon's book *Sex at Dusk*—a 364-page book written explicitly, and solely, to criticize it. One interesting point made about the latter book was published by Robin Hanson in 2012:

> A key question, to me, is what percentage of our forager ancestor kids were fathered outside pair-bonds. That is, what fraction of kids were born to mothers without a main male partner, or had a father different from that partner. This number says…a lot about how 'natural' are such things. Alas, none of these authors (of *Sex at dawn* and *Sex at dusk*) give a number, but my impression was that Saxon would estimate less than 20%, while the *Sex At dawn* authors would estimate over 50%. Even 20% would be consistent with a lot of human promiscuity…. I asked Saxon directly via email, however, and she declined to give a number – she says her main focus was to argue against *Sex At dawn*'s 'paternity indifference' theory [that humans (males) don't care which kids are theirs].

Saxon's book actually completely distorts empirical data, as when she refers to the sexuality of non-human primates and of our human ancestors. After reviewing data on great apes, she writes: "this even suggests an *alternative* breeding scenario for our common ancestor with the **gorilla**: monogamous pairs." Simply, there is no evidence at all to support such a statement. It is consensual that **orangutans** have a **dispersed type of sexual behavior**—they are often solitary, so it is difficult to accurately access the number of sexual partners, but they are clearly not sexually or socially monogamous, as noted in a 2015 paper by Dorus and colleagues. Extant gorillas are clearly **polygynous**: many females live around a single alpha male, although he is often "cheated on" by at least some of them. Both common chimpanzees and bonobos live mainly in **multimale-multifemale groups**, and paleontological studies of body size of individuals of our lineage strongly suggest that, at least during the first three million years after we split from chimpanzees, humans continued to have a pronounced sexual dimorphism, a feature usually correlated with both sexual and social polygamy. So, based on these consensual lines of empirical data, how can someone arguing to be "objectively guided" by empirical data state that such data indicate that the last common ancestor of gorillas, common chimpanzees, bonobos and early humans likely lived in "monogamous pairs"?

The same types of distortions also apply to the parts of *Sex at Dusk* focusing on hunter-gatherers, which are also crucial to understand the evolution of human sexuality. For instance, about **"pre-agricultural" groups**, Saxon spends more than 30 pages repeating sentences such as "the Cashinahua allow for discreet extramarital sex but public acknowledgement it is rare," "she…is expected to allow perhaps as many as 25 men to have sex with her…[but]…female orgasm does not occur," and so on. So many "buts": the reality is that none of the examples that she provides contradicts at all the point made by *Sex at Dawn* about them, that is, that they illustrate, indeed, cases of sexual and social non-monogamy. Primates that are **socially monogamous**, such as **gibbons**, have a few extra-couple affairs, that is why they are not truly **sexually monogamous**, as almost no mammalian truly is. But gibbons don't have rituals where a female has sequential sex with 25 males: having sequential sex with a huge number of partners clearly is out of any definition of social monogamy. In fact, Beckerman and Valentine's 2002 book *Cultures of Multiple Fathers*, to which Saxon refers many times in her book, clearly defends—as the term "multiple fathers" indicates—that these Amazonian societies are neither monogamous nor polygynous because females usually have sex with various males: that is, they have mainly a multimale-multifemale type of organization, exactly as defended in *Sex at Dawn*. This is made very clear in the official description of that 2002 book:

> *Cultures of multiple fathers* is the first book to explore the concept of **partible paternity**, the aboriginal South American belief that a child can have more than one biological father – in other words, that all men who have sex with a woman during her pregnancy contribute to the formation of her baby and may assume social responsibilities for the child after its birth. The contributors, all Amazonian ethnologists with varied anthropological backgrounds and arguably the world's experts on this little-known phenomenon, explore how partible paternity works in several **aboriginal societies** in the South American lowlands. Many findings in this book challenge long-held dogma in such fields as evolutionary psychology and evolutionary anthropology and sociology. For example, under some circumstances, children with multiple putative fathers have higher prospects for surviving than do children ascribed to only a single father. Among several ethnic groups, a strong case can be made for a pregnant woman's having a lover so that her child will have more than one father and provider.

Surprisingly, contrarily to the general tone and main thesis of *Sex at Dusk*, in chapters 5 and 6 of that book, Saxon recognizes that she agrees with previous authors in that the answer to the question of why **matrilocal societies** (societies in which a married couple resides with or near the wife's parents) such as the **Mosuo** "still have marriage when it is not necessary for reproduction or the economic division of labour," is related to the fact that humans "desire to both possess one's partner *and* to have multiple partners." That is, here Saxon is in agreement with the key idea defended in the present book: that each person, both women and men, tends to desire to have sex with many partners *and* that each of those many partners only have sex with them. And yes, this idea is indeed different from the ideas defended in *Sex at Dawn*, because that latter book suggested that jealousy is mainly a "social construct," rather than a "natural tendency" of humans. This is clearly the main scientific flaw of *Sex at Dawn*, although it needs to be stressed that such an idea about a "lack of **jealousy**" was not the most crucial idea defended in that book.

In his book *The Incurable Romantic*, Tallis writes, "why do people get jealous? If you love someone you should want them to be free and happy…true love knows no bounds; it releases the soul." But he notes that, as it is often the case, the roman poet and philosopher **Lucretius** "gets much closer to the truth when he warns us that the goddess of love has sturdy fetters…we are only free to be ourselves and that isn't very free at all." Tallis further notes that "**utopian communities** have adopted 'free love' as a guiding principle, but virtually all of them have dwindled or collapsed on account of group members reverting to [social] **monogamy**…the internet has opened up a channel of communication between young couples eager to explore a '**polyamorous**' lifestyle, yet many of them confess that overcoming jealousy is a major obstacle…couples who manage to maintain stable **'open' relationships** and raise children constitute only a tiny fraction of the general population". He concludes that "whenever social engineers or political visionaries have attempted to alter the structure of society, the **family unit** returns…our need to privilege a single, exclusive relationship and guard it jealously is clearly hardwired". However, again, this is not simply a matter of 'nature' versus 'nurture', because the extreme type of '**monogamous-jealousy**' that is so prevalent in today's Western popular culture is clearly also not 'natural', as Tallis recognized in subsequent parts of his book:

"some 20 to 40% of married heterosexual men admit to having had at least one extra-marital affair—as do 20 to 25% of heterosexual women…approximately 70% of dating couples cheat on each other…over half of the single population engage in 'mate poaching'—attempting to break up an existing committed relationship…from the perspective of evolutionary psychology, the human reproductive strategy is mixed, a judicious combination of pair bonding and opportunistic sex." The only aspect I would question is Tallis' use of the word "judicious," because I fully agree with Sapolsy's book *Behave* in that most other things we do are actually often rather "injudicious"—part of our characteristic ***Homo irrationalis*** component.

That is, a single person is often full of contradictions, not only regarding what he/she *wants* "naturally" versus how he/she is *supposed* to act in society—the main focus of *Sex at Dawn*—but even concerning *different things that he/she wants* "naturally." This applies precisely to the fact that both women and men usually "naturally" *desire* to have sex with many partners, but because we are also "naturally" territorial and jealous as apes tend to be, we have this constant paradox that permeates humans: (1) we have the power to be with many partners that are only with us, as kings and emperors had and many men still do in so many countries, or (2) we are polygamous and accept that our partners also have multiple partners and learn to live with jealousy, or (3) we try to force ourselves to be socially monogamous and accept that we cannot fulfill our desires to have sex with more than one person. Except the first option, which is ideal for the powerful ones but completely unfair to other people, options 2 and 3 will always involve some "natural" frustration to all involved, and this is actually a powerful demonstration of the *reality of the natural* world: *nothing is "ideal," things are just what they are, a chaotic, random, and contingent mix of trade-offs, in which for anything that can be deemed to be "good" according to some narratives, there is almost always an evolutionary "price to be paid."* As explained in a 2015 paper by Burton and colleagues, "the large size of the human brain at birth is one of the defining features of our species, and yet comes at a price…among the primates, humans have particularly difficult births, with high rates of maternal and fetal morbidity and mortality…approximately 287,000 maternal deaths occurred in 2010 worldwide, and complications during delivery, including obstructed labour, were significant contributors."

So, this is indeed a scientific valid criticism, by Saxon, of *Sex at Dawn*: Saxon provides many examples of **infanticide**, for instance in gorillas, where an alpha male gorilla will try to kill the offspring of a female he had sex with, if he gets the *impression* that the female had sex with, and thus could have offspring from, other male gorillas, a clear case of territoriality and of what humans often call jealousy. Such behaviors are widely seen in other primate taxa, and in other mammals. So, yes, our "natural jealousy" clearly plays a huge role in **human sexuality**, and should not be completely neglected in philosophical discussions and political and social agendas. This is precisely one of the main aims of the present book: to take into account our evolutionary history within such broader societal discussions. But the type of extreme jealousy that is so widespread in our cultures is sure *not* just "an hardwired part of our human nature," because it is bombarded to us since a very young age, within all the prevailing narratives of "romantic love" seen in paintings, books, and movies—including those for children, such as *Cinderella*, *The Mermaid*, and *Snow White*—that repeat over and over tales of "the only one," "if you love me truly you would not be with someone else," and "I will suffer and even kill myself if you are with others." In summary, while *Sex at Dawn* was indeed wrong about jealousy being *chiefly* a social construction, it is true that the pathological extreme type of "**monogamous-jealousy**" that is so rampant in our "modern" societies—to which Tallis refers in *The Incurable Romantic*—is indeed partially the product of Western social constructions, being much less extreme in many nomadic hunter-gatherer societies.

So, let's now discuss in more detail the first option listed above, used by the ones that have more power to avoid jealousy and still be able to be with many partners: *using their authority to oblige each of those partners to only have sex with them*. This was commonplace at the beginning of agriculture, as can be seen in the stories of the **Old Testament**, in which nearly all males had more than one wife: the wife was of course expected—that, is, often obliged—to be only with "her master," as women became seen, and literally treated, as the property of men. The same still applies nowadays in various countries such as Afghanistan, in which **polygyny** is prevalent because men use mostly tales not related to the Old Testament but to the Koran to justify the oppression of women, stressing the fact that each and every organized religion can be, and often is, used to oppress people at different moments in time. A similar but even more extreme resolution of the "**natural polygamous sexual desire *versus* natural jealousy paradox**" was applied in cases where a single ruler had a disproportionally huge power: the **Harem**, and its many analogues. While many people think immediately about the harems within older Islamic cultures—often to justify or a posterior confirm their biases and prejudices against Muslims—various of these analogues occurred in many regions of the globe: **Ashota**, the emperor of the **Mauryan Dynasty of India**, had an harem of around 500 women, while in **China** some emperors had **Hougongs** that had up to thousands of women, and the Aztec ruler **Montezuma II** had more than 400 concubines, as explained in Croutier's 1991 book *Harem – The World Behind the Veil*, and Singh's 2008 book *A History of Ancient and Early Medieval India*. So, clearly, the common denominator here is not the Islamic culture, but instead agricultural societies in which there is an enormous subjugation and oppression of women by men, something that was also the case in Western countries for most of their history but fortunately started to slowly change in the last centuries.

In fact, not only the existence, per se, of those societies in which men were/are able to "solve" this "natural paradox" by imposing **polygynous marriages** to women, but also the detailed study of what those women say illustrate the hugely important role played by jealousy in human societies, particularly in agricultural ones, and shows that jealousy is indeed not entirely a social construct. This is because the social constructs and norms that were mainly created by men but also propagated by women—for instance due to the **misogynistic role commonly played by mothers-in-law** in many of such **polygynous societies**—are based on narratives that state that this is "how life should be lived" and that it is actually "good for women" that will be "protected" by living in this way, and so on. But, despite that, this is not what most women report to be feeling, when they live such a type of live in places where such narratives are predominant: instead, they often report to be *very jealous and anxious*, particularly the "first-wife," which is of course the one that in the long term will spend less time, and have less sex, with the polygynous husband and that will see him with more new partners. Within the psychological studies done to women in such societies, a term that appears over and over is effectively the "**first wife syndrome**." Of course, this "syndrome" is not exclusively due to jealousy—as in many of those cases those women never truly desired/loved their husbands-, but also to other complex socio-cultural and politico-economic factors faced by those women. As explained in Al-Krenawi's 2013 paper "*Mental health and polygamy*":

> Historically, many factors have been identified that appear to perpetuate **polygamy**…men may have higher mortality rates than women because of disease, warfare, and the occupational dangers associated with hunting, ocean fishing, migrant labor, and other activities…. One can infer that the higher mortality rate of males may be responsible for an increase in polygamy. One study of the **Ngwa Igbo** [which is an agricultural group] in Nigeria has identified five basic reasons for men to practice polygamy; it allows the Ngwa husband: (1) have as many children as he likes; (2) heighten his prestige and boost his ego among his peers; (3) enhance his status within his community; (4) ensure sufficient working hands to perform the necessary farm work and other labor; and (5) satisfy his sexual urges. In the Middle East, one risk factor for **poor mental health** among millions of women may be found in the practice of polygamy. Although accurate data regarding its precise prevalence are not readily available, polygamous marriage is known to be a common family structure in the Middle East. One explanation for polygamy in the Middle East is embodied in **Islam** as a religion that permits a man to marry up to four wives. **Pro-polygamy Muslim thinkers** insist that men have to be fair to their wives.
>
> One aspect of fair treatment centers on spending **Al-Qaradawi**. The practical considerations revolving around polygamous families in **Muslim Arab society** are diverse. **Polygamous wives** may live together in the same house, or in separate households. A senior wife is defined as any married woman who is followed by another wife in the marriage. A "junior wife" is the most recent wife joining the marriage. This unique family structure forces cooperation between the wives in the household chores and the fields (in rural areas), while they are subject to the husband's authority and in constant competition over his love, attention and financial resources. Studies conducted in different countries have shown that polygamy can lead to **co-wife jealousy**, competition, and unequal distribution of household and emotional resources, and generate acrimony between co-wives and between the children of the different wives. They have also shown that polygamy is associated with **mental illness** (in particular, **depression** and **anxiety**) among women and children…[and] found a disproportionate number of women in polygamous marriages (mostly senior wives) among psychiatric outpatient and inpatient populations in Kuwait. A recent Turkish study found that the participants from polygamous families, especially senior wives, reported more psychological distress…first wives in polygamous families experience a major psychological crisis. Another finding is that women in polygamous marriages report **low self-esteem** and less **life satisfaction** than women in monogamous marriages….
>
> The present study is the first to examine the psychological…family function, marital satisfaction, life satisfaction and degree of agreement with the practice of polygamy among polygamous women with a control group from monogamous women in Syria…[the] sample consisted of 136 women, 64 from polygamous families and 72 from monogamous families. Sixty-two point five percent of the women from polygamous families were "senior wives" – their husbands' first wives; thirty-four point three were second wives and 3.2% were third wives. [Our] findings revealed that women in polygamous marriages experienced lower self-esteem, less life satisfaction, less marital satisfaction and more **mental**

> **health symptomatology** than women in monogamous marriages. Many of the mental health symptoms were different; noteworthy were elevated somatization, depression, hostility and psychoticism and their general severity index was higher. Furthermore, "first wife syndrome" was examined in polygamous families, comparing first with second and third wives in polygamous marriages. Findings indicated that first wives reported on more family problems, less self-esteem, more anxiety, more paranoid ideation, and more psychoticism than second and third wives.
>
> Polygamy's evident characteristic of competition and jealousy among co-wives is commonly observed within plural marriage communities. This seems predictable, as co-wives are likely to have very limited private time with the lone husband they share, and thus might vie for his attention and favor. In some polygamous communities, women's self-worthiness is linked to the number of children they bear and, therefore, having time with their husband is also critical to promote their status within the family and community. Studies showed that in certain contexts, jealousy between co-wives can escalate to intolerable levels, resulting in physical injuries sustained by the women, and **suicide** attempts amongst the women. The present research points out some concerns in relation to the degree of agreement with the practice of polygamy. The majority from both groups of women does not agree with polygamy. Only a small percent agree with the practice of polygamy under some circumstances, or agree. One important difference was that about 4.6% of the polygamous participants agree with the practice of polygamy, compared to 0% of their counterparts. Those women who practice it may seek to legitimate polygamy as a way of coping with the associated problems in their lives. Moreover, the notion of self-sacrifice has a cultural and political dynamic in the Arab culture, and the need to maintain a relationship for the sake of the children is a significant motivator for many women.

One interesting aspect of Al-Krenawi's 2013 paper is that it always uses the term **polygamy**, despite the fact that this term can be related to many types of phenomena, **polygyny** and **polyandry** being just two of them, as it is a **multimale-multifemale type of sexuality**, and so on. This is done by many authors, probably even realizing they are doing this, because they basically assume that polygamy equals polygyny by accepting a priori imaginary tales about women being less sexual. Moreover, by doing this, when such authors say they are comparing the mental illnesses of women and other items under "polygamy" versus under "monogamy," they are presenting a very distorted discussion of these issues, because they are actually comparing monogamy not with polygamy as a whole, but to the most oppressive and unfair type of polygamy imposed to women. So, it is completely normal that women that are obliged by men/cultural traditions to accept polygynous marriages are discontent with them, and suffer more mental illnesses than women that live within societies that tend to favor other models, such as the "monogamous marriage model" that is mostly imposed in Western countries or the multimale-multifemale type of polygamy displayed by many nomadic hunter-gatherer societies.

Let's now analyze a bit in more detail what is known about the models of sexuality displayed by different hunter-gatherer societies. An criticism that Saxon makes about *Sex at Dawn* is that "almost all peoples presented by Ryan and Jetha are settled horticulturalists or are otherwise not representative of pre-agricultural ancestors" and that they offer "no explanation…as to why marriage exists at all." Contrary to what is suggested by Saxon, *Sex at Dawn* does take into account marriage, following the explanation defended by most anthropologists and historians: that its origins in most, if not all human cultures had mainly to do with for instance economic—such as the distribution of material objects among people/their offspring—social—for example, bounding between families, bands or other groups of people—and/or teleological—religious/"purpose of women is," and so on—factors. It mainly had nothing to do, *originally*, with "true romantic love" or with "sexual desire" between the people involved, particularly the women, as explained in Coontz's book *Marriage, a History*.

A brilliant point raised in *Sex at Dawn* is the following: if monogamy and monogamous marriages were so "natural" to humans, why would we need threats of fire and brimstone and, in some cultures, even death to impose them to people? Eating is something clearly "natural" for our species, as is drinking water or sleeping, for instance. Do we need to have threats of fire and brimstone, and even to kill people, to oblige them to eat, drink water, or sleep? It is clear that in societies where polygynous marriages are common practice, a huge number of women do not agree with those types of marriages: they are *obliged to accept them*, by the misogynistic society in which they live. For instance, in the **Na of southwestern China**—said by some researchers to be the only society known in which marriage is not a significant institution—brothers and sisters live together, jointly raising, educating, and supporting the children of the sisters. What a coincidence that in a society where marriage is not *imposed* by peers/the society, instead of the type of pair-bonding that is said to be "natural" for our species by authors such as Saxon, one actually sees the type of organization predicted to be more "natural" by *Sex at Dawn*. In fact, as noted by Coontz, even when some Na couples do practice a more public relationship—for instance, the man comes to the woman's home earlier in the evening, more openly and more regularly than in the usual sexual affair—"the partners owe each other nothing," exactly as it would be predicted by *Sex at Dawn*.

I do agree that *Sex at Dawn* sometimes overemphasizes the similarity of nomadic hunter-gatherers with bonobos. But it should be stressed that they mostly do that when they are referring to sexual behavior, not to social organization as a whole. Still, studies of **body size** clearly indicate that **sexual dimorphism** between modern women and men is on average less prominent today than it was three million years ago, and than it is in modern common chimpanzees and bonobos, although it is clearly more prominent than in the socially monogamous gibbons. Moreover, as pointed by Saxon, a study of **promiscuity** and the primate **immune system** indicated that a higher white blood cell count is found in species where females mated with more males and that humans have counts more closely aligned with the polygynous **gorillas**, and secondarily with the socially monogamous **gibbons**. So, these empirical data seem to suggest that, in terms of our "true/ancestral human nature"—if it could be completely decoupled from our "nurture," what is not the case—our species is somewhere in between polygamous chimpanzees and socially monogamous gibbons. But these data do *not* suggest at all that we are truly at the level of gibbons. A main difference between us and gibbons is precisely that, as noted in *Sex at Dawn*, in humans the marriage/pair-bonding is often imposed to people via social norms/pressure, force, or even death threats. No field study of gibbons has shown that they need to be physically forced by older members of their families, or by threats of stoning, fire.

So, what applies to gibbons also applies obviously to common chimpanzees and bonobos, and this fact help us to answer Saxon's question about why pair-bonding is so common in humans but not in common chimpanzees: at least in part, it is surely precisely because they are not physically forced by their societies to do so. Let's go back to some of the reasons that anthropologists and historians often list to explain the origins of marriage: economic, social, and/or teleological reasons. About the last item, chimpanzees don't build and believe in complex teleological stories telling them that having a single partner is the "right thing to do" and that not doing that is a "sin." About economic reasons, it is clear that humans, even well before agriculture, were very different from chimpanzees concerning the possession of material goods, and I do agree that this is a point that is often neglected by people that try to over-romanticize hunter-gatherer societies, as if they had "nothing" and were completely "un-materialistic."

As explained in McBrearty and Brooks' 2000 and Driscoll and Thompson's 2018 papers, humans used stone tools for at least 2.6 million years ago, and employed projectile points, pigment processing, long distance exchange, bone tools, barbed points, notational pieces, microliths, beads, and even produced complex images at least some dozens of thousands years ago, much before the rise of agriculture (Fig. 5.7). We also know that human burials, as well as at least some

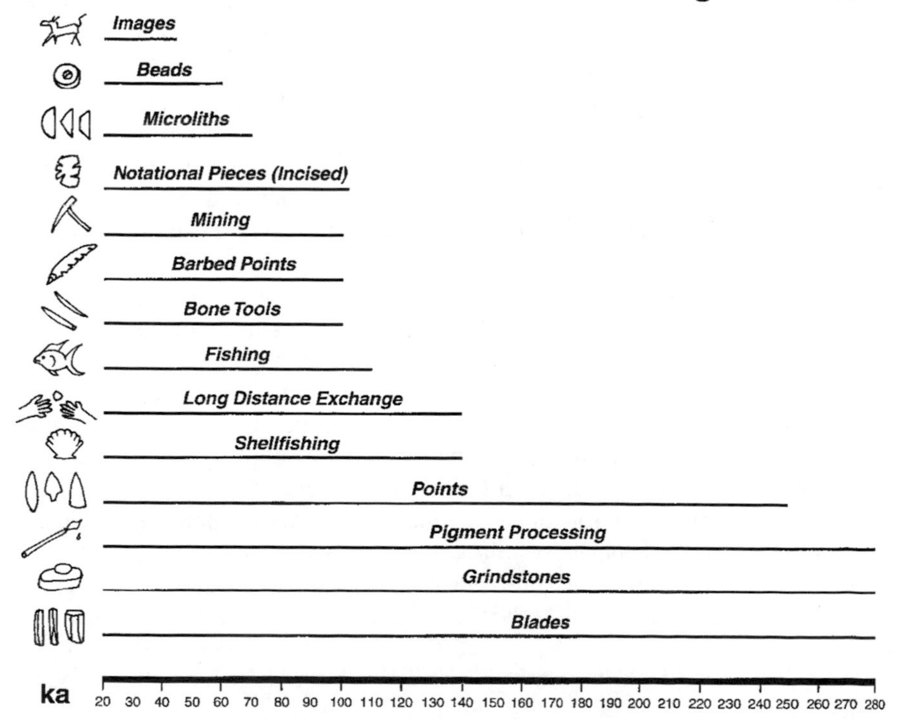

Fig. 5.7 Evolution of so-called modern behaviors in human evolution (ka means thousands of years ago)

kind of religious rituals, also appeared well before agriculture. So, taking into account these data, and the fact that marriage was, at least for millennia after the rise of agriculture, unequivocally more linked to materialistic/economic aspects and teleological narratives than to the "true will" of the individuals, it is not difficult to envisage that humans became different from chimpanzees concerning **pair-bonding and marriage** through "nurture" while retaining "natural" sexual desires similar to those of the **multimale-multifemale chimpanzees**. There is indeed plenty of evidence that in many hunter-gatherer societies, marriage was/is also mainly related to the exchange of material goods and/or to establishing connections between members of those societies.

One illustrative example is given in Coontz's *Marriage, a History*: "the **Bella Coola** [small indigenous group of Canada] and the **Kwakiult societies of the Pacific Northwest** provide a striking example of how establishing connections between kin groups sometimes took precedence over sexual or reproductive issues in determining marriage—if two families wished to trade with each other, but no suitable matches were available, a marriage contract might be drawn up between one individual and another's foot or even with a dog belonging to the family of the desired in-laws.

The 2016 book *Domestic Tensions, National Identities – Global Perspectives on Marriage, Crisis, and Nation*, edited by Kristin Cerello and Hanan Kholoussy, provides some interesting examples about these issues. It states that "states around the world…have sought to produce and promote certain types of **marital arrangements**, with the goal of maintaining *control* over their citizens/subjects…(also) turning men and women into husbands and wives, marriage has *designated* the way both sexes act in the world." One of the cases given in that book concerns the 1930s in colonial **Burma**: "at a time in which nationalist leaders were coming of age and the Burmese economy was struggling…resentment of Indian men's status in the colonial regime and access to financial resources ran high…the rapidly expanding press seized hold of this discontent and created a crisis, castigating the Burmese women who were willing to marry these men for sullying their race and religion…it thus became women's responsibility to marry the 'right' kind of man to guarantee the success of the Burmese national project." Similarly, "in China…after the communist came to power in 1949, the party's first piece of national legislation promoted marriage as a means of embodying **socialist values**, bringing the state more fully into the intimate lives of its citizens." Another example concerns "the long history of marriage in Nigeria…even though monogamous unions were the norm, pre-colonial Nigerians also sanctioned a wide range of marital practices, including **polygamy** – [including] polyandry…the community and ritually sanctioned encouragement of women to move from husband to husband, as they wished, while staying legally married to each – deity-to-human marriages, and woman-to-woman marriages…British colonial officials, as well as the post-independence Nigerian government, however, sought to quash this diversity in favor of a single acceptable model of heterosexual, male-headed unions."

Let's analyze in a bit more detail some of the examples about **hunter-gatherer groups** provided by Saxon in *Sex at Dusk*. She states: "in hunter-gatherers we often find groups with fluid membership mostly comprising mobile **nuclear family units** that are linked through **marriage** into a much larger network…looking at one hunter-gatherer culture, the **Hadza**…**serial monogamy** is the best way to describe the mating system though perhaps 20% of Hazda stay married to the same person their whole life…**divorce** is often due to the pursuit of **extramarital affairs.**" When asked "what happens if someone finds out his or her spouse has had an affair? 38% of men and women said the man would try to kill the other man, 26% said a woman would fight with the other woman, 20% said a man would leave is wife, and 13% said the woman would leave her husband." So, once again, this is another clear example in which both women and men may be "faithful to his/her spouse" not because they have no sexual desire for others, but instead because they would be abandoned, physically attacked, or even killed if they would do so. Apart from possible **social pressures** from the broader community, it does seem that **jealousy** might be playing a substantial role in these outcomes, as defended by Saxon. About the **Aka**, another hunter-gatherer African group, Saxon states that divorce is common among them, and that "in 64% of divorces the cause was the spouse sleeping with, searching for, or finding another mate."

The last example I will provide here from *Sex at Dusk* concerns again the**!Kung**, which Saxon cites from the book *Nisa – The Life and Words of a!Kung Woman*: "sex is also recognized as tapping some of the most intense and potentially explosive of human emotions – especially where **extramarital attractions** are concerned. In such cases, sex is considered outright dangerous: many affairs that become known lead to **violence**, which, in the past, sometimes resulted in death…therefore people that participate in such relationships are extremely careful and discreet…the best assurance against complications arising from love affairs is *not to be found out*." Of course, Saxon uses this text to criticize *Sex at Dawn*, but the reality is that this case instead clearly supports the main point of *Sex at Dawn*, because it refers directly to a!Kung woman that explains how such affairs are not *supposed* to happen and can even lead to death, but *still happen* frequently because women—as men—*did have* extramarital attractions. That is, such natural bodily *desires* and *attractions* are so strong that, despite the risk of death, they still occur very frequently, the only "solution" then being to make sure the sex affair is not known to the broader community. *Do as I say, not as I do*: a very common, longstanding motto concerning human sexuality, indeed.

As noted by Saxon, estimates by researchers that studied the **Aka**, between the ages of 18 and 45 years, indicate that married people have sex 2–3 times a week and 'do it' three times within each of those nights. These numbers are *much higher* than those reported for married people in almost any "modern" society. To give an example, cohort analyses of **sexual frequency** in a representative sample of 7483 people living in the U.S. in 1988 show that younger couples who have been married for less than 2 years tend to have sexual relationships an average of 2–3 times a week, whereas older couples and those who have been married for more than 2 years tend to engage in intercourse only 1.5 times per week, as explained in a 1995 paper by Call and colleagues. So, the average numbers for all the sampled married Aka between 18 and 45 years are only seen in the sampled U.S. that are *both* young *and* married for less than 2 years, that is in a very small proportion of the total married U.S. population. According to a 2016 book chapter by Theiss, "a variety of life events can also contribute to temporary conditions in a relationship that impede sexual contact…one obvious life event that interferes with sexual intercourse is **pregnancy** and **childbirth**…couples report engaging in intercourse four to five times per month during pregnancy, engage in no sex for the first month after the birth of the child, and return to levels of sexual activity similar to what they had experienced during pregnancy within the first year after birth."

Theiss adds: "in another study, new parents engaged in intercourse only one to two times per month for up to four years after the birth of the first child, often citing tiredness as the main barrier to physical intimacy". Moreover, in the country that is often said to be the 'technologically most developed one', Japan, there is on average a frequency of 45 intercourses per year, that is 0.9 times per week, with about only 27% reporting to have sex more than once a week, as noted by Sechiyama in *Tokyo Business today*. These numbers are consistent with those from a *Sexual Well Being Global Survey* involving 26,032 respondents from 26 countries—minimal age of 16, as summarized in a 2009 paper by Wylie: "two thirds (67%) of participants described having sex once a week, with people in Greece (89%) and Brazil (85%) having sex most often…sex happened the least for participants in Japan (38%)." So, what is striking is not only the fact that in Japan almost two thirds of people don't have any sex during a whole week: this also applies to one third of people all around the globe living within so-called modern societies, contrary to the much higher sexual frequencies almost always found within nomadic hunter-gatherer groups.

In *Marriage, a History*, **Coontz**, states:

> **Eskimo** couples often had **cospousal arrangements**, in which each partner had sexual relations with the other's spouse. In **Tibet** and parts of **India**, **Kahmir**, and **Nepal** a woman may be married to two or more brothers [**polyandry**], all of whom share sexual access to her…. The children of Eskimo cospouses felt that they shared a special bound, and society viewed them as siblings. Such different notions of marital rights and obligations made divorce and remarriage less emotionally volatile for the Eskimo than it is for most modern Americans. Among Tibetan brothers who share the same wife, **sexual jealousy** is rare. The expectation of mutual fidelity is a rather recent invention. Numerous cultures have allowed husbands to seek sexual gratification outside marriage. Less frequently, but often enough to challenge common preconceptions, wives have also been allowed to do this without threatening the marriage. In a study of 109 societies, anthropologists found that only 48 forbade extramarital sex to both husbands and wives…in some societies the choice to switch partners rests with the woman. Among the **Dogon** of **West Africa**, young married women publicly pursued extramarital relationships with the encouragement of their mothers. Among the **Rukuba** of **Nigeria**, a wife can take a lover at the time of her first marriage….
>
> Several small-scale societies in **South America** have sexual and marital norms that are especially startling for Europeans and North Americans. In these groups, people believe that any man who has sex with a woman during her pregnancy contributes part of his biological substance to the child. The husband is recognized as the primary father, but the woman's lover or lovers also have paternal responsibilities, including the obligation to share food with the woman and her child in the future. During the 1990s researchers taking life histories of elderly **Bari** women in **Venezuela** found that most had taken lovers during at least one of their pregnancies. Their husbands were usually aware and did not object. When a woman gave birth, she would name all the men she had slept with since learning

she was pregnant, and a woman attending the birth would tell each of these men: you have a child.... When Jesuit missionaries from France first encountered the North American **Montagnais-Naskapi** Indians in the early 17th century, they were shocked by the native women's sexual freedom. One missionary warned a Naskapi man that if he did not impose tighter controls on his wife, he would never know for sure which of the children she bore belonged to him. The Indian was equally shocked that this mattered to the Europeans: "you French people...love only your own children, but we love all the children of our tribe", he replied.

Clearly, the statements of Coontz—using both examples not cited and the same case studies cited by *Sex at Dusk* and *Sex at Dawn*—are much more in line to what is stated in the latter book. Actually, even concerning the topic that was not discussed in as much detail as it should have been in *Sex at Dawn*—**jealousy**—Coontz wisely shows that although jealousy is something natural that evolved from our primate ancestors' territoriality, the extreme importance given to it in our "modern" societies *also* depends on the teleological narratives and cultural norms that are prevalent within those societies. This is made clear by an illustrative example provided by her: "Tibetan brothers who share the same wife, sexual jealousy is rare...the *expectation of mutual fidelity* is a rather recent *invention.*" In other words, jealousy is natural, but if it is minimized within the cultural and teleological narratives that are predominant where you live, you will probably not give it so much importance as you do in countries in which the concepts of jealousy or 'honor' are over emphasized. This is another example of how "nature" and "nurture" are always deeply intermingled. Coontz further explains:

According to the protective or **provider theory of marriage**... –still the most widespread *myth* about the origin of **marriage** – ...women and infants in early human societies could not survive without the men to bring them the meat of woolly mammoths and protect them, from marauding saber-toothed tigers and from other men seeking to abduct them. But males were willing to protect and provide only for their 'own' females and offspring they had a good reason to believe were theirs, so a woman needed to find and hold on to a strong, aggressive male. One way a woman could hold a mate was to offer him exclusive and frequent sex in return for food and protection. According to the theory, that is why women lost the estrus cycle that is common to other mammals, in which females come into heat only at periodic intervals. Human females became sexually available year-round, so they were able to draw men into long-term relationships. In anthropologist Robin Fox's telling of the story "the females could easily trade on the male's tendency to want to monopolize (or at least think he was monopolizing) the females for matting purposes, and say, in effect, 'okay, you get the monopoly...and we get the meat'". The male willingness to trade meat for sex was, according to Fox, "the root of truly human society". Proponents of this protective theory of marriage claim that the **nuclear family**, based on sexual division of labor between the male hunter and the female hearth keeper, was the most important unit of survival and protection in the Stone Age. People in the mid-20th century found this story persuasive because it closely resembled the **male breadwinner/female homemaker family** to which they were accustomed.

The idea that in **prehistoric times** a man would spend his life hunting only for the benefit of his wife and children, who were dependent solely upon his hunting for survival, is simply a projection of 1950s marital norms onto the past. But since the 1970s other researchers have poked holes in the protective theory of marriage...they argued that...the origins of marriage lay not in the efforts of the women to attract protectors and providers but in the efforts of men to control the productive and reproductive powers of women for their own private benefit...some (researchers) denied that male dominance and female dependence came from us from our primate ancestors. Studies of actual human hunting and gathering societies also threw doubt on the male provider theory – in such societies, women's foraging, not men's hunting, usually contributes the bulk of the group's food. Nor are women in foraging societies tied down by child rearing. One anthropologist, working with an African hunter-gatherer society during the 1960s, calculated that an adult woman typically walked about twelve miles a day gathering food, and brought home anywhere from 15 to 33 pounds. A woman with a child under two covered the same amount of ground and brought back the same amount of food while she carried her child in a sling, allowing the child to nurse as the woman did her foraging. In many societies women also participate in hunting, whether as members of communal hunting parties, as individual hunters, or even in all-female hunting groups. Today most paleontologists reject the notion that early human societies were organized around dominant male hunters providing for their nuclear families…[instead] there is strong evidence that in many societies…[particularly] sedentary agriculturalists… – ***marriage** was indeed a way that men put women's labor to their private use.* Women's bodies came to be regarded as the properties of their fathers and husbands.

Indeed, the fact that men created narratives about marriage and "women's naturally tendency to do domestic tasks" as a way to put women's labor to their private use continues to have dramatic implications nowadays. Even today, in the twenty-first century, in the vast majority of countries, including so-called developed countries such as the Netherlands, U.S., Germany, and so on, women have much less leisure time than men, on average, as shown in Fig. 5.8, in which strikingly there is a single exception, Norway. In a nutshell, Coontz shows that the "man-the-provider" theory is a very old imaginary tale mainly based on **misogynistic biases**. This is also shown in many other books and specialized papers that provide extensive literature reviews about what empirical data tell us, including the influential 2009 book *Mothers and Others* of **Hrdy**, which explains that such outdated "man-made" tales include **Darwin**'s ideas:

From the outset, they [evolutionists] assumed…that (the) provider must have been her [the wife's] mate, as Darwin himself opined in *The descent of man and delection in relation to sex*. Indeed, it was the hunter's need to finance slow-maturing children, Darwin thought, that provided the main catalyst for the evolution of our big brains: "the most able men succeeded best in defending and providing for themselves and their wives and offspring", he wrote…it was the offspring of hunters with "greater intellectual vigor and power of invention" who were most like to survive". According to this logic, males with bigger brains would have been more successful **hunters**, better providers, and more able to obtain mates and thereby pass their genes to children whose survival was underwritten by a better diet. Meat would subsidize the long childhoods needed to develop larger brains, leading eventually to the expansion of brains from the size of an australopithecine's to the size of Darwin's own.

Thus did the '**hunting hypothesis**' morph into one of the most long-standing and influential models in anthropology…at the heart of the model lay a pact between a hunter who provided for his mate and a mate who repaid him with sexual fidelity so the provider could be certain that children he invested in carried at least half of his genes. This '**sex contract**' assumed pride of place as the "prodigious adaptation central to the success of early hominids"…[However] as it became apparent that among foragers (like the **!Kung**) plant foods accounted for slightly more calories than meat, researchers started paying more attention to female contributions…[also] when Frank Marlowe interviewed **Hazda** still living by hunting and gathering, he learned that only 36% of children had fathers living in the same group…a hemisphere away, among **Yanomano** tribespeople in remote regions of Venezuela

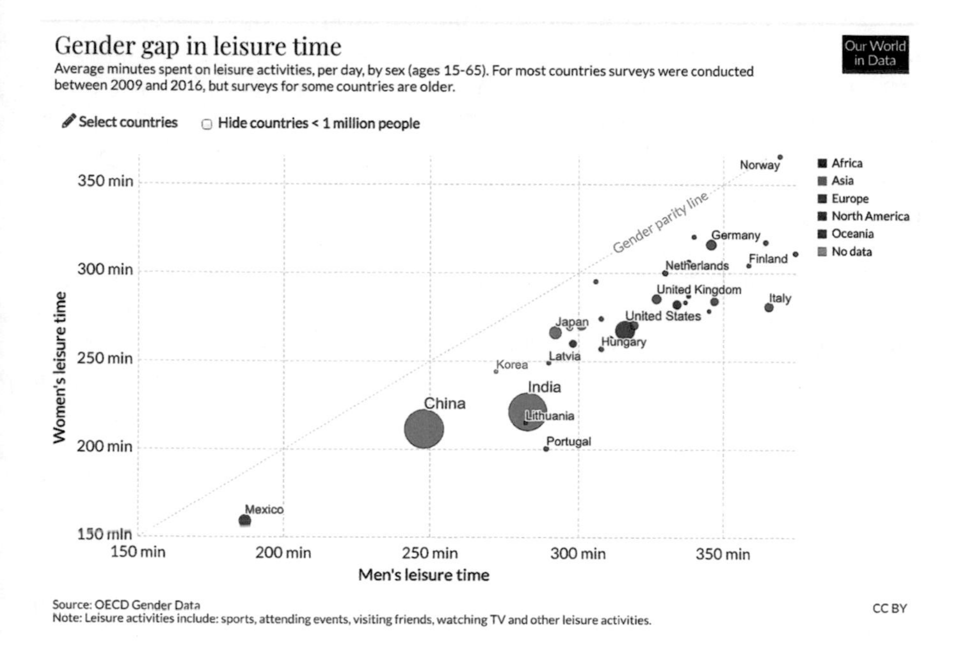

Fig. 5.8 An interesting way to look at how people spend their time globally is through the lens of gender. This figure shows gender disparities in time spent within 33 countries—women spend nearly three times more in unpaid care work compared to men, a whopping total of 1.1 trillion hours each year, which means a lot less leisure time, except in Norway, which is a very rare example. (Source: OECD Gender Data. Note: Leisure activities include: sports, attending events, visiting friends, watching TV and other leisure activities)

> and Brasil, the chance of a 10-year-old child having both a father and a mother living in the same group was 1/3, while the chance that a Central African **Aka** youngster between the ages of 11 and 15 was living with both natural parents was closer to 58%...pity the **Ongee** foragers living on the Andaman Islands: none of the 11- to 15-years-olds in that ethnographic sample still lived with either natural parent...When anthropologists reviewed a sample of 15 traditional societies, in 8 of them the presence or absence of the father had no apparent effect on the survival of children to age 5, provided other caregivers in addition to the mother were on hand in a position of help.

Basically, no living apes can be considered to be mainly "active hunters." **Chimpanzees** only hunt actively, and eat for instance meat of monkeys, a few occasions, and this only contributes to a very minor part of their diets. **Gorillas** and **orangutans** almost never eat meat in their natural environments. So, meat, or hunting, is not so crucial for most ape groups: females, and their kids, can live very well without them, without needing the males to "save them with their hunted meat." Moreover, **active hunting** has been only common in the last one or two million years of the six million years of our human lineage. Meat was very likely important before that time, but humans were mainly scavengers, a bit like hyenas are today, that is they ate meat more opportunistically, rather than by actively hunting and killing animals. This, of course, does not mean that **hunting meat** has not been crucial for many nomadic **hunter-gatherer societies** since active hunting started: clearly, it is, particularly for societies that depend hardly on eating animals like the **Nunamiut**, semi-nomadic people located in Inland Alaska. But in *most* nomadic hunter-gatherer societies, hunting actually contributes to less than 50% of the diet, as noted by Coontz and Hrdy and summarized in Kelly's 2013 book *The Lifeways of Hunter-Gatherers* (see Fig. 5.9). Moreover, even within those *few* groups in which hunting contributes more than 50% of the diet, sharing meat is usually very common, particularly in small nomadic hunter-gatherer groups, as noted by Kelly. So, even within those groups, one cannot say that if a father dies or leaves his group, the mother and kids will necessarily die from starvation: in most cases, they will still obtain at least some—often enough, as noted by Hrdy—meat from other hunters. In other words, even in those groups, one can say that the dependence is more on the *group as a whole*, more than on a specific "father-the-hunter-savior."

This also relates to the **power networks** and **diet types** that are prevalent within different groups of **hunter-gatherers**. As noted in Kelly's 2013 book, the **type of diet** will be a major determinant on "whether men's or women's foraging determines camp movement: **Agta** camp members, for example, discuss for hours or days whether to move, and foraging efforts of men *versus* women play a role in these debates." The "effective **foraging distance** for plants is shorter, in general, than it is for large game since many plant foods provide lower return rates than those of large game…since large game is usually procured by men, women's foraging

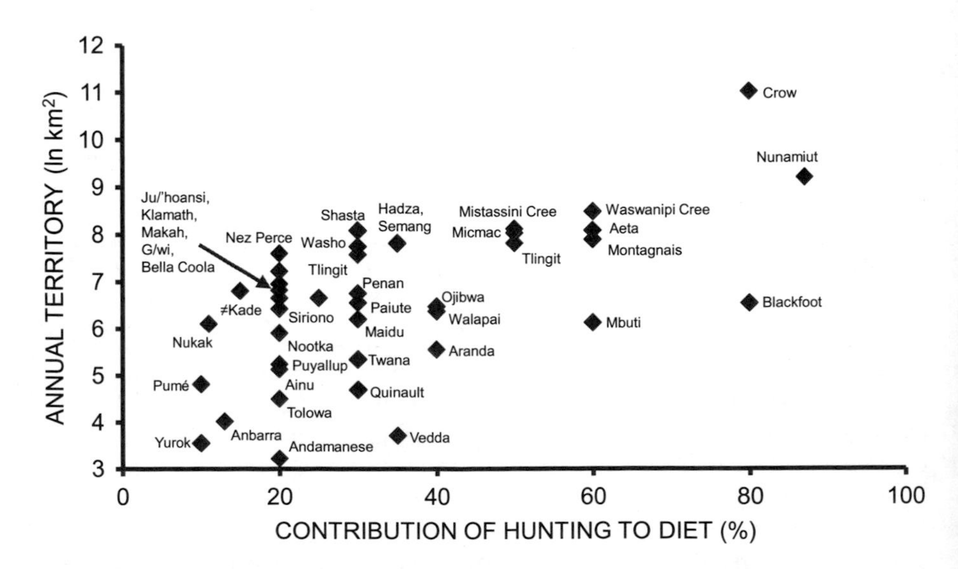

Fig. 5.9 The size of foragers' annual ranges plotted against the percent dependence on hunting—in general, as the dependence on hunting increases so does the size of the exploited territory. Overall, it is interesting to see how in the vast majority of non-agricultural hunting contributes less than 50% to their diet, contrary to the narrative of the "man-the-hunter-savior"

should normally determine when and where camp is moved." Therefore, "among the Agta, since hunting depends on mobile animals, it is not an important consideration [in determining moves]...men and women freely voice their opinion on residence change, but women, who must carry out the most gathering, have the final say." In fact, in numerous hunter-gatherer groups, such as within the so-called Central **African forest pygmies**, "in many cases, women are equal partners with men in...collective hunts," as explained by Bahuchet in the first chapter of the 2014 book *Hunter-Gatherers of the Congo Basin.* Such type of power relationships between gender are, therefore, generally very different from what happens in more **sedentary groups**, and particularly in **agricultural groups**. The **rise of agriculture** has resulted in an increased **subjugation of women**, who started to "work" a lot: in general, many hours per day, and doing harder tasks. This is something that people—either men or women—that criticize the so-called new trend of women going to work, instead of "staying at home as they should," completely neglect: apart from a few women that were married with rich men and thus did not work—for instance in the Victorian era, about which Hollywood movies are so obsessed with—the vast majority of women that lived in this planet never "stayed at home." A 2018 article published online by the World Economic Forum, entitled "*Women grow 70% of Africa's food – but have few rights over the land they tend*," explains that "studies show that women account for nearly half of the world's smallholder farmers and produce 70% of Africa's food...yet, less than 20% of land in the world is owned by women". It further points out that "over 65% of land in Kenya is governed by customary laws that discriminate against women, limiting their land and property rights...this means that women farmers have to access land through either their husbands or sons...sometimes these male family members move to the cities leaving women behind to tend the land – land they have no right to own, use as collateral or sell the output without consent from the men."

5.5 History of Marriage, Love, and Sex

If a man can possess a woman sexually – really possess – he won't need to control her ideas, her opinions, her clothes, her friends, even her other lovers. (Toni Bentley)

So, how can we infer, using empirical data, what was the predominant ancestral type of sexual organization of humans, since we split from chimpanzees about six million years ago, until the rise of agriculture, or **Neolithic Revolution**, about 10,000 years ago, that is during the vast majority—99%—of our human history? Within the biological sciences, there are three major ways of inferring behaviors of extinct individuals. A crucial one is to use the so-called outgroup method: to take into account what happens in the closest living relatives of the ingroup that is being studied. In this case, the ingroup is all humans—extinct and living—and the outgroup therefore is the **chimpanzees**, which include both the common chimpanzees (*Pan troglodytes*) and bonobos (*Pan paniscus*). About this point, there is no doubt at all: any biological anthropology textbook will tell you that in both these chimpanzee subgroups, each female and male tend to have multiple sexual partners of the opposite sex-and commonly also from the same sex, particularly in bonobos-not only during their whole life, but also during much shorter periods of time, including a single menstrual cycle, as noted in Dorus et al.'s 2005 paper. Another piece of evidence would be from paleontology and archeology, that is from data obtained from human fossils or archeological material. Unfortunately, that type of evidence is relatively scarce, but we do have data about **sexual dimorphism** for humans that lived millions years ago, based on their fossilized bones, such as the difference of body size between females and males, which is often a good approximation about the predominance of sexual or **social monogamy** versus polygamy—be it **polyandry**, **polygyny**, or a **multifemale-multimale type of organization**.

In socially monogamous primates such as **gibbons,** the body size of males tends to be somewhat similar to that of females, in part because there is not so much competition between males to get a female, nor of females to get a male, because for most of the time a single female lives with a single male, except when they "fool around." In contrast, in primates that have multimale-multifemale sexual "promiscuity" such as **chimpanzees**, and particularly within a polygynous type of organization such as that of **gorillas**, the males tend to be much bigger than the females, and to have other distinct traits such as bigger canine teeth, or crests in the skull, and so on, which they use to "show off"—and much less commonly to fight—against other males to compete for females. We know that until at least three millions of years or so, humans tended to have a sexual dimorphism similar to that seen in chimpanzees, as was the case in *Australopithecus afarensis* individuals such as the famous female Lucy and the males from that species. So, one can say that these two types of independent evidence—outgroup comparison and paleontology/archeology—strongly suggest that for a huge amount of time during our evolutionary history, we did not have a sexually monogamous—almost no mammal species have—nor a socially monogamous type of sexual organization. In fact, while in our species *Homo sapiens*—which only appeared about 250 thousand years ago or so—sexual dimorphism is not, on average, as high as it is in chimpanzees and in those humans that lived three millions of years ago, it is still significantly higher than it is normally—although there are some exceptions in the natural world of course—seen in typical socially monogamous species, such as gibbons.

The third type of evidence that can be used to discuss these topics concerns the study of human groups that are still living nowadays, but that can give us, somehow, an idea of how other human groups might have lived before the rise of agriculture. With this goal, most academics would refer to human groups that are living, or have lived until recently, a hunter-gatherer style of life, and particularly a nomadic one. This is because a non-agricultural lifestyle was predominant for six millions of years of human evolution, before the Neolithic Revolution. But, paradoxically, this third type of evidence is the one that has created more controversies between researchers, which tend to take different sides on the so-called **Hobbes versus Rousseau** debate. As noted above, this debate is often oversimplified. **Rousseau** was far from being someone that always defended the oppressed or that had what we now call "left-wing" ideas. For instance, he was as misogynistic as most men were in the eighteenth century, as can be seen in an excerpt of *Emile*: "The man should be strong and active, the woman should be weak and passive…the woman is made to please and be in subjection to man…if woman is made to please and be in subjection to man, she ought to make herself pleasing in his eyes and not provoke him to anger…she should compel him to discover and use his strength." More examples of Rousseau's misogyny are given in specialized works such as a paper published by Darling and De Pijpekamp's in 1994, in which they argue that "Rousseau's endorsement of male domination and his illiberal views of rape, punishment and the education of women have been seriously underestimated by…commentators who tend to produce expositions of his work that evade, ignore or marginalize this 'darker side' of his educational philosophy."

So, trying to go away from the oversimplifications, biases, and stereotypes of the "**Hobbes versus Rousseau's debate**," what can we say about the sexual and social organization of hunter-gatherers? The most succinct and adequate way to summarize in a single sentence the history of **marriage**, and its connection to—or, better say, mostly separation from—"**romantic love**" and "**sexual passion**," is to cite a brilliant sentence of Coontz' 2005 *Marriage, a History*: "*people have always loved a love story…but for the most of our past our ancestors did not try to live in one*". What Western anthropologists studying non-agricultural groups often named "marriage"—following the narratives commonly accepted in their countries of origin—does often not correspond at all to the Western concept of marriage. Among the **Warao**, a group living in the forests of Brazil, ordinary relationships are suspended periodically and replaced by ritual relations in which adults are free to have sex with whomever they like. It is likely that the famous **Brazilian carnival** has some historical links with such rituals, where "everything is allowed" for a day/few days. In the **Pirahã**, another indigenous group of Brazil, "marriage" is not allowed outside of their "tribe," so they have long kept their gene pool refreshed by permitting their women to sleep with outsiders, in order to avoid inbreeding. Many other examples of so-called marriage that apparently seem to involve some kind of *apparent* "**social monogamy**" but that actually do not involve at all a true **sexual monogamy** are provided in *Sex at Dawn*, and reported in many other books and specialized papers. A recent, fascinating book on this subject was published by Valentine and colleagues in 2017—*The Anthropology of Marriage in Lowland South-America: Bending and Breaking the Rules*—which, as its official description appropriately summarizes, reveals "that individuals in Amazonian cultures often disregard or reinterpret the marriage rules of their societies – rules that anthropologists previously thought reflected practice." The book considers "not just what the rules are but how people in these societies negotiate, manipulate, and break them."

In other words, what has been called by anthropologists as "marriage" in nomadic hunter-gatherers societies often has nothing to do with a "contract" in which both a woman and a man are supposed to be completely monogamous and, as it still happens in some agricultural states, in which women are somehow seen as a "property" of a men. Such a type of "contract" mainly begun, or at least became more widespread/strict, in sedentary societies and in particular since the rise of agriculture and organized religions: as Van Schaik and Michel's note in *The Good Book of Nature*, at that moment a married woman indeed begun to be mainly seen as the property of a man or his family, together with his house and his other material goods. So, while this type of "**property contract**" is still seen in many agricultural states, particularly those where a polygynous model is prevalent, how come that it changed so dramatically to the type of marriage that is now so commonly seen in Western countries? In Coontz's *Marriage, a History*, she explains that in "our Stone-Age ancestors…marriage spoke to the need of the larger group…it converted strangers into relatives and extended cooperative relations beyond the immediate family or small band by creating far-flung networks of in-laws." Then, "as civilizations got more complex and stratified…marriage became a way through which elites could board or accumulate resources…in Europe, from the early Middle Ages through the 18th century, the dowry a wife brought with her at marriage was often the biggest infusion of cash, goods, or land a man would ever acquire…finding a husband was usually the most important investment a woman could make in her economic future." In a way, one can thus say that this sounds a bit like the "whore woman-man the provider" idea of Saxon's reactionary *Sex at Dusk*. But there are key differences. As explained by Coontz, this crucial need, by a woman, to have a single man to take care of her precisely begun only after the rise of agriculture, and became particularly prominent within the U.S.' 1950s and 1960s model of "domestic women"—that is, it does not apply at all to 99% of our evolutionary history. Moreover, Coontz's sentence is "finding a husband…," not finding a sexual partner. Even nowadays in so-called civilized countries, including Western ones, some women make the economic investment of "finding a rich, often older" husband, but that does not mean at all that they will necessarily only *have sex* with him. These are two completely different things, as we will further discuss below.

Concerning this topic about women "finding their rich husband," or their "blue prince," it is so sad for me to now realize that movies that I saw—and much liked—when I was a child, such as the ***Little mermaid*** or ***Snow white***, were portraying women in such misogynistic ways. These and many other movies for kids tended to show women that just wanted to find a "prince" and be happy forever, as if that was the only goal within the life of a "good woman." Actually, many of the **Disney movies** we saw, and that many kids still see, are not only highly misogynistic but also extremely racist. This was implicitly recognized by the **Disney TV channels**, which now—after the ***Black Lives Matter*** protests in 2020—include a **content advisory notice for racism** for films such as ***Dumbo***, ***Peter Pan***, and ***Jungle Book***, stating: "this film includes negative depictions and/or mistreatment of people or cultures…these **stereotypes** were wrong then and are wrong now." The notice adds that, rather than removing the content, they "want to acknowledge its harmful impact, learn from it and spark conversation to create a more inclusive future together." As an example, the crows that teach **Dumbo** to fly are commonly described as having highly "exaggerated black voices"—what, in itself, is a stereotypical statement because obviously there is no such thing as a "black" voice, is the voice of a forest hunter-gatherer of Cameroon similar to that of an African-American in the U.S.? Moreover, the leader of those crows is called Jim, in an apparent reference to the racist **Jim Crow's segregation laws** that were used in the U.S. for decades.

Also, in the same movie, there is the ***Song of the Roustabouts***, which goes on like this—"we work all day, we work all night, we never learned to read or write, we're happy-hearted roustabouts," while showing African-Americans doing manual labor.

Contrary to the examples concerning such Disney movies, many people nowadays argue that the differences regarding how women and men consider they are *supposed*—versus how they *truly want or desire*—to behave are no longer dictated so much by culture, but rather by "true natural gender differences." This is because—they argue—our current popular culture is now mainly against gender stereotypes or even has a tendency to follow **feminist ideas**—an idea defended by many people, particularly reactionary men, the far right party *Vox* in Spain being an illustrative example of this. Such people seem to forget that kids are still bombarded from a very young age with not only those old animated movies but also newer ones that are equally misogynistic: most of them still are, indeed, with some notable exceptions, such as the movie ***Moana***. Regarding this topic, there was recently a controversy in my native country, Portugal, because in a TV ad, a parent called the daughter a "princess."

Feminist movements protested, and many people, including friends of mine, answered that this was an example of the "politically-correct" extremism of the **Me Too movement** and other similar "radical"—in their words—movements. However, this is a clear case where it is indeed correct to point out that, because of the still prevailing misogynistic idea that girls mainly just want to be "princesses," such an ad is indeed contributing to propagate such a wrong stereotype, and therefore the *status quo*. I do not see TV ads calling boys "princes." Why? And even if there were, the connotation would not be the same anyway, because a man is often a prince because he is the son of a king, not because the only and single thing he wants in life is to find, and marry with, a princess. Significantly, in what might seem a coincidence but is not, a few weeks after that controversy started in Portugal I went to Santiago de Chile—a city with very active and vocal youth movements—and I read, written in a graffiti in a Cultural Center, the sentence: "I am not your princess." Profoundly different countries, but similar teleological narratives and gender *Neverland* stereotypes being fought by a younger generation that no longer wants to follow them blindly, as their parents and grandparents so often did (see also Box 5.5).

Box 5.5: Self Illusion, Gender Stereotypes, and Child Development

In his book *The Self Illusion*, Hood explains:

> Although not cast in stone, **gender stereotypes** do tend to be perpetuated across generations. Many parents are eager to know the sex of their children before they are born, which sets up gender expectations such as painting the nursery in either pink or blue. When they eventually arrive, newborn baby girls are described mainly in terms of beauty, whereas boys are described in terms of strength. In one study, adults attributed more anger to a boy than to a girl reacting to a jack-in-the-box toy even though it was always the same infant. Parents also tend to buy gender-appropriate toys with dolls for girls and guns for boys. In another study, different adults were introduced to the same child wearing either blue or pink clothes and told that it was

either Sarah or Nathan. If adults thought it was a baby girl, they praised her beauty. If they thought it was a boy, they never commented on beauty but rather talked about what occupation he would eventually have. When it came to play, they were boisterous with the boy baby, throwing him into the air, but cuddled the baby when they thought it was a girl. In fact, the adults seemed to need to know which sex the baby was in order to play with them appropriately. Of course, it was the same baby, so the only difference was whether it was wearing either blue or pink. It is worth bearing in mind the association of the colour blue is only recent – a hundred years ago it would have been the boys wearing pink and the girls wearing blue. With all this encouragement from adults during the early months, is it any surprise that, by two years of age, most children easily identify with their own gender and the roles and appearances that they believe are appropriate? However, this understanding is still very superficial. For example, up until four years of age, children think that long hair and dresses determine whether you are a boy or girl. We know this because if you show four-year-olds a **Ken Barbie Doll** and then put a dress on the male doll, they think that he is now a girl. By six years, children's gender understanding is more sophisticated and goes over and beyond outward appearances. They know that changing clothes and hair does not change boys into girls or vice versa. They are already demonstrating an understanding of what it means to be essentially a boy or a girl. When they identify gender as a core component of the self, they will tend to see this as unchanging and foundational to who they and others are.

As children develop, they become more fixed in their outlook about what properties are acquired and what seem to be built in. For example, by six years, children think that men make better mechanics and women are better secretaries. Even the way parents talk to their children reinforces this generalized view of what is essential to gender. For example, parents tend to make statements such as 'boys play soccer' and 'girls take ballet' rather than qualifying the statements with 'some boys play soccer' or 'some girls take ballet'. We can't help but fall into the **gender trap**. Our interaction with children reinforces these gender divisions. Mothers tend to discuss emotional problems with their daughters more than with their sons. On a visit to a science museum, parents were three times more likely to explain the exhibits to the boys than to the girls. And it's not just the parents. Teachers perpetuate gender stereotypes. In mixed classes, boys are more likely to volunteer answers, receive more attention from teachers and earn more praise. By the time they are eight to ten years old, girls report lower self-esteem than boys, but it's not because they are less able. According to 2007 UK National Office of Statistics data, girls outperform boys at all levels of education from preschool right through to university. There may be some often-reported superior abilities in boys when it comes to mathematics but that difference does not appear until adolescence, by which time there has been ample opportunity to strengthen stereotypes. **Male brains** are different to **female brains** in many ways that we don't yet understand (for example, the shape of the bundle fibres connecting the two hemispheres known as the **corpus callosum** is different), but commentators may have overstated the case for biology when it comes to some gender stereotypes about the way children should think and behave that are perpetuated by society.

Stereotypes both support and undermine the **self illusion**. On the one hand most of us conform to stereotypes because that is what is expected from those in the categories to which we belong and not many of us want to be isolated. On the other hand, we may acknowledge the existence of stereotypes but maintain that as individuals we are not the same as everyone else. Our self illusion assumes that we could act differently if we wished. Then there are those who maintain that they do not conform to any stereotypes because they are individuals. But who is really individual in a species that requires the presence of others upon which to make a relative judgment of whether they are the same or different? By definition, you need others to conform with, or rebel against. Consider another universal self stereotype – that of **male aggression**. Why do men fight so much? Is it simply in their nature? It's an area of psychology that has generated a multitude of explanations. Typical accounts are that males need physically to compete for dominance so that they attract the best females with whom to mate, or that males lack the same negotiation skills as women and have to resolve conflicts through action. These notions have been popularized by the '**women are from Venus, men are from Mars**' mentality. It is true that men have higher levels of **testosterone** and this can facilitate aggressive behaviour because this hormone makes you stronger. But these may be predispositions that cultures shape. When we consider the nature of our self from the **gender perspective**, we are invariably viewing this through a lens, shaped by society, of what males and females should be. Males may end up more aggressive but surprisingly they may not start out like that. Studies have shown equal levels of physical aggression in one-year-old males and females, but by the time they are two years of age, boys are more physically aggressive than girls and this difference generally continues throughout development. In contrast, girls increasingly rely less on physical violence during conflicts but are more inclined to taunting and excluding individuals as a way of exerting their influence during bullying. Males and females may simply differ in the ways in which they express their aggression.

External events influence our choices in ways that seem to be somewhat out of our control. But what of the internal conflicts inside our heads? The self is a constructed web of interacting influences competing for control. To live our lives in society, we need to inhibit or suppress disruptive impulses, thoughts and urges. The drives of fleeing, fighting, feeding and fornicating are constantly vying for attention in situations when they are not appropriate. What of our reasoning and control when we submit to these urges? It turns out that the self-story we tell our selves can become radically distorted. In what must be one of the most controversial studies of late, Dan Ariely, he wanted to investigate how our attitudes change when we are sexually aroused. First, he asked male students to rate their attitudes to a variety of issues related to sex. For example, would they engage in unprotected sex, spanking, group sex and sex with animals? Would they have sex with someone they did not like or a woman over sixty? He even asked them whether they would consider spiking a woman's drink with drugs so that she would have sex with them. In the cold light of day, these men answered absolutely no way would they engage in these immoral acts. These were upstanding males who valued women and had standards of behaviour. Ariely then gave them $10, a copy of *Playboy* magazine and a computer laptop protectively wrapped so that they could answer the same questions again with one hand, while they masturbated with the other in the privacy of their dorm rooms. When they were sexually aroused something monstrous happened. These men were turned into animals by their passion. Ariely discovered these student Dr. Jekylls turned into veritable Mr. Hydes when left alone to pleasure themselves. They were twice as likely to say that they would engage in dubious sexual activities when they were sexually aroused. More worrying, there was a fourfold increase in the likelihood that they would drug a woman for sex! Clearly when males are thinking with their 'little brain', they tumble from their moral high ground, which they can usually maintain when they are in a non-aroused state. As Ariely put it, 'prevention, protection, **conservatism** and **morality** disappeared completely from their radar screen'. It was if they were a different person.

As noted in Saini's 2017 book *Inferior – How Science Got Women Wrong, and the New Research That's Rewriting the Story*, gender stereotypes, particularly those regarding "**women's inferiority**," are so deeply embedded in our societies that they are internalized by a large number of women themselves. She wrote: "in a study published in 2012, psychologist Corinne Moss-Racusin and a team of researchers at Yale University explored the problem of bias in science by conducting a study in which over a hundred scientists were asked to assess a résumé submitted by an applicant for a vacancy as a laboratory manager…every résumé was identical, except that half were given under a female name and half under a male name." When "they were asked to comment on these supposed potential employees, scientists rated those with female names significantly lower in competence and hireability…they were also less willing to mentor them, and offered far lower starting salaries." Strikingly, as recognized by the authors of the study, "the gender of the faculty participants did not affect responses, such that female and male faculty were equally likely to exhibit bias against the female student." That is, "**prejudice** is so steeped in the culture of science, their results suggested, that women are themselves discriminating against other women," concluded Saini. Going back to Coontz's book, she explains that since the Neolithic revolution "until the late 18th century…certainly, people fell in love…*sometimes* even with their own spouses…but most societies around the world saw marriage as far too vital and economic and political institution to be left entirely to the free choice of the two individuals involved, especially if they were going to base their decisions on something as unreasoning and transitory as love." Accordingly, as she notes, some centuries ago it was considered unwise, in many regions of the globe, to get married if one was "deeply in love":

> In ancient **India**, falling in love before marriage was seen as a disruptive, almost antisocial act…. The Greeks thought lovesickness was a type of insanity…. In the Middle Ages the French defined love as a "derangement of the mind"…. In the Chinese language the term *love* did not traditionally apply to feelings between husband and wife…it was used to describe an illicit, socially disapproved relationship…in the 1920s a group of intellectuals invented a new word for love between spouses because they thought such a radical new idea required its own special label. In Europe, during the 12 and 13th centuries, **adultery** became idealized as the highest form of love among the aristocracy…according to the Countess of Champagne, it was impossible for true love to "exert its powers between two people that are married to each other"…. Catholic and Protestant theologians argued that husbands and wives who loved each other too much were committing the sin of idolatry…. Although medieval Muslim thinkers were more approving of **sexual passion** between husband and wife…they also insisted that too much intimacy between husband and wife weakened a believer's devotion to God.

Coontz states that "only in the 17th century did a series of political, economic, and cultural changes in Europe begin to erode the older functions of marriage, encouraging individuals to choose their mates on the basis of personal affection and allowing couples to challenge of outsiders to intrude upon their lives." Then, "in the late 18th century, and only in Western Europe and North America…the notion of free choice and marriage for love triumph as a cultural ideal…people began to adopt the *radical new idea* that love should be the most fundamental reason for marriage and that young people should be free to choose their marriage partners on the base of love." According to her, "two seismic social changes spurred these changes in marriage norms…first, the spread of wage labor made young people less dependent on their parents for a start in life…second, the freedoms afforded by the market economy had their parallel in new political and philosophical ideas." **Enlightenment** "influential thinkers across Europe championed individual rights and insisted that social relationships, including between men and women, be organized on the basis of reason and justice rather than force…believing the pursuit of happiness to be a legitimate goal, they advocated marrying for love rather than wealth or status." According to Coontz, the **sentimentalization of the love-based marriage** in the eighteenth and nineteenth centuries and its **sexualization** in the twentieth century each represented a logical step in the evolution of this new approach to marriage. A dramatic illustration of that tendency happened in the 1950s, when "for the first time, a majority of marriages in Western Europe and North America consisted of a full-time homemaker supported by a male earner…also new in the 1950s was the cultural consensus that everyone should marry…for hundreds of years, European rates of marriage had been much lower…the baby boom of the 1950s was likewise a departure from the past, because birthrates in Western Europe and North America had fallen steadily during the previous hundred years."

By examining the pictures and Ads that were widely disseminated by the media in the 1950s and 1960s in the U.S., one can see that they played a crucial role within the rise of the obsession with the socially constructed idea of "**happy marriage**" and for the establishment of the cultural consensus about—or, I would say, the cultural imposition of—the extremely **romanticized idea of love**. Within such idea: (1) everybody should get married; and (2) husband and wife should not only be *always* tremendously in love, but they should also *always* have a huge desire for each other, and obviously for nobody else, *and* moreover should basically spend most of their leisure time, and be always happy during that time, with their spouse. That is, it is not only a concept of "**happy, highly romanticized, and highly sexualized marriage**" but also the radicalization of the notion of an "**anti-social marriage**," as will be discussed below. Marriage, as well as media Ads in general, became more and more sexualized in the twentieth century. But such sexualization was still following the **Victorian misogynistic paradigm** that the man is the main, proactive sexual player that "truly" wants sex, and wants it with many women. Such a narrative continues indeed to be followed, even within the twenty-first century, not only by a huge portion of the broader public, but also in many so-called scientific papers and books, such as *Sex at Dusk* and many similar evolutionary psychology pseudoscientific publications, as we have seen above.

Such misogynistic ideas about women being "less sexual" are social constructions that begun to be particularly prominent and popularized relatively recently in human history, namely, within Western countries, in the eighteenth and nineteenth centuries, chiefly during the Victorian era and the publication of Darwin's books. As Coontz wrote:

> Throughout the Middle Ages women had been considered the lusty sex, more prey to their passions than men. Even when **idealization of female chastity** began to mount in the 18th century…few of its popularizers assumed that women totally lacked sexual desire. Virtue

was thought to be attained through self-control; it was not necessarily innate or biologically determined. The beginning of the 19th century, however, saw a new emphasis on **women's innate sexual purity**. The older view that women had to be controlled because they were inherently more passionate and prone to moral and sexual error was replaced by the idea that women were asexual beings, who would not respond to sexual overtures unless they had been drugged or depraved from an early age. This cult of female purity encouraged women to internalize limits on their sexual behavior that 16th and 17th century authorities had imposed by force. Its result was an extraordinary **desexualization of women** – or at least of good women, the kind of woman a man would want to marry and the kind of woman a good girl would wish to be. Given the deeply rooted **Christian suspicion of sexuality**, however, the new view of women as intrinsically asexual improved their reputation. Whereas women had once been considered snares of the devil, they were now viewed as sexual innocents whose purity should inspire all decent men to control their own sexual impulses and based appetites. The cult of female purity offered a temporary reconciliation between the egalitarian aspirations raised by the **Enlightenment** and the fears that equality would overturn the social order….

Still, the new ideals of **marriage** and womanhood were more than simply a face-lift for **patriarchy**…women gained the moral right to say no to sex even though husbands continued to have legal control over their bodies. Furthermore, the cult of female purity was not…a one-way street…men were called upon to emulate this purity themselves…although they were thought to have strong sexual urges, these were seen as unfortunate impulses…animal passions…that had to be controlled and repressed. The **sentimentalization of marriage** made domestic violence much less acceptable as well. In addition, the unique moral influence accorded to mothers contributed to an expansion of educational opportunities for women. (However) for many women brought up with the idea that normal females should lack sexual passion, the wedding night was a source of anxiety or even disgust. In the 1920s, Katharine Davis interviewed 2200 American women, most of them born before 1890. Fully a quarter said they had initially been 'repelled' by the experience of sex. Even women who did enjoy sex with their husbands reported feeling guilt or shame about their pleasure, believing that 'immoderate' passion during the sex act was degrading. Of course many women *did* have sexual urges, and the struggle to repress them led to other problems…physicians regularly massaged women's pelvic areas to alleviate 'hysteria'…medical textbooks of the day make it clear that these doctors brought their patients to orgasm…in fact the **mechanical vibrator** was invented at the end of the 19th century to relieve physicians of this tedious and time-consuming chore.

Of course, this Victorian way of seeing women as "asexual," mainly promoted by men, brought huge problems for women but also created some problems for men: if "decent wives" are chiefly asexual, with whom could their husbands fulfill their "male-animal sexual instincts"? As noted in *Sex at Dawn*, they could of course "fool around," but if they would do so with "decent women…[they would be] threatening familial and social stability" anyway, so "they were expected to purge their lust with prostitutes…19th century philosopher **Arthur Schopenhauer** observed that 'there are 80000 prostitutes in London alone – and what are they if not sacrifices on the **altar of monogamy**'?" Indeed, prostitution, as alcohol, do provide further evidence against narratives about the supposed "naturalness" of human monogamy. The next time you hear—or say—"I was unfaithful because I drunk too much alcohol, *I was not myself*," just remind that person—or yourself—that actually she/he/you were in a way truly more "herself/himself/yourself" than in most other occasions. This is because it is known, scientifically, that alcohol generally leads to the "loosening up" of social norms and the disinhibition of "sexual impulses," thus often leading to higher sexual promiscuity and risky sexual behavior, as explained in a 2018 paper by Cottonham and colleagues.

Figure 5.11 shows another crucial, and related, point made in *Sex at Dawn*, and discussed in detail by Coontz in *Marriage a History*. This concerns the fact that in the 1950s and 1960s in the U.S., for the first time in the history of that country, a middle-class man often received a salary that was high enough to provide for him, his wife, and children. That is, the husband could do this without necessarily needing financial help from his parents/siblings. This, together with other dramatic social changes that occurred in the twentieth century, including the excessive romanticizing of love discussed above, led to a situation that was mainly unprecedented in human history: a "**nuclear family**" now became to be chiefly formed only by husband, wife, and their children. Of course, in occasions such as Thanksgiving in the U.S., and during vacations and so on, this "nuclear family" did often meet with some other members of the broader family. But more and more figures such as the one shown in Fig. 5.11 begun to be seen in the last decades: as can be seen there, even during Christmas one would often only see the father, mother, and children together.

This figure and Figs. 5.10 and 5.12 are also illustrative examples of the profound link between the type of **materialistic capitalism** so in vogue in the 1950s and 1960s in the U.S., and this new notion of a small "nuclear family," which in reality acted as the **crucial, "nuclear unit of consumption**." In general, I tend to not agree with **Marxist narratives**, particularly because they tend to be based on teleological tales that are disconnected from empirical scientific evidence (see for instance Boxes 2.6 and 7.11). But **Carl Marx**'s idea—obviously developed well before the 1950s and 1960s—that ultimately the kind of "nuclear family" shown in Fig. 5.11 would one day perform a critical, ideological function for capitalism was indeed not too far stretched. As noted by Coontz, this strong association between a reduced "nuclear family" and "home life," as opposed to the rest of the family and society, was chiefly the exaggeration and culmination of a concept that had its roots in the nineteenth century, when it started to be popularized, with people transferring "their loyalty from house (which before meant family *sensu lato*, or lineage) to home…the home begun to be seen as a sanctuary of domestic love…an oasis…a hallowed place…a quite refuge from the storms of life…only in the **sanctuary of home** do we find disinterested love…ready to sacrifice everything at the **altar of affection.**" Religious "as well as **secular moralists** came to view doing well for one's family as more important than doing good for society." The "**ideology of home and domesticity** [also] imposed new constraints on men's participation in the public sphere and curtailed many of men's traditional associations with other men…now their activities, apart from work and formal political occasions, began to center on their homes and the company of their wives and children…everyday meals were taken *en famille*, with the husband, wife and children eating alone." Even "when **honeymoons** to romantic places such as Niagara falls came into vogue, couples often took friends or relatives along for company…after 1850, however, the honeymoon increasingly became a time for couples to get away from others."

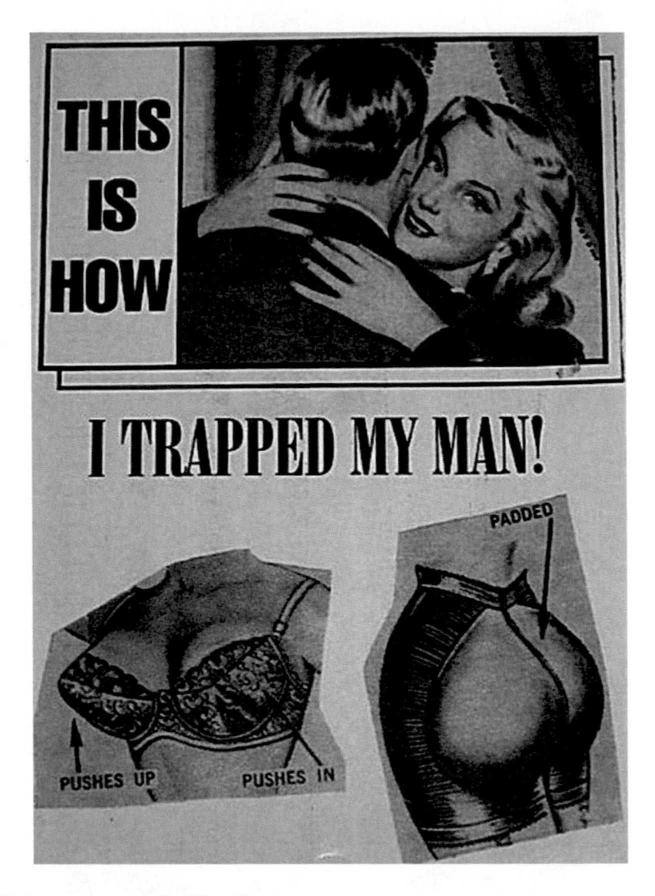

Fig. 5.10 1916 American ad for Woodbury soap

Fig. 5.11 Nuclear family celebrating Christmas

Fig. 5.12 Kenwood Chef Ad from the 1950s

In fact, this latter case is somewhat a historical reversion, because originally the "honeymoon" was, in some cultures, precisely also to get away from others, but due to a completely different—and surely less romantic—reason. As noted in Ackerman's *Natural History of Love*, "we think of the **honeymoon** as romantic days of sensual bliss under a tropic sky, but the original honeymoon had a more somber purpose…right after a groom captured or bought a bride, he disappeared with her for a while, so that her family and friends couldn't rescue her…by the time they found the couple, the bride would already be pregnant." A similar scenario occurs with the now highly romanticized "wedding ring." Ackerman explains: "**wedding rings** are very ancient…(made) of plain, strong metal, so it didn't break, which would have seemed a disastrous omen…naturally, there are romantic interpretations of the band—that it symbolizes harmony, unending love, and so on—but it originally served as a notice and a reminder that a woman was bond to her husband (who didn't have to wear a ring)." Another example of a current fairytale about romantic love that is contradicted by empirical data, in this case by physiological evidence, concerns a sentence that many have heard or read many times, either in real life or in movies or books: "if you really love me, you would not have sex with others." As noted by Charles Ryan in *The Virility Paradox*, "**testosterone** affects the desire for sex and the urge to reproduce…my observation is that it has relatively little to do with love," which is instead commonly linked to **oxytocin**.

He further notes: "when you remember that testosterone is the chemical of aggression, dominance, and reduced empathy, it makes perfect sense…it is rare indeed for my patients who undergo **testosterone-lowering therapy** to lose affection for their romantic partners…in fact, I've had many patients…who have fallen in love while taking these treatments." As he explains, "psychologists have determined that the failure of relationships is tied to higher testosterone levels…men with higher testosterone levels are more likely to get divorced, have extramarital affairs, and be violent with their spouses." Amazingly, this also applies to **women's testosterone levels**: "there's a negative correlation between the woman's testosterone level and the self-reported satisfaction of both members of the couple." He notes: "testosterone levels are 33% lower in fathers of newborns than in non-fathers, while ocytoxin levels are more than 25% higher…more oxytocin induces fathers to spend more time with their kids…higher levels of testosterone…can make fathers less likely to respond, or simply slower to respond, to the crying of their infant children." He therefore then cites Margaret Mead, who said "something like 'everyone should have three marriages: one for sex, one for kids, and one for companionship…all three can be with the same person'." He concludes that, while he suspects that "most people still at least try to find one person to fulfill all their needs, those needs, and the priorities given to them, are not only different from person to person, but they also might well change over time."

Going back to Coontz, she explains that "prior to the mid-19th century families rarely gathered together at holidays such as **Christmas**…that day had been a time to visit friends and neighbors and to greed a constant round of mummers, who came to the door dressed in costumes and expected to be offered food and alcohol." Also, "the doctrine…that men and women had innately different natures and occupied separate spheres of life [e.g. women being less sexual]…held back the inherently individualistic nature of the '**pursuit of happiness**' by making men and women dependent upon each other and insisting that each gender was incomplete without marriage." The apogee of this insistence was the 1950s and 1960s, when individuals were being hugely pressured—by their families, friends, politicians/ideologists and the media in general—not only to marry when they were young, but also to have children and chiefly live only with their spouses and those children under the concept of a "happy small nuclear family." The pressure was huge for men, as they needed to be sure they could find a job and provide for the whole "nuclear family": finally, in the 1950s and 1960s, the longstanding imaginary tales about "man-the-provider" became completely true. Coontz explain how in the U.S. "marriage was seen as the only culturally acceptable route to adulthood and independence…men who choose to remain bachelors were branded 'narcissistic', 'deviant', 'infantile', or 'pathological'…a 1957 survey in the USA reported that 80% of people believed that anyone who preferred to remain single was 'sick', 'neurotic', or 'immoral'." But, as often happened in human history, the pressure was even higher for women, as they were the ones that would in reality mainly contribute to the formation of the "nuclear family" by not only having kids but also raising them, while being mainly "passive" and "submissive" to the "chief of the family," including having sex with him, and him alone, in order to be able to raise her kids, following the "women-the-asexual" narrative (see Fig. 5.12).

As explained by Coontz, "**the idealization of married love** and the collapse of women's female networks left women more isolated and emotionally dependent than in the past…married women had the home and children, as they had in the 19th century, but they lacked the cultural support and the network of contacts that formed the separate sphere of 19th century women". This huge pressure, particularly the demand for people to always 'feel happy' within such an imposed model of life, as well as "the very features that promised to make marriage such a unique and treasured relationship [actually] opened the way for it to become an optional and fragile one." For instance, "if marriage was about love and lifelong intimacy, why would people marry at all if they couldn't find true love? What would hold a marriage together if love and intimacy disappeared?" She explains that her research shows that a major reason why marriage model begun to collapse in the 1970s was that before the 1970s, most people—particularly the women— could not yet afford to act on their aspirations for love and for personal fulfillment. Including, I would add, the fulfillment of what they truly desired, sexually. Interestingly, as she notes, this trend had started a bit earlier—before it was temporarily reverted within the model that became predominant in the 1950s and 1960s—in the late eighteenth century, when "the insistence that marriage be based on true love and companionship spurred some to call for further liberalization of divorce laws…the USA was simultaneously a world leader in embracing the ideals of marriage and…in divorce rates: between 1880 and 1890 it experienced a 70% increase in divorce." According to Coontz:

> Men [also] had their own complains about the typical family arrangement of the 1950s. In fact Barbara Ehrenreich argues that it was men, not women, who first revolted against the **male breadwinner marriage**…. Robert Lindner wrote that when a men tried to live up to all of society's expectations at work and home, he became 'a slave in mind and body'…. John Keats described the suburbs as 'jails of the soul'. In 1953 **Hugh Hefner** founded **playboy magazine** as a voice of revolt against male family responsibilities…[on the other hand] alongside lessons in femininity and homemaking, the women's magazines of the 1950s and 1960s [paradoxically also] nourished a 'discourse of discontent' by promoting intimacy and self-fulfillment as the purpose of marriage…it was reading about what marriage *ought to be* that many women saw what their own marriages weren't. As early as 1957 **divorce rates** started rising in the United States and several other countries. In fact, one of every three American couples who married in the 1950s eventually divorced…[however] until women had access to safe and effective **contraception** that let them control when to bear children and how many to have, there was only so far they could go in reorganizing their lives and their marriages…only in 1960 did a **birth control pill**, **Enovid**, become commercially available. The **contraceptive revolution** of the 1960s was a much more dramatic break with tradition than the so-called **sexual revolution**…for the first time in history any woman with a modicum of educational and economic resources could, if she wanted to, *separate sex from childbirth*…. By 1978 only 25% of Americans still believed that people who remained single by choice were 'sick', neurotic' or 'immoral', as most had thought in the 1950s…by 1979, 75% of the [U.S.] population thought that it was morally okay to be single and have children…the divorce rate more than doubled between 1966 and 1979. Although women still tend to be more eager than men to enter marriage, they are also more likely to become discontent once married…a survey conducted in the USA during the

> mid-1990s found that a majority of divorced wives said they were the ones who wanted out of marriage…fewer than 25% said their husbands had unilaterally wanted to get out of the marriage…a recent study of divorces that occur after age forty found that 2/3 were initiated by wives.

It is important to note that, well before the 1960s, another form of **sexual contraception** was in fact already available, the condom, although of course condoms did not allow women to have full power of deciding when and how—and how securely, and also with whom specifically—they would use such a contraception, contrary to the birth control pill. As explained in Holland's book *A Brief History of Misogyny*, **condoms** "first became available in London and Paris in the 17th century…through initially used as a prophylactic against venereal infection, the condom was soon functioning as a contraceptive device…[it] represented the first major step towards the transformation of sexual activity into a pursuit that was mainly, not just occasionally, recreational." Holland added: the "ability of women to protect themselves, and avoid pregnancy, challenged the biological determinism that lies behind so much misogyny…the anxiety that this creates, today as in the 17th century, is often disguised in moralizing that such protection makes women even more vulnerable to men's lusts…but it cannot hide the essential fear of women controlling their reproductive fate, thus achieving the autonomy that all misogynists dread."

As explained in Coontz's book, even in the 1950s—often called the "golden age of marriage in the West"—many people were discontent with marriage: a 1957 study of a cross section of all social classes found that only 47% of U.S. married couples described themselves as "very happy." Coontz cites Elaine Tyler May, whom defended the idea that in the 1950s "the idea of a 'working' **marriage** was one that often included day-to-day misery for one or both partners…a 1950 family that looked well functioning to the outside world could hide terrible secrets," such as **domestic violence**. As Coontz explains, "of women interviewed during the late 1950s and 1960s, even those who were content with their marriage almost always wanted a different life for their daughters." In fact, in another clear case that highlights the critical difference between *what one is supposed to do or to say* versus *what one truly wants and desires*—linked to "nurture" versus "nature," respectively, in a oversimplified way—she notes that "a 1962 Gallup poll reported that American married women were very satisfied with their lives…but only 10% of the women in the same poll wanted their daughters to have the same lives that they had." This is a typical problem that researchers face when they try to compare, let's say, happiness levels between married versus non-married, or religious versus non-religious, people, as there is a strong, and often unconscious, **link between self-reporting and cultural biases**. That is, if a person is married and/or religious, within a culture in which there is a strong social pressure to think that this is what is "normal" or "right," that person would often be more prone to self-report that she/he is "happy," because she/he was conditioned to believe that she/he is living life as "it should be lived."

Even in studies about **suicide** it is sometimes difficult to discern if the feeling that one is not living life "as it should be lived" might increase psychological distress, which is often mentioned together with the **absence of social integration** as some of the major risk factors for suicide after divorce. In a 2000 paper, Kposowa used the National Longitudinal Mortality Study of 1979–1989 to estimate the effects of marital status on death from **suicide**. This author concluded that divorced and separated persons were over twice as likely to commit suicide as married persons and suggested that the effect of **divorce** on **suicide risk** may be attributable to the absence of **social integration** and increased **psychological distress**. Interestingly, however, being single or widowed had no significant effect on suicide risk. Also, when data were stratified by sex, it was observed that the risk of suicide among divorced men was over twice that of married men, while among women there were no statistically significant differentials in the risk of suicide by marital status categories. Once again, such empirical data strongly contradict the narrative that women are the ones that most want, and need, to be married, with a single "man-the-provider." Historically, the opposite was almost always the truth: the "**agricultural" model of marriage**, and the other models of marriage constructed since then in so-called civilized countries, were instead created by, and for the advantage of, *men* (see Fig. 8.8). So, accordingly, and not surprisingly, men are still the ones that mainly want to keep such models and that logically tend to suffer more when such models are broken, as they indeed often do in cases of divorce. This is clearly attested, in a nutshell, precisely by the fact that during marriage women (1) are the ones that often feel more depressed and discontent and (2) the ones that tend to initiate divorce, while after divorce, men (3) are the ones that tend to become more depressed and (4) the ones that tend to suicide more. This pattern of married women being in general more depressed than married men is noted in Saxon's *Sex at Dusk*, when she discusses the 1972 book *The Future of Marriage* of the sociologist Jessie Bernard:

> Though men for centuries had railed against marriage it turned out to be very good for their psychological health and the majority of their needs. Men only had two complaints: their economic responsibilities and the constraints on their polygynous desires. The former, Bernard says, was being [after the decade of the 1960s] improved by women increasingly sharing this burden, and the latter by what appeared to be a greater tolerance for extramarital affairs, though she noted that extramarital affairs had historically been tolerated in men anyway, so it was not new. The big problem was to make marriage better for women…. Women suffered due to their low status as wives, their social isolation, and the burden of all the household chores and of having to minister to all the needs of a husband.

As noted in a 2013 paper by Clark and Georgellis, an extensive study based on a survey of 10,000 people in the UK between the ages of 16 and 60 years also showed that women are actually significantly more content than usual for up to 5 years following the end of their marriages, even more so than their own average or "baseline" level of happiness throughout their lives. Although men also felt slightly happier after receiving the decree absolute, the increase was much less marked. Moreover, as explained by Coontz, there are also far-reaching studies that show that "in states that adopted **unilateral divorce**, this was followed on average, by a 20% reduction in the number of married women committing suicide, as well as a significant drop in **domestic violence** for both men and women…[also] higher rates of marital separation lead to lower homicide rates against women." Regarding children, "although 75 to 80% of children recover well and function within normal ranges after divorce, **children from divorced families** have twice the risk of developing behavioral and emotional problems as children from continuously married families…but **children in high-conflict marriages** are often better off if their parents divorce than if they stay together…children also suffer when exposed to constant and chronic low-level friction in a marriage, such as parents not talking to each other, being critical or moody, exhibiting jealousy, or being domineering." This goes against a huge myth that is defended even nowadays by not only many lay people, but even by some so-called specialists: the myth that it is *always* better for kids to live with a married couple. About this subject, Coontz further notes that "sociologist Paul Amato estimates that divorce lowers the well-being of 55–60% of children…but it actually improves the well-being of 40–45%." This issue is also discussed in Browning's 2017 book *Fate of Gender*: "studies have *not* shown that compared to all other family forms, families headed by married, biological parents are best for children."

The intricate complexity of these topics, and the fact that there is clearly not "a single solution for all," is further noted by Coontz, who states that "a 3-year study of married couples in which one partner had mild hypertension found that in happy marriages, the blood pressure of the at-risk partner dropped when couples spent even a couple of extra minutes together…but for those who were unhappily married, a few extra minutes together raised the blood pressure of the at-risk spouse…having an argumentative or highly critical spouse can seriously damage a person's health, raising blood pressure, lowering immune functions, and even worsening the symptoms of chronic illnesses like arthritis." She additionally notes that "unhappy wives have higher rates of depression and alcohol abuse than single women…a bad marriage raises a woman's cholesterol readings and decreases her immune functioning…**unhappily married women** in their forties were more than twice as likely to have medical symptoms that put them at risk for heart attacks and strokes as happily married or never-married females…a long-term study of patients in Oregon even found that unequal decision-making power in marriage was associated with a higher risk of death for women." Indeed, many recent comprehensive studies are increasingly showing that the longstanding narrative that "a married life leads to an healthier life" has some huge flaws—see Luhmann et al.'s 2012, Tumin's 2017, and Kalmijn's 2017 works.

It is true that being married may increase the likelihood of surviving **cancer**, as well as decrease **heart disease** and lead to healthier blood vessels, as shown in a 2014 study of the American College of Cardiology with about 3.5 million people. One reason for this is that single people tend to be more **risk taking**, for instance they tend to go out at night more, in part to find new partners, or to be with friends, and this often involves the consumption of alcohol or eventually of tobacco, sleep less, and having less stable hours to sleep, and so on. However, on the other hand, the higher motivation of single people to find and attract potential mates often leads them to do on average more exercise, be less prone to gain extra weight, eat less fatty food, and so on, than married people, which is in turn related to having on average less incidence of a vast range of medical problems related to having too much weight or being too sedentary, and so on. The authors of these comprehensive studies also point out that, while most people nowadays tend to think that married people "have someone" and single people don't, this actually is not so simple, and often is even the opposite when one refers to larger social networks.

As made clear in Finkel's 2017 book *The All-Or-Nothing Marriage*, empirical studies plainly show that, overall, married people are far less likely than never-married people to see their parents, siblings, neighbors, and friends regularly (see Figs. 5.13, 5.14, and 5.15). This is a logical problem that resulted from the obsession about being almost all the leisure time with the "**small nuclear family**" that

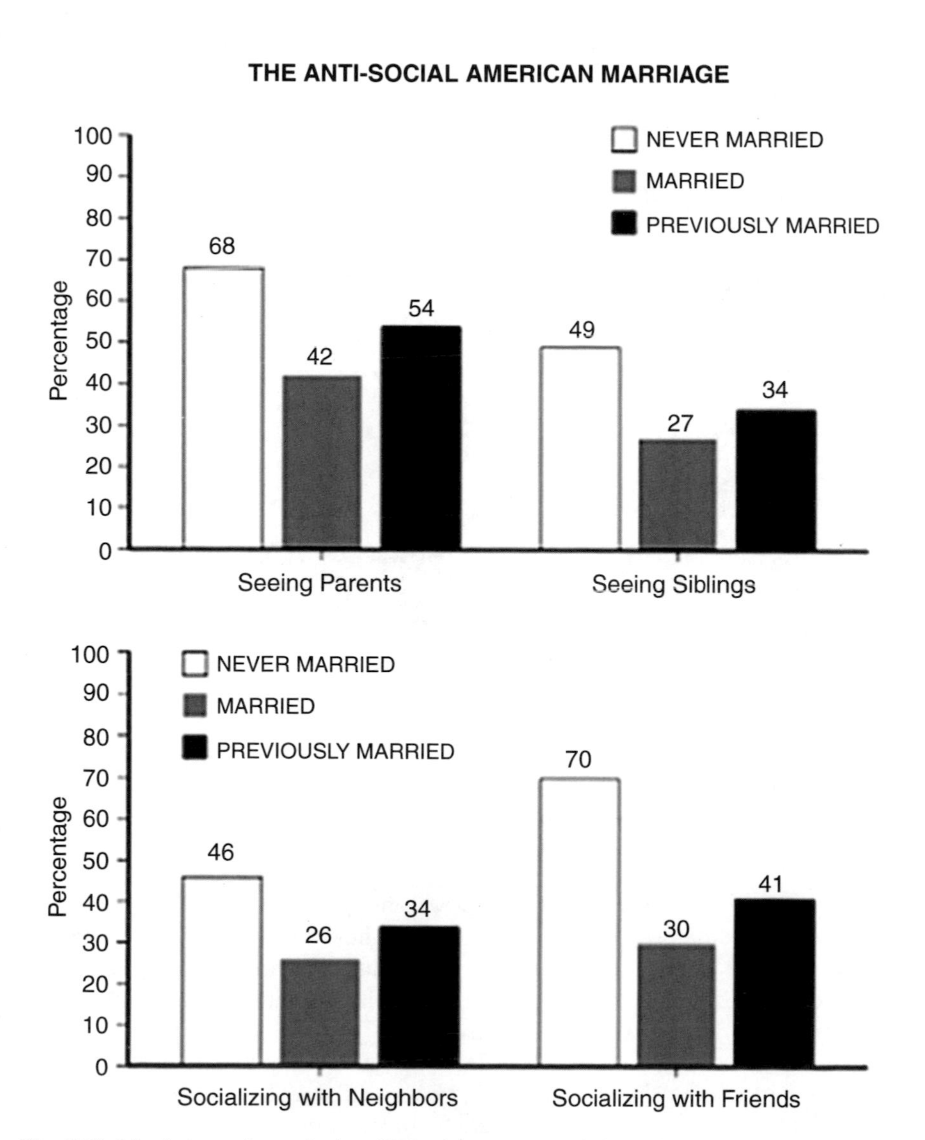

Fig. 5.13 Married people are far less likely than never-married people to see their parents, siblings, neighbors, and friends regularly

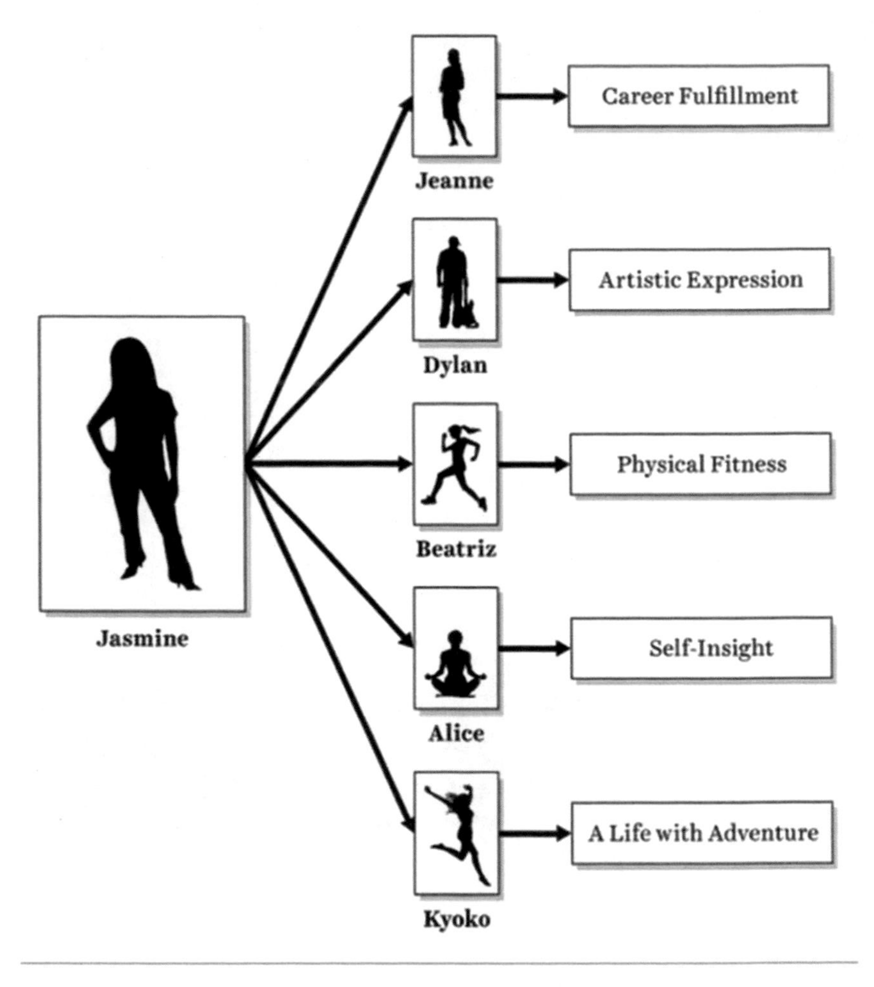

Fig. 5.14 Finkel's *All-Or-Nothing Marriage* theoretical example: Jasmine, who at 28 years was single and fulfilled in many different areas of life through a large social network

became particularly prevalent in the 1950s and 1960s, and still is nowadays. Therefore, concerning specifically the issue of **health**, although of course this depends on many factors, such as **genetics**, how "happy" a marriage truly is, and so on, one can say that, overall, the available date indicates that some items often associated with marriage nowadays, such as having a less risk taking life, seem to counter-balance others, such as taking less care of one's body weight, having less friends and smaller social networks in general, and so on. So, in conclusion, the existence of a huge number of recent, less biased, and more comprehensive studies contradicted two of the major myths of the longstanding narratives concerning

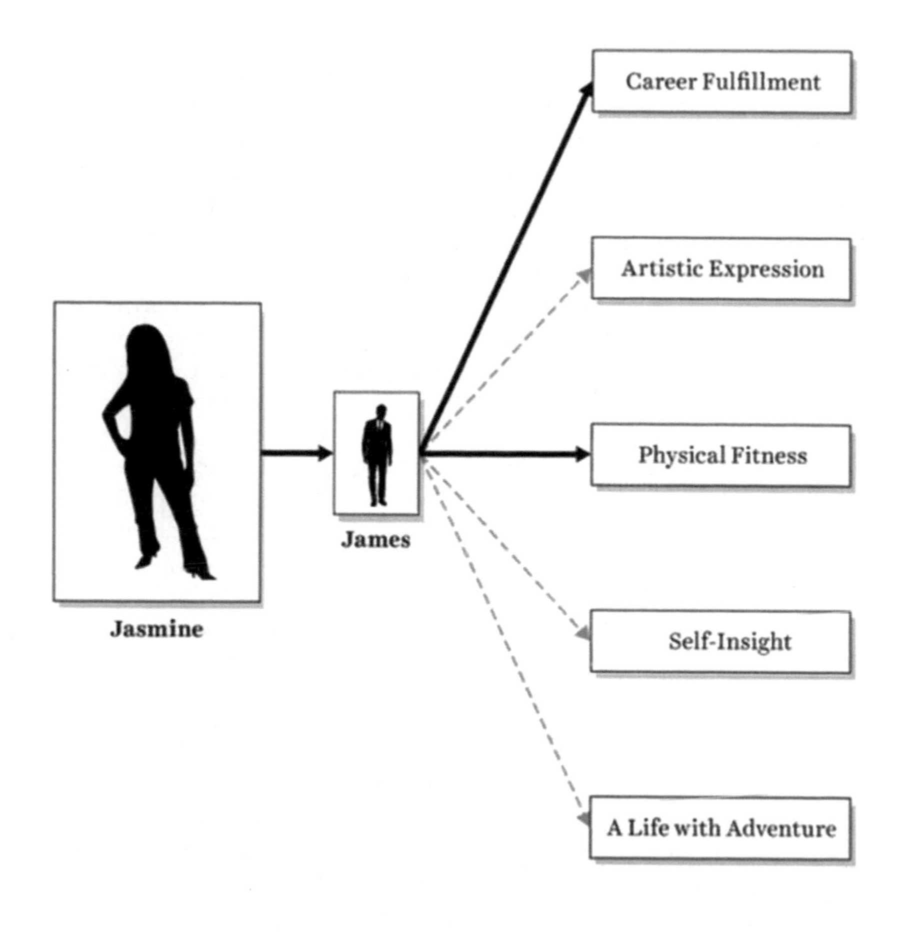

Fig. 5.15 Finkel's *All-Or-Nothing Marriage* theoretical example: Jasmine, 12 years later—now at 40 years—is married and mother, and because of her intense focus on her nuclear family, she is mainly fulfilled in only two of the many areas she was fulfilled at 28 years, the ones her husband is also interested in

marriage—that, on average, married people are healthier, and that the well-being of kids with married parents is higher than those with divorced parents—although it is importation to stress that such studies do not indicate that the opposite is necessarily true neither. As often, reality is just far more complex, with many nuances in between, but the fact that those two myths were contradicted surely allows people to take out a huge weight—and often guilt—from their shoulders: *in case you feel like just being single, or divorcing, you, and your kids, will not necessarily "pay a huge price" for that, in many cases actually you and them will have a much more rich, balanced, and even healthier—for both your body and mind—life, particularly if you are a women.*

So, in this sense, I completely agree with the main arguments made by Coontz in *Marriage, a History*, in particular in that the extremely **romanticized and anti-social type of marriage** constructed by the media that continues to be seen in **Hollywood movies** has an obvious, crucial problem: it is not "natural" at all in the sense that it is not based on any evident physiological or neurobiological trait of our evolutionary lineage, being as removed from reality as are **Disney movies** such as *Little Mermaid* or, at the other extreme, as are most porno movies nowadays. Such movies are **just-so stories** not grounded in any kind of scientific empirical data, being based instead on teleological narratives, for instance that the "purpose" of a woman is to find a prince or to raise a small nuclear family—as seen in many Disney movies—or to obey and please sexually and immediately every single man, or woman, or in some cases even non-human animals—as seen in some porno movies.

Moreover, humans are **highly social primates**: many hunter-gatherer groups, even nomadic ones, can have up to 100 people or more, which often see, interact, and communicate with each other almost on a daily basis. Also, with modern medicine, life expectancy is on average higher than it was a few centuries ago, for instance if one marries early, as one was *supposed to* under the social model of the 1950s and 1960s, and one dies at let's say 80, this would mean that one would live sometimes 60 years or so with the same person, spending a huge time within those six decades *only* with that person.

Now, suppose that you go to Africa, to undertake a long-term study of the also highly social baboons, which are commonly used as a model to understand the social networks of our ancestors millions of years ago. You check the data you compiled by studying the baboons, after some decades of field work, and you realize that all baboons are mainly interacting with each other, except *two of them*, a male and a female, which spent the vast majority of those two decades alone, just with each other. Let's say that this happened for 75% of that time, or even just 50%. What would be one of the first ideas that would come to the mind of any experienced anthropologist, primatologist, or veterinarian? Clearly, that there was something very wrong, likely even pathological, at a psychological or social level, with these two baboons, perhaps as the by-product of some huge traumatic experience, or some kind of hormonal imbalance leading to depression, and so on. *Well, this is exactly what a huge number of humans—that live in couples under the prevalent "**highly-romanticized, monogamous, anti-social model of marriage/love**" that we tend to see as "normal"—are forcing themselves to do*, for two, three, four and sometimes even more decades.

As noted in *Sex at dawn*, "if you ever doubt that human beings are, beyond everything, social animals, consider that short of outright execution or physical torture, the worst punishment in any society's arsenal has always been **exile**...having run short of empty places to exile our worst prisoners, we've turned to internal exile as our harshest punishment: **solitary confinement**...human beings are so desperate for social contact that prisoners almost universally choose the company of murderous lunatics over extended isolation." A related point is made in Finkel's *All-Or-Nothing Marriage* with an example that is theoretical—shown in Figs. 5.14 and 5.15—but that actually happens so often in relationships: Jasmine, who at 28 years was single and fulfilled in many different aspects of life through a large social network; 12 years later, now married and mother, because of her intense focus on her small nuclear family, she is only fulfilled in two of those aspects, the ones in which she shares interests with her husband. Socially constructed ideas that are still much in vogue—such as that one *should* "learn to love" our spouse, or that one should always be "flexible" and "adaptable," or that we are with our spouse because it was "meant to be," or that he/she is the "only love of our live"—put a huge pressure on the shoulders of both the husband and wife to be most of their time together. These social constructions are mainly **teleological narratives**, not only from religion—for instance, the "**sanctity of marriage**"—but also from just-so-stories about **cosmic purpose of life** and the related concept of "meant to be." In fact, many people that describe themselves as non-religious, particularly nowadays, believe in such teleological narratives about "meant to be," and thus in associated tales about "love of my life," despite the obvious fact that if there is no God, there is no planer, and thus no masterplan that is "meant to be fulfilled" by being with your "other half" "until death do you part."

Let's briefly discuss another complex equation in the life of couples, either married or not: **children**. This is a further example of how hard it has been, historically, for women to live within the dichotomy of what they are *supposed to do*—the narrative often being that their only "purpose" or "aim" is to have children—versus what they really *want* to do or what they truly *feel* about these issues. Only recently, this huge taboo subject begun to be discussed in so-called developed countries, and even now this remains one of the most sensitive issues for a lot of people, with people even attacking very harshly those very few women that start to publicly admit they do not *feel* as happy after having children as society tells them they *should be*. Various examples are given in a recent article—www.macleans.ca/regretful-mothers—by Anne Kingston, author of the book *The Meaning of Wife: A Provocative Look at Women and Marriage in the twenty-first Century*, which has the following subtitle: "in pushing the boundaries of accepted maternal response, women are challenging an explosive taboo – and reframing motherhood in the process." One of the examples concerns "Ami," who did *not want* to have kids, but ended up by having them and taking care of them 90% of her time:

> At first glance, 'Amy' is like many busy young moms – she's 34, lives in Alberta, works full-time and is devoted to her five-year-old. "I love my son with all my heart" she says… "my life revolves around this child." When discussing **motherhood**, however, Amy deviates from the maternal script: if she could make that choice over again, she says, she wouldn't. She never wanted – her husband did. Parenthood put an untenable strain on the marriage; her husband wasn't as involved as she wanted; they separated. Life is difficult, Amy reports: "our child has two homes and I'm still doing 90 per cent of it on my own." Amy's candour is part of a growing yet contentious conversation about parental regret, one primarily focused on mothers. Social media provides one hub, from the 9,000-member Facebook group "*I regret having children*". French psychotherapist Corinne Maier stoked an international firestorm and condemnation in 2008 with her manifesto *No kids: 40 good reasons not to have*

children; her two children left her "exhausted and bankrupt", and she couldn't wait for them to leave home, she wrote. In 2013, Isabella Dutton, a 57-year-old British mother of two grown children created furor with a *Daily Mail* essay headlined: "The mother who says having these two children is the biggest regret of her life". By 2018, however, Dutton and Maier are no longer freakish outliers; parental regret, or "the last parenting taboo" as it's dubbed in the media has been covered by everyone from the BBC to *Marie Claire* to *Today's*. The discussion has been stoked by the first scholarship on regret; Israeli sociologist Orna Donath thrust it into the spotlight with her 2015 book *Regretting motherhood: a study*, based on interviews with 23 Israeli women, all anonymous, aged 26 to 73, five of them grandmothers.

Unsurprisingly, women who express regret are called selfish, unnatural, abusive "bad moms" or believed to "exemplify the 'whining' culture we allegedly live in", as Donath puts it. One commenter called Dutton "an utterly miserable, cold-hearted and selfish woman." Even Donath has been savaged for her research: one critic suggested she be burned alive. Discussing **maternal regret** raises ethical dilemmas but is necessary, says Andrea O'Reilly, a professor at York University's School of Gender, Sexuality and Women's Studies and the author of 18 books on motherhood: "I understand the protection of children, but if you completely enforce that you have no mother voices telling their story, and you don't want that either." And what we're learning about regretful mothers upends binary thinking that **women who regret having children** must be neglectful or substandard parents: *it's motherhood these women regret, not the children*. Dutton expressed love for her offspring ("I would cut off my arm if either needed it"); it was maternal strictures she bristled against ("I felt oppressed by my constant responsibility for them"). In *Today's parent*, Augustine Brown called her children "the best things I have ever done" and assured readers she wasn't "a monster" before expressing conflicted feelings: "What I'm struggling with is that it feels like their amazing life comes at the expense of my own", she wrote, expressing remorse for "this life I wanted so badly and now find myself trapped in." "The more I feel [regret], the more I give them", a mother interviewed by Donath said about her two children.

Still, it's received as an affront to the "sanctity" of motherhood and the entrenched belief that the maternal instinct is innate and unconditional – despite ample historical evidence to the contrary. Pressure on women to have kids is intense, says Amy. "I work with a lot of girls and, if they haven't had kids, they're told, 'The clock is ticking'." But mothers voicing regret also signal something else: a larger groundswell of maternal reckoning, one Augustine Brown compares to the #MeToo campaign. "We still can't talk honestly about what it's like to live with those pressures and those sacrifices" she tells *Maclean's*. Response to her story was overwhelmingly positive, she reports, save one woman: "She told me I should see a therapist and that my children deserved a better mother." Yet dozens of women wrote to thank her for making them not feel so alone – or *abnormal*. Donath's book has also been a sensation: when published in Germany (an English translation arrived in 2017), #RegrettingMotherhood trended on that country's Twitter. Donath views suspicion over the existence of maternal regret as consistent with a traditional rejection of women expressing negative responses to motherhood, pointing to the reluctance to accept **post-partum depression** until the late 20th century; before then, mothers reporting **perinatal sadness** were dismissed as 'neurotic'.

So, what does the biological evidence available really tell us, for instance, concerning studies of other animals and of human hunter-gatherer societies? First of all, as noted by Donath, we know that **postpartum depression** is found in numerous other animals, including non-human primates, being therefore completely natural in this sense. So, it is factually inaccurate to consider that the mothers that have postpartum depression are necessarily "pathological" or "neurotic," as it was done until the late twentieth century. Second, in many hunter-gatherer groups, some mothers *do* abandon children: in some groups up to one child can be abandoned, within 100 newborns. Often this is related to factors such as the newborns having birth defects, but in reality we cannot truly be sure if some of those cases of abandonment have, or not, to do with postpartum depression, eventually—it is reasonable to assume that at least some of them might have.

In Kingston's online article, she notes that a 2002–2003 study by the U.S. Department of Health and Human Services found that 3% of parents disagreed with the statement: "the rewards of being a parent are worth it despite the cost and the work it takes," while a 2016 German study found that 8% of the 1200 parents polled said they would choose not to have children again. Based on what we know about our society and the social pressure imposed on people in general, it is very likely that these percentages are actually lower than the proportion of people that truly regret having children, because of the **social feeling of guilt** or, at least, of unease that men and particularly women may have when answering such questions, even anonymously, due to the huge pressure that society puts on them. This is further reinforced by the obvious complicatedness of children becoming aware of what their parents said and then thinking—in most cases, wrongly—that this means that their parents do not love them. Actually, one other related difficulty to answer in such a way is precisely the fact that some of the questions of these questionnaires can lead to confusion: for instance, concerning the mothers, it is not about **regretting having children**, but instead **regretting motherhood**.

In this sense, I completely agree with most authors cited above in that the "**regretful mom phenomenon**" is becoming more and more prevalent nowadays not only because at least some mothers can finally start saying what they *truly feel*—particularly with the rise of the "**me-too-movement**" and similar movements in recent years—but also because of the huge increase in time and effort that mothers dedicate, on average, to parenting, particularly in most Western countries, in the last decades. This might be surprising to you, as you can think that, now that there are less domestic mothers than there were in the 1950s and 1960s, mothers would have to work more hours and thus surely would tend to dedicate less hours per week, on average, to parenting. However, as Kingston noted, "time spent by parents with their kids has doubled in four decades, *The Economist* revealed: in an analysis of 11 wealthy countries, mothers spent an average of 104 minutes a day caring for children in 2012, up from 54 in 1965…men do less, but far more than they did in the past: 59 minutes a day, up from 16." This is an almost 100% increase for mothers, in just about five decades. As she also notes, "parenting standards have become far more draconian since…the 1980s and 1990s…[due to] a confluence of forces – the **rise of materialism**, **consumerism**, **neoliberalism** and **social media** – turning **parenthood** into a performance…parents now raise children in a far more difficult, competitive world and are pressured to do more with far less…expectations have been ramped up to such a point that standards are impossible to achieve." She further states that "the rise of '**intensive mothering**' has political implications; as Susan Douglas and Meredith Michaels wrote in their 2005 book *The mommy myth: the idealization of motherhood and how it has undermined all women*, an increasingly powerful conservative subculture is determined to 're-domesticate the women of America through motherhood'…."

As explained in Ackerman's *Natural History of Love*, the correlation between **parenting** and **divorce** is clearly seen since the social changes that occurred in the 1950s and 1960s linked to the notions of "small nuclear family" and "antisocial marriage," in which the mother did almost all the parenting without the help of her friends and of the broader families of her and her husband. She stated that "a study of an United Nations survey of marriage and divorce around the word showed that divorce usually occurs early in a marriage, during the couple's first parenting years…some couples do stay together and have other children, but *even more* don't." Together with the "nuclear-family anti-social marriage" (Figs. 5.11, 5.12, 5.13, 5.14, and 5.15) and the higher amount of time dedicated to work, the increasing parental obsession with kids—as compared to just some decades ago—further decreased the amount of time a married person spent with friends and relatives, as well as alone doing her/his own hobbies and having her/his own small pleasures of life (Fig. 5.16). As noted in Finkel's *All-Or-Nothing Marriage*, "according to one major study, the amount of time that childless Americans spent alone with their spouses declined from 35 to 26 hours per week from 1975 to 2003, with much of this decline resulting from an increase in hours spent at work…the decline for Americans with children at home was from 13 to 9 hours per week, with much of it resulting from an increase in time-intensive parenting" (Fig. 5.17).

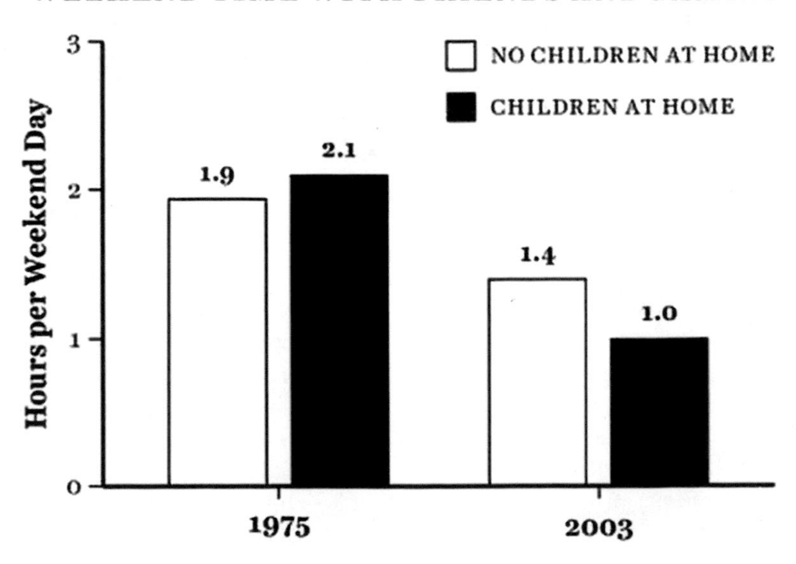

Fig. 5.16 Together with the "nuclear-family-anti-social" model of marriage and the higher amount of time dedicated to "work," the increasing parental obsession with kids—as compared to just some decades ago—further decreases the amount of time a married person spends alone with friends or relatives

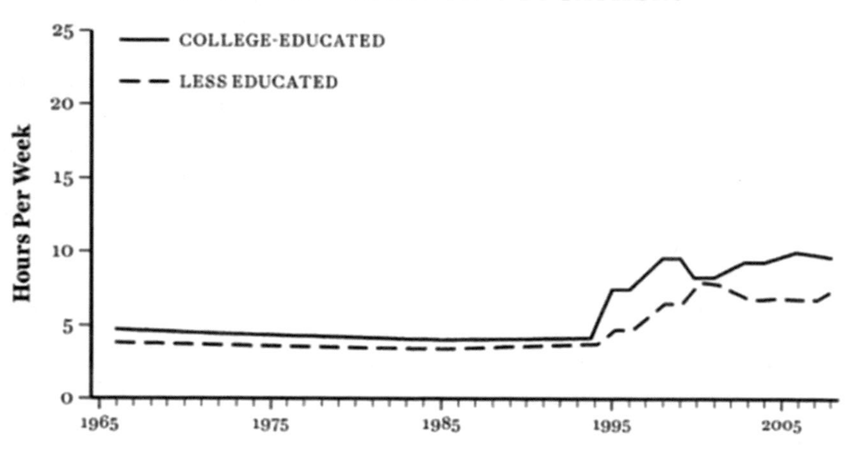

TIME SPENT PARENTING BY MOTHERS

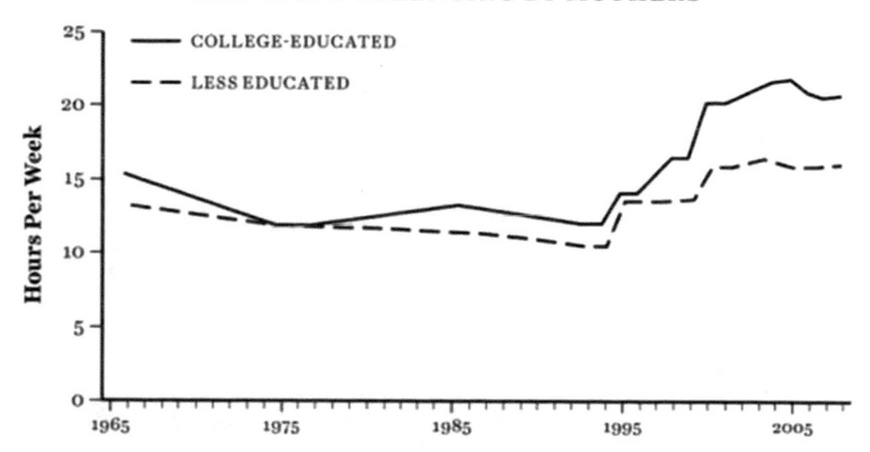

Fig. 5.17 Time spent parenting has significantly increased in the last decades, in general, all across so-called developed countries

He further explained that "the total amount of work new parents do (paid work + housework + child care) increases by an estimated 33.5 hours per week when the first baby arrives, with 63% of this increase absorbed by mothers…but even spouses without children are time-starved, in part because they work so much." When all these factors are put together, there is obviously an increase of stress, as Finkel wrote: "Americans' elevated stress levels result in part from the increasing difficulty of achieving work-life balance…whereas 24% of husbands and 23% of wives endured strong work-life conflict in 1980, the respective estimates had surged to 44% and 33% by 2000." These numbers are similar to the ones he showed in the image displayed in Fig. 5.18, the difference being that this image seemingly refers to men and women in general—that is, not only married ones—and that, in this image, women consistently have higher levels of stress. So, Finkel concluded, "perhaps we shouldn't be surprised that only 38% of mothers of young infants are highly satisfied in their marriage (in contrast to 62 percent of childless women)—or that the link between parenting and marital dissatisfaction has gotten stronger in recent decades."

As emphasized by Kingston, "a 2010 American Sociological Association study found that parents were more likely to be depressed than their child-free counterparts, and that people without kids were happier than any other group." As Feiler notes in his book *The First Love Story*, "one of the most persistent findings of social science is that having children is surprisingly harmful to parents, to their sense of happiness, and to their relationships with each other…children zap resources, decimate sleep, and reduce sex…childcare ranked 16 out of 19 on a list of things parents enjoy, behind even housework…83% of new mothers and fathers experience 'severe crisis', 90% suffer a decline in marital satisfaction." Taking all this into account, it is therefore not surprising that more and more people don't *want* to have kids, or at least want to delay having them. Also, it is not surprising that more and more people are questioning the current model of marriage, and even of monogamy per se. A 2016 survey from the polling firm *YouGov* posed this question to Americans: on a scale where 0 is completely monogamous and 6 is completely non-monogamous, what would your ideal relationship be? While 70% of people 65 years old or older

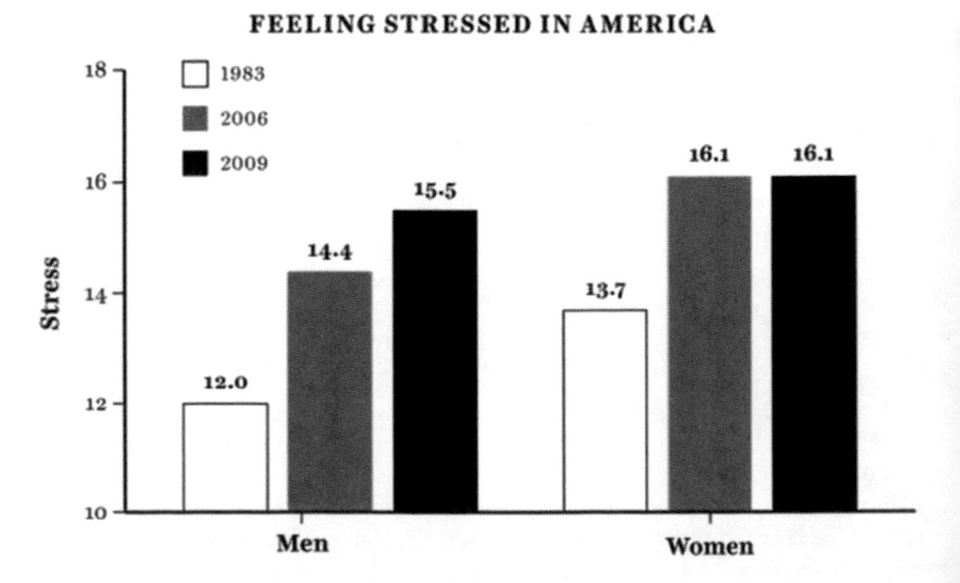

Fig. 5.18 Feeling stressed in the U.S.

"*believed*" in monogamy as an ideal, only 63%, 58%, and 51% of people 45–64 years old, 30–44 years old, and under 30 years old did so, respectively. Not only people increasingly doubt the "monogamy" ideal, but according to some studies those that are not applying it in their lives, for instance by being consensually non-monogamous, generally report to be more satisfied with, committed to, having more passion for, more trust in, and even less jealousy concerning, their partners, compared to those that try to apply it. Also, some studies have indicated that people in open relationships—even 55 year-olds, and older—are on average happier, healthier, and have sex more frequently, than the general population, as explained in a 2015 paper by Fleckenstein and Cox.

However, it is worthy to make three comments concerning these topics. First, some works have contradicted some of these latter studies, suggesting that there are in fact few differences in terms of relationship between individuals engaged in monogamy versus in consensual non-monogamous relationships, as explained in a 2017 paper by Conley and colleagues. Second, even if—as indicated by studies such as the one published by Fleckenstein and Cox in 2015—there is on average a higher frequency of sexual intercourse within people in consensual non-monogamous relationships, such behaviors are so complex that is often difficult to discern the cause from the effect. Are people having more sex because they are in such open relationships, or did they decide they would opt to be "open" because they had a higher sexual drive in the first place? Of course, even if this would be so, this would still be an important factor, as it would mean that people would have an a priori expectation that monogamous relationships would not be fulfilling for someone with a higher sexual drive. Be that as it may, many studies done on these subjects are indeed too simplistic. Third, it is obvious that even if research studies are indeed consistently showing that people that have children are, on average, less happy, and have less sex, than those that are childless, humans obviously cannot simply completely stop having kids, if not we would be extinct in a bit more than a century from now: happiness is surely not enough to survive within biological evolution.

A related, deeper question, is whether humans should have so many children, in a planet that is already overpopulated, and in which there are so many kids that lost their biological parents from wars, starvation, and other incidents, and that could be adopted. This is a difficult issue, because the prevailing narrative is that people "should have kids" and that if they don't have them, it is because they are "selfish." But in view of the high number of kids that are orphans, one may wonder about who is truly being "selfish": the people that decide to have "their own kids," which will propagate their own genes, or the people that decide either to not have kids to not overpopulate even more this planet or, in particular, to adopt orphans that desperately need someone to take care of, and love, them despite those kids not having their genes.

One thing is sure: in terms of what is "natural" versus what is not, having kids is clearly one of the most 'natural' things for sexual animals. So, the ultimate reason for having problems can't be humans having kids, in general. That is, it surely is "natural" that some people don't *want* to have them and that some women *want to* but then experience for instance postpartum depression, as it is completely normal that some people regret having kids many years after having them. However, many people that regret having kids clearly say that this is not about the children per se, or their love for them, but instead because of the "asphyxiation" they felt after having them within the social, economic, and cultural context of most so-called developed countries. These include: (1) the current obsession with parenting and being "the best parents ever"; (2) the present obsession with an extreme, Hollywood type of anti-social romantic love where social networks are highly decreased and thus parents spend less time with their relatives, friends, and colleagues and have less help from others to take care of their kids; and (3) the amount of time many people spend working, including even working at home extra hours, and so on (Figs. 5.13, 5.14, 5.15, 5.16, and 5.17).

All these three factors—and the increased stress as well as the overall higher dissatisfaction with both marriage and parenting that are related with/provoked by them—are moreover deeply linked with the type of **materialism** and **consumerism** that has become more and more prominent in most countries of the planet: love is monetized, parenting is monetized, children are monetized, and more work needs to be done to have money for all these things, plus the other million materialistic things that we, and our kids, "want." This is amplified by the fact that one cannot just be a "good" parent, or give the typical amount of gifts and pleasures to our kids because of the obsession and social pressure to be the best parents: one thus needs to give more than other parents, do more activities than other parents do with their kids, and so on, within our current "more-is-better-society."

So, what I am—and many others scientists are—arguing is obviously not that people should not have kids. Instead, we should combine the "positive" aspects we gained with so-called civilization, such as vaccination and other medical discoveries that hugely decreased child mortality, and the "positive" aspects that we lost either in the last decades—such as being less obsessed with being the "best parents ever" or with such an extreme version of romantic love—or in the last millennia since the rise of agriculture. These would include dedicating less time to "work" and more time to be surrounded by relatives and other people, which would moreover help enormously to take care of the children, and so on. The fact that the current obsession with, and huge increase of time dedicated to, parenting is indeed mainly a trait seen in "developed" countries is clearly reflected in our societies, from bank ads to social pressure and activities such as "baby showers" that nowadays often include a huge number of gifts being given to babies that were not even born yet, as well as in numerous movies or fiction and non-fiction books that have titles such as, literally, *A Woman Is a Woman Until She Is a Mother*—published by Prushinskaya in 2017. That is, looking back to Fig. 5.15, and thinking about what Jasmine will be when she reaches let's say 45 years and have one or two kids, one can say that it is not only that she gave up many of the things that she liked to do when she was 28 years (Fig. 5.15): no, "Jasmine" is not even "Jasmine" anymore, but simply the mother of "little Jas."

5.6 Misogyny, Religion, and Hypatia of Alexandria

> The history of men's opposition to women's emancipation is more interesting perhaps than the story of that emancipation itself. (Virginia Woolf)

A question that people often ask me in my talks is: "So, if women have all the capacities and abilities you listed in your talk, why is that men won more **Nobel prizes** and made more discoveries than women? Why were Da Vinci, Galileu, Newtown and Einstein all men?" The answer is very easy, actually, and may consist in a simple new question: if you would do this question about so-called black people and formula 1, or golf, just 20 years ago, asking me why all the best formula 1 drivers and golf players were 'white', this would also apply, right? But clearly this indeed applied 20 years ago not because "black" people were not good at all in those sports: it was because they often *lacked the opportunities* to be good at them—both sports are often played, and associated with, people that are "well off." We now know this for sure because in the last years Tiger Woods and Lewis Hamilton were particularly good at those two sports. Yes, some sports have a lot to do with some specific physical attributes—Olympic weightlifting is obviously normally not won by very skinny people with scarce muscle mass. But this does not apply to sports such as formula 1, golf, or chess, nor of course to Nobel prizes: in such cases, it is clear that, as shown by **Tiger Woods**, **Lewis Hamilton**, or **Marie Curie**, the fact that certain people of a so-called group A—remember that "whites" or "blacks" do not constitute natural, real groups biologically—won more prices or awards than certain people of a group "B" had chiefly to do with the narratives, social constructions, and even laws that lead to, and maintained, the segregation of those people "A" and "B."

So, while the distinction between women and men is biologically real, in terms of sex—although there are also many cases in between, biologically, as we all know—in terms of what we call gender it is in great part constructed. That is, on average, there are some significant differences between women and men, not only concerning their genitals and sexuality—women have a higher **orgasmic potential**—but also, yes, concerning some mental capacities—with women having higher performances, on average, than men in most tests. This stands in stark contrast with the so-called differences between "blacks" and "whites," not only because the distinction between them lacks any biological reality apart from the color of the skin but also because all living human groups diverged at the most about 250 thousands of years, when our species emerged—what corresponds to only 4 seconds of a "Earth's day" (see Fig. 2.6). Compare this with the evolution of **sexual beings**: in the fossil record of eukaryotes—organisms whose cells have a nucleus enclosed within a nuclear envelope, as we do—**sexual reproduction** appeared more than 1 billion years ago, and the **Y chromosome** that distinguishes males from females at the genetic level appeared at least 180 million years ago, that is 3000 times the amount of time that has passed since *Homo sapiens* first left Africa about some 60 thousand years ago. That is indeed a key reason explaining why the **biological difference between female and male sexes** is real and profound while the **differences between gender** are in great part socially constructed, and the **differences between so-called blacks, whites, yellows, and reds** are chiefly socially constructed.

So, we can now specifically address the question that people often ask me in my talks about the differences between genders: women have been subjugated by men since many millennia ago, particularly since the rise of **sedentism** and particularly of **agriculture** and because of that women could *not* do many tasks that were only allowed to men. Actually until only some decades ago in many "developed" countries, women could not even vote, or have their own research laboratories. Therefore, of course, they could *not* have the same amount of Nobel Prizes as men have. When I answer this in my talks, some people then ask me: "But why were they subjugated in the first place, if they are not mentally inferior?" This is a very common argument also raised by people about so-called other races: for them, the fact that Europeans colonized 'others', and not the other way round, shows that Europeans are "superior." Such arguments completely neglect that researchers and historians have already provided very strong hypotheses—based on sound empirical data and arguments—about why were the Europeans and not the "others" that did so, which have nothing to do with so-called innate superiority. Of course, the differences between genders are unlike from those between human societies originally living in different areas of the globe, in the sense that women and men from a similar society almost always co-existed physically. But this is precisely the point: contrary to the differences between Europeans versus the Mayans, the subjugation of women by men is, in this sense, actually much easier to explain, as it was mainly a **physical subjugation** combined with teleological man-made narratives that ultimately also lead to a legal, social, and economic subjugation.

That is, men partly, and chiefly, used one of the very few features in which they do have, *on average*, a biological advantage over women: **physical force**, the very same feature that alpha-male gorillas use to subjugate female gorillas and other male gorillas. Now, it is my time to ask: when there is a change of alpha-male in a gorilla group, for instance due to a fight, does the winner of the fight suddenly becomes "more intelligent" than the other? Of course not, the alpha male just became weaker physically, or its adversary became stronger physically, with time, showing that nothing is eternal, not even physical force: biological traits change with time, and depend on contingency, randomness, and on numerous very specific circumstances. I often provide in my talks an unfortunately very disturbing, but in my opinion appropriate, case study about humans themselves, to make this clear once for all. When Jews were physically living in **Nazi concentration camps**, did they do any kind of scientific research that was awarded later with a Nobel Prize? Of course not, because the Nazis *did not allow them* to use microscopes, laboratories, or any other type of scientific item that is imperatively needed in order to do scientific research to win such an award.

Does this mean that Nazis were necessarily intellectually superior to Jews (Figs. 4.5 and 9.17)? Of course not: a detailed review of the history of Nobel Prizes shows that, on average, people that define themselves as Jews tend to have a very high ratio of prizes won, per million people. So, this is what happened to women, not during a few years, but during the last millennia. In fact, the increased subjugation of women during the rise of agriculture allowed men to also create new forms of subjugation other than those directly related to the use of physical force, for instance, via the creation of religious teleological narratives, social norms, and stereotypes about woman being "more passive," or "less intelligent," or "physiologically weak." To mention here just some depressing empirical evidence, I will list here a few events that happened only after the Second World War: 1945, U.S.—Harvard Medical School admitted women for the first time; 1955, Qatar—First public school for girls; 1966, Kuwait—University education open to women; 1983, U.S.—Columbia College, Columbia University, allowed women to apply for admittance; 2016, Tibet—Women able to take the "geshe" Tibetan Buddhist academic exams for the first time.

Further examples of how the **physical subjugation of women** allowed men to then subjugate them in many other ways are given in the excerpts of Holland's outstanding book *A Brief History of Misogyny* that are included in Boxes 5.6–5.8. In that book, Holland argues that the type of **misogyny** we are nowadays mainly familiar with, particularly in the West, dates back as far as **ancient Greece** and **Judaea**.

Box 5.6: Pandora, Plato, Aristotle, and Misogyny

The exceptional book of Holland, *A Brief History of Misogyny*, provides the background to better understand the type of misogyny that is so familiar to us nowadays, particularly in the West or in regions of the globe that were highly influenced by **Western ideas**. This does not mean, of course, that the West is the only source of "post-agricultural" misogyny. Holland writes:

> It is hard to be precise about the origins of a **prejudice**. But if misogyny has a birthday, it falls sometime in the eighth century BC. If it has a cradle, it lies somewhere in the eastern Mediterranean. At around that time in both **Greece** and **Judaea**, creation stories that were to acquire the power of myth arose, describing the **Fall of Man**, and how woman's weakness is responsible for all subsequent human suffering, misery and death. Both myths have since flowed into the mainstream of **Western civilization**, carried along by two of its most powerful tributaries: in the **Jewish tradition**, as recounted in **Genesis** (which a majority of Americans still accept as true) the culprit is **Eve**; and in the Greek, **Pandora**. But in the history of misogyny, the Greeks also occupy a unique place as the intellectual pioneers of a pernicious view of women that has persisted down to modern times, confounding any notion we might still have that the rise of reason and science means the decline of prejudice and hatred. The **myth of Pandora** was first written down in the eighth century BC by **Hesiod**, a farmer turned poet, in two poems: '**Theogony**' and '**Works and Days**'. In spite of Hesiod's considerable experience as a farmer, his account of mankind's creation ignores some of the basic facts of life. The race of men exists before the arrival of woman, in blissful autonomy, as companions to the gods, 'apart from sorrow and from painful work/ free from disease'. As in the **Biblical account of the creation** of man, woman is an afterthought. But in the Greek version, she is also a most malicious one. **Zeus**, the father of the gods, seeks to punish men by keeping

from them the secret of fire, so that, like the beasts, they must eat their meat raw. **Prometheus**, a demi-god and the creator of the first men, steals fire from heaven and brings it to earth. Furious at being deceived, Zeus devises the supreme trick in the form of a 'gift' to men, 'an evil thing for their delight', Pandora, the 'all giver'. The Greek phrase used to describe her, 'kalon kakon', means 'the beautiful evil'. Since then, according to **Greek mythology**, mankind has been doomed to labour, grow old, get sick, and die in suffering.

One of the functions of **mythology** is to answer the sort of questions we asked as children, such as 'Why do the stars shine?' and 'Why did granddad die?' Myths also justify the existing order of things – both natural and social – and account for traditional beliefs, rituals, and roles. One of the beliefs most central to the Greek, and later the **Judaeo-Christian, traditions** was that man was fashioned by the gods, or God, separately from the creation of animals. (The persistence of this belief among conservative Christians is why Darwin's theory of evolution continues to meet with such resistance.) As well as burdening Pandora with responsibility for the mortal lot of man, the Greeks created a vision of woman as 'the Other', the antithesis to the male thesis, who needed boundaries to contain her. Most crucially, Greece laid the philosophical-scientific foundations for a dualistic view of reality in which women were forever doomed to embody this mutable, and essentially contemptible world. Any history of the attempt to dehumanize half the human race is confronted by this paradox, that some of the values we cherish most were forged in a society that devalued, denigrated and despised women. 'Sex roles that will be familiar to the modern reader were firmly established in the **Dark Ages** in Athens', wrote the historian Sarah Pomeroy. That is, along with Plato and the Parthenon, Greece gave us some of the cheapest sexual dichotomies of all, including that of 'good girl versus bad girl'. Having violent warrior divinities, however, is not necessarily an indication of a **misogynistic culture**. In the older civilizations the Greeks encountered, such as those of **Egypt** and **Babylon**, there was an abundance of war gods, but no equivalent of the Fall of Man myth. In **Mesopotamia**, the **Sumerian poem 'The Epic of Gilgamesh'**, which dates back to the third millennium BC, has a hero who like Prometheus aspires to rival the gods. **Gilgamesh** does so by seeking to share in their **immortality**; but women are not made the instrument of revenge by some vindictive deity seeking to punish man for challenging his mortal lot. Nor does Gilgamesh castigate women for being to blame for 'the lot of man'; the gods are to blame for our mortality.

Homer based both ***The iliad*** and ***The odyssey*** (the latter recounting the long journey home of **Odysseus**, one of the Greek kings) on material which dates back to the earlier dynastic period. In these works, women are generally portrayed sympathetically; they are complex and powerful, and among the most memorable characters in all literature. The end of this era was accompanied by a move from a pastoral to a labourintensive agricultural economy, one concerned about the conservation of property. Laws regulating women's behaviour and opportunities give the most graphic and pertinent examples of how **Hesiod's allegory of misogyny** became a social fact. Legally speaking, Athenian women remained children, always under the guardianship of a male. A woman could not leave the house unless accompanied by a chaperone. She seldom was invited to dinner with her husband and lived in a segregated area of the house. She received no formal education: 'Let a woman not develop her reason, for that would be a terrible thing', said the philosopher **Democritus**. Women were married when they reached puberty, often to men twice their age. Such a difference in age and maturity, as well as in education, would have enhanced the notion of women's inferiority. The husband was warned: 'He who teaches letters to his wife is ill advised: he's giving additional poison to a snake'.

Some have hailed **Plato** as the first feminist because in ***The republic***, his vision of **Utopia**, he advocated that women receive the same education as men. At the same time, however, his dualistic vision of the world represents a turning away from the realm of ordinary, mutable existence. This existence he held was an illusion and a distraction to be scorned by the wise man. It included **marriage** and procreation, lowly pursuits with which he identifies women. He himself never married, and exalted the 'pure' love of men for men higher than the love of men for women, which he placed closer to animal lust. His is a familiar enough dualism – identifying man with **spirituality** and woman with carnal appetites. But Plato gave it a kind of philosophic fire-power never seen before. Plato's ***Theory of forms*** is the philosophical basis for the Christian doctrine of **Original Sin**, in which the very act of conception is viewed as a falling away from the perfection of God into the abysmal world of appearances, of suffering and of death. It provided the allegory of Pandora and the Fall of Man with a powerful philosophical basis. Before this Fall, autonomous man lived in a state of harmony with God.

Aristotle has been described as one of the most ferocious misogynists of all time. His views on women take two forms: scientific and social. Although at times Aristotle was a precise observer of the natural world – his descriptions of various species impressed Charles Darwin – his observations of women were decidedly warped. As a sign of women's inferiority, he referenced the fact that they did not grow bald – proof of their more childlike nature. He also claimed that women had fewer teeth than men, about which Bertrand Russell is said to have commented: 'Aristotle would never have made this mistake if he had allowed his wife to open her mouth once in a while'. Aristotle introduced the concept of **purpose** as fundamental to science. The purpose of things, including all living things, is to become what they are. In the absence of any knowledge of genetics, or of evolution, Aristotle saw purpose as the realization of each thing's potential to be itself. In a sense, this is a materialistic version of Plato's *Theory of forms*: there is an Ideal Fish of which all the actual fishes are different realizations. The ideal is their purpose. When applied to human beings, notably to women, this has unfortunate but predictable results; it becomes a justification of inequality rather than an explanation for it. The most pernicious example is seen in Aristotle's theory of generation. This assumes different purposes for men and women: 'the male is by nature superior and the female inferior; and the one rules, and the other is ruled; the principle of necessity extends to all mankind'. Therefore, according to Aristotle, the male semen must carry the soul or spirit, and all the potential for the person to be fully human. The female, the recipient of the male seed, provides merely the matter, the nutritive environment. The male is the active principle, the mover, the female the passive, the moved. The full potential of the child is reached only if it is born male; if the 'cold constitution' of the female predominates, through an excess of menstrual fluid in the womb, then the child will fail to reach its full human potential and the result is female. 'For the female is, as it were, a mutilated male', Aristotle concludes.

Box 5.7: Teleology, Christianity, Adam and Eve, Homosexuality, and Misogyny

Holland's *A Brief History of Misogyny* also provides an interesting discussion on the links between the rise of **Christianity** and **misogyny**, which are obviously highly influential for about 30% of the people alive today, who are Christians, as well as for a huge number of people within the remaining 70%, due to the influence of Christianity and/or Western ideas on the planet's population as a whole:

> The rise of Christianity from an obscure sect to the world's dominant religion is a phenomenon unprecedented in human history. So too is the power and complexity of its misogynistic vision, which derives, essentially, from three sources. From the **Jews**, the early **Christians** took the Fall of Man myth, as well as the notion of sin and a profound sense of shame. Later, from the **Greeks**, they borrowed aspects of **Plato's dualistic philosophy** and **Aristotle**'s 'scientific' proofs of women's inherent inferiority. To this potent brew, Christianity itself contributed its central and unique tenet, that God has intervened in human history in the person of **Jesus Christ** to save mankind from death, sin and suffering, the evil effects of the fall from grace brought about by woman. The bitterest irony in this is that during Christianity's first three centuries, women were a key to its remarkable success, thanks to the fact that it gave them a kind of liberation unheard of in the ancient world. Unlike **Greek misogyny**, the Jewish version remained, as did Jewish religion, at the level of proverb, parable and practice. Instead of philosophy, the Jews had extensive commentary and interpretation of the sacred texts. But the similarities in both the creation and **Fall of Man** myths are clear. As in the **Greek myth**, in the Jewish tradition God creates the first man, **Adam**, as an autonomous being who lives a happy, contented existence in the **Garden of Eden**. His only communion is with the divine. **Eve**, like **Pandora**, is an afterthought. She is created from Adam's rib because God thought he required 'an help'. And, as with her Greek equivalent Pandora, Eve is disobedient, ignoring God's instruction not to eat the fruit from the **Tree of Knowledge**. 'The serpent did beguile me and I did eat', Eve confesses rather nonchalantly (Genesis, 3:13).
>
> The **God of the Old Testament** proves every bit as vindictive as **Zeus**. He tells Eve: 'And I will greatly multiply thy sorrow and thy conception; and in sorrow shalt thou bring forth children and thy desire shall be for thy husband and he shall rule over thee' (Genesis, 3:16). The message for Adam is clear. 'And I shall put enmity between thee and the woman', God tells him in what turns out to be a self-fulfilling prophecy (Genesis, 3:15). Along with sin came a sense of shame of the human body, something completely alien to the world of the **Greeks** and **Romans**. Shame strikes as the very first consequence of Eve's transgression: 'and the eyes of them both were opened, and they knew that they were naked; and they sewed fig leaves together and made themselves aprons' (Genesis, 3:7). Passing from the **Jewish tradition** into Christianity, shame took a firm grip on human sexuality. To a surprising extent, it has not yet relinquished this hold. It gave misogyny a new and destructive dimension. Linked to shame was the Jewish belief, which the Christians inherited as well, that sex was for procreation, not recreation. **Homosexuality** was forbidden, as was any wasteful spilling of man's seed, including sodomy, masturbation and oral sex. Not a drop could be spared from the business of begetting. Apart from the magnificent poetry of the Song of Solomon, the Old Testament is harsh and bleak in its attitude **to human sexuality**, and almost always hostile towards women. The God of the Old Testament sits gloomily aloft and alone, a brooder whose emotional range is usually restricted to jealousy and anger. The beauty of the beings he has created does not

usually fill him with pride, and never with desire. Unlike the divinities of Mount Olympus, he is devoid of love or even lust.

[But] what is most striking about the parables and proverbs attributed to Jesus, as recounted in the **Gospels**, is the absence of both misogyny and vengefulness. Women were among his first followers. We are told by Matthew: 'many women were there beholding afar off, which followed Jesus from Galilee, ministering unto him'. They had good reason to do so. Matthew also tells us (9:20–22) of a woman 'which was diseased with an issue of blood' who touched the hem of his garment. Jewish law had strict taboos on **menstruating women** as 'unclean', forbidding them contact with the male and entrance into the Temple, among others. In contrast, Jesus does not rebuke the bleeding woman but tells her, 'daughter, be of good comfort; thy faith hath made thee whole'. (28:55). In the Gospel according to **St John**, Jesus' disciples are said to have 'marvelled that he was talking with a woman'. (John: 4:57) In that Jesus was unique. None of the great Classical teachers/philosophers, nor the Jewish prophets who preceded him such as **John the Baptist**, gathered women followers about them to any significant extent. Because Christians held that every member of the faithful carries the spark of the divine in his or her soul, **infanticide** was forbidden, as was **abortion**. Since a majority of exposed infants were girls, this meant that gradually the proportion of females who were Christians began to rise. Women's numbers were further augmented in this new faith by its ban on abortion, which due to the dangers of the operation, killed many women and often rendered those who survived it infertile. In the ancient world, both in Greece and Rome, it was the man who, as head of the household, had the legal power to order a woman to have an abortion. Aristotle advocated it as a form of birth control. Evidence also shows that Christian women married later than their pagan contemporaries, so had better chances of surviving their first pregnancy. Nor were widows compelled to remarry, as was the common practice…. Christians were expected to marry for life, and infidelity was regarded as being as much of a sin for a man as for a woman. In this, Christianity leveled the moral playing field for women. Christian women were also less likely to be forced to marry, as Christians valued **virginity**. Traditionally, in the world of **Classical Antiquity**, men had been called on to resist the wiles of women. Now, for the first time, women were being told that they could reject men. Women were being offered a choice whether to marry or not. Since **marriage** was a perilous state, quite a few exercised that choice and opted for celibacy.

However [later]…at first sight, apparently, **St Paul** was an unimpressive and unattractive little man, with a 'big bold head', crooked legs, dark thick eyebrows that grew together and a large nose; hardly, one would think, a man to foment one of the great upheavals in the human psyche. But the letters of St Paul represent the beginning of a revolution in human sensibility of seismic proportions. In *Romans* he writes about his body: 'o wretched man that I am! who shall deliver me from this body of death? I thank God through Jesus Christ Our lord. So then with the mind I myself serve the law of God; but with the flesh the law of sin'. This is a declaration of war on the human body. And when a man declares war on himself, the first casualty is woman. It is a war that is still being fought. Though St Paul did not advocate that all Christians remain **celibate**, realizing that such a condition would have been incompatible with his ambitions to broaden the new faith's appeal, his bleak view of human sexuality as a necessary evil provided one justification for the Church's increasingly misogynistic vision. **Sanctity** was identified more and more with virginity. The rebellious body had to be put down, and like an enemy citadel, it was laid siege to with fasts, deprivations, and other punishments including, most importantly, abstinence from sex. The Greeks and Romans were taught it was necessary to master passions. But according to the Christian teacher **Clement of Alexandria** (circa AD 150–215) 'our ideal is not to experience desire at all'.

Box 5.8: Teleology, Sex, Misogyny, and Eastern Civilizations

In order to complement the information given in Boxes 5.2, 5.6 and 5.7, as well to avoid being too "Western centric," I will provide a few excerpts from Holland's *A Brief History of Misogyny* that show how sexuality and gender were historically seen in religions of Eastern civilizations. Basically, while sexuality per se was seen in very different ways within different cultures and religions, misogyny was commonly prevalent in all agricultural "civilizations":

> "**Hinduism** and **Buddhism** had developed in **India** over 1,000 years between 1500 BC and 500 BC, with **Taoism** and **Confucianism** in **China** emerging between the seventh and fifth centuries BC. Both civilizations retained traces of much earlier cultures, with what some have interpreted as matriarchal elements. In the earliest **Chinese creation myth**, for example, it was a goddess **Nu Wa** who moulded the human race from clay. Archaeological investigation of the earliest civilizations in the Indus valley reveals a plethora of terracotta figurines of naked women, and the later **Hindu pantheon** contains several powerful goddesses, including **Parvati**, **Durga**, **Sakti** and **Kali**" [see Fig. 9.23]. "Whatever conclusions we might draw from this about the status of women in these early societies, one thing is beyond doubt. Sexual and religious rituals in both civilizations recognized and at times exalted the role of women. Yet, alongside this was a profound contempt, especially noticeable in Confucianism, Hinduism and Buddhism.... The British and other Europeans were shocked, confused and fascinated by **Indian sexual attitudes** and behaviour. Indeed, in some Hindu and Buddhist sects, rituals of **orgiastic intercourse** were seen as the principal path to enlightenment, the way of escaping what the Mexican poet **Octavio Paz** has termed the 'dualistic trap'. [In fact] the great religions of the Eastern civilizations are profoundly different from **Christianity** in that they are not, essentially, philosophically or theologically oriented. Nor do they have a mission – a conviction that they are the holders of an absolute truth regarding the salvation of all mankind with a historical imperative to spread it. Instead their beliefs about the world and the human race's place in it have given rise to complex ethical systems in which ideas are ritualized. They are also completely ahistorical. That is, their beliefs have only personal, not historical consequences; their aim is to allow the individual to achieve happiness in this world (Hinduism, Taoism and Confucianism), or to escape suffering, most radically by extinguishing any sense of self (Buddhism). They do not share the missionary need of **Christians** and **Muslims** to convert or exterminate the unbeliever. That means that, unlike **Islam** and **Christianity**, their misogyny has largely been internal. But what Taoism, Confucianism, Hinduism and Buddhism do have in common with Christianity and Islam is their profound dualism in which the world is seen as being in a permanent state of tension, if not conflict, between body and spirit, self and nature, the one and the many, life and death, male and female, being and non-being.
>
> Except for Confucianism, which was less a religion than a code of etiquette and ethics, these Eastern religions shared a belief with Christians and **Platonists** that the world of the senses is fundamentally an illusion that prevents us from achieving a higher state of being. But unlike Christianity, they posited that dualism could be ended in this world through the practices of certain rituals. However, though the body was viewed as an obstacle to this goal, it was not held to be evil, a sign of our falling away from the divine as it is in Christianity. None of the Eastern religions had any concept equivalent to sin, which made the work of the first missionaries who

arrived in India and China in the seventeenth centuries extremely frustrating. Even in the most ascetic expressions of these beliefs, and both Buddhism and Hinduism produced traditions of holy men and monks who forswore this world for a life of contemplation and physical deprivation, **Puritanism** as the West understands it does not exist. Though scholars have linked **Eastern asceticism** with misogyny in Indian and Chinese societies, the impact this has on women's status remains full of contradictions. Indeed, in Taoism, as in the **Tantric versions** of Hinduism and Buddhism, the body, and in particular **sexual pleasure**, were viewed as a path to **immortality**. Among the practitioners of the Tantric disciplines, it was a release from the cycle of birth, death and **reincarnation**, a **path to Nirvana** in which the self is dissolved. In all these rituals, women played an essential role.

Taoism holds that the world is kept in balance between the interaction of two forces **yin** (female) and **yang** (male). This interaction gives rise to change, according to the ***I-Ching*** or ***Book of changes***. There are two keys to a long life. The first lies in the retention of semen – a belief found in many cultures around the world. The second key, held to be just as vital, is the imbibing of vaginal secretions. Taoists believed that while man produced a limited amount of his precious fluid, woman's supply was infinite. In China, it led to elaborate **sexual rituals**, the aim of which was to rouse the woman to **orgasm**, but not the man. Not surprisingly, **cunnilingus** [stimulation of the female genitals using the tongue or lips] was popular among the Chinese: 'the practice was an excellent method of imbibing the precious fluid', according to one authority. In a series of texts, known as *Bed treatises*, produced between the Sui and Ming dynasties (AD 581–1644), methods of retaining semen whilst absorbing as much of the female fluid as possible are outlined in minute detail. The ultimate aim was to unite the male and female fluids, obliterating sexual dualism, and achieving (it was believed) a kind of immortality. The treatises were eventually suppressed under the conservative **Qing dynasty** (1644–1912), along with the erotic novels of the **Ming years** (1368–1644), though some continued to survive on the Chinese black market.

Tantric Buddhism in India was a rebellion against the rigidity of the **Hindu caste system** and its religious rituals based on the belief in reincarnation (which held that a person's behaviour in this life determines his or her status in the next). **Tantrism's sexual rituals** were orgiastic, beginning with the banquet, in which food was eaten from the body of a naked woman lying face up; the devotees then had intercourse in public. They believed that through **sexual ecstasy** they could break free of the cycle of reincarnation and reach the state of **Nirvana**. One historian has compared **Tantrism** to the sexual revolution of the 1960s, its sexual permissiveness a challenge to moral, social and political authority. However, one does not have to go to the extremes of Tantric Buddhism to realize that Indian sexual practices differ from those of the West in their recognition of woman as a sexual being. From the **Kamasutra**, to the **Tantric rituals**, **Indian eroticism** sees the woman as an active participant, and the aim of both men and women is to give pleasure to each other. Likewise, among the Chinese, **sexual relations** between men and women were not dominated by a sense of sin or shame but by the need to manage desire and passion.

In the ***Confucian book of rites*** husbands are instructed that 'even if a concubine is growing old, as long as she has not yet reached 50, [you] shall have intercourse with her once every five days'. In that sense, a kind of sexual equilibrium is attained, which seems the very opposite to the **misogyny** that developed in the West and that tried to deny women their sexual nature. Yet, however much the recognition of **female sexuality** expressed itself both in the Indian and Chinese civilizations, it did not protect women from being treated with contempt in other ways. In the teachings of **Confucius** (551–479 BC), which dominated Chinese thinking for at least

2000 years, a complex ethical system was constructed, along with a precise etiquette to govern social relations. It was a **patriarchal system**, in which relations within the family reflected both the order of the cosmos, and the structure of the state. China was a **polygamous society** [polygynous] for much of its history – polygamy was only finally outlawed in 1912 with the collapse of imperial rule. It had a very large middle class, with most men possessing between three and a dozen wives and concubines. There were also luxurious establishments where the rich might visit courtesans. Although, in accordance with Confucian doctrine which aimed always at balance and order, the husband was expected to look after his wives and concubines' economic and sexual needs, in other ways women were treated with disdain. In India also, the voluptuous eroticism that exalted female sexuality coexisted with a host of discriminatory practices that lowered women's social status. In the Indian epic of the fifth century BC ***The mahabharata*** the birth of a daughter is hailed as a misfortune, and it is declared, 'women are the root of evils; for they are held to be light-minded'.... *The mahabharata* makes clear that traditionally Hinduism was especially fierce in its taboos against menstruating women. In some cases, a woman was whipped if she even touched a man while she was having her period."

Regarding the "three religions of the book"—Christianity, Judaism, Islamism—a particularly illustrative example that should be familiar to many of you concerns the case of **Maria Magdalena** (Fig. 5.19). Originally, according to the four canonical gospels, Magdalene traveled with **Jesus** and was a witness to his crucifixion, burial, and resurrection. That is, she clearly occupied a prominent role within the followers of Jesus. But, later, most earliest Church Fathers do not mention her at all, or only briefly. In contrast, **Ephrem the Syrian**, who lived in the fourth century, makes one of the first identifications of Magdalena as a redeemed sinner. Later, in the Middle Ages, she is conflated with Mary of Bethany and thus with a "sinful woman." The version of Mary Magdalene as a prostitute held on for centuries after **Pope Gregory the Great** made it official in a sixth-century sermon, although such a view was not adopted neither by Orthodoxy nor by Protestantism when those faiths later split from the **Catholic Church**. Only many centuries later, about just five decades ago, in 1969, this identification with a "sinful woman" was removed in the General Roman Calendar—but the view that she was a prostitute mainly continued to persist in popular culture, within many countries, until the present time. This example thus clearly illustrates how even one history can be dramatically changed through time, mainly by men, from Magdalena being one of the favorite—if not *the* favorite—follower of Jesus, to her being a repentant, penitent, prostitute.

Apart from contributing—together with the subjugation based on differences in physical force—to the exclusion of women in many tasks historically, man-made **misogynistic narratives** have also constantly minimized the historical importance of those women that were able to liberate themselves from the societal chains of subjugation, oppression, and prejudice. As always happens in history, such a minimization was not only done by political and religious leaders and the broader public, but even by scientists and, strikingly, by historians. One illustrative, and

Fig. 5.19 "The Penitent Magdalene" (c.1598), by Tintoretto

extremely absurd example of this, concerns one of the most renowned biologists and philosophers of all times, Aristotle, who was hugely misogynist and went all the way to even minimize the contribution of women for reproduction. Yes, you read correctly: for reproduction—the absurdities of *Neverland*'s world of unreality have no limit. As explained in the book *Biology and Feminism*, edited by Linn Nelson:

> **Aristotle** (384–322 BCE) was a logician, philosopher, physicist, and biologist.... But like many scientists, his views were in part informed by his historical and social context. For Aristotle, this context was **ancient Greece**, and fundamental differences between men and women, and males and females generally, were widely accepted and argued for. In brief, from at least **Plato** (428–348 BCE) forward, it was believed that men are superior to women in myriad ways. Like other philosophers of his time, Aristotle theorized about the differences between the sex/genders and sexes, largely in his biological research.... But he faced a conundrum putting together his theories concerning the "nature" of each sex/gender and his explanation of human reproduction. Women, Aristotle held, are "incomplete [or distorted] men", physically and intellectually inferior to them. And his explanation of sexual

reproduction was **androcentric** (i.e., "male centered"), apparently reflecting the view that women's role, in all areas of human endeavor, was of far less consequence than that of men. Aristotle maintained that women provide the space and necessary physical matter for a human to grow. Sperm contain the "form" of a tiny human male. Given this one would expect all babies to be male. But this is obviously not the case. Aristotle hypothesized that, while the form a man contributes is always male, female offspring result when a woman's womb is insufficiently warm – resulting in an incomplete or distorted child – that is, a female.

[More recently] in "*The energetic ggg*" an article written for the lay public and published in 1983, developmental biologists Gerald Schatten and Heidi Schatten argued that there were striking parallels between the **Grimm Brothers**' fairytale "**Sleeping Beauty**" and the then accepted account of **fertilization**. They challenged the account's portrayal of the egg as passive and dormant until "penetrated and activated by a sperm" based on observations they understood to indicate that the egg's role in fertilization was as active as that of sperm…. Based on their observations using scanning electron microscopy, and on their reevaluation of reports of observations others had earlier made, Schatten and Schatten maintained that "it is becoming clear that the egg is not merely a large yolk-filled sphere into which the sperm burrows to endow new life". Rather, they contended, "recent research suggests the almost heretical view that sperm and egg are mutually active partners". What Schatten and Schatten observed is that when the egg and sperm interact, the sperm does not "burrow into the egg"; rather the egg "directs" the growth on its surface of small finger-like projections (called "microvilli") to clasp the sperm so that it can draw the sperm into itself. Interestingly, Schatten and Schatten also noted that as early as 1895 E. B. Wilson had published photographs of sea urchin fertilization in which the egg's extension of microvilli to the sperm was visible. And there was more evidence to come that would support Schatten and Schatten's observations and interpretation of them.

In 1991, feminist anthropologist **Emily Martin**, who also compared the classic account of fertilization to "Sleeping Beauty" and maintained that scientists "had constructed a romance" between egg and sperm "based on stereotypical male-female differences", studied research into fertilization undertaken in the late 1980s and early 1990s in a lab at Johns Hopkins University. It had long been assumed that sperm used mechanical means to get through the zona (a thick membrane surrounding the ovum) and penetrate the egg. As Martin chronicles, earlier investigations had emphasized "the mechanical force of the sperm's tail" in enabling fertilization. To their great surprise, Martin states, investigators at Johns Hopkins Discovered…that the forward thrust of sperm is extremely weak, which contradicts the assumption that sperm are forceful penetrators. Rather than thrusting forward, the sperm's head was now seen to move back and forth. The sideways motion of the sperm's tail makes the head move sideways with a force that is ten times stronger than its forward movement. In fact, its strongest tendency, by tenfold, is to escape by attempting to pry itself off the egg. Given these observations, Martin noted, the scientists concluded that the egg traps the sperm and adheres to it so tightly that the sperm's head is forced to lie flat against the surface of the zona.

The same book, *Biology and Feminism*, also provides examples of even more "modern" and so-called scientific narratives that are plainly misogynistic and not based on empirical data at all, such as those about medicine included in Box 5.9, as well as others mentioned in the following excerpts of that book:

By 1978 when he published *On human nature* [a book intended for a general audience], [Edward] **Wilson** and other human **sociobiologists** had increasingly turned to explaining social behaviors they attributed to contemporary humans as adaptations selected for during the **Pleistocene** [often referred to as the 'Ice Age' including 'cave-men', this geological epoch lasted from about 2580000 to 11700 years ago]. In addition to proposing genetic bases for **xenophobia** and **war**, Wilson and other human sociobiologists devoted a good

Box 5.9: Links Between Misogyny, Science, Biases, and Medicine

In the book *Biology and Feminism*, edited by Lynn Nelson, it is stated:

> The changes the nineteenth century brought to medical views about women's biology and health were dramatic and multifaceted. First, differences between men's and women's organs and biological processes related to reproduction came to be viewed as significant and of consequence in relation to their "roles" and health. Second, women's organs and biological processes related to reproduction came to be understood as detrimental to **women's health**, if not pathological. Writing in 1900 about the "ravages" thought to be wrought by the "sexual storms" of **female puberty**, **menstruation**, **pregnancy**, and **childbirth**. The president of the American Gynecology Society used an analogy with shipwrecks to describe the dangers women encounter during each phase of their reproductive years. Third, medical experts came to view women's reproductive organs, her ovaries and uterus, as the source of disease and illness involving other organs, and as dictating that her role is that of wife and mother. Of course, this view of women's proper role was not new. What were new were the ideas that it is women's ovaries and uteruses that determine that role, and that their proper or improper functioning determines women's health or illness in every respect – even in terms of illnesses and diseases to which men are also subject. As one medical professor declared in 1870, it is "as if the Almighty in creating the female sex, had taken the uterus and built up a woman around it". Many physicians and psychologists emphasized women's ovaries as dictating all aspects of **femininity**, including women's psychology. One physician, speaking of women's ovaries and uteruses, declared "Women's reproductive organs are pre-eminent."
>
> We have seen that for many centuries menstruation was viewed as analogous to the kinds of purging of excess fluids and/or impurities that men purged by sweating or blood-letting. But, from the perspective of nineteenth and early twentieth century medicine, menstruation was a unique kind of biological process, a process that was debilitating if not pathological – putting women at high risk from the onset of menses until **menopause**. Women (at least those belonging to the middle or upper class), medical experts maintained, require extensive rest before and during their monthly periods. In 1870, the zoologist, Walter Heape, expressed a common view when he described menstruation as a "severe, devastating, periodic action…[that leaves behind] a ragged wreck of tissue, torn glands, ruptured vessels, jagged edges of stroma, and masses of blood corpuscles". Not surprisingly, Heape argued that menstruation requires medical treatment. "It would seem hardly possible", he maintained, that a woman could "heal satisfactorily without the aid of surgical treatment". Popular books advising women about menstruation sounded similar cautions.
>
> In the second half of the century, what Douglas-Wood describes as …"fumbling experiments with the female interior," gave way to surgeries, many of which were undertaken to deal with diseases in other organs as well as "**female personality disorders**". For a brief period in the 1860s, some physicians treated "**nymphomania**" and "**intractable masturbation**" by removing the **clitoris**, although many physicians disapproved of the surgery. The most common surgery for women's diseases, including "**personality disorders**", was removal of the ovaries. Ehrenreich and English note that in 1906, a leading American gynecological surgeon estimated that 150,000 women in the country had had their ovaries removed. In the last third of the century, physicians' arguments…came to reflect their understandings of the implications of **Darwinism**. Ehrenreich and English note that "civilization" was taken to explain why "the middle-class woman [was] sickly; her physical frailty went hand-in-white glove-hand with her superior modesty, refinement, and sensitivity". In con-

trast, working class women "were robust, just as they were supposedly 'coarse' and immodest". Assuming Darwin's general characterization of the significant sex/gender differences that evolution produced because of the greater selection pressures to which men were subject, it seemed, as feminists describe the issue, that in terms of the wealthy classes, that some came to believe that "men evolve and women devolve". Interests in maintaining the current "social order", as well as economics practices, also appear to have been factors in how poor and working class women were viewed. After all, someone had to do the work of scrubbing floors and other physical tasks; so, viewing poor and working class women as "robust", functioned to support the social order.

Feminists also cite growing interest on the part of middle class women in pursuing education and careers, and opposition against changes to **sex/gender roles**, as factors contributing to the perception of what might otherwise be viewed as normal and natural processes, such as puberty and menstruating, as requiring that women avoid intellectual activities. Arguments offered by medical experts against admitting women to college frequently cited the dangers to their reproductive organs. Among the most influential were arguments offered by Dr. Edward H. Clarke in his book, *Sex in education: or a fair chance for girls*. Clarke, a professor at Harvard, published the book in 1873 when pressure to admit women to that institution was at its height. As Ehrenreich and English chronicle, Clarke was opposed to women's admission. He appealed to what were common medical views, warning that women who engaged in strenuous mental activity – who studied in a "boy's way" – risked atrophy of their ovaries and uteruses, insanity, and sterility. Studying would cause their brains to drain energy and/or blood from their reproductive organs. Clarke's arguments persuaded many, even some successful women who came to believe education had harmed them; and some colleges cut back on the number of courses women could take in a year. A warning issued by R.R. Coleman, M.D. to women entering colleges or seeking to be admitted was representative. "Women beware"m he wrote, "you are on the brink of destruction…science pronounces that the woman who studies is lost". Clearly, at least some were using medical theories to reinforce notions about women's proper role and to prevent them from entering spheres that men had dominated.

deal of attention to proposing adaptations of what they assumed or argued to be **sex/gender differences** in mating strategies and behavior…of why men are "**promiscuous**" and "undiscriminating" and women are "coy" and "choosy"; of **rape**; and of societal practices and institutionalized norms that…. Wilson described as a universal "Double Standard" that allows men to engage in a range of activities frowned upon or outright denied to women…. [David] Barash offered the following argument in a book written for the lay public: "because men maximize their fitness differently from women, it is perfectly good biology that business and profession taste sweeter to him, while home and child care taste sweeter to women…while it may be true that it's "not nice to fool Mother Nature", it can be done…biology's whispers can be denied, but in most cases at a real cost…although women who participate [in work outside the home] may be attracted by the promise of "liberation", they are in fact simply adopting a male strategy while denying their own…. Cavalier female parenting is maladaptive for all mammals; for humans, it may be a socially instituted trap that is harmful to everyone concerned"…. He also suggested that rape may be a reproductive strategy: "rape in humans…is by no means as simple [as the rape among mallard ducks I have observed], influenced as it is by an extremely complex overlay of cultural attitudes…nevertheless, mallard rape and bluebird adultery may have a degree of relevance to human behavior...perhaps human rapists, in their own criminally misguided way, are doing the best they can to maximize their fitness...if so, they are not that different from the sexually excluded bachelor mallards", Barash notes.

It is remarkable that *despite* all these biases, prejudices, and narratives, and the historical subjugation of women to which they have contributed to, there is growing evidence showing that some women indeed managed to find a way to actually be far more relevant than what we are often told. And this applies to a wide range of intellectual tasks, including their involvement in something that was unheard of, until very recently: **manuscript production** in the European Middle Ages, as noted in a recent 2019 article by Radini and colleagues. One rather short, but beautifully illustrated and accessible book that I would recommend you to show to your kids and/or grownup friends and colleagues, as this is still so much needed precisely to fight against such just-so-stories, is **Rachel Ignotofsky**'s 2016 book *Women in Science – 50 Fearless Pioneers Who Changed the World.* Importantly, as it usually happened in human history, the mostly man-made narratives stating that women are passive players, not only concerning their sexuality but in most other aspects of life and in human evolution in general, were also widely disseminated by the use of art. This point was explained in **Stephanie Moser**'s also wonderfully illustrated 1998 book *Ancestral Images – The Iconography of Human Origins* (see also Box 3.7 and Figs. 3.4, 5.20, 5.21, and 5.22). As she explains, most images/displays about human evolution have historically focused on men, as if women were completely inert, or even totally inexistent, for millions of years. This point applies to three of the so-called most crucial moments of human evolution: **discovery of fire** (Fig. 5.20), **stone tool building and use** (Fig. 5.21), and **cave painting** (Fig. 5.22). Referring specifically to Fig. 5.22, she wrote: "in the...mural, featuring the **Cro-Magnons**, a male artist is seen adding the final touches to his depiction of a great beast on the cave wall...significant is the omission of women, which was an explicit instruction that Osborn gave to Knight (the painter)...this omission adds another dimension to the sexual division of labour in reconstructions, introducing the notion that women were not involved in areas of cultural achievement such as the production of art."

Of course, such **misogynistic narratives**, as most teleological tales, are self-contradictory, and not supported at all by empirical data. First, men can't have their cake and eat it too: they cannot be both "men the hunter-savior," "men the toolmaker," and "men the artist," all at the same time. If, as the misogynistic narratives suggest, they were the hunters, providers, and saviors of the whole group, being so brave that they go on for several kilometers a day to hunt huge dangerous animals in order to bring food to, and "save," everybody, including the "poor passive" women, when did they have time to invent fire, produce stone tools, and do cave painting? Does not seem easy to walk so many kilometers to hunt and, during those very long walks when they have to be particularly alert, still have the time to stop and the needed attention to invent all those things, particularly because all of them require being in a single place for a long time, particularly building stone tools and doing cave paintings. More logically, in such a narrative, would be that the ones that would travel less far away—to gather wild fruits and vegetables, a task that often

Fig. 5.20 "*Homo erectus* and fire"

involves walking a smaller distance, and that, we now know, did actually very likely usually provide the main bulk of diet for the entire group, as it does in most hunter-gatherer groups today—, the women, would be the ones that would have more time and be more often in the right place near the nomadic temporary settlements, to do such innovative and artistic tasks. In this sense, it is very likely that, *in the world of reality*, women have contributed hugely to new inventions, stone tool building, cave painting, and so on. Unless, of course, there were already narratives, social norms, and taboos against women doing such tasks, back then, hundreds of thousands, or even millions of years ago (some studies indicate that control of fire by humans might data back to 2 million years ago, or more). However, that does not seem very plausible, in face of the historical, archeological, anthropological, and biological evidence now available, at least not in the sense that it would apply to *all* these tasks, *continuously*, for *such a long time* of our evolutionary history, because in general in

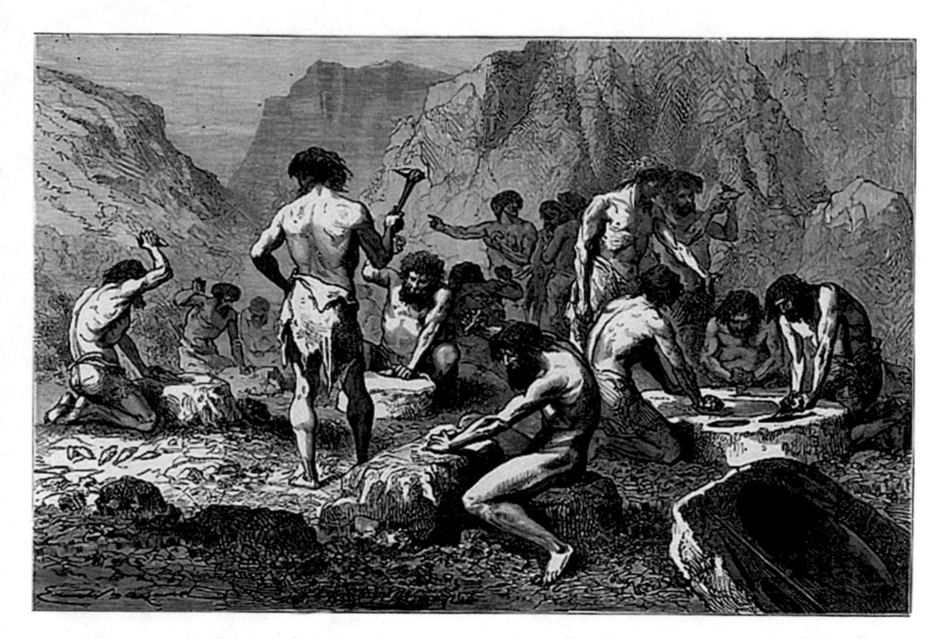

Fig. 5.21 "The earliest Manufacture and Polishing of Flints," by Emile Bayard for Louis Figuier's *L'homme primitif*, 1870

Fig. 5.22 "Cro-Magnon artists," by Charles Knight, 1924

nomadic hunter-gatherer groups, there is usually far less subjugation of women by men than in sedentary/agricultural societies.

In actual fact, several recent and less biased archeological studies are finding that many fossil skeletal remains that are found near places of **stone tool production** are from women, as would be expected from what we now know about the life of most nomadic hunter-gatherer groups, because moreover this precisely applies to some living groups that still produce stone tools. A 2003 paper published by Weedman and colleagues became tremendously widespread in the media across the globe as it was said to contain "shocking" news. As explained to the broader public in a press-release about that 2003 paper, published by the American website *ScienceDaily*:

> **Man the toolmaker**: the idea of men as stone tool producers may need some rechiseling, say University of Florida (UF) scientists who found women sometimes are the masters. The research among an Ethiopian group indicates stone tool working is not just a male activity, but rather that women probably had an active part in creating stone tools, one of the most ubiquitous materials found on prehistoric sites. "It really gives women a presence in the archaeological record and a chance for us to reflect upon a place in prehistory where women basically have been invisible", said **Kathryn Weedman**, a UF anthropology lecturer who led the National Science Foundation-funded research. There has always been this image of 'man the toolmaker' because it's generally perceived by the public, and many archaeologists, that males were the ones who made stone tools", said **Steve Brandt**, a UF anthropology professor and co-leader of the research team. "But we found that among one ethnic group, the **Konso of Ethiopia**, women dominate the activity". The Konso women create a stone tool called a scraper to clean animal hides to be made into bedding and clothing, he said. Stone tools are important because they were the first recognizable object people made, marking the beginning of the archaeological record dating back as early as 2.6 million years ago, Weedman said. Not until 5,000 to 10,000 years ago were pottery and metal tools introduced, she said. "Stone artifacts are critical for identifying a wide range of activities that will help us learn what life was like", she said. "Basically, they trace the evolution of human culture because, for better or worse, they are often the only things preserved".
>
> If people were found to be scraping antelope hides 100,000 years ago, for example, that might help tell us when they started making prepared clothing, Brandt said. The Konso project "is vitally important both in documenting how stone tools are made and used – most people who used stone tools have been dead for hundreds or thousands of years – and in the social context of their use", said Michael Shott, an anthropology professor at the University of Northern Iowa. The tradition of stone toolmaking continued into the 20th century in isolated parts of Africa, Australia and Siberia, but in the last couple of decades it has virtually disappeared as an everyday activity, except perhaps for the hide workers of Ethiopia, Brandt said. The project is unique because it provides evidence that women actively flake stone to produce tools, Brandt said. "This project changes our perspective dramatically because theoretically we can talk about **gender issues** – the role of men and women in ancient societies", he said. In the study described in the September/October issue of *Archaeology magazine*, the UF researchers identified 119 Konso hide workers who used flaked stone, glass or iron to scrape hides. Seventy-five percent of the hide workers were women, and most – 73 percent – were 40 or older, Weedman said. Members of this group are born into the hide-working profession and remain locked into it, Weedman said. The products they make have not yet been completely replaced by Western industrial products, she said. "No one is living in the stone age", Brandt said of today's stone toolmakers. "They're wearing Western clothes, they have radios, they may even know about 9-11", he said. "They make these stone tools because the tools work. Stone is still the superior material – they prefer it over glass and iron"."

Another, even more recent paper, published in September 2020, by Martínez-Sevilla and colleagues—entitled "*Who painted that? The authorship of schematic rock art at the Los Machos rockshelter in southern Iberia*"—provides the first direct evidence pointing out that women likely—probably, but not surely in the specific example examined in that paper—have also contributed to **cave painting**. As summarized for the broader public by Sam Jones in *The Guardian*, in an article in which he interviewed Martínez-Sevilla:

> One day, perhaps a little over 7,000 years ago, a man in his 30s and a younger companion dipped their fingers in ochre pigment and set about daubing the walls of a shallow cave in southern Spain with anthropomorphic, circular and geometric designs. Today, thanks to the fingerprints they left behind in the natural shelter of Los Machos in the province of Granada, researchers have been able to determine their sexes and ages. The study, carried out by experts from the University of Granada, Durham University and the Autonomous University of Barcelona, has established that the pictures were painted by a 36-year-old man and, most probably, by a woman a little less than half that age. Their findings, published in the journal *Antiquity*, suggest for the first time that both men and women took part in making rock art, and that it was a social, rather than an individual, act. Francisco Martínez Sevilla, a researcher at the University of Granada, said that while experts had long known about the 32 neolithic images painted on the slopes of the Cerro de Jabalcón some time between 5,500BC and 2,500 BC, the figures had only recently begun to yield some of their many secrets. "We looked at the number of fingerprint ridges and the distance between them and compared them with fingerprints from the present day", he told the *Guardian*. "Those ridges vary according to age and sex but settle by adulthood, and you can distinguish between those of men and women…you can also tell the age of the person from the ridges".
>
> While some of the fingers that painted the pictures belonged to a man aged at least 36, Martínez Sevilla and his colleagues cannot say for certain who his fellow artist was. There is a high likelihood that they were a woman aged under 20, or perhaps a juvenile male. "From our point of view, if there are two people taking part in the creation of this pictorial panel, it means it must have been a social, rather than an individual, act, as we'd thought until now…. It shows us that these manifestations of art were a social thing and not just done by one individual in the community, such as the shaman or whoever". Although Martínez Sevilla acknowledges that he and his colleagues will never know the relationship between the pair – nor the significance of the images they left behind – the enduring art speaks for itself. "The area where they are, and the fact that they haven't been changed or painted over, gives you the feeling that this was a very important place and must have had a really important symbolic value for this community", he said. As an archaeologist, added Martínez Sevilla, he tends to see the person behind the stone or the tool rather than just the object itself. "And when I look at these pictures, there's a bit of an emotional response because I see a person, many thousands of years ago, painting symbols or designs that would have meant something to them, or which would have been a way for them to express themselves, or identify the territory, or communicate socially".

Similarly, a paper that was published by Haas and colleagues, even more recently, in early November 2020, entitled "*Female hunters of the early Americas,*" argues that the "man-the-provider" narrative is wrong not only because by gathering wild plants women usually contribute more to the bulk of the diet eaten by the whole group but also because at least some women were moreover also involved, at least some times, in hunting, historically:

> Scholars generally accept that projectile points associated with male burials are hunting tools, but have been less willing to concede that projectile points associated with female

> burials are hunting tools…[our case study] presents an unusually robust empirical test case for evaluating competing models of gendered subsistence labor…. It is possible that the burial [studied by Haas and colleagues] represents a rare instance of a **female hunter** in a male-dominated subsistence field, but such an outlier explanation diminishes with the observation of 11 female burials in association with hunting tools from 10 Late Pleistocene or Early Holocene sites throughout the Americas, including Upward Sun River, Buhl, Gordon Creek, Ashworth Rockshelter, Sloan, Icehouse Bottom, Windover, Telarmachay, Wilamaya Patjxa, and Arroyo Seco 2. These results are consistent with a model of relatively undifferentiated subsistence labor among early populations in the Americas. Nonetheless, hunter-gatherer ethnography and contemporary hunting practices make clear that subsistence labor ultimately differentiated along sex lines, with females taking a role as gatherers or processors and males as hunters. Middle Holocene females and males at the Indian Knoll site in Kentucky were buried with atlatls in a respective ratio of 17:63, suggesting that big-game hunting was a male-biased activity at that time. Thirty percent of bifaces, including projectile points, are associated with females in a sample of 44 Late Holocene burials from seven sites in southern California. A similar trajectory may be observed in the European Paleolithic, where meat-heavy diets and absence of plant-processing or hide-working tools among Middle Paleolithic **Neandertals** would seem to minimize potential for sexually differentiated labor practices.
>
> Economies diversified in the Upper Paleolithic sometime after 48 ka, with increasing emphasis on plant processing and manufacturing of tailored clothing and hide tents creating new contexts for labor division. When and how such differentiated labor practices emerged from evidently undifferentiated ones require further exploration. Scholars have long grappled with understanding the extent to which contemporary gender behavior existed in our species' evolutionary past. A number of studies support the contention that **modern gender constructs** often do not reflect past ones. Dyble *et al.* show that both women and men in ethnographic hunter-gatherer societies govern residence decisions. The discovery of a **Viking woman warrior** further highlights uncritical assumptions about past gender roles. Theoretical insights suggest that the ecological conditions experienced by early hunter-gatherer populations would have favored big-game hunting economies with broad participation from both females and males. Such models align with epistemological critiques that reduce seemingly paradoxical tool associations to cultural or ethnographic biases. [Our case study] and the sum of previous archaeological observations on early hunter-gatherer burials support this hypothesis, revealing that early females in the Americas were big-game hunters.

However, one important distinction needs to be made: while there are no reasons to think that women were less involved than men—they might have been more—before the rise of agriculture concerning the discovery of fire, stone tool building and use, and cave painting—which are features chiefly found in humans—the scenario regarding hunting is very different. That is, women might have participated more in hunting than it has been assumed, but they clearly were not as prevalent in hunting as men, based on the evidence available not only about our evolutionary lineage but also concerning chimpanzees. In fact, in a 2017 paper by Gilby and colleagues, entitled "*Predation by female chimpanzees,*" the authors explained that the results "suggest that before the emergence of social obligations regarding sharing and provisioning, constraints on hunting by females did not necessarily stem from maternal care…instead, they suggest that a risk-averse foraging strategy and the potential for losing prey to males limited female predation [as is the often the case in chimpanzees]…sex differences in hunting behavior would likely have preceded the evolution of the sexual division of labor among modern humans."

Fig. 5.23 "Death of Hypatia in Alexandria"

I will end this Section by mentioning the particularly disturbing and tragic story of a women that was clearly an *active player*, in her personal life and within science in general, **Hypatia of Alexandria** (Fig. 5.23). Clearly, such an active role was not well taken, for numerous reasons, by most men, as it is often the case in agricultural states, as can be seen in excerpts written for the broader public, from the entry about her in the online *Ancient history encyclopedia* (www.ancient.eu/Hypatia_of_Alexandria):

> Hypatia of Alexandria (c. 370 CE – March 415 CE) was a female philosopher and mathematician, born in Alexandria, **Egypt** possibly in 370 CE. She was the daughter of the mathematician Theon, the last Professor at the **University of Alexandria**, who tutored her in math, astronomy, and the philosophy of the day which, in modern times, would be considered science. As the historian Deakin writes, "the most detailed accounts we have of Hypatia's life are the records of her death…we learn more about her death from the primary sources than we do about any other aspect of her life". She was murdered in 415 CE by a Christian mob who attacked her on the streets of Alexandria. The primary sources, even those Christian writers who were hostile to her and claimed she was a witch, portray her as a woman who was widely known for her generosity, love of learning, and expertise in teaching in the subjects of **Neo-Platonism**, mathematics, science, and philosophy in general. In a city which was becoming increasingly diverse religiously (and had always been so culturally) Hypatia was a close friend of the pagan prefect Orestes and was blamed by **Cyril**, the Christian Archbishop of Alexandria, for keeping Orestes from accepting the 'true faith'.
>
> She was also seen as a 'stumbling block' to those who would have accepted the 'truth' of Christianity were it not for her charisma, charm, and excellence in making difficult mathematical and philosophical concepts understandable to her students; concepts which contradicted the teachings of the relatively new church. Alexandria was a great seat of learning in

the early days of Christianity but, as the faith grew in adherents and power, steadily became divided by fighting among religious factions. It is by no means an exaggeration to state that **Alexandria** was destroyed as a centre of culture and learning by religious intolerance and Hypatia has come to symbolize this tragedy to the extent that her death has been cited as the end of the classical world. By all accounts, Hypatia was an extraordinary woman not only for her time, but for any time. Theon refused to impose upon his daughter the traditional role assigned to women and raised her as one would have raised a son in the **Greek tradition**; by teaching her his own trade. The historian Slatkin writes, "Greek women of all classes were occupied with the same type of work, mostly centered around the domestic needs of the family…women cared for young children, nursed the sick, and prepared food". Hypatia, on the other hand, led the life of a respected academic at Alexandria's university; a position to which, as far as the evidence suggests, only males were entitled previously. She never married and remained celibate throughout her life, devoting herself to learning and teaching. The ancient writers are in agreement that she was a woman of enormous intellectual power.

Deakin writes: the breadth of her interests is most impressive. Within mathematics, she wrote or lectured on astronomy (including its observational aspects – the astrolabe), geometry (and for its day advanced geometry at that) and algebra (again, for its time, difficult algebra), and made an advance in computational technique – all this as well as engaging in religious philosophy and aspiring to a good writing style. Her writings were, as best we can judge, an outgrowth of her teaching in the technical areas of mathematics. In effect, she was continuing a program initiated by her father: a conscious effort to preserve and to elucidate the great mathematical works of the Alexandrian heritage. This heritage was so impressive that Alexandria rivaled Athens as a jewel of learning and culture. From the moment of its founding by **Alexander the Great** in 332 BCE, Alexandria grew to epitomize the best aspects of civilized urban life. Early writers like Strabo (63 BCE-21 CE) describe the city as "magnificent" and the university was held in such high regard that scholars flocked there from around the world.

The great **Library of Alexandria** is said to have held 500,000 books on its shelves in the main building and more in an adjacent annex. As a professor at the university, Hypatia would have had daily access to this resource and it seems clear she took full advantage of it. In 415 CE, on her way home from delivering her daily lectures at the university, Hypatia was attacked by a mob of Christian monks, dragged from her chariot down the street into a church, and was there stripped naked, beaten to death, and burned. In the aftermath of Hypatia's death the University of Alexandria was sacked and burned on orders from Cyril, pagan temples were torn down, and there was a mass exodus of intellectuals and artists from the newly-Christianized city of Alexandria. Cyril was later declared a saint by the church for his efforts in suppressing paganism and fighting for the true faith. Hypatia's death has long been recognized as a watershed mark in history delineating the classical age of paganism from the age of **Christianity**. History is clear: Alexandria began to decline as Christianity rose in power and the death of Hypatia of Alexandria has come to embody all that was lost to civilization in the tumult of **religious intolerance** and the destruction it engenders.

5.7 Myths and Facts About Gender Roles and Differences

The evidence is clear: from the constitutional standpoint woman is the stronger sex. (Ashley Montagu)

The *Cambridge Dictionary* common definition of **misogyny** is: "feelings of hating women, or the belief that men are much better than women." Sedentism and in particular agriculture and the subsequent societal changes, including the origin of major organized religions, strongly contributed to many of the **misogynistic narratives** that are, unfortunately, still so prevalent nowadays. An interesting point is that when I was doing an extensive literature review for this chapter, I was more focused on the prevailing narratives about sex, love, and marriage. But, within the numerous papers and books I read, I found, once and once again, empirical evidence showing that women outperformed, on average, men in almost all the items discussed in those publications. This was striking for me. Not because the misogynistic narratives that are still so accepted nowadays tend to portrait exactly the opposite, as I know that they are just *Neverland* tales: it was instead because one would think that the distribution of so-called biological strengths would be more equal between genders, based on the evolutionary "trend-offs" that are so common in biological evolution. Of course, in order to discuss this subject in a comprehensive way, one would need to examine *all* the biological items that were studied, and compared, in women and men. Clearly, this was not the main goal of this book, and some people would probably even argue that it would better to not do this, as that could lead to the rise of discussions about the "superiority" of a certain gender, when one should instead focus on bringing genders, and all groups of humans in general, closer together.

But the fact that I *did not aim* at all do undertake such a comprehensive comparison is precisely, in a way, the main strength of this section, because within the biases that all scientists have—including myself, despite the fact that I try so hard not to—the cases I will refer to here are *not* the result of a priori planning or "cherry-picking." I obviously found features in which women perform, on average, worse than men for reasons that seem to be mainly biological—"nature"—rather than cultural. One example is the biological fact that women tend to be more prone to have **autoimmune diseases**, in general. This is precisely the type of cases that might well have to do with evolutionary trade-offs, being the "negative"—"too-much"—side of something that actually saves the lives of millions of women across the globe every year: within different dangerous infections that affect humans today, women tend to have a much more robust immune response to, and thus to die less as a result of, them (see also Box 5.10). Indeed, evolutionarily there is always a bit of tension concerning the immune system, because if it is "too weak" it might not be effective against germs and other organisms, but if it is "too strong," it might attack not only those outside organisms but our own cells: a further example that in the natural world nothing is "perfect" or the "ideal solution," everything depends on very specific circumstances. There are therefore many biological reasons proposed by researchers to explain this gender difference, and probably the reality is a combination of at least some of those reasons, so here I will just cite one of them, based on a recent study by Wilhelmson and colleagues, published in 2018. As summarized—for the broader public—in the website *ScienceDaily*, in a 2018 piece accordingly entitled "*New theory on why more women than men develop autoimmune diseases*":

> New findings are now being presented on possible mechanisms behind **gender differences** in the occurrence of **rheumatism** and other **autoimmune diseases**. The study, published in *Nature Communications*, can be of significance for the future treatment of **diseases**. "It's very important to understand what causes these diseases to be so much more common among women", says Asa Tivesten, professor of medicine at Sahlgrenska Academy,

> Sweden, a chief physician and one of the authors of the study. "In this way, we can eventually provide better treatment for the diseases". In autoimmune diseases, the immune system creates **antibodies** that attack the body's own tissue. Almost all autoimmune diseases affect women more often than men. The gender difference is especially great in the case of **lupus**, a serious disease also known as **systemic lupus erythematosus** or SLE. Nine out of ten of those afflicted are women. It has been known that there is a link between the male sex hormone **testosterone** and protection against autoimmune diseases.
>
> Men are generally more protected than women, who only have one tenth as much testosterone. Testosterone reduces the number of **B cells**, a type of **lymphocyte** that releases harmful antibodies. The researchers behind the study were trying to understand what the connection between testosterone and the production of B cells in the spleen actually looks like, mechanisms that have so far been unknown. After numerous experiments on mice and studies of blood samples from 128 men, the researchers were able to conclude that the critical connection is the protein BAFF, which makes the B cells more viable. "We have concluded that testosterone suppresses BAFF. If you eliminate testosterone, you get more BAFF and thereby more B cells in the spleen because they survive to a greater extent. Recognition of the link between testosterone and BAFF is completely new. No one has reported this in the past", says Asa Tivesten. The results correlate well with a previous study showing that genetic variations in BAFF can be linked to the risk of diseases such as lupus. That disease is treated with BAFF inhibitors, a medicine that has not, however, really lived up to expectations. "That's why this information about how the body regulates the levels of BAFF is extremely important, so that we can continue to put the pieces together and try to understand which patients should have BAFF inhibitors and which should not. Accordingly, our study serves as a basis for further research on how the medicine can be used in a better way".

These topics are being widely discussed while I am writing this book, during the **Covid-19** pandemic, because effectively within a same number of women and men infected with the virus, men are dying much more (Fig. 5.24). Not surprisingly, at the beginning, the media reported that they were "puzzled" by this, because it goes against the common *belief* that women are biologically "weaker" than men. Of course, the media should have known better, because there was already a huge amount of scientific data showing that women are commonly far more resilient in terms of health than men are, particularly concerning infectious diseases as they precisely have in general a much stronger **immune system**. But most media sources obviously neglected such studies, and instead preferred to create all the types of unscientific narratives, to explain what they designated as a "surprising, puzzling scenario." Obviously, when they interviewed the researchers that do know about this subject, those researchers explained that this pattern was not surprising at all.

Sabra Klein, a Johns Hopkins University Professor that is specialist on gender differences and infectious diseases stated, in an interview to the TV channel *France 24* aired the March 13, 2020, that in general women are indeed physiologically/medically tougher than men. **Garima Sharma**, also of Johns Hopkins University, who published a paper on sex differences in **Covid-19 mortality** with other colleagues, argued that one of the specific reasons that likely contributes to the physiological toughness of women in general, and against Covid-19 in particular, is the fact that women have a "backup" **X chromosome**: "X chromosomes contain a high density of immune-related genes, so women generally mount stronger immune responses." But that was clearly not the type of discussion that most of the media platforms—which unfortunately are not as objective as *France 24*—were *looking for*, or *wanted to engage in*: women toughest then men? So, accordingly, most of the media platforms continued instead to discuss another wide range of so-called theories about the "true reasons" leading to this "puzzling" pattern, that is, they continued to discuss just-so-stories made within the context of the longstanding man-made narratives and stereotypes that are still so prevalent nowadays.

A similar, but much older, emblematic case where reality "surprisingly" contradicted myths and stereotypes that were also propagated by the media, in this case about the "natural passivity of women," concerns the first time that women were allowed to vote in a parliamentary election in a currently existing independent country: namely, in **New Zealand**, in November 28, 1893. As proudly reported in the website *New Zealand History* (https://nzhistory.govt.nz/), this election occurred just 10 weeks after the governor, Lord Glasgow, signed the Electoral Act 1893 into law, after a longstanding, hard fight by women, including a third petition calling for them to be allowed to vote. Strikingly—but unfortunately not surprisingly, due to the reasons that have been explained so far—among the about 32,000 signatures that made part of this petition, just 21 are known to have been signed by men. When the date of the election was approaching, the newspaper *The Press* included an editorial

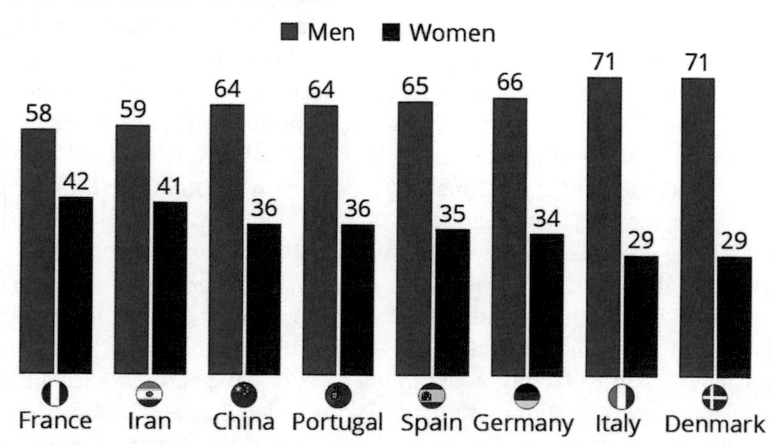

Fig. 5.24 Within the same number of infected women and men, many more men end up dying, than women. (Data as of March 27. Sources: Wall Street Journal, Global Health 50/50)

that called into question the "interest of women in voting," stating that the vast majority of women would not want to vote as they would naturally prefer to stay at home and do "their domestic tasks." Such a *belief* was completely destroyed by empirical facts. Despite the short timeframe for voter registration and the warnings from opponents of women's suffrage that "lady voters" could be harassed at polling booths, 109,461 women—about 84% of the adult female population—enrolled to vote in the election and, on polling day, 90,290 of them casted their votes, a turnout of 82% that was far higher than the 70% turnout among registered male voters.

When I talk with laypeople about differences between sexes, a comment I often hear is: "ohh, but I know a man, or a woman, that…." So I have to emphasize, once again, that these empirical studies usually refer to *averages*, as most scientific studies do. I also typically hear this comment when I refer to other items, such as about smoking normally affecting people's health: "ohh, but I know someone that smoked the whole life and lived more than 100 years." Yes, there are exceptions to—as it is often said, they actually confirm—the rule, but on *average* people that smoke live less that people that do not smoke, when all other factors are similar. Accordingly, if you know some women that are physically stronger than some men—many of them are—this does not mean that this is what happens *on average*. Apart from being more prone to have immune diseases and from having less physical force, one of the few other cases that seems to be potentially a valid case in which men *usually* biologically outperform woman concerns the capacity to handle specific **spatial tasks**, such as **three-dimensional orientation**. This topic was discussed in a 2018 study by Boone and colleagues, entitled "*Sex differences in navigation strategy and efficiency.*"

However, it should be noted that when one carefully reads the science behind some of these studies, one realizes that one needs to take them with a grain of salt, particularly the broader conclusions that they do, because at least some of their authors don't seem to able to escape from the prevailing *Neverland* narratives and cultural stereotypes. In a 2014 study by Oliveira-Pinto and colleagues about **olfactory capacities,** they show that there is a true sexual dimorphism in humans in the sense that females have, on average, more neurons and glial cells in the olfactory bulb than men do. In stark contrast, in the 2018 study by Boone and colleagues about navigation skills, the tests mainly concerned navigation in virtual environments, using computers. The problem is that, particularly at younger ages, males tend to be more used to computers because people tend to follow prevailing gender narratives and thus to give computers and videogames more often to boys than to girls. In fact, this possibility is addressed and recognized by Boone in an interview that accompanied the publication: "it is also possible that the sex difference in efficiency is due in part to facility with the interface or navigation in virtual environments, as men tend to spend more time playing video games." However, in those cases concerning men having "better 3D navigation skills," there are evolutionary reasons to think that this might be truly a "natural tendency," that is, that within exactly the same cultural conditions men can indeed be able, on average, to more easily learn how to navigate. This is because, as seen earlier, there is empirical evidence that for a long period of our evolutionary history as well as today in many nomadic societies, men often were/are more involved—with exceptions, of course—than women in hunting, an activity that frequently requires the learning/use of 3D navigational skills. However, more studies—ideally trying to minimize the possible influence of modern cultural stereotypes related to factors such as sons receiving more cars, computers and videogames than daughters—are needed to further analyze these topics in a comprehensive, unbiased way, as acknowledged by Boone.

So, what about the opposite cases, which are in *reality* far more numerous: the ones concerning traits in which women tend to biologically outperform men, *on average*? Before talking about them, one point should be made concerning the biases of scientific studies: the vast majority of the huge amount of studies showing cases in which women outperform men were surely published not because of such biases but *despite* them, because those biases almost always are related to man-made stories. An illustrative example is that empirical studies consistently show that women tend to experience more complex, elaborate, and intense **orgasms** than men, and to have **multiple orgasms** more frequently, *despite* the fact that for centuries many evolutionary biologists have argued that it is a "mystery" that women have any type of orgasms at all. As also seen earlier, scientific studies have also shown that women consistently outperform men in **smell sensitivity tests**, including those regarding scents related to **sexual arousal**, further contradicting the fairytale that women are "less sexual." Within the many other empirical case-studies in which women outperform men—see also Box 5.10 for further examples—one concerns reading **facial expressions**, as explained in Hrdy's book *Mothers and Others.* As noted in Shermer's 2004 *The Science of Good and Evil*, "to the extent that lie detection through the observation of body language and facial expressions is accurate (overall not very), women are better at it than men because they are more intuitively sensitive to subtle cues…in experiments in which subjects observe someone either truth telling or lying, although no one is consistently correct in identifying the liar, women are correct significantly more often than men." Clearly, this is something that is not taken into account within the recruitment of intelligence and police officers, which continue to be mostly men.

Shermer further added that "women are also superior in discerning which of two people in a photo was the other's supervisor, whether a male-female couple is a genuine romantic relationship or a posed phony one, and when shown a two-second silent video clip of an upset woman's face, women guess more accurately then men whether she is criticizing someone or discussing her divorce." Other examples concern the development of **complex cognitive language skills** and **fine motor skills** as well as **optimization of brain connections**: on average, girls tend to develop all these items faster than boys—see Boyatzis et al. 1993s, Labarthe's 1997, Hanlon et al.'s 1999, and Lenroot et al.'s 2007 works. Interestingly, in such studies of **brain development**, boys tend to develop, faster, areas of the brain involved in **spatial memory** and targeting, for instance they tend to perform better, at an earlier age, concerning **spatial-kinesthetic tasks** such as building with blocks or hitting a target with a ball. This point reinforces the idea that it does seem to be likely that these differences concerning 3D spatial tasks might be truly biological/physiological ones, although of course this is complicated by the fact that parents tend to give, to

Box 5.10: Gender Stereotypes, Health, and Male Research Biases

Saini wrote, in her 2017 book *Inferior – How Science Got Women Wrong, and the New Research That's Rewriting the Story*:

> We often think of males as being the tougher and more powerful sex. It's true that men are on average six inches taller and have around double the upper body strength of women. But then, strength can be defined in different ways. When it comes to the most basic instinct of all – survival – women's bodies tend to be better equipped than men's. The difference is there from the very moment a child is born. 'When we were there on the neonatal unit and a boy came out, you were taught that, statistically, the boy is more likely to die', explains Joy Lawn. Besides her academic research into child health, she has worked in neonatal medicine in the United Kingdom and as a paediatrician in Ghana. The first month following birth is the time at which humans are at their greatest risk of death. Worldwide, a million babies die on the day of their birth every year. But if they receive exactly the same level of care, females are statistically less likely to die than males. Lawn's research encompasses data from across the globe, giving the broadest picture possible of infant mortality. And having researched the issue in such depth, she concludes that boys are at around a 10 per cent greater risk than girls in that first month – and this is at least partly, if not wholly, for biological reasons. Thus, in South Asia, as elsewhere in the world, the mortality figures should be in favour of girls. The fact that they're not even equal, but are skewed in favour of boys, means that girls' natural power to survive is being forcibly degraded by the societies they are born into. 'If you have parity in your survival rates, it means you aren't looking after girls', says Lawn. 'The biological risk is against the boy, but the social risk is against the girl'. Elsewhere, **child mortality statistics** bear this out. For every thousand live births in subSaharan Africa, ninety eight boys compared with eighty six girls die by the age of five. Research Lawn and her colleagues published in the journal *Pediatric Research* in 2013 confirmed that a boy is 14 per cent more likely to be born prematurely than a girl, and is more likely to suffer disabilities ranging from **blindness** and **deafness** to **cerebral palsy** when he's at the same stage of prematurity as a girl. In the same journal in 2012 a team from King's College London reported that male babies born very prematurely are more likely to stay longer in hospital, to die, or to suffer brain and breathing problems.
>
> 'I always thought that it was physically mediated, because boys are slightly bigger, but I think it's also biological susceptibility to injury', says Lawn. One explanation for more boys being born preterm is that mothers expecting boys are, for reasons unknown, more likely to have placental problems and high blood pressure. Research published by scientists from the University of Adelaide in the journal *Molecular Human Reproduction* in 2014 showed that newborn girls may be healthier on average because a mother's placenta behaves differently depending on the sex of the baby. With female foetuses, the placenta does more to maintain the pregnancy and increase immunity against infections. Why this is, nobody understands. It could be because, before birth, the normal human sex ratio is slightly skewed towards boys. The difference after birth might simply be nature's way of correcting the balance. But the reasons could also be more complicated. After all, a baby girl's natural survival edge stays with her throughout her entire life. Girls aren't just born survivors, they grow up to be better survivors too. 'Pretty much at every age, women seem to survive better than men', confirms Steven Austad, chair of the biology department at the University of Alabama at Birmingham, who is an international expert on ageing. He describes women as being more 'robust'. It's a phenomenon so clear and undeni-

able that some scientists believe understanding it may hold the key to human longevity….

Austad is convinced that the difference is so pronounced, ubiquitous and timeless that it must mean there are features in a woman's body that underlie the difference. 'It's hard for me to imagine that it is environmental, to tell you the truth', he says. The picture of this survival advantage is starkest at the end of life. The Gerontology Research Group in the United States keeps a list online of all the people in the world that it has confirmed are living past the age of 110. I last checked the site in July 2016. *Of all these 'supercentenarians' in their catalogue, just two were men.* Forty six were women. 'It's a basic fact of biology', observes Kathryn Sandberg, director of the Center for the Study of Sex Differences in Health, Aging and Disease at Georgetown University in Washington, DC, who has explored how much of a role disease has to play in why women survive. 'Women live about five or six years longer than men across almost every society, and that's been true for centuries…first of all, you have differences in the age of onset of disease…so, for example, **cardiovascular disease** occurs much earlier in men than women…the age of onset of **hypertension**, which is high **blood pressure**, also occurs much earlier in men than women…there's also a sex difference in the rate of progression of disease…if you take chronic kidney disease, the rate of progression is more rapid in men than in women'. Even in laboratory studies on animals, including mice and dogs, females have done better than males, she adds. By picking through the data, researchers like her, Joy Lawn and Steven Austad have come to understand just how widespread these gaps are. 'I assumed that these sex differences were just a product of modern Westernised society, or largely driven by the differences in cardiovascular diseases', says Austad. 'Once I started investigating, I found that women had resistance to almost all the major causes of death'. One of his papers shows that in the United States in 2010, women died at lower rates than men from twelve of the fifteen most common causes of death, including cancer and **heart disease**, when adjusted for age. Of the three exceptions, their likelihood of dying from **Parkinson's** or stroke was about the same. And they were more likely than men to die of **Alzheimer's Disease**.

When it comes to fighting off infections from **viruses** and **bacteria**, women also seem to be tougher. 'If there's a really bad infection, they survive better…. If it's about the duration of the infection, women will respond faster, and the infection will be over faster in women than in men', says Kathryn Sandberg. 'If you look across all the different types of infections, women have a more robust immune response'. It isn't that women don't get sick. They do. They just don't die from these sicknesses as easily or as quickly as men do. One explanation for this gap is that higher levels of **oestrogen** and **progesterone** in women might be protecting them in some way. These hormones don't just make the **immune system** stronger, but also more flexible, according to Sabine Oertelt-Prigione, a researcher at the Institute of Gender in Medicine at the Charité University Hospital in Berlin. 'This is related to the fact that women can bear children', she explains. A pregnancy is the same as foreign tissue growing inside a woman's body that, if her immune system was in the wrong gear, would be rejected. 'You need an immune system that's able to switch from pro-inflammatory reactions to anti-inflammatory reactions in order to avoid having an abortion pretty much every time you get pregnant…the immune system needs to have mechanisms that can, on one side, trigger all these cells to come together in one spot and attack whatever agent is making you sick…but then you also need to be able to stop this response when the agent is not there any more, in order to prevent tissues and organs from being harmed'. The hormonal changes that affect a woman's immune system during pregnancy also take place on a smaller scale during her menstrual cycle, and for the same reasons. 'Women have more plastic immune systems...they adapt in different ways', says Oertelt-Prigione.

In 2011 health researcher Annaliese Beery at the University of California, San Francisco, and biologist Irving Zucker at the University of California, Berkeley, published a study looking into sex biases in animal research in one sample year: 2009. Of the ten scientific fields they investigated, eight showed a male bias. In pharmacology, the study of medical drugs, the articles reporting only on males outnumbered those reporting only on females by five to one. In physiology, which explores how our bodies work, it was almost four to one. It's an issue that runs through other corners of science too. In research on the **evolution of genitals** (parts of the body we know for certain are different between the sexes), scientists have also leaned towards males. In 2014 biologists at Humboldt University in Berlin and Macquarie University in Sydney analysed more than three hundred papers published between 1989 and 2013 that covered the evolution of genitalia. They found that almost half looked only at the males of the species, while just 8 per cent looked only at females. One reporter, Elizabeth Gibney, described it as 'the case of the missing vaginas'. When it comes to health research, the issue is more complicated than simple bias. Until around 1990, it was common for medical trials to be carried out almost exclusively on men. There were some good reasons for this. 'You don't want to give the experimental drug to a pregnant woman, and you don't want to give the experimental drug to a woman who doesn't know she's pregnant but actually is', says Arthur Arnold. The terrible legacy of women being given **thalidomide** for morning sickness in the 1950s proved to scientists how careful they need to be before giving drugs to expectant mothers. Thousands of children were born with disabilities before thalidomide was taken off the market. 'You take women of reproductive age off the table for the experiment, which takes out a huge chunk of them', continues Arnold. A woman's fluctuating hormone levels might also affect how she responds to a drug. Men's hormone levels are more consistent. 'It's much cheaper to study one sex...so if you're going to choose one sex, most people avoid females because they have these messy hormones...so people migrate to the study of males...in some disciplines it really is an embarrassing male bias'. This tendency to focus on males, researchers now realise, may have harmed women's health. 'Although there were some reasons to avoid doing experiments on women, it had the unwanted effect of producing much more information about how to treat men than women', Arnold explains. A 2010 book on the progress in tackling women's health problems, cowritten by the Committee on Women's Health Research, which advises the National Institutes of Health (NIH) in the USA, notes that **autoimmune diseases** – which affect far more women than men – remain less well understood than some other conditions: 'despite their prevalence and morbidity, little progress has been made toward a better understanding of those conditions, identifying risk factors, or developing a cure'.

boys, toys that are more related to navigation, building, or hitting targets, such as cars, jet planes, tool kits, building sets, weapons, balls, and so on. In contrast, parental biases do not seem to be connected to the traits that girls tend to developed earlier than boys, such as developing complex cognitive and fine motor skills and optimizing brain connections, because the toys stereotypically given to girls—Barbie dolls, easy-bake ovens, little ponies, tea sets, and so on—do not focus necessarily on any of those traits, well on the contrary, often.

In the *Fate of Gender*, Browning notes that "not only do little girls excel in preschools more quickly than little boys, but realms of data show that women's **mental agility** persists seriously longer than men's as everyone's life expectancy grows longer." A very interesting point made by Browning—which should be studied in further detail—is that the **better grades of women in school**, including at a young age, are not necessarily related to an inferior cognitive capacity of men. He notes: "researchers found after examining 5800 students from kindergarten through 5th grade…no particular evidence of inferior **cognitive capacity** among the boys…the boys were simply more fidgety, less attentive, and less 'eager to learn', and their teachers, stressed by larger class sizes, graded them down for their comportment—a track that followed them on through high school." On the other hand, he explains that "MRI imaging…[shows] that the **ventral prefrontal cortex**…well known to relate to **social awareness** and **interpersonal response**, was in fact about 10% larger in women than in men." However, the same researchers found that "little boys actually had (a) larger (ventral prefrontal cortex) than the little girls but that this *smaller* (size) correlated with greater **interpersonal activity** – exactly the opposite of what they found in the sixty adults." That is, "their investigations showed a far muddier portrait of what makes men masculine and women feminine, and as they age not only do their interests, behavior, and mannerisms change, but so too can their neural biology change." This example illustrates the very complex links between "nature" and "nurture," and the crucial links between them, development and epigenetics.

As put in *Fate of Gender*, "as neurologist Lise Eliot, author of *Pink brain, blue brain*, commented in a review of [the above] work, 'individuals gender traits – their preferences for masculine or feminine clothes, careers, hobbies and interpersonal styles – are inevitably shaped more by rearing and experience than by their biological sex…likewise, their brains, which are ultimately producing all this masculine or feminine behavior, must be molded – at least to some degree – by the sum of their experiences as a boy or a girl'…we are all of us both male and female, and the way we express our 'masculinity' and 'femininity' depends on the *circumstances* in which we find ourselves living – and moreover those experiences can alter our neurobiology and physiology." This is indeed the power of epigenetics, and what makes that discussions about "nature" versus "nurture," and in particular gene-centered ideas such as those of **Dawkins' *selfish genes***, are not only over simplistic, but often completely wrong. The power of plasticity and epigenetics, and particularly hormones, and how they change during lifetime, was also noted in Ryan's *The Virility Paradox*: "does it mean that **larger brains** make boys smarter than girls? Nope. In fact, having more **testosterone** and more active **androgen receptors** during adolescence may actually have a negative impact on brain function…tests on teenage male subjects…found…a significant higher rate of **depression** and **suicidal thinking** in boys with a combination of higher testosterone." Taking this into account, another item that can be added, in such a simplistic way, to the so-called natural/biological strengths of women, is that they are, on average, more resilient than men in terms of health in general, and that this difference starts at a very early age, as discussed in Box 5.10.

Actually, since just some weeks after gestation, when they are still in the wombs of their mothers, females tend to get less sick and have a quicker/stronger **homeostatic response** when exposed to conditions similar to those to which males are exposed. Accordingly, women tend to have a higher **life expectancy** in almost every country, including those where they are highly oppressed, subjugated, and neglected. Similarly, even in countries where education possibilities are far less available for women, they tend to be better students, on average, in most areas of knowledge, including mathematics. They also tend to do better in **multitasking tests**: this would be expected under even the prevalent gender stereotypes and narratives, because according to them women are *expected to* be "better at taking care of kids and at performing multiple domestic tasks." As noted in Browning's *The Fate of Gender* "most male brains tend to have as much as 7 times more 'gray matter' associated with concentration attention on a specific action or task while female brains can have ten times more white matter – associated with coordination between gray matter centers and, hypothetically, with women's alleged capacity to switch between tasks and to multitask, advantages that can be beneficial in **infant nurturing.**"

More surprising, for those following the prevailing man-made narratives and stereotypes, is that in *reality* women tend to be less emotional/stressed than men in the sense that, when exposed to similar types of stress, they tend to react in a less stressful and more pragmatic way, while men then to react in a more stressful and risk-taking way. Such empirical evidence is also considered to be "surprising" to most women I talk to: indeed, man-made narratives about women being more "emotional" are among the more internalized ones by women themselves, together with those about them being more "asexual." When I was in **Iran**, this was a recurring idea I heard from local women: that perhaps it is in fact better that men are politicians, or religious leaders, because if the country was to be ruled by women, they would tend to be more emotional and more influenced by "womanly things such as menstruation," and thus to take less "rational" decisions for the country. The brainwash done, and imposed, by men in the last millennia was indeed hugely powerful if women of a country such as Iran truly think that the country would be even worse and more chaotic than it already is right now, that its horrible answer to the Covid pandemic, the poverty of a huge proportion of its people, and the enormous amount of money and lives the country has spent in fighting wars would be even worse if Iran was led by women.

Compare what happened so far about **Covid-19** in Iran versus New Zealand, the country that probably reacted faster and better among the whole planet, with 0 cases reported for several months: a country led by a woman. And this is not an isolated case, the countries led by women in general had much better response to the pandemic than those led by men, particularly those led by "alpha-males" such as Putin, Trump, Bolsonaro, and so on. Actually, we have very detailed empirical data about this topic, which one more time deconstruct the myth of such misogynistic narratives: within the three states that reacted better to the pandemic according to analyzes of six different factors, two of them—including the very one on top—have governments led by women, **Taiwan** and **New Zealand**. In contrast, the five worse

(https://interactives.lowyinstitute.org/features/covid-performance/) 31 January 2021

RANK	COUNTRY	AVERAGE
1	New Zealand	94.4
2	Vietnam	90.8
3	Taiwan	86.4

RANK	COUNTRY/TERRITORY	AVERAGE
98	Brazil	4.3
97	Mexico	6.5
96	Colombia	7.7
95	Iran	15.9
94	United States	17.3

Fig. 5.25 Within the *Lowy Institute*'s analysis of how well or poorly states have managed the Covid-19 pandemic, two of the three states at the top, including the one on the very top, are led by women, Taiwan and New Zealand, while the five worse are all led by populist conservative—right wing/religious—men, four of them being particularly misogynistic: Trump, Obrador, Bolsonaro, and Khamenei

are all led by populist conservative—right wing/religious—men, four of them being particularly known for being very misogynistic: **Trump**, **Obrador**, **Bolsonaro**, and **Khamenei**, the latter being precisely the leader of **Iran** (Fig. 5.25). These are the *facts in the real world*, based on the analysis, by the **Lowy Institute**, of how well or poorly countries have managed the pandemic in the 36 weeks that followed their hundredth confirmed case of infection by Covid-19, based on a combination of factors such as confirmed cases, confirmed deaths, confirmed cases per million people, confirmed deaths per million people, confirmed cases as a proportion of tests, and tests per thousand people. As only about 29 states are led by women right now, this means that this barely applies to about 15% of the states worldwide, so the fact that two of the best three, including the very one on top, are led by women almost surely does not represent a mere random coincidence.

Unfortunately, similar examples of how the oppressed internalize the just-so-stories created by the oppressors are far too frequent in human history, as can be attested by an example that concerns girls and constructed gender differences—**Barbie dolls**, remember?—but is above all a profoundly sad illustration of the result of long-standing institutionalized racism in the U.S. (see Fig. 6.3). When African-American children are asked to choose between "white" and "black" Barbie dolls, or even when asked which of the two is "better," they tend to choose the "white" one. In a 2015 article that Browning wrote in the *California Magazine* of the *Cal Alumni Association of UC Berkeley*, he refers to other researchers that study gender differences, particularly concerning resilience. It is fascinating to see the reasons that the researchers cited by him—from very different backgrounds and fields of science—use to explain those gender dissimilarities, and how epigenetics non only indeed links the so-called nature and nurture, but also the so-called mind and body:

Andy Scharlach, a UC Berkeley professor of aging and director of its Center for the Advanced Study of Aging Services, says…[that] men, whatever age…tend to blather about safe topics circling over and around their tender anxieties, while women are ever more pragmatic and direct about the foibles and frustrations brought on by the passage of time. Over and over again, Scharlach's research has shown that women generally retain far more resilience as they age than men. **Biological difference and genetic inheritance** clearly play important roles in our health as we age, but resilience, an admittedly fuzzy concept, can also affect our biological response to stress, and therefore to both cardio illness and cognitive failure. Where does resilience come from? It begins very early, he says – often even before boys and girls learn to read – "you become resilient by dealing with small-scale stressors that you're able to learn from…women have many more opportunities to do that in their lives than men do, in part because they have more exposure to the stresses that come from being excluded from the privileges that come automatically to little boys; and that continues throughout women's lives as they carry different burdens and expectations from men…women still carry more child rearing responsibilities…they carry more of the emotional load in families…the gender biases that exist either beat you down, or you develop a sense of yourself and others as being OK". A second source of **female resilience** concerns what many sociologists have noticed in gender relations across the lifespan. Says Scharlach: "women develop richer **social networks** than men that are not as work bound, and not as sports bound, or activity bound".

Susan Folkman…spent the last decade of her professional life as a distinguished professor of medicine at the University of California, San Francisco. Her first major research concerned how men learned to take care of each other during the worst years of the AIDS epidemic. From there she looked at the kinds of people who learn and succeed in caregiving – especially as they age. "From a very early age", she notes, "boys are indoctrinated with the athletic metaphor: you don't give up…you keep going after that success…you fight for it…you don't take a second…you just fight harder…I don't think women are brought up with that metaphor". Nearly all the women understood threat and challenge as physical danger, something that had to be mastered, but more importantly, something that had to be understood. "I think all that's very embedded very early in males and females" she says. And there is something more profound at work: a marked difference in **spousal mortality rates** following the death of a partner. A man is much more likely to die in the first six months following the death of his wife than is a woman following the death of her husband. While there are slight differences concerning race and ethnicity in what's called the "widowhood effect", the greater death toll among men appears to be a worldwide phenomenon. Janice Schwartz, another gerontologist at UCSF, has also focused on gender gaps in health, longevity and caregiving. She is even more convinced of the negative consequences that result from conventional men's inability to form close friendships and strong social networks following retirement. Schwartz's research extended well beyond the upper-middle-class-territory of doctors, businessmen and technologists. She spent several years interviewing and following aging residents of a residential retirement trailer park north of San Francisco in Sonoma County, where she found the same behavior.

The Fate of Gender provides further fascinating details about studies on these topics:

Boys born prematurely are 1.7 times less likely to survive than premature girls…. A Canadian study of mothers living in a chemically polluted area found that male embryos were far less likely to survive through gestation than female embryos…. [Susan] Pinker wrote in her 2009 book *The sexual paradox* [that] cognitive and **attention deficit disorders**…were from 4 to 10 times more common in boys than in girls. Men develop **cardiovascular disease** on average 7 to 10 years earlier than women do. Men have strokes much more often than women – even though women strokes seem to be more severe. Cancer numbers are still worse: 4.5 to 5.5 times more throat and **mouth cancers** in men than in women and 3.3 to 1 for **bladder**

cancer; 2.3 to 1 for **lung cancer** deaths and about the same for **liver cancer**. The notion that males as the tougher and sturdier to the two human animals is highly suspect.

In a nutshell, it can be said that the available scientific data do show that "**gender roles**" in particular, and to a certain extent also "**gender differences**" in general, are mainly socially constructed. However, humans are not *only* the products of their culture. In contrast to "gender roles," sexual and physiological differences between males and females have a very deep evolutionary history, and accordingly, there are indeed some biological differences between men and women, not only anatomical ones concerning for instance the shape of their genitals, but also regarding *certain* mental abilities and responses to diseases.

Within our closest living relatives, the great apes, males contribute much less than females to child care. Archeological and paleontological data about our human lineage also indicate that men also tended to do so, for millions of years, and still do in most hunter-gatherer groups. Therefore, one can argue that for most of our human evolutionary history, if an adult mother died, or got very sick, or was exceedingly stressed, this could put in direct danger the life, or at least the well-being, of her children, even if other women would help taking care of those children, as they probably did, following Hrdy's concept of "mother and others." Based on empirical data collected from the study of nomadic hunter-gather societies, the same does not often apply, at least so directly, to the father, as in many of these societies a diseased or even a completely absent father does not often affect in a major way the well-being of his children. In other words, contrary to a sick or absent father, a mother being very sick, or dying, would very likely lead to her children being weak or sick, or even also dying, thus affecting directly their reproduction and/or survival. Therefore, in the long run, natural selection would tend to favor those cases in which the females were more resilient.

This means that, in a way, women were the subject of a kind of "**double selection**"—if something serious happened to a mother, it would often also strongly affect her children—while men did mainly experience a "**single selection**"—if something serious happened to a father, his children's well-being would not be so much affected usually. This idea of a double selection of females very likely does not apply only to humans but also to other primate and many other mammalian taxa in which the father does not contribute so much to taking care of the progeny. In fact, there are data showing that females tend to be also more resilient in various other mammalian taxa. This idea is also supported, in a way, by several examples given in Klarsfeld and Revah's 2003 book *The Biology of Death*, in which males of several species die before the females, in many cases just after reproducing with them—in some cases being literally killed, and even eaten, by them. In such cases, the females obviously need to live for some time in order to lay their eggs or deliver the progeny and, in many taxa, to then take care of them or help/produce others to do so. Some "**queen bees**" use the sperm of the dead males to produce, during *various years* after their death, the "**bee workers**" that are needed to keep the **beehives**. Another famous example is the **southern black widow**, *Latrodectus mactans*: the females of this venomous spider species occasionally eat their mate after reproduction (Fig. 5.26). As noted by Klarsfeld and Revah's, contrary to popular culture,

Fig. 5.26 "Black Widow" female spiders are much larger than—and often eat—the males

these females are actually *not* the champions of **erotic cannibalism**, but they do tend to live *on average* ten times longer than the males. The "black widow" female spiders not only are much larger than males, but also are the ones that are poisonous and that bite, and in addition sometimes kill and eat the males and live much longer than them. So, *Neverland*'s stories about "male biological superiority" are not only ridiculous but also not based at all in the *reality of the natural world* (see Fig. 5.27). They are just anthropocentric just-so-stories that tend to see humans as "the norm"—for instance, concerning the fact that human males tend to be physically stronger/larger than females, contrary to numerous other species—and, even worse, that are based in and try to impose the acceptance of wrong facts about our own lineage—for example, that human females are physiologically/medically weaker than males, while they are often stronger as it is the case in many other animals.

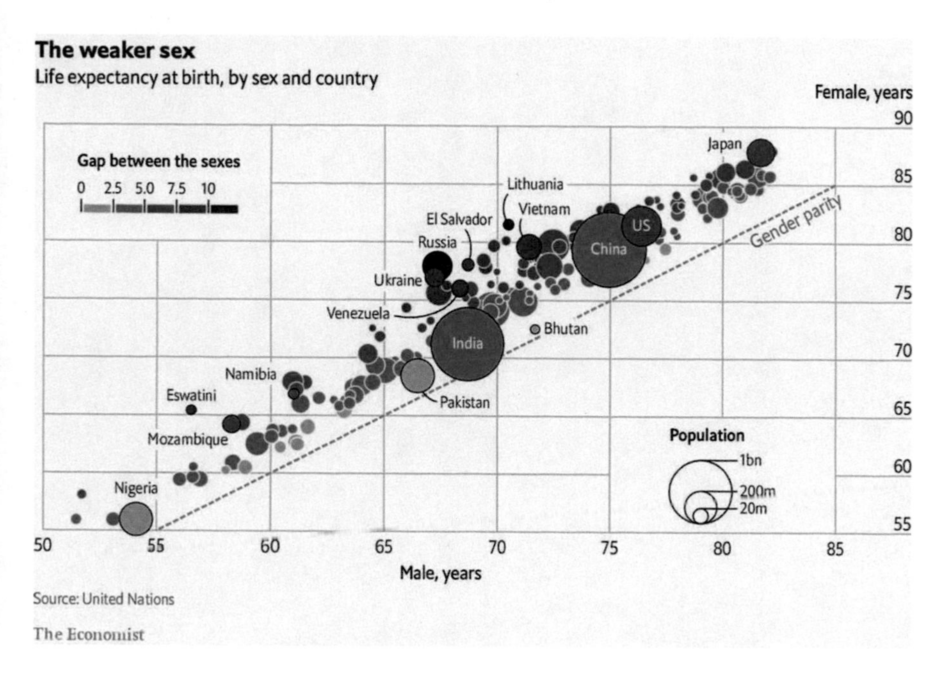

Fig. 5.27 The weaker sex: men tend to live less time than women in basically every country, including countries where women are particularly oppressed by them

Chapter 6
Darwin, Capitalism, Colonialism, and Beliefs

When the missionaries arrived, the Africans had the land and the missionaries had the bible...they taught us how to pray with our eyes closed...when we opened them, they had the land and we had the bible

(quote often attributed to Jomo Kenyatta, or to Archbishop Desmond Tutu, although it might have an earlier origin)

6.1 Darwin's Idolization, Darwinism, and Just-so-Stories

Idealizing Daddy is grand when you're five; it's crippling when you're twenty-five or thirty-five...for if you still believe in Daddy's miracles, you may not believe that you can make your own dreams come true...worse, you may not even be able to formulate them without his guidance (Victoria Secunda)

Many of the narratives that human beings have a propensity to create and believe tend to include **idealization of others**—be it Gods, saints, or just people such as loved ones, politicians, scientists, or actors. As noted in Tallis's *The Incurable Romantic*, idealization is a very common psychological phenomenon: "it simplifies the word in order to reduce the anxiety caused by inconsistency and troublesome complexities...*idealization* always incorporates a degree of *denial*, because in order to see someone as perfect we must deny the existence of their less favorable attributes." In the case of idealization of scientists, such denial often comprises neglecting the fact that every scientist has his/her own ego—that is, his/her own **sense of self-esteem or self-importance**—as well as his/her own biases and prejudices, which are related to the specific contingencies of life, such as when and where one is born, studies, lives, interacts with, and so on. In this Chapter I will pay special attention to the case of **Charles Darwin** because, paradoxically, on the one hand his ideas put in question the creation myths of most human societies—that is why Darwin and his evolutionary ideas were, and continue to be, hated by so many people—, but on the other hand many of his own ideas and of those that followed them are themselves powerful examples of how strong is the human tendency to create imaginary narratives. This is because

R. Diogo, *Meaning of Life, Human Nature, and Delusions*,
https://doi.org/10.1007/978-3-319-70401-2_6

Darwin could not completely escape *Neverland* himself, nor did most of the scientists that have defended his ideas, who tended to build themselves **idealization narratives** about Darwin—often recurring to the image of an "humble," "wise," "kind," "never-mistaken" bearded old naturalist that somewhat resembles that of a God—and/or to strictly defend all his evolutionary ideas or even some very extreme "adaptationist," "**struggle-for-existence**" versions of them. This is another profound paradox concerning Darwin: many of those that idealize him and defend extreme *Neverland* versions of his ideas are actually doing a disservice to Darwin and the huge and clearly more nuanced knowledge he had about the *reality of the natural world*. In fact, as noted in Delisle's 2017b book chapter entitled "*From Charles Darwin to the evolutionary synthesis*," "Darwinism" and "Neo-Darwinism" comprise multiple different ideas and ideologies, many of them having nothing to do with what actually Darwin meant in his writings. Therefore, in order to simplify the discussions presented here for a broader audience, when I use terms such as "**Darwinists**" or "NeoDarwinists" I am merely referring to the way scholars define themselves and/or are defined by other scholars.

The idealization of Darwin by a huge number of scientists is a particularly powerful example of the human tendency to build narratives because this has occurred, and continues to occur, *despite* the fact that, as will be shown below, the relatively few scholars that have been able to study in detail either the personal aspects and historical and geographic context of his life, or the validity of his ideas and/or of subsequent caricatures of them, have plainly shown that such idealizations and ideas are contradicted by factual data. That this inevitably happens in any case of idealization should be obvious to anyone, as noted in Tallis's *The Incurable Romantic*, but this obvious *fact* strikingly continues to be neglected by the majority of scientists and historians when it comes to talk about Darwin, contrary to what happens when they talk about almost any other scientist.

Many scholars continue to defend that we don't need—see Wray et al.'s 2014, Coyne's 2016, Stoltzfuz's 2017, Futuyma's 2017, and Gupta et al.'s 2017 works—any kind of new or even "**extended evolutionary synthesis** (see Fig. 5.6), contrary to what has been proposed recently in publications such as Pugliucci and Müller's 2010, Kull's 2014, and Laland et al.'s 2014, 2015, and 2016 works. This is because, according to the former scientists, what we have from Darwin, or from the subsequent "Modern Synthesis," is *enough*. As if all the thousands of evolutionary studies made in the last decades, and discoveries such as those concerning the **structure of the DNA**, the reading of the **genome of humans** and several other species, and the importance of **epigenetics** and the inheritance of characters acquired during life that do not imply changes in the genetic code, and so on, did not add anything at all that is relevant for how we comprehend the reality of the natural world. Such a reactionary position is indeed remarkable and unique, in contrast to what happened in other fields of science, such as physics. The information accumulated *after* the works of **Newton** has dramatically changed our understanding of the reality of the cosmos, and this is accordingly rightly recognized by the vast majority of scientists and historians.

Indeed, the type of idealization of—sometimes even veneration for—Darwin and/or his ideas, has no current parallel in other scientific disciplines also in the sense that it strikingly resembles that done by **religious fundamentalists**, who, faced with scientific discoveries contradicting the original narratives of the religious texts they believe in—for instance, earth not being the center of the universe—argue that these are just "minor" details that do not truly put those narratives in question. One of the more illustrative, extreme examples concerns Darwinian and Neo-Darwinian **evolutionary adaptationist ideas** that are "just-so-stories" *sensu* **Stephen Jay Gould**. As noted by Gould, when an adaptationist wants to "find out" the adaptative—that is, the "advantageous"—function of a certain biological feature, and hypothesizes that it is A, then B, then C, then D, even if these ideas are all contradicted, he/she would just state that this simply means that the "function" is not yet known, and then try E, F, G, H, and so on, instead of being at least open to the hypothesis that maybe there is no current adaptative function at all. As astutely put by Landau in his 1991 book *Narratives of Human Evolution*: "like the hand of Providence in the biblical account, natural selection justifies even where it fails to explain…what happens is not always 'right' or well understood, but it is 'fit'." Again, by doing this, adaptationists are actually paying a disservice to Darwin, because while Darwin defended that biological features often had a current adaptative function, he clearly was *not* a fundamentalist adaptationist: he explicitly wrote that *numerous features* displayed by an organism at a certain geological time and stage of development are *not* necessarily being "useful." In his 1859 ***On the Origin of Species***, he noted that **vestigial and rudimentary organs**, "or parts in this strange condition, bearing the stamp of inutility, are extremely common throughout nature." This is a further example of how "Darwinism" or "Neo-Darwinism" have further fascinating parallels with religion: like fundamentalists of a certain religion tend to be more extreme than the imaginary or real figures they venerate—would Jesus put bombs in abortion clinics?—, many self-designated "Darwinists" or "NeoDarwinists" have defended, and continue to defend, much more extreme versions of the ideas that were truly put forward by Darwin.

A typical argument used by "Darwinists" or "NeoDarwinists" defending such extreme ideas is to use the "straw-man" argument: that is, they say that "nobody really defends such ideas, you are using a 'straw-man' example." So, in order to clearly show we are *not* at all just creating theoretical "straw-men" examples of, for instance, **circular adaptationist reasoning**, in the Sections below I will provide well documented, specific examples of how such a reasoning is not only still frequently applied, but is also even explicitly defended as a valid research methodology by renowned and very influential scholars. Importantly, such a circular reasoning is deeply related to another common human feature associated to our obsession of creating just-so-stories and teleological narratives: our profoundly embedded tendency to seek for a "purpose." **Aristotle**—who is recognized by many historians and scientists as the very founder of the field formally known as biology—famously stated that nature "does nothing in vain"—a teleological notion that influenced Darwin and that continues to be highly influential today. As explained in Ruse's 1996 book, *Monad to Man – The Concept of Progress in Evolutionary Biology* and

2003 book, *Darwin and Design – Does Evolution Have a Purpose?*, this is indeed one of the most profound paradoxes concerning Darwin: he tried to avoid applying teleology in his theories, but was not successful in doing so.

Namely, he was successful in rebutting some of the oldest teleological narratives such as those used in organized religions since the rise of agriculture, but he promoted others, influenced by ideas of authors such as Aristotle and, more commonly, of writers that were particularly in vogue within the **Victorian era** and **Industrial revolution**, such as **Malthus**, as we will see below. Tragically, those teleological narratives that he promoted, or the way in which they were subsequently used by other authors, political leaders, and so on, were not mere historical anecdotes: they have strongly influenced not only biologists but scholars from various other areas of science such as anthropology (see Box 6.1), psychology and psychiatry, particularly evolutionary psychologists and psychiatrists (see Box 6.2), thus directly or indirectly influencing, and often directly affecting, the life of a huge number of people.

Box 6.1 Cultural and Human Behavioral Ecology and Adaptationism

In Kelly's 2013 book *The Lifeways of Hunter-Gatherers*, he provides a pertinent criticism of **adaptationism** as applied to the study of **hunter-gatherers** in particular, and anthropology in general, although the book does include, in some parts, views and arguments that are also adaptationist. The criticism he does is made during his brief account on the history of the disciplines commonly known as **cultural ecology** and **human behavioral ecology**. He explains that "**cultural ecological studies** tried to account for behaviors by showing how they were functionally linked to the acquisition of food in a particular region—example, how they improved foraging efficiency, reduced risk, or netted the highest returns as a continued reaction to the **racist claims** of unilinear evolution and that so-called primitive people acted out of superstition or stupidity rather than rational thought." He noted that "Omar Moore, for example, explained **scapulamancy**, the **Montagnais-Naskapi**'s use of burnt caribou scapulae to divine the direction of a hunt, as a way to randomize hunting excursions and avoid repeated hunts to one area…he argued that this made the most efficient use of time in an area where prey is widely scattered and mobile." This is in fact one of the main problems of adaptationists: that they are very uncomfortable with the existence of evolutionary mismatches, and by consequence with human behaviors that might actually be irrational and/or detrimental to the people displaying them. Therefore, they also tend to privilege narratives where a certain behavior is normally connected with life in a certain environment, for instance that some humans may have adopted agriculture because they lived in more fertile environments than the ones inhabited by hunter-gatherers. However, things are usually much more complex than that. One of Kelly's criticism of adaptationism is related to one of the most significant theoretical flaws of cultural ecology, which are:

(1) a **neofunctionalist concept of adaptation** and (2) an implicit reliance on group selection. By "neofunctionalist" we mean that cultural ecologists assumed that the "function" of behavior was to keep their society in balance with the environment. The term "adaptation" consequently came to refer to any behavior that seemed a reasonable way to maintain the status quo. Adaptation was seen as a state of being rather than what it is: a continual process of becoming. This led to an important tautology: behavior is adaptive because it exists – otherwise, it would not exist. But this **Panglossian view of life** held an important contradiction, for it assumed that if a behavior exists because it accomplishes a goal more effectively than other techniques or strategies, then, presumably, at some time those former techniques or strategies had existed. In this regard, cultural ecologists were like culture area theoreticians, in that they assumed that societies went through changes in the past but were, at the time of study, "best" adapted to their environment. It requires an unwarranted level of confidence to assume that societies had finally figured out adaptation just as anthropologists arrived on the scene.

The idea of group selection can be seen in cultural ecologists' view of **hunter-gatherer demography**. In the 1970s, anthropologists argued that hunter-gatherers held their population below carrying capacity, an environment's maximum sustainable population at a given technological level. In fact, conventional wisdom after *Man the hunter* was that hunter-gatherers maintain their population at only 20–30 percent of carrying capacity through a variety of cultural means, including infanticide, breastfeeding, and intercourse taboos. In so doing, they prevent overexploitation of food supplies and remain in balance with the environment. Hunter-gatherers, anthropologists thought, altruistically sacrifice their own reproductive interests (including, apparently, their own offspring) for the good of the population. The concept of **group selection** was especially important in explaining apparent self-limitations on reproduction. Anthropologists saw such limitations as a sacrifice performed for the benefit of the group. But they could also be a way to maximize individual reproduction by increasing the number of offspring raised to adulthood. **Ju/'hoan women**, for example, produce a child about every four years, a fact often used to demonstrate that hunter-gatherers intentionally hold their populations below carrying capacity to prevent resource overexploitation. Nicholas Blurton Jones, however, argued that Ju/'hoan women who produce a child every four years raise more children to adulthood than do women who have babies at closer intervals. Although their fecundity might also be limited by venereal disease, Ju/'hoan women may still be maximizing their reproduction; it just happens to be a fairly low maximum. Humans are animals, albeit cultural animals, and susceptible to the same evolutionary processes that govern the nonhuman world. Evolution is simply the differential persistence of variability over time, and adaptation describes the process of selection and differential reproduction. Cultural ecology, therefore, was not evolutionary in the Darwinian sense since its functionalist stance deemphasized the potential for competition between members of a group, the importance of natural selection, and hence the importance of behavioral variability within groups. Instead, it was a "theory of consequences", in which the end result, the consequence of adaptation, defined the process rather than vice versa. Like the culture area concept, cultural ecology did not specify how adaptive change occurs. When external circumstances changed, people seemed to decide that this or that way of doing things was better for the group. But the way in which these decisions were made was nebulous; and it was not clear what was meant by better (avoid extinction? increase tribal size? more offspring? stronger offspring? psychological satisfaction?). This produced some important paradoxes.

However, in reality **human behavioral ecology** is also plagued by **adaptationism**, as can be seen by the definition given by Kelly: "human behavioral ecology is less concerned with biology and more concerned with understanding how different human behaviors are adaptive within a particular environmental and social context." That is, it does not seem concerned so much with investigating *if* certain behaviors are adaptive—because it assumes a priori that they *are* adaptive—but rather makes a huge effort trying to explain how they are adaptive. As noted by Kelly, "the majority of behavioral ecologists, therefore, adhere to a "**weak sociobiological thesis,**" in which people tend to select behaviors from a range of variants whose net effect, on average, in a given social and ecological context is to maximize individual reproductive or inclusive fitness…it does assume that humans subconsciously evaluate the reproductive consequences of behaviors." However, are people that suicide themselves, that take drugs and are killed in an overdose, that drink alcoholic drinks continuously until they die *because* of that, really "subconsciously evaluating the reproductive consequences of their behaviors" all the time? Only a hard-core adaptationist could really say they are. Interestingly, Kelly does use some of these assumptions throughout his otherwise outstanding book, although of course in a much more nuanced, and informed, way.

Box 6.2 Design, Disorder, Adaptationism, and Evolutionary Psychology

The voluminous and very interesting 2012 book *Origin(s) of Design in Nature – A Fresh, Interdisciplinary Look at How Design Emerges in Complex Systems, Especially Life*, edited by Swan and colleagues, includes a very informative chapter by Adriaens titled "*Design and disorder – Gould, adaptationism and evolutionary psychiatry*." Basically, Adriaens shows that many of the narratives circulating within the field of **evolutionary psychiatry** might well be "just-so stories," as they assume a priori the dogmatic adaptationist tale that if a certain trait exists in even a very small subset of the population, and even if it is considered clinically to be a mental disorder, then it is because it is, or at least it was, surely an advantageous adaptation. Actually, many scholars go as far as to argue that even suicide can be often adaptive, as summarized in Bering's fascinating 2018 book *Suicidal – Why We Kill Ourselves*, while we know that in most "modern" cases of suicide this is clearly not the case. In Adriaens' chapter "*Design and disorder*," he explains:

> Most evolutionary psychiatrists…are tried and tested in the **adaptationist tradition**. Generally, there are two ways to be an adaptationist about **mental disorders**. First of all, some **evolutionary psychiatrists** disagree with mainstream **psychiatry** in suggesting that mental disorders are not disorders or dysfunctions at all, but adaptations. Thus, they have been spread over the population by natural selection because they confer some reproductive advantage to their bearers. The idea that some mental disorders may have some functional significance may seem outrageous, but such adaptationist hypotheses have been and are still being defended in the literature today, particularly in relation to depressive disorders. Hagen (in 1999), for example, has hypothesized that women affected by **postpartum depression** may signal that they are suffering an important fitness cost, either because they lack paternal or social support or because their newborn baby is in bad health. In this view, postpartum depression would be a bargaining strategy, enabling women to negotiate greater levels of investment from others. Similar hypotheses suggest that typical depressive symptoms, such as a loss of appetite and excessive ruminating, may have been designed by natural selection to signal yielding in a fierce social competition that cannot be won and to reconsider unfeasible ambitions and investments.
>
> The gist of these hypotheses is that **depression** is not a disorder, but a useful psychological mechanism that enables us to cope with the inevitable adversities of life, much like how fever enables us to fight bacterial infections and how coughing and sneezing help us to keep our airways clear. To my knowledge, [Stephen Jay] **Gould** has barely written anything about [mental] disorders. On a rare occasion, however, he does discuss the evolution of mental disorders, in a book review of **Sigmund Freud**'s posthumously published *A Phylogenetic fantasy*. **Freud**'s text is a rather rumbling attempt to examine 'how much the phylogenetic disposition can contribute to the understanding of the neuroses', particularly by linking up our ancestor's vicissitudes during and immediately after the last Ice Age with man's present day vulnerability to a series of mental illnesses. In Freud's view, for example, the disposition to phobia derives from our progenitors' useful fears when confronted with the privations of the Ice Age. Freud's just-so story confirms Gould's earlier claim that psychoanalysis is a textbook example of the pervasive influence of **recapitulationism** and **Lamarckism**. But there is more. In one of the last paragraphs of his review, Gould notes: 'I also deplore the overly adaptationist premise that any evolved feature not making sense in our present life must have arisen long ago for a good reason rooted in past conditions now altered. In our tough, complex, and partly random world, many features just don't make functional sense, period'.
>
> [There is] a second way of being an adaptationist about mental disorders. For convenience's sake, I will refer to such explanations as **mismatch explanations**. Mismatch explanations of mental disorders build on one of the central ideas in **evolutionary psychology** – another recent evolutionary discipline crucified by Gould for being 'ultra-adaptationist'. Evolutionary psychologists claim that our ancestral environment, i.e. the environment in which most of the evolution of our species took place, differs substantially from our modern cultural environment. Or, in the words of Tooby and Cosmides, 'our modern skulls house a stone age mind'. As a result, we are much better at solving the problems faced by our hunter-gatherer ancestors than the problems we encounter in modern cities. Evolutionary

psychiatrists often consider this mismatch to be the hotbed of many of today's mental disorders. Continuing Freud's example of phobia, they hold that such disorders mostly involve natural threats, such as snakes, spiders and heights. These threats were probably common in our ancestral environment, but they certainly aren't the most dangerous things in our contemporary environment. We do not fear guns the way we fear snakes, for example, even though guns pose a much greater threat to our fitness today than snakes do.

In Gould's view, however, there is no need to assume that currently maladaptive traits were once adaptive. Mental disorders, he suggests, may not have an evolutionary history at all, let alone a functional one: 'we need not view **schizophrenia**, **paranoia**, and depression as postglacial adaptations gone awry: perhaps these illnesses are immediate **pathologies**, with remediable medical causes, pure and simple'. In evolutionary psychiatry, Gould's solution is known as a breakdown explanation or medical explanation – a third category of evolutionary explanations of mental disorders. As Murphy notes, for example, both adaptationist and mismatch explanations seem to assume that 'none of our **psychopathology** involves something going wrong with our minds', while 'nobody should deny that our evolved nature suffers from a variety of malfunctions and other pathologies'. [Furthermore, many] claim that most evolutionary explanations of mental disorders, including trade-off explanations, smell of **panglossianism**: 'evolutionarily oriented mental health researchers, such as Darwinian psychiatrists and evolutionary psychologists, often go to torturous lengths to find hidden adaptive benefits that could explain the evolutionary persistence of profoundly harmful mental disorders such as schizophrenia or anorexia, but these accounts are often frustratingly implausible or hard to test'. When charging biologists (and, later on, philosophers of biology) with panglossianism, Gould and **Lewontin** did not only criticize their overly optimistic view of life but also their laziness in testing the predictions that follow from their hypotheses. Anyone can easily come up with stories about the function of, say, male baldness, being homesick or athletic skills, but there is an important difference between *just-so stories* and real science.

I should clarify that the aim of this Chapter is *not*—far from it—to criticize Darwin or all his ideas per se—I have a huge admiration for him, his life and travels, his persistence, and the brilliance of his idea of natural selection and how it helped to explain, and contribute to the acceptance of, the occurrence of **biological transformism**. Instead, by discussing subjects such as adaptationism, the notions of purpose and progress, and issues such as misogyny and racism in face of what Darwin wrote versus what the empirical data now available show, I want to call attention for a topic that in my opinion has not been as discussed as it should have been. Namely, the parallels between religious thinking and the inflexible—and sometimes unfalsifiable—way in which many scholars defended, and continue to defend, the validity of the ideas of Darwin and/or of his subsequent followers or minimize/minimized the huge social implications of such ideas and the completely inopportune metaphors used to disseminate them to the broader public.

6.2 Purpose, Struggle-for-Life, and Selfish Genes

There is not one big cosmic meaning for all; there is only the meaning we each give to our life, an individual meaning, an individual plot, like an individual novel, a book for each person (Anais Nin)

As explained in detail in my 2017 book *Evolution Driven by Organismal Behavior*, it is clearly not an historical accident that Darwin referred to gravity, and to **Newton's mechanism**, in the last sentence of his most prominent work, his 1859s *Origin*: "there is grandeur in this view of life, with its several powers…that, whilst this planet has gone cycling on according to *the fixed law of gravity*, from so simple a beginning endless forms most beautiful and most wonderful have been, and are being, evolved." By doing so, **Darwin** could in a way be seen as the Newton of biology—as noted in Ruse's 2018 *On Purpose*: "what spurred the move to natural selection was the strongly felt need to be the Newton of biology – to find a cause for the change." As Hoffmeyer put it in a 2013 book chapter, "Darwin created a perfectly ***externalist* theory**, a theory that seeks to explain the internal properties of organisms, their adaptations, exclusively in terms of properties of their external environments, natural selection pressures." However, as emphasized by Hoffmeyer, Darwin "was not a fundamentalist in his externalism, as were his followers (the **NeoDarwinists**) in the twentieth century, who thought they could get rid of organismic agency by enthroning the gene and seeing organisms as passive derivatives of genotypes." In reality, Darwin did distinguish between his notion of ("external") **natural selection** and "**sexual selection**" associated to the behavioral choices made by organisms of the very same species being selected, which broadly corresponds to a subset of "organic selection" *sensu* **Baldwin**—see Baldwin's 1895 and 1896a, b, c works, as well as discussions in Weber and Depew's 2013 and in my 2017a, b works. Another type of selection recognized by Darwin was "**artificial selection**," which mainly refers to the behavioral choices of humans concerning traits of other taxa, for instance during **domestication**.

However, the fact that Darwin's *Origin* paid much more attention to (external) "natural selection" and that for him such "natural selection" and "artificial selection" clearly contrasted with "sexual selection" does illustrate that he mainly emphasized the **passive role of organisms** in **biological evolution**. This is because both in his "natural selection" and "artificial selection" the organisms being selected are mainly passive players, powerless in face of the "selectors," that is in face of the external environment or of humans respectively. This contrasts with a much more logical grouping of these three "types" of selection: "external" natural selection *sensu* Darwin versus organic selection *sensu* Baldwin. The latter is mainly driven by the behavior of organisms themselves, rather than by the external environment, thus being in turn subdivided into "sexual" selection—driven by sexual behavior—and "artificial" selection—driven by the behavior of those taxa actively involved in the process of **domestication**. In other words, by establishing a parallel between planets impotently moved by the force of gravity and passive organisms selected by the external environment, and by combining it with an emphasis on the **Malthusian notion of "struggle-for-existence"** (see below), Darwin was in fact attributing a particularly powerful strength to his "external" natural selection.

In several occasions, such as a letter from Darwin to Lyell—18 June 1958, excerpt from Wetherington's 2011 book *Readings in the History of Evolutionary Theory*—, Darwin makes it very clear that this "struggle-for-existence" is not just a crucial part of his theory, the whole theory *depends* on it: "I explained to you here very briefly my views of 'natural selection' depending on the Struggle for existence." The vast majority of biologists nowadays—including myself—accept the importance of Darwin's "external" natural selection, as they accept the fact that there is "artificial" selection. But this acceptance does not mean that these are always, or even often, particularly strong phenomena that tend to lead to an optimal, or at least a suboptimal, current "design" due to a continuous, suffocating, struggle for life. Depending on the specificities of when, where and how it occurs, natural selection *sensu* Darwin can be very strong, or more relaxed, or very relaxed, as indicated by the frequent, and usually much neglected, occurrence of phenomena such as **morpho-etho-ecological mismatches**—that is, a mismatch between the anatomy of an organism, its behavior, and/or its ecology—and/or so-called **maladaptive behavior syndromes**, as explained in my 2017 book. The same applies to "artificial selection": there are cases in which it is quite intense—for instance occurring at each generation of the famous **Siberian domesticated foxes** described in Dugatkin and Trut's outstanding 2017 book *How to Tame a Fox*—while in others it is much more relaxed—for example, within some stray dogs or cats of villages of many parts of the globe.

Such **Neo-Darwinist views of evolution** defending the passive evolutionary role of organisms were defended by one of the most influential biologists in the last decades: **Richard Dawkins**, with his notion of "**selfish genes**" and famous quote that organisms are "no more and no less than survival machines." As explained in Noble's 2017 book *Dance to the Tune of Life*:

> The language of Neo-Darwinism and twentieth-century biology reflects highly reductionist philosophical and scientific viewpoints, the concepts of which are not required by the scientific discoveries themselves. In fact, it can be shown that, in the case of some of the central concepts of Neo-Darwinism, such as 'selfish genes'…no biological experiment could possibly distinguish even between completely opposite conceptual interpretations of the same experimental findings. There is no biological experiment that could distinguish between the **selfish gene theory** and its opposites, such as 'imprisoned' or 'cooperative' genes. This point was implicitly conceded long ago by Richard Dawkins in his 1982 book *The Extended Phenotype*, where he wrote 'I doubt that there is any experiment that could prove my claim'. [Such Neodarwinist] concepts therefore form a biased interpretive veneer that can hide those discoveries in a web of interpretation. I refer to a web of interpretation since it is the whole conceptual scheme of **NeoDarwinism** that creates the difficulty. Each concept and metaphor reinforces the overall mind-set until it is almost impossible to stand outside it and to appreciate how beguiling it is. Since Neo-Darwinism has dominated biological science for over half a century, its viewpoint is now so embedded in the scientific literature, including standard school and university textbooks, that many biological scientists may themselves not recognise its conceptual nature, let alone question incoherencies or identify flaws.

Some years ago, in his 2014 book *The Meaning of Human Existence*, **Edward Wilson** wrote a very interesting story about these issues. The story not only shows the fallacies of the "selfish gene" and the related obsession to focus exclusively on the theory of inclusive fitness and kin-selection, but also contains parts that are more personal, and that also show how Neo-Darwinists such as Dawkins often indeed behave more as closed-minded, dogmatic fundamentalist "believers" than as a scientist should in theory behave:

> At first I found the **theory of inclusive fitness**, winnowed down to a few cases of kin selection that might be studied in nature, enchanting. In 1965, a year after Hamilton's article, I defended the theory at a meeting of the Royal Entomological Society of London. Hamilton himself was at my side that evening. In my two books formulating the new discipline of **sociobiology**, *The Insect Societies* (in 1971) and *Sociobiology: The New Synthesis* (in 1975), I promoted kin selection as a key part of the genetic explanation of advanced social behavior, treating it as equal in importance to caste, communication, and the other principal subjects that make up sociobiology. In 1976 the eloquent science journalist Richard Dawkins explained the idea to the general public in his best-selling book *The Selfish Gene*. Soon kin selection and some version of inclusive fitness were installed in textbooks and popular articles on social evolution. During the following three decades a large volume of general and abstract extensions of the theory of kin selection was tested, especially in ants and extensions of the theory of kin selection was tested, especially in **ants** and other **social insects**, and purportedly found proof in studies on rank orders, conflict, and gender investment. By 2000 the central role of **kin selection** and its extensive inclusive fitness had approached the stature of *dogma*. It was a common practice for writers of technical papers to acknowledge the truth of the theory, even if the content of the data to be presented were only distantly relevant to it. Academic careers had been built upon it by then, and international prizes awarded.
>
> Yet the theory of inclusive fitness was not just wrong, but fundamentally wrong. Looking back today, it is apparent that by the 1990s two seismic flaws had already appeared and begun to widen. Extensions of the theory itself were growing increasingly abstract, hence remote from the empirical work that continued to flourish elsewhere in sociobiology. At the same time the empirical research devoted to the theory remained limited to a small number of measurable phenomena. Writings on the theory mostly in the social insects were repetitive. They offered more and more about proportionately fewer topics. The grand patterns of ecology, phylogeny, division of labor, neurobiology, communication, and social physiology remained virtually untouched by the asseverations of the inclusive theorists. Much of the popular writing devoted to it was not new but affirmative in tone, declaring how great the theory was yet to become. In 2010, the dominance of inclusive fitness theory was finally broken. After struggling as a member of the small but still muted contrarian school for a decade, I joined two Harvard mathematicians and theoretical biologists, Martin Nowak and Corina Tarnita, for a top-to-bottom analysis of inclusive fitness. Nowak and Tarnita had independently discovered that the foundational assumptions of inclusive fitness theory were unsound, while I had demonstrated that the field data used to support the theory could be explained equally well, or better, with direct natural selection – as in the sex-allocation case of ants just described. No fewer than 137 biologists committed to inclusive fitness theory in their research or teaching signed a protest in a Nature article published the following year. When I repeated part of my argument as a chapter in the 2012 book *The Social Conquest of Earth*, Richard Dawkins responded with the indignant fervor of a true believer. In his review for the British magazine Prospect, he urged others not to read what I had written, but instead to cast the entire book away, "*with great force*", no less.

It is effectively striking how Dawkins, who is now more known for his criticism of religion and religious fundamentalists—which, as explained earlier, is truly more a Western-centric criticism of Islam—, clearly behaves exactly like them, by urging people to cast books away—as if what they say is a "dangerous blasphemy," just because it goes against what Dawkins has stated in his "sacred" book *The Selfish Gene*. And, once again, as we can see, this is not only about Dawkins, or about using him as a "straw-man": instead, there are "straw-men" and "straw-women" everywhere, because as stated by Wilson, at least 137 biologists signed that letter. In fact, even authors such as Wilson, who now don't subscribe to Dawkins' selfish genes idea, often describe insect colonies using terms such as "growth-maximizing machines" formed by "cellular automata" whose operations can be portrayed using language of physical or computer science, as astutely noted in the excellent 2013 book edited by Henning and Scarfe, *Beyond Mechanism – Putting Life Back into Biology*. It is important to emphasize that the notion of evolutionary passivity of organisms within a suffocating "struggle-for-existence" is stressed in the most common current definition of natural selection as "the differential *survival* and reproduction of individuals due to the differences in phenotype," and particularly in the still prevailing Neo-Darwinist definition of evolution as "changes in allele frequencies within populations." The fact that these definitions are still the ones used in most biological and anthropological textbooks makes it difficult to accept that these are just "straw-men" examples, or that they just apply to what happened decades ago.

Among numerous examples I could provide here, I will refer to a few recent works that specifically focus on Darwin, "Darwinism" and "Neo-Darwinism" and that do show that there are indeed still a huge number of such "straw-men," in many fields of science. Depew, in his 2017 book chapter, states that in Bowler's 2013 book *Darwin deleted* "what counts as Darwinism is not far removed from what Gould called '**Darwinian fundamentalism**'…it is true that in recent decades gene-by-gene, trait-by-trait adaptationism, especially applied to animal and human behavior, passes as Darwinism's highest achievement, final justification, and hence defining mark." I completely agree that *Darwin deleted* provides illustrative examples both of the still high prevalence of adaptationism nowadays and, more importantly for the context of the present volume, of the **quasi-religious idealization of Darwin** and his ideas. As some religions argue that what is "good" can only be the product of God and what is "bad" has to be the work of the Devil, Bowler's book—which, it should be noted, is otherwise a fascinatingly profound work—somehow does the same for Darwin. Regarding the "positive" aspects related to Darwin's theory, such as the idea of natural selection, Bowler states that "no one else, not even **Wallace**, was in a position to duplicate Darwin's complete theory of evolution by natural selection."

So, what about the "negative" aspects that a few scientists—and of course many religious people, particularly creationists—relate to Darwin's ideas? Bowler does recognize that some scholars associate some of Darwin's ideas to "an outgrowth of **Victorian cutthroat capitalism** – social Darwinism was possible because the selection theory was actually modeled on the **ideology of competitive individualism**." He also recognizes that Darwin saw **Malthus**' "struggle-for-existence"—which "was a product of the **individualistic utilitarian ideology**...more individuals are born that can be fed, so many must die, and the result is competition for scarce resources"—as "the driving force of selection." He even subscribes to the point that I made above, that is, that "Darwin drew upon the Malthusian image of a world ruled by scarcity and struggle to promote his theory...he certainly modified that image by making struggle a creative force," also noting that some scholars have related "**social Darwinism**" and in particular a strong version of the notion of "struggle-for-existence" with "militarism, **racism**, or **eugenics**." Importantly, he moreover also recognizes that Darwin "may have highlighted the harsh implications of this image of nature" and that "Darwinism *was* involved, certainly in the promotion of the **heartless individualism of the mid-nineteenth century middle classes**, and less directly in the promotion of the later, very different models of '**progress through struggle**' – Darwin himself shared some of the concerns that drove social Darwinism." However, puzzlingly, after recognizing all these facts, and explaining that his aim is not "to absolve Darwinian from all responsibility" because there *is* empirical evidence showing that actually Darwin's "theory *was* used to justify" militarism, racism, or eugenics, Bowler then states, in the conclusion of his book—without providing plausible data to support such a statement—that "most of the effects that have been labeled as 'social Darwinism' could have emerged" without Darwin.

One needs to emphasize that, as pointed out by authors such as Bowler, it is true that the powerful "survival-of-the-fittest" metaphor, so much used by **white supremacists** such as the **Ku Klux Klan**'s (KKK) former **Grand Wizard David Duke**, or by Hitler, was originally coined by Spencer—*not* by Darwin. As explained in Black's outstanding book *War against the weak*:

> In the 1850s, agnostic English philosopher **Herbert Spencer** published *Social Statics*, asserting that man and society, in truth, followed the laws of cold science, not the will of a caring, almighty God. Spencer popularized a powerful new term: "**survival of the fittest**". He declared that man and society were evolving according to their inherited nature. Through evolution, the "fittest" would naturally continue to perfect society. And the "unfit" would naturally become more impoverished, less educated and ultimately die off, as well they should. Indeed, Spencer saw the misery and starvation of the pauper classes as an inevitable decree of a "far-seeing benevolence", that is, the laws of nature. He unambiguously insisted, "the whole effort of nature is to get rid of such, and to make room for better...if they are not sufficiently complete to live, they die, and it is best they should die". Spencer left no room for doubt, declaring, "all imperfection must disappear". As such, he completely denounced charity and instead extolled the purifying elimination of the "unfit". The unfit, he argued, were predestined by their nature to an existence of downwardly spiraling degradation. As social and economic gulfs created greater generation-to-generation disease and dreariness among the increasing poor, and as new philosophies suggested society would only improve when the unwashed classes faded away, a third voice entered the debate. That new voice was the voice of **hereditary science**.
>
> In 1859, some years after Spencer began to use the term "survival of the fittest", the naturalist Charles Darwin summed up years of observation in a lengthy abstract entitled *The Origin of Species*. Darwin espoused "natural selection" as the survival process governing most living things in a world of limited resources and changing environments. He confirmed that his theory "is the doctrine of **Malthus** applied with manifold force to the whole animal and vegetable kingdoms; for in this case, there can be no artificial increase of food,

> and no prudential restraint from marriage". Darwin was writing about a "natural world" distinct from man. But it wasn't long before leading thinkers were distilling the ideas of Malthus, Spencer and Darwin into a new concept, bearing a name never used by Darwin himself: **social Darwinism**. Now social planners were rallying around the notion that in the struggle to survive in a harsh world, many humans were not only less worthy, many were actually destined to wither away as a rite of progress. To preserve the weak and the needy was, in essence, an unnatural act.

However, it is also important to note that while **Spencer**'s ideas—including his 1850 book *Social Statics* and the notion of "survival-of-the-fittest"—were published years *before* Darwin's 1859 *Origin*, one cannot neglect the crucial fact that later in life Spencer—and his works—were in fact highly influenced by Darwin's ideas. The very fact that Darwin inflated the importance of the "struggle-for-existence" within biological evolution plainly shows that he was, obviously, just a human being with his own biases and prejudices, because his overgeneralization of this concept was moreover profoundly influenced from what he read—for instance, Malthus and Smith—and what he physically saw as "normal" in England, at that specific time, as explained in Wetherington's *Readings in the History of Evolutionary Theory*. As noted in that book, the "London Charles [Darwin]…settled in [after his travels] had added a million souls – numbering about 2.3 million…lighted factories could employ more people for longer hours…poverty increased…the unbelievable density of humanity – over four hundred people per acre in Greater London – brought the rampant disease, increased mortality, and accelerated reproduction so starkly described by **Adam Smith** and enumerated by **Thomas Malthus**." In other words, some of the most crucial aspects of Darwin's evolutionary theory were simply the result of *seeing London of the Victorian era as representing what generally happens in the planet as a whole*. That is, they were a direct product of Darwin's highly **Eurocentric views**, what is particularly striking because he wrote his 1859 book *after* traveling to, and observing in detail, so many other parts of the globe, including indigenous people living within areas with much lower population densities, or with more abundant natural resources.

Todes' 1989 book *Darwin without Malthus* discusses in detail this crucial fallacy of Darwin and the related obsession he had with the notion of a "struggle-for-existence." Todes explains that Russian biologists in general had no problem accepting Darwin's transformism and natural selection, but did have a huge problem with the "struggle-for-existence" metaphor and the related **capitalistic notions of individual selfishness**, due to obvious political—including socialism—and geographical differences. As he noted, in many parts of Siberia one would be lucky to see even a single animal, for hours and hours—as I can attest myself, after spending several days in a Tran Siberian train: so, where is the incessant competition between animal species? Personally, what I found particularly striking, after I went to the **Galapagos islands**, is that many people assume that Darwin's most important evolutionary ideas where chiefly based on what he observed on those islands, while in reality among all the numerous regions I have been in the globe, those islands are actually the place that conforms *less* to the notion of a "struggle-for-existence." Numerous scholars that have studied those islands in detail—see Jackson's 2016 book

Galapagos, a Natural History—pointed out that because the islands have plenty natural resources and many large animals have no natural predators, and were also not predated by humans until relatively recently, those large animals are in general relaxed and tame and do not display obvious signs of being afraid of other animals or even of humans.

Evolutionary psychology, **behavioral ecology**, and **evolutionary medicine** continue to be among the areas of science most plagued by adaptationism. A 2017 book chapter by Pigliucci provides several examples emphasizing this point. He states that "behavioral biologists…are still clinging to simplistic notions from **sociobiology** and evolutionary biology, which have long since been debunked…it's not the basic idea that behaviors, and especially human behaviors, evolve by natural selection and other means that is problematic…the problem, rather, lies with some of the specific claims made, and methods used, by evolutionary psychologists." There is probably not even a single human behavioral trait that has not been said to have crucial 'adaptative functions' in evolutionary psychology works. This even includes highly pathological traits (see Box 6.2) or phenomena such as suicide, as noted above, or deaths by a drug overdose: for adaptationists these cases *have to be adaptative because everything is*, so they just need to "find" their "true" adaptative functions. In Aubin et al.'s 2013 paper "*The evolutionary puzzle of suicide*," the abstract clearly summarizes this type of circular reasoning: "mechanisms of self-destruction are difficult to reconcile with evolution's first rule of thumb: survive and reproduce…however, evolutionary success ultimately depends on inclusive fitness…the **altruistic suicide hypothesis** posits that the presence of low reproductive potential and burdensomeness toward kin can increase the inclusive fitness payoff of self-removal…the **bargaining suicide hypothesis** assumes that suicide attempts could function as an honest signal of need…the payoff may be positive if the suicidal person has a low reproductive potential." These are of course just two of the current adaptationist hypotheses about suicide, the A and B of the series, there are also C, D, E, F, G, H, and so on (see Box 6.2).

In reality, *empirical data clearly show* that such adaptationist hypotheses don't apply at all to the vast majority of the about 8 hundred thousand **suicides** that happen in agricultural societies every year. Firstly, the members of society more emotionally attached to the people that kill themselves, such as their families/loved ones, are often those that suffer the most after the suicides take place, often suffering from a plethora of psychological problems and/or having very harsh lives after that event. Secondly, empirical data from clinical and psychological studies show that such suicides often have to do with phenomena such as hormonal unbalances, or particularly stressful/damaging previous experiences like being raped or having other traumatic experiences, or being bullied, or feeling very lonely, and so on. Therefore, saying that suicides are nothing more than a way to increase group selection is not only directly ignoring all the data available about those cases, but also a complete lack of respect towards those persons that took their lives and their loved ones, because it basically equates those that committed suicide with "destitute passive automata" that blindly obey to natural selection without knowing they are doing so, as if they were just some poor brainless robots obedient to the 'rules' of natural selection.

Thirdly, *we know* that societies that tend to be more individualistic and to promote individual selfishness tend to have more suicides, not the other way round. Also, as we will see below, the dramatic increase of sleep deprivation in "modern societies" might be related with the dramatic increase of suicides in such societies: part of the huge price such societies pay for their obsession with "work" and productivity. Accordingly, suicide is very rare in nomadic hunter-gatherer societies and almost—or completely, this is still the subject of controversy among scientists—inexistent within non-human highly social animals such as baboons. For those that will quickly try to absolve Darwin at any cost for the type of circular reasoning used in Aubin et al.'s 2013 paper about suicide, and argue that these latter adaptationist authors are just odd scientists that have nothing to do with—and were not influenced by—Darwin's writings, here's the sentence that they used immediately after their abstract, which reveals the main theoretical framework for their a priori assumptions: "*natural selection will never produce in a being any structure more injurious than beneficial to that being, for natural selection acts solely by and for the good of each…no organ will be formed for the purpose of causing pain or for doing an injury to its possessor – Charles Darwin*."

In *The world until yesterday*, Jared Diamond emphasizes that one cannot, indeed, reduce discussions of complex behaviors of hunter-gatherers to a simplistic adaptationist approach, and that one needs to at least consider the possibility of taking into account other types of approaches, namely:

> A second approach, lying at the opposite pole from that first approach (**adaptationism**), views each society as unique because of its particular history, and considers cultural beliefs and practices as largely independent variables not dictated by environmental conditions. Among the virtually infinite number of examples, let me mention one extreme case from one of the peoples to be discussed in this book, because it is so dramatic and so convincingly unrelated to material conditions. The **Kaulong people**, one of dozens of small populations living along the southern watershed of the island of New Britain just east of **New Guinea**, formerly practised the ritualized strangling of widows. When a man died, his widow called upon her brothers to strangle her. She was not murderously strangled against her will, nor was she pressured into this ritualized form of suicide by other members of her society. Instead, she had grown up observing it as the custom, followed the custom when she became widowed herself, strongly urged her brothers (or else her son if she had no brothers) to fulfill their solemn obligation to strangle her despite their natural reluctance, and sat cooperatively as they did strangle her. No scholar has claimed that Kaulong widow strangling was in any way beneficial to Kaulong society or to the long-term (posthumous) genetic interests of the strangled widow or her relatives. No environmental scientist has recognized any feature of the Kaulong environment tending to make widow strangling more beneficial or understandable there than on New Britain's northern watershed, or further east or west along New Britain's southern watershed. I don't know of other societies practising ritualized widow strangling on New Britain or New Guinea, except for the related **Sengseng people** neighboring the Kaulong. Instead, it seems necessary to view Kaulong widow strangling as an independent historical cultural trait that arose for some unknown reason in that particular area of New Britain, and that might eventually have been eliminated by natural selection among societies (i.e., through other New Britain societies not practising widow strangling thereby gaining advantages over the Kaulong), but that persisted for some considerable time until outside pressure and contact caused it to be abandoned after about 1957.

> Anyone familiar with any other society will be able to think of less extreme traits that characterize that society, that may lack obvious benefits or may even appear harmful to that society, and that aren't clearly an outcome of local conditions. Yet another approach towards understanding differences among societies is to recognize cultural beliefs and practices that have a wide regional distribution, and that spread historically over that region without being clearly related to the local conditions. Familiar examples are the near-ubiquity of monotheistic religions and non-tonal languages in Europe, contrasting with the frequency of non-monotheistic religions and tonal languages in China and adjacent parts of Southeast Asia. We know a lot about the origins and historical spreads of each type of religion and language in each region. However, I am not aware of convincing reasons why tonal languages would work less well in European environments, nor why monotheistic religions would be intrinsically unsuitable in Chinese and Southeast Asian environments. **Religions**, **languages**, and other **beliefs** and practices may spread in either of two ways. One way is by people expanding and taking their culture with them, as illustrated by European emigrants to the Americas and Australia establishing European languages and European-like societies there. The other way is as the result of people adopting beliefs and practices of other cultures: for example, modern Japanese people adopting Western clothing styles, and modern Americans adopting the habit of eating sushi, without Western emigrants having overrun Japan or Japanese emigrants having overrun the U.S.

However, adaptationist ideas are still prevailing not only in anthropology, but also within scholars that work with other groups of organisms, such as plants. An emblematic example of this concerns a monograph that I otherwise much enjoyed reading, Mancuso and Viola's 2015 *Brilliant green – the surprising history and science of plant intelligence*. These authors stated: "in the plant world as in the animals, no one does anything for nothing"—a direct reference to, and almost a copy of, Aristotle's "nature does nothing in vain," that remind us how the teleological ideas that this ancient Greek scientist and philosopher proposed more than two millennia ago continue indeed to be so imbedded in the minds of both the broader public and the scientific community. Another recent example is Walker's 2017 otherwise superb book *Why We Sleep*, in which he wrote "why would *Mother Nature* design this strange equation of **oscillatory phases of sleep**?" The inclusion of such a sentence in a non-fiction scientific book is troubling because (1) it is plagued of teleological terms, such as "why," "Mother Nature" and "design"; (2) it gives an idea of the natural world to the broader public that is false—and that Walker knows that is false—, according to which things are *designed by Mother Nature under an a priori masterplan*; and (3) it moreover employs a wrong **scientific methodology**, in which the ***adaptationist dogma*** is considered to be a self-evident truth that does not even need to be tested, that is, assuming a priori that the existence of oscillatory phases of sleep has to be a "good design."

An even more recent example is the book *The Science of Sin: Why We do the Things We Know We Shouldn't*. Published in 2018 by neurobiologist, writer and broadcaster Jack Lewis, the book has been widely discussed in numerous media and social media, including various TV shows, public talks and other sources that Lewis used to disseminate his ideas. Each of the chapters of that book is dedicated to one of the "***seven capital sins***," and in each of them Lewis repeats such a circular—and factually inaccurate—reasoning. In the chapter about "anger," he states: "based on the familial logic that we have used throughout this book…there is always *something* positive to be said about such horrible and malevolent behaviors, an aspect without which we could not be…if there were no benefits, anger would have been eliminated from our human genetic heritage since a long time ago." There are so many layers of scientific wrongness in this very short excerpt. Firstly, "anger" is not part of our genetic heritage.

There are no genes for "anger," as "anger" is obviously the result of a combination of a plethora of biological factors, including epigenetic ones. Secondly, such ideas completely exclude the possibility that at least some traits, even behavioral ones, that arose in the evolution of a certain group of organisms might just be neutral, while empirical data show that many features started by being neutral or even detrimental, for instance when they are the result of drift or part of an "**evolutionary package**" that also includes other traits. This point could apply very well to "**anger**," which could be, at least partially, a byproduct of the evolution of **hormones**, which play a plethora of crucial roles in the evolution and physiology of an endless number of species. Empirical data show that in humans, **testosterone** seemingly plays a crucial role in the arousal of certain behavioral manifestations related to **aggression** and that its levels tend to be higher in individuals with aggressive behavior or during the violent phases of some sports games. Thirdly, such ideas also totally leave out the prospect that at least some behavioral features that might have been advantageous at a certain period might later become neutral or even detrimental for millions of years, for instance due to social/cultural transmission: we know that this actually occurs very frequently in evolution as explained in my *Evolution Driven by Organismal Behavior* book. Fourthly, after repeating such a circular reasoning in each chapter of his *The science of sin*, Lewis then refers to cases of "brain dysfunctions," "malfunctions," or "injuries" that can be related to each of the "sins" discussed by him.

This is a further illustrative example of the type of logical inconsistency that plagues adaptationist just-so-stories: if a certain behavioral trait can only persist for a long time in an evolutionary lineage because it necessarily brings some current "benefits," as argued by Lewis, how can that trait be related, at the same time, with clear "brain dysfunctions," "malfunctions," or "injuries"? For instance, some of the behavioral traits discussed by him can be related to the occurrence of brain tumors. Of course, Lewis could argue that such a pathological biological relation is different from the type of phenomenon that would make a trait persist in a "healthy" population for millions of years. However, this does not solve the logical inconsistency of such **adaptationist stories**, because such cases show that a certain trait can just occur as the result of a pathology, or a "malfunction," without any previous, current or future advantages. Moreover, how can such dysfunctions, brain tumors, or other pathological conditions occur in humans, and in many other animals, since times immemorial? According to his **adaptationist way of thinking** such phenomena could only have been originated and persist during such a long time if they "always had *something* positive": this surely does not apply to brain tumors, unless Lewis and other adaptationists want to go as far as to argue that having a brain tumor is "advantageous" for humans.

Moving now from *Neverland*'s twilight zone to the reality of the natural world, such examples about a certain behavioral trait such as being aggressive occurring due to a brain cancer demonstrate the main fallacy of **adaptationist ideas**: they are both **reductionist** and **teleological**, focusing on a single specific trait—for instance "aggressivity," or "having a **cancer**" or "suicide"—and obsessively defending that the trait surely was/is advantageous to the animals having them. Such ideas neglect the fact that in the natural world everything is connected. No trait arises within a certain group of animals in a vacuum, by itself, independent of an endless number of anatomical, physiological, and behavioral features of those animals. Furthermore, those animals are also not living in a vacuum themselves, they live in a natural world with millions of other species. Some brain "malfunctions" that can dramatically change our behaviors can be produced by viral infections. This point, in turn, reminds us of one of the most important legacies of **Stephen Jay Gould**, who brilliantly deconstructed such adaptationist just-so-stories: that selection does not operate only at the level of the *individual*—for instance, of a single human being—but instead *also* at the level of communities, species, or cells, as attested by the fact that while for a person suffering from cancer the proliferation of cancerous cells is something "negative," for those cells themselves this is evidently something "positive." But, ultimately, if the cancer progression leads to the death of that person, that would be a "tragedy"—the complete termination—for both the individual and the cancerous cells, showing the absurdity of such adaptationist just-so-stories and of how we need to be careful when we use terms such as "evolutionary success".

Prum's 2017 book, *The Evolution of Beauty* cites numerous fascinating cases of quasi-religious, unfalsifiable adaptationist ideas. It stresses that, in particular, "contemporary **evolutionary psychology** has a profound, constitutive, often fanatical commitment to the universal efficacy of adaptation by natural selection…[which] is *the organizing principle* of the field…there is never any doubt what the conclusion of any evolutionary psychology study will be…the only question is how far the study will have to go to get there…this is a how faith-based scientific discipline operates – looking for new reasons, however inadequate, to maintain belief in a theory that has failed." He asks "where's the harm in this intellectual mission? What concerns me most is not merely that so much of evolutionary psychology is bad science…what's worse is that evolutionary psychology is beginning to influence how we think about our sexual desired, behavior, and attitudes." Sadly, the term "beginning to influence" does not reflect the whole truth: evolutionary psychology has *already* done a huge damage in the sense that it has influenced the way many scholars, and a vast part of the broader public, think about human evolution, in particular the way they continue to see women as "more passive," "less innovative," or "less sexual" than men.

Disturbingly, many adaptationists often try to *force* others to think like them and/or to apply their methodological flaws. Unfortunately, due to their huge influence in many fields, adaptationists are frequently indeed able to do this to others—for instance, during the process of peer reviewing, that is, when scientists review the studies of other scientists before those works can be officially published in so-called peer-reviewed journals. Among numerous disconcerting cases that have been revealed throughout the last decades, an illustrative, recent one was provided in Prum's *The Evolution of Beauty*. Notably, he submitted a paper to a peer-review journal about bird behavior, explaining that the data he collected indicated that a certain feature evolved "through arbitrary mate choice." As he notes, "the reviewers…argued…that I had not specifically rejected each of the many adaptative hypotheses that they could imagine…of course, this made it impossible to 'prove' my point, and I ultimately cut this section out of the manuscript in order to publish the paper." He thus asks: "how many of these adaptative hypotheses…would I have to test before I could conclude that any given display trait was arbitrary? When should I ever be done with this task? Even if I were able to test every adaptative explanation they could think of…their reasoning implied that I would have to test other hypotheses in order to satisfy other skeptical reviewers, and then others, *ad infinitum*." Further stressing the point that the numerous existing adaptationist "straw-men" are indeed dominant in many fields, he states: "I was trapped…the prevailing standard of evidence meant it would be impossible for me to ever conclude that any trait had evolved…arbitrarily."

Ruse's 2018 book *On Purpose* is a recent, clear case in point of how adaptationist "straw-men" are not only everywhere, but also that their ideas still continue to be seen as the "norm" within the scientific community, two decades after the beginning of the twenty-first century. He states: "*so where are we today in evolutionary thinking? Don't go away with the message that…biologists today are now questioning seriously what was labeled…the design-like nature of the world…in the world of organisms, adaptation is the norm – the hugely well-justified null hypothesis – and it is your task to make the contrary case if you wish…purpose thinking rules, and it is cherished…today's biologists use end-directed thinking and language when they are dealing with organisms*."

One needs to recognize that, however, something has started to change, although very slowly: firstly, in the end of the twentieth century, precisely mainly because of the strong criticism that Gould and his colleagues did of adaptationist just-so-stories; and more recently, when some authors—mainly some evolutionary biologists—have taken those criticisms one step further, expanding them to many other critical aspects of the **NeoDarwinian Evolutionary Synthesis**. In fact, some of these latter authors—including myself—have argued that we need a major change of mind-set: that is, an "**Extended Evolutionary Synthesis**." One very interesting historical point is that the empirical experimental data obtained in numerous recent studies have revived some of the ideas of one of the scholars that was often seen by Darwinists and Neo-Darwinists as one of the main—if not the major—"heretic": **Lamarck**. In reality, not only molecular and developmental studies, but also behavioral and ecological works are showing that types of **extra-genetic inheritance**, such as behavioral inheritance associated with **niche construction**, are crucial for evolution and widen the notion of heredity beyond genetics. It has been fascinating to assist to recent confrontations between scholars that write papers titled "*Lamarck rises from his grave*"—a 2017 paper by Wang and colleagues—versus papers titled "*We should not use the term Lamarckian*"—a 2019 paper by Speijer 2019. The latter paper literally states that "discussing examples of inheritance of acquired characteristics…is all fine as long as it is clear they do not embody alternatives to Darwinism, but illustrate the incredible versatility of natural evolution working in accordance with its basic assumptions…to paraphrase the most celebrated words from Darwin's magnum opus: 'by so simple a model endless forms most beautiful and most wonderful have been, and are being, explained'…so, stop using the 'L-word!'."

I don't know if there is a more illustrative example to show how there is indeed still often a quasi-religious veneration for Darwin/Darwinism, to the point of telling others scholars to not even do the "heresy" of simply using the name of its main "evil" heretic, and accordingly not even pronouncing that name, using instead just the capital letter L. This point clearly reminds me of a commonly used religious expression from the European Middle Ages: "don't speak of the devil…or he shall appear." And **Lamarck** did indeed re-appear, in the last decades. Actually, not only Lamarck: he is rising from his grave together with other main 'heretics', such as **Baldwin** and **Goldschmidt**. As explained in my 2017 book, Baldwin's idea of '**organic selection**' has been increasingly cited in recent years because of the growing empirical evidence accumulated in the last decades showing that organisms are not merely passive evolutionary players—they are often crucial active players, for instance in cases of **niche construction**. Regarding Goldschmidt, more and more authors are now showing that many of his ideas are also supported by empirical data, including his notion of "hopeful monsters" that was so discredited by Neo-Darwinists—see Weisbecker and Nilsson's 2008 and Dittrich-Reed and Fitzpatrick's 2013 papers.

The main reason why that notion was so discredited is that within a "struggle-for-existence" framework in which life is seen as a forceful, never ending "struggle, war, famine and death," organisms, including their anatomy, should in theory optimally "fit" the habitats in which they live. Therefore, a "macromutation" *sensu* Goldschmidt would in theory not be viable because any feature that is not optimal, or at least almost optimal, and that is not immediately directly related to the survival and/or reproduction of the organism is purged from existence. However, there are clear empirical examples of animals that are "hopeful monsters," many of them still living today, such as **chameleons**, as explained in detail in a recent paper that I wrote with two colleagues—Diogo et al. 2017b. Chameleons are '**hopeful monsters**' *sensu* Goldschmidt in the sense that their limbs have features that were described as "monstrosities"—more recently named "severe congenital malformations"—in humans, such as **syndactyly** and **zygodactyly**—that is, having some digits fused into "superdigits" and moreover being able to oppose such "**superdigits**" against one another, for instance to grasp tree branches. That is, for at least some time the ancestors of today's chameleons were able to survive *despite* those "monstrosities," and moreover were subsequently able to use those "**monstrosities**" in a way that allowed them to occupy specific niches in a very efficient way, for instance arboreal habitats.

This point therefore shows us another major problem with the "mode" of evolution emphasized in Darwin's *Origin*, in which Malthus' "struggle-for-existence" was considered to be *the* driving force of selection. This is because these and many other empirical examples show us, instead, that in at least some cases—for example when there are enough natural resources, or in ecologically relaxed environments where there are not so many predators—evolutionary changes—even neutral, or slightly detrimental ones—may prevail as long as the organism as a whole is "*good enough*" to survive and reproduce. From "*evolution is optimal or sub-optimal*," as defended by Darwin and, to a more extreme level, by NeoDarwinists—in particular adaptationists—*to* "*evolution is good enough*," as defended by a growing number of other evolutionary biologists, is probably *one of the most profound changes of evolutionary thinking in the decades*. Importantly, in a recent book published by me and some colleagues in 2016, we have empirically shown that such a "*good enough*" view of evolution applies to our own evolutionary lineage and our own bodies, contradicting the longstanding idea that our bodies are a "perfect machine." They are not: most of the structures of our bodies have a clearly non-optimal configuration for the type of life we had, which can only be understood when one takes into account both their evolutionary history and embryological development. That is why we chose the title of that 2016 book to be *Understanding human anatomy and pathology: an evolutionary and developmental guide for medical students*.

6.3 "Higher" or "Favored" Groups, Racism, and Capitalism

To prove women's inferiority, antifeminists began to draw not only, as before, on religion, philosophy and theology, but also on science: biology, experimental psychology and so forth. (Simone de Beauvoir)

Ruse has written extensively—for instance, in his 1996, 2003, and 2018 books—about the links between Darwin's works and two of the oldest and most obsessive traits of humans: the search for **cosmic purpose** and the **notion of cosmic progress**. What is particularly remarkable is how the concept of *cosmic progress* was, indeed also so prominent in Darwin's writings, because his ideas of natural selection should theoretically have contributed to finish this longstanding and clearly erroneous concept, once for all. The notion of "natural progress" was developed millennia ago, being for instance present in the non-evolutionary ideas of Aristotle, which were framed within the context of a *Scala Naturae*—"ladder of nature" or "great chain of being": from "lower" organisms to non-human animals, and then to humans, at the top (Fig. 3.2). Ruse's 2018 book *On Purpose* does a great work showing that, as the notion of purpose, the concept of progress does continue very much alive in science nowadays, being in fact probably stronger in the last centuries than ever before. This is because, among scientific circles, this concept begun to be even more prominent in the eighteenth century, the so-called **Age of Enlightenment** that so much influenced **Darwin** (see also Boxes 3.3 and 3.9).

Ruse emphasizes the parallel between the use of this concept in science and religious thinking: in "the Age of Enlightenment…thought and hope were actualized in the form of a formidable challenger: progress! No less end-directed, this was a philosophy of history that took the responsibility and control away from God and put it firmly in our hands…[however] to be honest, the two philosophies (Providence and progress) were frequently not all that different, and at times it is difficult to distinguish their ends." For those that might argue that saying that scientists nowadays follow such a notion of progress is just using a "straw-man" example, that such a view was only "commonly defended in the past," I will answer by providing a statement made by Ruse, which further provides an example of the "progress-religious belief" link:"many of today's leading evolutionists are quite open about their *beliefs* in biological progress…what is [even more] striking is how most evolutionists more or less take biological progress for granted—for all that they are prone to deny it when in public – and go on to argue from there."

The million-dollar question therefore is: if within Darwin's idea of natural selection organisms are mainly adapted to their *local* habitats, at a *specific time* in history, why did Darwin still refer so often to *general* "progress," "favored or preferred races," and to "higher taxa" as concepts that would apply to *all* geological eras and regions of the globe? He clearly did so in his *Origin*, from the very subtitle of the book—see below—to its very end, which finishes with the suggestion that, as put by Bowler in *Darwin Deleted*, "in the long run, natural selection would lead to the evolution of higher types of organization." As noted in Ruse's *On Purpose*, this is one of the few criticisms that historians of science and biologists dare to do about Darwin's works, because it is an obvious example of how not only some of his theories are probably wrong but are also self-contradictory. Ruse argues that Darwin recognized these inconsistencies in some parts of his books but that, for some reason, he was just unable to finish once for all with such teleological notions of cosmic purpose and progress, because either it was just too revolutionary for his epoch to do so, or he was actually personally inclined to accept those notions.

One case in which Darwin accepted these inconsistencies, and directly related some of his ideas with teleological narratives in general and religious ones in particular, is a sentence from his 1878 autobiography. Namely, he confessed that when he was writing his *Origin* he felt the "extreme difficulty or rather impossibility of conceiving this immense and wonderful universe, including man with his capacity of looking far backwards and far into futurity, as the result of blind chance or necessity…when thus reflecting I feel compelled to look to a *First Cause* having an *intelligent mind* in some degree analogous to that of man; and I deserve to be called a *Theist*." As summarized in Ruse's *On purpose*, "one feels a little as if Darwin is like Moses – he led his children to the Promised Land but never got there himself." One of the more illustrative examples showing that more and more writers are finally starting to recognize this, although very—too—slowly, concerns Delisle's 2017b and 2019 books, which stressed that "the first to challenge Darwin was Darwin himself in the *Origin*…the reader willing to go beyond Darwin's rhetoric encounters a book displaying at least five independent sets of issues or pictures…while some are squarely incompatible with one another, others are less than clearly related to each other" (see also Box 7.16).

Some of the arguments used by Darwin, and by some scholars that subsequently tried to deny that his works suffered from such theoretical inconsistencies, involved using the notion of "**arms races**," which according to them could explain the existence of a "general progress" in nature. That is, if a predator A and its prey B are co-evolving, and they respectively acquire "better" evolutionary "weapons" to hunt and to not be hunted, respectively, then we could explain how *both* A and B could get "better" with time. However, this argument is not convincing. First, even if that was the case, A and B would just be better at hunting B and not be hunted by A, respectively: they would not be "better" animals *as a whole*, evolutionary. Life is much more than a predator A being able to hunt or not hunt B—predators often hunt many different types of preys, and each prey is often eaten by many types of predators, and in addition both preys and predators have endless other types of ecological interactions such as competition, cooperation, having sex with others, taking care of their progeny and/or building nests if that applies, and so on. It would be like saying that humans living 1 million years ago were "better" as a whole just because they were "better" at escaping to a single predator, let's say leopards. So many other crucial aspects were relevant for humans at that time, such as using fire, doing and using stone tools, cooperation, migration, and so on—several of them likely having contributing to the evolution or our lineage as much, or more, than just being able to escape a single predator.

And what makes the "arms-races" argument even less convincing is that in most of the cases in which Darwin refers to "progress" or "higher clades," or "favored races" he is not even referring at all to preys or predators. Instead, the "highest" taxon of all for Darwin is our species, of course. This is another clear case of how Darwin was deeply influenced by teleological narratives and biases, as we all are: what a coincidence, that a member from one of the millions of species living in this planet would state that *his* species is *the* "highest" one. As summarized in Ruse's *On Purpose*, "Darwin was deeply committed to the **cultural ideology of progress** and to the belief in **biological progress**, something that ends not just with human beings but with Europeans, preferably **English capitalists.**" In fact, there are profound historical links between not only the **Age of Enlightenment** and the notion of progress, but also the **Industrial Revolution**, **capitalism**, and the **concept of individual selfishness** that was so prominent in Darwin's scientific ideas. For instance, "**Adam Smith** was important here [regarding the notion of progress], with his ideas of the importance of a division of labor and of the Invisible Hand making a virtue of individual selfishness." Or, as put in *Narratives of Human Evolution* by Landau, who shows that many elements of Darwin's works—particularly *The Descent of Man*—are indeed similar to imaginary narratives and tales used in **folklore** and **myth**: "the principle of natural selection, or 'struggle-for-existence', remains the chief agent of [Darwinian] evolution…it also explains events according to their consequences or 'final causes'…[it] may appear to operate in a teleological fashion, as though directed toward some overall design or purpose…Darwin…confesses that he does believe human evolution has been toward a preferred and higher state." As she further noted, in Darwin's *Descent*, as in other writings on human evolutionary history such as those of **Huxley** or **Haeckel**, "like most narratives, the story of human evolution is subject to an intrinsic '**teleological determinism**': elements are present not as they occur but as they contribute to the outcome of the story."

This is precisely the major flaw that many scholars, including Bowler in his *Darwin Deleted*, do when they refuse to recognize that Darwin's theories and metaphors did provide easy ammunition for the subsequent dissemination and political use of narratives about selfishness, racism and misogyny that became so popular and that were in actual fact used to justify a huge number of atrocities done since then. Yes, many of these narratives existed in the West before Darwin. But what was unique—and should never be forgotten—with Darwin is that the most eminent and influential biologist of all times not only *provided so-called scientific evidence to support such narratives*, but *did that explicitly in his published scientific works*, to the point of even including them in the original title of his most important book, the *Origin*. As noted above, nowadays when scholars refer to Darwin's 1859 book, they almost always refer to its title as *On the Origin of Species by Means of Natural Selection*, but the full title was actually *On the Origin of Species by Means of Natural Selection, or, the Preservation of Favored Races in the Struggle for Life*. As that 1859 book is one of the most cited and discussed books in human history, it seems difficult to accept that almost no scholar gets its original subtitle right: it is thus very likely that many scholars do this consciously—or unconsciously—within the prevailing context of idealization of Darwin and the biases it creates, trying to completely disconnect his works with the rise of **scientific racism** and **eugenics** in the last decades of the nineteenth and first decades of the twentieth centuries.

As will be discussed below, Darwin was seemingly "less racist" than many Westerners were at that time. He came from a family opposed to slavery and was appalled by the cruel treatment he saw meted out to slaves in South America on the Beagle voyage. However, there is a huge amount of historical data that unequivocally shows that he clearly and explicitly defended and "scientifically supported" both racist and Eurocentric ideas in his works, often using very powerful metaphors that became immensely widespread in popular culture and that *were directly used in the writings of leaders such as Hitler*, as we will see below. Therefore, one needs to at least acknowledge that in this regard he was particularly uncareful—clearly contrasting of how careful he was about other aspects of his scientific life, including the time he took to write and publish the *Origin*. Specifically, precisely because *he was aware* of how heated the discussions on racism, slavery and colonialism were at that time, both at a scientific and societal level, and precisely because he was against slavery, it does seems very uncareful, to say the least, to not realize that by explicitly "scientifically" supporting racist and Eurocentric ideas in his works he would provide easy ammunition to those defending racist and pro-slavery ideas.

What can be better for a racist, colonialist, misogynist, or slave-owner, to be able to defend his/her ideas by stating that even the most renowned "objective scientific expert" on the matter clearly stated, in his scientific books, that whites are superior, women are inferior, and so on? That this is a biological fact, "proven" by Darwin himself? Of course, I am not suggesting that Darwin was *directly* responsible for *everything* that happen subsequently to him regarding racism, misogyny, eugenics, and so on, that would be completely unfair and unrealistic. I am just saying that it is also unrealistic to argue that Darwin's life and works had *nothing* to do with that, and particularly to argue, as in Bowler's *Darwin deleted*, that racism, eugenics, and misogyny would likely have been even much worse in a world without Darwin.

In fact, I am not talking here just about theoretical historical speculations that cannot be tested: I am referring to a *huge amount of empirical, historical facts* that *show*, beyond any doubt, that Darwin's ideas and particularly the metaphors he created and disseminated were *directly and actively used by the leaders promoting these atrocities and genocides*. That is, we are talking about *facts* about what Darwin *did* write, about the increase of scientific racism and rise of eugenics *after* the publication of those writings, and about the explicit way **Nazis**, or members of the **KKK**, or other **white-supremacists**, directly used/continue to use what Darwin wrote to justify their racist ideas and the atrocities they did/are doing. I can understand that scholars can discuss whether Darwin was enormously uncareful, or very uncareful, or just a bit uncareful, when he wrote what he did, or whether racists could use the works of other scholars to support their views, although it is obvious that nothing can compare to have the most prominent biologist of all times directly stating that some human "races" or genders are "inferior" or "lower" than others.

However, there is *no doubt* that his ideas/metaphors *were used and continue to be used* by racists and white supremacists: to deny this is just denying facts, to keep living in *Neverland*'s world of unreality. To give just one unequivocal historical fact, among an endless number of others that I could provide here, **David Duke**, former **"Grand Wizard" of the KKK**—a group that mainly started 6 years after the publication of Darwin's 1859 *Origins* and that became progressively more prominent since then, until recently—explicitly wrote, in his 1998 book *My Awakening – A Path to Racial Understanding*: "Charles Darwin…demonstrated that principles of heredity combined with what he called, *Natural Selection*, had developed the exceptional abilities of mankind itself…his masterpiece, *Origin of Species* has a subtitle that expresses his whole idea in a nutshell: *The Preservation of Favored Races in the Struggle for Life* – preserving the Caucasian race is but a precondition for continuing its evolution to a higher level." It is enormously important to note that Duke precisely referred to the part of the title of Darwin's 1859 book that most scholars omit today—"*the preservation of favored races in the struggle for life*"—and that Darwin himself deleted in later versions of his *Origin* book, probably because he recognized how wrong and dangerous it was to have used it in the first place. Very dangerous indeed, as Duke's words remind us, in a particularly troubling and sad way, knowing the atrocities done and terror created by the KKK in the U.S. for such a long time.

In *Readings in the History of Evolutionary Theory*, Wetherington noted that "the idea that progress is a natural condition of the social order did not await Darwin for its expression…it was present at the Enlightenment…(Darwin's) natural selection simply gave it a sense of scientific authenticity." Indeed, even Bowler's *Darwin deleted*, which mainly defends narratives related to the idealization of **Darwin**, had to recognize that "evolutionism…offered a plausible (scientific) explanation of why some races might not have advanced as far as others up the scale leading from the ancestral ape…most of the Darwinians endorsed this way of thinking…with the notable exception of **Wallace**." This comment about Wallace is exceptionally significant, because it also contradicts one of the main lines of defense used by those idealizing Darwin: that **Darwin's racist and misogynistic writings** cannot be criticized per se because at that time everybody was misogynist and racist. The works of scientists such as Wallace, or Humboldt, clearly show that this is not necessarily so, because although they might have been racist and misogynistic in *some* aspects of their personal lives and their scientific works, in the overall their scientific writings have a very different, much less racist tone than those of Darwin. **Alexander Humboldt**, who died in 1859—precisely the same year when Darwin published the *Origin*—illustrates this point because he clearly had in general a much more positive view of the indigenous people he encountered in his fascinating travels than Darwin did (see below).

One example concerns his 1814 book *Views of the Cordilleras and Monuments of the Indigenous Peoples of the Americas*, as discussed in detail in a splendid book about Humboldt's life, travels, research, findings and writings that I strongly recommend: Wulf's 2015 *The Invention of Nature – The Adventures of Alexander von Humboldt, the Lost Hero of Science*. Humboldt was opposed to Trans-Atlantic slavery trade, as Darwin was, but in a marked contrast to Darwin he did indeed have, in the overall, a much more positive view of indigenous people, and in general did not fall into the trap of *believing* in a "progress" towards "civilization" or "higher people," and was accordingly also firmly opposed to **colonialism**. Of course, Humboldt should also not be idealized, he clearly had some less positive aspects regarding indigenous people, for instance as **Russel Wallace** he tended to neglect their contributions to "his scientific discoveries" in his writings. Still, the reality is that when leaders such as **Bolivar** fought for independence from Spain and against colonialism, they often referred *directly* to Humboldt as a main source of inspiration, and that by itself gives a lot of credit to Humboldt. In contrast, no South American "revolutionary" leader has directly stated that he was "inspired" by Darwin, nor would one think that they would have any reasons to do so, when we read—as we will, below—what Darwin *wrote* about indigenous people of South America, such as the Patagonians. Actually, the politicians that have used and continue to use the name of Darwin politically—such as **David Duke**—are often the political enemies of those that fight against colonialism, oppression, social injustices, and inequality. This point, per se, should make us think deeply about the social legacy of Darwin, or at least the way in which his works and metaphors became used, politically.

The idea that Darwin's writings and metaphors did provide easy ammunition for such politicians, as well as for racists, colonialists, slave holders and/or segregationists in general, was also strongly defended in one of the most extensive and well-documented books on the history of racist ideas in the U.S.—Kendi's 2016 *Stamped from the beginning*:

> [In the 1860s] it looked as if **polygenesis** had finally become mainstream. In actuality, the days of the notion of separately created human species were numbered. Another pernicious theory of the human species was about to take hold, one that would be used by **racist apologists** for the next one hundred years. In August 1860, polygenesist Josiah C. Nott took some time away from raising Alabama's first medical school (now in Birmingham). He skimmed through a five-hundred-page tome published the previous November in England. It had a long title, *On the origin of species by means of natural selection, or the preservation of favoured races in the struggle for life*. Nott probably knew the author: the eminent, antislavery British...biologist **Charles Darwin**. "The view which most naturalists entertain, and which I formerly entertained – namely, that each species has been independently created – is erroneous", Darwin famously declared. "I am fully convinced that species are not immutable". Recent discoveries were showing, he explained, that humans had originated much

> earlier than a few thousand years ago. Darwin effectively declared war on biblical chronology and the ruling conception of polygenesis, offering a new ruling idea: **natural selection**. In the "recurring **struggle for existence**", he wrote, "all corporeal and mental endowments will tend to progress towards perfection". Darwin did not explicitly claim that the **white race** had been naturally selected to evolve toward perfection. He hardly spent any writing time on humans in *The Origin of Species*. He had a grander purpose: proving that all living things the world over were struggling, evolving, spreading, and facing extinction or perfection. Darwin did, however, open the door for bigots to use his theory by referring to "civilized" states, the "savage races of man", and "half-civilized man", and calling the natives of southern Africa and their descendants "the lowest savages". Over the course of the 1860s, the Western reception of Darwin transformed from opposition to skepticism to approval to hailing praise. The sensitive, private, and sickly Darwin let his many friends develop his ideas and engage his critics. The mind of English polymath **Herbert Spencer** became the ultimate womb for Darwin's ideas, his writings the amplifier of what came to be known as **Social Darwinism**. In *Principles of Biology* in 1864, Spencer…[used] the iconic phrase "survival of the fittest". He religiously believed that human behavior was inherited. **Superior hereditary traits** made the "dominant races" better fit to survive than the "**inferior races**". Spencer spent the rest of his life calling for governments to get out of the way of the struggle for existence. In his quest to limit government, Spencer ignored the discriminators, probably knowing they were rigging the struggle for existence. Longing for *ideas to justify the nation's growing inequities*, American elites firmly embraced Charles Darwin and fell head over heels for Herbert Spencer.
>
> Charles Darwin's scholarly circle grew immeasurably over the 1860s, encircling the entire Western world. The *Origin of Species* even changed the life of Darwin's cousin, Sir **Francis Galton**. The father of modern statistics, Galton created the concepts of correlation and regression toward the mean and blazed the trail for the use of questionnaires and surveys to collect data. In *Hereditary Genius* (1869), he used his data to popularize the myth that parents passed on hereditary traits like intelligence that environment could not alter. "The average intellectual standard of the negro race is some two grades below our own", Galton wrote. He coined the phrase "nature *versus* nurture", claiming that nature was undefeated. Galton urged governments to rid the world of all naturally unselected peoples, or at least stop them from reproducing, a social policy he called "**eugenics**" in 1883. Darwin did not stop his adherents from applying the principles of natural selection to humans. However, the largely unknown codiscoverer of natural selection did. By 1869, British naturalist **Alfred Russel Wallace** professed that human spirituality and the equal capacity of healthy brains took humans outside of natural selection. Then again, as Wallace made a name for himself as the most egalitarian English scientist of his generation, he still professed European culture to be superior to any other. Darwin attempted to prove once and for all that natural selection applied to humans in *Descent of Man*, released in 1871. In the book, he was all over the place as he related race and intelligence. He spoke about the "mental similarity between the most distinct races of man", and then claimed that "the American aborigines, Negroes and Europeans differ as much from each other in mind as any three races that can be named". He noted that he was "incessantly struck" by some South Americans and "a full-blood negro" acquaintance who impressed him with "how similar their minds were to ours". On **racial evolution**, he said that the "civilized races" had "extended, and are now everywhere extending, their range, so as to take the place of the lower races". A future evolutionary break would occur between "civilized" Whites and "some ape" – unlike like the present break "between the negro or Australian and the gorilla". Both **assimilationists** and **segregationists** hailed *Descent of Man*. **Assimilationists** read Darwin as saying Blacks could one day evolve into **White civilization**; **segregationists** read him as saying Blacks were bound for extinction.

Let's thus now discuss Darwin's personal encounters with, and ideas about, non-Western people, which are deeply related with the links between science,

Fig. 6.1 Jemmy Button in 1831 and 1834

colonialism and the notion of "progress"? As explained in Ryan and Jetha's *Sex at Dawn*, in his travels Darwin recognized the clash between his **capital-based society** and his notion of **struggle-for-existence** and the related concept of **biological selfishness**, and what he saw as the natives' self-defeating kindness, writing: "nomadic habits…have in every case been highly detrimental…the perfect equality of all the inhabitants…will for many years prevent their civilization." Before Darwin's travels, **Malthus**, looking for an example of the world's most downtrodden "savages," had cited "the wretched inhabitants of Tierra del Fuego," who had been said by some earlier European travelers to be "at the bottom of the scale of human beings." When Darwin arrived to **Tierra del Fuego**, he agreed with Malthus, stating: "I believe if the world was searched, no lower grade of man could be found." **Captain Robert Fitzroy** of the **Beagle** had picked up three **Fuegians** on an earlier voyage, and took them to England to introduce them to the "highest" of civilizations, and then he returned them to Tierra del Fuego so they would serve as missionaries. But just a year later, in 1934, the huts and gardens that the British sailors built for the three Fuegians were empty: **Jemmy Button**, one of the three, appeared later and told the crew that he and the other two had reverted to their former way of living (see Fig. 6.1). Darwin, surprised and puzzled, wrote in his journal that he'd never seen "so complete and grievous a change" and that "it was painful to behold him."

Captain Fitzroy told Jemmy that he could take him back to England, but Jemmy answered that he had "not the least wish to return to England" as he was "happy and contented" with "plenty fruits," "plenty fish," and "plenty birdies." As a fascinating example of the Europeans' lack of understanding of and empathy towards the way of life, aspirations, and priorities of the indigenous people, Darwin could not comprehend how someone from the "lower grade of man" did not want to live a "highly civilized" life in London. So a way that Darwin used to try solving this "puzzle" was to suggest—in his *Voyage of the Beagle*—that Jemmy's unwieldiness to go back to London was probably due to the presence of his "young and nice looking wife." Like many adaptationists continue to create just-so-stories in order to support a priori assumptions based on wrong scientific ideas framed on the notion of cosmic purpose and progress, Darwin created a just-so-story in order to support his a priori assumptions based on wrong "alternative facts" framed on such notions. That is, in the mind of someone as Victorian and "civilized" as Darwin, who was *sure* that the "whites" living in England were a "highest," more "advanced" group that was "favored" by evolution, what other reason could a "savage" such as Jemmy have to not want to go back to London's "civilization," except for something as "primitive" as his love/desire for a "savage" woman?

In his 2001 book *Savage: The Life and Times of Jemmy Button*, Hazlewood provides more details about Darwin's interactions with, and ideas about, Jemmy and other indigenous people of Tierra del Fuego, which Darwin described as "the most abject and miserable creatures I anywhere beheld." After his first encounter with them, Darwin wrote to his sister Caroline: "an untamed savage is I really think one of the most extraordinary spectacles in the world…the difference between a domesticated and a wild animal is far more strikingly marked in man: in the naked barbarian, with his body coated with paint, whose very gestures, whether they may be peaceable or hostile are unintelligible, with difficulty we see a fellow-creature." Later, he remarked that he felt "quite a disgust at the very sound of the voices of these miserable savages." Regarding the moment of the farewell between Darwin and **Jemmy Button**, Darwin wrote in his diary "I am afraid whatever other ends this excursion to England produces, it will not be conducive to their happiness…they have far too much sense not to see the vast superiority of civilized over uncivilized habits, yet I am afraid to the latter they must return." So, here again, Darwin refers directly to the "superiority" of one group of people and their lifestyle, as if this was a timeless scientific fact. Such examples from Darwin's own writings are particularly critical for two main reasons. Firstly, they show that one cannot just excuse Darwin, as many scientists and historians tend to do, because he was "merely a product of his epoch." Other scientists living in the same epoch, when they encountered indigenous people during their travels, clearly reacted in a very different—undoubtedly much less racist—way, as did Humboldt and Wallace. Secondly, the extremely racist lines written by Darwin's also go directly against the main evolutionary idea defended by him in his *Origin*, according to which one can only talk about certain types of behavioral adaptations as a product of natural selection within a *specific* type of habitat, geographic place, and period of time. In the book *Biology and Feminism*, edited by Lynn Nelson, it is stated that:

> Some feminists also argue that, contrary to the way he is generally portrayed by historians of science, there are several respects in which Darwin was decidedly not "swimming upstream" – that is, he was not critically taking on prevailing sociopolitical or scientific views. He was, as we have seen, assuming the **gender stereotypes** of his day. In addition, some feminists and others point out that Darwin's model of natural selection – which involves waves of competition for scarce resources, and "winners and losers" – paralleled

> then current arguments for **capitalism**. So, too, Darwin assumed then current sociopolitical beliefs about race differences. Although an abolitionist, he appealed to differences between "the races" in brain size and intelligence…. Many [if not most] Darwin scholars recognize that the claims **Darwin** makes about sex/gender, sex, and racial differences that feminists criticize are in fact unsupported assumptions, assumptions that were characteristic of **Victorian England**.

The story of Jemmy Button is similar to a story reported in a 2013 article of BBC News entitled "*Return to the rainforest: a son's search for his Amazonian mother*." It concerns anthropologist Kenneth and a **Yonomamo woman** named Yarima. They eventually married, had children, went to live in the U.S., but then she went back to Amazonia, leaving her children in the U.S:

> But life in New Jersey was not working out for Yarima. It wasn't the weather, food or modern technology but the absence of close human relations. The **Yanomami** day begins and ends in the shapono, open to relatives, friends, neighbours and enemies. But Yarima's day in the US began and ended in a closed box, cut off from society. Other than Kenneth, no-one could communicate with Yarima in her own language and she had no means of speaking with her family back home. In Hasupuweteri, the men disappeared for a few hours in the day to go hunting, but husbands did not disappear all day, every day. Yarima would spend the day at home or roaming the shopping malls. Good also gave her video and sound recordings from Hasupuweteri that she would listen to over and over. Together with a co-writer, David Chanoff, Kenneth wrote his memoir, which was reviewed well, sold well and was translated into nine languages. He and Yarima became minor celebrities, appearing in People magazine three times. Articles appeared in newspapers with titles like *Americanization of a stone age woman* and *Two worlds: one love*. A 1992 film with National Geographic charted the family's first visit back to the jungle for almost four years. A five year-old David is seen squabbling with Vanessa over a heavy bunch of plantains, while baby Daniel is carried on Yarima's back in a sling attached to a headband, in the traditional Yanomami style. The film contains some joyful moments of Yarima showing off her children to her sister and going crab hunting again in the creeks, but it also captures her despondency. "They say I have become a nabuh," Yarima's translated voiceover tells us. "I live in a place where I do not gather wood and no-one hunts…the women do not call me to go kill fish…sometimes I get tired of being in the house, so I get angry with my husband…I go to the stores and look at clothing…it isn't like in the jungle…people are separate and alone…it must be that they do not like their mothers". A few months after the making of the film, on another return trip to Hasupuweteri, Yarima decided to stay.

The **Eurocentric evolutionary ideas** of Darwin, and of many of his Darwinian followers, were so prominent that even Bowler, in his *Darwin deleted* states: "I must concede that Darwinism did become involved with the culture of **imperialism**, providing a source of extremely effective rhetoric…the imperialists certainly used Darwinian terminology…and in a few cases they were (even) genuine scientific Darwinists." Further, "in the absence (as yet) of fossil hominids, modern savages were treated as equivalent of these primitive ancestors, and the physical anthropologists' alleged evidence of small brains and apelike features in the 'lowest' races was called in to confirm the link…. Darwin certainly contributed to this process in his *Descent of man*, and in Germany **Ernst Haeckel**"—often associated to the **Nazi ideology**—"built the idea that the human race show different levels of development firmly into his Darwinism." Moreover, while Bowler tried to minimize the links between Darwin/Darwinism and eugenics, he had to recognize that it was Darwin's cousin, **Francis Galton**, who began to argue for a "**eugenic program**, in effect a call to impose a mechanism of artificial selection on the human race…in a civilized society, we do not restrict the ability of people to have children, which means that even those with the lowest mental and moral capacities continue to breed…both Darwin and Galton worried that this might lead to degeneration." This program became popular in Britain, "and **sterilization programs** were introduced in a number of American states…the movement became particularly active in Germany, where the **Nazis** went beyond mere sterilization and began to exterminate those elements of society they wishes to suppress."

As wisely noted in Landau's *Narratives of Human Evolution*, one particularly striking aspect that is too often neglected in the literature is how "the real struggle in [Darwin's] *Descent of man* occurs not between animals and men but between humans of varying intellects." Darwin explicitly wrote: "for my own part I would as soon be descended from that heroic little monkey…or from that old baboon… – as from a savage who delights to torture his enemies, offers up bloody sacrifices, practices infanticide without remorse, treats his wives like slaves, knows no decency, and is haunted by the grossest superstitions." As Landau notes, "next to savages who are cruel and false, the European appears kindest and most faithful." Furthermore, as she explains, the problem is *not only* that Darwin had indirect links, and *direct* familiar connections, with the **eugenics movement**—including his cousin and his own son, as we will see below—, but also that Darwin's own writings strikingly resemble in a disturbing way those used by authors within that movement. In *The descent* he writes: "with savages, the weak in body or mind are soon eliminated; and those that survive commonly exhibit a vigorous state of health…we civilized men, on the other hand, do our utmost to check the process of elimination…we build asylums for the imbecile, the maimed and the sick…we institute poor laws…and our medical men exert their utmost skill to save the life of every one to the last moment…thus the weak members of civilized societies propagate their kind…no one who has attended to the breeding of domestic animals will doubt that this must be highly injurious to the race of man…it is surprising how soon a want of care, or care wrongly directed, leads to the degeneration of a domestic race; but excepting in the case of man himself, hardly anyone is so ignorant as to allow his worst animals to breed."

If someone would tell me that this was an excerpt of **Hitler's *Mein Kampf***, it would seem completely credible to me. But just because it is written by the so-often idealized and venerated Charles Darwin, then clearly this can't be as bad—nor as dangerous or influential—at it seems to be, it has to be simply a misinterpretation of his words: nothing coming from Darwin can be "bad," "wrong" or "dangerous," everybody living in *Neverland* "knows" that he was flawless. But in the *world of reality*, things are much more nuanced: although most authors do refer to Charles' direct familiar links to eugenicists such as his cousin Galton, for some reason they tend to neglect the fact that someone that *was raised by him*, his own son **Leonard Darwin**, also became a prominent actor within the **eugenics movement**. It is important to emphasize this often untold story about Darwin's family, because Leonard was predominantly involved with one of the most *negative versions of eugenics*, which was mainly initiated in the U.S. but was then exported to other parts of the globe such as Europe. One of the few authors that had no problem to discuss this neglected story in the broader context of the history of eugenics is **Edwin Black**, in his formidable book *Against the weak*:

> By 1912, America's negative eugenics had been purveyed to like-minded social engineers throughout Europe…hence the **First International Congress of Eugenics** attracted several hundred delegates and speakers from the United States, Belgium, England, France, Germany, Italy, Japan, Spain and Norway. Major **Leonard Darwin**, son of Charles Darwin and head of the EES [**UK's Eugenics Education Society**], was appointed congress president…. Leonard Darwin revealed his true feelings in a speech to the adjunct ***Cambridge University Eugenics Society***. "The first step to be taken", he explained, "ought to be to establish some system by which all children at school reported by their instructors to be specially stupid, all juvenile offenders awaiting trial, all ins-and-outs at workhouses, and all convicted prisoners should be examined by trained experts in mental defects in order to place on a register the names of all those thus ascertained to be definitely abnormal". Like his colleagues in America, Darwin wanted to identify not just the so-called unfit, but their entire families as well. Darwin emphasized, "From the Eugenic standpoint this method would no doubt be insufficient, for the defects of relatives are only second in importance to the defects of the individuals themselves-indeed, in some cases [the defects of relatives] are of far greater importance". British eugenicists were convinced that just seeming normal was not enough – the unfit were ancestrally flawed. Even if an individual appeared normal and begat normal children, he or she could still be a "carrier" who needed to be sterilized.

In Todes' *Darwin without Malthus* he explains that Darwin's notion of "struggle-for-existence" was said to be "most severe between the individuals of the same species, for they frequent the same districts, require the same food, and are exposed to the same dangers…. Darwin used the words 'struggle' and 'competition' interchangeably…the metaphor '**struggle for existence**', and in such phrases as 'the great battle for life' and the 'war of nature' contributed a certain rhetorical power to his argument." According to Todes, by sacrificing accuracy for eloquence, and proposing that within this struggle "death is generally prompt, and that the vigorous, the healthy and the happy survive and multiply," Darwin *did* gave *easy and powerful ammunition* for **eugenicists** around the globe. Similarly, in his 2014 chapter of the book *The Invention of Race*, Andreassen wrote that "Darwin's arguments about the survival of the fittest became central to theories about **racial hierarchies** and human development…many scientists began to see the different races competing against one another; the stronger and more intelligent would thrive, while the weaker and less intelligent races declined…[in other words] **racial Darwinism**." In Bethencourt's 2013 book *Racisms* he also shows the links between Darwin's ideas, capitalism, inequalities, racism, and the notion of "progress," and notes once again the inconsistencies between Darwin's writings about these subjects and some of the key evolutionary ideas defended by him:

> Darwin's…remark on the social system of the Fuegians shows again how reflections on the different stages of humankind and prevailing **racial constructions** were linked to the **issue of inequality**, which at the turn of the century was dealt with in a debate between **William Godwin** and **Thomas Robert Malthus** that is reflected here through reference to Malthus's assertion that only **self-interest motivates humankind**. For Darwin, the perfect equality among the individuals composing the Fuegian tribes had retarded their civilization. Peoples governed by hereditary kings were considered most capable of improvement, and among

races, the more civilized ones had the more sophisticated governments. *Darwin equated equality with baseness*: pieces of cloth given to the Fuegians were torn into shreds and distributed; no individual would be richer than the others. **Individual property**, the **notion of superiority**, and an **accumulation of power** were unthinkable in this tribal regime, yet for Darwin they were the sinews of improvement. The comparison between the "savages" and "barbarians" that Darwin met during his voyage around the world highlights his hierarchy. The **Fuegians** were placed at the bottom of the scale, along with the warlike cannibals and murderous **New Zealanders** (or **Maori**), **Australian aborigines** (skillful with the boomerang, spear, and throwing stick in climbing trees and methods of hunting, but feeble in mental capacity), and "wretched" **South African tribes** prowling the land in search of roots. They were all contrasted unfavorably with the relatively civilized **South Sea islanders** – the manners and even tattoos of the **Tahitians** were praised – and proficient **Eskimos**, with their subterranean huts and fully equipped canoes.

Darwin possessed an independent and acute mind, although for some of his observations he was indebted to **Captain Cook**'s journals. These observations reveal continuities in the descriptions of the peoples of the world, reminding us of the early accounts of **Native Americans** by **Columbus**, **Vespucci**, or **Caminha**, even though the detachment (and repugnance) concerning savages sounds even more pronounced after centuries of contact. The divergence was reinforced by the eighteenth century notion of civilization, and enhanced by the industrial revolution and progress in the comfort of daily life as well as the quality of transportation. The voyage of the Beagle was certainly more comfortable and safe than previous circumnavigations of the world, although the second trip made by Cook (1772–75) had been particularly successful in terms of a radical reduction in the loss of human lives. The filthiness of the native body along with the scanty clothes, diabolic body paintings and tattoos, constant warfare driven by revenge, cannibalism, cruelty, and absence of justice as well as the inferior local languages were not new topics; vehement disgust was an expression of the Europeans' projection of their own self-perception. Darwin's descriptions, however, represent the highest level then reached by travel accounts. They were attentive to habitat, housing, material culture, family structure, division of labor, and political specialization.

The claim that inequality was a source of **social improvement** lay at the core of contemporary debates between socialists and liberals; it shows that Darwin was aware of the major social and political discussions of his time. Nevertheless, Darwin's lack of empathy concerning the savages did not shake his abolitionist convictions. Darwin expressed his indignation when confronted with the daily cruelty toward slaves in Rio de Janeiro, where he saw instruments for their torture, heard the cries of slaves being punished, and intervened on various occasions to stop further suffering. He equated **slavery** with the moral debasement of a whole society; he protested against the idea of slavery as a tolerable evil, denouncing the way in which people were "blind[ed] by the constitutional gaiety of the negro"; and he refused the attempt to "palliate slavery by comparing the state of slaves with our poorer countrymen". Darwin raised a crucial issue that could be related to many other situations of oppression: "those who look tenderly at the slave owner and with a cold heart at the slave, never seem to put themselves into the position of the latter", concluding emotionally, "it makes one's blood boil, yet heart tremble, to think that we Englishmen and our American descendants, with their boastful cry of liberty, have been and are so guilty: but it is a consolation to reflect, that we at least have made a greater sacrifice, than ever made by any nation, to expiate our sins".... Scholars James Moore and Adrian Desmond attribute Darwin's more conservative stance in [his later book *The Descent of Man*]...to the hardening of attitudes in the 1860s.

But if we take one significant example, Darwin's remark concerning the **immorality of savages**, used to counter the Irish historian William Lecky's benevolent arguments, was in line with his observations of savages during his voyage on the Beagle. In terms of **eugenics**, though, I would agree with Moore and **Desmond**: Darwin quoted **William Greg**, **Wallace**, and **Francis Galton** on the failure of natural selection in civilized nations, as a result of vaccinations, poor laws, and asylums – medical care and social assistance for the less fortunate, which promoted the survival and propagation of the weaker members of society, leading to a "deterioration in the noblest part of our nature". Darwin blended eugenics with an essentialist approach to nations. He drew attention to another process of negative selection, produced by the **Spanish Inquisition** over centuries, which systematically excluded those people most ambitious in thought and action, and thus was responsible for long-term decline, while the emigration of the most energetic people to British America had produced the opposite outcome. But Darwin acknowledged that all civilized nations descended from barbarians, showing the possible improvement of savages through independent steps along the scale of civilization. He quoted anthropologist Edward Tylor, who in 1865 had published *Researches into the Early History of Mankind and Development of Civilization*, based on the idea of intellectual abilities shared by all groups of people and differences in social evolution resulting from education. Darwin explicitly rejected the idea of human being's decline: "to believe that man was aboriginally civilized and then suffered utter degradation in so many regions, is to take a pitiably low view of human nature", maintaining instead "that progress has been much more general than retrogression; that man has risen, though by slow and interrupted steps, from a lowly condition to the highest standards as yet attained by him in knowledge, morals and religion".

Perhaps even more pronounced than Darwin's racism—despite his very noble opposition to slavery, it has to be emphasized once again—was his **misogyny**, which also *did* have a huge influence in the history of **misogynistic ideas**, from the Victorian Era until the present time. Here I would like to analyze in particular the evolutionary idea that men are more prone to **polygamy** than women are, because men can "optimize"/"maximize" the number of their children by having many women, while women, when pregnant, have to wait at least 9 months to have another child. Apart from being based on the **misogynistic fallacy** of using **polygynous**—one male with various females—primates such as gorillas as a model for humans, this idea is also based on the **capitalist fallacy** of "more is better." That is, it assumes that the best "adaptative optimization" is always to have the higher number of descendants. As put by Landau, in Darwin's *Descent* it is argued that "success, in the long run, is measured by numbers of offspring." Apart from the fact that our closest relatives, the chimpanzees, have a "multimale-multifemale"—not a polygynous—type of sexual organization, such a reasoning is in itself paradoxical. This is because **Malthus**, who influenced Darwin so much, actually stated that a main problem of humanity would be **overpopulation**, leading to the "struggle-for-existence" that was so crucial, and exaggerated, in Darwin's writings. In human history, there were many cases of collapses of civilization that were indeed very likely mainly, or at least partially, related with overpopulation, which is also a problem faced by our planet nowadays, as explained in the brilliant 1999, 2005 and 2012 books of **Jared Diamond**.

This point also reflects another crucial problem in the way that Darwin, and particularly some of his subsequent Darwinian and Neo-Darwinian followers, saw evolution: mainly involving individual survival or reproduction, and therefore selfishness. In this specific case, under such a paradigm, for an individual human being it would indeed appear to be "better" to have as many children as possible. But in the long run if everybody would do that, this would obviously lead to overpopulation, overuse of resources, war and so on, and therefore be disadvantageous for the group as a whole. Accordingly, there are numerous examples of human societies that were and/or are **polyandrous**—one female with various males—, and many of them explicitly did/do so to avoid their "societal collapse." I mentioned above the example of **Canary Islands**, in which the **Guanches** took the political decision to change from a **multimale-multifemale model** to a **polyandrous model** with at least five husbands for a single wife precisely to try to control population growth and avoid wars because of the scarce natural resources of the islands. Other examples are given in Levine's 1998 book *The Dynamics of Polyandry*, Beckerman and Valentine's edited 2002 book *Cultures of Multiple Fathers*, or Cerello and Kholoussy's edited book *Domestic Tensions, National Anxieties – Global perspectives on Marriage, crisis, and Nation*. So, knowing that many other parts of the globe also have scarce natural resources, why would one assume as a dogma that a polygynous model would be universally "favored" everywhere and anywhere, within human evolution? This is another example of how a scientifically inaccurate "idea" that was made by, and for, men and subsequently said to be "supported" by scientist males, based on misogynistic biases and prejudices, then became accept by most people in Western countries and internalized even by many women (see Box 6.3).

Box 6.3 Darwin, Peacocks, Science, Biases, Misogyny, Marriage, Sex, and Just-so-Stories

In her book *Inferior*, Saini wrote:

> I'm holding three letters, all yellowing, the ink faded and the creases brown. Together they tell a story of how women were viewed at one of the most crucial moments of modern scientific history, when the foundations of biology were being mapped out. The first letter, addressed to **Darwin**, is written in an impeccably neat script on a small sheet of thick cream paper. It's dated December 1881 and it's from a Mrs **Caroline Kennard**, who lives in Brookline, Massachusetts, and a wealthy town outside Boston. Kennard was prominent in her local women's movement, pushing to raise the status of women (once making a case for police departments to hire female agents). She also had an interest in science. In her note to Darwin, she has one simple request. It is based on a shocking encounter she'd had at a meeting of women in Boston. Someone had taken the position, Kennard writes, that 'the inferiority of women; past, present and future' was 'based upon scientific principles'. The authority that encouraged this person to make such an outrageous statement was no less than one of Darwin's own books. By the time Kennard's letter arrived, Darwin was only a few months away from death. In her letter, Kennard naturally assumes that a genius like Darwin couldn't possibly believe that women are naturally inferior to men. Surely his work had been misinterpreted? 'If a mistake has been made, the great weight of your opinion and authority should be righted', she entreats. 'The question to which you refer is a very difficult one', Darwin replies the following month from his home at Downe in Kent. If polite Mrs Kennard was expecting the great scientist to reassure her that women aren't really inferior to men, she was about to be disappointed. 'I certainly think that women though generally superior to men [in] moral qualities are inferior intellectually', he tells her, 'and there seems to me to be a great difficulty from the laws of inheritance, (if I understand these laws rightly) in their becoming the intellectual equals of man'.
>
> It doesn't end there. For women to overcome this biological inequality, he adds, they would have to become breadwinners like men. And this wouldn't be a good idea, because it might damage young children and the happiness of households.

Darwin is telling Mrs Kennard that not only are women intellectually inferior to men, but they're better off not aspiring to a life beyond their homes. It's a rejection of everything Kennard and the women's movement at the time were fighting for. Darwin's personal correspondence echoes what's expressed quite plainly in his published work. In *The Descent of Man* he argues that males gained the advantage over females across thousands of years of evolution because of the pressure they were under in order to win mates. Male **peacocks**, for instance, evolved bright, fancy plumage to attract soberlooking peahens. Similarly, male lions evolved their glorious manes. In evolutionary terms, he implies, females are able to reproduce no matter how dull their appearance. They have the luxury of sitting back and choosing a mate, while males have to work hard to impress them, and to compete with other males for their attention. For humans, the logic goes, this vigorous competition for women means that men have had to be warriors and thinkers. Over millennia this has honed them into finer physical specimens with sharper minds.

Women are literally less evolved than men. 'The chief distinction in the intellectual powers of the two sexes is shown by man attaining to a higher eminence, in whatever he takes up, than woman can attain – whether requiring deep thought, reason, or imagination, or merely the use of the senses and hands', Darwin explains in *The Descent of Man*. The evidence appeared to be all around him. Leading writers, artists and scientists were almost all men. He assumed that this inequality reflected a biological fact. Thus, his argument goes, 'man has ultimately become superior to woman'. This makes for astonishing reading now. Darwin writes that if women have somehow managed to develop some of the same remarkable qualities as men, it may be because they were dragged along on men's coattails by the fact that children in the womb inherit attributes from both parents. Girls, by this process, manage to steal some of the superior qualities of their fathers. 'It is, indeed, fortunate that the law of the equal transmission of characters to both sexes has commonly prevailed throughout the whole class of mammals; otherwise it is probable that man would have become as superior in mental endowment to woman, as the peacock is in ornamental plumage to the peahen'. It's only a stroke of biological luck, he implies, that has stopped women from being even more inferior to men than they are. Trying to catch up is a losing game – nothing less than a fight against nature. To be fair to Darwin, he was a man of his time. His traditional views on a woman's place in society don't run through just his scientific works, but those of many other prominent biologists of the age. His ideas on evolution may have been revolutionary, but his attitudes to women were solidly Victorian.

Unconventional ideas can appear from anywhere, even the most conventional places. The township of Concord in Michigan is one of those places. In 1894…a middleaged schoolteacher from right here in Concord published some of the most radical ideas of her age. Her name was **Eliza Burt Gamble**. Gamble believed there was more to the cause than securing legal equality. One of the biggest sticking points in the fight for women's rights, she recognized, was that society had come to believe that women were born to be lesser than men. Convinced that this was wrong, in 1885 she set out to find hard proof for herself. She spent a year studying the collections at the Library of Congress in the US capital, scouring the books for evidence. She was driven, she wrote, 'with no special object in view other than a desire for information'. Although not a scientist herself, through Darwin's work Gamble realized just how devastating the scientific method could be. If humans were descended from lesser creatures, just like all other life on earth, then it made no sense for women to be confined to the home or subservient to men. These obviously weren't the rules in the rest of the animal kingdom…. But, for all the latent revolutionary power in his ideas, Darwin himself never believed that women were the intellectual equals of men. This wasn't just a disappointment to Gamble, but judging from her writing, a source of great anger. She believed that Darwin, though correct in concluding that humans evolved like every other living thing on earth, was clearly wrong when it came to the role that women had played in human evolution.

Her criticisms were passionately laid out in a book she published in 1894, called *The evolution of woman: an inquiry into the dogma of her inferiority to man.* Marshalling history, statistics and science, this was Gamble's piercing counter-argument to Darwin and other evolutionary biologists. She angrily tweezed out their inconsistencies and double standards. The peacock might have had the bigger feathers, she argued, but the peahen still had to exercise her faculties in choosing the best mate. And while on the one hand Darwin suggested that gorillas were too big and strong to become higher social creatures like humans, at the same time he used the fact that men are on average physically bigger than women as evidence of their superiority. He had also failed to notice, Gamble wrote, that the human qualities more commonly associated with women – **cooperation**, nurture, **protectiveness**, **egalitarianism** and **altruism** – must have played a vital role in human progress. In evolutionary terms, drawing assumptions about women's abilities from the way they happened to be treated by society at that moment was narrow-minded and dangerous. Women had been systematically suppressed over the course of human history by men and their power structures, Gamble argued. They weren't naturally inferior; they just seemed that way because they hadn't been allowed the chance to develop their talents. Gamble also wrote that Darwin hadn't taken into account the existence of powerful women in some tribal societies, which might suggest that the present supremacy of men now was not how it had always been…. It's hard to picture the directions in which science might have gone if, in those important days when Charles Darwin was developing his theories of evolution, society hadn't been quite as sexist as it was. We can only imagine how different our understanding of women might be now if Gamble had been taken a little more seriously. Historians today have regretfully described her radical perspective as the road not taken. In the century after Gamble's death, researchers became only more obsessed by sex differences, and by how they might pick them out, measure and catalogue them, enforcing the dogma that men are somehow better than women.

It is worthy to note that one can always argue that some people can even use "good" or "noble" ideas, or true scientific facts, to support racist, or Eurocentric, or misogynistic concepts, anyway, so perhaps this can apply to Darwin as well. That is, perhaps Darwin's ideas were just "noble," "good" and "correct" but "bad" people just used them in a "bad" way. But the crucial point is that those parts of the works published by Darwin and many of his subsequent Darwinian and Neo-Darwinian followers that were *directly* used by people such as **David Duke** or **Hitler** or other **white supremacists**—as well as by misogynists—were neither "good ideas" nor scientifically correct. Scientifically there is no such thing as "favored races" in evolution, or "higher taxa," or "cosmic purpose" or "natural progress," nor a suffocating, omnipresent, omnipotent and never-ending "struggle-for-existence" in which only one—in particularly the most selfish one, be it an individual or species, or a gene as argued by Neo-Darwinists such as Dawkins—can survive. Darwin was right in numerous parts of his writings, there is no doubt about it, but he was not right

100% of the time, no one is, and he was indeed wrong about those aspects discussed in this Section, and about other aspects that had to do with his **Eurocentrism**, **racism** and **misogyny**.

However, I want to stress again that my main is not to criticize Darwin per se, but rather to call attention to the remarkable parallel between **teleological narratives** such as those used in religious ideas and the inflexible—often unfalsifiable—way in which many defend Darwin's, Darwinist or NeoDarwinist ideas, in many cases even *after* those ideas were shown by empirical data to be plainly wrong. In fact, as I anticipate that **creationists** and so-called **intelligent designers** might use this book to criticize Darwin and evolutionary biologists in general, I have a note for them. The note is: I am above all criticizing the quasi-religious reasoning done by many scholars when they defend Darwin. Therefore, it would be a huge paradox—and also plainly wrong—to use the criticism of the type of circular reasoning used by Darwinians and NeoDarwinians to idealize Darwin in order to defend the type of religious circular reasoning used by creationists and intelligent designers.

Furthermore, the fact that I—an evolutionist—am providing *empirical scientific data* to test some of Darwin's ideas that are often venerated by many other evolutionists and, when those ideas are shown to be scientifically wrong by empirical data, that I then use those data to contradict and criticize that quasi-religious veneration, shows a major difference between scientists and creationists. Namely, scientists *can change even their core evolutionary ideas* in face of new empirical data. Yes, some scientists can—and many do—remain trapped for a long time in that quasi-religious type of *Neverland* thinking, but *at least some scientists* are able to escape from it by using the scientific method and testing and contradicting falsifiable theories. This contrasts with what is said to be the "truth" of the Bible, or the Koran. This is because religious people, and creationists in particular, don't test empirically their religious ideas as these ideas are actually not even falsifiable, at all. Well on the contrary, they often keep their ideas even if there is a huge amount of evidence contradicting them, for instance showing that humans evolved from apes, and apes from monkeys, and monkeys from other mammals, and mammals from other tetrapods, and tetrapods from fish, and that none of these animals, or any other organism, was created by God, or following God's masterplan.

6.4 "Savages," Colonialism, Slavery, and Neo-colonialism

> *Five centuries ago, with the 'discovery' of the so-called 'new world' what was really discovered was the true Spain herself, the reality of Western [colonizing, dominating] culture and the church as they were then (Ignacio Ellacuria)*

Expanding now the connection between subjects such as **colonialism** and **teleological narratives** beyond Darwin, Darwinism, and Neodarwinism, I will now provide examples of how colonialism was frequently, if not almost always, also "morally" defended by the use of such narratives and ideologies related to them. Europeans such as Columbus often *created tales* arguing that they were "helping" the "poor savages" of the non-European lands that they reached, helping to take them out of their "brutish" existence, and so on. A crucial point that needs to be made is however that those imaginary tales created by some Westerners, including by political and religious leaders, were *truly believed* and *internalized* by millions of Westerners. This critical point is often neglected by a huge number of historians and politicians, who, basing themselves on the false scientific premise that we are truly a sapient being—*Homo sapiens*—tend to focus only on the so-called rational explanations for why some countries colonized others. Many take this false premise to the extreme, arguing that colonialism, and basically almost everything else, is purely related to economic gains—using a particularly famous sentence, "[it's] the economy, stupid."

Yes, *in part* "it's the economy," but what is actually truly "stupid" is to think that it's just the economy. It is not: throughout the last millennia, we have numerous unequivocal examples showing that political and religious leaders were/are particularly good at *creating teleological tales* in order to "morally" justify atrocities that might allow them to have economic gain, as Columbus did (see Box 6.4). And why is that? Why do you think they do that, over and over? Because they have to convince the broader public, that is, to make lay people truly *believe* that doing such atrocities is either actually "morally" justified or is "good" for the broader public in general, or for "other" people, such as the "savages" in the case of colonialism. Political and religious leaders typically don't fight wars: they need people from the broader public to fight wars for them, and people will fight, or fight with more motivation, if they believe in *something*, for instance that they will also profit from those wars, or more often that they are combating for *something bigger than life*, *something transcendent*, such as for their God, or nation, or "common good."

Apart from being a scientist that obviously had to do an extensive literature research about these specific issues to write this book, I can attest this myself in a more personal way because, as a citizen of Portugal—a country that colonized regions of Africa, South America and Asia and was a crucial part of the **Trans-Atlantic slavery trade**—, I grew up being systematically told, including in public schools, that the Portuguese were "good colonizers" and even "good slave-owners," that they were truly "good-hearted" and "helped a lot" the "savages" of the places we colonized, something that is obviously historically not true (see Boxes 6.4 and 6.5). Why do basically all agricultural states use a huge amount of resources to undertake such a deep **inculturation**, if everything would be just "the economy, stupid"? Because, if people were not acculturated and did not believe in such imaginary *Neverland tales*, the leaders of modern "democratic" countries would have no way to engage in wars, colonialism, or genocides. Hitler alone would surely not be able to imprison millions of Jews, put them in trains, and keep and then kill them in the most atrocious ways in concentration camps: this was only possible because a huge number of Germans believed in his ideas and/or idealized him.

Compared to so-called democratic countries nowadays, particularly those that do not even have a mandatory military service, in early "civilizations" kings and religious leaders had it much easier to oblige people to fight, or to use a huge number of slaves to do that. But despite of this we do know, based on detailed historical records, that they still usually created imaginary narratives to try to convince the broader public that they were fighting for a "good cause" or for a God, or a King-God. So, *"it's much more than the economy, stupid"*: for instance, coming back to the issue of **colonialism**, in the specific case of those millions of Westerners that truly believed the tales made by people like *Columbus*—including the thousands that fought battles to enforce those tales' "noble ideas and goals"—, such tales often involved a combination of both the *acceptance of teleological narratives* related to cosmic purpose and progress *and* the *total incomprehension of and lack of empathy* towards the lifeways of "savages." However, one should emphasize that being highly irrational does not mean that we necessarily all *tend* to believe in bellicose, racist or misogynistic tales: we do tend to believe in teleological tales, but some of them are actually used against wars, oppression, racism and misogyny. A clear example is that some religious leaders, such as those of some specific religious orders, did intervene in favor of natives against the oppression of the colonialists, often following other types of teleological narratives such as religious ones, together with other, less "innocent"/altruistic reasons, as explained in Boxes 6.4 and 6.5.

Shermer's *The believing brain* provides a very interesting sidenote that emphasizes that the **colonization** of the New Word was often "morally" defended by the use of teleological narratives, and that the very first events that lead to that colonization were also deeply related to the **power of belief** and of **confirmation biases**:

> **Christopher Columbus**'s confidence in achieving a successful mission to the Far East by way of sailing west is a prime example of beliefs driving perceptions. His first voyage was premised on **Ptolemy's cartographical coordinates** for the length that the Euro-Asian continent extends east, as well as the overall circumference of the world, both of which were miscalculated to a degree perfectly in sync with Columbus's expectations. Thus, in one of the most prescient coincidences in the history of serendipitous discovery, after sailing a little more than 5,000 kilometers westward across the "Ocean Sea" (the Atlantic) on his maiden voyage, Columbus encountered land in the exact place where he had calculated the Indies would be, and thus he dubbed the people he engaged there "Indians". Why did Columbus not immediately realize he was not in Asia? Surely the flora and fauna and people he discovered were nothing at all like what **Marco Polo** had reported from his land excursions eastward from Europe where he had met the Great Khan and absorbed Asian culture. The answer can be found in the dual problem of *perception* and *cognition*, or *data* and *theory*. What threw Columbus off was coarse-grained data coupled with incorrect theory. Marco Polo's reports of Asia were sketchy at best, allowing ample wiggle room for interpreting New World data as Old World facts. Plus, there was no theory of a New World, so in Columbus's mind when he made first contact with the New World on that fateful day in October 1492, where else could he be *but* Asia?
>
> Because of the power of the paradigm to shape perceptions, Columbus's cognitive map told him what he was seeing. When his men dug up some common garden rhubarb, *Rheum rhaponticum* (used in pies), for example, the ship's surgeon determined that it was *Rheum officinale*, the medicinal Chinese rhubarb. The Native American plant gumbo-limbo was mistaken for an Asiatic variety of the mastic evergreen tree that yields resin used to make lacquer, varnish, and adhesives. The South American *nogal de pais* nut was classified as the Asian coconut, or at least what Marco Polo had described as such. Columbus deemed a plant with the aroma of cinnamon to be that valuable Asian spice. After first touching land in San Salvador, Columbus then sailed to Cuba, bringing with him some San Salvadorian captives to help with communications with the Cuban natives, who told him that there was gold to be found at "Cubanacan" – the middle of Cuba – which Columbus heard as "El Gran

Can," or the **Great Khan**. When Columbus touched down again in Cuba during his second voyage, he recorded his navigation along what he thought were the shores of the Mangi kingdom in southern China, which had been described by Marco Polo. And so it went for all four voyages to "the Indies," with Columbus never once doubting where he was, despite never meeting the Great Khan. *Such is the power of belief*. New data pouring in through old paradigms only reinforced his confidence that he was where he believed he was – on the eastern boundary of the Old World, not the eastern edge of the New World.

An illustrative example of the complex, interactive profound link between the economic interests of some *political leaders* in, and the teleological narratives that

Box 6.4 Beliefs, Confirmation Bias, Cannibals, and Western Atrocities

In his 2013 book *Racisms*, Bethencourt provides further interesting insights about Columbus, and in particular the atrocities done by Europeans, whom often called "cannibals" to "others" but ultimately became themselves accused of "cannibalism" by other people, precisely due to the atrocious acts they committed:

> It was **Christopher Columbus** who coined the noun **cannibal**. In the diary of his first trip to the Antilles (1492–93), Columbus mentioned that the natives of the main islands (Cuba and Hispaniola) feared certain tribespeople who ate human flesh and were supposedly hunting them from the southern islands. These **anthropophagi** were identified as *caribes*, or *canibes* as they were designated in Hispaniola, in contrast to the peaceful natives of the main islands, and were said to wear long lengths of hair attached to the back of a headdress of feathers plus be armed with bows and arrows. In a letter to Luís de Santángel, dated February 15, 1493, Columbus stated that the anthropophagi came from an island called caribe. Printed copies of the letter were made and the piece became a best seller, with two Spanish editions, nine Latin editions, three Italian editions, and one German edition, all in the space of four years (1493–97). The reference to eaters of human flesh did not go unnoticed. In his account of his second trip (1493–96), Columbus mentioned the cannibals (in Spanish canibales), transforming the word into what was to become an accepted noun, which he repeated in all his accounts and correspondence with European kings When Columbus arrived in the **Antilles**, the **Caribs** were settled farmers, fishers, hunters, and expert navigators who had spread from the northeastern coast of South America to the Caribbean islands and were driving away the **Arawaks**, who had previously settled there.
>
> Columbus may have reproduced the Arawak fear of the Carib as better equipped and more determined from a military point of view. Although Columbus praised the meekness of these people, who were supposed to be ideal recipients of the Gospel, he was also shocked by the "cowardice" of some Arawak communities. But the idea of anthropophagic practices served Columbus's own interests, clearly expressed in his account of the second trip. He suggested that the "**cannibals**" should be enslaved because they were infinite in number, and each of them would be worth three blacks from Guinea in strength and ingenuity (a "sample" group of Caribs was dispatched with the letter). In 1494, Columbus sent a letter to the Catholic kings in which he suggested that an investment in caravels, personnel, cattle, and tools could be paid for by the resulting enslavement of the cannibals. The argument was that freed from "that inhumanity", the cannibals would be the best slaves; their expertise as rowers could be put to good use in the galleys that Columbus intended to build for travels on the Caribbean Sea. In 1495, Columbus suggested enslavement of all Indians in a letter he wrote to the kings. Although he considered that the women of Hispaniola would not make good domestic **slaves**, he noted they would be valued as excellent workers on the land and in the manu-

facture of cotton fabrics. The assertion that the natives were extremely lazy (perezosos en grandísima manera) was put forward for the first time in another letter of the same year, without Columbus noticing that it would contradict his project for **enslavement**.

Columbus was familiar with the **Portuguese model of trips** to explore the west coast of Africa subsidized by the **slave trade**, the supply of which became the main purpose of those trips. He had lived in Lisbon and Madeira from 1476 to 1486, maintaining ties with the local elite through his marriage to Filipa Moniz, the daughter of a Portuguese captain of Porto Santo. Columbus claimed several times in his reports and even in the marginalia of his books that he had been in the Portuguese fort of Mina, in the Gulf of Guinea. Elmina, built in 1482, was a center for the gold and slave trades – a commerce that Columbus was trying to emulate in the Caribbean region. In 1495, the same year was trying to emulate in the Caribbean region. In 1495, the same year that Columbus made his project explicit, he sent a cargo from Hispaniola of five hundred slaves, who were received and sold in Seville by his factors Giannotto Berardi and Amerigo Vespucci. Queen Isabel temporarily suspended the auction, worried by its theological and political consequences – a compunction already expressed by the kings in their reply to a previous letter. In 1498–1500, Columbus insisted that the project should go ahead, adding brazilwood that he had seen on the coast of South America to the list of items to be traded. He drew attention to the fact that Castile, Portugal, Aragon, Italy, Sicily, the Canaries, and other islands "spent" a lot on slaves; those from this new source would be worth fifteen hundred maravedis each. The high death rate among enslaved Caribs, Columbus contended, was normal; the same had happened with the first black Africans and Canarians (**guanches**) sent to Iberia. If the project of enslavement of the Caribs was never implemented, it was basically for three key reasons: the theological and political doubts of the kings (after all, these natives were their new vassals); the absence of the institution of slavery in the region, which meant that the natives refused the oppression, and in many cases preferred to die; and the existence of an established maritime slave trade with West Africa (later with Central Africa) that could be diverted from Iberia to the Antilles. Meanwhile, the Arawaks and Caribs were decimated by war, displacement, and the disease brought by the Europeans, which made Columbus's project pointless…

[More than a century later], In a vivid account in his (1614) book *Peregrinação*, **Fernão Mendes Pinto** highlighted the Japanese curiosity and capacity for learning. In the early 1540s, in their first encounter with the Japanese, the Portuguese had left a harquebus as a gift for the governor of the island of Tanegashima. The governor had been extremely excited by the efficiency of the gun when he invited the Portuguese to go hunting, as there were then no firearms in Japan. When the Portuguese returned after six months, they discovered that the gun had been replicated; when they left the island there were already six hundred; when Mendes Pinto went back to Japan in 1556, he was told that across all the islands, a total of more than three hundred thousand harquebuses was to be found. Behind this high praise lay missionary euphoria about the possibilities offered by the Far East and a wish to attract more **evangelizers**. The elation soon cooled, although the Japanese experience proved to be extremely successful, with hundreds of thousands of conversions by the beginning of the seventeenth century. Success was cut short, however, by the persecution of Christians launched by the central authorities in 1614 and sealed by the expulsion of the Portuguese in 1639. In China, the Portuguese managed to maintain their position in **Macau**, although this was reduced to trade, and they accepted the small successes they had had with continental missions, even though recent work has challenged this view. Ironically, the Chinese accused the Portuguese of **cannibalism**, because they bought young slaves smuggled into China by the coastal population. The "barbarians of the south" were accused of the same awful unnatural crimes they attributed to the American Indians.

Box 6.5 Maps, Demons, Colonialism, and Slavery

Canizares-Esguerra's 2016 book *Nature, Empire and Nation – Explorations of the History of Science in the Iberian World* provides several examples of how the teleological narratives used to justify the "morality," or even the "obligatory mission" of the colonization of the New Word was elevated to a striking level of paranoiac detail and obsession:

> In Mexico…**demons** appeared so entrenched there that they had even shaped the landscape. In the 1690s, **Cristobal de Guadalajara**, a mathematician and cartographer from Puebla, realized after studying a copy of an early 17th century map of the rivers and lakes of the central valley of Mexico that the hydrographic contours of the valley represented the head, body, tail, wings, and legs of a **Satanic beast**. To counteract the **Aztec kingdom** of darkness, God had fortunately sent Spanish conquerors and the **Virgin Mary** to set the Amerindians free. The Virgin – who had appeared to a Nahua commoner, Juan Diego, in Tepeyac in 1531, leaving her image stamped on his cape – held the key to understanding Mexico's destiny. The image [see **Fig. 6.2**] had a virgin standing on the moon, surrounded by sun rays while eclipsing the sun, wearing a heavily starred blue shawl, and held up by an angel…. In 1648, Miguel Sanchez argued that the woman of the **Apocalypse**…is a prefiguration of **Our Lady of Guadalupe**, who had routed the kingdom of darkness of the **Aztecs**. Sanchez offered interpretations of every detail of the image: the moon underneath the Virgin represented her power over the waters; the Virgin eclipsing the sun stood for a New World whose torrid zone was temperate and inhabitable; the twelve sun rays surrounding her head signified Cortes and the conquistadors who had defeated the dragon; and the stars on the Virgin's shawl were the 46 good angels who had fought **Satan's army**.

However, one should also point out that some religious orders did intervene in favor of natives, although their reasons were not always completely "purely innocent" or altruistic, as explained in Bethencourt's *Racisms*:

> The enslavement of **Native Americans** was forbidden by the Spanish Crown in response to protests against the enormous scale of abuse during and after the conquest by the religious orders, mainly the **Dominicans**. That is why the Spanish authorities encouraged the slave trade from Africa as a way to ease the pressure to enslave Native Americans under the pretext of a just war – the one possible door left open by the legislation. The transatlantic slave trade began to function on an unprecedented scale, exceeding the traditional caravan trade to Muslim countries. But what needs to be highlighted is the crucial role played by the Crown's decision to transform converted Native Americans into subjects (or vassals), who could not be enslaved, while black Africans could be transported to perform the heaviest duties. We know that the percentage of Africans in the American population varied enormously with time and place, and ended at a low level in New Spain in the long term. Yet the image of black Africans as the lowest stratum, virtually without any rights – though this assertion has to be qualified – was certainly reinforced by this policy. At the same time, the Spanish Crown encouraged Catholic marriages between Spaniards and Native American elites, and did not forbid marriage between mixed races or castas, although illegitimate children were to be excluded from higher education and public office, while those marrying black women were excluded from the civil service in 1687. The **Catholic Church** played an undeniable role in moderating **colonial abuse**, not only through the sacrament of mar-

riage, but also through the promotion of confraternities among all strata of the population, including black slaves. The **Dominican confraternities of the rosary**, created for black people, whether slaves or freedpersons, intervened in major cases of **brutality against slaves** and helped them to pay for their manumission (emancipation). Spanish legislation favored manumission: there was no legal obstacle, and the variety of options was considerable – in the owner's will, through external work and payment by the slave, or in return for a contribution from a charitable institution. The condition of slaves was not the exclusive affair of the owner; the slave could legally denounce brutal treatment, require a change of master, and request the intervention of the king. These legal change of master, and request the intervention of the king. These legal possibilities were distant from daily practices, yet the structure of confraternities enabled a certain level of information and action.

The Portuguese case presented similar features, although the rights of Native Americans were never properly addressed: the king donated vast stretches of Brazil, declared as "my coast and land", from the coast to the interior, in order to attract **European colonists**, as if the land were totally empty of native rule. The fact that the colonists were dealing with seminomadic populations helped them to maintain an enormous level of native enslavement under cover of pursuing a just war. Dominican champions of natives' rights, such as **Las Casas**, were not active on the Portuguese side, but the **Jesuits** later performed this role, causing several of them to be summarily shipped back to Portugal by the angry colonists of São Paulo and São Luís do Maranhão, the major centers for hunting native slaves. Jesuits promoted **Indians rights**, obtained the support of the Crown, and in the long run managed to curb the practice of native enslavement by the colonists. The Jesuits were not entirely innocent, though; they wanted to control access to Native Americans, and managed an indigenous labor force within their own missions, where they also had sugar mills and **African slaves**. They compromised with the colonists in some areas, such as Maranhão, sending "their" Indians from the missions to temporarily undertake private and public labor.

are created by them about, as well as the scientific studies of, the colonies is given in Canizares-Esguerra's excellent book *Nature, Empire and Nation*:

> Determined to transform the viceroroyalties into colonies, the **Spanish Bourbons** turned to new sciences…**naturalists** sought to benefit the economy by identifying new products (dyes, **spices**, woods, gums, **pharmaceuticals**) or alternatives to already profitable staples from Asia. Spanish expeditions to the Andes, for example, put a premium on finding species of cloves and cinnamon to challenge the British and Dutch monopolies in the East Indies. The logic behind sending botanical expeditions to the New World was best expressed in 1777 by the architect of these policies, Casimiro Gomez Ortega, who assured: 'twelve naturalists…spread over our possessions will produce as result of their pilgrimages a profit incomparably greater than could any army of 100000 strong fighting to add a few provinces to the Spanish empire…These ideas also surfaced in Peru, another place that witnessed botanical studies sponsored by the Spanish Crown…. 'It seems', Unanue argued, 'that after having created the deserts of Africa, the flagrant and lush forests of Asia, and the temperate and cold climate of Europe, God made an effort to bring together in Peru all the productions he had dispersed in the three other continents…in this manner, God has sought to create a

> temple for himself worthy of his immensity'…Unanue speculated that certain local products were suitable substitutes for popular products currently monopolized by Spain's European rivals…. Mexico also found an avid audience for these kinds of microcosmic narratives…for Venegas, New Spain was 'the purse of Omnipotence; an Eden capable of *providing Europe* not only with precious metals but also with many of the noblest vegetables, roots, woods, fruits, gums, and balms'.

Within these excerpts, one part that is particularly relevant about how demented and teleological the **discourse of some of the colonialist leaders** often was is that they often stated that the New World was an Eden created in order to "*provide Europe*" with huge resources. As if the New Word was only made with the following cosmic purpose: first, to wait for billions of years until our species finally appeared; secondly, to then wait again for about a quarter of a million of years until the Europeans would "discover" it and take its richness to the "land of the chosen ones," Europe of course. These colonizers did not seem to spend too much time thinking deeply about why, in addition, God would have put in his own "Temple" the "brute" Native Americans to start with, nor why he then led them to create "civilizations" such as those of the Mayas and the Aztecs, which explored the richness of this Eden much before the Europeans "discovered" it.

I have seen, myself, how not only such imaginary stories can be *internalized* by the very same people that were enslaved and oppressed by those that created them (see Fig. 6.3), but also how additional tales—including **religious narratives**—can be *created* by those whose ancestors were colonized and enslaved, probably in an intent to try both to find a "cosmic purpose" for what happened and to see "the positive side" of something that was otherwise horrible for their ancestors. This happened when I was traveling in the Southern Caribbean Islands, and a local resident, descendant from African slaves, told me: "in a way, we were the *chosen ones*, because **slavery** and **colonialism** were part of God's masterplan to ultimately bring us here, to paradise, the Caribbean paradise." This teleological narrative clearly has the merit of trying as much as possible to see the glass half full, that is for sure. Be that as it may, what should be noted is that the use of teleological narratives by the colonialists was important within the "discovery" and colonization of lands and enslavement of people to work on those lands, as well as within the subsequent creation of new "nations" in those lands and the maintenance of their unity. Indeed, efforts to create and maintain a "nation," no matter where, when, and in what context that is done, almost always recur to teleological narratives that are often related to the notion of cosmic purpose and progress. Illustrative examples of this, as applied specifically to the Spanish colonies in the New World, are also given in Canizares-Esguerra's *Nature, Empire and Nation*, including in some of the outstanding figures cited and discussed by him in that book (see Figs. 6.2, 6.4, and 6.5):

> [In] one of the most important works of Mexican scholarship…. *Mexico a traves de los siglos* (1887), edited by the liberal **Vicente Riva Palacios**…the frontispiece to volume 1 presents a young woman crowned with olive branches and holding a pen and a book – **Fig. 6.4** – , standing for the Mexican republic and its hard-won liberal constitution. Surrounding

Fig. 6.2 Frontispiece to Cayetano de Cabrera y Quintero's *Escudo de armas de Mexico*: Lady of Guadalupe acts as a shield protecting Mexico from negative heavenly influences, such as those of the Aztec kingdom of darkness

the beautiful maid, symbols of Mexican identity and history lie scattered about: cactuses, ferns, and palms represent Mexico's various ecological habitats; pre-colonial objects signify the grandeur of the Amerindian past.... Behind the maiden hangs the seal of the Mexican nation, a purported Aztec depiction.... A narrative of radical change, however, unfolds parallel to the **evolutionary tale of progress**. Two brown-skinned **Amerindians**

Fig. 6.3 What systemic racism looks like

and a white-bearded Spanish conquistador stand next to a Mexican mestizo, whose deportment and costume differ as radically from the Amerindians' as from the Spaniard's. The narrative of Mexico as a mestizo civilization is reinforced by the posture of the white maiden, for the nation's visage is turned to the Amerindians…(it) argues that the nation has evolved from an ancient, grandiose yet barbarous society into a liberal polity, in which constitutional principles and the law rule at last. The book also suggests that no past historical period can be dismissed or belited, for a new mestizo Mexican civilization emerged out of the encounter of **Amerindians** and **Spanish conquistadors**. Mexico should stop running away from both its Amerindian and Spanish pasts…[the book] sets out to recover the colonial past from condescension of 19th century liberal intellectuals, who took it to be a period characterized by **medieval obscurantism**, **inquisitorial repression**, and **cruel Spanish domination**, an era to be forgotten rather than celebrated. The frontispiece of volume 2…seeks to make obvious the role of the Church in the crafting of the new mestizo nation – **Fig. 6.5** in the background stand the two volcanoes that overlook the valley of **Mexico**, Popocateptl and Ixtaccihuatl…the choice of image is deliberate, for the volcanoes had long been linked in the collective imaginary with a purported Aztec myth of *mestizage*.

In his 2019 book *Our Shadowed World*, Kirkham also directly connects teleological narratives with European colonialism:

Although anniversaries and centenaries come and go, the issues they raise and the controversies they generate tend to linger on. A case in point is the five hundredth anniversary of the discovery of America in 1992. Or at least that was the way innumerable articles and studies put it. But apart from the details of what took place in that fateful year of 1492, far more important and equally elusive is the perspective from which we must now view and evaluate them. Exactly what was "discovered"? For the Latin American writer Ignacio Ellacuría it was clear that what was "discovered" – in the sense of revealed or uncovered – was Europe itself and the nature of its civilization. Ellacuría did not mince words: "Thus, five centuries ago, with the 'discovery' of the so-called 'new world' what was really discovered was the true Spain herself, the reality of Western culture and the church as they were

Fig. 6.4 Frontispiece to volume 1 of *Mexico a traves de los siglos* (1887), edited by Vicente Riva Palacios

Fig. 6.5 Frontispiece to volume 2 of *Mexico a traves de los siglos* (1887), edited by Vicente Riva Palacios

then". In this perspective it was above all the conquistador, the dominator, who laid himself open to discovery. In reality it was what we now condescendingly call the Third World that discovered the First World in its most negative and truest aspects; the plunder and destruction of a continent was the harbinger of what Europeans had to offer the rest of the world in an epoch of discovery, **colonization**, and conquest – all the while passing themselves off as benefactors of the world.

This theme was taken up by another distinguished voice, a survivor of a U.S.-backed contra massacre, the Jesuit liberation theologian Jon Sobrino. In a powerful address delivered in Salford Cathedral titled "*500 Years: Reflections for Europe from Latin America*" he stated unequivocally that "in the reality of the South (or Third World), with all its poverty, injustice and death, the North (First World) can recognize itself, as in a reverse mirror-image, through what it has produced". This metaphor of a mirror – recalling those polished obsidian mirrors found in burial caches throughout the Americas – was also used by the Mexican writer Carlos Fuentes. In his quincentennial study, The Buried Mirror: Reflections on Spain and the New World, he challenges us to think about how our understanding of the world has been shaped and evolved in the face of the radically "other". We look into a mirror and see disconcerting and disturbing reflections. The European expansion, in this case led by the Spaniards, was driven by the insatiable quest for ever greater wealth and power at any cost, and always underpinned by a sense of divine legitimation. The ambiguity of this motivation was well expressed by the distinguished Austrian historian Friedrich Heer when he wrote, "the enthusiasm of the early Spaniards who first voyaged to the Americas was stirred and sustained by mythological and eschatological expectation, by hopes of finding Paradise, by fantasies of mounting a global crusade against Islam". The discovery of a new world was even placed on a par with the original act of creation and the incarnation, and the conquest of America was presented as an extension of the **Reconquista** – the crusade against Islam and paganism to create one catholic world order.

But regardless of such **ideological motives** as Christianizing the natives, the existence of **indigenous people** was seen as a means to an end; the primary goal was, as Sobrino noted, "to make the Spaniards rich, and later when that supply became exhausted, black Africans were enslaved so that they could in turn become instruments, just like modern sources of energy". The justification for all this took place in many and varied ways: ecclesiastically (the bull of **Pope Alexander VI** divided the new domains between Spain and Portugal), theologically (this was the providential reward for the reconquest of Spain from the Moors), politically (there were no legitimate owners of the lands), anthropologically (Indians were inferior and not even fully human), ethically (the evil and perverse customs of the Indians demanded suppression), and so on. But many and varied though these justifications may have been, the premise was always the same; namely, that the Europeans were going to stay in those lands and defend what was already held in possession in order to get rich. The few solitary voices raised in opposition, such as that of Bartolomé de las Casas and Francisco de Vitoria, were drowned out by the chorus of conquest…. [There was] a deeper mentality of arrogance and contempt that was already embedded in the minds of the **conquistadores**, five hundred years ago.

When **Columbus** stepped ashore on the Bahamas in 1492, he was welcomed by the **Taino people**, a hospitable multiethnic people who had inhabited many of the Caribbean islands for over a thousand years and had a well-ordered and peaceful society. These he summarily denounced as heretics and while women were raped he began mass burnings: at one **autodafe** eighty caciques (chieftains) were burned alive. After twelve years of genocidal butchery and disease brought by the Spanish, the islands had been depopulated. In such a case we may well ponder just who exactly were the savages. We may also ask why this should have happened: what was the origin of this behavior and mentality? Columbus seems to have had an **apocalyptic view of the world** and to have seen himself as chosen for a divinely inspired mission – the fulfilment of the single-minded dream of the Christian monarchs Ferdinand and Isabella for Christian uniformity and the reconquest of Muslim

> lands. The genocide (for this is what we would now call it) of the Taino people was the same treatment accorded to deviant or defiant groups across medieval Europe, from the so-called Cathars of Languedoc to the Prussians and Slavic peoples targeted by the great northern Teutonic crusades. This became the pattern for future European expansion. Ethnic or racial difference was not initially significant; the fact that people seemed to believe different things was sufficient indictment. The infamous instruction attributed to the papal legate Arnaud Amaury authorizing the massacre of the entire population of Béziers in 1209 sums it up: "kill them all…God will know his own". This is how a confessional civilization dealt with "otherness".

A recent, outstanding, comprehensive book about the history of **slavery**, **colonialism**, **racism,** and **prejudices** is Green's 2019 *A Fistful of Shells – West Africa from the Rise of the Slave Trade to the Age of Revolution*. The most original aspect of that book is that it is written from the perspective of Africans, as the *active players*, including concerning events related to horrible acts of slavery, thus going against the racist stereotypes that tend to see the "other" as an incapable, or at least as a mere passive, player. Contrary to such stereotypes, he reminds us that many cultural and artistic innovations or tendencies, including those often followed in the West, came—or were inspired—from African roots. Green wrote: "yet Africa also emerged in the twentieth century as a place of amazing, myriad, almost bewilderingly creative complexity…. African sculpture and masking inspired the emergence of **Cubism**, while the use across Africa of manufactured objects in religious shrines resonates in the readymades offered by **Marcel Duchamps** and **Andy Warhol** to the modern temples of Art." He further explained that "African musical traditions led to **Jazz**, **Blues** and **Soul**, to **Samba** and **Salsa**…. **African religious traditions** influenced the rise of **Evangelical Churches**, and the nature of shrines used by the **Afro-Catholic religions** in the New World such as **Candomblé in Brazil** and **Santería in Cuba** and New York." And this clearly does not apply only to the twentieth and twenty-first centuries:

> The landscape of West Africa is filled with relics of a past that few today in or outside Africa know much about. On the north bank of the Gambia River lie the **Wassu stone circles**, built sometime in the last few thousand years by cultures of which there is very little awareness today. In the region of Dô, in what is now south-central Mali, huge fields of tumuli lie scattered across a wide area, some of them 50 feet in diameter. The large walled fortress of **Loropéni**, perhaps dating from the seventeenth century, or perhaps earlier, now lies scattered in the bush of southern **Burkina Faso**. In southern Nigeria, earthen defences known as eredos, 33 feet tall and over 100 miles in length, and dating from the fourteenth century, are found in Ijebu. In many coastal and riverine regions, it is easy to come across enormous shellfish middens, piled up over the centuries by peoples whose names and beliefs have by and large been forgotten. For decades, outside a small circle of passionately dedicated scholars, these African pasts have suffered neglect. Yet they reveal ancient civilizations and a history whose relevance is absolutely contemporary. As early as the seventh century BCE, the **Nok culture** that grew around the plateau region of what is now central and northern Nigeria had developed agriculture and iron production. The settlement of **Jenne-jenò** in the inland delta of the Niger River had grown to a population of around 4,000 people by 400 CE, and had grown to as many as 26,000 by 800 CE. This growth was supported by rice production developed through iron tools smelted by local smiths. The iron ore was brought from around 30 miles away, while copper ornaments found in burial chambers probably came from much further afield, in the Sahara. Meanwhile, digs in the **Upper Senegal River**

Valley have shown a similar trade in copper artefacts by around 500–700 CE, where they were traded for cloth produced on spindle-looms.

For most historians, though, Africa has always been ‘outside history’. It is, after all, easy enough to dismiss something when you know little or nothing about it. Yet globalization came so early to many parts of Africa that one Chinese chronicle claims that ambassadors from the region of Ethiopia went to the Chinese Court around 150 BCE. It’s hard to imagine the Celts or the Jutes before the time of Christ doing the same. African trade connections expanded rapidly, especially after around the year 700 CE. By around 1000, **Madagascar** was linked to China through the trading town of **Kilwa**, located on an island off the coast of southern **Tanzania** and founded by a Persian sultan in the eleventh century. Many artefacts of Chinese porcelain found in recent excavations of Kilwa have confirmed the very extensive long-distance connections here from an early time. In **West Africa**, the pattern is similar. Early cave paintings from the era of Jenne-jenò reveal chariots with wheels, suggesting that this was a technology known in West Africa, either from long-distance trade to the Mediterranean or through local use. Analysis by archaeologists of the gold coins used in Tunis and Libya suggests a major change around the ninth century, when gold from the forest regions of what are now Ghana and Ivory Coast was dug out in large quantities and exported through networks of local traders. By the eleventh century, there were important mints in cities from **Siğilmāsa** in Morocco to a variety of cities along the Mediterranean coast, and the trans-Saharan trade from the states of West Africa influenced the commercial and cultural worlds of **Al-Andalus** in Spain.

If this surprises some readers, it is because ‘History’ as a subject has developed a rather selective memory over the years. There was a time when this was well known to many. One example is the Catalan Atlas, compiled by the Majorcan Jewish cartographer Abraham Cresques in around 1375. Here, the Emperor of Mali (rey Melli) sits enthroned with a sceptre and golden crown, dressed in elegant robes. In his right hand he extends a golden nugget to a North African trader, mounted on horseback, who emerges, his face wrapped in cloth, from the nomadic encampments of the Western Sahara. Across the Atlas Mountains, trade routes crisscross the desert towards North Africa, and some of them extend across the Mediterranean to the Iberian Peninsula. It is a powerful representation of the ways in which West African kings interacted with the Mediterranean worlds through the gold trade almost 650 years ago. It shows us how keen European rulers were to find out about Africa, and that, in fact, some of them already knew quite a bit about it. From his Majorcan home, Cresques designed the Atlas using information from travelers who knew both North Africa and the trans-Saharan trading routes. Cresques also built on the well-known pilgrimage of the Emperor of Mali, **Mansa Musa**, to Mecca in 1324-5; and he relied on longstanding trade routes linking Jewish communities in Saharan oases such as **Tuwāt** in Algeria and **Siğilmāsa** in Morocco with both **West African kingdoms** further south and Jewish communities in the Iberian worlds. Looked at like this, the Atlas is the product of centuries of cross-cultural exchange. Cities of the Mediterranean world such as Cairo, Lisbon, Seville and Tripoli did not impose themselves on West Africa and dominate their peoples; instead, West African and Mediterranean societies emerged like the Catalan Atlas, through trade and reciprocal exchanges.

In particular, Green’s book provides a very detailed account of how most regions of Africa were severely transformed with the rise of the **Trans-Atlantic slavery trade**, and with **colonialism**, with dramatic consequences that remain until today:

The transformations were, of course, not just political but also economic. The aforementioned trend of importing currencies was shifting by 1680. A vital factor in this negative economic trend was the place of **African labour** in shaping the economy of the West. The loss of labour through the trade in enslaved captives meant that surplus-value producing young men had their economic output placed within the growing European empires. This, combined with the export of gold discussed in the last chapter, and the import of currencies that lost their relative value over time – such as iron and cloth – placed West African political systems at a distinct economic disadvantage. Indeed, these imports added value to European economies as they boosted the European manufacturing base.... By the early seventeenth century, many of these changes were falling into place. The great empires changes were falling into place. The great empires had fragmented, replaced by smaller, centralized kingdoms in the creeks and forests of the Atlantic coastline of West Africa. The aristocracies of these kingdoms had developed a heavy dependence on the Atlantic trade to buttress their power. Thus, the local kings, and their representatives, assiduously monitored all ships that came to trade…

The place of alcohol consumption in **Greater Senegambian trade** at this early time already shows some of the changing dynamics; and even in 1995, when I visited the **King of Canogo** on the Bijagós Islands in **Guinea-Bissau**, it was still expected that I should bring a bottle of rum as a gift. This part of West Africa is certainly not alone in this, since alcohol (most often rum) was also long a lynchpin of the trade in Angola from the seventeenth century onwards, and alcohol must also be brought as a gift to meet local chiefs, or sobas, in Angola to this day. Alcohol was not only for elite consumption, however, but was also used for religious purposes, often replacing palm wine as a libation at spirit shrines. For this reason, alcohol use became associated with both **African kings** and with **African religious practice**, something that, as we shall see by the end of the book, would prove very significant. A classic oral narrative to exemplify this is that of **Kelefa Saane**, an epic of conquest and political change set in **Senegambia** in the early nineteenth century. In many accounts of Kelefa, his violent actions are associated closely with alcohol consumption: one historian, Sana Kuyate, described how Kelefa went from village to village, raising hell and selling inhabitants for slaves.

So it may be no accident that, alongside the sale of alcohol, Bautista Pérez also traded large numbers of knives and swords made in Portugal, stoking the sort of violence that appears in the epic of Kelefa. At one village, one oral historian says of Kelefa, 'the day they were leaving there, half the children of the village went to gather small sticks of firewood…he saw those children in the woods – a hundred children…he took all these and sold them all for alcoholic beverage'. Growing trade thereby promoted increased social hierarchies in **Greater Senegambia**. Those who had access to social and political capital could display this through new ornaments, through wearing imported cloths or by having a high time of it drinking imported rum; they could enforce this power, too, by buying an arsenal of guns and knives. With this increasing hierarchy came more authoritarian power structures. The political power to reorganize labour according to 'age-sets' (where field tasks were divided into groups of young people by age and gender), and to expand the agricultural production of provisions for the ships of the slave trade, was made possible by these new trading links. This meant that, initially, far from weakening West African political authority, a major consequence of this trade was the emergence of stronger centres of political power. Elites were able to charge tolls and, as mediators of the long-distance trade, develop into a powerful aristocracy.

European traders knew perfectly well that they could not have lived, worked and traded there without acknowledging African power. An early seventeenth-century account of the coast in Casamance describes how the peoples living there routinely robbed any ship that foundered and captured the crew for ransom. In nearby Cacheu, at a similar time, people would break into the houses of European traders by night, freeing slaves and stealing the ornaments from the local churches. By the 1660s, one priest described how there was no way to stop people setting Cacheu on fire, and every time that they wanted the Portuguese traders to do something, they would occupy the water sources of the town, rendering the Portuguese impotent. Further north, the dynamics were the same, and, by the 1680s, Jolof power over French traders near the estuary of the River Senegal was so clear that when one, De la Marche, stole some cotton and tobacco, he and the crew of his ship were set upon by a force of 500 sent by the Brac (King) of Waalo, killing Dela Marche and three of his companions. Thus, the beginnings of global trade as it affected West Africa would not just produce greater inequalities between West African societies and other parts of the world; it also produced greater economic and political inequalities within West Africa itself. A powerful trading class did emerge, but they did so alongside the growing power of kings. European traders had to adapt to this power, and those who could not fared badly. When **Capuchin missionaries** refused to bury someone in 1684 because of his multiple sexual partners, they were hauled out of the house, dragged by their beards and forced by arms to carry out the burial. If wars broke out between different kingdoms, the supply line in provisions to the trading ports could be completely stopped. In these cases, only those with good local connections did well, and these could come about only when foreign traders were prepared to adapt. As two Portuguese traders accused of cheating their African associates in the 1660s discovered, failure to adapt could be very dangerous, for 'the fury of the [Africans] was so great, and so many the arrows and darts that rained down on them and on us, that we were all in danger of death…with the [Africans] coming to seek them out again with arms [once the traders had hidden in the house of two missionaries], shouting loudly and shrilly and promising that no one was going to leave alive'.

Green further explained that:

The social transformations [in West Africa] needed to increase production of provisions for slave-trading ships were marked, with many of the provisions coming from his kingdom. Rice-growing boomed…in the following centuries. Indeed, the idea that 'West Africa could not feed itself' was the reverse of the truth for many centuries. Social transformations in agricultural work meant that rice-growing communities produced a substantial surplus, which was bought by the slave-trade ships. It was only during **Guinea-Bissau's War of Independence** from Portugal (1960–74) that ricegrowing went into decline, never really to recover; this cemented the country's current dependence on rice imports from China, which are bartered by farmers for the cashew nuts that constitute the country's major export. A sobering comment on these social transformations is that, in Senegambia, European traders were often assumed to be cannibals by West Africans. This notion was described by one sailor as early as 1455, and was a repeated trope long into the seventeenth century. And, indeed, across West Africa, from Cameroon and Angola to Senegambia, many different peoples believed in the cannibalism of the European slave traders; here was a mirror-image of the more famous trope, whereby Europeans discussed African 'cannibalism'. The Africans, however, had some justification, for in this case the slave trade was one that consumed people, devouring West and West-Central Africa's human resources.

Political violence was never far away, as the growth of a monetized economy was fuelled by this trade in captives. The process of physical and psychological violence, of course, began with the initial enslavement of peoples, through warfare between West African kingdoms, banditry, judicial enslavement for crimes such as adultery or witchcraft, or sale to clear debts, as described in the 1570s by André Álvares de Almada: 'the slaves which they have [on the Gambia River] and which they sell, they capture in wars and through judicial trials and kidnappings, because they go to steal them from one part or another'. Almada also described the way in which some of this occurred, discussing a river in Guinea-Bissau where there are 'large canoes in which there are thieves called Gampisas in the [Biafada] language. They are like bandits…they are so canny that if somebody comes from the bush inland, they pretend that they want to welcome and host them, and welcome them into their homes; and, having had them there a few days, they put the idea in their heads that they have some friends by the sea, and that they'd like to take them there so that they can meet them and enjoy each other's company; and going to the ships they sell them; and in this way they trick many people'. The inscribing of violence on to the bodies of the enslaved then continued at the ports of departure, where captives were branded with the mark of their 'owners'. Such violence was further exemplified through the account books of slave traders such as Bautista Pérez, who assessed captives whom they had purchased solely on the basis of their potential economic impact, describing infirmities such as 'swollen sides', cataracts and burns as 'losses [daños]'. The culmination of the violence was the Middle Passage itself, when potential losses from mortality were weighed up against profits to be gained from transporting the maximum number of captives. This reduction of human life to an economic equation was made explicit in the so-called 'books of the dead': account books that listed the losses incurred through the deaths of those enslaved captives who perished on the crossing to the Americas, and in which the branding marks of the owners were written clearly in the margins, alongside the simple names of the dead.

By the end of the seventeenth century, therefore, an economic system had been established across Greater Senegambia that favoured the emergence of small, powerful kingdoms at the expense of larger ones such as **Jolof**. The economy of the region was based around trade and exchange, and depended fundamentally on credit, in common with **slave-trading** economies throughout the Atlantic Basin, where, as soon as a credit source dried up, so did the trade. However, the weight of these currencies as set against global currencies was in decline, and thus **West African kings** did not have the economic power to resist the dumping of increasing volumes of cheaply manufactured goods. Nor did they want to, for these goods facilitated greater trade, and it was through control of these trading networks that their power had grown. The authority that went with ruling was one that the new trading patterns had tended to increase, entrenching the local aristocracies. This cycle of credit from the Atlantic economy created growth – as economic credit does – but it could be cashed out only through the violence involved in the increased export of enslaved captives.

The economic differential between Senegambian and Atlantic-world economies, therefore, grew throughout this period, as capital accumulation accelerated outside West Africa.... The fundamental dynamic at play across the whole region was of the concentration of political power, with the wealth derived from trade goods becoming a source of status and a marker of inequality. Increased trade did not, therefore, bring an equal benefit to the peoples of Greater Senegambia. It was a source of greater political violence, social insecurity and insubordination for many; whereas for the new elites, through control of trade, they were able to exercise and display their power. And resentment of the arbitrary power of the new aristocracies was beginning to emerge. The Atlantic trade was thus highly disruptive of existing institutions. In Greater Senegambia, the Jolof 'Empire' was fragmenting, while the **Empire of Kaabu** was rising; in **Sierra Leone**, during the sixteenth century, the **Kingdom of the Sapes** was overcome by migrants, known as the **Manes**, from the collapsing **Mali Empire**. Everything changed: when political systems were large scale and complex, they often fell apart; when they were smaller and weaker, they became stronger. The fundamental place of the Atlantic trade in West African political life was not, therefore, universal, but depended on pre-existing conditions. The unifying theme was the disruption that it had introduced.

Importantly, Greene deconstructed various misapprehensions, mostly regarding how Westerners have historically tended—and continue—to see slavery:

Yet, beyond the flowery prose [see in travel writings done by Westerners], the realities were grim; reproducing these external accounts alone is to reproduce a fantasy that was deliberately designed to conceal the truth. This is one of the reasons why the style and structure of this second part of the book are different from those of the first. [For instance] Cape Coast was not beautiful. It was the ultimate site for the acting out of the ugliest side of human nature – a place for the brutal exercise of male power. The English officers had multiple mistresses, taken from among their slaves, throwing those who refused into the 'punishment cell'. They were routinely 'punch-drunk' on imported Brazilian rum, sugar, water and limes (an early form of the **Brazilian caipirinha** and **Cuban mojito** cocktails). They presided over a regime of violence, their chapel and beautiful apartments sited above the foetid and grim realities of the slave dungeons, which housed 200 people at a time on a floor awash with their own excrement, vomit and death. Those who survived their imprisonment were marched along a tenebrous basement corridor lit by torches, counted through by spy-holes to make sure that no money was lost, despatched through the 'door of no return' and taken in lighters over the dangerous surf to the ships resting in the road of the castle, which would carry people chained to the New World – the lights of the ships bobbing by night, just as today the oil tankers do along the coast by **Lomé**, in Togo.

These realities of the many European slaveforts were not for 'public consumption' in the colonial capitals of London, Lisbon, Amsterdam and Paris. They appear not to have been discussed in 'polite society'. They were rarely written about in the many travel books on the region, which, of course, were calculated to appeal to the market, such as it was, in eighteenth-century sensibility. The violence that underwrote the expansion and accumulation of capital was hidden from European public view; but, in West Africa, it was in plain sight and remains so in view of the castle-prisons, and the memorial at the small town of **Assin Manso**, around 30 miles from the coast, which was the last stopping point for captives on their march to the coast. The English regime at Cape Coast castle was one of power and fear. Yet the English were also dependent on the people of the town of **Ooegwa**. English agents would travel by sea with their interpreters from Ooegwa to Ouidah, where they bought captives from the kings of Hueda – and, after the 1720s, from Dahomey. The captives were then brought back to Cape Coast, where they were kept in the castle dungeons, piled on top of one another, a thousand at a time, alongside those who had been sold by intermediaries from **Asante** and marched down to the coast. They were branded with the mark of their owners, after the commodification of their bodies had been completed. After some time in the dungeons, these enslaved persons were taken through the 'door of no return'.

The industrialization of this calculated programme to **dehumanize human beings** in the eighteenth century is one of the more significant episodes in human history. Its consequences in the Americas and Europe are, perhaps, well known; but its consequences in Africa are less so, and that is the focus of the second part of this book. As Africa globalized into the world, the world globalized into Africa. As increasingly authoritarian West African leaders profited from the Atlantic trade, they became ever more distant from their subjects in social and political terms. At the same time, their religious shrines became associated by some with the cruelty and corruption of the existing system, opening the door to the greater influence of, first, Islam, and then – in the nineteenth century – **Christianity**. The eighteenth century in Africa reveals one long period of struggle against the new political and religious power associated with the slave trade, which, in many places, was eventually overthrown. Yet successful revolutions do not always succeed in implementing restorative justice. In nineteenth-century West and West Central Africa, one type of inequality was replaced by another, as gender imbalances grew and new systems of captive labour were introduced to meet the needs of the 'legitimate' trade in tropical produce for European factories: palm oil, groundnuts and the like.

As Greene further noted:

Two core misapprehensions need to be addressed before going any further. First, slavery did not just exist in an Atlantic context, and is not just an Atlantic story. The **trans-Saharan trade** to North Africa and Egypt probably involved the forced migration of six million persons between around 800 CE and 1900; a further four million were sold into the Middle East and Indian markets via the Red Sea and the Indian Ocean. The trans-Atlantic slave trade database of slaveship voyages estimates that, between 1492 and 1866, 12,521,337 persons crossed the Atlantic Ocean as captives. The second core misapprehension is that slavery is exceptional to Africa. When talking of enslaved persons being a 'form of currency' in West Africa, the word 'slavery' can be problematic, for it carries a different meaning in different historical settings. Far from being the norm, the New World chattel slavery that enabled this to take place was the exception. Among the **Aztecs**, for instance, **slavery** was important for **religious rituals of sacrifice**; and among the Maya, while people could be enslaved for debts or could sell themselves into slavery to avoid starvation, chattel slavery did not exist, and – as among the Aztecs – war captives were usually sacrificed for religious reasons. In other words, the institution of slavery has not always been related to labour and economic value. In fact, economic value as it related to enslaved persons was largely a construct of the system of Atlantic slavery. The key to understanding slavery as it emerged in West and West-Central Africa from the fifteenth to the nineteenth centuries is not just its economic function, but its relationship to warfare, kinship and honour. Just as among **Native American** peoples, warfare often shaped how slavery was seen in many parts of Africa. Successful wars helped a society grow in size and strength. War captives could be incorporated as new members of an expanding society, with dependent status. By the seventeenth century in Angola, for instance, slaves were frequently described as 'captives' – just as they were among the **Bambara of Segu** (the jòn).

In the mid 1680s, the French colonist Michel Jajolet de La Courbe described a trade on the Senegal River in 'captifs' and did not use the word 'slave'. The use of the word 'captive' shows that the concept of 'slave' was, in fact, a rather different one, derived from Roman and then New World contexts and later imported to Africa. This fundamental relationship of dependence and warfare did, however, change over time. As the slave trade expanded hugely in the eighteenth century, so, too, did the capture of enslaved persons by **warfare**. Since slavery in West Africa created a class of dependent foreign aliens, it encouraged an expansion of warfare in order to create an ever larger servile class and greater social differentiation. From the Western viewpoint, thus, economic cycles of demand provoked the increase of enslavement in the eighteenth century; yet, from the West African viewpoint, this increase had more to do with the place of foreign dependants in society, and the ways in which these societies were transforming themselves as social hierarchies grew. That enslaved persons were most often outsiders is important, as it shows up the fallacy in the idea often put forward that it was 'Africans' who sold 'Africans' into the slave trade. This argument is usually developed as a way of alleviating discourses of Euro-American guilt, discourses that themselves emerge from the history of abolitionism. But it completely misunderstands identities in the seventeenth and eighteenth centuries, when…people did not see themselves as 'African' but rather as belonging to a specific lineage, kingdom and ritual community – just as people did not see themselves as 'Europeans' at the outset of this time, but rather defined themselves according to the style of Christian belief and nation.

An interesting piece of evidence of this comes from the Gold Coast in 1682, where the factor at Anomabu, Richard Thelwall, wrote that 'as concerning slaves, though the [Aboms] panyarrd [kidnapped] the Cormanteen people, yet they dare not sell them for they are all of one country'. Clearly, identity was not based on some common sense of belonging to an abstract continent, but on local and regional ties, ties that also determined who could and could not become enslaved. In sum, whether a person had uncles, aunts, brothers, fathers, mothers, sisters or cousins with inheritance and lineage rights shaped the social rights that they held. One anthropologist described well how it was those who were 'not integrated into the domestic community…who, having no relations of kinship, affinity or vicinity, were most vulnerable to capture'. It was not always the case that enslaved persons had no kin in the region where they lived, but it often was; and certainly the concept of 'slavery' cannot be separated from other forms of personal dependence that defined the claims that people might make on each other.

This important point about how many conflicts that are happening today still have so much to do with mistakes and atrocities done during colonialism, centuries ago, is also clearly made in Meral's 2018 book *How Violence Shapes Religion* (see Box 6.6).

Box 6.6 Colonial Mistakes, Religion, Conflicts, and Nigeria

One of the two main case studies provided in Meral's *How Violence Shapes Religion*, Nigeria—the other is Egypt—provides ample evidence of how colonial errors, occurred centuries ago, have influenced subsequent regional conflicts:

Adamolekun argues that the people of Nigeria's first contact with Christianity was in 1472 when Portuguese missionaries arrived in the Delta region but that until the close of the eighteenth century, Christianity did not gain any substantial ground. In fact, it was not really until late in the nineteenth century that Christian missionaries from Europe and North America began effective proselytism campaigns in the region, overlapping with both the advancement of colonial expeditions and the evangelical revival happening in missionary-sending countries. This resulted in various missionary groups, **Baptists** and **Anglicans** as well as European Catholic and North European Protestant groups, arriving in Nigeria. Methodist and Anglican mission boards were followed by Baptists and Roman Catholics, mostly active in the southern regions of Nigeria. Throughout the twentieth century, the Nigerian church saw rapid nationalization with local leadership, associations and theological education. Indigenous mission movements spread local churches independent of established historic churches. It was not until the later part of twentieth century that Christianity spread rapidly across northern Nigeria, both due to conversion from animistic religions and among **Hausa-Fulanis** as well as migration to the Middle Belt and north from other parts of **Nigeria**.... The journalist, Dame Flora Louisa Shaw, in an essay published by the Times on 8 January 1898, suggested that 'it may be permissible to coin a shorter title for the agglomeration of pagan and **Mahomedan States** which have been brought, by the exertion of the Royal Niger Company, within the confines of a British protectorate', which, she noted as 'Nigeria'. Managing such a complex territory and its peoples with limited military capacity and dubious trade agreements with local chieftains was no easy task for the British colonial engagement with Africa. Berman notes that colonial rulers, therefore, developed a 'divide and rule' strategy, which confined social activities into local administrative subdivisions, each containing a single culturally and linguistically defined community, directed by their indigenous structures and discipline processes, thus not only fragmenting the possibility of a united discontent across the land against the colonial rulers, but also drawing legitimacy and indirect control from traditional social structures...

In the late 1940s, the British colonial administration then divided the Nigerian territory into three administrative zones. The three zones – the North, East and West – implied the ruling power of the three ethnic groups that dominated those three geographical spreads, the **Hausa**, **Igbo** and **Yoruba**. This gave British rulers control over three major groups and indirect rule through their privileged position, which in return meant that the three major ethnic groups believed it was their destiny to dictate the future of Nigerian political, social and development. In fact, far from restricting the expansion of Muslim rule and Islam in the north, the colonial administration had 'deliberately prohibited Christian evangelization in Muslim areas', while Christian missionaries, schools and hospitals spread to other parts of Nigeria. The gradual accommodation of British rule by the Northern Muslim elite stood in stark contrast to the views expressed by the leadership of the **Sokoto Caliphate** at the time of British conquest.... The British faced difficulty in applying their indirect rule model to the Igbos, whose complex social and political systems were often misread, resulting in resentment and anger among them, as the colonial powers appointed people whom they thought to have traditional power, but in fact did not.

Ejiogu argues, that 'in the overall, while the **Hausa-Fulani** who cherished the preferential treatment that they received from the British in the course of colonial state building, saw themselves as rulers, the Igbo, the Yoruba and others became increasingly distrustful of the resultant supra-national state and its authority'. In addition, other ethnic groups who had local majorities but were regional numerical minorities feared ethnic domination by the three privileged groups, which resulted in tensions and rebellions.

For example, the northern **Fulani Muslim emirs** sought to control the Middle Belt region and the wide range of non-Muslim communities under their own traditional leaderships. Nnoli notes that 'inevitably, the non-Muslim majority of the territory mobilised against this injustice, domination and oppression. They were resentful of the emirs for imposing Islamic norms and religion on them and for relegating their own rulers to an inferior second class status'. For many Christians, 'at the local level the question of national independence was relatively unimportant to most of the politically minded Protestants…their main concern was to uphold their independence from the Muslim Fulani, not of the British…they generally saw the British as their best guarantee against Fulani rule'. The 1954 constitution formalized the power sharing of regional governments, giving them authority over education, administration, health and public works at the local levels, while creating a federal government which handled macro-economic policies, higher education, military, foreign policy and transportation. This created multiple layers of ethnic competition: among the three dominant ethnicities of the West (Yoruba), East (Igbo) and North (Hausa-Fulani) there was competition for control of the federal government. In return, these three dominant groups competed with ethnic minorities in their own local governments for power and access. Thus, national politics emerged as ethnic politics when the federal state was created. Colonial rule had not only united a diverse human geography under the umbrella it created, but also its governance shaped the way individuals and communities perceived and responded to emerging national space. As Berman observes in postcolonial states across Africa, 'the structures and practices of the colonial state, its demarcation of political boundaries and classification of people, as well as European expectations about African cultures and institutions, contained African political processes within the categories of 'tribe' and encouraged Africans to think ethnically'. Berman argues that it was 'within these intersecting social, cultural, economic and political processes that the social construction of modern African ethnicities has taken place – partially deliberate and intended, and partially as their unintended and unforeseen consequence'…the consequences of colonial rule continue to impact Nigeria and are directly linked to **ethno-religious violence** in the country.

For instance, after summarizing the history of Nigeria before its independence from British colonial rule (see Box 6.6), Meral explained that:

> Nigeria became independent from British "colonial rule on 1 October 1960 but faced major complications from the start. As Chief Obafemi Awolowo famously said: 'Nigeria is not a nation…it is a mere geographical expression…there are no Nigerians in the same sense as there are English, Welsh or French…the word Nigerian is merely a distinctive appellation to distinguish those who live within the boundaries of Nigeria from those who do not'. In fact, it was colonial rule that not only created the word Nigeria but also decided who was part of it and how they related to the created nation. Chinua Achebe captures the ironies of Nigeria's independence from the British colonial rule and yet dependence on its legacy in its emergence as a sovereign state in Nigeria's first national anthem: 'our national anthem, our very hymn of deliverance from British colonial bondage, was written for us by a British

woman who unfortunately had not been properly briefed on the current awkwardness of the word tribe…so we found ourselves on independence morning rolling our tongues around the very same trickster godling: though tribe and tongue may differ, In brotherhood we stand!'

It was a most ominous beginning. And not surprisingly we did not stand too long in brotherhood. In his study of **postcolonial African politics**, Berman asks a key question: 'what has happened to the colonial state and its links to patronage networks and ethnic development since independence?' He observed a paradox, that the answer is both 'very little' and also 'a very great deal' had changed. He sees a continuation of colonial power structures, bureaucratic authoritarianism and the use of local leaders to assert power and clientelism that bought allegiances along ethnic lines…. Indeed, independent Nigeria, too, has seen a paradoxical 'very little' but also 'a very great deal' changed. Since its independence, Nigeria has seen a vicious cycle of attempts to move to civilian rule paused by multiple military coups, suspending the normal functioning of the political system as well as maintaining a perpetual state of emergency in the governance of the country".

About the current situation in Nigeria, and in particular the links between violence, religion, and external influences, Meral stated:

The gradual transformation of indigene *versus* settler conflict into an **ethno-religious conflict** escalates the initial cause of violence to a much larger religious framework that moves beyond the politics of the Middle Belt to all of Nigeria and turns the local tension into a continuation of global developments involving apparently Christian and Muslim-majority nations. Ukiwo captures this dynamic while commenting on the September 2001 clashes that saw more than 1,000 fatalities in **Jos** alone. He notes that while 'the immediate cause of the riot was a dispute between a Christian woman and a Muslim outside a mosque, the riots have been linked to the appointment of a Muslim as head of the lucrative post of coordinator of the federal government's **Poverty Alleviation Programme**'. A prebendal state with weak rule of law also turns violence into a low risk opportunity for personal and communal gain…. A more organized and regular expression of this is the raids by Hausa-Fulani tribesmen and cattle herders across Plateau State.

There are numerous ongoing cases of Christian villages and cattle owners being attacked in rural areas, their properties and stocks looted, houses burnt and many left dead. Poor villagers often find themselves in further difficulties if they report their stolen cattle and file complaints. A nineteen-year-old boy interviewed at the outskirts of Jos still had fresh wounds from the attack on his house and farms two days earlier, which left his father dead, destroyed all of their property and took away more than 100 cattle. This last attack which completely destroyed the future of his family and caused the young man to be relocated continually with the fear of being hunted down was in response to an earlier raid by Hausa-Fulanis which the family reported to the police and pursued their case at the courts. They were verbally warned to withdraw their complaints by Hausas. If not, they would face the outcomes. The family did not back down. Their attackers, who are known by authorities and local communities, were not brought to justice. Christian Solidarity Worldwide recorded 50 dead, mostly women and children, in fourteen attacks on villages in Plateau State by Hausa-Fulani tribesmen during the first three months of 2011 alone.

Such raids are clearly organized against particular villages that are weaker and from different ethnic groups than those of the attackers, most of the time Christian, with a clear aim of stealing livestock. A pastor told me in an interview that this was not a religious issue, that 'there are economic undertones, but religion gives them a reason', thus they attack non-Muslims and see their looting as jihad and the legitimate spoils of war. The strong presence of the use of violence for ideological ends can be seen particularly in northern Nigeria and the Middle Belt region, especially among the myriad of Islamist elites, local networks and clerics. Glorified memories of the **Sokoto Caliphate** and the **Great Jihad** which had been launched to expand Islam to the rest of West Africa and Nigeria are still widely alluded to

> by Muslims, particularly in northern Nigeria. Far from being melancholic about a long lost kingdom that now seems too far away to restore, Muslims interviewed particularly in northern Nigeria held passionate beliefs that not only Muslim-majority states but the whole of Nigeria should be based upon Islam and the place of non-Muslims in Muslim-majority societies should be subjugated to provisions of **Shari'a**. These beliefs that refer to **traditional Islamic jurisprudence** about the place of minorities in a Muslim society and the place of Islam in governance for all provide strong frameworks for the use of violence, not simply legitimized by the outcomes of a prebendal state, but allowed, rewarded and at times demanded by God.

This connection between past and present should be obvious to most people, but strikingly a huge number of people in Western countries tends to—or tries to—forget or neglect this point. For instance, numerous people argue instead that the fact that many places in the Middle East and Africa continue to have such a high number of conflicts despite the fact that colonialism "finished" so long ago, can only mean that the peoples living in those areas of the globe—or at least their cultures—are truly "inferior."

Obviously, unfortunately colonialism is not over, far from it: we still face huge problems related to so-called neo-colonialism, concerning religion, science, economic interests, and many other items. In the sense that the governments of the former colonies are at least in theory a bit more "independent" from the former colonial powers, one can say that there are of course major differences between what happens now versus what occurred in colonial times. But the reality is that in a huge number of cases Westerners continue to have a huge influence, not only economic but also scientific, cultural, and religious, in countries that were previously colonized. To give just an example here, among many, I will refer to McAlister's excellent 2018 book *The Kingdom of God has No Borders – A Global History of American Evangelicals*. As explained by McAlister, **American Evangelicals** still have a huge influence on important cultural, political, and even legal aspects within various African countries in Africa, often using a combination of religious teleological narratives and economic incentives to exert such an influence. This point was also clearly illustrated in the multi-awarded 2013 movie *God loves Uganda*, which shows how U.S. evangelicals successfully influenced policies in Uganda with the aim to have this country officially punishing **homosexual acts** with life imprisonment or eventually even with the death penalty. Nowadays, Uganda has one of the most strict set of laws against homosexual acts: under its Penal Code, "carnal knowledge against the order of nature" between two males can lead to life imprisonment.

Chapter 7
Brains, Conspiracies, Witches, and Animal Abuse

> *War: a massacre of people who don't know each other for the profit of people who know each other but don't massacre each other.*
>
> (Paul Valery)

7.1 Brains, Behavior, Genes, and Religions

> *Alice laughed...'There's no use trying', she said, 'one can't believe impossible things'. 'I daresay you haven't had much practice', said the (Red) Queen...'When I was your age, I always did it for half-an-hour a day. Why, sometimes I've believed as many as six impossible things before breakfast'. (Lewis Carroll: Alice's Adventures in Wonderland)*

Shermer's 2011 *The Believing Brain* provides several illustrative examples of our tendency to try to look for a meaning and for meaningful patterns even when the data is actually meaningless. In his chapter about the afterlife, he noted that in a 2009 Harris Poll on **religious beliefs** among Americans, 71% answered that they believe in **soul survival**. Shermer listed six major reasons why so many people believe in **afterlife**, which are profoundly connected with the tendency of our brains to build teleological narratives and then accept them as true:

> 1. *Belief in the afterlife is a form of* ***agenticity***. In our tendency to infuse the patterns we find in life with meaning, agency, and intention, the concept of life after death is an extension of ourselves as intentional agents continuing indefinitely into the future.
>
> 2. *Belief in the afterlife is a type of* ***dualism***. Because we are natural-born dualists who intuitively believe that our minds are separate from our brains and bodies, the afterlife is the logical step in projecting our own mind-agency into the future without our bodies.
>
> 3. *Belief in the afterlife is a derivative of our* ***theory of mind***. We have the ability to understand that others have beliefs, desires, and intentions (we "read their minds") by projecting ourselves into the minds of others and imagining how we would feel. This theory of mind projection is another form of agenticity and dualism by which we can imagine the intentional minds of both ourselves and others as continuing indefinitely into the future.

R. Diogo, *Meaning of Life, Human Nature, and Delusions*,
https://doi.org/10.1007/978-3-319-70401-2_7

4. *Belief in the afterlife is an extension of our body schema*. Our brains construct a body image out of the myriad inputs from every nook and cranny of our bodies. When this single individual *self* is coupled with our capacity for agenticity, dualism, and theory of mind, we can project that essence into the future, even without a body.

5. *Belief in the afterlife is probably mediated by our left-hemisphere interpreter*. A second neural network that is likely integral for afterlife beliefs is the **left-hemisphere interpreter**, which integrates inputs from all the senses into a meaningful narrative arc that makes sense of both senseful and senseless data. Tie this process into our body schema, theory of mind, and dualistic agenticity and it becomes clear how easy it is to develop a plot in which we are the lead character whose meaning and importance is central to the story and whose future is eternal.

6. *Belief in the afterlife is an extension of our normal ability to imagine ourselves somewhere else both in space and time, including time immemorial*. Close your eyes and imagine yourself on the warm sands of a tropical beach on a beautiful sunny day. Where are you in this picture? Are you inside your skin looking out from your eyes at the crashing waves in the distance and children playing in the sand? Or are you above yourself looking down on your entire body as if there were a second you hovering overhead? For most people this thought experiment results in the second observational platform. This is called *decentering*, or imagining ourselves somewhere else from an Archimedean point beyond our body. In this same manner we envision ourselves in the afterlife as a decentered image removed from this time and space into an empyreal realm, the literal (and literary) dwelling place of God, the ultimate immortal and eternal agent.

As explained in Chap. 2, some of the six items listed above are questionable, but the important point I want to make here, and in that chapter, is to emphasize that at least some of these items are also found in other animals, for instance, **agenticity**. Therefore, the presence of such features very likely facilitated the subsequent origin of the belief in afterlife in human evolutionary history. What is interesting is that Shermer did *not* refer to *the* item that is most probably the more crucial one for the origin of the seemingly unique human belief in afterlife: the *awareness of the inevitability of death and randomness of both death and life*. In his chapter about the belief in God, Shermer notes:

According to Oxford University Press's *World Christian Encyclopedia*, 84 percent of the world's population belongs to some form of organized religion, which at the end of 2009 equals 5.7 billion people. That's a lot of souls. **Christians** dominate at around 2 billion adherents (with Catholics accounting for half of these), **Muslims** come in at a little more than a billion, **Hindus** at around 850 million, **Buddhists** at almost 400 million, and **ethnoreligionists** (**animists** and others in Asia and Africa primarily) make up most of the remaining several hundred million believers. Worldwide, there are about 10,000 distinct religions, each one of which may be further subdivided and classified. Christians, for example, may be apportioned among about 34,000 different denominations. Somewhat surprisingly – given that we are the most technologically advanced and scientifically sophisticated nation in history – America is among the most religious tribes of the species. A 2007 Pew Forum survey found that 92% of Americans believe in God or a universal spirit, 7% in **heaven**, 59% in **hell**, 63% that scripture is the word of God, 58% prey once a day, and 79% believe in **miracles**. Most striking to me…the **dualistic belief** that there must be something else out there is so pervasive that even 21 percent of those who identified themselves as **atheists**, and 55 percent who identify themselves as **agnostics**, expressed a belief in some sort of God or universal spirit. Such statistics stagger the imagination. Any characteristic that is this common in a species cries out for an explanation. Why do so many people believe in God? On one level…[this concerns]…**patternicity** and **agenticity**…God is the ultimate *pattern* that explains everything that happens, from the beginning of the universe to the end of time and everything in between, including and especially the fates of human lives. God is the ultimate intentional *agent* who gives the universe meaning and our lives purpose. As an ultimate amalgam, patternicity and agenticity form the cognitive basis of **shamanism**, **paganism**, **animism**, **polytheism**, **monotheism**, and all other forms of **theisms** and **spiritualisms** devised by humans.

Shermer then provides a brief summary of his personal take on the *evolution* of humans' belief in God (see also Box 7.1):

In his 1871 book *The descent of man*, **Charles Darwin** noted that anthropologists conclude that "a belief in all-pervading spiritual agencies seems to be universal; and apparently follows from a considerable advance in the reasoning powers of man, and from a still greater advance in his faculties of imagination, curiosity and wonder". What flummoxed Darwin about the universal nature of **religious beliefs** was how natural selection could account for them. On the one hand, he noted, "it is extremely doubtful whether the offspring of the more sympathetic and benevolent parents, or of those who were the most faithful to their comrades, would be reared in greater number than the children of selfish and treacherous parents of the same tribe. He who was ready to sacrifice his life, as many a savage has been, rather than betray his comrades, would often leave no offspring to inherit his noble nature". On the other hand, although Darwin was a strident proponent of restricting the range and power of natural selection to operate strictly at the level of the individual organism, he conceded that selection might also operate at the group level when it came to religion and between-group competition: "there can be no doubt that a tribe including many members who, from possessing in a high degree the spirit of patriotism, fidelity, obedience, courage and sympathy, were always ready to aid one another, and to **sacrifice** themselves for the common good, would be victorious over most other tribes; and this would be natural selection".

Picking up where Darwin left off, in my book *How we believe* I developed an evolutionary model of belief in God as one of a suite of mechanisms used by religion, which I define as *a social institution to create and promote myths, to encourage conformity and altruism, and to signal the level of commitment to cooperate and reciprocate among members of a community*. Around five thousand to seven thousand years ago, as bands and tribes began to coalesce into chiefdoms and states, government and religion co-evolved as social institutions to codify moral behaviors into ethical principles and legal rules, and God became the ultimate enforcer of the rules. In the small populations of hunter-gatherer bands and tribes with a few dozen to a couple of hundred members, informal means of behavior control and social cohesion could be employed by capitalizing on the moral emotions, such as shaming someone through guilt for violating a social norm, or even excommunicating violators from the group. But when populations grew into the tens and hundreds of thousands, and eventually into millions of people, such informal means of enforcing the rules of society broke down because free riders and norm violators could more readily get away with cheating in large groups; something more formal was needed. This is one vital role that religion plays, such that even if violators think that they got away with a violation, believing that there is an invisible intentional agent who sees all and knows all and judges all can be a powerful deterrent of sin. One line of evidence for this theory of religion can be found in human universals, or traits that are shared by all peoples. There are general universals, such as tool use, myths, sex roles, social groups, aggression, gestures, emotions, grammar, and phonemes, and there are specific universals, such as kinship classifications and specific facial expressions such as the smile, frown, or eyebrow flash. There are also specific universals directly related to religion and belief in God, including ***anthropomorphizing animals*** *and objects, general belief in the* ***supernatural****, specific* ***supernatural beliefs*** *and rituals about death, supernatural beliefs about fortune and misfortune, and especially* ***divination****,* ***folklore****,* ***magic****,* ***myths****,* and ***rituals***. Although such universals are not totally controlled by genes alone (almost nothing is), we can presume that there is a genetic predisposition for these traits to be expressed within their respective cultures, and that these cultures, despite their considerable diversity and variance, nurture these genetically predisposed natures in a consistent fashion.

A second line of evidence for the evolutionary origins of religion and belief in God can be found in anthropological studies of **meat sharing** practiced by all modern hunter-gatherer societies around the world. It turns out that these small communities – which can cautiously be used as a model for our own Paleolithic ancestors – are remarkably egalitarian. Using portable scales to measure precisely how much meat each family within the group received after a successful hunt, researchers found that the immediate families of successful hunters got no more meat than the rest of the families in the group, even when these results were averaged over several weeks of regular hunting excursions. Hunter-gatherers are egalitarian because individual selfish acts are effectively counterbalanced by the combined will of the rest of the group through the use of gossip to ridicule, shun, and even ostracize individuals whose competitive drives and selfish motives interfere with the overall needs of the group. Thus, a human group is also a moral group in which "right" and "wrong" coincide with group welfare and self-serving acts, respectively. Other hunter-gatherer groups employ supernatural beings and **superstitious rituals** to enforce fairness, such as the **Chewong people of the Malaysian rain forest** and the ritual *punen*, which is related to the calamities and misfortune that arise when you act too selfishly. In the Chewong world, the myth about **Yinlugen Bud** – a god who brought the Chewong out of a more primitive state by insisting that eating alone was improper human behavior – serves to ensure the sharing of food. When food is caught away from the village, it is promptly returned, publicly displayed, and equitably distributed among all households and even among all individuals within each home. Someone from the hunter's family touches the catch then proceeds to touch everyone present, repeating the word *punen*. Thus, both superstitious rituals and the belief in supernatural agents oversee the exchange process that reinforces **group cohesiveness**.

Box 7.1 Behavior, Genetics, and Religion

Shermer's 2011 *The Believing Brain* provides a fascinating summary on the links between behavioral genetics and religion:

> **Behavioral geneticists** attempt to tease apart the relative roles of heredity and environment on any given trait. Since there is variation in the expression of all traits, we are looking for a percentage of the variation accounted for by genes and environment, and one of the best natural experiments available for research are identical twins separated at birth and reared in different environments. One study of fifty-three pairs of identical **twins** reared apart and thirty-one pairs of fraternal twins reared apart…found that the correlations between **identical twins** were typically double those for **fraternal twins**, and subsequent analysis led…to conclude that genetic factors account for 41 to 47 percent of the observed variance in their measures of religious beliefs. Two much larger twin studies out of Australia (3,810 pairs of twins) and England (825 pairs of twins) found similar percentages of genetic influence on religious beliefs, comparing identical and fraternal twins on numerous measures of beliefs and social attitudes. They initially concluded that approximately 40 percent of the variance in religious attitudes was genetic. These researchers also documented substantial correlations between the social attitudes of spouses. Because parents mate assortatively (like marries like because "birds of a feather flock together") for social attitudes, offspring tend to receive a double dose of whatever genetic propensities may underlie the expression of such attitudes. When these researchers included a variable for assorta-

tive mating in their behavioral genetics models, they found that approximately 55 percent of the variance in religious attitudes is genetic, approximately 39 percent can be attributed to the nonshared environment, approximately 5 percent is unassigned, and only about 3 percent is attributable to the shared family environment (and hence to cultural transmission via parents). Based on these results, it would appear that people who grow up in religious families who themselves later become religious do so mostly because they have inherited a disposition, from one or both parents, to resonate positively with religious sentiments. Without such a genetic disposition, the religious teachings of parents appear to have few lasting effects.

Of course, genes do not determine whether one chooses Judaism, Catholicism, Islam, or any other religion. Rather, belief in supernatural agents (God, angels, and demons) and commitment to certain religious practices (church attendance, prayer, rituals) appear to reflect genetically based cognitive processes (inferring the existence of invisible agents) and personality traits (respect for authority, traditionalism). Why did we inherit this tendency? One line of research that may help answer this question is related to dopamine, which…is directly connected to learning, motivation, and reward. There may be a genetic basis to how much dopamine each of our brains produces. The gene that codes for the production of dopamine is called ***DRD4*** (**dopamine receptor D4**) and is located on the short arm of the eleventh chromosome. When **dopamine** is released by certain neurons in the brain it is picked up by other neurons that are receptive to its chemical structure, thereby establishing dopamine pathways that stimulate organisms to become more active and reward certain behaviors that then get repeated. If you knock out dopamine from either a rat or a human, for example, they will become catatonic. If you overstimulate the production of dopamine, you get frenetic behavior in rats and **schizophrenic behavior** in humans. Most of us have four to seven copies of the ***DRD4* gene** on chromosome eleven. Some people, however, have two or three copies, while others have eight to eleven copies. More copies of the *DRD4* gene translate into lower levels of dopamine, which stimulates people to seek greater risks in order to artificially get their dopamine fix. Leaping off of buildings, antennae, spans, or earth is one way to do it, although high-risk gambling in Las Vegas or Wall Street may also do the trick…[studies show that] people with high numbers on the risk-taking survey had more copies than normal of the *DRD4* gene.

The links between human evolution and the belief in Gods, and particularly the rise of organized religion and the belief in one single God—**monotheism**—are also discussed in detail in Van Schaik and Michel's fascinating 2016 book *The Good Book of Human Nature – An Evolutionary Reading of the Bible*. Their main idea is that the **Bible** was mainly a reflection of, and a way to cope with, the major changes that happened during the rise of agriculture, which has traditionally been seen by scholars as one of the "most amazing things" that happened in human history, but which an increasing number of authors now consider, based on a huge amount of empirical data, to have truly been—using Jared Diamond's words—"the worst mistake in the history of humanity." Van Schaik and Michel wrote:

> Most attention has usually focused on the progress this civilizing step brought with it. There's no doubt that the **Neolithic Revolution** set the stage for an unprecedented success story. Over the course of the last 10,000 years, *Homo sapiens*'s population has exploded from 4 million to nearly 8 billion individuals, and we have seen revolutionary **technological**

> **progress**. But very little has been said about the collateral damage of this demographic and economic success story. Archaeological excavations have shown that **violence** became a prominent part of everyday life; people became shorter, suffered more often from **starvation**, and died younger than their hunter-gatherer forebears. And with the **domestication of animals**, a number of **diseases** leaped from livestock to humans. Diseases such as the **plague**, **smallpox**, **measles**, **influenza**, and **cholera** emerged and quickly evolved [see Fig. 8.7]. **Tooth decay** exploded. At the same time, injustice and **repression** made themselves at home in the new societies, and women bore the brunt of the suffering. Over the course of the coming millennia, these various scourges would beset humanity like the **Four Horsemen of the apocalypse**. But there was no going back to the old way of life. The point of no return had long since come and gone. But still people needed to do something to come to terms with all of the epidemics and other disasters that befell them. The problems were too urgent, too life threatening to ignore. Had our only option been to wait for biological evolution to produce adaptations to these threats, humanity probably wouldn't have survived at all. It was time for our species to exercise its greatest talent – for culture to take over the reins. In order to gain the upper hand over misfortune, people began to look for the causes of all the catastrophes, violence, and **epidemics** so as to develop ways of protecting themselves from these dangers in the future. Cultural solutions were needed, and soon these efforts would result in a cultural "big bang".
>
> Back then – the span of millennia in which first chiefdoms, then states and finally advanced civilizations such as in **Mesopotamia** and **Egypt** developed – there were no independent fields of inquiry like science, medicine, law and religion. Instead there existed something of a primordial cultural soup. Slowly individual spheres of knowledge began to differentiate themselves. But all of them remained deeply influenced by religion, for belief in the rule of supernatural powers was woven into every aspect of life. Any type of misfortune was credited to wrathful spirits and gods. People began to formulate rules and other measures that would eventually develop into entire systems of crisis management aimed at soothing the gods' rage – in hopes of protecting them from diseases and catastrophes. Much of what we widely attribute to "religion" today began as part of this "cultural protection system." We are not trying to say that the **Bible** directly reflects humanity's **sedentarization**, the transition from a nomadic life to more permanent settlement. This would be difficult to imagine, for several thousand years lie between those events and the appearance of the written text. Much more important is that sedentism brought with it a number of problems that remained threatening for long periods. Some of them still pose a danger to this day – diseases are a good example. In sum, the Bible is humanity's diary, chronicling both the problems our ancestors faced and the solutions they came up with – among other things, how people dealt with the new cultural concepts of **property**, **patriarchy**, **monogamy** and **monotheism**.

The book *Sex at Dawn* notes that "the wealth of **civilization** is material…after reading every word of the **Old Testament**, journalist David Plotz was struck by its **mercantile tone**…'the overarching theme of the Bible', he wrote, 'particularly of **Genesis**, is real estate…God is…constantly making land deals (and then remaking them, on different terms)…it's not just land that the Bible is obsessed with, but also **portable property**: **gold**, **silver**, **livestock**'." However, although I consider Van Schaik and Michel's 2016 *The Good Book of Nature* to be an outstanding book and agree with many of its ideas, it should be noted that, paradoxically, some other ideas of that book are themselves the result of the human tendency to think teleologically. This is because those ideas were formulated by following an ***adaptationist framework***. I completely agree with them that many of the stories in the Bible were written as a way to talk about, and cope with, horrible events such as the rise of plagues, famines, and so on. However, this is very different from saying that all, or almost all, stories of the Bible brought an "adaptative advantage" to the people writing and/or believing in them. We now know that human evolution and life in general are much more complex than that: there are numerous examples in biology of mismatches between behavior and ecology, of detrimental behaviors, and of cases that are not optimal at all: in many cases, things are just "*good enough*" (Chap. 6). Accordingly, the writing of the **Bible** is very likely the result of a very complex intermix between a tentative to cope with the Neolithic revolution—as argued by Van Schaik and Michel—self-interest—for instance, used by men to subjugate and oppress woman—copying of previous epic tales—such as the ***Epic of Gilgamesh*** (Fig. 2.1)—irrational beliefs, and so on (see also Box 7.2). In fact, although they don't do this as often as they probably should, in a few parts of their book Van Schaik and Michel do recognize that some parts of the scriptures that might be seen as related to "adaptative reasons"—for example, related to medical aspects—were indeed also likely related to pure self-interest:

> The **Torah**'s authors were aware that **bodily fluids** are dangerous, even if they had no idea what made them so. Whenever they are involved, God seems exceptionally prone to meting out punishments. So you had better beware! By the same logic, all women's bodily fluids were under general suspicion. Even the birth of a child made a woman unclean – and the birth of a daughter made her unclean for twice as long as that of a son (sixty-six *versus* thirty-three days). Mere **menstruation** made her unclean for seven days (and anyone touching a menstruating woman remained unclean until the evening). Most importantly, any unusual instances of bleeding required additional reparations – an indication that the health risk posed was deemed greater than that of regular menstruation. It is important to stress that additional conscious or unconscious motivations may also have inspired these rules. An ethnographic comparison suggests that menstruating women might have been declared unclean for entirely different reasons. Among the **Dogon** of Mali in West Africa, for example, the 'ideology of menstrual pollution as the supernatural enforcement mechanism to coerce women' is instrumentalized to make women 'disclose their menses by going to the menstrual hut'. The result is an increase in 'coital frequency [with the spouse] around the time of ovulation', which in turn helps ensure the husband's paternity. The rules described in the Torah almost certainly helped to minimize the number of children a woman fathered by men other than her husband – an altogether unsurprising effect, given its patriarchal intentions.

7.2 Randomness, Lack of Control, and Conspiracy Theories

> *A credulous mind...finds most delight in believing strange things, and the stranger they are the easier they pass with him; but never regards those that are plain and feasible, for every man can believe such. (Samuel Butler)*

Although people often don't associate teleological narratives and conspiracy theories (see Fig. 2.11), the reality is that the latter are nothing more than a subset of the former, as are religions. Shermer wrote in his book *The Believing Brain*:

> Why do people believe in highly improbable **conspiracies**? I contend that it is because their pattern-detection filters are wide open, thereby letting in any and all patterns as real, with

little to no screening of potential false patterns. **Conspiracy theorists** connect the dots of **random events into meaningful patterns**, and then infuse those patterns with **intentional agency**. Add to those propensities the **confirmation bias** and the **hindsight bias** (in which we tailor **after-the-fact explanations** to what we already know happened), and we have the foundation for **conspiratorial cognition**. Examples of these processes can be found in…the **Freemasons**, the **Illuminati**…[and] the JFK assassination as a prime example. Add to these factors how compellingly a good narrative story can tie it all together – think Oliver Stone's *JFK* or Dan Brown's *Angels and demons*, both equally fictional – and you've got a formula for **conspiratorial agenticity**. The term **conspiracy theory** is often used derisively to indicate that someone's explanation for an event is highly improbable or even on the lunatic fringe, and that those who proffer such theories are most probably crackpots. Since conspiracies do happen, however, we cannot just automatically dismiss any and all conspiracy theorists *a priori*. So what should we believe when we encounter a conspiracy theory? What are some of the characteristics of a conspiracy theory that indicate that it is likely untrue?

1. There is an obvious pattern of connected dots that may or may not be connected in a causal way. When the Watergate conspirators confessed to the burglary, or Osama bin Laden boasts about the triumph of 9/11, we can be confident that the pattern is real. But when there is no forthcoming evidence to support a causal connection between the dots in the pattern, or when the evidence is equally well explained through some other causal chain – or through **randomness** – the conspiracy theory is likely false.

2. The agents behind the pattern of the conspiracy are elevated to near superhuman power to pull it off. We must always remember how flawed human behavior is, and the natural tendency we all have to make mistakes. Most of the time in most circumstances most people are not nearly as powerful as we think they are.

3. The more complex the conspiracy, and the more elements involved for it to unfold successfully, the less likely it is to be true.

4. The more people involved in the conspiracy, the less likely they will all be able to keep silent about their secret goings-on.

5. The grander and more worldly the conspiracy is believed to be – the control of an entire nation, economy, or political system, especially if it suggests world domination – the less likely it is to be true.

6. The more the conspiracy theory ratchets up from small events that might be true into much larger events that have much lower probabilities of being true, the less likely it is to be grounded in reality.

7. The more the conspiracy theory assigns portentous and sinister meanings and interpretations to what are most likely innocuous or insignificant events, the less likely it is to be true.

8. The tendency to commingle facts and speculation without distinguishing between the two and without assigning degrees of probability of factuality, the less likely the conspiracy theory represents reality.

9. Extreme hostility about and strong suspicions of any and all government agencies or private organizations in an indiscriminate manner indicates that the conspiracy theorist is unable to differentiate between true and false conspiracies.

10. If the conspiracy theorist defends the conspiracy theory tenaciously to the point of refusing to consider alternative explanations for the events in question, rejecting all disconfirming evidence for his theory and blatantly seeking only confirmatory evidence to support what he has already determined is the truth, he is likely wrong and the conspiracy is probably a figment of his imagination.

Such basic factors leading people to create and believe in imaginary conspiracy theories, such as **patternicity**, **agenticity**, and confirmation and hindsight biases were discussed in Chap. 2. Humans are in general very uncomfortable with

randomness and its crucial importance in our *lives*, and with the related feeling of lack, or loss, of control. A very recent, emblematic, and somewhat funny case of a conspiracy theory that illustrates very well this point concerns **Leonardo Di Caprio** and the **2017 Oscars**. As you might remember, conspiracy theories quickly became viral across the media about the fact that Warren Beatty and Faye Dunaway announced that the **movie *La La Land*** won the **Oscar for *Best Picture***, before we learned that the winner was actually the **movie *Moonlight***. What happened in the *word of reality* was very simple: Di Caprio read the envelope for the previous award, in which **Emma Stone** won the **Oscar for *Best Actress*** in *La La Land* and then Warren and Faye accidently mixed up the "Best Picture" envelope and the "Best Actress" envelope, reading "*La La Land*" from the latter. That is, this was a completely random mistake, a confusion between two envelopes, without any clear "purpose" or "purposeful agency" involved. However, humans, as other animals such as pigeons, have a major problem dealing with randomness and tend to see purposeful agents everywhere, and in addition—contrary to other animals—are obsessive storytellers.

So, within *Neverland*, things surely could not be so simple: Warren and Faye accidently mixing up two envelopes can't be the whole story. This story involves chance, lacks a purposeful agent, and furthermore is very boring, for such a storytelling irrational animal. It is a much better tale if we add a clear "agent," with a clear "purpose" and thus remove randomness from it: everything happens for a reason, right? So, in one of the most viral conspiracy theories the "purposeful agent" was said to be Di Caprio. According to this theory, the *agency* of Di Caprio was the outcome of his *purpose*, which was vengeance: he wanted to revenge for the fact that he only won the **Oscar for Best Actor** in 2016. This story is of course completely absurd because it claims that instead of revenging all those years in which he did *not* win the Oscar, Di Caprio decided to wait patiently to revenge *after* he got one. This conspiracy theory—as almost all imaginary conspiracy theories (see Fig. 2.11)—is a further powerful illustration of the fact that while we the *Homo fictus* are characterized by obsessively creating and believing in *a huge number of* imaginary stories about purpose, most of us mortals are particularly awful in what concerns the quality, plausibility, and logical consistency of the stories we create, because after all we are also a *Homo irrationalis*.

This is exactly what happened with the **Covid-19** pandemic. A human just eating a bat or any other non-human animal is accidently infected by a virus? That cannot be the whole story, it is not a good story, particularly because this one concerns the most uncomfortable thing that humans can face: our awareness of the inexorableness of death and arbitrariness of life and death. It suffices to say that, just mentioning examples from the U.S., the most well-known and discussed conspiracy theories tend to be indeed those that include something about death: **JFK**, **11/9**, **Covid-19**, and so on (Fig. 2.11). And, although you might not have noted this yet, those conspiracy theories dealing with death often share a similar, very revealing trait: the "purposeful agent" is someone from the same group that created the story, that is, of the *ingroup*. Why? Because it is particularly difficult for humans to think that they might, for instance, invest so much time and effort and resources in their lives, and in those of their children, and that after all that effort, after they found that amazing job at a top company located at the twin towers in New York, then completely unrelated people from an outgroup can come from other countries and take over some planes and crash them against those towers and bring an end to our lives. After all that effort, after everything we did, our lives, or those of our loved ones, are just randomly and suddenly lost for eternity.

So, while religious "stories" create a "purpose" using a supernatural agent—for instance, God took their lives because he loved them so much and wanted to have them in heaven with Him sooner —conspiracy theories create a "purpose" by often—not always, of course—assuming that there is a "purposeful agent" within the ingroup. In the specific case of 9/11, the *ingroup* has to be the U.S. government, or the CIA or the FBI: people living in that country are the active purposeful players, the ones in command. The idea that an outgroup can do so is particularly uncomfortable because it illustrates how in reality the ingroup lacks control: we cannot do nothing about it, or at the maximum we can do little to avoid it, within a chaotic, contingent, random world. By assuming that it was instead the CIA, FBI, and/or the American Government that did the masterplan for the 9/11 attacks, U.S. citizens are trying to bring agency to their ingroup and thus gain a false sense of control: they have control in the sense that they can, for instance, change their government through elections, each of them can contribute to do that, personally, while it is obviously much more difficult for them to personally intervene in a small meeting of a few people that are completely unrelated to them, in a small cave in a far away country such as Afghanistan.

Paradoxically, in what can seem to be a completely different imaginary scenario but actually follows exactly the same overall narrative and for the very same reasons, some Western people accept that the attack was planned in those caves in Afghanistan but say that the attack was mainly caused by the wrongdoing of Western countries. Both scenarios follow exactly the same pattern—and are equally racist, even if the latter one is often accepted by many people that deem themselves to be "liberal anti-racist" people—because the active, crucial agent of both scenarios is still the *ingroup*, that is, the Westerners. This racist narrative implies that other people and other cultures can never be the active players—they can only be either passive or re-active—and also gives a *false appearance of ingroup control*: if Western governments are the ones to blame because they gave no other choice to the poor members of **Al-Qaida**, then we the Westerners *can personally change that*, for instance, by voting those governments out of power or going to protest in the streets. The fact that such racist narratives are very often used to try to gain a false sense of control within a chaotic and highly complex world is unfortunately too often neglected in the literature and by the media and the broader public, as emphasized by Jonathan Freedland's *The Guardian* 2017 article "*Terror attacks are not just about us*":

> The debate about the causes of **terrorism**…one camp holds that the men who plant these bombs are driven by loathing for western values, for our freedom and permissive way of life, and especially for the liberty exercised by women…the other argues that the root cause is western foreign policy and our record of armed intervention in Muslim lands. Boiled down, it becomes a battle of who we are *versus* what we do. For one thing, foreign policy

clearly plays some role in these horrific events. Listen to the testimony of **Jomana Abedi**, sister of the **Manchester murderer**, who said of her brother: "He saw the explosives America drops on children in Syria, and he wanted revenge", before adding, rather chillingly: "Whether he got that is between him and God". Recall the posthumous video released by **Mohammad Sidique Khan**, ringleader of the **7/7 bombers**, in which he cast himself as an avenger for the **2003 invasion of Iraq**. And recall too the warnings of **Britain's security services**, who feared the Iraq war could lead to increased **radicalization**. Besides, such a stance has an appeal beyond the facts. It grants us a degree of control over these acts of catastrophe. It lets us think that we can bring an end to this horror, if only we change tack internationally. We can ensure there are no Manchester tragedies: it's up to us.

The trouble is, the link is not nearly so simple or direct. Talk to those who devote their lives to the study of **violent jihadism**, reading **ISIS's propaganda** and interviewing its devotees, and a different picture emerges. For one thing, it's not all about us. Most of jihadism's victims are other Muslims, in the Arab world or in Africa. When they murder and maim **Shia Muslims** by the hundreds, they're not doing that to punish western foreign policy. When ISIS set about the **massacre of Yazidi men** and the enslavement and mass rape of **Yazidi women** and girls, it wasn't revenge for western meddling in the Middle East. It takes an oddly **Eurocentric view** of the world to decide that this is a phenomenon entirely of the west's creation. The point is, this is an ideology that can rage against western inaction as much as action. When I spoke to Shiraz Maher, a senior research fellow at King's College London who studies radicalization up close, he put the problem concisely: "You're damned if you do, and damned if you don't." Maher suggests that western foreign policy often plays the role of a hook on which jihadis can hang a much larger set of ideological, and theological, motives. In his latest essay for the New Statesman, he quotes one British ISIS recruit he interviewed, who told him: "We primarily fight wars due to people being disbelievers. Their drones against us are a secondary issue". So it's not clear what a foreign policy designed to soothe rather than inflame jihadi opinion would look like – or that it would get you very far…Not against those who can regard an eight-year-old girl and her friends as "crusaders", worthy of death for the sin of dancing in a "shameless concert arena". Maybe it would be easier to bear if our fate was entirely in our hands, if a life of peace and calm beckoned if only we chose the right path. It would be a comfort, but a false one – for it would misunderstand the enemy we face.

So, how did people that engaged in conspiracy theories about **Covid-19** try to gain a sense of control? Doing exactly the same as we have seen above: assuming that the "purposeful agent" is from the ingroup, that is us humans, because it is hugely discomforting to recognize that an outgroup virus can come from nowhere and kill us. Just think about basically all the Covid-19 conspiracy theories you have heard of: the purposeful agents are humans—either Bill Gates, or the ones that did 5G, or the ones that want to kill old people, or the Chinese that produced the virus, and so on (see Fig. 2.11). This is exactly what we did with other viruses and bacteria involved in previous pandemics. During the **Black Plague** many conspiracy theories defended that the plague was created, or at least purposefully spread, by humans, such as the Jews, as will be explained below. When conspiracy theories don't involve death so directly, then yes, this rule-of-thumb can be broken more often, and for instance, people from a certain ingroup, such as Westerners, can say that people from the outgroup, such as Muslims, are the active purposeful players in the sense that they, for instance, conspire to come and populate the Western countries in order to take full control of them in the future. Such racist conspiracy theories are indeed very common, as can be seen in Fig. 2.11. This leads us to briefly talk about **politics** here, as political narratives and *Neverland* ideologies are hugely important for our daily lives and are very often related to beliefs, including numerous conspiracy theories, as pointed out in Shermer's *The Believing Brain*:

> Are you a political liberal or a conservative? If you are a liberal, I predict that you read the ***New York Times***, listen to progressive talk radio, watch **CNN**, hate **George W. Bush** and loathe **Sarah Palin**, adore **Al Gore** and revere **Barack Obama**, are pro-choice, anti-gun, adhere to the separation of church and state, are in favor of universal health care, vote for measures to redistribute wealth and tax the rich in order to level the playing field, and believe that global warming is real, human caused, and potentially disastrous for civilization if the government doesn't do something dramatic and soon. If you are a conservative, I predict that you read the **Wall Street Journal**, listen to conservative talk radio, watch ***FOX News***, love George W. Bush and venerate Sarah Palin, despise Al Gore and abhor Barack Obama, are pro-life, **anti-gun control**, believe that America is a Christian nation that should meld church and state, are against universal health care, vote against measures to redistribute wealth and tax the rich, and are skeptical of global warming and/or government schemes to dramatically alter our economy in order to save civilization. Although this cluster of specific predictions may not be a perfect match for any one person's positions, the fact that most Americans do fall into one of these two sets of attitudes indicates that even political, economic, and ***social beliefs*** form distinct patterns that we can identify and assess.

Some numbers given by Shermer are particularly fascinating, although not really surprising in face of what we have seen about *Neverland*, so far—one should however point out that some of Shermer's notes sometimes mix factual numbers with some of his own personal political opinions, so one should take them with a grain of salt:

> According to the National Opinion Research Center's General Social Surveys, 1972–2004, 44 percent of people who reported being "**conservative**" or "very conservative" said they were "very happy" *versus* only 25 percent of people who reported being "liberal" or "very liberal". A 2007 Gallup Poll found that 58 percent of **Republicans** *versus* only 38 percent of **Democrats** said that their mental health is "excellent". One reason may be that conservatives are so much more generous than liberals, giving 30 percent more money (even when controlled for income), donating more blood, and logging more volunteer hours. And it isn't because conservatives have more expendable income. The working poor give a substantially higher percentage of their incomes to charity than any other income group, and three times more than those on public assistance of comparable income. One explanation for these findings is that conservatives believe **charity** should be private (through nonprofit organizations) whereas liberals believe charity should be public (through government). Here we see a pattern of political party preferences grounded in different moral foundations, which we will explore below. One reason that liberals characterize conservatives in this manner may be the **liberal bias** of academic social scientists. To wit, a 2005 study by George Mason University economist Daniel Klein using voter registrations found that Democrats outnumbered Republicans by a staggering ratio of 10 to 1 among the faculty at the University of California-Berkeley and by 7.6 to 1 among the faculty at Stanford University. In the humanities and social sciences, the ratio was 16 to 1 at both campuses (30 to 1 among assistant and associate professors). In some departments, such as anthropology and journalism, there wasn't a single Republican to be found.
>
> The ratio for all departments in all colleges and universities throughout the United States, said Klein, is 8 to 1 Democrats over Republicans. Smith College political scientist Stanley Rothman and his colleagues found a similar bias in a 2005 national study: only 15 percent of professors describe themselves as conservative, compared to 72 percent who said they were liberal (80 percent in humanities and social sciences). A more nuanced nation-

wide study conducted in 2001 by UCLA's Higher Education Research Institute found that 5.3 percent of faculty members were far left, 42.3 percent were liberal, 34.3 percent were middle of the road, 17.7 percent were conservative, and 0.3 percent were far right. Comparing the extremes in this sample, there are seventeen times more far left liberals than far right conservatives. The bias appears even in law schools, where one would hope for a more balanced education in our future lawmakers. In 2005, Northwestern law professor John McGinnis surveyed the faculties of the top twenty-one law schools rated by U.S. News and World Report and found that politically active professors overwhelmingly tend to be Democrat, with 81 percent contributing "wholly or predominantly" to Democratic campaigns while just 15 percent did the same for Republicans. The liberal slant also appears to dominate many forms of the media. A 2005 study by UCLA political scientist Tim Groseclose and University of Missouri economist Jeffrey Milyo measured media bias by counting the times that a particular media outlet cited various think tanks and policy groups, and then compared this with the number of times that members of Congress cited the same groups. "Our results show a strong liberal bias: all of the news outlets we examine, except ***Fox News' Special Report*** and the ***Washington Times***, received scores to the left of the average member of Congress". Predictably, the ***CBS Evening News*** and the ***New York Times*** "received scores far to the left of center." The three most politically neutral media outlets were ***PBS's NewsHour***, ***CNN's NewsNight***, and ***ABC's Good Morning America***. Interestingly, the most politically centrist of all news sources was ***USA Today.***

These excerpts provide some interesting data per se, but the way they are discussed is not only a bit too personal but also too simplistic in general. This is because a left-wing person could easily argue that if let's say the number of liberals and conservatives is about the same in the U.S., the fact that most professions that require higher intellectual skills, such as being university professors, are dominated by liberals simply indicates that liberals in general have higher intellectual aptitudes, or at least are more prone to study and to further develop those aptitudes. They could moreover use what has happened in the last years in the U.S. since Trump's election as a further argument to support such a thesis: many conservatives are explicitly creating, using and/or following markedly **anti-intellectual narratives** and **anti-scientist narratives**, in which professors, journalists, researchers, and so on are seen as the "enemy." Of course, such arguments would also be too simplistic, because the fact that African-Americans and women were historically less represented in professions that require higher intellectual skills such as university professors had nothing to do with 'white' men having in general higher intellectual aptitudes. A huge number of factors can overcome pure **meritocracy**, such as the power of subjugation, oppression, negligence, and **social networks**—for instance, if people who have the power within, let's say, university communities, are white men, or liberals, they would tend to choose people like them to join their universities, and so on. And perhaps this is the most interesting part: the very same data can be used by different groups in completely different ways, and this is precisely the subject of the more appealing, and profound, part of Shermer's discussion on these subjects. This is because he refers to the third main component that defines our species: apart from the *Homo fictus* ability to create imaginary stories and the *Homo irrationalis* tendency to believe in them, we have a strikingly strong tribal *Homo socialis* nature, which is aggravated by our confirmation biases (see Box 7.2) and by its use and manipulation by political and religious leaders, or the 1%, to gain political, social, and/or economic advantages (see Figs. 4.4 and 9.5):

In their book *Partisan hearts and minds*, political scientists Donald Green, Bradley Palmquist, and Eric Schickler demonstrated that most people do not select a political party because it reflects their views; instead, they first *identify* with a political position, usually inherited from their parents, peer groups, or upbringing. Once they have made a commitment to that political position they choose the appropriate party and then follow the dictates of it. This is the power of **political belief**, and it shows in the very tribal nature of modern politics and the stereotypes of each tribe. In fact, research now overwhelmingly demonstrates that most of our moral decisions are grounded in automatic **moral feelings** rather than deliberatively **rational calculations**. We do not reason our way to a moral decision by carefully weighing the evidence for and against; instead, we make intuitive leaps to moral decisions and then rationalize the snap decision *after the fact* with rational reasons. Our moral intuitions – reflected in such **conservative-liberal stereotypes** – are more emotional than rational.

As with most of our beliefs about most things in life, our moral beliefs come first; the rationalization of those moral beliefs comes second. Over the years Haidt and his University of Virginia colleague Jesse Graham have surveyed the moral opinions of more than 118,000 people from over a dozen different countries and regions around the world, and they have found this consistent difference between **liberals** and **conservatives**: liberals are higher than conservatives on *harm, care*, and *fairness/reciprocity*, but lower than conservatives on *in-group/loyalty*, *authority/respect*, and *purity/sanctity*. Conservatives are roughly equal on all five dimensions: lower than liberals on the first and second but higher on the other three. The research by behavior geneticists on **identical twins** separated at birth and raised in different environments that found that about 40 percent of the variance in their religious attitudes was accounted for by their genes. These same studies also showed that about 40 percent of the variance in their political attitudes is due to inheritance. In his book *A conflict of visions*, economist Thomas Sowell argued that these two clusters of moral values are intimately linked to the vision one holds about human nature, either as constrained (conservative) or unconstrained (liberal). He called these the *constrained vision* and the *unconstrained vision*. Sowell showed that controversies over a number of seemingly unrelated social issues such as taxes, welfare, Social Security, health care, criminal justice, and war repeatedly reveal a consistent ideological dividing line along these two conflicting visions. Which of these natures you believe is true will largely shape which solutions to social ills you think will be most effective.

Box 7.2 Homo Irrationalis, Confirmation Bias, Prejudices, and Politics

Shermer's *The believing brain* provides illustrative examples of our **confirmation bias** based on data obtained from psychological and neurobiological studies:

> Confirmation bias…is the mother of all the **cognitive biases**, giving birth in one form or another to most of the other heuristics…no matter what the issue is under discussion (republicans and democrats), both sides are equally convinced that the evidence overwhelmingly supports their position. I'm sure it does because of the confirmation bias, or *the tendency to seek and find confirmatory evidence in support of already existing beliefs and ignore or reinterpret disconfirming evidence*. The confirmation bias is best captured in the biblical wisdom *Seek and ye shall find.* Experimental examples abound. In 1981, psychologist Mark Snyder tasked subjects

to assess the personality of someone whom they were about to meet, but only after they reviewed a profile of the person. Subjects in one group were given a profile of an introvert (shy, timid, quiet), while subjects in another group were given a profile of an extrovert (sociable, talkative, outgoing). When asked to make a personality assessment, those subjects who were told that the person would be an extrovert tended to ask questions that would lead to that conclusion; the introvert group did the same in the opposite direction. In a 1983 study, psychologists John Darley and Paget Gross showed subjects a video of a child taking a test. One group was told that the child was from a high socioeconomic class while the other group was told that the child was from a low socioeconomic class.

The subjects were then asked to evaluate the academic abilities of the child based on the results of the test. Even though both groups of subjects were evaluating the exact same set of numbers, those who were told that the child was from a **high socioeconomic class** rated the child's abilities as above grade level, and those who thought that the child was from a **low socioeconomic class** rated the child as below grade level in ability. This is a striking indictment of human reason but a testimony to the power of **belief expectations**. During the run-up to the 2004 presidential election, while undergoing a brain scan, thirty men – half self-described "strong" Republicans and half "strong" Democrats – were tasked with assessing statements by both George W. Bush and John Kerry in which the candidates clearly contradicted themselves. Not surprisingly, in their assessments of the candidates, Republican subjects were as critical of Kerry as Democratic subjects were of Bush, yet both let their own preferred candidate off the evaluative hook. Of course. But what was especially revealing were the neuroimaging results: the part of the brain most associated with **reasoning** – the ***dorsolateral prefrontal cortex*** – was quiescent. Most active were the ***orbital frontal cortex***, which is involved in the processing of **emotions**, and the ***anterior cingulate cortex*** – which is so active in **patternicity processing** and **conflict resolution**. Interestingly, once subjects had arrived at a conclusion that made them emotionally comfortable, their *ventral striatum* – a part of the brain associated with reward – became active. In other words, instead of rationally evaluating a candidate's positions on this or that issue, or analyzing the planks of each candidate's platform, we have an emotional reaction to conflicting data. We rationalize away the parts that do not fit our preconceived beliefs about a candidate, then receive a reward in the form of a neurochemical hit, probably **dopamine**.

7.3 Witches, Sexuality, Magic, and QAnon Conspiracies

Aside from traveling and teaching, there is hardly anything that Jesus did more often than cast out demons and heal the afflicted…to Christians he is the Son of God and the Lord and Savior of humankind; by occupation, he was primarily an exorcist and healer. (Jeff Levin)

The history of the so-called witches and the huge atrocities done against them (see Fig. 7.1) is deeply related to teleological narratives and related historical notions about human sexuality, morality, and religion, which are in turn profoundly associated with widespread misogynistic tales (see Box 7.3).

Fig. 7.1 Typical witch-burning in the sixteenth century, in this case of Elsa Plainacher at Vienna in 1583

Box 7.3 Teleology, Christianity, Misogyny, and Witches

Holland's book *A Brief History of Misogyny* provides an interesting introduction to the links between **Christianity**, **misogyny**, and the atrocities done to so-called 'witches', including **witch-burning** (see **Fig. 7.1**):

> At least three conditions conspired to create the emotional, moral and social context for the **witch-hunts**. First, the fourteenth century, which ushered them in, was, like the fifth century BC in Greece and the third century AD in the Roman Empire, a period of terrible calamities. **Plague** and war threatened to unhinge society. Fear and doubt caused people to view the world in a darker and more sinister light. Secondly, heretics real and imagined threatened a once seemingly all-powerful institution, the **Church**, and its claims to embody the absolute truth. Finally, **Christian society's deep-seated misogyny** provided the needed scapegoat in the form of woman. Just as centuries of **Christian anti-Semitism** provided the ideological grounds for the Nazi holocaust, so the long tradition of contempt for and dehumanization of women made the witch-hunts possible...The late Medieval mood of pessimism, mixed with doubt and fear, expressed itself in a way that would have a direct impact on the fate of women: the growth in interest in **demons**, a need to prove that they were real, and therefore that the **Devil** and his demons were abroad in the world. As the historian Walter Stephens summed it up, 'without proof of a devil, there can be no proof of

God'. The most convincing proof of the reality of demons would be their ability to interact with human beings. There is no more powerful and corporal form of interaction than sex. But to have sex, demons needed bodies. Many learned monks bent over ancient texts in bare cells burned the midnight oil pondering the corporality of demons; the great authorities **St Augustine** and **St Thomas Aquinas** were invoked for those who were in favour of **devilish embodiment**. Augustine had pointed to the pagan gods, who he believed were demons, and their fondness for raping and impregnating women as proof they could interact with humans. St Thomas Aquinas believed demons were the supreme, supernatural gender-benders. They could appear as females – **succubi** – and go about extracting semen from men. Then they would transform themselves into **male demons** or **incubi**, and impregnate women.

By the fourteenth century the arguments for the reality of demons had won crucial support at the highest levels of the Church. **Pope John XXII** (1316–1334) was obsessed with **witchcraft** and **heresy**; and he was a true believer in demons. It was during his long reign that for the first time in history a woman was accused of having sex with the Devil. In 1324, **Lady Alice Kyteler** of Kilkenny in Ireland earned that dubious distinction…under torture her maid Petronilla told the bishop how she acted as a go-between for the Devil and her mistress…when the Devil as lover first appears in history he does so in the form of three big, handsome black men…Petronilla said she saw with her own Lady Alice making love with them, sometimes in broad daylight. The **Inquisitors** have a simple explanation for why it is that nearly all witches are women: 'all witchcraft comes from **carnal lust**, which is in women insatiable', they write…'There are three things that are never satisfied, yea, a fourth thing which says not, it is enough; that is, the mouth of the womb…Wherefore for the sake of fulfilling their lust they consort even with the Devil'. They allege other faults in women that make them vulnerable to temptation, of course, including vanity, feeble-mindedness, talkativeness and credulity. But in the minds of the Inquisitors, women's greater carnality is the primary cause for witchcraft. Since presumably this fault identified as particular to women is not new, it might be asked why there are almost no reports of women copulating with the Devil before 1400, when the Church decreed making love to demons a capital crime? [Often, in the **Middle Ages**] the accused was imprisoned before being brought to trial, and while awaiting judgment, often for considerable periods of time, fed on a diet of bread and water. **Torture** was employed to extract confessions, and there was no appealing the sentence. The Inquisitor was prosecutor, judge and jury. Technically, the Church did not actually carry out the sentence of death, since it is forbidden to take life – it merely 'relaxed' its protection of the accused (if convicted). The victim was handed over to the civil authorities, who administered the punishment. The civil authorities, of course, could be certain to concur with the Inquisitor's findings.

Henry Kramer and James Sprenger sum up the Church's role in a chilling phrase when they speak of 'those whom we have caused to be burned'. The accused may be kept in a state of suspense by 'continually postponing the day of examination', the Inquisitors advise. If this does not make her confess 'let her first be led to the penal cells and there stripped by honest women of good character', in case she is concealing some instruments of witchcraft made 'from the limbs of unbaptized children'. It is then a good idea to shave or burn off all her hair, except in Germany, where shaving 'especially of the secret parts…is not generally considered delicate…and therefore we Inquisitors do not use it…' They are not so squeamish in other countries where 'the Inquisitors order the witch to be shaved all over her body'. In Northern Italy, the Malleus reports: 'the **Inquisitor of Como** has informed us that last year, that is, in 1485, he ordered forty-one witches to be burned after they had been shaved all over'. The unmistakable relish Kramer and Sprenger derive from stressing this detail betrays the underlying sadism. If the squalor of the prison and the humiliation of stripping and shaving, never mind the mounting terror as she awaits the coming torture, do not break her, the judge should 'order the officers to bind her with cords, and apply her to some engine of torture; and then let them obey at once but not joyfully, rather appearing to be disturbed by their duty'. Usually, the first instrument of torture applied was the strappado. Her hands are tied beneath her back. She is roped to a pulley and then yanked violently into the air, where she is jerked up and down until her shoulders are dislocated and her sinews torn. 'And while she is raised above the ground', the Inquisitors write with the detachment of civil servants, 'if she is being tortured in this way, let the Judge read or cause to be read to her the dispositions of the witnesses with their names, saying: 'See! You are convicted by the witnesses'.' If she is still obstinate, other tortures can be used. She might be burned with candles or with hot oil. Flaming balls of pitch might be applied to her genitals or gallons of water forced down her throat until she is bloated and the officers then beat her belly with sticks. She can be forced to sit on the witch's chair – a sort of narrow cage with clamps and a spiked seat. Thumbscrews, and other devices for crushing the legs and feet might be used. Some victims were held in irons so long in filthy conditions that they died of gangrene before coming to trial.

Between 1628 and 1631…in Germany, children as young as three and four were accused of **having sex with devils**. Children who had been convicted of attending the witch's Sabbat with their parents were flogged in front of the stake as their mother and father burned. Jean Bodin, the author of the 1580 treatise *De la démonomanie des sorciers*, writes 'children guilty of witchcraft, if convicted, are not to be spared, though, in consideration of their tender age, they may, if penitent, be strangled before being burnt'. Girls above the age of twelve, and boys over fourteen, were treated as adults…The **English Puritans** brought the fear of witchcraft with them to the New World. They brought with them too something of the Old World misogyny that was its inspiration. But the witch craze never caught on with the same ferocity in the colonies as it did in Europe. There were only two intense outbreaks – the first in Hartford, Connecticut (1662-1663) and the second and more infamous, in **Salem, Massachusetts**, for a few months beginning in December in 1691. In Hartford, thirteen were accused and four hanged, and in Salem, two hundred were accused and nineteen hanged. As in Europe, four-fifths of the victims were women; a half of the males who were accused were husbands or sons of witches. The conviction rate was far lower than that of the **European witch-hunts**. A more democratic system of justice prevailed, allowing those convicted to appeal to higher courts; the outbreaks endured for a far shorter period. The majority of the cases concerned **acts of possession**. There was only one instance, in 1651, of a woman accused of **going to bed with the Devil**, and she only did so when he appeared to her in the form of her lost child. At an official level, skepticism prevailed very rapidly. Within a generation of the Salem trials, a man and wife who accused one Sarah Spenser of **witchcraft** were sent to see a doctor in order to establish whether or not they were sane. Undoubtedly, one of the reasons that the persecution of witches in North America lasted for so short a time was the fact that Old World misogyny did not enjoy a completely successful transplant to the New World. The Puritan tradition shared something of the early Christians' belief in equality before the Lord. Women enjoyed a higher status in the colonies.

A book that provided an updated, exceptionally comprehensive review of the history of the fear and persecution of witches is Hutton's 2017 *The Witch – A History of Fear from Ancient Times to the Present*:

> There is little doubt that in every inhabited continent of the world, the majority of recorded human societies have believed in, and feared, an ability by some individuals to cause misfortune and injury to others by non-physical and uncanny ('magical') means: this has been the single most striking lesson of anthropological fieldwork and the writing of extra-European history. Speaking from anthropology, Peter Geschiere proposed that 'notions, now translated throughout Africa as "**witchcraft**", reflect a struggle with problems common to all human societies'. What is valuable about these insights is that they testify to the general truth that human beings traditionally have great trouble in coping with the concept of random chance. People tend on the whole to want to assign occurrences of remarkable good or bad luck to **agency**, either human or superhuman. It is important to emphasize, however, that malevolent humans have been only one kind of agent to whom such causation has been attributed: the others include **deities**, **non-human spirits** that inhabit the terrestrial world, or the **spirits of dead human ancestors**. All of these, if offended by the actions of individual people, or if inherently hostile to the human race, could inflict death, sickness or other serious misfortunes. Wherever they appear, these alternative beliefs either limit or exclude a tendency to attribute suffering to witchcraft.
>
> In addition, many societies have believed that certain humans have the power to blight others without intention to do so, and often without knowledge of having done so. This is achieved by unwittingly investing a form of words or a look with destructive power: in the case of malign sight, this trait has become generally known to English-speakers as 'the **evil eye**'. Belief in it tends to have a dampening effect on a fear of witches wherever it is found, which is mainly in most of the Middle East and North Africa, from Morocco to Iran, with outliers in parts of Europe and India. This is because it is thought to be part of the possessing person's organic constitution. Among a single people, the intensity with which witchcraft was feared could vary according to the kind of settlement in which people lived. The **Maya of the Yucatan Peninsula of Mexico** all hated witches equally in theory during the early twentieth century, but those in villages were rarely inclined to suspect anybody of being one, while the tension was much greater in towns: in the district capital of Dzitas during the 1930s, 10 per cent of the adult population were thought to have been either perpetrators or victims of witchcraft. Wim van Binsbergen, commenting on the complexities of belief in magic among Africans in 2001, could still conclude with regard to witchcraft that 'the amazing point is not so much variation across the African continent, but convergence'. Adam Ashforth, considering attitudes to destructive magic and its alleged perpetrators in the modern **Soweto township** near Johannesburg, decided that he had to use the terms 'witchcraft' and 'witch' because 'there is no avoiding them'.
>
> Gender is another worldwide variable, witches being, at different places within each continent, viewed as essentially female, or essentially male, or of both sexes in different proportions and according to different roles. It is fairly common, also, for societies to manifest a discrepancy between the gender of their **stereotypical witch** and that of the people whom they actually accuse. Those making the accusations are, likewise, normally female or male or both, according to the conventions of the culture to which they belong. In general, the comment made by Philip Mayer on Africans half a century ago holds good for human societies in general, that suspected witches and their accusers are people who ought to like one another but do not. To put it another way, as Eytan Bercovitch did after working in New Guinea, 'the witch is everything that people truly are as communities and individuals but would rather not be.' Suspicion of witchcraft has generally been one consequence of unmet social obligations. The circumstances under which that suspicion arises tend everywhere to be those of regular, close and informal relationships, especially those in confined and intense environments where it is difficult to express animosities in open quarrelling and

fighting: which is why, for example, in southern India accusations were never made between different social castes, as they never had intimate enough relations with each other. Although the consequences of allegations of witchcraft generally involved social groups, in essence they were generated by close personal relationships. In Godfrey Lienhardt's words, 'witchcraft is a concept in the assessment of relations between two people'. A belief in it is an aspect of face-to-face human encounters.

In his book, Hutton refers to the general belief that "*the witch can be resisted*". This is a major characteristic of most teleological narratives, from the belief in Gods to conspiracy theories, from stories about omens to tales about witches: to avoid the discomfort caused by randomness by trying to gain a *false feeling of control.* Earthquakes can't simply kill humans accidently, for no "reason": they are made instead by Gods or provoked by demons, sorcerers, or witches. So in *Neverland* we *can* pray to those Gods or give them human flesh as sacrifices, or resist the demons, or burn the witches, and therefore *gain* control over such events. The problem is that in the *world of reality* not only we can't have control over all those natural events, but moreover in many if not most of those cases, in order to try to gain control over something "bad" and random, the storyteller irrational animal has undertaken non-random atrocities to other humans and to other animals—often following directions from religious texts or even from written laws, such as it was done to sacrifice and burn and endless number of innocent people—that were far worse than the random events they were trying to avoid.

The fictional irrational stories created by the tribal self-entitled "sapient being" are indeed unique, not in the way they are often portrayed in *Neverland* but in what concerns their absurdity and the agonizing cruel way in which they are applied in the world of reality. Indeed, replacing random "bad" things with often much worse non-random "bad" things, that is, in a nutshell, one of the most profound legacies produced by our anxiety about our inevitable death. When we compare this point with what many scientists are doing now, in the world of reality, for instance, creating vaccines to prevent or reduce the harsh symptoms of infectious diseases provoked by bacteria or viruses such as **Covid-19** instead of creating imaginary political, racist, or misogynistic narratives to give us a fake sense of control over those infectious diseases, the difference is huge: random "bad" things replaced by non-random "good" things. Of course, this is a simplification, not only because some medical items, such as **thalidomide**, and in particular medical experimentation, have done very "bad" things as well (see Box 5.10), but also because science has also not been immune to teleological narratives, being actually plagued by it, particularly in the past. However, while this distinction is not so "black and white," and unfortunately *there was and still is*, a lot of gray in between, the *reality* is that a huge number of scientists are starting to slowly escape from *Neverland* narratives and indeed focusing more and more on creating ways to directly solve or reduce "bad" things instead of creating *Neverland* tales to give us a false sense of control over them. And this is indeed one of the many reasons for optimism concerning the possibility that despite our *Homo fictus, irrationalis et socialis* tendencies we might be able to escape from *Neverland* before it is too late, as will be discussed in Chap. 9.

Hutton gives the following examples concerning narratives about "resisting the witch:

> The belief that witches can be resisted by their fellow humans is also found worldwide, in the three main forms which it took in Europe. One of these was to protect oneself or one's dependants and property by using **benevolent magic**, which could turn away spells and curses; if the latter seemed to take effect, then stronger magic could be employed to break and remove the effects of **bewitchment**; and perhaps to make the witch suffer in turn. The second widespread remedy for bewitchment was to adjust the social relations that had created the suspicion of it. This could take the form of persuading or forcing the witch into removing the spell that she or he had placed, and so its destructive effects. Among the **Azande**, when a service magician or chief had decided that a malady was the result of bewitchment, then the next step would be to ask the alleged culprit to lift the spell. The third remedy was to break the power of the witch with a physical counter-attack, which could take the form of direct action, such as a severe **beating** or **murder**, or intimidation that ran the person concerned out of the neighborhood. In most societies, however, a formal and legal remedy was preferred to this sort of private action, by which the suspect was prosecuted before or by the whole community, and if found guilty was subjected to such punishment as it appointed. In many cases the identification of the culprit was assisted or carried out by the same kind of **magician** as that which provided counter-magic against **witchcraft**. Across much of the world, oracles and special rites were employed to find the guilty party when witchcraft was suspected. Once under suspicion, people were commonly forced to undergo an ordeal to demonstrate innocence or guilt.
>
> The traditional witch-finding society among the **Nupe of northern Nigeria** forced suspects to dig the ground with bare hands: if they bled, they were deemed to be guilty. The **Dowayo** would make them drink beer in which a poisonous sap had been mixed. A person who died or produced red vomit as a result was deemed guilty, while those who produced white vomit, and lived, were exonerated. Different forms of this poison ordeal were found across Central Africa, from Nigeria to Zambia and Madagascar, and its consequences depended on how toxic a potion was made. The **Lele** herded suspects into pens for testing, and the drink administered killed many of them. The same test was used in north-western New Guinea, where those who vomited the poison were declared guilty and put to death: as it was quite difficult to survive the poison without bringing it up, this was an ordeal heavily weighted against the person submitted to it. In Africa from Ghana to the islands off the Tanzanian coast, a chicken had its throat cut or was given poison in front of a suspect, whose guilt or innocence was determined by the final posture of the dying bird. The danger in which the accused was placed could be manipulated by deciding how many such postures counted as proof of innocence: in much of Nigeria during the 1940s and 1950s, the odds were heavily weighted against acquittal by the ruling that only one position did. A standard test for witchcraft on Flores, in the southern Indonesian island chain, was to have to pick a stone out of boiling water: the guilty would blister.
>
> Once a person was identified as a probable witch, **torture** was sometimes used to extract a confession: in India the **Dangs** commonly swung an accused person upside down over a fire. Across much of the rest of India and in Burma suspects were flogged with wood from a sacred tree. The **Navaho of the south-western United States** preferred to tie them up and starve them of food and shelter. How severe a penalty was imposed on those convicted of witchcraft depended both on local attitudes to it and the perceived extent of the damage done by the presumed witch. To societies that prescribed the death penalty for murder or other serious crimes against the person, it was logical to apply it to people convicted of inflicting death or ruinous damage by means of magic. Most peoples who have traditionally believed in witchcraft have killed at least some of those formally convicted of it. In communities that greatly feared witchcraft, the body counts achieved could be considerable. It was said that in pre-colonial days every village of the **Bakweri of Cameroon** had its witch-hanging tree. Among the **Pondo of South Africa**, the rate of execution ran at one per day

on the eve of the British conquest and this number did not include those who fled when accused, or were fined. A British official serving in India during the early nineteenth century estimated that about a thousand women had been put to death for alleged witchcraft on the northern plains during the previous thirty years: a rate of mortality far more serious than that caused by the more notorious local practice of sati, or widow-burning.

Box 7.4 Witches, Beliefs, "Traditional Societies," Colonialism, Economic Pressures, and Human Atrocities

In his 2017 book *The Witch*, Hutton explains:

> The rupturing of British rule over India in the rebellion of 1857 permitted a great witch-hunt, with lethal effects, to occur among the tribes of northern India. Before British colonialism arrived, the **Nyoro** allegedly burned many of their people alive as witches, while before the Germans conquered them, the **Kaguru** clubbed to death those convicted of witchcraft and left them to rot in the bush, and the **Pogoro** burned them alive. The **Greenland Inuit** cut the bodies of those executed into small pieces to prevent their spirits from haunting the living. Likewise, the **Northern Paiute** of what became Nevada and Oregon stoned convicted suspects to death and then burned the corpses. A Jesuit missionary working among the **Huron of Canada** in 1635 noted that they often murdered each other or burned each other alive on the testimony of dying men who accused the victims of having caused their fatal illness by magic. On Flores, the penalty for witchcraft before the Dutch conquest was to be buried alive, and this apparently occurred regularly. On another Indonesian island, Sulawesi, the **Toraja** people submitted accused witches to ordeals that allowed virtually no proof of innocence, and then beat them to death. Young boys were encouraged to participate in this to prove their courage. Before being ruled by the British, the tribes of what is now Botswana avenged deaths by presumed witchcraft either by allowing the bereaved relatives to kill the family of the suspected witch or by having the local chief try the suspects and execute those convicted: there were twenty-six such trials among the **BaNgwatetse** alone between 1910 and 1916. The former execution places of witches were still pointed out to British visitors to the region in the 1940s. The **Kaska**, who lived on the border between Canada and Alaska, had no concept of magical cures that could be used against witchcraft, and so the only known remedy was to deal with the witch, who in that society was usually thought of as a child. This belief led to persistent killings in the first two decades of the twentieth century, often by the families of the youngsters accused.
>
> Across the world, traditional peoples have often manifested the pattern of sudden upsurges in witch-hunting among populations hitherto or for a long time characterized by little of it. In general, people who have traditionally feared **witchcraft** tend to accuse neighbours of it much more frequently in times of economic pressure and/or of destabilizing economic, political and cultural change; but it is also true that such times do not automatically and necessarily produce an increase in accusations. When such an upsurge has occurred, it has tended to rebound on the social order in three different ways: to confirm the authority of the traditional leaders and society; to enhance the power of an individual member of the traditional elite; or to enable a new social group to seize authority. In Africa, Lobengula, king of the **Matabele**, Ranavalona, queen of the **Malagasy**, and Shaka, king of the **Zulus** are examples of nineteenth-century leaders who reinforced their **hereditary authority** by waging war on alleged witches. Shaka once summoned almost four hundred suspects to his court at once, and killed them all, while under Ranavalona about a tenth of her subjects were forced to endure the poison ordeal to test for witches, and a fifth of those

died. Lobengula presided over an average of nine to ten executions per month, mainly of relatively powerful men. In nineteenth-century North America the **Navaho** chief Manuelito executed more than forty of his political opponents on charges of witchcraft, and a generation earlier the **Seneca** chief Handsome Lake established himself as a religious leader by directing a persecution of it. Such figures sometimes used witch-hunting to defend traditional ways against innovation: in the eighteenth-century Ohio Valley the **Shawnee** prophet Tenskwatawa instigated it against Christian converts in his tribal confederacy. Political use of the mechanism could be deployed collectively as well as by particular rulers and prophets: thus, in the seventeenth century the north-eastern **Algonquian tribes of North America** made witchcraft accusations their main means to establish new territorial boundaries to service the fur trade developing with European settlers. On the other hand, some strongly based and long-established regimes chose to discourage witch-hunts as part of the demonstration of their authority. When a panic swept twelve provinces of China in 1768, that itinerant magicians were cursing people (especially male children) to death in order to enslave their souls, the imperial judges quashed the convictions imposed by local courts, although mobs had murdered some suspects before they could be arrested.

In Africa witch-finding movements were common in the colonial period, affecting much of the western and central parts of the continent, and functioning partly as a response to the prohibition or extreme modification of traditional trials for witchcraft by the European administrations. It is also possible that colonial rule, by shattering tribal institutions and moral codes, increased the instability in which a fear of bewitchment often flourishes. The **Lele** were caught up in no less than five witch-hunts between 1910 and 1952. Typically, they were conducted by young men who toured regions, crossing tribal boundaries and claiming the power both to detect witches and to render them permanently harmless. The latter process usually took the form of forcing suspects produced by communities to deliver up the materials with which they were supposed to work their magic, for destruction, and administering a drink or ointment to them, or a particular rite, which was supposed to remove their ability to bewitch. Likewise, in western India, the 'Devi' religious revival of the 1920s included the detection and banishment of witches from villages as part of its remit. Such movements originated from outside traditional structures of authority and custom but generally worked within them. Even under colonial rule, however, witch-hunters sometimes emerged who provoked a rejection and punishment of the familiar native elites or religions. The **Atinga** witch-finding cult in West Africa was conveyed by devotees of a single shrine in northern Ghana, who destroyed other traditional cult centres as they travelled. The **Nyambua**, their equivalent in Nigeria, denounced established chiefdoms as well as witches. Sometimes, also, such movements blended with anti-colonial feeling, or even with outright rebellion: the Maji Maji uprising against German rule in Tanganyika in 1905-1906 was led by a prophet who termed himself a 'killer and hater' of witches, and indeed ordered the death of anybody who refused the 'medicine water' that he administered to destroy evil magic.

This pattern has become much more common since the removal of European rule, as Africa has undergone programmes of self-conscious modernization that have produced major social change. Witch-hunting has often been prominent both in revolutionary movements which directly opposed and helped to end colonialism or white supremacy, and in the successor states, under native regimes, which emerged out of the former colonies. The groups of young men who attacked suspected witches in parts of the Transvaal during the 1980s were also those who led resistance to the system of **apartheid**, portraying the white government which both upheld apartheid and forbade witch-hunting, as the protector of witchcraft. After the establishment of black majority rule, they still found themselves marginalized by the new regime, and so continued their role as local defenders of their people, in the face of a largely alien central government, with persecution of witches still part of that role. Closer to the main centres of population in the new South Africa, in the **Soweto township**, the daily fear of witchcraft was reported as 'tremendous' by the early 1990s, and it was said that 'every older woman, especially if eccentric and unpopular, lives with the risk of being accused of witchcraft'. Among the **Mijikenda of the Kenya coast**, independence was followed by an upsurge of accusations and of violence against suspects, with tribal and national administrative leaders uniting to promote a particular healer as a witch-finder. From the 1970s direct and public accusations of witchcraft increased in Zambia, and with them the use of expert witch-finders, who were ubiquitous in rural areas by the 1980s. In the war of independence, which established native rule in Zimbabwe, the guerrillas assumed the traditional role of chiefs as witch-detectors, usually with the full support of local communities, and put those detected to death if those communities desired it. Unsurprisingly, the victims were often allies of the white government. After independence had been achieved in the country, during the early 1990s, a local hunt was conducted by a spirit medium obtained from a government-sanctioned National Traditional Healers' Association, who detected witches by making suspects step over his walking stick. Both sides in the Angolan civil war of the early 1990s, which followed the collapse of Portuguese rule, put alleged witches to death as an aspect of their attempts to enhance their popularity and claims to legitimacy; one tended to burn them alive and the other to kill them after making them dig their own graves. Refugees expressed outrage at the abuse of the activity, by targeting political opponents (and their children) as witches, but not at their execution. In those parts of the world in which native people were ruled for a time by European powers, a feature of the persecution of alleged witches was the manner in which selected features of **Christianity** were borrowed from the colonial rulers and integrated with traditional concepts of the witch. This was a natural enough process in Latin America, where for more than two centuries the ruling Europeans themselves feared witchcraft and outlawed all kinds of magic. Two parallel systems of witch-hunting thereby met and blended, with the early modern **European stereotype of witchcraft** as a form of **Devil-worship** infiltrating indigenous ideas and taking up permanent residence among them.

A particularly interesting part of Hutton's book is focused on the links between the atrocities caused or justified by the belief in and fear of witches and colonialism. In particular, he shows that this is one of the very rare cases in which, in a way, something "positive" can truly be attributed to Western colonialism because there were cases in which Westerners made laws or created rules to prevent the persecution and killings of "witches" in the colonized regions (see also Box 7.4). However, there are also of course a huge number of cases in which such laws and rules actually then led to catastrophic events, for instance, due to the breaking of social traditions and norms that were precisely created in the first place by the indigenous communities to minimize the type of conflicts that are often associated with the creation of tales about witchcraft. Moreover, highlighting the enormous complexity of and inter-connection between these issues, in many cases such conflicts within the indigenous communities were exacerbated by another type of colonialism—which can be defined as "**internal African colonialism**." Contrary to racist narratives, Western countries are not the only ones that have purposeful agents that subjugate and oppress and colonize others: in each continents, including obviously also Africa, internal colonialism has been done by indigenous people much before the arrival of Europeans, in particular by agricultural groups such as the Mayas, Bantus, and so on.

That is, as societies that were living a mainly agricultural, hierarchical type of life arrived to new regions of Africa and begun to dominate or at least to increasingly exert influence over nomadic hunter-gatherers that were already living in those regions, there was often a collapse of the existing social traditions and norms of such hunter-gatherer groups. In fact, a crucial point explained by Hutton is that although small, nomadic hunter-gatherer groups do of course often believe in witchcraft, they normally do not undertake **witch-hunting** at a large scale; the local conflicts are often controlled, to a certain extent—obviously with exceptions—at a local level. However, with the rise of **sedentism** and particularly **agriculture** the resulting increased inequality, economic pressures, and political and cultural changes often led to sudden upsurges in witch-hunting. That is, once again, against the ideas of authors such as **Hobbes**: it is not necessarily those that they call the "egalitarian anarchic nomadic savages" that do most of the atrocities related to witchcraft. Those atrocities are instead often the result of events related with the rise of "civilizations" characterized by inequalities and oppression and all the numerous social conflicts created by them (see Box 7.4).

Hutton also provides a far-reaching historical overview of the belief in witches and witchcraft, from ancient **Mesopotamia** and Greece (Box 7.5) to the rise of Christianity and the Middle Ages (Box 7.6), as well as within what is often designated as "early modern history" (Box 7.7).

About the Romans, Hutton points out that:

> The pagan Romans, both in their republican and imperial periods, were heavily influenced by Greek culture [see **Box 7.5**], and it is not surprising to find them embracing the same distinction between religion and magic; though it may equally be argued that the distinction concerned must have appealed to their own attitudes for it to have taken root. In the first century of the imperial period, and Christian era, both the playwright and philosopher **Seneca** and the scholar **Pliny** condemned magic as a wish to give orders to deities. Things become clearer only in the second century AD, when the work of the magus in general became equated with ***veneficium*** and with ***maleficium***, meaning the intentional causing of harm to others. By the third century, **Roman law codes** were adapting to this change, extending the *Lex Cornelia* to cover the making of love potions, the enactment of rites to enchant, bind or restrain, the possession of books containing magical recipes, and the 'arts of magic' in general. To own such a book now meant death for the poor and exile for the rich (with loss of property), while to practise **magical rites** incurred the death penalty, with those who offered them for money being burned alive. As the possession of books and the provision of commercial services were activities that could readily be proved objectively, these were relatively easy laws to enforce. Two major problems attend any attempt to understand the actual status of magic in the Roman world. One is that very little information

survives on how these laws were actually enforced; the other, that it was perfectly possible to conduct witch-hunts without having any law against magic itself, if the victims were accused simply of committing murder by magical means. It was recorded, centuries later, that in 331 BC an epidemic hit Rome, with high mortality, and over 170 female citizens, two of them noblewomen, were put to death for causing it with *veneficium*. This may have meant straightforward potions, as the first suspects, having claimed to be healers instead, were made to drink their alleged medicine and died, triggering the mass arrests.

The years 184 to 180 BC were also a time of epidemic disease in Italy, and much bigger trials were held in provincial towns, claiming over two thousand victims in the first wave and over three thousand in the second. Again the charge was *veneficium*, and it is impossible to tell whether this meant poisoning in the straightforward sense, or killing by magical rites, or a mixture. If the second or third sense of the word was what counted, and the reports are accurate, then the republican Romans hunted witches on a scale unknown anywhere else in the ancient world, and at any other time in European history, as the body counts recorded – however imprecise – surpass anything in a single wave of early modern trials. In addition, however, there are characters that have no parallel in **Greek literature**: women who habitually work a powerful and evil magic, using disgusting materials and rites and invoking underworld and nocturnal deities and spirits, and human ghosts. They appear in the later first century BC and continue into the later centuries of the empire. Such is **Horace**'s **Canidia**, a hag who poisons food with her own breath and viper's blood, has 'books of incantations', and enacts rites with her accomplices to manufacture love potions or blight those who have offended her.

It may fairly be wondered whether any of this was intended to be taken as seriously at the time as early modern demonologists were later to take it. These are, after all, literary inventions appearing in genres equivalent to romantic fantasy, Gothic fiction, satire and comedy. A major element of preposterous exaggeration was plainly present. On the other hand, such images – of potent magic worked by evil women – would not have been chosen had they not resonated to some extent with the prejudices and preconceptions of the intended audience. Kimberley Stratton has plausibly linked their appearance to a concern with the perceived **sexual licence** and **luxury** of Roman women in the same period, combined with an ideal of **female chastity** as an indicator of social stability and order. The image of the witch, in her view, emerged as the antithesis of this idealized and politicized version of female behaviour. However persuasive an argument, this still needs once again to take into account the likelihood that the image flowered so rapidly and luxuriantly because it was planted in soil made fertile for it. After all, the Romans who produced and consumed it had a historical memory of having put to death almost two hundred women in their city, centuries before, for having deliberately produced a major epidemic that claimed huge numbers of lives, by using *veneficium*. According to medical realities, all of them would have been innocent of this offence, and so their society would already have needed to believe in the capacity and will of women to commit it. An unknown number of women, perhaps the majority, would have been among the thousands of victims of the mass trials for the same crime in the 180s BC. Rome therefore already had a sense of wicked women as agents of murder and social disruption who used hidden means. Likewise, though with much more muted consequences both in social reality and in literature, it must be significant that when the Greeks conceived of divine or semi-divine figures who used dangerous magic, like Circe and Medea, these were female. It seems that cultures which had defined magic as an illicit, disreputable and impious activity, and in which women were excluded from most political and social power, such as the Greek and Roman (and Hebrew and Mesopotamian), were inclined to bring the two together into a single stereotype of the menacing Other. In the Roman case, however, the results, both practical and literary, were the most dramatic.

The ancient Romans were one people who possessed such a thought system, and in doing so tapped into another well-scattered aspect of human belief, the tendency to associate witches with owls. This family of birds has, after all, five features that people often find sinister: nocturnal habits, silent movements, predation, a direct stare and an ability to turn the head completely round. In the Native American languages of the **Cherokee** and the **Menominee** the word for the owl and the witch is the same, and the belief that witches could take the form of owls was found from Peru to Alaska. Even more widespread is the idea that owls, or humans in their shape, were responsible for the ubiquitous human tragedy of the sudden, unexpected and mysterious sickening and deaths of babies and small children. It was found among many North American peoples, but also in Central and West Africa and Malaya. This was…a feature of Roman culture, but as one corner of a complex of ideas spanning the Near East and Mediterranean, which also allows us some opportunity to penetrate the thought world of pagan Germany. It may be argued from all the data above that when magic is the subject under scrutiny, the ancient European world can indeed be divided into different regions, with contrasting attitudes and traditions. The Egyptians made no distinction between religion and magic, did not distinguish demons as a class of supernatural being, and had no concept of the witch figure. The **Mesopotamians** feared both demons and witches, and the **Persians** combined this fear with a division of the cosmos into opposed good and evil powers, the **Hittites** introduced it into high political life, and the **Hebrews** blended it with a belief in a single, good, deity with a single permissible cult. The **Greeks** (or at least some of them) made a distinction between religion and magic, to the detriment of the latter and some of its practitioners, but do not seem to have had an idea of witchcraft. The **Romans** made the same distinction, and accompanied it with a vivid concept of both witchcraft and witches, which extended to a criminalization of many forms of magic. The **Germans** feared a mythical sect of night-flying cannibal witches, which projected into real life, and criminal prosecution, a much more widespread mythology – found as far as Mesopotamia – about nocturnal **demonesses**.

Box 7.5 Witches, Mesopotamia, Hebrews, and Ancient Greece

In *The Witch*, Hutton explains:

> It was obvious to many early modern Europeans that their ideas and images of witchcraft were at least partly inherited from antiquity. The text that was most fundamental to their culture, the **Bible**, was itself ancient, and the authors of the **demonological texts** which supported witch prosecution quoted lavishly from it, and also from the Church Fathers. They also, however, included passages from pagan Greek and Roman authors: one of the most famous of such witch-hunters' guides, the ***Malleus maleficarum***, cited five of those; Henri Boguet's *Discours des sorciers* also had references to five; and Martin del Rio's *Disquisitiones magicae* drew on a grand total of twenty-nine. Creative writers were just as disposed to use such sources. Sometimes this process was implicit: the most famous witches in the whole of early modern literature, those who deal with **Macbeth** in **William Shakespeare**'s play, were derived originally from the ancient mistresses of prediction, the Fates or Norns, and in parts the chant they use seems similar to one composed by the pagan Roman poet **Horace**. The peoples of **Sumeria**, **Babylonia** and **Assyria** also believed in witches, in the classic sense of human beings, concealed inside their own society, who worked magic to harm others because they were inherently evil and associated with the demons which were the object of so much fear. The repertoire of the *āshipu* included many rites for undoing the harm that such people had wrought, while law codes prescribed death for those convicted of working such harm. The concern of the rites

to avert witchcraft, however, was always to remove the affliction and not to detect the witch: indeed, the rituals themselves were supposed to bring about the death of the witch at whom they were aimed. In any case, most misfortune was blamed on **angry deities**, **ghosts** or (of course) **demons**. **Mesopotamians** also believed in the **evil eye** (and the evil mouth, tongue and sperm), and thought it to be destructive of both people and their livestock. The stereotypical witch mentioned in the sources is assumed to be female, which seems to match the generally low status of women in Mesopotamian society and make witchcraft an assumed weapon of the weak and marginalized. This suggestion is borne out by the other kinds of people associated with the practice of it: foreigners, actors, pedlars and low-grade magicians. In the few cases of actual prosecutions for witchcraft, which span the whole period of the various Babylonian and Assyrian monarchies, the accused were all women.

[A] variation on the norm was found on the south-eastern fringe of the Mesopotamian world, among the **Hebrews**, who developed in the course of the first millennium BC an exceptional emphasis on one of their own gods, **Yahweh**, as the single deity whom they were henceforth permitted to honour. Spiritual power was therefore concentrated in the hands of priests and other holy men associated with Yahweh's cult, and this had an impact on attitudes to **magic**. The Hebrew Bible applauds wonder-working prophets who serve Yahweh, above all **Elijah** and **Elisha**, even when they deploy their powers as expressions of personal vindictiveness. It invests the objects of Yahweh's cult, especially his altar and the **Ark of the Covenant**, with intrinsic power, sometimes lethal. **Joshua**'s army stages an elaborate rite to draw on Yahweh's power to bring down the walls of **Jericho**, and the god himself tells **Moses** to make a bronze serpent to protect his people from snakebite. The ***Mosaic Law*** includes a ceremony to determine the guilt of a woman accused of adultery by making her drink water mixed with sacred texts and dust from the *Tabernacle floor*. All these could be termed ceremonies of a kind usually associated with magic, co-opted into the service of the official cult. The attitudes to magic expressed in it are not altogether coherent, but tend as before to credit outstanding holy men of the religion, now members of the official priesthood called rabbis, with the ability to work apparent miracles. These acts are always applauded, presumably as permitted and empowered by the true God, though this sanction for them is only sometimes made explicit. By contrast, anonymous women or heretics are treated as the natural practitioners of witchcraft, *keshaphim*, and are usually portrayed as being defeated by rabbis. Witches are not portrayed as having special looks or belonging to a special breed: they are just ordinary Jews, usually female, who have chosen to work harmful magic. In general, the Jewish literature of late antiquity rarely used magic as a polemical label for the religious practices of opponents, and attributed misfortune far more to the anger of the deity or the malice of demons than to witchcraft. None the less, it retained the traditional Mesopotamian belief in the existence of witches, who were generally presumed to be women.

The oldest European society from which evidence exists for attitudes to magic, including witchcraft, is the ancient Greek, which still has a far shorter recorded history than those of Egypt and Mesopotamia, extending back to the seventh or eighth centuries BC. By the fourth century at the latest the Greeks had developed their own distinctive set of beliefs, different again from any held in the great Near Eastern civilizations. One aspect of it was a distinction between religion and magic, often to the detriment of the latter, which was fundamentally that articulated at the opening of the present book, and indeed subsequently held by most Europeans until recent times. It first appears in a medical tract concerned with *epilepsy*, *On the sacred dis-*

ease, which has been dated to the years around 400 BC and so may well push back the distinction concerned into the fifth century. This opposes the disreputable use of spells and medicines which seek to compel divine beings, 'as if the power of the divine is defeated and enslaved by human cleverness', to the legitimate actions of people who only supplicate for divine aid. Shortly afterwards, the great Athenian philosopher **Plato** repeated it, attacking those who promised 'to persuade the deities by bewitching them, as it were, with sacrifices, prayers and incantations'. It seems, therefore, that by the central part of the classical age of **Greek civilization**, intellectuals, at least, were confidently articulating a matched pair of definitions that would become an enduring part of European culture. It has become something approaching a consensus among experts that…hostility to magic appeared in Greece in the fifth century BC, as one response to a number of developments. One of these was war with the **Persians**, which caused Greeks to define themselves more clearly against foreigners, and eastern foreigners in particular. Certainly the term **magos**, which gave rise to 'magic', was in origin the name for one of the official Persian priesthood, serving the **Zoroastrian religion**. What seems to be missing from this composite picture is witchcraft. There is no sense in any archaic or classical Greek text of hidden enemies within society who work **destructive magic** under the inspiration of evil. Plato called for the death penalty for any kind of magician who offered to harm people in exchange for financial reward, while those who tried to coerce deities, for any reason, should be gaoled. His targets, however, were service magicians offering morally and religiously dubious services in addition to the usual, theoretically benevolent, kind.

Furthermore, the fact that he needed to make this prescription perhaps indicates that no such laws already existed in his home city of Athens. There is no clear record of any trial of a person for working destructive magic in the whole of ancient Athenian history. The wider picture is equally enigmatic. Matthew Dickie has gathered hints that magicians were arrested and punished in Greek cities from the history of **Herodotus**, the drama of **Euripides** and a dialogue of Plato. He also points out, however, that such people were almost never practitioners of magic pure and simple, but doubled in other roles, such as priests, oracles or healers, so that their offences would be hard to match conclusively to magic. It is of a piece with these patterns that there seem to be no clear representations of witches in archaic or classical Greek literature. Two characters in **mythology** bear some resemblance to them, as powerful female figures who work destructive magic: **Circe** and **Medea**. Circe uses a combination of a potion and a wand to turn men into animals, and Medea uses **pharmaka** for various magical ends, including murder. Neither, however, is human, Circe being explicitly a goddess, daughter of the sun and a sea nymph, while Medea is her niece, product of a union between Circe's brother and either another ocean nymph or the goddess of magic, Hecate herself. Nor are they unequivocally evil, Circe becoming the lover and helper of the hero Odysseus once he overcomes her with the aid of the god Hermes, and Medea assisting and marrying the hero Jason. Medea certainly murders to help her beloved, and then again in an orgy of vengeance when he casts her off; but the attitudes of the Greek texts towards her remain ambivalent, and (like Circe) she escapes retribution for her actions. Both were to be immensely influential figures in later European literature, as ultimate ancestresses of many of its magic-wielding females.

Box 7.6 Witches, Christianity, Middle Ages, and Misogyny

Hutton's *The Witch* also focuses on the rise of **Christianity** and the **European Middle Ages**:

The **Christian religion** which underpinned the early modern witch trials combined the whole range of ancient traditions which individually established parts of a context for witch-hunting: **Mesopotamian demonology**; **Persian cosmic dualism**; a **Graeco-Roman fear of magic** as intrinsically impious; **Roman images of the evil witch**; and the **Germanic concept of night-roaming cannibal women**. Comments by respected scholars have not been lacking, indeed, to credit the Christian faith with an inherent propensity to encourage the **persecution of magicians**. Valerie Flint has argued that its institutionalized and monopolistic traits made it automatically into a state religion that demanded tighter control of human dealings with spirits, most of which became **evil** by definition. Richard Kieckhefer has pointed out that Christianity redefined magic in a totally new way, as the worship of false gods, alias **demons**. Michael Bailey has agreed, observing that Christians always posited a more fundamental distinction between religion and magic than that imagined by pagans and Jews. All this is correct, but there are two obvious features of the history of magic that provoke counterbalancing reflections. One is that the early **European witch trials** commenced a thousand years after the triumph of the new religion, raising the problem of why, if its ideology was so well suited to hunt witches, it took so long to do so. The second is that, as discussed earlier in this book, the pagan Roman Empire had proved perfectly capable of enacting a savage **code of laws against magicians**, based on wholly traditional attitudes, at precisely the same time as it was persecuting Christians with an equal brutality. It is, in fact, that legal and cultural context, of established and intense official hostility towards magic, that provides the reasons for the Christian perspective on the subject. It presented early Christians with an acute problem: that the miracles credited by them to their **Messiah** and his apostles could look like those promised by, or attributed to, **ceremonial magicians**. This charge was levied against them by some of their most effective pagan critics, such as **Celsus**, who wrote the first comprehensive attack on the new religion in the second century. The reply provided by the leading Christian theologian **Origen** became the standard one: **magicians** used **rites** and **incantations**, but true Christians only the name of **Jesus** and the words of the **Bible**, and a reliance on the power of their deity: a formula which plugged directly into the long-established Graeco-Roman distinction between religion and magic. Almost two hundred years later, **Augustine of Hippo** worked it up into its enduring form, which persisted through the Middle Ages: that the acts of magicians were accomplished with the aid of demons, whereas the miracles of Christian saints were made possible by the intervention of the one true God.

When dealing with ceremonial magic, however, and harm caused by magical means, the laws did little more than reinforce what had already been laid down under pagan emperors. This continuity would none the less still mask significant change if the existing laws against ceremonial magic, and all forms of **harmful magic**, were enforced more rigorously than before. An impression that this was indeed the case is provided in a series of famous passages by the fourth-century historian **Ammianus Marcellinus**. He recorded that following a law of **Constantius II**, in 358, which declared that magicians across the empire were enemies of humanity, anybody who wore an amulet to cure a disease, or passed a tomb after dark, was in danger of denunciation and execution. Being seen near a tomb was fatal because of suspicion that the person concerned was hunting for human body parts for use in **spells**. This wave of trials was followed by three more, at intervals between 364 and 371, under the brother emperors **Valentinian** and **Valens**. Those began by mainly affecting the

Roman senatorial class, but expanded over time to target commoners. Whole libraries were burned by their owners for fear they might be thought to contain magical texts. These persecutions affected both Rome itself and the eastern provinces, and torture was freely used to obtain evidence. **Ammianus** made clear that in most cases the pressure to prosecute came from the top, from emperors leading recently established and insecure dynasties and afraid of conspiracy: the charge of using magic had returned, for the first time in three hundred years, as a weapon in central politics. Medieval law codes, starting with those of the **Germanic kingdoms** which supplanted the western **Roman Empire**, continued to prescribe penalties for the deliberate working of harmful magic. If the harm done was serious, such as murder, then the penalties were as severe as those specified for doing equivalent damage by physical means; which is logical in societies, such as those in medieval Europe, which believed in the literal potency of spells and curses…

For two hundred years, learned authors in Western Europe conducted a debate over how far forms of the new complex magic could be assimilated into orthodoxy and be used for human benefit. By the early fourteenth century, however, the majority of them had swung firmly against such a rapprochement and reinstated the **Augustinian orthodoxy** that all magic was inherently demonic, whether its practitioners were conscious or not that they were working with demons. This development accompanied and overlapped with another, which was inspired largely by the appearance of widespread heresy in Western Europe between the eleventh and thirteenth centuries and the increasingly savage, and successful, Catholic counter-attack upon it, using crusade and inquisition as its main weapons: it was between 1224 and 1240 that burning came to be adopted as the standard mode of execution for heretics, as it had long been for magicians. They were routinely portrayed as devil-worshippers as part of that counter-attack, and this strategy encouraged an outbreak of political trials between 1300 and 1320 in which prominent individuals and organizations were accused of worshipping **Satan** in secret, and often ruined as a consequence. **King Philip the Fair of France** became the most ardent practitioner of this technique, using it to attack a bishop who was one of his own councillors, a pope and then the crusading order of the **Knights Templar**, and it was continued under his successor **Louis X**. In England, the bishop of Lichfield was accused in 1303, and suspicion of magic seems to have quickened at a local level, as a woman was banished from Exeter in 1302 for entertaining notorious magicians from South Devon, and in 1311 the bishop of London ordered measures to be taken to curb the growth of fortune-tellers. Both developments, the condemnation of magic and the escalation of political trials for devil-worship, were accompanied by a growing fear of the power of Satan in the world; which may itself have been generated, and was certainly reinforced, by the twin new threats posed by large-scale heresy and ceremonial magic. In this manner the scene was set for a direct and comprehensive attack on ceremonial magic, as demonic, launched by **Pope John XXII** between 1318 and 1326. He was already inclined to use the charge of malicious magic against personal opponents, having had a bishop burned as a result of it in 1317 and going on to deploy it again thereafter. In 1318 he appointed a commission to root out ceremonial magic from his own court at Avignon, and in the 1320s four other trials of alleged magicians were held in different parts of France, some directly encouraged by the pope: churchmen were accused in all of them, although sometimes assisted by lay practitioners. In 1326 John decreed that ceremonial magic had grown to the proportions of a plague, and excommunicated all concerned in it…

The most important feature of the concept of the satanic witch that appeared at the end of the Middle Ages was that it was new. This was, as shall be seen, fully acknowledged at the time of its appearance. In 1835 **Jacob Grimm**, as part of his pioneering work into the history of **Germanic folklore**, came up with a two-stranded explanation for its development. One strand, the more dynamic, consisted of the increasing concern of the medieval Catholic Church to purify the societies which it controlled, by identifying and eliminating heresy. This supplied the basis for the imagination of an organized sect of devil-worshipping witches, but Grimm also suggested that the forms which that imaginative creation took were conditioned by his second strand, popular beliefs inherited ultimately from the pagan ancient world. The record therefore shows that a belief in a conspiracy of **devil-worshipping magicians**, to harm other people, and especially to kill babies and children, appeared in the mid- to late 1420s at different points widely dispersed across a broad area, stretching in an arc from north-eastern Spain to central Italy. The single factor which can link them all is the preaching of friars who were co-operating in a campaign against popular heresy and unusually conscious of the danger posed by magic, as part of the resurgence in the prosecution of its practitioners which had commenced in Western Christendom in the 1370s. This seems to have ignited responses among the populace, amounting at times to panics, in particular places where the circumstances were propitious, perhaps because of unusual infant mortality and other misfortunes, and certainly where justice was in the hands of local secular lords and captains who were easily carried away by public feeling, in a period of political and economic instability. There was a clear connection between these responses and folk beliefs derived from ancient origins, but those did not so obviously derive from 'shamanistic' motifs of spirit-flight so much as the figure of the **child-murdering demoness**, the **Roman strix** and the **Germanic nocturnal cannibal woman**. The main part of the new construct, of a group of people who gathered secretly by night to worship the Devil, who appeared to them in animal form, was absolutely standard as an orthodox accusation against heretics in the high and later Middle Ages.

Box 7.7 Witches, 'Early Modern History', and European Atrocities

Hutton wrote in *The Witch*:

> The executions inspired by the new concept of the **satanic witch** lasted from (the) first known examples in the Pyrenees and at Rome in 1424 until the final one in Switzerland in 1782. Between those two dates between forty and sixty thousand people were legally put to death for the alleged crime of witchcraft, with the true figure more probably in the lower half of that range. This figure is, however, deceptive in two ways, for the trials were concentrated both in space and in time. They were found mostly in a zone extending across Northern Europe from Britain and Iceland to Poland and Hungary, and from the extreme north of Scandinavia to the Alps and Pyrenees. Furthermore, even within the region across which trials were relatively common, the new concept of the demonic witch proved to be a slow-burning fuse. During the fifteenth century it was confined mainly to the western Alps, northern Italy and Spain, the Rhineland, the Netherlands and parts of France; and does not seem to have claimed more than a few thousand victims at most. Between 1500 and 1560 this range did not much expand, and the overall number of trials seems to have decreased, before an explosion in the second half of the century. Most of the victims claimed by the early modern witch-hunts in fact died in the

course of a single long lifetime, between 1560 and 1640. Two factors may account for this. One is that it was the period in which the crisis in European religion ushered in by the **Reformation** came to a peak, and Catholic and Protestant engaged in a series of all-out contests. This sent the religious temperature to fever level in many places and individuals, and produced a greater willingness to perceive the world as a battleground between the forces of heaven and hell. The typical proponent of witch trials was a pious reformer, the age's equivalent to the Observant friars of the early fifteenth century, who wanted to purge society of wickedness and ungodliness; to such people the destruction of witches was usually only a single item on a list of measures to achieve an ideal Christian polity...

Averaged out across the Continent, about three-quarters of those tried were women, but this figure conceals major local variations. Likewise, the majority of victims were drawn neither from the wealthier ranks of society nor from the very poor, being ordinary peasants and artisans like their accusers, but, again, local experience threw up exceptions to this rule. If they conformed to a particular human type it was that of the bad neighbor, quarrelsome and inclined to curse and insult; yet very many were generally normal personalities who happened to have the wrong friends or enemies at the wrong moment. Religious identities were in general irrelevant to the matter, as the most intense regions of witch-hunting were Calvinist Scotland, Lutheran northern Norway, and some Catholic states in western and central Germany and in the Franco-German borderland. In most places the pressure to prosecute came from below in society, originating among the common people, but in a minority it was imposed by the rulers of the state concerned. Some of the worst rates and totals of execution were produced by 'chain-reaction trials', in which large numbers of people were arrested and forced to denounce yet more; but territories such as Lorraine, where one or two people were accused at a time, could still accumulate large overall death tolls. It is abundantly clear that personal and factional enmities, and political ambitions, often formed a context for accusations. The latter never seem, however, to have been merely a pretext for the resolution of such other tensions: rather, they were generated by very real fears of **bewitchment**. Witch trials did not seem to yield obvious and measurable benefits to communities that engaged in them...

The great majority of the early modern executions for witchcraft occurred between 1560 and 1640. This was also the period in which the regional inquisitions that defended the purity of the Roman Catholic religion in the Western Mediterranean basin, and which represented some of the most formidably efficient investigative and punitive machines in Europe, launched a determined attack on magical practices of all kinds. The results, however, have come to be recognized as remarkably mild: several thousand prosecutions for magic yielded at the very most five hundred death sentences. This was because of a general lack of a sense of danger from a satanic conspiracy, so that charges of collective devil-worship, and of pacts with Satan, were very rare. **Torture** was seldom used on those arrested, and there was little pressure on them to name accomplices: on the whole, witches were treated as ignorant folk deluded by the Devil, not as dangerous criminals. At Venice the **inquisitors** held over six hundred trials concerning magic between 1550 and 1650, about a fifth of which were for witchcraft, but most ended in acquittal and none in execution. Similarly, no executions are recorded in Sicily, and the **notorious Spanish Inquisition** managed to try more than five thousand people for using magic between 1610 and 1700, without burning any. The Portuguese one put one person to death for such an offence, although it regularly tried cases that concerned **magic** and sometimes prosecution of them rose to peaks...

The great influence of the Papacy and the Spanish upon the western Mediterranean lands in general explains why the other territories in the region followed the same trajectory in the same period. The new oversight and professionalism injected into the inquisitorial process by the foundation of central supervisory bodies seems in itself to have engendered a more rigorous and skeptical attitude towards accusations of **demonic witchcraft**, and a growing disposition to view even those people who confessed to dealings with the Devil as deluded and in need of redemption. This change then became a factor in regional power politics, as interventions to prevent credulous and destructive witch-hunting enabled the central tribunals to enforce their authority more effectively over the localities. Eventually a cautious attitude to accusations of witchcraft, and a programme of correction and not extermination for those convicted of attempting to work magic, became a matter of ethnic identity. Seventeenth-century Italians, in particular, could be surprised and horrified by the huge body counts being stacked up by witch-hunts in Northern Europe. The **Mediterranean inquisitions** remained forbiddingly effective machines for the persecution of magical practices, and even moderate punishments such as imprisonment, flogging and public penance would have been traumatic for those who suffered them.

None the less, they rescued a region representing about a quarter of Europe from the most concentrated and deadly period of the early modern witch trials. They seem to have done so, moreover, because of political and ideological developments among the religious elite, in which popular beliefs played only a supporting role, in certain places, and not a decisive one. What, however, of the core area of the early modern witch trials, where the majority of their victims perished: the German-speaking lands, the French-speaking parts of the Rhine and Moselle basins to their west and Poland to their east? It has already been noted that these had sprouted rich medieval popular traditions, such as those of the 'furious army', Holle and Perchte, which should have meshed easily with the concept of the witches' Sabbath. Modern folklorists, led by **Jacob Grimm**, uncovered a still flourishing lore of nocturnal spirits of this sort, with strong regional hallmarks. All this testifies to a prolific set of beliefs, grounded in the culture of ordinary people, which should have informed the nature of witch trials in the way in which **striges**, **wolf-riders**, **superhuman ladies** and **dream warriors** did further south; and yet most of the evidence suggests that it did not. In saying this, it is important once again not to forget the deeper perspectives. The image of the satanic witch that was transmitted to Northern Europe was based partly on an ancient concept, that of **the strix**, and the facility with which the Germanic cultural zone picked it up may well have owed much to its own ancient native tradition, of the cannibal witch who attacked all age groups rather than specifically children. This tradition might also help to explain why the majority of those accused in this region were women.

Unfortunately the believe in and fear of witches and witchcraft and the atrocities justified by/related to those beliefs, fears, and related teleological narratives are not just something "of the past." Far from being a mere curiosity for *history aficionados*, in many regions of the planet a huge number of people continue to have such beliefs and to do horrible atrocities because of, or at least justified by, them, as explained by Hutton:

> The process continued in Africa in the twentieth century under a very different colonial system, in which the official attitude to witchcraft was one of disbelief. Here the **Bible**, in early modern translations which affirmed a disapproval of witchcraft and ordered its suppression, often acted in its own right to confirm native beliefs: ironically, **Christianity** therefore had the effect of reducing the credibility of ancestor spirits and land spirits,

against which the missionaries preached, and so of producing a tendency to blame witches alone for uncanny misfortunes. In northern Uganda, the end of British rule was followed by a resumption of **witch-hunting** by chiefs, with considerable popular support. Suspects were tortured by being made to sit or walk naked on barbed wire, exposed to termite bites, beaten, made to drink their own urine, or having pepper put into their eyes. In 2007 the president of Gambia, **Yahya Jammeh**, sent a division of his personal bodyguard to join local police in rounding up over 1,300 suspected witches from one district of his country. They were taken to detention centres and dosed with a potion expected to remove their powers, which made many ill. Three years later a major hunt swept southern Nigeria, directed at children and driven by ministers from native Christian churches who offered to exorcize the accused and render them harmless. The young victims were often detained and tortured to induce them to confess, and then abandoned by their families after exorcism; and all this occurred despite the existence of a new national law forbidding accusations against children. By 2012 the panic regarding child witches had spread to Congo, and twenty thousand children were said to be living on the streets of the capital, Kinshasa, because they had been expelled from their homes.

By 2005 at least half a million people had emerged as self-proclaimed experts in dealing with the problems of **bewitchment** in South Africa alone. If **Christianity** had easily been assimilated into traditional beliefs regarding witches, and served to reinforce them, then so has modern technology. Indeed, as the anthropologist Adam Ashforth has emphasized, science has become the 'primary frame of reference' for interpreting witchcraft in some South African townships, as quantum physics, cell phones, digital imaging, cloning and artificial life are all more compatible with a **magical view of the universe** than that of the preceding machine age. Among the **Ambrym Islanders of Central Melanesia**, fear of witchcraft, and the homicides that it generated, had reached what was described as 'critical levels' in the late 1990s. By the 2010s other parts of Melanesia had become as severely affected, as a result of collapsing traditional social and cultural systems, declining health services, worsening poverty, and increasing lifestyle diseases and premature deaths. Violence against suspects was (as recently in southern Africa) mainly conducted by impoverished young men seeking to achieve value in the eyes of their communities, and was becoming more public as well as more extreme. In New Guinea a young woman was burned alive in 2013 in front of hundreds of onlookers, including police, and two other women publicly tortured and beheaded on Bougainville Island in the Northern Solomon archipelago. In 2014 two men were publicly hanged in a community hall in Vanuatu. Nor is legal action against witchcraft missing from the world outside Africa, above all in Islamic states. During the period between 2008 and 2012, laws against **magical practices** of all kinds were more strictly enforced in Afghanistan, the Gaza Strip, Bahrain and Saudi Arabia. In that period Saudi Arabia executed several people for such offences, mostly foreigners and mostly by beheading. A woman was murdered as a suspected witch in Gaza in 2010. The Saudi government trains employees not only as **witch-hunters** but in rituals to destroy the effects of witchcraft, while one recent president of Pakistan, **Asif Ali Zardari**, sacrificed a black goat nearly every day to ward off its effects. In Indonesia courts have become increasingly willing to try acts of magic as crimes, or at least as antisocial behaviour, judges often seeming to believe in it and their actions in doing so being popular.

It is possible to make a theoretical case that witch-hunting may, at least at times, serve a *positive social function*. In some contexts it may reinforce cultural norms, and so communal solidarity, by discouraging aberrant or antisocial behaviour. The identification of witchcraft with **jealousy**, **greed** and **malice** can serve to strengthen attachment to the countervailing virtues, and discourage the expression of animosity. It can be used to enforce economic obligations and reduce competition in favour of co-operation. In other contexts, it can be a midwife to change, in that anti-witchcraft movements have often legitimized or reinforced the power of new groups. Accusations have sometimes provided a means by which disempowered individuals, such as children or women, can attract attention and respect, and intimidate people normally in superior positions to them. They can articulate otherwise unspeakable fantasies, reveal and represent destructive impulses, and identify and express tensions within families and wider social groups, blasting away unsustainable relationships. Measures against presumed witchcraft have enabled humans to act purposefully in the face of adversity. It was for these reasons that an influential school of thought among anthropologists has held that witchcraft accusations functioned as instruments of social health rather than as symptoms of malfunction.

Others, however, have held a different opinion, and that is the one favoured here. It emphasizes that all these positive functions of belief in witchcraft have only acted to strengthen societies, or to enable them to adjust more effectively to changing circumstances, when the rate of accusation has been low and sporadic, and subjected to firm controls. In many cases this situation has not obtained, and suspicions and accusations have not resolved fears and hostilities, but aggravated them and represented obstacles to peaceful co-operation. At worst, they have torn communities apart and left lasting traumas and resentments, or greatly compounded the suffering consequent on adjustment to new economic and social developments. Most societies that have believed firmly in witchcraft have regarded it as a scourge and a curse, of which they have longed to be rid; but the only way in which they have been able to conceive of bringing about this happy result has been to destroy the witches. Such attempts have tended to reinforce vividly a consciousness of the threat from witchcraft, and so perpetuate fear of it, and make future witch-hunts likely, even if they have managed – often at a grim human cost – to reduce that which had existed at the moment.

It is truly disturbing that, while I was writing this book, there were so many news articles that provide awful cases in which the belief in or use of imaginary misogynistic stories about witches led to discrimination, subjugation, oppression, and even killings of numerous women—often **outcasts** within their societies—in so many different parts of the globe. In January 11, 2018, an article published by Yasmin in the *Scientific American*, entitled "*Witch hunts today: abuse of women, superstition and murder collide in India*" (see also Fig. 7.2), reported:

Men circled the three women, their fists wrapped around thick iron pipes and wooden sticks. The women huddled on the ground at the center of their village in the western Indian state of Gujarat and whimpered as the crowd gathered. Two young men had died in the village, and the women were being called **Dakan**, the Gujarati word for **witch**. They were accused of feasting on the young men's souls. Madhuben clutched her right upper arm. She had taken three blows from one of the pipes and was sure her bones were broken…her sisters-in-law, Susilaben and Kamlaben, covered their heads as wood and metal pounded their backs (the names of women targeted by witch hunts have been changed in this story, to minimize the risk of further assault or of jeopardizing pending legal cases) [see **Fig. 7.2**]. The attack on the trio, in Gujarat in 2014, was one of thousands of witch hunts that take place in India. More than 2,500 Indians have been chased, tortured and killed in such hunts between 2000 and 2016, according to India's National Crime Records Bureau. Activists and journalists say the number is much higher, because most states don't list witchcraft as a motive of murder. **Witch hunts** primarily target women and exploit India's **caste system** and **culture of patriarchy**. Men who brand women as dakan capitalize on deeply rooted **superstitions** and systems built on **misogyny** and patriarchy to lay blame on females. The accusations of sorcery are used to oust women from valuable land that men covet, in a region where flawed development plans have produced **agricultural failures**, say sociologists who study violence in India. Witches are also convenient explanations for rising **infant mortality rates** and deaths from **malaria**, **typhoid** and **cholera**.

The violence and accusations against Madhuben, Susilaben and their sister-in-law began in 2012. That year the three women found their male relatives routinely defecating in the plot of land where the sisters grew corn, lentils and peas. Almost half of India's households lack a toilet, according to the 2011 census, and many of those people defecate in the

Fig. 7.2 Superstition, imaginary stories, and witch-hunting continue to occur too frequently in the twenty-first century: this picture shows two of the thousands of women that continue to be attacked in witch-hunts, just in a single country, in this case in India

open. The sisters-in-law were upset with their male relatives' using their crops as a toilet. "I said to them, 'This is where we grow food…how are we supposed to deal with [human excrement] here?'" Susilaben says. This challenge to men in a culture where women are expected to be silent subordinates infuriated her family, she recalls. The men did not stop defecating on the land. Instead they turned on the women, beat them and ran them out of their home for 10 days. The situation worsened a year or so later, when two young men in their home became ill. One developed renal failure, the other cancer. Poor access to health care in the region meant the family was forced to take out loans and travel to neighboring towns for medical help. Money was scarce and stress was high. When the young men died, the sisters-in-law were accused of eating their souls and causing their premature deaths. And then the remaining men began a campaign to take their land. The plot where the women grew vegetables was fertile and in a prime location, at a four-way road junction in the village. That was the spot where they were beaten. Male relatives forced the sisters-in-law to sign a document saying they would hand over land ownership to the men. "We had no choice but to sign", Susilaben says. "They said they would kill us if we didn't give them our land". The farming land is now a series of roadside stores selling slippers, stationery, car parts and clothes. Battles over land and property are common starts to witch hunts, says Soma Chaudhuri, a sociologist at Michigan State University who studies **gender violence** in India. Chaudhuri says witch hunts and beatings provide an outlet for men living in poverty to vent frustrations over their own lack of power. "These rural communities are so marginalized and so oppressed, and they have no political resources and no avenues of protest. So what do people do when they're very frustrated? You look to your surroundings for an easy *scapegoat*. Women are that *scapegoat*."

It is perhaps even more troubling to see that imaginary stories about witches, child-murdering demonesses, and nocturnal cannibal women as inane as those mentioned in Boxes 7.6 and 7.7 continue to be recycled, over and over again, up to this point in time, in 2020. Distressing not because this is somehow surprising, but exactly for the opposite reason: it is just a further dramatic piece of evidence, within an endless number of them, showing how humans always were, and continue to be, immersed in the darkest dungeons of *Neverland*. An illustrative example of such a recycling of completely illogical old imaginary tales is the harebrained, increasingly viral, and now even politically powerful **QAnon conspiracy theories** (see Fig. 2.11). In this sense, this example effectively illustrates the point that conspiracy theories involving something about death tend to be particularly widespread and influential. To the point that two famous followers of these theories were recently elected to the U.S. House of Representatives, in November 2020. Yes, these politicians, as well many hundreds of thousands of people, *firmly believe today, in the twenty-first century,* in tales about **left-wing satanic sex traffickers** molesting and literally eating children. Such QAnon conspiracy theories are in a way a mix between old imaginary stories about child-murdering demonesses and about cannibal women, the only difference being that these theories are a bit less misogynistic in the sense that the satanic people are not only women but also men, but at the same time are more politically charged as the satanic men and women are mainly left-wing demoniac beings.

Apart from these differences, everything else is basically the same: QAnon theories illustrate once again that there is often a strong link between conspiracy theories and tales about witches or other people working for, or together with, demons or Satan himself. On the one hand these QAnon theories accomplish the main function of conspiracy theories: to try to give a false sense of control and order in a mainly random and chaotic world, by arguing that basically everything that happens, including viruses such as **Covid-19**, "happens for a purpose," and namely that the "bad purposeful agents" are humans, of course. So, those "bad human agents" can be controlled by us, humans: in the specific case of the QAnon theories, they can be solved by humans like **Donald Trump**. On the other hand, the QAnon theories try to accomplish another major function that most conspiracy theories aim to achieve:

to say that the "bad human agents" are precisely the people that the ones that created those theories don't like. In this case, as the QAnon theories are mainly created by far-right people and promoted by far-right political leaders, the "bad ones" are of course the "leftists," from **Hillary Clinton** to **Barack Obama**, from **Bill Gates** to **George Soros**. Within the thousands of articles summarizing these theories to the broader public, as well as their disturbing and worrying growing influence around not only the U.S. but numerous other countries, I provide here excerpts of one published November 24th, 2020, in the *CBS News* website, entitled "*What is the QAnon conspiracy theory?*":

> What started as a fringe movement among President Trump's supporters, confined to the shadier corners of the internet, has taken a mainstream turn. The **QAnon conspiracy theory** started on *4chan*, the bulletin board known for creating and spreading memes, but has moved to larger social media platforms. Facebook has taken action against QAnon groups and pages, while Twitter removed several thousand QAnon-linked accounts in 2020. The FBI has warned that **fringe conspiracy theories** like QAnon pose a growing domestic terrorism threat. What is the QAnon conspiracy theory? What do its followers believe? Those questions have become more difficult to answer as the movement has expanded since its inception in 2017. QAnon purports that America is run by a **cabal of pedophiles** and **Satan-worshippers** who run a global child sex-trafficking operation and that President Trump is the only person who can stop them. The information supposedly comes from a high-ranking government official who posts cryptic clues on ***4chan*** and the even more unfettered site ***8chan*** under the name "Q." That's the central gist of the theory. The rest is open to some degree of interpretation, which is necessary because Q's posts tend to read like riddles. But YouTube videos created by QAnon believers help fill in the gaps and create a storyline that's more-or-less comprehensible. QAnon exists as a kind of parallel history, in which a "deep state" took over decades ago. An all-encompassing theory of the world, it appears to tie together and explain everything from "**Pizzagate**" to ISIS to the prevalence of mass shootings and the JFK assassination. It claims the military, supposedly eager to see the deep state overthrown, recruited President Trump to run for president. But the deep state, which controls the media, quickly tried to smear him through "fake news" and unfounded allegations of **collusion with Russia**.
>
> It goes on to insist that despite the deep state's best efforts, however, President Trump is winning, and that Q is releasing sanctioned leaks to the public in order to galvanize them ahead of "The Storm," which is the moment when the deep state's leaders are arrested and sent to Guantanamo Bay. QAnon believers have called this process "The Great Awakening". The storm takes its name from President Trump's enigmatic comment from October 2018 about "the calm before the storm". Q began posting soon after and said that the storm Mr. Trump referenced is a coming series of mass arrests that would end the deep state forever. In QAnon lore, President Trump was secretly working with special counsel Robert Mueller to bring the deep state down, and the storm is a kind of **Judgment Day** in which the evildoers are punished and the faithful are redeemed. Q has repeatedly suggested that the storm would hit in the very near future and has even said certain people would be arrested at certain dates. When those dates come and go without any arrests, Q says that they needed to be delayed for one reason or another, but that President Trump still has the situation well in hand.

The article then explains that:

> Q's posts tend to be either vague or totally incomprehensible, but QAnon believers are more than happy to try and decipher them. Last year, for example, Q posted a photo of an unnamed island chain. Eager to divine the reasoning behind the post, QAnon adherents tried to "prove" that the photo must have been taken on Air Force One and thus that Q was

traveling with the president. The Q posts are known to the faithful as "breadcrumbs". The people who then try to figure out what they mean are called "bakers"…QAnon believers also spend a lot of time trying to figure out who in the government is a "white hat" Trump supporter and who is a "black hat" in league with the deep state…Q's identity…refers to Q-level clearance at the Energy Department. But who's behind the posts is anybody's guess…the QAnon faithful sometimes point to former national security adviser Michael Flynn and White House aide Dan Scavino as possibilities. Others believe it's Mr. Trump himself. Another theory is that John F. Kennedy Jr. faked his death and now posts on *8chan* as QAnon. On November 3 (2020), Election Day, *8chan* (now *8kun*) administrator Ron Watkins resigned from his post. Q did not post for the next week, raising questions about a connection. As the QAnon movement has migrated to more mainstream social media platforms such as Facebook and Twitter, it has developed new conspiracy theories that have helped subsume more followers. Many QAnon supporters believe that **President Kennedy** was set to reveal the existence of the secret government when he was assassinated. They also believe **President Reagan** was shot on the deep state's orders, and that all the presidents since he left office – with the exception of **President Trump** – have been deep state agents. QAnon believers have also latched onto other conspiracies, such as the **9/11 "truther" movement** and the Rothschild family owning the world's banks. Different QAnon followers identify with different conspiracies, though they all believe in the central conspiracy of **child sex trafficking rings** perpetrated by members of the **Democratic party**.

Most recently, followers have staged several #SaveOurChildren demonstrations. "It seems like they've hijacked the 'Save Our Children' movement, infiltrating it and putting their spin on it", says Daryl Johnson, who previously researched **right-wing terrorism** for the Department of Homeland Security. "Think about children and how vulnerable they are…the issue really tugs at the hearts of anybody…but they're linking it to their conspiracy theories, which are crazy and very dangerous". According to political science professor Joe Uscinski, who studies conspiracy theories, "the beliefs themselves are almost an incitement to violence…I mean, there isn't anything worse you can say about your political competitors than that they are **satanic sex traffickers** who molest and eat children". "It has a lot of properties that make it more like a cult", Uscinski said. At least 19 House Republican candidates who support or have elevated the QAnon movement were on the November ballot, according to tracking by Media Matters. Two QAnon supporters were elected to the United States House of Representatives: Marjorie Taylor Greene won her race for Georgia's 14th congressional district; Lauren Boebert won her race for Colorado's 3rd congressional district. QAnon spread from its fringe beginnings on *4chan* and *8chan* to larger social media platforms such as Facebook, Twitter, and YouTube. These platforms have faced increasing pressure to crack down on these accounts and groups, but have found it difficult to do so. "QAnon is not one organization that you can just cancel or remove", says CNET senior producer and CBSN correspondent Dan Patterson: "these organizations are heavily driven by algorithms, and these algorithms really favor engagement, which QAnon is really good at doing".

7.4 Teleological Tales, Violence, and Torture

As I walked out the door toward the gate that would lead to my freedom, I knew if I didn't have my bitterness and hatred behind, I'd still be in prison. (Nelson Mandela)

In his book *Better Angels*, Pinker gives several examples of how **extreme violence**, including **torture**, has been so prevalent for so long in most agricultural states. This is interesting because his book is mainly aimed to support Hobbes's view that life in such states is in general "better" and "less violent" than in non-state societies such as those of nomadic hunter-gatherers. Of course, Pinker is right about the fact that in many "developed" countries torture is much less prevalent, and particularly is much less promoted by those in power, than it was centuries ago, but he fails to then compare these two cases with what happened before "civilization," a type of failure that is very characteristic in various other sections of his book. Anyway, the important point is that as pointed out by Pinker the type of extreme violence—including **torture** (see Fig. 7.3)—undertaken in agricultural states was/is very often related to, or at least justified by, teleological narratives:

> I think even the most atrocity-jaded readers of recent history would find something to shock them in this display of **medieval cruelty**. There is **Judas's Cradle**, used in the **Spanish Inquisition**: the naked victim was bound hand and foot, suspended by an iron belt around the waist, and lowered onto a sharp wedge that penetrated the anus or vagina; when victims relaxed their muscles, the point would stretch and tear their tissues. The **Virgin of Nuremberg** was a version of the iron maiden, with spikes that were carefully positioned so as not to transfix the victim's vital organs and prematurely end his suffering. A series of engravings show victims hung by the ankles and sawn in half from the crotch down; the display explains that this method of execution was used all over Europe for crimes that included rebellion, witchcraft, and military disobedience. The Pear is a split, spike-tipped wooden knob that was inserted into a mouth, anus, or vagina and spread apart by a screw mechanism to tear the victim open from the inside; it was used to punish **sodomy**, **adultery**, incest, **heresy**, **blasphemy**, and "**sexual union with Satan**". The **Cat's Paw** or **Spanish Tickler** was a cluster of hooks used to rip and shred a victim's flesh. **Masks of Infamy** were shaped like the head of a pig or an ass; they subjected a victim both to public humiliation

Fig. 7.3 The inside of a jail of the Spanish Inquisition, with a priest supervising his scribe, while men and women are suspended from pulleys, tortured on the rack or burnt with torches

and to the pain of a blade or knob forced into their nose or mouth to prevent them from wailing. The **Heretic's Fork** had a pair of sharp spikes at each end: one end was propped under the victim's jaw and the other at the base of his neck, so that as his muscles became exhausted he would impale himself in both places…

Medieval Christendom was a culture of cruelty. Torture was meted out by national and local governments throughout the Continent, and it was codified in laws that prescribed blinding, branding, amputation of hands, ears, noses, and tongues, and other forms of mutilation as punishments for minor crimes. Executions were orgies of sadism, climaxing with ordeals of prolonged killing such as burning at the stake, breaking on the wheel, pulling apart by horses, **impalement** through the rectum, **disembowelment** by winding a man's intestines around a spool, and even hanging, which was a slow racking and strangulation rather than a quick breaking of the neck. **Sadistic tortures** were also inflicted by the Christian church during its inquisitions, witch hunts, and religious wars. Torture had been authorized by the ironically named **Pope Innocent IV** in 1251, and the order of **Dominican monks** carried it out with relish. As the Inquisition coffee table book notes, under **Pope Paul IV** (1555-1559), the Inquisition was "downright insatiable – Paul, a Dominican and one-time Grand Inquisitor, was himself a fervent and skilled practitioner of torture and atrocious mass murders, talents for which he was elevated to sainthood in 1712". Torture was not just a kind of rough justice, a crude attempt to deter violence with the threat of greater violence.

Most of the infractions that sent a person to the rack or the stake were nonviolent, and today many are not even considered legally punishable, such as heresy, blasphemy, **apostasy**, criticism of the government, gossip, scolding, adultery, and **unconventional sexual practices**. Both the Christian and secular legal systems, inspired by Roman law, used torture to extract a confession and thereby convict a suspect, in defiance of the obvious fact that a person will say anything to stop the pain. Torture used to secure a confession is thus even more senseless than torture used to deter, terrorize, or extract verifiable information such as the names of accomplices or the location of weapons. Nor were other absurdities allowed to get in the way of the fun. If a victim was burned by fire rather than spared by a miracle, that was taken as proof that he was guilty. A suspected witch would be tied up and thrown into a lake: if she floated, it proved she was a witch and she would then be hanged; if she sank and drowned, it proved she had been innocent. Far from being hidden in dungeons, torture-executions were forms of popular entertainment, attracting throngs of jubilant spectators who watched the victim struggle and scream. Bodies broken on wheels, hanging from gibbets, or decomposing in iron cages where the victim had been left to die of starvation and exposure were a familiar part of the landscape. Torture was often a participatory sport. A victim in the stocks would be tickled, beaten, mutilated, pelted with rocks, or smeared with mud or feces, sometimes leading to **suffocation**.

Pinker provides several examples of extreme violence—for instance, concerning **human sacrifice**, which is something also chiefly done in sedentary societies—that were explicitly related to, or justified by, ***superstition*** associated with *belief* in teleological narratives:

The most benighted form of institutionalized violence is human sacrifice: the torture and killing of an innocent person to slake a deity's thirst for blood. The **biblical story of the binding of Isaac** shows that human sacrifice was far from unthinkable in the 1st millennium BCE. The Israelites boasted that their God was morally superior to those of the neighboring tribes because he demanded only that sheep and cattle be slaughtered on his behalf, not children. But the temptation must have been around, because the Israelites saw fit to outlaw it in Leviticus 18:21: "You shall not give any of your children to devote them by fire to Molech, and so profane the name of your God." For centuries their descendants would have to take measures against people backsliding into the custom. In the 7th century BCE,

King Josiah defiled the sacrificial arena of Tophet so "that no one might burn his son or his daughter as an offering to Molech." After their return from Babylon, the practice of human sacrifice died out among Jews, but it survived as an ideal in one of its breakaway sects, which believed that God accepted the torture-sacrifice of an innocent man in exchange for not visiting a worse fate on the rest of humanity. The sect is called **Christianity**. Human sacrifice appears in **the mythology** of all the major civilizations. In addition to the Hebrew and Christian Bibles, it is recounted in the Greek legend in which **Agamemnon** sacrifices his daughter **Iphigenia** in hopes of bringing a fair wind for his war fleet; in the episode in Roman history in which four slaves were buried alive to keep Hannibal at bay; in a Druid legend from Wales in which priests killed a child to stop the disappearance of building materials for a fort; and in many legends surrounding the multiarmed **Hindu goddess Kali** [see **Fig. 9.23**] and the feathered **Aztec god Quetzalcoatl**. Human sacrifice was more than a riveting myth. Two millennia ago the Roman historian **Tacitus** left eyewitness accounts of the practice among Germanic tribes. **Plutarch** described it taking place in Carthage, where tourists today can see the charred remains of the sacrificial children. It has been documented among traditional Hawaiians, Scandinavians, Incas, and Celts. It was a veritable industry among the **Aztecs in Mexico**, the **Khonds in southeast India**, and the **Ashanti**, **Benin**, and **Dahomey** kingdoms in western Africa, where victims were sacrificed by the thousands. Matthew White estimates that between the years 1440 and 1524 CE the Aztecs sacrificed about forty people a day, 1.2 million people in all.

Human sacrifice is usually preceded by torture. The Aztecs, for example, lowered their victims into a fire, lifted them out before they died, and cut the beating hearts out of their chests. The **Dayaks of Borneo** inflicted death by a thousand cuts, slowly bleeding the victim to death with bamboo needles and blades. To meet the demand for sacrificial victims, the Aztecs went to war to capture prisoners, and the Khonds raised them for that purpose from childhood. The killing of innocents was often combined with other superstitious customs. Foundation sacrifices, in which a victim was interred in the foundation of a fort, palace, or temple to mitigate the effrontery of intruding into the gods' lofty realm, were performed in Wales, Germany, India, Japan, and China. Another bright idea that was independently discovered in many kingdoms (including Sumeria, Egypt, China, and Japan) was the **burial sacrifice**: when a king died, his retinue and harem would be buried with him. The **Indian practice of suttee**, in which a widow would join her late husband on the funeral pyre, is yet another variation. About 200,000 women suffered these pointless deaths between the Middle Ages and 1829, when the practice was outlawed. What were these people thinking? Many institutionalized killings, however unforgivable, are at least understandable. People in power kill in order to eliminate enemies, deter troublemakers, or demonstrate their prowess. But **sacrificing harmless children**, going to war to capture victims, and raising a doomed caste from childhood hardly seem like cost-effective ways to stay in power. In an insightful book on the history of force, the political scientist James Payne suggests that ancient peoples put a low value on other people's lives because pain and death were so common in their own. This set a low threshold for any practice that had a chance of bringing them an advantage, even if the price was the lives of others. And if the ancients believed in gods, as most people do, then human sacrifice could easily have been seen as offering them that advantage. "Their primitive world was full of dangers, suffering, and nasty surprises, including plagues, famines, and wars…it would be natural for them to ask, 'what kind of god would create such a world?'…a plausible answer was: a **sadistic god**, a god who liked to see people bleed and suffer". So, they might think, if these gods have a minimum daily requirement of human gore, why not be proactive about it? Better him than me.

A **sanguinary god** that hungers for indiscriminate human scapegoats is a rather crude theory of misfortune. When people outgrow it, they are still apt to look to supernatural explanations for bad things that happen to them. The difference is that their explanations become more finely tuned to their particulars. They still feel they have been targeted by

supernatural forces, but the forces are wielded by a specific individual rather than a generic god. The name for such an individual is a witch. Witchcraft is one of the most common motives for revenge among hunter-gatherer and tribal societies. In their theory of causation, there is no such thing as a natural death. Any fatality that cannot be explained by an observable cause is explained by an unobservable one, namely sorcery. It seems incredible to us that so many societies have sanctioned cold-blooded murder for screwball reasons. But certain features of human cognition, combined with certain recurring conflicts of interest, make it a bit more comprehensible. The brain has evolved to ferret out hidden powers in nature, including those that no one can see. Once you start rummaging around in the realm of the unverifiable there is considerable room for creativity, and accusations of sorcery are often blended with self-serving motives. Tribal people, anthropologists have shown, often single out despised in-laws for allegations of witchcraft, a convenient pretext to have them executed…In the 15th century two monks published an expose of witches called *Malleus maleficarum*, which the historian Anthony Grafton has called "a strange amalgam of Monty Python and Mein Kampf". Egged on by its revelations, and inspired by the injunction in Exodus 22:18 "Thou shalt not suffer a witch to live", French and German **witch-hunters** killed between 60,000 and 100,000 accused witches (85 percent of them women) during the next two centuries. The executions, usually by burning at the stake, followed an ordeal of torture in which the women confessed to such crimes as eating babies, wrecking ships, destroying crops, **flying on broomsticks** on the **Sabbath**, **copulating with devils**, transforming their demon lovers into cats and dogs, and making ordinary men impotent by convincing them that they had lost their penises. The psychology of **witchcraft** accusations can shade into other blood libels, such as the recurring rumors in medieval Europe that Jews poisoned the wells or killed Christian children during Passover to use their blood for matzo. Thousands of Jews were massacred in England, France, Germany, and the Low Countries during the Middle Ages, emptying entire regions of their Jewish populations.

Another type of tremendously violent acts explicitly related to, or justified by, teleological narratives are those against blasphemers, heretics, and apostates, as noted by Pinker:

Human sacrifice and witch-burnings are just two examples of the harm that can result from people pursuing ends that involve figments of their imagination…But the greatest damage comes from religious beliefs that downgrade the lives of flesh-and-blood people, such as the faith that suffering in this world will be rewarded in the next, or that flying a plane into a skyscraper will earn the pilot seventy-two virgins in heaven…the belief that one may escape from an eternity in hell only by accepting Jesus as a savior makes it a moral imperative to coerce people into accepting that belief and to silence anyone who might sow doubt about it. A broader danger of unverifiable beliefs is the temptation to defend them by violent means. People become wedded to their beliefs, because the validity of those beliefs reflects on their competence, commends them as authorities, and rationalizes their mandate to lead. Challenge a person's beliefs, and you challenge his dignity, standing, and power. And when those beliefs are based on nothing but faith, they are chronically fragile. No one gets upset about the belief that rocks fall down as opposed to up, because all sane people can see it with their own eyes. Not so for the belief that babies are born with original sin or that God exists in three persons or that Ali was the second-most divinely inspired man after Muhammad. When people organize their lives around these beliefs, and then learn of other people who seem to be doing just fine without them – or worse, who credibly rebut them – they are in danger of looking like fools. Since one cannot defend a belief based on faith by persuading skeptics it is true, the faithful are apt to react to unbelief with rage, and may try to eliminate that affront to everything that makes their lives meaningful. The human toll of the persecution of heretics and nonbelievers in medieval and early modern Christendom beggars the imagination and belies the conventional wisdom that the 20th century was an unusually violent era. Though no one knows exactly how many people were killed in these holy slaughters, we can get a sense from numerical estimates by atrocitologists…Between 1095 and 1208 Crusader armies were mobilized to fight a "just war" to retake Jerusalem from Muslim Turks, earning them remission from their sins and a ticket to heaven. They massacred Jewish communities on the way, and after besieging and sacking Nicea, Antioch, Jerusalem, and Constantinople, they slaughtered their Muslim and Jewish populations. Rummel estimates the death toll at 1 million. The world had around 400 million people at the time, about a sixth of the number in the mid-20th century, so the death toll of the **Crusader massacres** as a proportion of the world population would today come out at around 6 million, equivalent to the Nazis' genocide of the Jews…

Rummel estimates the death toll from the **Spanish Inquisition** at 350,000. After the **Reformation**, the **Catholic Church** had to deal with the vast number of people in northern Europe who became Protestants, often involuntarily after their local prince or king had converted. The **Protestants**, for their part, had to deal with the breakaway sects that wanted nothing to do with either branch of **Christianity**, and of course with the Jews. One might think that Protestants, who had been persecuted so viciously for their heresies against **Catholic doctrines**, would take a dim view of the idea of persecuting heretics, but no. In his 65,000-word treatise *On the Jews and their lies*, **Martin Luther** offered the following advice on what Christians should do with this "rejected and condemned people: First, …set fire to their synagogues or schools and…bury and cover with dirt whatever will not burn, so that no man will ever again see a stone or cinder of them…Second, I advise that their houses also be razed and destroyed…Third, I advise that all their prayer books and Talmudic writings, in which such idolatry, lies, cursing, and blasphemy are taught, be taken from them…Fourth, I advise that their rabbis be forbidden to teach henceforth on pain of loss of life and limb…Fifth, I advise that safe-conduct on the highways be abolished completely for the Jews…Sixth, I advise that usury be prohibited to them, and that all cash and treasure of silver and gold be taken from them and put aside for safekeeping. Seventh, I recommend putting a flail, an ax, a hoe, a spade, a distaff, or a spindle into the hands of young, strong Jews and Jewesses and letting them earn their bread in the sweat of their brow, as was imposed on the children of Adam. For it is not fitting that they should let us accursed Goyim toil in the sweat of our faces while they, the holy people, idle away their time behind the stove, feasting and farting, and on top of all, boasting blasphemously of their lordship over the Christians by means of our sweat. Let us emulate the common sense of other nations…[and] eject them forever from the country."

At least he suffered most of them to live. The **Anabaptists** (forerunners of today's Amish and Mennonites) got no such mercy. They believed that people should not be baptized at birth but should affirm their faith for themselves, so Luther declared they should be put to death. The other major founder of **Protestantism**, **John Calvin**, had a similar view about blasphemy and heresy: Some say that because the crime consists only of words there is no cause for such severe punishment. But we muzzle dogs; shall we leave men free to open their mouths and say what they please? …God makes it plain that the false prophet is to be stoned without mercy. We are to crush beneath our heels all natural affections when his honour is at stake. The father should not spare his child, nor the husband his wife, nor the friend that friend who is dearer to him than life. Calvin put his argument into practice by ordering, among other things, that the writer Michael Servetus (who had questioned the trinity) be burned at the stake. The third major rebel against Catholicism was **Henry VIII**, whose administration burned, on average, 3.25 heretics per year. With the people who brought us the **Crusades** and Inquisition on one side, and the people who wanted to kill rabbis. Anabaptists, and Unitarians on the other, it's not surprising that the **European Wars of Religion** between 1520 and 1648 were nasty, brutish, and long. The wars were fought, to be sure, not just over religion but also over territorial and dynastic power, but the religious differences kept tempers at a fever pitch. According to the classification of the military historian Quincy Wright, the Wars of Religion embrace the **French Huguenot Wars** (1562-94), the **Dutch Wars of Independence**, also known as the **Eighty Years' War**

(1568-1648), the **Thirty Years' War** (1618-48), the **English Civil War** (1642-48), the **wars of Elizabeth I in Ireland, Scotland, and Spain** (1586-1603), the **War of the Holy League** (1508-16), and **Charles V's wars in Mexico, Peru, France, and the Ottoman Empire** (1521-52). The rates of death in these wars were staggering. During the **Thirty Years' War** soldiers laid waste to much of present-day Germany, reducing its population by around a third. Rummel puts the death toll at 5.75 million, which as a proportion of the world's population at the time was more than double the death rate of **World War I** and was in the range of **World War II** in Europe. The historian Simon Schama estimates that the **English Civil War** killed almost half a million people, a loss that is proportionally greater than that in World War I.

7.5 Homo Irrationalis, War, Terrorism, and Mass Killings

Dear God, save us from the people who believe in you. (Anonymous, written on a wall in Washington DC after the terrorist attacks of 9/11)

The first paragraphs of an outstanding book published by **Catherine Nixey** in 2017 are, in my opinion, one of the most powerful beginnings of any non-fiction book that I have ever read:

> The destroyers came from out of the desert. **Palmyra** must have been expecting them: for years, marauding bands of bearded, black-robed zealots, armed with little more than stones, iron bars and an iron sense of righteousness had been terrorizing…Their attacks were primitive, thuggish, and very effective. These men moved in packs – later in swarms of as many as five hundred – and when they descended utter destruction followed. Their targets were the temples and the attacks could be astonishingly swift. Great stone columns that had stood for centuries collapsed in an afternoon; statues that had stood for half a millennium had their faces mutilated in a moment; temples that had seen the rise of the Roman Empire fell in a single day. This was violent work, but it was by no means solemn. The zealots roared with laughter as they smashed the 'evil', 'idolatrous' statues; the faithful jeered as they tore down temples, stripped roofs and defaced tombs. Chants appeared, immortalizing these glorious moments. 'Those shameful things', sang pilgrims, proudly; the 'demons and idols…our good Saviour trampled down all together' *…the 'triumph' of* ***Christianity*** *had begun.*

This excerpt is stupendously bright because it plays with our own **biases** and **stereotypes**, in particular within the current global context, illustrating that not only we are all biased but also that those biases depend on specifically where and when we live and a huge number of other specific circumstances. That is, many people could think that the above excerpt was referring to the most recent events involving **ISIS**—"***Islamic State of Iraq and the Levant***"—a terrorist militant group that follows a fundamentalist doctrine of **Sunni Islam**. But then we are stricken by the fact that this is not at all the case: the excerpt is referring to markedly similar events that happened millennia ago, which are the focus of her book, entitled *The Darkening Age – the Christian Destruction of the Classic World* (see also Box 7.8 and Fig. 7.4). Firstly, this example reminds us, once again, that human history, rather than "progress"—except of course in a few cases, such as the "progress" of some technology as that involved in the production of cars or computers—is instead mainly a repetition of very similar events.

Fig. 7.4 "The Triumph of Christianity," by Tommaso Laureti, c. 1585. As noted in Nixey's book *The Darkening Age*, "a god has been knocked from its pedestal by the triumphant cross of Christ and now lies broken on the floor…the bases of destroyed Greco-Roman statues were frequently used, once their statues had been removed, to mount Christian crosses"

Secondly, the excerpt also shows us how flawed racist narratives are, as they often use what happens in a specific region and time to build tales in which "others" are seen as being *always* "inferior." Many Europeans nowadays talk about Syrians or Iranians as if they were "inferior" because they have to leave their countries to come to "civilized" Europe, most of the times in a desperate way, risking their lives in boats that cross the Mediterranean Sea. However, if those Europeans would know a minimum of human history, they would be aware of the fact that millennia ago the Assyrian Empire, or the Persian empire, were the ones that were said to be more "developed," when compared to the so-called barbarian groups that were living in northern European regions by then. Moreover, they would also know that just some decades ago, during the **Second World War**, millions of Europeans had to migrate to the Middle East—including to the now ill-named Syrian city **Aleppo**—and that many of those Europeans were received in refugee camps that were said to even have kindergartens that "compared favorably with many in the United States". This is explained in an article titled "*During WWII, European refugees fled to Syria – here's what the camps were like*," published online in 2016 by *Public Radio International*:

> Since civil war erupted in Syria five years ago, millions of refugees have sought safe harbor in Europe by land and by sea, through Turkey and across the Mediterranean. Refugees crossed these same passageways 70 years ago. But they were not Syrians and they traveled in the opposite direction. At the height of **World War II**, the ***Middle East Relief and Refugee Administration*** (MERRA) operated camps in Syria, Egypt and Palestine where tens of thousands of people from across Europe sought refuge. MERRA was part of a growing network of refugee camps around the world that were operated in a collaborative effort by national governments, military officials and domestic and international aid organizations. Social welfare groups including the **International Migration Service**, the **Red Cross**, the

Near East Foundation and the **Save the Children Fund** all pitched in to help MERRA and, later, the **United Nations** to run the camps. In March 1944, officials who worked for MERRA and the International Migration Service (later called the International Social Service) issued reports on these refugee camps in an effort to improve living conditions there. The reports, which detail conditions that echo those faced by refugees today, offer a window into the daily lives of Europeans, largely from Bulgaria, Croatia, Greece, Turkey and Yugoslavia, who had to adjust to life inside refugee camps in the Middle East during World War II. Once registered, recent arrivals wound their way through a thorough medical inspection. Refugees headed toward what were often makeshift hospital facilities – usually tents, but occasionally empty buildings repurposed for medical care – where they took off their clothes, their shoes and were washed until officials believed they were sufficiently disinfected. Some camps even had opportunities for refugees to receive vocational training. At El Shatt and Moses Wells, hospital staff was in such short supply that the refugee camps doubled as nursing training programs for Yugoslavian and Greek refugees and locals alike.

A prominent nurse practitioner named **Margaret G. Arnstein** observed that students in the program were taught practical nursing, anatomy, physiology, first aid, obstetrics, pediatrics, as well as the military rules and regulations that governed camps. Because most had no formal education beyond grammar school, Arnstein noted that nursing curriculum was taught 'in simple terms' and emphasized practical experience over theory and terminology. The head nurses of the training program hoped they could eventually garner formal accreditation so that anyone who finished the program would be licensed to practice nursing after leaving the camps – at the time, nursing students in refugee camps were only able to treat patients because they were 'emergency nurses' operating by necessity in wartime. MERRA officials agreed that it was best for children in refugee camps to have regular routines. Education was a crucial part of that routine. For the most part, classrooms in Middle Eastern refugee camps had too few teachers and too many students, inadequate supplies and suffered from overcrowding. Yet not all the camps were so hard pressed. In Nuseirat, for example, a refugee who was an artist completed many paintings and posted them all over the walls of a kindergarten inside the camp, making the classrooms 'bright and cheerful.' Well-to-do people in the area donated toys, games, and dolls to the kindergarten, causing a camp official to remark that it 'compared favorably with many in the United States'. When they weren't working or going to school, refugees took part in various leisure activities. Men played handball and football and socialized over cigarettes – occasionally beer and wine, if they were available – in canteens inside the camp. Some camps had playgrounds with swings, slides and seesaws where children could keep themselves entertained, and camp officials, local troops, and Red Cross workers hosted dances and put on the occasional performance for camp residents.

Box 7.8 Athens AD 532, Christianity, and the Destruction of the Classical World

In Nixey's *The Darkening Age – the Christian Destruction of the Classic World* she emphasizes historical facts that are often not discussed as widely or profoundly as they should be, at least partially because the most disseminated and influential history books have been, for centuries, written by Christian westerners. In the first chapter of her book she focuses particularly on **Athens**, in AD 532, a key year for the destruction of the classic world:

> In AD 532, a band of seven men set out from Athens, taking with them little but works of philosophy. All were members of what had once been the most famous of **Greece's philosophical schools**, the Academy. The **Academy's philosophers**

proudly traced their history back in an unbroken line – 'a golden chain' as they called it – to **Plato** himself, almost a thousand years before. Now, that chain was about to be broken in the most dramatic way possible: these men were abandoning not just their school but the **Roman Empire** itself. Athens, the city that had seen the birth of **Western philosophy**, was now no longer a place for philosophers. This was no time for a philosopher to be philosophical. 'The tyrant', as the philosophers put it, was in charge and had many alarming habits. In **Damascius**'s own time, houses were entered and searched for books and objects deemed unacceptable. If any were found they would be removed and burned in triumphant bonfires in town squares. Discussion of religious matters in public had been branded a 'damnable audacity' and forbidden by law. Anyone who made sacrifices to the old gods could, the law said, be executed. Across the empire, ancient and beautiful temples had been attacked, their roofs stripped, their treasures melted down, their statues smashed. To ensure that their rules were kept, the government started to employ spies, officials and informers to report back on what went on in the streets and marketplaces of cities and behind closed doors in private homes. As one influential Christian speaker put it, his congregation should hunt down sinners and drive them into the way of salvation as relentlessly as a hunter pursues his prey into nets.

The savage 'tyrant' was **Christianity**. From almost the very first years that a Christian emperor had ruled in Rome in AD 312, liberties had begun to be eroded. And then, in AD 529, a final blow had fallen. It was decreed that all those who laboured 'under the insanity of **paganism**' – in other words Damascius and his fellow philosophers – would be no longer allowed to teach. There was worse. It was also announced that anyone who had not yet been baptized was to come forward and make themselves known at the 'holy churches' immediately, or face exile. And if anyone allowed themselves to be baptized, then slipped back into their old pagan ways, they would be executed. For Damascius and his fellow philosophers, this was the end. They could not worship their old gods. They could not earn any money. Above all, they could not now teach philosophy. For a while, they remained in Athens and tried to eke out a living. In AD 532, they finally realized they could not. They had heard that in the East there was a king who was himself a great philosopher. They decided that they would go there, despite the risks of such a journey. The Academy, the greatest and most famous school in the ancient world – perhaps ever – a school that could trace its history back almost a millennium, closed.

There was a lot of God, or at any rate of **Catholicism**, in my childhood. I suspect that, somewhere between monastery and world, their faith had died too. What never, ever died in our family, however, was my parents' faith in the educative power of the Church. And, in a way, my parents were right to believe this, for it is true. Monasteries did preserve a lot of classical knowledge. But it is far from the whole truth. In fact, this appealing narrative has almost entirely obscured an earlier, less glorious story. For before it preserved, the Church destroyed. In a spasm of destruction never seen before – and one that appalled many non-Christians watching it – during the fourth and fifth centuries, the Christian Church demolished, vandalized and melted down a simply staggering quantity of art. Classical statues were knocked from their plinths, defaced, defiled and torn limb from limb. Temples were razed to their foundations and burned to the ground. A temple widely considered to be the most magnificent in the entire empire was levelled. Many of the **Parthenon sculptures** were attacked, faces were mutilated, hands and limbs were hacked off and gods were decapitated. Some of the finest statues on the whole building were almost certainly smashed off then ground into rubble that was then used to build churches. Books – which were often stored in temples – suffered terribly. The remains of the greatest library in the ancient world, a library that had once held perhaps 700,000 volumes, were destroyed in this way by Christians. It was over a millennium before any other library would even come close to its holdings. Works by censured philosophers were forbidden and bonfires blazed across the empire as outlawed books went up in flames.

Attacks against the monuments of the 'mad', 'damnable' and 'insane' pagans were encouraged and led by men at the very heart of the Catholic Church. The great **St Augustine** himself declared to a congregation in Carthage that 'all superstition of pagans and heathens should be annihilated is what God wants, God commands, God proclaims!' St Martin, still one of the most popular French saints, rampaged across the Gaulish countryside levelling temples and dismaying locals as he went. In Egypt, **St Theophilus** razed one of the most beautiful buildings in the ancient world. In Italy, St Benedict overturned a shrine to Apollo. In Syria, ruthless bands of monks terrorized the countryside, smashing down statues and tearing the roofs from temples. The attacks didn't stop at culture. Everything from the food on one's plate (which should be plain and certainly not involve spices), through to what one got up to in bed (which should be likewise plain, and unspicy) began, for the first time, to come under the control of religion. Male homosexuality was outlawed; hair-plucking was despised, as too were make-up, music, suggestive dancing, rich food, purple bedsheets, silk clothes…The list went on. This is (thus) a book about the **Christian destruction of the classical world**. The Christian assault was not the only one – fire, flood, invasion and time itself all played their part – but this book focuses on Christianity's assault in particular. This is not to say that the Church didn't also preserve things: it did. But the story of Christianity's good works in this period has been told again and again; such books proliferate in libraries and bookshops. The history and the sufferings of those whom Christianity defeated have not been. This book concentrates on them.

One can say that our natural tendency to create and believe in teleological narratives, as well as our confirmation biases, and so on, are the *mother* of racism, misogyny, and ideology, and thus of phenomena such as terrorism and wars, but *ignorance* per se also plays a huge role, being the *father*. We all tend to create and believe in imaginary tales, we all are *Homo fictus et irrationalis*, yes. But it is also clear that if people would make an effort to read about and take into account scientific and historical facts they would *know* in advance that we have a tendency to create and believe imaginary stories and that available empirical data directly contradicts most of those stories. That is, knowledge about the reality of the natural world is by far the best antidote against *Neverland*'s world of unreality, to prepare people to not just blindly accept imaginary tales as "universal truths." Such a knowledge could thus have prevented, or at least minimized, some of the huge atrocities that have darkened, and continue to darken, human history, although as often things are not so simple, because many of the worst atrocities ever done, such as the Holocaust, actually actively involved some of the most "educated" members of Germany at that time, as we will see below. This complex topic is related to the also very complex question on whether the hate between Us versus Them is mainly related to our so-called tribal nature or to our tendencies to build and believe in teleological stories, or to both. We surely have a natural tendency to us *vs* them tribalism, as chimpanzees do, but our human tendency to have a "group" A fighting, killing people from, and in some cases even trying to commit genocide against people belonging to a "group" B that is *not* a true biological entity is surely in great part related to our tendency to build and believe in teleological stories.

When we define people from another religion as "others," or "leftist cannibals" as "them" as it is done in QAnon conspiracy theories, the "others" are clearly *not* a biological group. For instance, the group "Muslims" includes people from completely different regions and continents and with very different genetic backgrounds: it is not at all a true biological group but is instead defined by ideological, aesthetical, and/or theological factors directly associated with teleological narratives. The same applies to the "leftist cannibals" of the QAnon theories, as it can be easily perceived if we raise the following simple question: would chimpanzees, or for that matter any other non-human animal, be able to recognize who are the "leftist human cannibals" to which the QAnon theories refer to? If they saw Pelosi, Soros, Bill Gates, Biden, Trump, Marco Rubio, Bannon, and Pence, would they really know that the former four people are "leftist cannibals" and the latter four are not? Of course they would not be able to know that: actually even most of the 8 billion people alive today don't know that, only people specifically interested in U.S. politics and in conspiracy theories such as those of QAnon would know that, because there is no biological or even ethnic trait, no single physical trait, that would allow to differentiate the "good ones" versus the "leftists". Indeed, in Sapolsky's *Behave* he provides several examples showing that some people that are "supposed" to hate each other based on such ideologies and teleological stories are sometimes indeed able to understand that such tales are mainly completely arbitrary fictional social constructions. That is, by using factual data *and* thinking with their own heads—instead of just following the crowd as many ***Homo socialis*** do —humans do have the capacity to escape *Neverland* and ultimately to undertake extraordinary "noble" acts toward those they were "supposed" to hate, including war truces. Some examples of such acts involving a *recategorization* of Us *versus* Them, mentioned by **Sapolsky**, are:

> In the **Battle of Gettysburg**, Confederate general Lewis Armistead was mortally wounded while leading a charge. As he lay on the battlefield, he gave a secret Masonic sign, in hopes of its being recognized by a fellow Mason. It was, by a Union officer, Hiram Bingham, who protected him, got him to a Union field hospital, and guarded his personal effects. In an instant the Us/Them of Union/Confederate became less important than that of Mason/non-Mason. Another shifting of Thems also occurred during the Civil War. Both armies were filled with Irish immigrant soldiers; Irish typically had picked sides haphazardly, joining what they thought would be a short conflict to gain some military training – useful for returning home to fight for Irish independence. Before battle, Irish soldiers put identifying sprigs of green in their hats, so that, should they lie dead or dying, they'd shed the arbitrary Us/Them of this American war and revert to the Us that mattered – to be recognized and aided by their fellow Irish. Rapid shifting of Us/Them dichotomies is seen during **World War II**, when British commandos kidnapped German general Heinrich Kreipe in Crete, followed by a dangerous eighteen-day march to the coast to rendezvous with a British ship.
>
> One day the party saw the snows of Crete's highest peak. Kreipe mumbled to himself the first line (in Latin) of an ode by Horace about a snowcapped mountain. At which point the British commander, Patrick Leigh Fermor, continued the recitation. The two men realized that they had, in Leigh Fermor's words, 'drunk at the same fountains'. A recategoriza-

tion. Leigh Fermor had Kreipe's wounds treated and personally ensured his safety through the remainder of the march. The two stayed in touch after the war and were reunited decades later on Greek television. And…there is the **World War I Christmas truce**…the famed event where soldiers on both sides spent the day singing, praying, and partying together, playing soccer, and exchanging gifts, and soldiers up and down the lines struggled to extend the truce. It took all of one day for British *versus* German to be subordinated to something more important – *all* of us in the trenches *versus* the officers in the rear who want us to go back to killing each other. Thus **Us/Them dichotomies** can wither away into being historical trivia questions like the Cagots and can have their boundaries shifted at the whims of a census. Most important, we have multiple dichotomies in our heads, and ones that seem inevitable and crucial can, under the right circumstances, have their importance evaporate in an instant. But surprisingly, the same is shown by combatants who, instead of soaring, were cannon fodder, faceless cogs in their nation's war machine. In the words of a British infantry grunt serving in the bloodbath of trench warfare in World War I, 'at home one abuses the enemy, and draws insulting caricatures…how tired I am of grotesque Kaisers…out here, one can respect a brave, skillful, and resourceful enemy…they have people they love at home, they too have to endure mud, rain and steel'. Whispers of Us-ness with people trying to kill you.

These last sentences are particularly relevant for the subject of this book: the politicians and officers "in the rear who want us to go back to killing each other"; and "abuse the enemy, and draw insulting caricatures." This is because these sentences show that, even when politicians eventually do not believe in the Us *versus* Them teleological narratives that they build—having instead other reasons, such as economic ones, self-interest, and so on to use those narratives —they do *need* that the people that fight for them believe in those narratives. So, again, it's not at all "*just the economy, stupid*": economic reasons might be critical for a few politicians and high officers, but for many others leaders, such as **Hitler**, and more importantly for the vast majority of lay people directly involved in war battles or terrorist acts, teleological narratives and related ideologies are a crucial motivation for them to kill the "others." A tragic example that clearly illustrates this point is 9/11: the hijackers remained completely blinded by teleological narratives and radical ideologies from the very beginning of that day until the very end of their lives, when they shouted "Allah is the greatest…Allah is the greatest" just before crashing the planes. In this sense, one of the most powerful sentences I have ever read, due to both its depth and the particular time when it was written, is the one written by someone—nobody seems to know who did it—on a wall in Washington DC precisely after 9/11: "*Dear God, save us from the people who believe in you.*"

Not only most economists—obviously—but also most lay people, and even the vast majority of historians tend to minimize the fact that the *Homo fictus, Homo irrationalis, and Homo socialis* components are so crucial to understand acts such as terrorism (see also Boxes 7.9 and 7.10), wars, and many other terrible acts made by humans. As beautifully pointed out in books such as Taleb's 2010 masterpiece *The Black Swan – The Impact of the Highly Improbable*, historians tend to fall into the trap of **hindsight biases** and build *a posteriori* narratives that are often wrong in order to *rationalize* the decisions of politicians, military personnel, and lay people, as if humans are always perfectly rational: things can only be the result of a logical economic reason or geostrategic reason, and so on. Clearly, this makes no sense at all, because if humans were truly a "sapient animal," they would know how to avoid doing wars that kill millions of people over and over, or how to keep away from destroying the ecology of the planet where we live. As brilliantly showed in *The Black Swan*, most of the darkest events in human history were actually not predictable at all and were in great part related to mainly irrational reasons. Or does someone think that Hitler was truly a "perfectly rational" "sapient being," or that he planned the killing of millions of Jews and Gypsies, and people with congenital malformations (see below) for purely "logical reasons"? In *Stamped from the Beginning* Kendi stresses the direct link between Hitler's ideas and writings and the teleological notion of cosmic purpose, in a chapter where he also exposes the logical fallacies of similar racist ideas and their theoretical roots—such as the so-called IQ tests. Kendi also refers to one of the persons that mostly humiliated **Hitler** in a public event, **Jesse Owen**—a U.S. track and field athlete—who paradoxically—or perhaps expectedly, taking into account the history of the U.S.—later became mainly ignored in his own country:

> When Germany surrendered in the Great War, an embittered Austrian soldier sprinted into German politics, where he gained some cheers for his nasty speeches against Marxists and Jews. In 1924, **Adolf Hitler** was jailed for an attempted revolution. He used the time in prison…to write his magnum opus, ***Mein Kampf***. "The *highest aim of human existence* is…the conservation of race", Hitler famously wrote…[At that time] Eugenicist ideas also became part of the fledgling discipline of psychology and the basis of newly minted **standardized intelligence tests**. Many believed these tests would prove once and for all the existence of natural racial hierarchies. In 1916, Stanford eugenicist **Lewis Terman** and his associates "perfected" the IQ test based on the dubious theory that a standardized test could actually quantify and objectively measure something as intricate and subjective and varied as intelligence across different experiential groups. The concept of general intelligence did not exist. When scholars tried to point out this mirage, it seemed to be as much in the eye of the beholder as general beauty, another nonexistent phenomenon. But Terman managed to make Americans believe that something that was inherently subjective was actually objective and measurable. Terman predicted that the **IQ test** would show "enormously significant racial differences in general intelligence, differences which cannot be wiped out by any scheme of mental culture". Standardized tests became the newest "objective" method of proving Black intellectual inferiority and justifying discrimination, and a multimillion-dollar testing industry quickly developed in schools and workplaces. IQ tests were administered to 1.75 million soldiers in 1917 and 1918. *American Psychological Association* president and Princeton psychologist Carl C. Brigham used the results of the army intelligence tests to conjure up a genetic intellectual racial hierarchy, and a few years later, he constructed the SAT test for college admissions. White soldiers scored better, and for Brigham that was because of their superior White blood. African Americans in the North scored better than **African Americans** in the South, and Brigham argued that northern Blacks had a higher concentration of White blood, and that these genetically superior African Americans had sought better opportunities up North because of their greater intelligence…
>
> Hitler aimed to project the **supremacy of Aryan athleticism** through hosting the 1936 Summer Olympics. The disinterested Du Bois remained away from Berlin for much of August, but **Jesse Owens**, a little-known son of Alabama sharecroppers, made history at the games. He sprinted and leaped for four gold medals and received several stadium-shaking ovations from viewers, Nazis included. When Owens arrived back in the states to a ticker tape parade, he hoped he had also managed to change Americans' racist ideas. That was one race he could not win. In no time, Owens was running against horses and dogs to stay out

> of poverty, talking about how the **Nazis** had treated him better than Americans. If anything, Jesse Owens's golden runs deepened the color line, and especially the racist ideas of **animal-like Black athletic superiority**. **Racist Americans** refused to acknowledge the extraordinary opportunities Blacks received in sports like boxing and track, and the fact that a disciplined, competitive, and clever mind, more than a robust physique, was what set the greatest athletes apart. Instead, athletic racists served up an odd menu of anatomical, behavioral, and historical explanations for the success of Black sprinters and jumpers in the 1932 and 1936 Olympics. "It was not long ago that his ability to sprint and jump was a life-and-death matter to him in the jungle", explained University of Southern California legend Dean Cromwell, Owens's Olympic track coach. But Jesse Owens did not possess the "Negroid type of calf, foot and heel bone" that supposedly gave Blacks a speed advantage, Howard anthropologist **W. Montague Cobb** found in 1936. Since some track stars could pass for White, "there is not a single physical characteristic, including skin color, which all the Negro stars have in common which definitely classify them as Negroes". Cobb did not receive many admirers in a United States where people were convinced about the benefits of natural Black athleticism and biological distinctions. Almost everyone still believed that different skin colors actually meant something more than different skin colors.

I should note that the above excerpt touches me personally and deeply because it gives deserved credit to an African-American anthropologist that is too often neglected in history books, **W. Montague Cobb**. I am very proud to teach and do scientific research in the Historical Black University where he mainly did his work, **Howard University**, which has been and continues to be a key institution in the fight against racism, discrimination, and inequality in the U.S., being among the institutions that contribute the most to break the cycle of poverty in which many African-American families too often become trapped within this country. I am also particularly proud of actively collaborating with colleagues that have recently created, at this university, the *W. Montague Cobb Research Laboratory*, in particular its Director and close friend, an outstanding, visionary African-American Muslim female researcher that actively fights against ethnic, religious, and gender discrimination around the globe, **Fatimah Jackson**.

Box 7.9 Neuroscience, Sacred Values, Narrative Identities, and Finding One's Purpose

In their 2018 paper "*A Multilevel Social Neuroscience Perspective on Radicalization and Terrorism*", Decety and colleagues explain that, among the many factors that can lead people to commit **terrorist acts**, various of the most crucial ones are related to the construction of **fictional narratives**:

> In everyday life, humans construct rich narratives that represent causal connections between different events and so imbue these connections with meaning or purpose. Indeed, one of the ways in which humans understand themselves is by generating **autobiographical stories**, or **narrative identities**, that represent the self in past, present, and imagined future contexts. Coherent self-identities emerge gradually through adulthood from the integration of such narratives, which may themselves change over time, with social roles, psychological dispositions, motivations, and values. With the right psychological ingredients in place, narratives can make bad ideas appear attractive and good ideas wholly unappealing. Of particular relevance to the study of **radi-**

calization are **victimization narratives**, or narratives that frame individuals or social groups as the victims of intentional harm(s). Attempts to understand experiences of victimization, to assign these experiences meaning, and to reconcile these experiences with one's pre-existing identity may lead to the emergence of new narrative identities. In Palestinian and Dutch Muslim youths, for example, perceiving oneself or one's group as the victim of injustices was associated with greater support for violence and radicalism. It is thus not surprising that propaganda materials produced and distributed by **violent extremist organizations** in an effort to garner support and to mobilize recruits routinely employ such victimization narratives. They are a powerful means for transmitting emotions and values. Experiences of victimization through injustice may occur when one feels as though their sacred values or their group's sacred values have been violated. **Sacred values** are distinguishable from instrumental or material values in that they incorporate moral and ethical principles and, as such, are insensitive to trade-offs with other values Sacred values may interact with other aspects of individual and group identity to produce a willingness to make costly sacrifices in support of one's group or cause. For instance, in Moroccans living in neighborhoods associated with **militant jihad**, a willingness to engage in costly sacrifice for **Sharia** was especially pronounced in individuals who considered Sharia a sacred value.

Sacred values may also complicate attempts at conflict resolution – Israelis and Palestinians who were presented with hypothetical Israeli-Palestinian peace deals were more strongly opposed when these deals included material incentives but opposition decreased when adversaries were willing to make symbolic concessions relating to their sacred values. A recent **functional neuroimaging study** reported that sacred values, operationalized as values participants were unwilling to **sacrifice** in exchange for material gain, were associated with activation in **ventrolateral prefrontal cortex** and left **temporoparietal junction**. Interestingly, activation was not observed in typical value encoding regions which might be expected if individuals process sacred values in terms of utility. The authors interpreted the neural response as reflecting the encoding and retrieval of **deontic rules** [deontic rules are established by institutions that have the right or the power to introduce rules, to control whether or not the rules are kept, and to negatively sanction rule violations], which they argue suggests sacred values are concerned with what is right or wrong rather than with outcomes. Violations of **sacred, moral values** may trigger **disgust** and/or **anger responses** that may set the stage for **ideologically-motivated violence**. The 'significance quest' model provides a framework through which some victimization narratives, such as those constructed around **violations of sacred values**, can be understood as risk factors for radicalization. The significance quest refers to the universal need to make a difference, to be noteworthy, and to *find one's purpose*. On this view, the **radicalization process** requires the desire for significance, the identification of ideologically-motivated violence as the most appropriate means by which to achieve significance, and the devaluation of competing goals and desires. **Victimization** narratives may dampen feelings of personal significance and consequently lead to disconnections between individual narrative identities and 'master' societal narratives. Individuals who feel their life stories are at odds with societal narratives often seek understanding through interactions with individuals or groups that have had similar victimization experiences. These people or groups provide alternative 'master' narratives that may accord better with individuals' personal narratives. An individual with extreme views on **religious fundamentalism**, for example, might find the prospect of joining a violent extremist organization attractive if this organization shares their views, and if it can provide a means by which to achieve significance.

In his 2012 book *The World Until Yesterday*, Diamond also provides powerful examples of the profound links between teleological tales, such as those often associated with organized religions, and the justification of **wars**, which became exceptionally prominent with the rise of **sedentism** and particularly of **agriculture**. In fact, wars, which are one of the items that more quickly come to our mind when we talk about "uncivilized" acts, have been and continue to be deeply connected with what many continue to call "civilization," since its very beginning, as we will further discuss below: we can only hope that this deep connection will not remain until the very "end of civilization." Diamond writes:

> Another new problem faced by emergent chiefdoms and states, but not by the bands and tribes of previous history, involved **wars**. Because tribes primarily use relationship by blood or marriage, not religion, to justify rules of conduct, tribesmen face no moral dilemmas in killing members of other tribes with whom they have no relationship. But once a state invokes religion to require peaceful behavior toward fellow citizens with whom one has no relationship, how can a state convince its citizens to ignore those same precepts during wartime? States permit, indeed they command, their citizens to steal from and kill citizens of other states against which war has been declared. After a state has spent 18 years teaching a boy "Thou shalt not kill", how can the state turn around and say "Thou must kill, under the following circumstances", without getting its soldiers hopelessly confused and prone to kill the wrong people (e.g., fellow citizens)? Again, in recent as well as in ancient history, **religion** comes to the rescue with a new function. The **Ten Commandments** apply only to one's behavior toward fellow citizens within the chiefdom or state. Most religions claim that they have a monopoly on the truth, and that all other religions are wrong.
>
> Commonly in the past, and all too often today as well, citizens are taught that they are not merely permitted, but actually obliged, to kill and steal from believers in those wrong religions. That's the dark side of all those noble patriotic appeals: *for God and country*. It in no way diminishes the guilt of the current crop of **murderous religious fanatics** to acknowledge that they are heirs to a long, widespread, vile tradition. The **Bible's Old Testament** is full of exhortations to be cruel to heathens. Deuteronomy 20:10-18, for example, explains the obligation of the **Israelites** to practice genocide: when your army approaches a distant city, you should enslave all its inhabitants if it surrenders, and kill all its men and enslave its women and children and steal their cattle and everything else if it doesn't surrender. But if it's a city of the **Canaanites** or **Hittites** or any of those other abominable believers in false gods, then the true **God commands** you to kill everything that breathes in the city. The **book of Joshua** describes approvingly how Joshua became a hero by carrying out those instructions, slaughtering all the inhabitants of over 400 cities. The book of rabbinical commentaries known as the **Talmud** analyzes the potential ambiguities arising from conflicts between those two principles of "Thou shalt not kill [believers in thine own God]" and "Thou must kill [believers in another god]." For instance, according to some Talmudic commentators, an Israelite is guilty of murder if he intentionally kills a fellow Israelite; is innocent if he intentionally kills a non-Israelite; and is also innocent if he kills an Israelite while throwing a stone into a group consisting of nine Israelites plus one heathen (because he might have been aiming at the one heathen).
>
> In fairness, this outlook is more characteristic of the Old Testament than of the **New Testament**, whose moral principles have moved far in the direction of defining one's dealings with anyone – at least in theory. But in practice, of course, some of history's most extensive **genocides** were committed by **European Christian colonialists** against non-Europeans, relying for moral justification on the New as well as the Old Testament. Interestingly, among **New Guineans**, religion is never invoked to justify killing or fighting with members of an out-group. Many of my New Guinea friends have described to me their participation in genocidal attacks on neighboring tribes. In all those accounts, I have never

heard the slightest hint of any religious motive, of dying for God or the true religion, or of sacrificing oneself for any idealistic reason whatsoever. In contrast, the religion-supported ideologies that accompanied the rise of states instilled into their citizens the obligation to obey the ruler ordained by God, to obey moral precepts like the Ten Commandments only with respect to fellow citizens, and to be prepared to sacrifice their lives while fighting against other states (i.e., heathens). That's what makes societies of religious fanatics so dangerous: a tiny minority of their adherents (e.g., 19 of them on September 11, 2001) die for the cause, and the whole society of fanatics thereby succeeds at killing far more of its perceived enemies (e.g., 2,996 of them on September 11, 2001). Rules of bad behavior toward out-groups reached their high point in the last 1,500 years, as fanatical Christians and Muslims inflicted death, slavery, or forced conversion on each other and on the heathen. In the 20th century, European states added secular grounds to justify killing millions of citizens of other European states, but religious fanaticism is still strong in some other societies.

Box 7.10 White Supremacy, Timothy McVeigh, and Links to al-Qaida Narratives

In Younge's article "*White Supremacy Feeds on Mainstream Encouragement*" published in April, 2019, in *The Guardian*, he tells us about a mesmerizing story about the friendship of the **white supremacist terrorist Timothy McVeigh** and **Ramzi Ahmed Yousef**, who trained at **Al-Qaida** camps. The story is surprising in the sense that most people don't know about it, but as one often says, reality is often even more incredible than fiction, and taking into account what has been shown in the present book, such a link between the narratives of white supremacists and those of al-Qaida members is actually not surprising at all. This is because many teleological narratives used in different cultures or human groups or at different times are indeed strikingly similar to each other, being mainly a repetition of very old teleological tales about cosmic purpose and cosmic progress and related ideologies mixed with tribal notions of Us *versus* Them. Younge states:

> On 19 April 1995, Timothy McVeigh blew up the Alfred P Murrah Federal Building in Oklahoma City, killing 168 people and injuring another 684, in the deadliest act of domestic US terrorism to date. A **white supremacist**, among other things, he was radicalised by what he regarded as excessive federal government power, US foreign policy and a constellation of bigotries and inadequacies too numerous to mention. He was sentenced to death and placed in a maximum-security prison in Colorado. While he was there, McVeigh became good friends with Ramzi Ahmed Yousef, who was in the next-door cell. Yousef, who trained in **al-Qaida camps in Afghanistan**, had tried to blow up the World Trade Center in 1993. Born within four days of each other, here were two young men both unrepentant about their crimes and **ideologies**, even if McVeigh had ultimately been more successful in inflicting carnage and misery. He was executed in June 2001. Yousef would later state: 'I have never [known] anyone in my life who has so similar a personality to my own as his'. It is presumably in recognition of the unity of purpose between terrorists who attach themselves to Islam and those who embrace white supremacy that the **neo-Nazi terror group National Action** has called for a '**white jihad**'. Inadequate young men brimming with rage and brooding resentment, in pursuit of **moral certainty**, **doctrinal purity** and a desire to make their mark on a world in which they feel increasingly superflu-

ous and disoriented. They are as made for each other as Yousef and McVeigh. The central difference is in how these two complementary strands of political violence are understood and challenged.

White supremacy poses a serious and urgent threat to our political stability, social cohesion and general security. White people are being radicalised at an alarming rate and in disconcerting numbers. The number of far-right terrorists in British prisons tripled between 2017 and 2018. While the violence may come from the fringes, the encouragement comes from the centre. Only last week, Britain's counter-terrorism chief, Neil Basu, claimed far-right terrorists were being radicalised by mainstream newspaper coverage; only this week a cartoonist who portrayed refugees as vermin received a fellowship from the society of editors for his 50 years' service to the industry. A senior politician has described women in niqabs as looking like letterboxes; and Ukip, polling at 7%, appointed a notorious far-right activist, Tommy Robinson, as an adviser. There is of course a relationship between the brutal manifestations of bigotry and intolerance and the antiseptic talk of amendments and parliamentary procedure taking place around this stage of the **Brexit** negotiations. But it is contextual rather than causal. Brexit didn't create racism. It has exploited, leveraged, amplified, focused, suckled, fed and nurtured it. It has given many licence to say and do things that would previously be regarded as unacceptable. It has laid bare what was hidden and mainstreamed what was marginal. There were lots of reasons why people voted to leave the European Union, some of which were perfectly reasonable. But it would be more accurate to state that racism – particularly expressed in the form of nostalgia and **xenophobia** – helped to make Brexit possible, rather than the other way around. Moreover, this intensification of rightwing nationalism is global. It's just a few weeks since the **deadly shootings in the mosques in Christchurch**, New Zealand, and a few months since the **massacre at the synagogue in Pittsburgh**. We live in a period when the presence of **fascists in government in Europe** is no longer noteworthy and neo-Nazi demonstrators have an advocate in the White House. White supremacy, and the violence that it spawns, can no longer be lampooned as a postgraduate talking point. It is unpicking the now very fragile fabric that's keeping us all together. The number of rightwing extremists arrested in Europe nearly doubled between 2016 and 2017; there was a 74% increase in **antisemitic offences** in France last year; and the number of hate groups in the U.S. is the highest ever. According to a 2017 Washington Post/ABC poll roughly one in 10 Americans believes that holding Nazi views is acceptable – assuming very few of those were black or Latino, that's a lot of white people in a very dark place.

Were we to borrow the language directed against Muslims at this point, we would be talking about a 'death cult' deep in the **Caucasian psyche** and calling on moderate, law-abiding members of the white community to step up and speak out. There would be lofty lectures on the **Enlightenment** and liberal democracy and integration. We would be talking travel bans, rendition and muscular liberalism. The Home Office would revoke **Boris Johnson**'s citizenship on the basis that he was born in the US and his presence here is not conducive to the public good. The home secretary might treat a roomful of white people, as John Reid did Muslims in east London in 2006, to tips on parenting. 'Look for the tell-tale signs now, and talk to them before their hatred grows and you risk losing them for ever'. But we won't borrow that language, not only because it is patronising and racist, holding one group of people collectively responsible for the actions of just a few on the basis of their shared identity. We won't borrow it because we know it doesn't work. While police protection for MPs and more vigilance of hard-right groups might help in the short term, the reality is that the state can only contain a terror threat – it can't eliminate it. As

Northern Ireland showed us, political violence demands political solutions. At some stage that will require our political class to stop pandering to bigotry and start challenging it. It should be clear by now, if it was not before, that any concession that is made to bigots does not satisfy but emboldens them. That is, in no small part, how we got here. They need to call it out both in the chamber and on the doorstep. If it came from any other group they would have done it already. In the current climate they can no longer avoid it. Their lives may depend on it. They know that racism is the problem. They have yet to understand that anti-racism is the solution.

Fig. 7.5 General Custer on horseback with his U.S. Army troops in battle with Native American Lakota Sioux, Crow, Northern, and Cheyenne, at Little Bighorn Battlefield, June 25, 1876 (Little Bighorn River, Montana)

Another example, within the endless number of cases I could mention here, is Marshall's 2007 book "*The Day the World Ended at Little Bighorn,*" focused on the 1876 **Little Bighorn battle** (Fig. 7.5), as well as other battles between the U.S. army and **Native American groups** such as the **Lakota** and **Cheyenne**. Referring to the specific battle won by the latter groups, Marshall explained that "initial disbelief about the Lakota and Cheyenne victory stemmed from a pervasive Euro-American notion that natives were an inferior people." Clearly linking this vision of superiority with teleological narratives of "progress" and religion, he notes that "Euro-Americans considered natives to be uncivilized savages standing in the path of

Christian progress." Marshall explains in detail the numerous consequences of the Euro-American invasion, the most devastating one being "the loss of the primary resource on the northern plains: the bison herds…once seemingly numberless, the bison herds had been reduced to only a few thousand by the early 1870s…dire consequences of that reality were felt by all the Lakota, whose physical and spiritual survival depended on this endlessly useful animal."

This is because, "for generations, bison had been the source of food, clothing, utensils, toys, and weapons; every part of its body was utilized, from hooves to tail…at least twenty bison hides were used to construct one conical dwelling, the tipestola, or "pointed dwelling," more commonly known as the tipi…and its strength and independence were at the core of the spiritual symbiosis the Lakota shared with it…so long as the Lakota had bison to eat, they were strong and independent…its decline was both a practical and spiritual blow." The later process of "assimilation" of the natives by the Euro-Americans was clearly also related to/justified by teleological narratives, as explained by Marshall. Namely, it "was intended to destroy native culture by stopping its growth in the young…the premise was based on how America thought of itself, articulated by Judge Elmer Dundy's opinion in his decision in Standing Bear v. Crook in 1879, part of which stated: *On one side we have the remnants of a once numerous and powerful, but now weak, insignificant, unlettered, and generally despised race; on the, we have the representatives of one of the most powerful, most enlightened, and most Christianized nations of modern times*."

One important point that needs to be made is that of course different **Native American groups** also had wars between themselves, which were often related to/justified by teleological narratives and ideologies as well. It is indeed interesting that even those that consider themselves as "not racist" often neglect or omit those historical facts. They often try to do it to go against racist tales according to which the "savages" are brute and wild, but don't realize that actually by doing so they are falling into the trap of the "Euro-American superiority" fallacy. This is because within such a **Eurocentric narrative**, either agency is often not attributed to the "other," or the "other" is supposed to have only "basic," "brute" reasons to fight, not "higher" teleological ones. But there is vast empirical evidence showing that contrary to most nomadic hunter-gatherer groups, sedentary ones such as the Native American Lakota and Cheyenne often—and agricultural "civilizations" even more often—had wars, which were not only related to typical sedentary societal factors such as having higher population density, inequalities, and "valuable" private property, oppression, and so on, but also to beliefs and teleological just-so-stories. As noted by Marshall, "battles between enemy tribes on the northern plains did not occur as frequently as many historians believe…when they did, however, it was an opportunity for men to prove themselves…for the Lakota, demonstrating courage in the face of an enemy was the best way for a man to show himself worthy of leadership, on and off the battlefield." However, as he noted, "engaging in combat also had a higher purpose: warriors fought to protect their families and homes…it was considered the highest of callings and a sacred duty."

Another book that directly relates wars, as well as terrorist acts, to teleological narratives, including the *belief in the apocalypse*, is Kirkham's 2019 *Our Shadowed World*:

> **ISIS** is itself a product of **apocalypticism**. From the outset **Islam**, following the monotheistic tradition of the so-called **Abrahamic family of faiths** from which it springs, had an apocalyptic expectation of a threatened or promised ending. The contemporary clash of this tradition with modern Western civilization has created a new context for its manifestation. Ever since the dismemberment of the **Ottoman state and caliphate**, the Muslim world has suffered a deep sense of grievance, particularly over the arbitrary creation of the mosaic of modern Middle Eastern states (created to suit Western interests), which provides the context of contemporary conflict. As one radical preacher urged, "your father's Islam is what the colonizers left behind, the Islam of those who bow down and obey…our Islam is the Islam of combatants, of blood, of resistance". ISIS adherents believe in an imminent apocalypse that only the most devout will survive. Their attempt to restore a caliphate aims to create a space where believers – as well as nonbelievers – can kill and be killed and so earn martyrdom. Its values invert those of normal civilization: "you love life, we love death", its adherents taunt. As Roy says, this narrative "has the power to fascinate fragile individuals suffering from genuine psychiatric problems". Its certainties have great appeal to those adrift in the ambiguities of modernity or torn by the **cognitive dissonance** of conflicting values. Not only does it open up an opportunity for an increasing number of "lone wolves" to make a lasting affirmation of identity – like some character out of a novel by **Dostoyevsky** or Conrad – but it gives a rallying point for those newly converted to Islam to express their commitment. These converts to Islam in France, Belgium, and Britain as well as **born-again Muslims** who after living highly secular lives suddenly renew their allegiance, constitute the core of jihadists who operate beyond the traditional frameworks of religious organization in the pursuit of radicalism and death. Their bloodlust sets them apart as did the mark of **Cain** in the mythical past.
>
> Where civilization began, an apocalypse of planetary proportions now looms; the **four horsemen of destruction, war, famine, and death** are once again in the saddle. Many tremble at this descent into chaos, but the contemporary Middle East merely reprises the inexorable pattern of civilization; for as Lewis Mumford observed in his monumental study *The city in history*, the essence of civilization has always been the exertion of power in every form. The city became the paramount expression of this truth as an instrument of aggression, domination, and conquest. As **Plato** declared in *The laws*, "in reality every city is in a natural state of war with every other". The consequence of this thirst for power became expressed in empire, as with **Sargon**, **King of Akkad**, who was the first to create an empire, seeking to dominate all he beheld. Again, nothing expresses this state more clearly than ISIS in Syria. Though the rest of the world looks on aghast as ancient cities like Nineveh and Nimrud are utterly destroyed, they forget that these cities themselves represent and recapitulate centuries of destruction. The palaces built by Sennacherib bear witness to his power and his total annihilation of **Babylon** and other rivals: "the city and houses from its foundation to its top, I destroyed, I devastated, I burned with fire…the wall and the outer wall, temples and gods, temple tower of brick and earth, as many as there were, I razed…I made its destruction more complete than that by a flood". **Civilization** has always stood on the neck of the vanquished, and now ISIS apes this attitude and behavior, and, though perhaps unwittingly, *continues the story of civilization*.

One of the most recent and brilliant books discussing the history of brutality within our species is **Malesevic'** book *The Rise of Organized Brutality – A Historical Sociology of Violence*. Contrary to the ideas defended by Pinker and many other Hobbesian authors, Malesevic argues that violence has mainly increased in human history, in particular with the rise of **sedentism** and then of **agriculture** and the formation of the first states or "**civilizations**" (see Box 7.11). He shows that this was mainly related to the rise of leaders and the social organizations supporting them, such as bureaucratic and religious institutions. Furthermore, he also contradicts, using well-grounded empirical data, the widespread erroneous notion that wars are normally related to "rational" reasons or simply driven by "economic factors":

> Without intending to dispute the often observed finding that sharp class or status polarization and **inequalities** contribute to popular discontent, it is not self-evident why these factors matter more in some cases and have very little impact in others…**economics-centred explanations** of revolution, which also focus on economic disparities and deprivation, cannot provide plausible answers to the question of why popular discontent rarely translates into rebellion or revolutionary upheaval. In the same way, such theories cannot account for the nonexistence of **terrorism** in some of the poorest and most unequal societies in the world. Why is there so much terrorism in **Iraq**, **Pakistan** and **Afghanistan** and almost none in Namibia, Lesotho or Botswana, which top the list of the most unequal countries in the world? Although social injustice and economic inequality can under particular circumstances contribute substantially to the proliferation of violence, there is no reliable evidence that poverty, inequality and deep economic disparities by themselves cause a terrorist response. On the contrary, several important recent studies show the opposite trend, with an increase in living standards being positively correlated with involvement in and support for **terrorist activities**. There is also robust evidence demonstrating convincingly that rather than being uneducated and impoverished individuals, most terrorists come from the relatively affluent backgrounds. For example, **Al Qaeda** membership was disproportionally staffed by voluntary recruits from the upper and middle classes who were well educated, possessing science, medicine or engineering degrees from respectable universities. A similar pattern was established for other **Islamic radical groupings** involved in terrorism who in some important respects resemble their late-nineteenth and early-twentieth-century anarchist counterparts. Gambetta and Hertog (in 2009) have analysed the demographic and biographical details of over four hundred jihadists involved in violence and have found that a large majority had engineering degrees. Since terrorist activities entail a substantial degree of technological, organisational and communicational skills, it seems reasonable to assume that the educated middle classes are better equipped to take part in such activities. Nevertheless, this does not explain their motivations…
>
> **Wars** cannot be waged without effective **social organizations**. However, the historical record demonstrates that organisational power on its own is rarely enough to succeed on the battlefield. What has also been highly important is the ability of those who wage wars to justify their violent actions and if possible also mobilise a substantial degree of support for their cause. Hence successful **warfare** often entails the presence of persuasive **ideological doctrines**. To be deemed justifiable, wars have to be fought in the name of something important – to defend one's nation, to uphold the royal dynasty, to fulfill a **civilising mission**, to preserve the existence of a particular collective, to impose social justice, to ensure racial purity, to expand specific religious doctrine, to defend national territory and so on. The conventional views on the personal motivation of soldiers to take part in wars clash over the question of whether actors are motivated by 'greed', that is, the individual's self-interest, or 'grievances', that is, noneconomic motives such as identity, religion or other cultural factors. Hence Collier *et al.* (in 2009) analyse recent **civil wars** through the prism of an actor's material benefits, arguing that 'there is evidence…that where a rebellion is financially and militarily feasible it will occur'. In contrast, Cederman *et al.* (in 2013) insist that the available data indicate that the primary individual motivation in most civil wars stems from social grievances: 'political and **economic inequalities** following group lines generate grievances that in turn can motivate civil war'.
>
> While both of these factors certainly play an important part in individual motivation, there is a need for a nuanced sociological understanding that takes into account the com-

plexity of collective decision making, the changing historical dynamics of conflicts and geographical variation. For example, to account for the willingness of millions of British, French and German soldiers to fight and die in World War I, Michael Mann identifies five key reasons: the dominance of **militaristic culture** in the pre–World War I Europe; the youth's thirst for adventure as an attempt to escape the dull working- or middle-class life; the popularly shared belief, reinforced by mass media and government messages, that this was a war of self-defence and 'civilisation against barbarism'; and **local community pressure**, including the localised system of recruitment and the institution of regular pay and full employment generated by war…These factors were also present in World War II, although Mann argues that ideological commitment played a greater role in World War II when compared to World War I. However, even in this case, ideology mattered more for some than others, as he sharply distinguishes between the fascist regimes and the rest. More specifically, he accepts the view articulated by Bartov and others who see **Wehrmacht soldiers** as being more motivated by the **Nazi ideology** than the sense of a micro-, platoon-level comradeship associated with most other militaries. In this view, the fascist states were characterised by a substantial degree of ideological unity between frontline soldiers and political elites. Mann illustrates this with available research that analyses diaries and letters of Wehrmacht soldiers who adored Hitler and remained loyal until the end of the war.

But Malesevic notes that:

[However] Bartov's studies have been criticised for taking the letters of frontline Wehrmacht soldiers at face value while ignoring the fact that sincere emotional expressions or criticisms of the Nazi regime could not pass the censors and could even lead to the courts martial. The fact that **Nazis** and Japanese military authorities shot 20,000 soldiers each for desertion or showing cowardice on the frontline while only 146 U.S. soldiers were given death penalty is in itself a powerful indicator that coercion was more significant than ideological devotion. In a similar vein, Neitzel and Welzer's (in 2012) recent analysis of transcripts of secret recordings of German prisoners of war (POWs) indicate that the **microsolidarity** was a much more powerful source of individual motivation than any doctrinal principles. For example, key Nazi ideas such as folksgemeinschaft, 'global Jewish conspiracy' and 'Bolshevik promotion of Genetic inferiority' rarely if ever appear in the conversations of POWs. The authors demonstrate well that: 'as a rule German soldiers were not 'ideological warriors'…most of them were fully apolitical'. What really mattered is the sense of responsibility and duty towards one's comrades. This leads us the final point: the relationship between ideological power and microsolidarity. Saying that the bonds of microsolidarity often trump officially proclaimed ideological creeds does not suggest that ideology does not matter. On the contrary, in addition to coercive organisational might, **ideological power** is a cornerstone of much of social action. The point is that a great deal of ideological power does not stem from a set of uncompromising principles and beliefs. Instead, ideological processes work best when they are in perpetual motion and when they are able to amalgamate with other forms of social action. In other words, rather than assuming that ideology is an either/or singular phenomenon directly opposed to material interests or emotional, habitual or value rational action, it is much more fruitful to focus on **ideologisation** as a multifaceted process that blends with different forms of social action and taps into existing social relations. Hence ideological power is heavily dependent on the capacity to tie the diverse pockets of microsolidarity into a shared, wider, ideological narrative. In this context, **nationalism** has regularly proved to be the most potent social glue that brings together patches of microsolidarity into a society wide common narrative in which heterogeneous individuals and groups can instantly recognise their own personal experiences…

Since the backbone of social organizations is the coercive power, they are the main vehicles of violence. The popular views, reinforced by the dominance of the **Hobbesian paradigm** in social science, see human beings as intrinsically violent creatures. In this understanding, our primordial inclination to violence has been gradually constrained and overcome by the establishment of powerful and long-lasting institutions such as the state, civil society, capitalism or civilisation. The argument is that as such institutions gain in strength, they thwart conflicts, thus removing violence from the human everyday interactions and securing lasting peace and order. In contrast to this view, I argue that human beings are not naturally predisposed either for violence or for peace: we lack recognisable biological prerequisites for fighting, and we are not naturally prone towards living in very large congruous associations. Instead, it is in fact the development of complex social organizations that has given impetus to proliferation of violent acts. Hence, rather than stifling alleged genetically programmed aggressive impulses, social organizations create conditions for conflict and also foster expansion of violence among human beings. Organisational power has proved crucial for successfully mobilising the large groups of individuals as well as for maintaining those individuals in their specific roles. There would be no **wars**, **revolutions**, **genocides**, insurgencies or **terrorism** without durable social organizations. The bureaucratic entities are there to make sure that individuals pay taxes, levies or tributes that will finance violent conflicts; that they are available for recruitment when the organizations require fighters; that they are producing weapons, equipment and resources for future clashes; that they provide popular support for the use of violence in organisational causes; and that human beings comply with so many other organisational demands. It is the coercive might of organizations that has historically proven decisive for the popular compliance…While coercion and economic remuneration are important for establishing and making sturdy social organizations work, they are not sufficient in the long term. In addition, nearly all durable bureaucracies require also a more specific social glue that motivates its members and holds these entities together.

Throughout history, social organizations tended to rely on different cultural mechanisms to secure organisational legitimacy as well as to mobilise a degree of popular support. For example, composite kingdoms and pristine empires utilised religious principles, including the idea of the divine rights of rulers, to justify the existing social structure. In a similar vein, chartered companies such as **British East India** or **Dutch Verenigde Oostindische Compagnie** regularly deployed the idiom of civilising mission both to legitimate their **colonial expansion** and to justify their existence to their own employees. Today, nation-states rely on nationalism to maintain popular support as well as to mobilise public opinion for a particular course of action. Nevertheless, this is not to say that the premodern world required as much popular justification as it is needed today. While when waging wars or domestic persecutions rulers of pristine empires and early kingdoms were interested in acquiring as much support as they could muster, their target audience was usually very small – fellow aristocrats and the top clergy. As the majority of the largely peasant population were excluded from the body politic and military affairs and were traditionally perceived as inferior, almost subhuman, rabble, there was little organisational demand to justify political decisions. Hence the divine rights principle and other religious doctrines were really more proto-ideologies than fully fledged ideological doctrines. In contrast, the birth and expansion of modernity are associated with a much deeper ideological penetration. Once the ideals of humanity, equality, liberty and fraternity take centre stage and replace supernatural authorities as the dominant source of organisational legitimacy, ideological power becomes much more significant. **Ideological penetration** entails substantially increased **literacy rates**; mass printing of cheap and affordable books, pamphlets, popular magazines and newspapers; the development of public sphere; the existence of a military draft; enlarged urbanisation; and the existence of mass educational systems, among others; all of which help politicise ordinary citizens. Although ideological projects are cornerstones of modernity, as exemplified by the staggering rise and fall of **fascism**, **Nazism**, **Stalinism**, **Maoism**, **anarchism**, **classical liberalism** and many other modern -isms, **ideologisation** is particularly important in the context of violence.

Box 7.11 Democides, Genocides, Politicides, Marxism, and Nazism
As noted above, it is remarkable that, although Pinker's *Better Angels* is ultimately a defense of Hobbes' ideas that states, particularly those in which there is a strong political authority, are "better" at decreasing the level of violence and number of wars, most of the examples that he provides in that book regarding wars, and other horrendous atrocities related to teleological narratives and ideologies, concern the so-called civilized states, often with very powerful central leaders, including kings and dictators:

> Luard designates 1559 as the inception of the **Age of Religions**, which lasted until the **Treaty of Westphalia** ended the **Thirty Years' War** in 1648. Rival religious coalitions, often aligning with rulers according to the principle *Un roi, line loi, line foi* (**One king, one law, one faith**), fought for control of cities and states in at least twenty-five international wars and twenty-six civil wars. Usually **Protestants** warred against **Catholics**, but during **Russia's Time of Troubles** (an interregnum between the reign of **Boris Godunov** and the establishment of the **Romanov dynasty**), Catholic and Orthodox factions vied for control. The religious fever was not confined to **Christendom**: Christian countries fought Muslim Turkey, and Sunni and Shiite Muslims fought in four wars between Turkey and Persia. This is the age that contributed (to major) atrocities…The era broke new records for killing partly because of advances in military technology such as muskets, pikes, and artillery. But that could not have been the main cause of the carnage, because in subsequent centuries the technology kept getting deadlier while the death toll came back to earth. Luard singles out **religious passion** as the cause: *"It was above all the extension of warfare to civilians, who (especially if they worshipped the wrong god) were frequently regarded as expendable, which now increased the brutality of war and the level of casualties. Appalling bloodshed could be attributed to divine wrath. The duke of Alva had the entire male population of Naarden killed after its capture (1572), regarding this as a judgment of God for their hard-necked obstinacy in resisting; just as Cromwell later, having allowed his troops to sack Drogheda with appalling bloodshed (1649), declared that this was a "righteous judgement of God." Thus by a cruel paradox those who fought in the name of their faith were often less likely than any to show humanity to their opponents in war. And this was reflected in the appalling loss of life, from starvation and the destruction of crops as much as from warfare, which occurred in the areas most ravaged by religious conflict in this age.*" Names like the "Thirty Years' War" and the "**Eighty Years' War**", together with…never-equaled spike in war durations, tell us that the Wars of Religion were not just intense but interminable. The historian of diplomacy Garrett Mattingly notes that in this period a major mechanism for ending war was disabled: "as religious issues came to dominate political ones, any negotiations with the enemies of one state looked more and more like heresy and treason…the questions which divided Catholics from Protestants had ceased to be negotiable…consequently…diplomatic contacts diminished." It would not be the last time ideological fervor would act as an accelerant to a military conflagration. Of course, it all went horribly wrong. The **French Revolution** and the **French Revolutionary and Napoleonic Wars** caused as many as 4 million deaths, earning the sequence a spot in the twenty-one worst things people have ever done to each other…
>
> [Regarding genocides] so far I have tried to explain genocide in the following way. The mind's habit of essentialism can lump people into categories; its moral emotions can be applied to them in their entirety. The combination can transform

Hobbesian competition among individuals or armies into Hobbesian competition among peoples. But **genocide** has another fateful component. As Solzhenitsyn pointed out, to kill by the millions you need an **ideology**. Utopian creeds that submerge individuals into moralized categories may take root in powerful regimes and engage their full destructive might. For this reason it is ideologies that generate the outliers in the distribution of **genocide death tolls**. Divisive ideologies include Christianity during the **Crusades** and the Wars of Religion (and in an offshoot, the **Taiping Rebellion in China**); revolutionary romanticism during the **politicides** of the French Revolution; nationalism during the genocides in Ottoman Turkey and the Balkans; **Nazism** in the **Holocaust**; and **Marxism** during the purges, expulsions, and terror-famines in **Stalin's Soviet Union**, **Mao's China**, and **Pol Pot's Cambodia**. Why should utopian ideologies so often lead to genocide? At first glance it seems to make no sense. Even if an actual utopia is unattainable for all kinds of practical reasons, shouldn't the quest for a perfect world at least leave us with a better one – a world that is 60 percent of the way to perfection, say, or even 15 percent? After all, a man's reach must exceed his grasp. Shouldn't we aim high, dream the impossible dream, imagine things that never were and ask "why not"? **Utopian ideologies** invite genocide for two reasons. One is that they set up a pernicious utilitarian calculus. In a utopia, everyone is happy forever, so its moral value is infinite. Most of us agree that it is ethically permissible to divert a runaway trolley that threatens to kill five people onto a side track where it would kill only one. But suppose it were a hundred million lives one could save by diverting the trolley, or a billion, or – projecting into the indefinite future – infinitely many. How many people would it be permissible to sacrifice to attain that infinite good? A few million can seem like a pretty good bargain. Not only that, but consider the people who learn about the promise of a perfect world yet nonetheless oppose it. They are the only things standing in the way of a plan that could lead to infinite goodness. How evil are they? You do the math.

Democides are often scripted into the climax of an eschatological narrative, a final spasm of violence that will usher in millennial bliss. The parallels between the utopian ideologies of the 19th and 20th centuries and the **apocalyptic visions** of traditional religions have often been noticed by historians of genocide. Daniel Chirot, writing with the social psychologist Clark McCauley, observes: "***Marxist eschatology*** *actually mimicked* ***Christian doctrine****. In the beginning, there was a perfect world with no private property, no classes, no exploitation, and no alienation – the* ***Garden of Eden****. Then came sin, the discovery of private property, and the creation of exploiters. Humanity was cast from the Garden to suffer inequality and want. Humans then experimented with a series of* ***modes of production****, from the slave, to the feudal, to the* ***capitalist mode****, always seeking the solution and not finding it. Finally there came a* ***true prophet with a message of salvation****,* ***Karl Marx****, who preached the* ***truth of Science****. He promised redemption but was not heeded, except by his close disciples who carried the truth forward. Eventually, however, the proletariat, the carriers of the true faith, will be converted by the religious elect, the leaders of the party, and join to create a more perfect world. A final, terrible revolution will wipe out capitalism, alienation, exploitation, and inequality. After that, history will end because there will be perfection on earth, and the true believers will have been saved.*" Drawing on the work of the historians Joachim Fest and George Mosse, they also comment on **Nazi eschatology**: "*It was not an accident that Hitler promised a* ***Thousand Year Reich****, a millennium of perfection, similar to the thousand-year reign of goodness promised in Revelation before the return of evil, the great battle between good and evil, and the* ***final triumph of God over Satan****. The entire imagery of his Nazi Party and regime was deeply mystical, suffused with religious, often Christian, liturgical symbolism, and it appealed to a higher law, to a mission* decreed *by fate and entrusted to the* **prophet Hitler**".

The last point made in Box 7.11 is in line with what Epstein wrote in his 2010 bestseller *Good without God – what a Billion Nonreligious People Do Believe*: "**Hitler** is often erroneously labeled a secularist or atheist…[but] in carrying out the **Holocaust**, Hitler wrote, '*I am acting in accordance with the will of the Almighty Creator: by defending myself against the Jew, I am fighting for the work of the Lord*'…the Nazi army's belts were inscribed ***Gott mit uns!*** (*God is with us*)." Coming back to Pinker's *Better angels*, it further states:

> Certainly an **ideology of an afterlife** helps, as in the posthumous Playboy Mansion promised to the 9/11 hijackers. (Japanese kamikaze pilots had to make do with the less vivid image of being absorbed into a great realm of the spirit.) But modern **suicide terrorism** was perfected by the **Tamil Tigers**, and though the members grew up in Hinduism with its promise of **reincarnation**, the group's ideology was secular: the usual goulash of **nationalism**, **romantic militarism**, **Marxism-Leninism**, and **anti-imperialism** that animated 20th-century third-world liberation movements. And in accounts by would-be suicide terrorists of what prompted them to enlist, anticipation of an afterlife, with or without the virgins, seldom figures prominently. So while expectation of a pleasant afterlife may tip the perceived cost-benefit ratio (making it harder to imagine an atheist suicide bomber), it cannot be the only psychological driver. Using interviews with failed and prospective suicide terrorists, the anthropologist Scott Atran has refuted many common misconceptions about them. Far from being ignorant, impoverished, nihilistic, or mentally ill, suicide terrorists tend to be educated, middle class, morally engaged, and free of obvious psychopathology. Atran concluded that many of the motives may be found in **nepotistic altruism**. The case of the Tamil Tigers is relatively easy. They use the terrorist equivalent of file closers, selecting operatives for suicide missions and threatening to kill their families if they withdraw.
>
> Only slightly less subtle are the methods of **Hamas** and other **Palestinian terrorist groups**, who hold out a carrot rather than a stick to the terrorist's family in the form of generous monthly stipends, lump-sum payments, and massive prestige in the community. Though in general one should not expect extreme behavior to deliver a payoff in biological fitness, the anthropologists Aaron Blackwell and Lawrence Sugiyama have shown that it may do so in the case of **Palestinian suicide terrorism**. In the West Bank and Gaza many men have trouble finding wives because their families cannot afford a bride-price, they are restricted to marrying parallel cousins, and many women are taken out of the marriage pool by **polygynous marriage** or by marriage up to more prosperous Arabs in Israel. Blackwell and Sugiyama note that 99 percent of Palestinian suicide terrorists are male, that 86 percent are unmarried, and that 81 percent have at least six siblings, a larger family size than the Palestinian average. When they plugged these and other numbers into a simple demographic model, they found that when a terrorist blows himself up, the financial payoff can buy enough brides for his brothers to make his sacrifice reproductively worthwhile. Atran has found that **suicide terrorists** can also be recruited without these direct incentives. Probably the most effective call to **martyrdom** is the opportunity to join a happy band of brothers. Terrorist cells often begin as gangs of underemployed single young men who come together in cafes, dorms, soccer clubs, barbershops, or Internet chat rooms and suddenly find meaning in their lives by a commitment to the new platoon. Young men in all societies do foolish things to prove their courage and commitment, especially in groups, where individuals may do something they know is foolish because they think that everyone else in the group thinks it is cool. Commitment to the group is intensified by religion, not just the literal promise of paradise but the feeling of spiritual awe that comes from submerging oneself in a crusade, a calling, a vision quest, or a jihad. Religion may also turn a commitment to the cause into a sacred value – a good that may not be traded off against anything else, including life itself.

In their 2010 book *Causes of War* Levy and Thompson provide an interesting brief discussion about how decision-making at an individual level that leads to wars is often plagued by a mix of the typical storytelling of the *Homo fictus* and characteristic beliefs and biases of the *Homo irrationalis*:

> In contrast to **cognitive biases**, **motivated biases** derive from the emotional side of human beings, from their psychological needs, fears, guilt, and desires. People do not face up to information that makes them feel emotionally uncomfortable or that runs contrary to their goals, a pattern that some label '**defensive avoidance**'. Their beliefs about the world are often convenient rationalizations for their underlying political interests or unacknowledged emotional needs, and for the policies that serve those interests and needs. These 'motivated biases' are most likely to manifest themselves in decisions involving high stakes and important value tradeoffs. The stress inherent in these decisions often leads decision-makers to deny those threats and to deny the need to make tradeoffs between values. We begin our discussion of the individual sources of misperception with unmotivated biases, which until recently have received most of the attention in the literature since the 'cognitive revolution' in social psychology in the 1970s. One of the most important unmotivated biases involves the influence of an individual's prior *beliefs* on the ways in which s/he perceives and interprets information. The main hypothesis is that people have a strong tendency to see what they expect to see based on their prior *beliefs*. They tend to be more receptive to information that is consistent with their beliefs than to information that contradicts their beliefs. Thus there is a tendency toward selective attention to information. Another way of saying this is that *information processing tends to be more theory driven than data driven*. One consequence of the **selective attention** to information is a tendency toward premature cognitive closure. Instead of engaging in a complete search for information relevant to the problem at hand, there is a tendency to end the search for information after one's pre-existing views gain adequate support. These tendencies lead to the perseverance of beliefs beyond the point that the evidence warrants. These tendencies clearly violate the elements of a rational decision-making.
>
> Some kinds of beliefs are particularly resistant to disconfirmation by new information. A good example is the dual belief that the adversary is fundamentally hostile yet at the same time responsive to external threats and opportunities. Consider the '**inherent bad faith model**' of the adversary. This refers to situations in which people perceive aggressive actions by the adversary as reflecting the adversary's innate hostility, while perceiving conciliatory actions as reflecting the adversary's response to one's own resolute actions or perhaps strategic deception to induce complacency. Such beliefs are strongly resistant to change, because no matter what the adversary does its behavior reinforces one's mental model of the adversary. This can lead decision-makers to misinterpret conciliatory behavior by the adversary and consequently to miss good opportunities for conflict resolution. One explanation for the tendency to perceive apparently hostile actions by the adversary as reflecting its underlying hostile intentions is provided by the **fundamental attribution theorem**, a theory in social psychology that has received substantial support from the experimental evidence. The theory relates to the way people explain the behavior of others. Individuals have a tendency to interpret others' behavior, particularly behavior that they regard as undesirable, as reflecting dispositional factors rather than situational factors. If the adversary adopts hardline security policies, we tend to attribute those policies to the adversary's hostile intentions or evil character, not to a threatening environment (including our own actions) that might have induced such policies. One implication is that actors tend to underestimate the effects of the security dilemma. They minimize the extent to which apparently hostile behavior by the adversary might reflect a defensive reaction to the actor's own actions that the adversary perceives as threatening.
>
> This tendency to overestimate the adversary's hostility by attributing its behavior to its evil intentions rather than to a threatening environment is compounded by actors's tenden-

> cies to explain their own behavior in terms of situational factors rather than dispositional factors, which is the '**actor-observer discrepancy**'. While we attribute the adversary's hardline strategies to his hostile intentions, we attribute our own hardline strategies to external threats and to the need to defend ourselves. Moreover, since we believe that our own actions are defensively motivated, and since we assume that the adversary understands that, we interpret the adversary's hostile behavior as evidence that it must be hostile. This leads to mutually reinforcing negative feedback and often to an escalating conflict spiral. This set of judgments and responses was quite evident for both the United States and the **Soviet Union** during the **Cold War**. Soviet officials attributed high levels of US defense spending to **American ideological hostility** to the Soviet Union (and to the capitalist foundations of that hostility), and US officials emphasized the role of **communist ideology** underlying Soviet behavior. Each downplayed the effects of its own actions and other external pressures on the actions of the other…[An] emphasis on the emotions involved in ethnic conflict can be paired readily with Stuart Kaufman's stress on symbolism in generating warfare between ethnic groups…He establishes three preconditions for escalation to ethnic warfare: (1) a ***mythology justifying hostility*** between or among ethnic groups; (2) fears on the part of one or more ethnic groups that their existence is threatened; and (3) political opportunities to *mobilize the myths* and the threat perception. The conjuncture of the three preconditions creates an ethnic group security dilemma in which groups begin working harder at preserving their security and status. In the process of preserving their own welfare, their actions threaten the security and status of other groups. Political entrepreneurs can exploit these potentially escalatory situations to their own advantage, as well, by drawing attention to the myths and the threats in electoral rhetoric. The outcome is increased hostility among ethnic groups, which may only need some trigger to break out into civil war.

This is indeed a very exciting discussion. However, it should be noted that Levy and Thompson only dedicate a few pages to discuss these specific issues, in a book of about 250 pages, and moreover the discussion is almost completely focused on the motivation of leaders, and not of the "normal people" who actually do the fighting. This goes in line with what has been seen above: that these topics are indeed not often seen as so important within the literature about the causes of war and violence, further showing how Malesevic's book truly represents a much needed, timely, perceptive, and comprehensive take on such issues. This point is also evidenced by the fact that even authors such as Salposky—in the chapter "*War and Peace*" of his brilliant book *Behave*—often take for granted the "progressive" idea of Pinker's *The Better Angels* according to which there was a huge decrease of violence in the last centuries, contrary to what is argued in Malesevic's book. Salposky wrote:

> When it comes to our best and worst behaviors, the world is astonishingly different from that of the not-so-distant past. At the dawn of the nineteenth century, **slavery** occurred worldwide, including in the colonies of a Europe basking in the **Enlightenment**. Child labor was universal and would soon reach its exploitative golden age with the **Industrial Revolution**. And there wasn't a country that punished **mistreatment of animals**. Now every nation has outlawed slavery, and most attempt to enforce that; most have child labor laws, rates of **child labor** have declined, and it increasingly consists of children working alongside their parents in their homes; most countries regulate the treatment of animals in some manner. The world is also safer. Fifteenth-century Europe averaged 41 homicides per 100,000 people per year. Currently only El Salvador, Venezuela, and Honduras, at 62, 64, and 85, respectively, are worse; the world averages 6.9, Europe averages 1.4, and there are Iceland, Japan, and Singapore at 0.3. Here are things that are rarer in recent centuries: **forced marriages**, **child brides**, **genital mutilation**, **wife beating**, **polygamy**, **widow burning**, **persecution of homosexuals**, epileptics, albinos, **beating of schoolchildren**,

beating of beasts of burden, rule of a land by an occupying army, by a colonial overlord, by an unelected dictator, **illiteracy**, **death in infancy**, **death in childbirth**, death from preventable disease. Here are things invented in the last century: bans on the use of certain types of weapons, the World Court and the concept of **crimes against humanity**, the UN and the dispatching of multinational peacekeeping forces, international agreements to hinder trafficking of blood diamonds, elephant tusks, rhino horns, leopard skins, and humans, agencies that collect money to aid disaster victims anywhere on the planet, that facilitate intercontinental adoption of orphans, that battle global pandemics and send medical personnel to any place of conflict.

It needs to be pointed out that the aim of the present book is *not* to provide an overview of all the evidence published so far about the question on whether there was an increase—as defended by Malesevic—or a decrease—as defended by Pinker—of violence, in particular concerning wars and genocides, in the last centuries. Each of these authors seems to be right somehow, as both of them present empirical data to support their points, so probably in at least some cases they are not referring exactly to the same thing, or perhaps their differences concern the type of comparisons they do: for instance, as noted in Chap. 4 Pinker's comparisons between the number of crimes within 100,000 people in nomadic hunter-gatherer communities *versus* in major genocides are misleading. In fact, what is clear, and particularly important in the context of the present volume, is that there *was* undoubtedly an increase in terms of *organized* slavery, inequality, warfare, and oppression with the rise of **sedentism** and **agriculture** (see Box 7.12). But regarding the last centuries, I do agree with Pinker in that the rise of **humanism**—see section below—as well as the *current* existence of solid democratic governments in countries such as those of West Europe very likely contributed to a *regional* decrease of the *proportion* of people killed in **interpersonal violence**. In *The World Until Yesterday* Diamond argues that the existence of strong governments might indeed decrease—in at least some cases—one of the most prevalent vectors of violence and killings in hunter-gatherer societies: ***the vicious cycle of revenge***. The terms "**an eye for an eye**" or "**a tooth for a tooth**" refer to this cycle. This is a very old way of thinking, clearly expressed in the *Old Testament*. So, on the one hand, Pinker is right in the sense that "modern" strong central democratic governments *do have* legal ways to try to prevent, or at least minimize, this cycle of **personal/familial/small group revenge**.

In a very simplified way, if a person A kills a person B, then what typically follows is a legal case of "the state against person A," not a revenge case of "the family of person B against person A and/or his/her family." So commonly person A would end up going to jail and the family of person B would receive some kind of compensation—moral, and in some cases also monetary—the case being then seen as "closed." This is hugely different from the family of B applying "an eye for an eye" and killing person A or someone else of his/her family and then someone from this latter family revenging again against family B and so on (see Box 7.12). Apart from decreasing such cycles of individual/familial revenge, modern central governments should in theory also decrease violence related to local militias or other small criminal organizations *operating in those countries*, as clearly evidenced by the chaotic violence currently occurring in countries with very weak or even inexistent governments such as Mali, Yemen, Somalia, Libya, Central African Republic, South Sudan, and so on. In other words, within the so-called agricultural states, the beloved concept of "*Leviathan*" of Hobbes and Pinker might apply in such cases, in terms of decreasing *personal/familial/small group* revenge and violence: there is indeed available data supporting this idea.

However, on the other hand, Malesevic is right in the sense that, within agricultural states "Leviathan" not only increases but is actually often *the main promoter* of "big-group" violence, warfare, and genocides. For instance, in the last years there was tribal violence and also frequent terrorist attacks, plus cases of personal/familial revenge/jealousy killings, and so on, in Mali. As an example, *Acled*, an organisation that collates and analyzes data on conflicts, reported that more than 150 children were killed and 75 injured in violent attacks in the first 6 months of 2019, and that, including adults, nearly 600 Malian civilians were killed in those 6 months. But even within these particularly chaotic years in Mali, a country that is now often called a "failed state," this would mean about 1200 people killed in such violent attacks in a year. So in a very simplified way, that would mean let's say 144 thousand people since 1900, just as a very simplified, not factual thought experiment. Let's say, for that experiment, that the number is even higher, say 500 thousand people in 120 years, what of course would be a truly horrible number. But let's now compare that with the number of people killed in Germany/by Germans, which are part of a country that is often considered to be among the "most developed" in the planet and that undoubtedly had a very strong "*Leviathan*" government for some periods in the last century, such as **Hitler's Third Reich**.

The number is several millions of killed people, including in World War I, and particularly in World War II, and so on. So, which is more violent, in the overall, in the last 120 years? A so-called failed, third-world, poor country such as Mali, or a so-called highly developed, highly civilized, rich country such as Germany? Of course these are just two examples, among many that could be used. According to the World Health Organization website about interpersonal violence (https://apps.who.int/violence-info/), Mali and Germany had, in 2015, homicide rates of 10.8 *versus* 0.7 per 100,000 population, respectively. However, on the other hand "failed" countries do not have the means, in general, to efficiently disseminate narratives to, and then plan and execute, huge mass killings such as those done by Nazis, as Malesevic pointed out (see Figs. 4.5 and 9.17). So-called poor countries clearly can do this, yes, such as **Mao's China** did, or **Pol Pot's Cambodia**, or Rwanda, but all those countries had very strong, centralized governments and/or very strong, mobilized social networks that were involved in the genocides and/or mass killings: none of them is an example of a country with a very weak or "inexistent government." About these subjects, as well as Pinker's Hobbesian ideas about them, **Sapolsky** wrote in his book *Behave*:

> Pinker does something sensible that reflects his being a scientist. He corrects for total population size. Thus, while the eighth century's **An Lushan Rebellion** and civil war in **Tang**

dynasty China killed 'only' 36 million, that represented one sixth of the world's population – the equivalent of 429 million in the midtwentieth century. When deaths are expressed as a percentage of total population, **World War II** is the only twentieth-century event cracking the top ten, behind An Lushan, the **Mongol conquests**, the **Mideast slave trade**, the **fall of the Ming dynasty**, the **fall of Rome**, the deaths caused by **Tamerlane**, the **annihilation of Native Americans** by Europeans, and the **Atlantic slave trade**. Critics have questioned this – 'Hey, stop using fudge factors to somehow make World War II's 55 million dead less than the fall of Rome's 8 million'. After all, 9/11's murders would not have evoked only half as much terror if America had 600 million instead of 300 million citizens. But Pinker's analysis is appropriate, and analyzing *rates* of events is how you discover that today's London is much safer than was Dickens's or that some hunter-gatherer groups have homicide rates that match Detroit's. But Pinker failed to take things one logical step further – also correcting for differing durations of events. Thus he compares the half dozen years of World War II with, for example, twelve *centuries* of the Mideast slave trade and four centuries of Native American genocide. When corrected for duration as well as total world population, the top ten now include **World War II** (number one), **World War I** (number three), the **Russian Civil War** (number eight), **Mao** (number ten), and an event that didn't even make Pinker's original list, the **Rwandan genocide** (number seven), where 700,000 people were killed in a hundred days.

When one analyzes the specific horrendous events mentioned by Sapolsky, one can clearly see that not only they were indeed often done in/by states with strong governments/leaders but also that many are in fact directly and profoundly related to teleological narratives. To give just some obvious examples of atrocities directly associated with such narratives that we have already discussed in detail in the present volume, we can list World War II, annihilation of Native Americans by Europeans, Trans-Atlantic slavery trade, Mideast slave trade, Rwandan genocide, Mao's Marxist crusade, and so on. In fact, these are not merely cases crucially related to teleological tales: they are actually among *the most* emblematic examples of the huge importance of such tales—in particular those related to cosmic purpose and "progress," be it "biological progress," "racial progress," "political progress," or "social progress," such as in the case of **Marxist eschatology** (see Box 7.11).

Box 7.12 Teleological Tales, Agriculture, Sedentism, Symbolic Identities, Ritual Practices, and Warfare

In his 2017 paper "*Human Niche, Human Behavior, and Human Nature*" Fuentes reviews data contradicting Pinker's suggestion that warfare was more common, or deadly, in hunter-gatherers than it was after the rise of agriculture in various regions in the globe. In this sense, Fuentes is in line with most of the authors of the specialized papers and books I randomly—it is important to emphasize this—choose to read to write about these topics in the present book. Indeed, apart from referring to **sedentism, agriculture, inequality** and the concept of '**private property**' as key factors for the rise of warfare, Fuentes also adds the development of **symbolic identities** and **ritual**

practices related to **teleological narratives** about **group identity** and **ideology**—that is, he emphasizes the critical importance of not only our *Homo fictus et irrationalis*, but also of our *Homo socialis*, components for the increase of violence in human 'civilizations':

> Across the Pleistocene *Homo* **brain size** and complexity increased and an extreme **extended childhood period** and increasingly complex social structures developed. Such structures, and their concomitant ecological and technological outcomes, eventually altered the landscape of human evolution in especially distinctive ways, resulting in increasing group sizes, higher population densities and more diverse patterns of social interactions; human groups got larger and more structurally complex. One outcome of this increasingly complex social group living is the development of the human capacity for **warfare**. Warfare as a species-wide behavioural complex is distinctive to humans (for at least the last approximately 10-14000 years). Broadly accepted evidence of warfare coincides with a suite of particular demographic and behavioural patterns including **increased densities and group sizes**, **resource storage**, increased intra- and intergroup stratification and sedentism/agriculture. However, there is much disagreement about the origins of warfare and whether or not it has deeper roots in the **Pleistocene**; in short, whether prehistoric interpersonal violence can be considered warfare, a topic that is highly debated. Looking across the skeletal evidence of the genus *Homo* from the Pleistocene only 11 of the 447 sites, or approximately 2.5%, have fossils that show evidence of serious trauma. The overall database used to assess this pattern includes the remains from at least 2605 individuals, and of these only 58, or approximately 2%, show any evidence of traumatic violent injury. Thus, approximately 98% of the fossil evidence for the genus *Homo* between approximately 2 million years ago and approximately 15000 years ago show no sign of traumatic, lethal violence. In fact, given the current fossil datasets, there is no clear evidence of **coordinated intergroup lethal violence** in the Pleistocene fossil record and insufficient data to argue that inter-individual conflict was a major cause of mortality for most Pleistocene *Homo* populations.
>
> Recent reviews demonstrate that in the **Pleistocene–Holocene transition** (approximately 14000 to 8000 years ago) **violent trauma**, at what appears to be a group level, becomes more common in the archaeological record and that during the Holocene Neolithic (8-4000 years ago) substantially more human skeletal remains show signs of violent trauma. Individuals with identifiable violent injuries become increasingly common at Neolithic sites and this has been interpreted as the result of organized and lethal conflict between groups. The earliest solid evidence of intergroup lethal violence comes from two sites: **Jebel Sahaba in Northern Sudan** dating to approximately 14000 and 12000 years ago and **Nataruk**, a site west of Lake Turkana (Kenya), dating to approximately 9-10000 years ago. At Jebel Shaba roughly 40% of the 59 bodies are interpreted as showing evidence of traumatic violence. At Nataruk 27 individuals, and 12 full bodies were discovered. Ten of the 12 complete skeletons show signs of lethal violent trauma. A similar case comes from Ukraine and the sites of **Voloshkoe** and **Vasilyevka**, dating between 12000 and 10035 years ago. In each of these cases the humans killed were foragers in particularly rich local ecosystems. Interestingly, these sites also date to a period of rapid climate change. A suggestive hypothesis is that ***survival pressure*** and an ***inequality in access*** to the best locales and resources primed groups for violent conflict. However, despite these examples, most archaeological sites of the time period do not show signs of intergroup, lethal violence. Up until about 7500 years ago, clear evidence of larger scale, coordinated lethal violence between humans is relatively rare,

but by 6-7000 years ago there is a steady increase in the density of unambiguous evidence of coordinated, relatively **large-scale killing**. Numerous studies demonstrate that the emergence of more complex societies, **sedentism**, **increasing demographic pressures** and inequality are correlated with the appearance of evidence of intergroup lethal violence and warfare. While there has been some facile comparisons between these patterns and the general notion of **territoriality** and conflict in other organisms, the patterns, pace, characteristics, scale and intensity of the human expansion at this time period and its concomitant association with increased evidence of warfare is currently seen as a distinctive process relative to territorial conflict in other species.

When **agricultural settlements** become common in the Holocene they are accompanied by increasingly strong evidence for group identity, increased storage capacities, enhanced physical and social obligation to place, the potential for increased/more complex trading relationships between groups, and increased inequality. The expansion of the human niche to include **domestication** and sedentism created new ecologies, expanding the opportunities and incentive (pay-offs) for violence. **Hierarchies in status, wealth and power** and the control and management of larger surpluses of food, and the division of land and other protectable resources, created ecologies and altered patterns of gene flow and material exchange. These restructured the fitness implications of conflict behaviour, and **lethal violence**, and increased options and incentives for conflict, greed, distrust and violence. Bowles and Choi note the coincidence of sedentism, agriculture and storage practices with the emergence of **symbolic and behavioural processes** associated with the concepts of '**property**'. They argue that this produced specific patterns of ecological and ***symbolic inheritance*** and novel opportunities, and pay-offs, for collaboration and conflict between human groups. Key to the ability to conduct warfare that involves the use of specific tools, and has long temporal duration with multiple battles and outcomes between the two (or more) warring groups, is the capacity to develop an extreme sense of **shared community**, **social coordination** and a will to engage in highly **risky behaviour for the community**. How do human groups maintain such a sense of cohesion and coordination concomitant with the increased stratification of individuals and roles emerging in societies at this time? The tendency (or even the capacity) to engage in warfare is unlikely to be a specific trait, or even a suite of physiological or behavioural traits, that can be targeted via direct selection, as there is little, none or a negative correlation between participation in warfare and fitness (direct or inclusive) across populations where it has been assessed. One mechanism to facilitate the emergence of a human capacity to engage in warfare that is evident in the archaeological record (and in the ethnographic record) is the ***development of symbolic identities and ritual practices*** creating and reinforcing **group identities** (and **ideologies**), which can be a central feature in behavioural and symbolic inheritances, and social niche construction.

Given the core role of cooperation in human evolution, a sense of group identity that could be co-opted and deployed to get individuals to engage in warfare is likely to be very old in the human lineage, dating to at least the middle Pleistocene. However, increasing role differentiation in groups and the development of clans and lineages who leave material and symbolic evidence is much more recent…it is not until the terminal Pleistocene that we see evidence in archaeological sites of all of these variables coming together alongside sedentism, increasing stratification, consistent storage practices, and the emergence of specific symbolic and behavioural processes associated with the concepts of property. It appears that at some point in the terminal Pleistocene/early Holocene a suite of social, perceptual, behavioural and ecological facets of the human niche coincided in multiple regions resulting in a critical juncture in human history; the development of new human ecologies wherein inherited landscapes and materials involved **symbolic identities** and **structured ecosystems of ownership**, inequity and increasing group sizes. Add to this the emergence of institutionalized differences within and between groups and the increasing collective complexity manifest in the increasing specialization and diversity of societal roles and one can see that the template for a broader emergence of warfare is present. For most of human history lethal violence probably took the form of **homicides from revenge killings**, **fights over mates** and **domestic disputes**. In such disputes one, or a few, individuals were targeted. But the social, ecological and perceptual changes in human niches across the terminal Pleistocene and Holocene provided the context for the emergence of incentive and justification for group-level violence without identifying specific individuals as the targets. Humans made the mental shift from **individual-on-individual violence** to the possibility of perceiving another group as 'the enemy', creatively de-humanizing them.

Another very interesting book discussing these topics and the links between violence and religion in particular is Meral's 2018 *How Violence Shapes Religion*. I had the chance to see a talk by him in Washington DC, about this specific book, before I read it, and I particularly liked how he made an effort to avoid using ideas about these issues that are often repeated by the media and even by politicians, but that are too simplistic. The main idea of his book, as summarized in his talk, is that the available data he reviewed from the recent so-called religious conflicts in **Nigeria** and **Egypt** do not support the view that the role of religion in violence should be minimized, nor the opposite view that religion is mainly the "Satan" of violence, as sometimes suggested by authors such as Dawkins. As he explains, the view that with "progress"—whatever that means—we will see the decrease of religion, and thus of violence, clearly ignores the fact that warfare has been a major component of the life of human beings, in particular since the rise of agriculture. His idea is therefore that religion is just a way, but a *particularly powerful and effective one* indeed, to *justify* violence. I would add that we need to keep in mind two important points. Firstly, within religious teleological narratives, the ones related to **monotheism** are particularly powerful in justifying violence.

Secondly, **religious narratives** are just a subset of teleological narratives, so it is obvious that other types of teleological tales also contribute to justify violence. So, yes, simply blaming religion for the existence of *all* violence would be as simplistic as blaming soccer for the existence of violence between **soccer fans/hooligans**, without taking into account our tendency to engage in ***tribal Us versus Them violence***. The vast majority of soccer fans are not violent, and soccer is not a particularly violent sport neither. In fact, one can argue that at least some soccer hooligans would probably be—and often are—violent in another ways, such as when they undertake violent criminal activities, domestic violence, and so, particularly if they did not have at all the "escape" provided by soccer. However, having said that, one needs to recognize that there is an enormous difference between the power of using soccer versus religion as a *justification for violence*. Yes, soccer might be a *justification* for violence—including a combination of mechanisms similar to those used in religious conflicts as well as tribal behaviors, such as ingroup *versus* outgroup, us *versus* them, group cohesion, sense of belonging, "monotheism" in the sense that our club is the "only really good" or "pure" club, and so on. But soccer is obviously not as powerful as a justification for violence as some religions are, because religions are connected to many more daily-life events than just mere hours per week during or just before or after a soccer game. For highly religious people, religion is *everywhere*, *everytime*, within what one thinks about food, sex, dressing codes, abortion, vaccines, politicians, pandemics, and so on. An illustrative example is what is happening nowadays in the U.S.: religion is clearly much more *omnipresent* for most people and thus is commonly much more divisive than sports.

Conversely, as Meral rightly points out, while religion potentially has this huge power to help justifying violence before a conflict starts—for instance, Christians preparing a crusade—and obviously to keep fueling violence during a conflict, it can also be a powerful tool, at least in some cases, to *contain violence*, or to cope with traumatic acts resulting from it and/or to bring people from different sides of the conflict together, after the end of the conflict. Moreover Meral also accurately points out that religion clearly is not "dead" at all, contrary to what many intellectuals were predicting in the last centuries and particularly the last decades, based on the wrong assumption that we are truly a *Homo sapiens* and on Western-centric biases leading to use Europe as *the example* to discuss what is/will be happening in the globe as a whole. Such assumptions and biases are refuted by empirical data: for instance, if we consider the whole planet, the percentage of religious people is actually increasing, as explained above (Fig. 2.8). Meral also contradicts, with empirical evidence, some of the simplistic comparisons done by Hobbesian authors such as Pinker, regarding their ideas that there is much less religious and/or ethnic violence nowadays than decades or just years ago:

> The Pew Research Center notes that 'religious hostilities increased in every major region of the world', particularly in the Middle East and North Africa, with 33 per cent of the 198 countries surveyed by Pew having high religious hostilities in 2012, up from 29 per cent in 2011 and 20 per cent in 2007. Since some of the countries where there are social hostilities involving religion are among the most populous in the world, Pew calculates that the percentage of the world's population that live in countries with **religious hostilities** *went up* from 45 per cent in 2007 to 74 per cent in 2012. Subsequent Pew studies continued to record worrying levels of social hostilities involving religion across the world, noting the increase of use of violence or threat of violence to enforce religious norms in 16 countries in sub-Saharan Africa in 2015…While such studies suggest a recent intensification, violent conflicts involving religious actors and causes are not a new phenomenon. There have been widespread incidents of **ethno-religious violence** since the mid-twentieth century. Rapport notes that 'after World War II half of the internal struggles were ethno-religious; by the 1960s ethno-religious violence outstripped all others put together'. He estimates that some three-quarters of conflicts globally from 1960 to 1990 were instigated by **religious tensions**. Steve Bruce also claims that three-quarters of **violent conflicts** in the world had religious characteristics and argues that many who were involved in these conflicts 'explain or justify their causes by reference to their religion'. In his study of the State Failure Data Set, Jonathan Fox observes that 'throughout the 1960-96 period, religious conflicts constituted between about 33 per cent and 47 per cent of all conflicts'. The last ten years have seen further examples of this worrying trend with violent ethno-religious conflicts across Africa

and the Middle East, including in Sudan, Central African Republic, Egypt, Nigeria, Syria, Afghanistan and Iraq…It is, therefore, not a coincidence that the popular explanations on the relationship between religion and violence cited above contain declared and undeclared political visions, *teleologies of linear human progress* and *a priori beliefs* held about the place of religion and violence in the world. Incidentally, the horizon that all these views share is profoundly shaped by a particular form of European and North American modernity. This manifests itself as prejudices or assumptions in three critical areas that are central for this book: religion, violence and constructs of civilization.

In 1968, Berger [an influential sociologist] had famously stated that by the '21st century, religious believers are likely to be found only in small sects, huddled together to resist a worldwide secular culture'…[In 1999, the] reality pushed **Peter Berger** to abandon his earlier views on religions. He acknowledged that the idea that 'modernization necessarily leads to a decline of religion, both in society and in the minds of the individuals…has turned out to be wrong'. According to Berger, 'by and large, religious communities have survived and even flourished to the degree that they have not tried to adopt themselves to the alleged requirements of a secularized world'. The world remains as 'furiously religious as it ever was, and in some places more so than ever' and this automatically means 'that a whole body of literature by historians and social scientists loosely labelled '**secularization theory**' is essentially mistaken'. Berger presents two main reasons for this unseen outcome. First of all, since the modernization process has underlined 'taken for granted' realities which people held and left them in an uncomfortable place, religions that provide strong certainty have gained great appeal. Secondly, the **secular worldview** is located in an elite intellectual group, which excludes many people, who out of resentment choose to join religions that have a strong anti-secular bent.

The conditions of **World War II** could be vulnerable to being relativized as an extreme experience. In his bestselling book *The better angels of our nature: why violence has declined*, **Steven Pinker** argues that the world is less violent than ever before in human history. This is due to historical developments such as the containment of violence by modern states, the human rights movement, trade, advancement of human knowledge and **enlightenment ideals** as well as further development of human potential for empathy. Pinker makes grand philosophical assumptions that result in the relativization of the unprecedented levels of violence seen in the twentieth century and the 'civilized' world that embodies these 'advancements'. It is also problematic to use death rates from previous wars as the basis for arguing that comparatively lower mortality rates than from wars in previous eras mean a decline of human violence. Simply because this is no guarantee that the next century will not be bloodier than the current one. A staggering truth remains, as Hinton points out, 'during the twentieth century alone, sixty million people were annihilated by **genocidal regimes'**. Girard noted that what made global developments worrying today is not the intensity of the violence or comparisons of the number of dead today with that of yesterday, but 'that the unpredictability of violence is what is new'. In fact, new studies have demonstrated that the last decade has seen an increase in violent incidents globally. The report 'Global Peace Index 2015' released by the Institute for Economics and Peace (IEP) note that while it is in fact true that 'in Europe and in many other developed countries, homicide rates and other forms of interpersonal violence continue to drop and are at historic laws', there is a deterioration of peace across the world with some 86 countries seeing an increase in violence during the last eight years. Similarly, a study by the International Institute for Strategic Studies found that even though there were fewer wars, there has been an 'inexorable intensification of violence', citing some 56,000 fatalities in 63 active conflicts in the world in 2008, which increased to 180,000 fatalities in the world from 42 active conflicts in 2014. Norton-Taylor notes that the 'World Bank estimates that 1.2 billion people, roughly one-fifth of the world's population, are affected by some form of violence or insecurity'. What makes this cold fact all the more bitter is that the occurrence of violence and mass atrocities such as **ethnic cleansing** and **genocides** are 'neither particular to a specific race, class, or nation, nor rooted in any one ethnocentric view of the world'.

In the end of his book, based on his extensive discussion of the major case studies analyzed by him—Nigeria and Egypt—as well as many other cases studies examined in works of other authors, Meral gives examples of other, non-religious narratives, such as utilitarian, moralistic, or ideological ones that are employed "*to legitimatize murder in the name of a nation, a people group, a revolution or a political outcome*":

> It is clear that in both cases, religious beliefs, religious actors, institutions, sensitivities, social visions and mobilization channels play important roles [in conflicts]. Yet, what the comparison reveals is that the reality of the importance of religion and its interactions with social, economic and political spheres is dynamic and evolves and changes over time. Thus, rather than being an unmoved mover in the Aristotelian sense, religions themselves are impacted by macro and micro pressures, national and global developments and more often than not, rather than being the primary engine of politics and conflict, they themselves become impacted and guided by them. The context within which religions become a vital aspect of individual or communal responses and visions to better or correct the world or seek protections from the chaos and failures of states that are always in flux and failing to meet the needs of their populations is where the primary causes **of ethnoreligious violence** lies. Neither in Nigeria nor Egypt do we see **theological beliefs** or **imagined notions of civilizational identities** being the starting point of violent conflicts or even political tensions, but such beliefs and identities have developed in response to conflicts and tensions which are deeply local and contextual. In both countries we also see numerous religious initiatives to de-escalate conflicts and pursue peace, thus, while it is tempting to merely focus on religion and violence, religion's role in enabling the **breaking down of cycles of violence** is often overlooked. What is peculiar to observe in both countries is how global developments – whether from the 1970s wave of **Islamic revolution**, **ethnonationalism** and **separatism**, or the post-1990s **religious militancy** and post-2000s **Manichean battles between 'us' versus 'them'** – have directly impacted local conflicts or at times triggered new ethno-religious clashes and violence and fuelled deep local animosities. This demonstrates that not only the view that what we are witnessing in Africa and Middle East is an outcome of religions is not supported by two of the most cited cases, but that the view that these are somehow local expressions of a given global problem is also not correct. On the contrary, international developments and **globalized narratives of animosities** between **Islam** and **Christianity** directly impact local conflicts, which are created and sustained by local factors.
>
> A common scholarly approach that moves on from the failures of such popular explanations can be seen in the instrumentalist understanding of religions. As Stein explains, 'instrumentalism rejects the view that differences in religion are real causes of political conflicts' but recognizes 'that religion can play a part in violent conflict' as a 'tool used by self-interested elites to mobilize support and fighting power for conflict'. In fact, instrumentalism often stops decoding the relationship between religion and violence at the point of stating its use by conflicting parties. This tendency is inadequate for the interests of this book. The elite use of religious sensitivities, imageries and languages to achieve desired political outcomes does not explain fully why such religious sensitivities, imageries and languages are so powerful in moving human beings to risk their own lives. Similarly, it does not explain why it is that calls for or experiences of violence seem to find an effective space in religions at the first place and why and how local conflicts with religious characteristics are able to interact with and impact global developments. Thus, while helpful in explaining the use of religion, instrumentalist approaches fall short in exploring answers to questions pursued in this book beyond contributions to specific aspects of case analysis. As Burkert notes, 'blood and violence lurk fascinatingly at the very heart of religion'…

Meral further stated that:

In fact, religion in human history emerged from the **sacrificial rituals** that were utilized by primitive communities as a response to both human conflict and also to the violence of inhospitable natural conditions…Therefore, unlike the common argument that religion necessarily results in violence, historical studies on sacrificial systems point to an interesting dynamic whereby violence has actually given birth to religion, and religion is ultimately engaged in responding to violence, which is a deeper understanding of the relationship between the two than merely a focus on elite manipulation of religious sensitivities at particular conjunctures. Girard affirms this by saying that 'the sole purpose of religion is to prevent the recurrence of **reciprocal violence**'. In today's world, such a task is taken away from religions by the legal use of violence by sovereign states, but even contemporary forms of rule of law and enforcement mimic religious rituals throughout the court procedures and promise a final settlement of justice. Yet, the role the state assumes in today's world means neither the eradication of human potential for violence and human drives and social mechanisms that escalate risks of violence, nor that religions do not play a central role in the human experience of and addressing of violence.

As Clarke notes, 'most people who commit violent acts do so reluctantly and only after they have overcome internal constraints that would ordinarily make them feel guilty about harming others…when they do act violently, they do so in the *belief* that what they are doing is justifiable, all things considered'. Therefore, such an extraordinary call to inflict violence demands out-of-the-ordinary psychological and social processes to enable the individual to both process the trauma that the common order has been altered – making violence a viable if not the only option – but also provide a **moral suspension and authorization** to move beyond socially and religiously enforced inhibitions that forbid violence. This can be broken down into two crucial social mechanisms that are intrinsic to any deployment of violence, whether through its legal use by a state or by mobs, militants and crowds: ***legitimation*** and ***dehumanization***. Peter Berger defines legitimation as 'socially objectivated "knowledge" that serves to explain and justify the social order'. Acts, institutions and deployment of violence need an explanation, an answer to the question of 'why' and whether one has the 'right'. Legitimation is not simply limited to religion: countless **utilitarian**, **moralistic** or **ideological** attempts are used to **legitimatize murder** in the name of a nation, a people group, a revolution or a political outcome. As Steve Clarke points out, 'the religious generally justify their activities in much the same way as the secular and these justifications generally follow the same canons of logic as secular justifications. Religious arguments justifying violence are structurally similar to secular ones, but the religious are able to *feed many more premises* into those structures than the non-religious'.

That is why Berger argues that there is an important **relationship between legitimation and religion**, since 'it can be described simply by saying that religion has been the historically most widespread and effective instrumentality of legitimation'. This is because religion plays a key part with its unique role 'to "locate" human phenomena within a cosmic frame of reference…and since religious ideas are legitimized by virtue of emerging from a **transcendental source**, they are hardly subject to negotiation and compromise given their accepted supernatural origin'. Religion, with its ability to articulate why the other is not simply different due to the products of culture and language, but due to an eternal reason whereby the Other, with his false gods, is eternally separate from us, becomes far more effective in establishing differences and solidarities. Secondly, it does so through representation and effacing of the other in the languages of **enemies of God** – **evil**, **heresy**, **abomination**, **impure** – all of which serve to **dehumanize the other**. In northern Nigeria, we saw how minority Christians were regularly called names, such as arne – godless – and how in Egypt **Islamist media outlets** regularly used derogatory words for **Copts**. Once a person or a community is abstracted by such categories and language, we are no longer dealing with human beings. Thus, religion is not only able to provide grounding for why violence needs to occur, and why it is morally acceptable, but also why morality can be suspended and all theological provisions on the dignity and value of human life denied in pursuit of combating an 'evil'. One 'regrets' that bad things had to be done as a '**lesser evil**' in a '**just war**' in which violence deployed by us is 'different' in nature than that of our enemies.

Thus, even **nonreligious campaigns of violence** use religious descriptions about their enemies and their motivations and aims…It is only the effectiveness of modern policing and the rule of law that hides various levels of violence from the eyesight, whether it be domestic abuse or gang crime, creating the illusion that somehow people of the 'developed' world are less violent than those of 'developing' world. Yet, some of the world's most developed countries also have historically highest rates of incarceration, and their armed forces continually deploy brutal force across the world to assert national interests and foreign policy preferences. In fact, European and North American states have only been able to maintain the peace and order taken for granted by their citizens with their effective use of violence and its threat across the globe. Therefore, any argument that suggests the 'other' is somehow intrinsically predisposed to violence unlike 'us' is a mere illusion. *Thus, yes, there is an intrinsic relationship between religion and violence that runs deep, but not a causal one*. The absence of religion in the same contexts would not stop violence, but only push for some other all-encompassing framework to fill the role of religion, whether it is **nationalism**, **ethnic separatism**, **tribalism** or **communism**.

7.6 "Savages," Animal Abuse, and Humanism

True memories seemed like phantoms, while false memories were so convincing that they replaced reality. (Gabriel Garcia Marquez)

Teleological narratives, particularly those related to the notion of "**great chain of being**" or "**ladder of nature**," are directly related to the tales that are normally used to justify the use, and abuse, of other organisms by humans. Importantly, contrary to some ideas that are now so in vogue, this does not apply only to the Judeo-Christian traditions, but to almost all organized religions. Kasperbauer's 2018 book *Subhuman – The Moral Psychology of Human Attitudes to Animals* emphasizes the deep, direct ties between these topics, the related concepts of cosmic purpose and progress and of human "specialness" and "superiority" linked to our existential anxiety about death, and **animal abuse**:

Concern for animals seems to be at an all-time high. And not just for animals kept in our homes. Animals of all types are arguably treated better than ever before. It may thus seem that any comparison we make to animals primarily results in positive, not negative, judgments. Although tempting, this generalization fails to capture a wide range of attitudes about animals; care and concern do not tell the whole story…(there is) relevant empirical evidence on this issue…let's briefly consider some examples of how animals have been and continue to be evaluated negatively:

(1) *The Great Chain of Being*…the idea that there is a **hierarchical scale in nature** and humans' place within that scale. One common conception of humans' place in the natural scale…according to Lovejoy, is inherently paradoxical. Humans are "constitutionally discontented" with their relatively low rank in the scale, compared to angels and gods, but they also see this as "appropriate to [their] place in the scale". Humans constantly strive to ascend the **scale of nature** while knowing that they will inevitably fail. This conception of human beings would seem to have implications for conceptions of non-humans. One way of ascending the scale is to be less like animals, which are invariably beneath humans. But animals, like humans, are only filling their role in the natural order of things. So we would

expect animals to be denigrated if they threaten **human superiority** but not in order to help humans ascend the scale; pushing animals down does not lift humans up. Lovejoy's scale of nature suggests that animals will be viewed as mostly benign but threatening when they seem to be too close to humans on the natural scale.

(2) *Proximity Threats*...ascending the chain of being presents a symbolic, or psychological, threat to **humanness**. There are different ways that this threat can come about. In his classic analysis of animal metaphors, Edmund Leach (in 1964) proposed that denigration of certain animals was a result of their physical closeness to humans. It is well known that animal terms are used to insult and demean others. Leach suggested that the most common animal insults involved animals that encroached on human lives. **Pigs**, for instance, are commonly used in animal insults in many cultures (e.g., "filthy pig," "fat as a pig") and have historically been raised near or sometimes even inside of human homes, even when intended for slaughter. This physical breach of the human–animal boundary, Leach proposed, would have caused psychological discomfort. Pigs have also participated in other characteristically human activities, such as receiving human food, and are often seen as being social and having a family life, which would exacerbate the psychological threat. To deal with the threat, we put an extra stigma on pigs in order to reaffirm, with language and taboos, that they are separate from us. We alleviate our concerns with physical closeness by insisting on a fundamental metaphysical difference.

(3) *Mortality Salience*...the line of research that has explored the psychological threat of animals in greatest detail is known as '**terror management theory**'...[its] the basic assumption...as it applies to animals, is that animals remind us of aspects of our humanity that cause existential anxiety, particularly our mortality". "This is often referred to as '**animal reminder**' or '**mortality salience**.' Being reminded of one's mortality presents a psychological threat and disrupts normal psychological functions, so we have developed various strategies to suppress these feelings when they arise. For instance, Rozin and Fallon (in 1987) argue that, as a result of this threat, human beings' wish to avoid any ambiguity about their status by accentuating the **human–animal boundary**. An excellent illustration of how animals elicit mortality salience comes from Beatson and Halloran (in 2007). They presented participants with a prompt that made them think about death. They also presented an animal stimulus, which was a video of **bonobos having sex**. In one condition the researchers emphasized to participants how similar bonobo sex is to human sex. In another condition, the differences between the species' sex habits were emphasized. They also measured participants' **selfesteem** because increasing self-esteem is a well confirmed method for reducing the effects of mortality salience. In the condition where **bonobo sex** was seen as similar to human sex, those with low self-esteem evaluated animals more negatively, while those high in selfesteem evaluated animals more positively. These results indicate that bonobo sex elicits mortality salience, which presents a **psychological threat**. Reminding people of their animal nature caused negative evaluations, unless the level of self-esteem was sufficient to fend off the attendant **existential anxiety**.

As noted in Sproul's 1991 book *Primal Myths*, many **creation myths** of different cultures do share the notion "that human beings are superior to all other creatures and are properly set above the rest of the physical world by intelligence and spirit with the obligation to govern it." The first creation myth that she describes is that of the **Bushmen of Southern Africa**: "**Cagn** was the first being; he gave orders and caused all things to appear, and to be made, the sun, the moon, stars, wind, mountains, and animals...his wife's name was **Coti**...he had two sons, and the eldest was chief, and his name was **Cogaz**; the name of the second was **Gewi**...he was at that time making all animals and things, and making them fit for the use of men...." Such stories about "animal inferiority" are also very common in **children's narratives**—a further example of how we acculturate to *Neverland*'s tales from a very young age (see Box 7.13).

Box 7.13 Children, Categorization of the Animal World, and Cross-cultural Studies

In Kasperbauer's book *Subhuman*, he explains:

> Research into **children's classification of animals** was also inspired by Carey (in 1985). Carey argues that only with formal education do children learn that humans are animals, which raises the broader question of how children generally **categorize the animal world**. A Number of recent experiments have provided persuasive evidence that young children are reluctant to view humans as animals. For example…in one experiment, they told children that a dog and a bird could each be called a 'blicket.' They then presented pictures of other animals as well as humans and asked if they could also be called blickets. Neither 3- nor 5-year-olds were willing to classify humans as blickets, though they were for other animals (bees and ants). In a second experiment, humans were used as a model, such that children learned that humans and either a bird or a dog (depending on the condition) could be called blickets. Here, an age difference appeared: 5-year-olds extended the term 'blicket' to both humans and other animals (e.g., squirrel and deer), but 3-year-olds limited the term 'blicket' only to non-humans. This suggests that 5-year-olds have a concept of shared animal nature between humans and non-humans but 3-year-olds do not, which is consistent with Carey's (in 1985) suggestion that seeing humans as animals requires significant directed instruction.
>
> Further **cross-cultural evidence** for Carey's thesis comes from Leddon *et al.* (in 2012). They argue that children do not spontaneously classify humans as animals and are reluctant even with formal education instructing them to do so. From a review of previous studies, they concluded that children aged 6-9 categorized human beings as animals about 30% of the time. To test this further, they conducted a study with three different groups of 5- and 9-year-old children: an urban sample from Chicago; a rural sample from Shawano, Wisconsin; and a sample from the **Menominee**, a group of **Native Americans** living in rural Wisconsin. Menominee children provide an interesting test case because they are taught from a young age that humans originated from five different animals (bear, eagle, crane, wolf, and moose). Previous research also had found that Menominee 5-year olds would infer that properties possessed by animals would generalize to humans. However, in this study, 9-year-olds were willing to categorize humans as mammals but not as animals, and 5-year-olds would not categorize humans as either animals or mammals. Crucially, there were no statistically significant differences across the different communities, despite significant differences in education and experiences with animals. So despite significant differences in conceptions of humans' animal nature, young children from very different cultures may nonetheless share in the belief that humans should not be classified as animals.

This does not mean that all human cultures think that others animals are necessarily "inferior" and/or relate in the same manner with non-human organisms. **Animal *abuse*** is markedly more prominent in **agricultural societies** than in most **hunter-gatherer groups**, which mainly *use* animals in the sense that they hunt and eat them—and sometimes use their bones for decoration or kill them for sacrifices or magical acts. But even in the latter cases there is a huge difference with the type of animal abuse done in most "developed" countries because they do so *without affecting the normal way of life of those animals before killing them*. In general, the

Fig. 7.6 The self-proclaimed "sapient," "altruist," "moral," "superior," and "special" being at its best: Tess Thompson Talley, a hunter from Kentucky, said she was "conservation hunting" and showed the news crew around her home, showing off the custom-made gun case that she had made out of her black giraffe kill—"I have decorative pillows made out of him," she added, "and everybody loves them"

more disconnected humans are with nature—that is, the more we live in *Neverland*—the more we *think* that we don't depend on nature, so the more we tend to abuse animals and/or neglect the natural environments around us. This is tragically illustrated by the practice of **trophy hunting** done by so many people of "developed" countries, which deeply contrasts with the typical type of hunting done by hunter-gatherers. While the latter type is basically a *useful* practice to provide food for the community, the "modern" type of **trophy hunting** or **sport hunting** is mainly done for *human recreation*: for the *pleasure* of killing, and seeing suffer, another living beings (Fig. 7.6), some of them in danger of extinction.

Let's be realistic: such an atrocious act would be hardly defined by a truly neutral viewer—let's say, if a Martian would come to earth—as a "noble" feature of the "expanding moral circle" of "civilized" humans living in "developed" countries. Even less "civilized" is the fact that apart from killing animals for pleasure, those people then usually take their "trophies" to their far-away countries—something that is, furthermore, a tragic sign of how **post-colonialism** continues to be so prevalent nowadays—and then proudly display them in their houses, offices, or special "trophy rooms" so they are admired by other "modern civilized" people for their "successful hunt." First of all, saying that the "hunt" was "successful" is completely absurd, because this seems to imply that it was a very difficult task, as if two similar opponents were fighting. But this is not the case at all: just one of them had a weapon to shoot the other one, cowardly, at a distance, sometimes when the other one could not even see the "hunter." Secondly, in this sense this is not even a "hunt." If this would be done to a human, this would be qualified as "cold blooded murder":

that is, when someone deliberately kills someone else by carefully plotting the murder—not being driven by a sudden impulse or outburst while committing it—and then carrying it out in a completely unfeeling manner and fully aware of the nature of his/her actions and their consequences.

Even more disturbingly, both the "trophy rooms" and the selfies done by the "hunters" with the dead animals just after their cold blooded murder are a clear illustration of **animal subjugation by humans**—a reminder that humans are the "real bosses," the "special ones," in a terrifyingly similar way of what has been done in art for centuries (compare Figs. 7.6 and 3.6). Amazingly, despite of all these aspects, nowadays not only the "successful hunters" and the huge number of people close to/that admire them, but also a huge number of other people argue that "trophy hunting" could be "a good think" because it can bring cash for activities such as conservation efforts. Although the present book is not the place to discuss in detail this very complex subject, I would just say that, while such a way of thinking might eventually bring some pragmatic "positive" aspects in some cases, it does not sound logically coherent, at least in theory. This is because, if applied to humans, it would be the same as saying that we should legalize the killing of let's say some small kids by psychopaths as long as those psychopaths pay a lot of money to fund cops that help preventing other people from killing other kids in similar direct, or in other indirect, ways. Be that as it may, the links between trophy hunting, human irrationality, and our lack of respect for non-human beings and the related teleological narratives about how "special" we are, and how other animals were "put here for us to use/abuse," are clearly illustrated in an interview done with **Tess Thompson Talley**. She is the very proud woman shown in Fig. 7.6 after cowardly killing, just for pleasure, a black giraffe, as reported in a June 2019 article by Natasha Ishak in the website *allthatsinteresting.com*:

> A Kentucky hunter received backlash after she posted photos of her smiling over the black giraffe that she shot and killed. In a TV interview, she argued that she was contributing to conservation efforts…She said she was "**conservation hunting**". In the interview, Talley showed the news crew around her home, showing off the custom-made gun case that she had made out of her **black giraffe kill**. "I have decorative pillows made out of him", she added, "and everybody loves them". "He was delicious", Talley said of the giraffe while donning a white fur coat. "He really was. Not only was he beautiful and majestic, but he was good. We all take pictures with our harvest. It's what we do, it's what we've always done. There's nothing wrong with that." "It's a hobby, it's something that I love to do. It's conservation and this hunt in particular was a conservation hunt", Talley said. She argued that by hunting and killing these majestic animals, people like her are able to grasp a better appreciation for the animals. "Everybody thinks that the easiest part is pulling the trigger. And it's not", Talley tried to explain. "That's the hardest part. But you gain so much respect and so much appreciation for that animal because you know what that animal is going through. *They are put here for us*. We harvest them, we eat them".

Not only the trophy hunting itself and the way she poses in the pictures, but also the "sophistication" of her arguments, from how she used parts of the giraffe to make a gun-case and pillows that "everybody loves" to her "mission" to make people be "able to grasp a better appreciation for the animals," as well as her assertion that other animals are "put here for us," clearly show the self-proclaimed "sapient special" being at its best. If there were indeed a God, would a being that does such horrible things and justifies them in such unintelligent ways be really His "highest creation," made in His "image and likeness" using His "supreme intelligence"?

However, after seeing how "sophisticated" and "kind" to other animals "modern civilized" people can be, it is important to note that the idea that all nomadic hunter-gatherers are, in contrast, always "respectful" to and live in a "perfect equilibrium" with animals is also not completely right. Yes, they are, in general—with exceptions of course—much more connected with and thus respectful towards other animals and the non-human natural word in general. But they are not "perfect nobles." Such a view, defended by some, is the result of either imaginary tales created by hunter-gatherers—such as that hunted animals are willful participants who supposedly gain as much from the hunt as the hunters (see Box 7.14)—or of **Western romanticization of the "Noble Savage,"** as discussed in Diamond's 2005 book *Collapse*:

> Efforts to understand past collapses have had to confront one major controversy and four complications. The controversy involves resistance to the idea that past peoples (some of them known to be ancestral to peoples currently alive and vocal) did things that contributed to their own decline. We are much more conscious of environmental damage now than we were a mere few decades ago…To damage the environment today is considered morally culpable…Not surprisingly, **Native Hawaiians** and **Maoris** don't like paleontologists telling them that their ancestors exterminated half of the bird species that had evolved on Hawaii and New Zealand, nor do **Native Americans** like archaeologists telling them that the **Anasazi** deforested parts of the southwestern U.S. The supposed discoveries by paleontologists and archaeologists sound to some listeners like just one more racist pretext advanced by whites for dispossessing indigenous peoples. It's as if scientists were saying, "Your ancestors were bad stewards of their lands, so they deserved to be dispossessed". Some American and Australian whites, resentful of government payments and land retribution to Native Americans and **Aboriginal Australians**, do indeed seize on the discoveries to advance that argument today. Not only indigenous peoples, but also some anthropologists and archaeologists who study them and identify with them, view the recent supposed discoveries as racist lies. Some of the indigenous peoples and the anthropologists identifying with them go to the opposite extreme. They insist that past indigenous peoples were (and modern ones still are) gentle and ecologically wise stewards of their environments, intimately knew and respected Nature, innocently lived in a virtual ***Garden of Eden***, and could never have done all those bad things. As a New Guinea hunter once told me, "If one day I succeed in shooting a big pigeon in one direction from our village, I wait a week before hunting pigeons again, and then I go out in the opposite direction from the village." Only those evil modern First World inhabitants are ignorant of Nature, don't respect the environment, and destroy it.
>
> In fact, both extreme sides in this controversy – the racists and the believers in a past Eden – are committing the error of viewing past indigenous peoples as fundamentally different from (whether inferior to or superior to) modern First World peoples. Managing environmental resources sustainably has always been difficult, ever since *Homo sapiens* developed modern inventiveness, efficiency, and hunting skills by around 50,000 years ago. Beginning with the first human colonization of the Australian continent around 46,000 years ago, and the subsequent prompt extinction of most of Australia's former giant marsupials and other large animals, every human colonization of a land mass formerly lacking humans – whether of Australia, North America, South America, Madagascar, the Mediterranean islands, or Hawaii and New Zealand and dozens of other Pacific islands – has been followed by a wave of extinction of large animals that had evolved without fear of humans and were easy to kill, or else succumbed to human-associated habitat changes, introduced pest species, and diseases.

> Any people can fall into the trap of **overexploiting environmental resources**, because of ubiquitous problems…that the resources initially seem inexhaustibly abundant; that signs of their incipient depletion become masked by normal fluctuations in resource levels between years or decades; that it's difficult to get people to agree on exercising restraint in harvesting a shared resource; and that the complexity of ecosystems often makes the consequences of some human-caused perturbation virtually impossible to predict even for a professional ecologist. Environmental problems that are hard to manage today were surely even harder to manage in the past. Especially for past non-literate peoples who couldn't read case studies of **societal collapses**, ecological damage constituted a tragic, unforeseen, unintended consequence of their best efforts, rather than morally culpable blind or conscious selfishness. The societies that ended up collapsing were (like the **Maya**) among the most creative and (for a time) advanced and successful of their times, rather than stupid and primitive. Past peoples were neither ignorant bad managers who deserved to be exterminated or dispossessed, nor all-knowing conscientious environmentalists who solved problems that we can't solve today. They were people like us, facing problems broadly similar to those that we now face. They were prone either to succeed or to fail, depending on circumstances similar to those making us prone to succeed or to fail today. Yes, there are differences between the situation we face today and that faced by past peoples, but there are still enough similarities for us to be able to lean from the past.

Box 7.14 "Traditional Societies," Hunting, and Human–Animal Relations

Kasperbauer notes in his book *Subhuman*:

> Various communities of **indigenous people**, it has been argued, have very different conceptions **of human-animal relations**, according to which animals and humans are equals and must be treated accordingly. A core belief in **human-animal equality** has been identified across a wide variety of indigenous groups in different regions of the world. These views about equality typically derive from **spiritual beliefs about animals.** Animals are often seen as representatives of the spiritual world or as connected to individual spirits that communicate through animals. Nurit Bird-David (1999) argues that human–animal equality and the spiritual world that is intertwined with animals are not mere metaphors for many indigenous communities. Groups who hold this belief in equality see animals as agents that can be known just like other human beings and, because of this, treat animals as important subjects with which they can interact and do not use them solely for meeting their own, human, ends. Many different indigenous groups are said to have reciprocal relationships with animals, where each side is expected to benefit the other in some way. This too is typically driven by spiritual beliefs about animals. For example, people as diverse as the **Chewong of Malaysia**, the **Siberian Yukaghirs**, various **Amazonian communities**, and the **Rock and Waswanipi Cree of Canada** all believe that animals give themselves up to hunters who treat animal spirits appropriately. Interestingly, if you put yourself in this mindset of reciprocal relationships with animals, hunting is non-violent.
>
> Knight (2012) calls this the '**hunting-as-sharing thesis**' because animals are seen as willful participants who supposedly gain as much from the hunt as the hunters. In other words, as Nadasday (2007) argues, hunting represents a 'long-term relationship of reciprocal exchange'. Similarly, in describing the hunting behavior of the **Rock Cree in Canada**, who are expected to mimic animal behavior to achieve a successful

hunt, Brightman (1993) says, 'The event of killing an animal is not represented as an accident or a contest but as the result of a deliberate decision of the animal or another being to permit the killing to occur'. Hunting is non-violent in reciprocal relationships because animals must always present themselves by their own volition and must not be tricked by the hunter. As Tim Ingold (1994) puts it, 'coercion, the attempt to extract by force, represents a betrayal of the trust that underwrites the willingness to give…animals thus maltreated will desert the hunter, or even cause him ill fortune'.

[However] there is more antagonism in **indigenous attitudes toward animals** than anthropologists have generally acknowledged. The evidence supports this challenge to the very premise that the **human-animal relationship in indigenous communities** is a reciprocal one. Actually, beliefs about hunting-as-sharing are often used to justify wastefulness. The **Siberian Yukaghir**, for example, reportedly view the world of living animals as capable of being replenished by dead animals; as a result, prey are seen as an inexhaustible resource. Since they also view prey as animals that have chosen to offer themselves to the hunters, they kill every animal that presents itself so as not to disrespect the animal spirits. This results in significant waste. For instance, when Yukaghir slaughter elk they only take selected parts, leaving most of the carcass to rot. Willerslev (2007) suggests that this is the case for many indigenous groups because 'if a hunter is offered much, he must take much…failure to kill all the animals available is to put one's future hunting luck at risk'. Similar accounts have been described of other indigenous groups appearing unconcerned by unnecessary killing of animals that they claim are important parts of the spiritual world. Another source of support for Knight is that many indigenous communities do not seem concerned about animals' welfare. For example, Brightman (1993) describes the Rock Cree as concerned about animals stuck in snares only because it makes the meat taste bad. The animals' pain and suffering did not seem to matter.

They also seem to acknowledge that animals are taken against their will, despite the behaviors Cree must perform to convince animals to offer themselves to hunters. Moreover, many indigenous communities do not show high levels of concern for all animals. It is primarily prey animals that are viewed as engaging in reciprocal relationships with human beings. The Yukaghir, for example, reportedly view dogs as extremely helpful but still dirty and unworthy of respectful treatment, unlike elk, bears, and other animals that they hunt. In short, it's not clear what benefits animals actually receive from these relationships. They seem not reciprocal but thoroughly asymmetrical. Animals are treated well primarily to ensure future hunting success; much of the affection shown toward prey seems designed to assist in attracting and killing them. And respectful treatment is often applied only to the animals after they have died, in the form of rituals and tributes to the spirit world. As Brightman (1993) characterizes this phenomenon amongst the Rock Cree, 'Rituals of control occur before the hunt, while those emphasizing reciprocity and respect occur after'. A more cynical interpretation of the ethnographic literature is that these beliefs are merely used to justify killing animals. Humans and animals may indeed be seen as fundamentally the same, which may lead certain indigenous groups to treat animals differently, particularly in thinking of them as having value beyond their usefulness in a hunt. But these beliefs also lead to excusing behaviors: animals can be killed so long as sacrifices are made, animal suffering is unimportant because it is part of a broader spiritual process, and so on. At the very least, we should be suspicious that these behaviors result from caring attitudes toward animals. Animals are still seen as both predators and prey, they are still seen as threats, and even if certain indigenous groups do genuinely intend for animals to benefit from the relationship, the animals receive these benefits only by paying with their life.

Having seen above what Diamond wrote in *Collapse* and what Kasperbauer wrote in *Subhuman* (see Box 7.14), two important points need to be made. Firstly, discussions about "traditional societies" can be confused, and confusing, because the very term "traditional societies," as defined by many authors and often by the media and the broader public, is a Western-centric concept that includes an amalgam of "others" that is actually completely heterogeneous. This is because it includes nomadic hunter-gatherers, sedentary hunter-gatherers, and even sometimes the so-called lower agricultural groups. So, there are obviously major differences in the way in which each of those three main types of societies tends to treat animals, and moreover there is a huge diversity within each of those types of societies. Secondly, non-Western agricultural societies such as the Maya did a lot of animal abuse, as "modern" societies do, because that is what agriculture consists of: to dramatically change the way in which a huge amount of animals *lives*—from being wild and free to be domesticated and often having horrible, constrained lives subjected to a lot of abuses.

Amazingly, these two simple and well-documented points are often neglected by authors that defend reactionary, racist, and/or Hobbesian views about how "brute" the "savages" are and how there has been in the last centuries an "expanding moral circle" in "developed" countries. You might not believe this, but such authors actually refer to animal abuse as an example to reinforce the idea that in "developed" Western states life is much "better" than anytime in the past, not only for humans but also for…domesticated animals. Amazing, indeed, how far *Neverland* fairytales can go: an example of this is Pinker's *Better Angels*. A confessed Hobbesian, he writes: "I have demarcated the 18th century…to highlight the many humanitarian reforms that were launched in this remarkable slice of history…another was the prevention of **cruelty to animals**…in 1789 Jeremy Bentham articulated the rationale for animal rights in a passage that continues to be the watchword of animal protection movements today: 'The question is not *Can* they reason? nor *Can* they talk? but *Can* they suffer?'…beginning in 1800, the first laws against bearbaiting were introduced into Parliament…in 1822 it passed the ***Treatment of Cattle Act*** and in 1835 extended its protections to bulls, bears, dogs, and cats." Pinker then states: "like many humanitarian movements that originated in the **Enlightenment**, opposition to **animal cruelty** found a second wind during the **Rights Revolutions** of the second half of the 20th century, culminating in the banning of the last legal blood sport in Britain, the foxhunt, in 2005." Yes, there are effectively some laws protecting "animal cruelty" that did not exist some centuries ago, particularly concerning pets. However, there is no doubt, scientifically, that nowadays animals suffer more, both in proportion and in absolute numbers, than in any other time of this planet's history, at the hands of humans, and the fact that authors such as Pinker suggest otherwise is a further example of their Hobbesian biases as noted in Box 7.15.

Box 7.15 "Developed Countries," Pet Keeping, Moral Circle, Animal Abuse, Better Angels, and Animal Cruelty

In *Subhuman*, Kasperbauer states:

> The psychology of **dehumanization** indicates that we overlook the suffering of out-group members, particularly animals. This is part of seeing something as inferior to us. Seeing something as subhuman requires us to view its pain and suffering as less important than that of other human beings. However, the **expanding moral circle** [as argued for instance by Pinker] would seem to predict that harm to animals has come to be seen as morally problematic. We should expect that active harming of animals has decreased and active protection against harm has increased. But does the evidence support these claims? One type of evidence often cited in support of an expanding moral circle concerns the use of animals in laboratory research. Rowan and Loew's (2001) review of trends in the late 20th century found that use of animals in laboratory research had been steadily declining, after having peaked in the 1970s and 1980s (mainly in Britain and the United States). Similar trends have recently been reported for the United States. In 2014, 834,543 animals covered by the **Animal Welfare Act** were used in laboratory research, the lowest recorded since 1973. Similarly, the European Union's (2014) report on animal use found that the downward trend has continued, with 11.5 million animals used in 2011, down from around 12 million used in 2008. All of these statistics could be cited as evidence of an **expanding circle of moral concern**. It has been suggested, however, that the use of **laboratory animals** is vastly underreported. In the United States, the majority of animals used are rats and mice, which are not covered by the Animal Welfare Act and so do not figure into the numbers reported above. In the United Kingdom, around 90% of the animals used are mice and rats. A similar percentage, or more, is likely in the United States. A recent analysis by Goodman, Chandna, and Roe (2015) suggests that when mice in particular are included, it is clear that **animal use** in America is *increasing*. They found that over a period of 15 years, institutions that had received a significant amount of money from the National Institutes of Health had increased their use of vertebrates, from just over 1.5 million to just over 2.5 million. This trend may not reflect use by other institutions in the United States, but it does give reason to wonder whether there is expanded moral concern for animals. We should also consider the total number of animals being used. Even if the trend is downward, the European Union alone is using *over 11 million animals every year*. Good data on global trends are hard to find, but in 2005 it was estimated that over 100 million animals were used annually across 179 countries.
>
> We might also point to meat consumption as evidence of an expanding moral circle. In 2012, annual worldwide **meat consumption** was expected to reach 455 million tons by 2015. Though global meat consumption is increasing, the rate of meat consumption is declining. Overall meat consumption increased 2.6% from 1980 to 2006 but is expected to increase only 1.6% between 2005/2007 and 2030. Much of this decline will be seen in Western countries, which have traditionally been heavy consumers of animals reared in objectionable conditions. The overall amount of meat consumed in developed countries has actually already started to decrease. This could perhaps be cited as evidence in favor of an expanding moral circle. However, there are reasons to be skeptical. Although meat consumption is slightly decreasing in Western developed countries, this is more than made up for by increases in developing countries, especially in Asia. China, Brazil, and India, for instance, are all expected to significantly increase their rates of consumption through 2050. There are many causes of this, including decreasing costs of production, population growth, and malnourishment. These factors are unlikely to go away. Moreover,

agricultural intensification is increasing and projected to continue increasing. From 1997 to 2007, for example, world beef production increased by 1.2% a year, but total number of cattle increased by only .5%. This means that efficiency of cattle use improved by .7%. From an animal welfare perspective, it might seem that this is a good thing: a .7% increase in efficiency means that fewer animals are brought into existence purely to be killed for our consumption. There is a downside to this, however, in that intensification likely *entails worse lives for the animals*. Killing fewer animals is only an improvement if their lives are not made worse off.

Protection against harm to animals is still treated roughly along the lines of damage to property – namely, as a criminal act against the owner, not the animal. This might be changing, but there are still reasons to think it will be difficult for most legal traditions to incorporate animal interests and weigh them equally against those of humans. **Laboratory research, meat consumption**, and legal regulations might all be seen as only remotely connected to public perception of animals. If we want to know whether moral concern has expanded, one might think we should look at studies of what people actually say about animals. Herzog, Rowan, and Kossow (2001) argue that surveys conducted throughout the late 20th century indicate a definite improvement in attitudes toward animals among Americans. On pretty much every area of animal use, more people supported improved protection for animals at the end of the century than at any time in the previous 50 years. This trend seems to have continued. A Gallup poll conducted in May 2015 found that American attitudes toward animals improved between 2008 and 2015. Specifically, there was an increase from 25% to 32% in the percentage of respondents who said that animals 'deserve the same rights as people to be free from harm and exploitation'. There was a corresponding decrease in the number of people who said that animals can be used for human benefit, so long as they are not harmed or exploited. This suggests a shift in attitudes about whether animals can be used for human benefit – a seemingly significant moral change. More animals – and more types of animals – do seem to have moral status than ever before. We can make that conclusion based on many different studies, including Herzog *et al.*'s above.

This is not reflected in the Gallup poll, however. The number of people reporting that animals 'don't need protection' remained stable at 3% from 2003 to 2015. This makes it difficult to conclude that we are experiencing a moral expansion – that we should expect a continued increase in the number of animals seen as morally considerable. Past changes in attitudes toward animals seem to support the quantitative claim of the expanding moral circle, but recent changes do not. It's also unclear how we should interpret the increase in people who think animals cannot be used for human benefit (from 25% to 32%). **Pet-keeping**, for instance, might count as 'keeping animals for human benefit'. But it seems unlikely that a third of Americans object to pet-keeping. This raises the more general issue of how respondents understand the terms used. For instance, consider their interpretation of 'harm and exploitation': 94% of people thought animals deserved some degree of protection from harm and exploitation. However, another question in this poll found a much smaller percentage of people who expressed concern about the ways animals were treated as pets (68%), in zoos (78%), and as livestock (80%). These more pointed questions had not been asked previously, preventing any historical comparison. But the lower numbers should make us wonder whether people are applying a relatively narrow conception of harm and exploitation. A significant portion of respondents (20–30%) apparently does not think any harm or exploitation is involved in pet-keeping, captivity, and **intensive agriculture.** This too makes it difficult to conclude that current attitudes to animals are part of an expanding moral circle.

When very nuanced, well-informed authors—such as Kasperbauer in his book *Subhuman*—argue that not all the non-agricultural societies act as if there is a true human–animal equality, they often refer to *hunting*. That is, they refer essentially to "animal use," something that is done by *all* the organisms that eat animals. They don't refer to *animal abuse*, which refers to something abusive that, in my view, is *done for at least a substantial amount of time while the animal is alive*. The vast majority of hunter-gatherer societies, particularly the small, nomadic ones, don't do animal abuse and, as is also the generalized case with non-human predators, they would have no interest, or gain any advantage, in doing so anyway. The exception to this rule is when a certain nomadic group eventually abuses a living animal for some time during a sacrifice or magical ritual, as noted above. Or if they would take some time to skin animals alive to use their skin/furs, but this practice clearly does not seem to be so common in such groups as it is in at least some "developed countries," as documented in a truly horrifying—but much needed, to alert the broader public—way in the movie *Earthlings*. This is a 2005 documentary about humankind's dependence on, and the atrocities we make to, other animals, including five chapters: pets, food, clothing, entertainment, and scientific research. As shown in the movie, and well-documented in many other movies, books, and specialized papers, all "developed states," without any exception, *do* horrible acts of animal abuse, massively, and often employing extreme levels of cruelty (see Figs. 7.7, 7.8, and 7.9). For instance, we now have more "**animal factories**" than ever before, for massive production of meat, or milk, or eggs from animals that are often never free to simply walk a few meters, or see the sunlight: they are produced, from the very beginning until the very end of their lives, to just be used, abused, and killed by humans (see Fig. 9.6).

This goes in line with the point made by Kasperbauer in Box 7.15, when he notes that "though global meat consumption is increasing, the *rate* of meat consumption is declining…[but] killing fewer animals is only an improvement if their lives are not made worse off." Let's say that in a hypothetical country one century ago one cow was killed to feed 10 people for a day, while now one cow feeds 11 people a day, but in the former case the cow was free to walk in the grass during the whole day, and in the latter case she is trapped every single day of her miserable life as shown in the lower right of Fig. 9.6. Or, compare in particular the life of this latter cow with that of the ancestors of modern cows that were hunted by hunter-gatherers thousands of years ago, which had a completely wild life, just until the day they were killed, either by humans or by other animals. Killing fewer animals *in proportion* is not at all an "improvement" if the lives of those killed animals are more and more miserable.

Moreover, as noted above one of the problems of the comparisons made by Pinker and even by more nuanced authors such as Kasperbauer is that they tend to focus particularly on what happened in the last centuries, or the last millennia. However, if we do compare what happens nowadays in "developed" countries with what happened hundreds of thousands of years ago, or even with the few hunter-gatherer groups living today, would there be a lower rate of meat consumption per day, in our countries? As explained in Chaps. 4 and 5 and shown in Fig. 5.9, in many

Fig. 7.7 In his *Better Angels* book, Pinker talks about the "expanding moral circle" of humans, and how animals are "being better treated than ever." However, this is clearly another imaginary just-so-story of the storytelling animal, because empirical data clearly show otherwise: this is one of many thousands of pictures showing how animals are being treated *now* by the "sapient being." The atrocious *reality* is that humans are the only organisms that do this at a massive scale and to such extreme levels of cruelty. In fact, there are only thousands of pictures showing this kind of extreme abuse because most of the cases are not made public, being occulted from the broader public by the companies doing such atrocities and/or by other agents benefiting with them: this particular picture was only seen by the broader public due to a courageous undercover investigation done at the scientific research institution *Max Planck Institute for Biological Cybernetics*

Fig. 7.8 Another example of how animals are treated in the vast majority of today's countries by the so-called moral '*Homo sapiens*': experimentation with cosmetics to be sold by the company Victoria's Secret

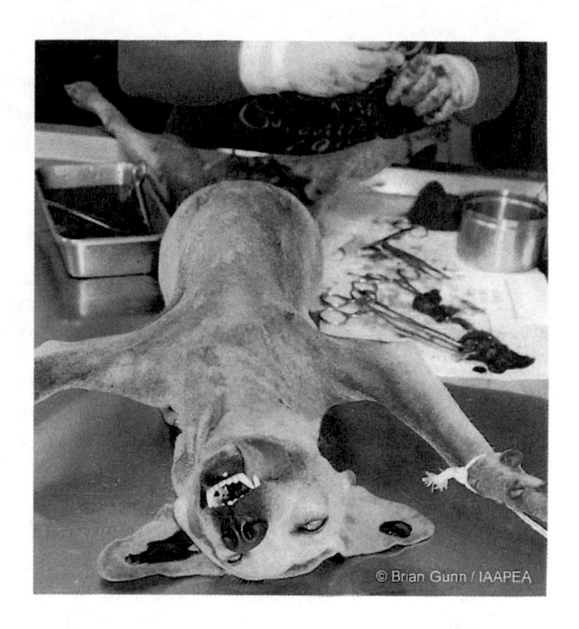

Fig. 7.9 A dog restrained in a brutal bloody experimental vivisection research testing lab: as noted in the caption provided by the IAA PEA—The International Association Against Painful Experiments on Animals—"atrocities are not less atrocities when they occur in laboratories and are called medical research"

nomadic hunter-gatherer groups the main part of the diet comes from the gathering of plants. Killing wild animals is not easy at all in general. This is precisely one of the numerous reasons why some groups of people begun to domesticate animals, thus starting the process of hugely decreasing *the quality of life of those animals*: it is much easier to kill—and to take milk, or eggs from—a non-free animal, of course. Even after the rise of agriculture in various regions of the planet, eating animal meat was seen as a luxury for most of the time, until relatively recently. This is exactly the opposite of what happens today in many Western countries, in which wealthier people or the growing number of vegetarians criticize "**cheap fast-food meat consumption**" and in which many health problems are due to an over-consumption of meat. Think about **Kentucky fried chicken**, **Macdonald's burgers**, and so on.

Furthermore, as wisely emphasized by Kasperbauer in Box 7.15—and often neglected by Pinker—animal abuse in "developed" countries does not concern only animal factories, very far from it. There are millions and millions of other animals that are produced, also from the very beginning until the very end of their lives, to simply be used and abused by humans for a plethora of other reasons, such as for medical and fundamental research. This includes truly horrifying experiments such as burning their eyes and skin, perforating their brains, paralyzing them, and so on (Figs. 7.7, 7.8, and 7.9). As the number of **animal experiments** increased enormously in the last centuries and continues to be done by an endless number of institutions nowadays in every single "developed" country, the fact that authors such as Pinker can go as low as to argue that animal abuse provides evidence for the concept of an "expanding moral circle" seems not only unscientific but a completely lack of respect—above all to the animals being abused, but also to the people that read/listen to their books/talks, as they are literally spreading factual inaccuracies. As stated in a 2012 article of the *Humane Society International*, already in that year—the numbers today are higher—it was estimated that *more than 115 million animals worldwide were used in laboratory experiments in a year*. Every single year, 115 million animals, basically produced to be trapped and tortured, over and over again, until the very end of their miserable lives. And such numbers are almost surely a huge underestimation, as explained in that article:

> The term "**animal testing**" refers to **procedures performed on living animals** for purposes of research into basic biology and diseases, assessing the effectiveness of new medicinal products, and testing the human health and/or environmental safety of consumer and industry products such as cosmetics, household cleaners, food additives, pharmaceuticals and industrial/agro-chemicals. All procedures, even those classified as "mild," have the potential to cause the animals physical as well as psychological distress and suffering. Often the procedures can cause a great deal of suffering. Most animals are killed at the end of an experiment, but some may be re-used in subsequent experiments. Here is a selection of common animal procedures: **Forced chemical exposure in toxicity testing**, which can include **oral force-feeding**, **forced inhalation**, injection into the abdomen, muscle, etc; Exposure to drugs, chemicals or infectious disease at levels that cause illness, pain and distress, or death; Genetic manipulation, e.g., addition or "knocking out" of one or more genes; **Ear-notching** and **tail-clipping** for identification; Short periods of physical restraint for observation or examination; Prolonged periods of physical restraint; **Food and water deprivation**; Surgical procedures followed by recovery; **Infliction of wounds**, burns and other injuries to study healing; **Infliction of pain** to study its physiology and treatment; **Behavioural experiments designed to cause distress**, e.g., **electric shock** or **forced swimming**; Other manipulations to create "animal models" of human diseases ranging from cancer to stroke to depression; **Killing by carbon dioxide asphyxiation**, **neck-breaking**, **decapitation**, or other means.
>
> What types of animals are used? Many different species are used around the world, but the most common include **mice**, **fish**, **rats**, **rabbits**, **guinea pigs**, **hamsters**, **farm animals**, **birds**, **cats**, **dogs**, **mini-pigs**, and non-human primates (**monkeys**, and in some countries, **chimpanzees**). It is estimated that more than 115 million animals worldwide are used in laboratory experiments every year. But because only a small proportion of countries collect and publish data concerning animal use for testing and research, the precise number is unknown. For example, in the United States, up to 90 percent of the animals used in laboratories (purpose-bred rats, mice and birds, fish, amphibians, reptiles and invertebrates) are excluded from the official statistics, meaning that figures published by the U.S. Department of Agriculture are no doubt a substantial underestimate. Within the European Union, more than 12 million animals are used each year, with France, Germany and the United Kingdom being the top three animal using countries. British statistics reflect the use of more than 3 million animals each year, but this number does not include animals bred for research but killed as "surplus" without being used for specific experimental procedures. Although these animals still endure the stresses and deprivation of life in the sterile laboratory environment, their lives are not recorded in official statistics.

No other species in this planet provokes such a sufferance to so many animals, as does our self-designated "moral," "rational," and even "altruistic" species. Lions are not at all "noble" animals that care about the suffering of zebras, when they kill them to eat, but they also do *not* create ways to make that suffering much higher, as many humans did, and still do, to themselves, for instance, during torture sessions that employ a huge number of tools and techniques specifically designed to inflict pain. Nor do the lions create narratives to be able to argue that the suffering of the zebras, or even the use of torture tools and techniques to make that suffering much worse, is justified by a "good cause," for instance, by arguing that zebras were "put in this planet" by the God of lions to be used and abused by them or, even worse, that by doing so lions are actually amazingly "altruistic" because that is ultimately "for the good of others" or of the "world." Only the *Homo irrationalis*, *fictus,* and *socialis* can live in such a world of *unreality* and be so cruel by creating and believing such completely absurd and scientifically untruthful tales. Even when people argue that in the case of animal medical research we are doing a "good thing" because this can potentially save human lives, one should ask: who or what gave us the right to abuse others so we can potentially improve our lives? As we have seen, 'whites' have argue exactly the same thing about - and done exactly the same type of atrocities with - 'non-whites', justifying that 'others' could be used and abused by them to 'develop' Western medicine (Fig. 3.19).

Outside of *Neverland*'s fantasies, we are actually nothing more than one mere species in a planet with millions of them: we are not special at all, there was no designer or "mother nature" that attributed more importance or specialness to humans than to other living organisms, these are just delusions. If let's say spiders would use millions of humans in cruel, painful experiments so just some spiders could have some pathologies temporarily mitigated—because, in the end, all living organisms die anyway, with or without medicine—would we say that the spiders were right in saying that this was for a "good," "moral," "altruistic" cause? When one only cares about the well-being of our ingroup, particularly when that implies inflicting enormous suffering and pain to the outgroups, this is not altruism: it is pure selfishness and cruelty. Actually, we don't need to make a hypothetical case with spiders, because the storyteller animal has already created similar stories that have been in vogue for decades now, about aliens coming to Earth to abduct and use humans in alien medical experiments. Those stories obviously considered that such a behavior by the aliens was a clear indication that they were selfish, "bad," cruel beings, and that humans should accordingly fight against them in every possible way. So, how can it be that when humans are the ones doing the very same thing to other beings, then suddenly they are the "good" ones? There is no way to justify cruelty, period.

Some people that are less concerned with such human "moral" constructs often argue, when they are asked about these issues, that the fact that humans are the ones doing the experiments with other animals precisely demonstrates that we are the "special" ones, and that we accordingly deserve to use and abuse other animals. Although this might sound as a valid point of view, if we think deeply about it we can see that it has huge flaws: basically under this logic, the group that arrives first to a position of dominance is allowed to do whatever atrocities it wants because it is the "special one." Such people seem to forget that exactly the same could be said to justify the horrible medical experiments done by Nazis that caused excruciating pain within a huge number of Jews and gypsies: they did "nothing wrong," they fully "deserved" to be able to do such experiments because they were the ones that were "able" to do them to Jews, not the other way round. This type of argumentation is disturbingly similar to that used, often in a distressingly joyful way, by scientists many decades ago, when they were doing one of the most cruel types of experiments on other animals—**vivisection** (Fig. 7.9).

Vivisection is commonly used to refer to the practice of **experimenting with live animals**, including the administration of drugs, infecting with diseases, brain damaging, blinding, and other types of very painful and invasive procedures (see Figs. 7.7 and 7.8). For a long time in human history, one of the most common types of vivisection was to open the body of non-human animals and, while they were still alive, to study their anatomy and physiology, for instance, how the blood flows, how some nerves can be blocked, and so on, often without any kind of anesthesia. I will not describe here in detail the type of pain that **vivisection** can inflict, because this would just be an horrendous description and because I think, or at least hope, that those people reading these lines have enough empathy to imagine how excruciating the pain would be. Therefore, I will just use an excerpt of Guerrini's book *Experimenting with Humans and Animals*, about a proclamation by the Russian physiologist **Elie de Cyon** in his *Methodik der physiologischen experimente und eivisectionen*, which shows that such experiments were often used to express the ideal aimed at by the leaders of the **vivisectionist school**. To put this highly enthusiastic, almost ecstatic declaration of Cyon—he seems to almost come to climax with it—in context, one needs to note that he died just about a century ago—in 1912—and that the types of vivisection he is referring to often included animals such as dogs (Fig. 7.9):

> The true vivisector must approach a difficult vivisection with the same *joyful excitement* and the same *delight* wherewith a surgeon undertakes a difficult operation from which he expects extraordinary consequences. He who shrinks from cutting into a living animal, he who approaches a vivisection as a disagreeable necessity, may very likely be able to repeat one or two vivisections, but he will never become an *artist in vivisection*. He who cannot follow some fine nerve-thread, scarcely visible to the naked eye, into the depths, if possible sometimes tracing it to a new branching, with joyful alertness for hours at a time; he who feels no enjoyment when at last, parted from its surroundings and isolated, he can subject that nerve to electrical stimulation…The *pleasure of triumphing* over difficulties held hitherto insuperable is always one of the highest **delights of the vivisector**. And the sensation of the physiologist, when from a gruesome wound, full of blood and mangled tissue, he draws forth some delicate nerve branch, and calls back to life a function which was already extinguished – this sensation has much in common with that which inspires a sculptor, when he shapes forth fair living forms from a shapeless mass of marble.

Furthermore, for those that might still defend that doing such atrocious experiments with Others—be them Jews, gypsies, people with Down syndrome, or other animals—is a "noble," justified "altruistic" act because it will ultimately benefit Us, there is a crucial aspect that needs to be emphasized. Namely, that many, if not most, non-human laboratory studies don't bring any practical benefit to humans, being instead a huge waste of human potential, time, money, and resources, what is of course nothing compared with the horrible sufferance they bring to millions of animals every year. Firstly, a huge number of such studies use "animal models" that are completely inappropriate for the type of pathologies being studied. Secondly, an endless number of such studies are not even "completed," or do not lead to "applicable results" at all. Thirdly, even worse—if there can be anything worse than this—a vast majority of the remaining studies causing horrible sufferance to millions and millions of animals every year concern either fundamental research about, or pharmaceutical research to test products for, items that have nothing to do with saving lives or curing diseases, being, for instance, related to things as superfluous as producing perfumes, makeup accessories, and other cosmetic products. A 2013 article published by the *Humane Society International* refers to one of these unnecessary items, cosmetics:

> Did you know that in many parts of the world, animals in laboratories are still suffering and dying to test cosmetics such as lipstick and shampoo? They have chemicals forced down their throats and dripped into their eyes and onto their shaved skin. It's the ugly secret of the beauty industry…We estimate that approximately 100,000-200,000 animals suffer and die just for cosmetics every year around the world. These are rabbits, guinea pigs, hamsters, rats and mice. While dogs and monkeys are never used to test cosmetics anywhere in the world, they are used to test other types of chemicals. Typically, **animal tests for cosmetics** include **skin and eye irritation tests** where chemicals are rubbed onto the shaved skin or dripped into the eyes of rabbits; repeated oral **force-feeding studies** lasting weeks or months to look for signs of general illness or specific health hazards, such as cancer or birth defects; and even widely condemned "**lethal dose tests**", in which animals are forced to swallow massive amounts of a test chemical to determine the dose that causes death.
>
> These tests can cause considerable pain and distress including blindness, swollen eyes, sore bleeding skin, internal bleeding and organ damage, birth defects, convulsions and death. Pain relief is not provided and at the end of a test the animals are killed, normally by **asphyxiation**, **neck-breaking** or **decapitation**. Almost without exception, companies have a choice about whether or not to test on animals. In the majority of cases, animal tests continue because some companies insist on developing and using "new" ingredients. These are ingredients that don't have existing safety data – because they're new! So new safety data has to be generated to satisfy the regulators before a product can go on sale, and that means new animal testing. Animal testing also continues in the cosmetics industry because of convention – that's the way it's always been done, animal tests are familiar even if they're flawed. Regulators, whose job it is to approve cosmetics for use, tend to be very conservative in their approach and can delay approve a product if the manufacturer provides safety data based on unfamiliar non-animal test methods. Some companies claim that they have to test on animals because they sell their products in countries like China where animal testing is still required by law for companies importing into the country. But this isn't really true. They have chosen to sell in China knowing that to do so will mean new animal testing.

One point I would like to make, and that might surprise some of you, is that the reasoning that humans "deserve" to do such cruel experiments to other animals because we are the ones that have the power to do so is very similar to that often employed by **humanists**, but applied in the opposite way. This is because humanists often accept teleological tales about how humans have a "cosmic" duty to forbid such cruel experiments precisely because we are "special" animals with a "unique mental ability." That is, both ideas are based in the longstanding teleological idea that humans are "special" beings: in fact, as it is often said, humanists tend to be in love with humanity, so basically this is one more recycled—more "modern"—version of our very old friend, the concept of "ladder of life" (see Fig. 3.2). However, although both ideas are based on the belief of such a factually inaccurate concept, personally I much prefer the way in which **humanists** and organizations such as the Humane Society International often use this concept *versus* the way it has been used by many people to justify atrocious types of abuses to other animals and/or to "Other humans". In other words, I am not at all criticizing what humanist organizations do in general. Well on the contrary, personally I profoundly admire many of them, and I directly contribute monetarily to help some of them, such as ***Save the***

Children—the real one that does an amazing work helping kids across the globe, not the "fake" one related to the **QAnon conspiracy theories**, of course. In this sense, I need to emphasize that one of the most interesting, and correct, contributions of Pinker's *Better Angels* is precisely that it is an ode to this type of "compassionate **humanism**." However, as explained in Harari's *Homo Deus*, this does not mean that historians or scientists should neglect the fact that humanism is also the result of the creation and believe in exactly the same type of teleological narratives that have been accepted since millennia ago and that were/are used by people to justify horrible atrocities. Particularly because recognizing this critical point also allows us to be aware of some important theoretical flaws and thus of potential misuses of humanism.

So, in a nutshell, this should be taken into account when we talk about **humanism**, or about **Enlightenment**, because (1) such narratives are, after all, part of a world of unreality, of *Neverland*, and (2) human history has clearly shown us how dangerous living in *Neverland*, and applying narratives about "human specialness," can be. These two points actually bring us back to a central topic already emphasized in the present book: that "there is only one religion, though there are a hundred versions of it"—humanism is precisely just a more recent version of it. This point is recognized in the very title of the 2010 bestseller *Good without God – What a Billion Nonreligious People Do Believe*, written by Epstein, an humanist chaplain at Harvard University. As noted in that book, "the **American Humanist Association** defines Humanism as a progressive lifestance that, without supernaturalism, affirms our ability and responsibility to lead ethical lives of personal fulfillment, aspiring to the **greater good of humanity**." *Greater good*—a teleological term used, and abused, in so many types of beliefs. Similarly, the way the word responsibility is often used by humanists is also teleological: who said we humans have a "responsibility" to do *greater good*? If there is no God, no designer, no masterplan, why do humanists—which include a huge number of self-defined "atheists"—talk in a way that clearly seems to indicate that we *do* have a cosmic "duty" and "responsibility"? And if we are not "made in the image of God," as many humanists such as Epstein rightly claim when they say "without supernaturalism," why would humans specifically, and not other animals, for instance, spiders, claim that they have such a cosmic "duty" to, let's say, "make a better world," or to "save the planet"? In his 2018 book *Seven Types of Atheism*, Gray also discusses these subjects, in a chapter titled "*Secular Humanism – a sacred relic*":

> For its followers the **religion of humanity** seems different from the religions of the past. Having repudiated monotheism, they imagine they stand outside the view of the world that monotheism expressed. But while they may have rejected monotheist beliefs, they have not shaken off a monotheistic way of thinking. The belief that humans are gradually improving is the central article of faith of modern humanism. When wrenched from monotheistic religion, however, it is not so much false as meaningless. For the **ancient Greeks and Romans**, history revealed no pattern other than the regular growth and decline of civilization – a rhythm not essentially different from those found in the natural world. There was no prospect of indefinite improvement. Judged by the standards of the time, civilization might improve for a while. But eventually the process would stall, then go into reverse. Rooted in the innate defects of the human animal, cycles of this kind could not be over-

come. If the gods intervened, the result was only to make the human world even more unpredictable and treacherous. Some ancient historians eliminated divine intervention in their accounts of events. Writing in the fifth century BC, the Greek historian **Herodotus** has the gods acting to punish wrongdoing (such as violation of temples), but there is no suggestion that they were interested in shaping the course of history.

Herodotus' successor **Thucydides** wrote history without recourse to any kind of divine intervention. His *History of the Peloponnesian war* records a succession of mishaps in which human will and reason are confounded by archetypal human flaws. Thucydides has been called the father of 'scientific history'. But for him there were no laws of history, only the fact of recurring human folly. A cyclical view of history was revived in Europe during the **Renaissance** by **Niccolò Machiavelli**. Rather than contesting **Christian belief**, the Florentine historian and adviser to princes stepped outside Christian ways of thinking. History was not a moral tale in which evil is punished or redeemed. A prince had to be ready to commit crime in order to protect the state. In order for virtue to survive, a ruler had to practise vice. Human goodness showed no tendency to increase over time. This view proved too uncomfortable to be adopted by Machiavelli's contemporaries, and it is one most secular thinkers have found intolerable. Until the rise of Christianity, a cyclical view of history was taken for granted by practically everyone. When in eighteenth-century Europe religion began to be replaced by secular creeds, the **Christian myth of history** as a redemptive drama was not abandoned but renewed in another guise. A story of redemption through divine providence was replaced by one of *progress through the collective efforts of humanity*. Nothing like this could have developed from polytheistic religions, which take for granted that human beings will always have disparate goals and values.

As explained by Gray, among the typical teleological narratives believed by humanists, the one about "cosmic progress" is clearly the most critical one. Strangely, while Gray correctly points out that this notion of "progress" is indeed one of the deepest myths of Christianity, he neglects the fact that this is mainly due to the influence exerted by the **ancient Greek notion of a "ladder of life**," and that this is why this notion is so profoundly entrenched in Western science and philosophy (see Box 7.16 and Chap. 3). However, despite neglecting this important historical fact, what Gray writes about Christianity and humanism does deserve to be taken into account and discussed here, as it highlights the links between the "**religion of humanity**," the **French Revolution**, **the Enlightenment**, and ideologies such as **Marxism**, **Communism**, and other other 'isms':

The modern faith in progress began with shifts in **Christian thinking**. Declaring itself superior to anything in the pagan or Jewish world, **Christianity** affirmed that a new order of things was open to everybody...With the **Reformation** and the rise of 'post-millennialism' in seventeenth-century **Protestantism**, this myth gave way to one that was more human-centred. The belief that evil would be destroyed in an **apocalyptic end-time** was supplanted by the conviction that evil could be slowly diminished in history. **Jesus** would still return and rule over the world, but only after it had been transformed by human effort. Emptied of its transcendental content, this Christian myth is the source of modern **meliorism** – the idea that human life can be gradually improved. Unlike the dominant view of history in the ancient world, which recognized improvement but accepted that what had been gained would over time be lost, the modern neo-Christian **belief in progress** asserts that human life can be made better cumulatively and permanently. The modern myth of progress came into being as a fusion of Christian faith with **Gnostic thinking**...while modern **meliorism** claims to be based in science, the idea that **civilization** improves throughout history has never been a falsifiable hypothesis. If it had been it would have been abandoned long ago. For those who believe in progress, any regression that may occur can only be a temporary halt in an onward march to a better world. Yet if you look at the historical record without modern prejudices you will find it hard to detect any continuing strand of improvement. The triumph of Christianity brought with it the near-destruction of classical civilization. Libraries and museums, temples and statues were demolished or defaced on a vast scale in what has been described as 'the largest destruction of art the world has ever seen'. Everyday life was hemmed in with unprecedented repression. While there was nothing in the pagan world of the liberal concern for individual freedom, pluralism in ways of life was accepted as a matter of course. Since religion was not a matter of belief, no one was persecuted for heresy. **Sexuality** was not demonized as it would be in the Christian world, nor gay people stigmatized. While they were subordinate to men, women were freer than they would be once Christianity had triumphed.

Today everyone is sure that civilization has improved with modern times. As we are forever being reminded, the medieval and early modern world was wracked by **wars of religion**. But faith-based violence did not fade away with the arrival of the modern age. From the **French Revolution** onwards, Europe and much of the world were caught up in revolutions and wars fuelled by secular creeds such as **Jacobinism** and **communism**, **Nazism** and **fascism** and a belligerently evangelical type of **liberalism**. In the twenty-first century a potent source of faith-based violence has emerged in **Islamist movements**, which blend ideas borrowed from **Leninism** and fascism with fundamentalist currents from within Islam. It is true that **slavery** and **torture** were flaws of pre-modern societies. But these practices have not disappeared. Slavery was reintroduced in the twentieth century on a vast scale in **Nazi Germany** and the **Soviet and Maoist gulags**. Slave auctions in the so-called Caliphate established by the **Islamic State** in parts of Iraq and Syria were advertised on Facebook. **Human trafficking** flourishes throughout much of the world. Torture has been renormalized. Banned in England in the mid-seventeenth century and in Europe by the Habsburg empress Maria Theresa in the late eighteenth, the practice was revived by the world's pre-eminent democracy when George W. Bush sanctioned it in the run-up to the invasion of Iraq. Instead of being left behind, old evils return under new names. No thread of progress in civilization is woven into the fabric of history. The cumulative increase of knowledge in science has no parallel in ethics or politics, philosophy or the arts. Knowledge increases at an accelerating rate, but human beings are no more reasonable than they have ever been. Gains in civilization occur from time to time, but they are lost after a few generations. When secular thinkers tell the history of humankind as a story of progress they flatter themselves that they embody the progress of which they speak. At the same time they confirm that their view of the world has been inherited from monotheism. It was only with the invention of Christianity that a history of humankind began to be told. Before that point, there was no universal history. Many stories were told – the story of the Jewish people, the Greeks, the Romans and multitudes of others.

Modern thinkers say that telling history as a story of all humankind marks an advance. But along with **Christian universalism** came a militant intolerance – a trait that Christianity transmitted to its secular successors. For neo-Christian believers any way of life that fails or refuses to fit into a story of progress can be regarded as sub-human, exiled to the margins of history and then consigned to extinction…the belief that humanity makes history in order to realize its full possibilities is a relic of **mysticism**. Unless you believe the species to be an instrument of some higher power, 'humanity' cannot do anything. What actually exists is a host of human beings with common needs and abilities but differing goals and values. If you set metaphysics aside, you are left with the human animal and its many contending ways of life…In *Beyond good and evil*, **Nietzsche** described Christianity as '**Platonism** for the masses' – an accusation that applies with equal or greater force to secular humanism. The faith that history has a built-in logic impelling humanity to a higher level is Platonism framed in historical terms. **Marxists** have thought of human development as being driven by new technologies and class conflict, whereas liberals have seen the growth of knowledge as the principal driver. No doubt these forces help shape the flow of events. But unless you posit a divinely ordained end-state there is no reason to think history has any overarching logic or goal. For **Plato** and **Plotinus**, history was a nightmare from which the individual mind struggled to awake. Following Paul and **Augustine**, the Christian Erigena made history the emerging embodiment of Logos. With their unending chatter about progress, secular humanists project this mystical dream into the chaos of the human world. An implacable enemy of Christianity, Nietzsche was also an incurably Christian thinker. Like the Christians he despised, he regarded the human animal as a species in need of redemption. Without God, humankind faced '**nihilism**' – life without meaning. But nihilism could be avoided if humans willed into being the meaning God had once secured. Only a few would ever be capable of this feat. It was these exceptional individuals – the supermen lauded in *Thus*

Box 7.16 "Religion of Science," Darwin, Racism, and Transhumanism

The third chapter of Gray's *Seven types of atheism*, entitled, "*A Strange Faith in Science*", provides clear examples of how belief and teleological narratives, particular about "progress," have been deeply embedded in the ideas/works of the most renowned scientists, including Darwin as we discussed in Chap. 6. Gray writes:

> In 1929, the Thinker's Library, a series of books published by the **Rationalist Press Association** in London to counter the influence of religion in Britain, produced an English translation of the German biologist **Ernst Haeckel**'s 1899 book *The riddle of the universe*…strongly hostile to Jewish and Christian traditions, Haeckel founded a new religion called **Monism**, which spread widely among intellectuals in central Europe. Among **Monist tenets** was a 'scientific anthropology' according to which the human species was composed of a hierarchy of racial groups, with white Europeans at the top. At the time '**scientific racism**' was not unusual in books promoting **rationalism**. Like Haeckel a proponent of a '**religion of science**', **Julian Huxley** joined Haeckel in promoting theories of innate racial inequality. In 1931, he wrote that there was 'a certain amount of evidence that the Negro is an earlier product of human evolution than the Mongolian or the European, and as such might be expected to have advanced less, both in body and in mind'. In the early twentieth century such attitudes were commonplace among rationalists. In his best-selling book *Anticipations* (in 1901) **H. G. Wells**, also a contributor to the Thinker's Library, wrote of a new world order ruled by a scientific elite drawn from the most advanced peoples of the world. Regarding the fate of 'backward' or 'inefficient' peoples, he wrote: "And for the rest, those swarms of black and brown, and dirty-white, and yellow people, who do not come into the needs of efficiency? Well, the world is a world, not a charitable institution, and I take it that they will have to go…it is their portion to die out and disappear."
>
> Huxley's '**evolutionary humanism**' asserted that if humankind was to ascend to a higher level, evolution would have to be consciously planned. Some religious thinkers followed Huxley in thinking in this way. A. N. Whitehead (1861-1947) and Samuel Alexander (1859-1938) developed a type of '**evolutionary theology**' in which the universe was becoming more conscious of itself – a process that would culminate in the emergence of a Supreme Being much like the God of monotheistic religion. The French Jesuit theologian **Pierre Teilhard de Chardin** (1881-1955) developed a similar view in which the universe was evolving towards an 'Omega Point' of maximal consciousness. Much of the **Enlightenment** was an attempt to demonstrate the superiority of one section of humankind – that of Europe and its colonial outposts – over all the rest. Evangelists for the Enlightenment will say this was a departure from the 'true' Enlightenment, which is innocent of all evil. Just as religious believers will tell you that 'true' Christianity played no part in the **Inquisition**, secular humanists insist that the Enlightenment had no responsibility

for the rise of modern racism. This is demonstrably false. ***Modern racist ideology** is an Enlightenment project*. **Racism** and **anti-Semitism** are not incidental defects in Enlightenment thinking. They flow from some of the Enlightenment's central beliefs. For **Voltaire**, **Hume** and **Kant**, European civilization was not only the highest there had ever been. It was the model for a civilization that would replace all others. The 'scientific racism' of the nineteenth and early twentieth centuries continued a view of humankind promoted by some of the greatest Enlightenment thinkers.

All these philosophies rely on an idea of evolution. There is a problem, however. As understood in Darwin's theory, the universe is not in any sense evolving towards a higher level. Thinking of evolution as a movement towards greater consciousness misses Darwin's achievement, which was to expel teleology – explaining things in terms of the purposes they may serve rather than the causes that produced them – from science. As he wrote in his *Autobiography*, "There seems to be no more design in the variability of organic beings, and in the action of natural selection, than in the course in which the wind blows." As **Darwin** makes clear in this passage, natural selection is a purposeless process. He did not always stick with this view, however. On the last page of *On the origin of species*, he wrote: "we may be certain that the ordinary succession by generations has never once been broken, and that no cataclysm has desolated the whole world…hence we may look with some confidence to a future of great length…and as natural selection works solely for the good of each being, all corporeal and mental endowments will tend to *progress to perfection*." In fact the theory of natural selection contains no idea of progress or of perfection. Darwin's inability always to accept the logic of his own theory is revealing. An eminent Victorian, he could not help believing that natural selection favoured 'progress to perfection'. Many less scrupulous thinkers have followed Darwin in this belief. But not only is it at odds with Darwin's account of evolution as a non-directional process. If evolution had a direction, human beings need not accept it. What if the world was evolving towards new forms of slavery? In that case, it would be better to resist evolution than go along with it. Theories of social evolution mirror the intellectual fashions of the time. If **Herbert Spencer** used evolutionary ideas to vindicate untrammeled **capitalism**, Haeckel and the early Julian Huxley used these ideas to bolster belief in **European racial superiority**. Others used them to prop up their political views. In a book co-authored with her husband Sydney Webb, *Soviet communism: a new civilization?*, first published in 1935, the sociologist Beatrice Webb – at one time Herbert Spencer's assistant – suggested that **Stalin's Russia** embodied the next stage of social evolution. In a famous essay *'The end of history?'* published in…1989, **Francis Fukuyama** was in no doubt that an evolutionary process was at work leading to the advance of 'democratic capitalism' throughout most of the world…Neither the Webbs' nor Fukuyama's prognostications have been borne out by events…

In his book *The singularity is near: when humans transcend biology* (2005), the futurologist Ray Kurzweil looks forward to an explosive advance in science that will enable humans to transcend the physical world and thereby escape death. It is telling that both Bernal and Kurzweil take the titles of their books from the **Bible** – Bernal's echoing a passage in the *New testament* (Letter to the Ephesians 2:1–3), Kurzweil's echoing **John the Baptist**'s *'The Kingdom of Heaven is near'* (Matthew 3:2). Knowingly or otherwise, both of them are suggesting that **transhumanism** is religion recycled as science. At bottom, the **transhumanist movement** is a modern variant of the dream of transcending contingency that possessed mystics in ancient times. Gnostics and disciples of **Plato** longed to be absorbed in a timeless Absolute, a refuge from the ugly conflicts of the human world. Transhumanism is a contemporary version of a modern project of human **self-deification**. One of the few to recognize this is the Israeli historian of science **Yuval Noah Harari**. In *Sapiens: a brief history of humankind*, first published in Hebrew in 2011, and *Homo Deus: a brief history of tomorrow* (2016), Harari suggests that the expanding powers that humankind is acquiring through the advance of science could end up bringing about human extinction. Using bioengineering and artificial intelligence, the human species will enhance its physical and mental capacities far beyond their natural limits. Eventually it will turn itself into God. "In the twenty-first century", Harari writes, "the next big project of humankind will be for us to acquire the divine powers of creation and destruction, and upgrade *Homo sapiens* into *Homo Deus*." Yet his history of the future rests on a confusion. 'Humanity' is not going to turn itself into God, because 'humanity' does not exist. All that can actually be observed is the multifarious human animal, with its intractable enmities and divisions. The idea that the human species is a collective agent, setting itself 'big projects' and pursuing them throughout history, is a humanist myth inherited from monotheism. As the post-human species that humans have created mutate and evolve, human history may indeed come to an end. The question remains why transhumanists find this prospect so appealing. Is it because they think humankind is only a channel for values that transcend the human animal, such as knowledge or information? But, unless you posit a Platonic heaven beyond the material universe, it is hard to know where these values are to be found. They are not features of the natural world.

spake Zarathustra – who would redeem humanity from a senseless existence. Nietzsche's *Übermensch* or superman played a Christ-like role.

In *The Rise of Organized Brutality – A Historical Sociology of Violence*, Masesevic argues that many scholars, as well as the media and broader public in general, tend to idealize the **Enlightenment**—that is, the movement that took place between the late seventeeth to early ninteenth century and that emphasized "reason," science, and **humanism**, as we have seen above. By doing so, they often tend to neglect that there are clear links, in various cases, between some of the ideas related to that movement and an increase of *violence*, and even of genocides, as a result of the type of racism and **dehumanization** that were produced by followers of such ideas (see also Boxes 3.9 and 7.16). Masesevic explains:

> Although the ideas of individual autonomy, popular sovereignty and equal moral worth of all human beings were articulated a long time before the **Enlightenment**, it was only after the concrete structural transformations, including the two famous revolutions, that such ideas started to become acceptable to most. Furthermore, and in contrast to **Pinker**, these ideas did not automatically pacify social order. Instead, these very ideas were inaugurated through some of the most violent episodes in human history. Not only were the **American and French revolutions** accomplished through a great deal of **postrevolutionary violence**, but once in motion these revolutionary ideas and practices quickly transformed into all-out warfare. Following the success of the revolution, the new **French Republic** was involved in numerous wars and crushing of domestic and foreign rebellions. Initially the focus was on obliterating the clerical and monarchist opposition in Vendée and Brittany (1793-1796), where much of the local civilian population were exterminated as the 'enemies of the free republic'. While the conflict is usually depicted solely as an ideological conflict, the initial uprising was motivated by the peasant rejection of the mass conscription

introduced by the new republic. Of course, ideology played an important role on the republican side as the revolt was interpreted to be a counterrevolution aimed at destroying the new republic. Consequently, the Committee of Public Safety made a conscious decision on 1 August 1793 to 'pacify' the entire region by killing all of its inhabitants. When the general in charge questioned the 'fate of the women and children', the Committee's order to him was 'eliminate the brigands to the last man, there is your duty'. The outcome of this deliberate policy of political cleansing was more than 160,000 locals killed out of the population of 800,000.

Since the revolutionaries saw themselves as being in the possession of the ultimate truth and absolute justice, they were motivated by this 'scientific' understanding of righteous zeal. For example, the new National Convention, a republican parliament in Paris, issued the Edict of Fraternity in November 1792 declaring that their aim was to 'export the French Revolution' with the promise of 'fraternity and help to all peoples who wish to recover their liberty'. The consequence of this policy was a gradual transformation of what was essentially a defensive war into an expansionist conquest – the **Napoleonic Wars** (1803-1815). These wars were historically distinctive in a sense that they brought together advanced military organization and society wide **ideological mobilisation**. The traditional scholarship tends to mythologise the military genius of **Napoleon** as being the decisive factor in the unprecedented victories of the revolutionary armies. However, recent analyses emphasise the organisational and ideological dimensions: the state's capability to recruit, train, arm, feed, clothe and coordinate huge numbers of ordinary recruits, some of whom were enthusiastic to fight but the majority of whom were mobilised through the combination of coercion and ideology. The ideological element was important as the soldiers tended to see themselves more and more as equal members of the abstract French nation. This is not to say that they joined the military because of a strong sense of French identity; rather, nationalism was in most cases a byproduct of military socialisation. However, what really matters is the ideological (and organisational) capacity and willingness of the French state to turn 'peasants into Frenchmen' through its military machine. The Napoleonic Wars inaugurated the new form of a mass-scale fighting resulting in a staggering number of casualties. Unlike previous European conflicts, where direct battlefield fatalities were generally small and more soldiers died from the disease or hunger, now the battlefields became the arena of mass death and destruction. For example, at the **Battle of Leipzig** (1813), 500,000 soldiers were involved, out of which 150,000 were direct fatalities...all these wars against old regime Europe between 1792 and 1815 cost the lives of well over 5 million Europeans.

He further directly links some of the ideas of the Enlightenment, and of humanism—particularly those concerning teleological narratives about "cosmic progress"—with **dehumanization** and violence:

Since the key ambition of the **modernist project** was to create the most rational, equal, free, just and truthful social order, any organized resistance to this ideal world could only be interpreted as an attempt to stop the *progress* and reestablish the old tyrannical ancient regime. Hence the revolutionaries tended often to understand their role in deeply Manichean terms: they saw themselves as the possessors of the ultimate truth and justice fighting against the forces of darkness who intend to reverse the (progressive) wheel of history. In such an environment, ideological zeal was perceived as a precondition for the ultimate victory. Moreover, to successfully delegitimize the representatives of the ancient regime and others who challenge the new order, it becomes critical to deny them a membership in the human race. More to the point, as the new modernist ideologies conceptualise all human beings as morally equal, it is not enough anymore to emphasise the opponent's moral failings. Instead, to fully delegitimise the enemy's actions, it is now crucial to deny any humanity to the enemy. While in the deeply hierarchical premodern world such categorisations were not necessary, as peasants were universally perceived to be inferior to their lords, in the modern context where everybody is of equal moral worth, deligitimisation inevitably slides in the direction of dehumanisation

If all humans are equal, then our despicable enemy can only be something less than human. The consequence of this ideological transformation was the justification of violence in new, universal terms. Hence the **mass-scale massacres** in regions that did not accept the legitimacy of the **French revolutionary government**, particularly in Vendée, were justified in strictly universalist terms. For example, one of the leaders responsible for the mass-scale violence of the revolutionary forces in Lyon stated that 'I am purging the land of liberty of these monsters according to the principle of humanity'. Thus the representatives of the ancient regime and their supporters are not the respected opponents but monsters that need to be purged in the name of humanity. This type of justification was also used by the revolutionary general **Francois Joseph Westermann**, who described his actions in a letter to the Committee of the Public Safety announcing that 'There is no more Vendée…according to the orders that you gave me, I crushed the children under the feet of the horses, massacred the women who, at least for these, will not give birth to any more brigands…I do not have a prisoner to reproach me…I have exterminated all…mercy is not a revolutionary sentiment'. In an almost identical way, the youngest leader of the **French Revolution**, Saint-Just, was a stringent proponent of the idea that revolution requires 'inflexible justice' and that 'whoever vilified or attacked the dignity of the revolutionary government should be condemned to death'. The direct result of this ***uncompromising ideological crusade*** was as many as four hundred thousand casualties of this brutal civil war. This **revolutionary radicalism** was equally visible in the **Haitian Revolution**, where the enemy was also dehumanised. For example, one of the revolutionary leaders, **Boisrond-Tonnerre**, stated that 'For our declaration of independence, we should have the skin of a white man for parchment, his skull for the inkwell, his blood for ink, and a bayonet for a pen!'. Since revolutions are usually premised on the idea that the new postrevolutionary order will bring about a society able to transcend the existing social conflicts, their ideological mobilisation inevitably stimulates **doctrinal radicalism**.

As these ideological blueprints are more often than not conceptualised as the ultimate truth delivering projects, buttressed either by scientific authority, **moral humanist absolutism** or both, then any attempt to stand in the way of such a noble endeavour can be interpreted as nothing less than deliberate evil. The possession of ultimate truth and absolute justice calls for righteous zeal and no restraint. Similarly, to establish a classless world of proletarian justice entails not only the 'expropriation of the expropriators' but also their ultimate destruction, as they are bourgeois stooges of the capitalist enemy nations. In the alternative modernist vista, the foundation of a new, better and racially pure world necessitates the implementation of genocidal policies against **malicious 'Judeo-Bolshevik' conspirators**. Hence those who oppose the creation or existence of such a perfected social order can be nothing other than 'parasites', 'leeches', 'rats' and 'monsters', and as such they do not deserve treatment equal to that of humans; they have to be destroyed. *Therefore, it is inclusivity, universalism and moral equality that paradoxically create the conditions for the greater dehumanisation of one's ideological enemy.* When there is an acknowledged and visible hierarchy between social groups, there is no great need to dehumanise the (already inferior) Other. It is only when all human beings are universally considered to be of equal moral worth that one needs to deploy an elaborate ideological apparatus to deny a particular group full membership in the human community. It is only in modernity that dehumanisation becomes necessary for both the justification of organized, interpolity violence and for popular mobilisation to participate in such violence. How else could one reconcile the idea of building a more liberal, democratic world that respects human rights and the dignity of all human beings while simultaneously deploying atomic bombs to kill hundreds of thousands of civilians? It is no accident that the destruction of **Hiroshima** and **Nagasaki** was justified through dehumanisation. When **President Truman** said that 'when you have to deal with a beast, you have to treat him as a beast', he was just reflecting the dominant perception of the Japanese shared by large sections of U.S. public. As various polls conducted at the end of the war show, most respondents enthusiastically supported bombings, with substantial numbers advocating dropping more atomic bombs and the extermination of all Japanese…

> The modern age differs from its premodern counterparts in hiding death, as both killing and dying are removed from the public eye. The animals we eat are killed in the closed and far away abattoirs, our old and sick die in hospices and hospitals, our morgues are removed from public view, we don't organise public hangings nor do we torture or burn people in town squares…For example, the number of **slaughtered animals** used for consumption in the United States alone is regularly over 25 billion per year. In 2011, 29 billion animals have been killed for food, including 23.1 million ducks, 38 million cattle, 109 million pigs, 256 million turkeys, 7.8 million chickens, 2.4. million rabbits, 14 billion fish and 40 billion shellfish. The world consumption of animal food has experienced staggering increase since 1860s, and it now amounts to over 63 billion animals killed per year. As Cudworth documents, 'at least 55 billion land-based non-human animals are killed in the farming industry per year'. Bourke vividly describes the enormous scale of animal destruction: 'In factories for killing pigs, "hog stickers" can slit over a thousand throats an hour…in the UK, twenty-eight animals are slaughtered for food every second – a total of over 883 million animals each year'…Animals are also used as the material for domestic uses, from soaps, dyes, lubricants, plastic, rubber, strings and fertilizers to cosmetics, adhesives and cleaning products. Although human beings could for the most part exist without the use of animal food, products and labour, *the foundations of all human civilisations have been built on the subjugation, exploitation and consumption of animals*…[Similarly] in times of war, the citizens of modern polities generally give tacit or explicit consent for the mass murder (often also removed from the public eye) of those who inhabit other polities. Moreover, it is regularly those who are the least affected by the calamities of war who often support the most extreme forms of violent retaliation. For example, **World War II** surveys of the British public clearly show that there was substantially greater support for the reprisal bombings of German cities among individuals living in areas unaffected by the Luftwaffe's aerial bombardment than by those who lived in the cities that were excessively bombed. Similarly, in the **wars of Yugoslav succession** it is the civilians (academics, journalists, university students and teachers) rather than frontline soldiers who regularly expressed the most extreme attitudes towards the despised enemy.

"*Therefore, it is inclusivity, universalism and moral equality that paradoxically create the conditions for the greater dehumanisation of one's ideological enemy*": Masesevic at his best—writing in a very concise, but eloquent and very astute way, that make us deeply re-think ideas that we had often taken for granted. In his outstanding 2016 book *Stamped from the Beginning* Kendi also deconstructs a plethora of myths and contradictions concerning the many so-called revolutionary ideas defended by numerous "Enlightenment intellectuals," such as **Thomas Jefferson**, as well as the hypocrisies between those ideas *versus* what those intellectuals actually did in practice not only within their personal lives but also within their political/public careers (see Box 7.17). Kirkham's 2019 book *Our Shadowed World* also criticizes the idealized way in which Enlightenment and the rise of humanism are so often portrayed in Western history books and textbooks:

> Since the Enlightenment humankind has increasingly come to accept the presence of an external world that is mechanical and indifferent to human presence. This has, of course, reinforced the need to re-create a dream of innocence, a single vision that will give hope, an ideology to serve as a pseudo or replacement religion. The very fragility of such a construct seems to engender a doubtdestroying fanaticism. An exasperated Austrian officer who was interrogating the young revolutionary terrorist **Felice Orsini** perceptively exclaimed, Orsini's nationalism had become "a religious monomania". Though modern tyrants are often deemed to be irreligious, this is far from the truth. Understanding the need for religion, both **Robespierre** and **Hitler** denounced atheism. Robespierre proclaimed the need of the

French people to place their trust in "the conception of an incomprehensible power, which is at once a source of confidence to the virtuous and of terror to the criminal". In return the *Chronique de Paris* reported, "There are some who ask why there are always so many women around Robespierre at his house…it is because this Revolution of ours is a religion and Robespierre…is the a priest at the head of his worshippers". And as Adam Zamoyski recounts in his study *Holy madness*, **nineteenth-century nationalism** acquired all the trappings of a replacement religion, with such secular messiahs as **Garibaldi in Italy** and **Kossuth in Hungary**. Ideology inexorably became a monomania! Ultimately, those in thrall to the tyranny of an absolutist doctrine do not imagine that human beings matter. Despite all the promises of better things to come, individuals are of no concern in the immediate present. The growth of human dispensability under modern tyrannies is staggering. When the Bastille fell, there were seven inmates; five years later over seven thousand filled the prisons of Paris to the bursting point, awaiting the guillotine. **Czarist Russia** had incarcerated several thousand political prisoners, but the **Soviet Union** held millions behind bars. During **Mao's Great Leap Forward** some forty million or more disappeared, but we will never know the true number because individuals simply did not matter, and nobody could be bothered to count. At a single site, the pits at Butovo near Moscow, over one hundred thousand bodies were disposed of. During **Stalin's purges** the overwhelming message of tyranny was clear: human life is worthless.

Box 7.17 Enlightenment, Racism, Thomas Jefferson's, and Hypocrisy

In *Stamped from the beginning*, Kendi noted:

> On june 7, 1776, the delegates at the Second Continental Congress in Philadelphia decided to draft an independence document. For years, European intellectuals like France's **Buffon** and England's **Samuel Johnson** had projected Americans, their ways, their land, their animals, and their people as naturally inferior to everything European. **Thomas Jefferson** [who was deeply influenced by 18th-century Enlightenment] disagreed. At the beginning of the *Declaration of Independence*, he paraphrased the Virginia constitution, indelibly penning: "all Men are created equal". It is impossible to know for sure whether Jefferson meant to include his enslaved laborers (or women) in his "all Men". Was he merely emphasizing the equality of White Americans and the English? Later in the document, he did scold the British for "exciting those very people to rise in arms among us" – those "people" being resisting Africans. Even if Jefferson believed all groups to be "created equal", he never believed the antiracist creed that all human groups are equal. But his "all Men are created equal" was revolutionary nonetheless; it even propelled Vermont and Massachusetts to abolish **slavery**. To uphold **polygenesis** and slavery, six southern slaveholding states inserted "All freemen are created equal" into their constitutions. Continuing the Declaration, Jefferson maintained that "Men" were "endowed by their creator with inherent and inalienable rights; that among these are life, liberty, & the pursuit of happiness". As a holder of nearly two hundred people with no known plans to free them, Thomas Jefferson authored the heralded American philosophy of freedom. What did it mean for Jefferson to call "liberty" an "inalienable right" when he enslaved people? It is not hard to figure out what Native Americans, enslaved Africans, and indentured White servants meant when they demanded liberty in 1776. But what about Jefferson and other slaveholders like him, whose wealth and power were dependent upon their land and their slaves? Did they desire unbridled freedom to enslave and exploit? Did they perceive any reduction in their power to be a reduction in their freedom? For these rich men, freedom was not

the power to make choices; freedom was the power to create choices. England created the choices, the policies American elites had to abide by, just as planters created choices and policies that laborers had to follow. Only power gave Jefferson and other wealthy White colonists freedom from England. For Jefferson, power came before freedom. Indeed, power creates freedom, not the other way around – as the powerless are taught. *Thomas Jefferson only really handed revolutionary license to his band of wealthy, White, male revolutionaries*. He criminalized runaways in the **Declaration of Independence**, and he silenced women…

Thomas Jefferson did propose a frontal attack on slavery in *Notes on the State of Virginia*, a plan he would endorse for the rest of his life: the mass schooling, emancipation, and colonization of Africans back to Africa. Jefferson, who enslaved Blacks at Monticello, listed "the real distinctions which nature has made", that is, those traits that he believed made free Black incorporation into the new nation impossible. Whites were more beautiful, he wrote, as shown by Blacks' "preference of them". He was paraphrasing Edward Long (and John Locke) in the passage – but it was still ironic that the observation came from the pen of a man who may have already preferred a Black woman. Black people had a memory on par with Whites, Jefferson continued, but "in reason [were] much inferior". He then paused to mask his racist ideas in scientific neutrality: "It would be unfair to follow them to Africa for this investigation. We will consider them here, on the same stage with the whites, and where the facts are not apocryphal on which a judgment is to be formed". On this "same stage", he could "never…find that a black had uttered a thought above the level of plain narration; never saw an elementary trait of painting or sculpture". "Religion", he said, "indeed has produced a Phyllis Wheatley; but it could not produce a poet"…With *Notes on the State of Virginia*, Thomas Jefferson emerged as the preeminent American authority on Black intellectual inferiority. Notes on the State of Virginia was replete with other contradictory ideas about Black people. "They are at least as brave, and more adventuresome" than Whites, because they lacked the forethought to see "danger till it be present", Jefferson wrote. Africans felt love more, but they felt pain less, he said, and "their existence appears to participate more of sensation than reflection"…

On July 15, 1787, eight-year-old **Polly Jefferson** and fourteen-year-old **Sally Hemings** reached Jefferson's Paris doorstep. Sally Hemings had come to Monticello as an infant in 1773 as part of **Martha Jefferson**'s inheritance from her father. John Wayles had fathered six children with his biracial captive Elizabeth Hemings. Sally was the youngest. By 1787, she was reportedly "very handsome, [with] long straight hair down her back", and she accompanied Polly to Paris instead of an "old nurse". As his peers penned the **US Constitution**, Jefferson began a sexual relationship with Sally Hemings. Her older brother James, meanwhile, was training as a chef in Paris to satisfy Jefferson's gustatory desires. Hemings was more or less forced to settle for the overtures of a sexually aggressive forty-four-year-old (Jefferson also pursued a married local Frenchwoman at the time). Jefferson pursued Hemings as he arranged for the publication of Notes in London. He did not revise his previously stated opinions about Blacks; nor did he remove the passage about Whites being more beautiful than Blacks. Jefferson had always assailed interracial relationships between White women and Black or biracial men. Before arriving in Paris, he had lobbied, unsuccessfully, for Virginia's White women to be banished (instead of merely fined) for bearing the child of a Black or biracial man. Even after his measure was defeated, even after his relations with Hemings began, and even after the relations matured and he had time to reflect on his own hypocrisy, Jefferson did not stop proclaiming his public position. "Amalgamation with the other color, produces degradation to which no lover of his country, no lover of excellence in the human character, can innocently consent", he wrote in 1814, after he had fathered several biracial children. Like so many men who spoke out against "amalgamation" in public, and who degraded Black or biracial women's beauty in public, Jefferson hid his actual views in the privacy of his mind and bedroom. In 1789, Jefferson had a front-row seat to the anti-royal unrest in Paris that launched the **French Revolution**. He assisted his friend the **Marquis de Lafayette** in writing the Declaration of the Rights of Man and of the Citizen, adopted in August, weeks before his departure. But while putting the starting touches on the French Revolution and the finishing touches on the **American Revolution**, Jefferson had to deal with a revolt from sixteen-year-old Sally Hemings. She was pregnant with his child, refused to return to slavery, and planned to petition French officials for her freedom. Jefferson did the only thing he could do: "He promised her extraordinary privileges, and made a solemn pledge that her children should be freed", according to an account Hemings told their son Madison. "In consequence of his promise, on which she implicitly relied, she returned with him to Virginia", Madison wrote in his diary. Hemings gave birth to at least five and possibly as many as seven children from Jefferson, a paternity confirmed by DNA tests and documents proving they were together nine months prior to the birth of each of Sally's children. Some of the children died young, but Jefferson kept his word and freed their remaining children when they reached adulthood…

From the perspective of the enslaved, the most profound instance of moral eminence was evolving in **Haiti**. Jefferson learned of the Black revolt on September 8, 1791. Within two months, a force of 100,000 African freedom-fighters had killed more than 4,000 enslavers, destroyed almost 200 plantations, and gained control of the entire Northern Province. As historian C. L. R. James explained in the 1930s, "they were seeking their salvation in the most obvious way, the destruction of what they knew was the cause of their sufferings; and if they destroyed much it was because they had suffered much". What Jefferson and every other holder of African people had long feared had come to pass. In response, Congress passed the *Fugitive Slave Act* of 1793, bestowing on slaveholders the right and legal apparatus to recover escaped Africans and criminalize those who harbored them. Thomas Jefferson, for one, did not view the **Haitian Revolution** in the same guise as the American or French Revolutions. "Never was so deep a tragedy presented to the feelings of man", he wrote in July 1793. *To Jefferson, the slave revolt against the enslavers was more evil and tragic to the feelings of man than the millions of African people who died on American plantations.* Jefferson would soon call General Toussaint L'ouverture and other Haitian leaders "Cannibals of the terrible Republic"…Then again, upwardly mobile Blacks seemed as likely to produce resentment as admiration. "If you were well dressed they would insult you for that, and if you were ragged you would surely be insulted for being so", one Black Rhode Island resident complained in his memoir in the early 1800s. It was the cruel illogic of racism. When Black people rose, racists either violently knocked them down or ignored them as extraordinary. When Black people were down, racists called it their natural or nurtured place, and denied any role in knocking them down in the first place…

Thomas' 1983 *Man and the Natural World – Changing Attitudes in England 1500–1800*—an exceptional book that is too often unnoticed, even by academics—provides a very informative analysis of the links between our obsession to create and believe in tales about "**human uniqueness**," the subjugation of other organisms, **animal abuse**, and the rise of **humanism**:

The justification for this belief...that there was a fundamental difference in kind between humanity and other forms of life...went back beyond Christianity to the Greeks. According to **Aristotle**, the soul comprised three elements: the nutritive soul, which was shared by man with vegetables; the sensitive soul, which was shared by animals; and the intellectual or rational soul, which was peculiar to man. This doctrine had been taken over by the medieval scholastics and fused with the Judaeo-Christian teaching that man was made in the image of God (Genesis i. 27). Instead of representing man as merely a superior animal, it elevated him to a wholly different status, halfway between the beasts and the angels. In the early modern period it was accompanied by a great deal of self-congratulation. Man, it was said, 'was more beautiful, more perfectly formed than any of the other animals'. Jeremiah Burroughes reminded his congregation that, when God saw his other works, he only said that they were 'good', whereas when he had made man he said 'very good': "Observe, it is never said 'very good' till the last day, till man is made." Even so, there was a marked lack of agreement as to just where man's unique superiority lay. The search for this elusive attribute has been one of the most enduring pursuits of Western philosophers, most of whom have tended to fix on one feature and emphasize it out of all proportion, sometimes to the point of absurdity. Thus man has been described as a political animal (Aristotle); a laughing animal (Thomas Willis); a tool-making animal (**Benjamin Franklin**); a religious animal (Edmund Burke); and a cooking animal (James Boswell, anticipating **Lévi Strauss**). What all such definitions have in common is that they assume a polarity between the categories 'man' and 'animal' and that they invariably regard the animal as the inferior.

In Tudor and Stuart England the long-established view was that the world had been created for man's sake and that other species were meant to be subordinate to his wishes and needs...theologians of the early modern period usually had no difficulty in arriving at a generally accepted synthesis. The **Garden of Eden**, they said, was a paradise prepared for man in which Adam had Godgiven dominion over all living things (Genesis, i. 28). At first, man and beast had cohabited peacefully. The humans were probably not carnivorous and the animals were tame. But with the Fall the relationship changed. By rebelling against God, man forfeited his easy dominance over other species. The earth degenerated. Thorns and thistles grew up where there had been only fruits and flowers (Genesis, iii. 18). The soil became stony and less fertile, making arduous labour necessary for its cultivation. There appeared fleas, gnats and other odious pests. Many animals cast off the yoke, becoming fierce, warring with each other and attacking men. Even domestic animals had now to be coerced into submission. Then, after the Flood, God renewed man's authority over the animal creation Henceforth men were carnivorous and animals might lawfully be killed and eaten...It is difficult nowadays to recapture the breathtakingly anthropocentric spirit in which Tudor and Stuart preachers interpreted the **biblical story**. For they did not hesitate to represent the world's physical attributes as a direct response to **Adam's sin**: 'Cursed is the ground for thy sake' (Genesis, iii. 17). It was only because of the Fall that wild animals were fierce, that obnoxious reptiles existed, and that domestic animals had to undergo blows in misery. 'The creatures were not made for themselves, but for the use and service of man', said a Jacobean bishop. 'Whatsoever change for the worse is come upon them is not their punishment, but a part of ours'. Despite the Fall, therefore, man's right to rule remained intact. He was still 'the Vicegerent and Deputy of Almighty God'. 'All the creatures were made for man, subjected to his government and appointed for his use'.

Box 7.18 Science, "Human Uniqueness," and "Uses and Virtues" of Plants

In his *Man and the Natural World*, Thomas notes:

> Meanwhile, the scientists and economic projectors of the seventeenth century anticipated yet further triumphs over the inferior species. For **Bacon**, the purpose of science was to restore to man that dominion over the creation which he had partly lost at the Fall, while **Robert Boyle** was egged on by his correspondent John Beale to establish what Beale called 'the empire of mankind'. To scientists reared in this tradition, the whole purpose of studying the natural world was 'that, Nature being known, it may be mastered, managed, and used in the services of human life'. As William Forsyth remarked in 1802, in a plea for the observation of caterpillars: 'it would be of great service to get acquainted as much as possible with the economy and natural history of all these insects, as we might thereby be enabled to find out the most certain method of destroying them.' The initial motive for the study of **natural history** was practical and utilitarian. **Botany** began as an attempt to identify the 'uses and virtues' of plants, primarily for medicine, but also for cooking and manufacture. It was the conviction that every part of the plant world had been designed to serve a human *purpose* which led Sir John Colbatch in 1719 to discover the medical use of mistletoe: 'It immediately entered into my mind that there must be something extraordinary in that uncommon beautiful plant; that the Almighty had designed it for farther and more noble uses than barely to feed thrushes or to be hung up superstitiously in houses to drive away evil spirits…I concluded, *a priori*, that it was…very likely to subdue…epilepsy.'
>
> **Zoology** was equally practical in its intentions. The Royal Society encouraged the study of animals with a view to determining 'whether they may be of any advantage to mankind, as food or physic; and whether those or any other uses of them can be further improved'. Centuries of selective breeding had already refined the stock of domestic animals, cows, sheep, chickens and pigeons, but many new possibilities were yet to be explored. Pigs, urged Sir William Petty, could be taught to labour; and, if their diet were changed, the flesh of domestic stock could be improved. In the nineteenth century the official purpose of the Zoological Society would be to acclimatize and breed new domestic animals. Animals, as the Rev. William Kirby put it in 1835, were of the deepest interest to everyone, because of their diversity, their beauty, 'but above all, their pre-eminent utility to mankind'. Plants were equally malleable. A large range of cultivated plants had been inherited from remote antiquity, but continuous breeding and experimentation opened new vistas. Agricultural writers described the great improvements which could be made by 'altering the species of such vegetables that are naturally produced, totally suppressing the one, and propagating another in its place.' A gardener declared in 1734 that man now had the power 'to govern the vegetable world to a much greater improvement, satisfaction and pleasure than ever was known in the former ages of the world'. In the conjectural history which became increasingly popular during the **European Enlightenment** of the eighteenth century, man's victory over other species was made the central theme. The true origin of human society, it was said, lay in the combination of men to defend themselves against wild beasts. Then came hunting and domestication. Man's crucial act, thought **Buffon**, was the taming of the dog. It led, agreed Thomas Bewick, to the conquest and peaceable possession of the earth. Without the camel, thought Herder, the deserts of Africa and Arabia would have been inaccessible, and, without the horse, the Europeans could never have conquered America. Lord Kames noted that, without the reindeer, Lapland would have been impenetrable. **Adam Smith** observed that crops and herds were the earliest forms of private property. 'Our toil is lessened,' pronounced Edward Gibbon, 'and our wealth is increased by our dominion over the useful animals'.

Thomas provides many interesting examples that illustrate how absurd the use of tales about "human uniqueness" and our "cosmic right" to use and abuse other animals can be (see also Boxes 7.18 and 7.19), including stories about the "innate instinct of obedience to humans" that God gave to other animals:

> The animals were less docile than they had been (before the Fall), but they had not all forgotten their duty…As Andrew Willet observed in 1605, there still remained 'a natural instinct of obedience in those creatures which are for man's use, as the ox, ass, horse'. 'Sometimes,' said Jeremiah Burroughes in 1643, 'you may see a little child driving before him a hundred oxen or kine this way or that way as he pleaseth; it showeth that God hath preserved somewhat of man's dominion over the creatures.' The instinct which brought fish in shoals to the seashore, noted the nonconformist divine Philip Doddridge a century later, 'seems an intimation that they are intended for human use'. The only purpose of animals, declared Thomas Wilcox, an Elizabethan, was to minister to man, 'for whose sake all the creatures were made that are made'. It was with human needs in mind that the animals had been carefully designed and distributed. Camels, observed a preacher in 1696, had been sensibly allotted to Arabia, where there was no water, and savage beasts 'sent to deserts, where they may do less harm'. It was a sign of **God's providence** that fierce animals were less prolific than domestic ones and that they lived in dens by day, usually coming out only at night, when men were in bed. Every animal was thus intended to serve some human purpose, if not practical, then moral or aesthetic. Savage beasts were necessary instruments of God's wrath, left among us 'to be our schoolmasters', thought James Pilkington, the Elizabethan bishop; they fostered human courage and provided useful training for war. Horse-flies, guessed the Virginian gentleman William Byrd in 1728, had been created so 'that men should exercise their wits and industry to guard themselves against them'. Apes and parrots had been ordained 'for man's mirth'. Singing-birds were devised 'on purpose to entertain and delight mankind'.
>
> The lobster, observed the Elizabethan George Owen, served several purposes in one: it provided men with food, for they could eat its flesh; with exercise, for they had first to crack its legs and claws; and with an object of contemplation, for they could behold its wonderful suit of armour…As for cattle and sheep, Henry More in 1653 was convinced that they had only been given life in the first place so as to keep their meat fresh 'till we shall have need to eat them'. As late as the 1830s the authors of the Bridgewater Treatises on 'God's goodness as manifested in the Creation' were still maintaining that all inferior species had been made to serve man's purpose. God created the ox and the horse to labour in our service, said the naturalist William Swainson; the dog to display affectionate attachment, and the chicken to show 'perfect contentment in a state of partial confinement'. The louse was indispensable, explained the Rev. William Kirby, because it provided a powerful incentive to habits of cleanliness. Vegetables and minerals were regarded in the same way. Henry More thought that their only purpose was to enhance human life. Without wood, men's houses would have been merely 'a bigger sort of beehives or birds' nests, made of contemptible sticks and straw and dirty mortar'; and, without metals, men would have been deprived of the 'glory and pomp' of war, fought with swords, guns and trumpets; instead there would have been 'nothing but howlings and shoutings of poor naked men belabouring one another…with sticks or dully falling together by the ears at fisticuffs'. Even weeds and poisons had their essential uses, noted a herbalist: for they exercised the industry of man to weed them out…Had he nothing to struggle with, the fire of his spirit would be half extinguished.

Box 7.19 Souls, "Universal Salvation," Animal Use, and Darwin

Thomas states in *Man and the Natural World*:

> [After] neither anatomy nor language nor even the possession of reason could any longer provide an indisputable barrier between them (humans) and the beasts…all that was left was the claim that man was the only religious animal, the sole possessor of an immortal soul…this was the only certain distinction between men and brutes, thought a late-seventeenth-century writer: 'other creatures seem in some measure to partake of reason, but not at all of religion.' A hundred and thirty years later, the physician William Lambe agreed. To deny that man belonged physiologically with the monkeys, apes and baboons was to display 'misplaced pride and an ignorant apprehension'. But in his noble part, his rational soul, man was 'distinguished from the whole tribe of animals by a boundary which cannot be passed'. Unfortunately, this allegedly uncrossable boundary was also the one whose existence was hardest to prove. For at a popular level religion had never been regarded as inaccessible to animals. **Protestant theologians** were contemptuous of medieval legends about **St Francis** preaching to the birds or **St Anthony of Padua**'s horse kneeling to receive the host. But many early modern farmers continued to regard their domestic animals, in the way the Jews had done before them, as essentially within the covenant. After all, the animals were supposed to rest on the **Sabbath**; in the nineteenth century it was even a question among some High Churchmen as to whether they should not also be made to starve on fast days. In the Victorian countryside on **Christmas Eve** the horses and oxen were rumoured to kneel in their stables and even bees gave out a special buzz. All animals were thought to have religious instincts. Classical authors taught that fowls had 'a certain ceremonious religion' and that elephants adored the moon. Such traditions were easily Christianized. Psalm 148 declared that all creatures praised the Lord, even 'beasts and all cattle; creeping things and flying fowl'. 'Let man and beast appear before him, and magnify his name together,' sang Christopher Smart. Some theologians, and many poets, regarded bird-song as a kind of hymnsinging. There are also hints of popular belief in something very close to the transmigration of souls. The souls of unbaptized children were vulgarly assigned a great number of animal resting-places: they became headless dogs in Devon, wild geese in Lincolnshire, ants in Cornwall, night-jars in Shropshire and Nidderdale. Fishermen sometimes regarded seagulls as the spirits of dead seamen.
>
> In such notions one can see a debased popular version of the doctrine of **metempsychosis** which had been taught by **Plato** and **Pythagoras**. Though condemned by all orthodox theologians, the notion had been intermittently espoused by medieval heretics, and it was revived by some of the **NeoPlatonists** of the **Renaissance**. By postulating the movement of the universal soul of the world into every kind of animate creation, it suggested that even beasts had the divine spark within them. It can be traced in some English seventeenth-century Platonist writing; the 'best of philosophers', thought **Henry More**, were not averse to conceding animals immortal souls. It is also clearly recognizable in the doctrine attributed to the Ranters, who allegedly held that 'when we die we shall be swallowed up into the infinite spirit, as a drop into the ocean, and so be as we were; and if ever we be raised again, we shall rise a horse, a cow, a root, a flower and such like.' Yet the roots of the idea that even animals might have an after-life were propagated by the theologians themselves, for the mortality of beasts was part of the curse with which Christ had come to do away. In Chapter 8, verse 21, of his *Epistle to the Romans* **St Paul** promised that 'the creature itself also should be delivered from the bondage of corruption into the glorious liberty of the children of God'. Medieval schoolmen had said that this meant only

men, together with creatures without life, such as the heavens and the elements. Some early Protestant writers, however, put forward the novel view, previously only held by a few isolated commentators, that by 'creatures' was meant all living animals, birds and plants; and in the century after the **Reformation** the text was subjected to a fascinating mixture of contradictory interpretation…a substantial proportion of commentators took the view that animals, like the rest of nature, would be restored to the perfection they had enjoyed before the Fall. Thorns, thistles and creatures engendered by putrefaction would disappear, leaving birds, beasts and useful plants to flourish in renewed perfection. This did not necessarily mean that every animal who had ever lived would be restored, merely that each species would be represented in heaven. But the notion that all animals were resurrected was not easily set aside. It was maintained by the early Protestant reformer John Bradford; and it was reiterated during the Civil War period by several radical believers in **universal salvation**.

Many contemporaries found it highly offensive to use the **Bible** to prove animal salvation. In his commentary on Romans Thomas Horton in 1674 found it necessary to refute 'such kind of persons who, from this present place of scripture, would very fondly and absurdly infer a resurrection of beasts. This,' he emphasized, 'does not follow from the text, neither has any other good foundation for it.' Yet in the later seventeenth century many otherwise orthodox clergy regarded the issue of animal immortality as entirely open. Samuel Clarke told an acquaintance that he thought it possible that the souls of brutes would eventually be resurrected and lodged in Mars, Saturn or some other planet, while the physician Dr. Charles Leigh thought there was 'a spiritual immaterial being' in all living creatures. Ralph Josselin dreamed in 1655 that Christ was born in a stable because he was 'the redeemer of man and beast out of their bondage by the Fall'. In the eighteenth century those who felt that **animal salvation** was at the very least a possibility included Bishop Butler, the clergyman William Whiston, the philosopher David Hartley and the writer Robert Wallace. Those who thought it highly probable or even a certainty included animal-lovers like John Hildrop and John Lawrence, Dissenters like Matthew Henry, and Methodists like Adam Clarke and John Wesley. The physician George Cheyne declared in 1740 that 'it seems utterly incredible that any creature…should come into this state of being and suffering for no other purpose than we see them attain here…there must be some infinitely beautiful, wise and good scene remaining for all sentient and intelligent beings, the discovery of which will ravish and astonish us one day.'

In the 1770s the Calvinist divine Augustus Toplady declared that beasts had souls in the true sense, adding that he had never heard an argument against the immortality of animals which could not be equally urged against the immortality of man. 'I firmly believe that beasts have souls; souls truly and properly so-called.' The idea of animal immortality seems to have made more headway in England than anywhere else at this period; and it was undoubtedly to pet-lovers that it made its greatest appeal. It was buttressed by arguments from scripture and by observation of the mental capacities of the animal in question. There were now many who felt that one had only to look into a dog's eyes (always his eyes) to settle the issue. They knew that cats and dogs could dream, which surely showed a spiritual quality, and they thought them capable of good and bad actions. Canine virtue, said the Quaker writer Priscilla Wakefield, was not very different from moral virtue. In Victorian times there were many whose religious faith in a just God was sorely tried by the official doctrine that pet animals were doomed to oblivion and that domestic animals had to suffer without hope of posthumous reward. The acceptance of evolution posed the dilemma more sharply, for if men had evolved from animals then either animals also had immortal souls or men did not. In 1816 the future Archbishop Sumner denounced all those writers 'who have taken an extraordinary pleasure in levelling the broad distinction which separates man from the brute creation'. A decade later, in 1827, the young **Charles Darwin** attended a meeting of the Plinian Society of the University of Edinburgh, where he heard a Mr. Grey read a paper 'in which he attempted to prove that the lower animals possess every faculty and propensity of the human mind'. When in 1871 Darwin published his *The Descent of Man*, he would himself argue not only that man and animals were descended from a common ancestor, but also that the mental difference between humans and the existing higher animals was only one of degree. Without the long history of pet-keeping in England and without the knowledge accumulated through centuries of experience of domestic animals, it is hard to believe that the author of *The Descent of Man* could have made his case in quite the way he did.

There are three aspects of Thomas' *Man and the Natural World* that I find particularly interesting. Firstly, he reminds us that teleological narratives are not only created, and believed by, religious people, or the "broader public," but also by scientists, economic projectors, politicians, and the so-called elite (Box 7.18). Secondly, he provides fascinating discussions on the existence of a "soul" in non-human organisms (Box 7.19). Thirdly, he shows that in *some respects* there was an "increased humanist moral circle" as authors such as Pinker defend, but that, paradoxically, humans continue to do huge atrocities to other animals (see also Box 7.15). Actually, most people in "developed" countries have actually no idea of what is truly being done to animals in "animal factories" and in biological or medical or cosmetic experiments. That is why movies such as *Earthlings* are so rare and so hard to watch. They are rare precisely because the people/organizations doing such horrible animal abuses, or benefiting from them, try as much as possible to not allow those images to reach the media and broader public, as noted above. Furthermore, most people don't truly make an active effort to try to see them anyway, because such documentaries are rare, yes, but they *do exist* and can be seen online if one really tries hard. But many people *prefer to be passive about this* because either they simply don't care or they don't want to be reminded that we are all accomplices of such cruel abuses.

I have actually attested this myself, during the few years I was helping teaching a Biological Anthropology course, during my second PhD, at George Washington University (GWU). For that class, I made the students watch *Earthlings* and then answer a small questionnaire. It should be noted that most students of GWU are well-educated young people, and that many of them probably would say they approve "humanist" ideas and actions, particularly regarding animal abuse. However, when they started to see the movie, various of them begun to feel uncomfortable and asked me to stop the movie, or if they could just go out and not answer the questionnaire they were supposed to fill after watching it. Some of them begun to cry, others said they were disgusted and even had nausea, and some went to the point to protest saying that they should not be required to see such a "disturbing" movie in a Biological Anthropology course. I answered them that I completely understood their point of view, as the movie is truly brutal, but I also pointed out that it was a documentary, so it was just showing the reality of what we truly do to animals, and that this is crucial to understand how humans are and how they relate to other animals, which is precisely what they were supposed to learn as part of that course. I also told them that using the word "disturbing" to classify the movie was perhaps not really appropriate, because what is truly disturbing is what humans do in the movie, not the movie itself, which was made by very brave people that took a huge risk to do something that many consider to be a true act of humanism. In a way, what might be really "disturbing" is that, after those people took such a risk to be able to produce such a revealing documentary to show us the type of atrocities that humans do to other animals, various students of a Biological Anthropology course that specifically included a component about our "place in nature," *don't want* to watch that documentary. To be able to escape from *Neverland*, humans will have to be able to accept reality, even if it is brutal, as it is in a huge number of cases, because human history is not only a fantastic case of biological evolution including fascinating technological discoveries, but also a journey that has involved and continues to involve, an enormous number of brutalities.

Think again about the people that recorded those brutal scenes of animal abuse displayed in *Earthlings*: they involved the hard work, and bravery, of a huge number of people that had to fake that they wanted jobs that they hated more than anything else, to watch such brutal scenes on a daily basis, often putting themselves at risk, in order to do undercover investigations to reveal that brutal reality to the broader public. So, refusing to see the movie is not only neglecting all the hard and brave work those people did and the risks they took, but also, above all, closing our eyes to the reality of **human animal abuse** specifically, and to the world of reality in general. This is precisely the issue discussed in the last pages of Thomas' superb *Man and the Natural World*, in which he reminds us of, and criticizes us for, such intolerable attitude of concealment. Concealment done not only by the people doing the atrocities and/or trying to keep them out of the public eye, but also by all the other people that don't simply care or that, as me—and *you*?—don't truly do something significant to pragmatically contribute to finish such brutal animal abuses once for all. Yes, it is very difficult to do something against this, because the whole system—well, the whole existence of agricultural states, of "civilizations"—is *based on animal abuse*. But let's say: if we were able to know, in *all cases*, which meat or eggs or milk come from animal factories, or which perfumes or other products come from companies that abuse animals—what is, per se, already hugely difficult within the system we live in—we could just decide to not buy such products anymore. In theory, if we would do this, at least these specific abusive practices would no longer bring economic gains to those factories and companies and thus would ultimately be highly reduced or even ultimately finished.

But of course, these topics are much more complex than this, for many reasons. For instance, most people don't have enough money to really have an option to buy more expensive meat or milk from, let's say, **free grass-feeding cows**, particularly because if nobody else bought products from "factory cows," the demand of free grass-feeding cows' milk and meat would be enormous, and the prices would rise exponentially. Moreover, in the world of reality it is not even sure that, even if everybody would care about this and had enough resources to buy such expensive food items, it would be possible to finish animal factories once for all, because there are simply too many people in the planet right now, to be fed exclusively with meat, eggs, milk, and so on from "free" domesticated animals. We have eight billion people living right now: this current problem of **overpopulation** is deeply linked to the colossal problem of animal abuse. Unless, of course, we would all become vegetarians, but this could be seen as unethical by those arguing that plants might also suffer—and actually some scientists have recently affirmed that they might indeed suffer, in some way that is very different from how animals suffer—or that without animal meat at least some humans can eventually tend to develop more health problems and so on. These are very murky waters, and we don't have the needed space to explore this topic in detail in the present volume, unfortunately. However, be that as it may, what is clear is that while we can in theory use animals to eat as many other animals do, abusing animals is something that should never exist if we were truly a "special," "sapient," "caring" animal. Moreover, the complex discussions above don't apply anyway to the millions of animals that are horribly abused to create perfumes and other hundreds of non-essential products: either if you are poor, or rich, or omnivorous, or vegetarian, there is no reason at all to continue buying such products, unless someone would argue that he/she is doing that to give jobs to the people working on such abusive companies. But this would really be missing the point here, because that can be said about basically anything, including the jobs that Nazis gave to the people taking Jews to concentration camps and gazing them, or the jobs that the Inquisition gave to people to torture and burn "witches" centuries ago.

In *Man and the Natural World*, Thomas remind us of the **paradox of humanism**, in which teleological narratives about humans being "special" organisms with a "cosmic duty" to make a "better world for all living beings" actually lead to laws, norms, and policies that end up, many times, by doing exactly the opposite—a much worse life for numerous non-human organisms. One example he gives is that when the **United Nations** and the **International Union for the Conservation of Nature**—two emblematic examples of the rise of humanism—defined, in 1969, the term "conservation," the definition used was: "*the rational use of the environment to achieve the highest quality of living for mankind*." As noted by Thomas, it is precisely the combination of our paradoxical attitude of concealment and the theoretical confusion created by such imaginary narratives that has prevented the huge cruelties that we do to other animals to be, in his words, "fully resolved." The problem, as he states, is that we cannot simply conceal the existence of such atrocities for ever, because this can lead to "ultimate consequences" that "we can only speculate" about, but that very likely can affect in a dramatic way not only non-human animals—as it has been the case since millennia ago and particularly since the rise of agriculture—but also humans and life in this planet as a whole. I will therefore end this section by quoting the last excerpts of his book:

> The embarrassment about **meat-eating** thus provides a final example of the way in which, by the end of the eighteenth century, a growing number of people had come to find man's ascendancy over nature increasingly abhorrent to their moral and aesthetic sensibilities.

This was the human dilemma: how to reconcile the physical requirements of **civilization** with the new feelings and values which that same civilization had generated. It is too often assumed that sensibilities and morals are mere ideology: a convenient rationalization of the world as it is. But in the early modern period the truth was almost the reverse, for, by an inexorable logic, there had gradually emerged attitudes to the natural world which were essentially incompatible with the direction in which English society was moving. The growth of towns had led to a new longing for the countryside. The progress of cultivation had fostered a taste for weeds, mountains and unsubdued nature. The new-found security from wild animals had generated an increasing concern to protect birds and preserve wild creatures in their natural state. Economic independence of animal power and urban isolation from animal farming had nourished emotional attitudes which were hard, if not impossible, to reconcile with the exploitation of animals by which most people lived. Henceforth an increasingly **sentimental view of animals as pets** and objects of contemplation would jostle uneasily alongside the harsh facts of a world in which the elimination of 'pests' and the breeding of animals for slaughter grew every day more efficient. Oliver Goldsmith wrote of his contemporaries that ‘they pity and they eat the objects of their compassion’.

The same might be said of the children of today who, nourished by a meat diet and protected by a medicine developed by animal experiments, nevertheless take toy animals to bed and lavish their affection on lambs and ponies. For adults, nature parks and conservation areas serve a function not unlike that which toy animals have for children; they are fantasies which enshrine the values by which society as a whole cannot afford to live. Of course most people in practice…retained their faith in the **primacy of human interests**, even if they lamented the effect of **material progress** on the natural world…it was not for the sake of the creatures themselves, but for the sake of men, that birds and animals would be protected in sanctuaries and wild-life parks. In 1969 the United Nations and the International Union for the Conservation of Nature defined ‘conservation’ as ‘the rational use of the environment to achieve the highest quality of living for mankind.’ The early modern period had thus generated feelings which would make it increasingly hard for men to come to terms with the uncompromising methods by which the dominance of their species had been secured. On the one hand they saw an incalculable increase in the comfort and physical well-being or welfare of human beings; on the other they perceived a ruthless exploitation of other forms of animate life. There was thus a growing conflict between the new sensibilities and the material foundations of human society. A *mixture of compromise and concealment* has so far prevented this conflict from having to be fully resolved. But the issue cannot be completely evaded and it can be relied upon to recur. It is one of the *contradictions* upon which modern civilization may be said to rest. About its ultimate consequences we can only speculate.

7.7 ‘Monsters’, Disabilities, and Mass Murder

I consider it useless and tedious to represent what exists, because nothing that exists satisfies me. Nature is ugly, and I prefer the monsters of my fancy to what is positively trivial. (Charles Baudelaire)

Asma 2009’s captivating book *On Monsters – An Unnatural History of Our Worst Fears* focuses on the long-standing fascination of humans with “monsters” and with what is “abnormal”—including people with **anatomical anomalies**, which are often called **congenital malformations** in biology and medicine. Numerous so-called monstrosity museums, or similar displays, have existed in the past, and continue to exist, such as—here in U.S.—the University of Michigan Museum of Art’s display "*Monsters and monstrosity*," the New York Metropolitan Museum of Art's "*Monstrosity and Otherness in medieval art*" exhibit, and the Mütter Museum of Philadelphia's "*Imperfecta*" recent display. Concerning books, some prominent examples published in the last decades are Leroi's 2003 *Mutants*, Blumberg's 2009 *Freaks of Nature*, and Bondeson's 1997 *A Cabinet of Medical Curiosities* and 2000 *The Two-Headed Boy*. As explained in DeSesso's 2019 review paper "*The arrogance of teratology: a brief chronology of attitudes throughout history*," while the discipline of **Teratology**—the study of the causes, mechanisms, and manifestations of congenital malformations—has "existed for about 60 years, there has been a deep interest in the causes of human malformations for millennia." As he noted, "absent the scientific method and acting on fervent beliefs that made sense to ancient/medieval populations, "mechanisms" were described and prognostications of future events were assigned to *terata* resulting in tragic (and unwarranted) sequelae" (see Boxes 7.20 and 7.21 and Figs. 7.10 and 7.11).

Box 7.20 "Monsters," Babylonian Omens, and Hindu Gods

DeSesso's 2019 review paper "*The arrogance of teratology*" provides a recent and very concise account about the history of ideas regarding malformations and their link with teleological narratives and stories about **Gods**, **demigods**, and even the **Devil**:

> During ancient times in Western culture, going back at least as far as **Hammurabi** (ca.1750 BCE), **congenital malformations** were known to be rare events. Observers noted their descriptions and paid attention to the ensuing historical events. Based on their observations, they believed that occurrences of malformations were **portents** of future events. As a result, **cuneiform clay tablets** were prepared by **Chaldeans** (a tribe related to ancient **Babylonians**) that listed specific malformations and the **predictions of future events**. Not only did congenital malformations have an impact on what ancient people believed the future held, but it is possible that the **malformed babies** may have influenced the depiction of some of the gods in their pantheon. For instance, is it possible that the image of the Greek god, **Janus** (god of beginnings and endings; gates; transitions) [see **Fig. 7.11**], who had two faces that looked toward past and future, could have been influenced by observation of…**conjoined twins** who are connected at the thorax and head in such a way that there are two equal faces on either side of the head. Another possible example can be found in the **Hindu religion**…
>
> Several of the gods are depicted as **multilimbed beings**. Examples include **Kali**, goddess of death [see **Fig. 9.6**]; **Durga**, protector of the universe; and **Ganesh**. Is it possible that the observation of **parasitic twins** influenced the concept? Another important concept that developed in antiquity related to **hybrids**. The early **Greco-Roman religions** worshipped gods with human attributes, who could interact with humans. When the interactions involved intercourse that resulted in the birth of a child, the child was a hybrid between a god and a mortal: a **demigod**. **Demigods** had human form but also had some traits transmitted to the child from the god-parent that made them 'better' than normal humans. Demigods included such notable individuals as **Hercules**, **Achilles**, **Asclepius**, **Helen of Troy**, **Aeneas**, and **Orion**. During the later portion of the ancient times, particularly in the **Greco-Roman era**, the fate

of surviving malformed babies was brutal by modern standards. In his treatise on *Politics* (at 1335b), **Aristotle** wrote "As to exposing or rearing the children born, let there be a law that no deformed child shall be reared". In the case of malformed infants, exposure was a euphemism for **infanticide**; thus, the infants were not directly murdered. Rather, they were abandoned and left for nature to take its course...

As the **Roman Empire** began to deteriorate, Europe and portions of the territories of Africa and Asia that bordered the Mediterranean Sea entered the **Dark Ages**. In Europe especially, intellectual curiosity ebbed. The human body was considered unclean. Normal bodily functions such as **menstruation** were considered impure and intercourse with a woman who was menstruating was thought to be causative in the birth of monsters. The prevailing thought throughout most of the time preceding the **Renaissance** was that the imperfect human body should be kept hidden. People believed that diseases and **plagues** (e.g., **The Black Death**, ca. 1343-1356) were due to **God's wrath**, which resulted from his displeasure with human imperfection and sinful activities. These thoughts carried over into the belief that **deformities in infants** were either a punishment from God visited upon the parents or the work of the devil. In the latter case, the ancient concept of **hybridization** was invoked to explain the genesis of deformities. The **devil** could take on the form of an animal and copulate with a woman resulting in the birth of a monster that had attributes of both its animal and human parentage. The result of such hybridization was the antithesis of the demigods of the ancient world, who were the result of procreation between mortals and a god. This thinking perfused into future eras. With the advent of the Renaissance, people's outlook on life improved. Science and art enjoyed a resurgence.

However, many deep-rooted, erroneous beliefs were slow to be purged from popular culture. In particular, the notion that **bestiality** (sexual contact with animals) could result in viable offspring remained strong well into the 17th century and spread from Europe to the New World. This belief had deadly consequences when it was wielded by leaders or persons in authority who applied 'justice' according to their own set of standards, without regard for logic, but based firmly on statements found in scripture. Even when individuals were not caught in the act, the evidence against them could include their offspring. According to the Danish physician and scientist, Thomas Bartholin...there was a case in 1638 of a young woman who was accused **of bestiality** for giving birth to an infant 'with the head of a cat.' This was likely a premature delivery of an **anencephalic fetus**. However, based on the 'evidence' of the fetus, the poor woman was convicted. In a complete miscarriage of justice, she was tied to a ladder and burned alive in the public square of Copenhagen. By the middle of the 17th century, a popular belief was that offspring developed from fully formed miniature versions of themselves that previously existed within their parents. One view (spermist) was that the father's semen was the source of the embryo and the mother's uterus served only as an incubator; another view (ovist) was that the woman's eggs had a completely formed individual that only needed stimulation by semen. This view fit well with the concept that malformations were the result of God's wrath and were punishments visited on the parents. The concept was espoused by some notable scientists of the time, such as **Marcello Malpighi**. Coincident with the rise in experimental science, physicians, and surgeons of the time were interested in obtaining 'curiosities' which they collected and displayed in museums known as 'cabinets.' While most of the curiosities related to tumors and various disease states, some of the curiosities were congenital malformations.

The notion has been around for millennia that if a pregnant woman experiences a great scare or a devastating loss during her pregnancy, she is at risk of having a child with some physical sign of the mother's experience. The physical sign of the maternal impression could be a birthmark or deformity. All of the 'proof' relative to such effects from **maternal impressions** is, of course, anecdotal. One such anecdotal case involved **Hippocrates of Kos** (460-370 BCE), the father of modern medicine. The following story was attributed to **St Jerome**. In ancient Greece, an aristocratic lady was accused of adultery by her husband. While there were no witnesses to her infidelity, the circumstantial evidence against her was fairly strong. Both she and her husband were white, but the child was dark-skinned. Shortly before the jury was about to render its verdict and sentencing, Hippocrates appeared in court of his own volition and motivation. He reminded the jury that the woman spent much of pregnancy abed in her room and that she had a picture of a Moor in her room. Consequently, she spent much of her time looking at that picture. In addition, the child bore a strong resemblance to the image in the picture. Thus, it was Hippocrates's opinion that these circumstances had created a 'maternal impression' that altered the child's skin color and overall body shape. Based on Hippocrates's widely acknowledged medical knowledge and reputation, the jury acquitted the woman. More than 2000 years later, William Dabney (1891) wrote a paper that attempted to separate facts from fantasy to determine if maternal impressions are a cause of malformations. He collected many anecdotal cases that alleged a causative role for maternal impressions. He excluded those cases wherein there was no specific stimulus or there was no specific period of gestation in which the stimulus occurred. One of his conclusions was that birth defects were observed only if the stimuli occurred during the first 2 months of gestation (which corresponds with the period of organogenesis in humans). Nevertheless, Dabney's reasoning and final conclusions were faulty. For one thing, Dabney believed that stimuli could include dreams (there is no way to verify when, or if, the dreams actually took place). While Dabney failed to follow the scientific method, his contemporaries were making great advances in the field. The prolific Scottish physician, J. W. Ballantyne (1861-1923) edited the first scientific journal devoted to birth defects: *Teratologia: A Quarterly Journal of Antenatal Pathology*. *Teratologia* first appeared in 1894 and provided a forum for physicians and scientists to report and discuss birth defects and their prognoses. In 1896, his treatise on the possible causes of **teratogenesis** appeared…the treatise thoroughly debunked the notion of a causative role for maternal impressions in the etiology of birth defects.

Our fascination with "monsters" is not only due to the fact that they are out of the norm and that they are therefore ideal to be incorporated by us humans, the storyteller animal, in our stories and fairytales. It is also due to the fact that within such stories, monsters are ideal characters for teleological narratives about "**cosmic progress**," "**cosmic punishment**," "**God's visitation**," or **morality**, thus also paving the way for easy discriminatory tales about the "other": thence the name of the MET exhibits, "*Monstrosity and Otherness in Medieval Art*." As famously stated in the 1549–1559 and 1662 editions of the *Book of Common Prayer*, namely in its *"Order for the Visitation of the Sick*," which was particularly popular in England and the rest of Europe for centuries: "Whatsoever your sickness is, know you certainly, that it is God's visitation." That is, if you are **disabled** and/or have any particular **malformation**, that is part of **God's masterplan**, probably to punish someone of your family, very likely because of your "mother's fault": such tales

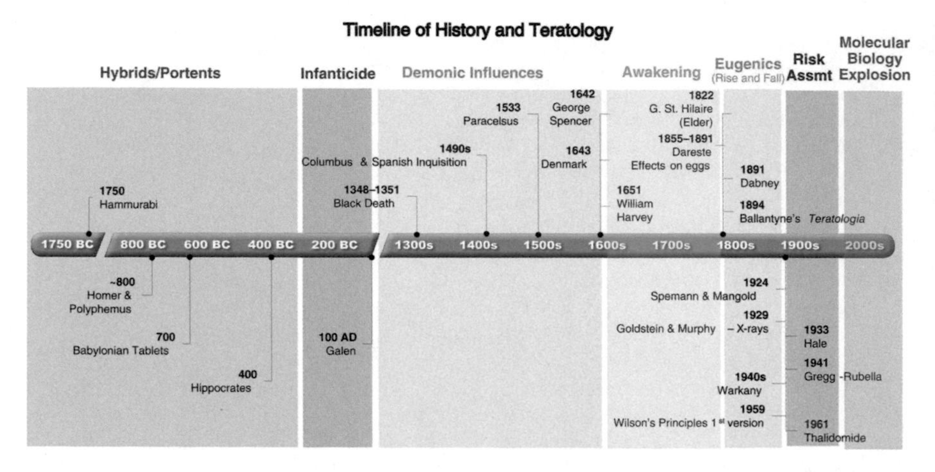

Fig. 7.10 As explained in DeSesso's 2019 paper "*The arrogance of teratology*," "this timeline displays selected events occurring over the past four millennia…the eras that are demarcated are arbitrary period that are useful for discussion…but are not meant to infer that were sudden changes in attitudes or that attitude were not changing during each era"—"assmt" is abbreviation of "assessment"

Fig. 7.11 As also noted in DeSesso's 2019 paper "*The arrogance of teratology,*" "Janus, the Greek god of beginnings and endings…transitions…and gateways…and after whom the month of January is named…he is often depicted with the face of an old man looking back towards the past and that of a young man gazing forward to the future…[it is] possible that the image…could have been influenced by observation of…conjoined twins"

indeed provide an easy ammunition to not only discriminate **people with malformations/disabilities** but also to blame others for their occurrence, particularly women or other groups that are subjugated/oppressed by those creating and disseminating such teleological tales. As explained in Asma's *On Monsters*, the term "***monster*** derives from the Latin word **monstrum**, which in turn derives from the root monere (to warn)...to be a monster is to be an **omen**...sometimes the monster is a display of **God's wrath**, a **portent of the future**, a **symbol of moral virtue or vice**, or an **accident of nature**...the monster is more than an **odious creature of the imagination**; it is a kind of cultural category, employed in domains as diverse as religion, biology, literature, and politics" (see Boxes 7.21, 7.22, and 7.23). He further explains that "as a *literal* creature, the monster is still a vital actor on the stage of **indigenous folk cultures**, and it's safe to say that even in our developed and otherwise secular world, the idea of a literal demon or devil still haunts the minds of many evangelical and mainstream Christians":

> The monster, of course, is a product of and a regular inhabitant of the imagination, but the imagination is a driving force behind our entire perception of the world. If we find monsters in our world, it is sometimes because they are really there and sometimes because we have brought them with us. Both the east and the west are rife with monsters of every stripe. **Demons**, **dragons**, **ghosts**, **wrathful Buddhas**, and **supernatural animals** occupy the theology, folklore, and daily rituals of religious cultures around the globe. The '**hungry ghost**' is a common creature in Asia. It usually represents a monstrous afterlife for a person who was gluttonous or greedy in this life; in the afterlife, the person is tortured by his insatiable hunger. These creatures, sometimes imagined with a giant stomach and a pinhole mouth or no mouth at all, continue to play an important role in Eastern cultures; Southeast Asia and China still have annual hungry ghost festivals. They are imaginative symbols of the frustrations of **hedonism** and the **doomed pursuit of pleasure**. The ancient story about monsters does not progress from crazy paranoia to cool-headed tolerance. Instead, superstition and rationalism shared territory, just as they do today. But it is still important to notice, even without the triumphal narrative, that cool-headed tolerance did evolve in the ancient world. The failure of the masses to adopt a scientific attitude toward abnormal beings does not diminish the impressive achievements of the rational minority who did. Anaxagoras's examination of a deformed ram's head is a good example to illustrate the uneasy simultaneity of **ancient mysticism** and **empiricism**. **Anaxagoras** developed a mechanical theory to explain the origin and motions of the heavenly bodies, suggesting that a powerful sifting force he called Nous, or Mind, slowly differentiated the material soup of the early cosmos. Presaging the materialism of later **atomist philosophers**, he argued that every physical thing had small bits of other substances hidden within it. So the transformations in nature that we see, such as growth and decay, are really the result of these invisible mechanical processes. He looked for predictable causes, rather than superstitious divinations, and his ideas demystified the natural world.
>
> One day **Pericles** heard that a monstrous ram had been born on one of his farms and he sent for the animal. When it arrived at his court, a crowd gathered around and studied the strange anomaly. The animal had only one horn growing from the center of its head. A revered fortune-teller named Lampon announced that the current political struggle between Pericles and his rival **Thucydides** would finally be resolved in Pericles' favor. The monstrous ram, found on Pericles' estate, was the auspicious sign indicating political victory. Lampon read the monster as a good omen. Anaxagoras, who was present for the spectacle, made a careful examination of the one-horned ram and then chopped its head in half. **Plutarch** reports: Anaxagoras, cleaving the skull in sunder, showed to the bystanders that the brain had not filled up its natural place, but being oblong, like an egg, had collected from all parts of the vessel which contained it, in a point to that place from whence the 'root of the horn took its rise.' In other words, he offered a scientific, causal explanation of the monster. A developmental glitch had produced a wonder. Apparently this bit of demonstrable empiricism won Anaxagoras a moment of respect and admiration from the many bystanders at the court. People could actually see with their own eyes the mechanical causes of the monstrosity. But this was a very short-lived triumph of reason over superstition,

because a brief time later Pericles did indeed prevail over his political rival and then Lampon the seer was the court darling all over again. This suggests, for one thing, that rational science was not exactly a juggernaut of truth, crushing the culture of superstition in its path. It also suggests that neither science nor superstition ever definitively rules out the other one. Explaining how a monster came to be monstrous, as Anaxagoras did, still failed to explain the monster's **purpose**. The purpose or **teleology of monsters** remained a vital concern for the ancients, and then the medievals, long after the mechanical explanations emerged.

Unlike many ancients who loved to speculate on the meaning or purpose of a particular monstrous birth, **Aristotle** concluded that monsters have no purpose or special meaning. To ascribe such meanings to natural accidents would be as wrongheaded as saying that a crack in the sidewalk is for the purpose of letting grass grow through. Monsters are just cases of biological bad luck and therefore don't require special explanations. Aristotle joins the other scientists in claiming that there is no additional purpose or portent in bizarre ram's horns (Anaxagoras) or seeming **centaurs** (Thales). All of Nature, according to Aristotle, should be understood in terms of purpose (teleology), such as when he says that an eye must be explained by its purpose of seeing and an acorn's purpose is the oak tree. But despite this framework, or rather because of it, there are no special purposes for monsters beyond the usual species-specific goals. A monster born of humans, no matter what it looks like, is a failed attempt to actualize a human essence. It is not a new species or a hybrid species or an alien creature or even a message from the gods. *It is just an anomalous or abnormal human being*. But Aristotle's demystification of monsters turned out to be a minority report, largely ignored by the ancient populace. Anomalous births continued to augur important revelations for superstitious Greeks and Romans.

Box 7.21 Hermaphrodites, Superstition, Fortune-Tellers, and Animal Sacrifice

In *On Monsters* Asma refers to the historical discrimination towards **hermaphrodite humans** (see also Fig. 7.12):

> **Prodiges** and **portents** were perceived to be everywhere in the ancient world. All of nature was sending **signals foretelling the future**. If one could read the signs properly, which was the job of augurs in Rome and **oracles** in Greece, then one could predict the fate of military campaigns, the health of marriages, or the prosperity of business ventures. The Romans practiced an art of **prophesy** that came down from the **Etruscans** and involved **reading the liver of a sacrificed animal**. The liver, thought to be the source of blood and life itself, was charted into subdivisions that corresponded to deities. Cults of **fortune-tellers** evolved an elaborate and **secret science of viscera interpretation**. The Roman historian **Livy** (59 BCE-17 CE) tells a typical story about the bad omen of a sacrifice in 90 BCE. The Roman consul Rutilius Lupus sacrificed an animal and 'failed to find the lobe of the liver among the organs; ignoring the omen he lost his army and was killed in battle.' In contrast, in 43 BCE **Caesar Augustus** sacrificed an animal on the eve of his military campaign against **Marc Antony**. Livy reports that 'the animal he sacrificed had twin sets of internal organs. Success followed him.' Like a missing liver lobe, the discovery of a **hermaphrodite** human was considered by most Romans to be very bad for the health of the state. Apparently the founder of Rome himself, **Romulus**, felt threatened by hermaphrodites and ordered them to be drowned upon discovery. The logic of this custom, as with many customs, is unclear.
>
> The classicist Carlin Barton suggests that hermaphrodites, with their ambiguous, unclassifiable sexuality, may have been simultaneously threatening to the increasingly rigid official Roman culture but also alluring and exciting to those Romans who felt repressed by the bureaucratic, authoritarian, and hierarchic mores of an expanding empire. The ambiguously gendered person did not conform to **tradi-**

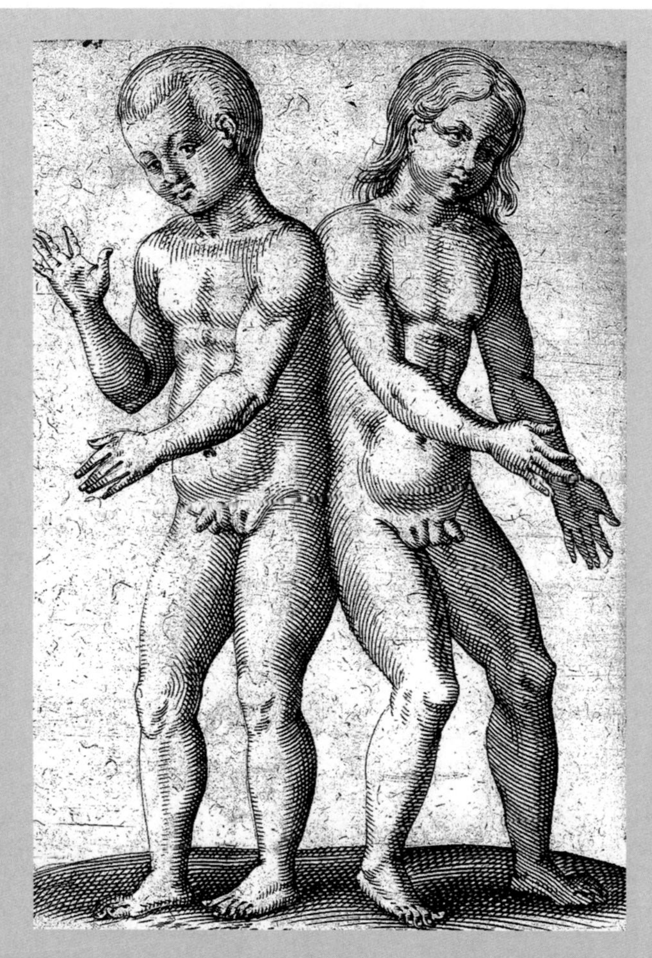

Fig. 7.12 Hermaphrodite twins joined at the back, born at Rorbach near Heidelberg, 1486, shown in *De Hermaphroditorum*, 1614. As noted in Asma's *On monsters*, "the practice of drowning hermaphrodites was extended to all seriously disabled children in the Roman Laws of the Twelve Tables: 'A father shall immediately put to death a son recently born, who is a monster, or has a form different from that of members of the human race'"

tional male or female parameters. Hermaphrodites, on this account, represented a dangerous freedom, in the same way that 'noble savages' must have done for Enlightenment-era urbanites. A more prosaic, and probably accurate, explanation is that monstrous offspring represented a terrible economic and energy burden on the family, and if they should make it to adulthood they would be a burden on the state as well. It's reasonable to expect laws and taboos to emerge in a society that reinforces the specific ecological survival needs of its families. This practice of drowning hermaphrodites was extended to all seriously disabled children in the *Roman Laws of the Twelve Tables*: 'A father shall immediately put to death a son recently born, who is a monster, or has a form different from that of members of the human race.' The hermaphrodite is a **liminal being.** Liminal comes from the Latin word limen, meaning 'threshold.' When you are on a threshold, you are neither inside nor outside but in between. Hermaphrodites, with their **ambiguous genitalia**, are in between the traditional categories of male and female. One sees, immediately I think, that the idea of a liminal being, something between categories, is a very useful way to think about many…('monsters'), not just hermaphrodites.

Griffins, with their ambiguous avian-quadruped shape, would qualify as liminal, as would **centaurs**, the **chimera**, the **Gorgons**, the **Minotaur**, and the **Hydra**. Mosaic beings, grafted together or hybridized by nature or artifice, reappear throughout the history of Western monsters as the **Golem**, **Frankenstein**'s creature, and **transgenic animals**. Even **zombies**, though not hybridized, are liminal monsters because they exist between the living and the dead. In short, **liminality** is a significant category for the uncategorizable. Livy chronicled many murders of hermaphrodites in the last two centuries before the Common Era. A short sampling of his very long list will suffice to demonstrate the common perception of hermaphrodites as monsters. In 133, 'in the region of Ferentium, a hermaphrodite was born and thrown into the river'; in 119 'a hermaphrodite eight years old was discovered in the region of Rome and consigned to the sea'; in 117 'a hermaphrodite ten years old was discovered and was drowned in the sea'; in 98 'a hermaphrodite was thrown into the sea.' Many other types of monsters are cited in Livy's encyclopedic history, including many unfortunate developmentally **disabled children**. A sad litany of abnormalities is offered as examples of bad omens, including conjoined twins and babies born with no hands and feet or too many hands and feet. Livy himself seems completely unmoved by any of these stories and recites them as though he's reading sports scores. His entry for 108 BCE reads, 'At Nursia twins were born to a freeborn woman, the girl with all members intact the boy with the following deformities; in front his abdomen was open, so that the uncovered intestine could be seen, and behind he had no anal opening; at birth he cried out once and died…the war against Jugurtha was carried on successfully.' But things seemed to be looking up for hermaphrodites, at least, by the time Pliny writes his *Natural history*. We find a refreshing tolerance developing toward hermaphrodites when he states, 'There are people who have the characteristics of both sexes. We call them hermaphrodites, the Greeks **androgyny**. Once considered portents, now they are sources of entertainment.' Some scholars see a teleological arc here. That is to say, the transition from **superstitious murder** of hermaphrodites to benign neglect and even amusement looks like progress. It looks like progress because it is **progress**, ethically speaking. But history and ethics don't always converge on the righteous path. The classicist Luc Brisson and the gender theorist Anne Fausto-Sterling both suggest that hermaphrodites suffered terribly in the early days of Roman law, but then rational progress ultimately created a more hospitable Rome for first-century hermaphrodites. But even while Pliny was assuring the reader that hermaphrodites were in the clear, so to speak, drownings continued. Monsters did not simply evaporate as **rational humanism** came on the scene. The fear of monsters hung on in the vast stretches of darkness, while the thin flare of rationality, possessed by a few elite philosophers, swept around the terrain, without much illumination or impact.

Box 7.22 "On Monsters and Marvels," "Mother's Fault," and Medicine

Asma explains in *On monsters* (see also Fig. 7.13):

The man who is usually credited with rescuing monsters from the melodramatic arena of spiritual and moral meaning is his predecessor, the French surgeon and scholar **Ambroise Paré** (1510-1590). His influence was not enough to effect a complete revolution in **monsterology**, and he himself was highly superstitious, but he paved the way

Fig. 7.13 Chas Eisenmann's photograph of Fedor Jeftichew (1864–1904), often ill-named "the Russian Dog-faced boy"

for future medical scientists to study birth anomalies. In the conceptual history of monsters he certainly represents a turn toward the more naturalistic explanation of extraordinary beings. Paré's book On *monsters and marvels* took a relatively empirical approach to monsters, preferring the collection and dissection of oddities rather than the pursuit of hearsay natural history. Monster races like the **Blemmyae** and **Cynocephali** were of little interest compared with human monstrosity. Paré's *On monsters* is really a transitional work, steeped in the superstitions of the day but struggling to extricate itself from dead-end research avenues. One finds all the usual ingenuousness about **unicorns**, sea creatures, and such, but also an attempt to put some monster legends to rest. Perhaps the most surprising cause in Paré's list is the fifth item, the imagination. Paré follows his ancient predecessors (i.e., **Hippocrates**, **Aristotle**, and **Empedocles**) in upholding a theory about the role of the *mother's imagination* at the moment of conception and in early gestation. If a woman in coitus is exposed to frightening or disturbing or simply strong imagery, through either the senses or memory, the offending image may be impressed on her offspring. Paré accepts the reality of a physiological process, one that begins as a disturbing sense impression and ends with a distorted fetus. He offers a few cases to illustrate his point, some of which strain his own credulity and some that seem quite credible to him. Undermining his own embryonic empiricism, he cites the authorities of old. He tells of **Queen Persina of Ethiopia**, who with **King Hidustes** mysteriously produced a white baby 'because of the appearance of the beautiful **Andromeda** that she summoned up in her imagination, for she had a painting of her before her eyes during embraces from which she became pregnant.' Likewise we are told of a girl who was born as furry as a bear. Her unfortunate state was the result of her mother's having looked too intensely at the image of **Saint John** [the Baptist] dressed in skins, along 'with his [own] body hair and beard, which picture was attached to the foot of her bed while she was conceiving.'

A more contemporary example is offered in Paré's story of a baby born in France in 1517 with the face of a frog. When asked what the cause of this monster might be, the father of the child explained that his wife had been ill with a fever and had taken the curative advice of her friend. The folk cure required the wife to carry a frog in her hand until the frog died, at which point she would be cured of the fever. 'That night she went to bed with her husband, still having said frog in her hand; her husband and she embraced and she conceived; and by the power of her imagination, this monster had thus been produced'...Of course, Paré was not a secular humanist, and **devilish demons** were quite real for him. He did not have an ironic or literary response to the demonic. 'Satan's actions', he says, 'are supernatural and incomprehensible, surpassing the human mind, [it] not being able to explain them, any more than [it can] the magnet which attracts iron and makes the needle turn.' But he goes on to say that we should not fall into a general skepticism about the 'principles and reasons of natural things.' The human mind may not be up to the challenge of **supernatural spiritual riddles**, but let us not give up, he seems to suggest, our attempts to grasp the natural world. Here we find an inconspicuous boundary marker in Paré's thinking: the **natural monsters** are appropriate subjects for medical study, but the **supernatural monsters** exist in a domain that cannot be penetrated properly by science. The proper response to this latter domain is prayer and piety, not scientific exploration. This important concession to the Church, reiterated by most scientists of the following century, helped to create an autonomous domain for previously forbidden explorations of nature.

Unfortunately, the disturbing links between teleological narratives and the **mistreatment of people with malformations** are not just something of the past. It is true that there were significant changes in the last centuries in the way that, in general, people look at and care for those with disabilities, as correctly noted in Pinker's *Better Angels*. But there are of course exceptions, many of them, and among them there are some truly horrific ones, which are probably the most horrendous ones ever done in human history. To give an example, which is unfortunately not as discussed as it should be, within—and probably also because of—the endless number of dreadful atrocities done by **Nazis**, is the way they dealt with people with disabilities. It is estimated that close to 250,000 disabled people were murdered under the Nazi regime: to my knowledge, this atrocious event, just about 7 decades ago, is by far the biggest and quickest mass killing of people with disabilities in human history. And it started well before disabled people were sent by the Nazis to **concentration camps**: between 1939 and 1941, a huge number of them were killed under the "**euthanasia program**" called "**mercy killing**." The shockingly disturbing details are summarized in the *Holocaust Encyclopedia* of the U.S. Holocaust Memorial Museum. Although I understand that it is very distressing to read these details, it is important to include some of them here because of their historical importance and precisely because they are too often ignored, because by closing our eyes to reality and remaining in our comfortable *Neverland* bubble will just lead to the repetition of such atrocities, over and over, in the future:

> The **Euthanasia Program** was the systematic murder of institutionalized patients with disabilities in Germany. It predated the genocide of European Jewry (the **Holocaust**) by approximately two years. The program was one of many radical eugenic measures which aimed to restore the racial 'integrity' of the German nation. It aimed to eliminate what **eugenicists** and their supporters considered 'life unworthy of life': those individuals who – they believed – because of severe psychiatric, neurological, or physical disabilities represented both a genetic and a financial burden on German society and the state. In the spring and summer months of 1939, a number of planners began to organize a secret killing operation targeting disabled children. They were led by **Philipp Bouhler**, the director of **Hitler**'s private chancellery, and **Karl Brandt**, Hitler's attending physician. On August 18, 1939, the Reich Ministry of the Interior circulated a decree requiring all physicians, nurses, and midwives to report newborn infants and children under the age of three who showed signs of severe mental or physical disability. Beginning in October 1939, public health authorities began to encourage parents of children with disabilities to admit their young children to one of a number of specially designated pediatric clinics throughout Germany and Austria. In reality, the clinics were children's killing wards. There, specially recruited medical staff murdered their young charges by lethal overdoses of medication or by starvation. At first, medical professionals and clinic administrators included only infants and toddlers in the operation. As the scope of the measure widened, they included youths up to 17 years of age. Conservative estimates suggest that at least 5,000 physically and mentally disabled German children perished as a result of the child "euthanasia" program during the war years. "Euthanasia" planners quickly envisioned extending the killing program to adult disabled patients living in institutional settings. In the autumn of 1939, Adolf Hitler signed a secret authorization in order to protect participating physicians, medical staff, and administrators from prosecution. This authorization was backdated to September 1, 1939, to suggest that the effort was related to wartime measures.
>
> The Führer Chancellery was compact and separate from state, government, or **Nazi Party** apparatuses. For these reasons, Hitler chose it to serve as the engine for the 'euthanasia' campaign. The program's functionaries called their secret enterprise "T4." This code-name came from the street address of the program's coordinating office in Berlin: Tiergartenstrasse 4. According to Hitler's directive, Führer Chancellery director Phillip Bouhler and physician Karl Brandt led the killing operation. Under their leadership, T4 operatives established six gassing installations for adults as part of the "euthanasia" action. Using a practice developed for the child "euthanasia" program, in the autumn of 1939 T4 planners began to distribute carefully formulated questionnaires to all public health officials, public and private hospitals, mental institutions, and nursing homes for the chronically ill and aged. The limited space and wording on the forms, as well as the instructions in the accompanying cover letter, combined to give the impression that the survey was intended simply to gather statistical data. The form's sinister purpose was suggested only by the emphasis placed upon the patient's capacity to work and by the categories of patients which the inquiry required health authorities to identify. The categories of patients were: those suffering from **schizophrenia**, **epilepsy**, **dementia**, **encephalitis**, and other chronic psychiatric or **neurological disorders**; those not of German or "related" blood; the criminally insane or those committed on criminal grounds; those who had been confined to the institution in question for more than five years. Secretly recruited "medical experts", physicians – many of them of significant reputation – worked in teams of three to evaluate the forms. On the basis of their decisions beginning in January 1940, T4 functionaries began to remove patients selected for the "euthanasia" program from their home institutions. The patients were transported by bus or by rail to one of the **central gassing installations** for killing.

Within hours of their arrival at such centers, the victims perished in **gas chambers**. The gas chambers, disguised as shower facilities, used pure carbon monoxide gas. T4 functionaries burned the bodies in crematoria attached to the gassing facilities. Other workers took the ashes of cremated victims from a common pile and placed them in urns to send to the relatives of the victims. The families or guardians of the victims received such an urn, along with a death certificate and other documentation, listing a fictive cause and date of death. Because the program was secret, T-4 planners and functionaries took elaborate measures to conceal its deadly designs. Even though physicians and institutional administrators falsified official records in every case to indicate that the victims died of natural causes, the "euthanasia" program quickly become an open secret. There was widespread public knowledge of the measure. Private and public protests concerning the killings took place, especially from members of the **German clergy**. Among these clergy was the bishop of Münster, **Clemens August Count von Galen**. He protested the T-4 killings in a sermon August 3, 1941. In light of the widespread public knowledge and the public and private protests, Hitler ordered a halt to the Euthanasia Program in late August 1941. According to T4's own internal calculations, the "euthanasia" effort claimed the lives of 70,273 institutionalized mentally and physically disabled persons at the six gassing facilities between January 1940 and August 1941. Hitler's call for a halt to the T4 action did not mean an end to the "euthanasia" killing operation. Child "euthanasia" continued as before. Moreover, in August 1942, German medical professionals and healthcare workers resumed the killings, although in a more carefully concealed manner than before. More decentralized than the initial gassing phase, the renewed effort relied closely upon regional exigencies, with local authorities determining the pace of the killing.

Using drug overdose and lethal injection – already successfully used in child "euthanasia" – in this second phase as a more covert means of killing, the "euthanasia" campaign resumed at a broad range of institutions throughout the Reich. Many of these institutions also systematically starved adult and child victims. The Euthanasia Program continued until the last days of World War II, expanding to include an ever wider range of victims, including geriatric patients, bombing victims, and foreign forced laborers. Historians estimate that the **Euthanasia Program**, in all its phases, claimed the lives of 250,000 individuals. Persons with disabilities also fell victim to German violence in the German-occupied east. The Germans confined the Euthanasia Program, which began as a racial hygiene measure, to the Reich proper – that is, to Germany and to the annexed territories of Austria, Alsace-Lorraine, the Protectorate of Bohemia and Moravia, and the Warthegau in former Poland. However, the Nazi ideological conviction which labeled these persons "life unworthy of life" also made institutionalized patients the targets of shooting actions in Poland and the Soviet Union. There, the killings of disabled patients were the work of SS and police forces, not of the physicians, caretakers, and T4 administrators who implemented the Euthanasia Program itself. In areas of Pomerania, West Prussia, and occupied Poland, SS and police units murdered some 30,000 patients by the autumn of 1941 in order to accommodate ethnic German settlers (Volksdeutsche) transferred there from the Baltic countries and other areas. SS and police units also murdered disabled patients in mass shootings and gas vans in occupied Soviet territories. Thousands more died, murdered in their beds and wards by SS and auxiliary police units in Poland and the Soviet Union. These murders lacked the ideological component attributed to the centralized Euthanasia Program. Planners of the "**Final Solution**" later borrowed the gas chamber and accompanying crematoria, specifically designed for the **T4 campaign**, to murder Jews in German-occupied Europe. T4 personnel who had shown themselves reliable in this first mass murder program figured prominently among the German staff stationed at the Operation Reinhard killing centers of Belzec, Sobibor, and Treblinka. Like those who planned the physical annihilation of the European Jews, the planners of the Euthanasia Program imagined a racially pure and productive society. They embraced radical strategies to eliminate those who did not fit within their vision.

Fig. 7.14 The exhibition "The Miracle of Life" at Kaiserdamm in Berlin, 1935

That is, the teleological narratives embedded in such a "vision" of a "**racially pure" society**, used to justify such killings, were similar to those used to justify the genocide of Jews and gypsies. They involved the combination—as so often done in **eugenic movements**—of the longstanding narrative of "progress" with more recent evolutionary notions of "struggle-for-life" and "survival of the fittest" related, in this specific case, to the "improvement of the **Aryan race**" (see also Box 7.23). In the image shown in Fig. 7.14, which was part of the 1935 "*The miracle of life*" shown at the Kaiserdamm Hall in Berlin displaying different types of Nazi propaganda, the narrative is the following: you should be scared, because if we allow that "lesser individuals" have more children than us, the "superior Arian ones," our whole group, will gradually disappear. So, something has to be done, and *now*. And unfortunately something terrible was indeed done, then, by the Nazis, to individuals that were deemed to be "inferior" because they had disabilities, or were from "non-Aryan races," or believed in a certain God, or had "hybrid sexuality," or for whatever reason that did not conform to the tales created by the *Homo fictus et irrationalis* to feel "special" or "superior" within the "progress" of the "ladder" of life.

One particularly irrational, paradoxical, and incoherent point made in the type of narratives displayed in Fig. 7.14 is that if the ingroup—in this case, the "Aryans"—is the "chosen one," the "special favored race," the "superior breed," why does it always end up by losing any natural type of competition, unless the ingroup does something horrible to prevent such a natural course? Nazis argued that they had to kill Jews, gypsies, homosexuals, or people with disabilities because otherwise they would be "naturally" outcompeted by them: so they seem to be assuming that actually Nazis were "naturally inferior," not the other way round. Similarly, current **white supremacy** or **far-right political movements**, both in Europe and the U.S., defend the existence and/or construction of walls to separate "Them" from the "Others"—be it Mexicans, or Africans, or people from the Middle East—because otherwise those "Others" would just come to their cities and naturally outcompete them by "taking their jobs," "filling their universities," "having more kids," and so on. So, once again, the self-designated "superior whites" actually seem to think they are instead "poor whites" that have no change of naturally competing with "Others." In fact, this logically incoherent way of thinking is clearly attested by the fact that white supremacists, and many of the far-right wing parties they are associated with, tend to portrait themselves as the "victims," the ones "oppressed by the deep state," the "silent majority," and so on, in a way that remind us small Chihuahuas that are very noisy and bark aggressively to other dogs because in reality they are hugely terrified when they see bigger and stronger dogs.

Box 7.23 Colonialism, "Modernism," Social Engineering, and the Holocaust

In *The Rise of Organized Brutality*, Malesevic explains with his characteristic perceptiveness:

> Most **genocidal action** in the **imperial colonies** was undertaken indirectly through the destruction of the native habitat; the **spread of European diseases**; and the introduction of new, often parasitic, animals and crops. In this way, the populations of the Americas and Australia and some parts of Africa and Asia were completely decimated. In regions of North America, where the majority of European settlers moved, 90 per cent of the native population died. In Australia, by 1920s only 20 per cent of the original Aboriginal population survived. As Mann points out, the **pre-Columbian native population** of what is today the United States was somewhere between 4 to nine million, while 'in the US Census of 1900 there were only 237,000 Indians, a loss of 95%'. The speed and scale of social destruction were most visible in California, where the 1849 gold rush intensified mass slaughter: 'By 1860, after 10 years of statehood, Californian Indians numbered only 31,000 – an 80 percent loss rate over only 12 years! The **Third Reich** also lasted 12 years and killed 70 percent of European Jews'…Hence, the more comprehensive definition of **genocide** has to incorporate this type of indirect violent action…Tilly's famous quote about the inherent link between war making and state making in premodern Europe applies even more to colonial expansion, as the organisational capacities of both imperial states and private colonial corporations have largely been built on top of the bodies of their colonial subjects. The direct consequence of this organisational development were millions of native deaths: it is estimated that by the beginning of the twentieth century the indigenous populations of the Americas has declined by more than 80 per cent. While in other parts of the world these figures may not be as stark, there is no doubt that the colonial experience had extremely violent outcomes. Although the institution of genocide emerged very late in human history, the organisational ingredients for its creation were largely forged throughout the centuries of violent **colonialism**.

In *Modernity and the Holocaust*, Bauman offered the first sustained modernist sociological account of genocide. Although his focus was almost exclusively on the **Nazi extermination** of European Jews, his intention was to articulate a more general argument about the modern foundations of genocidal projects. For Bauman, the inherent modernity of the **Holocaust** stems from its engineering ambitions and ideological blueprints aimed at bringing about a completely new world. Although the **Nazi utopia of racial hierarchies** and **eugenic purity** is often perceived as a fluke aberration and a regression from *Europe's enlightened march forward*, for Bauman the Nazi project too was a child of the **Enlightenment**. He argues that unlike the premodern pogroms and *ad hoc* massacres of Jews associated with the morbid rituals and individual instances of brutal rage, the '**Final Solution**' was a systematic programme aimed at implementing a blueprint for a new world. Whereas the premodern pogroms were simple expressions of individual rage and hate, modern genocides are ideologically articulated and instrumentally executed exercises in social engineering…In this context, the genocidal apparatus is deployed in a similar way to how a gardener intentionally uses bleach or vinegar to destroy grubby garden weeds: as Jews, Roma Gypsies, Slavs, homosexuals and the disabled did not fit into this **Nazi racial utopia**, they had to be removed in the same way weeds are cleansed to create a perfect garden. Hence the Holocaust is a direct byproduct of an Enlightenment-induced ambition to create a perfect social order. For Bauman, this modernist obsession is well reflected in the execution of the Final Solution, which was accomplished relying on conventional bureaucratic mechanisms and routines. The same principle of instrumental rationality that operates in private corporations, modern factories or state administration was utilised in the extermination camps of the Nazi regime: 'Rather than producing goods, the raw material was human beings and the end product was death, so many units per day marked carefully on the manager's production charts…the chimneys, the very symbol of the modern factory system, poured forth acrid smoke produced by burning human flesh…The brilliantly organized railroad grid of modern Europe carried a new kind of raw material to the factories'…

As political conflicts intensify in modernity, elaborate ideological means are deployed to **dehumanise one's enemy**. In **Goebbels**'s cinematography and radio broadcasts, ordinary individual German citizens of Jewish heritage are transformed into 'the swarms of disease ridden vermin and parasites'; in **Stalin**'s **Soviet Union**, moderate landowning peasants (i.e. kulaks) were depicted as 'leeches' and 'rats' bent on destroying the socialist homeland; and in the '**Hutu power**' controlled **radio station Libre des Mille Collines** and the influential **magazine Kangura**, **Tutsis** were described as cockroaches that needed to be exterminated. While delegitimisation of one's enemy is not a historically new phenomenon, what is distinct about modern contexts is the availability of sophisticated organisational and technological mechanisms to generate and distribute propagandistic messages and images. More importantly, only in modernity do such dehumanising images make popular sense as they become integrated into the wider ideological narrative. And this leads us to the second consequence: unlike the protoideologies which are usually confined to relatively small groups of individuals and which mostly appeal to a single social strata, ***modern ideologies*** (**nationalism**, **socialism**, **liberalism**, **conservatism**, **religious fundamentalism**, etc.) draw their mass support base from a variety of social strata. For example, while nationalism was a doctrine that attracted a small number of intellectuals and members of the upper-middle classes in early-eighteenth century, by the end of the twentieth century this belief system has become a dominant ideology of its age. This widening of an ideological support base is crucial for the proliferation of genocidal projects, as genocides ultimately grow out of ideologies that have a substantial degree of popular support. This is not to say that modern individuals easily consent to mass murder. On the contrary, in Enlightenment-infused modernity, human life is deemed much more precious than in the previous historical epochs. However precisely because life is so valued, it becomes critical to delegitimize and destroy all organizations and people associated with the imminent threats to, what ideologues perceive to be, a direct road to human happiness. Simply put, as modern ideological projects devise grand vistas of perfected social orders, the realisation of such ultimate societal goals fosters extreme intolerance towards opponents of such projects. Once a particular ideological outlook is taken to be the ultimate truth, any challenge to such a project is regularly interpreted as a malicious attempt to prevent the fulfilment of a noble goal. This way of thinking opens the door for the use of the most extreme measures to implement a particular goal: annihilating all ethnic minorities to build an ethnically uniform folksgemeinschaft; destroying all kulaks and Western spies to create a perfect **communist utopia**; sterilising genetically inferior populations to produce a eugenically perfected human race; etc....

There is a popular perception that **genocidaires** are inherently sadistic individuals who enjoy torturing and killing other human beings. However, without denying the fact that some killers suffer from various mental illnesses which might destroy their sense of empathy, there is abundance of empirical evidence indicating that the overwhelming majority of genocide perpetuators do not suffer from any mental disorders. Taking into account that **genocides** usually involve a large number of perpetuators and that serious mental illnesses affect only a small section of any general population, it seems highly unlikely that mental disorders play any significant role in such killings. For example, the Rwandan gacaca courts have tried at least 120,000 individuals accused of being directly involved in genocidal acts. No serious case could be made that such a large number of people were affected by mental disorders...Step-by-step **radicalization** regularly transpires in the context of large-scale warfare, when future genocide perpetuators are already organisationally and ideologically caged in the particular doctrinal project. The outbreak of war contributes to radicalization in a variety of ways: the propagandistic discourses deployed to delegitimise the enemy foster **dehumanisation**; increased battlefield casualties relativise the universalist principles that hold all human life sacred, thus opening up the possibility that killing others does not breach moral norms; and deteriorating living conditions with an increased sense of fear foster a greater lack of empathy. In such an environment, radicalization presents ordinary individuals with difficult choices: to embrace extremist doctrines, to reject such acts and risk punishment or to ignore the actual social reality and pretend that such violent events do not take place. As both Fulbrook and Browning demonstrate, most 'ordinary men' in 1940s Germany were not initially supportive of extreme violence against enemy civilians. However, once they joined the **Nazi bureaucratic machinery** (including the various armed forces), they gradually became more accepting of such extremist acts. Browning shows how some members of Reserve Police Battalion 101 disliked killings of Jewish women and children and would initially shoot in the air, would vomit after the murderous actions and would apply for transfer to other units. Nevertheless, he also shows how killings of civilians gradually became normalised and how many 'ordinary men' embraced their role as killers. What is crucial in such a step-by-step radicalization is the sense of attachment to one's **microgroup**.

Chapter 8
"Progress," Morality, and "Good" and "Evil"

With or without religion, good people would tend to behave well and bad people would do evil things, but the peculiar contribution of religion throughout history has been to allow 'good people' to do evil things.

(Steven Weinberg)

8.1 Morality, "Progress," Revolutions, and Health

A mind is its own place, and in itself...can make a Heav'n of Hell, and a Hell of Heav'n. (John Milton)

Almost any person has made questions about what is moral and immoral, and numerous philosophers and literally thousands of publications have focused on this topic. The aim of this chapter is *not* to provide a comprehensive or extensive review on the history of these questions and on complex issues such as the links between **morality**, **consequentialism**, **deontology**, **utilitarianism**, the **philosophy of happiness**, and so on (see for instance Box 8.1). Instead, here I will provide a brief discussion on some of those issues that are directly related within the main topics covered in the present book and the information provided in the previous chapters. In order to do so, I will start with what Vaas stated in his 2009 book chapter, which we partially discussed mentioned in Chap. 2:

> Although the truth of **religious faith** cannot be demonstrated (or even need not and must not be demonstrated, as many believers claim, for otherwise it would not be **existential belief** anymore), it is often said that **religiosity** and **religion** are at least *useful*. This might be true for gaining power, wealth or consolation and in the restricted context of biological evolution if religiosity enhances **reproductive fitness**. Thus religious belief and behavior seems to be a profitable illusion like the assumption of a strong (libertarian) kind of **free will**. But, of course, it would be a logical and **naturalistic fallacy** to infer truth or moral values from this. Usefulness is not equal to *truth* nor is it an ethical accolade (besides, exponential human reproduction might, in the not too distant future, even destroy the biosphere due to overpopulation and its many devastating effects). So it might not be surprising that religions are widespread, because: 1) for many people, especially the desperate,

R. Diogo, *Meaning of Life, Human Nature, and Delusions*,
https://doi.org/10.1007/978-3-319-70401-2_8

religion is a source of hope or relief – a drowning man will clutch at a straw (even if it is just a **self-delusion** or the manipulating promise of others) – religious doctrines can motivate believers to persevere and sometimes even to change power relationships; 2) religions strengthen social support and propagate rapidly with the reproduction of their adherents, who often indoctrinate their offspring in early childhood; and 3) religions, if sufficiently well established, are used and enforced by the potentates, not only among their followers, but also outward (**proselytization**). These three factors interact and depend crucially on social and ecological boundary conditions. From this perspective, though it is a very crude sketch, one can understand why religions are so common: it is due to a kind **of quasi-Darwinistic self-organization** – including meme competition – and independent of whether or not religiosity is an adaptation. It is often said, "If God does not exist, everything is permitted" …meaning that **morality** requires religion as a source or justification of **moral values**. This is clearly not the case, however; moreover, religious persons do not behave ethically better than non-believers, as many studies have shown. And even if religion was useful or advantageous in the past, it might be harmful and detrimental nowadays or in the future. The main problem is that **ideological dogmas** – and there are not only religious ones! – claiming to own absolute, infallible truths, can and often did motivate people to **dehumanize** and debase others.

I see "*usefulness*," "*goodness*," and "*truth*" (see Box 8.1) as very different concepts that have often been confused or linked in an inappropriate way. For instance, within **biological evolution**, what non-human organisms do has nothing to do with being "morally right" or "morally wrong." **Cuckoos** have been evolutionarily successful by undertaking a type of niche construction that lead them to evolve parasitic-like features such as laying their eggs in the nests of other bird species so they are incubated by the foster parents, who rear the young cuckoos. If something similar was done by humans, probably most people would say that those that tricked the foster parents did something "morally wrong." But it does not seem to make sense saying that the cuckoos, or any other organisms with parasitic-like features, are doing something that is morally wrong and that they should change their behavior, accordingly. The cuckoos are living and evolving, and are actually an example of evolutionary success, and nobody—in particular a God that does not exist in the world of reality—can tell them that they are "bad animals" because of this: within the reality of the natural world, *things are just what they are*. People often react to this example by noting that humans are different from cuckoos precisely because humans have **consciousness**, a **theory of mind**, and have mental capacities that allow them to display moral features related to fairness and justice. However, none of these aspects truly only applies to humans. Empirical data clearly show that animals such as **chimpanzees** have some type of theory of mind, consciousness and display features that in humans are considered to be related to "fairness" and "justice." This topic was briefly discussed above, and I provide more details about it in my 2017 book *Evolution Driven by Organismal Behavior*, and if you are interested to know more about it I strongly recommend you to read De Wall's 2016 *Are We Smart Enough to Know How Smart Animal Are?*

Box 8.1: Morality, Ethics, Utilitarianism, and Philosophy of Happiness
In his 2016 book *The big picture – on the origins of life, meaning, and the universe itself*, theoretical physicist **Sean Carroll** – not to be confused with the biologist Sean Carroll, also discussed in the present volume – explains:

> Philosophers find it useful to distinguish between **ethics** and **meta-ethics**. Ethics is about what is right and what is wrong, what moral guidelines we should adopt for our own behavior and that of others. A statement like 'killing puppies is wrong' belongs to ethics. Meta-ethics takes a step back, and asks what it means to say that something is right or wrong, and why we should adopt one set of guidelines rather than some other set. ***Poetic naturalism*** [see **Box 9.11**] has little to say about ethics…but it does have something to say about meta-ethics, namely: our **ethical systems** are things that are constructed by us human beings, not discovered out there in the world, and should be evaluated accordingly. To help with that kind of evaluation, we can contemplate some of the choices we have when it comes to ethics. Two ideas serve as a useful starting point: **consequentialism** and **deontology**. At the risk of vastly oversimplifying thousands of years of argument and contemplation, consequentialists believe that the moral implications of an action are determined by what consequences that action causes, while deontologists feel that actions are morally right or wrong in and of themselves, not because of what effects they may lead to. 'The greatest good for the greatest number', the famous maxim of **utilitarianism**, is a classic **consequentialist way of thinking**. 'Do unto others as you would have them do unto you', the *Golden Rule*, is an example of deontology in action. Deontology is all about rules. (The word 'deontology' comes from the Greek deon, for 'duty,' while 'ontology' comes from the Greek on, for 'being.' Despite the similarity of the words, the two ideas are unrelated.) Greene has studied volunteers hooked up to an MRI machine while being asked to contemplate various moral dilemmas. As expected, contemplation of 'personal' situations (like pushing someone off of a bridge) led to increased activity in areas of the brain that are associated with emotions and social reasoning. 'Impersonal' situations (like pulling a switch) engaged the parts of the brain associated with cognition and higher reasoning. Different modules within ourselves spring to life when we're forced to deal with slightly different circumstances. When it comes to morality, the unruly parliament that constitutes our brain includes both deontological and consequentialist factions. Sticking someone inside an imposing medical scanner and asking them to consider philosophical thought experiments might not tell us much about how that person would actually react in the situation described. The real world is messy – are you sure you could stop the trolley by pushing that guy off the footbridge? – and people's predictions about how they would act in stressful situations aren't always reliable.
>
> Consequentialism and deontology aren't the only kinds of ethical systems we can consider. Another popular approach is **virtue ethics**, which traces its roots back to **Plato** and **Aristotle**. If deontology is about what you do, and consequentialism is about what happens, virtue ethics is about who you are. To a virtue ethicist, what matters isn't so much how many people you save by diverting a trolley, or the intrinsic good of your actions; what matters is whether you made your decision on the basis of virtues such as courage, responsibility, and wisdom. Virtue sounds like a good thing to strive toward. Like consequentialism and deontology, it's an ostensibly attractive moral stance. Sadly, all of these attractive approaches end up offering different advice in important cases. How should we decide what ethical system to abide by? What kind of **morality** shall we construct? There is no unique answer to this question that applies equally well to all persons. Perhaps the most well-known

approach to ethics is the consequentialist theory of utilitarianism. It imagines that there is some quantifiable aspect of human existence, which we can label 'utility,' such that increasing it is good, decreasing it is bad, and maximizing it would be best of all. The issue then becomes how we should define utility. A simple answer is 'happiness' or 'pleasure,' but that can seem a bit superficial and self-centered. Other options include 'well-being' and 'preference satisfaction'. A…challenge for utilitarianism was offered by philosopher Robert Nozick: the 'utility monster,' a hypothetical being with incredibly refined sensibilities and an enormous capacity for pleasure. At face value, standard utilitarianism might lead us to think that the most moral actions are those that keep the utility monster happy, no matter how sad that might make the rest of us, because the monster is so incredibly good at being happy. Relatedly, we could imagine technology progressing to the point where we could place people in machines that would render them immobile, but generate in their brains maximal feelings of happiness or preference satisfaction or a feeling of flourishing or whatever other utility measure we dreamed up. Should we work toward a world where everyone is hooked up to such machines?

If we were to accept that morality is constructed, (one could say that) individuals will run around giving in to their worst instincts, and we would have no basis on which to condemn obviously bad things like the Holocaust. After all, somebody thought it was a good idea, and without objective guidance how can we say they were wrong? We might hope, in the spirit of **Kant**, that simple logical requirements of internal consistency would lead every rational person to construct the same moral rules, even starting from slightly different initial feelings. But that hope seems slim indeed. If (people)…act on their impulses in ways that bring harm to others, we should respond as we actually do in the real world: by preventing them from doing so. When criminals refuse to be deterred, we put them in jail. As a practical matter, the worries associated with **constructivism** are somewhat overblown. Most people, in most circumstances, want to think of themselves as doing good rather than evil. It's not clear what operational benefit would be gained by establishing morality as an objective set of facts. Presumably we envision a person or group who was relatively rational, but disagreed with us about morality, whom we could sit down with over coffee and convince of the mistake they were making. In practice the recommended strategy for a constructivist would be essentially the same: sitting down and talking with the person, appealing to our common moral beliefs, attempting to work out a mutually reasonable solution. ***Moral progress*** is possible because most people share many moral sentiments; if they don't, reasoning with them wouldn't help much no matter what.

[Regarding the ***philosophy of happiness***] we live at a time when the search for happiness has taken center stage as never before. Books, TV shows, and websites are constantly offering pointers about how to finally achieve and sustain this elusive and sought-after state of being. If only we were happy, everything would be okay. Imagine a drug that would make you perfectly happy, but remove any interest you might have in doing anything more than simple survival. You would lead a thoroughly boring treadmill of a life, from the outside – but inside you would be blissfully happy, romping through imaginary adventures and always-successful romantic escapades. Would you take the drug? Think of **Socrates**, **Jesus**, **Gandhi**, **Nelson Mandela**. Or **Michelangelo**, **Beethoven**, **Virginia Woolf**. Is 'happy' the first word that comes to mind when you set out to describe them? They may have been – and surely were, from time to time – but it's not their defining characteristic. The mistake we make in putting emphasis on happiness is to forget that life is a process, defined by activity and motion, and to search instead for the one perfect state of being. There can be no such state, since change is the essence of life. Scholars who study meaning in life distinguish between synchronic meaning and diachronic meaning. Synchronic meaning depends on your state of being at any one moment in time: you are happy because you are out in the sunshine. Diachronic meaning depends on the journey you are on: you are happy because you are making progress toward a college degree. If we permit ourselves to take inspiration from what we have learned about ontology, it might suggest that we focus more on diachronic meaning at the expense of synchronic. The essence of life is change, and we can aim to make change part of how we find meaning in it. At the end of the day, or the end of your life, it doesn't matter so much that you were happy much of the time… We have aspirations that reach higher than happiness. We've learned so much about the scope and workings of the universe, and about how to live together and find meaning and purpose in our lives, precisely because we are ultimately unwilling to take comforting illusions as final answers.

Interestingly, when I referred to some of the empirical examples provided in De Wall's book to my colleagues in the 2017 *Human Enhancement* meeting organized in Lisbon by philosophers, some of them—including the influential and in my opinion particularly bright bioethicist **John Harris**—argued that such cases—even those found in chimpanzees—can at the maximum be seen as an example of **protomorality**. This has been a typical reaction of humans when empirical data contradict **human exceptionalism**: concerning this specific example, surely chimpanzees cannot have a "true morality," because "everyone knows" that only the "special" self-designated "sapient being" can truly have morality. It should be noted that for a long time humans classified themselves, scientifically, in a separate biological kingdom with no other living beings because of our *belief* that we were the only beings displaying features A, B, C, D, and so on. As noted in my anatomical works with **Bernard Wood**—such as Diogo and Wood 2011, 2012, 2013—the more empirical data is acquired, the less inclusive is the taxon including only humans: after our "own" kingdom, we were placed alone in an order, then in a family, and now just in a genus, *Homo*.

Various researchers nowadays argue that we should not even be alone in a genus, because our *Homo sapiens* species should be placed in a same genus together with all the other fossil human species plus all living and fossil species of chimpanzees, including the common chimpanzees—*Pan troglodytes*—and bonobos—*Pan paniscus*, as discussed in Jared Diamond's 1991 book *The Rise and Fall of the Third Chimpanzee*. One of the features that most researchers resisted to accept, for a long time, to be also present in non-human animals was *culture*. Even when clear examples of culture in other animals were started to be described in detail in scientific publications—including the one about **potato washing in Japanese macaques**—many researchers continued to argue that these were just examples of "protoculture." In fact, one of the few remaining behavioral features that does seem to be truly unique for humans is precisely one that is key to understand why we create such just-so-stories about our supposed specialness, as discussed in Chap. 2: although we cannot be completely sure if *all* other animals truly lack this ability, the data available do seem to indicate that humans are the only animals that are fully aware of the inevitability of death.

Some other fascinating topics covered in that 2017 *Human Enhancement* meeting also relate to subjects discussed in the present volume. One is that, even nowadays, **ethics** and in particular **bioethics** about **gene editing and "human enhancement"** continue to be often rooted in teleological notions of the "sacred." For instance, data shown in the meeting indicate that religious people are normally more skeptical about gene editing and "human enhancement" techniques than atheists are. If one thinks that humans are made "in the likeness of God," then techniques that might in any way change them would logically be generally perceived more negatively. Similarly, **anti-abortion movements** are more prominent within religious groups. An interesting question was raised in the meeting: is the notion that the human genome is sacred and thus that it should not be modified by gene editing not actually similar to the eugenics idea that the "pure white race" is sacred and should not be "lost" due to mixing with other so-called races? Because teleological narratives were used to justify so many atrocities in human history, including those promoted/defended by eugenicists, I think that one should avoid as much as possible to have bioethical decisions being based/rooted in such narratives: whenever is feasible, they should mostly take into account what is shown by scientific empirical data. For example, when one talks about the "human genome" as sacred, it is as if our genome has been always the same, fixed, while in reality genomes obviously have changed and continue to change. Also, many bioethical discussions *are* based on a priori assumption that "genes are our essence," while the available empirical data shows that genes are just part of the multiple layers and factors that one needs to take into account to have a comprehensive understanding of human evolution and biology, as explained above.

So, when people say that humans "should be protected from science"—a stereotype often seen in popular science fiction books and movies, and lately getting particularly prominent within a substantial portion of the population of various countries, I would answer that the most important and urgent thing we need to do is to protect—or better said, liberate—humans, science, and bioethics from imaginary teleological tales. This is because many topics included in philosophical discussions on "**human enhancement**" and bioethics are indeed related to out-dated tales about "human progress" and "human complexity," such as the notion that humans are more anatomically complex than animals such as other primates. This notion is simply wrong in terms of most types of comparisons we can do, for instance, regarding the total number of anatomical structures such as muscles and bones. Anatomically, humans are instead "simplified mammals," with fewer muscles and bones than most other primates, as shown in works done by my colleagues and me, such as Diogo and Wood 2011, 2012, 2013. Based on these data, perhaps we should think if, contrary to the "progress-based" idea that "human enhancement" needs to involve an increase of the number of structures of our body—for instance, including robotic ones—one would need instead to simplify our bodies even more, in order to expand what has been occurring in our evolutionary lineage since we split from chimpanzees.

As seen in the previous chapters, such notions of progress are deeply related to the concept of "economic growth" and the idea that "more is always better," as if let's say an economic growth of 0 is the worst thing ever. These notions are related to what is beautifully said in a famous quote by the **Red Queen**, in Lewis Carroll's *Alice's Adventures in Wonderland*: "*it takes all the running you can do, to keep in the same place*." Although Carroll was seemingly not referring to, or supporting in any way, such notions with that story, teleological tales about "progress" are used, over and over, to create this *Neverland* delusion that we can't never stop, that we need to "produce" and "grow" more, and thus we can't bring to a halt using more and more resources from the planet. The problem is that, as wisely stated in various tales by **Confucians**, something seen as "good" now can have—and normally does have—dark consequences later. For instance, when humans were "finally capable" of inventing cars, this then led to a massive production and use of cars, which in turn led to an unprecedented level of pollution in our planet, which consequently led to an increase of various human diseases and global warming, and so on. As summarized in a 2016 paper by Jiang and colleagues, entitled "*Air pollution and chronic airway diseases: what should people know and do?*":

> According to the **World Health Organization** (WHO) report in 2008, 1.3 million deaths were estimated to be related to **ambient air pollution** globally. The figure became 3.7 million in 2012, which was nearly tripled. Two million deaths were attributable to the effects of **household air pollution** in 2008. This number also increased as nearly doubled (4.3 million) according to the latest report based on 2012 data by WHO recently. More than two million premature deaths each year were related to air pollution. Globally, seven million deaths were attributable to the joint effects of household and ambient air pollution in 2012. Air pollution has impact on most of the organs and systems of human body. Air pollutants can induce and aggravate diseases like **cardiocerebral vascular disease**, **ischemia heart disease**. Air pollution even has adverse effects on nervous system, digestive system, and urinary system. Long-term ambient air pollution exposure was reported to increase all-cause mortality. Air pollution is the cause and aggravating factor of many respiratory diseases like **chronic obstructive pulmonary disease**, **asthma**, and **lung cancer**. Struggle against air pollution seems to be a longtime task for both developed countries and developing countries, especially China.

Conditions such as chronic obstructive pulmonary disease and asthma are among the major "pre-conditions" that put us at particular risk within the **Covid-19** pandemic we are experiencing right now, as I write these lines. As are many other conditions that are mainly also a by-product of our obsession with "progress," and in particular of the so-called **Industrial Revolution**. For instance **obesity**, which is partially related to very low levels of exercise, together of course with the mass production of highly caloric food, and so on. Sitting down many hours per day, particularly just after eating, is also associated with an increased risk of **inflammation**. Within other animals, inflammation typically plays a critical role in fighting against **infectious germs**, but chronic inflammation in humans might lead to many **inflammatory diseases**, which in turn increase the risk of many other conditions, such as some cancers, **atherosclerosis**, **periodontitis**, and **hay fever**. A very brief list of long-term conditions that physicians associate with inflammation include **asthma**, **chronic peptic ulcers**, **tuberculosis**, **ulcerative colitis**, **Crohn's disease**, **sinusitis, active hepatitis, and rheumatoid arthritis**, **dermatitis**, **systemic lupus**, and **allergic reactions**. Inflammation may also contribute to a wide range of other chronic diseases, such as **type 2 diabetes** and **heart disease**, which are also extremely common, and serious. As noted in a 2017 paper by Daneshmandi and colleagues, entitled "*Adverse effects of prolonged sitting behavior on the general health of office workers*":

> The participants worked in a sitting position for an average of 6.29 hours during an 8-hour working shift. It was found that women sat longer than men (6.47 *versus* 6.07 hours/day, respectively). Our study also revealed that the participants had an average exercise time of 2.16 hours per week. The results showed that 48.8% of office workers did not feel comfortable with their workstations and 73.6% were exhausted during their working day. In addition, 6.3% of the studied workers suffered from **hypertension** and 11.2% reported **hyperlipidemia**…neck (53.5%), lower back (53.2%) and shoulder (51.6%) symptoms were the most prevalent problem among the office workers in the past 12 months. The results of our statistical analysis indicated that prolonged sitting times among office workers could have an effect on exhaustion during the working day, job satisfaction, hypertension (blood pressure above 140/90 mmHg), and symptoms in the shoulders, lower back, thighs, and knees of office workers…. Other studies have shown that reducing one's energy expenditure and the lack of localized excitation-contraction of muscles that results from a prolonged sitting position can cause **suppression of lipoprotein lipase activity**. The activity of lipoprotein lipase is critical for the attraction of triglycerides and the production of high-density **lipoprotein cholesterol**. Prolonged sitting additionally reduces **insulin secretion**, interferes with the uptake of blood glucose by skeletal muscles and may also increase **proinflammatory cytokines**, which are associated with the development and progression of many **cardiovascular disorders.**

As noted in Chap. 4, in *The Origins of Human Disease* McKeown discussed the main changes concerning health and disease during the **agricultural revolution**. About what he called as "diseases of affluence" in industrialized countries, he wrote:

> The transition to agriculture from hunting and gathering, the first of the two major changes in conditions of human life, occurred in a period of a few thousand years. By leading to the predominance of **infectious diseases** as causes of sickness and death, it had a large effect on health and a considerable effect on **population growth**. The second change, to industry, is taking place in a much shorter period, at most a few hundred years. Its effects on health and population growth are even more profound, chiefly as a result of the decline of the infections and their replacement by **non-communicable diseases**…. For almost the whole of his existence man lived in an environment that was relatively constant in its impact on health, and to which he was well adapted through natural selection. The expansion of populations under agriculture exposed him to new hazards which led to the predominance of the infections; but with this exception many of the determinants of health changed little before the nineteenth century. Most people still led an active outdoor life under conditions not too far removed from those of their nomadic ancestors. But with **industrialization** and the transfer from rural to urban life, changes have been rapid and profound, affecting not only basic influences such as food, exercise and patterns of reproduction, but also many aspects of behaviour and the environment to which people are exposed….
>
> Boyden discussed ways in which conditions of life have departed from those for which our hunter-gatherer genes have prepared us, and concluded that 'the majority of the disorders of which people complain in Western society are **disorders of civilization**, in the sense that they would have been rare or non-existent in primeval society.' This was perhaps the first clear statement of this important idea, which was later considered by others under such headings as Western diseases and diseases of affluence…. Many aspects of physical and mental health were affected by the deterioration of working and living conditions. Factories

were built without regard for the health of those who worked in them, and conditions of employment were for all practical purposes uncontrolled. If we need to be reminded that our claim to progressive social policies is of recent origin, it is only necessary to recall that a hundred and fifty years ago it was possible to exploit paupers, to use female and child labour for work in mines, to force a child of six to do manual work for fourteen hours a day for six days a week, and to expose workers to the risk of industrial disease or accident without obligation to compensate them or their dependants should they fall sick or die. The domestic environment was also bad. Back-to-back houses of poorest type were hastily erected and are still to be seen in the slum property of industrial towns.

Before the eighteenth century nearly everyone had access to clear air; it was reported recently that even in Switzerland one must now go above 6000 feet to find it. The pollution of air and water results from the discharge of domestic and **industrial effluents**. The main sources of **atmospheric pollution** are the combustion of solid fuel (coal and coke) and of petroleum products (kerosene, diesel oil and motor spirit). The use of nuclear power does not cause **atmospheric pollution**, but it leads to the discharge of waste products into the sea and introduces hazards from radiation and from the possibility of an occasional major disaster such as from the **nuclear reactor at Chernobyl** near Kiev in 1987.... [Regarding diet] Braudel tells us that the revolution in the making of bread occurred between 1750 and 1850. The practice of sifting flour to remove the bran was practised from the fourteenth century or earlier in France; but white bread was a rarity and a luxury, eaten by only about 4 per cent of the European population. From the middle of the eighteenth century, however, wheat gradually replaced other cereals in many countries, and bread was increasingly made from flour that had the bran removed.

Nevertheless the change in **bread making** occurred slowly, and it was only from about 1850 that most people in Europe had white bread. To the extent that lack of fibre contributes to the occurrence of digestive and other diseases, it has been a common influence for a little more than a hundred years. In relation to health, probably the most important change in diet was increased consumption of fat, derived chiefly from dairy products. They were not taken in the vegetarian Far East, and although milk, eggs and cheese were eaten in some European countries, because of difficulties in production and preservation, consumption was limited. We can date the increase in the amount of fat mainly to the widespread use of **dairy products** from the beginning of the present century, when **refrigeration** and **pasteurization** made it possible to handle milk and its products safely. **Sugar** is an ancient food, and was in use in India and China in the eighth century. But its extensive production began in Brazil in the sixteenth century, and from that time it was taken as a food rather than as a medicine. Nevertheless, for some time it remained a luxury, and until the eighteenth century there were large areas of Europe where it was unknown. Its widespread use is therefore quite recent, and consumption at present high levels has occurred only in this century, greatly advanced by the promotion of **refined sugar** and **processed foods**. For centuries there has been a world-wide trade in **salt**, which was essential for the preservation of meat and fish. But it had numerous other uses, in preservation of fats and vegetables (from the eighteenth century) and to make food more palatable to those who had acquired a taste for salt, a feature of both the European and the Eastern cooking traditions. It is said that in Europe about 20 gms per person was consumed daily, so that considerable amounts of salt have been taken by some people for quite a long time, although others, more deprived or perhaps more fortunate, have had little or none.

McKeown further explained:

There is also little doubt about the time when the physical demands of life were greatly reduced. At all times there were people, in what would have been regarded as privileged positions, who limited their movements, generally by relying on others to do their work for them. In both urban and rural life physical demands were reduced by mechanization from the eighteenth century, but the widespread reduction has resulted from the **introduction of**

the automobile since 1900.... What are now regarded as **illicit drugs** have been available since ancient times, but there has undoubtedly been a large increase in their production and distribution in the last three centuries. **Opium**, for example, was often consumed in the West, particularly in Turkey, and in the East it spread from India to China and the East Indies. According to Braudel, 'the great turning point came about 1765, just after the conquest of Bengal, when a monopoly of poppy fields was established to the advantage of the East India Company.' There has been an even greater increase in the present century, and illicit drugs are now produced and distributed widely throughout the world.... [About **cancer**] on the basis of data from 28 mainly industrialized countries, the World Health Organization reported that from 1960 to 1980 male deaths from cancer (all types) increased by 55 per cent, or 40 per cent when a correction is made for the increasing age of the populations. In the same period mortality from cancer of the lung (age corrected) increased by 78 per cent for males and 80 per cent for females. There was also an increase in breast cancer (by 43 per cent), and indeed in all the other types examined with the exception of cervical and stomach cancers, which showed relatively small but consistent decreases. This change is particularly significant in Japan, where **stomach cancer** is the most common form of the disease. In recognition of the environmental and behavioural origins of cancer, and of the appearance of those influences in the Third World, the World Health Organization predicted that 'there will be an epidemic of cancer in the majority of the developing world by the year 2000'. In this context we should also note the remarkable differences in disease death rates in different countries. For example, age-adjusted death rates of breast and **colon cancer** in many parts of the world are less than one fifth of the rates in the United States. Willett and MacMahon concluded that these differences cannot be accounted for by variation in genetic predisposition.

[Another piece of evidence showing that there is a mismatch about our way of life and the biology of our body concerns the] change of environment. A racial group which has changed its environment and associated ways of life exhibits the disease pattern of the population with which it shares its environment rather than that of the population with which it shares its genes. This experience has been particularly striking in cancer, whose frequency differs widely by type and country. Japanese migrants in Hawaii, for example, had the high stomach cancer rates seen in Japan, but the rates had fallen in their children. There was a sharp increase in the risk of **breast cancer** for Japanese women living in the San Francisco Bay Area. For cancers of the stomach, intestinal tract and lung, rates for Polish migrants to the United Kingdom were intermediate between the levels of people living in Poland and England and Wales... Perhaps the most persuasive evidence that the non-communicable diseases are essentially new is the observation that they are rare in populations which have retained their traditional way of life, but begin to appear when they change to the western lifestyle. In *Western Diseases* Trowell and Burkitt brought together reports by 34 contributors who described their experience of changes in the pattern of disease in several countries as westernization occurs. There are four main lines of evidence, not all equally secure, (a) Until recently many of the non-communicable diseases now predominant in the West were uncommon or absent in **hunter-gatherers** and **peasant agriculturalists**, (b) When these populations change from their traditional ways of life to those of developed countries, they begin to exhibit the western pattern of disease, (c) The incidence of some of the diseases has declined in western populations which have reversed certain features of their lifestyle to bring it closer to that of peasant agriculturalists, (d) Of the multiple influences responsible for the western pattern of disease, Trowell and Burkitt considered that dietary changes are probably the most important. The evidence assembled on the first two points is impressive. Before 1940, in Africans of Kenya and Uganda, blood pressure did not rise with age and essential hypertension was rarely seen; it is now a common disease. Obesity was almost unknown in 1930...today 'the towns of East Africa contain many fat upper class Africans and their leaders seen on television are often grossly obese'..

> In Kenya in the 1930s, diabetes was rare in Africans but not in Europeans and Indians; there are now large diabetic clinics in all town hospitals. **Cerebrovascular disease** was the first **arterial disease** of clinical significance to emerge in Africans. Before 1948 a case due to hypertension was rarely or never seen; in 1970 it was the commonest cause of death in a large series of neurological patients in a Ugandan hospital. Coronary artery disease is the last major cardiovascular western disease to appear – the first clinical reports of cases were made quite recently in Uganda (1956) and in Kenya and Tanzania (1968). And it is said that '**coronary thrombosis** has begun only recently to emerge in Zimbabwe Africans and angina remains a rare disease'. On the basis of observations of this kind from many parts of the world, Trowell and Burkitt prepared a provisional list of Western diseases. In addition to those already mentioned, it includes **gallstones**, **varicose veins**, **constipation**, **appendicitis**, **diverticular disease**, **haemorrhoids**, **cancers of the bowel**, chest and lung and **dental caries**. While there may be differences of opinion about the acceptability of some of the conditions as Western diseases, the general conclusion that they are appearing in developing countries where formerly they were rare is not in doubt.

I was particularly impressed by the obesity levels that a substantial proportion of people had when I went to various small cities of Utah. I could not avoid to compare these obesity levels with the scientific/historical descriptions of the in general much healthier day-a-day lives of the **Native Americans** that were living in those regions centuries ago before the arrival of Europeans. Of course, one can argue that the people that live in small cities of Utah nowadays tend to have a greater **life expectancy** than did those Native Americans centuries ago. However, one should not forget that life expectancy in non-industrial societies was often substantially decreased by high rates of child mortality: as will be discussed below, we now know that in many of those societies if people did not die until they were 15 or 20 years old, they could then have very healthy, long lives, until 70, 80, or even older ages, often with a general quality of life that was much higher on average than that of people living in many small cities of Utah nowadays. Clearly, on average, adult Native Americans had lower levels of obesity, higher levels of sexual activity, fewer heart and mental diseases, a higher connection with nature and level of physical activity, a more egalitarian distribution of goods, a less stressful type of life, and a much more healthy, fresh, and less processed food, as powerfully shown in Fig. 8.1.

So, when I walked in those small cities, I wondered: is this really all that we humans can do with all the "revolutions" we had and the fascinating scientific discoveries that were done in our history, such as the invention of electric light, cars, air conditioning, TVs, phones, cell phones, and so on? Is this all we can achieve? Those cities were always mainly deserted, at any moment of the day, basically because people tend to remain inside buildings and, in the very brief moments they need to move from building to building to eat, go to doctors, or to work, they often go by car, even when the distances separating those buildings are minimal. To tell the truth, many of those cities are not very appealing so one does not feel like walking there anyway, as they mostly consist of a large, central street which is actually part of the main road that connects them, and often have no main squares or fountains, or beautiful parks to walk, nor mindful places to sit down and read in the street

Fig. 8.1 The paradox of "progress"—case 1: one picture truly has more value than a thousand words. Work "Daily Bread", by photographer Gregg Segal: he asked kids from different regions/cultures to make detailed lists about all the food items they consumed during a week, and in the end of the week he photographed them around all the items they listed, to show differences between "industrialized" and "non-industrialized" societies. On the left: Kawakanih Yawalapiti, 9 years old indigenous girl, Alto Xingu, Mato Grosso, Brazil. On the center: Henrico Valia, 9 years old boy, Brasília city, DF, Brazil. On the right: Ademilson dos Santos, 10 years old boy from the Quilombola Kalunga community of Vão de Almas, Goiás, Brazil

(see Fig. 8.2). Moreover, many so-called restaurants are food chains, such as MacDonald's, Subway, and so on, and therefore the food served there consists of high-caloric items, without natural fresh juices, and so on. Is this really the result of "progress," is this truly a "more fulfilling" or healthy—"physically" and "mentally"—life than the one many Native Americans had centuries ago in those very same regions?

If one would want to go deeper with such comparisons, to really understand this in a broader biological and medical context, one could expand them to include not only **non-industrialized societies** versus industrialized ones, but also nomadic versus sedentary hunter-gatherer societies as explained above, as well as comparisons between the health of individual groups within each of those types of human societies versus those of domesticated *and* also of non-domesticated animals. For instance, McKeown's *The Origins of Human Disease* points out that the type of **non-communicable diseases** that have been particularly prevalent in humans since the industrial revolution, such as diabetes, hypertension, and cardiovascular diseases, also occur to some extent in domesticated animals but are very rare in wild non-human animals:

> To throw light on the problems of **human disease**...what is needed for other animals is evidence, particularly quantitative evidence, of their disease experience when living in their natural habitats. Unfortunately this information is rarely available. The observations that have been made are chiefly on animals domesticated, in zoos or, if free living, affected in various ways by the man-made environment; disease reports on **wild animals** are usually

Fig. 8.2 The paradox of "progress"—case 2: a typical, rather uninteresting, small city in Utah, Ogden: like many others, with a lot of restaurants and food chains, where cars can be easily seen while almost no people is seen walking on the street, and where there are no beautiful places to seat or read near fountains or picturesque squares. Such small towns illustrate the point that the intention of those that built and/or "live"—perhaps better said, that exist—in such towns is more to "survive," that is to reproduce, eat and "work," and obviously to "consume," rather than to promote a healthy, mindful environment with pleasant green areas to walk or cultural events to enjoy, which are indeed so often neglected in most of such small U.S. cities—and even bigger ones, just think about most of the areas of a city such as LA

based on single or few specimens and without knowledge of the size of the populations from which they are drawn. But although it is not possible to make estimates of disease frequency or to assess accurately the relative importance of different causes of death, some general conclusions have been drawn. First, many ecologists are agreed that 'in populations of wild animals, disease is of rather *secondary importance* in population control, and only becomes of significance when population pressure becomes so heavy that the animals lose their natural resistance to **parasites**.' Second, the disease which occurs is predominantly due to infection caused by parasites; indeed symposia on disease in free-living wild animals have been devoted almost exclusively to **infectious diseases**. Third, although **non-communicable diseases** do occur in animals affected by the man-made environment, in the wild they are very uncommon, particularly in **primates**. Because of its bearing on the problems of human health, the evidence on which the last conclusion is based is of great interest, particularly in relation to the most common causes of death in developed countries, **cardiovascular disease** and **cancer**. There have been scattered reports of **sclerotic vascular lesions** in wild animals, but only the pig, the whale and possibly some nonhuman primates have been observed to develop a 'spontaneously' progressive disease which resembles **human atherosclerosis**. In primates, systematic studies have been made in baboons, rhesus monkeys and squirrel monkeys. They suggest that **hypertension** and atherosclerotic changes (not affecting the cerebral or cardiac vessels) are sometimes seen in monkeys used in experiments and kept in regular cages over long periods; they are rarely seen in animals kept in open-air cages and 'there is no direct reference in the literature to the possibility of atherosclerosis developing in monkeys in their natural habitats'. The complications of atherosclerosis – **thrombosis**, **aneurysm** and **myocardial infarction** – are very uncommon, and only five cases of **coronary occlusion** caused by **infarcts** have been reported for **non-human primates**. From their extensive study of the pathology of **Rhesus monkeys**, Lapin and Yakovleva concluded that 'coronary insufficiency is rightly considered a disease seen only in man'.

So far as it goes, the evidence suggests that **spontaneous tumours** of free-living wild animals are also rare. Knowledge of tumours, as indeed of most diseases of wild animals, is derived from observations on animals kept in the unnatural environments of breeding farms and zoos. Even under such conditions tumours, both benign and malignant, are very uncommon in primates and they seldom metastasize. Fiennes noted that before 1972 'only some 200 spontaneous tumours had been described in the literature, in spite of the large numbers of monkeys and apes that have been kept in zoos and research establishments'. Moreover, in other mammals, unlike man, the tumours which do occur are believed to be mainly viral in origin. The same is said to be true of **plant cancers**. Writing more generally of non-communicable diseases in primates, Lapin and Yakovleva concluded that 'the diseases that are most common in man, i.e. **malignant neoplasms**, **rheumatism**, cardiovascular diseases, they either occurred seldom or not at all in monkeys'. At first sight the conclusion that disease is of secondary importance in population control appears to be inconsistent with many well authenticated examples of high mortality from infectious diseases of wild animals. In his review of the literature Holmes noted several references to large-scale epizootics of disease. In waterfowl, among the most intensively studied vertebrates, some were caused by viruses, such as duck viral enteritis in South Dakota where 40,000 mallards died; others were due to bacteria, such as **avian cholera** in Missouri, where 1100 snow geese died in a single night, or **botulism** which killed 4-5 million ducks in 1952; still others were from fungi, blood protozoa or nematode infections. Many more examples could be cited in vertebrates, such as the dramatic panzootic of rinderpest that swept through Africa and killed vast numbers of cattle and wild ungulates. In spite of this evidence, many population biologists are not convinced that infection plays a large part in the regulation of animal numbers, particularly in vertebrates. Although it is sometimes a proximate cause of death, it appears to be a consequence of population size rather than a determinant of it; parasites are generally rare in low density populations.

One particularly interesting comparison concerns rates of suicide: although it is difficult to have accurate data about mental health and **suicides** among human hunter-gatherer groups that were/are not affected directly or indirectly by industrialized societies, the reality is that within the huge number of anthropological works covering many dozens of thousands of hours spent within several groups one very rarely reads about cases of suicide. As noted in Bering's 2018 book *Suicidal*, this does not mean that there are no cases of suicides in hunter-gatherer societies, but they do seem to be much less frequent, on average, than they are in most "developed" countries. For instance, in a 2013 study by Long and colleagues they stated: "the **Mla Bri** are a small group of **nomadic hunter-gatherers** living in northern Thailand who since the 1990s have begun to settle in semi-permanent villages…settlement has resulted in improved health conditions in terms of malaria eradication, infant mortality, and other common indicators…however, there have also been suicides in a society which had apparently previously experienced none…[likely] associated with the rapid social change the group has experienced during the transition from nomadic hunter-gatherers to semi-settled status."

Importantly, **suicides** are not only more prevalent in “civilized” countries than they are in hunter-gatherer societies: within the former, “suicide rates are higher in developed nations than in less prosperous ones,” as recognized by Bering in *Suicidal*. As noted above, and discussed more in detail below, the dramatic increase of sleep deprivation might be related with the dramatic increase of suicides in such societies. This is one more piece of evidence that can seem to be surprising for those that believe in tales of “progress” but that, when we take into account the empirical facts discussed so far in the present book, is actually not surprising at all. Malesevic’s *The Rise of Organized Brutality* provides another remarkable piece of evidence, which is often neglected by the media and broader public as suicide continues to be a huge, overlooked taboo within “developed countries.” Namely, in such countries there are more people that kill themselves than people that are killed by others:

> According to all available statistics, global **suicide rates** have increased by 60 per cent over the past forty-five years, with more than 1 million individuals dying as a result of suicide each year. The suicide rates in most countries are substantially higher than **rates of homicide**, with the global suicide rate at 16 per 100,000 people. Furthermore, in most societies male suicide rates are often five or more times higher than male homicide rates: Lithuania 70.1 compared to 7 (in 2004); Guyana 70.8 compared to 17 (in 2012), South Korea 41.7 compared to 0.9 (in 2012) or the United States 20.7 compared to 4. Although suicide is conventionally excluded from statistics on **violent deaths**, there is no doubt that this extreme form of self-harm is a product of **changing social conditions**. Since Durkheim’s (1952) early studies on altruistic, anomic and egoistic suicide, sociologists have become well aware that organisational and ideological pressure combined with the sense of emotional and moral responsibility towards one’s family and friends has significant impact on suicide rates. Just as with homicides and war casualties, suicide too is first and foremost a social phenomenon. In this context, the continuous rise in suicide rates can be interpreted as a product of organisational and ideological development where this type of violence increases at the expense of homicides and other forms of violent action.

8.2 “Noble Savages,” Technological Development, and Loneliness

> *Bacteria are not evil because they make us sick, they are highly successful organisms, spreading themselves far & wide…earthquakes that kill people are not, in and of themselves, evil, they are just a natural event, without purpose or intention. (Michael Shermer)*

In *The Science of Good and Evil*, Shermer notes that humans can be both “good” and “bad” depending on the *circumstances*. After all, we are all the product of a chaotic, random, contingent evolutionary process, so our behaviors depend a lot of the specific conditions, and for humans particularly the type of social organizations within which we live. Although Shermer mostly refers to examples about sedentary hunter-gatherer groups such as the **Hohokam of southern Arizona** and tends to neglect the huge societal differences between those groups and nomadic hunter-gatherer ones—as so many authors unfortunately tend to do, see Chap. 4—some of the points raised by him about the enormous importance of specific circumstances are worthy of note:

In 1670, the British poet **John Dryden** penned this expression of humans in a state of nature: "I am as free as Nature first made man/When wild in woods the noble savage ran." A century later, in 1755, the French philosopher **Jean-Jacques Rousseau** canonized the **noble savage** into Western culture by proclaiming, "nothing can be more gentle than him in his primitive state, when placed by nature at an equal distance from the stupidity of brutes and the pernicious good sense of civilized man". From the **Disneyfication of *Pocahontas*** to Kevin Costner's ecopacifist Native Americans in *Dances with Wolves*, and from post-modern accusations of corrupting modernity to modern anthropological theories that indigenous people's wars are just ritualized games, the noble savage remains one of the last epic creation myths of our time. Within this myth lies the antithesis of the myth of pure evil, and that is the myth of pure good. The latter is just as detrimental toward a deeper understanding of human moral nature as is the former. The evidence from all the human sciences overwhelmingly supports the view that humans are good and bad, cooperative and competitive, selfish and altruistic. The potential for the expression of both moral and immoral behavior is built into human nature. How, when, and where such behaviors are expressed depends on a host of variables. But the myth of the noble savage extends far beyond what Rousseau envisioned and is still embraced today by many scientists, academics, and social commentators, in what I call the **Beautiful People Myth** (BPM).

In a fascinating 1996 study, for example, University of Michigan ecologist Bobbi Low used the data from the Standard Cross-Cultural Sample to test the hypothesis that we can solve our ecological problems by returning to the mythological Beautiful People's attitudes of reverence for (rather than exploitation of) the natural world, and by opting for long-term group-oriented values (rather than short-term individual values). Her analysis of 186 hunting-fishing-gathering (HFG) societies around the world showed that their use of the environment is driven by ecological constraints and not by attitudes, such as **sacred prohibitions**, and that their relatively low environmental impact is the result of low population density, inefficient technology, and the lack of profitable markets, not from conscious efforts at conservation. Low also showed that in 32 percent of HFG societies, not only were they not practicing conservation, environmental degradation was severe; again, it was limited only by the time and technology to finish the job of destruction and extinction. Extending the analysis of the BPM to other areas of human culture, UCLA anthropologist Robert Edgerton surveyed the anthropological record and found clear evidence of **drug addiction**, **abuse of women** and children, **bodily mutilation**, **economic exploitation** of the group by political leaders, suicide, and mental illness in indigenous preindustrial peoples, groups not contaminated by Western values (allegedly the source of such "sick" behavior).

Anthropologist Shepard Krech analyzed a number of Native American communities, such as the **Hohokam of southern Arizona**, and discovered that a large-scale irrigation program led to the salinization and exhaustion of the Gila and Salt River valleys, ultimately triggering the collapse of their society. Krech says that even the reverence for big game animals we have been led to believe was ingrained into the world-view of America's indigenous peoples is a myth. Many, if not most, Native Americans believed that common game animals such as elk, deer, caribou, beaver, and especially buffalo are replenished through **divine physical reincarnation**. Game populations bounced back after successful hunts not because Native Americans made it happen through ecological veneration, but because they believed the gods willed it. Given the opportunity to overhunt big game animals, Native Americans were only too willing to do so. One of the most poignant examples of this is the famous "Head-Smashed-In" buffalo kill site in southern Alberta, Canada. I had an opportunity to visit **Head-Smashed-In** (the name alone belies the Noble Indian myth). It is a most dramatic site. Standing on the edge of the cliff, one looks down upon a thirty-foot-thick deposit of buffalo bones that reflects five thousand years of Native American mass hunting. Looking back away from the cliff, one sees a vast and expansive V-shaped valley in which the hunters ambushed and drove their game for tens of miles. The terrain is on a slight decline toward the cliff, so these massive animals built up so much speed that upon reaching the cliff they were unable to stop themselves. They tumbled over, one after another, until there were so many carcasses that most were left unused. Buffalo populations were ultimately stable not because of a Native American conservation ethic, but because they simply did not have the numbers and technology to drive these big game animals into extinction. Other species were not so fortunate. The evidence is now overwhelming that woolly mammoths, giant mastodons, ground sloths, one-ton armadillo-like glyptodonts, bear-sized beavers, and beefy saber-toothed cats, not to mention American lions, cheetahs, camels, horses, and "many other large mammals, all went extinct at the same time that Native Americans first populated the continent in the mass migration from Asia some 15,000 to 20,000 years ago. The best theory to date as to what happened to these mammals is that they were overhunted into extinction.

Shermer then provides a broader discussion on topics such as what is "noble," "good" and "evil" and notes that people which would otherwise probably have lived completely "normal"—and perhaps even so-called noble—lives often easily became "monsters" within a specific circumstance—the so-called banality of evil:

The real-world example of **Nazi Germany**, in some sense, serves as a historical experiment. At the **Nuremberg Trials** following the war, psychologist G. M. Gilbert was assigned to study the men imprisoned for committing these "crimes against humanity". He discovered that not only were the perpetrators well-cultured and highly educated, they tested out two to three standard deviations above average on a **standardized intelligence test** used at the time, the **Wechsler Adult Intelligence Scale**. Where the average IQ is 100, Reichs-Commissioner Seyss-Inquart tested at 141, Reichsmarschall Hermann Goring at 138, Reich Chancellor Franz Von Papen at 134, Poland Governor-General Dr. Hans Frank at 130, Foreign Minister Joachim von Ribbentrop at 129, Hitler's architect and Reichsminister of Armaments Albert Speer at 128. The prison psychiatrist Douglas Kelley, after his evaluations, offered this observation about the moral character of the leading Nazis: "As far as the leaders go, the Hitlers and the Görings, the Goebbels and all the rest of them were not special types". Their personality patterns indicate that, while they are not socially desirable individuals, their like could very easily be found in America. **Neurotic individuals** like **Adolf Hitler**, suffering from **hysterical disorders** and **obsessive complaints**, can be found in any psychiatric clinic. And there are countless hundreds of similar ones, thwarted, discouraged, determined to do great deeds, roaming the streets of any American city at this very moment. No, the Nazi leaders were not spectacular types, not personalities such as appear only once in a century. They simply had three quite unremarkable characteristics in common – and the opportunity to seize power. These three characteristics were: "overweening ambition, low ethical standards, and a strongly developed nationalism which justified anything done in the name of Germandom."

This leads us to a brief discussion about "goodness," "rightness," "usefulness," and "truth." For instance, when we say that a certain way of life is more connected to "nature" or to what is "natural," does this necessarily means this way of life is "better," or "good," or "right," or "truthful?" I shave once every week and wash my teeth with an industrialized toothpaste more than once a day: chimpanzees and gorillas surely do not do this, nor did our ancestors within the first millions of years since we split from chimpanzees. However, I think that an important point needs to be made: I do *not* wash my teeth with an industrialized toothpaste to find or accomplish a supposed "cosmic purpose of life," nor to conform to any cosmic teleological narrative, particular a narrative that says that I have to "fight against my bodily desires." I do it simply because empirical studies have shown that by doing so people will, in average, be more likely to reduce dental problems that otherwise can lead to pain, infections, and so on. In fact, I know how important is to do so precisely because I know that there was a major change in the last centuries that makes that things are much different from what they "naturally" were back then: we now consume much more sugar, and thus are much more prone to have such dental problems. Moreover, cleaning my teeth is mainly neutral, in the sense that it does not hurt others—unless the production of the toothpaste involves animal abuse, but as far as a I know this does not apply to the brand I use. This contrasts, for instance, with what happens when people impose by force monogamy to others, by using teleological narratives, what has led to the subjugation of women, burning of people, and so on.

We can take these ideas to another level, and compare, in a direct, empirical way, the way of life of some nomadic hunter-gatherer groups and some typical patterns of the lifestyle of people that live in a city that is seen, by many people, as the technologically most "developed" town in the planet nowadays, Tokyo (see Fig. 9.6). Of course, there are all kinds of people living in Tokyo, with very different lives, as there is also a huge variation within the behavior of individuals of even a single nomadic hunter-gatherer group. Furthermore, some of the behavioral features that are more commonly seen among Japanese people are specific to that country, and not exclusively the consequence of technological "development": this applies, for instance, to the historical importance given, in a generalized way, to shame and honor in Japan. However, having said this, as a rough and very simplified approximation that needs to be done in order to undertake such type of comparisons—because otherwise one would never be able to compare two groups of humans—and to make them less vague and more objective, we will focus here specifically on very specific and well studied cases concerning Tokyo for which we have specific empirical numbers. For instance, several studies have stressed the very high *average rates* of depression and suicides among people in Japan and in Tokyo specifically, as well as the huge amount of hours dedicated, on *average*, to paid work in this city. Some of these studies were discussed in detail in October 2017, when the media extensively covered the case of **Miwa Sado**, a young Japanese woman that died because she worked too much and slept too little. The extensive media coverage was precisely due to the fact that this was not at all a random event, but instead something that confirms the existence of a huge number of "**workaholics**" in Japan's biggest cities. In the study shown in Fig. 8.8 including 33 countries from different continents, Japan—which is not shown in the figure—was indeed the one in which people spent, in *average*, more time per day in paid work: 326 minutes, that is approximately 5.5 hours. As explained by McCurry in a 2017 article in *The Guardian* about the tragic story of Miwa Sado, there is even a term commonly used in Japan for **death from overwork**, "**karoshi**":

> **Japan** has again been forced to confront its **work culture** after labour inspectors ruled that the death of a 31-year-old journalist at the country's public broadcaster, NHK, had been caused by overwork. Miwa Sado, who worked at the broadcaster's headquarters in Tokyo, logged 159 hours of overtime and took only two days off in the month leading up to her

death from heart failure in July 2013. A labour standards office in Tokyo later attributed her death to karoshi (death from overwork) but her case was only made public by her former employer this week. Sado's death is expected to increase pressure on Japanese authorities to address the large number of deaths attributed to the punishingly long hours expected of many employees. The announcement comes a year after a similar ruling over the death of a young employee at Dentsu advertising agency prompted a national debate over Japan's attitude to work-life balance and calls to limit overtime. Matsuri Takahashi was 24 when she killed herself in April 2015. Labour standards officials ruled that her death had been caused by stress brought on by long working hours. Takahashi had been working more than a 100 hours' overtime in the months before her death. Weeks before she died on Christmas Day 2015, she posted on social media: "I want to die". Another message read: "I'm physically and mentally shattered". Her case triggered a national debate about Japan's work practices and forced the prime minister, Shinzō Abe, to address a **workplace culture** that often forces employees to put in long hours to demonstrate their dedication, even if there is little evidence that it improves productivity…. More than 2,000 Japanese killed themselves due to work-related stress in the year to March 2016, according to the government, while dozens of other victims died from **heart attacks**, **strokes** and other conditions brought on by spending too much time at work. According to the white paper, 22.7% of companies polled between December 2015 and January 2016 said some of their employees logged more than 80 hours of overtime each month – the level at which working hours start to pose a serious risk to health. Research shows that Japanese employees work significantly longer hours than their counterparts in the US, Britain and other developed countries. Japan's employees used, on average, only 8.8 days of their annual leave in 2015, less than half their allowance, according to the health ministry. That compares with 100% in Hong Kong and 78% in Singapore.

If we take a step back and think carefully about this issue, that someone can die "from overwork" is not only unheard of, but also almost impossible to be understood and probably even imagined—and rightly so—for a small nomadic hunter-gatherer group. This also applies to one of the other major problems that we face because of this "modern" obsession about "progress," "working," and "producing": **sleep deprivation**. As explained in Matthew Walker's fascinating 2017 book *Why We Sleep*, this is one of the most neglected problems of our "developed" societies, because **sleep** is related to a plethora of hugely important health items such as restocking our **immune system**, fine-tuning our **metabolism**, regulating our **appetite**, improving learning, mood, and energy levels, regulating **hormones**, helping prevent **cancer**, **Alzheimer's**, and **diabetes**, slowing the effects of **aging**, increasing **longevity**, and so on. As noted by him, apart from the enormous price we pay at an individual level because of sleep deprivation, we also pay huge societal prices. For instance, hundreds of thousands of people die every year because they fall asleep in cars: actually more people die because of this than because people drive under the effect of alcohol or drugs. The question is, thus: why has nothing been done about this for such a long time? As pointed out by Walker, one of the main reasons for this concerns the over simplistic teleological ***adaptationist way of thinking*** that is often used to discuss topics concerning evolution and medicine (see Chap. 6).

Namely, scholars and doctors were unable to properly study and understand the evolutionary origin and physiology of sleep for such a long time—and we still know too little about these topics—because they were focusing almost exclusively on a single "purpose" of sleep, while sleep is related to the well-being of a large number of body organs and physiological processes. A second major reason has to do with the *Neverland* ***dualistic separation of mind*** **versus** ***body*** that is so prominent in Western countries—**Descartes** being a major example of it—and of ***emotion*** **versus** ***rationality***. As noted above dualism is still palpable in many fields of knowledge, despite the fact that it has been contradicted by various empirical studies in the last decades, including precisely studies about sleep. As explained by Walker, not only something such as sleeping—or the lack of it—has a huge impact in numerous body organs, but also **dreaming** per se mollifies painful memories and creates a virtual reality space in which the brain melds past and present knowledge to inspire creativity. That is, "mind" and body, reality and unreality, rationality and irrationality, all mixed in one, when we sleep and dream. In this sense, it is fascinating to see that the *world of unreality* seems to be evolutionarily useful not only for humans to cope with the anxiety created by an inevitable death but also for other reasons within the huge number of animals that sleep and dream, including ourselves, as many of them spend even more than 8 hours a day unawake in such a type of "surreal" mixed world of reality and unreality.

Well, in reality the notion that humans spend in average 8 h per day sleeping is nowadays itself a *Neverland* tale, because in most "developed countries" a huge number of people spends much less than 1/3 of the day sleeping, due to our obsession with "progress." As astutely put by Walker, "unfortunately human beings are the only species that voluntarily deprives itself from sleep despite the fact that *this does not represent any benefit.*" Yes, there is *no* benefit at all for *our* health and well-being, well on the contrary, but still this pattern has become the trend and is now a troubling societal problem within "developed" countries. Apart from our longstanding obsession with *Neverland* tales about cosmic progress, this disturbing trend is also due to the fact that political and religious leaders, CEOs and the 1% are particularly efficient in using those tales for their own **political or economic gains**. By using such tales to make self-domesticated humans sleep even fewer hours there will be more "work," more "productivity," and thus the more money the "system," and consequently the leaders, CEOs and the 1%, will make, even if that will imply the burnout, exhaustion, depression, and even deaths due to overwork or suicide of a huge number of people: that is not their problem. In fact, as put by Walker, one of the desires of many within the "system," such as those in the U.S. military, is that one day they can reduce the amount of sleep of self-domesticated humans even to 0 hours per day for at least some periods of time: for instance, soldiers during battles. That is why the U.S. government spends so much money every year to finance studies about sleep deprivation in non-human animals that are able to dramatically reduce the amount of sleep for several weeks, such as the **crowned white sparrow**. This is a clear reminder that "civilization" has involved, and continues to involve, having a huge number of people working, or fighting, or giving their whole lives, for the dreams of others.

This is also why the prevailing socio-economic "system" has truly done nothing about the huge number of deaths caused by sleep deprived drivers, while "developed" countries have laws against driving under the influence of alcohol and drugs, despite the fact that the latter is often less deadly than driving in a sleep-deprived mode. It simply has to do with productivity: who cares that a few millions of people die because of that, if the remaining billions of people that sleep less work much more and thus are more productive for the "system?" "Developed" countries are particularly tuff against what they call **"illicit" drugs** and **alcoholism**—and I am not saying this should not be the case, as they can also cause a plethora of horrible health problems—because drug-addicts and alcoholics tend to be among the less productive people. In fact, in most textbooks and websites, one of the most commonly emphasized problems of **alcohol use disorder** is that it can cause harm in social and professional aspects of life as there could be failure to complete major obligations and responsibilities at work, school, or home due to repeated alcohol use. So, while there are also now many pharmaceutical approaches that aim to *decrease* the desire to consume alcohol, the "system" and the pharmaceutical companies are *increasingly* investing money and resources to *create more and more* "licit" drugs, food and drink products, and pills that allow the self-domesticated humans to *work a lot more* despite being sleep-deprived. These include, among others, drinks with 2, 3 or even more shots of coffee, energetic drinks, and medications to make people feel "more alert," "less tired," and "less sleepy," so they could *produce more*. Indeed, as pointed out by Walker, the most consumed "licit" drug in "civilized" countries is caffeine, being the only one "that we offer without hesitation to our children and teenagers," with negative consequences when this is done in excess and/or when this is done to make them sleep less, do more "work" or "homework."

As Walker points out, empirical studies in teenagers have demonstrated a clear link between sleep deprivation and suicidal ideas, suicide attempts and, tragically, completed suicides: another huge price our societies pay for our obsession with "work" and productivity. This topic about "licit" drugs such as coffee and tobacco, "illicit" drugs, and their link with productivity is astutely discussed by **Pier Vincenzo Piazza** in his book *Homo Biologicus*. The **self-domesticated *Homo servus*** can't stop, he/she has to produce, to follow the norms, to obey his "masters," exactly as domesticated cows can't stop producing more and more milk and meat for the human "masters" that domesticated them. With a crucial difference, tough: domesticated cows are obliged by humans to do so, while most humans do this because they self-domesticate themselves by believing *Neverland* imaginary stories made and/or used by others (see Fig. 9.6). In this sense, within this "system," what scientists such as Walker are trying to do—such as "trying to get doctors to 'prescribe' hours of sleep to patients" as the easier, cheaper, and most effective way of fighting the huge individual and societal problems created by sleep deprivation—will surely be seen as an heresy by the "masters," leaders, CEOs, military commanders, and pharmaceutics companies.

One crucial clarification needs to be made here: when I talk about the "system," I do not mean a group of people that conspire to control the world, and keep the *status quo*, the top 1%, and so on. This book is about deconstructing imaginary stories, including conspiracy theories, so clearly I would never subscribe to conspiracy theories such as those concerning the ***Illuminati***, which as noted above are fictional tales about people that conspire to control world affairs by masterminding events and planting agents in government and corporations, and so on. Such conspiracy theories are as absurd as—and actually are in some way included in some of—the **QAnon conspiracy theories**. So, I am instead referring to a highly complex, interactive socio-political system that, because of its intrinsic characteristics, tends to perpetuate itself. Obviously, some people that were not part of the 1% when they were younger, such as Bill Gates, became part of the 1% during their lives: something that actually shows that theories about the ***Illuminati*** make no sense. However, when you get to the 1%, or simply become a millionaire, it is much easier to keep being so, than to become part of the middle class or poor again. A clear example, among endless others, of how the system tends to favor those with more money concerns the housing market. If you are rich, and you buy a house paying upfront, you do not need to spend a huge amount of money paying interest to the bank in a mortgage.

However, those that are particularly poor to start with, and have much less money to pay upfront, will have a relatively high interest, that is, in the long term instead of paying only X money to buy the house as the very rich, they need to pay $X + Y$, the Y referring to the interest and being usually a substantial amount of money. This system was not created by a group of 2 or 3 heartless *Illuminati* that met together and decided to be "bad" and "explore the poor of the world." Instead, the system created itself, slowly, and now almost everybody in "modern civilization" is acculturated to see such a system as normal, while in fact if we truly think deeply about it, it is a very obscene and unfair system in which the very rich pay much less than the very poor do for exactly the same item. Or, as put by **Robert Frost**, "*a bank is a place where they lend you an umbrella in fair weather and ask for it back when it begins to rain.*" In a nutshell, this is thus what I mean by "system": something that often includes such absurd rules, laws or costumes and that was *not* master-planned, being rather the product of a combination of specific temporal and local circumstances and chaotic, random and contingent events, but that starts to be seen as "normal" and thus perpetuated via teleological narratives, fictional tales, acculturation, social norms, and ultimately even written laws.

Another topic that has been the subject of much discussion in recent years in "highly developed" countries such as Japan, and in particular within people living in big cities such as Tokyo, concerns the fact that those people are doing, less and less, another basic human pleasure that is also related to a plethora of health benefits: **having sex**. As noted in Chap. 5, Japan has one of the **lowest rates of sexual activity** in the planet, and there is an increasing number of people that join groups that are totally "against" or "disgusted by" sexual intercourse and even sexual desire per se, arguing for instance that sex is repulsive or a waste of time. Indeed, "**sex disgust syndrome**" is another term now being commonly used in Japan, even among couples. Of course, one individual or a few persons being repulsed by sex can occur sporadically anywhere but, once again, it would be unthinkable to hear this term being so *commonly* employed within a small nomadic hunter-gatherer group, or in any non-human sexual organisms for that matter, because sex is obviously one of the key pillars within the evolution of such organisms. Disturbingly, while so many Japanese people from big cities argue that having sex, or even dating, or socializing in general, is a "waste of time," a huge number of them spend an enormous amount of time seeing **porno movies** or playing "**virtual sex**" games and other similar media, which are growing exponentially in Japan (see also Box 9.4 and Figs. 9.11 and 9.12, for further examples of "modern" virtual reality). That is, this is not *only* a question of "lack of time"—which is itself a very strange term that only makes sense in a culture of "work" and "progress," because obviously every single day has basically the same time since we split from chimpanzees millions of years ago: so, one cannot "gain time" or "lose time" or "waste time," time passes by continuously no matter what we do.

Instead, this is also a matter of preferences imposed to the *Homo socialis* following teleological tales about "progress" and productivity and about "resisting" and "fighting" against anything that remind us that we are animals with decaying bodies. The crucial role played by these tales and our deep immersion in *Neverland*'s world of unreality, illusion and delusion was tragically shown by what happened to a South Korean couple that met online and then married and had a sickly premature baby: they decided to continue to spend dedicating a huge amount of time to their **online lives** in a café near their apartment, playing a game in which they raised a **virtual baby**, letting their own flesh-and-blood baby to die of severe dehydration and malnutrition (see Box 9.4). They had no time to take care of their own *real* child because they *preferred* to take care of their *Neverland*'s virtual baby.

Unfortunately, this is not all: there are more, many more, disturbing facts within the *world of unreality* of "highly-developed" countries. For instance, Tokyo is also particularly reputed for the remarkably small size of an enormous number of apartments in which individuals, couples, or even whole families live due to the very high population density within the city. I have seen this by myself, in a very unpleasant way, when I went some years ago to Tokyo to visit a friend and unwisely—and unknowingly—my girlfriend and me accepted to stay at his so-called apartment. It was too late when we discovered that the whole "apartment" where the three of us were supposed to sleep, eat, rest, and socialize was the size of a room of a typical townhouse of the city where we lived, Washington DC. You could be thinking: well, but this is likely because your friend was not "well-off," so that can happen in every city. But that is actually not the case: my friend is a "well-off" professor at a Tokyo university. What happens—and this is the most troubling part—is that this is something that actually happens very often in Tokyo and that is seen not only as something "normal" but even as something "cool." As explained in an article in the *All about Japan's* website, entitled "*Tiny Tokyo apartments are surprisingly popular,*" for many Tokyo inhabitants that work until they are completely devastated, the "living in such atrociously small apartments" (see Fig. 8.3) attitude is increasingly seen as something "customary" and actually "good":

> Japanese apartments, in the eyes of the international community, are notoriously small. It's all relative, though. What might looks like a small living space by U.S. or European standards could feel pretty spacious to a lot of people who grew up in Japan. Of course, there are apartments that Japanese people think are small too…it's just that compared to other countries, Japan's small apartments are downright tiny. Japanese real estate company

Fig. 8.3 The paradox of "progress"—case 3: "living"—again, better said, existing—in extremely tiny apartments is seen not only as 'normal' but also being increasingly sold as something 'cool' or 'popular' in Tokyo

Spilytus manages the **Ququri** line of apartments, and even in Japan, it's hard to find a much cozier room than these. Like most apartments in Japan, Ququri's units are measured in jo, a traditional Japanese unit of area that's equal to the size of one tatami mat. Ququri's apartments all offer just three jo of studio room floor space, which works out to 4.64 square meters (49.94 square feet). Even still, pretty much all of Ququri 1,200 rooms, split across multiple buildings around Tokyo, are being rented. They're especially popular with people in their 20s and 30s, who make up about 90 percent of the tenants. They're not only being used as bachelor bunkers, either, as roughly 40 percent of Ququri residents are women. So what's the appeal? Location, for one. Ququri offers apartments close to some of Tokyo's busiest downtown districts, such as Shinjuku, Shibuya, Ebisu, and Nakameguro.

Not only do those provide access to entertainment and cultural events, but they're all also major rail hubs, making for short, easy commutes to offices and schools in the city (Spilytus estimates about 60 percent of residents are office workers, and another 30 percent students). Another advantage is the low cost.… Obviously, a certain level of commitment to a minimalist lifestyle is a plus, but each room also has a loft that measures about four jo (6.19 square meters). Most residents choose to use the loft as their sleeping space, laying out a futon or compact mattress, which lets them use all of the actual floor space for non-sleeping furniture. Each apartment has an indoor washing machine hookup, unlike many micro-apartments in Japan where you have to leave your machine outside in the elements. Other luxuries not found in all tiny Japanese apartments: cooking space and a private bathroom and shower. Spilytus reports high satisfaction among Ququri's residents, many of whom quickly adapt to the size of their homes. "The convenience store down the street is basically my refrigerator", said one, when asked how he manages not having a big pantry…. Granted, if you're the kind of person who needs tons of space, some of these living conditions probably look more like a cell than an apartment. But for those who want to live with no wasted space, in a highly personalized and cozy environment, Ququri, which means "cocoon" in Greek, seems to be just the right size.

"**Ququri,**" another term used in Japan to which we will probably need to get used, in our "development" towards a future of "technological progress." So far, we have talked about a huge number of Tokyo inhabitants having too little sleep, followed by a huge amount of hours of work, and then being completely exhausted to socialize, flirt, and even to have sex, thus preferring to drag themselves at night in unspeakably small apartments. But what about the time they spend *between* these amazingly pleasant, healthy, and mindful—using imaginary terms based on *Neverland*'s tales, of course—activities? For instance, what about transportation? To be sure, Japan has one of the most "developed" systems of public transportation in the planet. For instance, metros are almost always "on time" in Tokyo, and one can plan and buy tickets in advance very easily, using many types of electronic devices, in order to not "waste time" or "have to talk with other people," such as ticket sellers. However, two points need to be made. Firstly, despite its huge population density, Tokyo is a huge city, what shows again one of the main problems, and trends, of "progress": the Greater Tokyo Area is said to be the most populous metropolitan area in the world, with more than 37 million residents as of 2020.

This is almost four times the whole population of the country where I was born, Portugal, compressed into a single metropolitan area. So, this means that although the public transportation is amazingly efficient, when people go to work they often have to spend a huge amount of time in public or private transportation. Secondly, because Tokyo is so densely populated, its public transportation is often terribly crowded (see Fig. 9.6). Unfortunately, I have also seen this with my own eyes, because those few days I spent in Tokyo my friend took us from place to place by metro: it was full of people, a huge number of them making no effort to talk, or even to look, to other humans, looking instead attentively to a very small machine that is called cell-phone but that they almost never use to call others—not only in the metro but also in the streets as well, I guess because that involved "real talking" with "real people"—or just looking to an empty space as if they were zombies. As domesticated cows in "animal factories" often move like passive zombies from where they are stuck most of their lives to where they will be killed, without any reaction or sign of pleasure, also a huge number of people in big cities behaves like passive zombies that have interiorized that public transportation is merely a way of going to, or coming from, work. As the main function of domesticated cows is to give milk and meat to their "masters," the function of a huge proportion of self-domesticated humans is to work for the dreams of others: *they don't work to live, they live to work* (see Fig. 9.6).

It is important to emphasize that I am *not* criticizing Japanese people per se, and much less people that live in Tokyo specifically. Firstly, although I would not like to live in such a big city, I did actually enjoy a lot those days in Tokyo, particularly because it was a unique experience in my life and precisely because I was not there to work, but mainly to visit my friend, visit museums, try to observe and melt in a new culture, and eat amazing food. It was truly a fascinating "lost in translation" learning scenario in many ways, and I loved the amazing sushis, tempuras, udon noodles, ramens, sakedons, and sweet sakis, as well as the splendidly beautiful, rich, elegant Japanese art. Secondly, as explained above, my criticisms about some aspects of "progress" are not directed to specific individuals, but above all to the *circumstances* and *socio-political systems* in which they live. So, when one looks at Fig. 9.6, one knows that if those very same individuals were living in let's say a small island in Japan or a small nomadic group in the Amazon, they would very likely have a completely different attitude and pose, such as being more social, having more sex, working less, being more relaxed and mindful, and so on. Exactly in the same way that if the domesticated cows shown in Fig. 9.6 were let free in a small island of Japan with a lot of grass and no humans, they would have completely different lives. That is actually one of the main flaws of racist ideas and of ideologies of Us versus Them: they forget that above all, everything depends on the specific *circumstances* and on both *the imaginary tales believed by* and the *system imposed to* human individuals.

Moreover, as I also noted above, I do not criticize all aspects related to "progress": I would have to be hypocrite to do so, as I am writing a book, using a computer, and in that trip to Japan I went there by plane, and so on. I love to read books, go to movies, and would not like to live my whole life in the middle of a forest in Congo, as I would not like to live my whole life in a big city such as Tokyo. For instance, apart from the many aspects that I liked a lot, while I was in Japan, there are many outstanding aspects of that country, such as the fact that in 2018 it was the top leader in the planet regarding **life expectancy**. As will be explained below, this has a lot to do with a more relaxed life in non-urban areas, particularly in smaller Japanese islands, but it is true that it is also related to a more general trend seen in the country as a whole, combining a fantastic **health system** with an overall mainly **healthy diet**. As the country is an island itself, a substantial portion of the diet of many people is based on fish and sea food, including many raw products as those included in sushi. So, in a way, the main "secrets" of Japan's very high life expectancy rates are associated with the peculiar mixture of a highly technological medical system—for instance, amazingly efficient to prevent the death of children—and a more traditional way of living. In other worlds, medication, vaccines, and so on play a huge role, but they cannot be the main factor of Japan's exceptionally high life expectancy because they also apply to many other "developed countries" that have much lower numbers regarding life expectancy. In fact, the main problem of Japan, which is already seen in cities such as Tokyo is that traditional ways to cook and eat are now being seriously threatened by "progress" and the consumption of processed foods and/or fast foods such as Macdonald's burgers and Kentucky's fried chicken. Together with the increased levels of stress, suicide, over-working, and numerous other aspects of life that are unhealthy for the body in general and the "mind" in particular, that is precisely why the life expectancy is lower in big cities such as **Tokyo** than in islands such as **Okinawa**, as we will see below.

Another example of a trend that is clearly not healthy, mindful or pleasant and that is quickly growing in Japan's big cities—and which is also related to a passive self-domesticated attitude towards life in general—is the **Hikikomori phenomenon** (Fig. 8.4). This exceptionally troublesome tendency is profoundly linked to both **social isolation** and the **pressures of "modern life**," as summarized in the textual description of a very interesting January 2019 *France24Reporters* documentary entitled "*Japan's modern-day hermits – the world of the hikikomori*":

Fig. 8.4 The paradox of "progress"—case 4: as a huge number of young Japanese people, the young man show in this picture was living as a *hikikomori* when the picture was taken, in 2004

> In Japan *half a million people* live **isolated in their bedrooms**, unable to face the outside world…since April 2018, the Japanese government has been conducting a nationwide study in a bid to fully understand this strange phenomenon…. Once limited to young people, it now affects the whole of Japanese society…most hikikomori are under 40, but this **extreme form of isolation** increasingly affects older people…older hikikomori sometimes have no relatives or friends to support them and remain cut off from society for longer periods…their ultimate fear is to die alone…*every year in Japan*, some 30000 people pass away without a loved one by their side – arguably an even more worrying phenomenon.

As explained in the documentary itself, this phenomenon is now also increasing in other "developed" countries such as France, because those countries share similar patterns: **increased stress levels** and **amplified physical isolation**, associated with an increased use of/access to TV, videogames, and so on. And, ironically, this phenomenon is also linked to the use of the so-called social media—or better said, antisocial medial—which amplifies the social pressure to be—or better said, to "seem to be"—"successful" and "always happy," to "be someone" within the teleological purpose dictated by a materialistic system and **work culture**. Indeed, as also noted in the documentary, and recognized by the Hikikomori people interviewed in it, in most cases the isolation starts as a reaction to an enormous pressure made by the society, including their own families. Another somewhat related troublesome increasing trend in Japan is that many elders want to go, or at least are "happy" to stay in, jails. Relatively, Japan has a low incarceration rate, but in proportion it has one of the higher rates for **incarcerated elders** (Fig. 8.5). This disturbing tendency—which, again, would be unthinkable in most nomadic hunter-gatherer societies, in which elders are very often intellectually active, highly respected and socially integrated individuals—has been reported in a January 2019 *BBC news* article entitled "*Why some Japanese pensioners want to go to jail*":

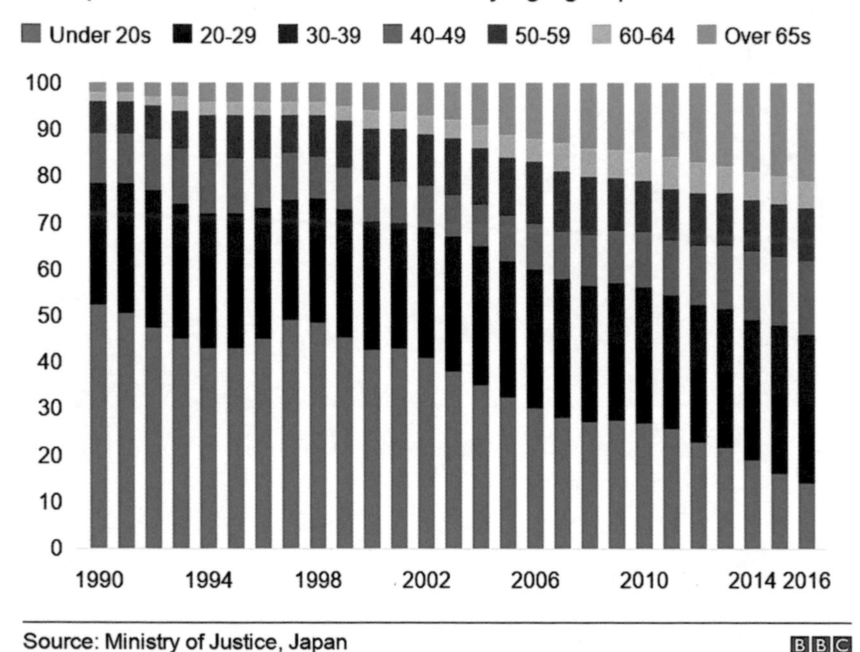

Fig. 8.5 The paradox of "progress"—case 5: older Japanese pensioners often want to go to jail. (Source: Ministry of Justice, Japan)

Japan is in the grip of an elderly crime wave – the proportion of crimes committed by people over the age of 65 has been steadily increasing for 20 years. At a halfway house in Hiroshima – for criminals who are being released from jail back into the community – 69-year-old Toshio Takata tells: "I reached pension age and then I ran out of money. So it occurred to me – perhaps I could live for free if I lived in jail…so I took a bicycle and rode it to the police station and told the guy there: 'Look, I took this'". The plan worked. This was Toshio's first offence, committed when he was 62, but Japanese courts treat petty theft seriously, so it was enough to get him a one-year sentence. Small, slender, and with a tendency to giggle, Toshio looks nothing like a habitual criminal, much less someone who'd threaten women with knives. But after he was released from his first sentence, that's exactly what he did. "I went to a park and just threatened them. I wasn't intending to do any harm. I just showed the knife to them hoping one of them would call the police. One did." Altogether, Toshio has spent half of the last eight years in jail. He points out an additional financial upside – his pension continues to be paid even while he's inside. "It's not that I like it but I can stay there for free," he says. "And when I get out I have saved some money. So it is not that painful." Toshio represents a striking trend in Japanese crime. In a remarkably law-abiding society, a rapidly growing proportion of crimes is carried about by over-65s. In 1997 this age group accounted for about one in 20 convictions but 20 years later the figure had grown to more than one in five – a rate that far outstrips the growth of the over-65s as a proportion of the population (though they now make up more than a quarter of the total).

And like Toshio, many of these elderly lawbreakers are repeat offenders. Of the 2,500 over-65s convicted in 2016, more than a third had more than five previous convictions.

Michael Newman, an Australian-born demographer with the Tokyo-based research house, Custom Products Research Group points out that the "measly" basic state pension in Japan is very hard to live on. In a paper published in 2016 he calculates that the costs of rent, food and healthcare alone will leave recipients in debt if they have no other income – and that's before they've paid for heating or clothes. In the past it was traditional for children to look after their parents, but in the provinces a **lack of economic opportunities** has led many younger people to move away, leaving their parents to fend for themselves. "The pensioners don't want to be a burden to their children, and feel that if they can't survive on the state pension then pretty much the only way not to be a burden is to shuffle themselves away into prison," he says. The repeat offending is a way "to get back into prison" where there are three square meals a day and no bills, he says. Newman points out that **suicide** is also becoming more common among the elderly – another way for them to fulfill what he they may regard as "their duty to bow out".

The director of "With Hiroshima", the rehabilitation centre where I met Toshio Takata, also thinks changes in Japanese families have contributed to the elderly crime wave, but he emphasizes the psychological consequences not the financial ones. "Ultimately the relationship among people has changed. People have become more isolated. They don't find a place to be in this society. They cannot put up with their loneliness" says Kanichi Yamada, an 85-year-old who as a child was pulled out of the rubble of his home when the atomic bomb was dropped on Hiroshima. Toshio's story about being driven to crime as a result of poverty is just an "excuse", Kanichi Yamada suggests. The core of the problem is his **loneliness**. And one factor that may have prompted him to reoffend, he speculates, was the promise of company in jail. It's true that Toshio is alone in the world. His parents are dead, and he has lost contact with two older brothers, who don't answer his calls. He has also lost contact with his two ex-wives, both of whom he divorced, and his three children. There are classes for older offenders…(one of them) begins with a karaoke rendition of a popular song, The Reason I was Born, all about the meaning of life. The inmates are encouraged to sing along. Some look quite moved. "We sing to show them that the real life is outside prison, and that happiness is there," Yazawa says. "But still they think the life in prison is better and many come back." Michael Newman argues that it would be far better – and much cheaper – to look after the elderly without the expense of court proceedings and incarceration. "We actually costed a model to build an industrial complex retirement village where people would forfeit half their pension but get free food, free board and healthcare and so on, and get to play karaoke or gate-ball with the other residents and have a relative amount of freedom. It would cost way less than what the government's spending at the moment," he says. But he also suggests that the tendency for Japanese courts to hand down custodial sentences for petty theft "is slightly bizarre, in terms of the punishment actually fitting the crime".

8.3 Affluence, Abundance, Growth, and Inequity

Two men came to a hole in the sky. One asked the other to lift him up…. But so beautiful was it in heaven that the man who looked in over the edge forgot everything, forgot his companion whom he had promised to help up and simply ran off into all the splendour of heaven. (Inugpasugjuk, from an Iglulik Inuit prose poem)

Smith and colleagues published in 2010 an interesting study using Ginis—a measure commonly used by economists to infer **inequality**—in which they estimated that **Ginis ranges** in general from 0.2 to 0.5 in hunter-gatherer groups, with an average of 0.25. This average is much lower—that is, there is much less inequality—than that of the vast majority of agricultural states, including almost all democratic "developed" countries nowadays: in fact, a **Ginis value** of 0.25 was only found, in a 2007 list, in the country that had the *lowest value* in the planet, **Denmark**. What about Japan? The World Bank estimate for **Japan** in 2008 was 0.32. In this sense Japan is doing also particularly well, with less inequality than most countries—states such as South Africa and Namibia have values above 0.6 for instance. But this still means that the most technologically "developed" country in the planet has on *average* substantially more inequality than the *average* of hunter-gatherer groups in general, and thus *a tremendously higher disparity than small nomadic hunter-gatherer groups in particular*. What about the so-called biggest economy in the planet, the U.S.? In 2007, its Ginis was 0.41. As noted by Smith and colleagues, "to the extent that our measures for this set of foragers are representative, wealth inequality" in hunter-gatherer groups is far from a state of "primitive communism"—as argued by some defenders of **Rousseau**'s **Noble Savage**—but is still obviously "very low by current world (agricultural states) standards." We should think deeply about this: after millennia doing so-many "revolutions"—agricultural, industrial, scientific, technological—"progressive" changes, and fighting so hard for "economic growth"—and according to Hobbesian authors such as Pinker, with an "expanding moral circle" occurring in the last centuries—*we are now in a situation in which even the "technologically more developed" and the "economically most powerful" countries of the planet are still much more unequal, on average, than our ancestors were for millions of years before the rise of* ***sedentism*** *and* ***agriculture***. Clearly, there is something wrong with this tale of "improvement" and "progress." The fallacies of *Neverland*'s prevailing fairytales about "progress" and "economic growth" are plainly shown by a recent report of *Oxfam*, which states that **income inequality** has reached a new global extreme: disturbingly, *just 62 individuals now hold the same wealth as the bottom half of humanity, while the number was 80 in 2014 and 388 in 2010.*

Is the fact that in a planet of about 7.6 billion people just 62 individuals have the same wealth as about 3.8 billion of people truly a reflection of what authors such as Pinker call the "civilizing process"—particularly when there is almost no inequality among small nomadic hunter-gatherer societies, as Pinker himself recognizes? It seems very difficult to argue that *we* are "better" nowadays at a societal or community level, unless the "we" refers exclusively to those 62 individuals, of course. As explained in Chap. 4, it was the new type of circumstances created by the rise of **sedentism** and particularly of **agriculture** that brought opportunities for political and religious leaders to start creating a system based in social inequities, and to actively create a plethora of teleological tales, practices, social norms, and laws—including those related to oppression, wars, and slavery—to enforce such a system.

That is, to make people *believe* that such a system of inequity was "natural/normal"—for instance, created by God—or, for those that did not believe in such just-so-stories, to oblige them to accept them anyway by the use of force in order to maintain a *status-quo* is which, for the first time in millions of years of human evolution, the 1% made the 99% *work for them* and *for their dreams* (see Figs. 9.5

and 4.4). This is precisely what happened in all the "**early civilizations**" that are so often praised in historical books. From **Egypt** to Mesopotamia, from the **Indus Valley** to China, from **Aztecs** to **Mayans**, and from **ancient Greece** to **ancient Rome**, all of them were characterized by **huge social inequities**, **colossal famines**, **never-ending wars**, and **massive use and abuse of slaves or so-called 'inferior casts'**. It is therefore truly puzzling for me why all the praise for, and the use of the name "civilizations" to designate, such early agricultural states, because in theory "civilized" means things like "having an advanced or humane culture or society" or "being polite or refined."

One of the most detailed—and brilliant—books about this topic is **Walter Scheidel**'s *The great leveler: violence and the history of inequality from the stone age to the twenty-first century*. Based on an incredible amount of historical and economic empirical data, Scheidel states that within those societies in which there was a rise of agriculture and "civilization," the **socio-economic system** almost *always* tended to lead to an increase of inequities. The major exceptions have been what he calls the "**four horsemen of levelling**": **mass-mobilization warfare**, **transformative revolutions**, **state collapse**, and **catastrophic plagues**. In fact, in contrast to the narratives that one is so acculturated to *believe*—which have been repeatedly defended by many scholars, such as Pinker—stable Western societies are not levelers, but instead just follow the same trend of rising inequalities. Even stable Western democracies, which are the so-called—by Westerns that live in them, obviously—as the "best of all types of socio-political government," one mainly sees the same general trend, with very few exceptions. As show by Scheibel, even during the last decades of stability and absence of war in Western European and North American countries, inequality has risen significantly within many countries, with the 1% having a much higher percentage of the total "wealth" than they had in the 1980s, in most of those countries. For instance, within the U.S., they had 8.2% in 1980, 13% in 1990, and 17.5% in 2010, with the lowest value in the last half century being in 1973, when it was 7.7%. As noted by Scheibel, although these numbers may vary within different studies and economic markers, the general trend seems to be clear.

This topic is directly related to the concept of "work": clearly, there was an enormous increase regarding the *average* of hours dedicated to "work" during and after the transitions to sedentism and agriculture. The more the 1% want to make more gains or to fulfill their dreams—be it building pyramids in **ancient Egypt** or **Trump towers** in big "modern" cities, so they are remembered after they die—the more the 99% have to work. And, of course, we are not referring only to paid work, as in "early civilizations" a huge amount of work was made by people that did not receive any remuneration, such as slaves. Today the 1% continues to use the very same type of teleological narratives to maintain the *status-quo*, such as using "hard work is a virtue" tales and inculturation processes. Such processes are just rebranded by employing new names, such as the "trickle-down economics" so in vogue in countries such as the U.S., or the "work culture" so in vogue in countries such as Japan. As shown in Fig. 8.8, specific circumstances—including the place and culture in which one is born—are indeed a critical factor for the number of hours that each individual works, on *average,* clearly illustrating that narratives and inculturation do play a major role here.

As illustrated in that figure, in most countries people spend an *average* of three to more than 5 hours per day, just in paid work—Japan reaching the 5.5 hours per day, as noted above. Moreover, that figure refers to an average including *all* people, from 15 to 64 years old. As most people in those countries that are 15 or 16 or 17 don't tend to work full time, and some people above 60 tend to work less, this means that in average people that tend to work full time from let's say 25–60, work even much more than 5.5 h per day, in countries such as Japan. Furthermore, the list of course includes many people that don't have a "paid work," such as religious leaders, other groups with privileges, and obviously people with disabilities, and so on. However, one could of course also argue that this is a major "noble" point about "developed" countries: that a huge number of people, such as disabled people, or retired people, don't have paid jobs and still receive money from the government, until the very last moments of their lives. Yes, I completely agree that this is one of the most important social achievements of "developed" countries, there is no doubt about that. However, a few points have to be clarified. For instance, retired people are *not* just "receiving" a "noble" charity monthly income given altruistically by "civilized" governments or the 1%: in the vast majority of countries those people have paid *themselves* for a substantial part of that income, during each of the countless years they have "worked," when they were working a huge number of hours per day mostly—with exceptions, of course—for the dreams of others and massive gains of CEOs and the 1%. Clearly, in the vast majority of cases, the amount of money those people received during their whole lives for their "work"—both when they were "working" or when they retired—is in general much less that the amount of money that *their* work generated for the 1% and the CEOs: if that was not the case, we would not have reached a situation in which, for the first time in history, *just 62 individuals now hold the same wealth as the bottom half of humanity*.

As noted above, empirical data from many fields of knowledge—including **neuro-economic evidence**—clearly show that—contrary to the Hobbesian fairy-tales of authors such as Pinker—in terms of "work," inequalities, and **greediness**, our "***Better Angels***" were indeed our nomadic hunter-gatherer ancestors. As explained by neurobiologist Jack Lewis in his 2018 book *The Science of Sin*, the huge levels of greediness displayed by so many people in "developed" countries are very likely a very recent phenomenon in human evolution, which requires a lot of "*a posteriori* rationalization" and "thinking," as put by him. He describes a very interesting study done at Harvard University in which people had to undertake a series of different tasks, including choosing to be greedy or to cooperate. When they had to decide in a very quick way, without really having the time to "think about it so much," they indeed tended to choose to cooperate, and in general they were relatively fair to others. However, when they were given time to ponder carefully the different options they could take, they instead tended to take much more selfish and greedy actions. He argues that this study and many other similar studies show that "instinctively" we tend to be more fair and less greedy. One of those other studies used electric stimulation to interfere with the regions of the **dorsolateral prefrontal cortex**, which is said to be related to some of our greedy decisions: when this area of the brain was magnetically deactivated, people tended to be less money-hungry and more fair.

This actually makes sense evolutionarily, because while some primates such as gorillas are very hierarchical, our closest relatives the chimpanzees tend to be much less hierarchical and to display "fairness" behaviors as explained in Chap. 2. They might do that because they need to do so, within the type of social networks often formed in chimpanzee communities, or because they just have a "natural" "innate" tendency to do so, or because of both: it would be interesting to do detailed studies about this, similar to those that have been done in humans. The idea that our "modern" levels of greediness are a very recent phenomenon in human evolution also makes evolutionary sense because chimpanzees don't seem to be able to build and believe in the type of complex teleological stories that are often crucial to rationalize a posteriori the inequalities and unfairness of human "civilization." In fact, it is important to stress that those stories are not only about how "work is a virtue" because that is what God wants or the only way one can be "productive to society" as noted above, but also about how we—the 99%, not the kings, CEOs or the1% of course—should not be greedy because "avarice" is one of the seven capital sins, as explained in Lewis' *The Science of Sin*.

What do the available empirical data tell us about the way in which non-agricultural societies spend time? As discussed in Chap. 4, there are huge differences between different **hunter-gatherer groups**, and it is therefore impossible to put forward even a decent estimate of "work hours" per day—many scholars use this term "work," but obviously this has nothing to do with our "modern" "paid work"—within those groups as a whole, particularly due to the huge differences between nomadic and sedentary hunter-gatherer societies. In his well-documented 2017 book *Affluence Without Abundance*, Suzman discusses this subject based on the observations he did of the southern Africa's **San peoples** for nearly a quarter of a century, as well as on an extensive literature review. He explains that one of the most detailed measures of "work hours" per day within these groups was made by **Richard Lee**, who, "struck by the apparent lack of effort that went into the food quest" in the**!Kung**—which often call themselves **Ju/'hoansi** (Fig. 8.6)—decided to analyze how much time they dedicated to get food. Lee "established that on average…healthy adults worked 17.1 h per week on food colleting, with that number skewed upward by hunting trips, which almost always took up much more time than gathering excursions…for women, the workweek rarely exceeded 12 h. Lee's survey also revealed that the Ju/'hoansi ate well…adults consumed on average over 2300 calories of food each day…this is more or less the recommended caloric intake for adults according to the World Health Organization." This idea of "**affluence without abundance**" was also supported by Suzman's own observations of the San. These numbers are impressive, because they mean that the *total average*—including both women (12 h per week) and men (17.1 h per week of men)—is 14.55 h per week, so basically just 2 h per day for the sampled!Kung adult, healthy population.

Fig. 8.6 Ju/'Hoansi-San women, bushwalk

That is, less than half the *average* of most countries, particularly "highly-developed" countries such as Japan, as seen above. So, as we did about inequality above, this begs once again a key question: how can it be that after all the "progress" and the "revolutions" that were supposed to be "beneficial" because our "work" would be replaced or at least decreased by animals—**agricultural revolution**—or factories—**industrial revolution**—or computers/robots—**technological revolution**—we actually have much less leisure time than our ancestors did for millions of years before these "revolutions?" This fairytale about "development," "growth," and "improvement" also just does not add up (see also Box. 8.2).

Of course, authors that are more reactionary and/or defend Hobbesian views often make a way to turn the empirical data upside down and try to "show things from a different prism," as in their minds there is no way that their a priori dogmatic assumption that "savages" had a "brutish" life can be wrong. So, for instance, in Saxon's *Sex at dusk*, she refers to Kelly's 1995 book *The Foraging Spectrum*, stating: "one problem Kelly notes is in the definition of the word 'work'...only time spent acquiring food in the bush was counted as work, not the time spent processing food, making tools, childcare, carrying water, collecting firework or cleaning...more accurate estimates of time spent foraging and processing food demonstrates that, while there is variation across hunter-gatherer societies, some hunter-gatherers spend seven, eight, or even more hours a day 'working'." Well, if we would apply this logic to "developed" countries—which clearly is almost never applied, as illustrated in Fig. 8.8—then we would also count buying food, processing food, childcare, housework and many other activities as "work." If we would do this, then basically the whole time we are awake would be considered to be related to "work," with the exception of 3–5.30 hours a day for "pure leisure" (see Fig. 8.8).

In "developed" countries, the only thing that we are truly bringing to our home after "paid work" is the money we receive for doing it, nothing else. At least nomadic hunter-gatherers bring food directly to their temporary settlements after they finish "working." But we, if we are, let's say, in the center of a "highly developed city" such as Tokyo, we probably need to spend at least half an hour, if not much more, to commute from the working place to our home, then more time to go to stores or supermarkets to buy the food, waiting in line to pay for it, and then bring the food home. Then, we need to process it—as hunter-gatherers do—to take care of our children—and individually we spend much more time on average doing so then hunter-gatherers do, as explained in Chap. 5—then eventually organizing and cleaning a bit the house, and so on. Moreover, another aspect that is not directly related to "work," but that authors such as Saxon forget to mention when they discuss the links between how time is spent, leisure, well-being, and health, is that, disturbingly, in many "developed" countries a substantial proportion of the few hours dedicated to the so-called total leisure time is spent in activities such as seeing TV or listening to radio (Fig. 8.8). In other words, this leisure time is often actually used to increase even more a sedentary, often detrimental, and many times anti-social type of life trapped in a world of unreality, in this case literally in *Neverland*'s world of TV. Notably, and tellingly, this applies in an extreme way to the "economically stronger" country in the planet, the U.S., in which people spend in average more than half of their "total leisure time" seeing TV or listening to radio (148 out of 292 minutes) (see also Boxes 4.2, 8.2–8.4).

Box 8.2: Hunter-Gatherers and the 'Modern' Obsession to 'Work Hard' and 'Create New Wealth'

Suzman's 2017 book *Affluence without abundance* provides, in my opinion, one of the most clear, less-biased, and well-documented discussions about the links between **affluence** and **abundance**, and the lessons we can learn from them, and from hunter-gatherers in general, and nomadic ones in particular:

> In the winter of 1930, **Keynes** was understandably preoccupied with the depression that was strangling the life out of European and American economies and the collapse of his personal fortune in the stock market crash the preceding year. Perhaps to persuade himself of the ephemeral nature of the crisis, he published an optimistic essay titled "*The economic possibilities for our grandchildren*". The future to which

Keynes's wings flew him was an **economic Canaan**: a promised land in which **technological innovation**, **improvements in productivity**, and **long-term capital growth** had ushered in an age of "economic bliss". An era in which we are all able to satisfy our material needs by working no more than fifteen hours in a week and in which we are liberated to focus on more profound joys than money and wealth accumulation. While Keynes was uncertain as to whether humanity would be able to easily adjust to a life of **leisure**, he was convinced that, save for war or cataclysm, this reality would come to pass in the time of his grandchildren. "I would predict", he wrote, "that the standard of life in progressive countries one hundred years hence will be between four and eight times as high as it is today." Keynes was right about improvements in **technology** and **productivity**. The U.S. Bureau of Labor Statistics tells us that labor productivity in the United States saw a fourfold increase between 1945 and 2005. But Keynes was wrong about the fifteen-hour week. While average working hours have declined from around forty hours per week in Europe and America to between thirty and thirty-five hours per week in the last fifty years, the drop has been much slower than the rise in individual productivity. Given the increases in labor productivity in the United States, the modern American worker should be able to enjoy the same standard of living as a 1950s worker on the basis of a mere eleven hours of productive effort a week. But Keynes was prescient about this too. He anticipated that there would be a lag between improvements in productivity and technology and its translation into fewer working hours. For him, the biggest obstacle to overcome was our *instinct to work hard* and *to create new wealth..*

[But] he failed to anticipate our capacity to consume whatever new things our increased productivity enabled us to create. He also underestimated quite how far people would go to create work when – in material terms, at least – there was none to do. Keynes was also unable to predict the environmental costs of **humankind's obsession with work** or, for that matter, his own inadvertent role in ensuring the ascendance of a global economic model focused myopically on **capital growth** and the **ever-quickening cycle of production**, **consumption**, and disposal that it spawned. Perhaps Keynes would have had a better sense of the scale of this problem – and of its genesis – had he realized that hunter-gatherers, the least economically developed of all the world's peoples, had already found the economic promised land that he dreamed of and that the **fifteen-hour working week** was probably the norm for most of the estimated two-hundred-thousand-year history of biologically modern *Homo sapiens*. But Keynes was a creature of his time. He could not have known something that would only be revealed some thirty years after his death. To him, the idea that primitive people with no interest whatsoever in labor productivity or capital accumulation and with only simple technologies at their disposal had already solved the "economic problem" would have seemed preposterous.

In many ways the secret of their [the **Khoisan**'s] success, and the endurance of their way of life, was based on their having reached a form of dynamic equilibrium with the broader environment, a balance between its relative stability and harshness. The evolutionary success of Khoisan, in other words, was based not on their ability to continuously colonize new lands, expand and grow into new spaces, or develop new technologies, but on the fact that they mastered the art of making a living where they were. That Khoisan remained in one region for such a long period of time may well also have helped ensure the survival of much of Africa's megafauna into the modern era. The expansion of modern *Homo sapiens* across the planet is associated with the waves of mass extinction that took place in Europe, Asia, Australia, and the Americas over the past hundred millennia. North America, South America, and Australia all witnessed the disappearance of nearly 80 percent of their large mammal species in the period following the arrival of modern *Homo sapiens*. Sub-Saharan

Africa, by contrast, only saw the extinction of two out of forty-four large mammal genera. It is hard to imagine that the globalized economy symbolized by **da Gama's voyage** will endure anywhere near as long as the Khoisan managed to survive by hunting and gathering. Some of the more doom-laden forecasts of climate change suggest that it is already too late – that an important threshold has already been passed and that we should all be channeling our inner Noah and building arks of one sort or another. But then again the promise of apocalypse has been built into most organized religions that evolved since the Neolithic, because while the transition to farming incrementally increased productivity and population size, it drastically *reduced the quality of life* for most people and introduced them to a whole range of perils unimaginable to most hunter-gatherers, like **viral epidemics** from livestock and **mass famines** when harvests failed. Or maybe our willingness to contemplate apocalypse expresses a suppressed sense that the same qualities that have enabled our species to exert such an extraordinary impact on the world around us hold within them the seeds of their own destruction.

After writting the excerpts show in Box 8.2, Suzman pointed out that he is completely aware of the often too romanticized view that some anthropologists have put forward regarding so-called traditional societies. Taking this into account, he then provides examples from his own observations of the **Ju/'hoansi**, including very interesting reflections on the professed "nurture *versus* nature debate":

Foraging Ju/'hoansi don't animate their environment like the **Mbuti**. They also don't talk about animal spirits or speak of conscious, living landscapes. Rather, they describe their environment's providence in more matter-of-fact terms: it is there and it provides them with food and other useful things, just as it does for other species. And just as importantly, even if they consider their environment to be provident, they don't think of it as "generous" – firstly because it can sometimes be austere, and secondly because Ju/'hoansi do not think of their environment as a "thing" capable of agency. Rather, they describe it as a set of relationships between lots of different things capable of agency – plants, insects, animals, people, spirits, gods, and weather – that interact with one another continuously on what Ju/'hoansi called the "earth's face." The idea that nature should be "pristine" and unpolluted by the detritus of human life is, of course, a relatively new one. In most complex societies, after all, "natural" or "wild" places are still considered to be dangerous or at the very least inhospitable. And while it is true that most societies conceptualize humans as something distinct from the natural world, foraging societies like the Ju/'hoansi simply do not. To them everything in the world is natural and everything cultural in the human world is also cultural in the animal world, and "wild" space is also domestic space. So while Ju/'hoansi consider the litter to be an irritation, few see it as pollution – at least in the way the tourists do. To most of them, the litter is no more offensive than the leaves that fall from the trees in the Kalahari autumn or the broken baobab seedpods that litter the soil near the Holboom.

Many of the foods Ju/'hoansi gather come into season in a neatly staggered sequence, meaning that at almost any particular time of year something is in season. Paradoxically, the period just after the first rains fall and when the Kalahari becomes a sea of green is considered the leanest time of the year. This is the period when hunting is hardest, as animals are able to move farther away from waterholes, sustaining themselves from puddles, seasonal waters, and in many cases the moisture content of newly germinated grasses. It is also the time of year when most of the food plants the Ju/'hoansi rely on begin a new seasonal cycle of growth, with the result that their tubers or fruits are too small or immature to make good eating. Ju/'hoansi complained a great deal about hunger at these times. They also grumbled about the additional effort involved in gathering enough to get by. Lee [the anthropologist that first wrote about these subjects, see above] didn't extend his detailed nutritional survey to these months, but over successive years he and other researchers measured people's weight variations at different times of the year. This revealed a mixed picture. Some researchers suggested that Lee's data was unrepresentative and that Ju/'hoansi typically experienced noticeable nutritional stress during lean seasons. Others suggested that Lee was right and that Ju/'hoansi suffered only minimal seasonal weight variations. When evidence like this points in two different directions, then the simplest solution is to assume that they are both right. Rain is so variable in the Kalahari that the difference between a good year and a bad year can be extreme. Some years are much leaner than others. In lean years Ju/'hoansi inevitably had to work much harder to get enough food, and when the energy they expended hunting and gathering exceeded the nutritional value of the foods they procured, they lost weight and condition and women's reproductive cycles were interrupted.

But focusing on this short-term variability obscures the most important point raised by Lee's work. This is simply that Ju/'hoansi were not captive to an interminable food quest and were content to expend no more effort than was strictly necessary to meet their basic short-term needs even in the toughest months. In other words, even when the food quest was difficult, Ju/'hoansi never lost faith in the abundance of their environment. And just as importantly, when foods were superabundant, as was often the case, Ju/'hoansi did not wallow in their plenty or try to maximize short-term benefits by gorging themselves. Instead, they ate to the point of satisfaction and appreciated the fact that the daily food quest hardly took any effort at all. Some Ju/'hoansi took the view that the farmers struggled to control their appetites because of all the alcohol they consumed. After all, Ju/'hoansi also had a hard time keeping a lid on their basic desires after a few drinks, and the farmers did drink an awful lot of beer and brandy. Others took the view that it was because most white people were simply incapable of controlling their appetites, their tempers, and their sexual desires. A few Ju/'hoansi reached the conclusion that it was a cultural matter and that greed was something that was taught in "cities." They noted that the few Ju/'hoansi who had been given fancy jobs in the capital after independence fattened up so fast that they may as well have been bitten by puff adders.

As explained by Suzman, the fact that nomadic hunter-gatherer peoples such as the!Kung tend to "work" to get the food they need, and *not* to get a lot of materialistic items they don't need, is not related to them being "Noble Savages": it has a lot to do with the *circumstances* and the way they live. Of course, nomadic hunter-gatherers did, and do, a huge number of practical innovations to adapt to climate changes and so on: they are *active players*, as we saw above. But it is obviously less likely that, in general, they would be obsessed in producing or collecting new "materialistic goods" that they would then need to carry every time they move to a new place. This topic was discussed in a recent article by Spencer in *The Guardian Weekly*, which had a fitting title—"*Long before tech bros, Silicon Valley had a highly developed society*"—and questioned Western concepts such as the "chain of being," "cosmic progress," "civilization," and "technology":

There was a point in time, before **colonization**, when the San Francisco Bay Area was dominated by a people with a way of life and philosophy that did not revolve around technology or technological improvement. Five hundred years ago, this swath of northern California was populated by the **Ohlone peoples**…so rich in plant and animal life was this region that the Ohlone were able to survive without farming or animal domestication; indeed, western explorers, when they eventually arrived, were amazed at the quantity of wild animal life. The Ohlone lived off acorns from all the different varieties of oaks, blackberries and gooseberries, chia, shellfish and the roots of many plants. They hunted squirrels, rabbits, elk, bear, whale, otter and seal. They did not 'farm' in the western sense of the word, though they had a complex knowledge of how to use controlled burns to cultivate plant and animal food sources. Their laissez-faire social relationships were alien to the hierarchy-obsessed **Spanish missionaries**, who commented that 'in their pagan state no superiority of any kind was recognized'. Likewise, the Ohlone lived in a **communal society** – which vaguely resembled a gift economy – that shocked the missionaries. 'They give away all they have…[and] whoever reached their dwelling is at once offered the food they possess', one missionary said. There was no obvious form of government. Status and competition were not important to the Ohlone; generosity and family were. This led early missionaries, who were subject to powerful European governments, to conclude that the Ohlone lived in 'anarchy'. The Ohlone peoples had a very different relationship with animals than the Europeans. Predators like foxes, bobcats, mountain lions and coyote were plentiful, yet coexisted peacefully with the Ohlone. 'Animals seem to have lost their fear and become familiar with man', said Frederick William Beechey, an English captain. It has been suggested that as the European colonizers openly hunted and killed easy game over several generations, animals adapted to the presence of gun-toting hunters and learned to keep their distance. 'We take it entirely for granted that animals are naturally secretive and afraid of our presence', wrote historian Malcolm Margolin, 'but for [the Ohlone] who lived here before us, that was simply not the case.'

In the late 18th century, the newly-arrived Spanish quickly set up missions in California, and began forcibly taking Ohlone subjects into the missions – ostensibly to convert them. Yet the Ohlone were held against their will and forced to labor for the Spanish, who separated men and women, lashing and hitting them when they refused to act as the missionaries pleased. One firsthand account describes the Spanish missions as indistinguishable from slave plantations. In addition to violence against the Ohlone, the missionaries brought **measles** and other diseases with them, which killed many Ohlone independently. Various **epidemics** in the 1790s killed hundreds at Mission San Francisco and Mission Santa Clara. And over the course of the 19th century, the native population of California dropped from an estimated 310,000 to 100,000. This mirrors what was happening in the rest of North America: there were an estimated 10 million American Indians living 'north of Mexico' when **Columbus** arrived, a number that eventually fell to less than one million. As the Spanish established their Missions, they also imposed their technological ideals on the land. By 1777, Mission Santa Clara had a farming and livestock operation that included pigs, chickens, goats, roosters, corn and wheat, mostly nonnative species. Despite re-shaping the landscape to their technological whims, the missionaries were surprised at how the Ohlone continued to 'nourish themselves' on acorns, trout, and other wild harvests. The Spanish did not understand why the Ohlone did not have reverence for their 'superior' systems. 'For one who has not seen it, it is impossible to form an idea of the attachment of these poor creatures for the forest', wrote Basque missionary Fermín Francisco de Lasuén. '[Outside the Mission] they are without a roof, without shade, without food, without medicine, and without any help. Here they have all of these things to their heart's content. Here the number who die is much less than there. They see all this, and yet they yearn for the forest.' It was unfathomable to missionaries like Lasuén that the Ohlone might prefer a world without the rigid hierarchies and controlling attitude towards nature that the Europeans possessed.

The differences between the Ohlone and the Spanish ways of life reveal the contradictions inherent to our present-day idea of 'technology'. To borrow the Silicon Valley business-speak of today, who possessed more advanced **technology**? The Ohlone or the Spanish? Who was more innovative? The deep knowledge of the maintenance of the landscape, and the communal lifestyles enjoyed by the Ohlone, meant that the Bay Area remained in a relatively stable ecological state for a thousand years. The incursion of the colonizers disrupted this; they imposed their technological whims and their agricultural logic on the landscape and enslaved and exploited the Ohlone. You can no longer survive in the Bay Area on acorns and wild trout and blackberries, as the Ohlone did; much of the plant and animal life has been extirpated to make way for Western civilization. Hence, the notion that the Spanish were more 'advanced', technologically-speaking, is arguable. As I write this in 2018, I am reminded of a recent news story about a newly-released consumer product called the Juicero. The Juicero is a $400 so-called 'juicer' whose parent company is backed by $120m in investment capital, including money from Google. It is a wifi–enabled juicer that connects to the internet to inform you of your juice's origin as you drink. Despite being termed a juicer, it doesn't really juice anything; you can't drop a carrot, apple or orange inside it. It can only make juice by wringing out proprietary, pre-sealed packages shipped by the company to consumers. A mini-scandal erupted after a Bloomberg reporter discovered that one could use one's hands to wring juice out of the proprietary juice packs, and fill a glass with juice much faster than the machine can. Shortly thereafter, the company ran out of money and shut down. Human hands are not generally thought of as particularly high-tech. But in this case, they were, from a technological standpoint, superior to the $400 Juicero.

The Juicero saga attests to the fact that sometimes technology doesn't make us more advanced, or intelligent, or make our lives better or faster at all. Sometimes it merely makes us dependent on new, more resource-intensive systems, while casting aside those that are incompatible with so-called economic logic. While the overall number of Spanish missionaries in California had been small, American settlers began arriving in droves in the 1840s. The American settlers were cruel and genocidal towards the remaining **Native Americans**, perhaps more so than the Spanish or Mexicans. As one historian wrote, the American incursion into California marked 'one of the last human hunts of civilization, and the basest and most brutal of them all'. There are still some alive today who personally knew those involved in the slow-moving genocide of the Ohlone and other native groups. My grandfather, still living, who grew up in the Bay Area in the 1920s, recalls as a child hearing adult men talking about heading out to the woods to hunt (read: murder) American Indians. It is not as distant an era as we might think. So it went that over the course of a century, the indigenous Ohlone peoples – and other Native Americans of California – were killed and displaced as the Spanish, Mexicans and later Americans, rebuilt California into an edifice of Western society.

Spencer wrote "as I write this in [the end of] 2018." So, I will add two even more recent illustrative examples, from 2020. One concerns a March *The Guardian* article by Harvey, entitled "*One in five Europeans exposed to harmful noise pollution.*" As explained in the article, such an exposure to harmful levels of **noise pollution** will very likely increase in the next decade, with traffic being the biggest culprit. Excessive noise can cause physical and **mental illness**, and is associated with higher levels of **heart disease**, **stress,** and **sleeplessness**—about 12,000 **premature deaths** are caused by noise in Europe each year, according to the European Environment Agency (EEA). Obviously, **excessive noise** is another item with which hunter-gatherer societies do not have to worry about. The other example concerns the **Covid-19 pandemic**. The fact that a mere non-living organism has killed millions of people from the "special superior species" has led to a reaction, both by the media and the broader public, that clearly shows the human discomfort with randomness and the related obsession to search for a "meaning" for such a huge loss of human lives. Even many philosophers and scientists can't avoid to attribute a "purpose" to the pandemic.

For people that don't like capitalism, the virus is "telling us" that we absolutely need to change our economic system. For people that don't like globalization, the virus is "indicating" that we do need strong borders and even walls. For others, the virus is clearly a way for "Mother Earth" to oblige us to pollute less, in order to restore the planet's "equilibrium." Still for others, the virus is a plan of God to remind us to not be selfish, or greedy, or whatever comes to their mind. Such stories, on turn, further remind us how people that declare themselves as "non-believers" are almost always actually also believers, they just replaced "God" by "Mother Nature" or "Mother Earth" (see Fig. 2.10). This is because, within the *reality of the natural world*, the pandemic obviously has no "purpose" at all because no supernatural being planned it. Instead, within the *circumstances* of human evolution the rise of sedentism and agriculture lead to an increase of population density and thus to the rise of pandemics, which have been numerous since then and will naturally continue to occur in the future. For instance, the "**Antonine Plague**" affected Mesoptamia and the Roman empire from 165 to 180 AD and is supposed to have killed up to five million people, the "**Plague of Justinian**" is supposed to have killed up to half of the population of Europe in 541–542, and the "**Black Death**" probably killed up to 75–200 million people in 1347–1351 (Fig. 8.7; see also Box 8.3). There is nothing mysterious about these pandemics, if ones takes into account what we know about the reality of life and evolution in this planet.

Right now, while I write these lines in January 2020, many people are anxiously waiting to take the **Covid-19 vaccines** that were developed in the last months, in a record time, by scientists. In theory, this should have given a lesson to the storytelling animal: it is not praying to God, or shamanism, or Nirvana, or using crystal balls or the "vital force of energy," or the "unique power of the mind" that helped us to develop vaccines, which are right now the only way we have to potentially finish this very tragic pandemic. It was instead the *objective use of scientific trial-and-error methodology*. However, and not surprising due to everything we have seen so far in this book, the power of *Homo fictus et irrationalis*, of confirmation biases and of our tendency to try avoiding cognitive dissonance at all costs, are much more authoritative than logical objective scientific ideas, as attested by the fact that now that we do have vaccines available, many are thanking God for that. Even Trump, who was at that time president of the U.S. and had to please the religious voters that voted on him, said that the vaccines were a "miracle." But, if God is omnipotent, why would he create the virus to kill millions of humans, and then wait almost a year and about 1.5 million deaths before he decided to do a "miracle" to end up our suffering. Why did he not do the miracle some months before, at least? Does he not care about those 1.5 million people, and their families and loved ones? Religious people often say that this is a way of God to test our faith: but if this would truly be so, what a cruel, sadistic, horrible, narcissistic God that would be, to kill 1.5 billion human beings and impose a huge sufferance to their families just to test if we still like him after that.

In the meanwhile, while God does his "miracles" in *Neverland*, in the world of reality almost all the numerous scientists that were involved in the development of those vaccines that will save the lives of dozens of millions of people continue to be completely unknown to—and not acknowledged by—the vast majority of the storytelling animals. In fact, there is now a huge number of people that not only does not

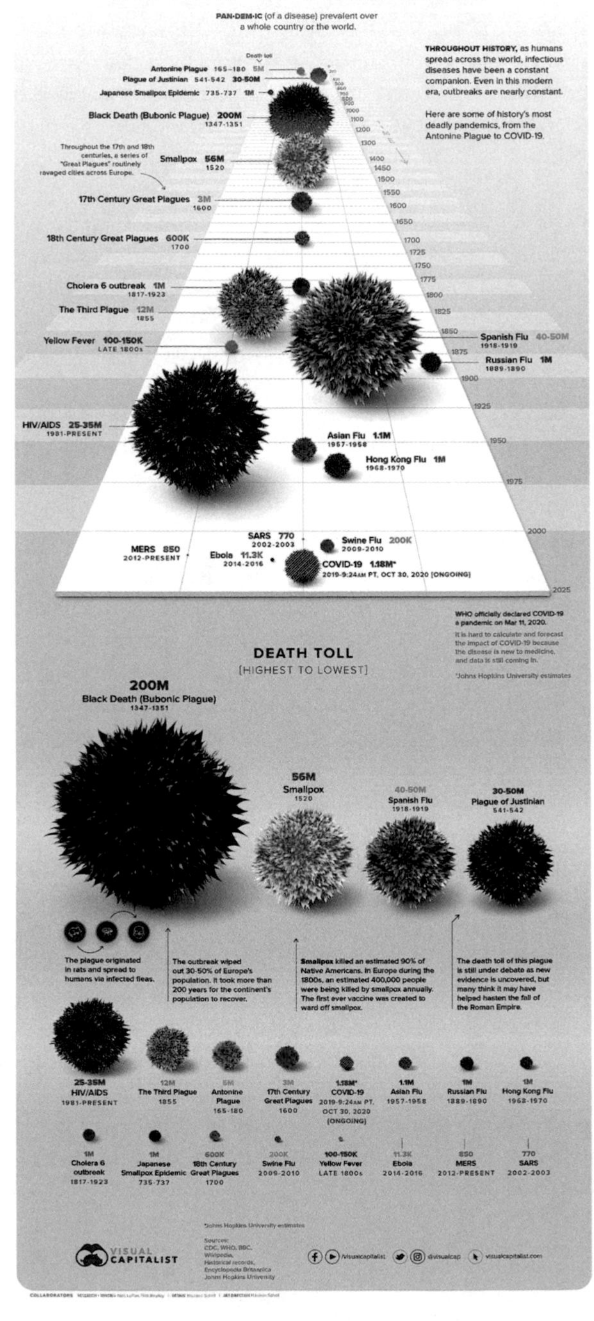

Fig. 8.7 History of pandemics, with worldwide covid-19 death toll updated until October, 30, 2020

recognize the amazing work done by those scientists, but that actually criticize them—for instance, using conspiracy theories that they are part of a "masterplan" made by Bill Gates, or Soros, to oblige us to take their demoniac, harmful vaccines—or doubt that the vaccines they created, and tested so carefully, are actually safe. To the point that in some countries, such as France, right now, as I write these lines, *most people*—I will repeat, more than 50% of French people—say they don't want to be, or are afraid of being, vaccinated. This is in great part due to the growing power of **anti-vax movements** and/or of **conspiracy theories** in the social media (see also Fig. 2.11).

To give an illustrative example about scientists being unrecognized, may I ask you: do you know who is **Maurice Hilleman**? Very likely, you don't. Me, a professor that teaches at a University college of medicine, I had never heard his name, before writing this book, what shows how not only the media and popular culture, but also science and education in general, have indeed huge problems and weird priorities. Completely unknown to 99.99% of the people nowadays, Maurice was *the* key person within the development of many vaccines that saved and continue to save the life of about eight million people—or even more—every single year. In other words, humans have not been able to escape *Neverland* not only because of our natural tendency to create and believe imaginary stories—which is amplified by our *Homo socialis* culture and inculturation—but also because most humans simply don't have any desire to even try to escape it anyway. That is why scientists such as Hilleman, who in their quest to know the *reality* about viruses, infections, and how they can be prevented/minimized have saved so many millions of lives, are infinitesimally less known worldwide than an endless number of singers, actors, fiction writers, movie producers, religious leaders, and defenders of conspiracy theories. While one can say that a huge proportion of people in the U.S. have probably heard about Kim Kardashian, only very few know about Hilleman. It suffices to say that even current or former presidents of countries such as U.S., Brazil, Russia, Turkey, Philippines, and numerous other ones, believe in and propagate conspiracy theories, even the most irrational and absurd ones such as those concerning **9/11** or **QAnon**. Two firm believers of **QAnon conspiracy theories** are now sited in the U.S. Congress, and many others politicians and celebrities believing in those theories are often interviewed on a daily basis and seen by millions of people in TV channels such as *Fox News*.

So, in order to at least try to contribute to change this *Neverland status quo*, let's learn together what Hilleman did, so more people in the planet will hopefully know one day at least some aspects of his research and life, as they show us, once again, how life is both random and contingent. This is because not only some of his discoveries were related to completely random infections, such as that of his own daughter, but also because he almost died when he was a kid. If that would have happened, it is likely that some **vaccines** that we now take for granted would have not been developed yet, or would have been developed much later, the consequence thus being that millions of people living nowadays would not be alive. I am citing here a brief excerpt from the website about a beautiful documentary that was made about him, entitled "Hilleman: a perilous quest to save the world's children"—which I strongly recommend you to see (https://hillemanfilm.com/dr-hilleman):

> Dr. Maurice Hilleman is considered by many to be the **father of modern vaccines**. Over the course of his career, he developed many of the vaccines that are routinely recommended for children today. By the end of his career, Dr. Hilleman had prevented **pandemic flu**, combined the **measles-mumps-rubella** vaccines (MMR), developed the first vaccine against a type of **human cancer**, and much more. It is likely that Dr. Hilleman's work has saved more lives worldwide than any other scientist in history. Maurice Ralph Hilleman was born August 30, 1919, in Miles City, Montana. Maurice's twin sister and mother died soon after his birth. His mother's dying wish was to have his sister buried in her arms and to have Maurice raised by his aunt and uncle, Bob and Edith. They lived nearby and had no children. Living near his siblings and biological father, Gustav, gave Maurice a different experience than that of his older siblings. His adoptive father, Bob, was more open-minded than Gustav. This likely allowed more room for Maurice's curiosity and intellect to flourish. But, he craved the approval of his biological father. This likely drove Maurice's lifelong goal to succeed. Even as a young boy, Maurice showed a strong aptitude for science. The writings of Charles Darwin were particularly interesting to the future scientist. Maurice graduated from high school in 1937, and was awarded a scholarship to Montana State College (now Montana State University). He majored in chemistry and microbiology and graduated in 1941 at the top of his class.
>
> He furthered his education at the University of Chicago where he studied **chlamydia**, a common sexually transmitted disease. In the early 1940s, chlamydia was believed to be caused by a virus. But through his doctoral work, Maurice determined that chlamydia was actually caused by a bacterium, which meant that it could be treated with **antibiotics**. Maurice and one of his professors also taught the first course ever offered in **virology** at the University of Chicago. In 1943, he married his hometown sweetheart, Thelma Mason. Upon completion of his PhD in 1944, Dr. Hilleman started working at a company called E.R. Squibb & Sons where he developed his first vaccine. The vaccine protected against **Japanese encephalitis** (JE) virus: the most common cause of **encephalitis** in the world. Infection with JE virus can lead to swelling of the brain and death. The vaccine was used to protect U.S. troops during World War II. In 1949, Dr. Hilleman became the chief of **respiratory diseases** at the Walter Reed Army Institute of Research in Washington, D.C. His project was to study **influenza virus**. During this time, Dr. Hilleman made discoveries critical to our understanding of influenza. First, he observed that the influenza virus underwent changes. Sometimes, the changes were small, but other times they could be dramatic. Later, the small changes became known as **antigenic drift**, and the sudden, major changes became known as **antigenic shift**. Antigenic shift can result in **pandemics** because almost no one in the world is immune to the significantly changed virus. Second, Dr. Hilleman realized in 1957, that an influenza pandemic had started in Hong Kong. He was the first person in history to predict a pandemic.
>
> The result was that he created a vaccine before the virus arrived in the U.S. Close to 70,000 deaths in the U.S. occurred as a result of the pandemic. Public health officials estimate that the number of deaths in the U.S. could have reached 1 million had Dr. Hilleman's vaccine not been available. For this effort, he was awarded the Distinguished Service Medal from the American military. At the end of 1957, shortly after his daughter Jeryl Lynn was born, Dr. Hilleman began working for Merck & Co. to oversee their vaccine research and development. He remained at Merck for the rest of his career. Sadly, his wife, Thelma, died in 1963. Dr. Hilleman married Lorraine Witmer the following year. They had a daughter, Kirsten, in 1965. While at Merck, Dr. Hilleman developed vaccines to protect us from **chickenpox**, **hepatitis A**, **hepatitis B**, **pneumococcus**, **meningococcus**, **measles**, **mumps**, and **rubella**. He was also the first person to combine viral vaccines when he created the MMR vaccine. With one shot, children could be protected against three diseases (measles, mumps, and rubella). The story of the mumps vaccine is unique because Dr. Hilleman iso-

lated the virus from his daughter, Jeryl Lynn, when she contracted mumps in 1963. By weakening the mumps virus he had obtained from Jeryl Lynn, he was able to make a safe and effective mumps vaccine. The same strain of mumps virus is used to make the **mumps vaccine** today. It is called the **Jeryl Lynn strain**. When he created the **hepatitis B vaccine**, Dr. Hilleman became the first person to develop a vaccine against a virus that **causes liver cancer** in people. While at Merck, Dr. Hilleman worked with colleagues to create two versions of the **hepatitis B vaccine**. Dr. Hilleman retired from Merck in 1984 at the age of 65 as this was the retirement age mandated by company rules. In the years after his retirement, he continued to contribute to science as a consultant and mentor to those still working to stop infectious diseases from harming or killing people around the world. For example, during this time, Dr. Hilleman served as an advisor to the World Health Organization (WHO). In 1988, President Ronald Regan awarded him the National Medal of Science, the highest science award given in the U.S. Colleagues note that Dr. Hilleman was always more concerned with preventing disease than being recognized for his efforts. Many people remain unfamiliar with Dr. Hilleman's name, despite the fact that his work has directly impacted their lives. Dr. Hilleman's work is estimated to save about eight million lives every year.

One crucial aspect about this discussion about Hilleman is that it illustrates that humans tend to give much more importance to people and jobs related to the creation of the *world of unreality* than to people and jobs associated with the study of reality, a fact discussed by authors such as **Pier Vincenzo Piazza**, in his recent book *Homo biologicus*. Even in the very rare cases in which scientists become part of popular culture, such as *Charles Darwin*, this is in great part because the storytelling animal has somehow fantasized their lives to make them more "spicy," more *Neverlandish*, for instance idealizing them or even converting those scientists into **infallible deities**. The reality is that Darwin, or Hilleman, are very far from being ideal, no human or non-human organism is. For instance, on the one hand Hilleman used and abused a huge number of non-human animals in order to develop vaccines, although on the other hand at least in his case those procedures allowed to save millions of human lives every year, that is the painful animal experiments that he did were not merely done to develop non-essential cosmetic products, for instance.

Hilleman himself recognized, later in his life, that he felt compelled to "pay back to chickens" the fact that he did use and abuse so many of them: according to him, that is why he created the first **vaccine against a virus that caused cancer** in chickens. However, such an idea only makes sense in an anthropocentric context, because those vaccines were applied only to *domesticated chicken that are used and abused by humans anyway*, not to birds living in the wild. That is, those vaccines were mainly beneficial, ultimately, to humans, who could thus eat more chicken and chicken eggs due to the decrease of the rates of mortality within domesticated chicken. Contrary to the tales that the storytelling animal loves to create, the natural world is indeed complicated: things are not "black" or "white," "good" or "bad," there are no "flawless heroes": "beautiful smart" dolphins kill fish to survive, "creative cuckoos" put their eggs on nests done by members of other species, some invertebrates kill and eat their mates after having sex and reproducing with them (Fig. 5.26). Things are just what they are.

Still about this topic concerning scientific innovations and **vaccines**, another aspect that needs to be pointed out, and that will be further discussed in the Sections below, is that vaccines are one of the key factors explaining differences of *average* **life expectancy** between "developed" countries versus **hunter-gatherer societies**. This is because vaccines dramatically reduced child mortality in the last decades, although it should also be noted that many of the vaccines created by scientists such as Hilleman concern infectious germs that became particularly prevalent—and that have mainly lead to pandemics—after the rise of agriculture, as explained above. Another key factor for the recent decrease of child mortality, for exactly the same reason, concerns the development of **antibiotics** and other types of medication that help us fight against a wide range of harmful germs, although overmedication and the exaggerated use of antibiotics have also lead to some serious "modern" health problems such as an increase of **autoimmune diseases** that lead to the death of millions of people around the globe, as explained above. In a way, it can be said that contrary to the use of vaccines, which are basically harmless in the vast majority of cases, the use of antibiotics has saved hundreds of millions of lives, particularly of children, but its overuse has also decreased the quality of life of many people living nowadays.

This is aggravated by the fact that exactly the same happens to the domesticated animals that provide a huge number of products we consume on a daily basis, such as milk, yogurts, meat, eggs, and so on: those animals themselves are forced to take a huge number of antibiotics—which can then be passed on to our bodies when we eat them—in order to increase their survival within the horrible conditions in which they live. Indeed, one should note that even in the wild, non-human animals such as mammals, including our closest relatives the chimpanzees—in particular the younger and the older ones—tend to be particularly vulnerable to infections by germs and parasites.

So, apart from the general increase of pandemics in agricultural societies, and the lack of vaccines and medication such as antibiotics in hunter-gatherer societies, what do the available empirical data tell us about other differences concerning health and well-being between such societies? Suzman's book *Affluence Without Abundance* provides some interesting points about this topic. However, before you read the below excerpt, and in order to not be confused with what is shown in Fig. 5.9, I needed to clarify two points. Firstly, as explained in Chap. 4, **caloric contribution to a diet** is not necessarily the same as the **total contribution to a diet**. To give an example, if you eat a Cheesecake Factory Chocolate Tower Truffle Cake every day, and eat a total of about 2500 calories a day, that means that this 1679-calories cake contributes 2/3 to your daily caloric consumption, but it clearly does not contribute 2/3 to your total diet, as a diet is much more than just counting the number of calories. This is because a diet that is minimally healthy needs to include a plethora of different items, including fibers, proteins, fats, antioxidants, vitamins, iron, and so on. Accordingly—and this is the second point—although meat has indeed a lot of calories and proteins the major contribution to the healthy diets of most hunter-gatherer societies comes not from hunting but from the plants that are by and large gathered by women, as explained above (Fig. 5.9). We thus need to take this into account, when reading the following lines from Suzman (see also Box. 8.3):

When compared to other well-researched **hunting and gathering societies**, foraging **Ju/'hoansi** were among the most moderate meat eaters. Excluding the almost entirely carnivorous arctic peoples, most other hunting and gathering societies acquired around two-thirds of their caloric intake from animal and fish products. **Australian Aboriginals** in Arnhem Land, for example, acquired 77 percent of their nutrition from animal products; the **Aché in Paraguay**, 70 percent; and the **Khoisan**'s closest genetic and linguistic relatives outside of southern Africa, the **Hadza** of Tanzania, 48 percent. Isotopic analysis of Paleolithic hominid collagen tissues suggests that these higher proportions were the norm throughout the history of modern *Homo sapiens*. Contemporary wisdom regarding the perils of **meat eating** suggests that eating as much meat as hunter-gatherers would be an evolutionary cul-de-sac. But it clearly wasn't, a fact now latched onto eagerly by contemporary advocates of '**Paleo diets**'. The science of understanding the relationship between what we eat and who we are has a long way to go. And it would not surprise anyone if the official advice on eating meat and animal products were to be reversed yet again in the near future on the basis of new evidence. There is also no real clarity on why hunter-gatherers have had such low incidences of **cardiovascular disease** despite their meat-heavy diets. It certainly isn't genetic. Among the handful of urban Ju/'hoansi and other **San** who have had enough money to eat what they want, when they want, bulging waistlines are the norm and **heart disease** has cut many of these lives tragically short. The most persuasive suggestion is that meat eating at the levels hunter-gatherers do in the context of their *broader diets* that include **high fiber and antioxidant intakes** combined with hardly any **carbohydrate-rich foods**, has probably made the difference. Exactly why it has made a difference, though, is uncertain. Perhaps it has something to do with the fact that they had to spend so little time working.

Box 8.3: Sedentism, Agriculture, Affluence, and Wellbeing

Suzman's 2017 book *Affluence without abundance* provides an interesting discussion on **sedentism, agriculture, affluence, religion** and **health**, which contradicts the idea that the rise of agriculture was some kind of 'natural', 'inevitable' step towards 'progress' and a 'better life' in the so-called 'chain of being'. Instead, it reinforces the idea—as do many other lines of evidence compiled in the last decades by historians, anthropologists and biologists, as noted in Chap. 4—that the **rise of agriculture** was likely the byproduct of a *local worsening of the quality of life*. This latter phenomenon in turn led to a subsequent further decrease of wellbeing in 'early civilizations'—for instance **famine** and **pandemics** killing millions of people (Fig. 8.7)—and a profound change of mindset regarding the way in which the broader public thinks about life, in 'civilized' societies:

> It is…likely that this transition to agriculture was pushed along by a far greater emphasis on **food storage** by hunter-gatherers, whose confidence in their environment's providence would have been severely dented by climatic upheavals. Whereas they were once certain that their environment would always be able to deliver something of use, they must have started to worry that if they had food today, there was no guarantee that there would be food the following day, week, month, or year. In circumstances like these, it is hard to sustain an **immediate-return economy**, and they would have gone to much greater efforts than they had in the past to preserve or store foods for later consumption. This practical measure alone would have started

to transform the way hunter-gatherers experienced and understood time, because in storing surplus foods their economic efforts became increasingly focused on meeting future as much as present needs. But over longer periods of time than a single seasonal cycle, **farming societies** were much more likely than hunter-gatherers to suffer severe, recurrent, and enduring famines. Hunting and gathering is a low-risk way of making a living. Hunter-gatherers hedge their bets by relying on many different potential food sources and so can capitalize on an environment's own dynamic responses to periodic droughts, floods, and other climatic anomalies. The 125 different edible plant species utilized by **Ju/'hoansi** in Nyae Nyae all have slightly different seasonal cycles, respond differently to different weather conditions, and occupy specific environmental niches. Some are more drought-resistant than others, some are more responsive to excessive rainfall, some cope better with exceptional cold, and others with exceptional heat. This means that when the weather proves unsuitable for one set of species, it is likely to benefit another. When you add hunting into the mix, this hedge becomes even more successful. While a severe drought will drastically reduce the total yields of some of the most important food plants in the Kalahari, it also usually makes hunting and scavenging easier, because animals' conditions decline, they become less alert, and they congregate closer to permanent water points or scarce food resources. Thus, among the foraging Ju/'hoansi, meat typically made up a larger proportion of people's diets during "lean months" than in good ones.

But in **agricultural societies** that depend on only a few staples, when there is not enough rain or rivers run dry, then harvests inevitably fail. And if a harvest fails and insufficient contingency has been made by setting aside food from the previous season or by having a sufficiently sophisticated exchange network to procure food from elsewhere, then famine is inevitable. This doesn't only apply to droughts. Most farmed plant species tend to be quite fussy and sensitive to **pests**. They only do well when they have just the right amount of rain at precisely the right time, just the right soil, and just the right amount of sunshine. Too much or too little of any of these may not necessarily result in a famine, but it will lead to a disappointing harvest. An infectious viral disease like **measles** with a mortality rate of nearly 90 percent among cattle populations with no immunity, the rinderpest scythed its way through the herds of southern and eastern Africa's pastoralists during the 1890s.... Livestock perishing in biblical numbers as a result of plagues like the rinderpest was not the only viral peril farmers faced. Living in such close proximity to increasingly intensively farmed animals meant that some **livestock diseases** adapted to human hosts. Our historic love affair with beef brought us **tuberculosis** and measles, and our hankering for bacon and chicken wings ensures that every once in a while we are hit with terrifying new strains of **influenza**. These diseases hit early Neolithic populations particularly hard, because they were on the whole poorly nourished compared to hunter-gatherers. This was not only due to occasional poor harvests but because their diets tended to be dominated by only one or two usually carbohydrate-rich crops that produced systemic **vitamin and mineral deficiencies**.

The vulnerability of agricultural societies to famine, disease, and natural disasters meant that the expansion of the Neolithic was punctuated by catastrophic **societal collapses**. The aggressive expansion of Neolithic peoples is revealed in the archaeology and genetic history of early Neolithic societies as well as more recent documented history. Comparisons of DNA extracted from the skeletons of Europe's early farmers with the DNA of Europe's various hunting and gathering peoples shows that farming did not spread because hunting and gathering populations were persuaded to adopt agriculture by their more productive neighbors. Instead it shows that hunting and gathering populations were typically displaced by aggressive farmers seeking new lands. Genetic histories of modern European populations point to just such a series of catastrophic collapses that coincided first with the Neolithic expansion through central Europe around 7500 years ago and then later with the Neolithic expansion into northwestern Europe about 6000 years ago. These collapses may have been caused by disease. The **mortality rates** associated with them were between 30 and 60 percent, a proportion roughly equivalent to mortality rates associated with the **Black Death** that stalked Europe in the fourteenth century. But the early Neolithic populations in Europe were small and spread out. This would have made it hard for **plagues** to be transmitted widely. Rather, it's more likely that these deaths came about as a result of unsustainable farming practices, an overreliance on a small number of plant and animal species, and a couple of years of unfortunate weather – perhaps a sustained drought, too cold a winter, flooding, or a combination of all three.

It is no coincidence that agricultural metaphors pepper the holy texts of the great religions. **Judeo-Christian texts** in particular are littered with parables and stories in which fields, farmers, domestic animals, harvests, and shepherds take center stage. And as if to remind us of the fears associated with agricultural disaster, the eventual fate that awaits humankind according to the **Christian Bible** is a harvest in which the "wheat of the blessed will be gathered into God's storehouse and the chaff of the damned cast into eternal fire". Where hunter-gatherers considered themselves to be part of their environments, farming societies saw their environments – or at least parts of them – as something separate from themselves, something manipulable. Once humans conceptually separated themselves from their environments in this way, it made sense that they started to reorganize and reclassify the world around them in terms of their ability to exert their control over it. To them, all the world was a "wild" and "natural" and often dangerous space when left to its own devices. And while farmers recognized that they depended on their ability to harness natural forces, they also took the view that where nature intruded unbidden into domesticated spaces, it became a pest. Unwanted plants growing in a plowed field became weeds, and unwanted animals partial to a farmers' grains or preying on his livestock were declared vermin. The conceptual separation farming communities made between the natural/wild and the human/cultural worlds was so widespread that for a long period social anthropologists believed it was a human universal. Over and above this, farming also makes numerous secondary demands on people's time. So when work in the fields is done, tools need to be made and mended, and farm structures built or maintained. And on top of all that, farmers need to perform ordinary everyday tasks, like looking after infants, preparing food, and maintaining their homes. Just as importantly, because farming generates harvests seasonally, storage vessels and systems also need to be built and other foods have to be preserved for consumption until the next harvest.

The almost universal belief now that **hard work is a virtue** is perhaps the most obvious of the **Neolithic Revolution**'s many social, economic, and cultural legacies. There are few societies on earth where work is not considered as fundamental a part of our humanity as our desire to reproduce or our need for companionship. In many societies it defines who we are and almost everywhere it dominates politics. In advanced economies, the airwaves are packed with the rhetoric of politicians and ordinary people alike invoking the virtues of "strivers" and "working families" and decrying the laziness of "shirkers" and "freeloaders." And in most less-developed economies, consultants and experts of all kinds spend their energy developing policy briefs and grand plans to create jobs. Almost everywhere **full employment** remains an ideal for politicians of all shades, among whom fears of rising unemployment invoke the specter of electoral defeat. It is no wonder, then, that John Maynard Keynes believed that our desire to solve the "economic problem" was not only "expressly evolved by nature" but was the sum of "all our impulses and deepest instincts." He just failed to recognize that economics became a problem only with the transition to agriculture and that our preoccupation with solving this problem was a consequence of our ancestors' having created it in the first place. If the agrarian equation between hard work and prosperity is an enduring legacy of the Neolithic Revolution, it is not the only one. For the additional demands of labor, coupled with increased productivity, laid the conceptual foundations of an economic model that inspired the likes of **Da Gama** and **Dias** to sail south (from Europe) and still shapes our ongoing preoccupation with productivity and trade. It did so by making hard work into a virtue and transforming time into a commodity, objects into assets, and systems of exchange into commerce.

Suzman's *Affluence Without Abundance* also provides fascinating discussions on broader topics such as **philosophy of life** and well-being that are related with topics such as **money** and affluence, **Keyne's economic thoughts**, and **Marx's communist utopia** (see also Boxes 8.1–8.4):

Old /Engn!au was the only **Ju/'hoan** I ever met to propose a theory of sorts about where money came from…. Unlike his other stories, the elements of this one – that money is created by magic; that it is acquired through trickery and deceit; that it inspires greed, violence, fear, possessiveness, and anger; and that it often appears to come from assholes and is covered in shit – do not require tremendous reinterpretation to find meaning in it…it is the only one of the many stories of the *First Times* that /Engn!au and others told me that resonates powerfully with life beyond the Kalahari. I suspect that this is because it deals with an element of their recent experience that is now near universal…. For most of us living in the world's richest countries, our absolute needs are almost universally met, and if resources were more evenly distributed among the population, they could arguably be met several times over. We are adequately nourished and live in warm homes packed with all sorts of enterprising gadgetry and comforts. And all of this stuff is imported or produced by the roughly one in ten of us who are employed in agriculture or manufacturing. The rest of us expend our productive and creative energies in the ever more expansive services sector, leaving some to wonder whether there is any point at all to what they do. As much as it is easier for some people to blame globalization, immigration, or any number of fantastical conspiracies for the decline in manufacturing jobs, the truth is that increased productivity and technological advancement are the real culprits..

Keynes was unusual among economists in his view that our productive instincts are secondary to our spiritual ones. Most other economists consider work to be the elemental particle of human sociality and economics the science of interpreting and manipulating the ever more complex forms arising out of these interactions. This view of human nature, which underwrites the free-market capitalism championed by Keynes's critics, also lay at the heart of **Karl Marx**'s critique of the free market. Marx, like generations of economists before and since, believed that human nature was to spontaneously and creatively produce in a manner that is conducive to social and individual satisfaction. To him the urge to produce was the essence of humanity, and his anxieties about **capitalism** had their roots in his belief that capitalism robbed people of the profound fulfillment that came from producing things. Marx's communist Utopia, in contrast to Keynes's post-labor Utopia, was one in which everyone continued to work but was liberated to seek a more profound fulfillment from their work by owning the "means of production." The evidence of hunting and gathering societies suggests that both Marx and **neoliberal economists** were wrong about human nature: we are more than capable of leading fulfilled lives that are not defined by our labor. But if this is so, then why is it proving so hard for humans to embrace abundance the way hunter-gatherers did? In part it is because hunter-gatherers' "primitive affluence" was neither a mind-set nor the economic expression of any particular ideology: there is no "manifesto of **primitive communism**." Their economic perspective was anchored in, among other things, their confidence in the providence of their environment, a hunter's empathy for his prey, an immediate-return economy, and indifference to the past and the future, and reaffirmed by social relationships shaped as much by jealousy as affection. It is also because there is another, more fundamental obstacle in the path to achieving Keynes's vision. For the hunter-gatherer model of primitive affluence was not simply based on their having few needs easily satisfied; it also depended on no one being substantially richer or more powerful than anyone else. If this kind of **egalitarianism** is a precondition for us to embrace a post-labor world, then I suspect it may prove to be a very hard nut to crack..

Namibia has an exemplary record of governance since independence. There have been problems, but these have often been more a function of capacity and resources shortfalls rather than bad intentions. Yet Namibia, like its neighbors South Africa and Botswana, remains among the top five most unequal countries in the world. This inequality is not, as has so often been the case in the world's "developing" economies, the result of the actions of a corrupt, kleptocratic class so much as a function of the nature of economic growth. But the net result is a massive concentration of available wealth in relatively few hands and a pronounced and obvious underclass making up half the population, with the **Bushmen** sitting at the bottom of the pile. If in becoming *New Time people* (which is part of the **Ju/'hoansi** beliefs) Ju/'hoansi have accepted that their lives are shaped by the unpredictable eddies and currents of an ever-changing world, then they might take some comfort in the idea that we are on the cusp of a new age in which we will no longer be hostage to the economic problem and in which the productive mind-set that the **Neolithic Revolution** nurtured will no longer be fit for purpose. Doing so will require that we learn to be more like the Ju/'hoansi's immediate ancestors, embrace the affluence we have created, and recognize value in things other than our labor. Working a whole lot less might well be a good place to start. And it may well be that ***millennials*** – a group in the first world who have known nothing but abundance and who seem increasingly inclined to seek out work that they love rather than persuade themselves to learn to love the work they find – will lead the way in doing this.

"Working a whole lot less might well be a good place to start…and it may well be that *millennials*…will lead the way in doing this": in a way, this idea of Suzman seems indeed to start gaining momentum in the last years, at least to a certain degree, as we will see in Chap. 9. To finish this Section, I will provide one of the most emblematic examples of how human "progress" can both be related to fascinating discoveries and to a decrease of quality of life and an increase of health problems, further illustrating that in evolution there is nothing just "good" or "bad," but instead an interminable series of trade-offs: the example concerns **space discovery**. This is because this is the *most extreme example* of a mismatch between the type of life our human and non-human primate ancestors had for dozens of millions of years versus the type of niche we are asking our bodies to adapt to in an extremely short geological time: first to highly populated cities, then to industrialized states, later to highly technological countries, and in the near future to life in space, inside space machines and/or in other planets. As explained in Davis' book *The Beautiful Cure*:

The consequences of disrupting our body clock are especially evident when magnified by the extreme conditions of space. As the **International Space Station** whizzes around the earth at around 17,000 miles per hour, astronauts are in sunshine for forty-five minutes, then darkness for forty-five minutes; sixteen days whizzing by for every one of ours on earth. A survey of sixty-four **astronauts** on **space-shuttle missions**, and twenty-one astronauts on the **International Space Station**, showed that most take **drugs** to help them sleep. Blood taken from space-station astronauts several times over a six-month period showed that, by all kinds of measures, their **immune systems** were in disarray. Many types of immune cells were redistributed in the body, activation thresholds had shifted and **T cells** had become less responsive. As far as we know, nobody has developed **cancer** or an **autoimmune disease** in space. **NASA** has its own rule that astronauts should not have their lifetime cancer risk raised by more than 3% by their work. Contrary to popular belief, though, astronauts do experience medical problems in space. These tend not to be from recent **infections**; since the earliest missions, precautions have been taken to prevent such occurrences…Instead, it is relatively common for a virus that is already present in the astronaut's body but which is dormant to be reactivated – in a similar way that the chicken-pox virus can reactivate later in life to cause shingles – probably because the astronaut's immune system is no longer able to keep it under control. Reactivations of all kinds of viruses (**cytomegalovirus**, **Epstein–Barr virus** and **herpes viruses**) have been documented in astronauts on both short- and long-duration missions. As far as we're aware, this has not led to anyone developing clinical problems in space – in other words, while the virus has become active and multiplied, the astronaut displays no symptoms of illness – but medical privacy or other rules about space flight might prevent disclosure of such a problem even if it did occur.

As well as reactivation of latent viruses, several crew members of the space station have developed a **skin rash**. In one case where blood samples were analysed, the presence of a rash correlated with changes in their immune system, including reduced functionality of their T cells and altered levels of **cytokines** in their blood. For this astronaut, the rash coincided with itchy watery eyes and sneezing, indicating an **allergic reaction**. This almost certainly developed from a disruption to the immune system caused by space flight. The astronaut had never had any of these problems on earth before and symptoms cleared up within days of the astronaut returning to earth. Symptoms peaked at the same time as **stressful events** during the mission, including just after a spacewalk, which fits with the idea that stress tends to make allergic reactions worse. Allergies in space are not rare. **Antihistamines**, used to counteract the effects of allergies, are the second most-taken medicine in space after **sleeping pills**. In at least one case, supplies of antihistamines ran out and more had to be sent up in the next scheduled space-shuttle docking. So for long-term space missions, the prospect of allergies, reactivation of latent viruses and possibly the development of autoimmune disease or cancer, are serious concerns. Brian Crucian, in his 'dream job' as NASA's lead scientist for all things immunology, thinks this could be a problem for, say, a trip to Mars. But, he says, it's hard to say if it's any more of a problem than the many other effects space flight has on the body: in addition to changes to the immune system, it undergoes **bone loss**, **muscle loss**, **cardiovascular problems**, **impaired vision** and **psychological stress**. *In short, we are not built for space*. The human body has evolved to fit our environment. It is tuned to the level of gravity felt at the earth's surface, the twenty-four-hour cycle of day and night, the way we interact socially, and so on. If there is ever a realistic plan for humans to settle elsewhere in the **solar system**, our **immune system** and many other bodily systems will need to be tricked into thinking we haven't left home.

To be clear, I am not criticizing at all **space exploration**, well on the contrary. Personally, I am a huge fan of it, because it involves scientific discovery and has already contributed, and will surely contribute even much more, to our understanding of the fascinating cosmos in which we live, and thus of the *world of reality* (see Chap. 9). I am not suggesting, at all, that we should not innovate or that we should not change at all our way of life and should live as our ancestors lived millions of years ago. Such a way of thinking makes no sense at all, because all organisms have changed their way of life in a way or another, and are still doing so: the ancestors of whales were land tetrapods, and the ancestors of those tetrapods were fish. Biological evolution never stops. What I am trying to emphasize, instead, is that our obsession with "progress" is often rushing things in a way that is completely different from what naturally happened with all other animals: our species only has about 200 or 300 thousands of years, and we want to go from African habitats to other planets, and our bodies can't keep with that. It would be like saying to the land ancestor of whales: OK, be ready now, we will drop you in the deep ocean. Simply, that land ancestor would just quickly drown in water, as it would not have had millions of years to develop fin-like upper limbs and so on.

So what I mean is that we should not be rushing all the time, for instance we can surely try to go to other planets, and discover fascinating details about them and the cosmos in general, but that does not mean that we should all go and live to those planets before we find reasonable healthy ways to inhabit them. That is, when **space exploration** will "revolutionize" the daily-lives of humans, decades or centuries from now, I hope humans will reflect deeply, without rushing, about the *pros and cons*—the *evolutionary trade-offs*—concerning the quality of life and health, and try to balance the cons with knowledge that we have accumulated from evolving in this planet for millions of years. Clearly, that kind of balance and knowledge has been neglected in the last millennia with all the major "revolutions," due to our obsession about "progress" and about discarding our "animal" roots and bodily needs and pleasures by portraying them as "inferior traits" or even as sins. We should just not forget—and in particular not be obsessed in forgetting—our evolutionary past: we should remember that now, after millions of years of evolution, whales live on the sea, but still they *need* to go up from time to time to breed oxygen, as their land ancestors did for hundreds of millions of years, and as we also still need to.

Box 8.4: Dopamine, Rewards, Wanting 'More', and Economic Growth
Sapolsky's 2017 book *Behave* provides examples of fascinating links between the biology of our body, our obsession with wanting 'more and more', and **teleological narratives**, including those about an **after-life**:

> In typical mammals the **dopamine system** codes in a scale-free manner over a wide range of experience for both good and bad surprises and is constantly habituating to yesterday's news. But humans have something in addition, namely that we invent pleasures far more intense than anything offered by the natural world. Once, during a concert of cathedral organ music, as I sat getting gooseflesh amid that tsunami of sound, I was struck with a thought: for a medieval peasant, this must have been the loudest human-made sound they ever experienced, aweinspiring in now-unimaginable ways. No wonder they signed up for the **religion** being proffered. And now we are constantly pummeled with sounds that dwarf quaint organs. Once, **hunter-gatherers** might chance upon honey from a beehive and thus briefly satisfy a hardwired food craving. And now we have hundreds of carefully designed commercial foods that supply a burst of sensation unmatched by some lowly natural food. Once, we had lives that, amid considerable privation, also offered numerous subtle, hard-won pleasures. And now we have drugs that cause spasms of **pleasure** and dopamine release a thousandfold higher than anything stimulated in our old

drug-free world. An emptiness comes from this combination of over-the-top non-natural sources of reward and the inevitability of habituation; this is because unnaturally strong explosions of synthetic experience and sensation and pleasure evoke unnaturally strong degrees of habituation. This has two consequences. First, soon we barely notice the fleeting whispers of pleasure caused by leaves in autumn, or by the lingering glance of the right person, or by the promise of reward following a difficult, worthy task. And the other consequence is that we eventually habituate to even those artificial deluges of intensity. If we were designed by engineers, as we consumed more, we'd desire less. But our frequent human tragedy is that the more we consume, the hungrier we get. More and faster and stronger. What was an unexpected pleasure yesterday is what we feel entitled to today, and what won't be enough tomorrow.

Schultz's group has shown that the magnitude of an anticipatory dopamine rise reflects two variables. First is the size of the anticipated reward. A **monkey** has learned that a light means that ten lever presses earns one unit of reward, while a tone means ten presses earns ten units. And soon a tone provokes more anticipatory dopamine than does a light. It's "This is going to be great" versus "This is going to be great." The second variable is extraordinary. The rule is that the light comes on, you press the lever, you get the reward. Now things change. Light comes on, press the lever, get the reward…only 50 percent of the time. Remarkably, once that new scenario is learned, far more dopamine is released. Why? Because nothing fuels dopamine release like the "maybe" of intermittent reinforcement So dopamine is more about anticipation of reward than about reward itself. Time for one more piece of the picture. Consider that monkey trained to respond to the light cue with lever pressing, and out comes the reward; as we now know, once that relationship is established, most dopamine release is anticipatory, occurring right after the cue. What happens if the post–light cue release of dopamine doesn't occur? Crucially, the monkey doesn't press the lever. Similarly, if you destroy its **accumbens**, a rat makes impulsive choices, instead of holding out for a delayed larger reward. Conversely, back to the monkey – if instead of flashing the light cue you electrically stimulate the **tegmentum** to release dopamine, the monkey presses the lever. Dopamine is not just about reward anticipation; it fuels the goal-directed behavior needed to gain that reward; dopamine "binds" the value of a reward to the resulting work. It's about the motivation arising from those **dopaminergic projections** to the PFC [**prefrontal cortex**] that is needed to do the harder thing (i.e., to work). In other words, dopamine is not about the happiness of reward. It's about the happiness of pursuit of reward that has a decent chance of occurring.

This is central to understanding the nature of motivation, as well as its failures (e.g., during **depression**, where there is inhibition of dopamine signaling thanks to stress, or in anxiety, where such inhibition is caused by projections from the **amygdala**). It also tells us about the source of the frontocortical power behind willpower. In a task where one chooses between an immediate and a (larger) delayed reward, contemplating the immediate reward activates **limbic targets of dopamine** (i.e., the **mesolimbic pathway**), whereas contemplating the delayed reward activates frontocortical targets (i.e., the **mesocortical pathway**). The greater the activation of the latter, the more likely there'll be gratification postponement. One more complication: these studies of temporal discounting typically involve delays on the order of seconds. Though the dopamine system is similar across numerous species, humans do something utterly novel: we delay gratification for insanely long times. No warthog restricts calories to look good in a bathing suit next summer. No gerbil works hard at school to get good SAT scores to get into a good college to get into a good grad school to get a good job to get into a good nursing home. We do something even beyond this unprecedented gratification delay: we use the **dopaminergic power** of the **happiness** of pursuit to motivate us to work for rewards that come *after we are dead* – depending on your culture, this can be knowing that your nation is closer to winning a war because you've sacrificed yourself in battle, that your kids will inherit money because of your financial sacrifices, or that you will spend eternity in paradise. It is extraordinary neural circuitry that bucks temporal discounting enough to allow (some of) us to care about the temperature of the planet that our great-grandchildren will inherit. Basically, it's unknown how we humans do this. We may merely be a type of animal, mammal, primate, and ape, but we're a profoundly unique one.

8.4 Rousseau, Hobbes, Biases, and "Civilization"

Celia's child, about four months old, died last Saturday the 12th...this is two negroes and three horses I have lost this year. (Davin Gavin)

I have noted above that the myth that most nomadic hunter-gatherers died/die in their 30s or 40s is being deconstructed by an increasing number of empirical studies. This myth was constructed in great part due to two main factors. One is obviously the blind acceptance of narratives about "cosmic progress" and related notions about the "ladder of life" and Hobbesian statements about the "nasty, brutish, and short" life of "primitives." The other is the longstanding and still widespread confusion between the terms **life expectancy at birth**, **median lifespan**, and **modal lifespan**. These terms have completely different meanings. In a very simplified way, lifespan is the number of years that a person lives, and thus median lifespan is the *average* lifespan of people within a certain group. Let's say that 40% of a population die in infancy, before being 1 year old, and the remaining 60% die at 70: the median lifespan would be 42.

In contrast, modal lifespan refers to the age at which the greatest frequency of people in a given dataset died: so, in the same case, this would be 70 years old. Life expectancy is also based on averages, but refers to the number of years that someone is expected to live from a specific starting point, changing as people grow older and face special risks at it is happening right now during the Covid-19 pandemic, and so on: in a very simplified way, life expectancy is more similar to median lifespan than to modal lifespan, as the former ones are particularly influenced by child mortality, while this does not apply to the modal lifespan of many nomadic hunter-gatherer societies. This is because, as explained above, in these latter groups there are often substantial rates of **child mortality**, but then within those people that remain alive during those critical stages of development and reach early adult life—let's say, 15 years old or so—a large proportion of them might live for many decades more, often having a high quality of life and health until they are very old, being, for instance, highly mobile and participating in various activities that are crucial for the whole group.

In fact, the modal lifespan of many hunter-gatherer groups was calculated in some detailed cross-cultural examinations, which have consistently indicated that in numerous groups it can be up to 70 years, with groups such as the **Tsimane of Bolivia** having modal lifespans of up to 78 years as explained in Gurven and Kaplan's 2007 paper "*Longevity among hunter-gatherers: a cross-cultural examination.*" These numbers don't just completely contradict the still-prevailing Hobbesian myths about the "short" life of "brutish savages": they are truly impressive, because such modal lifespans are substantially higher than those of most agricultural states for most of the time since the rise of agriculture, until very recently. For instance, the modal lifespan for the total population of Sweden—which is one of the Nordic countries that were so often seen by many racists, including scientists, as the "pinnacle" of the "ladder of life"—in the 1850s was about 70 years according to Canudas-Romo's 2010 paper "*Three measures of longevity.*" That is, it was 8 years below that of the Tsimane, a group that those same racists would designate as "brute" and "primitive." In contrast to those racist ideas, an increasing number of recent empirical studies are showing that Tsimane people that are 70 or even 80 are on *average* much healthier, in many ways, than people that are 70 or 80 in the vast majority of "developed" countries nowadays. For instance, one of those studies became quickly widespread within the globe precisely because it made the media, broader public and even many scientists reconsider the Hobbesian myths they had *believed in* for such a long time. A March 2017 *Guardian* article entitled "*Tsimané of the Bolivian Amazon have world's healthiest hearts*" summarized the results of the study for the broader public:

> A high carbohydrate diet of rice, plantain, manioc and corn, with a small amount of wild game and fish – plus around six hours' exercise every day – has given the **Tsimané people of the Bolivian Amazon** the healthiest hearts in the world. It may not be a life that everyone would choose. The Tsimané live in thatched huts with no electricity or modern conveniences. Their lives are spent on hunts that can last for over eight hours covering 18km for wild deer, monkeys or tapir and clearing large areas of primal forest with an axe, as well as the gentler pastime of gathering berries. But as a result of this pre-industrial lifestyle, the Tsimané have hardly any hardening of the arteries. **Heart attacks** and strokes, the biggest killers in the US and Europe, are almost unknown. The study published in the Lancet medical journal and being presented at the American College of Cardiology conference shows that an 80-year-old Tsimané man has the vascular age of an American in his mid-50s. Researchers, who investigated the lifestyles of the Tsimané and checked out their arteries with CT scanners, say that there are lessons for those of us who live sedentary lives in urban areas and eat packaged foods. "This study suggests that coronary atherosclerosis [hardening of the arteries] could be avoided if people adopted some elements of the Tsimané lifestyle, such as keeping their LDL **cholesterol**, **blood pressure** and **blood sugar** very low, not smoking and being physically active", said senior cardiology author Dr Gregory S Thomas from Long Beach Memorial Medical Centre in the US. "Most of the Tsimané are able to live their entire life without developing any **coronary atherosclerosis**. This has never been seen in any prior research. While difficult to achieve in the industrialized world, we can adopt some aspects of their lifestyle to potentially forestall a condition we thought would eventually effect almost all of us."
>
> Coronary atherosclerosis is the build-up of plaque in the arteries leading to the heart, which slows the blood flow and can cause blood clots – which may in turn lead to a heart attack. The researchers found that almost nine out of 10 of the 705 Tsimané adults who took

> part in the study had no risk at all of heart disease; 13% had a low risk and only 3% – 20 individuals – had moderate or high risk. Even in old age, 65% of those aged over 75 had almost no risk and only 8% (four out of 48) had a moderate to high risk. By contrast, in the US, a study of more than 6,800 people found that half had moderate to high risk – five times as many as among the Tsimané people – and only 14% had no risk of heart disease at all. In the Tsimané population, heart rate, blood pressure, cholesterol, and blood glucose were also low. The study suggests that genetic risk is less important than lifestyle. "Over the last five years, new roads and the introduction of motorised canoes have dramatically increased access to the nearby market town to buy sugar and cooking oil", said Dr Ben Trumble, of Arizona State University, US. "This is ushering in major economic and nutritional changes for the Tsimané people". Those whose lifestyle is changing have higher cholesterol levels than others who stick to hunting and fishing. Senior anthropology author Prof Hillard Kaplan, from the University of New Mexico, said the loss of subsistence diets and lifestyles could be classed as a new risk factor for vascular ageing. "We believe that components of this way of life could benefit contemporary sedentary populations", he said. Tsimané people are more likely to get infections than those in the US, but even so, he said, "they have a very high likelihood of living into old age". The researchers cannot yet say whether diet or the active lifestyle is the more important component, said Kaplan, but they want to go on to investigate that by following those of the community whose lifestyles change with exposure to the town. "My best guess is that they act and they interact", he said. And it could be as much the foods that the Tsimané do not eat that gives them healthy hearts as the food that they do. Their diet is high in **unrefined carbohydrates** (72%) with about 14% protein and it is very low in **sugar** and in **fat** – also 14%, which amounts to about 38g of fat a day including 11g of **saturated fat**. "In the evolutionary past, fat and dense energy in the form of sugar were in short supply", Kaplan said.

As noted above, a crucial topic, when one talks about the difference between life expectancy and median and modal lifespans, is **child mortality**. As I will discuss this topic in Sect. 8.5, for now I will just stress that although there is a clear correlation between being a "developed" country and having in general lower rates of child mortality, the correlation is not necessarily related to **GDP (gross domestic product)**. For instance, the U.S. is the country that spends the most on health, per capita, but has one of the highest **Infant Mortality Rates** within the "developed" countries. Similarly, it is important to emphasize that the countries that have a higher proportion of centenarians are not at all those with a higher GDP. For instance, as of November 2019, the six countries that appear more consistently in different lists about "top-centenarian countries" were: sixth Italy, 31.5 centenarians per 100,000, then France in fifth with 32.1, Thailand with 35.9, Spain with 37.5, my own country Portugal with 38.9, and then Japan in first place with 48.2, with the estimate average for all the countries in the globe being 6.2 centenarians per 100,000. The fact that three of the four southern Mediterranean countries that are often pejoratively called as "PIGS"—for Portugal, Italy, Greece, and Spain—as well as another Mediterranean country, France, are among the six countries listed as having the highest centenarian ratios clearly goes, once again, against the commonly accepted Anglo-Saxon notions of "progress." It also goes against tales about "working being a virtue" or "improvement via economic growth," because, as seen in Fig. 8.8, three of the "PIGS," and France, are listed—among the countries shown in that figure—as the ones in which people spend *less* time doing "paid work." This cannot be—and factually *it is not*—just a coincidence. Particularly because the very high rate of

How do people spend their time?

Averages of minutes per day from time-use diaries for people between 15 and 64.

Our World in Data

	Paid work	Sleep	Other unpaid work (Care work, volunteering)	Housework & shopping	Personal care	Eating & drinking	TV & Radio	Seeing friends	Other leisure	Total leisure
China	315 mins	9 hours 2 mins	56	123	52	100	127		78	228 mins
Mexico	302 mins	8 hours 19 mins	84	202	58	77	66	44	62	172 mins
South Korea	288 mins	7 hours 51 mins	70	89	90	117	102	42	114	258 mins
Austria	280 mins	8 hours 18 mins	65	145	55	79	109	82	101	292 mins
India	272 mins	8 hours 48 mins	44	160	75	84	61	73	119	253 mins
Canada	269 mins	8 hours 40 mins	81	139	52	65	109	53	116	278 mins
Portugal	259 mins	8 hours 26 mins	52	176	58	112	114	44	83	241 mins
USA	251 mins	8 hours 48 mins	96	122	57	63	148	44	100	292 mins
New Zealand	241 mins	8 hours 46 mins	89	134	42	80	124	69	108	301 mins
UK	235 mins	8 hours 28 mins	95	133	58	79	133	47	125	305 mins
Ireland	231 mins	8 hours 11 mins	132	118	42	75	85	49	178	312 mins
Poland	229 mins	8 hours 29 mins	77	160	57	91	122	45	119	286 mins
Germany	224 mins	8 hours 18 mins	71	141	55	95	118	61	152	331 mins
Netherlands	218 mins	8 hours 23 mins	68	133	65	114	113	73	130	316 mins
Turkey	217 mins	8 hours 35 mins	88	138	50	118	124	68	94	286 mins
Norway	201 mins	8 hours 12 mins	100	103	56	79	129	57	183	369 mins
Denmark	200 mins	8 hours 9 mins	73	154	52	119	123	81	124	328 mins
Finland	200 mins	8 hours 28 mins	104	136	52	81	118	55	158	331 mins
Belgium	194 mins	8 hours 33 mins	52	149	53	99	131	50	158	339 mins
Greece	187 mins	8 hours 20 mins	45	141	57	128	137	56	148	341 mins
Spain	176 mins	8 hours 36 mins	89	141	51	126	129	51	136	316 mins
France	170 mins	8 hours 33 mins	39	151	107	133	114	55	124	293 mins
Italy	149 mins	8 hours 33 mins	70	162	68	127	104	65	154	323 mins

Education: In school & study

Data source: OECD Time Use Database, Gender Data Portal. For most countries surveys were conducted between 2009 and 2016, but surveys for some countries are older.
OurWorldinData.org – Research and data to make progress against the world's largest problems.

Fig. 8.8 How people spend their time within 33 countries

centenarians living in Japan has much more to do with extremely high rates seen in rural areas or islands such as Okinawa than with what happens in highly technological and populated cities such as Tokyo. Basically, in islands like Okinawa for many people life continues to be, in many ways, "as it was back in the old days," as is precisely also the case in the rural regions and islands of Mediterranean countries such as **France**, **Portugal**, **Italy,** and **Spain** in which the rates are also extremely high. Just to give an example, in Okinawa there are, in average, 40 times more centenarians than the average of the rest of Japan.

This has to do with a plethora of factors including the much more relaxed and traditional type of life, the fact that on average people in such islands consume diets with less calories than those consumed in more "modern" cities, and so on. This has been discussed in detail in numerous empirical studies, and hundreds of popular books that try to better understand aging in our species to try to solve once for all the main obsession of humans since we became aware of the inevitability of our death: such studies have the aim of literally *making death avoidable*. The *Homo fictus et irrationalis* is tired of being anxious and suffering for millions of years with this obsession, and wants to become a *Homo Deus*, as pointed out by Harari, in order to stop this anxiety once for all. The paradox is that all the studies about centenarians made so far have shown that those humans that live longer do not do so because of "progress" and "technology," but instead because the places where they live are precisely not as 'developed' or technologically 'advanced' as places such as New York or Tokyo. Instead of obsessively trying to look for an hyper-technological *Homo Deus*, we are now understanding that in many ways it is those living "back in the old days," including those that were deemed to be "savages" such as the Tsimane, that can actually give us important clues about how to "cheat" aging. One of the most popular books about these issues is Buettner's 2012 *The blue zones, second edition—9 lessons for living longer*. These **Blue Zones**—which have a very high ration of centenarians—are, indeed, chiefly "lay-back" places that are not at all those typically described as the most "developed" or "technological" ones: for instance, **Ikaria in Greece**, **Costa Rica**, **Sardinia in Italy**, and precisely **Okinawa in Japan**. For example, concerning Sardinia, factors that are listed are the **importance of community**, associated with **lower rates of depression**, **suicide,** and **stress**—exactly the opposite of what happens in hyper-technological Tokyo—of **celebrating elders**—also contrary to what happens in Tokyo—and of being relaxed—something that also hardly applies to Tokyo. They also include simply taking a walk, or drinking a glass or two of red wine daily and laughing with friends—both these factors being associated with reduced stress and increased social interactions.

That is, factually, all these factors in many ways correspond to features that are often seen in most nomadic hunter-gatherer groups, except taking wine per se. But obviously similar actions, related to the reinforcement of social interactions, are also frequently done by such groups, the difference being that instead of taking wine they take other drinks or foods or undertake other practices. It is fascinating to note that most—if not all—these features were mainly seen as "primitive" or at least as "backward," by most influential scholars in Europe in the nineteenth century and beginning of the twentieth century, such as those from UK and Germany. Actually, their criticism did not only concern the "primitive" features of "savage" hunter-gatherers, but also the "backward traits" of Southern Europeans, for instance. The **Mediterranean siesta**—sleeping in the afternoon for a while—and the type of relaxed, "lay-back," and communal lifestyle typical of many rural areas and islands of the "PIGS" were commonly seen by those northern European scholars as "lazy," "inefficient," and "unproductive," and the **Mediterranean olive-oil-based type of diet** was frequently seen as "oily" or "greasy." Obviously, scientific studies have been contradicting, one by one, all these racist and xenophobe imaginary tales in the last decades. For instance, the Mediterranean diet was inscribed in 2013 on the ***Representative List of the Intangible Cultural Heritage of Humanity*** and is now considered one of the most—probably *the* most—healthy type of diet within "developed" countries. Also, it is now widely accepted that the siesta brings numerous health benefits, particularly after walking for a short time after lunch. A 2007 study by researchers at the Harvard School of Public Health and Greek scientists involving 23,000 adults has shown that Greeks who took regular 30-min **siestas** were 37% less likely to die of **heart disease** over a 6-year period than those who never napped.

However, I want to go even deeper into this topic, because despite the recent openness to recognize the value of and the lessons we can learn from other types of way of life, traditions, and diets, there are still many biases and prejudices concerning how these subjects are discussed in the media and even in books such as that of Buettner. For instance, why does the media in general, or Wikipedia, or many of those popular books, don't refer to the amazingly high centenarian rates of countries such as **Dominica**? When I went there, many locals proudly told me that they are one of the countries in the planet with a higher rate of **centenarians**. However, I had never encountered information about this fact within the numerous books, specialized papers, documentaries, podcasts, and numerous other sources I had read, seen, and heard until then. Nor was I able to find—after the locals told me this—Dominica listed in the vast majority of the "centenarian-lists" that are easily found in the web. So I decided to do the calculations myself. And the locals were completely right: Dominica had, in 2019, a ratio of 33.8 centenarians per 100,000 people, being therefore even above Italy and France, just below Thailand, Spain, Portugal, and Japan, using the numbers that I mentioned above.

Unfortunately, one of the key reasons why Dominica does not come in such lists, or is discussed by the main media, top Websites, and popular books is not mysterious at all: it is because the a priori ideas and imaginary tales that are still so prevailing nowadays do not even allow most Westerners to think that Dominica *could possibly* be one of the countries in the planet with a higher rate of centenarians. One thing is to start having a change of mindset and recognizing that the Mediterranean "PIGS," or even Costa Rica, have "Blue Zones of centenarians"—this goes against various longstanding narratives about "progress," but still countries such as Portugal and Spain are nowadays often considered to be "developed" countries by most Westerners, even if a bit "backward" for many of them. But Dominica? A country that for many people of such "developed" countries is just one of the "shit-hole countries"—paraphrasing the words of the former U.S. president, Trump—of the Caribbean region, with a population that mainly consists of "black" descendants from former slaves, how can it possibly have a higher rate of centenarians than most "hyper-developed" countries such as Germany, England, and even the very country of Trump, the U.S.? This is unthinkable. It contradicts the notion of "progress," as well as the deepest expectations of most people of such "developed" countries, be them overtly racist or not. Particularly because moreover Dominica is especially vulnerable to hurricanes, its hospitals and health system are clearly not among the most "developed" in the planet, and its GDP per capita is one of the lowest within the Eastern Caribbean states, which in general have low GDPs. In other words, Dominica fulfills all the typical items to be considered a "non-developed" country and, indeed, according to the official report of **IMF**'s ***World Economic Outlook Database*** of April 2019, it is included in the group of "developing/emerging" ones, which are defined as countries "with a less developed industrial base (industries) and a low *Human Development Index* (HDI) relative to other countries."

However, despite—or better said in at least some ways precisely *because* of—being "undeveloped," the fact is that Dominica does have a higher ratio of centenarians—33.8—than that of almost all the countries in the planet, something that should make us reconsider the appropriateness of using terms such as "Human Development Index." Indeed, 33.8 is almost five times higher than in so-called highly developed countries such as South Korea (7.7), four times higher than

Netherlands (10.4), more than the double than Norway (13.1), Austria (16.1), Denmark (16.1), and Switzerland (16.6), and about 1.5 times higher than the United Kingdom (21.5) and the Unites States (22). As a picture can be more valuable than a thousand words, I think the best answer I can give you to complement everything that was said above and explain the so-called puzzling paradox of Dominica's centenarians is to ask you to look carefully at Fig. 8.9. I took these pictures in Dominica and other small East Caribbean countries—some of them also have very high ratios of centenarians, although of course they are also neglected in most lists and discussions about centenarian for the very same reasons that Dominica is. The picture I took in Dominica is the one in which I am with the manager of the hotel where I stayed, who did a dinner with local items just for me, completely for free, as I arrived a bit late from the airport. Basically, these pictures show a **laid-back type of life** in which many people often have several hours of leisure per day in which they walk near, or are inside the sea, often talking with other people. Note also the very low levels of air pollution and the very fresh—and in general, very healthy—food that is constantly available, including a huge diversity of fruits and fish.

During such travels, I often wonder about this: the countries with higher GDPs should in theory have people that would be able to get the most fresh and healthy

Fig. 8.9 Pictures from my 2018 summer trip to Dominica—in one of the pictures I am standing with the amazingly friendly and kind manager of the hotel in which I stayed in that country—and other small East-Caribbean countries. Many of them—including Dominica—have extremely high ratios of centenarians among their populations

food, to have more time for leisure, to do more physical activity, and thus to have a higher quality of life in general, but often, very often indeed, what truly happens is exactly the opposite. How can it be that in the U.S., the "biggest economy" in the planet, there is: one of the highest ratios of **obese people** in the planet; a huge number of **suicides** and of people with **depression**, **anxiety,** and/or problems related to extreme **loneliness**; a vast proportion of people eating food that is mostly non-fresh—often heated in microwaves— and so unhealthy; a troubling scarcity of fresh fruit juices; and so on? According to a 2011 report by the ***National Center for Health Statistics (NCHS)***, the rate of **antidepressants use** in the U.S. among people aged 12 and older increased by almost 400% between 1988 and 2008, with **antidepressants** being the third most common medical prescription taken by people in that country between 2005 and 2008. However, it should be noted that saying that people in the richest countries don't necessarily have the highest quality of life, food, and leisure, is *not at all* the same than stating that people in the poorest countries have the highest quality of life. This is clearly not the case. Actually, such imaginary narratives about the "happy poor" are often constructions made by people living in rich countries to feel good with themselves and with the fact that they do nothing at all to help those "happy poor". In fact, within all the countries I have been, I have never heard someone from the poorest countries saying they are happy with that, or because of that, or that they want that. The poorest countries tend to have millions of refugees, or millions of people that starve or die with diseases that have been already eradicated in "developed countries," or that have to face terrible traumas and wars. Yemen is, currently, a particularly disturbing example of this.

The above discussion leads us to a key point that is often neglected in comparative analyzes about **life expectancy** and **median and modal lifespans** between different "developed" countries or between them and sedentary and nomadic hunter-gatherer societies: none of them is an ideal measure to infer **quality of life**. I will provide here a very extreme theoretical example, just to illustrate this point more clearly. If I take a sample from a nursing home of a city in a "developed" country such as the U.S., it is very likely that a large proportion of those people sampled will be more than, let's say, 70 years old, and that a substantial proportion of them will live at least 15 years more. So, the modal lifespan in this case is a bit higher than that of a group such as the Tsimane. However, I think nobody would argue—unless one would be very biased—that *most people* that are 70 within the Tsimane do not have a higher quality of life, on average, than the people living in many U.S. nursing homes. This is because we actually *know*, based on empirical data, that older Tsimane people do tend to be healthier than people with the same age living in the U.S., as noted above. This is particularly so because nowadays a substantial part of the population of "developed" countries has to endure, for many years or even decades, a plethora of chronic non-communicable diseases that were not so common centuries ago and particularly before the rise of industrialization (Chap. 4).

A very sensitive, but crucial, point that relates to this discussion is that, apart from the huge number of people of "developed" countries that commits **suicide** there is also a substantial number of people in those countries that express an **euthanasia** wish but that are not allowed to die as they want to. This is attested by the fact that there are major movements in many of those countries to try to legalize euthanasia, and many countries already did, but most countries in the planet still don't. I am completely in favor or it, I must note: as long as one ensures that the person that feels that she/he no longer wants to live consistently confirms that this is so for at least a few times during let's say several weeks, so we are sure that this is not just a temporary feeling. In this sense, euthanasia, with all the related "moral" and legal complications related to it, is indeed one of the most powerful, and excruciating, illustrations that having long lifespans and a high quality of life can be—and often indeed are, in at least some 'developed countries'—very different things. Because this topic is so delicate, it is still a major subject of taboo nowadays, particularly because in *Neverland*'s world of unreality people often cannot face—and in a certain way, stand—anything concerning death, as explained in **Atul Gawande**'s 2014 book *Being Mortal—Medicine and What Matters in the End*. Gawande stresses that our "modern" society is so obsessed about "progress" and, accordingly, about increasing life expectancies and modal lifespans at all costs with technological innovations and "state-of-the art" medication and practices, that it often fails to realize that numerous people that have their lives extended at such a cost actually *don't want that*, precisely because they are enduring a very low—actually, often very horrible and painful circumstances without a true—quality of life.

An even more extreme example that we could provide—perhaps the most extreme of all—concerns those people that are in **coma** for many years. I profoundly admire the technology that was developed and that allows such people to be alive and, in many cases, to recover after some years or even decades: literally, those people would be dead if they were living in hunter-gatherer societies or in the Amazon with the **Tsimane**. *As long as the people that are saved do want to be saved—that is, as long as their euthanasia wishes are fully respected—any life saved is a major milestone*. I definitively admire the strength of people that undertake and endure such harsh circumstances of life, as I admire the scientists that worked so hard to develop treatments and technologies to allow the survival of people that would otherwise no longer be with us. My own mother, who is right now here next to me, would not be alive if scientists would not have invented pacemakers, so I am forever grateful to the scientists and doctors that contributed to develop such keystone innovations, as well as to the surgeons that placed the pacemaker in her body, and the extremely friendly and caring nurses that took care of her in the days before and after the surgery was done. The only point I want to make is that, if let's say a "developed" country would one day be able to prevent most deaths by achieving a way to keep people in coma before they die, waiting for a future cure, and let's say, that would increase the modal lifespan of people of that country to 100 years, this clearly would had nothing to do with a general increase of the quality of life.

Well on the contrary: that increase would only be possible by achieving a way to keep alive people with an extremely low quality of life, in coma. But, again, this theoretical example about coma is a very extreme one, as many medical procedures, such as those involving the placement of **pace-makers** that give me the privilege to be sitting next to my mother, do not imply at all a decrease of quality of life. Actually the pace-maker not only allowed her to be alive and us to enjoy her presence since the surgery was done—and hopefully for many years to come—but it also actually *increased her quality of life*. In those frequent cases where her heart would naturally—"nature"—go astray, the pacemaker—a product of human cultural innovations, so in a way of "nurture"—activates and prevents this to happen. So, she no longer have moments of extreme tiredness, dizziness or faintness, what makes us think again about how delicate and complex is the longstanding discussion about "nature *versus* nurture": naturally, without human culture and innovation, my mother would no longer be here with me. So, this is a reminder for those that might want to criticize this book for being against science and innovation, or for wanting that humans go "back to the jungle." It is not at all about that, it is about combining some amazing achievements of "civilization" such as the development of vaccines and pace-makers with some fascinating aspects of the lifeways of nomadic hunter-gatherers, such as the lack of slavery or the more healthy type of life they do and food they eat. Above all, this book is *an ode to science*, as the scientific method is precisely the most powerful and efficient weapon we have right now to fight **Captain James Hook** and thus to escape the dungeons of *Neverland*.

This discussion thus brings us back to the so-called **Hobbes versus Rousseau** debate, and therefore also to **Pinker**'s *Better Angels*. We have seen above how a huge amount of data contradict many of the ideas defended in that book, which does include some interesting scientific and historical information but then portraits it in an extremely biased, Hobbesian way.

Pinker writes, for instance: "in an article…called "*The Decline of Elite Homicide,*" the criminologist Mark Cooney shows that many **lower-status people**—the poor, the uneducated, the unmarried, and members of minority groups—are effectively stateless." I am not putting in question the data provided by Cooney: I am instead amazed by the use, by Pinker, of the term "lower-status" people to refer to minority groups, or unmarried people. Does everybody, in the twenty-first century, need to be married, or to not be African American, or Native American, in order to not be considered "lower-status people?" Another example of a very awkward statement made by Pinker is when he states that one of the reasons for the increase of crime rates in the U.S. in the 1960s was likely related to the sexual liberation of women. Of course, Pinker needed to "make up" some weird idea, truly anything that would explain that increase, because it goes against his idea that there is a "progress" towards less crime, less violence, and so on in the last centuries/decades. So, he uses one of the oldest tactics: he blames women. But the way in which he uses that tactic is paradoxically, *in itself*, a powerful demonstration that there is no "cosmic progress" or "expanding moral circle" as he is obsessively trying to convince us in his book, because the book plainly shows how people like him continue to be as reactionary and misogynistic as as so many 'white' men were in the 1950s.

Namely, about the "fault" of women within the increase of the crime rates in the 1960s in the U.S. he wrote "together with self-control and societal connectedness, a third ideal came under attack (in the 1960s): **marriage** and **family life**, which had done so much to domesticate **male violence** in the preceding decades…the idea that a man and a woman should devote their energies to a **monogamous relationship** in which they raise their children in a safe environment became a target of howling ridicule." That is, the "fault" was mainly of women because they forgot that one of the key functions they *have* to perform in life is to "domesticate male violence." Under such a *Neverland* **reactionary** and **misogynistic scenario**, the idea is that women became so "selfish" that they started to do horrible egotistical things such as to be with whoever they wanted, to divorce from men that they did not love, to have their own jobs and money, and so on. Pinker makes this even more obvious in a subsequent sentence: "the 1960s **decivilizing process** affected the choices of individuals as well as policymakers…many young men decided that they ain't gonna work on Maggie's farm no more and, instead of pursuing a respectable family life, hung out in all-male packs that spawned the familiar cycle of competition for dominance, insult or minor aggression, and violent retaliation, the **sexual revolution**, which provided men with plentiful sexual opportunities without the responsibilities of marriage, added to this dubious freedom." For Pinker, the sexual revolution of the 1960s was a "dubious" event, and only people that followed the 1950s model of family life can be truly considered to be "respectable" and, obviously, of "high status."

Pinker further added that "the mixture was as combustible in the inner city as it had been in the cowboy saloons and mining camps of the Wild West, this time not because there were no women around but because the women lacked the bargaining power to force the men into a civilized lifestyle." Here it is, again: not only the part about the main function of women being to "domesticate" men, but also a new component—if you don't follow the 1950s model of family life, you are not "respectable" and cannot even be called a "civilized" person. You are as "bad" as the "nasty, brutish" uncivilized cave-man to which Hobbes referred to. Fortunately, Pinker's imaginary, groundless tale finishes with a happy ending. It had to, because Pinker, as Hobbes, base their ideas on very old Western teleological tales about "progress" and "ladder of nature," so they have to end on a "high note": very high on the ladder, literally. Therefore, he states: "so how can we explain the recent crime decline [in the last decades]? I think two overarching explanations are plausible: the first is that the **Leviathan** got bigger, smarter, and more effective…the second is that the **Civilizing Process**, which the **counterculture** had tried to reverse in the 1960s, was restored to its **forward direction.**" So, here it is: the "villains"—that is, the **counterculture movement**, including sexual liberation and the fight for the **rights of women** and minorities within the U.S.—had "tried to reverse the **Civilizing Process**," but fortunately the "good guy"—that is, the force of "cosmic progress"—was stronger and thus made things go back to the "forward direction" to which the cosmos is moving since time immemorial. Concerning the creative imagination put on such just-so-stories, as well as their total disconnection to reality—there is nothing such as "forward direction" within the cosmos—such "bigger-than-life" stories made-up by Pinker rival those of the ***Epic of Gilgamesh*** about "**cosmic purpose**," written thousands of years ago. To finish this Section, I will provide an excerpt from Kirkham's 2019 book *Our Shadowed World*, in which he refers to some very interesting historical notes as well as personal ideas about these long-standing quests for, and imaginary teleological tales about, cosmic purpose and cosmic progress that continue to be so prominent today. Although some of his personal ideas should be taken with a grain of salt, such as that about a "Green revolution," I do think it is worthy to read this excerpt, as they further show how recent narratives such as Marx's "**march of history**" and Pinker's "forward direction" mainly use, indeed, ideas recycled from very old teleological narratives and religious texts:

> **Zoroastrianism** [was] the first of the great religions to understand the world in terms of a moral order and a teleological conclusion. Previously the world had been viewed through the medium of **myth**, of heroic conflicts that subdued chaos and ensured the natural cycles of fertility. **Zarathustra** proclaimed a moral order under the direction of one supreme being (**Ahura Mazda** – the **God of Light**), who directed all things to a final end, a day of judgment when books would be opened, the record of all past actions proclaimed, and our lives evaluated by a final decision that would stand forever. Thereafter history would increasingly be viewed as periodic, progressive, even predestined, and having a final END. But all these expectations, like those original hopes of **John the Baptist** for a unilateral **divine intervention**, came to nothing. **Utopian dreams** of an eschatological ending remained just that, dreams – or, as the perceptive and radical social commentator **Karl Marx** put it, an opiate of the people…. Crossan's work helps in distinguishing between a rhapsodic and impossible **utopia** (Greek for "notplace"/nowhere) and a possible eutopia ("good-place"/ this place), which it is within our power to construct: "**Eutopia** imagines a social world of universal peace, a human world of nonviolent distributive justice where all get a fair and adequate share of God's world". It is exactly this imminent or realized eschatology, of human willingness to act for the common good starting here and now, that is the crucial element of a transformed or transfigured world. In the nineteenth century such an understanding of apocalyptic expectation was beginning to take shape under a new name, **Communism**, a new specter that now stalked the cities of Europe and proved a lightning rod for the growing resentment of the newly urbanized masses of impoverished workers. As Eric Hobsbawm has written, "Around 1840 European history acquired a new dimension: the 'social problem', or (seen from another point of view) potential social revolution". A new, purely secular millenarianism now haunted Europe, and **theology** gave way to **ideology**, transforming old utopian and **apocalyptic dreams** into **revolutionary social demands**. Marx provided the new **millenarianism** with a "scientific" basis grounded on an economic understanding of the inexorable *march of history*. Nothing could now alter mankind's final destiny.
>
> But once again it didn't happen. By the end of the twentieth century that dream was also in tatters, and its Soviet delivery system had collapsed. For all his brilliance Marx had overlooked one simple reality: nature. The environmental indifference and destructiveness of **Soviet-style Communism** was monumental and catastrophic. Vast industrial and agribusiness plans – like the cotton plantations of central Asia that desiccated the Aral Sea – proved apocalyptic in a manner and degree quite contrary to expectation. It was an environmental recklessness that was finally epitomized by the nuclear disaster at **Chernobyl** – a name ironically meaning "wormwood" – that inevitably drew attention to the passage in Rev 8:10, in which a star called Wormwood fell from heaven and poisoned rivers and springs so that the people died. But **Soviet communism** was not alone in its environmental destructiveness. **Western capitalism** proved equally rapacious. In complete disregard of its optimistic ideology, its exploitation of the finite and fragile resources of the natural world turned out to be unsustainable. It took the new prophetic voice of Rachel Carson – a woman's voice at last! – in the face of abusive skepticism to alert the world to an impending ecological apocalypse. Ecologists such as Thomas Berry challenged thunderous disbelief to emphasize that "the basic disruption of all the basic life systems of earth has come about within a culture that emerged from a **biblical-Christian matrix**". This was not accidental in a culture and belief system that viewed the earth as having been created for the benefit and delight of humanity. Just as theology once gave way to ideology, now ideology had to make way for ecology. The coming of the **Green Revolution** seemed to confirm what two

German academics, Rotteck and Welcker, had written in 1842 of the "proletarian revolution" that replaced the conflicts of the feudal world: "No major historical antagonism disappears or dies out unless there emerges a new antagonism". Ecocide, the shadow side of economic growth, had now become the apocalyptic specter stalking the modern secular milieu. The two specters that now in particular overshadow the world are population growth and climate change – two seemingly insoluble crises that in turn reflect a disconnect of people and governments from the natural world.

Here, then, we see a pattern of events closer to home than we might like to think. As Patrick Geddes has pointed out, each historic civilization begins with a living urban core, the polis, and ends in a shattered ruin – a necropolis, a city of the dead, of fire-scorched ruins, empty workshops, heaps of meaningless refuse, the population massacred or driven into slavery. Mumford cites mighty Rome as the perfect example of how one civilization after another, having achieved power and centralized control, fails to reach "an organic solution of the problem of quantity". In fact, on reading his description of Rome's breaking point – the suffocating numbers, rising rents, and deteriorating housing conditions; the overexploitation of resources – one could almost mistake it for the London news, or for news from many great contemporary cities. As he concludes, "When these signs multiply, Necropolis is near, though not a stone has yet crumbled After each collapse, apocalyptic as it may have been to those involved, new beginnings were always made somewhere else". So what if anything is different now? The answer is devastatingly simple: all preceding collapses were regional events initiated by localized factors – there was always "somewhere else". But today's global society has many features of a single global civilization and is threatened by such global phenomena as climate change, ocean acidification, collapsing ecosystems, and exponentially expanding overpopulation. There is no longer a "somewhere else". Archaeologist Ronald Wright warns that we have become victims of a "progress trap" of our own making. *Our belief in progress has, he argues, hardened into an ideology*, "a **secular religion**, which like the religions that progress has challenged, is blind to certain flaws in its credentials". This was always evident, but little noticed, even at the outset when the drainage of land and cultivation of crops made the first towns of Sumer possible. But these very technological developments in time created salt pans that ruined the land, destroyed the crops, and starved the cities – a process that was among the first installments of the "progress trap". Since then the human species has become infinitely more inventive, expansive, and powerful. Wielders of flint knives could be overpowered by those with bronze or iron swords, who in turn would fall to those with muskets and canons; now the whole of humanity can be exterminated with atomic weapons. This may be a logical progression and material progress of sorts, but the final step is one too far…when the progress trap is sprung one last time.

8.5 Child Mortality, Unlimited Wants, and our Planet

When I look back on all these worries, I remember the story of the old man who said on his deathbed that he had had a lot of trouble in his life, most of which had never happened. (Winston Churchill)

Two papers published in 2019 in the journal *Nature Human Behavior* have provided some interesting information about subjects that we have discussed in the Sections above, such as the links between "work," "leisure," affluence, and the rise of agriculture. As noted in a "news and views" article by Reyes-Garcia discussing the original research presented in in one of those two papers, published by Dyble and colleagues:

When considering the evolution of human societies, the adoption of **agriculture** is often seen as a way to escape from an arduous **foraging lifestyle**, the foraging-to-farming transition being associated with increased food availability and fertility, despite the decline in dietary breadth and overall health. In a seminal work four decades ago, Sahlins argued that the transition from foraging to agriculture also came at the cost of more working and less **leisure time**. While the thesis has been enthusiastically embraced by many anthropologists, only a handful of researchers have put it to empirical tests. This gap is surprising because, despite many other differences, all humans are equal in that we have 24 h per day, and the way we spend them can be considered an objective measure of livelihoods. So the question remains: did humans give up their free time to embrace agriculture? In a new study, Dyble and colleagues use data from a contemporary society of politically egalitarian **hunter-gatherers**, the **Agta** from the northern Philippines, to put the hypothesis to empirical test. In his inspiring book, which continues to be mandatory reading in most anthropological courses, **Sahlins** proposed that the foraging way of life was "the **original affluent society**" in which "all the people's wants are easily satisfied". Sahlins argued that hunter-gatherer societies are affluent because they meet their needs with what is available to them and desire little else. Sahlins argument was based on the idea that, while foragers did not enjoy material wealth, they were able to subsist with few hours of work per day and enjoyed large amounts of leisure time. This hypothesis has found much acceptance among anthropologists aiming to counter the idea hunter-gatherer societies are primitive, but, as mentioned, it has rarely been tested. Moreover, the scarce empirical research on the topic has indicated that Sahlins's argument might not apply universally, as there are important variations in foraging and farming populations in issues such as food processing times or cyclical fluctuations in food availability. Moreover, interpretations of what is considered 'affluence', 'work' and 'leisure' time are context-dependent, making interpretations made by the Western eye misleading.

The Agta are particularly suitable for examining whether the adoption of agriculture is associated to the amount of time devoted to leisure because, although they continue to largely rely on foraging, they are increasingly engaging in agriculture and other forms of non-foraging work. Using a unique dataset with almost 11,000 individual spot observations collected in ten camps, Dyble and colleagues compare use of time across camps that vary in their relative engagement in foraging *versus* non-foraging out-of-camp work. The authors find that individuals who live in camps that generally engage more in non-foraging work spend more time working and have less leisure time than their peers who live in camps where foraging is more frequent. The result is largely explained by changes in women's use of time, as women in their sample seem to significantly increase their out-of-camp work as camps move away from foraging. This study provides compelling evidence that greater engagement in farming and other non-foraging work is associated with increased out-of-camp work time and, consequently, *decreased leisure time*, thus apparently confirming Sahlins' intuition. However, the interpretation of these findings should be done with caution…[for instance] extrapolation from data collected among contemporary foragers to explain **foraging-to-farming transitions** in prehistory is contentious. Challenges of interpretation aside, Dyble's work remains one of the few empirical tests of Sahlins' hypothesis. Future work in this line could further test the ecological impact of Sahlins' theory. Sahlins argued that forager societies enjoyed more leisure time because they had limited "wants". In the current situation of **ecological crisis** and **overexploitation of resources** linked to **unlimited wants** for material consumption, Sahlins' argument seems to have a rabid relevance. Perhaps we should be looking back to prehistoric forager societies to learn how to limit our material wants and, in passing, recover some free time.

Having unlimited wants in a planet that has limited resources *does* seem to be a very dangerous, irrational idea. This topic lead us to finally discuss one of the most sensitive subjects that we will address in the present book: **child mortality**. By being related to a plethora of emotions, it is very difficult to address this subject in a purely "objective" and encompassing way, particularly because moreover any comprehensive discussion of it involves taking into account several layers of complexity, and various long-term implications. For instance, on the one hand clearly most people would agree that it is "morally right"—or, as humanists would say, our "duty"—to reduce **infanticide** (see also Box 8.5) and **child mortality**. I have a son myself, and I cannot even think about what I would feel if I were to lose him—it is something horrific, unimaginable. Accordingly, the decrease of child mortality and infanticide have been among the major examples used by scholars such as Pinker to defend their Hobbesian ideas about "progress" and "civilization," because **hunter-gatherer societies** tend to display much higher child mortality rates than "developed countries," as explained above. As noted in Hrdy's book *Mothers and Others*: "the best available data for **Hadza**, **Ju/'hoansi**, or **Aka** foragers indicate that 40–60% of children in these populations—and more in bad times—died before 15." Forty to 60% is truly a lot, surely a number that no "modern" society would tolerate nowadays. In fact, as noted in Klarsfeld and Revah's 2003 book *The Biology of Death*, humans from "developed" countries have lower initial rates of mortality than any other animals studied so far: just about 0.02%, compared to about 0.2% for African elephants, 2% for dogs, 1% for female mice, and 7% for some fish species. So, there is no doubt that in this sense this is an amazing—probably the higher, "morally"—achievement of *science and medicine*. This need to be emphasized: the merit of medicine and science was tremendous because this achievement was attained *not* due to *Neverland*'s imaginary stories and tales about Gods or Mother Nature, but instead *despite* of those stories.

However, on the other hand, only someone very biased would refuse to accept that it was precisely the huge decrease of mortality rates—particularly of children but also of many adults—in the last centuries and principally in the last decades related to medical discoveries and innovations such as vaccines, antibiotics, and pace-makers that lead, in great part, to the two major threats faced by us and by the planet in general right now: *overpopulation*, and thus the *overuse of its limited resources* (see Fig. 9.6). Not surprisingly, such biased authors do exist: many of them, including some scholars, dissociate overpopulation from current problems such as global warming, overuse of resources, and so on, while still others even go all the way to argue that overpopulation has nothing to do with the decrease of child mortality that occurred in the last centuries. However, such arguments have been strongly contradicted by countless empirical works, such as those summarized in Diamond's excellent book *Collapse*, or specialized papers such as Shelton's 2014 article "*Taking exception – reduced mortality leads to population growth: an inconvenient truth.*" Shelton did not use the term "inconvenient" because the results he presents are not sound or even consensual, but instead precisely because some biased authors are desperately trying to deny this obvious truth. This is because by doing so, one would be able to avoid the **cognitive dissonance** created by the fact that something we all praise—decrease of child mortality—is related to the biggest problem faced by us and many other species—Earth's ecological collapse. However, the world of reality and the natural world don't care about comfortable tales, or happy-ending imaginary stories, and the fact is that these two items are almost surely related, indeed. As noted in Shelton's 2014 article, the comforting tale that overpopulation and child mortality might not be associated is mainly part of an Western-centric view of life, based on a trend that is basically only seen in Western countries and very few other states in the last centuries, in which both **child mortality** and **fertility rates** decreased enormously. In the vast majority of other countries, this is not at all the case, as explained by Shelton:

> Of course, the situation among modern developing countries varies and is different from that in Europe a century ago. For one thing, child mortality rates have typically declined much more rapidly in developing countries. And modern communications have fueled rising aspirations for many. But notably, substantial mortality declines in a number of countries, especially in Africa, have not yet been followed by appreciable declines in fertility. A prime example is **Nigeria**. Despite declines in infant mortality over many years, total fertility has persisted at about 6 children per woman…for fertility levels to decline, women and couples must have good *means* to control their fertility, in addition to motivation…. For most of human history, global **population growth** was extremely slow, because mortality and fertility levels were in fairly close equilibrium. But recent times have taken us rapidly to 7 billion and counting. As demonstrated in the classic work of Thomas McKeown, *The modern rise of population,* the only plausible explanation is declines in mortality. Consider, there are only 3 possible determinants of population change – fertility, migration, and mortality. Fertility may sometimes have increased marginally but, overall, certainly not appreciably; and migration is net zero for the planet, with mostly some out-migration for most developing countries. That leaves only mortality decrease as the primary explanation for the profound increase in population. Moreover, reduced *child* mortality plays a huge role. Deaths to children under 5 typically account for at least half of all deaths in pre-transition societies, and child mortality declines have been dramatic. In addition, child survival contributes to population "momentum" because most of those surviving children will eventually have children themselves. Thus, not only does reducing mortality contribute to rapid population growth, it is the *predominant* cause, notwithstanding the partial virtuous cycle that reduced child mortality may partially help over time to reduce fertility levels.

As it has been stressed throughout the present volume, in the natural world there is truly nothing completely "good" or completely "bad," even if in the particular case of child mortality we all want to believe that this is something totally "good," with no "negative" consequences or implications. But the decrease of child mortality and of mortality in general may indeed ultimately lead to the worse thing that can happen to our species: its total extinction, which will necessarily imply a huge increase of child mortality to levels never seen in the 6 millions of years we have been in this planet. If this would happen, then of course one—not us, as we would not be here, but some other organism—could then ask a profound philosophical question such as: was it truly a "good" thing to reduce child mortality in the first place from X% to Y%, if they would have known this would then lead to a subsequent increase to 100%, in which all children, as well as teenagers, adults, and so on, would die leading to the extinction of the whole *Homo sapiens* species? The answer is not simple at all: do this mean we should let children die, while we have a way to save them? Would it not make much more sense to have less kids, for instance? If two parents have two kids, then they are not increasing the world population, even if child mortality is reduced to 0%. One important point, related to this topic, is that most people in "developed" countries tend to think that the high child mortality occurring in most hunter-gatherer societies is a result of them (1) not having modern medicine and (2) **abandoning/letting die newborns** that are in some way "defective" or at least a "burden," reflecting (3) the "true ape/primate nature" of our deep ancestors, including the last common ancestor of chimpanzees and humans. However, only items 1 and 2 are correct: point 3 is an erroneous assertion based on our tendency to think that humans necessarily have a so-called higher morality than other animals. In reality, as noted in Hrdy's book not only point 3 is wrong, but point 2 should also be taken with a grain of salt. This is because abandonment is actually very rare in hunter-gatherers and in many current agricultural states newborns are abandoned—a substantial number of children are orphans for this reason, the critical difference with hunter-gatherer societies being that many of them are then taken care of by orphanages, boarding schools, residential treatment centers, group homes or foster care—or even let to die. This happens not only when parents have no resources to take care of their children, but also if they consider that they have the "wrong" sex—think for instance about what happened in China under the rule of "one child only." Hrdy explains:

> Many mammalian mothers can be surprisingly selective about which babies they care for. A mother mouse or prairie dog may cull her litter, showing aside a runt; a lioness whose cubs are too weak to walk may abandon the entire litter with no attempt to nudge them to their feet, carry them or otherwise help. Some mammals (and this includes humans) even discriminate against healthy babies, if they happen to be born the 'wrong' sex. But not Great Ape or most **primate mothers**. No matter how deformed, scrawny, odd, or burdensome, there is no baby that a wild ape mother won't keep. Babies born blind, limbless, or afflicted with cerebral palsy – newborns that a hunter-gatherer mother would likely abandon at birth – are picked up and held close. If her baby is too incapacitated to hold on, the mother may walk bipedally or tripedally as to support the baby with one hand…[in] a particularly extreme case (within **Japanese macaques**), a newborn with neither hands nor legs…his mother carries him everywhere and holds him up to nurse when he can't reach her nipple. Had local people not fed these **monkeys**, the mother would not have been able to constantly assist her handicapped infant to stay aboard and still remain fed and safe herself. But there is no question that she would try. Maternal devotion in the human case is more complicated…*more conditional*…a newborn perceived as defective may be drowned, buried alive, or simply, wrapped in leaves and left in the bush within hours of birth. In most traditional hunter-gatherer societies, abandonment is [however] rare, and almost always undertaken with regret. It is an act no woman wants to recall…. Back when the **!Kung** still lived as nomadic hunter-gatherers, the rate of abandonment was about one in one hundred live births. Higher rates were reported among people with strong sex preferences, as among the pre-missionized Eipo horticulturalists of highlands New Guinea. 41% of live births in this group resulted in abandonment, and in the vast majority of cases the abandoned babies were newborn daughters whose mothers hoped to reduce the time until a song might be born.

That is, contrary to the idea that most people would have a priori, based on tales created by the storytelling animal, **maternal devotion** is actually *more conditional* in humans than in many other primate groups. So, now, things are turned up-side-down, for those authors that defend the idea of "progress" by associating it to a supposed "expanding moral circle," such as Pinker. That is, who is more "civilized," a langur monkey mother that does not abandon her newborn, even if it is already a desiccated corpse (Fig. 8.10), or some people in human hunter-gatherer societies

Fig. 8.10 Old World monkey and non-human ape mothers are not known to discriminate their progeny based on infantile attributes and, moreover, many times continue to carry the corpse of a dead baby for days, as this young langur monkey mother is doing

that abandon their newborns if they are "defective" or a "burden," or some people from "modern" countries such as China that did abandon newborns because they were from the "wrong sex?" On the one hand this stresses, once again, that evolution is highly dynamic, and there is no profound "big divide" between humans and other animals: the abandoning of the progeny by some mothers is a feature shared by some humans—seen in many hunter-gatherer groups and some "modern countries"—and many mammals but not by numerous other non-human mammals such as many primates. In contrast, the non-discrimination based on a baby's particular attribute such as sex is something shared by some animals—non-human primates, particularly apes—and some humans, particularly those in Western countries nowadays (see also Box 8.5). On the other hand, this discussion also puts in question romanticized views of "nature" that are now so much in vogue, such as the notion that humans are "always worse" than other animals. This is because **common chimpanzees** can also sometimes attack "Others" in a particularly violent—and sometimes deadly—form of **tribalism**, and lionesses many times leave their weak cubs without any "attempt to nudge them to their feet, carry them or otherwise help them." In turn, some so-called noble aspects that are often displayed by hunter-gatherers and thus seen as an illustration of Rousseau's "Noble savage," such as their striking **egalitarianism**, can actually be not just a mere reflection of our "true ape nature," but also the result of an active cultural effort that they do to enforce **social equity** (Chap. 4). Biological evolution is complicated, and accordingly the truth is often between the extremes, being therefore probably a combination of some aspects of both Rousseau's views and Hobbes' ideas. In fact, such a "midway, both good and bad" nuanced view is exactly the type of view defended in Hrdy's book, when she points out that most hunter-gatherer societies had/have both a high child mortality *and* a high adult quality of life often associated to a remarkable philosophy of life:

> Over generations children would have watched with dismay as half or more of their siblings and cousins died at young ages. Yet by definition, individuals who *did* survive would have done so surrounded by others who cared for and shared with them. This endowed them with a personal confidence notably different from that of many modern people who grow up in environments with more available resources but less caring. People with French and German agricultural ancestors like my own are more likely to have been reared to beware of strangers. Many of us were put to bed with folktales about the world 'outside over there', a scary place peopled by impoverished widows, cruel stepmothers, hungry orphans, and unwanted children who lived surrounded by a dangerous forest where malign creatures – wolves and witches – lurked. To an **Mbuti** child, the forest is not so much dangerous as nurturing – it is a benignly encompassing mother-figure. Such a child is taught to be at least initially (until encountering information to the contrary) curious rather than fearful of outsiders.

Box 8.5: Infanticide and the Complex Links Between Biological Evolution, Morality, and Religion

Kunz's 2009 chapter of the book *The biological evolution of religious mind and behavior* discusses mesmerizing cases exemplifying the very complex links between **biological evolution, morality**, and **religion**:

> A multiplicity of examples gathered by cultural anthropologists and sociologists…show how extensive the discrepancy can be between the statements made by humans and their behavior, and also between cultural values and actual behavior. This can be shown by the example of handling **infanticide**. Its rejection…becomes a part of the life-affirming orientation as provided by **Christianity**. Infanticide is a very common phenomenon in humans and animals. Children who were born to an unfavorable, or in an unfavorably felt time, were and are subject to the high risk of being put to death. Also, children living in bad health conditions were (and still are) frequently killed by their parents. From the perspective of **genetic fitness maximization**, this behavior can be quite adaptive, since these parents receive the opportunity to concentrate their resources on the remaining children with a better chance for survival. From the time of the Middle Ages up to modern times, infanticide was very common in Europe. With the rise of **Christianity** in Europe, the practice of infanticide came under pressure, since it became regarded as a sin…(leading to the) question as to whether it is adaptive from the perspective of the culture-and-reproductive-success approach to kill children if this leads to a larger number of grandchildren. Or is it adaptive to follow the laws of Christian religion to raise children, who have only a small chance of survival? The answer to these questions was supplied by the Christians themselves. Many Christians of medieval Europe and later times found some ways of avoiding Christian values. One example is the practice of interpreting some children as being a changeling. Sickly babies were labeled as nonhuman **demons** or as **goblin babies**, which were exchanged for the actual human

infants. This type of infant became an enfant change in France, a Wechselbalg in Germany, and a fairy child or changeling in England. Moreover, even in the context of the practice of wet-nursing, Christians found an opportunity for infanticide…in eighteenth-century Paris 95% of all babies were nursed by a wet-nurse. Fifty percent of these infants were born into the middle class, in which infants were mostly raised by a wet-nurse. However, instead of having their children raised by a wet-nurse living in their vicinity, many mothers decided to give them to distant living nurses in the countryside. This was a special form of infanticide, because most of these children died. In this case, neither the ideational aspects of religion thus led to adaptive behavior, nor does the result of the behavior of these Christians conform with the values of their religion. They found a way to symbolically reconcile the practice of infanticide with **Christendom** without changing their practical behavior.

Religion is thus often the exact opposite of a guide to adaptive behavior. Religion requires (reproductive) offerings of the believers, which can be used as cooperative signals to other persons and/or the community. Furthermore, many conceptions of universal religions are not accepted despite their uselessness (and often even because of their teachings that are absolutely not realizable), but rather because of their uselessness. Useless features are perfect ingroup/outgroup signals…. In a permanently changing world, those must be particularly stable ideational aspects of culture, which do not exert the behavior of humans or influence it to only a small degree. This is because behavior is exposed to the permanent pressure of (genetic and) cultural selection processes. Only for this reason can religious conceptions be transported over thousands of years. This runs alongside the fact that many believers in their daily lives not only pay little attention to their religion, but often do not know major parts of it. Therefore, the everyday behavior of humans not only differs within one religion from region to region, but also substantially within one society. As an intermediate result, it can now be stated, that religions can contain information which leads to adaptive behavior. The reason, however, lies above all in the fact that religions illustrate human knowledge, values, desires, and consciousness of conflicts during the time religions were formed or created. Apart from this, some religious aspects do not have any influence on behavior, others, on the other hand, contribute decisively to life style. These, however, do not compellingly benefit inclusive fitness, such as the rules concerning infanticide or the hostility of Christianity toward education and science in historical times, for example. This is cause for the assumption that the guidance to adaptive behavior does not represent the core of religious tasks.

The point made above in one of the excerpts of Hrdy's book—that in general the practice of infanticide in hunter-gatherers has been highly exaggerated in the literature—has also been supported in many other detailed works. For instance, Kelly's 2013 book *The Lifeways of Hunter-Gatherers* refers to a 1979 study by Howell in which the **infanticide rates** were very low when tabulated from informant interviews—for instance only 1.2% for the Dobe **Ju/'hoansi**—and to a 1987 work by Morales in which within a sample of 34 hunter-gatherer groups only less than half practiced infanticide. Moreover, as explained by Kelly, one of the main reasons why something "good" such as the decrease of infanticide has happened when some nomadic hunter-gatherer groups became sedentary, and in particular when they developed agriculture, was the increase of something "bad": of **child labor** (Chap.

4). The fact that there is nothing "good" or "bad" in nature per se has been defended by various other scholars, as explained in Chap. 2, as for instance by Shermer in his book *The Science of Good and Evil*:

> If there were no humans there would be no evil. **Earthquakes** that kill people are not, in and of themselves, evil. A shift between two tectonic plates that causes the earth to make a sudden and dramatic movement cannot possibly be considered evil outside the effects such an earthquake might have on the humans living near the fault line. It is the effects of the earthquake on our fellow humans that we judge to be evil. Evil as a physical concept requires human evaluation of a behavior and its effects on humans. As such, **bacterial diseases** cannot be inherently **evil**. By causing humans to sneeze, cough, vomit, and have diarrhea, bacteria are highly successful organisms, spreading themselves far and wide. As their human hosts, we may label the effects of a disease as evil, but the disease itself has no moral existence. Good and evil are human constructs. Which is not to say that a person is not morally responsible for his or her choices and their effects.

Let's now focus on the differences concerning sexual relationships and attitudes about sex in hunter-gatherer groups versus "highly-developed" countries such as Japan, which is the country we have been using in this Chapter as our key comparative case study. One enthralling example provided in Hrdy's *Mothers and Others* concerns the *belief* in "**partible paternity**" shared by many Amazonian societies—which, as explained in Chap. 5, involves sex between a woman and many men, that is, **polyandry**. As noted in that chapter, the book *Sex at Dawn* used this example to argue that such Amazonian societies reflect our "natural chimpanzee instincts" of displaying a **multimale-multifemale** type of sexual organization. However, although it is true that humans don't seem to be "naturally monogamous"—contrary to most religious narratives, Hollywood romantic tales, fairytales and scientific just-so-stories—with respect to this specific case it needs to be noted that within these Amazonian societies such a behavior clearly does not involve only "natural instincts." That is, it also entails "nurture," in the sense that such a behavior is profoundly related with cultural narratives and beliefs, in particular with such a *belief* in "partible paternity." The main difference between such beliefs and those believed by most members of organized religions is that these beliefs are a bit more "natural," if one can say this, because they do *not* involve teleological narratives about "*fighting against* our savage nature" and "going against our natural **polygamy**." Instead, those Amazonian societies have no problems *accepting* that "natural" polygamy and accordingly elaborate imaginary stories that are probably in part aimed to reduce the societal tensions provoked by another "natural" feature, **jealousy** (see Chap. 5). Hrdy writes:

> Given what a powerful emotion **sexual jealousy** is, **polyandrous liaisons** are a risky strategy, dangerous for all concerned. But widely help beliefs about 'partible paternity' help ease some of this tension. In these cultures. semen from every man a woman has sex with in the months before her infant is born supposedly contributes to the growth of her fetus, resulting in chimeralike composite young sired by multiple men. Each possible father is subsequently expected to offer gifts of food to the pregnant woman and to help provide the resulting child…. It is presumably with the ultimate goal of promoting child survival under perilous conditions that **costumary rituals among South American tribes** like the **Canela** or the **Kulina** provide publicly sanctioned ways for mothers to pick up an **extramarital provisioner**. When they find themselves 'hungry for meat', Kulina women order men to go

> hunting. On their return, each woman selects a hunter other than her own husband as a partner. At the end of the day men return in a group to the village, where the adult women form a large semicircle and sign **erotically provoking songs**…asking for their 'meat'. The men drop their catch in a large pile in the middle of the semicircle, often hurling it down with dramatic gestures and smug smiles, after which the women scramble to grab a good sized portion. After cooking the meat and eating, each women retires with the man whom she selected as her partner for the sexual tryst…. In [such] part of the world where one father was unlike to suffice, lineages that invented and retained beliefs about partible paternity proved best adapted to persist and pass on these customs to subsequent generations. People have converged upon ideological solutions functionally similar to the physiological solutions that in other cooperatively breeding animals evolved through natural selection. Without actually producing genetic chimaeras (as females of various non-human taxa sometimes do), women give birth to children that men *believe* to be chimeras.

As also noted in Chap. 5, within the **Aka**, an African hunter-gatherer group, between the ages of 18 and 45 married people have sex 2–3 times a week and average sex three times on each of those nights—numbers that are much higher than those reported for married people in almost any "developed" country. So, now, what about the data available about sex frequency in the "pinnacle of technology," Japan? The title of a 2014 *Tokyo Business Today* article by Sechiyama, a University of Tokyo professor that teaches a course about gender theory, provides a clear clue about the answer: the title is "*Japan, the sexless nation.*" He refers to the ***Sexual Wellbeing Global Surveys*** that are done by **Durex**, a leading **condom manufacturer**. He explains that although these surveys can have inconsistencies—for instance biases and inaccuracies concerning self-reporting by the participants as they are conducted on the Web—it is obvious that Japanese have, on average, the lowest sexual frequency within the 41 countries included in the 2005 survey: a frequency of only 45 intercourses per year, that is just 0.9 times per week. He notes that in the 2011 survey, which included 37 countries, "Japan is the only country where the percentage of people who are not happy with their sex life is higher than that of those who are." That 2011 survey also "included the percentage of people who have sex more than once a week, and unsurprisingly, Japan was in last place at 27%."

Sechiyama also refers to another survey, conducted by NHK, Japan's national public broadcasting organization, in 1999 based on random sampling, entitled "*Survey on **sexual behavior** and **sexual awareness** of Japanese people.*" He states that "this survey was targeted at people who are in their 20s, 30s, and 40s, and have partners (whether married or not)…since there is a possibility that each person's definition of "sexless" is different, in this survey, "sexless" had been defined as having sex less than once a month…according to that definition, the percentage of people in their 30s who are "**sexless**" is 19%, even higher than that of people who are in their 40s (16%)." He points out that "although this result contradicts the results from the survey conducted by Durex (27% have sex more than once a week), it is important to note that while Durex's survey targeted anyone above the age of 18, NHK's survey only targeted people who have partners." In summary, compared to the hunter-gatherer Aka couples aged 18–45 that have in average sex two or three times a week, with an average of 6–9 intercourses per week in total, in the "most technologically developed" country in the planet, Japan, within the couples of a similar age range 16–19% don't even have a single intercourse in a whole month. As also noted in Chap. 5, such numbers are consistent with those from a ***Sexual Well Being Global Survey*** involving 26,032 respondents—minimal age of 16—from 26 countries: as summarized in a 2009 paper by Wylie, "two thirds (67%) of participants described having sex once a week, with people in Greece (89%) and Brazil (85%) having sex most often…sex happened the least for participants in Japan (38%)." So, it is indeed remarkable that not only in Japan almost two thirds of people don't have any sex during a whole week, but also that this also applies to one third of people at a global scale, within "modern agricultural" societies. In other words, people from these "modern" societies have not only less leisure time, they also less sex, and thus less sexual pleasure than most nomadic hunter-gatherers do. *Less leisure, sex, pleasure and physical activity, and more work, obesity, hypertension, diabetes, heart diseases, anxiety, stress, suicides, warfare, slavery, inequity, discrimination, depression, oppression, misogyny, and so on: undoubtedly, an extraordinary list of "achievements" of "progress."*

The negative correlation between **sexual activity** and the type of stress and pressure felt by so many people within the "modern, fast-paced, workaholic, materialistic, productivity-centered lifestyle" of numerous "developed" countries and particularly their bigger cities—Tokyo being indeed a particularly illustrative example of this—is very well documented in the scientific literature. It is also consensually accept by a substantial number of sexologists and other specialists dealing with these topics—see, for instance, Bodenmann et al.'s 2006 and 2010, Kinsberg and Janata's 2007, Hamilton and Meston's 2013, and McCool-Myears et al.'s 2018 works. For instance, Hamilton and Meston's 2013 paper "*Chronic stress and sexual function in women*" explains that a type of **chronic psychosocial stress** that is often very damaging not only for reproduction but also for **sexual arousal** and activity is the "accumulation of small stressors that are *constantly or frequently* present, such as deadlines that never seem to be met, traffic, or financial worries," which are of course traits that particularly apply to the daily-life in "modern developed" cities:

> **Chronic stress** has been linked to problems with **female reproduction** both in human and non-human animals in numerous studies…many women experience hectic daily lives in which they are regularly exposed to small, **chronic stressors**…. In the research literature, chronic psychosocial stress is generally defined as either 1) major life events that induce an extended period of stress, such as a death in the family, or 2) as the accumulation of small stressors that are constantly or frequently present, such as deadlines that never seem to be met, traffic, or financial worries…. Kanner *et al.* were the first to find that these small stressors, which they called daily hassles, have more of a negative effect on health than the more severe but less common stressors. The negative effects of **daily hassles** have been replicated in several studies across multiple health domains, including a survey study on the effects of daily hassles and major life events on **sexual function**. It was found that daily hassles, not major life events, were related to **sexual difficulties**. The link between **chronic daily stressors** and sexual function has also been examined in additional survey research which found that for women, higher levels of chronic daily stressors were related to higher levels of **sexual problems** lower levels of **sexual satisfaction**. To our knowledge, only one study has found a positive relationship between stress and sexual functioning. The sample in this study included a large portion of unemployed men and women, and the researchers found

> that daily hassles stress combined with unemployment actually resulted in higher levels of desire; however this did not translate into higher levels of sexual activity. In addition to survey studies of the relationship between chronic stress and sexual function, there has been one laboratory-based study that examined the relationship between chronic stress and women's sexual function. In this study, women completed a daily stressor checklist while in the lab. Based on a median split of their scores on the checklist, researchers categorized participants as having high or low levels of chronic daily stress. Those in the high stress group had lower levels of **genital arousal** (measured by **vaginal photoplethysmography**), but not **psychological arousal** (measured continuously) than those in the low chronic daily stress group when shown an erotic film. This laboratory study added to the survey-based findings that chronic stress is linked to lower levels of sexual responding. While the evidence of the negative effects of chronic stress on sexual function is fairly consistent, the mechanisms involved in this relationship are not well defined, particularly in humans.
>
> Chronic stressors, whether they are daily stressors or major life events, increase an individual's allostatic load (i.e., the body's response to the accumulation of chronic or recurring stressors). In turn, the individual's physiological systems need to change and adapt in an attempt to maintain allostasis, which is defined as maintaining stability under changing demands. Increased chronic stress leads to higher levels of cortisol, which can cause harmful effects when elevated over extended periods of time. There has been extensive research (primarily in animal models) demonstrating how hormones released from the **hypothalamic-pituitary-adrenal (HPA) axis** in response to stress can interfere with hormonal secretion from the **hypothalamic-pituitary-gonadal (HPG) axis**, which is involved in the control of reproduction and sexual response. Briefly, **glucocorticoids** (i.e., **cortisol** in humans) released from the **adrenal gland** inhibit the HPG through interfering with the release of **gonadotropin releasing hormone (GnRH)**, **luteinizing hormone (LH)**, and **follicle stimulating hormone (FSH)** at the **hypothalamic and pituitary levels**. This model or similar disruptions of the HPG axis by **hormones** released from the HPA axis has been demonstrated in several species. A reduction in gonadotropin release results in less production of **gonadal steroids**, such as **testosterone** and **estradiol**, both of which have been shown to have facilitatory effects on women's genital arousal and may be involved in subjective arousal as well. Chronic stress is also linked to increases in **sympathetic nervous system (SNS) activity** that are harmful to health over time (e.g., **increased blood pressure**), which can inhibit blood flow to all areas of the body over the long term. Inhibited blood flow to the genitals, by definition, interferes with genital arousal. In contrast, acute, moderate increases in SNS have been shown to facilitate women's genital arousal, while inhibition of SNS activity impairs genital arousal. With chronic stress, there will not be an acute spike of SNS activity that has shown to be beneficial for women's arousal, so any differences in SNS activity related to chronic stress will likely correspond with decreases in sexual arousal. Psychologically, stress can interfere with sexual activity through both emotional and cognitive changes that distract the individual from focusing on sexual cues. These psychological responses can distract the focus of the participant toward the stressful stimuli and away from the appropriate stimulus. This would be relevant to any stress or **anxiety-related stimuli** participants experience that draws focus away from sexual stimuli and towards negative outcomes…. Distraction from sexual cues has been shown to have deleterious effects on both genital and subjective arousal in women.

Browning's book *The Fate of Gender* also discusses several problems related to stress and a "modern" lifestyle: "rather than leading to a new and more flexible balance between home and work, the **Internet** and the magic toys that accompany it appear to have led to even more reported stress in family relations—up 74% over the pre-Internet, pre-dual-income households era…the highest levels of stress showed up in employees between age 30 and 44, widely understood to be the period most important for childcare and job performance evaluation…stress-related

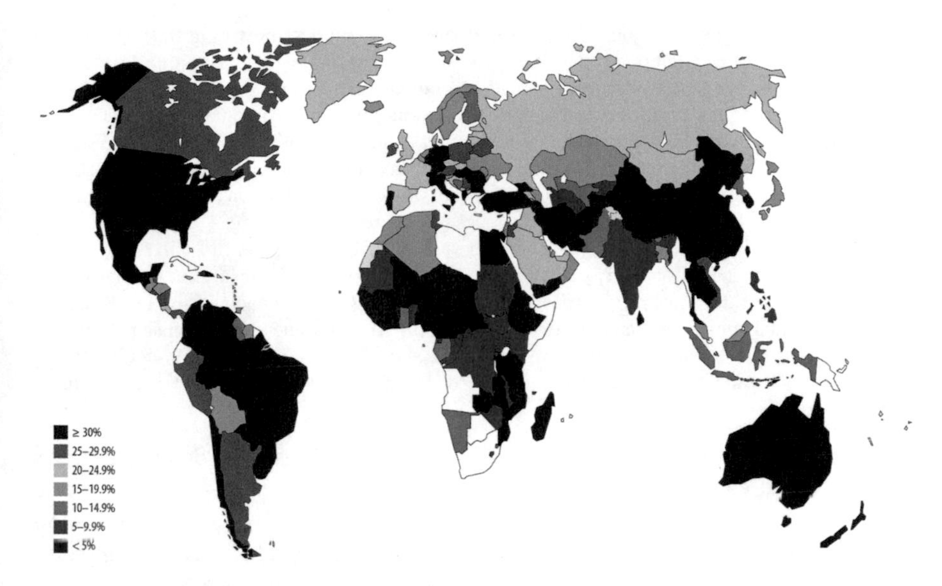

Fig. 8.11 Caesarean section rates by country. Note that since 1985 the international healthcare community has considered the "ideal rate" for caesarean sections to be between 10 and 15%, a percentage that is much smaller than the one seen nowadays in most so-called developed countries

disability claims have also continued to rise…3/4 of physician visits in America have become directly stress-related, ranging from neuroskeletal pain and distortion to binge eating and drinking to cardiovascular problems." Coming back to subjects related to sexuality and fertility, as well as health and well-being, and to examples of how some features of "modern societies" might be seen as "good or bad" depending on the context and the way in which they are done, or how often they occur, a very interesting example concerns the medical practice called ***Caesarean Section*** or **C-section**. This practice can be considered to be "good" because in many cases it saves the life of the child or of the mother or both. But it can also be seen as "bad" because it is increasingly being employed in an exaggerated way in many "developed" countries, as shown in Fig. 8.11. The overuse of this practice can lead to major potential health risks for the mother—including her death, as we will see below—and to a plethora of potential long-term health problems for the newborns. This issue was discussed in detail by authors such as Betran and colleagues, whom, in their 2016 paper entitled "*The increasing trend in caesarean section rates: global, regional and national estimates: 1990-2014,*" wrote:

> A caesarean section (CS) is a life-saving surgical procedure when certain complications arise during pregnancy and labour. However, it is a major surgery and is associated with immediate **maternal and perinatal risks** and may have implications for future pregnancies as well as long-term effects that are still being investigated. The use of CS has increased dramatically worldwide in the last decades particularly in middle- and high-income countries, despite the lack of evidence supporting substantial maternal and perinatal benefits

with CS rates higher than a certain threshold, and some studies showing a link between increasing CS rates and poorer outcomes. The reasons for this increase are multifactorial and not well-understood. Changes in maternal characteristics and professional practice styles, increasing malpractice pressure, as well as economic, organizational, social and cultural factors have all been implicated in this trend. Additional concerns and controversies surrounding CS include inequities in the use of the procedure, not only between countries but also within countries and the costs that unnecessary caesarean sections impose on financially stretched health systems. The lowest rates of CS are found in Africa (7.3%) and more specifically in Western Africa (3%). The highest rates of CS are found in Latin American and the Caribbean (40.5%) and South America is the subregion with the highest average CS rates in the world (42.9%). Countries with the highest CS rates in each region are Brazil (55.6%) and Dominican Republic (56.4%) in Latin America and the Caribbean, Egypt (51.8%) in Africa, Iran and Turkey in Asia (47.9% and 47.5%, respectively), Italy (38.1%) in Europe, United States (32.8%) in Northern America, and New Zealand (33.4%) in Oceania.

Since 1985 the international healthcare community has considered the "ideal rate" for caesarean sections to be between about 10 and 15%, while some recent studies are suggesting it is about 19%: this means that, even if we would use this this latter number as a reference, the medically "ideal rate" would be two, or in some cases even almost three, times lower than the rate occurring nowadays in some of the countries cited just above. As noted in Doucleff's 2018 article "*Rate of C-sections is rising at an alarming rate*":

Such high rates are due mainly to an increase of **elective C-sections**, says Salimah Walani, the vice president of global programs at *March of Dimes*, a U.S. maternal and child health organization. "The procedure is done when it is not really necessary or indicated," she says. Then the surgical procedure can do more harm than good for moms and babies, Walani says. For a mom, an elected C-section can raise the chance of death by at least 60 percent, and in some circumstances as much as 700 percent, several studies have reported. And it increases a woman's risk of life-threatening **complications during childbirth**, such as bleeding, **uterine rupture**, **hysterectomy** and **cardiac arrest** by about fivefold. This risk rises even further in subsequent deliveries. For babies, C-sections raise the chance of **obesity** and **autoimmune diseases** later in life. When the procedure occurs before 39 weeks, an early birth increases the infant's risk of **respiratory problems**. So what's driving the global rise of C-sections? It's likely three factors working together: financial, legal and technical, says Holly Kennedy, a professor of midwifery at the Yale School of Nursing: "as an obstetrician told me…'you're going to pay me more [to do a C-section], you're not going to sue me and I'll be done in a hour'," Kennedy says. When it comes to C-sections, there seems to be an optimal rate that provides the most benefit to women and babies. Doctors are still debating what that optimal rate is – and it probably depends on the location. The **World Health Organization** suggests it lies between 10 and 15 percent, while a more recent study found it is a little higher, around 19 percent. North America and Western Europe are well above this optimal rate, with 32 percent and 27 percent of babies in 2015 delivered by C-section, respectively. The only region with a higher rate than North America is Latin America and the Caribbean, where 44 percent of all deliveries were C-sections in 2015. To bring these rates down, hospitals need to pay doctors equally for **vaginal births**, a team of researchers write in a commentary. At the other end of the spectrum, sub-Saharan Africa is still struggling to give moms access to C-sections when required. Across this region, the C-section rate has changed very little since 2000, hovering right around 5 percent. So many moms around the world end up with less-than-optimal care when it comes to C-sections: it's either too little too late, or too much too soon.

So, what about comparisons between **violence** and in particular **warfare**, within nomadic hunter-gatherer versus "modern" societies? This topic was partially discussed in Chaps. 4 and 5, in which I explained that although Hobbesian authors such as Pinker argue that in general there was/is a higher frequency of violent deaths per capita in hunter-gatherer groups, even those authors recognize that nomadic hunter-gatherer societies had/have a relatively low number of organized wars, compared to agricultural states. According to authors such as Hrdy, this also applies—and is probably at least partially related to—differences concerning the degree in which such societies have "aversion" to—and/or think they can use, manipulate or explore—"others." She wrote in *Mothers and others*:

> One of the most dangerous things that could ever happen to a **common chimpanzee** would be to find himself suddenly introduced to another group of chimpanzees. A stranger risks immediate attack by the group's same sex-members. Now think about **Christopher Columbus**' arrival in the Bahamian Islands, his first landfall in the New World. To greet his ship, out came **Arawak islanders**, swimming and paddling canoes, unarmed and eager to greet the newcomers. Lacking a common language, they proceeded to proffer food and water, as well as gifts of parrots, balls of cotton, and fishing spears made from cane. Something similar to what happened to **Captain Cook** on his arrival in the Hawaiian islands: "the very instance I leaped ashore …[people] brought me a great many small pigs and gave us without regarding whether they got anything in return. European sailors were amazed by such spontaneous generosity, although Columbus simply found Arawak naive. Columbus's description of first contact parallels those of Westerners with the Bushmen and other pre-Neolithic peoples: when you ask for something they have, they never say no…to the contrary, they offer to share with anyone". But Columbus, himself a product of Europe's long post-Neolithic traditions, had different ideas: "they do not bear arms, and do not know them, for I showed them a sword…and [they] cut themselves out of ignorance…they would make fine servants…with fifty men we could subjugate them all and make them do whatever we want", the explored noted in his log".

However, once again one should not fall into the trap of over-idealizing non-agricultural societies, because some of them were actually also very violent, particularly the larger and most sedentary ones (Chaps. 4 and 5). For instance, authors such as Saxon describe—in *Sex at Dusk*—some cases of warfare:

> In the **Hiwi** 36% of all adult killings were due to **warfare** and **homicide**…due to either **competition over women**, reprisals by jealous husbands (or both their wives and their wives' lovers), or reprisals for past killings…. Ryan and Jetha (the authors of *Sex at Dawn*) tell us about a study of the **Waorani of Equador** which showed that they were free of most diseases and had no evidence of health problems such as hypertension, heart disease, or cancer. What they don't add is that…42% of all population losses were actually caused by Waorani killing other Waorani…. When the Waorani found invaders, they speared them. Their reputation for ferocity was earned by violence against each other as well as outsiders…(in total) 64% of a small (Waorani) population that was spread over a vast area dying due to in-group violence and warfare. Is the absence of 'modern' diseases really the only thing we needed to know about the Waorani? Apart from anything else it is certainly interesting that the 36% who lived amongst but survived this violence do not develop the stress-related illnesses we might expect.

Precisely in order to not fall into the murky waters in which many have fallen, and that are often plagued by ideology, biases, and prejudice, I think the most important point that needs to be made is that the different types of violence seen in different societies seem to be much more related to their type of social organization, such as the existence of organized institutions that might help preventing interpersonal violence and/or promoting, planning, and undertaking genocides, as we have seen above. In this sense, at least about the **African slave trade**, Pinker did recognize, in *Better Angels*, that:

> The African slave trade in particular was among the most brutal chapters in human history. Between the 16th and 19th centuries at least 1.5 million Africans died in transatlantic slave ships, chained together in stifling filth-ridden holds, and as one observer noted, "those who remain to meet the shore present a picture of wretchedness language cannot express." Millions more perished in forced marches through jungles and deserts to slave markets on the coast or in the Middle East. Slave traders treated their cargo according to the business model of ice merchants, who accept that a certain proportion of their goods will be lost in transport. At least 17 million Africans, and perhaps as many as 65 million, died in the slave trade. The slave trade not only killed people in transit, but by providing a continuous stream of bodies, it encouraged slaveholders to work their slaves to death and replace them with new ones. But even the slaves who were kept in relatively good health lived in the shadow of flogging, rape, mutilation, forced separation from family members, and summary execution.... Many politicians and preachers defended **slavery**, citing the Bible's approval of the practice, the inferiority of the African race, the value of preserving the southern way of life, and a paternalistic concern that freed slaves could not survive on their own.

In fact, when one thinks deeper about the so-called **Rousseau versus Hobbes** debate, after everything we have seen and discussed so far, one can understand that one of the major differences between the two sides of the debate—or better say between the way in which they are often oversimplified and used nowadays—is that the ideas defended by Hobbesian authors such as Pinker are not only against the "Noble Savage." They are also, and chiefly, the endorsement of a "total assimilation" of people from any society to the culture in which people like Pinker live, that is, to the "Western modern way of living." Basically, for the Pinkers of this world, *all* humans should "assimilate" into Western culture and to its "expanding moral circle," this is the *only way* to achieve a global "moral progress" and become "fully civilized"—in other words, to reach the final step of the "ladder of life." Indeed, just think about the very use of the term "expanding moral circle": this term clearly illustrates that such authors are truly endorsing a "complete assimilation" into their culture because "morals" are obviously the result of sociocultural constructions, being different within different societies and also within the same society in different times as noted above. Therefore, as it is clearly factually inaccurate to talk about "religious progress," or "storytelling progress," it is equally inaccurate to talk about "moral progress" or a "expanding moral circle," as if the "morals" of a certain specific culture or society are "superior" or more "advanced" than those of others groups, or eternal or universal truths. All these fallacies become clear in the following excerpt of Shermer's 2015 book *The Moral Arc*, which is basically a repetition of many of the ideas—and actually often includes very similar data and graphs—presented in Pinker's 2011 *The Better Angels:*

> The philosopher Peter Singer too was ahead of the [expanding moral] curve when he published *The Expanding Circle* in 1981, anticipating the developments in the sciences of **evolutionary psychology** and **evolutionary ethics** that unfolded in the 1990s and 2000s and

> out of which a science of morality could be developed. Singer makes the case for reason and science as providing rational arguments for why we should value the interests of X as much as we value our own interests, with X being racial minorities, gay people, women, children, and now animals.

The inconsistencies of such arguments—specifically concerning such erroneous notions of "moral progress"—can also be seen in this excerpt of the same book:

> A case can be made that our improved ability to reason abstractly is the result of the spread of scientific thinking – that is, science in the broader sense of reason **rationality empiricism** and sense of reason, **rationality**, empiricism, and skepticism. Thinking like a scientist means employing all our faculties to overcome our emotional, subjective, and instinctual brains to better understand the true nature of not only the physical and biological worlds, but the social world (politics and economics) and the moral world (abstracting how other people should be treated) as well. That is, the moral arc of the universe may be bending, in part, because of something like a **Moral Flynn Effect**, as Pinker calls it. Pinker says "the idea is not crazy", but I would go farther. I claim that our improvement in abstract reasoning generally has translated into a specific improvement in **abstract moral reasoning**, particularly about other people who are not our immediate kith and kin. Evolution endowed us with a natural tendency to be kind to our genetic relations but to be xenophobic, suspicious, and even aggressive toward people in other tribes. As our brains become better equipped to reason abstractly in such tasks as lumping dogs and rabbits together into the category of "mammal", so too have *we* improved in our capacity to lump blacks and whites, men and women, straights and gays into the same category of "human".

For instance, the sentence "*we* improved in our capacity to lump blacks and whites, men and women, straights and gays into the same category of 'human'": to whom does the "we" refer to? What about the votes of hundreds of millions of "we" people for racist—and often also homophobic—politicians such as Trump, Bolsonaro, Le Pen, and so many others, which have led to the victory of some of those politicians in recent elections? What about the millions of "we" people that are participating actively in the recent rise of white supremacy movements in the U.S. and many other countries? What about the millions of "we" people that have voted for and contributed to the rise of political parties such as *Vox* in Spain, which has both racist and misogynistic tendencies? What about the "we" police officers that killed George Perry Floyd Jr., an African American man, during an arrest, asphyxiating him for a long time in front of everybody, as if they were doing nothing special: they all lived in the U.S., so clearly they are part of Shermer's and Pinker's "we." Have those cops really gained an "improved capacity to lump blacks and whites into the same category of 'human'"? In this sense, the highly biased ideas about "moral progress" of these authors strikingly contrast with what is written by authors such as **Jared Diamond**, who are scientist that have done serious, well-grounded field studies in remote places of the globe, and that have a proper background, and personal experience, to write about not only the "we" but also about humans in a broader way, with fewer biases and stereotypes and preconceived ideas. Indeed, it is not a coincidence, or the mere result of similar political views, that the ideas presented in let's say Diamond's *The World Until Yesterday* are so similar to those presented in let's say Scott's *Against the grain* or Suzman's *Affluence Without Abundance*, and to ideas that are becoming more and more consensually accepted by authors that did spend countless hours of research and/or field work about/direct exposure to the lifeways of "others," instead of being focused exclusively on the "we," on "our culture" and "our civilization." For example, in *The World Until Yesterday* Diamond states that "with the **transition to agriculture**, the average daily number of work hours increased, nutrition deteriorated, infectious disease and body wear increased, and lifespan shortened…conditions deteriorated even further for urban proletariats during the **Industrial Revolution**, as work days lengthened, and as hygiene, health, and pleasures diminished." Regarding **health**, and specifically the **consumption of salt**, Diamond compared, based on empirical data, hunter-gatherer societies and "developed" countries, including "highly-technological" ones such as Japan, exactly as we have been doing throughout this Chapter for so many other items:

> While there are many different chemicals falling into the category termed "salts" by chemists, to laypeople "salt" means **sodium chloride**. That's the salt that we crave, season our food with, consume too much of, and get sick from. Today, salt comes from a salt-shaker on every dining table and ultimately from a supermarket, is cheap, and is available in essentially unlimited quantities. Our bodies' main problem with salt is to get rid of it, which we do copiously in our urine and in our sweat. The average daily salt consumption around the world is about 9 to 12 grams, with a range mostly between 6 and 20 grams (higher in Asia than elsewhere). However, for dozens of other traditional hunter-gatherers and farmers whose daily salt intake has been calculated, it falls below 3 grams. The lowest recorded value is for Brazil's **Yanomamo Indians**, whose staple food is low-sodium bananas, and who excrete on the average only 50 milligrams of salt daily: about 1/200 of the salt excretion of the typical American. A single **Big Mac hamburger** analyzed by Consumer Reports contained 1.5 grams (1,500 milligrams) of salt, representing one month's salt intake for a Yanomamo, while one can of chicken noodle soup (containing 2.8 grams of salt) represents nearly two months of Yanomamo salt consumption. **High blood pressure** (alias **hypertension**) is among the major risk factors for **cardiovascular diseases** in general, and for **strokes**, **congestive heart disease**, **coronary artery disease**, and **myocardial infarcts** in particular, as well as for **Type-2 diabetes** and **kidney disease**. Salt intake also has unhealthy effects independent of its role in raising blood pressure, by thickening and stiffening our arteries, increasing platelet aggregation, and increasing the mass of the heart's left ventricle, all of which contribute to the risk of cardiovascular diseases. Still other effects of salt intake independent of blood pressure are on the risks of stroke and stomach cancer. Finally, salt intake contributes indirectly but significantly to **obesity** (in turn a further risk factor for many **non-communicable diseases**) by increasing our thirst, which many people satisfy in part by consuming **sugary high-calorie soft drinks**.
>
> The population that I already mentioned as having the world's lowest recorded salt intake, Brazil's Yanomamo Indians, also had the world's lowest average blood pressure, an astonishingly low 96 over 61. The two populations with the next two lowest salt intakes, **Brazil's Xingu Indians** and **Papua New Guinea Highlanders** of the Asaro Valley, had the next two lowest blood pressures (100 over 62, and 108 over 63). These three populations, and several dozen other populations around the world with traditional lifestyles and low salt intakes, showed no increase in blood pressure with age, in contrast to the rise with age in Americans and all other Westernized populations. At the opposite extreme, doctors regard Japan as the "land of apoplexy" because of the high frequency of fatal strokes (Japan's leading cause of death, five times more frequent than in the United States), linked to high blood pressure and notoriously salty food. Within **Japan** these factors reach their extremes in northern Japan's Akita Prefecture, famous for its tasty rice, which **Akita farmers** flavor with salt, wash down with salty miso soup, and alternate with salt pickles between meals. Of 300 Akita adults studied, not one consumed less than 5 grams of salt daily (three months

of consumption for a Yanomamo Indian), the average Akita consumption was 27 grams, and the most salt-loving individual consumed an incredible 61 grams—enough to devour the contents of the usual 26-ounce supermarket salt container in a mere 12 days. That record-breaking Akita man consumed daily as much salt as an average Yanomamo Indian in three years and three months. The average blood pressure in Akita by age 50 was 151 over 93, making hypertension the norm. Not surprisingly, Akita's frequency of death by stroke was more than double even the Japanese average, and in some Akita villages 99% of the population died before 70.

Regarding **diabetes**, Diamond wrote:

Western diets that are high in sugar and in **sugar-yielding carbohydrates** are to diabetes as salt is to hypertension. Compared with the diet to which our evolutionary history adapted us, they (many modern supermarket products) differed in their much higher content of sugar and other carbohydrates (71% to 95% instead of about 15% to 55%) and much lower protein and fiber content. Many types of natural experiments...illustrate the role of environmental factors in diabetes.... One such type of natural experiment involves the rise and fall of diabetes prevalences accompanying the rise and fall of Western lifestyle and affluence in the same population. In Japan, graphs against time of diabetes prevalence and economic indicators are parallel, down to details of year-to-year wiggles. That's because people eat more, hence they risk developing more diabetes symptoms, when they have more money. Diabetes and its symptoms decline or disappear in populations under starvation conditions, such as French diabetes patients under the severe food rationing imposed during the 1870–1871 siege of Paris. Groups of **Aboriginal Australians** who temporarily abandoned their acquired sedentary Western lifestyle and resumed their traditional vigorous foraging reversed their symptoms of diabetes; one such group lost an average of 18 pounds of body weight within seven weeks. (Remember that **obesity** is one of the leading risk factors for diabetes.) Decreases in diabetes symptoms and in waist circumference were also noted for Swedes who for three months abandoned their very unMediterranean Swedish diet (over 70% of calories from sugar, margarine, dairy products, alcohol, oil, and cereals) and adopted instead a **Mediterranean diet** typical of slim Italians.

Developing countries that have recently been growing more affluent and Westernized have correspondingly been growing more diabetic. In first place stand the eight **Arab oil-producers** and newly affluent island nations that now lead the world in national diabetes prevalences (all of them above 15%). All Latin American and Caribbean countries now have prevalences above 5%. All East and South Asian countries have prevalences above 4% except for five of the poorest countries, where prevalences remain as low as 1.6%. The high prevalences of the more rapidly developing countries are a recent phenomenon: India's prevalence was still below 1% as recently as 1959 but is now 8%. Conversely, most sub-Saharan African countries are still poor and still have prevalences below 5%. These proofs of an environmental role in diabetes are illustrated by the tragedies of the two peoples with the highest rates of diabetes in the world: **Pima Indians** and **Nauru Islanders**. To consider the Pimas first, they survived for more than 2000 years in the deserts of southern Arizona, using agricultural methods based on elaborate irrigation systems, supplemented by hunting and gathering. Because rainfall in the desert varies greatly from year to year, crops failed about one year in every five, forcing the Pimas then to subsist entirely on wild foods, especially wild jackrabbits and mesquite beans. Many of their preferred wild plants were high in fiber, low in fat, and released glucose only slowly, thereby constituting an ideal **antidiabetic diet**. After this long history of periodic but brief bouts of starvation, the Pimas experienced a more prolonged bout of starvation in the late nineteenth century, when white settlers diverted the headwaters of the rivers on which the Pimas depended for irrigation water. The result was crop failures and widespread starvation. Today the Pimas eat store-bought food. Observers who visited the Pimas in the early 1900s reported obesity to be rare and diabetes almost non-existent. Since the 1960s, obesity has become widespread among

> the Pimas, some of whom now weigh more than 300 pounds. Half of them exceed the U.S. 90th percentile for weight in relation to height. Pima women consume about 3160 calories per day (50% over the U.S. average), 40% of which is fat. Associated with this obesity, Pimas have achieved notoriety in the diabetes literature by now having the highest frequency of diabetes in the world. Half of all Pimas over age 35, and 70% of those at ages 55 to 64, are diabetic, leading to tragically high occurrences of blindness, limb amputations, and kidney failure.

Therefore, Diamond argues, on the one hand, that there are many features that are typical of nomadic hunter-gatherer that we should take into account in discussions about how to improve our health and well-being, such as those concerning their quality of life and time dedicated to leisure, levels of physical activity, egalitarianism, diets based on fresh products and healthy lifestyles, slow pace of life and low levels of stress, and so on. As noted by Hillard Kaplan about the fact that almost nine out of 10 of 705 Tsimané adults who took part in a study had no risk at all of heart disease—while in the U.S. a study of more than 6800 people found that half had moderate to high risk: "we believe that components of this way of life could benefit contemporary sedentary populations." Yes, our sedentary "modern" societies surely could benefit with such components, which are moreover not so difficult to apply: eating less fatty and sugary foods, doing more exercise, and so on. On the other hand, Diamond's analysis of the empirical data also indicates that there are surely many features typical of many "developed" countries that most people would deem to be "positive," such as the prevention of revengeful violent acts and thus of interpersonal violence, the huge amount of resources used to take care of disabled or very sick people, and so on, exactly as it has been stressed in the present volume. *So, in a nutshell, this is basically one of the key take-home messages of the present Chapter: a much more nuanced view about different types of human societies. A view that recognizes "positive" aspects within two markedly different ways of living in particular—nomadic hunter-gatherer groups and "developed" countries—and that therefore is much more close to a "the virtue is in the middle" idea. A view that emphasizes that no single group or society or way of life is "completely good," or is the one having "morals" or "more morals" or "moral progress" or an "expanding moral circle," or being "better" or "superior" than others, or a "model" that needs to be imitated by others, or to which others should assimilate.*

Chapter 9
Towards a Fulfilling Life in This Splendid, Non-Purposeful, Planet

The world is a mirror of infinite beauty, yet no man sees it

(Thomas Traherne)

9.1 Nature, Nurture, Inculturation, and Self-Domestication

What was God doing before creating the world? He was preparing Hell for those who seek to scrutinize deep mysteries

(Saint Augustine)

In Peterson's 2001 excellent book *Being Human—ETHICS, Environment, and Our Place in the World* she describes two main opposing "modern" positions concerning the "nature *versus* nurture" debate: the **constructionist view** of authors such as the anthropologist Clifford Geertz and the **naturalistic view** of scholars such as the sociobiologist **Edward Wilson**. As she noted, both views tend to fall into the teleological trap of considering humans to be "special beings":

> The anthropologist Clifford Geertz is one of the most influential and articulate defenders of the constructionist position. Geertz rejects the notion of a **universal human nature**, either as reality or as heuristic device. We do not need and should not want, he writes, a "lowest-common-denominator view of humanity" that holds that "what it means to be human is most clearly revealed in those features of human culture that are universal rather than those that are distinctive to this people or that." The only kind of people, Geertz insists, are distinctive ones, and they are distinctive all the way to the core, not merely in their surface adornments. Thus Geertz decisively rejects the Enlightenment quest for a "common" human nature beneath the trappings of culture. He summarizes by saying, "Men unmodified by the customs of particular places do not in fact exist, have never existed, and most important, could not in the very nature of the case exist." More concisely he states, "There is no such thing as a human nature independent of culture." Geertz defines this dependence as a biological fact: humans are unable to develop or to survive in isolation from public symbol systems – in short, from culture. This is because humans, in contrast to all other species, have only very general instincts or innate behaviors. This allows for great plasticity…the *unfinished character* of human nature makes culture both possible and necessary. This *biological incompleteness*, in turn, stems from the distinctive character of human evolution, which, Geertz insists, is not only cultural and physical but simultaneously cultural and physical.

R. Diogo, *Meaning of Life, Human Nature, and Delusions*,
https://doi.org/10.1007/978-3-319-70401-2_9

"Humans, in contrast to all other species, have only very general instincts or innate behaviors": this factually inaccurate sentence illustrates the long-standing obsession of the storytelling animal to think that he/she is not at all as other animals are, he/she is much more than that. No empirical study has ever shown that humans have "only very general instinct or innate behaviors," or that such instincts or behaviors are much more reduced in our species than in let's say bonobos and common chimpanzees. Moreover, there is nothing in biology such as "unfinished characters" or " biological incompleteness," because there are no cosmic goals to be "finished" or "masterplans" to be "accomplished." *There are no end goals in evolution, things are just what they are: all organisms that inhabit this planet, including non-living viruses, are obviously biologically complete, in the sense that they exist and are functional.* Regarding the parts about humans being "simultaneously cultural and physical," simultaneously "nature and nurture," as well as the statement "men unmodified by the customs of particular places do not in fact exist, have never existed, and most important, could not in the very nature of the case exist," those parts are of course scientifically true. But this is a "lapalissade," an obvious truth, because it applies not only to humans but also to all organisms that have any kind of culture. Actually, the very last statement applies even to non-living beings such as viruses, because they have adapted specifically to certain host species, so it is indeed also scientifically correct to say that "viruses/parasites unmodified by the particular places they inhabit do not in fact exist, have never existed, and most important, could not in the very nature of the case exist." Peterson, with her characteristic depth, recognizes such flaws in Geertz' ideas, when she compares his constructionist view with the naturalistic view of Wilson and explains that Geertz's assertions—and also Descartes' claims as she explains—about "human exceptionalism" versus "lower animals" are indeed highly problematic:

> Wilson argues that many of these challenges, along with more sweeping rejections of biological imagery of origins of human nature, stem from a misunderstanding of **gene-culture coevolution** that confuses it with 'rigid genetic determination.' It is possible, Wilson insists, to combine a conviction that **biological evolution** (genes, epigenetic rules) helps determine human behavior and a sophisticated understanding of complex, tortuous, and multiple links between genes and culture. These differences become clearer when we look at the ways Wilson and Geertz respond to the paradoxical fact that, as Wilson puts it, "At the same time that culture arises from human action, human action arises from culture." Both acknowledge the truth of the paradox, but their efforts to resolve it reveal fundamental disagreements. Geertz answers the paradox with a straightforward assertion of ***human exceptionalism***: humans, and only humans, 'create themselves,' becoming 'artifacts' of the environment they themselves have created. This claim…suffers from serious internal contradictions. While Geertz and hard constructionists claim to avoid defining human nature by some essential feature, in the end they merely redescribe the ***vital principle***. For Geertz, *human exceptionalism* results from a unique biological and evolutionary incompleteness that requires humans to invent themselves, as individuals and as a species. Our unique need and ability to finish ourselves, Geertz explicitly states, make us human. Our 'desperate' dependence on culture 'is what we really have in common.' Ultimately, lack becomes the essential character of humanity, or, as Carl Esbjornson puts it, "absence" replaces "essence". In addition to the internal contradictions of hard constructionism, Geertz's position suffers from a failure to take seriously the biological grounds of human behavior, the reality of biological universals as well as cultural particularities.

Wilson points to an alternative way of resolving the paradoxical relationship between culture and human action and, thus, of rethinking ***human distinctiveness***. He notes that countless species, from termites to elephants, help construct the environments in which they live [**niche construction**]. There exist, in other words, various forms of reciprocity between environment and behavior, from the relatively simple ways that certain insects create the kind of soil in which they need to live to the infinite complexity of human societies. While human culture is by definition unique, Wilson argues, the underlying principle of gene-environment coevolution is the same. This argument is tied to another revealing area of difference to learn, such as speaking a language or making and using tools, distinguish us among animals. However, we share with other species the more general or underlying fact that we are prepared to learn some things and not others. Cats, for example, are prepared to learn to hunt small rodents, and songbirds are prepared to learn particular tunes. However, the fulfillment of these innate potentials, like the acquisition of language by human children, requires external conditions or triggers, which range from intensive training by parents to environmental factors such as sufficient territory. [Instead] Geertz radically distinguishes humans from '*lower animals*', whose behavior is largely determined by 'genetic sources of information' and constrained within 'much narrower ranges of variation, the narrower and more thoroughgoing the lower the animal.' Despite his sharp critiques of **Enlightenment rationality** on some topics, on this point Geertz echoes **Descartes**: other animals all have 'instructions coded in their genes and evoked by appropriate patterns of external stimuli: physical keys inserted into organic locks.' Geertz's mechanistic and reductive portrayal of nonhuman species contrasts sharply, as it is meant to, with the endless possibilities of human behavior. It also clashes with Wilson's claims of continuity among humans and other animals. Wilson argues that all animals, including humans, are innately prepared to learn some behaviors and predisposed to avoid others. He calls this 'prepared learning,' a subset of epigenetic rules of behavior.

No single feature can define what it means to be human, and no single principle can explain all the different aspects of humanness, including cultural diversity and biological universals. One attempt at this pluralism is the '**nonreductive physicalism**' articulated by various philosophers, theologians, and scientists. While these thinkers differ among themselves, they share a commitment to biological explanations of human behavior and life generally without restricting all explanation to biology. **Francisco Ayala**...concludes that "the proclivity to make ethical judgments, that is, to evaluate actions as either *good or evil*, is rooted in our (biological) nature, a necessary outcome of our *exalted intelligence*". A number of writers, for example, have explored the links between bodies and emotions, arguing that mind-body dualisms parallel dualistic conceptions of rationality and feeling. This issue is tied to the larger claim, central to the ethical anthropology I sketch here, that the human body is an integral whole. In this light, the body, and specifically the brain, provides the indispensable frame of reference for what **Antonio Damasio** calls "the neural processes we experience as the mind". Our body, not some external or supernatural reality, constitutes the reference point and yardstick for our constructions of the world and our sense of subjectivity. Damasio summarizes this: "The mind exists in and for an integrated organism; our minds would not be the way they are if it were not for the interplay of body and brain during evolution, during individual development, and at the current moment. The mind had to be first about the body, or it could not have been".

It is indeed striking how, in the twenty-first century, so-called modern scientists continue to use terms such as "human exceptionalism" or "lower animals." Similarly, it is astonishing how even "modern" evolutionary biologists and philosophers such as Ayala continue to use terms such as the "exalter intelligence of humans": common definitions of "exalter" are "glorifier" or "magnifier" and, according to the biblical definition of "exalt," this can also be interpreted as "raise in rank," thus directly evoking the longstanding and factually inaccurate teleological notion of "**ladder of life**." Regarding the part in which Peterson states "these thinkers…share a commitment to biological explanations of human behavior and life generally without restricting all explanation to biology," there is often some confusion that relates directly to a key misunderstanding about the so-called nature versus nurture debate. Humans are part of nature, of life in this planet, so clearly they are part of "biology," as biology is precisely the study of life as a whole. That is, both what we call "nurture" and "nature" have biological explanations, because humans and other animals do *not* transcend life, they are just part of it. It is important to finish once for all with this misunderstanding, with this false dichotomy between humans and other organisms, with this notion that contrary to other animals "human nature" cannot be discussed by biologists, but only by social scientists and fields of "humanities," as discussed in Chap. 1. If we can, and do, study the behavior of other organisms within the natural sciences, including *social non-human animals* such as chimpanzees or ants, why can't we study human behavior within natural sciences as well? Doing that would be, a priori, to exclude humans, as a subject of study, from nature, from all other types of life in this planet, something that is obviously completely wrong and that would contradict everything we know about biological evolution.

In this sense, it is interesting that the author that Peterson cites, in the above excerpts, as an example of how we can *indeed* explain, biologically—without the needed to invoke any "mystery" or type of "transcendence"—the existence of the "mind" is my fellow countryman **Antonio Damasio**. As argued by Damasio, the mind simply exists in an integrated organism, as the natural result of the interplay of the various organs of the body, including the brain—I should note that I rephrased Damasio's "interplay of **body and mind"** because this is misleading, as there is obviously no true dichotomy between "body" and "brain," as he is of course fully aware. Indeed, these issues are precisely the main focus of Damasio's new 2021 book *Feeling & Knowing—Making Mind Conscious*, which explains that **feeling**—particularly related to our internal organs, such as the guts, including pain and pleasure—is one of the most crucial aspects of consciousness. As he elegantly put it, "*we are puppets in the hands of pain and pleasure, occasionally liberated by our creativity.*" In this sense, the scientifically accurate data provided and discussed by Damasio contradict many of the futuristic ideas that are being put forward about the "nature *versus* nurture" debate in fields such as **cybernetics**. For instance, **Kevin Warwick**—an influential cybernetics expert at the University of Reading—stated, in the context of a supposed dichotomy "between brain *versus* mind," that "the physical body that we've got is perhaps not suitable anymore…we could do with something better" (*Daily Mail Reporter*, 2013). **Transhumanists** (see Box 7.16) often defend similar ideas, for instance that there is a compatibility between the **human mind** and computer hardware, with the theoretical implication that **human consciousness** may someday be transferred to alternative media, a type of "mind uploading," as explained in Moravec's 1990 book *Mind Children*.

What about the existence of "biological universals," which constructionists tend to see with skepticism, as noted by Peterson? In the way they are defined by Peterson, it is clear that there *are* "biological universals" in humans: for instance, we need to have ways to put water and other nutrients in our bodies, as all animals do—either by drinking or eating, or using more recent practices such as those employed for people in coma, for instance. After my lectures about these issues, people from various types of cultures and backgrounds often comment, in the Q&A, that we are surely "more than animals," "more than just biology," that we can "transcend our bodies," for instance by praying to a God or by doing meditation. I answer them with a simple example: indeed, there are studies indicating that meditation can potentially lead to a *momentaneous* decrease in metabolism, breathing rate, heart rate, and blood pressure, but even if you pray and meditate all the time, if you don't drink any liquids and eat any food for let's say 3 months, you will surely die. This is because we are just living organisms that, as any other being, have a plethora of different evolutionary constraints. This leads us to the main focus of Peterson's *Being Human* book: the links between ecology and our place in the world. Namely, she discusses some of the **"Mother Nature"** and **"Gaia"** **"New Age"** ideas that are being recycled in the last decades (see Chap. 2), as well as what is often designated as "Deep Ecology." Interestingly, although she supports some of these recycled longstanding teleological ideas, she also shows that, paradoxically, even people that *say* that "humans are just part of nature" can actually be, in practice, quite **anthropocentric**, egoistical, or even imperialist:

> We need to reconceive our self-interest as tied to the interest and well-being of all of life, which will lead to ecological restraint as a sort of enlightened selfinterest. The concept of enlightened or expanded self-interest is central to **Deep Ecology**, particularly as articulated by Arne Naess, and is embodied in John Seed's claim that to say 'I am protecting the rain forest' really means 'I am part of the rain forest protecting myself.' For Deep Ecologists, environmental responsibility is a form of expanded selfinterest rather than a moral obligation imposed from outside. Some critics argue that the expanded self of Deep Ecology does not necessarily lead to **ecologically superior behavior**. According to Val Plumwood, Deep Ecologists believe that "once one has realized that one is indistinguishable from the rain forest, its needs would become one's own." However, Plumwood argues, "there is nothing to guarantee this – one could equally well take one's own needs for its." She contends that the "expanded self is not the result of a critique of egoism…rather, it is an enlargement and an extension of egoism…it does not question the structures of possessive egoism and self-interest; rather, it tries to allow for a wider set of interests by an expansion of self." The expanded self recognizes others morally only to the extent that they are incorporated into the self and their differences denied. Simply being in relation to an ecosystem or seeing its good as attached to one's own does not guarantee that one will preserve that ecosystem. An expanded self might, in fact, become an even stronger reason to exploit nature, if its good is seen as inseparable from one's own narrowly conceived selfinterest. Perhaps the rainforest with which I identify wants me to have teak garden furniture, for example. Expansion, in other words, might unleash an imperial, not an enlightened, self. In light of this danger, the notion that morality is irrelevant, that we should just expand our Selves and then trust enlightened self-interest to fix things, falls short.

A crucial example regarding the so-called nature versus nurture debate that also illustrates how we cannot completely discard our evolutionary history and constraints and thus completely neglect, or go against—or "transcend"—all our bodily needs and desires is related to our sexuality. Let's go back to what I wrote in the Preface about **Saint Jerome** and Fig. 1, and discuss this topic together with the

Fig. 9.1 The "Temptation of St. Anthony," by Lovis Corinth

story of the **temptation of Saint Anthony**. This is a key story in **Christian scripture** that has been represented countless times in art: to make a long story short, many painters show how **Saint Anthony** is beset by **female naked demons**, and try as much as he can to "resist" their sexuality. One striking detail of the painting shown in Fig. 9.1 illustrates a key point made in the present book: that teleological narratives are often related to the aim of separating humans from other animals, in particular to avoid thinking that, as other organisms, our moment will come and we will just die, without any possible "salvation" or "after-life." In this painting, the association between the **sexual desire** for the female Demon and our "natural, wild, ape instincts" is emphasized by the addition of chimpanzees to the scene. As noted by Tallis in *The incurable romantic*, we cannot, however, blame only religion for attaching guilt to sex. I completely agree: it is something much more deep, as even people that are said to be atheists display behaviors that will lead children to logically think that having sex is something "bad," or at least something one should feel "ashamed of." For instance, since we are small kids, our parents close the door of the room when they are "doing it." They don't close the doors when they are eating, or talking, or seeing TV, but they do so when they are defecating or having sex, as they would do so if a father was for instance beating his wife: as if defecating or having sex was as "morally" wrong, or "disgusting," as domestic violence. Actually, for centuries this was indeed the case, and it still is nowadays, too often, in many countries, such as India or Pakistan. Centuries ago in the West, or today in Pakistan, it is surely much more common to see a husband beat his wife in public than to see him having sex with her in front of other people. This is also part of inculturation, it is not only about learning in school: that is, the association for the kids is clear—either sex is "*bad*" or at least it is something that *should not* be seen, that would cause embarrassment or one to feel ashamed. As Tallis explains: "the problem exists, most probably, because of a mismatch…we are constantly struggling to reconcile the contradictory parts of our totality, constantly trying to negotiate compromises…it is confusing being a rational animal, a creature that can derive pleasure from the transcendental complexities of a Mozart symphony and anal-oral contact…how do these two identities fit together"?

A disturbing example of how such longstanding teleological narratives can be very dangerous and damaging for those that believe in them and in particular for those that are abused by them, concerns the "passion for Christ/God" and **clerical celibacy** within the **Catholic church**: only unmarried men are ordained to the episcopate, and with some exceptions also to **priesthood**. As noted in the ***Code of canon law***—Catholic Church 1983 —although even the married may observe **abstinence from sexual intercourse**, the obligation to be celibate is seen as a consequence of the obligation to observe *perfect and perpetual* continence for the "sake of the **Kingdom of heaven,**" clerical celibacy being seen as "a special gift of God by which sacred ministers can more easily remain close to Christ with an undivided heart, and can dedicate themselves more freely to the service of God and their neighbor." Some of you might be arguing: OK, this is clearly a teleological narrative that is an extreme example within the "nurture *versus* nature" debate, but what is really the problem with it, can't priests do what they want? As noted in Chap. 2, some of these "extreme efforts," such as priests "resisting natural sexual temptations" probably contribute to the unity of the ingroup as a whole—in this case of the Catholic church—and may therefore ultimately bring some kind of advantages to that ingroup via group selection. So if priests *say* that they *want* to do so, who are we to criticize them? The problem is that as clearly shown by factual historical evidence and by many trials that are occurring right now and cases that are being widely discussed in the media, a huge number of priests clearly does *not* simply do that. This delicate, and disturbing, topic is related to what we discussed throughout this book: *saying something* that others want/expect to ear—in this case, priests saying to the Catholic church, and the society in general, that they will "resist" the temptations of sex—is completely different from what people often *truly want or desire,* and *ultimately end up doing*. That is precisely why so many catholic priests ended up by having sex after all, in many cases with children (Fig. 9.2).

Sexual abuse of children was done by an *enormous number* of catholic priests, as has been recently recognized by the Catholic church itself. For instance, in February 2016 **Pope Francis** lead a special summit on **clerical sexual abuse**, and it is now recognized that priests have done this for centuries, in basically every single region of the globe where there are catholic churches. That is, such atrocious abuses were done to an endless number of kids, resulting in massive psychological impacts throughout their whole lives. Some could argue: but many adults that do not undertake **celibacy vows** also sexually abuse kids, we all know that there are many **pedophiles**, including many atheist ones. But again this is confusing what happens in a few cases versus what is *an average*: the massive scale of the cases already reported about catholic clerical sexual abuse are just the tip of the Iceberg, as most of these cases were left unreported by the molested children or their families, and many of

Fig. 9.2 Clerical sexual child abuse

those that were reported were silenced by the Catholic church, by society or even by the children's families. So, if we take into account the countless number of cases that are known *despite* of that, the evidence available clearly indicates that there is, on *average*, a much higher proportion of **child sexual abuse** within clerics that undertook celibacy vows than within the general population. In fact, although there are obviously cases of sexual child abuse in the **protestant clergy**, it is interesting to note that one of the central disputes between **Catholics** and **Protestants** since the sixteenth century was precisely over the role of marriage. In general, protestants tended to argue that clergy should be allowed to marry, because **clerical celibacy** would only encourage priests to keep concubines and seduce their parishioners. As noted in Coontz's *Marriage, a history*, many protestants argued that Catholics were wrong to call **marriage** a necessary evil or a second-best existence to celibacy, because marriage was a "glorious estate."

That is, apart from factors such as the enormous power given to clerics by religious narratives, the very act of doing **celibacy vows** aiming to "resist natural wild sexual instincts" ironically—and tragically—led to carrying out abominable **sexual abuses** against defenseless children. In other words, by trying to "resist" a completely "natural," nonharmful type of sexuality, these clerics ended up by undertaking an extremely harmful type of sexual behavior that is completely "unnatural"—in the sense that we don't see any group of adult apes so consistently and frequently sexually abusing their youngsters. We can't simply go completely against natural evolutionary constraints that date back to millions or even billions of years all of the sudden: we cannot simply "try to resist" eating, defecating, or breathing during our whole lives. This might be possible in *Neverland*'s bizarre imaginary world, but within the reality of the natural world things do not work that way, at all. Our ancestors, us, and any other animals had, and will always have, certain "bodily" needs and desires, and that is nothing to be ashamed of, well on the contrary—it is an extraordinary example of the vastness, richness, and deep history, as well as the unity, of life in this planet.

However, amazingly, this simple fact is confused over and over again by countless authors, particularly about sexuality, which confuse what *people do or say they want to do*, with their natural bodily needs and desires, as if "nurture" simply did not existed, as if culture, inculturation, and society played no role in how people behave or say they want to behave. For instance, they would argue that because a huge number of priests don't have sex with adults, that shows that is what they "truly want naturally." Yes, many priests don't have sex with adults, but in great part that is because one of the main aims of their lives is precisely to try to "resist" what they see as our "bodily, *natural*, evil" instincts, due to peer pressure, social pressure, and the power of teleological narratives: that is, even priests themselves recognize, by using terms such as "resist sexual temptations," that having sex with other adults *is indeed natural.* Strikingly, such reasoning mistakes are very typical even within works of some scientists, as seen for instance in an otherwise very interesting 2019 review paper by Schacht and Kramer's, entitled "*Are we monogamous? A review of the evolution of pair-bonding in humans and its contemporary variation cross-culturally*":

> **Testis size** is another commonly used metric of mating system as it indicates, generally, **female multiple mating**, such that large testis relative to body size is positively correlated with the frequency of females mating with multiple males simultaneously. Adjusting for body size, human testes are smaller than would be predicted, and, when compared to our closest living relatives, are considerably smaller than those of chimpanzees. Together this provides evidence of relatively low rates of sex outside of a pairbond. However, human testes are somewhat larger than those of other **monogamous primates**, leading some to argue that this hints at a measure of **extrapair copulation** not expected in a monogamous species. Yet studies employing genetic methods find that rates of non-paternity are low among humans (2%) when compared to those of socially monogamous birds (20%) and mammals (5%), casting doubt on claims of relatively high rates of extrapair engagement in human males compared to males in other monogamous species.

That is, the crucial biological evidence discussed in the paper to infer what seems to be our "ancestral" or "natural" tendency—if there were no social pressure, group control, oppression, and so on—is that: "**human testes** are somewhat larger than those of other **monogamous primates**, leading some to argue that this hints at a measure of **extrapair copulation** not expected in a **monogamous species.**" However, as this biological evidence about the size of human testes goes against the authors' a priori assumption that humans are "monogamous," they quickly discard it by arguing that *in practice* humans have lower rates of non-paternity compared to other socially monogamous, and also polygamous, animals. As we saw above, in *practice* only a minority of catholic priests have sex with adult women and men, but that clearly does not mean that it is "natural" for humans to not have sex with other human adults. Using the flawed logic employed by Schacht and Kramer, one could also say that because in *practice* the vast majority of **African Americans** were slaves some centuries ago, this means that they are "**natural slaves**." Or, to refer to another horrible practice that we have also discussed above, in *practice* the **clitoris** of a huge number of girls is cut in countries such as **Somalia** and **Egypt**, so one could say that this is "natural." None of these things is "natural": instead, they emphasize, once again, that what the **self-domesticated members of our**

“civilizations” *do* is often—too often indeed—not what they *want* to do nor what they would *tend* to do if they were free from the impositions of their “masters”—either specific groups or the society as a whole—be it through force as in the **Trans-Atlantic slavery trade**, or through peer pressure as in the cases of **female genital mutilation**, or via the power of the church and of **teleological narratives**, as in the case of **cleric celibacy**.

This discussion thus take us to one of the most crucial questions of the present book, which was introduced in Chap. 2, partially discussed in Chap. 8, and finally addressed, taking everything we saw so far in this book, in this Chapter: *would we therefore not be ‘better off’ if we were able to free ourselves from such ‘masters’, such narratives, such beliefs, such a bizarre world of unreality of Neverland, or do we really ‘need’ them after all*? On the one hand, by taking into account what we have seen in the previous chapters, there is no doubt that most atrocities done since the rise of agriculture were done in the name of, or at least justified by, such narratives and beliefs. On the other hand, as we have also seen, beliefs, and religion in particular, do play some important functions, particularly (1) help coping with our **anxiety about the inevitability of death**, (2) being a **“crisis management system”** that probably saved, and continues to save, millions of lives particularly after the rise of agriculture, and (3) improving health/decreasing stress by means of finding a “purpose” or “meaning” and thus giving a false sense of “order” and “control,” and so on. Apart from the many empirical works I cited in Chap. 2 illustrating each of these latter three points, one that I strongly recommend for you to read is a book published just a few months before I wrote these lines: Levin’s 2020 *Religion and medicine*. I do agree with some scholars in that Levin, who recognizes that he is religious, seems sometimes to be pushing too far the idea that **religion** was—and particularly that it *continues to be*—crucial for the development of **medicine**. However, some of the examples provided by him are factually indisputable, for instance about not only the history of, but also the current status of and practices done at, **hospitals** related to **religious institutions**:

> If one doubts that **Christian churches**, collectively, constitute a **religion of healing**, one need only observe the extensive denominational branding of **medical care institutions**. So **many hospitals**, **healthcare systems**, **medical care organizations**, and provider practices incorporate the name of a respective Christian denomination in their formal corporate name or that of particular buildings or structures. Consider today how many medical centers are branded as Catholic, Lutheran, Baptist, Methodist, Presbyterian, Episcopal, Adventist, Mennonite, LDS, and so on. Within Roman Catholicism, religious orders own and operate community-based hospitals, regional academic medical centers, and healthcare facilities of almost every type, often reflected in the names of these institutions as well. This is not an exclusively Christian phenomenon: **Jewish-branded hospitals** exist in the United States, as well as **Buddhist hospitals** throughout Asia, **Hindu hospitals in India**, and **Muslim hospitals** throughout the Middle East. Since the founding of the earliest hospitals in the West by the major **Abrahamic traditions** nearly a millennium ago, religious movements have been at the forefront of providing medical care to the public. The establishment and continued operation of religiously led healthcare systems speaks to their dual vocation or purpose, of “mending bodies, saving souls”, reflective of the prophetic, pastoral, and charitable mission of institutional religions. Specifically, the presence of religiously branded hospitals, clinics, and care facilities in most communities speaks to a ubiquitous under-

standing that God's love can and must be externalized, through the agency of religious institutions, to meet worldly needs of human beings, including and especially health and healthcare needs.

With the rise of **scientific medicine** in the nineteenth century, the dynamics of religiously operated medical care institutions were altered. Historians have described how the **evolution of medicine** from speculative art to **empirical science** led to an inevitable **secularization** of medical theory and **laicization** of medical care delivery. Increasingly, as well, the religious value system of these institutions has been preempted by federal mandates governing practice standards. Since expansion of these programs in the United States with the advent of **Medicaid** and **Medicare**, it has been "often difficult to distinguish religious from secular hospitals – except perhaps for their names and, in Catholic institutions, their distinctive policies related to reproduction". Now, even the latter point of religious identity may be obsolete, in light of ongoing legal debate over the nondiscrimination mandate within the **Affordable Care Act** that continues to adjudicate competing claims of religious liberty (on the part of Catholic providers) and access to care (on the part of patients). The largest religiously operated medical care provider in the United States is the **Adventist Health System**. It is among the largest nonprofit healthcare systems in existence, with nearly eighty thousand employees working at forty-six hospital campuses in nine states. In its corporate statement on identity and values, foremost among the core principles is its "Christian Mission", defined as service "in harmony with Christ's healing ministry". This principle is reinforced in its institutional mission statement: "Extending the Healing Ministry of Christ". Results of a recent study of providers sampled system-wide suggests that the Adventist Health System has been successful in maintaining its corporate vision in the face of secularizing trends that have challenged other religiously based healthcare systems. The study concluded that "a significant number of Adventist Health System providers and staff favor engaging in spiritual practices with patients", and do so, including praying with patients, sharing religious beliefs, and encouraging more active religious participation among patients. There is also considerable provider support for a screening spiritual history at intake, not yet a system-wide standard. As the Adventist Health System has grown into a multi-institutional corporate entity, it has shifted away from the distinctively eclectic health and healing-related principles of Ellen G. White, nineteenth-century founder of the **Seventh-day Adventist Church**. These included strict dietary practices, such as vegetarianism, as well as aversion to drug therapy and promotion of light and water therapy.

Since mid-century, "the distinctiveness of **Adventist medicine** lay less in its therapies than in its customs, motives, and philosophical justification". To the extent that it has succeeded in maintaining these values in the changing healthcare environment, the Adventists have been more effective in facing such challenges than Catholic and other **Protestant healthcare systems**. The presence of a desire to serve, individually and corporately – and concomitants that such a value mandates, as far as service to others – can be found in the vision and mission statements of hospitals across the religious spectrum, not just among Christian-owned institutions that use such language explicitly, such as Seventh-day Adventists. In *Radical loving care*, Erie Chapman, founder of the **Baptist Healing Trust**, advocates building what he terms Healing Hospitals. He offers the following description: "A Healing Hospital is about loving service to others…it is about recognizing something that has increasingly been forgotten in a flood of the complex technology and magic-bullet drugs that now dominate America's hospitals…a Healing Hospital is not built with bricks and mortar. It is built with people who have Servant's Hearts, or a passion to serve, and who know that the fundamental relationship between caregiver and patient can be understood as a Sacred Encounter…it can be created in any healthcare setting where leaders and staff join together in a new commitment to what we call **Radical Loving Care** – creating a continuous chain of caring light around each and every patient". This beautiful statement remains an ideal, not yet fully realized in any large healthcare system that I am aware of at present. But it is a wonderful ideal. Who would not want to experience the patient role in such an institution if, heaven forbid, one were facing hospitalization? Significantly, it is a faith-based vision. The language Chapman uses to describe his vision is unlikely to appear in the mission statement of a secular or public hospital, which might be required by law to eschew such wording and even associated underlying concepts.

As stated by Levin, probably no **hospital**, in any country, be it associated to religious institutions or not, truly engages in "loving care." I have seen this myself, with the death of my paternal grandmother, for instance. Basically, the last days of her life were spent, as they are for the vast majority of humans in Western countries, in a room of an hospital, with some visits from her family but most of the time being alone or surrounded by medical staff that she hardly knows and that therefore don't truly love her, despite the fact that they try to do their best—and are almost always amazing at that, I have a huge admiration and respect for them. Basically, she was just one more terminal patient, waiting to die: when that happen, she liberated her bed to be used by the next terminal patient, and so on, over and over, in a hospital room in which, as it happens in the vast majority of hospital rooms, almost everything is at least a bit repelling. Namely, they smell "disease," they have no paintings, there is usually no music, the beds are often uncomfortable, the food is healthy but soulless. The question is, if we spend such a huge amount of money in **healthcare**, **medication**, and so on, why don't we spend just a bit more money to make those rooms more appealing? The answer is simple: because all the major "actors" within the prevailing health system would not really benefit much with it: that is, the hospitals, the pharmaceutical companies, the insurance companies, the government, and so on. The ones that *would* benefit a lot are just the individual "people" that stay in those rooms, and their families and loved ones that visit them, as well as the medical staff that has to spend several hours during many days a week in such characterless rooms for a great part of their working lives. Just think about this: when you go to a spa, in general you feel good, that is why you even pay to go there. There are often beautiful paintings, candles, ambience music, you feel calm, you feel cared about, sometimes it truly seems almost a small "paradise on earth." So, why don't we try to mirror at least some of these features in **hospital rooms** or **nursing homes**, so very old or very sick people would take advantage of a more cozy, comfortable and appealing—and thus also mentally healthy—atmosphere during the very last years, months, or days of their life, of the *only* life that they will ever get to live?

This reminds us of **Claude Levi-Strauss**' ideas: that things that are often seen as 'noble', such as writing and education, have been mainly used by 'civilizations' "*to favor rather the exploitation than the enlightenment of mankind*". Of course hospitals and medicine in general contribute to the well-being and survival of several millions of lay people every year, so they do benefit the general public for sure, although it is clear that as noted above the mental well-being of those people that are in hospitals is mainly neglected by the health system, because the pharmaceutical companies, insurance companies, and so on don't have any direct economic interest in changing that. But the point I want to make is that, going deeper into what Levi-Strauss argued about **education** and **inculturation** in "civilized" societies, we can see that both in "early civilizations" and in our "developed" countries, such societies don't only tend to neglect mental health and well-being when one is close to death: instead, this basically applies to many of the other most important moments of one's life. Actually, this might well be one of the major reasons leading to the apparent paradox of why the more "developed" and "richer" countries such as South Korea get, the higher the suicide rate, as noted in Harari's *Homo Deus* (see Chap. 9). That is, if one truly things deeply about this, one can realize that basically "civilized" societies in a way neglect a series of crucial items regarding birth, growing up, and how to live in general, to eat healthy, to have sex, and how to die.

This is particularly evident concerning education/inculturation, which is often seen as the "opus magnum" of Western "civilization." I spent more than 20 years in the educational system, from pre-school to the end of university. I did not learn anything about what to do, or what to expect, before and after the birth of my own child, nor about how to take care of a newborn and then educate a child and a teenager, as a father. Basically, people have sex, and ooops, a kid is coming: now you are a father, good luck with it. Then, when we become fathers or mothers, just a few weeks before or after the birth, we do have some "advice," and if we are lucky, some "parental lessons" for a few hours, within those weeks, but it is clearly far too less, particularly because within "developed" countries and the "nuclear family" model, there is less and less opportunity to be with our own parents during those important weeks, who could teach us how they took care of us and/or help us taking care of our own kids/their grandchildren. Same thing applies to sex, in general. There is basically no sexual education in school. Even in those few cases in which there is, it mainly focuses—as do the media, most textbooks, and society in general—on sex in a very biased, and almost always factually inaccurate, way, moreover ignoring even basic things such as providing updated summaries of all the major sex-related diseases and what one should do to prevent them, and so on.

For instance, despite being a scientist and having therefore read some superficial information about the ***human papillomavirus***, I only got to *really* know the details about it when a physician told me that I had it myself, and only that very same day, when I was already 44 years old, I learned that there is actually a vaccine available to prevent it, which could be taken only before one is 45 years old. If I had known that much before, I would surely have taken that vaccine—fortunately, in my case, the virus it type 6, which is one of the less problematic types, but still some people can get major problems with that strain, and if it was type 16 that would be much worse, leading to the death of a huge number of people around the globe. So, the question is: why did I had to memorize the name of each and every Portuguese kings, most of them people that died several centuries ago and that were completely irrelevant within a global historical context, when it would have clearly been more important for *my life and well-being*, and for those of any member of the general public, to know more about sex, the papillomavirus, and other sexual diseases, which continue to kill millions of people around the planet every single year? Again, the answer is simple: because education is still, in great part, not for the well-being of the general public, but for inculturation, to keep the *status quo*—in this case, to see the grandness of Portugal, to learn to respect and obey to kings or "modern" politicians, and so on. I will talk more about this below, but before that, I just ask you to compare this lack of care and of information about the most important moments and transitions of your life, with the key importance that such transitions have within many hunter-gatherer societies. Think for instance about their "**initiation rituals**," which can take weeks, or even months, and can apply not only to the "initiation" of sexual activity and to get people prepared for the subsequent moment when one becomes a father or mother, but also to growing up in general. Such rituals are truly major "**coming of age" lessons and celebrations**, often involving—and importantly, being assisted by—a large part of the community. Such lessons and celebrations became completely lost in most "modern" countries, with exception to a few examples such as a major celebration of the 16th anniversary of girls and boys in some South and Central American countries and some Latino communities within the U.S.—the "**Dulces dieciséis**" or "**Sweet Sixteen"** in English.

Apart from not learning about sex, sexual diseases, or how to educate/deal with newborns and small children, I also did not learn in school about how to eat and details about healthy food, or natural anti-inflammatory food items, and so on. I did, as many kids do, learn about the typical "food pyramid" in a very basic way, but this is an old-dated concept that does not even take into account empirical data showing, for instance, that a huge component of the health of our guts depends actually on the **microbiome**—that is, the germs that live there. I did not learn for instance that the exaggerated use of **antibiotics**—which was a common practice by then, when I was in school—could affect that microbiome, and thus make it more difficult to digest many of those vegetable items that are displayed in that "food pyramid," as well as leading to unbalances that can make it more likely to have infections of the digestive system by the so-called bad germs such as ***Helicobacter pylori***. One could argue that we now know much more about these topics than we did back then, but most of these basic principles were already know by scientists at that time. But they were not—*and actually they continue to not be, even in 2020*—taught to kids or teenagers in schools, in general. Another clear example is that obesity is nowadays one of the leading medical concerns in Western countries, such as the U.S., affecting more and more kids at younger and younger ages. However, numerous schools in the U.S. often don't teach how to truly prevent this: for instance, the notion that it mainly suffices to do some exercise is completely wrong—changing diets and reducing food items with a lot of sugar, or drinks such as Sprite or Coca Cola, is also extremely important. In fact, not only most schools don't teach that, but they are themselves among the main culprits of the current obesity pandemic, by providing horrible kinds of food items to kids, including such highly caloric food items and artificial drinks. Why would schools do that, contributing directly to a huge plethora of major medical problems? Because within the current socio-economic system they often care less with the well-being of the kids they are supposed to teach and care about, than with doing business with big companies such as **Subway**, **McDonald's**, **Dunkin's**, **Pizza Hut**, **Burger King**, **Wendy's**, **Taco Bell**, **Coca Cola,** and **Pepsi Cola**: *Inculturation*, indeed—"*to favor rather the exploitation than the enlightenment of mankind.*"

These topics were brilliantly discussed in **Naomi Klein**'s pulsating, well-documented and widely read 1999 book *No Logo—No Space, No Jobs, No Choice*, who presented a huge amount of factual evidence showing that in the past decades brands and fast-food chains have indeed "muscled" their presence into the school system. Basically many of the most **unhealthy, caloric, sugary, and salty food items** being marketed in the U.S. can be easily found in most U.S. schools. Many people tell me that there is nothing wrong with that, particularly those defending **capitalism**, or the notion of "**self-regulated markets**," which is a very weird and factually wrong *Neverland* metaphor, because obviously markets are not living organisms with an internal homeostasis, so clearly there is no way they can "self-regulate themselves," within the reality of the natural world. According to those people, books such as *No Logo* miss a crucial point: that if kids don't want to eat those unhealthy products, they have the choice to do so, nobody points a gun at them: that is precisely "the power of and main difference between liberal capitalism *versus* communism," they often say. However, if they had truly read *No logo*, they would know that the book precisely showed, empirically, that this argument is completely flawed—that is why Klein included "*no choice*" in its subtitle.

For instance, in schools, and basically everywhere else in the U.S. the prices of those unhealthy foods are very low when compared to much more healthy, fresh, and organic products—just to give an example, in the organic store just next to my apartment, one or two organic apples are often more expensive than a whole **McDonal's Happy Meal Box** for kids. Another example that you all can easily test by yourselves: try to get a **fresh fruit juice** at the airport, the next time you fly within the U.S. Even a mere **fresh orange juice**—but I mean a real one, when oranges are truly squeezed in front of you. If you are extremely lucky, you will be able to find one, after trying to find it around the airport for a long time. But in a huge number of airports, you will not find one: "*no choice.*" Now, try to get a fake "freshly squeezed orange juice" at any airport, and you will get it very easily, in a few minutes or even seconds, such as **Coca cola's *Minute maid*** or a similarly caloric, sugary, and basically chiefly artificial drink. Such orange juices, contrary to what many parents *believe*, have nothing fresh, and are instead among the most caloric and dangerous things kids can drink, because by having almost no fibers, the high amount of sugar is very easily and quickly absorbed by the body. But the big companies, and amazingly even some people serving food and drinks at restaurants, will make anything possible to miseducate the parents, and convince them that those drinks are healthy, and "fresh": they, and the system in general in countries such as the U.S. care mainly about the money they can make, not about the health of your children and if they get obese. At least until obesity will start to cause the government to spend a huge amount of money, in health care and so on, as smoking tobacco did: then yes, perhaps the system will finally care about your kids, or better said, about spending so much money taking care of them if they will become obese adults plagued by several types of health problems.

Now, compare this again with most **hunter-gatherer nomadic societies**, in which eating and drinking healthy and fresh food and liquids is the norm (see Fig. 8.1 and Chap. 8). In such societies, knowledge about such healthy food items, and about how to obtain them in nature through hunting and/or gathering, is often obtained from the parents or the broader community via a direct, active, interactive, engaging, mindful, and healthful type of social transmission (Fig. 9.3). Most kids in

Fig. 9.3 Different ways of learning: in nomadic hunter-gatherer societies kids normally learn in a much more direct, active, engaging, interactive, and thus in general more mindful and healthful, way

countries such as the U.S. spend a huge part of their lives in **school**—sometimes up to 7 to 8 hours a day, mainly in a sitting position that is obviously not the most healthy posture for humans neither—mostly learning things they will never truly need to know or use when they will be adults. Even in the case of a scientist as myself, I only learned throughout my education, including during university, a maximum of 5%, not even that much, of the knowledge I truly use nowadays, for instance when I write this book. Of course, part of those 5% include knowing how to write and read, which is fundamental for almost everything we do nowadays, but the reality is that we learn how to write and read in the very first years of school, which are by far the most important ones in that sense.

After those first years, a countless number of things that we learn will be useless later in life and/or are part of an indoctrination into our 'great' countries and cultures and political systems. The educational "system" in Portugal considered that it was crucial for me to learn things like the name of all regions of Portugal, or of the capitals of each and every Portuguese districts, or their rivers, or completely irrelevant wars and battles that happened near those rivers centuries or even millennia ago. This by itself is a clear example of how most things we learn in school is mere inculturation: they refer to the culture and historical events of "our great country," to things that obviously no kids from any other country would be forced to learn or would almost surely never care about. Instead, they will be forced to learn similar futile things, but applied to their "great nations," such as the names of all their past kings, rivers, and an endless number of things that they will never need to remember and that accordingly they will mostly obviously forget, with a few exceptions. In a way, having kids learning how "great" their countries, or cultures, or 'races' are is one of the first steps within the acculturation of tribal teleological Us versus Them narratives.

As explained above, I am not arguing that "modern schools" should just be closed, for ever and ever. I am completely aware that being able to go to a "**modern school"** is something that many kids of the so-called second or third world countries would *wish they had the opportunity to do*, and would absolutely benefit immensely by doing so, as it is one of the most efficient ways to break the ***cycle of poverty*** in which a huge number of families is trapped, in such countries, as noted above. But that is precisely the point: this cycle of poverty is the *result* of the type of system we are living in, particularly since agriculture and the rise of "early civilizations" and the related rise of **inequity** that accompanied those transitions. So, yes, for an individual domesticated sheep that is part of a herd being handled by a shepherd, when that sheep is trapped in such a vicious cycle of **domestication**—in the case of humans, of **self-domestication—control**, and **dominance**, it is better indeed to follow the herd and the shepherd. But this does not mean that this is *always* the right thing to do, particularly for the whole population of sheep, in the long term. Indeed, within the natural world, it was precisely because some non-human animals made the mistake of starting to get used to, comfortable with, trust, and thus follow and then obey to humans millennia ago that ultimately they ended up by being domesticated, used, and abused by us (see Figs. 7.9 and 9.4).

Therefore, yes one can say that it is better for a kid in Yemen to good to school and learn things even if a great part of them are and will always be useless to him/her, because in reality what is truly most important is not what we learn in school but instead being able to put in a CV, or say in an interview, that one has done that, that one is an "acculturated productive/successful" individual within that "system": that is what will ultimately help him/her to get a job and "work" and eventually, hopefully break the cycle of poverty so characteristic of "civilized" states. That is, the *reality* is that not only the **cycle of poverty** of many families in Yemen is the result of that same *system* they desperately want their kids to be acculturated to, but

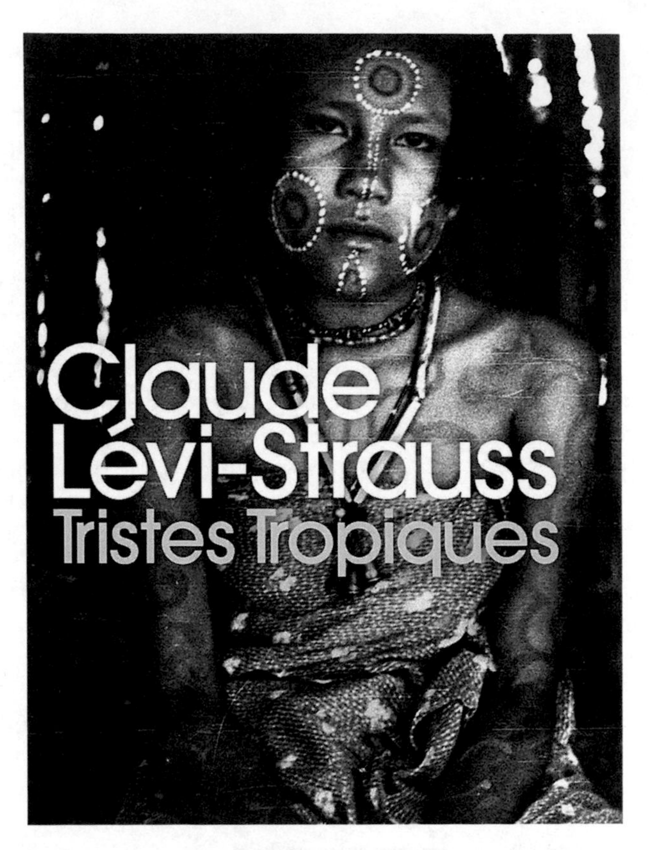

Fig. 9.4 One of the most beautiful and respectful front covers published in a non-fictional book: front cover of the English version of Levi-Strauss' *Tristes tropiques*, showing a Caduveo indigenous girl made ready for her puberty rites, photographer by him in Brazil

the system itself precisely *needs* more and more kids to receive that type of inculturation given in schools so it survives trough the continuation of the *status-quo*, by learning the name of former kings/leaders, admiring them, venerating powerful people, and ultimately *obeying* to them and to authority. Let's be honest, and deconstruct what truly happens in such a cycle of poverty: we all know that in Yemen, Central Africa, or Guatemala, and many other countries, a huge number of kids will go to school, and yes, a few of them will be able to break the cycle of poverty if they show they have been successfully acculturated to that system. However, the truth is that the *vast majority* of them will not be able to do so: they will remain poor, as their parents were, and their grandparents, and all their other ancestors for millennia, since the rise of agriculture and "early civilizations." Of course in many Western countries, and even several non-Western ones, there is now a growing middle class: that has been a major change in recent history and needs to be highly praised. But sometimes we tend to be so Western-centric that we forget that this does not apply to most countries in the globe, in which a huge number of people continue to be as miserable as their ancestors were since they became "civilized."

Indeed, as a friend once told me, "oppression, inequity and control will obviously always persist within 'civilizations', because we need to have someone to clean our toilets, nobody would do it voluntarily, right?" One recent example that clearly illustrates this fact, and further illustrates that schools are part of, and needed for, our current political and economic system, concerns the fact that during this **Covid-19 pandemic** most governments made everything they could to keep schools open, despite the obvious risk for the students and teachers, and their families, loved ones, and friends, and the broader population in general. For example, in the school in which my brother teaches, near Pacos de Ferreira in Portugal, a substantial number of teachers *and* students got infected, but amazingly that school did not close for months, until Portugal became the country in the whole planet with higher rates of deaths and infections because of the virus. Then yes, the government was finally forced to close all schools, including the one where my brother teaches. Of course, the main argument used by the government was not that the schools had to be open so the kids could learn, because obviously they could learn online at home, as was done since the very start of the pandemic in many universities, including the one where I teach in the U.S., Howard University. I am not saying that online teaching is ideal for kids, I am just saying that this was *not* the main argument used by most governments and by economists to keep schools open.

So, what was the reason why daycares, pre-schools, and schools with kids and teenagers had to be open, putting in risk the lives of so many people and literally contributing to the death of thousands of people in my own country? Because parents had to go to "work," physically or online from home, and they would not be able to do so—or would be less *productive*—if the kids stayed at home. That is, our own governments had to recognize, once for all, that-apart from acculturation-one of the main functions of schools in "developed" countries is to liberate their parents so they can "work" for and be "productive" to the system. And if that puts in risk the lives of millions of people with a pandemic? Who cares: *the show must go on*, if not the "economy will stop," the "system would collapse," and the non-essentials would not have "essentials" to clean the streets and their houses, to produce items and deliver them food, and so on, and above all the "masters"—the 1%, leaders, and CEO's—would not have people to work for their dreams. Indeed, the "essentials" are, not surprisingly, the ones that tend to have lower salaries, and do "works" such as cleaning toilets, streets, supermarkets, airports, and so on, as well as cleaning the apartments and houses of the non-essentials or the villas of the 1% (see Fig. 9.5). Another clear example concerns what we call "**law enforcement and police**." For instance, it is well known, and now widely discussed because of the rise of the **black lives matter movement**, that one of the earlier forms of policing in the Southern colonies that later became part of the U.S. were **slave patrols** that had the main goal of preventing **slave rebellions** and **enslaved people** from escaping. Again, not a "noble" function related to the *"enlightenment of mankind,"* but instead related to the preservation of the *status quo* within a system—one more—of *"exploitation,"* in this case truly the worse type of human exploitation that the storytelling

Fig. 9.5 Far-right Portuguese politician Andre Ventura—or it could be a religious leader or a CEO or anybody from the 1%—using *Neverland* stories to "divide-and-conquer" the 99% by making them fearful/against the "others"—a longstanding distraction tactic so the 99% don't realize that, in the *world of reality*, the leaders and the 1% love such tales as they help them maintain the status-quo and continue or increase the huge inequalities of the system

animal has ever created, justified, and undertaken. A similar point was made in Malik's 2014 book *The Quest for a Moral Compass* about the so-called civilized institutional religions such as **Christianity**, or "ancient civilizations" such as the **Roman empire**—and how they often became ultimately truly related with enforcing authority, usually recurring to teleological narratives—when he referred to **Saint Augustine**:

> By the time that **Augustine** and **Pelagius** were locked in doctrinal combat over **sin and salvation**, Christianity had moved from being a fringe, persecuted faith to being one of power and authority, the official religion of the Roman Empire. The key moment was **Emperor Constantine**'s Edict of Milan in 313 legalizing Christian worship. Constantine did more than end persecution. He converted to **Christianity** and became a patron for the faith, supporting it financially, promoting Christians to high office, and building an extraordinary number of basilicas. When between 324 and 330 Constantine transformed **Byzantium**, an ancient Greek town on the Bosporus, into his new imperial capital, Constantinople, he turned it into a Christian city, building churches within the city walls but no pagan temples. The edict of Thessalonica in 380 made Christianity the official state religion of the Roman Empire. The combination of fractured communities and a deep desire for scriptural authority gave rise to a characteristic feature of Christianity: the ferocious doctrinal disputes that of ten rent the faith, and the denunciation of opponents as 'heretics' whose heresy of ten had to be expunged not simply through excommunication, but sometimes through torture and death. Increasingly, too, there was an insistence on placing faith above reason. 'Let us Christians prefer the simplicity of our faith, which is the stronger, to the demonstrations of human reason' wrote **Basil of Caesarea**, an influential fourth-century

theologian and monastic. 'For to spend much time on research about the essence of things would not serve the edification of the church'. Theologians did not reject reason as such. Rather, they insisted that reason had to be subordinate to faith. 'Unless you believe, you will not understand', as Augustine put it, borrowing a phrase from Isaiah. Reason was not only subordinate to faith, it was, as in many classical cultures, seen also as exclusive to the elite, who alone could be entrusted with it. *For the masses, faith was the cement of obedience.* Faith, wrote **Origen** (185–245), perhaps the first great Christian theologian, is 'useful for the multitude', a means of teaching 'those who cannot abandon everything and pursue a study of rational argument to believe without thinking out their reasons'. Faith, in other words, no longer meant a state of surrender or openness to Revelation but rather an absolute trust in the Church hierarchy, which alone possessed the reason to discern God's meaning. Faith had become the means of **enforcing authority**.

Augustine, while also drawn to asceticism, had developed a theology of authority and order. 'It is in the natural order of things', he preached, 'that women should serve men, and children their parents, because this is just in itself, that the weaker reason should serve the stronger'. As with family, so with society. It was given by nature for the lower orders to serve the upper orders, and for all to serve the emperor. Slavery, too, was 'ordained as a punishment by that law which enjoins the preservation of the order of nature, and forbids its disturbance'. War, repression and the torture even of innocent men were all acceptable to compel obedience and to secure order. But while the rulers of a society could take punitive action to defend social peace, individuals had no such right. In Augustine, the theologian John Rist observes, 'the powers of ordinary citizens are almost non-existent'. **Plato** and **Aristotle**, Rist adds, who themselves worried about the mob and feared for social peace, nevertheless 'would have shuddered at such an empty concept of citizenship'. In the space of four centuries, Christianity had transformed itself from a faith for the dispossessed to a 'religion fit for gentlemen', as the historian Diarmaid MacCulloch has aptly described the imperial Church. **Jesus** insisted that 'it is easier for a camel to go through the eye of a needle than for a rich man to enter into the kingdom of God'. Augustine suggested that 'the poor could act as heavenly porters to the wealthy, using their gratitude to carry spiritual riches for their benefactors into the next life'.

As also pointed out by Malik:

Today **Rousseau** is viewed as, at best, naively eccentric, at worst dangerously deluded. The idea of the **'noble savage'**, for which Rousseau is perhaps best known, is portrayed as a romantic celebration of **primitivism**. The concept of the 'general will', by which Rousseau meant the authority to which individuals within a collective must accede, is often seen as paving the way to totalitarianism. In fact, Rousseau was far subtler in his arguments than modern critics allow. Though indelibly associated with the concept of the noble savage, Rousseau neither used the phrase nor believed in the idea…. Born in Geneva in 1772, the son of a watchmaker, Rousseau was brought up a Calvinist but converted to Catholicism in his teenage years. Moving to Paris, he became friends with leading Encyclopaedists including **Diderot**, d'Alembert and **Voltaire**. Rousseau found himself increasingly alienated from their easy optimism, and drawn towards a darker view, not of human nature, such as that possessed by most pre-Enlightenment thinkers, but of civilization, which the philosophes had seen as the tool for human betterment…. Like earlier thinkers, Rousseau believed that social life emerges as humans come to recognize the value of co-operation. The creation of society also leads, however, to the institutionalization of private property in which Rousseau finds the source of inequality, oppression and enslavement. Imagine, Rousseau wrote, '*The first man who, having fenced off a plot of ground, took it into his head to say this is mine and found people simple enough to believe him'. What crimes, wars and horrors would the human race have been spared, he wondered, by 'someone who, uprooting the stakes or filling in the ditch', had 'shouted to his fellow men: beware of listening to this imposter…you are lost if you forget that the fruits of the earth belong to all and the earth to no one'*!

As I have referred above to our self-domestication, let's analyze in more detail this subject, which is now widely mentioned within discussions about "human nature." The idea that humans domesticated themselves is not new: for instance, some decades ago the biologist **Konrad Lorenz** popularized the term "Verhausschweinung," which literally means that "modern" urban life makes Western people similar to domesticated pigs. Of course, the **human self-domestication** hypothesis that is currently accepted by many scientists does not correspond directly to Lorenz's notion of "**Verhausschweinung,**" and some of its details are still the subject of much discussion and controversy, as explained in a 2019 paper written by my close colleague and friend **Sanchez-Villagra** and by anthropologist van Schaik, entitled "*Evaluating the self-domestication hypothesis of human evolution.*" Basically, as they explain, in the current "modern" context "human self-domestication" is often invoked to explain human evolution, arguing that selection for reduced aggression on animals undergoing domestication provides a model for selection favoring **prosocial behaviors** in humans and for a set of seemingly independent features, such as changes in facial anatomy and so on. However, it is important to stress that some aspects of the hypothesis currently defended by many scholars have several flaws, such as the a priori assumption that hunter-gatherers are in general more violent than "civilized" people and the non-recognition that there are strikingly different levels of warfare in small nomadic versus large sedentary hunter-gatherer societies, as explained above. In this sense, a major problem with the "self-domestication" hypothesis as currently defined is that it still follows the longstanding "ladder-of-life" teleological framework of the "good-civilization-narrative" that has been seriously put in question by a huge number of works from other fields, recently (see Chaps. 4 and 8).

That is, many authors discussing our self-domestication absolutely "want" to see it as something that was necessarily "good," because it is related to "sedentism" and "civilization" and that of course "has" to be "good": that was what we were told in school, so how could it possibly be wrong? But if we take a step back and discard *Neverland*'s inculturation for a moment, and accordingly take into account what the available empirical data truly show us—that is, that such long-standing "civilization-progress" narratives are factually wrong—then we can perceive that self-domestication has instead been a major contributor to the increase of warfare, subjugation, and oppression since the rise of "civilization." Indeed, it contributed to a situation in which a huge number of people, mostly living in horrible conditions, have fought and killed, and many times loss their lives, for the sake of a few "rich masters," by obeying them blindly, exactly as domesticated donkeys do work for their human "masters," and sometimes *die working for them*, as an increasing number of people do in countries such as Japan. So, how can we, the human "masters," make dunkeys work, and sometimes die working, for us? Because (1) if the donkeys don't do what we want we punish them; (2) the donkeys were somehow "acculturated" to think that what they "do" is normal, in the sense that they can "see" that all the other donkeys do the same, as their parents did, and their grandparents, and so on; and/or (3) simply because that is what they are "told to do" by us. As we have seen before, exactly the same three ways have been used to self-domesticate humans, which begs the question: outside of *Neverland*, would anyone seriously argue that by doing that, a domesticated donkey has a "higher quality of life" than let's say a wild worse or a wild zebra?

Some people could argue against this idea, and say for instance that in some cases domesticates "benefited" from the domestication process, for instance when wolfs became closer and closer to humans and this ultimately led to the **evolution of dogs**. However, firstly, this would only apply to dogs, and perhaps to a lesser extent to cats, and surely not to the vast majority of other domesticated animals that we domesticate, use, abuse, and eat, such as pigs, cows, chicken, and so on. Secondly, contrary to what Western-centric people tend to *believe*, the life of dogs in a huge number of countries, such as China, can be truly horrible, as they are often trapped in wet markets and animal factories, often without being even able to move at all, exactly as we do with pigs, chicken, cows, and other animals in our Western animal factories (see Fig. 9.6). Again, would someone seriously argue that such dogs that are even unable to move, and are just waiting to be killed and then eaten by humans, are "better off" than wild wolves? Actually, millions of other dogs are also trapped in other types of cages in "developed" countries, including Western ones: those in pet stores, or dog shelters, where dogs can wait for months, and even years, to be sold or adopted, so those cages will then become available to have more dogs that will be entrapped for months or years, and so on. Moreover, even those dogs that are "lucky enough"—within a human perspective—to be bought or adopted, are often plagued by a plethora of **"modern" diseases** that wolfs don't

Fig. 9.6 The self-domesticated "sapient being": pictures of Tokyo metro illustrating what the self-proclaimed *Homo sapiens*—or *Homo Deus* as proclaimed by scholars such as Harari—call "progress." *Homo Deus*, or instead *Homo servus*—an irrational being that self-domesticated and self-enslaved himself through the creation and belief of imaginary narratives of "purpose" and "progress" by blindly obeying to the social norms and obligations that resulted from them

have: often exactly the same "modern" diseases that we have in "developed countries" that most hunter-gatherers don't have (see Chap. 4). Actually, apart from such "modern" diseases, many dog breeds selected by humans in the last centuries or decades have **chronic diseases** from the very beginning to the very end of their lives. For example, **bulldogs** tend to have a **chronic respiratory problem** named "**brachycephalic obstructive airway syndrome**." This is the result of a process in which dogs and humans are no longer both active players of the domestication process. Instead, at a moment in time the human "masters" started to be the only active players, selecting the dog breeds for their own benefits or according to their desires, even the most weird ones, such as those causing severe obstructive airway problems and many other types of chronic diseases within the selected dogs.

Another interesting and often also neglected key point about domestication is that a typical trait of domesticates—with some exceptions, of course—is the decrease of brain size. This point should be taken into account in discussions about human self-domestication, because such a trait does seem to match the type of "**mechanization"**—or as some scholars call it, the "**stupidification"**—that became a major part of life for a huge proportion of humans with the rise of **agriculture**, and was then further augmented within the "industrial revolution." Actually, some scholars argue that such a process of "stupidification" was even amplified some decades ago with the invention of the "modern opium of people," that is, of TV: sitting for ours in front of it, passively, thus moreover often getting fat and even more unhealthy by doing that, in a very similar way to what happens to the mainly passive, fat and unhealthy domesticated pigs or cows of our "animal factories."

Having said that, one of the major tenets of the human self-domestication hypotheses defended by many scholars is unquestionable: humans can live nowadays in big cities with a very high level **of inter-personal tolerance** and a low level of **inter-personal violence**, at least in what concerns physical aggression. For instance, in Copenhagen—a city with more than 1.3 million people—most days there is not even a single murder. This is indeed a remarkable feature of some "developed" countries, particularly Western European ones, as well as many Asian ones such as Japan, about this point there is no doubt that Hobbesian authors such as **Pinker** are completely right, particularly when one compares this to the huge levels of violence—and wars—that were so common and widespread in "early civilizations." However, one should note that many "developed" countries, including the economically most powerful one—the U.S.—continue to display very high crime rates. Moreover, one cannot simply say that humans are "more docile" in the last centuries because those centuries precisely included the most deadly genocides ever—the Holocaust, the Rwandan genocide, the Armenian genocide, and so on—and the most atrocious occurrence in human history: the Trans-Atlantic slavery trade. All these terrible atrocities have involved massive planning, massive violence, massive inhumanity, and massive and complete lack of empathy for the "Other." Furthermore, another critical incongruence of the "self-domestication-less violence-more docile" hypothesis that is accepted by many scholars is that according to them this process started to occur hundreds of thousands of years ago, much earlier than the rise of agriculture, something that does not match with the unquestionable fact that there was an enormous rise of slavery, warfare, and oppression during the rise of agriculture and "early civilizations."

In summary, although empirical data do point out that there has been a self-domestication of humans, that process does not correspond to the way it is often portrait by many scholars and in the media in recent years. Firstly, it might have started hundreds of thousands of years ago, but was clearly augmented during the rise of agriculture, "civilizations," and organized religion, which have been the three major players within this process, since then until this very day. As a consequence, as we know that the rise of those **"three historical horsemen of apocalypse"** was clearly associated with an increase of the **"three societal horsemen of apocalypse"** mentioned just above—that is slavery, warfare and oppression of "Others," which include "other" groups, "other" sexual preferences, and "other" genders—this means that self-domestication indeed was not, and continues to not be, necessarily "good" or "docile" (see Figs. 3.14, 3.15, 3.19, 5.23, 7.1, 7.5, 7.6, 7.7, 7.8, 7.9, 7.14, 8.4, 9.2, 9.6, 9.7, 9.8, 9.9). So, this begs another question: although obviously humanity as a whole has been playing a role in its self-domestication, who have been, and continue to be, the most active players promoting it, speeding it up, and benefiting with it, in other words, who are the main "masters" within that process of domestication? Perhaps the major clue comes precisely from the analysis of the

Fig. 9.7 Lynching of Elias Clayton, Elmer Jackson, and Isaac McGhie in Duluth: on June 15, 1920, these African-American circus workers, suspects in an assault case, were taken from jail and lynched by a white mob of thousands in Duluth, Minnesota, because there were rumors that six African Americans had raped and robbed a 19-year-old woman, although a physician who examined her found no physical evidence of rape

Fig. 9.8 One of the most horrible aspects related to the creation of teleological narratives about *purposeful* imaginary supernatural beings, about what they *want from us*: sacrifices done by humans to humans, provoking excruciating pain to people that were not only alive but also conscious about what would be happening to them, without any kind of anesthesia

three 'historical horsemen of apocalypse' and the moment they arose. For instance, within the horrific stories about warfare, slavery, oppression, and violence in "early civilizations," such as those mentioned in the Old Testament, it seems to be clear that those that benefited the most were the kings, leaders, slave owners, the religious leaders, the richest ones, and so on—in other words, the elites. In fact, as explained above—and as will be seen below when I discuss Harari's *Homo Deus*, which puts forward ideas that are similar to those I am defending in this Section—those elites would have never been able to have so much money, food, power and so many material goods without having their domesticated servers—the 99%—to pay taxes to them and to fight, die, starve, work, suffer, plant, and collect crops, and built palaces for them either blindly, willingly, or forcibly (see Figs. 4.4 and 9.5).

A publication that discussed these topics in some detail is Wrangham's 2018 paper "*Two types of aggression in human evolution*":

> **Proactive aggression** involves a purposeful planned attack with an external or internal reward as a goal. It is characterized by attention to a consistent target, and often by a lack

Fig. 9.9 Humans can be, depending on the circumstances and biosocial and political systems in which they live, both the most oppressive and cruel—as seen for instance in such images showing horrible tortures or atrocious acts such as mistreating, starving and killing kids in concentration camps—and, although to a much lesser extent, the most altruistic and caring animals—as seen for example with the care given in some countries to people with Down syndrome, which can have a higher quality of life than ever before and than any non-human animals with similar syndromes would normally have

of emotional arousal. Aggressors normally initiate action only when they perceive that they are likely to achieve their goals at an appropriately low cost. Examples include bullying, stalking, ambushes, and premeditated homicides, whether by a single killer or a group. By contrast, **reactive aggression** is a response to a threat or frustrating event, with the goal being only to remove the provoking stimulus. It is always associated with anger, as well as with a sudden increase in **sympathetic activation**, a failure of **cortical regulation**, and an easy switching among targets. Examples are **bar fights** arising from mutual insults and **crimes of passion** immediately after the discovery of infidelity. Note that the term "reactive aggression" refers to the nature of the aggressive act rather than the reason for acting aggressively. According to this definition acts of revenge are not necessarily reactive and in fact are unlikely to be so, given that revenge typically involves planning.

As explained by Wrangham, the existence of these two types of aggression is well supported by neurobiological and experimental data. I do agree that, compared to chimpanzees, humans living in "civilizations" do seem, in general, to have a decreased reactive aggression—as illustrated by the example of Copenhagen, above—and an increased proactive aggression. That is, chimpanzees are reported to plan "proactive" killings of chimpanzees from other groups through the formation of "patrols," but still this cannot be compared to the careful planning and execution, by humans, of genocides and mass killings of millions of people. Having said that, it should be noted that some of the broader implications of Wrangham seem a bit dubious, such as his proclamation that he "solved the **Rousseau-Hobbes debate"** because the two would be right according to his thesis. **Rousseau** would be right in that humans "naturally" have very low reactive violence, as is the case with nomadic hunter-gatherers, while **Hobbes** would be right in that humans "naturally" have, compared to other primates, very high proactive violence. However, such a statement misses a key point about that debate, which is mostly centered in the "brutish" state of humans, as seen in "savages," versus the "civilized" state of humans, as seen in "modern countries/societies." Wrangham's thesis misses this point because he is precisely one of the many authors that is instead referring to changes that supposedly happened much earlier than, and thus that do not apply at all to, the transition between nomadic hunter-gatherer "savages" and agricultural "civilized" societies. For instance, he recognizes that "in comparison with both chimpanzees and bonobos, frequencies of fighting in small-scale human societies are very low…quantitative data on rates of human fighting conforms to the conclusions of ethnographers who uniformly stress the peaceful tenor of daily life within small-scale societies."

Of course, this latter statement - about 'savages' having very low fighting rates and a very 'peaceful' lifestyle in general - is completely right as we have seen above, but it would surely make Hobbes roll over in his grave. But Wrangham could argue, as he did, that although **Hobbes** was wrong about this, within Wrangham's thesis he would be right in that humans "naturally" have, compared to other primates, very high proactive violence, as noted just above. However, the reality is that interestingly, and rather oddly, Wrangham does not refer at all to the levels of proactive aggression of nomadic hunter-gatherers versus "civilized humans." That is, he seems to be assuming a priori that the high level of proactive aggression that humans display in "early civilizations" and "modern societies" was already present before the rise of "civilizations," and thus that nomadic hunter-gatherers are more "proactively aggressive" than common chimpanzees, but he does not present compelling data to support this idea, particularly because he actually refers to nomadic hunter-gatherers as having a "peaceful life" in general as noted above. In fact, what seems to be really unique for humans in terms of "proactive aggression" is the amount of massive killings and/or atrocities that they can do in a proactive, planed way, such as the Trans-Atlantic slavery trade, or the Holocaust, what applies above all to "civilized" societies, as empirically shown in the numerous studies that we have discussed throughout this book. So this would *not* support at all Hobbes' ideas, well on the contrary, it would support Rousseau's ideas because such massive atrocities were only possible within an "Hobbesian" system, a **Leviathan**, that is, with strong centralized governments, systemic oppression, massive bureaucratic planning, and so on. In this sense, this latter idea, defended in the present book and in Harari's *Homo Deus*, is much more in line with the ideas put forward in a very recent book that focuses precisely on human self-domestication: Hare and Woods' 2020 *The survival of the friendliest*. These authors argue that the behavioral and physical tools used for keeping reactive aggression at low levels *despite* the rise of bigger agglomerations such as agricultural villages or cities, through the control by leaders and the use of force and the "rule of law" and of narratives about Gods, "meant to be," "morals," and "progress," were exactly the same tools that were systematically used to oppress, enslave, and torture people from other groups since then-in other words, to undertake proactive aggression at levels never seen before the rise of agriculture, in humans or in any other species.

Specifically, Hare and Woods argue that there is a main behavioral difference between humans and other animals, which is profoundly related to the central topic of the present group: as other social animals, humans can be "friendly" to other humans because of **familiarity**—such as "we saw each other since we were young," and so on. However, contrary to most other social animals humans can also be "friendly" to other humans because of an *identity* that is socially constructed by using complex teleological narratives. For instance if you are an U.S. citizen and travel to Mongolia, and you see someone that you never saw before but that happens to also have a U.S. passport, you immediately tend to feel more "related" with and thus to react in a more friendly way to that person. This "identity" sense of "familiarity" can be logical, as you could assume that such a person knows more about your own "culture," for instance about U.S. politics, NBA, burgers, and so on, to use just some typical stereotypes of what is being "American." However, such stereotypes are themselves just a social construct, so not only you are not truly "familiar" with that person in a literal way, but that person might not even share at all your "culture"—for instance, if they are Latino-Americans, they might care much more about soccer than about NBA, or about Tacos al pastor than about burgers, to use again some constructed stereotypes. As argued by Hare and Woods, this ability to be "friendly" with—and feel greater empathy towards—strangers based on such a constructed notion of "identity"—either real or much more often, completely imaginary—was likely crucial, and would precisely explain why we can live in cities of millions of people without having more inter-personal violence. However, the price paid because of this is that the very same narratives and imaginary stories that were/are often used to create this sense of "ingroup identity and cohesion" are indeed the very same ones that are employed to exclude, kill, or plan the genocide of "others" that are not part of that ingroup. That is, because those "others" were not "chosen by your God," or are not from the U.S., or are not heterossexuals, and so on. This explains why humans are the only animals that have planned and committed genocides—proactive aggression—of millions of people, particularly by using narratives and fictional tales involving the **dehumanization** of "others," which lead to an extreme level of *both inclusion and exclusion*: those "others" not only believe in a different God, or are from different "races" and so on, *they are not even from our human species* (see Fig. 9.10).

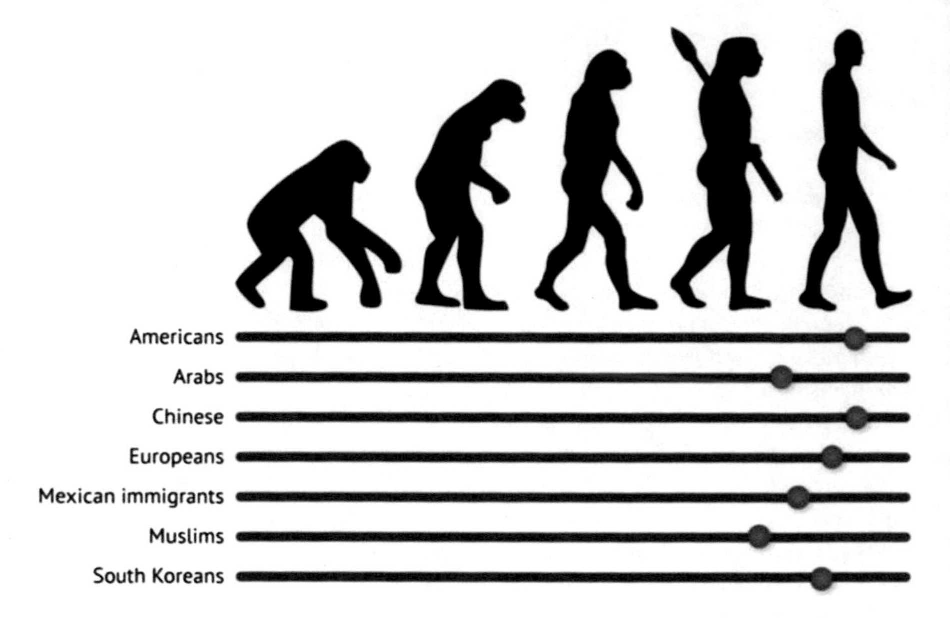

Fig. 9.10 The "ascent of humans", related to the notion of "ladder of life" discussed in the present volume, has been used by Kteily and colleagues to measure dehumanization. Scores are provided using a slider scale ranging from 0 to 100, with 0 corresponding to the left side of the image (that is, quadrupedal human ancestor) and 100 corresponding to the right side of the image ("full" modern-day human). This popular image, which is of course not a scientifically accurate representation of human evolution but instead of a factually wrong 'chain of being'—is—precisely because of this—indeed useful to test the acceptance, by the broader public in different countries and social contexts, of narratives of "linear progression" that are so widespread in humans and that often lead to dehumanization, for instance Americans versus others, or Palestinians versus Israelis, and so on. This specific picture disturbingly shows, for example, how U.S. people that were included in a research study considered that "others," such as Arabs, Mexican immigrants, and Muslims, are not "fully" humans

One clear example of **self-domestication** is given in a recent book by Jack Lewis, and actually in its very title: *The science of sin: why we do the things we know we shouldn't*. Of course, as Lewis is completely aware, within the natural world there are no "sins," so saying that we "know we shouldn't" do them, particularly when some of those "sins" refer to things as natural as having a strong desire for sex—**"luxury"**—is something weird, coming from a neurobiologist. In fact, the whole book, which is otherwise interesting, is plagued by such logical inconsistencies as well as by teleological narratives that are so typical of *Neverland*, including highly adaptationist ones (see Chap. 6). For instance, in the chapter focused on "luxury" Lewis makes an effort—rightly so—to show how wrong some "moral judgments" are, but then he quickly falls again into *Neverland*'s trap, as for instance when he refers to Tiger Woods or Michael Douglas. Namely, he states that one should probably *not* consider the many sexual affairs of people like them as a "sexual dependency" medical condition, but then says that those affairs should be seen instead as "**moral weakness**." Is it a moral weakness to have, or to want to have, a lot of sex? If that is so, then most hunter-gatherers would be considered to be miserably weak, morally, as on average they have more sexual intercourses per week than **Michael Douglas** or **Tiger Woods** very likely had in the vast majority of the weeks of their adult lives. Lewis states that the closest thing to "**sexual dependency**" that has been more widely studied by neurobiologists is **compulsive sexual behavior**, in the sense that both conditions can affect the "capacity of living a normal and productive life." This example plainly illustrates the type of mechanisms that reinforce our self-domestication: first one defines what is "normal"—in this case, having a reduced amount of **sexual desire**, intercourse, or whatever—and then one delineates the "price" paid for not being "normal." In this case, the price is paid by both the individual and the society as a whole, because the individual is not "productive" within that society. As **domesticated donkeys** that work for us non-stop, or domesticated cows, the *Homo servus* is *not* supposed to "fool around," to have too much sex, too much fun, and so on, as that distracts his/her main function: to serve its "masters" and be productive (see Fig. 9.6). Actually, Lewis explicitly states what is the *specific societal price* that is often paid by those that dare to commit the sin of having too much sexual desire in "civilized" societies: "this type of *unpleasant* behavior frequently leads to divorce, or affects one's reputation"—including affecting one's job, as it happened to Tiger Woods—"or even results in legal problems."

Such a process of self-domestication could, in theory, indeed substantially affect, in the long term, the **sexual desire of humans** at a broader evolutionary scale. That is, if by having too much sex I am more likely to lose my job and/or romantic partners, have legal problems, and basically be seen as a "looser" or a "weirdo" by society, this could surely affect my ability to find someone that will want to have kids with me. Of course, this was not so much a problem when men were just able to force—for instance, physically or economically—women to have sex with them and there were no condoms and birth control pills, as attested by the fact that **Genghis Khan** is said to have 12 million descendants nowadays. But in 2020 in Western countries such as the U.S., who wants to have, and take care of, kids with a so-called pathological "looser" or "sicko" that constantly commits the atrocious sin of wanting "too much sex" and is moreover unproductive to society? It is much better to have kids with *"normal" and productive* people such as those shown in Fig. 9.6. Although this is just a theoretical example of one of numerous potential mechanisms that could be involved in a self-domestication process, it does show how the "system" and self-domestication can indeed be self-preserved or even augmented without involving any a priori master-planning by any *Illuminati* or political or religious leaders, or the 1%. That is why it is precisely so difficult to escape from the system, and particularly from *Neverland* as a whole, which is the product of an endless number of such "fictional systems," as explained by Harari in *Homo Deus* (see below).

We actually have a vast amount of historical and empirical data showing that such a process of self-domestication *did happen* in the last millennia, in at least some cases. For instance, the "technological most developed" country in the planet, Japan, which in many ways is indeed the most self-domesticated state in the globe as noted above (see Fig. 9.6), does have, in average, the lowest frequency rates of sexual intercourse globally. Moreover, there is also factual evidence indicating that this trend is likely further increasing right now, before our eyes. For example, because people in "developed" countries tend to have much less sex than hunter-gatherers do, and thus than our hunter-gatherer ancestors almost surely also did for millions of years, one of the resulting consequences is that this is somehow "compensated" by people by watching more and more **pornographic movies**. In the long-term, this in turn tends to make people to be less likely to desire to, or even to *be able* to, have sex, as explained by Lewis. He cites various studies showing that seeing a lot of porn indeed tends to lead to a long-term decrease of **sexual arousal**, and might often also lead to **sexual dysfunction**—something probably related to **dopamine**, in a somewhat similar way to what happens with some drugs. That is, after seeing so much porn, particularly the type of "*Neverland* super sex" usually displayed in such movies, the brain begins to want also "super" things, or even "super-duper" ones: more intense, more shocking, and so on. So, in a way, this is "**super-sized" self-domestication**, because this cycle may end up in a scenario in which sex is chiefly to be watched and consumed, at a massive scale, *not to be done*. Obviously, people that see more porn are, on average, those that have less sex—a huge number of them don't actually have *any* sex at all, and while this cycle is perpetuated, and in particular more people fall into it, the porno industry will become bigger and bigger. Just to give an example, as of 2006, according to the *Internet Filter Review*, every second, $3075.64 was being spent on **pornography**, 28,258 Internet viewers were viewing pornography, 372 Internet users were typing adult search terms into search engines, and every 39 minutes a new pornographic video was being made in the U.S. And, of course, the numbers are immensely higher nowadays, 14 years later. This is how a "system/business" that is clearly not "natural"—in the sense that no non-human animals or hunter-gatherer humans have a **porno industry**—and that is extremely absurd—instead of having sex and pleasure for free a huge number of people actually pay a lot of money to see others doing so—not only perpetuates itself but growths up every single year. There is hardly a more illustrative—symbolically and literally—example of how the *Homo fictus et irrationalis* is intensively immersed in *Neverland*'s surreal world of absurdity.

Indeed, Lewis relates this **vicious cycle of pornography** and **unnaturalness** with the work of Nobel Laureate **Nikolaas Tinbergen** about **egg preferences of gulls**: after being continuously exposed to very colorful artificial "super" eggs, the gulls ended up by preferring those fake eggs than their own real ones, thus putting at risk their own progeny, and thus their own survival in the long term. Many other studies have shown similar cases in which a "**supernormal stimulus"**—such as *Neverland*'s "super porn sex" or "super virtual reality"—hijacks the **natural stimulus**—in this case, to have sex naturally as other primates do. So, here we can clearly see the profound connection between our self-domestication process, our disconnection with the natural world, and our deeper immersion in ***Neverland***. First, something completely natural—**sexual intercourse**—begun to be seen as "sinful" and not "productive," so it became to be done less and less frequently, being then increasingly "replaced" by the *consumption* of more and more porn via TVs and computers bought and porn websites paid with the resources obtained from "working" in "productive" jobs. In turn, this ultimately lead, in many cases, to the increasing hijacking of natural sexual intercourse by ***Neverland*****'s "super" porn sex** and/or **by *Neverland*'s virtual sex** or even more dramatically by ***Neverland*****'s completely virtual lives** (see Figs. 9.11 and 9.12). In a nutshell, the type of people shown in Figs. 9.6 and 9.11 are an emblematic example of what the deep immersion in *Neverland* leads to: our *Homo fictus, irrationalis et socialis* obsessions ultimately also created, after the rise of sedentism/agriculture, the ***Homo servus*** and ***Homo Consumericus***. And, although this was obviously *not* planned a priori by the rulers, religious leaders, elites, CEOs, and the 1%, the truth is that they surely don't mind at all about—and often are very happy to be the main promoters of and contribute to—a scenario in which they can make use of such domesticated, dedicated servers and consumers.

That many of the items we were told to be "noble" features of "civilizations" are not necessarily the result of a path to the *"enlightenment of mankind"* but instead to use, abuse or oppress a huge number of people also becomes evident when one understands that since the origin of schools there were several methods, including painful ones, to punish and "correct" students if they were not learning what—or in the way in which—the teachers and/or leaders wanted them to learn. In Scott's first (1938) and second (1959) editions of the book *The History of Corporal Punishment—a Survey of Flagellation in its Historical, Anthropological and Sociological Aspects*, he wrote:

Fig. 9.11 David Pollard and Amy Taylor in "real" life and their "fictional selves" Dave Barmy and Laura Skye in the virtual online world *Second Life*

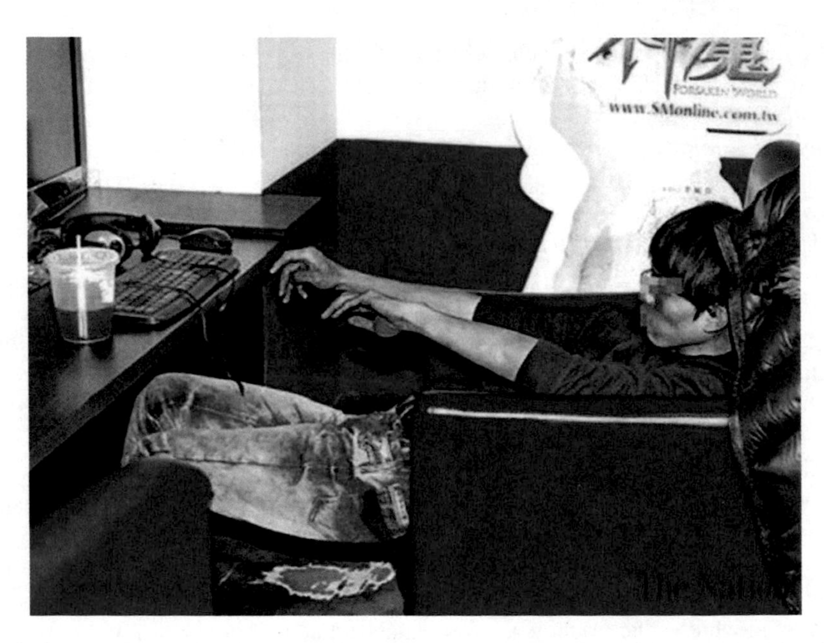

Fig. 9.12 Corpse of a 23-year-old man surnamed Chen, in a New Taipei City Internet cafe, where he was found dead

With the widespread popularity of **whipping** as a form of punishment for transgressors, and as a means of deterring others from committing crime in adult life, it was only to be expected that the whip should rank as an admirable instrument for the **correction of children**. Solomon's dictum: "He that spareth the rod hateth his son; but he that loves him chastises him betimes"; and his admonition: "withold not correction from the child; for, if thou beatest him with the rod, he shall not die. Thou shalt beat him with the rod, and shalt deliver his soul from hell", were acted upon to the letter by parents all over the world; and the maxim "Spare the rod and spoil the child" was accepted and considered to constitute full and complete justification for **flagellating children** of both sexes right through the ages until comparatively recent times – so recent, indeed, that within the memories of those of the present day who have, at any rate, reached middle age, must be vivid recollections of the sting of the birch or the cane. In the olden days boys and girls both, of working-class parentage, were flogged by their parents at home, and by their employers at work; while the children of the aristocracy received their floggings at the hands of their governesses or private tutors, and later at school. Even so long ago as the days of **ancient Greece**, pretty nearly a couple of thousand years ago, if history does not lie, the schoolmaster used the birch as an instrument of correction. **Homer** was flogged by his tutor; so was **Horace**; and so no doubt were all those who went to school at all. Indeed, it seems to have ranked as a universal corrective in all countries; this whip, or stick, or other analogous fustigating instrument. Even the teachers of religion, and the clergy, seem to have called upon the birch to help ram their arguments home. The monk **Udalric**, writing in 1087 in the *Coutumes de Cluny*, says: "At prayers, if the children sang badly or fell asleep, the prior or master will strip them to their shirts and flog diem with osiers or specially prepared cords". In China, whipping was customary in every school until **Confucius** put an end to it.

From the days when schools were first established until the beginning of the present century, the birching of boys has been inseparable from the discipline of nearly every school in Great Britain. Occasionally the **floggings** were so severe as to cause death. Thus, in *The Percy Anecdotes* there is an account of the trial of a schoolmaster named Robert Carmichael, in 1699, for the murder of one of his scholars. According to the evidence, Carmichael gave the boy three successive beatings "and in rage and fury, did drag him from his desk, and beat him with his hand upon the head and back with heavy and severe strokes, and after he was out of his hands he immediately died". An examination of the body revealed stripes on the back and thighs, from which much blood had issued, and marks on the head; and the jury found the beating which the boy received to be the cause of death. Carmichael was sentenced to seven stripes and banishment from Scotland for life.... The reason for the abandonment of birching, which occurred about the middle of the nineteenth century, was mainly due to the growth of **Victorian prudery**. It was held, and held most strongly, that the exposure of the naked fundament was indecent and immoral, particularly so in the case of girls. The alternative of whipping on the back, or on the loins or the shoulders, was recognized to be much too dangerous a practice. The **Puritans** were between the devil and the deep sea – in the end, though with some reluctance, they were obliged to abandon the use of the birch altogether.

That one of the reasons for the abandonment of birching had to do with the **obsession with sexual purity** in the **Victorian age** is emblematic: it shows that, within the fictional world created by the storytelling animal, sometimes new imaginary narratives are the only way to stop atrocious practices that were justified at first by the use of previous ones. And this, in turn, also shows that the assertion of Hobbesian authors that what they call "the expanding moral circle" in Western countries is mainly related to "rational" reasons of, and decisions made by, their leaders or the general public, is obviously wrong in some, if not most, cases. Scott's book bring us back to the example given in the **Preface** of the present book, shown in Fig. 9.1: how humans self-flagellate themselves—both literally and symbolically—because of the narratives they create and ultimately start to believe in. In his discussions of **self-flagellation**, he explains:

We are compelled to fall back upon the need which so often occurs in the case of **religious fanatics**...of finding some means of repressing the worldly cravings which arise irresistibly in their minds; hence the popularity of **self-torturing** in many and devious ways, of which, in ancient times, **flagellation** was one of the most widespread. The belief in the efficacy of the voluntary submission to pain or suffering or humiliation, as a means of expiation for a sin or transgression committed against God or the Church, was firmly established; and, indeed, to this day, is an integral part of many varieties of religion. Penance looms largely in the **Catholic faith**, it ranks as the fourth of the seven sacraments. It was this firm belief which let the leaders of the Churches, in those ancient days, go so far as to whip themselves, or to suffer whipping at the hands of their disciples, to wear sackcloth next to their skin, to crucify their own flesh, to fast for long periods, to parade about in rags and filth, to humiliate themselves in a hundred different ways. It was, too, this self-same firm belief which caused them, whenever they happened to be beset with **temptations**, which was a frequent occurrence, to disperse such longings by **self-punishment** and **self-humiliation**. One must not overlook the fact that in many cases the priests genuinely believed that self-punishment, being a form of **sacrifice**, would propitiate the god they worshipped. This provides one of the explanations of all forms of asceticism; from the **chastity** of **Roman Catholic priests** to the extreme **self-tortures** practised by the **yogis of Tibet** and the **fakirs of India**. Also, and often coincident with this propitiation of their god, the arousing of the sympathy or compassion of the public, which, inevitably, is connected with any form of martyrdom, was no doubt in the minds of those indulging in self-flagellation.

It was undoubtedly by these and other (true or apocryphal) analogous practices that the saints of old established and retained their reputations. There are for the finding many

> revealing instances. Thus in *Lives of the Saints Canonized in 1839*, in a reference to Saint Liguori, it is stated that he flagellated himself so severely that "one day his secretary had to burst open the door – and snatch the discipline out of his hands, fearing lest the violence with which he scourged himself might cause his death". And, according to he same authority, Saint Pacificus was accustomed to scourge himself to such an extent "as to fill all those with horror who heard the whittlings of the lash, or saw the abundance of blood which he had shed during the flagellation". Then, too, there was the example set by the Biblical heroes. Saint Paul, revered of all associated with the Christian religion, was trotted out as a believer in and a practitioner of self-flagellation. "I keep under my body and bring it into subjection". (I Cor. ix. 27.) Here, if ever man did, he stands self-confessed. And we read in **Psalms**: "For all day long have I been plagued and chastened every morning".... Sex entered largely into the matter, fornication being one of the major sins against the dictates of the Churches. Self-punishment of various kinds were favourite methods adopted by the early saints to subdue sexual thoughts and cravings.

So, humans first create *absurd imaginary stories* to *demonize sex*—something that is completely *normal in the natural world*, and is moreover *highly pleasurable* and often also *very healthy*—and then accordingly try to "subdue" it by doing something that is not only completely absent in non-human animals but also *extremely painful*: self-flagellating themselves. Do you imagine any animal, other than the weird storytelling creature, let's say a dolphin, or a tiger, or a fish, self-flagellating itself to "subdue its sexual desire?" If, let's say, an alien would see humans doing this, and compare such a bizarre, illogical, irrational behavior with what all other sexual species do, do you think that the alien would think that our species is the "rational one?" Well on the contrary, I guess. And, for those that might argue that these types of weird behaviors such as self-punishment, and particular self-flagellation, were only displayed by humans long ago, in the Middle Ages, when humans were "less civilized," I can point out several examples showing that this is obviously not so. Very likely, you don't need me to do this as you almost surely have seen some examples yourself, in TV, as they are so widespread globally, year after year: for instance, during the **Ashura religious festival** within the **Shia community** in Iraq, Pakistan, and many other countries (Fig. 9.13). Many other examples of **corporal punishment** are provided in Scott's books:

> An example of a **superstition** which prevailed through the ages...is the notion prevalent among sailors of ancient days that **whipping** the passengers would prevent a storm. In the *Satyricon of Petronius* is related how Encolpus and Giton were flagellated with this express purpose in view. Thus: "It was resolved among the mariners, to give us each 40 stripes, in order to appease the tutelsir deity of the ship...no time in consequence is lost; the furious mariners set upon us with cords in their hands, and endeavour to appease the Deity by the effusion of the meanest blood; as to me I received three lashes, which I endured with Spartan magnanimity". **Seneca** made a general statement which influenced many early writers on diseases and their treatment. He said: "Medicine begins to have an effect on insensible bodies when they are so handled as to feel pain". Also he recommended whipping as a specific in the treatment of fever. Others followed in his tracks, and soon those afflicted with such widely divergent malaises as **lockjaw** and **smallpox**, **rheumatism** and **bowel troubles**, found themselves soundly flogged for their pains. According to Mercusialis, **Galen** was not alone in advising whipping as a means of inducing the putting on of flesh. Many physicians prescribed the same course, and for centuries **slave-dealers** were accustomed to flog their captives with the express purpose of increasing their plumpness and consequently their market value. According to Kisch, in **ancient Greece**, it was customary

Fig. 9.13 Shiite Muslims flagellate themselves during the Islamic month of Muharram, which marks the seven-century martyrdom of Prophet Mohammad's grandson Imam Hussein who was killed in a battle in Karbala in Iraq in 680 AD

for the woman who was not blessed with a child during the first years of her marriage to pay a visit to the **Temple of Juno** in Athens, there to be cured by one of the priests of Pan, of her sterility. To this end she was ordered to strip to the buff, prostrate herself belly downwards, in which position she was flagellated by the priest with a whip made of goat's hide..

Superstitions die hard, whether they are connected with **religion** or with **medicine**; and for this reason we need experience little surprise that many of these ideas, crude as they were, survived through the centuries. One Bartholin, writing in 1669, says: "Among the Insubres, as I have proved in my *Cento of Histories*, the dead foetus is extracted from the mother by compressing the belly strongly, or striking it with wooden or steel balls. I have observed that boys, and men too, have been cured of pissing in bed by whipping". As a remedy for sexual impotence in men and sterility in women **flagellation** enjoyed a great reputation during many centuries of the Christian dispensation. In these directions, Meibomius…was a great believer in its powers; so, too, was the Abbe Boileau. Millingen, as comparatively recently as 1839, wrote at length on the virtues of flogging in the treatment of **disease**, upholding the theories of the ancients. He says: "Flagellation draws the circulation from the centre of our system to its periphery. It has been known in a fit of ague to dispel the cold stage. Galen had observed that horse dealers were in the habit of bringing their horses into high condition by a moderate fustigation; and therefore recommended this practice to give embonpoint to the lean. **Antonius Musa** treated a **sciatica** of **Oetairus Augustus** by this process. **Elidaeus Paduanus** recommended flagellation or **urtication** when the eruption of **exanthematic diseases** is slow in its development. Thomas Campanella records the case of a gentleman whose bowels could not be relieved without his having been previously whipped. **Irritation of the skin** has been often observed to be productive of similar effects. The erotic irregularities of **lepers** is well authenticated; and various other

cutaneous diseases, which procure the agreeable relief that scratching affords, have brought on the most pleasurable sensations…. The effect of flagellation may be easily referred to the powerful sympathy that exists between the nerves of the lower part of the spinal marrow and other organs"…

The **Old Testament**, which ranks as the greatest catalogue of acts of cruelty and persecution that a **sadistic God** ever revelled in, betrays all the evidence one needs of the widespread practice among the **Hebrews** and the **Egyptians** of **whipping** or **flogging** as a form of punishment for all manner of crimes and misdemeanours. According to the laws of **Moses**, up to forty strokes of the rod could be given, the exact number varying with the nature of the offence and the whim of the judge. Thus: "If there be a controversy between men, and they come unto judgment, that the judges may judge them, then they shall justify the righteous, and condemn the wicked. And it shall be, if the wicked man be worthy to be beaten, that the judge shall cause him to lie down, and to be beaten before his face, according to his fault, by a certain number. Forty stripes he may give him, and not exceed: lest, if he should exceed, and beat him above these with many stripes, then thy brother should seem vile unto thee" (Deuteronomy xxv. 1-3.)…. **Jesus** himself had recourse to the rod on at least one occasion. Thus: "And the Jews' passover was at hand; and Jesus went up to Jerusalem, and found in the temple those that sold oxen and sheep and doves, and the changers of money sitting: and when he had made a scourge of small cords, he drove them all out of the temple, and the sheep, and the oxen; and poured out the changers' money, and overthrew the table" (John ii. 13-15).

"The notion prevalent among sailors of ancient days that whipping the passengers would prevent a storm": again, can you imagine animals such as dolphins, or whales, whipping—or, to make this example more realistic as they have no hands, biting—other members of their own species to "prevent a storm?" This is an extremely powerful example of living in a world of *reality* versus a world of *unreality*. As explained by Scott, the worse kind of corporal punishment, in terms of its widespread and magnitude, was that related with something also completely irrational—**racism**, either of the "cultural/epigenetic" type as often done by the Romans, or the more recent and even more horrible "innate/supremacist type":

Civilization has much to answer for. Its history is punctuated with the most horrible and most revolting cruelties that the mind of man can conceive. In no form has mankind's penchant for persecution shown itself more flagrantly than in the treatment of those of its kind whom circumstances have placed totally within its power, beyond hope of rebellion, resistance or escape. The treatment meted out to **slaves**, since the days of the Romans to within the memory of the living, will remain for all time a damning indictment of **Christianity** and **paganism** alike; an ineradicable blot upon the history of so-called civilization. The writings of **Horace**, **Plautus**, **Juvenal**, **Petronius**, **Terance**, **Ovid**, **Martial** and others, all provide testimony most abundant as to the universality with which the aristocracy of the **Roman Empire** flagellated their slaves. So much so was this the case that the whip itself became the emblem of slavery. The owner of a slave was vested with the power that goes with absolute possession. He owned the slave body and soul. The human creature ranked with the horse, the cow, or the dog, to be kicked and beaten at the whim of its master. Slave he was from birth to death. He was whipped for any and every crime or misdemeanour; he was whipped for the sins of omission as well as commission; he was whipped often enough to provide amusement for his master's guests. Frequently, very frequently, the slave died, either under the whip, or as a direct result of the punishment inflicted. For there would appear to have been no limit to the fiendish cruelty of the Roman aristocracy. They exercised their brains in devising means of intensifying the severity of the whippings. Not content with the scarification inflicted by the terrible flagellum, to the leather thongs they tied nails, bones and leaden weights.

After seeing all the above examples, we can now come back to one of the key discussions of the present Chapter, about the price paid for and/or the benefits gained by our tendency to create and believe imaginary stories and our social obsession to "fit in." Similarly—and related—to the way beliefs in general and religion in particular do bring some clear benefits concerning various aspects of life, under certain circumstances, our *Homo socialis* component also brings us a mixture of different benefits. For instance, studies show that having people next to us, or talking with them, touching them, or even just knowing them or knowing that someone cares about us, can have huge psychological benefits, reduce stress, and lead to multi-organ benefits, as well as boost the immune system and thus be able to fight infections harder, and so on. As explained in Davis' *The beautiful cure*:

> In a **fever**, these **cytokines** and **hormones** act on a region of the brain called the **hypothalamus**. In response, the hypothalamus signals for the body to produce another hormone, **noradrenaline**, which constricts blood vessels in the body's extremities and triggers brown fat cells to burn up energy and produce heat (the specialist job for this type of fat cell), as well as acetylcholine, which acts on muscles to cause shivering, for example, all of which serves to increase the body's temperature. The hypothalamus also controls our feelings of **hunger**, **thirst** and **sleep**, as well as more complex emotions such as seeking closeness with others and our **sex drive**. Because of this, as well as feeling sleepy and **losing appetite**, secretions from immune cells affect all sorts of behaviours and **emotions**. Although this is not very well understood in detail, our immune system undoubtedly shapes our moods and feelings. Some of this might just be a chance outcome of the way in which hormones and cytokines are interconnected, but some of this is likely to have evolved for a reason. There's an advantage in, for example, seeking comfort from others who may care for you when ill. Music, it seems, is not the only food of love; caring affections can be fired up by the chemical reaction of immune cells detecting germs. Broadly, the **immune system** and our **nervous system** are in constant dialogue, each affecting the other through the body's flux of cytokines and hormones. Many hormones affect our immune system, including the **sex hormones oestrogen** and **testosterone**, but it is stress hormones that have the greatest impact. We all know what stress is, though it's hard to define. It can be as all-encompassing as a fever or as fleeting as butterflies in the stomach. What is clear is that stress can have major effects on our health, because of its connection with the immune system. Reducing stress may boost immunity, for example.

We can thus go one step further, and combine our tendency to build and believe in imaginary stories—*Homo fictus et irrationalis*—and our propensity to like to be with other people—*Homo socialis*—in an example that might seem out of context here, but that actually illustrates a crucial point. The example concerns **imaginary friends**, which is a well-known, and studied, psychological phenomenon. If a child has an imaginary friend, psychologists and the broader public tend to consider it is "OK" because this is "normal"—in the sense that it is *frequent* within our Western societies. And also because empirical studies have suggested that having such "friends" may be encouraging for children, provide them motivation, increase their self-esteem, and lead them to undertake more "moral" decisions, as discussed in Hoff's 2004 paper "*A friend living inside me – the forms and functions of imaginary companions.*" These benefits are strikingly similar to those gained by human adults, if they believe in, and pray/talk with **imaginary Gods** (Chap. 2). However, when adults have imaginary friends that are not religious deities, most people consider that this is something "abnormal" or even "pathological." A main reason for this is that it is *much less frequent* to find adults with imaginary friends than to find adults that pray/talk to deities or children with such friends. However, empirical studies suggest that adults with unreal friends don't have a greater risk of developing pathologies such as **psychosis** or **schizophrenia**: they are just more likely to have common forms of **hallucinations**, as explained by Davis and colleagues in a 2010 paper. Such examples illustrate, once again, that what is seen as "normal"—or for that matter as "good" or "bad" or pathological—in a certain group or society is defined by the rulers, the elites, social norms, or "majority-rules," depending on the situation.

This leads us to a million-dollar question: if in general religious people in countries such as the U.S. seem to have better health, less stress, take less drugs, drink less alcohol, and be less depressed when faced with terminal diseases, then why is the present book telling us to escape from *Neverland*? When humans first became aware of the inevitability of death, the evolutionary answer that was more successful in terms of survival and reproduction was to be a believer, after all. However, a huge amount of time has passed, and a huge number of dramatic changes, have occurred since then, and moreover we have a tool that our ancestors did not have back then: historical data. Namely, we *now know* that while their full immersion in *Neverland* brought them evolutionary advantages at that moment, in the long term *Neverland* has led not only to a huge number of atrocities but also to a scenario in which our species is literally at a high risk of extinction due to an ecological collapse of the planet. An example of how things have changed in the last centuries is that there is nowadays a higher number of people than ever before that despite of not believing in Gods still are able to have amazingly happy, healthy, and long lives and feel completely fulfilled. In fact, today if people don't believe in Gods or on a "cosmic purpose," they might be a plethora of different things. For instance, there are a lot of **pessimistic nihilists** and **pessimistic existentialists**, but there are also, **optimistic existentialists**, and each of these different ways of seeing/reacting to life affects differently the health of our body as a whole, including our mental health. As explained in Chap. 2, many atheists do tend to feel "empty" or "lost," and that is probably a major reason why the vast majority of humans became to be believers since times immemorial, and also why nowadays the rising number of people that say they don't believe in God is being accompanied by a rising number of people that believe in other stuff, such as many New Agers tend to do (Fig. 2.10). However, there are now millions of atheists that have a more positive view of life and that do not feel "empty" or "lost" and consequently don't "need" to believe in New Age religious or spiritual stuff to be less stressed, less depressed, and so on (see Box 9.1).

Box 9.1: Sartre, Dawkins, Bering, and Cosmic Purpose

The vast majority of humans are, in a way or another, believers, but an increasing number of people in Western countries is starting to not *need* to believe in teleological narratives about a **'cosmic purpose'**. As explained in Bering's *The instinct belief*, in those countries more and more people have been able to deconstruct such narratives and to realize that *life just happens, it is just what it is, and that this is precisely what makes life and the cosmos particularly fascinating*:

> This (belief) is nonsense, said Sartre. In reality, we simply come to exist as individuals, just as beads of condensation form on a glass of water or spores of mold appear on old bread. And if there is no God, as **Sartre** believed, then metaphysical meaning – applied to the individual's raison d'être, as well as to life itself – is only a mirage. But Sartre cautions us not to fall into the Christian trap of seeing this startling truth of God's nonexistence as being reason to experience a crumbling sense of despair. Rather, says Sartre, we should rejoice in this divine absence, because now we are free to define ourselves as we please. That is to say, because God hasn't fettered any of us with a particular function in mind, selfishly obligating us to preordained tasks in this fleeting existence of ours, we've no legitimate grounds to stew over our incorrigible and immovable fates. Instead, our purpose is entirely our own affair: *we decide who we are, not God*.... God doesn't endow each man with an 'essence' – or prewritten, underlying purpose – said Sartre...purpose is a human construct. [Similarly] in an interview with Salon magazine conducted shortly after the release of his book [*The God delusion*], Dawkins was asked, "What is our purpose in life?". He responded, "*If you happen to be religious, you think that's a meaningful question...but the mere fact that you can phrase it as an English sentence doesn't mean it deserves an answer...those of us who don't believe in a god will say that is as illegitimate as the question, why are unicorns hollow? It just shouldn't be put. It's not a proper question to put. It doesn't deserve an answer*"

Bering then gives his own opinion on this issue:

> To see an inherent purpose in life, whether purpose in our own individual existence or life more generally, is to see an intentional, creative mind – usually God – that had a reason for designing it this way and not some other way. If we subscribe wholly and properly to Darwin's theory of natural selection, however, we must view human life, generally, and our own lives, individually, as arising through solely nonintentional, physical means. This doesn't imply that we are 'accidents', because even that term requires a mind, albeit one that created by mistake. Rather, *we simply are*. To state otherwise, such as saying that you or I exist for a reason, would constitute an obvious category error, one in which we're applying **teleo-functional thinking** to something that neither was designed creatively nor evolved as a discrete biological adaptation. Yet owing to our theory of mind, and specifically to our undisciplined **teleological reasoning**, it is excruciatingly difficult to refrain from seeing human existence in such intentional terms. To think that we are moral because morality works in a mechanistic, evolutionary sense is like saying that we are moral because we are moral; it's unfulfilling in that it strips the authority away from a God that created us to act in specific ways because *He knew best*, and He would become disappointed and angry if we failed to go along with His rules for human nature. But peeling back the cognitive illusion of the purpose of life...gives us our first glimpse into the question: has our species' unique cognitive evolution duped us into believing in this, the grandest mind of all? So far, the answer is clearly 'yes'.

I often answer, to those that ask me, after my talks, "how can one have a positive attitude towards life if one does not believe in God," by saying that what is truly "depressive" is to believe that I am just part of a "designed script," of a "masterplan," like an actor of a play that is obliged to say exactly the words that the scriptwriter wrote, over and over, every single night. In this sense, my answer is indeed very similar to that provided by authors such as Sartre (Box 9.1). One important point is that although all living humans are *Homo fictus, irrationalis et socialis*, including obviously nomadic hunter-gatherer groups with their cosmologies and teleological narratives, what we know about these latter groups shows us that their daily-lives are *not* so immersed in *Neverland* as are those of most agricultural societies in general, and particular of "developed" countries. This was recognized by the first Europeans that contacted American indigenous groups, who criticized those groups by arguing that they "live according to nature," "without laws or faith," which are paradoxically terms that now tend to sound as markedly progressive. This is attested by **John Lennon**'s "Imagine" idealistic, utopian lyrics: no private property, no king, no boundaries, obeying "nobody," and so on. Of course, the descriptions of Europeans were exaggerated and, as we have seen above, their arguments were moreover part of a racist and factually wrong way to see the natives as "primitive" and "passive," because all those native groups obviously had beliefs and often had very strong social norms that were *actively* imposed to enforce egalitarianism, and so on. But it is true that, *despite* those beliefs and norms, nomadic hunter-gatherers in general are much more connected to the reality of the natural world, as they depend on it much more directly, on a daily-basis. One of the reasons why Europeans tended to exaggerate their descriptions and criticism of indigenous groups was to pave the way to "morally" justify their colonization and evangelization. For instance, they said they were hugely violent and that they usually practiced **cannibalism** and **human sacrifices** and had **wars** all the time, as explained in Bergreen's excellent and captivating book *Over the Edge of the World—Magellan's Terrifying Circumnavigation of the Globe*:

> Discussing the region's [Brazil] **indigenous tribes**, **Vespucci** wrote out of his own experience: "I tried very hard to understand their life and customs because for twenty-seven days I ate and slept with them". He assembled a disturbing if tantalizing picture of the Indians whom **Magellan** and his crew would encounter in Rio de Janeiro: "They have no laws or faith, and live *according to nature.* They do not recognize the immortality of the soul, they have among them no private property, because everything is common; they have no boundaries of kingdoms and provinces, and no king! They obey nobody, each is lord unto himself".... More troubling, the Indians practiced **cannibalism** and **human sacrifice** in the course of their battles. "They are a warlike people, and among them is much cruelty", he warned. "Nor do they follow any tactics in their **wars**, except that they take counsel of old men; and when they fight they do so very cruelly, and that side which is lord of the battlefield bury their own, but the enemy dead they cut up and eat. Those whom they capture they take home as slaves".... Vespucci's Indians were most likely representatives of the vast network of **Guaraní tribes**. At the time of Magellan's arrival, there may have been as many as 400,000 Guaraní Indians, grouped by dialects. They occupied huge regions of South America extending all the way to the Andes, and lived communally in huts sheltering about a dozen families each; polygamy was not unknown to them, but it was not common. They were short – rarely more than five feet tall – and, by European standards, stout. The men wore a simple G-string and occasionally a headpiece made of feathers; the women were fully clothed. They were adept at pottery, wood carving, and skillful in their weapons of choice: the bow and arrow and the blowgun.

9.2 Homo Irrationalis, Socialis, and Fictus in Neverland

The Truth and the Lie meet one day. The Lie says to the Truth: 'It's a marvellous day today'! The Truth looks up to the skies and sighs, for the day was really beautiful. They spend a lot of time together, ultimately arriving beside a well. The Lie tells the Truth: 'The water is very nice, let's take a bath together!' The Truth, once again suspicious, tests the water and discovers that it indeed is very nice. They undress and start bathing. Suddenly, the Lie comes out of the water, puts on the clothes of the Truth and runs away. The furious Truth comes out of the well and runs everywhere to find the Lie and to get her clothes back. The World, seeing the Truth naked, turns its gaze away, with contempt and rage. The poor Truth returns to the well and disappears forever, hiding therein, its shame.
(19th century legend)

After everything we saw in this book, we now realize that there is a profound paradox concerning humanity: within all the endless types of organisms living in this planet, it is the self-proclaimed "*Homo sapiens*," and particularly the one living in self-proclaimed "developed countries" that commonly lives a daily life influenced by completely imaginary—and mostly unwise and illogical—stories, completely immersed in a *world of unreality*. We claim to be the only ones with a unique "mental capacity" allowing us to be fully aware of, and understand, the *truth* about both the organisms and the inanimate objects of the cosmos. Whether we proclaim ourselves to be "atheist," "religious," "believer," "non-believer," "spiritual," "romantic," "agnostic," or "humanist," or any combination of these, the vast majority of humans believe in at least some kind of such imaginary tales. Of course, it is true that no animal "sees" or "perceives" the habitat around it as it *really* is, as explained in Hoffman's 2019 book *The Case Against Reality: Why Evolution hid the Truth from Our Eyes*. However, it is also true that in general the behavior of the vast majority of non-human organisms is mainly guided, in their daily-lives, by biological drives, desires, functions, and behavioral choices that are deeply rooted in the *world of reality*. Apart from the very few that seemingly display some kind of apparent "mystic behaviors," like chimpanzees (Chap. 2), the norm is that non-human animals do things such as eating, drinking water, hunting or escaping, having sex, running or resting, taking care of their descendants or eating the progeny of other animals, and so on.

Now, compare this with the daily life of billions of humans whom are driven by, and thus do, things like going to a church, mosque, synagogue, or other types of temples to pray to fictitious religious deities, reading about imaginary "zodiac signs or zodiac animals" in local newspapers, discriminating against, oppressing or even killing other humans because of invented tales about "races" or "nations" or "castes," and so on. As summarized by Edwardes in his 2019 book *The Origins of Self—an Anthropological Perspective*, "unlike humans, other apes do not share unreal versions of the world (versions of the world that are "real" simply because people agree they are "real," even though they know they are not real—like the value of Bitcoin)." Furthermore, as put by Hood in his superb 2013 book *The Self Illusion—How the Social Brain Creates Identity*, we create stories and narratives not only about imaginary deities and cosmic purposes, but even about our own identity: "who we are is *a story of our self*—a *constructed narrative* that our brain *creates*" (see Boxes 9.2, 9.4, 9.6 and Fig. 9.14). *Neverland* is not only our evolutionary niche, our special

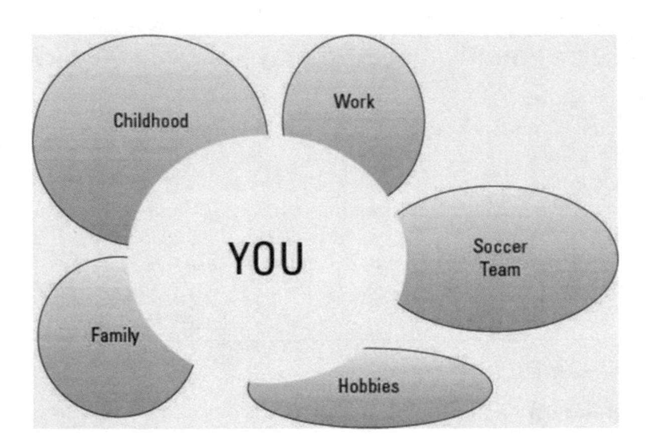

Fig. 9.14 The self-illusion: as put by Hood in his 2013 book *The Self Illusion*, "who we are is a story of our self – a constructed narrative that our brain creates"

habitat: it is also a part of our own constructed self. Or, as elegantly put by Gottschall in his 2012 book—from which I adapted the terms "*Neverland*'s world of unreality" and "*Homo fictus*," as well as one of the most beautiful pictures illustrating how we are indeed creatures of story (see Fig. 9.15)—entitled *The storytelling animal—how stories make us human*:

> ***Homo fictus*** (**fiction man**)...the great ape with the **storytelling mind**. You might not realize it, but you are a creature of an imaginative realm called *Neverland*. **Neverland** is your home, and before you die, you will spend decades there. If you haven't noticed this before, don't despair: story is for a human as water is for a fish – all-encompassing and not quite palpable. While your body is always fixed at a particular point in **space-time**, your mind is always free to ramble in lands of make-believe. And it does. Children...delight in stories and start shaping their own pretend worlds as toddlers. Story is so central to the lives of young children that it comes close to defining their existence. What do little kids do? Mostly they do story. It's different for grownups, of course. We have work to do. We can't play all day. In James Barrie's play *Peter Pan* (1904), the Darling children adventure in Neverland, but eventually they get homesick and return to the real world. The play suggests that kids have to grow up, and growing up means leaving the pretend space called Neverland behind. But Peter Pan stays in Neverland. He won't grow up. And in this, we are all more like **Peter Pan** than we know. We may leave the nursery, with its toy trucks and dress-up clothes, but we never stop pretending. We just change how we do it. Novels, dreams, films, and fantasies are provinces of Neverland. Humans are *creatures of story*, so story touches nearly every aspect of our lives. Archaeologists dig up clues in the stones and bones and piece them together into a saga about the past. Historians, too, are storytellers. Some argue that many of the accounts in school textbooks, like the standard story of **Columbus's discovery of America**, are so rife with distortions and omissions that they are closer to **myth** than history. Business executives are increasingly told that they must be creative storytellers: they have to spin compelling narratives about their products and brands that emotionally transport consumers. Political commentators see a presidential election not only as a contest between charismatic politicians and their ideas but also as a competition between conflicting stories about the nation's past and future. Legal scholars envision a trial as a story contest, too, in which opposing counsels construct narratives of guilt and innocence – wrangling over who is the real protagonist. We live in Neverland because we can't not live in Neverland. *Neverland is our nature. We are the storytelling animal. Neverland is our evolutionary niche, our special habitat.*

Fig. 9.15 "Literature," by James Koehnline's, 2007

So, this leads us again to the central question of the present Chapter: OK, humans are indeed storytelling animals, we live in *Neverland*, but what is wrong with that? It is clearly a more appealing world, that is why Peter Pan precisely decided to stay in it, after all, right? After all, would you prefer to have the fascinating live of Peter Pan in *Neverland* or the boring life of the other kids that were with him after they returned to the houses of their parents? Well, the problem is precisely that in the world of *reality* we know that the stories and narratives that the human-Peter Pans have created and believed in for millennia have led to, or at least justified, atrocities that *no other species has done*. None has encaged and abused and even enslaved so many other animals (Figs. 7.6, 7.7, 7.8, 7.9, 9.6), as well as members of its own species (Figs. 3.14 and 3.15); produced weapons explicitly designed to kill as many as possible; tenaciously produced and sold drugs that created addiction and destroyed the life of so many millions of their own kind; carefully planned and undertook bloody, horrible, mass 'religious human sacrifices' provoking excruciating pain to their kind (Fig. 9.8); created such a huge inequality among members of the same species; done atrociously painful experiments to or burned or tortured those of the same species (Figs. 3.19, 5.23 and 7.1); done so much arm to themselves, including **self-flagellation** (Figs. 1 and 9.13); and polluted so much the planet, putting their own species, and numerous other species, at risk of extinction. These are the facts, plain, and crude, about **Peter Pan** and the world of *Neverland*: at first sight, it appears to be a magical, wonderful, and comforting place, but in reality its dungeons are horribly ugly, plagued by atrocities and by a huge number of **Captains James Hook** that subjugate, oppress, kill, and even genocide people just because they are a bit different from or don't obey to him. After all, that is one of the major reasons why almost all kids wanted to go back to their homes, except Peter Pan precisely because he was already profoundly immersed, and irreversibly lost, in *Neverland*: that is precisely what we should avoid—what this book is trying to help to avoid. Because, specially for a self-proclaimed "sapient" being that is convinced to be made "in God's image" and to have a "unique" brain and "morality," what we humans have done in the last millennia is not only a truly horrifying historical record: it is literally unsustainable. As emphasized in a brilliant quote from the environmental organization **Greenpeace** that used a metaphor somewhat similar to that shown in Fig. 2.6, "the Earth is 4.6 billion years old…let's scale that down to 46 years…we've been here for 4 hours…our industrial revolution began 1 minute ago…in that time we've destroyed more than 50% of the World's Rain Forests…*this isn't sustainable.*"

As often one image is more powerful than 1000 words, I think the dangerous *Neverland* world of unreality created by the *Homo fictus, irrationalis et socialis* is plainly, and disturbingly, displayed in Fig. 9.7. Without the fictional teleological narratives, without our inability to discern between them and reality, and without the social pressure to not only accept but also undertake—and being proudly photographed, even smiling, during—such violence and madness, these type of horrible events would not be possible. One more horrible illustration of this is shown in Fig. 9.8: a **human sacrifice**, a practice we briefly mentioned above. A description of how painful such a practice can be was given in Tiesler and Oliver's 2020 paper *Open Chests and Broken Hearts*. Importantly, before you read it, we need to remind ourselves that these practices were done to people that were not only *alive* but were also almost always *non-anesthetized* and many times still *conscious of what was happening* and *what exactly was going to happen*:

> In the ancient world, **human sacrifice** was a supreme **religious expression**. The term '**religious human sacrifice**' is understood here as the act of killing a human being for the sake of collective well-being, which is consecrated by the divine. Its study provides broader insights into native notions of cosmic functioning and worship, the ritual enactment of **myths**, and religious politics. **Heart sacrifice** is a pervasive theme in pre-Hispanic Mesoamerican imagery and was repeatedly recounted by living witnesses during early colonial times. Benavente or Motolinía describes a common procedural sequel of **heart sacrifice** among the Aztecs, which has been attested to over and over again by other chroniclers: 'So the sacrificer comes, which is not a minor office among them, and with a stone knife, which cuts as if it was made from iron but is as big as a big knife, and in less [time]

than it takes to cross oneself, he sticks the knife into the chest, opens it, and pulls out the hot and beating heart. In this instance the supreme pontiff takes it and smears the blood on the mouth of the principal idol; and without further ado, he takes this blood and throws it towards the sun, or some star if it is night time; and they spread [the blood] on the mouths of the other stone or wooden idols, and on the door cornice of the chapel where the main idol is.' In addition to the heart and its blood, such practices could target segments from the head and the extremities, either with their flesh or defleshed to the ligaments and bone. Violent **chest penetration** of a still-living victim is bound to leave traces in the form of torsion fractures and sharp-force trauma. This evidence allows inferences of instruments, positioning of the victim, and **sacrificial altars**, along with the anatomical location of trunk penetration and manner of extraction.

[For instance in] chest opening from beneath the chest (subdiaphragmatic thoracotomy) the anatomic feature most frequently cut to open the chest was in all likelihood the sulcus below the left rib cage. This part of the trunk is readily visible from the front and does not require cutting through bone to access the heart from beneath. This was likely the access point used during **Aztec mass sacrifices** and referred to as the 'usual sacrifice', as the Spanish friars once called it. Torquemada Friar Diego Durán specifies that while the **sacrificial killings** lasted for hours or sometimes several days in a row, the leaders of the Triple Alliance would physically tire and need to pass their duty on to other priests. By the end, a large number of men had been ritually slaughtered by heart extraction. Procedurally, the best exposure is obtained when the body is stretched on its back face up over a raised block with the maximum backward curving flexure at midchest level. In order to ensure this very position in practice, the victim's extremities were extended and held down by four helpers, while a fifth practitioner would push the throat back and toward the ground, either by hand or with the help of a collar or yoke. Thus immobilized on a rigid support, the victim was struck with the sacrificial flint knife from above. The knife probably entered the middle of the trunk and was pulled out toward the left, cutting right below the left lower rib line. This scenario resonates with the chroniclers, who refer to this act as 'opening the trunk up,' 'breaking it open,' 'splitting it in two,' or even more explicitly, 'lifting the chest high by cutting the trunk' and leaving it 'partitioned in two,' just like 'a broken pomegranate'. The best exposure of the heart is reached from beneath when the cut is extended across the trunk. The produced opening may have resembled depictions from Piedras Negras, with the trunk visually divided into a lower and upper portion. Once accessed from below, the subsequent ritual removal of the heart may have been carried out by introducing a sickle-shaped knife with which to pull out the organ before cutting the anchoring vessels, as the image in the Atetelco murals from Early Classic Teotihuacan seems to suggest. An additional method, described still by living witnesses surrounding European contact, consists of grasping and pulling the heart with one hand while cutting it off the anchoring vessels with the other, either with the same penetrating knife or a second cutting implement.

Box 9.2: Self-Illusion, Social Brain, Identity, and the Storytelling Animal

Hood states, in *The self illusion* (see also Fig. 9.14 and Boxes 9.4, 9.5 and 9.6):

> Understanding that the **self** could be an illusion is really difficult. It may be one of the most, if not the most, difficult concept to accept. Our self seems so convincing, so real, so us. But then again, many aspects of our experiences are not what they seem. Take the most lucid experience that you are having right now as you read these words. As your eyes flit across the page, your visual world seems continuous and rich but you are actually only sampling a fraction of the text one bit at one time, rarely reading all the letters in between. Your **peripheral vision** is smeared and colourless, yet you could swear that it is perfectly clear just like the centre of your visual field.

There are two **blindspots**, the size of lemons at arm's length, just off-centre from your field of view that you do not even notice. Everything in your visual world is seamless and unbroken, yet your visual world is blacked out for a fraction of a second between eye movements. You are not made aware of any of these imperfections because our brain provides such a convincing cover story. The same deception is true for all human experience from the immediacy of our perception to the contemplation of inner thoughts, and that includes the self. Today, the findings from contemporary brain science have enlightened the nature of the self. Some of the most compelling evidence that the self depends on the brain comes from studies of unfortunate individuals who have suffered some form of brain damage either through aging or accident. Their personalities can be so radically changed that, to those who knew them, they become a different person. At the other end of the spectrum, many deliberately alter their brains temporarily with a variety of drugs that affect its workings. Whether by accident, disease or debauchery, these studies show that if the brain is damaged, the person is different. If taking drugs that change functioning alters the brain, the person behaves and thinks differently. So who we are depends on our brain. However, we are not just our brains in isolation. One of the messages that I wish to relay here is that each brain exists in an ocean of other brains that affect how it works.

The second major discovery is that there is no centre in the brain where the self is constructed. The brain has many distributed jobs. It processes incoming information from the external world into meaningful patterns that are interpreted and stored for future reference. It generates different levels and types of motivations that are the human drives, emotions and feelings. It produces all sorts of behaviours – some of them automatic while others are acquired through skill, practice and sheer effort. And then there is mental life. Somehow, this 1.5 kg lump of tissue inside our skull can contemplate the vastness of interstellar space, appreciate Van Gogh and enjoy Beethoven. It does this through the guise of a self. But the sense of self that most of us experience is not to be found in any one area. Rather it emerges out of the orchestra of different brain processes like a symphony of the self. Our brain constructs models of the external world. It can weave experiences into a coherent story that enables us to interpret and predict what we should do next. Our brain simulates the world in order to survive in it. This simulation is remarkable because much of the data that needs processing are corrupted. And yet, our **brain** fills in missing information, interprets noisy signals and has to rely on only a sample of everything that is going on around us. We don't have sufficient information, time or resources to work it all out accurately so we make educated guesses to build our models of reality. That working-out includes not only what's out there in the external world but also what is going on in the internal, mostly unconscious workings of our mind. *Who we are is a story of our self – a constructed narrative that our brain creates.* Some of that simulation is experienced as conscious awareness that corresponds to the **self illusion** that the average person in the street reports. At present we do not know how a physical system like the brain could ever produce those non-physical experiences like the conscious self. In fact, it is turning out to be a very hard problem to solve. We may never find an answer and some philosophers believe the question is misguided in the first place.

Dan Dennett also thinks the self is constructed out of narratives: 'Our tales are spun, but for the most part, we don't spin them; they spin us.' There is no self at the core. Rather it emerges as the 'centre of a narrative gravity'. (As) an illusion created by the surrounding elements…take the context away, and the square disappears…in the same way, the self is an illusion created by our brain" [see Fig. 9.14]."One dramatic case in which the brain science backs up the claim of true separated selves comes from a recent German DID patient who after fifteen years of being diagnosed as blind gradually regained sight after undergoing psychotherapy. At first, only a few of the personalities regained vision, whereas others remained blind. Was the patient faking? Not according to the electrical measurements recorded from her visual cortex – one of the early sensory processing areas in the brain. When her personality was sighted, electrical activity was normal over this region but absent when the patient was experiencing a **blind personality**. Somehow, the parts of her brain that were generating the multiple personalities were also switching on and off the activity of the visual part of the brain. This finding is beyond belief – literally. To believe that you are blind is one thing, but to switch off parts of the lower level functioning sensory processing areas of your own brain is astounding. Somehow, the network of connections that operates further upstream in the brain to deal with complex concepts, such as the self and personality, can control earlier basic processing input relay stations downstream in the brain.

Occasionally, we get a glimpse of the illusions our brains create. We may mishear a comment, bump into things or mistakenly reach for a shadow that looks graspable. This happens when we misinterpret the physical world. The same mistakes also happen in our personal world – the world that our self occupies. We reinterpret our failures as successes. We think we are above average on good attributes and not like others when it comes to behaving badly. We sometimes do things that surprise us or at least surprise others who think they know us well. This is when we do things that seem inconsistent with the story of our self. We say, 'I was not myself' or 'It was the wine talking' but we still retain a belief that we are an individual, trapped in our bodies, tracing out a pathway through life and responsible for our thoughts and actions. These influences work from the very beginning. Proportionally, humans spend the greatest amount of time in childhood compared to any other animal. This is not only so that we can learn from others, but also so we can learn to become like others. Becoming like others and getting on with them involves creating a sense of who we are – a participating member of the human species. This development of the self emerges across childhood as the interplay between the modelling brain, constructing stories from experience, and the influences of other people. This does not mean that we are blank slates at birth and that babies are not individuals. Anyone who has raised children or ever encountered identical twins knows they can think and behave differently right from the very beginning even though they are raised in the same environment. Our dispositions vary from one individual to the next, a legacy of our genetic inheritance, no doubt. However, we all share a common goal to become part of the human race through our social interactions and that can only take place when people construct a sense of self.

That process of **constructing the self** does not end with childhood. Even as adults we are continually developing and elaborating our self illusion. We learn to adapt to different situations. Sometimes we even describe our self illusion as multifaceted as if we have the work self, the home self, the parent self, the political self, the bigoted self, the emotional self, the sexual self, the creative self and even the violent self. They seem to be almost different individuals but clearly there is just one body. We seem to switch effortlessly between these different selves but we would be wrong to think that there is an individual doing the switching. That's part of the illusion. There is not one self or multiple selves in the first place. Rather, it is the external world that switches us from one character to another. This idea that we are a reflection of the situations is sometimes called the *looking-glass self* – we exist as the reflection of those around us. Initially as infants, we are bundles of self-interested activity but evolution has pre-programmed our self to emerge and attend to others. Our greatest influence during childhood moves from the immediate family that looks after our needs to the competitive world of young children. We learn to interpret, predict, anticipate and negotiate in the playground. Gradually over late childhood and adolescence we increasingly elaborate the narrative of who we are and eventually strike out to become a character differentiated from those who shaped us. For many adults, adolescence marks the turning point at which we 'discover' our true self. We use groups, possessions, tastes, politics and preferences to create the self – an individual that is different. At least, that is the story of self-formation in the West; other cultures provide a different framework that shapes a different type of self. Even hermits and outcasts from society are defined by their rejection of the principles that the rest of us accept. But whether we are distancing our self from the herd, or ingratiating our self as part of the herd, it is the existence of others that defines who we are. If the self is largely shaped by those around us, what does that mean for our everyday lives? When other people screw up it's because they are stupid or losers but when I screw up it's because of my circumstances. The self illusion makes the fundamental attribution error an easy fallacy to accept. Also putting all the blame on the individual self is tantamount to excusing all the policies that create inequality in our society. Maybe it's time to redress this imbalance by rethinking success or failure not so much as issues of the self alone, but more of society in general. Knowing that the self is an illusion cannot stop you thinking that it exists and even if you succeed, as **Buddha** and **Hume** did, then maybe it is best not to try in the first place. *But knowledge is power.* Understanding that the self is an illusion will help to reconcile the daily inconsistencies that you may experience in the way you think and behave. We are all too quick to notice how others can be manipulated, but we rarely appreciate how our own self is equally under the influence and control of others. That is something worth knowing and watching out for.

After what was said above, someone could argue: well, yes, no other animal does the kind of horrible stuff you listed above for humans, but some of them also do very awful things, to other species, or even to members of their own, other than just killing them to eat them, or parasitizing them. Yes, this is obviously true. As noted throughout this book, the non-human natural world is surely not "good" or "pure" or "kind," as some New Age people tend to argue, as if humans are the source of all "evil." As **Charles Darwin** famously wrote in a letter to the ground-breaking botanist **Asa Gray** about the existence of God, there is "too much misery" in the natural world that he compulsively studied in so much detail during his whole life. He stated: "*with respect to the theological view of the question…this is always painful to me…. I am bewildered…. I had no intention to write atheistically, but I own that I cannot see as plainly as others do, and as I should wish to do, evidence of design and beneficence on all sides of us…there seems to me too much misery in the world…. I cannot persuade myself that a beneficent and omnipotent God would have designedly created the Ichneumonidae [a parasitoid wasp family] with the express intention of their feeding within the living bodies of caterpillars.*" Actually, some animal species do have some kind of "**warfare**," and a few of them even have something somewhat similar to **slavery**, or use other beings in a way that is similar to how we do it in agriculture. For instance, this is done by some **eusocial animals** such as **ants**, as pointed out in **Edward O. Wilson**'s 2014 book *The Meaning of Human Existence*:

Eusociality stands out as an oddity in a couple of ways. One is its extreme rarity. Out of hundreds of thousands of evolving lines of animals on the land during the past four hundred million years, the condition, so far as we can determine, has arisen only nineteen times, scattered across insects, marine crustaceans, and subterranean rodents. The number is twenty, if we include human beings. This is likely to be an underestimate, perhaps a gross one, due to sampling error. Nevertheless, we can be certain that the number of originations of eusociality was relatively very small. Once attained, advanced social behavior at the eusocial grade found a major ecological success. Of the nineteen known independent lines among animals, just two within the insects – ants and termites – globally dominate invertebrates on the land. Although they are represented by fewer than twenty thousand of the million known living insect species, ants and **termites** compose more than half of the world's insect body weight. *Slavery* is widespread in ants of the north temperate zone. It starts when colonies of the slave-making species conduct raids on other ant species. Their workers are shiftless at home, seldom engaging in any domestic chore. However, like indolent Spartan *warriors* of ancient Greece, they are also ferocious in combat. In some species the raiders are armed with powerful sickle-shaped mandibles capable of piercing the bodies of their opponents. During my research on **ant slavery** I found one species that uses a radically different method. The raiders carry a hugely enlarged gland reservoir in their abdomen (the rear segment of the three-part body) filled with an alarm substance. Upon breaching the victim's nest, they spray large quantities of the pheromone through the chamber and galleries. The effect on the defenders of the allomone (or, more precisely, pseudo-pheromone) is confusion, panic, and retreat. They suffer the equivalent of our hearing a thunderously loud, persistent alarm coming from all directions. The invaders do not respond the same way. Instead, they are attracted to the pheromone, and as a result they are able easily to seize and carry away the young (in the pupal stage) of the defenders. When the captives emerge from the pupae as adults, they become imprinted, act as sisters of their captors, and serve them willingly as slaves for the rest of their lives.

The most complex societies of all ant species, and arguably of all animals everywhere, are the leafcutters of the American tropics. In lowland forests and grasslands from Mexico to warm temperate South America, you find conspicuous long files of reddish, medium-sized ants. Many carry freshly cut pieces of leaves, flowers, and twigs. The ants drink sap but don't eat solid fresh vegetation. Instead, they carry the material deep into *their nests*, where they convert it into numerous complex, spongelike structures. On this substrate *they grow a fungus*, which they do eat. The entire process, from collection of raw plant material to the final product, is conducted in an assembly line employing a sequence of specialists. [Regarding] leafcutter nests, [they] are constructed as a giant air-conditioning system. Channels near the center accumulate exhausted, CO2-laden air heated by the *gardens* and the millions of ants living on them. As the air is warmed, it moves by convection through openings directly above. At the same time fresh air is pulled into the nest through openings to channels located around the periphery of the nest. The advanced **superorganisms** of ants, bees, wasps, and termites have achieved something resembling *civilizations* almost purely on the basis of instinct.

This excerpt shows how inaccurate it is to assign terms such as "good" or "bad" to cases of biological evolution, for instance saying that eusociality is "good" or "noble," as many do and as even Wilson somewhat seems to imply by the use of the word "achieved"—a term commonly defined as successfully bringing about or reaching a desired objective or result by effort, skill, or courage. For instance, when he states that some **non-human superorganisms** "have *achieved* something resembling civilization." Someone could similarly—and also erroneously—use the term "bad" to argue exactly the opposite: that eusociality is "bad" because it can lead more easily to "civilized" things like warfare, slavery, using others, changing more the surrounding environment, and so on, as such phenomena precisely often require many organisms of a same group doing stuff together. It would be very hard for a more solitary animal like a leopard to have a group of "slaves," or fight in wars, or do agriculture, for instance. Regarding specifically what Wilson wrote about the resemblances between the type of organization of some **eusocial organisms** and what we call human "civilization," some of them are truly impressive. However, there is a key difference: in those non-human species to which Wilson refers, "civilized" acts such as wars and slavery are commonly done to members of other species, while within humans we do these atrocities to members of the same species. Moreover, it is also interesting to notice how Wilson—a naturalist who praises so much science and its objectivity—also paradoxically uses so often teleological terms that have no empirical support. For instance, he writes "the definitive part of the long *creation story* evidently began with the *primitive Homo habilis* two million years ago…prior to the habilines the *prehumans had been animals.*" 'Pre-humans' is a teleological term that has no place in evolutionary biology, as it suggests that not only 'humans' were ought to exist but also that they are the 'real think', that 'pre-humans' only existed so we could exist. Moreover, stating that "pre-humans were animals" is also awkward, because *all* humans are obviously animals, there was and there will be no moment in history in which we will rise the '**ladder of life**' and no longer be animals.

One could think that this was just a typo made by Wilson, but that is not the case because in a different part of his book, he wrote: "don't get me wrong…. I am not implying that we are driven by instinct in the manner of animals". Actually, this sentence has two major errors, because even if he had written "in the manner of *other* animals," this would still be a wrong assertion because other animals such as chimpanzees, birds, and so on also have culture, as we have seen above. Wilson further stated: "one of the lineages, the direct antecedents of *Homo sapiens*, won the *grand lottery* of evolution…we are a very special species, perhaps the *chosen species*…the *payout* was civilization based on symbolic language, and culture, and from these a gargantuan power to extract the nonrenewable resources of the planet – while cheerfully exterminating our fellow species…[however the] chance of eventually *reaching* the human level (could) at each step have sharply declined." So much teleology, so many factual inaccuracies, in a book written in the twenty-first century by a very prominent, and otherwise very bright biologist.

Another paradoxical aspect of Wilson's book is that even if we were to accept the type of comparisons and teleological arguments used by him, humans would still not be the most "special" animals at the top of the "ladder." This is because he refers to measures such as the total number of individuals of a certain group, or their total "body weight" or biomass, or the number of living species within a certain taxon. In terms of the latter measure, humans would actually be one of the evolutionarily less successful groups, because within six millions of years of evolution, we are now left with a single species of living humans. Even our closest relatives, the great apes, which also have a very low number of living species in general and are moreover often endangered in the wild in great part as a result of what we are doing to the planet—another example of our evolutionarily unsuccessfulness—have more living species than us. Chimpanzees have 2 species, gorillas 2 species, and orangutans 3 species. One could argue: well, but there are now about 7 to 8 billion humans, and contrary to great apes we are clearly not endangered, well on the contrary. Well, the first assertion is right, but the second is not, because precisely as a result of the enormously high number of humans living today and in particular of the average huge **ecological footprint** of a vast proportion of them, particularly those of "developed" countries, we are truly entering a very dangerous critical **ecological point of no return regarding climate change**. Moreover, even if someone knowing this would still aim to use the existence of 7 to 8 billion people as an argument to defend that humans are the most "special" and "successful" species, this would be a flawed argument anyway because numerous species—for instance of insects—have a much higher total number of individuals. And in addition they do that without endangering the Earth's whole ecological balance as we do, obviously because the ecological footprint of each of those individuals has nothing to do with the absurd individual average within people living in our "developed" countries.

A third potential type of "measure" that could be used to substantiate the "ladder of life" notion that our species is biologically the most "special" and "important" one would be total **biomass**. But this argument would also not stand in face of the available evidence. Let's say that a neutral viewer—an hypothetical "Martian"—came to earth and looked at our planet before landing, from space: humans, or primates—or all animals for that matter—would hardly be seen as the "dominant ones," because humans make up just about 1/10,000 of earth's total biomass, while plants represent about 80%. That is why our planet, when seen from space, is mostly blue (oceans) and green (plants). Alternatively, if that neutral observer landed on Earth and studied in detail all the organisms of the planet, perhaps it/he/she would consider that **prokaryotes**—microscopic single-celled organisms such as **bacteria** and **cyanobacteria** that do not have a distinct nucleus with a membrane nor other specialized organelles—are the ones that truly dominate life on Earth, not eukaryotes such as animals and plants. As noted by Noble in his 2017 book *Dance to the Tune of Life*, "appearances are deceptive…in fact, on two major counts, duration and quantity, the prokaryotes easily dominate life on Earth…they may have done so alone for one billion years or so from the period of origin of cellular life, which may have been around 3.5 billion years ago…there are microfossils from that long ago…(while) the earliest eukaryotic fossils date from 1.7 billion years ago." He further noted that "the prokaryotes also dominate in quantity…there are vastly more of them than there are of us and our fellow eukaryotic creatures…it is estimated that the total quantity of carbon locked up in prokaryotes is similar to the amount locked up in all plants on Earth." And "there is another sense, too, in which we, and the whole domain of eukaryotes, are the smaller domain…we and other animals and plants are made from prokaryotes, and we are still totally dependent on them…the cells in our bodies, like all eukaryotic cells, arose from symbiotic association." In summary, scientifically and objectively, there is no empirical argument to justify our anthropocentric narrative that humans are biologically more "special" or "resilient" or "strong"—or the "chosen ones" as argued by Wilson—within the natural world: sadly the only thing in which we might be truly unique is that we will very likely be the ones that will eventually destroy a huge number of other forms of life in this planet.

Another example showing that we are not "special" or particularly "resilient" or "strong" organisms concerns our **immune system**. As explained above, humans are particularly prone to have infectious diseases and, particularly those living in "developed countries," to also be attacked by their own immune systems. In *The Beautiful Cure*, Davis pointed out that contrary to humans many other animals, such as some insect species, don't seem to suffer from opportunistic infection at all, having apparently some kind of specially potent **immune defense**. In contrast, the self-designated "superior" humans spend trillions of dollars, lose millions of lives, and dedicate thousands of hours and huge resources trying to scientifically find a solution, without much success so far, for a problem that those "inferior" insect species don't even seem to have. In fact, even Wilson contradicts, in his book, the very ideas about human "specialness" that he had defended in other parts of the same book, by pointing out for instance that humans are actually "extremely limited" and even "one of the poorest" species concerning our **senses**:

> Of all the continua mapped by science, the most relevant to the humanities are the senses, which are extremely limited in our species. **Vision** is based in *Homo sapiens* on an almost infinitesimal sliver of energy, four hundred to seven hundred nanometers in the electromagnetic spectrum. The rest of the spectrum, saturating the Universe, ranges from gamma rays trillions of times shorter than the human visual segment to radio waves trillions of times longer. Animals live within their own slivers of continua. Below four hundred nanometers, for example, butterflies find pollen and nectar in flowers by the patterns of ultraviolet light reflected off the petals – patterns and colors unseen by us. Where we see a yellow or red blossom, the insects see an array of spots and concentric circles in light and dark. Healthy people believe intuitively that they can hear almost every sound. However, our species is programmed to detect only twenty to twenty thousand hertz (cycles of air compression per second). Above that range, flying bats broadcast ultrasonic pulses into the night air and listen for the echoes to dodge obstacles and snatch moths and other insects on the wing. Below the human range, elephants rumble complex messages in exchanges back and forth with other members of their herd. We walk through nature like a deaf person the streets of New York, sensing only a few vibrations, able to interpret almost nothing.
>
> Human beings have one of the poorest **senses of smell** of all the organisms on Earth, so weak that we have only a tiny vocabulary to express it. We depend heavily on similes such as 'lemony' or 'acidic' or 'fetid.' In contrast, the vast majority of other organisms, ranging in kind from bacteria to snakes and wolves, rely on odor and taste for their very existence. We depend on the sophistication of trained dogs to lead us through the olfactory world, tracking individual people, detecting even the slightest trace of explosives and other dangerous chemicals. Our species is almost wholly unconscious of certain other kinds of stimuli without the use of instruments. We detect electricity solely by a tingle, a shock, or a flash of light. In contrast, there exist a variety of freshwater eels, catfish, and elephant-nose fish, confined to murky water where, deprived of vision, they live instead in a galvanic world. They generate charged fields around their bodies with trunk muscle tissue that has been modified by evolution into organic batteries. With the aid of electric shadows in the pattern of charges, the fish avoid obstacles around them, locate prey, and communicate with others of the same species. Yet another part of the environment beyond the reach of humans is Earth's magnetic field, used by some migratory birds to guide them during their long-distance journeys.

This remind us of what happened with Darwin's *Origin* (Chap. 6): as Darwin's book, Wilson's book sometimes seem to have been written by two different authors, one being a "rational, objective, naturalistic scientist" and the other a "teleological storyteller," emphasizing again *Neverland*'s immense power. I will therefore just provide here a last excerpt from Wilson's book, because it is about an issue that is particularly relevant for the topics discussed in the present chapter in particular and the whole book in general, namely our chief *Homo socialis* component and its links to **tribalism**, **racism,** and **religious bigotry**:

> Evolutionary biologists have…searched for the *grand master* of advanced social evolution, the combination of forces and environmental circumstances that bestowed greater longevity and more successful reproduction upon the possessors of *high* social intelligence. Two competing theories of the principal force have been in contention. The first envisions **kin selection**: individuals favor collateral kin (relatives other than offspring), making it easier for altruism to evolve among members of the same group. Complex social behavior can evolve when group members individually reap greater benefits in numbers of genes passed to the next generation than losses from their **altruism**, averaged through their behavior toward all members of the group. The combined effect on the survival and reproduction of the individual is called inclusive fitness, and the explanation of evolution by it is called the theory of **inclusive fitness**. In the second, more recently argued theory (full disclosure: I am one of the modern version's authors), the grand master is **multilevel selection**. This formulation recognizes two levels at which natural selection operates: individual selection based on competition and cooperation among members of the same group, and **group selection**, which arises from competition and cooperation between groups. Group selection can occur through violent conflict or by competition between groups in the finding and harvesting of new resources. Multilevel selection is gaining in favor among evolutionary biologists because of recent mathematical proofs that kin selection can operate only under special conditions that rarely if ever exist. Also, multilevel selection is easily fitted to all of the known real animal cases of **eusocial evolution**, whereas kin selection, even when hypothetically plausible, can be fitted less well or not at all. Are human beings intrinsically *good* but corruptible by the forces of *evil*, or the reverse, innately *sinful* yet redeemable by the forces of good? Are we built to pledge our lives to a group, even to the risk of death, or the opposite, built to place ourselves and our families above all else? Scientific evidence, a good part of it accumulated during the past twenty years, suggests that we are both of these things simultaneously.
>
> Each of us is inherently conflicted. I am convinced after years of research on the subject that multilevel selection, with a powerful role of group-to-group competition, has been a major force in the forging of advanced **social behavior** – including that of humans. In fact, it seems clear that so deeply ingrained are the evolutionary products of group-selected behaviors, so completely a part of the contemporary human condition are they, that we are prone to regard them as fixtures of nature, like air and water. They are instead idiosyncratic traits of our species. Among them is the intense, even obsessive interest of people in other people, which begins in the first days of life as infants learn particular scents and sounds of the adults around them. Research psychologists have found that all normal humans are geniuses at reading the intentions of others, whereby they evaluate, proselytize, bond, cooperate, gossip, and control. Each person, working his way back and forth through his social network, almost continuously reviews past experiences while imagining the consequences of future scenarios. Social intelligence of this kind occurs in many **social animals**, and reaches its highest level in chimpanzees and bonobos, our closest evolutionary cousins. A second diagnostic hereditary trait of human behavior is the overpowering *instinctual urge to belong to groups in the first place*, shared with most kinds of social animals. To be kept forcibly in solitude is to be kept in pain, and put on the road to madness. A person's membership in his group – his tribe – is a large part of his identity. It also confers upon him to some degree or other a sense of superiority. When psychologists selected teams at random from a population of volunteers to compete in simple games, members of each team soon

came to think of members of other teams as less able and trustworthy, even when the participants knew they had been selected at random. All things being equal (fortunately things are seldom equal, not exactly), people prefer to be with others who look like them, speak the same dialect, and hold the same beliefs. An amplification of this evidently inborn predisposition leads with frightening ease to ***racism*** and ***religious bigotry***. Then, also with frightening ease, good people do *bad things*. I know this truth from experience, having grown up in the Deep South during the 1930s and 1940s.... Within groups selfish individuals beat altruistic individuals, but groups of altruists beat groups of selfish individuals. *Or, risking oversimplification, individual selection promoted sin, while group selection promoted virtue*.

The last statement is very interesting, because we do know that **altruism** is indeed mainly related to group selection, and that in at least some circumstances groups of altruists in theory can beat groups of selfish individuals. This is actually one of the major arguments used to explain the existence of altruism, particularly among evolutionary biologists that still have a "struggle-for-life" view of life in this planet (Chaps. 5 and 6). One example that is often given to support this idea is that if everybody was selfish and just did what they wanted, there would be no division of labor, so no bread makers, weapon-makers, soldiers, and so on: such groups would not be able to win fights against another group with such professions. And, in part, we know from human history that more organized and hierarchical groups such as the **Roman empire** won many wars against groups of "barbarians" because the latter tended to lack professional soldiers and mercenary troops. However, I am not sure we should say that **Roman emperors**, professional soldiers, and **mercenary troops** were an example of "virtue" or were "altruistic," particularly because they were often paid or obliged to fight. Moreover, as explained by Wilson, **group selection** is often also related to **tribalism**, which can lead—using his own terms—"good people" to do "bad things" such as **racism** and **religious bigotry**. So, contrary to Wilson's oversimplification, I would instead say that *reality* is far more complex, with group selection potentially leading to traits such as altruism and cooperation, but also to features such as discrimination, warfare, racism and bigotry, exactly as defended in Hare and Woods' *The Survival of the Friendliest*.

The fact that *reality* is indeed far more complex, and thus often more fascinating, than fiction is, and that humans are just a paradoxical, mostly irrational, and enthusiastically creative and social by-product of a combination of random, contingent and chaotic evolutionary natural events, was beautifully and eloquently stated by **Ann Druyan**, when she was asked about the death of her ex-husband, **Carl Sagan**:

When my husband died, because he was so famous and known for not being a believer, many people would come up to me – it still sometimes happens – and ask me if Carl changed at the end and converted to a belief in an **afterlife**. They also frequently ask me if I think I will see him again. Carl faced his death with unflagging courage and never sought refuge in **illusions**. The tragedy was that we knew we would never see each other again. I don't ever expect to be reunited with Carl. But, the great thing is that when we were together, for nearly twenty years, we lived with a vivid appreciation of how brief and precious life is. We never trivialized the meaning of death by pretending it was anything other than a final parting. Every single moment that we were alive and we were together was miraculous – not miraculous in the sense of inexplicable or supernatural. We knew we were beneficiaries of chance.... That pure chance could be so generous and so kind.... That we could find each other, as Carl wrote so beautifully in Cosmos, you know, in the vastness of space and the immensity of time…. That we could be together for twenty years. That is something which sustains me and it's much more meaningful…. The way he treated me and the way I treated him, the way we took care of each other and our family, while he lived. That is so much more important than the idea I will see him someday. I don't think I'll ever see Carl again. But I saw him. We saw each other. *We found each other in the cosmos, and that was wonderful.*

Ann's outstanding and optimistic answer indeed shows how *reality*, and a view of the world based on empirical scientific data, can certainly be wonderful and also poetic—Carroll literally used the term ***poetic naturalism*** to refer to such a realistic view of life in his 2016 book *The Big Picture*, as noted in Box 9.11. Talking about Carl Sagan—whom we will discuss in more detail below—leads us to analyze another major difficulty preventing the storytelling animal to escape *Neverland*, which is too often neglected in discussions about nature, the cosmos, and reality: that many aspects of the *world of reality* often collide with "common sense." As pointed out by Carroll, one reason for this is precisely that the senses of humans are in general very feeble when compared to those of most other mammals, which are, themselves, not ideal at all to grasp the true complexity of the cosmos (see Fig. 9.16). A major evolutionary cause for this mismatch about what the senses seem to indicate and the reality of the cosmos is that each group of animals, such as termites, lions, or dolphins, obviously don't "need"—that is, there is *no* selective pressure for them to—know about things such as black holes, in their daily-lives. Some examples concerning the very complex, fascinating and "weird"—to our senses, of course—cosmos we live in are given in Rovelli's enlightening 2017 book *Reality is Not What it Seems*:

Say I am on **Mars** and you are here. I ask you a question and you reply as soon as you've heard what I said; your reply reaches me a quarter of an hour after I posed the question. This quarter of an hour is time that is neither past nor future to the moment in which you've

Fig. 9.16 The often "non-sense" *world of reality*: against "common sense," the Earth actually turns around the Sun because space-time around the Sun is curved, somewhat as a bead that rolls on the curved wall of a funnel

replied to me. The key fact about nature that **Einstein** understood is that this quarter of an hour is inevitable: there is no way of reducing it. It is woven into the texture of the events of space and of time: we cannot abbreviate it, any more than we can send a letter to the past. It's strange, but this is how the world happens to be. As strange as the fact that in Sydney people live upside down: strange, but true. One gets accustomed to the fact, which then becomes normal and reasonable. It is the structure of space and time that is made like this. This implies that it makes no sense to say of an event on Mars that it is taking place 'just now', because 'just now' does not exist. The present is like the **flatness of the Earth**: an illusion. We imagined a flat Earth because of the *limitations of our senses*, because we cannot see much beyond our own noses. Had we lived on an asteroid of a few kilometers in diameter, like the **Little Prince**, we would have easily realized we were on a sphere. Had our brain and our senses been more precise, had we easily perceived time in nanoseconds, we would never have made up the idea of a 'present' extending everywhere. We would have easily recognized the existence of the intermediate zone between past and future. We would have realized that saying 'here and now' makes sense, but that saying 'now' to designate events 'happening now' throughout the universe makes no sense. It is like asking whether our galaxy is 'above or below' the **galaxy of Andromeda**: a question that makes no sense, because 'above' or 'below' has meaning on the surface of the Earth, not in the universe. There isn't an 'up' or a 'down' in the universe. Similarly, there isn't either always a 'before' and an 'after' between two events in the universe. The resulting knitted structure that space and time form together…is what physicists call '**space-time**'…we are not contained within an invisible, rigid scaffolding: we are immersed in a gigantic, flexible mollusc (the metaphor is Einstein's). The Sun bends space around itself, and the Earth does not circle around it drawn by a mysterious distant force but runs straight in a space that inclines. It's like a bead which rolls in a funnel: there are no mysterious forces generated by the centre of the funnel, it is the curved nature of the funnel wall which guides the rotation of the bead. Planets circle around the Sun, and things fall, because space around them is curved [see **Fig. 9.16**]. A little more precisely, what curves is not space but space-time – that space-time which, ten years previously, Einstein himself had shown to be a structured whole rather than a succession of instants.

But within this equation there is a teeming universe. And here the magical richness of the theory opens up into a phantasmagorical succession of predictions that resemble the delirious ravings of a madman but which have all turned out to be true. Even up to the beginning of the 1980s, almost nobody took the majority of these fantastical predictions entirely seriously. And yet, one after another, they have all been verified by experience. Let's consider a few of them. it is not only space that curves: time does, too. Einstein predicts that time on Earth passes more quickly at higher altitude, and more slowly at lower altitude. This is measured, and also proves to be the case. Today we have extremely precise clocks, in many laboratories, and it is possible to measure this strange effect even for a difference in altitude of just a few centimeters. Place a watch on the floor and another on a table: the one on the floor registers less passing of time than the one on the table. Why? Because time is not universal and fixed, it is something which expands and shrinks, according to the vicinity of masses: the Earth, like all masses, distorts space-time, slowing time down in its vicinity. Only slightly – but two twins who have lived respectively at sea-level and in the mountains will find that, when they meet up again, one will have aged more than the other. This effect offers an interesting explanation as to why things fall. If you look at a map of the world and the route taken by an airplane flying from Rome to New York, it does not seem to be straight: the airplane makes an arc towards the north. Why? Because, the Earth being curved, crossing northwards is shorter than keeping to the same parallel.

The distances between meridians are shorter the more northerly you are; therefore, it is better to head northwards, to shorten the route. Well, believe it or not, a ball thrown upwards falls downwards for the same reason: it 'gains time' moving higher up, because time passes at a different speed up there. In both cases, airplane and ball follow a straight trajectory in a space (or space-time) that is curved. **Quantum mechanics** brings probability to the heart of the evolution of things. This *indeterminacy* is the third cornerstone of quantum mechanics: the discovery that chance operates at the atomic level. While **Newton's physics** allows for the prediction of the future with exactitude, if we have sufficient information about the initial data and if we can make the calculations, quantum mechanics allows us to calculate only the probability of an event. This absence of determinism at a small scale is intrinsic to nature. An electron is not obliged by nature to move towards the right or the left; it does so *by chance*. The apparent determinism of the macroscopic world is due only to the fact that the **microscopic randomness** cancels out on average, leaving only fluctuations *too minute for us to perceive in everyday life*. All of this sounds like a tale told by an idiot, full of sound and fury, signifying nothing. And yet, instead, it is a glance towards reality. Or better, a glimpse of reality, a little less veiled than our blurred and banal everyday view of it. A reality which seems to be made of the same stuff our dreams are made of, but which is nevertheless more real than our clouded daily dreaming.

This applies to an endless number of other aspects of reality, including biological ones. For instance, I hear many scientists saying that they don't understand why so many people "still don't accept evolution and believe in Gods and **creationism.**" Well, there are many reasons for this, including of course the fact that accepting evolution goes against all the creation myths and teleological narratives about cosmic "purpose" that we are so prone and also acculturated to believe since times immemorial. However, another major, and much less discussed, reason is simply that our senses truly don't "see," "hear," "touch," "taste," or "smell" evolution on a daily-basis, or even during a lifetime. We don't "see" any clear example in which a certain animal, or group of animals, that we encountered when we were younger then becomes part of a completely new species, when we got older. And surely we don't "see," within a lifetime, a completely new group of animals appearing in front of our eyes. So, one could argue that what we "see" in the fossil record is just the result of "fake evidence" built by some atheists to trick us with "alternative facts" so we don't believe in God and His creation. One does not believe what one does not "see," that skeptic theist could argue—ignoring of course the key point that he or she have obviously also never seen God, or angels, or demons, and still they "believe" in them.

This type of discussions have indeed been recurrent in human history, one of the most illustrative examples being the supposed **flatness of the Earth**: that is how the planet seems to be, from our common points of view on Earth, although we now *know* this is an illusion. This shows how science and technology are the most powerful tools against *Neverland* beliefs and why, accordingly, religious fundamentalists from any religion tend to be against scientific knowledge—the *Taliban* being one of the most recent clear examples of that. Actually, in a different way but related to some similar reasons, a growing number of people in Western countries— particularly within the current explosive context of the rise of **"fake news"** and **anti-intellectualism** in some of those countries, such as the U.S.—are coming back to the notion that Earth is flat. Paraphrasing once again Chesterton's brilliant quote, when people "stop believing in God they don't believe in nothing; they believe in anything." And one of the major reasons for believing in anything, including that our planet is flat—apart from the power and attraction of *Neverland*'s "alternative facts," "fake news," and conspiracy theories—is precisely because reality is often not what it "seems": as the saying goes, there is much more than meets the "eye." It is indeed an enormous paradox, and a profound irony of life, that illusions such as the flatness of the Earth, or the homogeneity and continuity of our "self" (Boxes 9.2, 9.4, 9.5 and 9.6) "seem" more real that *reality* itself. This fact is sometimes invoked by many **postmodernists** and **constructionists** to argue that there is no *reality* at all, that everything depends on our "senses." But such a constructionist assertion is not only factually inaccurate, but also perhaps even more anthropocentric and arrogant that some of humans' most hardcore longstanding "ladder of life" and "made-in-the-image-of God" teleological narratives, because it basically means that what humans cannot sense is merely not true. Real objects, events and organisms—such as the sun, meteorites, and bacteria—were in this cosmos much before humans came to be (Fig. 2.6) and will very likely continue to exist much after we will be gone, so obviously the reality of their existence does not depend at all of the feeble senses of the storytelling anthropocentric animal.

Coming back to how we sense—or, better said, often are not able to sense—biological evolution, let's say that a child was born and put in a remote island and never had any contact with a human since then. Almost surely that child would think that there is no biological evolution, and that this planet is flat: that is "common sense." This is a crucial point: we have been stuck in *Neverland* mainly because we are irrational and obsessively social storytelling animals that create, believe or are forced to pretend to believe or conform to fictional tales. But the fact that there is often a huge **evolutionary mismatch**—more than in many other animals as seen above—between our senses and the reality of the cosmos does not help at all our escape from—actually, often contributes to our even deeper immersion within—that created world of unreality. That is, to escape from *Neverland* we have to escape not only from our created beliefs but also, to some extent, to common sense. This point was also made in Kirkham's 2019 book *Our Shadowed World*, in which he explained that "perhaps the decisive innovation of the **Enlightenment** was insistence on the simple recognition of facts, or what we now call data…when **Galileo** tried to persuade his inquisitors to look through his **telescope** at the moons of Jupiter—something they refused to do on the grounds of diabolical deception—he was merely seeking to challenge authority with fact." He noted that "looking through what is often regarded as perhaps the most monumental scientific work of all time, ***Newton's Principia Mathematica***, one is struck by the vast collection of observational detail in factual tables…the same is true of the records of **Lavoisier**'s experiments in chemistry and **Dalton**'s daily notes on meteorology." And added that "it would be the factual observation of nature – the hallmark of the **European scienza nuova** – that would gather increasing momentum after the time of Galileo…it would lead so often **to counterintuitive theories** that pitted the true understanding of reality against **common sense** and **common beliefs.**" And this is precisely another fascinating point about the *reality* of the natural world: that it includes such evolutionary mismatches, being full of "errors"—actually, mostly *based* on them, such as unreliable copies of genes or behaviors—unpredictable, within the dazzling strange cosmos described just above. As underlined in Rovelli's *Reality is not what it seems*:

> **Science** works because, after hypotheses and reasoning, after intuitions and visions, after equations and calculations, we can check whether we have done well or not: the theory gives predictions about things we have not yet observed, and we can check whether these are correct, or not. This is the power of science, that which grounds its reliability and allows us to trust in it with confidence: we can check whether a theory is right or wrong. This is what distinguishes science from other kinds of thinking, where deciding who is right and who is wrong is usually a much thornier question, sometimes even devoid of meaning. When **Lemaître** defends the idea that the universe is expanding, and **Einstein** does not believe it, one of the two is wrong; the other right. All of Einstein's results, his fame, his influence on the scientific world, his immense authority, count for nothing. The observations prove him wrong, and it's game over. An obscure Belgian priest is right. It is for this reason that scientific thinking has power. The sociology of science has shed light on the complexity of the process of scientific understanding; like any other human endeavour, this process is beset by irrationality, intersects with the game of power and is affected by every sort of social and cultural influence. Nevertheless, despite all of this, and in opposition to the exaggerations of a few **postmodernists**, **cultural relativists** and the like, none of this diminishes the practical and theoretical efficacy of scientific thinking. Because in the end, in the majority of cases, it is possible to establish with clarity who is right and who is wrong. And even the great Einstein could go on to say (and he did so), 'Ah…. I made a mistake!' Science is the best strategy if we value reliability.
>
> There is always, in this world, someone who pretends to tell us the ultimate answers. The world is full of people who say that they have *The Truth*. Because they have got it from the fathers; they have read it in a *Great Book*; they have received it directly from a god; they have found it in the depths of themselves. There is always someone who has the presumption to be the ***depository of Truth***, neglecting to notice that the world is full of other depositories of Truth, each one with his own real Truth, different from that of the others. There is always some prophet dressed in white, uttering the words, 'Follow me, I am the true way.' I don't criticize those who prefer to believe in this: we are all free to believe in whatever we want. Maybe, after all, there is a grain of truth in the joke reported by St Augustine: '*What was God doing before creating the world? He was preparing Hell for those who seek to scrutinize deep mysteries.*' But these deep mysteries are precisely the 'depths' in which **Democritus**…invites us to seek the truth [see **Box 9.7**]. For my part, I prefer to look our ignorance in the face, accept it and seek to look just a bit further: to try to understand that which we are able to understand. Not just because accepting this ignorance is the way to avoid being entangled in superstitions and prejudices – but because to accept our ignorance in the first place seems to me to be the truest, the most beautiful and, above all, the most honest way. The world is more extraordinary and profound than any of the fables told by our forefathers. I want to go and see it. To accept uncertainty doesn't detract from our sense of mystery. On the contrary: we are immersed in the mystery and the beauty of the world. The world revealed by **quantum gravity** is a new and strange one—still full of mystery, but coherent with its simple and clear beauty.

I guess by now you might have made this question to yourself: but, wait, is the present book arguing that we should remove *everything* that is imaginary, from our cultures? No, not at all. Our *Homo fictus* storytelling creative minds are not a problem, well on the contrary, as long as we don't *believe* in the fictional stories we create. I love to read and see movies about imaginary stories such as **Lewis Carroll**'s

Alice's Adventures in wonderland, **James Barrie**'s ***Peter Pan***, and **Charles Dickens**' ***Christmas Carol***: such fascinating stories, such outstanding imagination and creativity, are deeply absorbing and can be very fulfilling. Accordingly, I profoundly admire their authors: in terms of pure genius and creativity, Lewis Carrol and Charles Dickens are among the people I have a higher regard for. Moreover, it is important to note that, even within the reality of the natural world, many non-human animals display **play behaviors**, for instance when wolf or dog puppies pretend to be "fighting" with each other, as human toddlers often do (see Fig. 9.17). But the point is that when those puppies, and those toddlers, are pretending to fight, they *are aware* that the fight is not real: we can show that empirically by comparing those behaviors with those displayed in real fights—the stress levels, the type of injuries, and so on. Similarly, we have empirical scientific data showing that mice can't talk as we do, so we *know*, empirically, that **Mickey Mouse** can't be real. However, that does not happen with other imaginary tales that are as far removed from reality, and as contradicted by empirical evidence, as the existence of Mickey Mouse is: for instance, that a virgin had a child that was the son of God is not at all more plausible, scientifically, than a mouse talking like a human.

As explained in Chap. 2 our beliefs are related to a series of features that were already displayed by many other animals before we even existed, such as **patternicity**, **agenticity,** and **superstition**. As also noted just above, other animals also have **pretend play**, but as far as we know when other animals display play behaviors they are aware of the reality of the situation: if the puppies would lose this control of reality, or their mothers, or other members of their species, that could be potentially detrimental, obviously. But, because of all the reasons discussed in the present book—in particular the emergence of our awareness of the inexorableness of death—there was a key moment in which we started to lose that control of reality. We slowly became more and more immersed in *Neverland* by actually start believing in the imaginary stories we built, in particular those concerning teleological tales about "cosmic purpose." While wolf puppies that are engaged in playing *pretend* to have "angry" or "bad" adversaries, humans started to truly believe they had them: the "monsters," the "witches," "other" races or "casts," those with a different gender or type of sexual preference or religion, or the blasphemers, or people with malformations, and so on (Fig. 9.17). **Captain James Hook** was no longer a fictional adversary of **Peter Pan** within an imaginary ***Neverland***. He became a real, and omnipresent, one: he might be anywhere, you have to be careful and ready to face him, to fight against him, to defeat him (Fig. 9.5). Scientifically, this is a profoundly fascinating evolutionary story, but it is also an ultimately profoundly tragic occurrence in human history. And, unless we become fully aware of the enormous price we ended up paying because of this, and are able to escape once for all from the heavy handcuffs of *Neverland*, from being continuously and anxiously afraid of and obsessed with "bad villains" and "others" and using those fears, anxieties and obsessions as justifications to discriminate, oppress, enslave, torture, burn, kill, or genocide "Them," we will remain enchained in *Neverland*'s most deep, obscure, and dangerous dungeons.

Fig. 9.17 Two very different ways of using imaginary tales: pretend play within dog puppies and human kids versus horrendous acts of violence within adult storytelling beings: the bottom left figure displays a Holocaust photo in which a Jew in Ukraine is kneeling in front of a mass grave as a Nazi officer points a gun to his head moments before shooting him, while the bottom right figure shows a Christian attacking a Muslim with a knife after a series of lynching sprees undertaken by both Christian and Muslim mobs in Central African Republic

9.3 No Need for Better Angels, Economic Fairytales, Nor New Delusions

Somehow we've weathered and witnessed a nation that it isn't broken, but simply unfinished. We, the successors of a country and the time where a skinny Black girl descended from slaves and raised by a single mother can dream of becoming president only to find herself reciting for one. And yes, we are far from polished, far from pristine, but that doesn't mean we are striving to form a union that is perfect. We are striving to forge our union with purpose.
(Amanda Gorman, at Inauguration of U.S. President Biden, 2021)

In biological evolution, **over-specialized animals** that repeat over and over certain behaviors even when they become highly detrimental within the environmental reality of when and where they live, often become entrapped in **evolutionary dead endings** that usually lead to **extinctions**. An example of such an **evolutionary behavioral-ecological mismatch** that I gave in my 2017 book *Evolution Driven by Organismal Behavior* concerns **Giant pandas**. Bamboo shoots and leaves continue to make up more than 99% of their diet, despite the fact that such food items are seriously threatened by factors related to **overpopulation**, **deforestation**, **economic**

growth, and the so-called technological developments in China, putting their survival as a species at risk: they are now officially classified as a "vulnerable" species (see also Box 9.3). In this sense, it seems that our "modern" societies are lacking the evolutionary plasticity to stop the vicious cycle leading them to the behavioral-ecological mismatch of being more and more immersed in *Neverland* and thus less connected to the reality of the natural world, putting our whole species and the Earth's ecology at risk by doing so. Our obsession to create new teleological narratives to replace old ones is well-exemplified in Boxes 9.3 and 9.4 as well as in a book that, as discussed above, is very interesting but ultimately also falls into the storytelling trap: Peterson's 2001 *Being human*. In the first chapters, Peterson focuses on **non-Western teleological narratives** such as those related to **Asian Buddhist and Taoist traditions** (Box 9.5) and to **Native American traditions** (Box 9.7) as an example of how we can eventually contribute to the "construction of a coherent ethical framework" and of "modern" narratives with the potential to contribute to the "survival of global ecosystems" (see also Box 9.6).

In the present book I do also argue that we should try to incorporate into our "modern" ways of life many features typical of nomadic hunter-gatherer societies, such as egalitarianism, eating fresh food, having less taboos about sexuality, and so on. However, contrarily to Peterson, I am exclusively referring to *what those societies do* in their daily-lives within the reality of the natural world: I am *not* referring at all to the imaginary tales *they believe in*. This point is also a major difference between the present book and the otherwise fascinating books of **Edmund Wade Davis**, a brilliant cultural anthropologist, ethnobotanist, and photographer that does truly amazing work with, and is a passionate defender of, non-Western people, and that tends to suggest that we would be "better off" by incorporating into our Western societies their beliefs. I don't agree with this specific point because I don't think that to escape *Neverland* we need to believe in even more imaginary stories (see also Boxes 9.3 and 9.4). Actually, we have seen how many recycled old beliefs are permeating for instance New Agers in recent years, and the results have not been promising at all. For example, many of them are now even more immersed in *Neverland* than their parents ever were, contributing to the rise of the **anti-vax movement** as well as many other types of anti-scientific movements such as those denying the reality and/or danger of the **Covid-19 pandemic**, thus putting in risk not only their lives but, selfishly, the lives of their loved ones, of friends, of colleagues, and of millions of other people. Simply and plainly, it is as wrong to say that Christian imaginary tales are "superior," as was defended by most Westerners for a long time, as to say that Native American fictional stories are "better" as it is now so much in vogue within so many New Agers and many other people, including numerous scholars. Humans tend to be highly dichotomic, moving very quickly from one extreme—being Western-centric—to the other—being anti-Western—forgetting, once again, that normally reality lies between extremes. The *reality* is that all imaginary tales—animistic ones about a tree talking to us, Christian ones about a virgin having a child, Disney ones about Mickey Mouse talking as humans do, or QAnon conspiracy theories—are equally factually wrong, being nothing more than *Neverland*'s fictional stories.

Box 9.3: Neoliberalism, Social Democracy, Falsehoods, and Grounding Our Lives in Reality

Some of the points made by Monbiot, in a *The Guardian Weekly* article published in September 2017 entitled "*It's time to tell a new story if we want to change the world*", are in line with some of the ideas defended in the present book, particularly those about grounding our lives and decisions in *reality*. However, paradoxically, Monbiot can't escape *Neverland*'s trap-the arguments are plagued by teleology, including his proposed solution: we need to "build" a new "narrative" (see also Box 9.4). Monbiot seems to imply that one cannot escape from *Neverland* and that the only way to do a 'better world' is therefore to navigate its dungeons in a different way:

> Is it reasonable to hope for a better world? Study the cruelty and indifference of governments, the disarray of opposition parties, the apparently inexorable slide towards climate breakdown, the renewed threat of **nuclear war**, and the answer appears to be no. Our problems look intractable, our leaders dangerous, while voters are cowed and baffled. But over the past two years, I have been struck by four observations. The first observation is the least original. It is the realisation that it is not strong leaders or parties that dominate politics as much as powerful **political narratives**. The political history of the second half of the 20th century could be summarised as the conflict between its two great narratives: the stories told by **Keynesian social democracy** and by **neoliberalism**. First one and then the other captured the minds of people across the political spectrum. When the **social democracy story** dominated, even the Conservatives and Republicans adopted key elements of the programme. When neoliberalism took its place, political parties everywhere, regardless of their colour, fell under its spell. These stories overrode everything: personality, identity and party history. This should not surprise us. Stories are the means by which we navigate the world. They allow us to interpret its complex and contradictory signals. We all possess a **narrative instinct**: an innate disposition to listen for an account of who we are and where we stand. When we encounter a complex issue and try to understand it, what we look for is not consistent and reliable facts but a consistent and comprehensible story. When we ask ourselves whether something 'makes sense', the 'sense' we seek is not rationality, as scientists and philosophers perceive it, but narrative fidelity. Does what we are hearing reflect the way we expect humans and the world to behave? Does it hang together? Does it progress as stories should **progress**? A string of facts, however well attested, will not correct or dislodge a powerful story. The only response it is likely to provoke is indignation: people often angrily deny facts that clash with the narrative 'truth' established in their minds. *The only thing that can displace a story is a story. Those who tell the stories run the world.*
>
> I came to the second, more interesting, observation with the help of the writer and organiser George Marshall. It is this. Although the stories told by social democracy and neoliberalism are starkly opposed to each other, they have the same narrative structure. We could call it the *Restoration Story*. It goes like this: disorder afflicts the land, caused by powerful and nefarious forces working against the interests of humanity. The hero – who might be one person or a group of people – revolts against this disorder, fights the nefarious forces, overcomes them despite great odds and restores order. Stories that follow this pattern can be so powerful that they sweep all before them: even our fundamental values. For example, two of the world's best-loved and most abiding narratives – The ***Lord of the Rings*** and the ***Narnia series*** –

invoke values that were familiar in the middle ages but are generally considered repulsive today. Disorder in these stories is characterised by the usurpation of rightful kings or their rightful heirs; justice and order rely on their restoration. We find ourselves cheering the resumption of autocracy, the destruction of industry and even, in the case of Narnia, the triumph of divine right over secular power. If these stories reflected the values most people profess – democracy, independence, industrial 'progress' – the rebels would be the heroes and the hereditary rulers the villains. We overlook the conflict with our own priorities because the stories resonate so powerfully with the narrative structure for which our minds are prepared. *Facts, evidence, values, beliefs: stories conquer all.*

The social democratic story explains that the world fell into disorder – characterised by the **Great Depression** – because of the self-seeking behaviour of an unrestrained elite. The elite's capture of both the world's wealth and the political system resulted in the impoverishment and insecurity of working people. By uniting to defend their common interests, the world's people could throw down the power of this elite, strip it of its ill-gotten gains and pool the resulting wealth for the good of all. Order and security would be restored in the form of a protective, paternalistic state, investing in public projects for the public good, generating the wealth that would guarantee a prosperous future for everyone. The ordinary people of the land – the heroes of the story – would triumph over those who had oppressed them. The **neoliberal story** explains that the world fell into disorder as a result of the collectivising tendencies of the overmighty state, exemplified by the monstrosities of **Stalinism** and **nazism**, but evident in all forms of state planning and all attempts to engineer social outcomes. **Collectivism** crushes freedom, **individualism** and opportunity. Heroic entrepreneurs, mobilising the redeeming power of the market, would fight this enforced conformity, freeing society from the enslavement of the state. Order would be restored in the form of **free markets**, delivering wealth and opportunity, guaranteeing a prosperous future for everyone. The ordinary people of the land, released by the heroes of the story (the **freedom-seeking entrepreneurs**) would triumph over those who had oppressed them. Then – again with Marshall's help – I stumbled into the third observation: the narrative structure of the **Restoration Story** is a common element in most successful **political transformations**, including many **religious revolutions**. This led inexorably to the fourth insight: the reason why, despite its multiple and manifest failures, we appear to be stuck with neoliberalism is that we have failed to produce a new narrative with which to replace it.

You cannot take away someone's story without giving them a new one. It is not enough to challenge an old narrative, however outdated and discredited it may be. Change happens only when you replace one story with another. When we develop the right story, and learn how to tell it, it will infect the minds of people across the political spectrum. **Political renewal** depends on a new political story. Without a new story that is positive and propositional, rather than reactive and oppositional, nothing changes. With such a story, everything changes. The *narrative we build* has to be simple and intelligible. If it is to transform our politics, it should appeal to as many people as possible, crossing traditional political lines. It should resonate with deep needs and desires. It should explain the mess we are in and the means by which we might escape it. And, because there is nothing to be gained from spreading falsehoods, it must be firmly grounded *in reality*.

Box 9.4: Virtual Worlds, Second Life', and New Delusions

In his book *The self illusion*, Hood gives us disturbing examples of what the life of the storytelling animal can be in the near future, when new narratives and new illusions within the world of *Neverland* are created, as is the case with the **virtual online world** named **Second Life** (see also Figs. 9.11 and 9.12):

> The **Internet** can become addictive and it can also be dangerous, especially in the case of immersive gaming where individuals can play for hours in **fantasy worlds**. In 2010, **South Korea** had a greater proportion of its population online than any other nation (81 per cent of forty-six million). Most Koreans spend their online time in **internet cafés** that provide fast but cheap connections. This can have devastating consequences. Many of them develop serious medical conditions related to hours of online activity at the cost of offline inactivity. Their joints swell up. They develop **muscular pain**. Sometimes it's others that get hurt. In the same year, a South Korean couple who met online married in real life, but unfortunately had a sickly premature baby. But then they decided to continue their lives online in the café across the road in a game where they raised a virtual baby. *They only returned to the house once a day to feed their own real baby. This lack of care meant that their own child eventually died of severe dehydration and malnutrition.* Undoubtedly, this is an extreme case and many children raised in poverty are neglected but it highlights the compulsion of the Web. I recently hosted a highly educated academic family visiting from the United States and after the initial social conversation and exchange of anecdotes over dinner; we soon dispersed to check our email, Facebook and other online lives. It was not only the adults in the group, but the children as well. At one point, I looked up from my laptop and saw everyone else in the room silently immersed in their own Web. *Whereas we once used to compartmentalize our lives into the working day and time with the family, the Web has destroyed those boundaries forever.* Most of us are connected and we like it that way. Just like ***drug addiction***, many of us get withdrawal symptoms of **anxiety** and **irritability** when we are denied our Web access.
>
> [So], what do you do if you are unemployed, overweight and living off benefits with no prospect of escaping the poverty trap? Since 2003, there has been another world you can live in – a world where you can get a second chance. This is Second Life, a virtual online world where you reinvent your self and live a life among other avatars who never grow old, have perfect bodies, never get ill, have fabulous homes and lead interesting lives. **David Pollard** and **Amy Taylor** are two individuals who separately wanted to escape the dreariness of their mundane lives. Both of them lived in Newquay, a seaside resort in southwest England that has become a Mecca for drunken teenagers who come in their hordes to party away the summer. The town is far from idyllic and I would imagine living there, without a job and prospects, must be depressing. To escape the drudgery, David and Amy (who initially met in an online chatroom) joined Second Life where they became Dave Barmy and Laura Skye "[see **Fig. 9.11**]." Dave Barmy was in his mid-twenties, six foot four, slim, with long dark hair, and was a nightclub owner who lived in a sprawling villa. He had a penchant for smart suits and bling. In reality, David Pollard was forty, overweight at 160 kg, balding and living off incapacity benefits in a bedsit. He wore T-shirts and tracksuit bottoms. Laura Skye was an equally exotic character. She was also in her mid-twenties, a slim six foot with long, dark hair, living in a large house. She liked the country and western look of tight denim blouses and boots. In reality, Amy Pollard was an overweight, five-foot-four redhead who was also living off benefits. The contrast between reality and fiction could hardly have been greater"[**Fig. 9.11**]." When the couple met online as Dave Barmy and Laura Skye, they fell in love and married in Second Life. But they also met up in real life with Amy moving in with

David in Newquay. After two years, they married for real – just like the Korean couple. However, as in real life, that's when things started to go wrong. Laura (Amy) suspected Dave was playing around in Second Life so she hired a virtual detective to check up on her virtual husband. At one point, she discovered Dave Barmy having sex with a call girl in the game. In real life, David apologized and begged for forgiveness. The final straw came when Amy caught her real husband in front of the computer in their small flat watching his avatar cuddling affectionately on a couch with another Second Life character, Modesty McDonnell – the creation of Linda Brinkley, a fifty-five-year-old twice-divorcee from Arkansas, USA. Amy was devastated. She filed for divorce on the grounds of unreasonable behaviour even though Dave had not actually had sex or an affair in real life. Soon after, Dave proposed to Modesty online and in real life even though the couple had never met.

When the world discovered that a couple was divorcing on the grounds of make-believe unreasonable behaviour, the press flocked to Newquay. However, in what can only be described as reality imitating art, imitating reality, the Cornish couple initially declined to give interviews and would not answer the door. Then something very odd happened. Two enterprising journalists from the South West News hit on the bright idea of going into Second Life to secure an interview. From their offices miles away in Bristol, Jo Pickering and Paul Adcock created virtual ace reporters 'Meggy Paulse' and 'Jashly Gothley' to seek out Dave Barmy and Laura Skye for an interview. Jo still works on South West News and she told me that she had the idea after speaking to a colleague who had been using avatars to attend online courses. As Meggy Paulse, Jo found Laura Skye in Second Life. She told me that the online Laura Skye was much more approachable and confident than the real life Amy. Eventually Meggy Paulse persuaded Amy to logoff and go downstairs and open the door to speak to the reporters camped on her doorstep. They eventually got their story. Jo explained that Amy had felt that the betrayal online was far worse than betrayal in real life, because both she and David had created these perfect selves and still that was not good enough. In real life, we are all flawed and often put up with each other's weaknesses, but in Second Life there were supposed to be no weaknesses. That's why the online betrayal hurt. As Jo says, '*She had created this perfect version of herself – and even that wasn't good enough for him.*'

Some argue that one of the main uses of the Internet is **for sex**. A 2008 survey of more than 1280 teenagers (13–20 years) and young adults (20–26 years) revealed that one in five teenagers and one in three young adults had sent nude or semi-nude photographs of themselves over the internet. One online dating site, Friendfinder.com, estimates that nearly half of its subscribers are married. Either they are looking for new partners or the opportunity to flirt. Probably one of the most remarkable cases was US Army Colonel **Kassem Saleh**, who had simultaneously wooed over fifty women online and made marriage proposals to many of them despite the fact that he was already married. The ease and speed of the Web, as well as the perceived dissociation and distance from reality, lead to an escalation of brazen activity. This can easily slide into **moral indiscretions** that are *unregulated by social norms compared to real life*. Just like bullying, the apparent anonymity, distance and remoteness of being online allows us to not be our self as we would behave in the real world.

An example, discussed in detail in Peterson's *Being human*, of how very ancient teleological narratives continue to be propagated and followed by a huge number of people, are the **Asian Buddhist and Taoist traditions**. Her discussion is briefly summarized in Box 9.5, which shows how some of the ideas of those ancient traditions are in line with some very recent discoveries in the fields of psychology, anthropology, and neurobiology. This fact stresses an interesting, and important, point: contrary to the teleological narratives associated to those traditions—and for that matter linked to any teleological narratives as all of them are obviously factually wrong as there is no cosmic purpose in the universe—some of the non-teleological ideas related to them, as to other types of religion/philosophy of life, might be factually accurate, or at least not so far removed from reality (see also Box 9.6).

For instance, the Buddhist notions that "the profoundly relational character of the person leads to the crucial concept of no-self (*anatta*)," that "what we call "I," or "being," is only a combination of mental and physical aggregates, which are working together interdependently in a flux of momentary change within the law of cause and effect," are strikingly similar to the current scientific notion of **self-illusion** (see Boxes 9.2, 9.4, 9.5, 9.6 and Fig. 9.14). Also, empirical data from various areas are showing that there is no "linear progression" of life as it is often defended in Western countries (see Fig. 3.2) and are accordingly contradicting the "more linear, unidirectional understanding of causality [often seen] in Western traditions," thus being more in line with some Asian traditions that tend to view in general life as more cyclic, complex, and interconnected, and with the Buddhist notion that "mutual causality views reality as a dynamic interaction of mutually conditioning events."

Box 9.5: Taoism and the Similarities Between Buddhism and 'Modern' 'Self-Illusion' Concepts

In her book *Being human*, Peterson writes:

> Western thinking (has) shed the notion that one unique element – be it the soul, the rational mind, language, or culture – defines and sets our species apart from other animals and the whole natural world. This conviction about humanity's discontinuity with the rest of creation is far from universal, though. A wide range of ways to conceive human nature exists, within and outside Western culture…[for instance in] **Buddhist and Taoist traditions**. Related to the *Four Noble Truths* is the central Buddhist idea that, as Walpola Rahula puts it, "there is nothing permanent, everlasting, unchanging and eternal in the whole of existence." Things are not merely characterized by flux and change, **Buddhism** insists; everything actually is flux and change. Thus "*there is no unmoving mover behind the movement*. It is only movement. It is not correct to say that life is moving, but life is movement itself." The illusion that humans can halt or freeze this constant flow of change and movement is the root of *attachment or desire*, which in turn is the cause of suffering (***dukkha***). Unenlightened humans want things, such as relationships, possessions, or patterns in their lives, to remain constant. When inevitably they change, the result is pain and confusion. As notions of causality are central to philosophy generally, the conception of mutual causality or dependent co-origination is central to **Buddhism** and constitutes its major innovation over both Western and traditional **Indian (Vedic) philosophies**. In contrast to the linear, unidirectional understanding of causality in these traditions, mutual causality views reality as "a dynamic interaction of mutually conditioning events," in which there is no prime cause or unconditioned absolute to which occurrences can be traced in linear fashion. Ideas about mutual causality and interdependence also ground Buddhist ideas about human nature. Macy writes,

"Where all is process, so is the self, which by that token is neither categorically distinct from others nor endowed with any changeless essence." In contrast to the widespread assumption that the self is an entity that has experiences from which it is distinct, in Buddhism the self is "not separable from its experience nor isolable as an agent from the thinking, saying, and doing we attribute to it." Like reality as a whole, in Buddhist thought, persons are "not alone but thoroughly relational, and the grounds for a relational nature must be found within man's own nature and not in something external, to which he must react on a one-to-one basis." The profoundly relational character of the person leads to the crucial concept of *no-self* (***anatta***). Just as there is no permanent, unchanging substance in all the cosmos, there is no separate, eternal soul or ego as commonly conceived in **Western religious, philosophical, and psychological traditions**.

Rahula summarizes: "What we call 'I,' or 'being,' is only a combination of mental and physical aggregates, which are working together interdependently in a flux of momentary change within the law of cause and effect". The doctrine of no-self should not be understood as the opposite of the Hindu or Western concepts of atta (***atman***), or soul. Instead, anatta reflects and underlines the larger Buddhist challenge to ego-based ways of viewing the world, the radicalness of interdependence. As Macy writes, in Buddhism "the belief in a permanent, separate self is a fundamental error: engendering **greed**, **anxiety**, and **aggression**, it is an illusion basic to the suffering we experience and which we inflict on others." Although Buddhism rejects the notion of a **substantial self** and the **spirit-matter dualism** behind Western ideas of the soul, *it does not deny a special place to humans*. In Buddhism, the core of human distinctiveness lies in the claim that "*a man and only a man can become a Buddha*". Every man has within himself the potentiality of becoming **Buddha**, if he so wills it and endeavors. Because humans, and only humans, can practice meditation and become enlightened, Buddhism places great value on the good fortune of **human rebirth**. This valuation of human life, however, is far from absolute. Buddhism sees all sentient beings as fundamentally similar, in their urge to avoid pain and experience wellbeing. While birth as a human is highly valued, it is also a birth into "that vast universal web of interdependence in which what relates beings to each other is much more fundamental than what divides them into species." Rather than feeling superior to other beings, we should feel compelled to protect them from harm, because we know how much we do not want to suffer and how closely related all beings are. In Buddhism, then, "the preciousness of human birth is in no way due to human rights over other forms of life, for a human being was and could again be other forms of life." The unique value of human rebirth makes sense only in the larger context of the core doctrine of interdependence, which, as Gary Snyder puts it, teaches "mdesty in regards to human specialness." A number of writers agree with and have elaborated on, with variations, this idea that the key to Buddhist environmental ethics is the end of separation between an "I" and the "world." The extended self and the ethical correlate that one acts for others not out of duty but out of an enlightened self-interest help illuminate some possible weaknesses in Western approaches to environmental ethics and even social ethics more generally.

The emphasis in Western, particularly **post-Enlightenment**, ethics on rights and obligations rests on an assumption that a person's self-interest is inherently opposed to the interests of other persons (or beings) and of any larger human or nonhuman whole. Acting on behalf of others is perceived as "**altruism**" which suggests sacrificing one's own good for another's benefit. Underlying these assumptions is a definition of persons as not only separate from but antagonistic to each other and to larger (social or natural or both) communities. For Macy and other Buddhist thinkers, this anthropological assumption is fundamentally wrong. Instead, they begin with the Buddhist conviction that humans, like all beings, are inextricably constituted by and caught up in a web of mutual interdependence. In this perspective, one's own interests can never be cleanly or constructively separated from the interests of other beings or of the whole. The false dichotomy between the self and others lies at the root of mistreatment of the physical environment, other species, and other persons. Realizing that we do not stop at the borders of our skin leads to the logical conclusion that we should try to avoid harming other parts and wholes, since we are intimately related to them all. Some observers have charged that Buddhism, in stressing **spiritual practices**, such as **meditation**, which aim at **enlightenment**, is ***world denying***. In response, Buddhists often insist that there is an integral relationship between spiritual and ethical practices. The goal of enlightenment, they contend, is resolutely this-worldly, since **nirvana** (*Sanskrit, nirvana*) is not a separate realm but rather this world free of desire and thus of suffering. Spiritual practices are not distractions from ethical concerns. Rather, they are the only way to free the mind from egoism and desire and to cultivate both the virtues and the actions that Buddhists should bring to ethical problems. Buddhist practice begins "with the impulse to purify the mind and cultivate one's own sense of self, through a sense of the self's interdependence with a network of all other beings, to a sense of affection and love for all existence." The Vietnamese Buddhist monk **Thich Nhat Hanh** summarizes: "Meditation is not an escape from life…but preparation for really being in life." In this perspective, meditation, mindfulness, and other elements of spiritual training and self-cultivation lead directly to appropriate ethical ideas and practices.

A relational view of reality is crucial not only to Buddhism but also to understandings of self and nature in many other Asian traditions of thought, including the major indigenous belief systems of China, **Confucianism** and **Taoism**. While Taoism and Confucianism differ and even conflict in significant ways, they share key concepts and values, including an emphasis on the continuity of the universe and the relatedness of all beings in it. Within this context, Confucianism is more "humanistic", stressing the ethical and philosophical dimensions of human society and relationships. Taoism de-emphasizes or relativizes the role of humans in the cosmos, emphasizing instead the value of "flowing with" the Tao, usually defined as the "way" of all things, a sort of law of nature, but also suggesting, in Roger Ames's words, the "natural environment of any particular". Taoism's ultimate goal is harmony with this law or environment, rather than, as in Confucianism, the perfection of human character and society. **Wu-wei**, usually translated as nonaction or as effortless action, represents the achievement of this harmony with all entities and dimensions of the natural and social worlds. Taoism assumes that the underlying reality of all life is continuity of being, in which all things are part of the same larger whole and thus interrelated and interdependent. Further, there is no outside creator that brought them into being. While in Taoism humans' responsibility to care for nature cannot be understood as dominion, Taoism does suggest some form of ***human exceptionality***. Like Buddhism, Taoism views humans as the most sentient and complex beings in the universe. In Taoism, humans "are made of the same psychophysiological stuff that rocks, trees, and animals are also made of" but humans are unique insofar as "our consciousness of being human…enables and impels us to probe the transcendental anchorage of our nature". Humans have a distinctive spiritual capacity to enlarge and deepen their care for the universe, but this is neither automatic nor absolute. Further, it brings not only privileges but also, and especially, responsibilities. (However) Yi-Fu Tuan contends that traditional Chinese attitudes toward nature are "quiescent" and "adaptive", leading not to harmony but to passivity, even fatalism, in relation to nature. [Also\ Stephen Kellert concludes, based on extensive survey evidence, that Japanese culture values not wild nature but rather nature that humans have transformed in particular ways, nature as "garden," as aesthetic object. The idealized perception of nature that predominates in Japan, Kellert argues, includes little empirical understanding of or interest in nonhuman species or ecosystems.

Mindfulness, as well as **meditation**, a major focus of Buddhism and some other Asian traditions, is also something that is not only being more and more in vogue in the West, but also that is being shown, empirically, to be able to ameliorate, in at least some cases, the health and well-being of people to a certain extent (Chap. 2). However, it should also be noted that there is in fact some ambiguity, and there are often some inconsistencies, within some of the narratives of Buddhism. For instance, concerning the links between meditation, **mindfulness**, and the "root of *attachment or desire*, which in turn is the cause of suffering," as described by Peterson. The present volume shows, empirically, that if we are able to deconstruct ourselves from stories about the past or future, about "purposes," "goals," and so on, and concentrate instead on the *now* and on the *here* of the fascinating cosmos and planet in which we live, this could indeed avoid the causes of much suffering and anxiety. So I would clearly agree with the part about "attachments" being a main cause of suffering, **anxieties** and frustrations, moreover because attachment refers not only to both the past and future but also to toxic relationships and ideas of "meant to be," as well as to material possessions, and so on. However, this does not apply to "desire" in general, because although the term "desire" can be applied to let's say the "desire" to have more money, or more gold or more cars, it is much more commonly applied to wishes that are more momentaneous, such as one's desire, right now, to swim in the beach, or to have sex with someone. So, this does seem to be a key contradiction, because mindfulness should be about focusing on living *in the moment*, to enjoy it, to feel our bodies, to feel the pleasure it can give us, or the pleasure of swimming in the beach, or to listen to sound, or to taste a delicious coffee, and many other types of such simple and "this-worldly" things.

As recognized by Peterson, this is indeed a criticism that many scholars and philosophers do concerning **Buddhism**, which in at least some ways does emphasize instead the "negative" side of such "bodily desires," and moreover focuses a lot on "bodily problems" such as sickness, aging, and dying, much more than it often does on aspects such as the pleasures of sex, food, playing, and so on. In fact, despite the fact that some of the broader ideas of Buddhism are somewhat supported by scientific data, many of its more specific ideas have indeed more to do with an overall pessimistic view of earthly life, often reinforcing the typical taboos of agricultural states and institutionalized religions against sex, the naked body, and so on. This is an interesting point that seems to be often forgotten by many New Age people that idealize Buddhism, as well as by scholars such as **Edmund Wade Davis**. As explained by Malik in *The Quest for a Moral Compass*—a book that discusses and compares ideas from all around the globe—recently "there has been a tendency…especially in the West, to overplay the rational and humanistic quality of Buddhism…[but] at its core **Buddhism** is a doctrine of salvation." The **Buddha**, he explains, "did not view ethics as a means of building the good life on this Earth, but rather as a means of escaping the *bad life* of this Earth…his teachings embody an *intensely pessimistic* view of the world as a place of unremitting hurt and disappointment." That is, "suffering without end in a futile round of rebirths after rebirths – that is the fate of most mortals…escape comes through **nirvana.**"

So, more than "denying" the material world, I would say that in at least some aspects Buddhism has instead a very negative view of it. It is surely not a coincidence that one of the most important stories of Buddhism is that **Siddhartha** saw, when he left the palace, "the sufferings of life": "bodily" sickness, "bodily" aging, and "bodily" death. In a region with such a richness and diversity of natural beauty as the one where **Siddhartha** is said to have lived—near the actual border between southern Nepal and northern India—he could have focused on an endless number of positive aspects of the material world, such as the beauty of trees, the sound of rivers, the pleasures of sex, the laugher of kids playing in the street, the songs of birds, the blue tones of the sky, the endless number of forms of the clouds, and so on. But he focused instead obsessively on "the sufferings of life"—very similar to how fundamentalist Christians of medieval Europe argued that one should not focus on "this miserable life on Earth," but instead on heaven's paradise. Without entering into more details–because clearly this Section is not the place to provide a comprehensive history of Buddhism and its ideas–I would simply do the following statement. As we have seen in the above Chapters, most teleological narratives created in agricultural societies are not focused at all on the beauty of non-human life nor on the "bodily needs and desires" of humans, because such societies were mainly built on the basis of domesticating, using, and abusing non-human animals and of using imaginary narratives to separate them from our "special" species. In this sense, the teleological narratives of Buddhism are indeed not "better" or "worse" than those of most other religions: they are basically very similar to, and often actually even more negative than, many of them. Another piece of evidence showing that not only Buddhism, but also **Taoism**, just follow longstanding trends related to our obsession with the inevitability of death and thus the construction of and believe in stories of human specialness, concerns Buddhism's tales about "human rebirth" and a "*special place* of humans," and the fact that Taoism also "suggests some form of *human exceptionality*," as recognized by Peterson.

Box 9.6: 'Self-Illusion', Sociopathy, Schizophrenia, and Mental Disorders

As explained in Edwardes' 2019 book *The origins of self – an anthropological perspective*, the '**self-illusion**' that we create seems to be at least partially linked to **language cognition**, in the sense that we apparently need to have certain adequate linguistic tools to be able to think about our life experiences as an **evolving narrative**, that is to build our '**narrative self**':

> There is no negotiation toward meaning for **sociopaths** because there is only one relevant meaning to the universe – that of the undifferentiated self. And criticism does not matter, because there is no self-model to criticize, only a projected model or appearance that can be adjusted to meet current needs. If someone does not like the self you are projecting, project a different self. To quote **Groucho Marx**, 'Those are my principles, and if you don't like them…well, I have others'. It is possible to surrender to the multiple selves by losing the capacity to control which single self-model dominates at any one time. This is the problem that **schizophrenics** face, with

the different self-models competing in the **conscious mind**, rather than being policed by the **subconscious mind** and presented to the conscious mind one at a time. In this case, my inconsistencies do not represent a failure or fraying of the integrated model of my selfhood, they really are the fraying of my **selfhood**. This collapsing selfhood causes the boundaries between actuality, reality and virtuality to blur even more than usual, hence typical schizophrenic symptoms include **hallucinations** and **delusions**. The lack of a cohesive self-model also affects the self that the person can project, causing breakdowns in their social relationships. **Schizophrenia** is a particularly interesting 'solution' to the **manyselves dilemma**, because it seems to be a by-product of having language: the condition has been linked to language in several ways. First, **schizophrenia** is linked to **dysphasia**, or the **loss of communicative competency**; and it also seems to affect **phonology**, leading to flat-toned speech.

Second, schizophrenia has been shown to be implicated in the language and **social-modelling areas of the brain**. Radanovic et al. (in 2013) discovered a link between formal **thought disorder** (a diagnostic criterion for schizophrenia) and **language impairment**. The severity of both impairments was correlated with deficits in the **left superior temporal gyrus** and the **left planum temporale**, both areas in a Statistically Standard Brain (SSB) implicated in language; and in the **orbitofrontal cortex**, which is implicated in modelling for **decision-making**, including **social calculus modelling**. Pu et al. (in 2017) identified correlated deficits in the anterior part of the **temporal cortex**, the **ventro-lateral prefrontal cortex**, the **dorso-lateral prefrontal cortex** and **frontopolar cortex** areas of the brain – all areas implicated in both **social cognition** and language production. It seems that schizophrenia is somehow involved in the neural connections between social cognition and language cognition..

Basically, the *Narrative self* is the model we make of our life experiences as an evolving story – a stitching-together of the various **Episodic selves** in such a way that they can be viewed as aspects of a single self. Where the Episodic selves, being self-models, are differentiated (a series of models instead of an integrated single model), the Narrative self is an integrated meta-model. As the Narrative self is a product of the migration of selfhood, from the Social self through the self-model and then the Episodic self, *it is more virtual than real*; and yet it is the self we most often call on to define our me-ness. What is it that makes this self so attractive as a model of me? The answer appears to be that the Narrative self provides the individual with a sense of unity and purpose. It establishes the two cognitive concepts that having a self is supposed to enact: the concept of the single me and the concept of the continuous me. Although we cannot know it from the inside, this is what the Actual self seems to be from the outside: an entity delimited in both space and time; but, within those limits, a single integrated entity…[For instance] when it comes to efforts to recreate a humanlike experience in machine form, scientists working in artificial intelligence have a clear understanding of the need for a Narrative self. For Pointeau and Dominey, the Narrative self is a necessary tool for sharing plans, and it allows individuals to negotiate toward meaning in joint enterprises. These authors equipped their ***iCub* robot** with an AutoBiographical Memory (ABM), which is a simulation of a Narrative self, and showed that, when the ABM is linked to language, plans and activities could be negotiated between the robot and the trainer.

With the discussions above, we come full circle to a key point made at the beginning of the present book: they show how the type of accounts about different cultures and about "humanity" as a whole, such as that presented in Peterson's book and, very frequently, in social sciences and humanities in general, often fail to recognize the importance of biological factors and causes and our evolution as a whole.

For instance, Peterson argues that certain Asian traditions have certain specific traits A or B that are not so commonly seen in let's say Christian or Jewish or Muslim transitions. However, while this is surely true concerning such specific traits, regarding broader subjects such as those concerning the way they see bodily needs and desires, these latter three traditions are very similar to Asian traditions such as **Buddhism**, as they are about issues such as "human specialness" in Asian traditions such as **Taoism**, or about the use of misogynistic tales in Asian traditions such as **Confucianism**. In fact, in all regions where agriculture has been used, be it in Asia, the Americas—for instance by Mayans—Africa—for instance by Bantus—and the Middle East, there was an increase in inequity, slavery, oppression, and misogyny as compared to nomadic hunter-gatherer societies, for the reasons explained in Chaps. 4 and 5, many of them being indeed biological ones.

That is, the type of **biosocial organization** that early agricultural societies tended to adopt, in great part related to biological aspects such as the plantation of crops, agglomeration of domesticated animals, having those crops and animals as "private property," and so on, ultimately lead to the subjugation of women. Exactly in the same way that the biosocial organization of herbivorous **gorillas** and their behavioral choices in African dense forests lead to a scenario where alpha male gorillas tend to dominate female gorillas, while this did not happen within the type of social organization of **bonobos**. This also applies to **warfare**, or **genocides**, or **slavery**, contrary to many stories that one reads over and over in the media and in a substantial number of books and even specialized papers published by authors from the social sciences and humanities, arguing for instance that Asian traditions are generally "more peaceful" than European ones, or than African ones, or vice versa. These researchers forget that not everything is about the "mind," or even culture: a person alone cannot create a system of slavery, or fight a war, or oppress millions of women: clearly, Hitler could not have done the **Holocaust** alone, no matter how mentally disturbed or racist he was. He needed other people—a whole **bio-socio-political system**—to do so: so, if humans were, biologically, not a highly social animal, derived from African apes that were also social, **Hitler** would never had planned and be able to undertake the killing of millions of other humans.

For example, regarding fairytales about Asians being in general less "violent," the empirical reality is that some of the major atrocities, as those done in Cambodia under **Pol Pot's regime** and currently in **North Korea**, and the one leading to one of the highest—if not *the* highest—loss of human lives in a shorter period of time in human history—**Mao's regime**, with just the famines created by it killing up to 55 million people according to some studies—happened in Asia. And right now, in Myanmar, a country in which not only most people are Buddhist but also in which **Buddhism** has an enormous influence within the bio-socio-political system, there are Buddhist monks openly defending the mass killings of the **Muslim Rohingya** minority. How do they justify this? Exactly with the very same type of teleological narratives that have been so prevalent in agricultural societies and states everywhere else: for example, about how "special" their ingroup is versus how "horrible" and "immoral" the "others"—in this case, **Muslims**—are, or how such a mass killing is "meant to be," and so on. Simply, no group of humans, or of other organisms, is just "mentally good" or "bad," or "moral" or "immoral," everything depends on the *specific circumstances*, which are related to a huge number of factors and their interactions, most of them being related directly to biological and evolutionary features. Plainly, both the **biological determinists** that argue that doing A or B is "just our nature," as if we were mere robots pre-programmed to do A at time X and B at time Y, and the **social constructivists** that argue that is just "nurture," as if every child was born in a *tabula rasa*, are wrong.

So, let's continue analyzing Peterson's *Being human*, as a case study for discussing such topics, which are key within the context of this chapter. Peterson provides some examples to describe what she defines as "**Native American worldviews**": she refers, for instance, to an agricultural society, the **Navajo**, and to a society that has consisted mainly of hunter-gatherers, the Alaskan **Koyukon** (Box 9.7). A major argument of Peterson's book is that "native" people that lived a long time in a same place tend to feel more deeply connected to that place and thus to respect more its ecology. Despite obviously recognizing differences between these two groups, she therefore does amalgam them in that broader definition of a "Native American worldview." The fact that she does so is probably also partially related to the fact that Peterson, writing her book in 2001, was not as aware as we are nowadays, based on a huge compilation of empirical case studies and comparative data, that the major difference—in general, there are of course exceptions—regarding the subjugation and mistreatment of other animals, as well as of women and "others," arose mainly within the transitions from small nomadic to large sedentary, and in particular to agricultural, societies. That is, I am not challenging her argument that there are some "shared themes in Native American traditions." For instance, both the Navajo and the Koyukon have teleological narratives stating that humans are special. As she notes, "according to Koyukon tradition, humans are distinctive in their possession of a soul that differs from animal spirits."

What I am saying is that, as explained above, most aspects of the so-called culture—or "nurture"—of a certain group are in reality mainly related to specific circumstances that have a lot to do with the bio-socio-political organization, and again one of the major differences in that regard concerns being an agricultural society, such as the **Navajo** versus being an hunter-gatherer society, such as the Alaskan **Koyukon**. That is, one should be extremely careful to amalgam the "others" into a same group, just because they are "Native Americans," saying that there is a "Native American worldview," because while one might be doing this to go against racism towards those groups, such simplistic views often actually provide easy ammunition to racists, by defining those "others" as a mainly homogeneous group. Native Americans have created hugely dissimilar societies by themselves, from completely different small nomadic ones to very diverse large sedentary hunter-gatherer ones, to huge agricultural ones, such as the Mayans. So, while we might eventually find one, or a few, traits that eventually are truly shared by *all* Native American groups, it seems indeed extremely difficult that there is truly a single, homogenous, compact "Native American worldview," as racist people such as Columbus argued centuries ago.

Box 9.7: 'Native American Traditions', Animism, Anthropomorphism, and Ecology

In *Being human*, Peterson writes:

> The **Koyukon people** live in the northwestern interior of Alaska. Their language is part of a family called **Athapaskan**, widespread in many variants throughout northwestern North America and even, as a result of long-ago migrations, found in parts of the southwestern United States. Athapaskan people have been in the region now occupied by the Koyukon for at least a thousand years. The Koyukon first came into contact with Europeans – in their case Russians – in 1838, although Western items such as iron pots and tobacco and Western diseases such as smallpox had already reached the interior Koyukon lands as a result of trade with coastal **Inuit (Eskimo) people**. Most Koyukon people today still subsist, as they have for centuries, primarily by hunting, especially large animals like caribou, moose, and bear and, to a lesser extent, by fishing. The Koyukon also trap furbearing animals, mainly beaver but also rabbits, foxes, wolves, and wolverines. Richard Nelson...finds in the traditional Koyukon way of life and ideas about nature and selfhood a powerful resource, or at least inspiration, for environmental ethics: "Traditional Koyukon people live in a world that watches, in a forest of eyes...a person moving through nature – however wild, remote, even desolate the place may be – is never truly alone...the surroundings are aware, sensate, personified...they feel...they can be offended...and they must, at every moment, be treated with proper respect...all things in nature have a special kind of life, something unknown to contemporary Euro-Americans, something powerful." This view of the natural world as personified, conscious, and demanding of human respect is grounded on a close relationship not only between human and nonhuman realms but also between the natural and the **supernatural**. This relationship means that just as human actions affect the natural world, so events in nature are shaped by **spiritual forces**. These rules, the Koyukon believe, stem from the relations between humans and other beings in the "Distant Time." During that time, the subject of most of **Koyukon oral history**, "the animals were human," meaning they had human form, lived in human society, and spoke human language...some of these protohumans died and became animal or, in a few cases, plant beings, still possessing "a residue of human qualities and personality traits." Thus the Koyukon identify a host of shared characteristics among species, primarily behavioral, which are not those usually emphasized by Western taxonomists. For example, ravens' wit, ego, genius, love of play, and guile make them similar to humans; and like people, wolves cooperate in pursuit of prey and share the spoils.
>
> In this context of shared characteristics, common origins, and constant mutual influence, the distinction between humans and other animals lacks the clarity and absoluteness it possesses in dominant Western models. Further, explains Nelson, the similarity between humans and other species "derives not so much from the animal nature of humans as from the human nature of animals": their emotions, personalities, and capacity for communication both among themselves and with humans. This view contrasts with the dominant Western perspective, which sees any attribution of shared traits between humans and other animals as **anthropomorphism**, the extension of essentially human qualities to "lesser" beings. For the Koyukon, shared qualities of both human and nonhuman animals stem from shared origins. Such qualities are not the original or exclusive property of humans. According to Koyukon tradition, *humans are distinctive in their possession of a soul that differs from animal spirits*. However, humans' role is not to dominate but rather to serve a natural universe that is nearly omnipotent and demands propitiation for humans to survive. Because spiritual power is everywhere in nature, a moral system must guide human

behavior toward the nonhuman world. They also prohibit caging or otherwise harming or humiliating animals for human amusement. All these specific rules harmonize with what Nelson identifies, in a later work, as the Koyukon's "fundamental canons of restraint, humility, and respect toward the natural world." As a direct result of their conservationist practices, and thus as a result of their ethic, even today the Koyukon environment functions much as it would if no humans inhabited it, Nelson contends. While the Koyukon have not left the north woods completely unaltered, few outsiders can perceive the transitions from utilized to unutilized environment. This contrasts strikingly with Western ideas about the inevitable consequences of human habitation of wilderness. As Nelson writes, "The fact that Westerners identify this remote country as wilderness reflects their inability to conceive of occupying and utilizing an environment without fundamentally altering its natural state…but the Koyukon and their ancestors have done precisely this over a protracted span of time."

The **Navajo** or **Dine** (or **Dineh**) people constitute a much larger population than the Koyukon – in fact, theirs is the largest native nation in the contiguous United States. Most of the approximately two hundred thousand Navajo people in the United States today live within a twenty seven-thousand-square-mile reservation spanning parts of Arizona, New Mexico, and Utah. This is part of an ancestral homeland at which they probably arrived between 1300 and 1500 c.e., a remnant group of Athapaskan hunters (thus connected, remotely, to Koyukon culture). Through stories, rituals, and other aspects of everyday life, as Schwarz explains, "Navajo people are taught from earliest childhood to consider phenomena such as the earth, sky, sun, moon, rain, water, lightning, and thunder to be living kin." Thus Navajo stories…that "the land and everything on it are alive, permeated by a life essence of mingled air, light, and moisture…each 'thing' in this system, including landforms, individual plants, animals, natural forces, and so forth, has its own immortal humanlike 'inner form' (Holy Person), which interacts with the other 'things' in the system…there are also other Holy People, who move freely about the earth and sky." [Regarding the] shared themes in Native American traditions…"the notion that nature is somewhere over there while humanity is over here, or that a ***great hierarchical ladder*** exists…is antithetical to tribal thought…the **American Indian** sees all creatures as relatives (and in tribal systems relationship is central), as offspring of the Great Mystery, as cocreators, as children of our mother, and as necessary parts of an ordered, balanced, and livingwhole".

For native peoples such as the **Huaorani** the fluidity of species boundaries does not meant that they view humans as being identical to other animals. Rather, these cultures recognize differences but interpret and evaluate them differently than do dominant European cultures. This is often difficult for non-native peoples to understand: Western cultures generally insist that the difference between humans and other species must be absolute and, further, must posit **human superiority**. Attempts by some to understand species differences in other ways often draw charges that they have erased the idea of difference altogether. Such accusations may reflect, more than anything, the narrowness of established Western models of thinking about both personhood and nonhuman animals. In contrast, many indigenous cultures acknowledge that difference exists – not only between humans and other species but also among nonhuman species – without automatically positing superiority and without perceiving shared features as threats. Perhaps long-term shared inhabitation of a particular ecosystem, along with the mutual dependence and the knowledge that it engenders, enables natives to perceive a continuum of shared traits rather than opposition. These common qualities are expressed in many cultures in stories of common origins, human-animal communication, and animal-human metamorphosis. In this light, qualities like social organization, consciousness, spirit, or personhood do not become less desirable or less fitting for humans just because they are evidently shared with other species.

The fact that while authors such as Peterson are trying to contradict Hobbes' ideas about "savages" by defending "Native American traditions" might indirectly actually contribute to reinforce racist dichotomies that have been used to divide "us"—Westerners—and "them"—Native Americans—is made clear, when she cites a statement by another author, Nelson. The statement is: "even today the Koyukon environment functions much as it would if no humans inhabited it…this contrasts strikingly with Western ideas about the inevitable consequences of human habitation of wilderness…the fact that Westerners identify this remote country as wilderness reflects their inability to conceive of occupying and utilizing an environment without fundamentally altering its natural state…but the Koyukon and their ancestors have done precisely this over a protracted span of time." In fact, the true difference here is not between "Westerners" and "Native Americans," but between a small hunter-gatherer society such as the Koyukon, and agricultural societies in general, because obviously no agricultural society can keep the environment as "it would if no humans inhabited it." Saying that the Navajo and Koyukon, or the African Baka "forest pygmies" and agricultural Bantu, are different from Western agricultural societies because of their "ecological practices" makes no sense at all: agriculture is in itself, by definition, an extreme form of using and abusing, and literally modifying, a huge number of plants and animals.

As emphasized in the present volume, scholars such as Pinker and Dawkins defend the huge importance of **humanism**, which is profoundly related to teleological narratives about cosmic "purpose" or "duty" and also about "cosmic progress," such as the notion of a "ladder of life." The fact that humanism is deeply associated to such narratives about "progress" often creates a paradox for **humanists** such as Peterson that try to extol non-Western societies, including nomadic hunter-gatherer ones, as astutely pointed out by Harari in *Homo Deus*. For instance, although Peterson does not agree with the notion of a "linear, progressive ladder of life" and praises the Navajo and Koyukon for also not seeing the world in that way, in some aspects of *Being human* she does employ a humanistic philosophy of "progress," although in a different way than what is often done by Pinker. For example, apart from referring a lot to teleological terms such as "our common duty," she writes the following about the Koyukon people, whom, as explained above, have teleological narratives affirming that humans are "special" and different from the "rest of nature," as all living societies do:

> Western observers have often called these cultures and their ideas 'animistic', singling out their attribution of agency to nonhuman others as a defining and inferior ('primitive') feature. Perhaps it is possible to recover the term animism, to define it not as a derogatory term indicative of a foolish, naive, and even sinful misreading of the natural world but rather as descriptive of a careful and *realistic reading*. If many Westerners find it difficult to take seriously indigenous perspectives such as **animism**, taking nonhumans seriously seems nearly impossible.

I completely agree with Paterson in that Westerns should not consider that their imaginary teleological narratives are 'superior', or less 'primitive', than the ones of the Koyukon people. My reason for this is that, empirically, all such narratives are equally factually inaccurate. Therefore, there is a major problem with the second part of Peterson's argument, according to which we could see such animistic stories as a "realistic reading." As explained above, many humanists now fall exactly into the same trap, as **Edmund Wade Davis** does: by wanting to praise, and even to defend the life and existence of non-Western groups—something I completely agree with, and hope I am contributing to, with the present book—they then defend exactly the same type of fairytales that have justified the colonization, subjugation and oppression of those groups and put them at risk of extinction, in the first place. Two wrongs don't make a right: just because Westerns oppressed and devastated those groups and justified those horrible acts by saying that Western Gods and imaginary stories were the "real ones," one should not now go to the other extreme and say that the deities and fictional stories of non-Westerns, being them Native Americans or Asians, are the "real ones."

None of them are real: they are all fictional *Neverland* stories that can, within certain specific circumstances, be extremely dangerous, as indeed were the Mayan myths that were used to justify the sacrifice a huge number of humans centuries ago (Fig. 9.8) and are the just-so-stories used by Buddhist monks to openly defend the mass killings of the **Muslim Rohingya**, nowadays. Having millions of people firmly and blindly believing in any imaginary story, be it Western, Native American, Asian or African, is always hugely dangerous because, to paraphrase Voltaire, "those who can make you believe absurdities, can make you commit atrocities." In a nutshell, the fact that nomadic hunter-gatherers don't undertake genocides, don't have slaves, don't have huge wars, and don't destroy the whole Amazonian ecosystem is principally not related to their imaginary stories. Instead, it is simply because, *despite* of those imaginary stories, those small groups don't have enough individuals to form the type of socio-political organization that has the needed bureaucratic capacity, division of labor, and population density to effectively use those stories to justify, promote and convince others to do such atrocities, or to force others to do them. There is nothing mysterious about this: this happens to *all* small nomadic groups of any other social mammals, such as wolves or chimpanzees, and that is precisely why none of those groups does such atrocities, not because wolves or chimpanzees have "noble" imaginary stories. No "good" imaginary stories, no "bad" imaginary, no fairytales: *it is just what it is*.

This is actually, in a way, recognized by Patterson, when she states that "misguided efforts to force a certain goal, or a deluded conviction that a particular society will realize utopia, can lead to intolerance and even genocide…in religious terms, this reflects the arrogance of thinking that a given project has a divine mandate, that one is doing God's will." Still, she then argues that "despite the risk," we do need a "new story that will heal, guide, and *discipline us,*" something that sounds—in face of what we know about human history and have seen in the present book—very risky indeed, because an endless number of previous atrocities have precisely been justified by imaginary narratives that had the aim to "guide and discipline." Therefore, creating and believing even in more, and newer, teleological narratives does not seem to be the solution. My question is: why don't we try to, let's say, keep the balance of Earth's ecosystems simply by arguing that those ecosystems and life on Earth in general are fascinating per se, instead of having to lie to ourselves and to others with "newer" imaginary stories saying that we have a "cosmic duty to so" or about fictional deities such as the **Pachamama**? Apes, for instance, are mesmerizing animals, and fighting against deforestation not only will help keeping such mesmerizing animals alive but also the ecosystems of the forests where they live in balance. Is not about "them" versus "us," the "special" humans "saving" the 'poor apes' because we have an imaginary 'cosmic duty' to do so, it is about *all* life in the planet being fascinating. But for some reason, authors such as Peterson seem to think that we can never ever go out of *Neverland*, so instead they want to defend "others," or "apes," by just creating newer and more intense Neverland fairytales. Indeed, she clearly states that we *should* build newer narratives in order to accomplish our common "duty" to "make a better world," and that such narratives should be constructed from an amalgam of not only "Native American traditions" and "Asian traditions," but also of older Western Christian narratives, as well as other cultures/traditions. In the last part of her book she does a very muscular defense of **teleology** and **teleological narratives** in general and of the urgent need for **new utopian narratives** in particular:

> Some Christian thinkers echo [Edward] Wilson's vision of evolution as a ***sacred narrative***. Most influential, perhaps, is Thomas Berry's insistence that ethics requires a narrative and that the narrative most suited for our situation is evolution: "It's all a question of *story*. We are in trouble just now because we do not have a good story. We are in between stories. The old story, the account of how the world came to be and how we fit into it, is no longer effective. Yet we have not learned the *new story*." Traditional **Christian and Jewish narratives** provided a meaningful framework in the West for a long period, Berry explains, but the biblical story is now "dysfunctional in its larger social dimensions." We require a different narrative that offers the meaning and guidance that religion previously supplied. For this "we must begin where everything begins in human affairs – with the basic story, our narrative of how things came to be, how they came to be as they are, and how the future can be given some satisfying direction. We need a story that will educate us, a story that will heal, guide, and *discipline us*." Darwinian evolution generates a very powerful narrative. It has the potential to describe our origins, as well as the stages through which we have passed and the operating principles that brought us to where we are now. It also tells us that change is continuous and how change works. However, it does not, cannot, tell us where the changes *will lead*. Natural selection is not goal oriented, and viewing evolution as moving toward ever-higher goods seriously distorts scientific understandings of evolution. In this sense, the evolutionary epic differs decisively from the narratives of **Christianity**, **Marxism**, and other *teleological models*, which describe not only where we came from and how history works but also where we will end up, at least if we act as we ought. The ending helps tie together the beginning and middle of the story and provides its ethical efficacy. A projected ending, a vision of where we can and *ought to wind up*, is one of the most powerful motivators of ethical practice, especially in the face of risks and uncertainties. This is evident, for example, in stewardship ethics, which conceptualize proper care of nature as part of the story of humanity's relationship with God, destined to end with redemption. As Jim Cheney argues, "To contextualize ethical deliberation is, in some sense, to provide a narrative or story, from which the solution to the ethical dilemma emerges as a fitting conclusion." The absence of a clear picture of the future in the evolutionary epic weakens evolution's capacity to provide a motivational, as well as cognitive, basis for a lived ethic.
>
> For some thinkers, however, the **nonteleological nature of evolution**'s epic makes it a better possibility for grounding a **moral vision**. *Utopian narratives*, such as those of Christianity and Marxism, can lead to dangerous consequences, as postmodernist critics in particular have emphasized. (The fact that stories require endings, and big stories require

big endings, may be a primary reason that postmodernism rejects metanarratives so decisively.) Critics point to several problems with **utopias**. One problem is simply that they are implausible. At the beginning of the twenty-first century, how can any sensible person believe in the realization of an ideal human society? Quite plainly, we cannot get there from here. However, this critique is not really the most important one for ethics, since utopian narratives often influence even those who know they are fantastical. Two other ethical and political problems of utopian thinking are especially relevant here. First, misguided efforts to force a certain goal, or a deluded conviction that a particular society will realize utopia, can lead to intolerance and even genocide. In religious terms, this reflects the arrogance of thinking that a given project has a divine mandate, that one is doing God's will. A second danger is the mirror image of this hubris, and it rests on the same foundations. This is the risk of falling into resignation and what Sharon Welch calls cultured despair, of thinking that since everything we might realize will fall short of perfection, then we should not try for anything. In rejecting utopias, contemporary critics relinquish not only the power to motivate repressive movements but also the capacity to inspire *progressive ones*. Images of a desirable future generate much of the compelling power of many religious and political movements. ***Utopian hopes*** can encourage people to transform their present lives so that they more closely resemble the better lives they imagine. Without such a vision of where change might lead, critique alone rarely motivates sustained effort to improve present conditions. Even when utopian visions lead to a sense of not belonging, of just passing through, that very alienation can sometimes provide a powerful political or ethical motive, if a desirable alternative is also presented. For example, images of the river Jordan and the promised land in **African American Christianity** have inspired generations of creative and persistent activism. Utopian dreams also provide compelling standards by which to evaluate concrete movements and projects. None may match the ideal, but some come closer than others. Put in other terms, religion's "already, not yet" carries more ethical and political force than **postmodernism**'s "always already".

Peterson then concludes: "while we need to take seriously the dangers of *utopian narratives*, we may not have to discard narratives and utopias entirely…*despite the risks*, without utopian or utopian hopes ethical and political action often lacks meaning and almost always lacks the enthusiasm necessary for long-term commitment…visions of how things *ought to be* provide a goal towards which transformative action can aim, without needing guarantees of reaching the goal." One of the most striking aspects of these excerpts, written by an otherwise very sophisticated, knowledgeable writer, in a book published in the twenty-first century, is that they are more plagued by teleological terms and ideas than many of the texts of scholars of **ancient Greece** such as **Socrates** or **Aristotle**, written more than two millennia ago. It is for instance remarkable to see Peterson's frustration with the fact that "evolution does not, cannot, tell us where the changes *will lead*…the absence of a clear picture of the future in the evolutionary epic weakens evolution's capacity to provide a motivational, as well as cognitive, basis for a lived ethic." Most humans continue to think that the *reality*—including the facts about biological evolution—does not make a "good story," a reminder of how difficult it will be to escape *Neverland*.

This is particularly so because teleological discourses resonate extremely well with a major portion of the general public, as illustrated by the case study discussed in Leeman's book *The Teleological Discourse of Barack Obama*. Obama's recurrent use of teleological ideas, both in his speeches and books, was very likely a crucial component within his two victories in the presidential elections. There is no doubt that humanists such as **Obama** and those that he mostly admires have contributed to outstanding accomplishments, such as the possibility for woman and African-Americans to vote, the decrease of child labor, the end of slavery in the West and, specifically concerning Obama, the landmark changes leading to what is nowadays often called the "**Obamacare.**" A clear reminder of such accomplishments, which at the same time are an emblematic example of the power and beauty of such **humanist teleological narratives**, but also of how such narratives are indeed so prevailing in 2021, was recently given by one of the most outstanding speeches and performances I have seen in my life, by **Amanda Gorman**. She is a 22-year-old African American who became the youngest person to ever read at a presidential inauguration. She read her poem "*The hill we climb*" at the inauguration of President Biden, to a crowd that precisely also included Obama. "Climbing" is a term commonly used in humanist teleological narratives, being often also used by Obama himself: it resembles the concept of *scala naturae*, and is actually related to it, because as noted above both the **Enlightenment movement** and the **humanist movement** are deeply related to the teleological notion of "cosmic progress." I will just refer here to a very short excerpt of Gorman's poem, to illustrate how it also includes several other notions and terms typical of teleological narratives, such as "unfinished"—as if there is a cosmic masterplan that we have the "duty" to fulfill—and of course, the most central concept within teleological tales, "*purpose*":

> Somehow we've weathered and witnessed a nation that it isn't broken, but simply unfinished. We, the successors of a country and the time where a skinny Black girl descended from slaves and raised by a single mother can dream of becoming president only to find herself reciting for one. And yes, we are far from polished, far from pristine, but that doesn't mean we are striving to form a union that is perfect. We are striving to forge our union with *purpose*.

9.4 Medicine, Wellbeing, Science, and Fake News

> *I am, somehow, less interested in the weight and convolutions of Einstein's brain than in the near certainty that people of equal talent have lived and died in cotton fields and sweatshops.*
> *(Stephen Jay Gould)*

Most people agree that **medicine** is probably one of the most altruistic things done by humans, in the sense that those undertaking medical practices don't often have a close or familial relationship with the patients they are treating. A very well known story illustrating this, that is often repeated in the social media, was told by anthropologist Margaret Mead, in an interview:

> When asked by a student about what she considered to be the first sign of civilization in a culture, the student expected Mead to talk about fishhooks or clay pots or grinding stones. But no. Mead said that the first sign of civilization in an ancient culture was a femur (thighbone) that had been broken and then healed. Mead explained that in the animal kingdom, if you break your leg, you die. You cannot run from danger, get to the river for a drink or hunt for food. You are meat for prowling beasts. No animal survives a broken leg long enough for the bone to heal. A broken femur that has healed is evidence that someone has taken

time to stay with the one who fell, has bound up the wound, has carried the person to safety and has tended the person through recovery. Helping someone else through difficulty is where civilization starts, Mead said.

Although I completely agree with its main point, and I appreciate its good intentions, I have to say that the last part is a bit too anthropomorphic and perhaps unnecessary. Because yes, medicine as defined by us is likely uniquely found in humans, but "helping" others clearly is not a feature that only applies to our lineage, and much less to human "civilizations." Many other animals, special social ones, help each other. A termite alone clearly cannot build a termite nest, and moreover there are empirical studies showing that some ants do also help other ants that were injured in "ant wars" (see above) and then take care of them. Even non-human organisms from *different species* have been reported to help each other. Furthermore, as everything in life, things are not white or black, and surely not "good" or "bad." For instance, concerning the altruist character of human medicine per se, one needs to point out that since early times those doing medical practices, or pretending to do them—that is, including shamans or charlatans—tend to be, on average, "better off" than the rest of the population. In fact, more recently, the rise of **capitalism** has led to a situation in which, in countries such as the U.S., the explosive mixture of the very **high salaries of physicians** and the clearly non-altruistic—sometimes actually the opposite, a truly vampire-like parasitic—character of most **pharmaceutical and insurance companies** lead to a situation in which the most poor, and even the most sick are often unable to have decent, or even any type, of day care. For example, if they don't have the money to pay **health insurance**, or if they do but the **insurance companies** atrociously use their "pre-existing conditions" to make it much more difficult for them to be covered, or reimbursed more fairly. That is why **Obamacare** was so important, as noted above, although obviously it did not change the whole **health system in the U.S.**, which is indeed extremely capitalistic and often very heartless, with a huge number of U.S. citizens dying from health complications that could be easily avoided if they had more money. This issue has been widely discussed in the U.S. as you probably know, so here I will provide just a short excerpt of an article entitled "*The problem of doctors' salaries,*" published in 2017 in the *Politico* website by Dean Baker, whom was then a co-director of the think tank *Center for Economic and Policy Research*:

The United States pays more than twice as much per person for **health care** as other wealthy countries. We tend to blame the high prices on things like drugs and medical equipment, in part because the price tag for many life-saving drugs is less than half the U.S. price in Canada or Europe. But an unavoidable part of the high cost of U.S. health care is how much we pay doctors – twice as much on average as physicians in other wealthy countries. Because our doctors are paid, on average, more than $250,000 a year (even after malpractice insurance and other expenses), and more than 900,000 doctors in the country, that means we pay an extra $100 billion a year in doctor salaries. That works out to more than $700 per U.S. household per year. We can think of this as a kind of doctors' tax. Doctors and other highly paid professionals stand out in this respect. Our autoworkers and retail clerks do not in general earn more than their counterparts in other wealthy countries. Most Americans are likely to be sympathetic to the idea that doctors should be well paid. After all, it takes many years of education and training, including long hours as an intern and resident, to become a doctor. And people generally respect and trust their doctors. But they likely don't realize how out of line our doctors' pay is with doctors in other wealthy countries. However, as an economist, I look for structural explanations for pay disparities like this. And when economists like me look at medicine in America – whether we lean left or right politically – we see something that looks an awful lot like a cartel. The word "cartel" has some bad connotations; most people's thoughts probably jump to OPEC and the 1970s crisis caused by its reduction in the supply of oil. But a cartel is not necessarily completely negative. It means that the suppliers of a good or service have control over the supply. This control can be used to ensure quality, as is the case with many agricultural cartels around the world. However, controlling supply also lets the cartel exert some control over price.

In the United States, the supply of doctors is tightly controlled by the number of **medical school** slots, and more importantly, the number of **medical residencies**. Those are both set by the Accreditation Council for Graduate Medical Education, a body dominated by physicians' organizations. The United States, unlike other countries, requires physicians to complete a U.S. residency program to practice. This means that U.S. doctors get to legally limit their competition. As a result, U.S. doctors receive higher pay, and like anyone in a position to exploit a cartel, they also get patients to buy services (i.e., from specialists) that they don't really need. There are two parts to the high pay received by our doctors relative to doctors elsewhere, both connected to the same cause. The first is that our doctors get higher pay in every category of medical practice, including general practitioner. If we compare our cardiologists to cardiologists in Europe or Canada, our heart doctors earn a substantial premium. The same is true of our neurologists, surgeons, and every other category of medical specialization. Even family practitioners clock in as earning more than $200,000 a year, enough to put them at the edge of the top 1 percent of wage earners in the country. The other reason that our physicians earn so much more is that roughly two-thirds are specialists. This contrasts with the situation in other countries, where roughly two-thirds of doctors are general practitioners. This means we are paying specialists' wages for many tasks that elsewhere are performed by general practitioners. Since there is little evidence of systematically better outcomes in the United States, the increased use of specialists does not appear to be driven by medical necessity.

In recent years, the number of medical residents has become so restricted that even the American Medical Association is pushing to have the number of slots increased. The major obstacle at this point is funding. It costs a teaching hospital roughly $150,000 a year for a residency slot. Most of the money comes from Medicare, with a lesser amount from Medicaid and other government sources. The number of slots supported by **Medicare** has been frozen for two decades after Congress lowered it in 1997 at the request of the **American Medical Association** and other doctors' organizations. Furthermore, Medicare exerts little control over the fields of specialization in the residency slots it supports, largely leaving this up to the teaching hospitals, which have an incentive to offer residencies in specialties from which they can get the most revenue per resident…. There are enormous obstacles to any effort to reduce the pay of doctors. The restrictions that limit competition and keep physicians' pay high are mostly obscure and not even understood by many policy wonks. Any efforts to change them in ways that seriously threaten doctors' pay will encounter massive opposition from a very powerful political lobby. Furthermore, doctors generally enjoy a great deal of respect in society, and Americans tend not to think of their high salaries as part of the health care cost problem. But if we want to stop paying a $100 billion premium for health care that doesn't make us healthier, we're going to need to overcome those political barriers. Getting U.S. health care costs down is a herculean task; getting doctors' pay in line is a big part of the solution. It's time we broke up the doctor cartel…. The fact that most people like their doctors will make the effort harder. Most of us like our letter carriers too, but that doesn't mean they should make $250,000 a year.

However, one needs to remember that in most "developed" countries—the situation is often very different from that of the U.S., not only because physicians tend to receive less money in average but also because the insurance and pharmaceutical companies have many more restrictions in terms of applying such vampire-like practices. Unfortunately, this is not recognized by most people in the U.S., in great part because its politicians—with some notable exceptions, such as **Barack Obama**, and more recently people such as **Bernie Sanders**—as many physicians and insurances and pharmaceutical companies don't want this to be part of a well-informed discussion within the broader public. But outside of the U.S. this is well known, and I can attest myself that in most European countries almost all people, including the poorest and surely the most sick, can have access to high quality medicine. That is, in such countries, medicine and the health system can indeed be considered as an example of the "best" and more "noble," things we can do as humans. To give just a current example, my father, who lives in Portugal, will do a colonoscopy in just 2 days, and the amount of money he will ultimately pay for that procedure will be a few Euros. In contrast, I did a colonoscopy a few months ago in the U.S. and despite the fact that I have a very good health insurance in comparison to most people in the U.S. because I teach at a College of Medicine—at Howard University—I still paid several hundreds of dollars from my own pocket. And without any health insurance, it would be much more, almost surely above 1000 dollars. The very same health practice, but very different political and economic contexts and thus enormously different prices, stressing that the main differences between people among different societies or even among different "developed" countries depend mainly on the *specific circumstances* and socio-political context (Fig. 9.9).

It is also important to note that contrary to Hobbesian narratives about "brutish savages," there are empirical case-studies showing that even without having basically any material goods and medical tools, some hunter-gatherers have provided, and continue to provide, care for the very sick and even for highly disabled people. For instance, an empirical work that seriously puts in question such narratives that before "civilizations" people did not care for their children, or the weak, or the disabled, was published recently—in 2020—by Halcrow and colleagues, with the title "*Care of infants in the past: bridging evolutionary anthropological and bioarchaeological approaches*":

> Western thought discerns **medical care** as sophisticated and complex…[However] there has been a recent surge in modelling the implications of **care provisioning for people with serious disabilities** in the bioarchaeological and paleopathological literature…[for instance, work by Tilley and colleagues] has provided essential consideration of social responses and compassion towards people with long term care needs from **disabilities** and other health-related conditions…. Tilley provides a multiple step system within her model of care to determine whether people in the past needed long-term health-related care and to interpret the nature of that care within a biosocial context. Presenting a case study of an individual with paleopathological evidence for **quadriplegia** in prehistoric Viet Nam, Tilley and Oxenham (2011) argue that the survival and good mental health of a person with a serious disability necessitated the provision of long-term, skilled and consistent care, likely involving multiple group members, including the allocation of food/a special diet, water, shelter, bedding, a hazard-free environment, help with eating and drinking, and man-

> aging hygiene (removal of wastes, bathing)…. [Similarly, regarding childcare] some paleo-anthropological studies in the Upper Paleolithic already explore aspects of **mortuary behavior** to ascertain information on social responses to **infant loss**, familial relationships, and social age. Recently, there has been an increased research interest in grief and emotion from the archaeological context, and part of this research is starting to consider community members' responses to **infant and fetal death**. [In the past] the purported marginalization of fetuses along with infants in the archaeological record, including location and simplified mortuary treatment has led some scholars to interpret that they were of little concern beyond immediate family members…[however] considering evidence for intense grief after miscarriage and infant death starts to challenge the notion that their loss was of little consequence.

The point that hunter-gatherers cared not only about the very sick, but even the dead, should have been obvious to everybody since a long time ago, because we have countless lines of evidence since decades ago showing that **mortuary behavior and rituals** date back hundreds of thousands years ago, much before the rise of agriculture (Chap. 2). If by then people spent the time, effort, and often even the very few material values that they had to undertake **burials**, putting dead bodies—of people that were likely sick and weak for a substantial amount of time before dying –in certain specific positions, with certain objects, and/or with certain types of decorations, this obviously means that they cared a lot about sick, weak, very old, and dead people. Tilley and Oxenham, in the 2011 paper that was cited in the excerpt above, also made it clear that the fact that many people—including some scholars—have not acknowledged this obvious fact for so long is related to biases that led people to *overlook the available data* or, in some cases, to even *actively ignore those data* when they became aware of them:

> Archaeology has *overlooked* prehistoric health care provision as a specific focus for analysis, and a valuable source of information on past behaviours is being *ignored*. A 'bioarchaeology of care' analytical framework…illustrated through application to the case of M9, is being developed to address this deficit. Between 3700-4000 years ago in northern Vietnam a young man survived for approximately 10 years with disabilities so severe he would have been dependent on assistance from others for every aspect of daily life. Paralysed from the waist down and with at best very limited upper body mobility, the skeletal remains of *Man Bac Burial 9* (M9) provide evidence of a pathological condition difficult to manage successfully in a modern medical environment. In a subsistence Neolithic economy the challenges to health maintenance and quality of life would have been enormous, yet M9 lived with minimally **paraplegia** and maximally **quadriplegia** from childhood into his third decade. M9's survival reflects *high quality, continuous and time-consuming care within a technologically unsophisticated prehistoric community*…[This work and] the perspective it offers should enrich more general analyses [about prehistoric health care]…[for instance] one other instance of survival with such extensive paralysis is known; that of an individual from Hokkaido, dating to around, 3500 years before present.

However, while it is obvious that hunter-gatherers do/did take care of the sick, disabled, and even the dead, it is obvious that they simply did not have the needed resources to treat people in coma, or the needed pacemakers, and so on, as we have nowadays. In that sense, some "developed" countries arose to a situation in which they go to a level of care for people that are extremely sick—either "physically" or "mentally"—or disabled, that has no parallel in the rest of the natural world. For instance, while the seemingly increasing occurrence, in such countries, of a huge number of kids with very problematic conditions such as cancer and immune diseases might well be due precisely to some aspects of "modern life," as seen above, the reality is that kids with any type of harsh medical conditions would not receive such a level of care, and surely not have the same survival rates, within hunter-gatherer societies. This applies as well to older people, which in hunter-gatherer societies tend to be healthier and be more respected and integrated in the broader community as noted above, but clearly do not receive the same amount of medical care. I don't think that anyone, even the most anti-Hobbesian or "anti-progress," or more extremist New Age people, would argue that the existence of machines and medical procedures that help millions of people of "developed" countries that would otherwise not be able to be cured, or to move, or to survive, is a "bad" thing, *as long as they do want to survive*, as emphasized above. This also applies to people with so-called mental disabilities, such as Down syndrome for instance, whom can now have a much higher quality of life in many "developed" countries than they ever had. Several of them go to schools, with a massive amount of people being involved in the establishment of either special schools or of special conditions that allow them to be included in schools frequented by the broader community. And many have jobs that allow them to receive their own money and thus to be a bit more independent, for example (Fig. 9.9).

In summary, while it is true that the increase of life expectancy in "developed" countries does not necessarily mean that there was a general increase of quality of life *for most people*—very likely it did not, in many cases, as explained above—there is no doubt that for those people that were born with, or that developed, very serious diseases and incapacities, they never had the change to have such a quality of life as they can have in "developed" countries nowadays. As noted in Sect. 8.5, humans from such countries have lower initial rates of mortality than any other animal studied in detail so far: just 0.02%, compared to 0.2% for African elephants, 2% for dogs, 1% for female mice, 7% for some fish species. So this is indeed an amazing accomplishment of *science and medicine*, despite of all the very damaging and dangerous *Neverland* imaginary stories that humans—including scientists and physicians—have built and believe in. The problem is, of course, that the natural world is made of trade-offs, and we are now paying a huge price for this, in terms of overpopulation and the carnage of the planet's resources, as also explained in that Section. In fact, the pattern discussed just above applies not only to people with health problems or disabilities, but also to people that have their material possessions, or their lives in general, put at risk by natural events such as cyclones—which, on the other hand, might be part of a trade-off of some aspects of our "modern" lives, such as those leading to global warming—or earthquakes or tsunamis: such people are also often helped in many "developed" countries. Even people that decide to risk their own lives for the sake of it, such as those that try to climb the Himalayas or the Alps, will normally be helped, if they face major problems and their life is at risk, often involving huge and very expensive operations that include the use of helicopters and so on. In many countries, such rescue missions are paid by the governments and, consequently, ultimately by all of us, the taxpayers, and still people accept this: to pay some money to help particularly irresponsible people in moments of need.

A different point about **medicine** that is also important for the context of the present book and that concerns particularly what was discussed in the previous section, is that it contradicts the view of authors such as Peterson, who in her book *Being Human* defended—as many other scholars, politicians, and philosophers now do—that we need a "new teleological narrative to replace the old ones." Because that is not, and has never truly been, the case with medicine. As shown in this book, most of the great achievements done in medicine were produced by *removing*, or *after* we *removed*, teleology from it. For instance, when scientists and physicians started to no longer consider that diseases were the result of a punishment by God or of witchcraft, or part of a 'cosmic purpose', or an omen. They did not replace all these imaginary tales with yet another imaginary teleological stories. Well on the contrary: nowadays the major dangers faced by medicine and its outstanding contributions, such as vaccines, are precisely the return of **teleological narratives** and **conspiracy theories** as those now so widely disseminated for instance by **anti-vaxers**, or of **falsehoods** about "**miraculous oils**" or "**healing crystal materials**" also now so in vogue again. While I was writing this book, a fascinating *Nature* paper about these topics and the **Covid-19 pandemic**, published by Johnson and colleagues and entitled "*The online competition between pro- and anti-vaccination news*," presented a series of results that are deeply troubling (see Fig. 9.18). As summarized in their abstract, "distrust in scientific expertise is dangerous…opposition to vaccination with a future vaccine against **Sars-CoV-2**, the causal agent of **Covid-19**, for example, could amplify outbreaks, as happened for measles in **2019**…homemade remedies and falsehoods are being shared widely on the Internet, as well as dismissals of expert advice." They analyzed a global pool of around three billion **Facebook** users and revealed a "multi-sided landscape of unprecedented intricacy that involves nearly 100 million individuals partitioned into highly dynamic, interconnected clusters across cities, countries, continents and languages…although smaller in overall size…[such] **anti-vaccination** clusters manage to become highly entangled with undecided clusters in the main online network, whereas pro-vaccination clusters are more peripheral" (Fig. 9.18). Importantly, their theoretical framework "reproduces the recent explosive growth in **anti-vaccination views**, and predicts that *these views will dominate in a decade*." In other words, they predict that by 2030 most undecided people that will try to get information about vaccines within the social media will very likely follow/believe in the **anti-vax narratives**. This can put in danger not only those people and their families and in particular their children, but the whole population in cases of infectious diseases such as measles or Covid-19. In fact, it takes a single, badly-informed, *Neverland* selfish mother or father that decides that a kid will not be vaccinated for, let's say, measles, to put in risk a lot of kids in the school attended by that kid, and the kids that know those kids, particularly those with debilitating medical conditions. This is precisely why an increasing number of outbreaks of infectious diseases such as measles, which were almost unheard off in "developed" countries such as Germany, just a few years ago, have been happening in such countries in recent years.

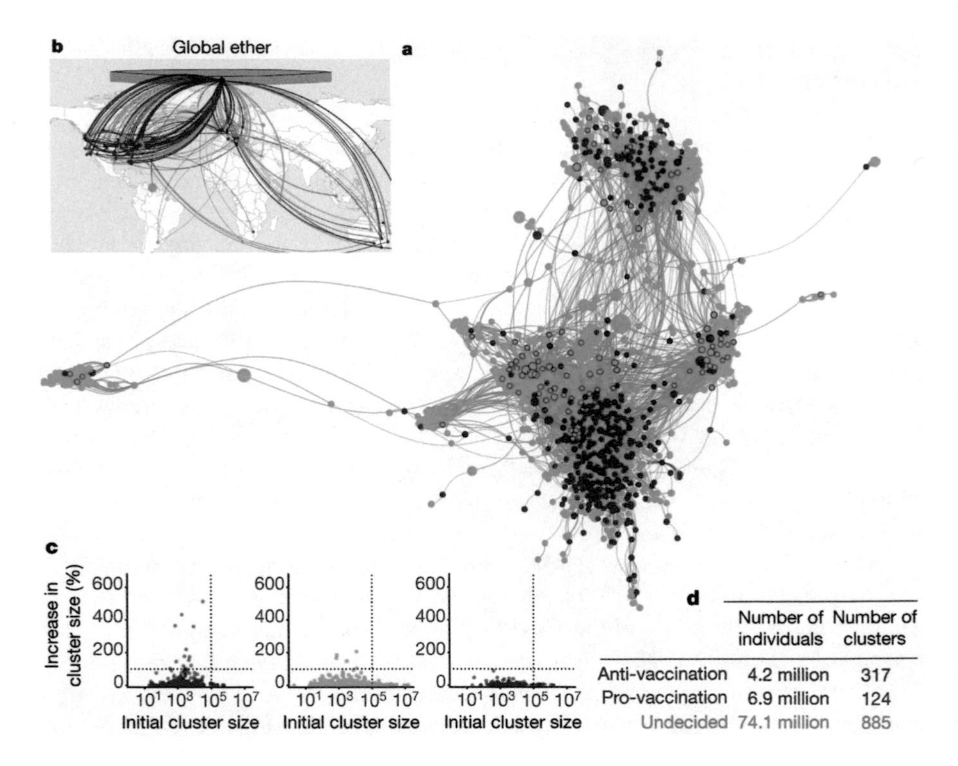

Fig. 9.18 A truly disturbing picture, and the enormous price we can pay for this in the near future: paper published in 2020 in *Nature* by Johnson and colleagues predicting that in 2030 most undecided people that will try to get information about vaccines in the social media will likely follow anti-vax narratives, thus putting in danger not only their relative ones but the whole population in general. (**a**) Snapshot from 15 October 2019 of the connected component in the complex ecology of undecided (green), anti-vaccination (red) and pro-vaccination (blue) views comprising nearly 100 million individuals in clusters (pages) associated with the vaccine topic on Facebook; the color segregation is an emergent effect (that is, not imposed); cluster sizes are determined by the number of members of the Facebook page; black rings show clusters with more than 50% out-link growth; each link between nodes has the color of the source node. (**b**) Global spread of *figure a* for a small number of clusters; the "global ether" represents clusters that remain global (grey). (**c**) Anti-vaccination clusters have a stronger growth in cluster size; each colored dot is a node; data are from February–October 2019. (**d**) Anti-vaccination individuals are an overall numerical minority compared with pro-vaccination individuals; however, anti-vaccination individuals form more separate clusters

As explained by Johnson and colleagues, one of the major reasons leading to such a troublesome current scenario is that "anti-vaccination clusters offer a wide range of potentially *attractive narratives* that blend topics such as safety concerns, conspiracy theories, and alternative health and medicine, and also now the cause and cure of the Covid-19 virus…this *diversity in the anti-vaccination narratives* is consistent with other reports in the literature…by contrast, pro-vaccination views are far more monothematic." So, should we do as suggested by Peterson and try instead to "build an attractive narrative" to counteract those narratives that are being spread by the anti-vax movement? I clearly don't think we should, because this would be like counteracting medieval imaginary narratives about witches used centuries ago in Germany with Nazi fictional stories about Jews used decades ago: we all know very well what happened as the result of this (Fig. 4.5). The major difference between science and theology is that although scientists can be—and often are—biased, as explained in the present book, science has in the overall been very successful in contradicting imaginary stories based on teleological narratives. The Earth is the center of the universe, God or witches cause diseases, HIV pandemics is a way to punish homosexuals: all of them have been completely contradicted by using the scientific method and gathering empirical evidence about what truly happens in the natural world and the cosmos (see also Box 9.9). In this sense, taking into account the absurdity of the fairytales created and believed by the *Homo fictus, irrationalis et socialis*, and all the destruction and atrocities that we have caused and/or justified, one can say that, despite contributing to many of such dangerous tales, in the overall science has indeed been one of the few *candles in the dark* since the rise of human 'civilizations,' as put by Carl Sagan (see below).

For instance, about **Covid-19**, scientists and physicians are not interested—and should not be—in "building new teleological narratives," but instead in studying in detail the virus—its molecular details, how it spreads to the human body, how does it affect it, and how to defeat it. Compare that with what happened for instance with the **Black Plague**, centuries ago, at a time when medicine was deeply influenced by teleology, and when there was therefore not such a candle in the dark within the darkness of *Neverland*'s dungeons. In addition to the much higher number of deaths caused then by the plague per se (Fig. 8.7), there was also a huge numbers of killings of members of certain groups—such as Jews—because of the paranoia and fears that were created by *Neverland*'s imaginary teleological tales and conspiracy theories. The Jews, or the sinners, or the non-believers, or the witches, they were to blame, so kill them all. Among the numerous books focusing on these subjects, I would recommend Kelly's 2006 *The great mortality: an intimate history of the Black Death, the most devastating plague of all time*. With Covid-19 there was a similar increase of stories blaming or attitudes against the Chinese in some countries, or against other groups of humans, yes, but there were no killings because of that, and despite the huge number of people that died so far because of the pandemics—2.2 million so far, in January 2020—there is no comparison with the number of people that died of disease because of the black plague, despite the fact that the world's population is much higher nowadays (Fig. 8.7). In fact, many historians argue that the black plague was precisely the major catalyzer for the decreasing influence of teleology in science and medicine in the last centuries. According to them, after such a huge number of deaths and the fact that nothing could be done to avoid them, people finally started to realize that **praying to the Gods**, or **killing Jews**, or **using magical tricks**, did not save them at all. Accordingly, scientists and health workers begun to realize that a medicine inundated by teleology failed terribly to understand and address what was truly happening. Therefore, arguing that scientists and health professionals should now—in the twenty-first century—start again recurring to teleological narratives completely misses the point (see also Box 9.9).

These discussions about **Covid-19** and the **anti-vax movement** lead us to revisit an issue that is key for the present book and that we partially addressed in Chap. 8, about the differences between truth, "goodness" and "usefulness." One point that many authors make, concerning such discussions, is that if beliefs, including those that are part of religion, are "useful" or "evolutionary adaptive," then there is nothing "bad" with them, despite the fact that they don't reflect the *reality* of biological evolution. Namely, one could argue: in a life without cosmic purpose and without an after life, maximizing the time an individual is here, or at least "feeling in the best way possible" while he or she is here, should be seen ultimately as a very "good" thing, actually as the only thing that matters, right (see Box 9.8)? One answer to that argument is: for a specific person A at a specific moment B, yes, that might be "good," but if everybody would just happily believe in imaginary stories, and not care at all about knowing the *reality of the natural world and the cosmos*, then that person would probably not even have the option to believe in such stories or feel optimistic during, let's, say a cancer treatment. This is because there would be no available cancer treatment to start with, or the Covid-19 pandemic would very likely have killed 10 or 100 times more people than it did so far, as happened with the Black Plague (Fig. 8.7). But one could then argue: OK, but without science life is still possible, and surely enjoyable, because while other animals do help themselves in many ways, they don't have vaccines or chemotherapy, and most of them have very healthy lives in general, anyway. My answer would be: yes, they do, but there are two main differences. Firstly, those animals do *not* create and believe in complex teleological tales or conspiracy theories defending that they should do risky things, or take unhealthy products, either to "cure" themselves or to please any religious deities. Nor do they take the huge amounts of other very unhealthy products that humans nowadays tend to take, such as drugs, alcoholic drinks, or tobacco. Secondly, and related to the latter point, people living in agricultural societies have a huge number of diseases that are not commonly seen in other animals, and since the so-called "industrial revolution" there was moreover a huge increase of the so-called modern diseases. As a result, people living in "developed" countries have to face a much higher number of different medical problems and, as seen now with the current Covid-19 pandemics and to a lesser extent also with **Sars** and **Mers** some years ago, of "new" pandemics during a lifetime than the vast majority of other animals have.

Box 9.8: Alzheimer and the Potential Medical Benefit of Teleological Stories

In Chap. 2 I provided various examples of potential benefits of believing in teleological narratives, even when such narratives are completely falsified by empirical data. Within the context of the topics discussed in the present chapter I provide one further—and last—example, concerning the belief on a 'cosmic meaning in life' (but see Box 9.9 for an opposite example). Namely, in a 2021 paper of *Archives of General Psychiatry*, entitled "*Effect of purpose in life on the relation between Alzheimer disease pathologic changes on cognitive function in advanced age*", Boyle and colleagues explain:

> ***Purpose in life***, the psychological tendency to derive meaning from life's experiences and possess a sense of intentionality and **goal directedness**, is a related psychological factor and component of well-being that has long been hypothesized to be associated with positive health outcomes. Indeed, studies (mostly cross-sectional) have shown that purpose in life is associated with cognitive and psychological health in elderly persons, but the neurobiologic basis of the beneficial effect of purpose in life remains unknown. In recent years, systematic examination has shown that purpose in life is associated with a substantially reduced risk of incident **Alzheimer Disease** (AD), **mild cognitive impairment**, disability, and death. Furthermore, purpose in life is associated with a reduced risk of incident disability and death. Initial evidence suggests that purpose in life may also be modifiable, rendering it a potential treatment target. To date, however, we are not aware of any study that has examined the neurobiologic basis of the protective effect of purpose in life. In this study, we sought to extend these findings by examining the neurobiologic basis of the protective effect of purpose in life on cognition. Thus, we tested the hypothesis that higher levels of purpose in life reduce the deleterious effects of AD pathologic changes on cognition in older persons. Participants were 246 persons from the Rush Memory and Aging Project, a longitudinal, epidemiologic, clinicopathologic study of aging. We found that higher levels of purpose in life reduced the deleterious effects of AD pathologic changes on cognitive function. This protective effect was observed for a global measure of the changes as well as a more molecularly specific measure of tangle pathologic changes. In addition, the **protective effect of purpose in life on cognition** persisted after controlling for various potentially confounding variables. Moreover, higher levels of purpose in life reduced the deleterious effect of AD pathologic changes on the rate of cognitive decline. These findings suggest that purpose in life provides neural reserve by protecting against the harmful effects of AD pathologic changes on cognitive function in elderly persons.
>
> Purpose in life is a complex and multifaceted trait like construct, and it is likely that purpose in life works via complex mechanisms to provide reserve. The ability to *find meaning in life's experiences* and develop a sense of direction and intentionality requires self-reflection, synthesis of diverse experiences into a narrative, awareness of one's role and potential within the broader context, establishment of goals and priorities, and focus. Furthermore, having a sense of purpose is thought to generate motivation to behave in ways consistent with one's purpose and work toward goals. Purpose in life is related to aspects of **psychological health**, including **happiness**, satisfaction, **personal growth**, and **better sleep**, and associations between purpose in life and aspects of personality (ie, **neuroticism**, extraversion, and **conscientiousness**), as well as depressive symptoms, have been reported. Although few studies have examined the extent to which purpose is related to engagement in health-promoting behaviors, results of a meta-analysis showed that a higher level of purpose in life was associated with **better health**, everyday competence, social integration, participation in the labor force, and socioeconomic status among middle-aged and older persons. Purpose in life also is associated with better treatment outcomes for persons with addiction. Taken together, the available data suggest that persons with higher levels of purpose tend to be goal-oriented and resilient, and their active pursuit of goals likely enhances the strength and efficiency of neural systems. Furthermore, although purpose in life may be most beneficial in aging (when cognitive and other resources are diminishing), elderly persons who report

higher levels may have acquired over their lifespan an expanded repertoire of behaviors that facilitate **neurocognitive development**. Although one could speculate from this that purpose in life may somehow prevent the accumulation of AD pathologic changes, we did not find evidence of a direct association with the changes. This may indicate that, instead of preventing the accumulation of pathologic changes, purpose in life contributes to the development of efficient neural systems that allow one to maintain cognition even in the face of accumulating characteristics of AD.

Another possibility is that purpose in life may reduce the association of AD pathologic features with cognition by helping to invoke compensatory processes in the face of accumulating damage. The current study was motivated in part by a clinicopathologic study that showed that social networks modified the association of AD pathologic characteristics with cognition. Given that the development of social networks includes brain regions not involved in traditional aspects of cognition, such findings may indicate that social cognitive brain regions are brought online to maintain cognition in the face of accumulating AD pathologic changes in regions that support traditional cognitive abilities. It is noteworthy that higher levels of purpose in life are associated with less negative affect, more positive social relations, and better sleep and other health outcomes. Although controlling for covariates such as neuroticism, **social networks**, and depressive symptoms did not affect our findings, it seems intuitive that the noncognitive abilities that allow some people to readily derive meaning from life's experiences and persist through challenging events might also provide reserve by increasing the availability of social cognitive or other **noncognitive neural networks** that can help preserve cognition even as AD pathologic changes accumulate. Notably, the beneficial effect of purpose in life was most evident in analyses with tangles as compared with amyloid. Previous studies have shown that tangles are more strongly associated with cognitive function than amyloid. This generally was true in our data, and the modifying effects may have been most evident in analyses with tangles because of their relatively stronger association with cognition. Similar differential effects were reported in 2 studies; one of these showed that processing resources protected against the deleterious effects of AD pathologic changes on other cognitive systems, particularly in analyses of tangles, and the other 10 showed a similar effect with social networks. These findings likely suggest that tangles are a major driver of cognitive impairment in old age and that the association of tangles with cognition is the predominant beneficiary of factors that provide reserve.

One of the aspects that particularly impressed me when I was reading about the life of **Maurice Hilleman—**whom we mentioned above–is a powerful point that he made about the reality of *Neverland*'s world of unreality. In the last years of his life, when the **anti-vax movement** was growing, he astutely stated that, paradoxically—and tragically—this phenomenon was directly related to the successful vaccines that scientists such as him had developed during decades. Firstly, many of the young people promoting, or believing in, that movement would indeed not be alive today, or would not have the strength to fight so hard against vaccines, if scientists like him had not dramatically decreased child mortality and disease in the last decades, by the millions. Secondly, it is precisely because of that huge decrease of child mortality and disease that almost no anti-vaxes have ever seen, in the "developed" countries which are chiefly those in which the movement is stronger, small children suffering, and dying, with Polio, Measles, Mumps, and so on. That is, by not directly seeing such a huge loss of lives and massive suffering with their own eyes, anti-vaxes don't "*see*" the *need* for vaccines: why to take vaccines when we are young, if one can "see" that everything is OK? This is a one of the most crucial point to answer the theoretical question mentioned above, about what is the problem of an individual believing in imaginary tales that makes him/her feel "good." The answer therefore is: apart from the wars, racism, misogyny, tortures, burning of people, oppression, and so on that were promoted or at least justified by such tales, there is indeed a huge problem. Namely, that believing in factual inaccuracies such as that vaccines are "bad"—or made by Bill Gates and his "demons," a theory currently disseminated by many anti-vaxes about **Covid-19** vaccines—can put at risk the well-being and health of the planet's population as whole. In the case of Covid-19, this is happening just before our eyes: people that don't want to take the vaccines are not only dying more in proportion—for instance, in countries such as Israel, this is happening to many **ultra-orthodox Jews**—but are also jeopardizing the possibility of achieving herd immunity against the virus in a shorter term, what has huge health, medical, societal and economic implications (see also Box 9.9).

The belief in factual inaccuracies and conspiracy theorics (see Fig. 2.11) had actually already caused a huge number of Covid-19 related deaths even *before* we had vaccines, from the very beginning of the pandemic. In the U.S. a substantial part of people, particular those more influenced by the alternative facts and fake news propagated by many right-wing politicians in TV Channels such as *Fox*, radios, websites, and social media, did believe in a series of conspiracy theories about the virus, including that everything was just a hoax. That was totally disastrous, not only for those people, but also for the rest of the population, because this allowed the virus to propagate even faster than those alternative facts as a huge portion of the population refused to take any measures of social distancing, or to use masks. As a result the U.S. was, by far, the country in the planet with both more infections and more deaths, with almost half a million deaths in just less than a year since the pandemic arrived there. This is a powerful reminder that the price for believing in imaginary stories is ultimately not only detrimental for the *population* as a whole, be it because viruses can propagate more easily, or because people are racist and misogynistic, or do mass killings of Jews, Gypsies, and disabled people, but can also be hugely harmful to the very individuals that believe in those stories. That is, in many cases believing can bring some comfort, reduce depression or the consumption of alcohol and drugs, and so on, but in many others it can lead to tragic consequences such as when people are part of sects that promote mass suicides, or auto-flagellate themselves (Fig. 9.13), or do not accept life-saving medical practices because they are "forbidden" by their religious texts, or do not take vaccines because of conspiracy theories, among an endless number of other devastating decisions. One of such countless tragic events was reported in a 24th of August, 2020, *BBC News* online article entitled "*Man who believed virus was hoax loses wife to Covid-19*":

> Brian Lee Hitchens and his wife, Erin, had read claims online that the virus was fabricated, linked to 5G or similar to the flu. The couple didn't follow health guidance or seek help when they fell ill in early May. Brian recovered but his 46-year-old wife became critically

ill and died this month from heart problems linked to the virus. Brian spoke to the BBC in July as part of an investigation into the human cost of **coronavirus misinformation**. At the time, his wife was on a ventilator in hospital. Erin, a pastor in Florida, had existing health problems – she suffered from asthma and a sleeping disorder. Her husband explained that the couple did not follow health guidance at the start of the pandemic because of the **false claims** they had seen online. Brian continued to work as a taxi driver and to collect his wife's medicine without observing social distancing rules or wearing a mask. They had also failed to seek help as soon as possible when they fell ill in May and were both subsequently diagnosed with Covid-19. Brian told BBC News that he "wished [he'd] listened from the beginning" and hoped his wife would forgive him. "This is a real virus that affects people differently. I can't change the past. I can only live in today and make better choices for the future", Brian explained. "She's no longer suffering, but in peace. I go through times missing her, but I know she's in a better place." Brian said he and his wife didn't have one firm belief about Covid-19. Instead, they switched between thinking the virus was a hoax, linked to 5G technology, or a real, but mild ailment. They came across these theories on *Facebook*. "We thought the government was using it to distract us" Brian explained, "or it was to do with 5G." But after the couple fell ill with the virus in May, Brian took to *Facebook* in a viral post to explain that he'd been misled by what he'd seen online about the virus. "If you have to go out please use wisdom and don't be foolish like I was so the same thing won't happen to you like it happened to me and my wife", he wrote.

"Please use wisdom"—and factual data—"and don't be foolish"—or fooled: the key question is indeed why so many humans are not using wisdom and factual data, being instead so easily fooled, and the aim of the present book is precisely to not only analyze this question in detail and answer it using empirical evidence, but also to try to contribute to change this, once for all. Because while it is true that we can't "change the past," we surely can, and urgently need to, "make better choices for the future"—choices based on real information, within the *world of reality*. Another story that came in the press a few weeks ago—the fifth of January, 2021—illustrates, perhaps in an even more dramatic way, how believing in conspiracy theories can be dangerous to both the believer *and* the community as a whole, particularly in the context of the highly-connected, globalized world of today. As reported by *NBC news*:

A pharmacist accused of trying to destroy hundreds of doses of **coronavirus vaccine** is a conspiracy theorist who believed the medication wasn't safe, Wisconsin authorities alleged Monday. The man, Steven Brandenburg, 46, was ordered held in lieu of $10,000 bond by Ozaukee County Circuit Court Judge Paul Malloy during a brief appearance. Police in Grafton, about 20 miles north of Milwaukee, arrested Brandenburg, a pharmacist with Advocate Aurora Health, on Thursday after 57 vials of the **Moderna vaccine** appeared to have been spoiled. Police said Brandenburg took the vaccine doses from a refrigerator and left them out for 12 hours, possibly rendering them useless. Each vial contained 10 doses; in total, the material was worth $8,550 to $11,400, according to a probable cause statement by Grafton police Detective Sgt. Eric Sutherland. Brandenburg is an "**admitted conspiracy theorist**", and he "told investigators that he believed that **Covid-19 vaccine** was not safe for people and could harm them and change their DNA", Sutherland wrote. "He admitted this was an intentional act".

Another article that make us think even more deeply about the dichotomy between the world of *reality* of those scientists that create vaccines or the health workers that provide them versus those immersed in *Neverland* was published in

December 2020 in *The Times of Israel*. Entitled "*American Jewish doctor shocked to find Nazi tattoos on COVID-19 patient,*" it reported:

> A Jewish doctor working with **coronavirus patients** in California shared his shock about the moment he saw **neo-Nazi tattoos** on the body of a severely ill man he was treating. As his team – which included a Black nurse and a respiratory specialist of Asian descent – prepared the man to be intubated, Taylor Nichols said on Twitter he spotted the **Nazi tattoos**. "The **swastika** stood out boldly on his chest. **SS tattoos** and other insignia that had previously been covered by his shirt were now obvious to the room", he tweeted Monday. "We all saw. The symbols of hate on his body outwardly and proudly announced his views. We all knew what he thought of us. How he valued our lives", said Nichols, who was later interviewed about his experience by various media outlets. Nichols talked about the conflicting emotions he felt, after months of battling the disease and seeing patients die, while living in isolation to avoid contaminating loved ones, constantly in fear of falling ill himself. "Unfortunately, society has proven unwilling to listen to the science or to our pleas. Begging for people to take this seriously, to stay home, wear a mask, to be the break in the chain of transmission", he said. Nichols said the man – whom he described as older and heavy set, his teeth lost to years of methamphetamine abuse – had begged him to save his life. "Don't let me die, doc", he said, according to Nichols. The man was admitted to the hospital near Sacramento in the middle of November, already "clearly working hard to breathe. He looked sick. Uncomfortable. Scared. I reassured him that we were all going to work hard to take care of him and keep him alive as best as we could", said Nichols, admitting he had asked himself how the man might have acted had the roles been reversed. "For the first time, I recognize that I hesitated, ambivalent. The pandemic has worn on me", he said. "And I realize that maybe I'm not ok", he said. Nichols later told the San Francisco Chronicle that when he saw the hate symbols tattooed on the man's body, "I didn't feel compassion for him in that moment". And he told ABC News he wondered "how much he would have cared about my life if the roles were reversed". It "really made me double down and look into myself and extend that compassion towards him". Nichols said he's "faced these situations countless times since medical school.... The swastikas. The racist patients. Every single time I feel a bit shaken, but I went into this job wanting to save lives". Nichols said he did not know if the patient with the Nazi tattoos had died or not, but said he had done everything he could to save his life before moving on to the next patient.

Real life is indeed often more fascinating—and in this case more tragic—than fiction. Examples similar to those about anti-vaxes above are those concerning individuals that say that they don't "believe" in **global warming**, or that it has nothing to do with human activities, or that scientists are just trying to build "fake-news" so left-wing politicians or ecologists destroy their jobs and give new jobs to the "others." In these cases, contrary to the anti-vaxes, those people will not suffer more than the rest of population by believing in such untruths, but the huge societal price paid by such a belief is even more enormous than that concerning the whole Covid-19 pandemic. That is, when seen in such a long-term *and* global context, such a belief indeed results in a self-centered, selfish way of thinking and behaving that puts in risk the whole human species and the life of a huge number of other species. In a 2007 paper by Pratarelli and Chiarelli, "*Extinction and overspecialization: the dark side of human innovation,*" these authors argue that we are so stuck—particularly after the rise of agriculture and "civilizations"—in a persistent, monotonic behavioral obsession with teleological stories, as well as related notions of "progress," "technological development," and "economic growth," that is dragging us to a scenario in which **technology**, **consumption patterns**, **overpopulation,** and **pollution** are being critically putting in risk the global ecosystem. As long as a huge part of the population of this planet continues to feel "good" by believing in the factually wrong idea that this is not truly happening, and to be comfortably immersed in *Neverland*, we are indeed rapidly getting closer and closer to the point of non-return for climate action, and thus to irreversibly hurting human existence in the only *real land* we inhabit, the marvelous planet Earth.

Box 9.9: Anti-Vaccination and the Potential Medical Disadvantages of Teleology

Opposite to the example provided in Box 9.8, here I provide a further case—also concerning the 'search for cosmic meaning'—that shows the dangers of believing in teleological narratives. It refers to a paper titled "*Risk perception and communication in vaccination decisions: a fuzzy-trace theory*", published in 2012 by Reyna:

> Strident **anti-vaccine messages** are attempts to predict or explain adverse outcomes, and to link them to vaccinations. As the fastest growing source of health information, the internet has enormous potential to spread such messages. The internet and social media can amplify the perceived frequency of adverse events as rare events can be quickly shared around the globe. According to **fuzzy-trace theory**, such anti-vaccine messages are expected when people do not understand **vaccination** (which is widespread) and when mysterious adverse events occur in close contiguity to vaccination. The ***search for meaning*** and the tendency to interpret events – to connect the dots – provides a powerful impetus to generating strident anti-vaccine messages under the right conditions. As long ago as 1948, Skinner demonstrated that pigeons would connect their own arbitrary behaviors (that happened to occur at the moment of delivery) to a food reward, even when food delivery was random. This so-called **superstitious behavior** is also evident in humans, for example, when baseball players continue to wear a lucky hat or use a lucky bat in the hope of recreating home runs. Connecting events that merely co-occur randomly is a rote or verbatim strategy because it does not depend on understanding (e.g., inferring a causal mechanism). Thus, individuals with very low levels of causal knowledge are likely to engage in superstitious behavior, much like Skinner's pigeon "[see Chap. 2]," connecting vaccinations to the adverse events that might follow them. However, the fuzzy-processing preference suggests that most adults will attempt to understand associations rather than connect them arbitrarily. For example, when events occur randomly, people will test hypotheses about why the events occurred in order to predict future occurrences. They perceive illusory correlations, seeing a relation even when none exists. Thus, they are better able to detect non-random patterns when they occur, but their performance is woefully inadequate when events are actually random.
>
> Events such as **autism** are increasing in frequency, and appear connected to vaccinations because they manifest around the same age that children receive vaccinations and because anti-vaccination messages 'explain' their co-occurrence. As Downs et al. (in 2008) showed, official communications, such as those from the Centers for Disease Control, are 'spare' and seem cryptic to those who lack background knowledge, whereas anti-vaccine communications 'tell more coherent stories, supported by narrative explanations'. In other words, anti-vaccination messages

attempt to create a highly coherent gist, but official sites often do not. Because of the drive to extract meaning, the widespread lack of knowledge about vaccination creates fertile ground in which misleading 'explanations' can take root. Hence, in addition to low knowledge, strident anti-vaccination messages are predicted when: (a) specific ideas have a priori plausibility (that the government would deliberately infect people with a dread disease; that authorities are untrustworthy) and when (b) adverse outcomes occur that are poorly understood (e.g., autism, **multiple sclerosis**, and **fibromyalgia**). Regarding b, mysterious adverse events, such as diseases whose causes are unknown, are a 'meaning threat', challenging what Albert Camus called our 'wild longing for clarity'. Anti-vaccine messages that make sense of unexplained events and associations, that satisfy that longing for clarity, are apt to diffuse more rapidly through the internet and social media.

Regarding a, plausibility has been shown to increase the uptake of suggestions in **false memory studies**, for example, presumably because plausible events that never happened are consistent with the gist of events that did happen. Belief bias also colors logical inference, making believable conclusions easier to infer than unbelievable ones. Therefore, conclusions that might seem **irrational** (that MMR causes autism, that **Obama** was not born in the U.S., or that humans never landed on the moon) will achieve greater uptake among subgroups who hold prior beliefs that make such conclusions plausible. Drawing on beliefs and plausibility, gist is extracted at the level of words or numbers (individual items), across lists of items or inferences across sentences, and at the level of whole narratives. At the level of narrative, the gist representation is a coherent story about causality. Hence, the tendency to jump to conclusions, to connect the dots, is a cornerstone of normal cognition. However, some people jump to conclusions more readily; **paranoid delusions** represent an extreme point along a continuum of inference and explanation. For example, Peters and Garety (in 2006) showed that in a simple random draws task with two jars (one jar with a ratio of 85 black beads to 15 orange beads and the other with the opposite ratio), patients diagnosed with paranoid delusions required significantly fewer draws to jump to the conclusion about which jar was being sampled, compared to non-clinical subjects. Depressed patients required significantly more draws than non-clinical subjects to jump to the conclusion about which jar was being sampled. These results again illustrate that jumping to conclusions is a natural outgrowth of seeing meaning and patterns, even in random events. Carried too far, this natural tendency can become irrational, as in delusional patients. Likewise, anti-vaccination messages fill in a vacuum of knowledge, and nature abhors a vacuum, which in some instances lead to conspiracy theories and other apparently irrational beliefs.

9.5 Epicurus, Lucretius, Sagan, Gould: A Splendid Non-Purposeful Cosmos

It is far better to grasp the Universe as it really is than to persist in delusion, however satisfying and reassuring
(Carl Sagan)

As explained in McMahon's 2006 book *Happiness—a History*, one important point that needs to be made in discussions about what is a fulfilling life is that, as it happens when a certain individual says that he/she feels "good" or "happy" about something in particular, the notion that one is "happy" is, in itself, chiefly the result of **social construction**. That is, it is true that, as explained in Damasio's 2021 book *Feeling & Knowing—Making Mind Conscious*, "feeling" is something much more ancestral, in biological evolution, than human social constructions. An insect can surely "feel pain" if a certain part of its body that is innervated was cut, as a dog can "feel" pain or discomfort if it has a cancer in its body. But when a person says to another that he/she is "feeling" in a way A or B, and particularly if that statement has to do with broader and more intangible things like feeling "happy," "positive," or "fulfilled," such statements are indeed hugely influenced by what is considered to be "normal" within his/her society, which is in turn related to the teleological narratives believed by the other members of that society, at a specific moment in time and geographical context. To give a simple example, it is much more likely to hear a woman saying to a man that she is "aroused" in a bar at night in the city of Paris than in a public store during the day in Cabul, despite the fact that each of those women might have exactly the same sexual desire for the man she is talking with. In particular about **happiness**, McMahon cites **Rousseau**: "the closer to his natural condition man has stayed, the smaller is the difference between his faculties and his desires, and consequently the less removed he is from being happy." This contrasts with what happens in "modern societies," in which, according to McMahon, "in the race to fulfill present needs, we continually created new ones, resulting in a disturbing phenomenon…or, as stated by Rousseau, in learning to desire, [we] have made [ourselves] the slaves of [our] desires." That is, as explained by Rousseau, unfelt by "men of earliest times, lost to the enlightened men of later times, the happy life of the golden age was always a state foreign to the human race, either because it went unrecognized when humans could have enjoyed it or because it had been lost when humans could have known it…as soon as man's needs exceed his faculties and the objects of his desire expand and multiply, he must either remain eternally unhappy or seek a new form of being from which he can draw the resources he no longer finds in himself." A beautiful image that McMahon used in his book to illustrate this issue is shown in Fig. 9.19.

The fact that different notions of happiness are related to different cultures and social norms and ways of seeing the world was shown empirically in a very broad cross-cultural study published by Gardiner and colleagues in a 2020 paper entitled "*Happiness around the world: a combined etic-emic approach across 63 countries.*" As explained by them to the broader public, in the press release about that paper:

> A worldwide comparison of two tests that measure people's **happiness** highlights the importance of a country's cultural context and suggests that a test developed in Japan may currently be a better tool for cross-cultural research than a U.S.-developed test. Most **studies of happiness** have used tests developed in Western countries, which typically prioritize individuals' independence. However, studies on happiness in Eastern countries emphasize **connectedness** with others, or **interdependence**. Non-Western tests of happiness have emerged in recent years, but it is unclear how the two concepts of happiness might generalize beyond the Eastern and Western countries typically examined in happiness research. To explore the cross-cultural applicability of Eastern versus Western concepts of happiness, Gardiner and colleagues employed two happiness tests: one developed in Japan that emphasizes interdependence and a U.S.-developed test that emphasizes independence. As part of

Fig. 9.19 "Chasing After Happiness": Edmund Youngbauer's "Die Jagd Nach Dem Gluck," late nineteenth century. In a way, this painting can be used to also symbolize the illusion of "happiness" or "fortune" in *Neverland*'s *world of unreality*: the man is desperately trying to reach the woman, which displays a socially constructed image of "fortune"—both physically and symbolically, including of richness, as symbolized by the crown—ignoring everything else, including diseases or other hazards—death is riding along with him and there is a dangerous break in the bridge—and/or the wellbeing of others, even to the point of hurting/pushing/damaging them—the girl being trampled under his horse. Everything is a delusion, as symbolized by the bubble that is about to burst in the Twilight-Zone of *Neverland*

a larger, overarching research project, 15,358 college or university students from 63 countries across six continents volunteered to complete a survey that included both tests. Analysis of the survey data and country-specific factors showed that the test emphasizing interdependence was more reliable in countries that are more culturally similar to Japan, such as other East Asian countries, while the independence-focused test was more reliable in countries more similar to the U.S., such as Western European countries. The study also found that, while the interdependence-focused test had slightly lower overall reliability than the independence test, it was more consistently reliable across countries. Therefore, the interdependence-focused test may be a stronger research tool for **cross-cultural comparison**. Both tests showed lower reliability in African and Middle Eastern countries, suggesting the need to develop tests of happiness that are more universal. The authors add: "The way researchers currently assess happiness is typically using a **Western-biased measure** that assumes one's own happiness is largely independent of others, but we find that this viewpoint is by no means universal."

This discussion about the notion of happiness, and in particular how Westerns tend to think about it chiefly in a Western-centric way, leads us to the key theme of this last Section of the present book, which comes full circle to issues discussed in its first chapters. Namely, in this Section, and to follow up on the discussion done in the previous Sections about the fact that many people, including various scholars, argue that we "need new narratives," I hope to be able to show that reality can be, by itself, amazingly *fascinating and fulfilling*: reality as *it is here and now*, not as an imaginary "ought to be." The enormous diversity of species in this planet (Fig. 9.20), and the immensity of the universe, and its unpredictable mix of random, contingent, and chaotic events, that is what is *really*—figuratively and literally—breathtaking. No teleological imaginary stories, not even the most creative and engaging ones, such as the Bible, the Koran, the Talmud, the Vedas, or the *Epic of Gilgamesh*, are able to reflect the depth, complexity, vastness, diversity, and unpredictability of the cosmos. Actually, they are over-simplifications of it, because at the moment they were written most of what we know today, about the natural world and the cosmos, was not known. People back them did not know about black wholes, nor about bacteria and viruses, nor did they know that organisms evolved and that we are closely related to animals as strange as colugos, or about the fishes that live in the deep sea.

Moreover, to those people that feel the urge to "transcend"—that is, to go "beyond"—this mesmerizing reality about our planet and the universe as a whole, be it by reaching "nirvana," by praying to imaginary Gods, or by experimenting a "shamanic trance," I would point out exactly the same: the vast majority of them don't know, or never even tried to know, about 10% of the fascinating details about the natural world. In fact, one paradoxical point about this topic is that empirical data show that indeed those that proclaim to have such a urge to go "beyond" this "simple material world" tend to be, in average, those that truly know very little about it. Even scientists such as Carl Sagan, who knew immensely more than the general public about the cosmos, only knew an infinitesimal part of the marvels of the universe, so how can someone that only knows a very tiny fraction of what he did have such a "urge" to go "beyond" it? Contrary to a saying that is now so in vogue within New Agers, it is not their curiosity that leads to their even deeper immersion into *Neverland*'s world of unreality: it is mainly their ignorance about, or even fear to know, what is truly out there, to leave the tempting but fake comfort of *Neverland*. It is as if I arrive to a country to which I have never went, let's say Australia, and after just seeing the airport where I arrived, without having seen its amazing beaches, and corals, and cities, and people, and wildlife, I say: ohhh, I need to "transcend" Australia, to go beyond it, either going to another country or to a virtual or Second Life country. As a wanderer and a wonderer, I am intellectually open to anything, and I have nothing against trying new experiences, either if they are "real" or imaginary, I have felt what it "feels" like to be "out of this world," for instance when one is deeply drunk. But I don't feel the *need* to constantly or continuously escape from the reality of life on Earth and of the cosmos, I want instead to continue discovering and enjoying this life, here and now, to enjoy this moment in *real* life, to mindfully walk in a real forest, swim in the sea with real fishes, taste a real pasta *al dente* in Italy, drink an exquisite real Geisha coffee in Panama, to have sex with a real person, run in the beautiful streets of a real city such as Washington DC. Precisely because I *know* that I will not be able to enjoy all those *real* things for ever: one day my life will just stop, my heart and brain will cease working, and there

Fig. 9.20 Amazing, breathtaking diversity of life within this splendid planet

will be no after life. We only have this fascinating real life, so better to fully enjoy it here and now, before it is too late: **Carpe Diem**, indeed.

Coming full circle to the beginning of the present book, it is interesting to note that many of the issues we discussed in it, such as that the main reason that lead humans to build and belief in teleological narratives was precisely the awareness of the inevitability of death, were precisely what lead to the conception of the term Carpe Diem. This term—known by most people but unfortunately so rarely applied in practice—can be translated to English as "pluck the day" or "seize the day," and was used by the epicurean Roman poet **Horace** (65–8 BC) to express the idea that we should precisely enjoy *this life* while we can. This idea was defended by the **ancient atomists**, including **Leucippus** and **Democritus** and later **Epicurus** (Fig. 9.21) and his epicurean followers—apart from Horace, one of the most famous ones was **Lucretius** (see Box 9.10). Remarkably, without knowing most of the facts we now know about the non-human organisms that inhabit this planet and the cosmos as a whole, they had a way of understanding life, millennia ago, that is much closer to *reality* than that proclaimed in the teleological stories followed by the vast majority of people living nowadays, including many scholars. Earlier in this volume I explained some of the main ideas of the **epicureans**, and pointed out how fortunately in the last decades more and more scholars are starting to fully recognize the intelligence and broadness of their ideas and to disseminate them to the broader public. One striking fact about human history is that, if more people had been able to embrace Epicurus' ideas more than 2 millennia ago instead of continuing to engage in the creation of more and more teleological stories and becoming increasingly immersed in *Neverland*, humans could likely have avoided at least some of the most horrible atrocities since then, to ourselves, to other animals, and to life in this planet in general (see Box 9.10). In his 2020 book *Heaven and Hell—a History of the Afterlife*, Erhman wrote the following about the profound wisdom of the epicureans:

> The **fear of death** for many people in antiquity differed from the terrors of torment or horrors of actual nonexistence experienced by so many in the West today. It was instead the dread of losing out on everything a full life has to offer, everything that makes living pleasant. For many ancients there was indeed a kind of non-tortured existence after death, but it was bleak, dreary, and completely uninteresting. Some ancient philosophers found such views of postmortem blessings and curses very disturbing and disruptive – not for themselves personally but for people at large. There was a strong minority position that maintained that tales of the **afterlife**, and the beliefs based on them, were damaging to a person's well-being, since they corresponded to *no reality*. In this alternative view, the horrors of the afterlife in particular were pure fictions that not only terrorized innocent people but forced them to behave in ways contrary to their health and happiness. Of those who held such skeptical views, none was more important than the Greek philosopher **Epicurus** (341-270 BCE). Throughout history Epicurus has had a completely undeserved reputation as a **hedonist**, interested only in promoting physical pleasure. This in fact is a mischaracterization of his views. Like many philosophers in antiquity, Epicurus was interested in knowing how a person could lead the best life with the greatest amount of ***happiness***. It is true that, in his view, the happiest life was one that avoided pain and promoted ***pleasure***. But Epicurus argued for the simple pleasures: moderate food and drink, good friends, intelligent discussions on important and compelling topics. Happiness also requires people to understand

Fig. 9.21 Epicurus (341–270 BC), an ancient Greek philosopher that is an emblematic example of the term "being ahead of his time" or, better said, it was perhaps the storyteller animal world that was frozen since then, remaining entrapped in *Neverland*. Fortunately, there are now some changes in framework that indicate that at least some *Homo sapiens* start to understand and apply some of Epicurus' ideas about life and the non-purposeful, and amazingly fascinating cosmos

what it means to be human and *not to allow baseless and irrational fears to overwhelm their mental lives*. No fear, for Epicurus, is more irrational than the fear of death, based as it is on a profound misunderstanding of what it means to be human, specifically about what it means to have a soul. Epicurus firmly believed that the soul is a corporeal entity, made up of a kind of matter. It consists of a large number of fine particles dispersed throughout the body. Only when the soul is united with the body is sense perception possible. When at death the soul separates from the body, its atoms are simply dispersed into the air. At that point, the body, lacking its soul, can no longer feel anything. But neither can the dissipated and therefore no-longer-existing soul. Since a departed and therefore dispersed soul no longer exists, it cannot be rewarded or punished. *It simply disappears*. Or as he writes to a man named Menoeceus, in one of the few letters that is preserved: "*Get used to believing that death is nothing to us…for all good and bad consists in sense-experience, and death is the privation of sense-experience…hence a correct knowledge of the fact that death is nothing to us makes mortality of life a matter for contentment, not by adding a limitless time [to life] but by removing the longing for* ***immortality***."

Not many philosophers in antiquity were persuaded by Epicurus's views. In some ways, the deeply rooted human sense that this life cannot be all there is proved too strong. So far as we know, humans have always imagined there must be life beyond. Possibly, in part, that is because individual humans have always – as long as they have been able to think – known nothing other than existence, making it very difficult indeed to imagine a never-experienced state of nonexistence. But, for whatever reason, the understanding of death that made such brilliant sense to Epicurus did not catch on, either among professional thinkers or the population at large. There were some notable exceptions, however, the most famous of whom appeared in Roman circles over two centuries later: Epicurus's latter-day disciple **Lucretius** (circa 98-55 BCE). Unlike Epicurus, for whom we have only a few scant literary remains, Lucretius has bequeathed to us an entire philosophical work, openly and proudly indebted to the views of the one he considered the greatest philosopher of all time. The book, called *On the Nature of Things*, tries to accomplish nothing less than to explain the *nature of reality*. In it Lucretius develops a theory that may sound remarkably prescient. Everything in the world, all that we experience and do not experience, is made up of atoms that have come together in chance combinations over infinite amounts of time as they run into each other in infinite reaches of space. We ourselves are the products of matter, time, and chance. As such, we will eventually dissipate as our atoms dissolve their connections. Dissolved with them will be not only our bodies, which obviously disappear eventually, but also o our souls. In many ways, Lucretius's entire treatise on the atomic basis for all reality is meant to accomplish a specific aim: to dispel the fear of death and destroy any foolish notions of life beyond the grave. As he says at one point in the book, he seeks to "drive out neck and crop that fear of Hell which blasts the life of a person from its very foundations, sullying everything with the blackness of death and the leaving no pleasure pure and unalloyed."

So, here it is again: we scientists don't need new—or for that matter also old—teleological narratives to make the world of *reality* appealing to the broader public, or to make people not be depressed by it, and so on. Instead of using such narratives, what is critical—against "alternative facts" and "fake-news" or conspiracy theories about vaccines—is to talk about the mesmerizing *reality* of the natural world and the cosmos, or about more specific topics such as how vaccines are truly done, what is the science behind them, or how they are in general very safe, particularly when compared to the risk of not taking them. It is crucial to do this in a way that *accurately and emphatically transmits* that fascination to the broader public, without recurring to lies or fictional stories. That is, in a way that is not perceived by the broader public as "boring" or "nerdish" or "distant" or "monochromatic" or "professorial." We know this can be successfully done, because this has been done in the past: illustrative examples of how this *was*, and should *be*, done are authors and science communicators such as the late **Carl Sagan** and the late **Stephen Jay Gould**. As noted above, the **series *Cosmos*** was widely seen, and the books of Gould were widely read, by—and clearly fascinated the—general public of a huge number of countries and of different cultures. Not because they included teleological narratives, well on the contrary, as both Sagan and Gould are renown for their anti-teleological views. It was instead because these authors had the empathy to care about what the broader public would be thinking or would consider more engaging, and also to be humble enough to communicate in such an engaging way to them without using a complex jargon and terms that could only be understood by very few intellectuals or scientists. As explained in Chap. 1, both Sagan and Gould were often criticized by their peers—particularly by those that considered themselves to be "erudite" or "elite" thinkers—precisely for writing popular books in such a simple and engaging way. But these two splendid scientists persisted despite of that criticism, and were indeed highly successful in communicating to the broader public an endless number of enthralling details about the *reality* of the cosmos and of the natural world. In this sense, they allowed hundreds of millions of people to escape, at least for a bit—or at least to not be so deeply immersed in—*Neverland*, and to gain the needed weapons—for instance, knowledge about reality, history and the scientific method—to further do this by themselves and also to teach those weapons to others.

A case study that is both interesting and distressing and that puts together three key issues that we have discussed in this chapter was provided by **Antonei Csoka**—him and me are faculty members of the same department at **Howard University**—in a 2015 paper entitled "*Innovation in medicine: Ignaz the reviled and Egas the regaled.*" Firstly, the paper shows how dangerous it is to have so many scientists enclosed in their "erudite" castles and thinking that communicating with the broader public, or discussing ideas with it, or learning from it, or from "non-erudite" people such as midwives, is something that would put in question their "specialness," their "authority," and "their egos." Secondly, it shows how this arrogance, together with the fact that so many scientists have been—and some still are—so influenced by teleological tales and in particular by those about "progress" and thus about the obsessive need to "create more, innovate more," results in a particularly hazardous combination in which those scientists are removed both from the broader public and from reality itself. Thirdly, the paper clearly shows, once again, that being removed from reality is never truly just a "free-ride": ultimately, in the vast majority of cases, if not all, when a group of people *believes* in fictional stories, this indeed can be very harmful, either on the short- or the long-term, be it at a psychological, societal, political, and/or ecological level. Teleological narratives, the obsession with progress, strong egos, lacking humbleness, and the wanting of connections to the real world—both figuratively concerning the detachment from the broader public and literally regarding the disentanglement from what the factual evidence truly shows—is indeed a very perilous mixture. As explained by Csoka:

> Mainstream medical science is constantly presented with **innovative concepts** and potential breakthroughs, a continuous stream of ideas collectively pushing the frontiers of human knowledge ever closer to a hypothetical technological singularity. Many ideas are validated and adopted, but others are discarded and ridiculed, perhaps to be recycled and reinvented at a later time. Why do some ideas blossom and take root, while others wither and fail? What are the hallmarks of robust innovation? The most obvious criteria, whether or not an innovative idea is actually true, false, uplifts, or degrades the human condition, may actually have little to do with its initial reception. Rather, the overall character, mood and skepticism of the professional audience can have a sizable impact on the response and reaction to an innovation, and of its ultimate adoption or rejection. One of the most dramatic examples of the mutable nature of the receptive climate and its influence over innovative progress involves two European physicians, Doctors Moniz and Semmelweis. **Egas Moniz** was born

Antonio Caetano de Abreu Freire on November 19, 1874 in Avanca, Portugal. Dr. Moniz was a neurologist who pioneered work in cerebral angiography as a tool to identify and localize cerebral lesions such as brain tumors. However, his name has gone down in history for the development and performance of prefrontal **leucotomy** (**lobotomy**), for which he was awarded the Nobel Prize in physiology and medicine in 1949. Dr. Moniz pursued experiments with angiography that paved the way for his signature "innovative" surgical technique. The procedure involved making incisions that destroyed connections between the prefrontal region and other parts of the brain. Psychiatric patients, mostly suffering from **schizophrenia**, were subjected to experimental trial and error until success was achieved with the aid of a device Moniz developed for the procedure called a "leucotome". This medical instrument was modeled after an ice pick, and was inserted into the tear duct and driven through the bone of the skull with a mallet. The international medical community responded by lavishing praise and accolades upon Dr. Moniz and his "groundbreaking" lobotomy technique.

Contrast the innovation of Dr. Moniz and the stunningly positive reception it received with the story of Dr. **Ignaz Philipp Semmelweis** (1818–1865) and his seminal innovation involving hygiene. He was a Hungarian obstetrician who gained some limited credit early in his career for work at the Vienna General Hospital (Allgemeines Krankenhaus) in the labor and delivery wards. Throughout history, childbirth has carried a degree of risk. Disease and illness were commonly associated with the birth of a child, especially within the first 3 days after birth, and often progressing with rapid and devastating results. Unfortunately, while advances in medicine saw the advent of hospitals to assist in the birthing process, understanding of diseases and their transmission had not yet caught up. Sanitation was virtually nonexistent, and male medical doctors and surgeons were replacing female midwives in the hospital setting. These changes alone drove the mortality rate for childbirth skyrocketing. In hindsight, it is clear that traditional midwives who tended one woman at a time were able to maintain a fairly clean and sterile environment, while their male counterparts were notorious for wearing bloody and soiled clothing from patient to patient. The startlingly high death rate of women who delivered their babies with the assistance of doctors and medical students was noticed and tracked by Dr. Semmelweis. He observed that women who were assisted by midwives experienced relatively safe deliveries and a much lower death rate. He had read theories regarding **septicemia** published by other forward-thinking physicians…upon considering their observation with his own, Semmelweis accurately deduced that the higher death rates in the women treated by doctors and medical students resulted from a fever; that this fever seemed to be communicable between people and/or objects (in other words that it was contagious); that the doctors and medical students who came directly from other areas of the hospital without washing were the mode of communication; and that the simple activity of hand washing might resolve the problem. To his credit, he took decisive action by instituting the practice of hand washing in a chlorinated lime solution before and after attending patients – the first institutionalized attempt at **antiseptic hygiene** for medical personnel. The mortality rates for the women in his clinics dropped by 90 %. [However] not only were his findings largely rejected, but Semmelweis' colleagues publically ridiculed and demonized his work.

Technology itself plays a critical role in ideological reception. The innovation of Dr. Semmelweis was quite simple, and did not require any specialized skills or training. No tools were required. Only hand washing, an easy procedure that anyone could accomplish at trivial expense and little effort but with revolutionary consequences. And not just in obstetrics, but with universal applications. Perhaps one reason for the downfall of Semmelweis' ideas was his identification of physicians as the culprits. He put the blame squarely on the doctors themselves, and told them that they literally held the solution in their own hands. His innovation required that they take persona responsibility for maintaining the sanitation of hospital and patient by keeping their own hands clean, and correctly observed that failure to implement this simple activity would render physicians themselves

> responsible for infecting and killing patients. He was quite vocal in laying blame on his fellows, and this lack of social diplomacy made him very unpopular. To the point, his innovation was ignored by most of his contemporaries. Therefore, the rejection of his innovation was due in large part to the arrogance of the medical community as a whole, many of whom chose to continue working in ignorance with their egos intact rather than admit their own direct role in the problem. In stark contrast, Dr. Moniz…came up in the **post-industrial age**, a new and enlightened era that supported **progressive thinking** and **cutting-edge innovation** by minds such as Tesla, Rutherford and Einstein. The time was ripe for exploration in all fields of science, including medicine. Unlike Semmelweis, Moniz's innovation did not hold physicians accountable for any wrongdoing, nor did it require an overhaul of the medical profession. Rather it involved a new surgical technique to be practiced exclusively by highly skilled professionals, correlating its usage with cachet rather than stigma. Furthermore, the innovation of Dr. Moniz focused on a patient cohort consisting of the mentally ill, a pitiful group whose treatment was clearly in need of medical advancement and whose plight was in no way caused by doctor intervention (prior to treatment). Rather, the patients themselves were to some extent held in contempt, and the blame placed on their own shoulders.
>
> Dr. Moniz died in 1955, less than a decade after being regaled for his innovative pursuits and winning the Nobel Prize. Although confined to a wheelchair after being shot by a patient in 1949, he enjoyed professional celebrity for his innovation. However, with the passage of time, his techniques and devices have come to be viewed with disdain. In fact, lobotomies have attained infamy in the horror genre. The only purpose left for leucotomes nowadays are as hideous, gruesome and sideshow-esque collectibles. Conversely, Dr. Semmelweis died alone, humiliated and reviled. Frustrated to the limit, he suffered a mental breakdown and was committed to an asylum in 1865, where he died within days after being severely beaten by guards. Years passed until Louis Pasteur established germ theory and proved the accuracy of the Semmelweis data. The innovation of Semmelweis, washing of hands with disinfectant, was borne out as a pivotal advancement in scientific thought. Not only do surgeons today scrub before and after surgery, but the garments they wear are called "scrubs", a term which has entered into vernacular usage as a result of Semmelweis' insight. In addition to countless departments and academic institutions bearing his name, Dr. Semmelweis has been honored with his likeness on the Austrian 50 Euro gold coin, anno 2008. Perhaps the most moving and meaningful legacy of Semmelweis is the attitude of curiosity and open-mindedness our current age enjoys, open to interpret innovative ideas from all angles, and not reject a new idea out of hand simply because it contradicts the established zeitgeist; this would render one guilty of a "Semmelweis reflex"…. Indeed, ultimately, the decision to condone or condemn rests with peers.

We scientists own something to the broader public, not because of any fictional "cosmic duty," but because we often receive public funds—partially paid by them, the taxpayers—to do our research, particularly when we have grants from institutions such as NSF or NIH in the U.S., as I do. Therefore, we need to be humble enough to go down from our castles full of "erudite scientists" and strong egos and from the 5-star hotels where an endless number of scientific meetings take place. We should go down of our pedestal and meet and communicate with the broader public. The way in which many "elite" scientists behave, and many things they do, such as staying in such 5-star hotels to attend such meetings, often paying those very expensive hotels with grant money that comes in part from taxpayers, reminds me many people I know in Washington DC that work for the World Bank. When they go to countries such as Benin, in which a huge amount of people live with less than 1 dollar a day, they often spend dozens of thousands of dollars flying in first class and then sleeping for a week or two in the few 5-star hotels available in those countries. There, they often just have 1 or 2 days to do "field work"—a term that sounds as if they were going to study some non-human animals in a jungle—that is, to go to small villages to meet the people they are supposed to "help."

As pointed out in Malik's 2014 book *The Quest for a Moral Compass* (see also Box 9.10):

> Epicurus was born around 342 BCE into a family of Athenian expatriates on the Aegean island of Samos. In 306 he took up residence in a house just outside the walls of **Plato's Academy** in Athens, which became a kind of philosophical commune. Though he apparently wrote more than three hundred books, all that survive of his writings are three letters and two sets of maxims. Much of what we know of Epicurus, we know from the sketch provided by **Diogenes Laertius**, the famous third century biographer, in his Lives of the Eminent Philosophers. The starting point of Epicurus' philosophy was the attainment of pleasure, which he described as "the beginning and end of the blessed life". He was not, however, a hedonist in the way that we might understand it. He and his commune lived more like Mahatma Gandhi than Mick Jagger; they abstained from sex, partied little, and survived mainly on a diet of bread and water. A Saturday night knees-up at Epicurus' commune was clearly an occasion to lift the spirit. "I spit on luxurious pleasures", Epicurus wrote, "because of the inconveniences that follow them". The wise man would not only abstain from sex but would "not fall in love" or "marry and rear a family". Nor would he partake in any form of public life. It was best to avoid both fame and power, Epicurus thought, for these created enemies. And "since the attainment of great wealth can scarcely be accomplished without slavery to crowds or to politicians", so "a free life cannot obtain much wealth". **Pleasure** was not about self-gratification but about the elimination of pain, by which Epicurus meant not just physical pain but also fear, worry, passion and envy. Pain was everywhere, in every desire, every temptation, every act of consumption. It was a notion of pleasure that could only make sense in a stormy, turbulent era when the idea of withdrawal to a safe harbour might have seemed the most precious of luxuries. "Our one need", as Epicurus put it, "is untroubled existence".

Box 9.10: Leucippus, Democritus, Epicurus, Lucretius, and the Joy of Life

In Rovelli's 2017 *Reality is not what it seems—the journey to quantum gravity*, he explained:

> From time immemorial, or at least since humanity had left written texts which have come down to us, men had asked themselves how the world had come into being, what it was composed of, how it was ordered, and why natural phenomena occurred. For thousands of years they had given themselves answers which all resembled one another: answers which referred to elaborate stories of **spirits**, **deities**, imaginary and **mythological creatures**, and other similar things. From **cuneiform tablets** to **ancient Chinese texts**; from **hieroglyphic writing in the Pyramids** to the **myths of the Sioux**; from the most **ancient Indian texts** to the **Bible**; from African stories to those of aboriginal Australians, it was all a colourful but basically quite monotonous flow of Plumed Serpents and Great Cows, of irascible, litigious, or kindly deities who create the world by breathing over abysses, uttering 'Fiat lux', or emerging out of a stone egg. Then, at Miletus, at the beginning of the fifth century before our era, **Thales**, his pupil **Anaximander**, **Hecataeus** and their school find a different way of looking for answers. This immense revolution in thought inaugurates a new mode of

knowledge and understanding, and signals the first dawn of scientific thought. The Milesians understand that by shrewdly using observation and reason, rather than searching for answers in fantasy, ancient myths or religion and, above all, by using critical thought in a discriminating way it is possible to repeatedly correct our world view, and to discover new aspects of reality which are hidden to the common view. It is possible to discover the new. According to legend, **Heracles** descended to Hades from Cape Tenaro. Hecataeus visits Cape Tenaro, and determines that there is in fact no subterranean passage or other access to Hades there and therefore judges the legend to be false. This marks the dawn of a new era. This new approach to knowledge works quickly and impressively. Within a matter of a few years, Anaximander understands that the Earth floats in the sky and the sky continues beneath the Earth; that rainwater comes from the evaporation of water on Earth; that the variety of substances in the world must be susceptible to being understood in terms of a single, unitary and simple constituent, which he calls apeiron, the indistinct; that the animals and plants evolve and adapt to changes in the environment, and that man must have evolved from other animals. Thus, gradually, was founded the basis of a grammar for understanding the world which is substantially still our own today.

Together…two thinkers have built the majestic cathedral of ancient atomism: **Leucippus** was the teacher…. **Democritus**, the great pupil who wrote dozens of works on every field of knowledge, and was deeply venerated in antiquity, when people were familiar with these works. The idea of Democritus's system is extremely simple: the entire universe is made up of a boundless space in which innumerable atoms run. This is the weave of the world. This is reality. Everything else is nothing but a by-product, random and accidental, of this movement and this combining of atoms. The infinite variety of the substances of which the world is made derives solely from this combining of atoms. Just as by combining the letters of the alphabet in different ways we may obtain comedies or tragedies, ridiculous stories or epic poems, so elementary atoms combine to produce the world in its endless variety. The metaphor is Democritus' own. *There is no finality, no purpose, in this endless dance of atoms.* We, just like the rest of the natural world, are one of the many products of this infinite dance. The product, that is, of an accidental combination. Nature continues to experiment with forms and structures; and we, like the animals, are the products of a selection which is random and accidental, over the course of eons of time. Our life is a combination of atoms, our thoughts are made up of thin atoms, our dreams are the products of atoms; our hopes and our emotions are written in a language formed by combinations of atoms; the light which we see is comprised of atoms which bring us images. The seas are made of atoms, as are our cities, and the stars. It's an immense vision; boundless, incredibly simple and incredibly powerful. Plato and Aristotle were familiar with Democritus's ideas, and fought against them. They did so on behalf of other ideas, some of which were later, for centuries, to create obstacles to the growth of knowledge. Both insisted on rejecting **Democritus's naturalistic explanations**, in favour of trying to understand the world in finalistic terms believing, that is, that everything that happens has a purpose; a way of thinking that would reveal itself to be very misleading for understanding the ways of nature or in terms of good and evil, confusing human issues with matters which do not relate to us.

I often think that the loss of the works of Democritus in their entirety is the greatest intellectual tragedy to ensue from the collapse of the old classical civilization. Take a look at the list of his works in the footnote; it is difficult not to be dismayed, imagining what we have lost of the vast scientific reflections of antiquity. We have been left with all of **Aristotle**, by way of which Western thought reconstructed itself, and nothing by Democritus. Perhaps, if all of the works of Democritus had survived, and nothing of Aristotle's, the intellectual history of our civilization would have been better…. But centuries dominated by monotheism have not permitted the survival of **Democritus's naturalism**. The closure of the ancient schools such as those of Athens and Alexandria and the destruction of all the texts not in accordance with Christian ideas were vast and systematic, at the time of the brutal anti-pagan repression following from the edicts of **Emperor Theodosius**, which, in 3901 declared that **Christianity** was to be the only and obligatory religion of the empire. **Plato** and Aristotle, pagans who believed in the immortality of the soul or in the existence of a Prime Mover, could be tolerated by a triumphant Christianity. Not Democritus.

But a text survived the disaster, and has reached us in its entirety. Through it, we know a little about ancient atomism and, above all, we know the spirit of that science. It is the splendid poem *De rerum natura* (*The Nature of Things*, or *On the Nature of the Universe*), by the Latin poet **Lucretius**. The beauty of the poem lies in the sense of wonder which pervades the vast atomistic vision. There is a sense of luminous calm and serenity about the poem, which comes from understanding that there are no capricious Gods demanding of us difficult things, and punishing us. There is a vibrant and airy joyfulness. And there is a serene acceptance of the inevitability of death, which cancels every evil and about which there is nothing to fear. For Lucretius, religion is ignorance; reason is the torch that brings light…the rediscovery of *De rerum natura* had a profound effect upon the Italian and **European Renaissance** and its echo resounds, directly or indirectly, in the pages of authors ranging from **Galileo to Kepler**, and from **Bacon** to **Machiavelli**. The **Catholic Church** attempted to stop Lucretius: in the **Florentine Synod of December 1516** it prohibited the reading of Lucretius in schools. In 1551, the **Council of Trent** banned his work. But it was too late. An entire vision of the world which had been swept away by medieval Christian fundamentalism was re-emerging in a Europe which had reopened its eyes. It was not just the **rationalism**, **atheism** and **materialism** of Lucretius that were being proposed in Europe. It was not merely a luminous and serene meditation on the beauty of the world. It was much more: it was an articulate and complex structure of thinking about reality, a new mode of thinking, radically different from what had been for centuries the mind-set of the Middle Ages. The medieval cosmos so marvelously sung by **Dant**e was interpreted on the basis of a hierarchical organization of the universe which reflected the hierarchical organization of European society: a spherical cosmic structure with the Earth at its centre; the irreducible separation between Earth and heavens; finalistic and metaphorical explanations of natural phenomena. Fear of God, fear of death; little attention to nature; the idea that forms preceding things determine the structure of the world; the idea that the source of knowledge could only be the past, in revelation and tradition…. There is none of this in the world of Democritus as sung by Lucretius. There is no fear of the gods; no ends or purposes in the world; no cosmic hierarchy; no distinction between Earth and heavens. There is a deep love of nature, a serene immersion within it; a recognition that we are profoundly part of it; that men, women, animals, plants and clouds are organic threads of a marvelous whole, without hierarchies. There is a feeling of deep universalism in the wake of the splendid words of Democritus: 'To a wise man, the whole earth is open, because the true country of a virtuous soul is the entire universe'.

Taking into account what we have seen in the paragraphs above and in Box 9.10, it is not surprising that the epigraph of **Claude Levi-Strauss**' influential book *Tristes Tropiques* was drawn from *De rerum natura*, a work written by Epicurus's latter-day disciple **Lucretius**' that precisely refers to a world in perpetual change and movement without an overlying cosmic purpose: *nec minus ergo ante haec* *quam tu cecidere, cadentque*. Levi-Strauss' book was originally published in French in 1955, and both that version and the subsequent English version have front covers that are, in my opinion, among the most beautiful and respectful ever published in scientific books: a wonderful, considerate tribute to the indigenous people of Brazil that he encountered during his travels (see Fig. 9.4). In the book, Levi-Strauss wrote:

> The notion of travel has become corrupted by the notion of power. No longer can travel yield up its treasures intact: the islands of the South Seas, for instance, have become stationary aircraft-carriers; the whole of Asia has been taken sick; shanty-towns disfigure Africa; commercial and military aircraft roar across the still virgin but no longer unspoilt forests of South America and Melanesia.... The great civilization of the West has given birth to many marvels; but at what a cost! As has happened in the case of the most famous of their creations, that atomic pile in which have been built structures of a complexity hitherto unknown, the order and harmony of the West depend upon the elimination of that prodigious quantity of maleficent by-products which now pollutes the earth.... I understand how it is that people delight in travel-books and ask only to be misled by them. Such books preserve the illusion of some thing that no longer exists, but yet must be assumed to exist if we are to escape from the appalling indictment that has been piling up against us through twenty thousand years of history. There s nothing to be done about it: civilization is no longer a fragile flower, to be carefully preserved and reared with great difficulty here and there in sheltered corners of a territory rich in natural resources: too rich, almost, for there was an element of menace in their very vitality; yet they allowed us to put fresh life and variety into our cultivations. All that is over: humanity has taken to monoculture, once and for all, and is preparing to produce civilization in bulk, as if it were sugar-beet. The same dish will be served to us every day.... Now that the Indians masks have been destroyed...(photographic) albums have taken their place. Perhaps our readers hope, by the intermediacy of these colour-plates, to take on something of the Indian charms? To have destroyed the Indians is not enough the public may, indeed, not realize that the destruction has taken place and what the reader wants is to satisfy, in some sort, the cannibal-instincts of the historical process to which the Indians have already succumbed..
>
> I should have liked to live in the age of real travel, when the spectacle on offer had not yet been blemished, contaminated, and confounded; then I could have seen Lahore not as I saw it, but as it appeared to Bernier, Tavernier, Manucci There s no end, of course, to such conjectures. When was the right moment to see India? At what period would the study of the Brazilian savage have yielded the purest satisfaction and the savage himself been at his peak? Would it have been better to have arrived at Rio in the eighteenth century, with Bougainville, or in the sixteenth, with Lery and Thevet? With every decade that we travelled further back in time, I could have saved another costume, witnessed another festivity, and come to understand another system of belief. But I am too familiar with the texts not to know that this back ward movement would also deprive me of much information, many curious facts and objects, that would enrich my meditations. The paradox is irresoluble: the less one culture communicates with another, the less likely they are to be corrupted, one by the other; but, on the other hand, the less likely it is, in such conditions, that the respective emissaries of these cultures will be able to seize the richness and significance of their diversity.
>
> The alternative is inescapable: either I am a traveler in ancient times, and faced with a prodigious spectacle which would be almost entirely unintelligible to me and might, indeed, provoke me to mockery or disgust; or I am a traveler of our own day, hastening in search of a vanished reality. If we want to correlate the appearance of writing with certain other characteristics of **civilization**, we must look elsewhere. The one phenomenon which has invariably accompanied it is the formation of cities and empires: the integration into a political system, that is to say, of a considerable number of individuals, and the distribution of those individuals into a hierarchy of **castes** and classes. *Such is, at any rate, the type of develop-*

> *ment which we find, from Egypt right across to China, at the moment when* **writing** *makes its debuts; it seems to favour rather the exploitation than the enlightenment of mankind.* This **exploitation** made it possible to assemble workpeople by the thousand and set them tasks that taxed them to the limits of their strength: to this, surely, we must attribute the beginnings of architecture as we know it. If my hypothesis is correct, the primary function of writing, as a means of communication, is to facilitate the enslavement of other human beings. The use of writing for disinterested ends, and with a view to satisfactions of the mind in the fields either of science or the arts, is a secondary result of its invention and may even be no more than a way of reinforcing, justifying, or dissimulating its primary function.

One point that needs to be made is that I avoided to read *Tristes tropiques* before I wrote this Chapter, because I thought that it would be particularly relevant for the topics discussed in it. So, it was striking to realize, after I read that book, that its ideas are so conspicuously similar to many of the points made in the above Chapters. Levi-Strauss always made it clear, in his works and interviews, that he never argued that what people call "traditional societies" are "superior" or "better" than "modern" ones. Instead, he explained, as I did throughout this volume, that both of them are plagued by teleological narratives and social norms, and argued that each human culture has some traits that can be considered "positive" and others that can be seen as "negative": above all, this depends mainly on the circumstances, not on the individuals themselves being "good" or "bad." This is *not* cultural relativism, as some proclaim. In fact, the discussions and empirical data provided in the above Chapters and Sections refute that proclamation by showing that when one objectively compares different aspects of the lifeways of different cultures and societies, *using empirical data*, some so-called negative traits, such as infanticide and child mortality, are more typically seen in nomadic hunter-gatherer societies, while others, such as slavery or warfare, are more prevalent in agricultural "civilizations." So, not relativism, but instead scientific knowledge, based on the compilation of empirical facts and discussions about what those facts tell us about the *reality* of life in this planet: in a way, "**experimental philosophy**" or "**scientific philosophy,**" as explained in Chap. 1. Obviously, Levi-Strauss exposes the ideas that are strikingly similar to some of those made in the previous Chapters in a much more beautiful and elegant way than I do, for instance stating:

> It will eventually become plain that no human society is fundamentally good: but neither is any of them fundamentally bad; all offer their members certain advantages, though we must bear in mind a residue of iniquity, apparently more or less constant in its importance, which may correspond to a specific inertia which offers resistance, on the level of social life, to all attempts at organization. This may surprise the habitual reader of travel-books, who delights in hearing of the barbarous customs of this people or that. But these superficial reactions are soon put in their place, once the facts have been correctly interpreted and re-established in a wider perspective. Take the case of **cannibalism**, which is of all savage practices the one we find the most horrible and disgusting. We must set aside those cases in which people eat one another for lack of any other meat as was the case in certain parts of Polynesia. No society is proof, morally speaking, against the demands of hunger. In times of starvation men will eat literally anything, as we lately saw in the **Nazi extermination-camps**. There remain to be considered what we may call the positive forms of cannibalism those whose origins are mystical, magical, or religious. By eating part of the body of an ancestor, or a fragment of an enemy corpse, the cannibal hoped to acquire the virtues, or perhaps to neutralize the power, of the dead man. Such rites were often observed with great discretion, the

vital mouthful being made up of a small quantity of pulverized organic matter mixed, on occasion, with other forms of food. And even when the element of cannibalism was more openly avowed, we must acknowledge that to condemn such customs on moral grounds implies either belief in a bodily resurrection, which would be compromised by the material destruction of the corpse, or the affirmation of a link between body and spirit, and of the resulting dualism. These convictions are of the same nature as those in the name of which ritual cannibalism is practised, and we have no good reason for preferring the one to the other all the more so as the disregard for the sanctity of death, with which we reproach the cannibal, is certainly no greater, and indeed arguably much less, than that which we tolerate in our European anatomy (dissection) lessons.

But above all we must realize that certain of our own usages, if investigated by an observer from a different society, would seem to him similar in kind to the cannibalism which we consider uncivilized. I am thinking here of our judicial and penitentiary customs. If we were to look at them from outside it would be tempting to distinguish two opposing types of society: those which practise cannibalism who believe, that is to say, that the only way to neutralize people who are the repositories of certain redoubtable powers, and even to turn them to one's own advantage, is to absorb them into one's own body. Second would come those which, like our own, adopt what might be called ***anthropoemia*** (from the Greek emein, to vomit). Faced with the same problem, they have chosen the opposite solution. They expel these formidable beings from the body public by isolating them for a time, or for ever, denying them all contact with humanity, in establishments devised for that express purpose. In most of the societies which we would call primitive this custom would inspire the profoundest horror: we should seem to them barbarian in the same degree as we impute to them on the ground of their no-more-than-symmetrical customs. Societies which seem to us ferocious may turn out, when examined from another point of view, to have their humane and benevolent sides. Take the Plains Indians of North America: they are doubly significant first because some of them practised a moderated form of cannibalism, and second because they are one of the few primitive peoples who were endowed with an organized police force. This force, which also had to mete out justice, would never have imagined that the punishments accorded to the guilty could take the form of a severance of social links. An Indian who broke the laws of his tribe would be sentenced to the destruction of all his belongings his tent and his horses. But at the same time the police became indebted to him and were required, in fact, to compensate him for the harm he had been made to suffer. This restitution put the criminal, once again, in debt to the group, and he was obliged to acknowledge this by a series of gifts which the entire community including the police would help him to get together. These reciprocities continued, by way of gifts and counter-gifts, until the initial disorder created by the crime and its punishment had been completely smoothed over and order was once again complete. Not only are such customs more humane than our own, but they are more coherent.

As **Claude Levi-Strauss**, also **Carl Sagan** stood in the shoulders of **Epicurus** and **Lucretius**, referring to the latter in his 1997 book *The Demon-Haunted World—Science as a Candle in the Dark*. That book is a masterpiece due to the way in which Sagan meticulous dismantles teleological narratives, and to its passionate defense of science and of science communication and the use of *multidisciplinary empirical data*—not "new narratives"—to fight against such narratives and pseudoscience. Above all, the book is truly a fascinating ode to the *world of reality*, which was clearly an obsession for Sagan in the end of his life. Indeed, Sagan seemingly was fully aware of what was going to happen—and *is happening*—around the globe, including the U.S. That is, the rise of **skepticism about science and scientists**, of so-called **anti-intellectualism**, of **New agers** and their old-beliefs and recycled teleological narratives, of **pseudoscience**, of **"fake news"** and of **"alternative facts"**:

If we long to believe that the stars rise and set for us, that we are the reason there is a Universe, does science do us a disservice in deflating our conceits? Science may be hard to understand. It may challenge cherished beliefs. When its products are placed at the disposal of politicians or industrialists, it may lead to weapons of mass destruction and grave threats to the environment. But one thing you have to say about it: it delivers the goods. Not every branch of science can foretell the future – paleontology can't – but many can and with stunning accuracy. If you want to know when the next eclipse of the Sun will be, you might try magicians or mystics, but you'll do much better with scientists. They will tell you where on Earth to stand, when you have to be there, and whether it will be a partial eclipse, a total eclipse, or an annular eclipse. They can routinely predict a solar eclipse, to the minute, a millennium in advance. You can go to the witch doctor to lift the spell that causes your pernicious anemia, or you can take vitamin B12. If you want to save your child from polio, you can pray or you can inoculate. If you're interested in the sex of your unborn child, you can consult plumb-bob danglers all you want (left-right, a boy; forward-back, a girl – or maybe it's the other way around), but they'll be right, on average, only one time in two. If you want real accuracy (here, 99 per cent accuracy), try amniocentesis and sonograms. Try science. Think of how many religions attempt to validate themselves with prophecy. Think of how many people rely on these prophecies, however vague, however unfulfilled, to support or prop up their beliefs. Yet has there ever been a religion with the prophetic accuracy and reliability of science? There isn't a religion on the planet that doesn't long for a comparable ability – precise, and repeatedly demonstrated before committed skeptics – to foretell future events. No other human institution comes close.

But **superstition** and **pseudoscience** keep getting in the way…providing easy answers, dodging sceptical scrutiny, casually pressing our awe buttons and cheapening the experience, making us routine and comfortable practitioners as well as victims of **credulity**. These are all instances of pseudoscience. They purport to use the methods and findings of science, while in fact they are faithless to its nature – often because they are based on insufficient evidence or because they ignore clues that point the other way. They ripple with gullibility. With the uninformed cooperation (and often the cynical connivance) of newspapers, magazines, book publishers, radio, television, movie producers and the like, such ideas are easily and widely available. Far more difficult to come upon…are the alternative, more challenging and even more dazzling findings of science. Pseudoscience is easier to contrive than science, because distracting confrontations with reality – where we cannot control the outcome of the comparison – are more readily avoided. The standards of argument, what passes for evidence, are much more relaxed. In part for these same reasons, it is much easier to present pseudoscience to the general public than science. But this isn't enough to explain its popularity. Naturally people try various belief systems on for size, to see if they help. And if we're desperate enough, we become all too willing to abandon what may be perceived as the heavy burden of scepticism. Pseudoscience speaks to powerful emotional needs that science often leaves unfulfilled. It caters to fantasies about personal powers we lack and long for (like those attributed to comic book superheroes today, and earlier, to the gods). In some of its manifestations, it offers satisfaction of spiritual hungers, cures for disease, promises that death is not the end. It reassures us of our cosmic centrality and importance. It vouchsafes that we are hooked up with, tied to, the Universe. Sometimes it's a kind of halfway house between old religion and new science, mistrusted by both. At the heart of some pseudoscience (and some religion also, New Age and Old) is the idea that wishing makes it so. How satisfying it would be, as in folklore and children's stories, to fulfill our heart's desire just by wishing. How seductive this notion is, especially when compared with the hard work and good luck usually required to achieve our hopes. The enchanted fish or the genie from the lamp will grant us three wishes – anything we want except more wishes. Who has not pondered – just to be on the safe side, just in case we ever come upon and accidentally rub an old, squat brass oil lamp – what to ask for?

> The ancient Ionians were the first we know of to argue systematically that laws and forces of Nature, rather than gods, are responsible for the order and even the existence of the world. As **Lucretius** summarized their views, 'Nature free at once and rid of her haughty lords is seen to do all things spontaneously of herself without the meddling of the gods.' Except for the first week of introductory philosophy courses, though, the names and notions of the early Ionians are almost never mentioned in our society. *Those who dismiss the gods tend to be forgotten.* We are not anxious to preserve the memory of such sceptics, much less their ideas. Heroes who try to explain the world in terms of matter and energy may have arisen many times in many cultures, only to be obliterated by the priests and philosophers in charge of the conventional wisdom, as the Ionian approach was almost wholly lost after the time of **Plato** and **Aristotle**. With many cultures and many experiments of this sort, it may be that only on rare occasions does the idea take root. In *The genealogy of morals*, **Friedrich Nietzsche**, as so many before and after, decries the 'unbroken progress in the selfbelittling of man' brought about by the scientific revolution. Nietzsche mourns the loss of 'man's belief in his dignity, his uniqueness, his irreplaceability in the scheme of existence'…[but for me] to discover that the Universe is some 8 to 15 billion and not 6 to 12 thousand years old [instead] improves our appreciation of its sweep and grandeur; to entertain the notion that we are a particularly complex arrangement of atoms, and not some breath of divinity, at the very least enhances our respect for atoms; to discover, as now seems probable, that our planet is one of billions of other worlds in the Milky Way galaxy and that our galaxy is one of billions more, majestically expands the arena of what is possible; to find that our ancestors were also the ancestors of apes ties us to the rest of life and makes possible important – if occasionally rueful – reflections on human nature. For me, *it is far better to grasp the Universe as it really is than to persist in delusion, however satisfying and reassuring.*

It is indeed remarkable to think that millennia ago a minority of people defended a key idea defended in the present book, but that mainly become lost in history for such a long time: that teleological tales, particularly those related to the inevitability of death and the related creation of stories about cosmic purpose and progress and human specialness, have indeed be used to "terrorize innocent people." As well as to "force them to behave in ways contrary to their health, *happiness* and *pleasure,*" by making "*baseless and irrational fears to overwhelm their mental lives.*" As put by **Lucretius**, one should "dispel the fear of death and destroy any foolish notions of life beyond the grave" in order to "drive out neck and crop that fear of Hell which blasts the life of a person from its very foundations, sullying everything with the blackness of death and the leaving no pleasure pure and unalloyed." Lucretius was however not completely immune to the tentacles of *Neverland*: he defended a few imaginary teleological tales, for instance that human history was, in a certain way, directed towards a natural "progress" (see Box 3.7). But, still, he and other Epicureans were, in many ways, among the thinkers that were more near to escaping *Neverland.* They realized that a major problem faced by the *Homo fictus, irrationalis et socialis* is that the imaginary and often absurd narratives created by him/her are often reinforced by the bombardment of an endless number of societal myths, rules, and norms since a very young age, thus not only leading to huge *societal abuses and atrocities* but usually also to a high cost paid *at an individual level.* Basically, most people blindly follow those rules and norms, created by *others*, often centuries ago and for reasons that don't have nothing to do with the context in which we live today, such as not eating pork, or eating only kosher food.

It is, however, important to note that the feeling that one needs to "fit in," which is often referred to as **behavioral conformism** in scientific discussions, is also found in many other animal species, including our closest living relatives, the apes, and monkeys. A very interesting empirical experimental example of that—reviewed in De Wall's 2016 outstanding book *Are We Smart Enough to Know How Smart Animal Are?*—concerns monkeys and shows how **conformism** often tends to lead, in the long term, to **evolutionary mismatches**, exactly as it has led in human history. As we have seen above, both the habitats of this planet and biological evolution per se are highly dynamic, so even a trait that was advantageous at a certain time, or a neutral trait acquired as a by-product of another feature that was advantageous at that time, often becomes detrimental after a long time. In that study, researchers gave **vervet monkeys** living in a game reserve various open plastic boxes with maize corn. There were always two boxes with two colors of corn, blue and pink: one color was good to eat while the other was laced with aloe, making it repulsive for the monkeys. Depending on which color corn was palatable, and which was not, some groups choose to eat blue, and others pink. After some time, researchers removed the distasteful treatment—that is, from that moment on, the blue and pink food would taste exactly the same. Researchers then waited for infants of each group to be born and new males to immigrate from neighboring areas to the group.

The results clearly showed how strong behavioral conformism was in those groups: all adults continued to stuck to their acquired preference, despite the fact that both foods now tasted exactly the same. In other words, they never even discovered the improved taste of the alternative color because they *never even tried it*: that is, they did not adapt to the new ecological conditions. With one exception—an infant whose mother was so low in rank, and that was very hungry—all 27 newborn infants learned to eat only the locally preferred food, not touching the other color, even though it was freely available and just as good as the other. However, there was no advantage at all, well on the contrary, as both colors were now palatable and thus, *in reality*, a useful source of nutrient for the monkeys. In summary, such a **behavioral social conformism** somehow contributes to a *world of unreality*, in the sense that a certain group would only eat food of a certain color, and reject the food with the other color, as if they currently lived in a reality where only one food was not repulsive—something that was only the truth previously, and for a very short amount of time: *it was not the reality of here and now*. The results of this study go in line with what I said above about **New Agers**: *curiosity "kills" not the cat, but the believer*.

Additional results of the same study showed the even higher impact of behavioral social conformism: a small group of male immigrants that moved between groups ended up by starting then to eat the color that was being preferred by the monkey group to which they were transferred to, even if they arrived from a group with the opposite preference. In this sense, this case showed how even something similar to "culture" or "**social identity**" can easily be changed by **social conformism** and the **need to "fit in**." One could argue that this is actually "good," to be more **socially tolerant**, and don't just stick with one's **cultural identity**: in this case, that small group of monkeys did not see preferences of "others" as something "bad," instead adapting to them. Yes, that is true. However, the fact is that in *both cases*—the monkeys that stayed in a group and continued to only eat the food color preferred by that group, and the monkeys that changed groups and started to eat the food color preferred by the new group—there is a kind of extremism, intolerance, and surely of unreality. This is because the most logical scenario, the one that would truly make sense within the *world of reality*, would be to eat *both* food colors, which at that time were already again, *factually*, both pleasant and a valuable source of nutrient for the monkeys. The social conformism displayed by the monkeys in this study does indeed resemble some examples found in humans, such as the fact that many Jews continue to only eat Kosher food. Moreover, it also shows a key, and extremely dangerous, difference between humans and other animals, which is at the core of the present book. Humans not only blindly follow the crowd to "fit in" as monkeys do—*Homo socialis*—but create in addition complex narratives to further reinforce such "cultural norms," for example saying that "God does not want you to eat food A" (*Homo fictus*), and then start to believe in such narratives even if they do not reflect the *true reason* why the cultural preference was adopted in the first place (*Homo irrationalis*). An emblematic illustration of this, discussed in Chap. 5, concerns the fact that in Western countries boys are now often associated with the blue color—that is, the way they dress, the material possessions they have, and so on—and girls with the pink color. That just has nothing to do with reality at all, either now or in the past, it was a completely random behavioral choice that started for no **adaptive reason** and that, despite the fact that almost nobody actually knows why and when it started, persists for completely illogical reasons, due to social conformism linked to *Neverland*'s **imaginary narratives linking gender and color preferences**.

In the previous Chapters, we have seen how the obsession of the *Homo socialis* to "fit in" leads to cases in which even if an individual—or for that matter a whole group of individuals—does not truly believe on the imaginary narratives and resulting social norms and rules made by the *Homo fictus et irrationalis*, he/she will still tend to follow them in order to not be seen as an "outcast." Importantly, this is not only merely because the *Homo socialis* wants to *feel* "part of a group," but also because being seen as an "outcast" can put one's live in risk. History tragically have shown us this, over and over again: the **stoning to death of "adulterers"** in countries such as Afghanistan nowadays is, in that sense, not different from the **burning of witches** in medieval Europe. In actual fact, one of the key elements about the imaginary stories about **witches**, and both the fear and fascination that they produce, is precisely because witches are almost always seen as "**outcasts**" that do *not blindly follow the social norms and rules of the societies* in which they live. That is, they have **agency**, they are not passive *Homo servus*: they actively "think with their own heads," and that is mainly why they do *not* follow the crowd or obey to other humans. That is something that the socio-political "system" created by human "civilizations" can't stand: thence, they tend to live alone, in remote places, either because they actively want or because they actively escaped from the villages or cities precisely because they were seen as outcasts and could be at great risk. And this show us the reverse, and very dangerous, coin of the *Homo socialis* component, which is the one that is deeply related to our *Homo servus* tendencies: that not only he/she is usually obsessed to fit in but, when a few "Others" are exceptions to that rule, they usually feel discomforted, and often also afraid of Them.

The *Neverland* reasoning created and believed by the storytelling *Homo socialis* within the context of an acculturated "civilized" crowd is: if that person is not, and does not want to be, a *Homo servus*, surely it is because he/she is up to no good, this must have to be the work of the devil, or of evil spirits. So, accordingly, the crowd often threatens and/or tries to oblige those outcasts to conform to the norm, to become "good normal productive servers," but if they resist, then the general public and the "system" as a whole start to discriminate, abuse, torture, or even kill those outcasts. As we all know, the imaginary stories about "witches" have indeed led, and continue to lead to a lesser extent (Fig. 7.2), to the discrimination against and subjugation, oppression, and the killing of a countless number of outcasts. The tragic conclusion is: even those that eventually try to avoid to fall into the trap created by the explosive combination of the *Homo fictus, irrationalis, socialis et servus* are often promptly, and aggressively, reminded that they should *not* even attempt to do so.

A particularly timely, and very tragic example of this, which often has horrible long-term consequences for the women to which it is done, is **female genital mutilation** (FGM), a procedure that was already mentioned in Chap. 5. This practice is one of the most atrocious examples of how far humans are detached from the world of reality. If I asked you: do you think that wolves, or birds, or spiders, cut parts of their own bodies that not only give them *a lot* of pleasure but that are also involved with their own reproduction, resulting in *excruciating pain* and in a plethora of long-term physiological problems? My guess is that you would answer me: no, of course not, such animals are not so stupid, not even spiders. And you would be completely right: there are reports of **self-injurious behavior** in some non-human animals, but they are often related to stress, usually seen in domesticated animals or other animals being abused by humans. And while a few cases do concern wild animals, they often concern particularly stressful, peculiar conditions and are moreover very different, and surely at a much lower scale, than the very widespread, horrible, and profoundly absurd practices done by the self-proclaimed "sapient being." Indeed, apart from creating and believing in totally unintelligent stories about how "good" cutting the genital parts of females is, it is well known that in the last decades this practice is so prevalent in places such as Africa and the Middle East chiefly because of social conventions, norms and pressure, related precisely with the fear of being excluded and becoming an outcast. As emphasized in a 2011 progress report of the World Health Organization (WHO) entitled "*An update on WHO's work on female genital mutilation*":

> Estimates based on survey data suggest that in Africa 91.5 million girls and women aged 10 years and above have been subjected to the practice. Of these, 12.4 million are between 10 and 14 years of age. In most cases, the procedure is carried out on girls under the age of 15 years, although obtaining data on FGM prevalence in that age group poses several methodological challenges. In some communities, and in some situations, women are subjected to FGM later in life; including when they are about to be married, or after marriage, during

pregnancy and after childbirth, or when their own daughters undergo the procedure. Most women who have experienced FGM live in one of the 28 countries in Africa and the Middle East – nearly half of them in just two countries: Egypt and Ethiopia. Countries in which FGM has been documented include: Benin, Burkina Faso, Cameroon, Central African Republic, Chad, Cote d'Ivoire, Djibouti, Egypt, Eritrea, Ethiopia, Gambia, Ghana, Guinea, Guinea-Bissau, Kenya, Liberia, Mali, Mauritania, Niger, Nigeria, Senegal, Sierra Leone, Somalia, Sudan, Togo, Uganda, United Republic of Tanzania and Yemen. The prevalence of FGM ranges from 0.6% to 98% of the female population. Incidences of FGM have been documented in some other countries, including India, Indonesia, Iraq, Israel, Malaysia, Thailand and the United Arab Emirates, but no national estimates have been made. In addition, the practice of FGM and its harmful consequences also concerns a growing number of women and girls in Europe, North America, Australia and New Zealand as a result of international migration. The exact number of women and girls living with FGM in Europe is unknown, but is estimated to be around 500 000, and 180 000 girls are estimated to be at risk of being subjected to the practice. The most common short-term consequences of FGM include severe pain, shock caused by pain and/or excessive bleeding (**haemorrhage**), difficulty in passing urine and faeces because of swelling, oedema and pain, as well as infections. Death can be caused by haemorrhage or infections, including tetanus and shock…[Regarding] long-term health risks…the most common complications are dermoid cysts and abscesses. **Chronic pelvic infections** that can cause chronic back and pelvic pain, and repeated urinary tract infections have been documented in both girls and adults. A recent WHO-led study showed that FGM is associated with increased risk for complications for both mother and child during childbirth. Rates of **caesarean section**…were both more frequent among women with FGM compared with those without FGM. In addition, there was an increased probability of tearing and recourse to **episiotomies**. The risk of birth complication increases with the severity of FGM. FGM of the mother is also a risk factor for the infant…the study found significantly higher death rates (including stillbirths) among infants born from mothers who have undergone FGM than women with no FGM…. FGM can also lead to negative psychological consequences. Documented effects include **posttraumatic stress disorder**, **anxiety**, **depression**, and **psychosexual problems**. A recent study shows that women who have undergone FGM may be more likely than others to experience **psychological disturbances** (psychiatric diagnosis, suffer from anxiety, **somatization**, **phobia** and **low self-esteem**).

Research has shown that **sexual problems** are also more common among women who have undergone FGM. Women with FGM were found to be 1.5 times more likely to experience pain during sexual intercourse, experience significantly **less sexual satisfaction** and they were twice as likely to report that they did not experience sexual desire…. Further surgery is usually necessary later in women's lives when infibulations must be opened to enable sexual intercourse and further again in childbirth. In some countries this is followed by re-closure (**reinfibulation**), and hence the need for repeated **defibulation** later. Urinary and menstrual problems are not uncommon, particularly prior to defibulation at first marriage. For many women sexual intercourse is painful during the first few weeks after sexual initiation, as the **infibulation** must be opened up either surgically or through penetrative sex. The male partner can also experience pain and complications. FGM is also associated with infertility…. Evidence suggests that the more tissue is removed, the higher the risk for infection. The continuation of FGM in a practising community is motivated by a complex mix of interlinked **sociocultural factors**, which vary from region to region, within single countries, between and even within practicing communities. FGM is generally practised as a matter of **social convention**, and is interlinked with **social acceptance**, **peer pressure**, the fear of not having access to resources and opportunities as a young woman and to secure prospects of marriage. Therefore individuals' actions are interdependent on those of others. This social convention is connected to different concrete **sociocultural perceptions**, most of which are linked to local **perceptions of gender**, sexuality and religion. With regards to gender, there is often a perception that women's bodies need to be 'carved' to become fully female. Often the **clitoris** is seen as 'male-like' organ that needs to be removed to ensure **pure femininity**. FGM is also frequently associated with **sexual morality**, and the perception of the clitoris as the origin of sexual desire. Hence the clitoris is removed in order to reduce women's sexual drive, in the belief that this will improve the prospect of **premarital virginity**, and **marital fidelity**, and to ensure 'decent behaviour'. The practice is often also linked to a **ritual marking of the coming of age** and **initiation to womanhood**. In many communities FGM is often perceived to be prescribed by the locally common religion, which includes Islam, Christianity and traditional faith systems.

Perhaps the most brilliant discussion I have ever read about the hugely important, and dangerous, link between the *Homo socialis* and the *Homo fictus et irrationalis* and how it feeds the *Homo servus*, is given in Malesevic' *The Rise of Organized Brutality*:

As scientific knowledge is cold and rational it cannot bind members of particular social organizations together. As Gellner argues powerfully: '**social cohesion** cannot be based on truth…truth butters no parsnips and legitimises no social arrangements'. As pure knowledge is factual, dry and coldly rational, it cannot in itself bind human beings nor provide comfort. The truth does not set you free; explanations are inevitably cold, blunt and heartless. Telling a simple truth to a small child that there is no **Tooth Fairy** or **Santa Claus** will make her more knowledgeable, but it is bound to hurt her feelings. While **Marx** and **Durkheim** saw alienation and anomie, respectively, as modern processes that lead away from one's true self, it is really truth that alienates, not illusions. Unlike religion, culture or ideology, which provide human beings with meanings and communal warmth, scientific truth is chillingly unsentimental. The sincere believers who attend a **religious ceremony**, tribesmen who take part in the **rain dance** or co-nationals who experience the shiver in their spines when the **national anthem** is played forge an intensive sense of belonging not by way of truth but through the **shared untruth**. Once one realises that dancing does not bring rain, that there is no god or that nationalist rituals are recent inventions, one attains knowledge at the expense of communal solidarity. The cold, rational truth brings enlightenment, but the price is very high: solitude, emotional deprivation and the lack of meaning. Hence, as human beings are first and foremost emotional creatures, all social organizations require potent and cohesive social glue to keep them together. Much of this glue is generated through the centrifugal ideologisation which supplies ideational ingredients for doctrinal organisational cohesion. Nevertheless, this is never a smooth and uncontested process but is regularly riddled with tensions, as bureaucratic and ideological principles clash and collide. To resolve or bypass these inherent tensions, the most efficient social organizations have managed to link organisational and ideological processes with **microsolidarity**…. It is difficult to image how networks of microsolidarity can be linked with bureaucratic units without centrifugal **ideologisation**. In this sense, all complex, longlasting social organizations tend to utilise specific ideological discourses to integrate large numbers of people. When successful, centrifugal ideologisation helps to bridge this huge gap between the **hyperrationality** of bureaucracy and intimacy of family and friendship. However, to accomplish this difficult task, ideologisation has to penetrate the hubs of microsolidarity and glue them to the organisational scaffold. This is usually achieved over long periods of time after many years of attempting to project the image of organizations as those resembling one's family and friends.

In some instances, the effort is made to develop ideological narratives that directly subsume networks of microsolidarity. For example, both **ethnonationalism** and some **religious fundamentalisms** embrace the kinship metaphors and refer to the actual or potential members of their organizations as 'brothers', 'sisters', 'sons' or 'daughters'. Thus both **ISIS** and **Al Qaeda** address their constituencies in such familial terminology, insisting that all Muslims are brothers and sisters, part of the great **umma**. Similarly, the newspapers and websites of **Basque nationalist organizations** refer to the Basque population as 'sons and

> daughters of the Basque land'. The ambition here is to represent a particular bureaucratic social organization (i.e. ETA, ISIS) as resembling an extended family. When social organizations are successful in projecting this kinship image, they are in position to attract a degree of strong emotional attachment that individuals usually reserve only for their closest friends and family. In this context, loyalty to organisational goals is understood in terms of the moral responsibility towards one's family members: if I do not work towards fulfilling these goals, I will disappoint my brothers and sisters or bring shame on my family. This link among ideologisation, **bureaucratisation** and microsolidarity is most apparent in the context of organizations that are the principal purveyors of violence. As soldiers, police officers, **paramilitaries**, **terrorists** and **revolutionaries** are regularly involved in the violent encounters, they need to know that their actions are legitimate and morally acceptable. This means that their respective organizations have to devise and implement effective and believable ideological mechanisms capable of bringing together organisational aims and micro-level attachments. Since ordinary human beings, as individuals, are not particularly comfortable with the use of violence, it is paramount that when such violent episodes occur that they are interpreted through the prism of a specific ideological frame. For example, the battlefield experience of soldiers who fought in the trenches of World War I was dependent not only on the capacity of a military organization to force individuals to shoot and kill other human beings but also on their ability to establish and successfully disseminate an ideological narrative which explains and justifies this experience. Only when soldiers recognised that they were fighting for a noble cause were they willing to kill and die for such a purpose. For military organizations to achieve this type of mass recognition, it was necessary to integrate the wider nationalist ideology with the sense of moral responsibility that soldiers expressed for their family, friends and close neighbours back home. Hence when this noble cause is articulated in the language of life preservation of those who are dearest to us, then ideology, organization and microsolidarity become successfully fused.

Another exceptional book that brilliantly demonstrates, within the modern context of a "developed" country such as the U.S., the huge *individual price* often paid by the *Homo irrationalis*, *fictus,* and *socialis* is **Arlie Hochschild**'s 2016 *Stranger in their Own Land*. She went to live in the South of the U.S. to try to understand the paradoxes of people that, according to sound empirical data, would in theory benefit at various levels if some specific so-called left-wing policies would be implement in the regions where they live, and importantly that often *were aware of that*, but still voted otherwise. The book is brilliant because it does not pretend that it is always more rational to vote for democrats in then U.S. That would make no sense, and would be just playing the game of "divide to conquer," as explained above: statistically, it is impossible that all republicans or right wing people, or alternatively all democrats or right-wing people, are always wrong or "bad." So, why did people vote in ways that would make them more poor, or the villages/cities where they live more polluted, and so on, including people that were fully aware of this fact? As she reports, their reasons chiefly had to do with **religious narratives** (*Homo fictus*), with what is typically done by most other people living in the same Southern regions of the U.S. (*Homo socialis*), and/or with the prioritizing of other items, such as honor or anger, over the items that they deemed to be the ones that would make their lives "better off" if democrats were to win (*Homo irrationalis*). As explained in Chap. 3, one of the most consistent findings of recent psychological, neurological, and even physiological—related for instance with the feeling of "disgust"—studies is that one of the major reasons why people tend to divide themselves into "conservative" versus "liberal" groups is related to the prioritizing of different "gut-feelings" and items associated for example with honor, order, or keeping stability/the status quo (right-wing) versus social equity and change (left-wing). Accordingly, Hochschild's book plainly contradicts the idea that most voters chiefly decide their vote in a rational way—the "sapient being" false narrative—and/or that "it's just the economy, stupid"—the *Homo economicus* fairytale. As explained by Harari in *Homo Deus*, this idea is one of the major imaginary theoretical fallacies made by what he calls "liberal democracies," which are therefore often "surprised" when people like Trump, or Bolsonaro, or Putin, win elections. Clearly, in the world of reality someone as prepared as Hillary Clinton should *in theory* be voted to be president of the U.S., rather than someone as unprepared, misogynistic, and racist such as Trump. But contrary to what "liberal democracies"—which, to be clear, are still by far the less atrocious and oppressive socio-political "system" ever created within agricultural states—tend to portrait, voters mostly live in *Neverland*, and there anything can happen, such as a mouse talking like humans, a virgin having a child, or Trump being president of the most powerful country in the planet. As explained by Hochschild's in *Stranger in Their Own Land*:

> Lee had worked at hard, unpleasant, dangerous jobs. He had loyally followed company orders to contaminate an estuary. He'd done his company's moral dirty work, taken its guilt as his own, and then been betrayed and discarded himself, as a form of waste. The most heroic act of Lee's life had been to reveal to the world a company's dirty secret, and to tell a thousand fishermen furious at the government that companies like PPG were to blame. Yet over the course of his lifetime, Lee Sherman had moved from the left to the right. When he lived as a young man in Washington State, he said proudly, "I ran the campaign of the first woman to run for Congress in the state". But when Lee moved from Seattle to Dallas for work in the 1950s, he shifted from conservative **Democrat** to **Republican**, and after 2009, to the **Tea Party**. So while his central life experience had been betrayal at the hands of industry, he now felt – as his politics reflected – most betrayed by the federal government. He believed that PPG and many other local petrochemical companies at the time had done wrong, and that cleaning the mess up was right. He thought industry wouldn't "do the right thing" by itself. But in the role of counterweight, he rejected the federal government. Indeed, Lee embraced candidates who wanted to remove nearly all the guardrails on industry and cut the EPA. The Occupational Safety and Health Administration had vastly improved life for workmen such as Lee Sherman – and he appreciated those reforms – but he felt the job was largely done. In the life of one man, Lee Sherman, I saw reflected both sides of the *Great Paradox* – the need for help and a principled refusal of it. As a victim of toxic exposure himself, a participant in polluting public waters, hating pollution, now proudly declaring himself as an environmentalist, why was he throwing in his lot with the anti-environmental Tea Party? Not because the Koch brothers were paying him to, at least directly. Lee was putting up Tea Party lawn signs for free. Still, his source of news was limited to ***Fox News*** and videos and blogs exchanged by right-wing friends, which placed him in an echo chamber of doubt about the EPA, the federal government, the president, and taxes. Indeed, Tea Party adherents seemed to arrive at their dislike of the federal government via three routes – through their religious faith (the government curtailed the church, they felt), through hatred of taxes (which they saw as too high and too progressive), and through its impact on their loss of honor.
>
> Most of what polluted the bayou sank to the bottom of it – mercury, heavy metals, **ethylene dichloride** (EDC), and **chlorinated dioxins**. So at first the danger lay mainly there. But when the U.S. Army Corps of Engineers twice dredged the nearby ship channel to ease the passage of commercial ships, "they scooped the toxic sludge from the bottom and

> pasted it on the banks right and left, without marking where they put it", Harold tells me. So now the Arenos can't trust the banks either. That was a decision of the U.S. Army Corps of Engineers, the federal government, I noted. What about stricter regulation of the polluters? I ask, wondering if the Arenos had voted for political candidates who pushed for cleaning the mess up or, like Lee Sherman, had not. "Stricter regulation would be good", Harold replies. "We're not against industry", Annette clarifies. "We were happy when industry came. It brought jobs. We were glad for Harold to get one. But for decades now, they've done nothing to clean up the bayou or compensate us to move". Like other friends and family, the Arenos are Republican and had voted in the presidential election of 2012 for **Mitt Romney**. "He's a big business guy, of course", Harold explains. "If he were here he'd be having friendly visits with the CEOs of the companies around here. He wouldn't be cleaning up the mess". But Harold and Annette speak with a mildness of manner, a flatness of voice, that makes me sense I am inquiring into an area of life in which they'd mostly given up interest. "*We vote for candidates that put the **Bible** where it belongs*", Harold adds. "We try to be right-living, clean-living people, and we'd like our leaders to live that way and believe in that, too". Before settling on Romney in the 2012 election, they had favored the former senator from Pennsylvania, **Rick Santorum**. The Arenos disapprove of "greedy corporations" stepping on the little guy. "Oil interests tried to suppress the development of the electric car", Annette adds. Agreeing, Harold says: "Republicans stand for big business. They won't help us with the problems we've got here." But Republicans put **God** and family on their side and "we like that…the Scripture says **Jesus** wants us to be about his Father's business", Annette says. Their faith had guided them through a painful loss of family, friends, neighbors, frogs, turtles, and trees. They felt God had blessed them with this courage to face their ordeals, and they thanked Him for that. "I don't know what people do if they don't know Him", Annette adds. For the Arenos, religious faith has moved into the very cultural space in which politics might have played a vital, independent role. Politics hadn't helped, they felt, and the Bible surely had…. People on the right seemed to be strongly moved by three concerns – taxes, faith, and honor. Lee Sherman was eager to lower his taxes, the Arenos to protect their **Christian faith**. Added to these basic motives were certain personal wishes: Lee, who had borne the guilt of polluting public waters and been cheated by a dishonest official at a tax office, wanted to feel vindicated. The tax office was corrupt, and taxes themselves were connected to dishonesty, he felt. One didn't know where they went or for what. The Arenos shared Lee's concern, but added another personal wish. Given their extended ordeal and the importance of God and the church in getting through it, they felt a powerful drive to place themselves in spiritually guided hands. For both Lee and the Arenos, at issue in politics was trust. It was hard enough to trust people close at hand, and very hard to trust those far away; to locally rooted people, Washington, D.C., felt very far away. Like everyone I was to talk with, both also felt like victims of a frightening loss – or was it theft? – of their cultural home, their place in the world, and their honor.

Humans in a nutshell: "their cultural home" (*H. fictus*), "their place in the world" (*H. socialis*), "and their honor" (*H. irrationalis*). Regarding the type of empirical data that Hochschild presented to show how these people were truly voting against their own individual interests, or those of the places in which they live—for instance, concerning **pollution**—she directly contrasted the teleological narratives often accepted/employed by them versus the reality of those data:

> If the power elite want to forget about pollution, and if they impose structural amnesia on a community, you need an omnipotent mind to remember how things once were. You needed, the Arenos felt, God. He remembers how it was. He knows what was lost. If the federal government was committed to a multicultural America that dimmed the position of the **Christian church**, it was getting in the way of that church, diminishing the importance of God, and it was God who had enabled them to survive their terrible ordeal. To Derwin –

who, having brought lunch to his parents, is packing up to go – the solution to Bayou d'Inde lies far beyond power, politics, or science. A devoted believer in the rapture, as are his parents, Derwin describes the approach of the "**End Times**". Quoting from the ***Book of Revelation***, he says, "The earth will burn with fervent heat". Fire purifies, so the planet will be purified 1,000 years from now, and until then, the devil is on the rampage, Derwin says. In the ***Garden of Eden***, "there wasn't anything hurting your environment. We'll probably never see the bayou like God made it in the beginning until He fixes it himself. And that will happen pretty shortly, so it don't matter how much man destroys".... Lee and the Arenos had played different roles in the pollution of Bayou d'Inde, but each recognized the other as a victim. They'd become good friends. In 2012, all three were watching speeches by Republican presidential candidate **Mitt Romney**. He wouldn't help the country clean up dirty rivers, they thought, but as an opponent to the right to abortion, he was for "saving all those babies" – and that seemed to them the more important moral issue on which they would be ultimately judged. Harold walks me to my car. I get in, open my window, and fasten my seat belt. "We're on this earth for a limited amount of time", he says, leaning on the edge of the window. "But if we get our souls saved, we go to Heaven, and Heaven is for eternity. We'll never have to worry about the environment from then on. That's the most important thing. I'm thinking long-term".

When I got home, I discovered an answer – a startling 2012 study by sociologist Arthur O'Connor that showed that residents of red states suffer higher rates of industrial pollution than do residents of blue states. Voters in the twenty-two states that voted **Republican** in the five presidential elections between 1992 and 2008 – and who generally call for less government regulation of business – lived in more polluted environments. Residents in the twenty-two **Democratic states** that generally favor stricter regulation, he found, live in cleaner environments. This would be discouraging news for my **Tea Party** friends. My Berkeley-based research assistant, Rebecca Elliott, and I asked one further question: was it just red states that were correlated to higher rates of pollution, or the counties of red-leaning individuals within any given state? We looked at the relationship between political views and pollution. For one we went to publicly available data on the EPA website. There we found scores for each county in the nation reflecting risk of exposure to pollution (Risk-Screening Environmental Indicators, or RSEI scores). These measure the amount of chemical release, the degree of its toxicity, and the size of the exposed population. It is the best measure we have of citizen exposure to pollution. This we linked with a second source of information – individual opinions recorded in the well-established General Social Survey. We were studying the link between what people believed about the environment and politics, and their actual risk of exposure to pollution linked to the county they lived in. If, in 2010, you lived in a county with a higher exposure to **toxic pollution**, we discovered, you are more likely to believe that Americans "worry too much" about the environment and to believe that the United States is doing "more than enough" about it. You are also more likely to describe yourself as a strong Republican. There it was again, the *Great Paradox*, only now it applied to my keyhole issue: environmental pollution across the entire nation.

Of course, the "*great paradox*" can only be seen as a paradox when we continue to be blind to the fact that instead of an objective "sapient being," we are merely a highly biased and subjective *Homo fictus*, *irrationalis et socialis*. If we realize this once for all, then we can easily understand that deciding something based on what the *Book of Revelation* says instead of what the empirical data shows is not surprising at all: this is what billions of humans have done for millennia. This discussion leads us to come full circle regarding the discussion of the long-standing so-called **nurture versus nature** and **Hobbes versus Rousseau** debates. This issue was discussed in much detail in **Levi-Strauss**' *Tristes tropiques*, in which he explained how such debates are too often oversimplified. He noted that **Rousseau** was obviously not referring to "good savages" versus "bad civilized people" but precisely to the fact that humans pay in general a very high price—that is, both those from non-agricultural *and* "developed" societies—for their obsessive *Homo socialis* mania. That is, for being so obsessed to "fit in"—or to oblige others to do so—and thus for blindly obeying—or being accordingly obliged to obey—to the social "order," norms and obligations ensuing from longstanding imaginary narratives created by *others* about how things are "meant to be." Levi-Strauss wrote:

> Rousseau is much decried these days; never has his work been so little known; and he has to face, above all, the absurd accusation that he glorified the state of Nature for its own sake. (That may have been **Diderot**'s error, but it was never Rousseau's.) What Rousseau said was the exact contrary; and he remains the only man who shows us how to get clear of the contradictions into which his adversaries have led us. Rousseau, of all the philosophes, came nearest to being an anthropologist. He never travelled in distant countries, certainly; but his documentation was as complete as it could be at that time and, unlike **Voltaire**, he brought his know ledge alive by the keenness of his interest in peasant customs and popular thought. Rousseau is our master and our brother, great as has been our ingratitude towards him; and every page of this book could have been dedicated to him, had the object thus proffered not been unworthy of his great memory. For there is only one way in which we can escape the contradiction inherent in the notion of position of the anthropologist, and that is by reformulating, on our own account, the intellectual procedures which allowed Rousseau to move forward…. He it is who showed us how, after we have destroyed every existing order, we can still discover the principles which allow us to erect a new order in their stead. There is no risk of his confusing the state of Nature with the state of Society; he knows that the latter is inherent in mankind, but that it brings evils with it, and that the question to be solved is whether or not these evils are themselves inherent in that state. We must go beyond the evidence of the injustices of abuses to which the social order gives rise and discover the unshakable basis of human society.
>
> In that myth-minded age, Man was no more free than he is today; but it was his humanness alone which kept him enslaved. As he had only a very restricted control over Nature, he was protected, and to a certain degree emancipated, by the protective cushion of his dreams. As and when these dreams turned into knowledge, so did Man's power increase; this gave us, if I may so put it, the upper hand over the universe, and we still take an immense pride in it. But what is it, in reality, if not the subjective awareness that humanity is being progressively more and more sundered from the physical universe? The great determining factors in that universe are no longer acting upon us as redoubtable strangers; rather is their operation not now through the intermediacy of thought, as they colonize us in the interests of the silent world whose agents we have now become? Rousseau was probably right when he held that it would have been better for our happiness if humanity had kept to the middle ground between the indolence of the primitive state and the questing activity to which we are prompted by our *amour-propre*. That middle state was, he said, the best for Man; and only some ill-boding turn of events could have caused us to leave it. That turn of events was found in the development of **mechanical civilizations**, a phenomenon doubly exceptional in that it was first, unique, and second, belated…. Rousseau's solution is eternal and universal. Other societies may not be better than our own; even if we believe them to be so we have no way of proving it. But knowing them better does none the less help us to detach ourselves from our own society. It is not that our society is absolutely evil, or that others, are not evil also; but merely that ours is the only society from which we have to disentangle ourselves. In doing so, we put ourselves in a position to attempt the second phase of our undertaking: that in which, while not clinging to elements from any one particular society, we make use of one and all of them in order to distinguish those principles of social life which may be applied to the reform of our own customs, and not of those of societies foreign to our own. In relation to our own society, that is to say, we stand in a posi-

tion of privilege which is exactly contrary to that which I have just described; for our own society is the only one which we can transform and yet not destroy, since the changes which we should introduce would come from within.

Right now, as I write these lines, I am literally seeing, before my eyes, an example of "fitting in" and obeying to rules that make no biological sense at all, being instead related to absurd and arbitrary teleological narratives, particularly religious ones. Thousands of people are right now either in the sand or water on this sunny day at this beautiful Portuguese beach called "Praia da Rocha." All of them, without exception, are wearing something to cover their genitals, and in addition almost all the women are wearing something to cover their breaths. This, of course, comes from the religious obsession to try to disconnect humans from "nature": we are not "animals," we have "decency," we should cover parts of our bodies that will lead to "temptation." Importantly, such stories are followed not only by Jewish, Christian, and Muslim people here in Europe, but also by many others as well. Actually, a substantial part of the younger people that are at the beach right now, following those absurd social norms, are not religious. Basically, in almost all beaches in the planet people follow exactly the same "rules," be them spiritual believers, spiritual non-believers, or non-spiritual non-believers. This fact teaches us two extremely critical lessons. Firstly, as it usually happens with the imaginary stories humans believe in, the fairytales that lead to these specific "social norms" are plagued by logical inconsistencies. Why would an omnipotent God make genitals and women's breasts to start with, if they are so "evil" or "not to be seen?" Secondly, the fact that even young non-religious people mostly follow such norms blindly, without even thinking about why they are doing so, shows us that—contrary to what we try to convince ourselves—most things we do in life are not really the result of a deep, personal introspection. Instead, even daily-life things we do, such as going to buy a bikini, wearing it, taking it to the beach, and so on, basically follow longstanding rules that others have created millennia ago.

In fact, many of such absurd rules have been incorporated into laws, which is the most extreme case of acculturation. This includes most democratic "developed" countries, which claim to have separated law from religion, but in reality continue to enforce laws that were originally related to religious stories, such as laws against "nudity," laws of "decency," and so on. The question is: why would you be obliged, in a twenty-first century democratic "developed" country, to cover your genitals in a beach or any other place? What is the logical rationale for doing so? But an amazing thing about the acculturated self-domesticated *Homo servus* is precisely that he/she is not used to stop and think about such critical questions, he/she is much, much better at simply following them blindly. If people did stop and think, they should be able to easily understand how those rules are completely arbitrary: if they had been created at a different moment in time, or by different people or cultures, the regions of the body that would be seen as "evil" would have been very different. During long periods of time a huge number of **Christian people** followed the idea that women should cover their hair—if they did not do this, this could lead to "temptation"—a practice still followed by many **Christian women** today—mostly orthodox ones—as well as by most **Muslim women** and married **Hasidic Jewish**

women. Moreover, some people, such as many Afghan men, argue that women should cover not only the hair, but also the whole face, and even the whole body, including the fingers. So, such cultural practices are as arbitrary as the fact that Donald Duck almost always uses a red bow tie: it could be yellow, it could be green, it could be whatever would come to the mind of the person that created it in the first place, at that particular day, time, and location when that random artistic decision was made. In a more crude way, this topic remind us of a brilliant point made by **Thomas Malcolm Muggeridge**: "*never forget that only dead fish swim with the stream.*"

These topics were discussed in Edwardes' 2019 book *The Origins of Self—an Anthropological Perspective*, in which he stressed the importance of what he calls "*emic facts,*" which are not "true scientific facts" but instead the result of the unique combination of our *Homo fictus, irrationalis et socialis* characteristics:

> Emic facts form the basis of most human cultures. Humans are outstandingly good at creating and enforcing **emic facts**, basing them on agreement *rather than evidence*. This becomes unsurprising if we accept the proposition that human communication is itself based on emic facts: our languages work not because there is a special 'language mechanism' inside each of us, but because we are able to negotiate toward meaning. Like language, exchanging social models also requires negotiation toward meaning: we each accept and use the unverified models of the social relationships of others when they are offered to us, even though we know them to be emic opinions and not etic facts. This acceptance of the opinions and beliefs of others about others unlocks all kinds of useful linguistic and modelling tricks and devices, such as referencing non-current events, referencing possible but not yet actual events, and using shared imagination. It also has an effect on how we model ourselves. Social modelling gives me access to what other people think (or say they think) about me, allowing me to build a model of myself as a social being. This social self-model is an emic fact, a thirdperson representation of my self as an entity in my social calculus; but I can treat it as an etic model inasmuch as it represents an objective view of me as an other; it's the best understanding of my self available to me. While human culture is an outcome of **human socialisation** and language, it nonetheless generates yet another, and very different, selfmodel. With its emphasis on emic facts, human culture presents me with an ideal model of *what an individual should be in the particular culture in which I find myself*. The ***Cultural self*** is based on the emic social expectations of others rather than their mostly emic social knowledge – and, as I am a member of the same culture, they are probably expectations that I (my social self-model) have about myself.
>
> This emphasis on emic facts in human culture creates a very odd inversion in the social strategies of our species. Like eusocial animals, we have societies with high levels of organisation, complexity, cooperation, individual specialisation, task-sharing and **self-sacrifice**; but where the eusocial lifestyle involves a physical culture bound by genetic imperatives, human society is governed by symbolic culture. **Eusocial societies** work because of the high level of relatedness in a nest and the fact that there are few fertile females – usually only one per nest. The only way for the sterile nest members to get their genes into the future is to protect the queen, their mother, and her fertile offspring, their sisters and brothers. This means that the range of cultures possible is severely limited, because they have to be based on etic realities, not emic beliefs. In contrast, the high levels of organisation, complexity, cooperation, individual specialisation, task-sharing and self-sacrifice in human cultures are all generated *emically, through group expectations and shared beliefs*. The correspondences between the needs of eusocial and human societies help to explain why humans seem to have adopted a pseudoeusocial social system; but, where reliance on etic facts makes the cultural range available to eusocial animals extremely small, the human

range of cultures, based on emic facts, is bewilderingly large: *any set of shared beliefs can become the basis for a culture*. An outcome of relying on emic facts is that, whereas eusocial cultural systems are stable and durable, human cultures are *vulnerable to collapse and elimination when key beliefs are challenged*. There is little durability in human cultures, which tend to last only hundreds of years rather than the millions of years for eusocial animals.

As further explained by Edwarnes:

The existence of **fairy stories**, **moral tales** and **mythic systems** in almost all human cultures indicates that sharing exemplars of **cultural morality** is common, while the recorded age of some of these stories indicates that they represent an ancient human tradition. However, the emic nature of the cultural facts in these stories means that their worth is negotiable, unlike the etic social facts that social selfmodelling provides. The cultural self-model is not a product of how I am seen or how I see myself; it reflects how I believe I should be. It does not provide accuracy, *but it does provide acceptability*…. Human history is full of comments promoting and praising this level of self-awareness: the **Oracle at Delphi** had the maxim 'Know thyself' over its entrance; the fifth-century BCE philosopher **Lao Tzu**, the founder of **Taoism**, said 'He who knows others is wise; he who knows himself is enlightened'; and Pythagoras said 'No one is free who has not obtained the empire of himself', to which **Socrates** added, 'True wisdom comes when we know how little we know about life, ourselves, and the world around us'. In **Hindu doctrine**, knowing your eternal self, or atman, is the route to enlightenment…. Yet, as we have seen, the only conscious representation of our self that we have available is what we have cobbled together from the social and cultural models offered to us by others. When we model our self we are modelling ourself from the outside looking in; but our unmodelled selfhood is imposed on the world from the inside looking out. The selfknowledge that language allows us is not reflexive but reflective, creating an image of our self that is recognisable and acceptable to others, and which we can then use to define and refine our self-model. This externalised vision of our self is, therefore, not so much what we are but *what others believe and want us to be*; it is a model of the socialised and **enculturated self**, *a representation we advertise or aim for* rather than actually are. Social and cultural self-modelling means that we are constantly trying to *meet expectations imposed on us by others*: human socialisation requires us to see self-promotion and hubris as vices, while human culture promotes humility, self-effacement and modesty as virtues. We also cannot avoid comparing our own third-person modelled selfhood with the other third-person models in our social calculus, and with the ideal self that symbolic culture imposes on us – all of which means that *we are always finding ourselves wanting*. Indeed, the human cultural view, at least in the West, seems to be that everyone is incomplete and improvable: whatever we are, *we could be better*. However, these individually unfit strategies are precisely what make us successful as a pseudo-eusocial species, as they encourage high levels of organisation, complexity, cooperation, *individual specialisation, task-sharing and self-sacrifice*..

A *Cultural self*, like the Social selves, is a model offered to the individual by others; but, unlike the **Social selves**, it is a virtual self. It is *a model of an ideal individual in this particular culture*, explicitly *the ideal self that the individual can be*. A culture usually has many ideal models, differentiated by gender, role, lineage, age group and any other way that the culture divides up its population. For instance, the **Hindu caste system** is based largely on gender and lineage, and it delimits not just the range of roles possible for an individual, it dictates how they are treated, whom they can marry, what they can eat and even what or whom they can touch. There are four main castes: priests, warriors, owning professions and labouring professions – a pattern repeated in internally specialist societies across the world. Unlike most other systems, however, the four Hindu castes are formally subdivided into sub-castes (in other cultures this level of differentiation is usually informal); but, like many other systems, there is also a formal gender-based differentiation, further limiting life choices. The caste system is a powerful engine for ensuring that life goes on regardless of who is in charge; but this also allows one ruling class to be replaced by another relatively seamlessly, without affecting the day-to-day functioning of the society. After seizing power in India, the **Delhi Sultanate**, the **Mughal Empire** and the **British Raj** all re-emphasised the caste system *to retain control over the populace*. The Hindu caste system is one of several historical systems that modern, global, pluralistic societies are breaking down; but *a socially differentiated system nonetheless remains an important feature of most cultures today*.... It is in the Cultural self that dispassionate **selfsacrifice begins**, so the Cultural self *is the key to a large number of our anxieties and self-doubts*.

In conclusion, Edwarnes agrees with Claude Levi-Strauss in that language, and writing in particular, have been key within the construction of our "cultural self" via the creation of and belief in narratives, or "emic facts," that ultimately are used to transmit *what others want us to be* and thus to ensure the continuation of high levels of organization, individual specialization, task-sharing and "self-sacrifice," particularly since the rise of agriculture and "civilization." This is a crucial point: all humans are a combination of *Homo fictus, irrationalis et socialis*, but this combination only became particularly explosive, to the point of leading the acculturated *Homo servus* to "self-sacrifice" their own freedom and even their own lives or to be enslaved—mentally or physically—after the rise of large sedentary groups and principally of agriculture. As seen above, in egalitarian small nomadic hunter-gatherer groups usually *no one can force others to do their will*. Therefore, while people living in such groups are not "free spirits" in the sense that they do have to conform to strict norms that are often created by, and for the well-being of, the group as a whole, they normally don't need to obey to arbitrary rules created by a single individual that are detrimental to the well-being of most people within that group. In contrast, a key aspect of hierarchical agricultural "early civilizations" is that *most people*, particularly those on the "bottom," mainly worked—and literally in many cases, lived their whole lives—for the will/dreams of a very small group of people: the kings, the pharaohs, the religious leaders, the elites, or the 1%.

When we think about the fact that the film and book industry makes billions of dollars, and that the vast majority of the books that are bought, and movies that are seen, are completely fictional, one can understand two major points—one commonly acknowledged, the other often neglected. The first is that, apart from our obsession to create imaginary stories, most people that go to **fictional movies** or read **fictional books** do so in order to precisely "disconnect" with the *reality of their daily "modern" lives*, as it is often said. However, the daily-lives of most people of "modern" countries are already vastly disconnected from *reality*, to start with. That is, despite the fact that most people are deeply immersed, 24/7, in *Neverland*, that does not seem to be enough: they want even more unreality, a "super-duper" version of *Neverland*, as attested by the fact that a huge number of them spend various hours per day in front of a TV, or reading or seeing fictional books or movies (see Fig. 8.8). **Hitchcock** is reported to have said something like this: if you are someone cleaning your house and working hard all day, at night when you go out, do you really want to go to the cinema to see a movie about someone cleaning a house and working hard the whole day? Of course not, you want to see a fictional movie to escape from such a *Homo servus* life: like *The birds*, or *The man that knew too much*, or *Vertigo* for instance. Moviemakers know very well our obsession with, and attraction for, **super-duper unreality**, and this take us to the second point, the often neglected one. If we accept that fictional movies, and books, are often an escape from the harsh daily-lives that a huge number of humans have in agricultural states, even in the so-called developed countries, why do we accept such hard lives as a "normality," to start with? Why would someone spend hours cleaning houses that often have "too much stuff," when in reality we don't really need most of that stuff? Why would someone think that it is "normal" to spend one third of a full day, five times per week, or even more time, working so hard, particularly doing something that he or she does not really enjoy doing, as so many humans accept to do? No other animal does that. Neither did humans for 6 millions of years, nor do most hunter-gatherers living today.

I now understand that I always had a huge tendency—since I can remember—to enjoy fresh food, walking in forests, being surrounded by other animals, and so on. But only when I was writing this book I realized the connection between that personal tendency and the key topics we have discussed so far. Most kids living in big cities nowadays have never seen or played with a living chicken, despite the fact that they had already eaten them an endless number of times. Most adults living in such cities have not snorkeled in seas and seen fishes living in their natural environments, but they probably have spent much time in swimming pools, which are nothing more than non-natural agglomerations of water put in a excavated hole or a prefabricated cubicle. We all use paper bills that have no real value as "valuable money," and now there are even "**bitcoins**," which are described in the web as "a **cryptocurrency** invented in 2008 and that started to be used in 2009 when its implementation was released as open-source software." This is a description that any person not living in agricultural states, and even a huge number of people living in such "developed" countries, would consider a completely abstract, intangible thing. A huge number of people living in big cities go to the gym to "run," but actually that "running" is done at exactly the same place, above a machine, often looking at a screen, while fewer people do indeed *run* in front of a beautiful lake, or river, or beach. Most "civilized" people eat processed foods frequently, while very few of them have eaten fresh food directly coming from a forest, or river (see Fig. 8.1). As a personal anecdote, I can tell you that, after having experienced thousands of restaurants of "developed" countries—I normally don't eat at home, as I spend less time there as possible—including some highly rated by *Michelin*, the best potatoes I tasted in my life were sold in the street of a village in the end of nowhere in Benin, and the best fish I have experienced was a fish that was taken from the river and grilled right on the spot also in a very small village in the same country.

The huge paradox of "civilization" and *Neverland*'s disconnect with the natural world is that some of the fictional movies that many people pay to see in "developed" countries, or fictional stories they read in the books they buy, are about people such as those that gave me that fish for free, and sold me those amazing potatoes in villages of Benin, or those shown on the left and right sides of Fig. 8.1. They see those movies and read those books because they want to "escape," yes: they want to escape the completely unnatural daily-lives they live in big cities, such as precisely the one shown in the center of the same figure. But if, by doing this, they seem to recognize that the images on the left and right sides of Fig. 8.1 are more appealing—or at least that the food items shown there are fresher and healthier—why do they continue to follow blindly, without questioning, the conventions, and norms, and way of life, or the very same socio-political "system" that creates, maintains, and aggravates the type of scenario that is shown in the image on the center of that figure?

Surely, there is a different way, "a middle way" where normally lies the "virtue," as it is often said. A way to be able to discover more realities about the amazing cosmos without most people remaining completely ignorant about these realities and even forgetting others that were commonly known previously, such as knowing how a living chicken looks like, or how it behaves, or at least to know that it is this animal that ultimately ends up in the **chicken Mcnuggets** that people eat at **Mcdonald's**. Or a way to not destroy the planet, or to not have people getting more and more obese, or overmedicated, or depressed, or to not have more and more suicides and people living a life of isolation and loneliness. Carl Sagan, and authors such as Jared Diamond, defended that such a different way is not only possible, but achievable, and urgently needed in face of what we have already done to the planet, and I completely agree with them. If we are indeed a "sapient" being, we have to be able to do so. I think that perhaps a better way to summarize the discussion provided in these two last Chapters, in particular Chap. 8 in which I used Tokyo, the capital of the "most technologically developed" country as a main comparative case study—is to look closely at Fig. 9.6. When one carefully analyzes such pictures of the Tokyo metro, and the unhealthy and uncomfortable way people are overcrowded, inside a machine deep down the soil, far from sun exposure, from fresh air, from rivers and from trees or birds or natural sounds, looking continuously to another machine—the cell phone—that constantly stresses them by reminding them about how "busy" they are and all the countless things they "need" to do, one cannot stop wondering: is this truly the supposedly "noble' result of the allegedly "progress" of a purportedly "sapient-being" that is said to be made in the image and likeness of God? Or of the *Homo Deus*, to use the term of employed by **Yuval Noah Harari**?

As you can see, I waited until the very last Chapter of this book to talk about Harari, who is considered to be one of the most bright thinkers of the twenty-first century. And, after reading his last book, *Homo Deus*—published in 2017—I do agree that his style of writing, and his broader discussions, are phenomenal. However, one should note that the book has three very different parts, which even seem to have been written by three different people, or at least under very different moods. The beginning of the book is ultra optimistic, very Hobbesian, being even in a way potentially disrespectful to hundreds of millions of people. The second part is very sober, and informative, and "in the middle," very similar to the present book in that regard, and defending various key ideas also defended in the present book, in another powerful example of convergence of intellectual ideas, as I just read that book when I started to write this last Chapter, as I did with **Claude Levi-Strauss**'

Tristes tropiques. The third part is more speculative, futuristic, and pessimistic about "progress," and in that sense a bit more Rousseauian, if one can say so.

An illustrative and very tragic empirical case that happened in the planet during the last 12 months and that plainly shows the type of tone used in the first part of *Homo Deus*, and how its ideas and predictions are in a way a reflection of human arrogance, concerns the **Covid-19 pandemic**. Harari wrote that "[We are now] in a healthy, prosperous and harmonious world…we know very swell what we have to do to avoid starvation, pandemics, and war – and, in general, we are successful…many people think that something like the Black plague can happen again…but we have good reasons to think that, in a race between germs and doctors, the latter will be faster." Well, what *truly happened in the world of reality* is that not only we *were not* able to avoid the pandemic, but we died like flies with it—a mere non-living organism was responsible for the loss of about two million lives, officially, in just a year: this means 166.6 thousand deaths per month, 5.5 thousand deaths per day, 231 one deaths per hour. And most scientists think that the real numbers are almost surely much higher, because many countries did very few tests, so a lot of people dying in them at that time were not counted as Covid-19 deaths. It is indeed very likely, taking into account the total difference between the people that died in the globe in those 12 months versus the 12 months before, that up to 3 or even four million people died because of Covid-19 in that 12-month period. If this is so, this means that *for a whole year, at least one person has died per second, because of Covid-19*. Moreover, the Word Health Organization predicted that 2021 will have even more deaths, globally, than 2020. A "healthy, prosperous and harmonious" world, indeed: it is very difficult to understand how Harari could have said such things and used these specific terms, in the first part of the book, particularly knowing how bright he is, and comparing the tone used in that first part with the much more sober, and dark, tone he used in the other two parts.

The statements Harari makes in that first part of *Homo Deus* about **famines**, **wars**, **genocides,** and **politics** are as inaccurate as those about pandemics. To say that humans "solved" starvation is not only completely factually wrong, but is purely an enormous lack of respect for the dozens of millions of people that starve and die of hunger in the globe every single year. "Even before Covid-19 hit, 135 million people were marching towards the brink of **starvation**…this could double to 270 million within a few short months," stated David Beasley, head of the **UN World Food Programme**.

But suddenly, in the second part of his book, Harari does completely change the tone, moving from an Hobbesian *Neverland* to the *world of reality*, using a much more humble and respectful tone, not only about humanity as a whole, but about the notion of "progress" in particular. That tone is indeed very similar to the one used in the present volume, including discussions and empirical data that do have a lot in common to those provided here, for instance in the excerpt he refers to suicides and "progress," in which he even refers to Epicurus as we have been doing in this Chapter (see also Fig. 9.22):

Fig. 9.22 What "progress" looks like: the suicide of a K-pop star in South Korea has brought the issue to national attention and lead to the country to make suicide pacts a criminal offence, in an effort to reverse one of the world' highest suicide rates

> **Epicurus** was apparently on to something. Being happy doesn't come easy. Despite our unprecedented achievements in the last few decades, it is far from obvious that contemporary people are significantly more satisfied than their ancestors in bygone years. Indeed, it is an ominous sign that despite higher prosperity, comfort and security, the **rate of suicide** in the developed world is also much higher than in traditional societies. In **Peru, Guatemala**, the **Philippines** and **Albania** – developing countries suffering from poverty and political instability – about one person in 100,000 commits suicide each year. In rich and peaceful countries such as Switzerland, France, Japan and New Zealand, 25 per 100,000 take their own lives annually. In 1985 most South Koreans were poor, uneducated and tradition-bound, living under an authoritarian dictatorship. Today **South Korea** is a leading economic power, its citizens are among the best educated in the world, and it enjoys a stable and comparatively liberal democratic regime. Yet whereas in 1985 about nine South Koreans per 100,000 killed themselves, today the annual rate of suicide has more than tripled to thirty per 100,000.

Harari subsequently also puts in question the notion of "progress" and Pinker's concept of "**expanding moral circle**," when he discusses the links between **organized religion, agriculture**, our disconnection with the natural world, the notion of "ladder of life," and topics such as animal abuse, slavery, oppression, and private property, in an excerpt that deeply resonates with the ideas defended in Chaps. 4, 7 and 8:

> The **Bible**, along with its belief in human distinctiveness, was one of the by-products of the **Agricultural Revolution**, which initiated a new phase in **human-animal relations**. The advent of farming produced new waves of **mass extinctions**, but more importantly, it created a completely new life form on earth: **domesticated animals**. Initially this development was of minor importance, since humans managed to domesticate fewer than twenty species

of mammals and birds, compared to the countless thousands of species that remained 'wild'. Yet with the passing of the centuries, this novel life form became dominant. *Today more than 90 per cent of all large animals are domesticated.* Alas, domesticated species paid for their unparalleled collective success with unprecedented individual suffering. Although the animal kingdom has known many types of pain and misery for millions of years, the Agricultural Revolution generated completely new kinds of suffering, that only became worse over time. To the casual observer domesticated animals may seem much better off than their wild cousins and ancestors. **Wild boars** spend their days searching for food, water and shelter, and are constantly threatened by lions, parasites and floods. **Domesticated pigs**, in contrast, enjoy food, water and shelter provided by humans, who also treat their diseases and protect them against predators and natural disasters. True, most pigs sooner or later find themselves in the slaughterhouse. Yet does that make their fate any worse than the fate of wild boars? Is it better to be devoured by a lion than slaughtered by a man? Are crocodile teeth less deadly than steel blades? What makes the fate of domesticated farm animals particularly harsh is not just the way they die, but above all the way they live. Two competing factors have shaped the living conditions of farm animals from ancient times to the present day: human desires and animal needs. Thus humans raise pigs in order to get meat, but if they want a steady supply of meat, they must ensure the long-term survival and reproduction of the pigs. Theoretically this should have protected the animals from extreme forms of cruelty. If a farmer did not take good care of his pigs, they would soon die without offspring and the farmer would starve. Unfortunately, humans can cause tremendous suffering to farm animals in various ways, even while ensuring their survival and reproduction. The root of the problem is that domesticated animals have inherited from their wild ancestors many physical, emotional and social needs that are redundant on human farms. Farmers routinely ignore these needs, without paying any economic penalty. They lock animals in tiny cages, mutilate their horns and tails, separate mothers from offspring and selectively breed monstrosities. The animals suffer greatly, yet they live on and multiply..

All agricultural religions – **Jainism**, **Buddhism** and **Hinduism** included – found ways to justify human superiority and the exploitation of animals (if not for meat, then for milk and muscle power). They have all claimed that a **natural hierarchy of beings** entitles humans to control and use other animals, provided that the humans observe certain restrictions. Hinduism, for example, has **sanctified cows** and forbidden eating beef, but has also provided the ultimate justification for the **dairy industry**, alleging that cows are generous creatures, and positively yearn to share their milk with humankind. Humans thus committed themselves to an '**agricultural deal**'. According to this deal, *cosmic forces* gave humans command over other animals, on condition that humans fulfilled certain obligations towards the **gods**, towards nature and towards the animals themselves. It was easy to believe in the existence of such a cosmic compact, because it reflected the daily routine of farming life. **Hunter-gatherers** had not seen themselves as superior beings because they were seldom aware of their impact on the ecosystem. A typical band numbered in the dozens, it was surrounded by thousands of wild animals, and its survival depended on understanding and respecting the desires of these animals. Foragers had to constantly ask themselves what deer dream about, and what lions think. Otherwise, they could not hunt the deer, nor escape the lions. Farmers, in contrast, lived in a world controlled and shaped by human dreams and thoughts. Humans were still subject to formidable natural forces such as storms and earthquakes, but they were far less dependent on the wishes of other animals. A farm boy learned early on to ride a horse, harness a bull, whip a stubborn donkey and lead the sheep to pasture. It was easy and tempting to believe that such everyday activities reflected either the natural order of things or the will of heaven. It is no coincidence that the **Nayaka of southern India** treat elephants, snakes and forest trees as beings equal to humans, but have a very different view of domesticated plants and animals. In the Nayaka language a living being possessing a unique personality is called mansan. When probed by the anthropologist

> Danny Naveh, they explained that all elephants are mansan. 'We live in the forest, they live in the forest. We are all mansan…. So are bears, deer and tigers. All forest animals'. What about cows? 'Cows are different. You have to lead them everywhere'. And chickens? 'They are nothing. They are not mansan'. And forest trees? 'Yes – they live for such a long time'. And tea bushes? 'Oh, these I cultivate so that I can sell the tea leaves and buy what I need from the store. No, they aren't mansan'. We should also bear in mind how humans themselves were treated in most agricultural societies. In **biblical Israel** or **medieval China** it was common to whip humans, enslave them, torture and execute them. Humans were considered as mere **property**. Rulers did not dream of asking peasants for their opinions and cared little about their needs. Parents frequently sold their children into **slavery**, or married them off to the highest bidder.

Harari's statements about the links between agricultural states, inequality, oppression, the use and abuse of power, politics and our *Homo socialis* inculturation are also strikingly similar to those made in Chaps. 4 and 8:

> History provides ample evidence for the crucial importance of large-scale cooperation. Victory almost invariably went to those who cooperated better – not only in struggles between *Homo sapiens* and other animals, but also in conflicts between different human groups. Thus **Rome** conquered **Greece** not because the Romans had larger brains or better toolmaking techniques, but because they were able to cooperate more effectively. Throughout history, disciplined armies easily routed disorganised hordes, and unified elites dominated the disorderly masses. In 1914, for example, 3 million Russian noblemen, officials and business people lorded it over 180 million peasants and workers. The Russian elite knew how to cooperate in defence of its common interests, whereas the 180 million commoners were incapable of effective mobilisation. Indeed, much of the elite's efforts focused on ensuring that the 180 million people at the bottom would never learn to cooperate. In order to mount a revolution, numbers are never enough. **Revolutions** are usually made by small networks of agitators rather than by the masses. If you want to launch a revolution, don't ask yourself, 'How many people support my ideas?' Instead, ask yourself, 'How many of my supporters are capable of effective collaboration?' The Russian Revolution finally erupted not when 180 million peasants rose against the tsar, but rather when a handful of communists placed themselves at the right place at the right time. In 1917, at a time when the Russian upper and middle classes numbered at least 3 million people, the **Communist Party** had just 23,000 members. The communists nevertheless gained control of the vast **Russian Empire** because they organised themselves well. When authority in Russia slipped from the decrepit hands of the tsar and the equally shaky hands of Kerensky's provisional government, the communists seized it with alacrity, gripping the reins of power like a bulldog locking its jaws on a bone. The communists didn't release their grip until the late 1980s. Effective organisation kept them in power for eight long decades, and they eventually fell due to defective organisation..
>
> In the late 1980s the **Soviet Union** withdrew its protection and the communist regimes began falling like dominoes. By December 1989 Ceauşescu [the Romanian leader] could not expect any outside assistance. Just the opposite – revolutions in nearby countries gave heart to the local opposition. The Communist Party itself began splitting into rival camps. The moderates wished to rid themselves of Ceauşescu and initiate reforms before it was too late. By organising the Bucharest demonstration and broadcasting it live on television, Ceauşescu himself provided the revolutionaries with the perfect opportunity to discover their power and rally against him. What quicker way to spread a revolution than by showing it on TV? Yet when power slipped from the hands of the clumsy organiser on the balcony, it did not pass to the masses in the square. Though numerous and enthusiastic, the crowds did not know how to organise themselves. Hence just as in Russia in 1917, power passed to a small group of political players whose only asset was good organisation. The **Romanian Revolution** was hijacked by the selfproclaimed National Salvation Front, which was in fact

a smokescreen for the moderate wing of the Communist Party. The Front had no real ties to the demonstrating crowds. It was manned by midranking party officials, and led by Ion Iliescu, a former member of the Communist Party's central committee and one-time head of the propaganda department. Iliescu and his comrades in the National Salvation Front reinvented themselves as democratic politicians, proclaimed to any available microphone that they were the leaders of the revolution, and then used their long experience and network of cronies to take control of the country and pocket its resources. In communist Romania almost everything was owned by the state. Democratic Romania quickly privatised its assets, selling them at bargain prices to the ex-communists, who alone grasped what was happening and collaborated to feather each other's nests. Government companies that controlled national infrastructure and natural resources were sold to former communist officials at end-of-season prices while the party's foot soldiers bought houses and apartments for pennies. **Ion Iliescu** was elected president of Romania, while his colleagues became ministers, parliament members, bank directors and multimillionaires. The new Romanian elite that controls the country to this day is composed mostly of former communists and their families. The masses who risked their necks in Timişoara and Bucharest settled for scraps, because they did not know how to cooperate and how to create an efficient organisation to look after their own interests…. A similar fate befell the **Egyptian Revolution of 2011**. What television did in 1989, Facebook and Twitter did in 2011. The new media helped the masses coordinate their activities, so that thousands of people flooded the streets and squares at the right moment and toppled the **Mubarak regime**. However, it is one thing to bring 100,000 people to **Tahrir Square**, and quite another to get a grip on the political machinery, shake the right hands in the right back rooms and run a country effectively. Consequently, when Mubarak stepped down the demonstrators could not fill the vacuum. Egypt had only two institutions sufficiently organised to rule the country: the army and the **Muslim Brotherhood**. Hence the revolution was hijacked first by the Brotherhood, and eventually by the army. The Romanian ex-communists and the Egyptian generals were not more intelligent or nimble-fingered than either the old dictators or the demonstrators in Bucharest and Cairo. Their advantage lay in flexible cooperation. They cooperated better than the crowds, and they were willing to show far more flexibility than the hidebound Ceauşescu and Mubarak.

A particularly interesting excerpt of that second part of Harari's *Homo Deus* is about how the three components that I call in the present book as "*Homo fictus, irrationalis, et socialis*" combine to form a "third level of reality"—which he designates as the "intersubjective level"—that plays a critical role in our deep immersion in *Neverland*:

People find it difficult to understand the idea of 'imagined orders' because they assume that there are only two types of realities: **objective realities** and **subjective realities**. In objective reality, things exist independently of our beliefs and feelings. **Gravity**, for example, is an objective reality. It existed long before **Newton**, and it affects people who don't believe in it just as much as it affects those who do. Subjective reality, in contrast, depends on my personal beliefs and feelings. Thus, suppose I feel a sharp pain in my head and go to the doctor. The doctor checks me thoroughly, but finds nothing wrong. So she sends me for a blood test, urine test, DNA test, X-ray, electrocardiogram, fMRI scan and a plethora of other procedures. When the results come in she announces that I am perfectly healthy, and I can go home. Yet I still feel a sharp pain in my head. Even though every objective test has found nothing wrong with me, and even though nobody except me feels the pain, for me the pain is 100 per cent real. Most people presume that reality is either objective or subjective, and that there is no third option. Hence once they satisfy themselves that something isn't just their own subjective feeling, they jump to the conclusion it must be objective. If lots of people believe in **God**; if **money** makes the world go round; and if **nationalism** starts **wars** and builds empires – then these things aren't just a subjective belief of mine. God, money and nations must therefore be objective realities. However, there is a **third level of reality**: the *intersubjective level*. **Intersubjective entities** depend on communication among many humans rather than on the beliefs and feelings of individual humans. Many of the most important agents in history are intersubjective. Money, for example, has no objective value. You cannot eat, drink or wear a dollar bill. Yet as long as billions of people believe in its value, you can use it to buy food, beverages and clothing. If the baker suddenly loses his faith in the dollar bill and refuses to give me a loaf of bread for this green piece of paper, it doesn't matter much. I can just go down a few blocks to the nearby supermarket. However, if the supermarket cashiers also refuse to accept this piece of paper, along with the hawkers in the market and the salespeople in the mall, then the dollar will lose its value. The green pieces of paper will go on existing, of course, but they will be worthless.... The value of money is not the only thing that might evaporate once people stop believing in it. The same can happen to laws, gods and even entire empires. One moment they are busy shaping the world, and the next moment they no longer exist. **Zeus** and **Hera** were once important powers in the Mediterranean basin, but today they lack any authority because nobody believes in them. The **Soviet Union** could once destroy the entire human race, yet it ceased to exist at the stroke of a pen.... It is relatively easy to accept that money is an ***intersubjective reality***. Most people are also happy to acknowledge that ancient **Greek gods**, evil empires and the values of alien cultures exist only in the imagination. Yet we don't want to accept that our God, our nation or our values are mere fictions, because these are the things that give meaning to our lives. We want to believe that our lives have some objective meaning, and that our sacrifices matter to something beyond the stories in our head. Yet in truth the lives of most people have meaning only within the network of stories they tell one another. Meaning is created when many people weave together a common network of stories. Why does a particular action – such as getting married in church, fasting on **Ramadan** or voting on election day – seem meaningful to me? Because my parents also think it is meaningful, as do my brothers, my neighbours, people in nearby cities and even the residents of far-off countries. And why do all these people think it is meaningful? Because their friends and neighbours also share the same view. People constantly reinforce each other's beliefs in a self-perpetuating loop. Each round of mutual confirmation tightens the ***web of meaning*** further, until you have little choice but to believe what everyone else believes. Yet over decades and centuries the web of meaning unravels and a new web is spun in its place. To study history means to watch the spinning and unravelling of these webs, and to realise that what seems to people in one age the most important thing in life becomes utterly meaningless to their descendants...

Other animals may also imagine various things. A cat waiting to ambush a mouse might not see the mouse, but may well imagine the shape and even taste of the mouse. Yet to the best of our knowledge, cats are able to imagine only things that actually exist in the world, like mice. They cannot imagine things that they have never seen or smelled or tasted – such as the US dollar, Google corporation or the European Union. Only Sapiens can imagine such chimeras. Consequently, whereas cats and other animals are confined to the objective realm and use their communication systems merely to describe reality, Sapiens use language to create completely new realities. During the last 70,000 years the intersubjective realities that Sapiens invented became evermore powerful, so that today they dominate the world. Will the chimpanzees, the elephants, the Amazon rainforests and the Arctic glaciers survive the twenty-first century? This depends on the wishes and decisions of intersubjective entities such as the European Union and the World Bank; entities that exist only in our shared imagination. No other animal can stand up to us, not because they lack a soul or a mind, but because they lack the necessary ***imagination***. Lions can run, jump, claw and bite. Yet they cannot open a bank account or file a lawsuit. And in the twenty-first century, a banker who knows how to file a lawsuit is far more powerful than the most ferocious lion in the savannah.

Harari's notion of ***intersubjective reality***, together with his statements about how a few leaders, or the 1%, are able to condemn millions or billions of people to subjugation, poverty and even starvation, leads us to one of the key points I want to make in the present book, about one of the biggest prices we pay for living in *Neverland*'s agricultural "civilizations." Namely, about how very few people are able to use our *Homo fictus, irrationalis et socialis* components to create a *"divide to conquer"* world of unreality in order to *force millions of others to do their will*, something that would be impossible outside the context of the extreme type *of intersubjective reality* existing within hierarchical agricultural "civilized" states and without the invention of writing, including the **religious texts** of **organized religions** as astutely emphasized by Harari and **Claude Levi-Strauss**. The **"Arab spring"** provides another recent powerful example of how this *"divide to conquer"* strategy continues to be as efficient today as it has been since millennia ago, and of how "intersubjective reality" usually plays a major role within this strategy, together with the use of **teleological tales** about **Gods**, narratives about Us versus Them, and the manipulation of information and disinformation (see Figs. 4.4 and 9.1). The example concerns Bahrain, as explained in a January 30th, 2020, review article in the Portuguese newspaper *Publico*:

> Having been one of first revolts of 2011, it was quickly forgotten for the rest of the world…the protests in **Bahrain** marked a turning point in the so-called Arab Spring: on the one hand, the pro-democratic movement in the archipelago of 1.3 million inhabitants was the first to succumb to repression; on the other hand, it was there that the wave of protests began to be seen in the context of a **sectarian concept**. The only revolt of 2011 to reach a **Gulf monarchy** happened in the one where the majority of the population is a **Shiite Muslim**, ruled by a **Sunni minority** family who managed to present the manifestations as if they were an uprising of the Shiite population. **Saudi Arabia**, which sees itself as the main **Sunni power** of the region and tends to observe the world around the threat of the "Shiite crescent", helped. The first major protests happened in February. In March, about 1500 Arabian military Saudi Arabia and the United Arab Emirates arrived in the country to help repression. "What the regime did was to unify the Sunni block…they told them that if they did not leave the street, they were helping Shiites to come to power and that when that happened, they would take revenge on all Sunnis…it worked", summarizes Joost Hiltermann, director of the think tank International Crisis Group. The Al-Khalifa family, who govern the archipelago since 1783, was also "skilled in the use of repression", says the analyst. "The repression was effective without being overly violent…it was very hard at first, but then it calmed down…. Bahrain is a small society, almost everyone knows each other…there were Sunnis and Shiites in the demonstrations, as there were communists, liberals, left-wing people, Islamists, academics, doctors, artists…there were no differences, everyone had legitimate claims", recalls the human rights activist Sayed Yousif Al-Muhafdah. "Unfortunately, they scared the Sunnis, they told them that if the Shiites reached the power, they were going to lose jobs, benefits". The regime knew how to promote fear. "We only have a channel, the television of Bahrain: presenters said that the Shiites were attacking the Salmaniyah hospital, kidnapping Sunni doctors…and they cried…it was all fake, of course", he says. It was on March 17th, one of the worst days of repression, when the hospital Salmaniyah was surrounded by tanks and military that did not let the injured enter or doctors leave. "Imagine the impact: presenters of the only channel to say such stuff…many within the Sunni community believed…they lied because they wanted to 'divide to conquer' and so far we have paid the consequences. Many Sunnis are still afraid", he describes. To underline the dimension of the threat, the king said the Shiites were controlled by Iran and that terrorist cells were preparing attacks.

And so it was: in the *world of reality* the Al-Khalifa family kept a hugely unequal system of power by using *Neverland*'s "alternative facts," "fake news," and teleological narratives to "divide and conquer," as it has done since it governs the archipelago, almost 250 years ago. The above excerpt, as well as the second part of Harari's *Homo Deus*, thus bring us back to our discussion about *Homo socialis*, and specifically about how *"intersubjective reality"* can be so powerful in agricultural states and particularly in "developed" countries. As noted by Harari, "animals such as wolves and chimpanzees live in a dual reality…on the one hand, they are familiar with objective entities outside them, such as trees, rocks and rivers…on the other hand, they are aware of subjective experiences within them, such as fear, joy and desire." In contrast, "Sapiens live in triple-layered reality…in addition to trees, rivers, fears and desires, the Sapiens world also contains stories about money, Gods, nations and corporations…as history unfolded, the impact of Gods, nations and corporations grew at the expense of rivers, fears and desires." One of the most potent examples illustrating how the vast majority of the self-domesticated *Homo servus* self-sacrifice even their desires, as well as pleasures, for the sake of others, and of the *system's "intersubjective reality,"* is the work of **Bronnie Ware**. Her work was summarized by Susie Steiner, for the broader public, in a February 2020 *The Guardian* article entitled *"Top five regrets of the dying"*:

> Bronnie Ware is an Australian (palliative) nurse who spent several years working in palliative care, caring for patients in the last 12 weeks of their lives. She recorded their dying epiphanies in a blog called Inspiration and Chai, which gathered so much attention that she put her observations into a book called ***The Top Five Regrets of the Dying***. Ware writes of the phenomenal clarity of vision that people gain at the end of their lives, and how we might learn from their wisdom. "When questioned about any regrets they had or anything they would do differently", she says, "common themes surfaced again and again." Here are the top five regrets of the dying, as witnessed by Ware:
>
> 1. *I wish I'd had the courage to live a life true to myself, not the life others expected of me*. "This was the most common regret of all. When people realise that their life is almost over and look back clearly on it, it is easy to see how many dreams have gone unfulfilled. Most people had not honoured even a half of their dreams and had to die knowing that it was due to choices they had made, or not made. Health brings a freedom very few realise, until they no longer have it."
>
> 2. *I wish I hadn't worked so hard*. "This came from every male patient that I nursed. They missed their children's youth and their partner's companionship. Women also spoke of this regret, but as most were from an older generation, many of the female patients had not been breadwinners. All of the men I nursed deeply regretted spending so much of their lives on the treadmill of a work existence."
>
> 3. *I wish I'd had the courage to express my feelings*. "Many people suppressed their feelings in order to keep peace with others. As a result, they settled for a mediocre existence and never became who they were truly capable of becoming. Many developed illnesses relating to the bitterness and resentment they carried as a result."
>
> 4. *I wish I had stayed in touch with my friends*. "Often they would not truly realise the full benefits of old friends until their dying weeks and it was not always possible to track them down. Many had become so caught up in their own lives that they had let golden friendships slip by over the years. There were many deep regrets about not giving friendships the time and effort that they deserved. Everyone misses their friends when they are dying."

5. *I wish that I had let myself be happier*. "This is a surprisingly common one. Many did not realise until the end that happiness is a choice. They had stayed stuck in old patterns and habits. The so-called 'comfort' of familiarity overflowed into their emotions, as well as their physical lives. Fear of change had them pretending to others, and to their selves, that they were content, when deep within, they longed to laugh properly and have silliness in their life again."

What's your greatest regret so far, and what will you set out to achieve or change before you die?

Coming back to *Homo Deus*, Harari makes a desperate, urgent plead—similar to the plead that has been done throughout the present book—to humans to "strive to distinguish fiction from reality," in order to escape from *Neverland* before it is too late:

When examining the history of any human network, it is therefore advisable to stop from time to time and look at things from the perspective of some real entity. How do you know if an entity is real? Very simple – just ask yourself, 'Can it suffer?' When people burn down the temple of **Zeus**, Zeus doesn't suffer. When the euro loses its value, the euro doesn't suffer. When a bank goes bankrupt, the bank doesn't suffer. When a country suffers a defeat in war, the country doesn't really suffer. It's just a metaphor. In contrast, when a soldier is wounded in battle, he really does suffer. When a famished peasant has nothing to eat, she suffers. When a cow is separated from her newborn calf, she suffers. This is reality. Of course suffering might well be caused by our belief in fictions. For example, belief in national and religious myths might cause the outbreak of war, in which millions lose their homes, their limbs and even their lives. The cause of war is fictional, but the suffering is 100 per cent real. *This is exactly why we should strive to distinguish fiction from reality*. Fiction isn't bad. It is vital. Without commonly accepted stories about things like money, states or corporations, no complex human society can function. We can't play football unless everyone believes in the same made-up rules, and we can't enjoy the benefits of markets and courts without similar make-believe stories. But the stories are just tools. They should not become our goals or our yardsticks. *When we forget that they are mere fiction, we lose touch with reality*. Then we begin entire wars 'to make a lot of money for the corporation' or 'to protect the national interest'. Corporations, money and nations exist only in our imagination. *We invented them to serve us; how come we find ourselves sacrificing our lives in their service*? In the 21st century we will create more powerful fictional stories and more totalitarian religions than ever before…*being able to distinguish between what is fiction and what is real and between what is religion and what is science will thus be more difficult, but also more important than anytime in the past.*

This notion that we are at a pivotal, defining moment in which the very worse of *Neverland* might actually be knocking at our doors—due to a combination of the increasing prevalence of **disinformation on social media**, technology, **virtual and alternative realities**, and **addictive computer algorithms**—has been subscribed by a person that has suffered the most, both personally and politically, due to imaginary tales and **conspiracy theories**: **Hillary Clinton**. She is one of the most hated persons in the U.S., particularly—but not exclusively, it is important to note—within republicans. Still, in a recent interview, she calmly and astutely analyzed the reasons that lead dozens of millions of people to believe such tales and theories about her—many of them literally defending the idea that she is a **satanic blood drinker**—and to accordingly detest her so much. She gave that interview for a *New York Times* article by Michelle Goldberg, entitled "*QAnon believers are obsessed with Hillary Clinton – she has thoughts*":

> QAnon is the obscene apotheosis of three decades of Clinton demonization. It's other things as well, including a repurposed version of the old **anti-Semitic blood libel**, which accused **Jews** of using the **blood of Christian children** in their rituals, and a cult lusting for mass public executions. According to the F.B.I., it's a domestic terror threat. But **QAnon** is also the terminal stage of the national derangement over Clinton that began as soon as she entered public life. "It's my belief that QAnon really took off because it was based on Hillary Clinton", said Rothschild. "It was based specifically on something that a lot of ***4chan*** dwellers wanted to see happen, which was Hillary Clinton arrested and sort of dragged away in chains". I was curious what Clinton thinks about all this, and it turns out she's been thinking about it a lot. "For me, it does go back to my earliest days in national politics, when it became clear to me that there was a bit of a market in trafficking in the most outlandish accusations and wild stories concerning me, my family, people that we knew, people close to us", she told me. The difference is that, even if ***Fox News*** or Rush Limbaugh spread demented lies about the Clintons, there was no algorithm feeding their audience ever-sicker stuff to maximize their engagement. For most ordinary people, there were no **slot machine-like dopamine hits** to be had for upping the ante on what might be the greatest collective slander in American history. Looking back to the 1990s, it's easy to see **QAnon's antecedents**. In "Clinton Crazy", a 1997 *New York Times Magazine* story, Philip Weiss delved into the multipronged subculture devoted to anathematizing the first couple.... The people Weiss wrote about targeted both Clintons, but there was always a special venom reserved for Hillary, seen as a **feminist succubus** out to annihilate traditional family relations. An attendee at the 1996 Republican National Convention told the feminist writer Susan Faludi, "It's well-established that Hillary Clinton belonged to a **satanic cult**, still does". Running for Congress in 2014, Ryan Zinke, who would later become Trump's secretary of the interior, described her as **"the Antichrist"**.... Trump himself called Clinton **"the Devil"**.
>
> For Clinton, these **supernatural smears** are part of an old story. "This is rooted in **ancient scapegoating of women**, of doing everything to undermine women in the public arena, women with their own voices, women who speak up against power and the **patriarchy**", she said. "This is a **Salem Witch Trials** line of argument against independent, outspoken, pushy women. And it began to metastasize around me". In this sense...is just a particularly disgusting version of misogynist hatred she's always contended with.... QAnon took it several steps farther.... To my surprise, Clinton thought Greene's [one of the republication politicians sitting in congress that deeply believes **QAnon conspiracy theories**] passive account of her own radicalization wasn't entirely absurd. "We are facing a mass addiction with the effective purveying of **disinformation on social media**", Clinton said. "I don't have one iota of sympathy for someone like her, but the algorithms, we are now understanding more than ever we could have, truly are addictive...and whatever it is in our brains for people who go down those **rabbit holes**, and begin to inhabit this **alternative reality**, they are, in effect, *made to believe*". Clinton now thinks that the creation and promotion of this alternative reality, enabled and incentivized by the **tech platforms**, is, as she put it, "*the primary event of our time*". Nothing about QAnon or **Marjorie Taylor Greene** is entirely new. Social media has just taken the *dysfunction* that was already in our politics, and rendered it *uglier than anyone ever imagined.*

In a nutshell, from everything we have seen so far in the this book, we are indeed at a crucial, unique moment in human history. On the one hand, there is a huge number of people that are increasingly immersed in the rabbit holes of *Neverland*, to a level that is far higher "than anyone ever imagined." This is the "*primary event of our time,*" as astutely put by Hillary, and those that don't understand—or don't want to recognize—this, and continue to proclaim that everything is just OK, or is "just about the economy, stupid," clearly don't understand humans and human history. However, on the other hand—and without wanting to sound overoptimistic—there is also an increasing awareness of this fact, as attested by the rising number of publications about this topic by scholars such as Harari, as well by how some of those publications quickly became so influential and popular within the general public, many of them being best sellers. That is, a huge proportion of people are thirsty for *Neverland*'s stories, as they have been since times immemorial, but a substantial segment of the population is also thirsty for change, to escape from the world of unreality and modify the *status quo* by using their awareness about the vicious cycle of *Neverland*. Such an increasing dichotomy, between those such as *Marjorie Taylor Green* that are digging 24/7 within *Neverland*'s rabbit holes and those such as Hillary Clinton that are able to understand both the social and neuronal—that is, in a oversimplified way, both the "nurture" and "nature" components involved in the—mechanisms that lead to those rabbit holes, might seem terrifying in the first place. The U.S. is deeply divided, as are many countries of the planet, right now, and this might be a troubling sign of something horrible to come, as many historians have pointed out. However, such dichotomies and extremisms are also typical within pivotal moments that have changed history: within major societal changes, such as that leading to the end of slavery in the U.S., there is often a number of people that react very strongly to them, who are precisely often called "reactionaries," as noted above. So, yes, these are very risky and potential dangerous times, but on the other hand they also provide a unique opportunity to escape once for all from the ***Neverland*** **rabbit holes** that humans begun to dig by creating and believing in teleological narratives when they became aware of the inevitability of their death, which then became huge ***Neverland dungeons*** during the rise of agriculture and organized religion, and later became immense ***Neverland black Holes*** with the rise of **industrialization** and more recently of **virtual and alternative realities** and **addictive logarithms**. In fact, this is not really an unique opportunity: it is a necessity. Very likely—and not wanting to sound too dramatic, now—this might well be our *last* opportunity, before we attain a point of no return, concerning the well-being of our "minds"—Marjorie Taylor Greene being a disturbingly worrying illustration of this—and bodies as well as of the members of countless species in this planet.

In the above sections we have seen some signs that this opportunity might indeed be knocking at our doors, so we better open them before they close for ever, making the rabbit holes, dungeons, and black Holes of *Neverland* completely inescapable. For instance, we saw that there is a growing recognition, in the last decades, by not only scientists but importantly also by a substantial portion of the general public, about the fact that humans are domesticating themselves—a phenomenon directly related to our increased detachment from the *world of reality* (Fig. 9.6). Moreover, an increasing number of authors are showing that this phenomenon is, in turn, very likely related to the fact that humans of "civilized" agricultural states display levels of discrimination, oppression, and exclusion towards "others" that have almost no parallel in other species. The recognition of these facts, and its public discussion—for instance in popular books such as Hare and Woods' *The Survival of the Friendliest*—as well as the undertaking of several projects and public discussions about **dehumanization** in many different regions of the globe (see Fig. 9.10), are examples of the recent major change of mindset concerning these subjects. Particularly when we compare such examples with what a substantial part of scientists were writing, doing, and promoting just a century ago, such as the rise of the **eugenics movement**, the propagandizing of **scientific racism and misogyny**, the scientific defense of **colonialism**, and so on.

Another major change of framework that allow us to be reasonably optimistic about the future of our planet and the possibility of escaping *Neverland* was emphasized recently by a leader of an indigenous movement, **Ailton Krenak**. In his latest book he provides a sober, deep, broader, and mostly non-teleological analysis of a series of broader issues discussed in the present volume, from the **Covid-19 pandemic** to the notions of "cosmic purpose," "progress," "civilization," "good," or "useful," and human "specialness," as well as their links to inequality, misogyny and our "modern obsession with work." This is also an emblematic example of how the voices and faces of "others"—of minority groups, women, homosexuals, and so on—are much more visible in the public sphere nowadays, compared to what was happening a century ago. A Brazilian writer, journalist, and philosopher, Ailton Krenak was forcibly separated from his people—of **Krenak ethnicity**, of which only 130 individuals are left today—at age 9. A defender of the preservation of the natural world, he does not make the common erroneous and anthropocentric assumption that **Covid-19** is a "message" that nature delivered to humans. Instead, in the 2020 book entitled *A vida não é útil*—something like "*Life is not useful,*" in English—he argues instead that the flawed way in which humans reacted to Covid-19 has a lot to do with "our denial of the events that Nature produces." He states that self-designated "civilized" people such as Europeans "cannot live with the idea of living aimlessly…they think that work is the reason for their existence…they enslaved so much "others," that now they need to enslave themselves…as if becoming "civilized" was our destiny…they can't stop and experience life as something that is simply part of a marvelous world…this is their religion: the ***religion of civilization.***" I am translating here an excerpt of an interview destined to the broader public that he gave for the December 28, 2020, edition of the Portuguese newspaper *Publico* about that book, in which he defends ideas that are very similar to some of the thoughts defended by **Epicurus** millennia ago:

> The virus, that act of Nature, exposed the weaknesses of what we know as **Humanity**, at a time when everything seemed possible thanks to the apparently unprecedented development of the technique. "There is no need for a complex war system to extinguish Humanity: it can be extinguished as easily as mosquitoes in a room after an insecticide is applied. *We are useless here*". The author rejects, however, the idea that Nature "communicates" with human beings through punishments or benefits – it is moreover a frivolous claim by men to humans to attribute anthropogenic attributes or dispositions to Nature, suggests Krenak. But there is something to be taken from this event, as from any tragedy. The activist recalls the idea that the **Lisbon earthquake of 1755** would have meant a "divine warning". "But there was also an enlightened debate, discussing that the city did not foresee the possibility of a seismic event. Who knows, in the future, science may explain that *[the pandemic] is not a warning from Nature, but the result of our stupidity, of our denial of the events that Nature produces*", he says. In *A vida não é útil*, Krenak makes a short but incisive question

> about our way of life, showing that everything is connected and, if nothing is done, everything will flow into a horrible future, a "**mediocre humanity**". The voracity of the consumption of the resources offered by the Earth is the product of a claim on the property of the planet that is not alien to the condition of **gender**, with **divine authorization**. "It is the **masculine gender** authorized by a **God**, masculine also, without a doubt, to make the Earth a platform for his experiences of government, of war, of **subjecting life to a practical objective**, which is this idea of going out to work", defends the philosopher. In a few previous situations, this idea of humans as agents of utility was so evident, measured by their capacity to produce, even in the face of an existential threat, as in the one we live in [in the pandemic]. Brazilian society, in particular, has spent the past few months engaged in a heated debate – one more – over whether "going out to work" is an inalienable right or a sinister obligation.
>
> This, too, troubled Krenak. "Work is an instrument of exercising authority. Whoever summons to work does not work", observes the activist, who compares the summonses so that labor "normality" is restored to warlike orders. "It looks like those generals who call for war, sit down and watch the soldiers die, while clapping their hands." Krenak speaks in the country where the **pandemic** was the lamp that illuminated the dark corners of **inequality**. The virus began to enter through the rich who returned in February and March from their holidays in Europe, where domestic workers were waiting for them in their homes, a battalion of millions of women, especially black women, who make up one of the most unprotected groups in Brazilian society, while they are the foundation of that same society. The first fatal victims of the Brazilian pandemic were employed as domestic workers. Supermarket workers followed, drivers working for digital applications, workers, then field workers, Native Americans, prisoners. Public hospitals were on the verge of rupture, with health professionals themselves being overthrown by the disease. At the same time, the poor among the poor endured huge queues every month, rain or shine, waiting for the famous 600 reais – less than one hundred euros, which meanwhile went up to half – given by the Government to prevent the pandemic if hunger joined. All of this in the name of "perpetuating that ***divine mandate that man should rule the Earth***", says Krenak. "It is obvious: to dominate the Earth, he needs many slaves, and the world of work has managed to solve this equation in a very clever way. It does not say that you are a slave, it says that you are a worker", maintains the activist. But Krenak refuses to dress up as a pessimistic loser and prefers to gather all the little signs that help him to conclude that something is changing. He sees people leaving cities and their jobs full of "utility" in exchange for a simpler life, in communion with the land. He sees people abandoning cars, radically altering their diet. And he believes that these steps go beyond what he calls the "sustainability myth", sold by major brands under the motto of social responsibility. One of the developments that gives him hope is the tendency for more and more millionaire heirs to families who have made a fortune in the industry or in the financial market to give up that money and go for something really different. "I'm not just talking about hippies who leave town with no experience to live in the country. I am talking about wealthy young people who occupy decision-making positions on the [management] board of their companies and who have decided that they will not continue this *Russian roulette* [of the 'modern' world]", he explains.

An additional change of framework that has also been occurring in the last decades, and that surprisingly might play a critical role in our escape from *Neverland*, concerns the rise of another type of new generation: not of people, but of ***Artificial Intelligence*** (AI). In a very simplified way, AI is commonly defined as "intelligence" demonstrated by non-living **machines**. At first sight, one could think: well, AI is, above all, an example of our total immersion in *Neverland*. After all, we are talking about *artificial* intelligence, right? Well, that is true. But on the other hand, precisely because they *are* artificial, non-living machines should in theory *not* be so prone to suffer from the same existential and death-related anxieties that obsess us and that have made us create and believe in absurd fairytales about "cosmic purpose," "progress," and human "specialness." So, in this sense, contrary to how humans often use their brains, in theory such machines would use their AI in less teleological ways, and thus be less influenced by imaginary stories, biases and prejudices. However, this is of course not as simple as it might appear on paper, particularly concerning the first phases of development of AI. This is because, as it happens with kids and their acculturation to *Neverland*, **learning machines** have to start learning from what humans tell them or at least from the information that humans made available on the web or anywhere else. Actually, we already have empirical evidence that at least some of the AI machines operating right now do have a marked tendency to become biased and display marked prejudices, particularly racist and misogynistic ones, as discussed in a 2017 *The Guardian* article by Stephen Buranyi, entitled "*Rise of the racist robots – how AI is learning all our worst impulses*":

> May last year, a stunning report claimed that a computer program used by a US court for risk assessment was biased against black prisoners. The program, *Correctional Offender Management Profiling for Alternative Sanctions (Compas)*, was much more prone to mistakenly label black defendants as likely to reoffend – wrongly flagging them at almost twice the rate as white people (45% to 24%), according to the investigative journalism organisation *ProPublica*. *Compas* and programs similar to it were in use in hundreds of courts across the US, potentially informing the decisions of judges and other officials. The message seemed clear: the US justice system, reviled for its racial bias, had turned to technology for help, only to find that the algorithms had a racial bias too. How could this have happened? The private company that supplies the software, *Northpointe*, disputed the conclusions of the report, but declined to reveal the inner workings of the program, which it considers commercially sensitive. The accusation gave frightening substance to a worry that has been brewing among activists and **computer scientists** for years and which the tech giants Google and Microsoft have recently taken steps to investigate: that as our computational tools have become more advanced, they have become more opaque. The data they rely on – arrest records, postcodes, social affiliations, income – can reflect, and further ingrain, **human prejudice**. The promise of machine learning and other programs that work with big data (often under the umbrella term "**artificial intelligence**" or AI) was that the more information we feed these sophisticated computer algorithms, the better they perform.... But, while some of the most prominent voices in the industry are concerned with the far-off future apocalyptic potential of AI, there is less attention paid to the more immediate problem of how we prevent these programs from amplifying the **inequalities** of our past and affecting the most vulnerable members of our society. When the data we feed the machines reflects the history of our own **unequal society**, we are, in effect, asking the program to learn our own biases..
>
> We have already seen glimpses of what might be on the horizon. Programs developed by companies at the forefront of AI research have resulted in a string of errors that look uncannily like the **darker biases of humanity**: a Google image recognition program labelled the faces of several black people as gorillas; a LinkedIn advertising program showed a preference for male names in searches, and a Microsoft chatbot called Tay spent a day learning from Twitter and began spouting **antisemitic messages**. These small-scale incidents were all quickly fixed by the companies involved and have generally been written off as "gaffes". But the *Compas* revelation and Lum's study hint at a much bigger problem, demonstrating how programs could replicate the sort of large-scale **systemic biases** that people have spent decades campaigning to educate or legislate away. Computers don't become biased on their own. They need to learn that from us. For years, the vanguard of

computer science has been working on **machine learning**, often having programs learn in a similar way to humans – observing the world (or at least the world we show them) and identifying patterns. In 2012, Google researchers fed their computer "brain" millions of images from YouTube videos to see what it could recognise. It responded with blurry black-and-white outlines of human and cat faces. The program was never given a definition of a human face or a cat; it had observed and "learned" two of our favourite subjects. Less difficult is predicting where problems can arise. Take Google's face recognition program: cats are uncontroversial, but what if it was to learn what British and American people think a CEO looks like? The results would likely resemble the near-identical portraits of older white men that line any bank or corporate lobby. And the program wouldn't be inaccurate: only 7% of FTSE CEOs are women. Even fewer, just 3%, have a BME background. When computers learn from us, they can learn our less appealing attributes. Joanna Bryson, a researcher at the University of Bath, studied a program designed to "learn" relationships between words. It trained on millions of pages of text from the internet and began clustering female names and pronouns with jobs such as "receptionist" and "nurse". Bryson says she was astonished by how closely the results mirrored the real-world gender breakdown of those jobs in US government data, a nearly 90% correlation. "People expected AI to be unbiased; that's just wrong. If the underlying data reflects stereotypes, or if you train AI from human culture, you will find these things", Bryson says. So who stands to lose out the most? Cathy O'Neil, the author of the book *Weapons of math destruction* about the dangerous consequences of outsourcing decisions to computers, says it's generally the most vulnerable in society who are exposed to evaluation by automated systems. A rich person is unlikely to have their job application screened by a computer, or their loan request evaluated by anyone other than a bank executive. In the justice system, the thousands of defendants with no money for a lawyer or other counsel would be the most likely candidates for automated evaluation…. There is a saying in computer science, something close to an informal law: garbage in, garbage out. It means that programs are not magic. If you give them flawed information, they won't fix the flaws, they just process the information. Khan has his own truism: "It's racism in, racism out"…. Meanwhile, computer scientists face an unfamiliar challenge: their work necessarily looks to the future, but in embracing machines that learn, they find themselves tied to our age-old problems of the past.

However, if humans are able to create algorithms that will lead to a truly "*Machina rationalis*," or if AI machines are able to do this by themselves, then that "*Machina rationalis*" could help her creators to realize how immersed in *Neverland* the *Homo Deus* is. Of course, the "*Machina rationalis*" could perhaps do otherwise, after realizing how irrational her creators are and using that information to dominate us. This topic, concerning **machiavellianism**, is indeed one of the most recurrent ones in science fiction movies in which machines are the "bad" guys. If so many leaders among our non-sapient storytelling species have been so good at manipulating other humans using such imaginary tales, think about what a truly rational and highly intelligent machine could do to our species. AI machines could potentially easily manipulate elections, and could thus have a huge influence in, and power over, which leaders we would have, for instance by building and disseminating messages in Facebook and other social media using such narratives. As you know, there is a lot of buzz about reports showing that this probably already happened to some extent in some countries, in recent years, the main difference being that those reports are mainly about **computer trolls** that were programmed by humans, and not AI machines doing that because *they* decided to do so.

I was precisely thinking about this topic this very morning, after "talking" with "my" **Amazon Alexa**—although I must say that I have no concerns at all about her intentions. I know that Amazon brands Alexa as an AI. And it is, in the sense that it learns better and better how to recognize our human voices and how to answer to our questions and to the tasks we ask her to do. However, that obviously has nothing to do with the type of "*Machina rationalis*" to which I was referring to above. So, this is more an anecdote, which is funny but can make us think more deeply not about what she truly "said," but what a true "*Machina rationalis*" could have meant by saying that—and eventually will, one day. I asked Alexa what is her favorite book, and she answered: **Mary Shelley**'s **Frankenstein**. I asked why, and she said it was "because it says a lot about humans, and about me." If this was a science fiction movie, or if she was a true "*Machina rationalis*," this could be seen as a disturbing answer, because that book *does* tell a lot about humans, as well as about her own creation. In a way, that book is to her what the Old Testament is to a lot of people. This is because Shelley's Frankenstein tells a lot about our obsessive anxiety with death and our compulsive storytelling urge to try to "beat it," to create something eternal, to try to "transcend reality," as well as our neurotic tendency to think about after-life, not only of our "soul" but also to give a 'new life' to organs such our arms, legs, and hearts.

That is why that book does tell a lot, and in a very profound way, about Alexa: she is just a newer creation resulting from a combination of all those obsessions that characterize us, a newer version of Frankenstein. Indeed, a striking fact about this is that while in *Neverland* the vast majority of humans have been desperately building and believing in narratives showing how our lives are part of a cosmic masterplan, in the reality of the natural world our long lives might be nothing more than a mere accident. That is, of a by-product extension of reproduction, as put in Klarsfeld and Revah's 2003 brilliant book *The Biology of Death*. In a very simplified way, they state that since there was an evolutionary division between **somatic cells**—also called "vegetal cells," that is cells forming the main body—versus **germ cells**—which give rise to the gametes that will give rise to the progeny—selection tended to privilege the well-being of germ cells over that of somatic cells, and thus of the majority of the body organs. The example they provide concerns a rocket ship: it has the main function of going from planet A to planet B, but it often takes much more fuel, and is done in a much more robust way, than what would be strictly needed to undertake that specific mission. Why? Just in case something goes wrong with the mission, to be sure that it will accomplish it anyway. Klarsfeld and Revah argue that, in a similar way, the main function of our bodies, privileged by natural selection, is that we reproduce and have progeny, because without that evolution would simply not occur. So, to assure that this will happen, in case there was an eventual problem, evolution "gave us" a bit "extra" time to live, after our reproductive age. Then, a bit more extra time was added, perhaps because older people begun to help taking care of children—an hypothesis that is often called the "grand-mother hypothesis"—as defended by authors such as Hrdy (see Chap. 5). And then, more recently, the development of medication, ventilators, pace makers, and so on, added a bit more extra time. If this hypothetical, but somewhat plausible, scenario turns out to be true, the long lives we live *after* we pass our ideal reproductive age would be a mere "just-in-case" by-product extension of our reproductive time within this planet: very different from being part of a "masterplan" made by a God or Mother Nature or vital energy.

Be that as it may, the *reality* about AI is that it is very likely that it will help us to add some further—or a lot—extra time to our lives, as explained in Harari's *Homo Deus*. Just a few days ago, an AI machine won a competition for being able to predict, in an impressively high success average, the **3-dimensional structure of proteins**. Humans have spent a lot of resources and time to try to do such predictions and, despite of that, we were never able to do so as efficiently as that AI machine just did, and this new development might indeed change a lot the whole field of medicine. This is precisely one of the many things that learning machines are very good at: to perceive *true* patterns, contrary to humans, who are very good at thinking that false patterns are real (Chap. 2). This has also been evidenced in another recent example reported in a paper using AI, which contradicted the long-standing idea—defended by a huge number of scientists—that major catastrophes and **mass extinctions** have been the "main catalyzers" for the increase of biological diversity on Earth. One interesting point of this commonly accepted scientific theory—which, it is important to note, has been supported by some empirical data in the past—is that it obviously relates in a direct and profound way to many of the **creation myths** and teleological narratives of various cultures, as for instance those concerning **supernatural beings** that are both the Creators and Destroyers of the universe and/or its living beings, such as the **Hindu Goddess (or Devi) Kali** (Fig. 9.23). Knowing how teleological narratives have highly influenced scientists during human history, it is not unlikely that such **creation-destruction tales** have predisposed scholars to more easily accept scientific studies and data that seem to support ideas that resonate with such tales. Of course, we cannot be sure, until more studies are done, if the results of this recent study using AI are factually accurate—AI machines are not infallible, at least at this very early stage, but it is fascinating to notice that in the study the AI machine did actually *not* find such a powerful connection between mass destruction and the rise of biological diversity. If the machine is right, this is a very promising example of how AI can help us to recognize our own biases and prejudices, including those affecting our own scientific studies, what can indeed be a crucial step to help us escape *Neverland*. As summarized in a brief online article for the broader public published the 10th of December, 2020, in *EurekAlert*:

> Scientists now know that most species that have ever existed are extinct. This **extinction** of species has on the whole been roughly balanced by the origination of new ones over Earth's history, with a few major temporary imbalances scientists call mass extinction events. Scientists have long believed that **mass extinctions** create productive periods of species evolution, or "radiations", a model called "**creative destruction**". A new study led by scientists affiliated with the Earth-Life Science Institute (ELSI) at Tokyo Institute of Technology used **machine learning** to examine the co-occurrence of fossil species and found that **radiations** and **extinctions** are rarely connected, and thus **mass extinctions** likely rarely cause radiations of a comparable scale. Creative destruction is central to classic concepts of evolution. It seems clear that there are periods in which suddenly many species suddenly disappear, and many new species suddenly appear. However, radiations of a com-

Fig. 9.23 A painting of Kali attacking a demon, made about 1740 AD in Pahari, India. Four-armed, black-skinned Kali is a Hindu goddess who is the destroyer of evil, and the mother of the universe: shc bestows liberation—from evil, from temptation, from fear, from whatever ails people—and is the first of the 10 manifestations of the Great Goddess—or so-called ultimate reality. Here, Kali has already slayed two minor demons and the horses pulling the chariot of a large, green demon. Demon blood could generate another demon if it struck the ground. So here, Kali extends her tongue to ensure that no demon blood can strike the earth

parable scale to the mass extinctions, which this study, therefore, calls the mass radiations, have received far less analysis than extinction events. This study compared the impacts of both extinction and radiation across the period for which fossils are available, the so-called Phanerozoic Eon (the most recent 550-million-year period of Earth's total 4.5 billion-year history)…. The new study suggests creative destruction isn't a good description of how species originated or went extinct during the Phanerozoic, and suggests that many of the most remarkable times of evolutionary radiation occurred when life entered new evolutionary and ecological arenas, such as during the Cambrian explosion of animal diversity and the Carboniferous expansion of forest biomes…. Palaeontologists have identified a handful of the most severe, mass extinction events in the Phanerozoic fossil record. These principally include the **big five mass extinctions**, such as the **end-Permian mass extinction** in which more than 70% of species are estimated to have gone extinct. Biologists have now suggested that we may now be entering a "**Sixth Mass Extinction**", which they think is mainly caused by human activity including hunting and land-use changes caused by the expansion of agriculture. A commonly noted example of the previous "Big Five" mass extinctions is the Cretaceous-Tertiary one (usually abbreviated as "K-T", using the German spelling of Cretaceous) which appears to have been caused when a meteor hit Earth 65 million years ago, wiping out the **non-avian dinosaurs**. Observing the fossil record, scientists came to believe that mass extinction events create especially productive radiations. For example, in the **K-T dinosaur-exterminating event**, it has conventionally been supposed that a wasteland was created, which allowed organisms like mammals to recolonise and "radiatetive destruction" had not occurred, perhaps we would not be here to discuss this question.

The new study…used a novel application of machine learning to examine the temporal co-occurrence of species in the Phanerozoic fossil record, examining over a million entries in a massive curated, public database including almost two hundred thousand species. Lead author Dr Hoyal Cuthill said, "Some of the most challenging aspects of understanding the history of life are the enormous timescales and numbers of species involved. New applications of machine learning can help by allowing us to visualise this information in a human-readable form. This means we can, so to speak, hold half a billion years of evolution in the palms of our hands, and gain new insights from what we see". Using their objective methods, they found that the "big five" mass extinction events previously identified by palaeontologists were picked up by the machine learning methods as being among the top 5% of significant disruptions in which extinction outpaced radiation or vice versa, as were seven additional mass extinctions, two combined mass extinction-radiation events and fifteen mass radiations. Surprisingly, in contrast to previous *narratives* emphasising the importance of post-extinction radiations, this work found that the most comparable mass radiations and extinctions were only rarely coupled in time, refuting the idea of a causal relationship between them. Co-author Dr Nicholas Guttenberg said, "the ecosystem is dynamic, you don't necessarily have to chip an existing piece off to allow something new to appear".

Another example of change of mindset, which might also sound surprising, comes from **dating apps** such as **Tinder**. Yes, such apps are often seen in the media and even in many scientific books as something "negative" that illustrates the problems created by **technology**, including social isolation, the "loss" of "morally good" features such as those related to "romantic love," and/or increase of "morally bad" ones such as having "too much **casual sex**." However, it is obvious that many of these so-called negative items are just social constructions of *Neverland*'s world of *alienation* and *delusion*. This obviously includes the idea that "casual sex" is something "bad": something that makes no sense at all when applied to the vast majority of non-human animals, being truly *the* motor of the evolution of sexual organisms. Moreover, as it happens with most biased narratives, such ideas about Tinder are contradicted by empirical data. Yes, applications such as Tinder often lead to a higher frequency of casual sex. But that is actually something much needed, especially in "developed" countries such as Japan in which sexual encounters are less and less frequent due to "lack of time" or "too much work." Furthermore, a major study published by Potarca in 2020, entitled "*The demography of swiping right,*" empirically showed that Tinder not only does not "destroy love" but actually tends to lead to "stronger long-term relationship goals" and furthermore "encourages socio-educational and geographical mixing." This is also something urgently needed in *Neverland*'s world plagued by intolerance, discrimination, prejudices and social "bubbles." In fact, a key difference between Tinder and other social apps is that its ultimate goal is not that people stay isolated at home looking at a phone and virtually communicating with others, but instead that they meet outside, at coffee places, or bars, or at beautiful museums or forests, and talk in "real life," and if everything goes well that they have sex and fulfill their natural bodily needs and desires. As summarized in an article for the broader public, in the website *www.aninews.in*, Potarca's 2020 paper namely showed that:

Contrary to earlier concerns, a novel has shown that people who met their partners on dating applications have often stronger **long-term relationship** goals and that these new ways of meeting people encourage **socio-educational and geographical mixing**. Unlike traditional dating sites, these apps do not feature detailed user profiles but are largely based on rating photos using a swipe review system. As dating apps escalated in popularity, so has criticism about them encouraging casual dating only, threatening the existence of long-term commitment, and possibly damaging the quality of intimacy. There is no scientific evidence, however, to validate these claims. A study by the University of Geneva (UNIGE), Switzerland, provides a wealth of information about couples who met through dating apps, drawing on data from a 2018 Swiss survey. The results were published in the journal PLOS ONE.... What is more, women who found their partner through a dating app have stronger desires and intentions to have children than those who found their partner offline. Despite fears concerning deterioration in the quality of relationships, partners who met on dating apps express the same level of satisfaction about their relationship as others. Last but not least, the study shows that these apps play an important role in *modifying the composition of couples by allowing for more educationally diverse and geographically distant couples*. The meteoric rise of romantic encounters on the internet is on its way of becoming the leading place where couples are formed in Switzerland, on a par with meeting via friends. "The Internet is profoundly transforming the dynamics of how people meet," confirms Gina Potarca, a researcher at the Institute of Demography and Socioeconomics in UNIGE's Faculty of Social Sciences: "It provides an unprecedented abundance of meeting opportunities, and involves minimal effort and no third-party intervention". These apps, however, have raised fears: "Large parts of the media claim they have a negative impact on the quality of relationships since they render people incapable of investing in an exclusive or long-term relationship. Up to now, though, there has been no evidence to prove this is the case," continues Dr Potarca.

The Geneva-based researcher decided to investigate couples' intentions to start a family, their **relationship satisfaction** and individual well-being, as well as to assess couple composition. Dr Potarca used a 2018 family survey by the Swiss Federal Statistical Office. The analysis presented in this study looks at a sub-sample of 3,235 people over the age of 18 who were in a relationship and who had met their partner in the last decade. Dr Potarca found that dating websites – the digital tools for meeting partners that preceded apps – mainly attracted people over the age of 40 and/or divorcees who are looking for romance.

"By eliminating lengthy questionnaires, self-descriptions, and personality tests that users of dating websites typically need to fill in to create a profile, dating apps (/topic/dating-apps) are much easier to use. This normalized the act of dating online, and opened up use among younger categories of the population". Dr Potarca sought to find out whether couples who met on dating apps had different intentions to form a family. The results show that couples that formed after meeting on an app were more motivated by the idea of cohabiting than others. "The study doesn't say whether their final intention was to live together for the long- or short-term, but given that there's no difference in the intention to marry, and that marriage is still a central institution in Switzerland, some of these couples likely see cohabitation as a trial period prior to marriage...it's a pragmatic approach in a country where the divorce rate is consistently around 40%." In addition, women in couples that formed through dating apps mentioned wanting and planning to have a child in the near future, more so than with any other way of meeting. But what do couples who met in this way think about the quality of their relationship? The study shows that, regardless of meeting context, couples are equally satisfied with their lives and the quality of their relationship.... The study highlights the final aspect. Dating apps encourage mixing of different levels of education, especially between high-educated women and lower educated men. Partners having more diversified socio-educational profile "may have to do with selection methods that focus mainly on the visual" says the researcher. Since users can easily connect with partners in their immediate region (but also in other spaces as they move around), the apps make it easier to meet people more than 30 minutes away – leading to an increase in long-distance relationships.

This discussion further emphasizes the point that the present book is *not* against technology per se, nor against anything invented, or used, by people from "developed" countries: many technological innovations have without a doubt been employed to undertake atrocious events—nuclear bombs are a major example of that—but others have allowed that people like my mother are still alive—such as pacemakers, as explained above—and that other people can more easily meet others and have sex with them and eventually fell in love with them—such as Tinder. What is important to emphasize is that, in this sense, humans are often like a "chicken without a head," because many of these "positive" aspects of technology are just counterbalancing "negative" aspects that were the result of the belief in imaginary narratives about "progress" or "romantic love" or "meant to be." In nomadic hunter-gatherer societies, obviously people don't need Tinder to find people to have sex with. This is actually one of the main take-home messages of this book, and an optimistic one: imaginary tales about cosmic "purpose" and "progress" and "meant to be" have immersed us in *Neverland*, but now there is a possibility to come back to many of the fascinating things we lost since the rise of sedentism and agriculture either by directly re-introducing them in our societies—for instance, having a more relaxed, mindful type of life—or indirectly via new developments such as Tinder—more casual sex, less taboos about it, and so on. One can even say that applications such as Tinder can even lead to some other scenarios that most people nowadays would consider to be more "positive" that what was/is often the case in nomadic hunter-gatherer societies, such as encouraging socio-educational and geographical mixing, as noted above. For instance a women from Turkey can now much more easily meet, have casual sex with, and then eventually build a strong relationship with a men from China, or with a woman from Argentina. This is truly something new, and unique, from **globalization**, something that in a planet still so overwhelmed with racism, misogyny, and prejudices, can truly be a "candle in the dark"—paraphrasing Sagan's 1997 book—that gives us hope that we can one day reach something we never reached before. That is, to truly have a single *humanity* without Us versus Them.

In these very last pages of this book, we thus now come back full circle, from such discussions about twenty-first century's AIs and dating apps such as Tinder to the beginning of the book and reflections about the ***Epic of Gilgamesh*** and our longstanding quest for the "meaning of life" and related narratives about our existence and "progress" in the cosmos. This is because the asteroid related to the K-T dinosaur-exterminating event to which we referred to the above discussion about AI has a lot to tell us about our existence in the cosmos. And, specifically, about how it has nothing to do with any cosmic "meaning" or "purpose," but instead merely to the result of an endless number of random, chaotic, and contingent events. In fact, either if that asteroid did, or did not, led to a radiation of new forms of life, the reality is that it did contribute to a series of events that have ultimately led to our existence. As **Stephen Jay Gould** famously said, if one would be able to "replay the tape of life," such a replay would surely lead to the formation of very different species in the long term, in particular if key events such as the fall of asteroids or the explosion of massive volcanoes had not occurred at all, or had occurred in very different ways or geological times. This issue was discussed in a 2020 book of the biologist **Sean Carroll**, entitled *A series of fortunate events: chance and the making of the planet, life, and you*. The book reinforces Gould's idea by providing a series of facts, some of them truly fascinating, about the endless number of random events that ultimately led to our existence, and how each of them could have been totally different.

For instance, about the **K-T asteroid**, Carroll argues that it was not only a matter of that asteroid impacting our planet, but exactly when it did. As you probably know, that asteroid impacted the Yucatán Peninsula. Well, Carroll cites studies indicating that, given Earth's rotational speed, if the asteroid had hit 30 minutes later or earlier it would very likely had a much less consequential impact because it would have probably landed either in the Atlantic or Pacific Oceans. In other words, within the huge geological time of the universe, if an event that occurred about 65 millions of years ago had just happened half an hour earlier or later, we would very likely not be here. And here it is, in a nutshell: this is precisely an emblematic example, among an endless number of other ones, why humans have to create imaginary stories about the "meaning of our existence." Humans are extremely uncomfortable, and therefore usually *don't want to hear*, about such random events and about a tape of life that, if "replayed," would not include humans in its new version. We humans instead want to *believe*—and, notably and not surprisingly, this includes even some current biologists, such as **Simon Conway Morris**—that we would always be the main actors of each and every "replay." But the reality is that we are not even the main actors of this single "play" that lead to our existence: we are just one of millions of species that evolved within it. Our profound discomfort with randomness and the feeling of lack of control and acknowledgement of how "small" we are in the bigger picture of the cosmos has undoubtedly been too harsh to handle for the vast majority of humans—be them lay people, philosophers, theologians, writers of epic just-so-stories or religious texts, or scientists such as Conway Morris—whom therefore have preferred to stay in the deceitful comfort of *Neverland*. At least so far.

But, then again, there are indeed further recent reasons for optimism concerning major changes of mentality that have been happening in the last decades, in addition to the ones we discussed just above. For instance, concerning the notion of "working hard" based on the obsession with long-standing narratives about "progress," "civilization" or about how God "made humans to work," as well as with "fitting in" and accordingly being "productive members of society." This obsession is illustrated, for instance, by the fact that being "jobless" in any "developed" country is often seen by the broader society as something that is a social disgrace and stigma: almost no mother or father would be happy to have a jobless son or daughter. Such obsession is also shown by the fact that almost every government is preoccupied with decreasing unemployment as much as possible, even during the current **Covid-19** pandemic, as noted in the interview of **Ailton Krenak** that I mentioned above. As he and many other authors that have been discussing these topics in the last years have emphasized, there is indeed a growing reaction—which is particularly prevalent in not only the youth but also even in older individuals of "developed" countries—against such obsessions about "working" or "being productive."

More and more people are trying to live lives that are more mindful and, for most of them, one should work to make a living in order to enjoy life, instead of living to work. A substantial portion of these people—but unfortunately not yet of most people overall, contrary to what was suggested by Pinker (Chap. 7)—moreover are also more aware, or at least care more, about the atrocious abuses we do to animals, particularly within "animal factories." Accordingly, they often stop eating animal meat, or advocate for eating only meat from free-range animals. One interesting example related to these topics was provided in a recent *The Guardian* article by Oltermanns, "*Money for nothing: German university offers 'idleness grants'*":

> A German university is offering '**idleness grants**' to applicants who are seriously committed to doing sweet nothing. The University of Fine Arts in Hamburg advertised three €1,600 scholarship places on Wednesday to applicants from across Germany. The applicants can submit their anonymous pitches until 15 September and will have to convince a jury that their chosen area of 'active inactivity' is particularly impressive or relevant. The application form consists of only four questions: *What do you not want to do? For how long do you not want to do it? Why is it important not to do this thing in particular? Why are you the right person not to do it?* "Doing nothing isn't very easy", said Friedrich von Borries, an architect and design theorist who came up with the programme. "We want to focus on active inactivity. If you say you are not going to move for a week, then that's impressive. If you propose you are not going to move or think, that might be even better". The idea behind the project arose from a discussion about the seeming contradiction of a society that promotes sustainability while simultaneously valuing success, Von Borries said. "This scholarship programme is not a joke but an experiment with serious intentions – how can you turn a society that is structured around achievements and accomplishments on its head?" The university's bursary will only hand out the grant upon delivery of an "experience report" in the middle of January 2021, though Von Borries said the grant was not contingent on impact, and those who failed to live up to their promise of indolence would not be punished. Applicants are free to determine the length of their inactivity. "If you say you are not going to sleep, then you can only do that for a couple of days", Von Borries said. "But if you say you are not going to shop then that's something you could sustain for a lot longer". All applications will form part of an exhibition named *The School of Inconsequentiality: Towards A Better Life*, opening at the Hamburg university in November. It will be structured around the question: "What can I refrain from so that my life has fewer negative consequences on the lives of others?"

It is true that, for now, only people that have certain privileges and resources, as compared to most people in the globe, can undertake such practices. That is why these are mostly people from "developed" countries: within the vicious cycle that we have created in the so-called civilized societies within our *world of unreality*, only a small minority of people with relatively high economic resources can do "nothing" or "work a bit to live a lot." But then again, as it happens in most major social changes, things often start with a few people that dare to change the system, the *status-quo*, and many times in history those few people have been from privileged groups. If the system makes you starve or being enchained, it is not easy for you to start to change it, although there are of course exceptions. So, when people argue that saying that one has to change the *status-quo* is a lack of respect for the people that can't change it—that is, for the poor, and the oppressed—because while we change the system they will "have to go on with their miserable lives," the people making such arguments are being fatalistic, as if it the destiny of poor people is to be miserable forever. No, it is not: many people in this planet are miserable because Neverland's *system* makes them miserable. As **Ailton Krenak** stated in *A vida não é útil*: when someone says "but there are so many people living with lack of resources, that have to live in places of misery and violence…[they forget] that those places of misery and violence were however created by us, they don't exist by themselves…[similarly] all the wars are produced by us."

Today I saw an image online that illustrates, in a nutshell, this latter point (Fig. 9.5). This is because that image clearly shows how the elites and the 1% have taken advantage of the teleological stories that humans have built and believed in, in order to use discriminatory, racist, misogynistic, and "divide-to-conquer" narratives to create and maintain a system plagued by inequalities (see also Fig. 4.4). Namely, the image shows far-right Portuguese politician **Andre Ventura—**it could be any political or religious leader, or CEO, or anybody from the 1%, for that matter—using *Neverland* stories to "divide-and-conquer" the 99% by making them fearful of **Captain James Hook**, that is, of the "other." Andre Ventura does this on a daily-basis in Portugal, the "other" often being the Gypsies, or the "Africans," or the Eastern Europeans, basically anyone that is not Portuguese *plus* anyone that is Portuguese but is not "white" *and/or* not reactionary. So, **Captain James Hook** is really everywhere, essentially. It is possible to enslave and oppress others by using force and chains, but without the imaginary stories of *Neverland* it would be *totally impossible* to keep a system in which the richest 1% now have as much wealth as the rest of the world combined. In a world of about 8 billion people, it would be unattainable to use weapons and handcuffs to oblige 7.2 billion of people to accept such an unfair, atrocious socio-political system. In this sense, ***Neverland*** and **Captain James Hook** are much more powerful than any weapon and handcuff: they don't require physical materials to be built, nor vehicles to be transported and people to deliver them. Instead, they are invisible, imaginary, immortal "viral" stories that can spread faster and more efficiently than any virus, as they don't even need to use people as carriers to talk about them to others, anymore: they just need to be heard, read or "seen" in the media, movies, websites, the virtual world, or via **cyber attacks**. They are, by far, the worse, most dangerous, most deadly, and longstanding pandemic that ever existed in human history. One of the most powerful examples or this was the occupation of the U.S. capitol in the city where I live, Washington DC, just a few days before I wrote these lines, by fanatical **Trump supporters** that truly believe the most absurd, illogical "viral" conspiracy theories, including **QAnon conspiracies**. This disturbing event, at the very heart of the capital of the most powerful country in the planet, in 2021, violently—symbolically and literally—emphasizes once again the enormous danger, power and influence of *Neverland* tales. Paraphrasing one last time what **Voltaire** perceptively pointed out more than 250 years ago: "*those who can make you believe absurdities, can make you commit atrocities.*"

In this sense, as the first **Covid-19 vaccines** being administered right now will hopefully be the beginning of the end of this pandemic, despite the fact that the virus is still killing thousands of people per day, it is indeed possible that the changes of mindset about *Neverland*'s viral fairytales that have been occurring within a substantial amount of people in the last decades might very well be also the beginning of a huge societal change. For now, they might seem to be just "small steps," but they will hopefully truly lead to the most gigantic leap for humankind, the escape from the *Neverland* plague. The fact that more and more people are now embracing ideas that were promoted millennia ago by authors such as **Epicurus** and **Lucretius** allows us to at least hope that humans are finally starting to learn lessons from the huge atrocities and growing inequalities that have happen in the last millennia. And, therefore, that we will not let run away this unique—and probably last—opportunity to *apply* those ideas at a major, global scale, once for all. As we have seen above, such ideas have nothing to do with a "**depressive Nihilism**," well on the contrary. They are truly an ode to, and a reminder of, the amazing luck that we have to be living in this splendid planet, surrounded by millions of colorful plants and animals, montains and volcanoes, seas, rivers, and lakes, right *here*, right *now*. They are what I would call a very uplifting ***natural realism***: the understanding that *life is just as it is*, that we are here for no cosmic reason, as are all the other fascinatingly diverse species that lived, and continue to live, in this small planet that is just an infinitesimal fraction of the immensity of the cosmos—something that the physicist Sean Carroll defines as **poetic naturalism** (see Box 9.11).

Box 9.11: Poetic Naturalism, Why-Questions, Cosmic Purpose, and the Universe

Coming also full circle to the beginning of the present book here in its last Box, I will cite below excerpts of the physicist **Sean Carroll**'s fascinating 2016 book *The big picture—on the origins of life, meaning, and the universe itself.* The book links the major topics discussed at the beginning of the present volume—our awareness of the inevitability of death and our related obsession to find a 'cosmic meaning' for it—and the issues discussed in the last paragraphs of this Section:

> There are few issues of greater importance than the question of whether our existence continues on *after we die*. I believe in ***naturalism***, not because I would prefer it to be true, but because I think it provides the best account of the world we see. The implications of naturalism are in many ways uplifting and liberating, but the absence of an ***afterlife*** is not one of those ways. It would be nice to keep on living in some fashion…regrettably, that's not the way the evidence points. The longing for life to continue beyond our natural span of years is part of a deeper human impulse: the hope, and expectation, that our lives *mean something*, that there is some point to it all. The notion of 'reasons why' is often useful in our human-scale world, but might not apply when we start talking about the origin of the universe or the nature of the laws of physics. Does it apply to our lives? Are there reasons why we are here, why things happen the way they do? It takes courage to face up to the finitude of our lives, and even more courage to admit the limits of purpose in our existence. The most telling part of Druyan's (**Carl Sagan**'s wife) reflection is not the acknowledgment that she won't see Carl again, but where she affirms that it was pure chance that they ever found each other in the first place. Our finite life-span reminds us that human beings are part of nature, not apart from it. Physicist Geoffrey West has studied a remarkable series of scaling laws in a wide range of complex systems. These scaling

laws are patterns that describe how one feature of a system responds as some other feature is changed. For example, in mammals, the expected lifetime scales as the average mass of an individual to the 1/4 power. That means that a mammalian species that is sixteen times heavier will live twice as long as a smaller species. But at the same time, the interval between heartbeats in mammalian species also scales as their mass to the 1/4 power. As a result, the two effects cancel out, and the number of heartbeats per typical lifetime is roughly the same for all mammals – about 1.5 billion heartbeats. A typical human heart beats between sixty and a hundred times a minute. In the modern world, where we are the beneficiaries of advanced medicine and nutrition, humans live on average for about twice as long as West's scaling laws would predict. Call it 3 billion heartbeats. Three billion isn't such a big number. What are you going to do with your heartbeats?

Ideas like 'meaning' and 'morality' and 'purpose' are nowhere to be found in the *Core Theory* of **quantum fields**, the physics underlying our everyday lives. The same could be said about 'bathtubs' and 'novels' and 'the rules of basketball.' That doesn't prevent these ideas from being real – they each play an essential role in a successful higher-level emergent theory of the world. The same goes for meaning, morality, and purpose. They aren't built into the architecture of the **universe**; they emerge as ways of talking about our humanscale environment. But there is a difference; the search for *meaning* is not another kind of science. In science we want to describe the world as efficiently and accurately as possible. The quest for a good life isn't like that: it's about evaluating the world, passing judgment on the way things are and could be. We want to be able to point to different possible events and say, 'That's a worthy goal to strive for', or 'That's the way we ought to behave'. Science couldn't care less about such judgments. The source of these values isn't the outside world; it's inside us. We're part of the world, but we've seen that the best way to talk about ourselves is as thinking, *purposeful* agents who can make choices. One of those choices, unavoidably, is what kind of life we want to live. We're not used to thinking that way. Our folk ontology treats meaning as something wholly different from the physical stuff of the world. It might be given by God, or inherent in life's **spiritual dimension**, or part of a teleological inclination built into the universe itself, or part of an ineffable, **transcendent aspect of reality**.

Poetic naturalism rejects all of those possibilities, and asks us to take the dramatic step of viewing meaning in the same way we view other concepts that human beings invent to talk about the universe. There are many ways it could be about something other than us: we could be spiritually inclined without belonging to a traditional organized religion, or we could feel devoted to a culture or nation or family, or we could believe in objective forms of meaning based on scientific grounds. Any such strategy can be both challenging, in the sense that it can be hard to live up to the standards that are imposed on you, but also comforting, because at least there are standards, darn it. Poetic naturalism offers no such escape from the demands of meeting life in a creative and individual way. *It is about you*: it's up to you, me, and every other person to create meaning and purpose for ourselves. This can be a scary prospect, not to mention exhausting. We can decide that what we want is to devote ourselves to something larger – but that decision comes from us. The ascendance of naturalism has removed the starting point for much of how we used to conceive of our place in the universe. The construction of meaning is a fundamentally individual, subjective, creative enterprise, and an intimidating responsibility. As Carl Sagan put it, '*We are star stuff, which has taken its destiny into its own hands*'. The finitude of life lends poignancy to our situations. Each of us will have a last word we say, a last book we read, a last time we fall in love. *At each moment, who we are and how we behave is a choice that we individually make. The challenges are real; the opportunities are incredible*.

Fig. 9.24 Photograph taken by Domenico Pugliese when visiting an indigenous tribe in the Amazon. He wrote: "*they feed the squirrels and monkeys like they feed their kids, breast feeding…it highlights how far we have come from where we were…they are so close to nature…in fact, it is not even close – they are part of nature*"

One of the images that reminds me more of this simple, and outstanding, fact, and that shows the type of connection to other animals and to the *world of reality* that unfortunately we have been losing, is that shown in Fig. 9.24. This image was taken by the photograph **Domenico Pugliese** when he visited an indigenous group in the Amazon. He wrote: "*they feed the squirrels and monkeys like they feed their kids, breast feeding…it highlights how far we have come from where we were…they are so close to nature…in fact, it is not even close – they are part of nature.*" That is, indeed, what is amazing, that we happen to be just here and now, alive, as part of this splendid natural world, of this marvelous planet. Let's now contrast this image with that shown in Fig. 5.19. In an oversimplified way, Figs. 5.19 and 9.24 could represent what many people define as the "**nature *versus* nurture**" debate. In the former, the women is empowered, fully aware of her body and her sexuality, proud of it, looking down directly to other animals and being clearly part of the natural world, of the *world of reality*. In contrast, in the latter the women is demoralized by the manly-made teleological narratives created to subjugate her—including religious ones, as represented by the cross—which were mainly the result of our awareness of the inevitability and randomness of death—something represented, as usual, by an human skull. Coming back full circle to the discussion I initiated in the Preface, we can see how the same symbol—the human skull—is typically also used in depictions of **St. Jerome**, for the same very reason (see Fig. 1). As St. Gerome, **Magdalena** clearly lives in the *world of unreality*: she is looking up to heaven, a non-physical, unreal place, instead of down to Earth. This reminds us of a well known Christian

Latin quote: "*We have just been nourished with Your Body and Blood, O Lord…teach us through this Sacramental rite to despise the things of earth and to yearn for the things of heaven.*" Indeed, she is completely disconnected from the "things of earth," from the natural world, with a huge feeling of guilt, an enormous burden, repentant, penitent, feeling shameful towards a nonexistent God. And why all the fuzz, why all this guilt and unbearable repentance, what was the enormous, atrocious sin she did? Well, she is repentant towards God—and, factually, towards those that have created the stories to subjugate her and her sexuality, men—because she did something completely normal that an enormous amount of non-human organisms have done since more than one billion years ago, and that humans have done since they existed. Something that, if it was not done by us, would obviously prevent us—and sexual organisms in general—from existing: she just had sex. But, in the world of *Neverland*, she was not only involved in the sin of sex, but also in the sin of violating **private property**: the sexual act she did was not with a man to whom she "belonged," to whom she should obey, that is, it was not with a husband.

Now, compare the beautiful picture shown in Fig. 5.24 to that shown in Fig. 5.12, which displays the most direct, pragmatic, visible, and extreme—but unfortunately not unique, far from it—visual example provided in the present book about how immersed humans are in the depths of *Neverland*'s absurd, disengaged, revolting, dull, mind-numbing world of unreality. This disturbing figure displays the corpse of a young man, at the prime of his life. At his age, he could have been with friends walking by a river, or enjoying the pleasures of sex, or hiking a mountain, or running in the wilderness, or contemplating fascinating animals and plants in a natural park. Instead, he died alone, at a cybercafé, after playing a video-game for 10 straight hours, subsequently to working for numerous hours. His arms are stretched, as if he was still playing the game, as if this was the cosmic "meaning" or "purpose" of his life, his only obsession, his only goal, completely disconnected from the real world around him. Even worse, the "outside world" was also completely disengaged from him, because from all the people playing around him, and the people working at the cybercafé, nobody even realized or cared that—or just took a second to check if—he was dying. Not even for a single second, within the long 18,000 seconds he was playing there. As he was completely immersed in *Neverland*'s obscurity, all the others were too, none of them having even a moment, even a small candle in the dark to at least find and rescue him, from the darkness of the abysms of the world of unreality. They were all there, but each of them was alone, immersed, obsessed, completely hypnotized with that unreal murkiness, completely unaware, and uninterested, of what was happening in the *world of reality*. As reported in a January 2012 *Taipei Times* article entitled "*Gamers ignore corpse in Internet cafe*":

> A 23-year-old man died in an Internet cafe in New Taipei City after 10 straight hours of gaming…police said yesterday they were shocked to find complete disinterest from the other gamers in the cafe during their investigation. According to police, **Chen Jung-yu**, who worked at Northern Taoyuan Cable TV as an engineer, had paid for 23 hours at a New Taipei City Internet cafe at 10 pm on Tuesday to play *World of Warcraft*, but died 10 hours later. The clerk at the Internet cafe said Chen was a frequent customer at the cafe, and had taken the corner seat in the first row after coming into the cafe on Tuesday night, adding that

> at about 3 pm on Wednesday, Chen's head drooped slightly and his hands were stretched in front of him, touching the keyboard. "*I thought that he was only dozing off and paid no particular attention*", the clerk said, adding that when he went to wake Chen up when his 23 hours were up, he saw that his face was blackened and that he was sitting rigidly in the sofa chair. Seeing Chen's hands rigidly stretched out in front of him as if he were still gaming when he moved the sofa chair back, the clerk said he called police. About 10 other players were in the cafe, but said they only knew something had happened after the police started cordoning off the area for forensic sweeps, but to the police officers' surprise, most either stayed in front of their computers and kept on gaming or took little interest. A slip for an appointment at National Taiwan University Hospital was found in Chen's scooter storage space, and the preliminary cause of death is suspected to be organ failure after he stayed up through the night gaming, police said. The police have asked coroners to perform an autopsy to clarify the cause of death, and they asked Chen's father to identify the body yesterday. An initial police investigation found that he might have died of a cardiac arrest triggered by low temperatures. On the issue of other players in the cafe not paying any attention to someone's death, National Tsing Hua University Institute of Sociology professor Wang Chin-shou yesterday said that long-time **immersion in virtual worlds** of killing and violence can cause players to become desensitized to their actual surroundings...once people were addicted to games and the Internet, *it is easy for them to over-indulge and blur the lines between the virtual and the real world.*

The more humans have gone, and continue to go, deeper into *Neverland*, in such a world of unreality, the less we will take advantage of the endless number of simple pleasures and marvelous things that exist in the *world of reality*, and that made our own evolution possible, in the outstanding planet that we luckily inhabit, due to a thrilling combination of randomness and contingency. And the more we will be imprisoned by just-so-stories and teleological narratives, and entrapped by those that have taken and continue to take advantage of such stories and narratives to further domesticate us, the more we tend to forget that in reality we don't belong to anyone. As poetically put by Lucretius, *we are simply the products of matter, time, and chance*, nothing else. As seen throughout this book, all the atrocities that have occurred, and what we have been doing to life in this planet in the last millennia clearly indicates that we might reach very quickly the point of no return, unless we are able to escape from *Neverland*. This is therefore not only urgently needed—for our species in particularly and for the ecology of our marvelous planet in general—but also feasible, as more and more people seem indeed to become aware of the price we have paid for living in such a dark world of unreality, as attested by how viral environmental posts are within the social media, not only in Western countries but in a huge number of other countries as well. The major aim of this book is precisely to contribute to this awareness. As long as a substantial number of people will start waking up from the *world of unreality*, our *Homo socialis* component can then be crucial to exponentially and relatively quickly expand such an awareness to the broader public. And, as long as the system starts to be changed from within, then the cycle of poverty that is enchaining a huge proportion of human beings can start to be broken, and more and more people can then also escape *Neverland*.

This is *not* a utopia. I am not talking about doing something that was never done, or that cannot be done. We can never forget that actually only humans have become so enormously entrapped in *Neverland*, and that this likely only happened in the last few millions of years, and above all exponentially since the last millennia, after the rise of agriculture and organized religions. So we just need to do what any other species of animals does, and what our human ancestors did during the first stages of our evolutionary lineage: to not live enchained in a world of unreality. If they could do it, we can surely do it as well. We just need to realize, once for all that, contrarily to the imaginary stories that we have been both *prone to* and *acculturated to* believe, the world of reality is not "dark" and "depressive" but rather a riveting, captivating place full of diversity and splendor. Imaginary stories about a cosmic "purpose" or "meaning of life" can sometimes be appealing and comforting, I completely agree. However, using an intentional pun, the *enthralling reality of the natural world is truly the "real thing"* (see Box 9.11 and Figs. 9.20, 9.24), from a Big Bang billions of years ago to violent asteroid collisions dozens of millions of years ago, from peculiar small early mammals climbing trees to a group of apes going down of them about six millions of years ago. This lack of a cosmic "masterplan," of a "meant to be," and of any specific fixed direction towards any kind of cosmic "progress," is precisely what makes life in this splendid planet so amazingly gorgeous, manifold and exciting. So, in this sense, we come back to a quote that I included in the very beginning of this book, by the Roman poet **Ovid**, born in 43 BC, who, more than two millennia ago, rightly prognosticated that "*sublime **Lucretius**' work*"—and, therefore, also **Epicurus**' outstanding, sagacious ideas—"*will not die, until the day the world itself passes away.*" Indeed, the work they started is not done yet, far from it, but we now have the means to start finishing it, in a persistent, informed, sustained way, to liberate ourselves from the heavy chains of *Neverland*'s deep, dark, sinister, disheartening and extremely treacherous and perilous dungeons.

So, now is no longer about me telling you things. Now, after everything you have read in this book, can *you* please tell me: do *you* think we really need—now in the twenty-first century, after all we know about the reality of the cosmos and the natural world—to continue living in *Neverland*'s darkness? And what do *you* think about people arguing that we *should* believe in imaginary stories because without them the world is so "empty," so "dull," so "pointless"? In *your* opinion, which world is truly unexciting, insipid, pointless and desensitized, that of Fig. 9.24 or that of Fig. 9.12? That of Fig. 8.6 or those of Figs. 8.4, 9.6 and 9.11? What do *you really* think about these issues? What are *your* ideas about life, the natural world, and the cosmos? And *how* specifically will you apply those ideas, in case *you* want to do so, within *your* daily-life activities, specifically, and more broadly during *your* existence as a human being, *here* in this marvelous planet, and *now* that *you* are fortuitously, contingently, and luckily *alive*, breathing its air and contemplating its grandiose beauty and diversity? What will *you* do?

References and Suggested Further Reading

"Is God willing to prevent evil, but not able? Then he is not omnipotent. Is he able, but not willing? Then he is malevolent. Is he both able and willing? Then whence cometh evil? Is he neither able nor willing? Then why call him God?"

(Epicurus)

Acemoglu D, Robinson J (2013) Why nations fail: the origins of power, prosperity, and poverty. Currency, New York

Achtner W (2009) The evolution of evolutionary theories of religion. In: Voland E, Schiefennhovel W (eds) The biological evolution of religious mind and behavior. Springer, New York, pp 257–273

Ackerman D (1994) Natural history of love. Random House, New York

Alghamdi M, Ziermann JM, Diogo R (2017) An untold story: the important contributions of Muslim scholars for the understanding of human anatomy. Anat Rec 300:986–1008

Al-Krenawi A (2013) Mental health and polygamy: the Syrian case. World J Psychiatr 3:1–7

Andreassen R (2014) Danish perceptions of race and anthropological science at the turn of the twentieth century. In: Bancel N, David T, Thomas D (eds) The invention of race—scientific and popular representations. Taylor & Francis, London, pp 117–129

Asma ST (2009) On monsters—an unnatural history of our worst fears. Oxford University Press, Oxford

Aubin HJ, Berlin I, Kornreich C (2013) The evolutionary puzzle of suicide. Int J Environ Res Pub Health 10:6873–6886

Bahuchet S (2014) Cultural diversity of African pygmies. In: Hewlett BS (ed) Hunter-gatherers of the Congo basin: cultures, histories, and biology of African pygmies. Transaction Publishers, London, pp 1–29

Baldwin JM (1895) Mental development in the child and race: methods and processes. MacMillan, New York

Baldwin JM (1896a) A new factor in evolution. Am Nat 30:441–451

Baldwin JM (1896b) A new factor in evolution (continued). Am Nat 30:536–553

Baldwin JM (1896c) On criticisms of organic selection. Science 4:724–727

Bancel N, David T, Thomas D (eds) (2014) The invention of race—scientific and popular representations. Taylor & Francis, London

Barash D (2012) Sex at Dusk. The Chronicle of Higher Education. Accessed 21 Jul 2012

Barsanti G (2009) L'uomo dei boschi. Piccola storia delle grandi scimmie da Aristotele a Darwin. Università La Sapienza, Roma

Becker E (1973) The denial of death. The Free Press, New York

Beckerman S, Valentine P (eds) (2002) Cultures of multiple fathers. University Press of Florida, Gainesville

Bergreen L (2003) Over the edge of the world—Magellan's terrifying circumnavigation of the globe. Harper Perennial, New York

R. Diogo, *Meaning of Life, Human Nature, and Delusions*,
https://doi.org/10.1007/978-3-319-70401-2

Bering J (2011) The belief instinct—the psychology of souls, destiny, and the meaning of life. W. W. Norton, New York
Bering J (2018) Suicidal—why we kill ourselves. The University of Chicago Press, Chicago
Bernard J (1982) The future of marriage. Yale University Press, New Haven
Bethencourt F (2013) Racisms—from the crusades to the twentieth century. Princeton University Press, Princeton
Betran AP, Ye J, Moller A-B et al (2016) The increasing trend in caesarean section rates: global, regional and national estimates: 1990–2014. PLoS One 11:e0148343
Black E (2003) War against the weak—eugenics and America's campaign to create a master race. Four Walls Eight Windows, New York
Blackiston D, Silva Casey E, Weiss M (2008) Retention of memory through metamorphosis: can a moth remember what it learned as a caterpillar? PLoS One 3:e1736
Blair KL, Cappell J, Pukall CF (2018) Not all orgasms were created equal: differences in frequency and satisfaction of orgasm experiences by sexual activity in same-sex versus mixed-sex relationships. J Sex Res 55:719–733
Blumberg MS (2009) Freaks of nature: what anomalies tell us about development and evolution. Oxford University Press, New York
Blume M (2009) The reproductive benefits of religious affiliation. In: Voland E, Schiefennhovel W (eds) The biological evolution of religious mind and behavior. Springer, New York, pp 117–149
Blumenbach F (1804) De l'unite du genre humain. Allut, Paris
Bodenmann G, Ledermann T, Blattner D et al (2006) Associations among everyday stress, critical life events, and sexual problems. J Nerv Ment Dis 194:494–501
Bodenmann G, Atkins D, Schär M et al (2010) The association between daily stress and sexual activity. J Fam Psychol 24:271–279
Boesch C (1992) New elements of a theory of mind in wild chimpanzees. Brain Behav Sci 15:149–150
Boetsch G, Blanchard P (2014) From cabinets of curiosity to the "Hottentot Venus": a long history of human zoos. In: Bancel N, David T, Thomas D (eds) The invention of race—scientific and popular representations. Taylor & Francis, London, pp 185–194
Bondeson J (1997) A cabinet of medical curiosities. W. W. Norton, New York
Bondeson J (2000) The two-headed boy, and other medical curiosities. Cornell University Press, Ithaca
Bonner JT (2013) Randomness in evolution. Princeton University Press, Princeton
Boone AP, Hegarty M, Gong X (2018) Sex differences in navigation strategy and efficiency. Mem Cognit 46:909–922
Bowler PJ (1987) Theories of human evolution—a century of debate, 1844–1944. John Hopkins University Press, Oxford
Bowler PJ (2013) Darwin deleted. University of Chicago Press, Chicago
Bowler PJ (2017) Alternatives to Darwinism in the early twentieth century. In: Delisle RG (ed) The Darwinian tradition in context: research programs in evolutionary biology. Springer, New York, pp 195–218
Boyatzis C, Chazan E, Ting CZ (1993) Preschool children's decoding of facial emotions. J Gen Psychol 154:375–382
Boyle PA, Buchman AS, Wilson RS et al (2012) Effect of purpose in life on the relation between Alzheimer disease pathologic changes on cognitive function in advanced age. Arch Gen Psychiatry 69:499–505
Brattain M (2007) Race, racism, and anti-racism: UNESCO and the politics of presenting science to the postwar public. Am Hist Rev 112:1386–1413
Brooks RC, Griffith SC (2010) Mate choice. In: Westneat DF, Fox CW (eds) Evolutionary behavioral ecology. Oxford University Press, New York, pp 416–433
Browning F (2017a) Survival secrets: what is about women that makes them more resilient than men. In: California Magazine. Cal Alumni Association UC, Berkeley. Accessed 29 Apr 2018

Browning F (2017b) The fate of gender: nature, nurture, and the human future. Bloomsbury, New York

Brune M (2009) On shared psychological mechanisms of religiousness and delusional beliefs. In: Voland E, Schiefennhovel W (eds) The biological evolution of religious mind and behavior. Springer, New York, pp 217–228

Buettner D (2012) The blue zones—9 lessons for living longer, 2nd edn. National Geographic, Washington, DC

Buklijas T, Gluckman PD (2013) From evolution and medicine to evolutionary medicine. In: Ruse M (ed) The Cambridge Encyclopedia of Darwin and Evolutionary Thought. Cambridge University Press, Cambridge, pp 505–514

Burton GJ, Moffett A, Keverne B (2015) Human evolution: brain, birthweight and the immune system. Philos Trans R Soc Lond B Biol Sci 370:20140061

Butler FP (2012) Evolution without Darwinism—the legacy of Stephen Jay Gould. CreateSpace, New York

Call V, Susan S, Pepper S (1995) The incidence and frequency of marital sex in a national sample. J Marriage Fam 57:639–652

Camper P (1778) Account of the organs of speech of the Orang Outang. Philosophical Transactions LXIX, London

Camper P (1782) Natuurkundige Verhandelingen van Petrus Camper over den orang outang; en Eenige Andere Aap-Soorten. Erven P. Meijer en G. Warnas, Amsterdam

Camper P (1791) Dissertation sur les variétés naturelles qui caractérisent la physionomie des hommes des divers climats et des différens ages: suivie de réflexions sur la beauté, particulièrement sur celle de la tête: avec une manière nouvelle de dessiner toute sorte de têtes avec la plus grande exactitude. Chez H. J. Jansen, Paris

Canizares-Esguerra J (2006) Nature, empire and nation—explorations of the history of science in the Iberian world. Stanford University Press, Stanford

Canudas-Romo V (2010) Three measures of longevity: time trends and record values. Demography 47:299–312

Carroll S (2016a) The big picture—on the origins of life, meaning, and the universe itself. Dutton, New York

Carroll SB (2016b) The Serengeti rules: the quest to discover how life works and why it matters. Princeton University Press, Princeton

Carroll SB (2020) A series of fortunate events: chance and the making of the planet, life, and you. Princeton University Press, Princeton

Carter CS, Perkeybile AM (2018) The monogamy paradox: what do love and sex have to do with it. Front Ecol Evol 6:202

Casserius I (1600–1601) De Vocis Auditus Que. Organis Historia Anatomica, Ferrariae, Venice

Catholic Church (1983) Code of canon law, Latin-English edition. Catholic Church, Vatican

Cavanaugh J (2018) The prosocial paradox: unraveling oxytocin's role in monogamous relationships. PhD dissertation, University of Nebraska at Omaha

Cerello K, Kholoussy H (eds) (2016) Domestic tensions, national identities—global perspectives on marriage, crisis, and nation. Oxford University Press, New York

Church of England (1844) The Book of common prayer: printed by Whitchurch, March 1549; commonly called The first book of Edward VI. William Pickering, London

Clark AE, Georgellis Y (2013) Back to baseline in Britain: adaptation in the British household panel survey. Economica 80:496–512

Cloquet J (1821–1831) Anatomie de 'homme. Charles-Philibert, Paris

Cole FJ (1975) A history of comparative anatomy—from Aristotle to the eighteenth century. Dowe Publications, New York

Conley TD, Matsick J, Valentine B et al (2017) Investigation of consensually non-monogamous relationships: theories, methods and new directions. Persp Psychol Sci 12:205–232

Coontz S (2005) Marriage, a history: how love conquered marriage. Penguin Books, New York

Corbey RHA (2005) The metaphysics of apes: negotiating the animal-human boundary. Cambridge University Press, Cambridge
Corbey RHA, Theunissen B (eds) (1995) Ape, man, apeman: changing views since 1600. Leiden University, Leiden
Cottonham DP, Madson MB, Nicholson BC et al (2018) Harmful alcohol use and alcohol-related sex expectancies as predictors of risky sex among African American female college drinkers. J Ethn Subst Abuse 17:389–400
Coyne J (2016) Why do some scientists always claim that evolutionary biology needs urgent and serious reform? Blogpost. https://whyevolutionistrue.wordpress.com/2016/12/26/why-are-scientists-always-saying-that-evolutionary-biology-needs-urgent-and-serious-reform/
Crews F (2017) Freud: the making of an illusion. Metropolitan Books, New York
Croutier AL (1991) Harem: the world behind the veil. Abbeville Press, New Work
Csoka AC (2016) Innovation in medicine: Ignaz the reviled and Egas the regaled. Med Health Care Philos 19:163–168
Cunningham A (1997) The anatomical Renaissance: the resurrection of the anatomical projects of the ancients. Scolar Press, Aldershot
Cuvier G (1797) Tableau elementaire de 'histoire naturelle des animaux. Bandouin, Paris
Daily Mail Reporter (2013) Who needs a body anyway? Experts are now investigating whether we could one day live as just a BRAIN on far away planets. Daily Mail. Accessed 14 Aug 2013
Damasio A (1994) Descartes' error—emotion, reason and the human brain. Putnam Publishing, New York
Damasio A (2021) Feeling & knowing—making mind conscious. Pantheon, New York
Daneshmandi H, Choobineh A, Ghaem H et al (2017) Adverse effects of prolonged sitting behavior on the general health of office workers. J Lifestyle Med 7:69–75
Darling J, De Pijpekamp MV (1994) Rousseau on the education, domination and violation of women. Br J Educ Stud 42:115–132
Darwin C (1859) On the origin of species by means of natural selection, or, the preservation of favored races in the struggle for life. J. Murray, London
Darwin C (1871) The descent of man, and selection in relation to sex. J. Murray, London
Daston L, Park K (1998) Wonders and the order of nature, 1150–1750. Zone Books, New York
Davis DM (2018) The beautiful cure—the revolution in immunology and what it means for your health. The University of Chicago Press, Chicago
Davis PE, Webster LAD, Fernyhough C et al (2019) Adult report of childhood imaginary companions and adversity relates to concurrent prodromal psychosis symptoms. Psychiatry Res 271:150–152
Dawkins R (1976) The selfish gene. Oxford University Press, Oxford
De Queiroz A (2014) The monkey's voyage—how improbable journeys shaped the history of life. Basic Books, New York
De Wall F (2016) Are we smart enough to know how smart animal are? W. W. Norton, New York
De Wall F (2019) Mama's last hug: animal emotions and what they tell us about ourselves. W. W. Norton, New York
Decety J, Pape R, Workman CI (2018) A multilevel social neuroscience perspective on radicalization and terrorism. Soc Neurosci 13:511–529
Delisle RG (2007) Debating humankind's place in nature, 1860–2000: the nature of paleoanthropology. Pearson Prentice Hall, Upper Saddle River, NJ
Delisle RG (ed) (2017a) The Darwinian tradition in context—research programs in evolutionary biology. Springer, New York
Delisle RG (2017b) From Charles Darwin to the evolutionary synthesis: weak and diffused connections only. In: Delisle RG (ed) The Darwinian tradition in context: research programs in evolutionary biology. Springer, New York, pp 133–167
Delisle RG (2019) Charles Darwin's incomplete revolution—the origin of species and the static worldview. Springer, New York

Depew DJ (2017) Darwinism in the twentieth century: productive encounters with saltation, acquired characteristics, and development. In: Delisle RG (ed) The Darwinian tradition in context: research programs in evolutionary biology. Springer, New York, pp 61–68
DeSesso JM (2019) The arrogance of teratology: a brief chronology of attitudes throughout history. Birth Defects Res 111:123–141
Diamond J (1991) The rise and fall of the third chimpanzee. Hutchinson Radius, London
Diamond J (1999) Guns, germs, and steel: the fates of human societies. W. W. Norton, New York
Diamond J (2005) Collapse: how societies choose to fail or succeed. Viking Press, New York
Diamond J (2012) The world until yesterday: what we can learn from traditional societies? Penguin Books, New York
Dibble HL, Aldeias V, Goldberg P et al (2015) A critical look at evidence from La Chapelle-aux-Saints supporting an intentional Neandertal burial. J Archaeol Sci 53:649–657
Diogo R (2010) Comparative anatomy, anthropology and archaeology as case studies on the influence of human biases in natural sciences: the origin of 'humans', of 'behaviorally modern humans' and of 'fully civilized humans'. Open Anat J 2:86–97
Diogo R (2017a) Evolution driven by organismal behavior—a unifying view of life, function, form, trends and mismatches. Springer, New York
Diogo R (2017b) Etho-eco-morphological mismatches, an overlooked phenomenon in ecology, evolution and Evo-Devo that supports ONCE (Organic Nonoptimal Constrained Evolution) and the key evolutionary role of organismal behavior. Front Ecol Evol 10:3389
Diogo R (2018a) Links between the discovery of primates and anatomical comparisons with humans, the chain of being, our place in nature, and racism. J Morphol 279:472–493
Diogo R (2018b) Where is, in 2017, the Evo in Evo-Devo (Evolutionary Developmental Biology)? J Exp Zool B 330:15–22
Diogo R (2019) Sex at Dusk, Sex at Dawn, selfish genes: how old-dated evolutionary ideas are used to defend fallacious misogynistic views on sex evolution. J Soc Sci Humanit 5:350–367
Diogo R, Abdala V (2010) Muscles of vertebrates—comparative anatomy, evolution, homologies and development. Taylor & Francis, Oxford
Diogo R, Wood B (2011) Soft-tissue anatomy of the primates: phylogenetic analyses based on the muscles of the head, neck, pectoral region and upper limb, with notes on the evolution of these muscles. J Anat 219:273–359
Diogo R, Wood B (2012) Comparative anatomy and phylogeny of primate muscles and human evolution. Taylor and Francis, Oxford
Diogo R, Wood B (2013) The broader evolutionary lessons to be learned from a comparative and phylogenetic analysis of primate muscle morphology. Biol Rev 88:988–1001
Diogo R, Potau JM, Pastor JF et al (2010) Photographic and descriptive musculoskeletal atlas of Gorilla—with notes on the attachments, variations, innervation, synonymy and weight of the muscles. Taylor & Francis, Oxford
Diogo R, Potau JM, Pastor JF et al (2012) Photographic and descriptive musculoskeletal atlas of gibbons and siamangs (Hylobates)—with notes on the attachments, variations, innervation, synonymy and weight of the muscles. Taylor & Francis, Oxford
Diogo R, Potau JM, Pastor JF et al (2013a) Photographic and descriptive musculoskeletal atlas of chimpanzees (Pan)—with notes on the attachments, variations, innervation, synonymy and weight of the muscles. Taylor & Francis, Oxford
Diogo R, Potau JM, Pastor JF et al (2013b) Photographic and descriptive musculoskeletal atlas of orangutans (Pongo)—with notes on the attachments, variations, innervation, synonymy and weight of the muscles. Taylor & Francis, Oxford
Diogo R, Pastor JF, Hartstone-Rose A et al (2014) Baby Gorilla: photographic and descriptive musculoskeletal atlas of the skeleton, muscles and internal organs—including CT scans and comparisons to other gorillas and primates. Taylor & Francis, Oxford
Diogo R, Ziermann JM, Linde-Medina M (2015) Is evolutionary biology becoming too politically correct? A reflection on the scala naturae, phylogenetically basal clades, anatomically plesiomorphic taxa, and "lower" animals. Biol Rev 90:502–521

Diogo R, Bello-Hellegouarch G, Kohlsdorf T et al (2016a) Comparative myology and evolution of marsupials and other vertebrates, with notes on complexity, Bauplan, and "Scala Naturae". Anat Rec 299:1224–1255
Diogo R, Noden D, Smith CM et al (2016b) Learning and understanding human anatomy and pathology: an evolutionary and developmental guide for medical students. Taylor & Francis, Oxford
Diogo R, Molnar JL, Wood B (2017a) Bonobo anatomy reveals stasis and mosaicism in chimpanzee evolution, and supports bonobos as the most appropriate extant model for the common ancestor of chimpanzees and humans. Sci Rep 7:608
Diogo R, Guinard G, Diaz R (2017b) Dinosaurs, chameleons, humans and Evo-Devo-Path: linking Étienne Geoffroy's teratology, Waddington's homeorhesis, Alberch's logic of 'monsters', and Goldschmidt hopeful 'monsters'. J Exp Zool B 328:207–229
Diogo R, Shearer B, Potau JM et al (2017c) Photographic and descriptive musculoskeletal atlas of bonobos, with notes on the attachments, variations, innervation, synonymy and weight of the muscles. Springer, New York
Dittrich-Reed DR, Fitzpatrick BM (2013) Transgressive hybrids as hopeful monsters. Evol Biol 40:310–315
Dorus S, Evans PD, Wyckoff GJ et al (2004) Rate of molecular evolution of the seminal protein gene SEMG2 correlates with levels of female promiscuity. Nat Genet 36:1326–1329
Doucleff M (2018) Rate of C-Sections is rising at an alarming rate, report says. NPR, 12 October 2018 online
Driscoll CA, Thompson JC (2018) The origins and early elaboration of projectile technology. Evol Anthropol 27:30–45
Dugatkin L, Trut L (2017) How to tame a fox {and build a dog}. University of Chicago Press, Chicago
Duke D (1998) My awakening: a path to racial understanding. Free Speech Press, Covington
Dyble M, Thorley J, Page AE et al (2019) Engagement in agricultural work is associated with reduced leisure time among Agta hunter-gatherers. Nat Hum Behav 3:792–796
Edwardes MPJ (2019) The origins of self—an anthropological perspective. UCL Press, London
Eldredge N (2014) Extinction and evolution: what fossils reveal about the history of life. Firefly Books, Toronto
Eldredge N, Gould SJ (1972) Punctuated equilibrium: an alternative to phyletic gradualism. In: Schopf TJM (ed) Models in paleobiology. Freeman, Cooper, San Francisco, pp 82–115
Engelmeier H (2016) Der Mensch. Der Affe. Böhlau Verlag, Köln
Epstein GM (2010) Good without God—what a billion nonreligious people do believe. Harper, New York
Erb CM (1998) Tracking King Kong—a Hollywood icon of world culture. Wayne State University Press, Detroit
Erhman BD (2020) Heaven and hell—a history of the afterlife. Simon & Schuster, New York
Fábrega H (1997) Evolution of sickness and healing. University of California Press, Berkeley
Fabrici G (1600) De Formato Foetu. Embryo Project Encyclopedia (2008-08-27), ISSN, 1940–5030
Feiler B (2017) The first love story—Adam, Even and US. Penguin Press, New York
Figes O (2007) The whisperers—a private life in Stalin's Russia. Picador, New York
Fine C (2017) Testosterone Rex—myths of sex, science and society. W. W. Norton, New York
Finkel EJ (2017) The all-or-nothing marriage—how the best marriages work. Penguin Press, New York
Fleckenstein JR, Cox DW (2015) The association of an open relationship orientation with health and happiness in a sample of older U.S. adults. Sex Relat Ther 30:94–116
Fox CW, Westneat DF (2010) Adaptation. In: Westneat DF, Fox CW (eds) Evolutionary behavioral ecology. Oxford University, New York, pp 16–32
Freedland J (2017) Terror attacks are not just about us. The Guardian. Accessed 26 May 2017
Frey U (2009) Cognitive foundations of religiosity. In: Voland E, Schiefennhovel W (eds) The biological evolution of religious mind and behavior. Springer, New York, pp 229–241
Fuentes A (2017) Human niche, human behaviour, human nature. Interface Focus 7:20160136

Futuyma DJ (2017) Evolutionary biology today and the call for an extended synthesis. Interface Focus 7:20160145
Gardiner G, Lee D, Baranski E et al (2020) Happiness around the world: a combined etic-emic approach across 63 countries. PLoS One 15:e0242718
Gawande A (2014) Being mortal—medicine and what matters in the end. Metropolitan Books, New York
Gilby IC, Machanda ZP, O'Malley RC et al (2017) Predation by female chimpanzees: toward an understanding of sex differences in meat acquisition in the last common ancestor of Pan and Homo. J Hum Evol 110:82–94
Goncalves A, Susana C (2019) Death among primates: a critical review of non-human primate interactions towards their dead and dying. Biol Rev 94:1502–1529
Goodall J (1988) In the shadow of man. Houghton Mifflin, Boston
Gottschall J (2012) The storytelling animal—how stories make us human. Houghton Mifflin Harcourt, New York
Gould SJ (1981) The mismeasure of man. W. W. Norton, New York
Gould SJ (1996) Full House: the spread of excellence from Plato to Darwin. Belknap Press, Cambridge
Gould SJ (2002) The structure of evolutionary theory. Harvard, Belknap
Gray J (2013) The silence of animals: on progress and other modern myths. Farrar, Straus & Giroux, New York
Gray J (2018) Seven types of atheism. Penguin Books, London
Green T (2019) A fistful of shells—West Africa from the rise of the slave trade to the age of revolution. The University of Chicago Press, Chicago
Greenblastt S (2017) The rise and fall of Adam and Eve. W. W. Norton, New York
Greenblatt S (2011) The swerve—how the world became modern. W. W. Norton, New York
Groves C (2008) Extended family: long lost cousins. A personal look at the history of primatology. Conservation International, Arlington
Guedron M (2014) Panel and sequence: classification and associations in scientific illustrations of the human races (1770–1830). In: Bancel N, David T, Thomas D (eds) The invention of race—scientific and popular representations. Taylor & Francis, London, pp 60–67
Guerrini A (2003) Experimenting with humans and animals—from Galen to animal rights. The Johns Hopkins University Press, Baltimore
Guerrini A (2015) The Courtiers' Anatomists—animals and humans in Louis XIV's Paris. The University of Chicago Press, Chicago
Gupta M, Prasad NG, Dey S et al (2017) Niche construction in evolutionary theory: the construction of an academic niche? J Genet 96:491–504
Gurven M, Kaplan H (2007) Longevity among hunter-gatherers: a cross-cultural examination. Popul Dev Rev 33:321–365
Haas R, Watson J, Buonasera T et al (2020) Female hunters of the early Americas. Sci Adv 6:eabd0310
Haeckel E (1868) Die Natürliche Schöpfungsgeschichte. Georg Reimer, Berlin
Haeckel E (1870) Die Natürliche Schöpfungsgeschichte, 2nd edn. Georg Reimer, Berlin
Haeckel E (1874) Anthropogenie oder Entwickelungsgeschichte des Menschen. In: Keimes- und Stammesgeschichte. Engelmann, Leipzig
Haeckel E (1887) The history of creation, or the development of the earth and its inhabitants by the action of natural causes. Appleton, New York
Halcrow S, Warren R, Kushnick G et al (2020) Care of infants in the past: bridging evolutionary anthropological and bioarchaeological approaches. Evol Hum Sci 2:e47.
Hamilton LD, Meston CM (2013) Chronic stress and sexual function in women. J Sex Med 10:2443–2454
Hanlon H, Thatcher R, Cline M (1999) Gender differences in the development of EEG coherence in normal children. Dev Neuropsychol 16:479–506
Hannon E, Lewens T (eds) (2018) Why we disagree about human nature. Oxford University Press, Oxford

Hanson R (2012) Sex at Dusk v. Dawn. Overcoming Bias (Blog). http://www.overcomingbias.com/2012/08/sex-at-dusk-v-sex-at-dawn.html. Accessed 30 Aug 2012

Harari YN (2017) Homo Deus—a brief history of tomorrow. Harper, New York

Hare B, Woods V (2020) The survival of the friendliest—understanding our origins and rediscovering our common humanity. Random House, New York

Harvell LA, Nisbett GS (eds) (2016) Denying death—an interdisciplinary approach to terror management theory. Routledge, New York

Haught JE (2000) Science and religion in search for of cosmic purpose. Georgetown University Press, Washington, DC

Hazlewood N (2001) Savage: the life and times of Jemmy Button. Thomas Dunne Books, New York

Henning BG, Scarfe AC (eds) (2013) Beyond mechanism: putting life back into biology. Lexington Books, Lexington

Hewlett BS (ed) (2014a) Hunter-Gatherers of the Congo basin: cultures, histories, and biology of African pygmies. Transaction Publishers, London

Hewlett BS (2014b) Hunter-gatherer childhoods in the Congo basin. In: Hewlett BS (ed) Hunter-Gatherers of the Congo Basin: cultures, histories, and biology of African Pygmies. Transaction Publishers, London, pp 245–275

Hirsch AR (1998) Scentsational sex: the secret to using aroma for arousal. Element Books, New York

Hochschild AR (2016) Stranger in their own land—anger and morning on the American right. The New Press, New York

Hoff EV (2004) A friend living inside me—the forms and functions of imaginary companions. Imagin Cogn Pers 24:151–189

Hoffman D (2019) The case against reality: why evolution hid the truth from our eyes. W. W. Norton, New York

Hoffmeyer J (2013) Why do we need a semiotic understanding of life. In: Henning BG, Scarfe AC (eds) Beyond mechanism: putting life back into biology. Lexington Books, Lexington, pp 147–168

Holland J (2012) A brief history of misogyny—the world's oldest prejudice. Constable & Robinson, London

Hood B (2013) The self illusion—how the social brain creates identity. Oxford University Press, Oxford

Hood RW, Hill PC, Spilka B (2009) The psychology of religion—an empirical approach. The Guilford Press, London

Hoquet T (2014) Biologization of race and racialization of the human: Bernier, Buffon, Linnaeus. In: Bancel N, David T, Thomas D (eds) The invention of race—scientific and popular representations. Taylor & Francis, London, pp 17–32

Hoßfeld U (2010) Ernst Haeckel. Biographienreihe absolute. Orange Press, Freiburg

Hoßfeld U (2016) Geschichte der biologischen Anthropologie in Deutschland—Von den Anfängen bis in die Nachkriegszeit, 2nd edn. Franz Steiner Verlag, Stuttgart

Hrdy SB (2009) Mother and others—the evolutionary origins of mutual understanding. Belknap Press, Cambridge

Huber E (1931) Evolution of facial musculature and expression. The Johns Hopkins University Press, Baltimore

Humboldt A (1914) Views of the cordilleras and monuments of the indigenous peoples of the Americas. In: Kutzinski VM, Ette O (eds) Views of the Cordilleras and monuments of the indigenous peoples of the Americas: a critical edition, 2012. University of Chicago Press, Chicago, pp 1–370

Huneman P, Walsh DM (eds) (2017) Challenging the modern synthesis—adaptation, development, and inheritance. Oxford University Press, Oxford

Hutton R (2017) The witch—a history of fear from ancient times to the present. Yale University Press, New Haven

Huxley TH (1863) Evidence as to man's place in nature. Williams and Norgate, London

Jablonka E, Lamb MJ (2005) Evolution in four dimensions—genetic, epigenetic, behavioral, and symbolic variation in the history of life. MIT Press, Cambridge
Jackson MH (2016) Galapagos—a natural history. University of Calgary Press, Calgary
Janson HW (ed) (1952) Apes and Ape lore in the Middle Ages and the Renaissance. Warburg Institute University of London, London
Jeffery AJ, Shackelford TK, Zeigler-Hill V et al (2018) The evolution of human female sexual orientation. Evol Psychol Sci 5(1):1–16.
Jiang XQ, Mei XD, Feng D (2016) Air pollution and chronic airway diseases: what should people know and do? J Thorac Dis 8:E31–E40
Johnson MR (2005) Aristotle on teleology. Clarendon Press, Oxford
Johnson NA, Lahti DC, Blumstein DT (2012) Combating the assumption of evolutionary progress: lessons from the decay and loss of traits. Evol Educ Outreach 5:128–138
Johnson NF, Velásquez N, Restrepo NJ et al (2020) The online competition between pro- and anti-vaccination views. Nature 582:230–233.
Kaiser D (2011) How the hippies saved physics—science, counterculture, and the quantum revival. W. W. Norton, New York
Kalmijn M (2017) The ambiguous link between marriage and health: a dynamic reanalysis of loss and gain effects. Soc Forces 95:1607–1636
Kasperbauer TJ (2018) Subhuman—the moral psychology of human attitudes to animals. Oxford University Press, Oxford
Kauffman SA (2010) Reinventing the sacred: a new view of science, reason, and religion. Basic Books, New York
Kelly RL (1995) The foraging spectrum: diversity in hunter-gatherer lifeways. Smithsonian Institution Press, Washington
Kelly J (2006) The great mortality: an intimate history of the Black Death, the most devastating plague of all time. Harper Collins, New York
Kelly RL (2013) The lifeways of hunter-gatherers—the foraging spectrum. Cambridge Press, Cambridge
Kendy IX (2016) Stamped from the beginning—the definitive history of racist ideas in America. Nation Books, New York
Kevles DJ (1995) In the name of eugenics—genetics and the uses of human heredity. Harvard University Press, Cambridge
Kingsberg SA, Janata JW (2007) Female sexual disorders: assessment, diagnosis, and treatment. Urol Clin N Am 34:497–506
Kirkegaard EOW (2019) Race differences: a very brief review. Mankind Quart 60(2):142–173
Kirkham D (2019) Our shadowed world—reflections on civilization, conflict, and belief. Cascade Books, Eugene
Klarsfeld A, Revah F (2003) The biology of death: origins of mortality. Cornell University Press, New York
Klein N (1999) No logo—no space, no jobs, no choice. Picador, Toronto
Kposowa AJ (2000) Marital status and suicide in the National Longitudinal Mortality Study. J Epidemiol Community Health 54:254–261
Krenak A (2020) A vida não é útil—ideias para salvar a humanidade. Objectiva, Sao Paulo
Kteily N, Bruneau E, Waytz A et al (2015) The ascent of man: theoretical and empirical evidence for blatant dehumanization. J Pers Soc Psychol 109:901–931
Kühl HS, Kalan AK, Arandjelovic M et al (2016) Chimpanzee accumulative stone throwing. Sci Rep 6:22219
Kuklick H (ed) (2008) A new history of anthropology. Blackwell, Oxford
Kull K (2014) Adaptive evolution without natural selection. Biol J Linn Soc 112:287–294
Kunz J (2009) Is there a particular role for ideational aspects of religions in human behavioral ecology? In: Voland E, Schiefennhovel W (eds) The biological evolution of religious mind and behavior. Springer, New York, pp 89–104

Kvarnemo C (2018) Why do some animals mate with one partner rather than many? A review of causes and consequences of monogamy. Biol Rev 93:1795–1812

Labarthe JC (1997) Are boys better than girls at building a tower or a bridge at 2 years of age? Arch Dis Child 77:140–144

Lagerkvist U (2005) The enigma of ferment—from the philosopher's stone to the first biochemical Nobel prize. World Scientific, Hackensack

Lahti DC (2009) The correlated history of social organization, morality, and religion. In: Voland E, Schiefennhovel W (eds) The biological evolution of religious mind and behavior. Springer, New York, pp 67–88

Laland KN, Odling-Smee J, Turner S (2014) The role of internal and external constructive processes in evolution. J Physiol 592:2413–2422

Laland KN, Uller T, Feldman MW et al (2015) The extended evolutionary synthesis: its structure, assumptions and predictions. Proc R Soc Lond B 282:20151019.

Laland K, Matthews B, Feldman MW (2016) An introduction to niche construction theory. Evol Ecol 30:191–202

Landau M (1991) Narratives on human evolution. Yale University Press, New Haven

Lee RB (2018) Hunter-gatherers and human evolution: new light on old debates. Ann Rev Anthropol 47:513–531

Leeman RW (2012) The teleological discourse of Barack Obama. Lexington Books, New York

Lenoir T (1982) The strategy of life—teleology and mechanics in nineteenth-century German biology. The University of Chicago Press, Chicago

Lenroot RK, Gogtay N, Greenstein DK et al (2007) Sexual dimorphism of brain developmental trajectories during childhood and adolescence. Neuroimage 36:1065–1073

Leroi AM (2003) Mutants: on the form, varieties and errors of the human body. Harper Collins, London

Leroi AM (2014) The lagoon: how Aristotle invented science. Bloomsbury, London

Levin J (2020) Religion and medicine—a history of the encounter between humanity's two greatest institutions. Oxford University Press, Oxford

Levine N (1998) The dynamics of polyandry: kinship, domesticity, and population on the Tibetan border. University of Chicago Press, Chicago

Levi-Strauss C (2011) Tristes tropiques. Penguin Books, New York

Levy JS, Thompson WR (2010) Causes of war. Wiley-Blackwell, New York

Lewis HS (2001) Boas, Darwin, science, and anthropology. Curr Anthropol 42:381–406

Lewis J (2014) Egalitarian social organization: the case of the mbendjele BaYaka. In: Hewlett BS (ed) Hunter-gatherers of the Congo basin: cultures, histories, and biology of African pygmies. Transaction Publishers, London, pp 219–243

Lewis J (2018) The science of sin: why we do the things we know we shouldn't. Bloomsbury Sigma, New York

Lieberman P (1991) Uniquely human: the evolution of speech, thought and selfless behavior. Harvard University Press, Cambridge

Lindholm M (2015) DNA dispose, but subjects decide -learning and the extended synthesis. Biosemiotics 8:4431–4461

Linnaeus C (1735) Systema naturae. Laurentius Salvius, Stockholm

Long M, Long E, Waters T (2013) Suicide among the Mla Bri Hunter-Gatherers of Northern Thailand. J Siam Soc 101:156–176

Lovejoy AO (1936) The great chain of being: a study of the history of an idea. Harvard University Press, Cambridge

Luhmann M, Hofmann W, Eid M et al (2012) Subjective well-being and adaptation to life events: a meta-analysis. J Pers Soc Psychol 102:592–615

Lupo K, Ndanga AJ-P, Kiahtipes C (2014) On Late Holocene population interactions in the Northwestern Congo Basin: when, how and why does the ethnographic pattern begin. In: Hewlett BS (ed) Hunter-gatherers of the Congo Basin: cultures, histories, and biology of African Pygmies. Transaction Publishers, London, pp 59–83

Mah K, Binik YM (2001) The nature of human orgasm: a critical review of major trends. Clin Psychol Rev 21:823–856

Mah K, Binik YM (2002) Do all orgasms feel alike? Evaluating a two-dimensional model of the orgasm experience across gender and sexual context. J Sex Res 39:104–113

Malesevic S (2011) The rise of organised brutality—a historical sociology of violence. Cambridge University Press, Cambridge

Malik K (2014) The quest for a moral compass—a global history of ethics. Melville House, Brooklyn

Malik A, Ziermann JM, Diogo R (2017) An untold story in biology: the historical continuity of evolutionary ideas of Muslim scholars from the eighth century to Darwin's time. J Biol Educ 52:3–17

Mancuso S, Viola A (2015) Brilliant green: the surprising history and science of plant intelligence. Island Press, Washington, DC

Marshall JM (2007) The day the world ended at Little Bighorn—a Lakota history. Penguin Books, London

Martin D (1984) Primate origins and evolution. Chapman and Hall, London

Martínez-Sevilla F, Arqués M, Jordana X et al (2020) Who painted that? The authorship of Schematic rock art at the Los Machos rockshelter in southern Iberia. Antiquity 94:1133–1151

Mayr E (1976) Evolution and the diversity of life: selected essays. Harvard University Press, Cambridge

McAlister M (2018) The kingdom of God has no borders—a global history of American evangelicals. Oxford University Press, New York

McBrearty S, Brooks AS (2000) The revolution that wasn't: a new interpretation of the origin of modern human behavior. J Hum Evol 39:453–563

McClelland BA (2006) Slayers and their vampires: a cultural history of killing the dead. University of Michigan Press, Ann Arbor

McCool-Myers M, Theurich M, Zuelke A et al (2018) Predictors of female sexual dysfunction: a systematic review and qualitative analysis through gender inequality paradigms. BMC Womens Health 18:108

McMahon DM (2006) Happiness—a history. Grove Press, New York

McShea DW (2012) Upper-directed systems: a new approach to teleology in biology. Biol Philos 27:663–684

Meijer MC (2014) Cranial varieties in the human and orangutan species. In: Bancel N, David T, Thomas D (eds) The invention of race—scientific and popular representations. Taylor & Francis, London, pp 33–47

Meral Z (2018) How violence shapes religion—belief and conflict in the Middle East and Africa. Cambridge University Press, Cambridge

Minelli A (2009) Forms of becoming—the evolutionary biology of development. Princeton University Press, Princeton

Monsó S, Osuna-Mascaró AJ (2020) Death is common, so is understanding it: the concept of death in other species. Synthese

Montagu MFA (1943) Edward Tyson, M.D., F.R.S., 1650–1708. Mem Am Philos Soc 20:1–488

Moravec H (1990) Mind children. Harvard University Press, Harvard

Morris D (2013) Monkey. Reaktion Books, London

Moser S (1998) Ancestral images—the iconography of human origins. Cornell University Press, Ithaca

Nee S (2005) The great chain of being. Nature 435:429–429

Nelson LH (ed) (2017) Biology and feminism—a philosophical introduction. Cambridge University Press, Cambridge

Nisbet R (1980) History of the idea of progress. Basic Books, New York

Nixey C (2017) The Darkening age—the Christian destruction of the classic world. Macmillan, London

Noble D (2006) The music of life: biology beyond the genome. OUP, Oxford

Noble D (2017) Dance to the tune of life—biological relativity. Cambridge University Press, Cambridge

Nour NM (2008) Female genital cutting: a persisting practice. Rev Obstet Gynecol 1:135–139

Öberg KG, Hallberg J, Kaldo V et al (2017) Hypersexual disorder according to the hypersexual disorder screening inventory in help-seeking Swedish men and women with self-identified hypersexual behavior. Sex Med 5:e229–e236

Odling-Smee FJ, Laland KN, Feldman MW (2003) Niche construction—the neglected process in evolution (Monographs in population biology 37). Princeton University Press, Princeton

Oliveira-Pinto AV, Santos RM, Coutinho RA et al (2014) Sexual dimorphism in the human olfactory bulb: females have more neurons and glial cells than males. PLoS One 9:e111733

Omland KE, Cook LG, Crisp MD (2008) Tree thinking for all biology: the problem with reading phylogenies as ladders of progress. Bioessays 30:854–867

Panese F (2014) The creation of the 'negro' at the turn of the nineteenth century: Petrus Camper, Johan Friedrich Blumenbach, and Julien-Joseph Virey. In: Bancel N, David T, Thomas D (eds) The invention of race—scientific and popular representations. Taylor & Francis, London, pp 48–59

Pavlicev M, Wagner G (2016) The evolutionary origin of female orgasm. J Exp Zool B 326:326–337

Perrault C (1676) Memoires pour servir a l'histoire naturelle des animaux. De l'Imprimerie Royale, Paris

Persaud TVN (1984) Early history of human anatomy: from antiquity to the beginning of the modern era. Charles C Thomas, Springfield

Persaud TVN, Loukas M, Tubbs RS (2014) A history of human anatomy, 2nd edn. Charles C. Thomas, Springfield

Peterson A (2001) Being human—ethics, environment, and our place in the world. University of California Press, Berkeley

Piazza PV (2019) Homo Biologicus: Comment la biologie explique la nature humaine. Albin Michel, Paris

Pigliucci M (2017) Darwinism after the modern synthesis. In: Delisle RG (ed) The Darwinian tradition in context: research programs in evolutionary biology. Springer, New York, pp 94–104

Pigliucci M, Müller GB (eds) (2010) Evolution—the extended synthesis. MIT Press, Cambridge

Pinker S (2011) The better angels of our nature: why violence has declined'. Penguin Books, New York

Potarca G (2020) The demography of swiping right—an overview of couples who met through dating apps in Switzerland. PLoS One 15:e0243733

Pratarelli ME, Chiarelli B (2007) Extinction and overspecialization: the dark side of human innovation. Mankind Quart 48:83–98

Pringle P (2008) The murder of Nikolai Vavilov—the story of Stalin's persecution of one of the great scientists of the twentieth century. Simon & Schuster, New York

Pruetz JD, LaDuke TC (2010) Reaction to fire by savanna chimpanzees (*Pan troglodytes verus*) at Fongoli, Senegal: conceptualization of "fire behavior" and the case for a chimpanzee model. Am J Phys Anthropol 141:646–650

Prum RO (2017) The evolution of beauty: how Darwin's forgotten theory of mate choice shapes the animal world—and us. Anchor Books, New York

Prushinskaya A (2017) A Woman is a woman until she is a mother: essays. MG Press, Des Plaines

Puchner M (2017) The written world: the power of stories to shape people, history, civilization. Penguin Random House LLC, New York

Pugliese D (2016) Awa: Alto Turiacu. Lens culture. Accessed 24 Dec 2016

Puts DA, Dawood K, Welling LLM (2012) Why women have orgasms: an evolutionary analysis. Arch Sex Behav 41:1127–1143

Radini A, Tromp M, Beach A et al (2019) Medieval women's early involvement in manuscript production suggested by lapis lazuli identification in dental calculus. Sci Adv 5:eaau7126

Ramsey G, Pence CH (eds) (2016) Chance in evolution. The University of Chicago Press, Chicago

Reiss JO (2009) Not by design: retiring Darwin's watchmaker. University of California Press, Berkeley
Rendu W, Beauval C, Crevecoeur I et al (2014) Evidence supporting an intentional Neandertal burial at La Chapelle-aux-Saints. PNAS 111:81–86
Reyes-García V (2019) Did foragers enjoy more free time? Nat Hum Behav 3(8):772–773
Reyna VF (2012) Risk perception and communication in vaccination decisions: a fuzzy-trace theory approach. Vaccine 30:3790–3797
Reynaud-Paligot C (2014) Construction and circulation of the notion of 'race' in the nineteenth century. In: Bancel N, David T, Thomas D (eds) The invention of race—scientific and popular representations. Taylor & Francis, London, pp 87–99
Richards RJ (2008) The tragic sense of life: Ernst Haeckel and the struggle over evolutionary thought. University of Chicago Press, Chicago
Richert RA, Smith EI (2009) Cognitive foundations in the development of a religious mind. In: Voland E, Schiefennhovel W (eds) The biological evolution of religious mind and behavior. Springer, New York, pp 181–204
Rigato E, Minelli A (2013) The great chain of being is still here. Evol Educ Outreach 6:18
Riva A, Orrù B, Pirino A et al (2001) Iulius Casserius (1552–1616): the self-made anatomist of Padua's golden age. Anat Rec 265:168–175
Rossano M (2009) The African interregnum: the 'where', 'when', and 'why' of the evolution of religion. In: Voland E, Schiefennhovel W (eds) The biological evolution of religious mind and behavior. Springer, New York, pp 127–141
Rovelli C (2017) Reality is not what it seems—the journey to quantum gravity. Riverhead Books, New York
Ruse M (1996) Monad to man: the concept of progress in evolutionary biology. Harvard University Press, Cambridge
Ruse M (2003) Darwin and design—does evolution have a purpose? Harvard University Press, Cambridge
Ruse M (2013) From organisms to mechanisms—and halfway back? In: Henning BG, Scarfe AC (eds) Beyond mechanism: putting life back into biology. Lexington Books, Lexington, pp 409–430
Ruse M (2018) On purpose. Princeton University Press, Princeton
Ryan C (2018) The virility paradox: the vast influence of testosterone on our bodies, our minds, and the world we live in. BenBella Books, Dallas
Ryan C, Jetha C (2010) Sex at dawn: how we mate, why we stray, and what it means for modern relationships. Harper Collins, New York
Sagan C (1997) The demon-haunted world—science as a candle in the dark. Ballantine Books, New York
Saini A (2017) Inferior—how science got women wrong, and the new research that's rewriting the story. Beacon Press, Boston
Sánchez-Villagra M, van Schaik CP (2019) Evaluating the self-domestication hypothesis of human evolution. Evol Anthropol 28:133–143
Sapolsky RM (2017) Behave—the biology of humans at our best and worst. Penguin Press, New York
Saxon L (2012) Sex at Dusk: lifting the shiny wrapping from Sex at Dawn. CreateSpace, New York
Schacht R, Kramer KL (2019) Are we monogamous? A review of the evolution of pair-bonding in humans and its contemporary variation cross-culturally. Front Ecol Evol: 2019.00230
Scheidel W (2017) The great leveler: violence and the history of inequality from the stone age to the twenty-first century. Princeton University Press, Princeton
Schmitt S (2004) Histoire d'une question anatomique: la repetition des parties. Museum National d'Histoire Naturelle, Paris
Schueler GF (2005) Reasons & purposes—human rationality and the teleological explanation of actions. Clarendon Press, Oxford

Scott GR (1938) The history of corporal punishment—a survey of flagellation in its historical anthropological and sociological aspects. T.W. Laurie, London
Scott GR (1959) The history of corporal punishment—a survey of flagellation in its historical, anthropological and sociological aspects, 2nd edn. Luxor Press, London
Scott JC (2017) Against the grain—a deep history of the earliest states. Yale University Press, New Haven
Sechiyama K (2014) Japan, the sexless nation. Tokyo Business Today. Accessed 19 Dec 2014
Shanahan T (2017) Selfish genes and lucky breaks: Richard Dawkins' and Stephen Jay Gould's divergent Darwinian agendas. In: Delisle RG (ed) The Darwinian tradition in context: research programs in evolutionary biology. Springer, New York, pp 31–36
Shelton JD (2014) Taking exception—reduced mortality leads to population growth: an inconvenient truth. Glob Health Sci Pract 2:135–138
Shermer M (2004) The science of good and evil—why people cheat, gossip, care, share, and follow the golden rule. Times Books, New York
Shermer M (2011) The believing brain—from ghosts and gods to politics and conspiracies, how we construct beliefs and reinforce them as truths. Times Books, New York
Shermer M (2015) The moral arc—how science makes us better people. St. Martin's Griffin, New York
Singer C (1959) A history of biology to about the year 1900. Abelard-Schuman, London
Singh U (2008) A history of ancient and early medieval India: from the Stone Age to the twelfth century. Pearson Education, New York
Skinner BF (1948) 'Superstition' in the pigeon. J Exp Psychol 38:168–172
Smith RJ (2016) Freud and evolutionary anthropology's first just-so story. Evol Anthropol 25:50–53
Smith EA, Hill K, Marlowe F et al (2010) Wealth transmission and inequality among hunter-gatherers. Curr Anthropol 51:19–34
Sommer M (2015) Evolutionäre Anthropologie zur Einführung. Junius, Hamburg
Sommer V, Vasey PL (eds) (2006) Homosexual behavior in animals—an evolutionary perspective. Cambridge University Press, Cambridge
Sorenson J (2009) Ape. Reaktion Books, London
Sproul B (1991) Primal myths—creation myths around the world. HarperCollins, New York
Stevens B (2016) Nihilism—a philosophy based on nothingness and eternity. Manticore Press, New York
Stoltzfus A (2017) Why we don't want another "synthesis". Biol Direct 12:23
Sugie NF (2017) When the elderly turn to petty crime: increasing elderly arrest rates in an aging population. Int Crim Just Rev 2017:19–39
Sugiyama Y (2017) Sex-biased dispersal of human ancestors. Evol Anthropol 26:172–180
Sultan SE (2016) Organisms & environment—ecological development, niche construction, and adaptation. Oxford University Press, Oxford
Suzman J (2017) Affluence without abundance—the disappearing world of the Bushmen. Bloomsbury, New York
Swan L, Gordon R, Seckbach J (eds) (2012) Origin(s) of design in nature—a fresh, interdisciplinary look at how design emerges in complex systems, especially life. Springer, New York
Taleb NN (2010) The black swan—the impact of the highly improbable, 2nd edn. Trader House Trade Paperback, New York
Tallis F (2018a) Women are more prepared for love. In: O Publico, pp 14–15. Accessed 28 May 2018
Tallis F (2018b) The incurable romantic: and other tales of madness and desire. Basic Books, New York
Thagard P (2010) The brain and the meaning of life. Princeton University Press, Princeton
Theiss JA (2016) Frequency of sexual relations in marriage. In: Shehan C (ed) Encyclopedia of family studies. Wiley-Blackwell, Hoboken, pp 1–5

Thomas K (1983) Man and the natural world—changing attitudes in England 1500–1800. Oxford University Press, Oxford
Tiesler V, Olivier G (2020) Open chests and broken hearts: ritual sequences and meanings of human heart sacrifice in Mesoamerica. Curr Anthropol 61:168–193
Tilley L, Oxenham M (2011) Survival against the odds: modeling the social implications of care provision to seriously disabled individuals. Int J Paleopathol 1:35–42
Todes DP (1989) Darwin without Malthus. Oxford University Press, Oxford
Trüper H, Chakrabarty D, Subrahmanyam S (eds) (2015) Historical teleologies in the modern world. Bloomsbury, New York
Tulp NP (1641) Observationes Medicae. Vivie, Leiden
Tumin D (2018) Does marriage protect health? A birth cohort comparison. Soc Sci Quart 99:626–643
Turner JS (2000) The extended organism—the physiology of animal-built structures. Harvard University Press, Cambridge
Turner JS (2007) The tinkerer's accomplice: how design emerges from life itself. Harvard University Press, Cambridge
Turner JS (2013) Biology's second law: homeostasis, purpose and desire. In: Henning BG, Scarfe AC (eds) Beyond mechanism: putting life back into biology. Lexington Books, Lexington, pp 183–204
Turner JS (2016) Homeostasis and the physiological dimension of niche construction theory in ecology and evolution. Evol Ecol 30:203–219
Turner DD (2017) Paleobiology's uneasy relationship with the Darwinian tradition: stasis as data. In: Delisle RG (ed) The Darwinian tradition in context: research programs in evolutionary biology. Springer, New York, pp 333–352
Tuttle RH (ed) (1975) Primate functional morphology and evolution. Aldine, Chicago
Tyson E (1699) Orang-Outang sive Homo sylvestris, or the anatomy of a pygmie compared to that of a monkey, an ape and a man. T. Bennet, London
UNESCO (1950) U.N.E.S.C.O. on race. Man 50:138–139
UNESCO (1951) U.N.E.S.C.O.'s new statement on race. Man 51:154–155
UNESCO (1952) U.N.E.S.C.O.'s new statement on race. Man 52:9
Vaas R (2009) Gods, gains and genes—on the natural origin of religiosity by means of bio-cultural selection. In: Voland E, Schiefennhovel W (eds) The biological evolution of religious mind and behavior. Springer, New York, pp 25–49
Valentine P, Beckerman S, Ales C (2017) The anthropology of marriage in lowland South-America: bending and breaking the rules. University Press of Florida, Gainesville
Van Arsdale A (2017) Human evolution as a theoretical model for an extended evolutionary synthesis. In: Delisle RG (ed) The Darwinian tradition in context: research programs in evolutionary biology. Springer, New York, pp 105–130
Van Schaik C, Michel K (2016) The good book of human nature—an evolutionary reading of the bible. Basic Books, New York
Van Wyhe J, Kjaergaard PC (2015) Going the whole orang: Darwin, Wallace and the natural history of orangutans. Stud Hist Philos Sci C 51:53–63
Varella D (2017) Prisioneiras. Companhia das Letras, Sao Paulo
Varki A (2009) Human uniqueness and the denial of death. Nature 460:684
Veracini C, Teixeira DM (2016) Perception and description of New World non-human primates in the travel literature of the fifteenth and sixteenth centuries: a critical review. Ann Sci 74:25–65
Vesalius A (1543) De humani corporis fabrica libri septem. Ex officina Joannis Oporini, Basel
Vinicius M (2012) Modular evolution: how natural selection produces biological complexity. Cambridge University Press, Cambridge
Voland E (2009) Evaluating the evolutionary status of religiosity and religiousness. In: Voland E, Schiefennhovel W (eds) The biological evolution of religious mind and behavior. Springer, New York, pp 9–24

Voland E, Schiefennhovel W (eds) (2009) The biological evolution of religious mind and behavior. Springer, New York
Wagner A (2014) Arrival of the fittest: solving evolution's greatest puzzle. Oneworld Publications, London
Walker W (2017) Why we sleep—unlocking the power of sleep and dreams. Simon & Schuster, New York
Wang Y, Liu H, Sun Z (2017) Lamarck rises from his grave: parental environment-induced epigenetic inheritance in model organisms and humans. Biol Rev 92:2084–2111
Washington HA (2006) Medical apartheid: the dark history of medical experimentation on black Americans from colonial times to the present. Anchor Books, New York
Weatherford CB (2017) Hearts and Minds: how the Doll Test opened schoolhouse doors. Southern Quart 54:164–168
Weber BH, Depew DJ (eds) (2003) Evolution and learning: the Baldwin effect reconsidered. MIT Press, Cambridge
Weisbecker V, Nilsson M (2008) Integration, heterochrony, and adaptation in pedal digits of syndactylous marsupials. BMC Evol Biol 8:160
West-Eberhard MJ (2003) Developmental plasticity and evolution. Oxford University Press, Oxford
West-Eberhard MJ (2004) Ryuichi Matsuda: a tribute and a perspective on pan-environmentalism and genetic assimilation. In: Hall BK, Pearson RD, Müller GB (eds) Environment, development and evolution: toward a synthesis (The Vienna Series in Theoretical Biology). A Bradford Book, Cambridge, MA, pp 109–116
West-Eberhard MJ (2007) Dancing with DNA and flirting with the ghost of Lamarck. Biol Philos 22:439–451
West-Eberhard MJ (2014) Darwin's forgotten idea: the social essence of sexual selection. Neurosci Biobehav Rev 46:501–508
Westneat DF, Fox CW (eds) (2010) Evolutionary behavioral ecology. Oxford University Press, New York
Wetherington RK (2011) Readings in the history of evolutionary theory. Oxford University Press, Oxford
White C (1799) An account of the regular gradation in man. C. Dilly, London
Wibowo E, Wassersug RJ (2016) Multiple orgasms in men—what we know so far. Sex Med Rev 4:136–148
Wilhelmson AS, Lantero Rodriguez M et al (2018) Testosterone is an endogenous regulator of BAFF and splenic B cell number. Nat Commun 9:2067
Wilson EO (2014) The meaning of human existence. W. W. Norton, New York
Wolfe SE, Tubi A (2018) Terror management theory and mortality awareness: a missing link in climate response studies? WIREs Clim Change 2018:e566
Wrangham RW (2018) Two types of aggression in human evolution. PNAS 115:245–253
Wray GA, Hoekstra HE, Futuyma DJ et al (2014) Does evolutionary theory need a rethink? No, all is well. Nature 514:161–164
Wulf A (2015) The invention of nature—the adventures of Alexander von Humboldt, the lost hero of science. Alfred A. Knopf, New York
Wylie K (2009) A global survey of sexual behaviours. J Family Reprod Health 3:39–49
Yasmin S (2018) Witch hunts today: abuse of women, superstition and murder collide in India. Sci Am. Accessed 1 Nov 2018

Figure Credits

As children tremble and fear everything in the blind darkness, so we in the light sometimes fear what is no more to be feared than the things children in the dark hold in terror

(Lucretius)

Fig. 1 (image freely available at https://commons.wikimedia.org/wiki/File:Jacopo_da_Ponte_-_St_Jerome_-_WGA01442.jpg)

Fig. 2.1 (image freely available at https://upload.wikimedia.org/wikipedia/commons/7/7a/British_Museum_Flood_Tablet.jpg)

Fig. 2.2 (image modified from Kühl et al. 2016)

Fig. 2.3 (image freely available, credit: https://creativecommons.org/licenses/by/4.0/)

Fig. 2.4 (image freely available from https://commons.wikimedia.org/wiki/File:ALS_clinical_picture.png)

Fig. 2.5 (image freely available at http://www.italianrenaissance.org/michelangelo-creation-of-adam/)

Fig. 2.6 (image freely available at https://www.pinterest.com/pin/493707177879097106/visual-search/?x=16&y=16&w=530&h=671)

Fig. 2.7 (image freely available at https://www.pewforum.org/2017/04/05/the-changing-global-religious-landscape/pf_17-04-05_projectionsupdate_fertility640px/)

Fig. 2.8 (modified from Blume 2009, for more details see that work)

Fig. 2.9 (image freely available at https://www.pewforum.org/2017/04/05/the-changing-global-religious-landscape/pf_17-04-05_projectionsupdate_changepopulation640px/)

Fig. 2.10 (image freely available at https://www.pewresearch.org/fact-tank/2017/09/06/more-americans-now-say-theyre-spiritual-but-not-religious/)

Fig. 2.11 (image freely available at https://imgur.com/r/Infographics/XY5PAvd; credit Abbie Richards/@tofology)

Fig. 3.1 (image freely available at https://upload.wikimedia.org/wikipedia/commons/b/bb/Psalter_World_Map%2C_c.1265.jpg)

Fig. 3.2 (due to its antiquity, this figure has no copyright; adapted from https://www.dreamstime.com/ascent-life-vintage-engraving-engraved-illustration-earth-man-image-162956266)

Fig. 3.3 (image freely available at https://upload.wikimedia.org/wikipedia/commons/7/74/Lucas_Cranach_the_Elder_-_The_Close_of_the_Silver_Age_%28%3F%29_-_Google_Art_Project.jpg)

R. Diogo, *Meaning of Life, Human Nature, and Delusions*,
https://doi.org/10.1007/978-3-319-70401-2

Fig. 3.4 (modified from Moser's 1998 book *Ancestral images—the iconography of human origins*)

Fig. 3.5 (modified from Diogo 2018a)

Fig. 3.6 (modified from Diogo 2018a)

Fig. 3.7 (due to its antiquity, this figure has no copyright; freely available and adapted from https://www.pinterest.com/pin/371617406729501758/)

Fig. 3.8 (due to its antiquity, this figure has no copyright; freely available and adapted from https://uk.pinterest.com/pin/292733466299986614/)

Fig. 3.9 (due to its antiquity, this figure has no copyright; freely available and adapted from https://commons.wikimedia.org/wiki/File:Nicolaes_Tulp_1641_3de_capvt_lvi_satyr.JPG)

Fig. 3.10 (due to its antiquity, this figure has no copyright; freely available and adapted from http://flashbak.com/wp-content/uploads/2016/08/Tyson-chimpanzee-cowper.jpg)

Fig. 3.11 (due to its antiquity, this figure has no copyright; freely available and adapted from https://archive.org/details/beitrgezurnh00burm)

Fig. 3.12 (due to its antiquity, this figure has no copyright; freely available and adapted from https://upload.wikimedia.org/wikipedia/commons/8/8a/Hoppius_Anthropomorpha.png)

Fig. 3.13 (due to its antiquity, this figure has no copyright; freely available and adapted from http://gallica.bnf.fr/ark:/12148/bpt6k1054470t/f179.item)

Fig. 3.14 (freely available and adapted from https://upload.wikimedia.org/wikipedia/commons/b/b1/A_Pair_of_Broad_Bottoms.jpg)

Fig. 3.15 (freely available and adapted from https://www.theguardian.com/world/2015/jun/03/the-man-who-was-caged-in-a-zoo)

Fig. 3.16 (modified from Diogo 2018a)

Fig. 3.17 (due to its antiquity, this figure has no copyright; freely available and adapted from Huber 1931)

Fig. 3.18 (for more details, see text and https://en.wikipedia.org/wiki/File:US_Race_Household_Income.png, where this figure is freely available)

Fig. 3.19 (modified from Washington 2006)

Fig. 3.20 (freely available from https://snl.no/James_Watson)

Fig. 4.1 (adapted from Diamond's 2012 book)

Fig. 4.2 (modified from Kelly 2013)

Fig. 4.3 (modified from Kelly 2013)

Fig. 4.4 (figure freely available and adapted from www.facebook.com/permalink.php?story_fbid=471045994265375&id=105766454126666)

Fig. 4.5 (figure made by the author, from pictures freely available from pixabay.com)

Fig. 4.6 (adapted from Diamond's 2012 book)

Fig. 5.1 (figure freely available and adapted from https://en.wikipedia.org/wiki/Polygyny#/media/File:Legality_of_polygamy.svg)

Fig. 5.2 (figure freely available and adapted from https://ifstudies.org/blog/is-monogamy-unnatural)

Fig. 5.3 (figure freely available and adapted from https://www.statista.com/chart/13668/where-babies-are-born-outside-of-marriage/)

Fig. 5.4 (figure freely available and adapted from https://ec.europa.eu/eurostat/statistics-explained/index.php?title=File:Crude_marriage_and_divorce_rates_in_the_EU,_1965%E2%80%932019_(per_1_000_inhabitants)_May_2021.png)

Fig. 5.5 (figure freely available and adapted from https://ifstudies.org/blog/who-cheats-more-the-demographics-of-cheating-in-america)

Fig. 5.6 (modified from Laland et al. 2015)

Fig. 5.7 (modified from McBrearty & Brooks 2000)

Fig. 5.8 (text and figure freely available from https://www.visualcapitalist.com/how-people-spend-their-time-globally/)

Fig. 5.9 (modified from Kelly 2013)

Fig. 5.10 (figure freely available and adapted from https://en.wikipedia.org/wiki/History_of_advertising#/media/File:1916-skin-touch-soap-ad.jpg)

Fig. 5.11 (figure freely available and adapted from https://www.devostock.com/stock-photo/devostock-xmas-retro-christmas-family-61885.html)

Fig. 5.12 (figure freely available and adapted from https://www.flickr.com/photos/methodshop/7599554838)

Fig. 5.13 (adapted from Finkel 2017)

Fig. 5.14 (adapted from Finkel 2017)

Fig. 5.15 (adapted from Finkel 2017)

Fig. 5.16 (adapted from Finkel 2017)

Fig. 5.17 (adapted from Finkel 2017)

Fig. 5.18 (adapted from Finkel 2017)

Fig. 5.19 (figure freely available and adapted from https://upload.wikimedia.org/wikipedia/commons/4/40/Tintoretto_-_Penitent_Magdalene_-_Google_Art_Project.jpg)

Fig. 5.20 (figure freely available and adapted from Moser 1998)

Fig. 5.21 (figure freely available and adapted from Moser 1998)

Fig. 5.22 (figure freely available and adapted from https://upload.wikimedia.org/wikipedia/commons/2/27/Font-de-Gaume.jpg)

Fig. 5.23 (figure freely available and adapted from https://upload.wikimedia.org/wikipedia/commons/e/ea/Mort_de_la_philosophe_Hypatie.jpg)

Fig. 5.24 (figure freely available and adapted from https://www.statista.com/chart/21345/coronavirus-deaths-by-gender/)

Fig. 5.25 (figure freely available and adapted from https://interactives.lowyinstitute.org/features/covid-performance/)

Fig. 5.26 (figure freely available and adapted from https://en.wikipedia.org/wiki/Latrodectus#/media/File:Latrodectus_hesperus_Berkeley,_California.jpg)

Fig. 5.27 (figure freely available and adapted from https://ourworldindata.org/life-expectancy)

Fig. 6.1 (figure freely available and adapted from Hazlewood 2001)

Fig. 6.2 (figure freely available and adapted from Canizares-Esguerra 2016)

Fig. 6.3 (figure freely available and adapted from Weatherford 2017)

Fig. 6.4 (figure freely available and adapted from Canizares-Esguerra 2016)

Fig. 6.5 (figure freely available and adapted from Canizares-Esguerra 2016)

Fig. 7.1 (figure modified from Hutton's 2017 book *The witch—a history of fear from ancient times to the present*)

Fig. 7.2 (figure freely available and adapted from Yasmin 2018)

Fig. 7.3 (figure freely available and adapted from https://wellcomecollection.org/works/mfygkhgm)

Fig. 7.4 (due to its antiquity, this figure has no copyright; adapted from https://commons.wikimedia.org/wiki/File:Tommaso_Laureti_-_Triumph_of_Christianity_-_WGA12505.jpg)

Fig. 7.5 (figure freely available and adapted from https://commons.wikimedia.org/wiki/File:Custer%27s_last_charge.jpg)

Fig. 7.6 (figure and text adapted from freely available article at https://everipedia.org/wiki/lang_en/tess-thompson-talley)

Fig. 7.7 (photo taken by Cruelty Free International and Soko Tierschutz, freely made available by Cruelty Free International)

Fig. 7.8 (photo freely made available by PETA)

Fig. 7.9 (photo by Brian Gunn/IAAPEA, freely made available by IAAPEA at https://www.animalexperimentspictures.com/photo/797)

Fig. 7.10 (figure modified from DeSesso 2019)

Fig. 7.11 (figure modified from DeSesso 2019)

Fig. 7.12 (figure freely available and adapted from https://wellcomecollection.org/works/kkj466nz?wellcomeImagesUrl=/indexplus/image/L0007589.html)

Fig. 7.13 (figure freely available and adapted from https://upload.wikimedia.org/wikipedia/commons/e/ed/Fedor_Jeftichew_portrait.jpg)

Fig. 7.14 (figure freely available and adapted from https://upload.wikimedia.org/wikipedia/commons/c/c3/Bundesarchiv_Bild_102-16748%2C_Ausstellung_%22Wunder_des_Lebens%22.jpg)

Fig. 8.1 (modified from Diogo 2017)

Fig. 8.2 (photo freely available from https://upload.wikimedia.org/wikipedia/commons/4/49/Downtown_ogden.jpg)

Fig. 8.3 (photo freely available from https://upload.wikimedia.org/wikipedia/commons/5/5f/Geekhouse_Higashinihonbashi.jpg)

Fig. 8.4 (photo and caption freely available online at https://upload.wikimedia.org/wikipedia/commons/8/8c/Hikikomori%2C_Hiasuki%2C_2004.jpg)

Fig. 8.5 (adapted from Sugie 2017)

Fig. 8.6 (adapted from Suzman 2017)

Fig. 8.7 (adapted from figure freely available from https://www.visualcapitalist.com/history-of-pandemics-deadliest/)

Fig. 8.8 (figure freely available from https://www.visualcapitalist.com/how-people-spend-their-time-globally/)

Fig. 8.9 (pictures from my 2018 summer trip to Dominica and other Caribbean islands)

Fig. 8.10 (figure adapted from Hrdy 2009)

Fig. 8.11 (adapted from figure freely available from Betran et al. 2016)

Fig. 9.1 (figure freely available and adapted from https://www.wikiart.org/en/lovis-corinth/the-temptation-of-saint-anthony-1897)

Fig. 9.2 (figure freely available and adapted from https://www.justiceinfo.net/en/45133-sexual-abuse-church-map-justice-worldwide.html)

Fig. 9.3 (images freely available and adapted from https://upload.wikimedia.org/wikipedia/commons/1/1f/Hadzabe_Hunters.jpg and https://theirworld.org/news/china-one-child-policy-meant-many-missed-out-on-school)

Fig. 9.4 (figure adapted from Levi-Strauss 2011)

Fig. 9.5 (figure freely available from https://escolapt.files.wordpress.com/2021/01/chegano.jpg)

Fig. 9.6 (pictures freely available from pixabay.com)

Fig. 9.7 (freely available and adapted from https://en.wikipedia.org/wiki/Duluth_lynchings)

Fig. 9.8 (figure adapted from Rovelli 2017)

Fig. 9.9 (figures freely available and adapted from https://pixabay.com/photos/kids-sports-summer-baby-boy-2639869/, https://upload.wikimedia.org/wikipedia/commons/a/a1/Hra_na_klav%C3%ADr.jpg, https://uago.at/-Fctn, https://www.newtimes.co.rw/rwanda/jewish-german-communities-rwanda-mark-holocaust)

Fig. 9.10 (figure adapted from Kteily et al. 2015)

Fig. 9.11 (adapted from Hood 2013)

Fig. 9.12 (adapted from Hood 2013)

Fig. 9.13 (freely available and adapted from https://en.wikipedia.org/wiki/Mourning_of_Muharram#/media/File:Grief_of_yore_and_gore-Muharram_procession_in_Hyderabad._01.jpg)

Fig. 9.14 (adapted from Hood 2013)

Fig. 9.15 (adapted from Gottschall 2012)

Fig. 9.16 (figure adapted from Rovelli 2017)

Fig. 9.17 (figures freely available and adapted from https://pixabay.com/photos/playing-puppies-young-dogs-790638/, https://www.theatlantic.com/photo/2013/12/christians-muslims-clash-in-central-african-republic/100652/, https://www.timesofisrael.com/amazon-removes-shirts-with-famous-photo-of-nazi-executing-jew/ and https://pixabay.com/photos/kids-games-pillow-fight-family-5026259/)

Fig. 9.18 (figure modified from Johnson et al. 2020)

Fig. 9.19 (figure modified from McMahon's 2006)

Fig. 9.20 (adapted from Diogo 2017)

Fig. 9.21 (picture from Thomas Stanley's 1655 *The history of philosophy*, freely available from https://commons.wikimedia.org/wiki/File:Epicurus_in_Thomas_Stanley_History_of_Philosophy.jpg)

Fig. 9.22 (pictures freely available from https://upload.wikimedia.org/wikipedia/commons/c/cc/Kim_Jong-hyun_performs_at_MBC_%22Turn_Up_the_Radio%22_in_September_2014_02.jpg)

Fig. 9.23 (figure and text freely available from https://commons.wikimedia.org/wiki/File:Kali_attacking_a_demon_-_Cleveland_Museum_of_Art_(28483606155).jpg)

Fig. 9.24 (adapted from Pugliese 2016)

Index

Tomorrow, and tomorrow, and tomorrow…creeps in this petty pace from day to day…to the last syllable of recorded time…and all our yesterdays have lighted fools…the way to dusty death…out, out, brief candle! Life's but a walking shadow, a poor player…that struts and frets his hour upon the stage…and then is heard no more…it is a tale told by an idiot, full of sound and fury, signifying nothing.

(Shakespeare, in Macbeth)

R. Diogo, *Meaning of Life, Human Nature, and Delusions*,
https://doi.org/10.1007/978-3-319-70401-2

Index

R. Diogo, *Meaning of Life, Human Nature, and Delusions*,
https://doi.org/10.1007/978-3-319-70401-2

D

F

H

I

N

Q

T

Printed in the United States
by Baker & Taylor Publisher Services